MasteringBIOLOGY™ Your gateway to the most widely-used online biology self-study materials and more...

With your purchase of a new copy of Campbell/Reece *Biology*, Eighth Edition, you should have received a **Student Access Kit for MasteringBiology™**. This kit contains instructions and a code for accessing this dynamic website. Your Student Access Kit looks like this →

DON'T THROW YOUR ACCESS KIT AWAY!

If you did not purchase a new textbook or cannot locate the Student Access Kit, you can purchase your subscription to MasteringBiology™ online with a major credit card. To do so, go to www.masteringbio.com. Identify the exact title and edition of your textbook when prompted.

INSTRUCTORS:

Look for your "Instructor Access Kit" that you received with your book. Contact your Pearson Benjamin Cummings sales representative if you need assistance.

What is MasteringBiology™?

MasteringBiology™ offers:

1. The **Study Area** for studying on your own or in a study group
2. **MasteringBiology™ Assignments** that your instructor may require
3. **myeBook,** a customizable eBook (see other side) myeBook

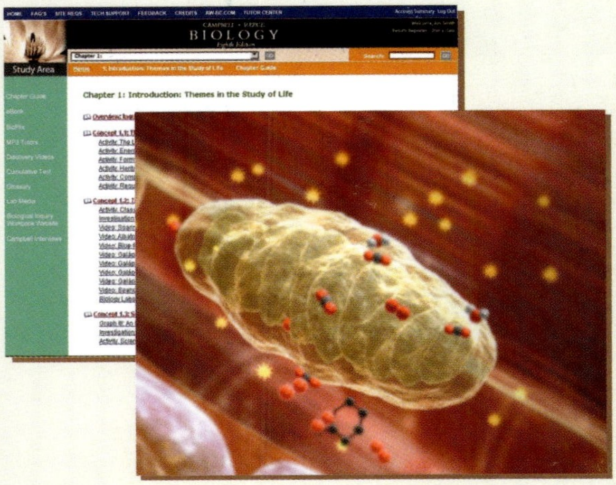

The **Study Area** provides state-of-the-art BioFlix™ 3-D Animations and student tools plus MP3 Tutors, Discovery Videos, Activities, Investigations, Practice Tests, myeBook, and much more (see other side).

MasteringBiology™ Assignments may include Tutorials, BioFlix™ Tutorials, Activities, GraphIt!, Multiple Choice Chapter Quizzes, and other Multiple Choice Questions required by your instructor. MasteringBiology™ Tutorials coach you with feedback specific to your needs and offer hints when you get stuck, optimizing your study time and maximizing your learning.

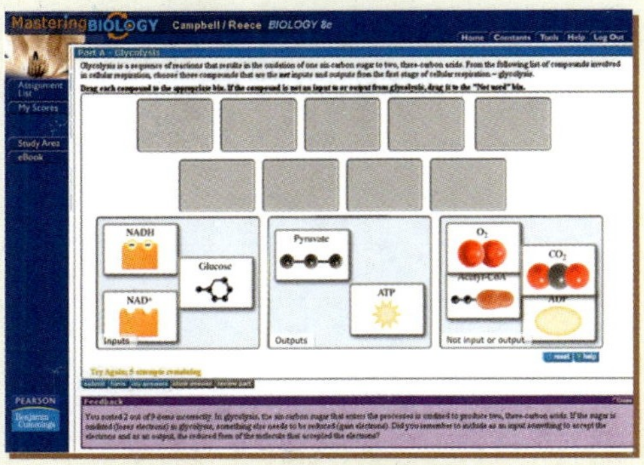

Features of the MasteringBiology™ Study Area for Campbell/Reece *Biology*, Eighth Edition

CHAPTER GUIDE

Review each chapter efficiently and effectively through a focus on Key Concepts. BioFlix™ 3-D Animations and student tools, MP3 Tutors, Discovery Videos, Videos, Activities, Investigations, GraphIt!, and the eBook are linked directly to each numbered Key Concept in the chapter. The Chapter Guide also includes Quizzes, Practice Tests, Word Study Tools, and Art.

BIOFLIX™ 3-D ANIMATIONS AND STUDENT TOOLS

BioFlix™ cover the most challenging biology topics with 3-D, movie-quality animations and carefully constructed student tutorials, labeled slide shows, study sheets, and quizzes. Topics include tour of an animal cell, tour of a plant cell, membrane transport, cellular respiration, photosynthesis, mitosis, meiosis, DNA replication, protein synthesis, water transport in plants, how neurons work, and muscle contraction.

MP3 TUTORS

These MP3 Tutor sessions carefully walk you through "audio tours" of 22 key biological topics.

VIDEOS

29 video clips from the Discovery Channel™ cover topics such as fighting cancer, antibiotic resistance, and introduced species. An additional 153 videos bring biology to life, including 65 Cell Biology Videos.

ACTIVITIES

Explore 212 interactive activities, including animations, virtual labs, and review exercises.

INVESTIGATIONS

Experiment, investigate, and analyze 55 case studies that develop scientific thinking skills.

GRAPHIT!

Learn how to build and interpret graphs with 11 graphing exercises.

PRACTICE TESTS AND QUIZZES

Assess your understanding with three different multiple-choice quizzes for each chapter: an online version of the Self-Quiz from the book, an Activities Quiz with graphics that tests your understanding of the media Activities, and a Practice Test that covers the entire chapter. You may also take Practice Tests on several chapters at once by using the Cumulative Test feature. Tests and Quizzes are graded automatically, and most questions have hints and feedback for right and wrong answers.

ART

View and print art from the textbook, available with and without labels.

WORD STUDY TOOLS AND AUDIO GLOSSARY

Flashcards, Word Roots, Key Terms, and an Audio Glossary will help you improve your biology vocabulary. Includes new audio pronunciations for every term.

MYE-BOOK

With **myeBook**, you can search, take notes, highlight text, bookmark pages, click on a term to see the definition, and link to Practice Tests, BioFlix™ Animations, and more. Also, instructors can write notes that their students can view.

CAMPBELL INTERVIEWS

Learn how scientists think and why they love what they do. Access 64 in-depth interviews with eminent biologists from all eight editions of Campbell/Reece, *Biology*.

BIOLOGY

Campbell / Reece
Urry / Cain / Wasserman / Minorsky / Jackson

Volume 1
Custom Edition

Taken from:
Biology, Eighth Edition
by Neil A. Campbell, Jane B. Reece, Lisa A. Urry, Michael L. Cain,
Steven A. Wasserman, Peter V. Minorsky, and Robert B. Jackson

Custom Publishing

New York Boston San Francisco
London Toronto Sydney Tokyo Singapore Madrid
Mexico City Munich Paris Cape Town Hong Kong Montreal

Cover Art: Courtesy of Corbis

Taken from:

Biology, Eighth Edition
by Neil A. Campbell, Jane B. Reece, Lisa A. Urry, Michael L. Cain, Steven A. Wasserman, Peter V. Minorsky, and Robert B. Jackson

Published by Benjamin Cummings
San Francisco, California 94111

This special edition published in cooperation with Pearson Custom Publishing.

Printed in the United States of America

10 9 8 7 6 5 4 3 2 1

2009140052

CG

Pearson
Custom Publishing
is a division of

PEARSON

www.pearsonhighered.com

ISBN 10: 0-558-23064-4
ISBN 13: 978-0-558-23064-7

Brief Contents

About the Authors

Neil A. Campbell

Neil Campbell combined the investigative nature of a research scientist with the soul of an experienced and caring teacher. He earned his M.A. in Zoology from UCLA and his Ph.D. in Plant Biology from the University of California, Riverside, where he received the Distinguished Alumnus Award in 2001. Neil published numerous research articles on desert and coastal plants and how the sensitive plant (*Mimosa*) and other legumes move their leaves. His 30 years of teaching in diverse environments included general biology courses at Cornell University, Pomona College, and San Bernardino Valley College, where he received the college's first Outstanding Professor Award in 1986. Neil was a visiting scholar in the Department of Botany and Plant Sciences at the University of California, Riverside. In addition to his authorship of this book, he coauthored *Biology: Concepts & Connections* and *Essential Biology* with Jane Reece. Neil died shortly after the initial planning of this revision, but his legacy continues in *BIOLOGY*, Eighth Edition.

Jane B. Reece

Lead author Jane Reece, Neil Campbell's longtime collaborator, has participated in every edition of *BIOLOGY*—first as an editor and contributor, then as an author. Her education includes an A.B. in Biology from Harvard University, an M.S. in Microbiology from Rutgers University, and a Ph.D. in Bacteriology from the University of California, Berkeley. Before migrating to California from the Northeast, she taught biology at Middlesex County College and Queensborough Community College. At UC Berkeley, and later as a postdoctoral fellow in genetics at Stanford University, her research focused on genetic recombination in bacteria. Besides her work on *BIOLOGY*, she has been a coauthor on *Biology: Concepts & Connections*, *Essential Biology*, and *The World of the Cell*.

For the Eighth Edition, Jane Reece is joined by five coauthors whose contributions reflect their biological expertise as scientific researchers and their teaching sensibilities gained from years of experience as instructors.

Lisa A. Urry

Units 1–3 (Chapters 2–21) and Chapter 47

Lisa Urry is a professor at Mills College in Oakland, California, and was a major contributor to the Seventh Edition of *BIOLOGY*. After graduating from Tufts University with a double major in Biology and French, Lisa completed her Ph.D. in Molecular and Developmental Biology at MIT. Following postdoctoral appointments at Harvard Medical School, Tufts University, and UC Berkeley, she began teaching at Mills College, where she currently holds the Letts-Villard Professorship and serves as Chair of the Biology Department. She has published research articles on various topics involving gene expression during embryonic development. Her current research interest is in sea urchin development. Lisa is also deeply committed to promoting opportunities for women in science education and research.

Michael L. Cain

Units 4 and 5 (Chapters 22–34)

Michael Cain is an ecologist and evolutionary biologist currently at Bowdoin College. Michael earned a joint degree in Biology and Math from Bowdoin College, an M.Sc. from Brown University, and a Ph.D. in Ecology and Evolutionary Biology from Cornell University. After postdoctoral work in plant ecology at the University of Connecticut and molecular genetics at Washington University in St. Louis, Michael went on to teach general biology, ecology, and evolution in a diverse range of settings, including Carleton College, New Mexico State University, and Rose-Hulman Institute of Technology in Indiana. Michael is the author of dozens of scientific papers on topics that include foraging behavior in insects and plants, long-distance seed dispersal, and speciation in crickets.

Steven A. Wasserman

Unit 7 (Chapters 40–46, 48–51)

Steve Wasserman is a professor at the University of California, San Diego. He earned his A.B. in Biology from Harvard University and his Ph.D. in Biological Sciences from MIT. Since a postdoctoral sojourn at UC Berkeley, where he investigated topological transformations of DNA, he has focused on regulatory pathway mechanisms. Working with the fruit fly *Drosophila*, he has contributed to the fields of embryogenesis, reproduction, and immunity. As a faculty member at the University of Texas Southwestern Medical Center and UC San Diego, he has taught genetics, development, and physiology to undergraduate, graduate, and medical students. He has also served as the research mentor for more than a dozen doctoral students and nearly 40 aspiring scientists at the undergraduate and high school level. Steve has been the recipient of distinguished scholar awards from both the Markey Charitable Trust and the David and Lucille Packard Foundation. He recently received the 2007 Academic Senate Distinguished Teaching Award for undergraduate teaching at UC San Diego.

Peter V. Minorsky

Unit 6 (Chapters 35–39)

Peter Minorsky revised Unit 6 for the Sixth and Seventh Editions of *BIOLOGY* and is a professor at Mercy College in New York, where he teaches evolution, ecology, botany, and introductory biology. He is also the science writer for the journal *Plant Physiology.* He received his B.A. in Biology from Vassar College and his Ph.D. in Plant Physiology from Cornell University. After a postdoctoral fellowship at the University of Wisconsin at Madison, Peter taught at Kenyon College, Union College, Western Connecticut State University, and Vassar College. He is an electrophysiologist who studies plant responses to stress and is currently exploring the possible effects of geomagnetism on plant growth.

Robert B. Jackson

Unit 8 (Chapters 52–56)

Rob Jackson is a professor of biology and Nicholas Chair of Environmental Sciences at Duke University. He directed Duke's Program in Ecology for many years and is currently the Vice President of Science for the Ecological Society of America. Rob holds a B.S. in Chemical Engineering from Rice University, as well as M.S. degrees in Ecology and Statistics and a Ph.D. in Ecology from Utah State University. He was a postdoctoral scientist in Stanford University's Biology Department and an Assistant Professor at the University of Texas, Austin. Rob has received numerous awards, including being honored at the White House with a Presidential Early Career Award in Science and Engineering from the National Science Foundation. He has published a trade book about the environment, *The Earth Remains Forever,* and a children's book of poetry about biology and animals called *Animal Mischief.* His second children's book, *Not Again,* will be published in 2008.

Preface

by Jane Reece

Much has changed in the world since the completion of the previous edition of *BIOLOGY*. In the realm of the biological sciences, the sequencing of the genomes of many more species has had deep ramifications in diverse areas of research, providing new insights, for example, into the evolutionary histories of numerous species. There has been an explosion of discovery about small RNA molecules and their roles in gene regulation and, at the other end of the size spectrum, our knowledge of Earth's biodiversity has expanded to encompass hundreds of new species, including parrots, monkeys, and orchids. And during the same period, biology has become more prominent than ever in our daily lives. The news is filled with stories about the promise of personalized medicine, novel cancer treatments, the possibility of producing biofuels with the help of genetic engineering, and the use of genetic profiling in solving crimes. Other news stories report climate change and ecological disasters, new drug-resistant strains of the pathogens that cause tuberculosis and parasitic infections, and famine—crises in the world around us that are posing new challenges for biologists and their allies in the other sciences. On a personal level, many colleagues and I have missed our inspiring friend, the late Neil Campbell, even as our commitment to leadership in biological education has grown. Our changing world needs biologists and a scientifically literate citizenry as never before, and we are committed to working toward that goal.

The New Coauthors

The Seventh Edition of *BIOLOGY* has been used by more students and instructors than any previous edition, remaining the most widely used college textbook in the sciences. With the privilege of sharing biology with so many students comes the responsibility of improving the book to serve the biology community even better. For that reason, Neil would have been delighted to see that this Eighth Edition fulfills our decade-long goal of expanding the author team. As biological discoveries proliferated, Neil and I realized that it was becoming harder than ever to make judicious decisions about which biological concepts are most important to develop in depth in an introductory textbook. We needed an author team with first-hand expertise across the biological spectrum, and we wanted coauthors who had honed their teaching values in the classroom. Our new coauthors—Lisa Urry, Michael Cain, Steve Wasserman, Peter Minorsky, and Rob Jackson—represent the highest standards of scientific scholarship across a broad range of disciplines and a deep commitment to undergraduate teaching. As described on pages iv–v, their scientific expertise ranges from molecules to ecosystems, and the schools where they teach range from small liberal arts colleges to large universities. In addition, both Lisa and Peter, as major contributors to earlier editions, had prior experience working on the book. The six of us have collaborated unusually closely, starting with book-wide planning meetings and continuing with frequent exchanges of questions and advice as we worked on our chapters. For each chapter, the revising author, editors, and I together formulated a detailed plan; subsequently, my own role involved commenting on early drafts and polishing the final version. Together, we have strived to extend the book's effectiveness for today's students and instructors, while maintaining its core values.

Our Core Values

What are the core values of this book? They start with getting the science right but then focus on helping students make sense of the science. Below I highlight our longtime values and describe how they've been put into practice in the Eighth Edition. You can see examples of many of the book's features in "To the Student: How to Use This Book" (pp. xiv-xix).

Accuracy and Currency

Getting the science right goes beyond making sure that the facts are accurate and up-to-date. Equally important is ensuring that our chapters reflect how scientists in the various subdisciplines of biology, from cell biology to ecology, currently view their area. Changes in the basic paradigms in various biological fields may call for us to reorganize some chapters and even create new ones in a new edition. For example, a new Chapter 21 discusses genomes and their evolution, and neurobiology is now covered in two chapters (Chapters 48 and 49), one focused on the cellular level and one at the organ system level. On pages ix–x, you can read more about new content and organizational improvements in the Eighth Edition.

A Framework of Key Concepts

The explosion of discoveries that makes biology so exciting today also threatens to suffocate students under an avalanche of information. Our primary pedagogical goal is to help students build a framework for learning biology by organizing each chapter around a small number of "Key Concepts," typically three to six. Each chapter begins with a list of its Key Concepts, a photograph that raises an intriguing question, and an Overview section that addresses the question and introduces the chapter. In the body of the chapter, each Key Concept serves as a numbered heading for a major section, in which the prose and pictures tell a more detailed story. At the end of each concept section, Concept Check questions enable students to assess their understanding of that concept before going on to the next concept. Students encounter the Key Concepts one last time when they reach the Chapter Review at the end of the chapter; the Summary of Key Concepts restates them and offers succinct explanatory support in both words and summary diagrams—new to this edition.

Active Learning

Increasingly, instructors tell us that they want their students to take a more active role in learning biology and to think about biological questions at a higher level. In the Eighth Edition, we provide several new ways for students to engage in active learning. First, the Concept Check questions in this edition build in difficulty, and each set now ends with a new "What if?" question that challenges students to integrate what they have learned and to think analytically. There are also questions accompanying selected figures within the text; each of these questions encourages students to delve into the figure and assess their understanding of its underlying ideas. And new "Draw It" exercises in every chapter ask students to put pencil to paper and draw a structure, annotate a figure, or graph experimental data. In addition to appearing regularly in the Chapter Review, a "Draw It" question may show up in a Concept Check or figure legend. Finally, the website that accompanies the book features two especially exciting new student tools, both of which focus on biology's toughest topics: MasteringBiology tutorials and BioFlix 3-D animations and tutorials. These are described on page xx.

Evolution and Other Unifying Themes

Together with *BIOLOGY*'s emphasis on key concepts, a thematic approach has always distinguished our book from an encyclopedia of biology. In the Eighth Edition, as previously, the central theme is evolution. Evolution unifies all of biology by accounting for both the unity and diversity of life and for the remarkable adaptations of organisms to their environments. The evolutionary theme is woven into every chapter of *BIOLOGY*, and Unit Four, Mechanisms of Evolution, has undergone a major revision. In Chapter 1, the other unifying themes have been streamlined from ten to six. And throughout the book, these themes are now referenced more explicitly in Key Concepts and subheadings. The former themes of "scientific inquiry" and "science, technology, and society" continue to be highlighted throughout the book, not as biological themes but as aspects of how science is done and the role of science in our lives.

Integration of Text and Illustrations

We regard text and illustrations as equal in importance, and starting with the First Edition, have always developed them simultaneously. The Eighth Edition has a number of new and improved figures, with the increased use of a more three-dimensional art style where it can enhance understanding of biological structure. At the same time, we avoid excess detail, which can obscure the main point of the figure. We have also improved our popular "Exploring" Figures and have added more (see the list on p. xii). Each of these large figures is a learning unit that brings together a set of related illustrations and the text that describes them. The Exploring Figures enable students to access dozens of complex topics very efficiently. They are core chapter content, not to be confused with some textbooks' "boxes," which have content peripheral to the flow of a chapter. Modern biology is challenging enough without diverting students' attention from a chapter's conceptual storyline.

Telling the Story at the Right Level

Whether in pictures or prose, we are committed to explaining biology at just the right level, and we've continued to use Neil's "quantum theory of teaching biology" as a touchstone. According to this idea, there are discrete levels at which a concept can be successfully explained, and a successful explanation must avoid getting "stuck between levels." Of course, most seasoned instructors have independently recognized this issue, also known as the "too much–too little" problem. The author team has drawn upon both scientific expertise and teaching experience to tell the story of biology at an appropriate level.

The Importance of Scientific Inquiry

Another of our core values is our belief in the importance of introducing students to the scientific way of thinking. In both lecture hall and laboratory, the authors and many of our colleagues are experimenting with diverse approaches for involving students in scientific inquiry, the process by which questions about nature are posed and explored. Special features in the textbook and in inquiry-based supplements make this edition of *BIOLOGY* more effective than ever in helping instructors convey the process of science in their courses.

Modeling Inquiry by Example

Every edition of *BIOLOGY* has traced the history of many research questions and scientific debates to help students appreciate not just "what we know," but "how we know," and "what we do not yet know." In *BIOLOGY*, Seventh Edition, we strengthened this aspect of the book by introducing "Inquiry" Figures, which showcase examples of experiments and field studies in a format that is consistent throughout the book. Each of these inquiry cases begins with a research question, followed by sections describing the experiment, results, and conclusion. Complementing the Inquiry Figures are "Research Method" Figures, which walk students through the techniques and tools of modern biology.

In the Eighth Edition, we have added many more Inquiry Figures; there is now at least one in every chapter and often more (see the list of Inquiry Figures on pp. xii–xiii). Each

Inquiry Figure now ends with a "What if?" question that requires students to demonstrate their understanding of the experiment described. We have also expanded the usefulness of the Inquiry Figures in another important way: In response to feedback from many instructors, we now cite the journal article that is the source of the research, providing a gateway to the primary literature. And the full papers for nine of the Inquiry Figures are reprinted in *Inquiry in Action: Interpreting Scientific Papers*, by Ruth Buskirk and Christopher Gillen. This new supplement, which can be ordered with the book for no additional charge, provides background information on how to read scientific papers plus specific questions that guide students through the nine featured articles.

Learning Inquiry by Practice

BIOLOGY, Eighth Edition, encourages students to practice thinking as scientists by tackling the "What if?" questions in the Concept Checks and Inquiry Figures (and occasional figure legends), as well as the "Scientific Inquiry" questions in the Chapter Review. Many of those in the Chapter Reviews ask students to analyze data or to design an experiment.

The supplements for the Eighth Edition build on the textbook to provide diverse opportunities for students to practice scientific inquiry in more depth. In addition to *Inquiry in Action: Interpreting Scientific Papers*, these include new editions of several other supplements that can be made available without cost. One is *Biological Inquiry: A Workbook of Investigative Cases*, Second Edition, by Margaret Waterman and Ethel Stanley; another is *Practicing Biology: A Student Workbook*, Third Edition, by Jean Heitz and Cynthia Giffen. You can find out more about these and other student supplements, both print and electronic, on pages xx–xxiii.

The BIOLOGY *Interviews: A Continuing Tradition*

Scientific inquiry is a social process catalyzed by communication among people who share a curiosity about nature. One of the many joys of authoring *BIOLOGY* is the privilege of interviewing some of the world's most influential biologists. Eight new interviews, one opening each unit of the textbook, introduce students to eight of the fascinating individuals who are driving progress in biology and connecting science to society. And in this edition, each unit of the text includes an Inquiry Figure based upon the research of the unit's interviewee; for example, see Inquiry Figure 2.2, on page 31. The interviewees for this edition are listed on page xi.

A Versatile Book

Our book is intended to serve students as a textbook in their general biology course and also later as a useful tool for review and reference. *BIOLOGY's* breadth, depth, and versatile organization enable the book to meet these dual goals. Even by limiting our scope to a few Key Concepts per chapter, *BIOLOGY* spans more biological territory than most introductory courses could or should attempt to cover. But given the great diversity of course syllabi, we have opted for a survey broad enough and deep enough to support each instructor's particular emphases. Students also seem to appreciate *BIOLOGY's* breadth and depth; in this era when students sell many of their textbooks back to the bookstore, more than 75% of students who have used *BIOLOGY* have kept it after their introductory course. In fact, we are delighted to receive mail from upper division students and graduate students, including medical students, expressing their appreciation for the long-term value of *BIOLOGY* as a general resource for their continuing education.

Just as we recognize that few courses will cover all 56 chapters of the textbook, we also understand that there is no single correct sequence of topics for a general biology course. Though a biology textbook's table of contents must be linear, biology itself is more like a web of related concepts without a fixed starting point or a prescribed path. Diverse courses can navigate this network of concepts starting with molecules and cells, or with evolution and the diversity of organisms, or with the big-picture ideas of ecology. We have built *BIOLOGY* to be versatile enough to support these different syllabi. The eight units of the book are largely self-contained, and, for most of the units, the chapters can be assigned in a different sequence without substantial loss of coherence. For example, instructors who integrate plant and animal physiology can merge chapters from Unit Six (Plant Form and Function) and Unit Seven (Animal Form and Function) to fit their courses. As another option, instructors who begin their course with ecology and continue with this top-down approach can assign Unit Eight (Ecology) right after Chapter 1, which introduces the unifying themes that provide students with a panoramic view of biology no matter what the topic order of the course syllabus.

Our Partnership with Instructors

A core value underlying all our work as authors is our belief in the importance of our partnership with instructors. Our primary way of serving instructors, of course, is providing a textbook that serves their students well. In addition, Benjamin Cummings makes available a wealth of instructor resources, in both print and electronic form (see pp. xx–xxiii). However, our relationship with instructors is not a one-way street. In our continuing efforts to improve the book and its supplements, we benefit tremendously from instructor feedback, not only in formal reviews from hundreds of scientists, but also via informal communication in person and by phone and e-mail. Neil Campbell built a vast network of colleagues throughout the world, and my new coauthors and I are fully committed to continuing that tradition.

The real test of any textbook is how well it helps instructors teach and students learn. We welcome comments from the students and professors who use *BIOLOGY*. Please address your suggestions to me:

Jane Reece, Pearson Benjamin Cummings
1301 Sansome Street, San Francisco, CA 94111
E-mail address: JaneReece@cal.berkeley.edu

New to the Eighth Edition

This section provides just a few highlights of new content and organizational improvements in *BIOLOGY*, Eighth Edition.

UNIT ONE The Chemistry of Life

New examples make basic chemistry more engaging for students, including an explanation of why steam can burn your skin in Chapter 3, the structures of the enantiomeric medications ibuprofen and albuterol in Chapter 4, and information on trans fats in Chapter 5. A new Inquiry Figure in Chapter 3 relates acidity to the emerging global problem of ocean acidification and its effects on coral reefs. The new Inquiry Figure in Chapter 5 shows Roger Kornberg's 3-D model of the RNA polymerase–DNA–RNA complex, work for which he won the 2006 Nobel Prize in Chemistry.

UNIT TWO The Cell

The judicious addition of recent research includes updated coverage of the sensory roles of primary cilia in Chapter 6, new developments regarding the membrane model in Chapter 7, and Paul Nurse's Nobel Prize–winning work on the cell cycle in Chapter 12. Chapter 11 now ends with a section on apoptosis, formerly in Chapter 21. New Inquiry Figures in this unit describe research on the role of microtubules in orienting cellulose in cell walls (Chapter 6), allosteric regulators of enzymes (Chapter 8), ATP synthase (Chapter 9), yeast cell signaling (Chapter 11), and a cell cycle regulator (Chapter 12).

UNIT THREE Genetics

Chapter 14 now includes "Tips for Genetics Problems." In Chapter 15, sex linkage is discussed directly after the discussion of the white-eye trait in Morgan's fruit flies. Chapter 16 covers replication of the bacterial chromosome and the structure of the eukaryotic chromosome (including a new Exploring Figure), formerly in Chapters 18 and 19, respectively.

We have reorganized Chapters 18–21 with the dual aims of telling a more coherent story and facilitating instructors' coverage of molecular genetics. Regulation of gene expression for both bacteria and eukaryotes is now consolidated in Chapter 18, which also includes a concept section on the crucial role of small RNAs in eukaryotes. We have streamlined material on the genetic basis of development (formerly in Chapter 21), and included it in Chapter 18, where it provides the ultimate example of gene regulation. Chapter 18 ends with a section on the molecular basis of cancer (previously in Chapter 19), to demonstrate what happens when gene regulation goes awry. Material on bacterial genetics in Seventh Edition Chapter 18 has been moved to other chapters within the genetics unit and to Chapter 27 on prokaryotes.

Chapter 19 now covers only viruses (from Seventh Edition Chapter 18), giving this chapter the flexibility to be assigned at any point in the course. Chapter 20 continues to cover biotechnology, but genome sequencing and analysis have been moved to Chapter 21. Cloning and stem cell production are now in Chapter 20. Newly explained techniques include the screening of an arrayed library, BAC clones, Northern blotting, RT-PCR, and *in situ* hybridization. The explosion of discoveries about genomes and their evolution led us to develop a chapter devoted to this subject, the new Chapter 21. This chapter consolidates new material with topics from Chapters 19–21 of the Seventh Edition.

UNIT FOUR Mechanisms of Evolution

Our revision emphasizes the centrality of evolution to biology and the breadth and depth of evidence for evolution. New examples and Inquiry Figures present data from field and laboratory studies and reveal how scientists study evolution.

Chapter 22 discusses how evolution can be viewed as both a pattern and a process, and introduces three key observations about life that are explained by evolution: the match between organisms and their environments (adaptation); the shared characteristics (unity) of life; and the diversity of life. This discussion serves as a conceptual anchor throughout Units Four and Five.

Chapters 24 and 25 have been significantly reorganized. Chapter 24 is now more tightly focused on speciation, enabling better pacing of this highly conceptual material. A new concept section explores hybrid zones as natural laboratories for studying speciation. Chapter 25 focuses on macroevolution, incorporating topics formerly in Chapters 24 and 26, such as the correlations between Earth's geologic and biological history. But the primary storyline concerns what we can learn from the fossil record about the evolutionary history of life. New text and figures explore how the rise and fall of dominant groups of organisms are linked to large-scale processes such as continental drift, mass extinctions, and adaptive radiations. Coverage of evo-devo has been expanded.

Phylogenetic trees are introduced earlier, in a new section on "tree-thinking" in Chapter 22. This material supports students in interpreting diagrams before studying phylogenetics more fully in Chapter 26.

UNIT FIVE The Evolutionary History of Biological Diversity

A new Chapter 26, Phylogeny and the Tree of Life, introduces the unit. Extending material formerly in Chapter 25, it describes how evolutionary trees are constructed and underscores their role as tools for understanding relationships, rather than facts to be memorized. New sections address common misconceptions in interpreting trees and help motivate students with practical applications.

Chapter 27 has a new concept section on prokaryotic reproduction, mutation, and recombination (formerly in Chapter 18). This unifies the coverage of prokaryote biology and supports students in developing a fuller understanding of these microorganisms.

Throughout Unit Five, along with updating the phylogenies of various groups of organisms—introducing, for example, the "supergroup" hypothesis of eukaryotic phylogeny (in Chapter 28) —we have found new opportunities to use the study of phylogeny as an opportunity to illustrate the iterative nature of the scientific process. We aim to help students stay focused on the big picture of why biologists study evolutionary relationships. Each chapter also now includes an Inquiry figure that models how researchers study organisms and their relationships. At the same time, in each chapter we highlight the key roles that various organisms play in the biosphere as well as their applied importance for humans.

UNIT SIX Plant Form and Function

Revisions to this unit draw more attention to the experimental basis of our understanding of plant biology. New examples include recent progress toward identifying the flowering "hormone" (Chapter 39). Featured in new Inquiry Figures are experiments demonstrating, for example, that trichomes affect insect feeding (Chapter 35) and that informational molecules transported through the symplast affect plant development (Chapter 36).

In Chapter 36, now titled Resource Acquisition and Transport in Vascular Plants, a new first concept section explores how architectural features of plants facilitate resource acquisition, helping students relate the transport of water and nutrients to what they learned in Chapter 35 about plant structure and growth. Another new concept section, on symplastic transport, discusses recent insights into changes in plasmodesmata shape and number and the transmission of electrical and molecular signals throughout the plant.

This unit now has more examples of practical applications of plant biotechnology. For instance, Chapter 37 discusses how genetic modification has increased the resistance of some plants to aluminum toxicity and has improved the flood tolerance of rice crops. Chapter 38 elaborates on the principles of plant breeding and incorporates a new section on genetic engineering of biofuels.

UNIT SEVEN Animal Form and Function

An evolutionary perspective more strongly pervades this unit, underscoring how environment and physical laws shape adaptations across animal groups. Each chapter now includes at least one Inquiry Figure; together, these figures highlight the wide range of methodologies used to study animal physiology, including several experiments using molecular biology techniques students studied earlier in the book.

Chapter 40 has been revised and reorganized to highlight functional relationships at all levels of organization in animal bodies; thermoregulation serves as an extended example throughout the chapter. Chapter 43, The Immune System, has been extensively revised. For instance, we now contrast recognition of pathogen class in innate immunity with antigen-specific recognition in adaptive immunity, helping overcome the common misconception that recognition is absent in innate immunity.

We have divided the former nervous system chapter into two, enabling us to better pace difficult material and highlight dynamic current research by focusing first on cellular processes in Chapter 48, and then on nervous system organization and function in Chapter 49. Chapter 50 rounds out the discussion of nervous system function by examining sensory and motor mechanisms. This sequence leads naturally into Chapter 51 on animal behavior (formerly in Unit Eight), which ties together aspects of genetics, natural selection, and physiology, and provides a bridge to the ecology unit.

UNIT EIGHT Ecology

This unit, which now includes Chapters 52-56, incorporates many new examples that demonstrate a range of methods and scales of study. For example, a new figure in Chapter 52 describes a large-scale field experiment in which researchers manipulated precipitation levels in forest plots, while new Research Method figures describe determining population size using the mark-recapture method (Chapter 53), using molecular tools to measure diversity of soil microorganisms (Chapter 54), and determining primary production with satellite data (Chapter 55). By building on earlier units, we hope to demonstrate how ecology represents a fitting capstone to the book.

We provide more microbial examples and more aquatic ones, from diverse locations around the globe. For instance, Chapter 52 now discusses the importance of salinity in determining the distribution of aquatic organisms, and Chapter 54's coverage of the intermediate disturbance hypothesis includes a new figure on a quantitative test of the hypothesis in New Zealand streams. The unit highlights the great relevance of ecology to society and to students' lives. A new concept section in Chapter 54, for example, discusses how community ecology helps us understand pathogen life cycles and control disease.

Interviews

UNIT ONE

The Chemistry of Life

Deborah M. Gordon

Stanford University

UNIT TWO

The Cell

Paul Nurse

Rockefeller University

UNIT THREE

Genetics

Terry L. Orr-Weaver

Massachusetts Institute of Technology

UNIT FOUR

Mechanisms of Evolution

Scott V. Edwards

Harvard University

UNIT FIVE

The Evolutionary History of Biological Diversity

Sean B. Carroll

University of Wisconsin-Madison

UNIT SIX

Plant Form and Function

Patricia Zambryski

University of California, Berkeley

UNIT SEVEN

Animal Form and Function

Masashi Yanagisawa

University of Texas Southwestern Medical Center

UNIT EIGHT

Ecology

Diana H. Wall

Colorado State University

Featured Figures

Exploring Figures

Inquiry Figures

*The original research paper and study questions are provided in *Inquiry in Action: Interpreting Scientific Papers.*

Research Method Figures

To the Student: How to Use This Book

Focus on the Key Concepts

Each chapter is organized around a framework of 3 to 6 **Key Concepts** that will help you stay focused on the big picture and give you a context for the supporting details.

Before you begin reading the chapter, get oriented by reading the **list of Key Concepts**, which introduces the big ideas covered in the chapter.

Each **Key Concept** serves as the heading for a major section of the chapter.

25

The History of Life on Earth

▲ **Figure 25.1 What does fossil evidence say about where these dinosaurs lived?**

KEY CONCEPTS

25.1 Conditions on early Earth made the origin of life possible
25.2 The fossil record documents the history of life
25.3 Key events in life's history include the origins of single-celled and multicelled organisms and the colonization of land
25.4 The rise and fall of dominant groups reflect continental drift, mass extinctions, and adaptive radiations
25.5 Major changes in body form can result from changes in the sequences and regulation of developmental genes
25.6 Evolution is not goal oriented

OVERVIEW

Lost Worlds

Visitors to Antarctica today encounter one of Earth's harshest, most barren environments. In this land of extreme cold where there is almost no liquid water, life is sparse and small—the largest fully terrestrial animal is a fly 5 mm long. But even as early antarctic explorers struggled to survive, some of them made an astonishing discovery: fossil evidence that life once thrived where it now barely exists. Fossils reveal that 500 million years ago, the ocean waters surrounding Antarctica were warm and teeming with tropical invertebrates. Later, the continent was covered in forests for hundreds of millions of years. At various times, a wide range of animals stalked through these forests, including 3-meter-tall predatory "terror birds" and giant dinosaurs such as the voracious *Cryolophosaurus* **(Figure 25.1)**, a 7-meter-long relative of *Tyrannosaurus rex*.

Fossils discovered in other parts of the world tell a similar, if not quite as surprising, story: Past organisms were very different from those now alive. The sweeping changes in life on Earth revealed by fossils illustrate **macroevolution**, the pattern of evolution over large time scales. Specific examples of macroevolutionary change include the origin of key biochemical processes such as photosynthesis, the emergence of the first terrestrial vertebrates, and the long-term impact of a mass extinction on the diversity of life.

▲ *Cryolophosaurus* skull

Taken together, such changes provide a grand view of the evolutionary history of life on Earth. We'll examine that history in this chapter, beginning with hypotheses regarding the origin of life. The origin of life is the most speculative topic of the entire unit, for no fossil evidence of that seminal episode exists. We will then turn to the fossil record and what it tells us about major events in the history of life, paying particular attention to factors that have helped to shape the rise and fall of different groups of organisms over time.

CONCEPT **25.1**

Conditions on early Earth made the origin of life possible

The earliest evidence of life on Earth comes from fossils of microorganisms that are about 3.5 billion years old. But when and how did the first living cells appear? Observations and experiments in chemistry, geology, and physics have led scientists to propose one scenario that we'll examine here. They hypothesize that chemical and physical processes on early Earth, aided by the emerging force of

When you finish reading a concept section, check your understanding using the **Concept Check questions** at the end of the section. If you can answer them (see Appendix A to check your work), you're ready to move on.

Each set of Concept Check questions builds in difficulty, ending with a **"What if?" question** that asks you to apply what you've learned. Work through these questions on your own or in a study group—they're good practice for the kinds of questions you might be asked on an exam.

CONCEPT CHECK 53.1

1. One species of forest bird is highly territorial, while a second lives in flocks. Predict each species' likely pattern of dispersion, and explain.
2. **DRAW IT** Each female of a particular fish species produces millions of eggs per year. Draw and label the most likely survivorship curve for this species, and explain your choice.
3. **WHAT IF?** As noted in Figure 53.2, an important assumption of the mark-recapture method is that marked individuals have the same probability of being recaptured as unmarked individuals. Describe a situation where this assumption might not be valid, and explain how the estimate of population size would be affected.

For suggested answers, see Appendix A.

Practice thinking like a scientist

To succeed in biology, you need to do more than just memorize facts—you need to understand how scientists go about asking and investigating questions. In each chapter, you'll find one or more **Inquiry Figures** that highlight how researchers designed an experiment, interpreted their results, and drew conclusions.

Scientific Inquiry questions in the Chapter Review provide more opportunities to practice inquiry skills, including developing hypotheses, designing experiments, and analyzing real research data.

SCIENTIFIC INQUIRY

8. **DRAW IT** The nitrogen-fixing bacterium *Rhizobium* infects the roots of some plant species, forming a mutualism in which the bacterium provides the plant with nitrogen, and the plant provides the bacterium with carbohydrates. Scientists measured how well one such plant species (*Acacia irrorata*) grew when its roots were infected by six different *Rhizobium* strains.

(a) Graph the data. (b) Interpret your graph.

Rhizobium strain	Plant mass (g) after 12 weeks of growth
1	0.91
2	0.06
3	1.56
4	1.72
5	0.14
6	1.03

Source: J. J. Burdon, et al., Variation in the effectiveness of symbiotic associations between native rhizobia and temperate Australian *Acacia*: within species interactions, *Journal of Applied Ecology*, 36:398-408 (1999).

Note: In the absence of *Rhizobium*, after 12 weeks of growth, *Acacia* plants have a mass of about 0.1 g.

▼ Figure 2.2 Inquiry

What creates "devil's gardens" in the rain forest?

EXPERIMENT Working under Deborah Gordon and with Michael Greene, graduate student Megan Frederickson sought the cause of "devil's gardens," stands of a single species of tree, *Duroia hirsuta*. One hypothesis was that ants living in these trees, *Myrmelachista schumanni*, produce a poisonous chemical that kills trees of other species; another was that the *Duroia* trees themselves kill competing trees, perhaps by means of a chemical.

To test these hypotheses, Frederickson did field experiments in Peru. Two saplings of a local nonhost tree species, *Cedrela odorata*, were planted inside each of ten devil's gardens. At the base of one, a sticky insect barrier was applied; the other was unprotected. Two more *Cedrela* saplings, with and without barriers, were planted about 50 meters outside each garden.

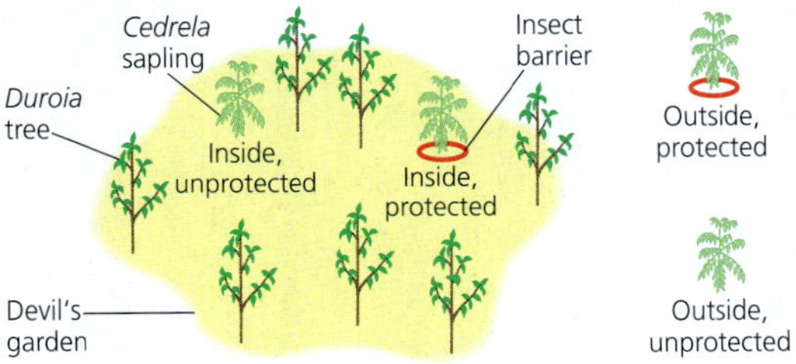

The researchers observed ant activity on the *Cedrela* leaves and measured areas of dead leaf tissue after one day. They also chemically analyzed contents of the ants' poison glands.

RESULTS The ants made injections from the tips of their abdomens into leaves of unprotected saplings in their gardens (see photo). Within one day, these leaves developed dead areas (see graph). The protected saplings were uninjured, as were the saplings planted outside the gardens. Formic acid was the only chemical detected in the poison glands of the ants.

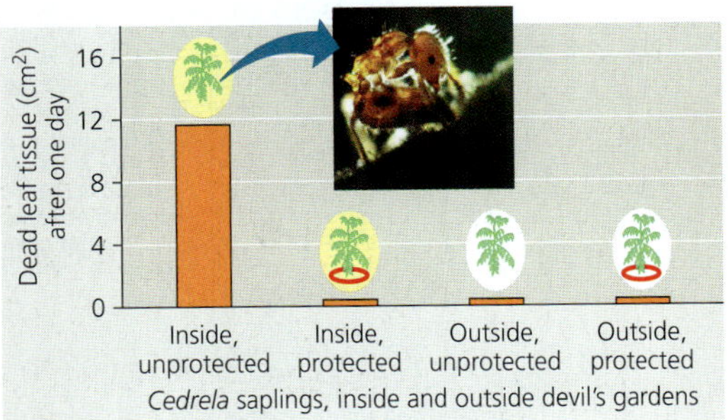

CONCLUSION Ants of the species *Myrmelachista schumanni* kill nonhost trees by injecting the leaves with formic acid, thus creating hospitable habitats (devil's gardens) for the ant colony.

SOURCE M. E. Frederickson, M. J. Greene, and D. M. Gordon, "Devil's gardens" bedevilled by ants, *Nature* 437:495–496 (2005).

Inquiry in Action Read and analyze the original paper in *Inquiry in Action: Interpreting Scientific Papers*.

WHAT IF? What would be the results if the unprotected saplings' inability to grow in the devil's gardens was caused by a chemical released by the *Duroia* trees rather than by the ants?

Some Inquiry Figures invite you to read and analyze the original research paper in its complete form. You can find the journal article, along with a worksheet guiding you through it, in the separate book ***Inquiry in Action: Interpreting Scientific Papers*** (ISBN 978-0-321-53659-4 | 0-321-53659-2).

After exploring the featured experiment, test your analytical skills by answering the **"What if?" question**. Suggested answers are provided in Appendix A to help you gauge your understanding.

As you read the text, study the art and photos

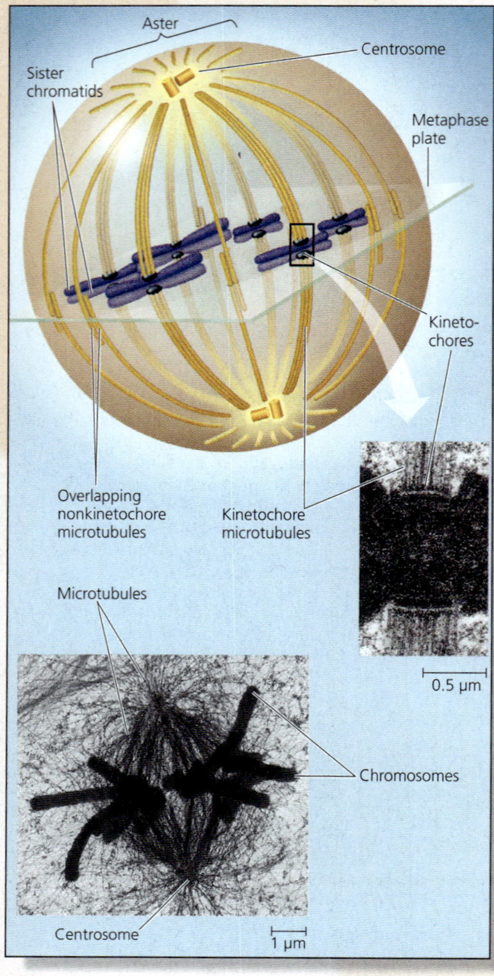

In selected illustrations, a **three-dimensional art style** helps you visualize biological structures.

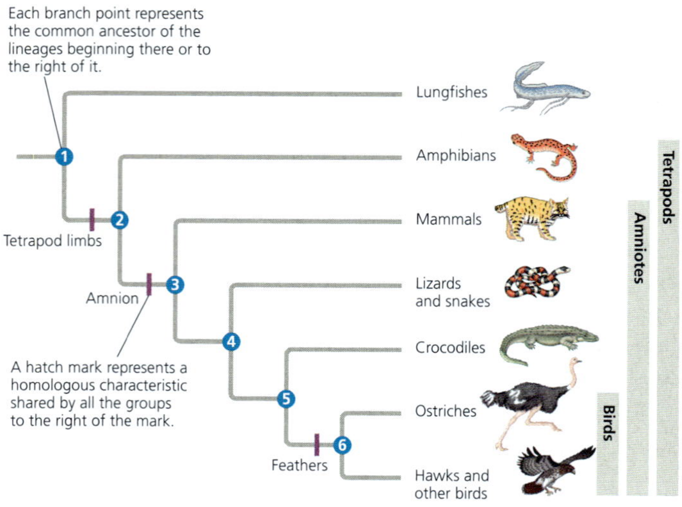

▲ **Figure 22.19 Tree thinking: information provided in an evolutionary tree.** This evolutionary tree for tetrapods and their closest living relatives, the lungfishes, is based on anatomical and DNA sequence data. The purple bars indicate the origin of three important homologies, each of which evolved only once. Birds are nested within and evolved from reptiles; hence, the group of organisms called "reptiles" technically includes birds.

? *Are crocodiles more closely related to lizards or birds? Explain your answer.*

The book's **phylogenetic trees** will prepare you to analyze the evolutionary tree diagrams you may encounter in the scientific literature. This figure walks you through some of the conventions of the phylogenetic trees you will see throughout the evolution and diversity units.

"Guided Tour" figures include concise explanations that walk you through complicated diagrams.

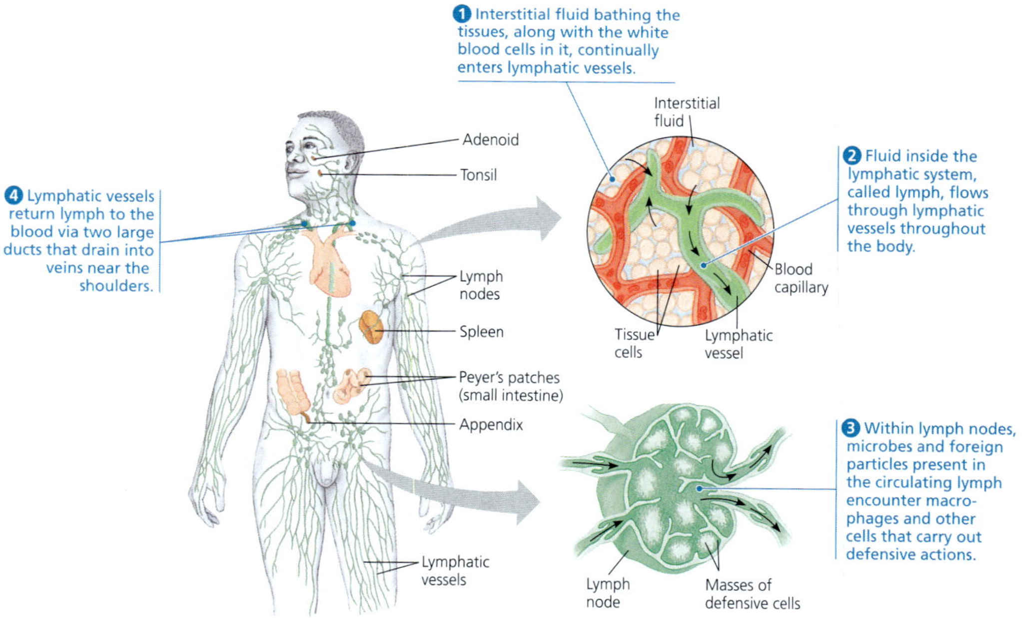

▲ **Figure 43.7 The human lymphatic system.** The lymphatic system consists of lymphatic vessels, through which lymph travels, and various structures that trap "foreign" molecules and particles. These structures include the adenoids, tonsils, lymph nodes, spleen, Peyer's patches, and appendix. Steps 1–4 trace the flow of lymph.

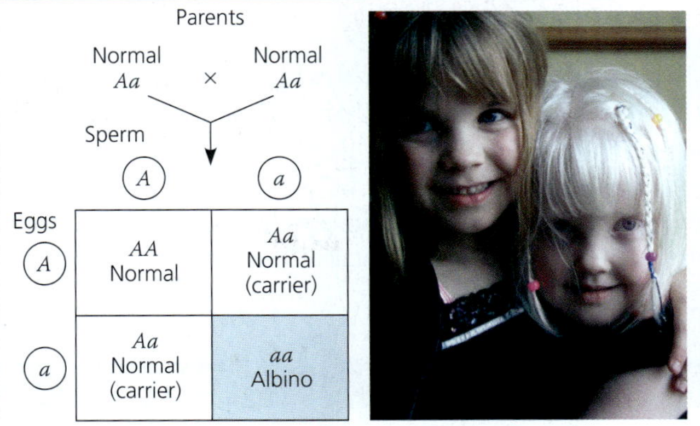

▲ **Figure 14.16 Albinism: a recessive trait.** One of the two sisters shown here has normal coloration; the other is albino. Most recessive homozygotes are born to parents who are carriers of the disorder but themselves have a normal phenotype, the case shown in the Punnett square.

? *What is the probability that the sister with normal coloration is a carrier of the albinism allele?*

Figure Legend questions prompt you to delve into a figure's content to make sure you understand it.

By integrating text and visuals, **Exploring Figures** help you learn more efficiently.

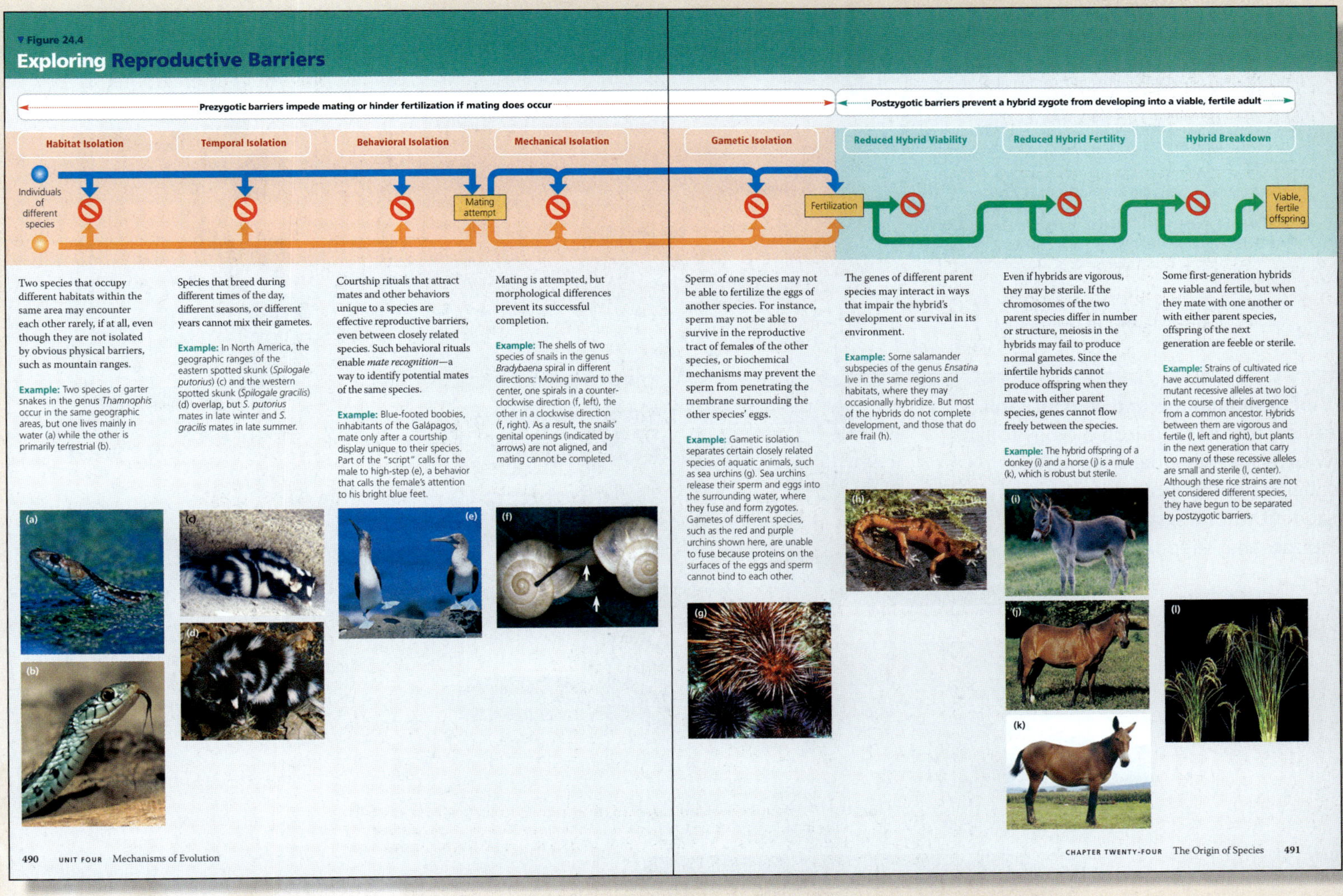

▼ Figure 24.4

Exploring Reproductive Barriers

Habitat Isolation: Two species that occupy different habitats within the same area may encounter each other rarely, if at all, even though they are not isolated by obvious physical barriers, such as mountain ranges.

Example: Two species of garter snakes in the genus *Thamnophis* occur in the same geographic areas, but one lives mainly in water (a) while the other is primarily terrestrial (b).

Temporal Isolation: Species that breed during different times of the day, different seasons, or different years cannot mix their gametes.

Example: In North America, the geographic ranges of the eastern spotted skunk (*Spilogale putorius*) (c) and the western spotted skunk (*Spilogale gracilis*) (d) overlap, but *S. putorius* mates in late winter and *S. gracilis* mates in late summer.

Behavioral Isolation: Courtship rituals that attract mates and other behaviors unique to a species are effective reproductive barriers, even between closely related species. Such behavioral rituals enable *mate recognition*—a way to identify potential mates of the same species.

Example: Blue-footed boobies, inhabitants of the Galápagos, mate only after a courtship display unique to their species. Part of the "script" calls for the male to high-step (e), a behavior that calls the female's attention to his bright blue feet.

Mechanical Isolation: Mating is attempted, but morphological differences prevent its successful completion.

Example: The shells of two species of snails in the genus *Bradybaena* spiral in different directions: Moving inward to the center, one spirals in a counter-clockwise direction (f, left), the other in a clockwise direction (f, right). As a result, the snails' genital openings (indicated by arrows) are not aligned, and mating cannot be completed.

Gametic Isolation: Sperm of one species may not be able to fertilize the eggs of another species. For instance, sperm may not be able to survive in the reproductive tract of females of the other species, or biochemical mechanisms may prevent the sperm from penetrating the membrane surrounding the other species' eggs.

Example: Gametic isolation separates certain closely related species of aquatic animals, such as sea urchins (g). Sea urchins release their sperm and eggs into the surrounding water, where they fuse and form zygotes. Gametes of different species, such as the red and purple urchins shown here, are unable to fuse because proteins on the surfaces of the eggs and sperm cannot bind to each other.

Reduced Hybrid Viability: The genes of different parent species may interact in ways that impair the hybrid's development or survival in its environment.

Example: Some salamander subspecies of the genus *Ensatina* live in the same regions and habitats, where they may occasionally hybridize. But most of the hybrids do not complete development, and those that do are frail (h).

Reduced Hybrid Fertility: Even if hybrids are vigorous, they may be sterile. If the chromosomes of the two parent species differ in number or structure, meiosis in the hybrids may fail to produce normal gametes. Since the infertile hybrids cannot produce offspring when they mate with either parent species, genes cannot flow freely between the species.

Example: The hybrid offspring of a donkey (i) and a horse (j) is a mule (k), which is robust but sterile.

Hybrid Breakdown: Some first-generation hybrids are viable and fertile, but when they mate with one another or with either parent species, offspring of the next generation are feeble or sterile.

Example: Strains of cultivated rice have accumulated different mutant recessive alleles at two loci in the course of their divergence from a common ancestor. Hybrids between them are vigorous and fertile (l, left and right), but plants in the next generation that carry too many of these recessive alleles are small and sterile (l, center). Although these rice strains are not yet considered different species, they have begun to be separated by postzygotic barriers.

(a) (b) (c) (d) (e) (f) (g) (h) (i) (j) (k) (l)

490 UNIT FOUR Mechanisms of Evolution

CHAPTER TWENTY-FOUR The Origin of Species 491

Review what you've learned

At the end of the chapter, the **Chapter Review** focuses you again on the main points of the chapter, with a Summary of Key Concepts, practice questions, and references to other study tools.

Go to the **Study Area** at **www.masteringbio.com** for BioFlix 3-D animations and student tools plus MP3 Tutors, Discovery Channel Videos, Practice Tests, Activities, Investigations, an eBook, and more.

Chapter 9 Review

MEDIA Go to the Study Area at **www.masteringbio.com** for BioFlix 3-D Animations, MP3 Tutors, Videos, Practice Tests, an eBook, and more.

SUMMARY OF KEY CONCEPTS

CONCEPT 9.1

Catabolic pathways yield energy by oxidizing organic fuels (pp. 162–167)

- **Catabolic Pathways and Production of ATP** To keep working, a cell must regenerate the ATP it uses. The breakdown of glucose and other organic fuels is exergonic. Starting with glucose or another organic molecule and using O_2, aerobic respiration yields H_2O, CO_2, and energy in the form of ATP and heat. Cellular respiration includes both aerobic and anaerobic respiration; the latter uses another electron acceptor instead of O_2, but also yields ATP.
- **Redox Reactions: Oxidation and Reduction** The cell taps the energy stored in food molecules through redox reactions, in which one substance partially or totally shifts electrons to another. The substance receiving electrons is reduced; the substance losing electrons is oxidized. During cellular respiration, glucose ($C_6H_{12}O_6$) is oxidized to CO_2, and O_2 is reduced to H_2O. Electrons lose potential energy during their transfer from organic compounds to oxygen. Electrons from organic compounds are usually passed first to NAD^+, reducing it to NADH. NADH passes the electrons to an electron transport chain, which conducts them to O_2 in energy-releasing steps. The energy released is used to make ATP.
- **The Stages of Cellular Respiration: *A Preview*** Glycolysis and the citric acid cycle supply electrons (via NADH or $FADH_2$) to the electron transport chain, which drives oxidative phosphorylation. Oxidative phosphorylation generates ATP.

MEDIA
BioFlix 3-D Animation Cellular Respiration
Activity Build a Chemical Cycling System
Activity Overview of Cellular Respiration

CONCEPT 9.2

Glycolysis harvests chemical energy by oxidizing glucose to pyruvate (pp. 167–169)

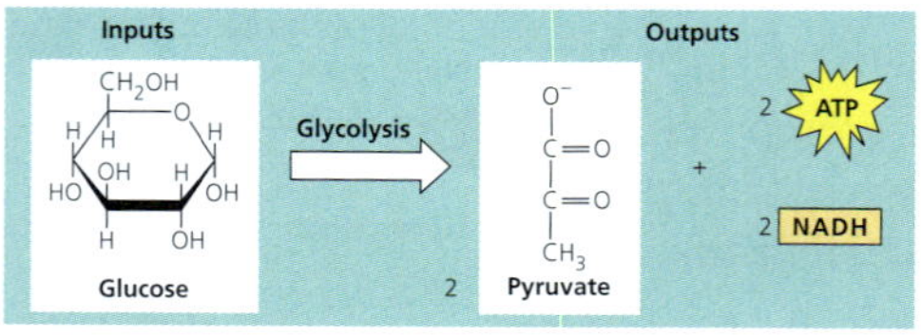

MEDIA
MP3 Tutor Cellular Respiration Part 1—Glycolysis
Activity Glycolysis

CONCEPT 9.3

The citric acid cycle completes the energy-yielding oxidation of organic molecules (pp. 170–172)

- In eukaryotic cells, the import of pyruvate into the mitochondrion and its conversion to acetyl CoA links glycolysis to the citric acid cycle. (In prokaryotic cells, the citric acid cycle occurs in the cytosol.)

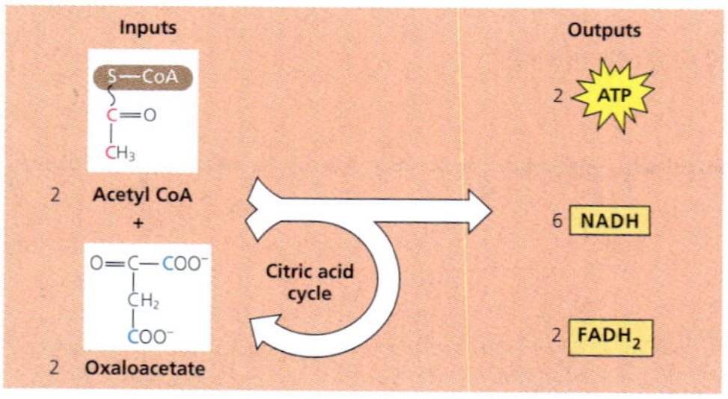

MEDIA
Activity The Citric Acid Cycle

Summary Figures present key information in a visual way, serving as "one-stop shopping" study tools.

MP3 Tutors explain some of the most difficult biological concepts in simple terms. A tutor that goes anywhere you go, anytime.

BioFlix animations and student tools review the most difficult biology topics with 3-D, movie-quality animations, labeled slide shows, carefully constructed student tutorials, study sheets, and quizzes that support all types of learners.

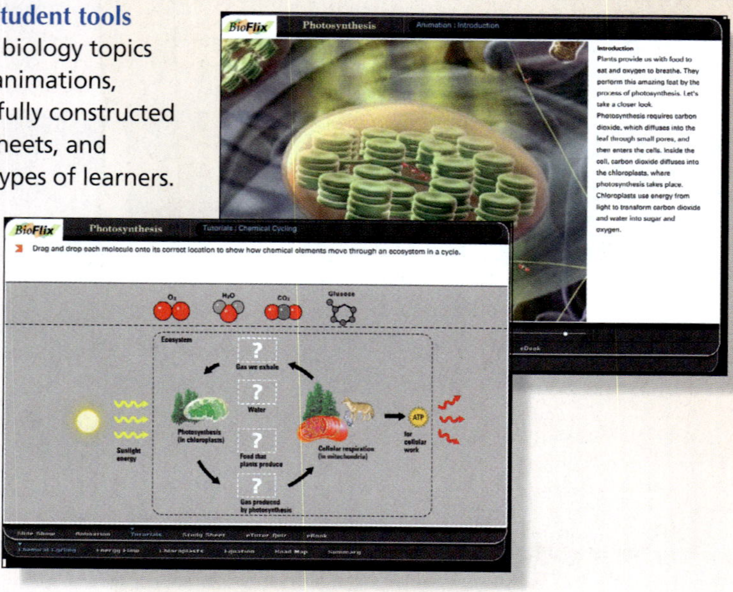

TESTING YOUR KNOWLEDGE

SELF-QUIZ

1. What is the reducing agent in the following reaction?

   $$\text{Pyruvate} + \text{NADH} + \text{H}^+ \rightarrow \text{Lactate} + \text{NAD}^+$$

   a. oxygen
   b. NADH
   c. NAD^+
   d. lactate
   e. pyruvate

2. The *immediate* energy source that drives ATP synthesis by ATP synthase during oxidative phosphorylation is
   a. the oxidation of glucose and other organic compounds.
   b. the flow of electrons down the electron transport chain.
   c. the affinity of oxygen for electrons.
   d. the H^+ concentration gradient across the inner mitochondrial membrane.
   e. the transfer of phosphate to ADP.

3. Which metabolic pathway is common to both fermentation and cellular respiration of a glucose molecule?
   a. the citric acid cycle
   b. the electron transport chain
   c. glycolysis
   d. synthesis of acetyl CoA from pyruvate
   e. reduction of pyruvate to lactate

4. In mitochondria, exergonic redox reactions
   a. are the source of energy driving prokaryotic ATP synthesis.
   b. are directly coupled to substrate-level phosphorylation.
   c. provide the energy that establishes the proton gradient.
   d. reduce carbon atoms to carbon dioxide.
   e. are coupled via phosphorylated intermediates to endergonic processes.

To start preparing for a test, take the **end-of-chapter Self-Quiz**.

"Draw It" exercises ask you to put pencil to paper and draw a structure, annotate a figure, or graph experimental data. Check your work by turning to the suggested answers in Appendix A.

11. DRAW IT Trace the uptake of water and minerals from root hairs to the vessels in a root, following a symplastic route and an apoplastic route. Label the routes.

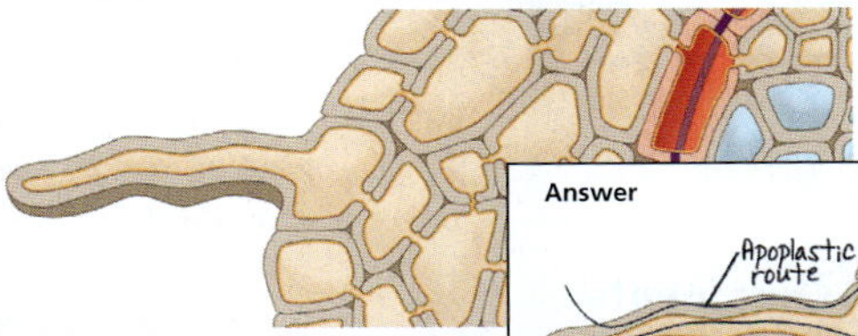

Answer

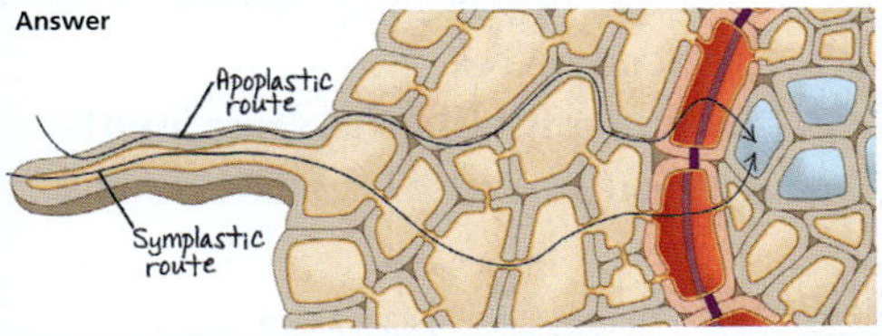

10. DRAW IT The graph here shows the pH difference across the inner mitochondrial membrane over time in an actively respiring cell. At the time indicated by the vertical arrow, a metabolic poison is added that specifically and completely inhibits all function of mitochondrial ATP synthase. Draw what you would expect to see for the rest of the graphed line.

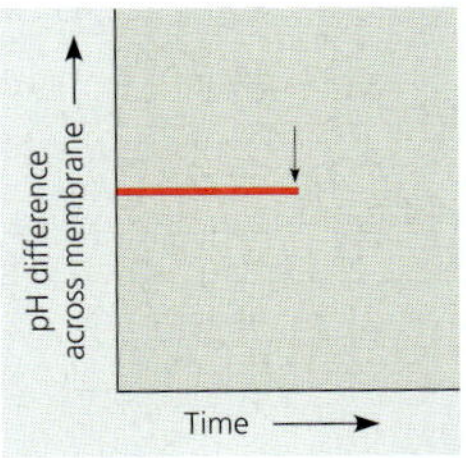

For Self-Quiz answers, see Appendix A.

MEDIA Visit the Study Area at **www.masteringbio.com** for a Practice Test.

For more **practice test questions** visit the Study Area at **www.masteringbio.com**

EVOLUTION CONNECTION

11. ATP synthases are found in the prokaryotic plasma membrane and in mitochondria and chloroplasts. What does this suggest about the evolutionary relationship of these eukaryotic organelles to prokaryotes? How might the amino acid sequences of the ATP synthases from the different sources support or refute your hypothesis?

SCIENTIFIC INQUIRY

12. In the 1940s, some physicians prescribed low doses of a drug called dinitrophenol (DNP) to help patients lose weight. This unsafe method was abandoned after a few patients died. DNP uncouples the chemiosmotic machinery by making the lipid bilayer of the inner mitochondrial membrane leaky to H^+. Explain how this causes weight loss.

SCIENCE, TECHNOLOGY, AND SOCIETY

13. Nearly all human societies use fermentation to produce alcoholic drinks such as beer and wine. The practice dates back to the earliest days of agriculture. How do you suppose this use of fermentation was first discovered? Why did wine prove to be a more useful beverage, especially to a preindustrial culture, than the grape juice from which it was made?

Biological Inquiry: A Workbook of Investigative Cases Explore fermentation further in the case "Bean Brew."

The **essay questions** in the Chapter Review give you practice writing about biological topics and making connections between the content of different chapters, as you may be asked to do in class discussions or on exams.

Supplements

Supplements for the Student

MasteringBIOLOGY™

www.masteringbio.com

NEW! **MasteringBiology™** offers:

1. The **Study Area**
2. **MasteringBiology™ Assignments**
3. **myeBook**, a customizable eBook

Study Area

Media assets in the Study Area include BioFlix™, MP3 Tutors, Discovery Videos, Cell Biology Videos, Activities, Investigations, GraphIt!, Quizzes, Practice Tests, Cumulative Test, Word Roots, Key Terms, Flashcards, Audio Glossary, Art, myeBook, and the Campbell/Reece Interviews.

In the **Study Area,** a convenient **Chapter Guide** correlates the student media to the Key Concepts of each textbook chapter.

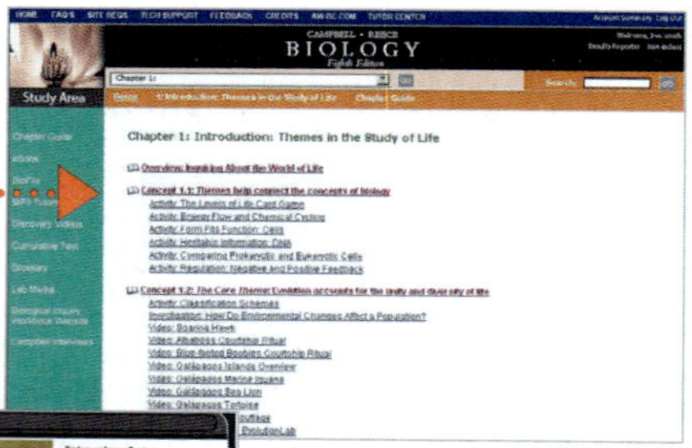

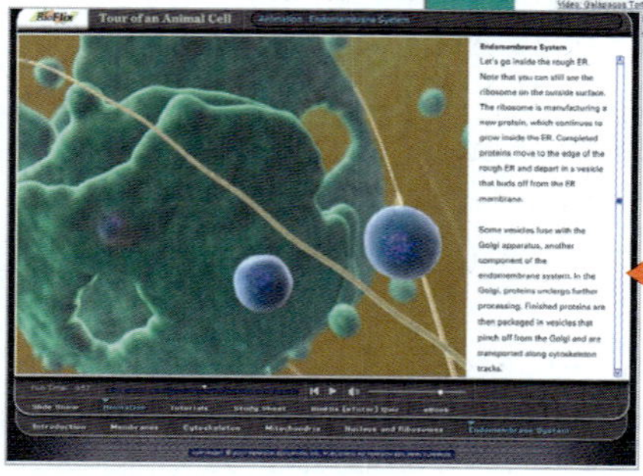

NEW! ***BioFlix*™**

The student versions of **BioFlix™** offer engaging 3-D animations plus labeled slide shows, basic tutorials, study sheets, and quizzes.

Discovery Channel™ Videos bring biology to life.

Discovery CHANNEL

- **MP3 Tutors** can be played from the website or downloaded onto personal MP3 players for studying anytime.
- **NEW! Audio pronunciations** are included for every term in the Audio Glossary, Key Terms, and Flashcards.

MasteringBiology™ Assignments

MasteringBiology™ offers assignable in-depth Tutorials, BioFlix™ Tutorials, Activities, GraphIt!, Chapter Quizzes with ten questions per chapter, and thousands of additional Multiple Choice Test Bank Questions.

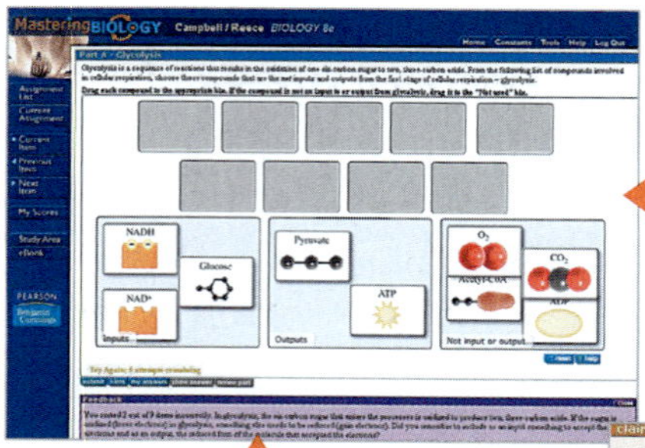

The MasteringBiology™ **Tutorials** guide students through difficult concepts with a series of interactive questions that build on each other.

If a student answers incorrectly, the student is given specific **feedback** that addresses the student's misconceptions.

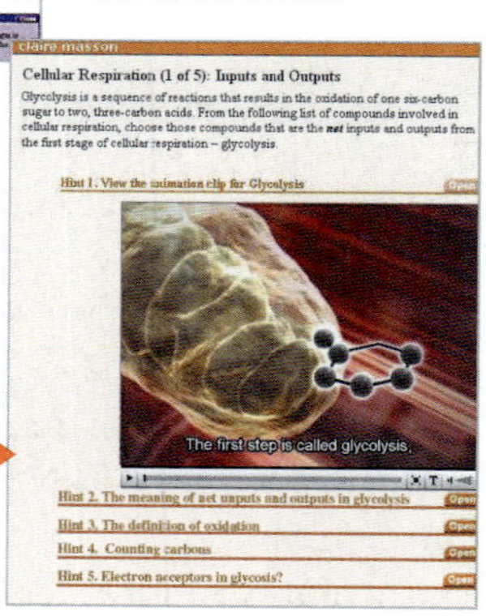

Students can view **hints** for background information, including animations, and additional assistance.

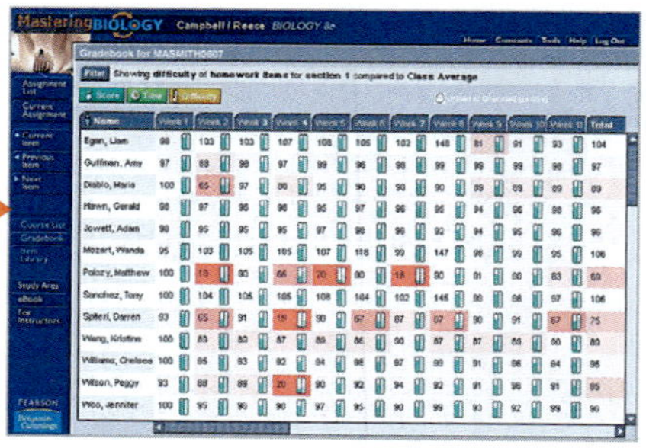

A powerful **Gradebook** automatically records student scores. Shades of red instantly reveal students in trouble or areas of difficulty for the class. Instructors can assign questions to encourage students to read the chapter before class and then adjust their lectures to address student misconceptions. Grades can be exported to another course management system or Excel™, or grades can be imported into the MasteringBiology™ Gradebook.

For more information about media for students, see the card inserted at the beginning of this book and the **Access Kit for MasteringBiology™**, which is included with every new copy of *BIOLOGY*, Eighth Edition.

NEW! **Inquiry in Action: Interpreting Scientific Papers**
(978-0-321-53659-4/0-321-53659-2)
Ruth Buskirk, University of Texas, Austin, and Christopher M. Gillen, Kenyan College
Selected Inquiry Figures in the Eighth Edition direct students to read and analyze the complete original research paper. In this new supplement, those articles are reprinted and accompanied by questions that help students analyze the article.

Student Study Guide for *Biology,* Eighth Edition
(978-0-321-50156-1/0-321-50156-X)
Martha R. Taylor, Cornell University
This popular study guide helps students extract key ideas from the textbook and organize their knowledge of biology. Exercises include concept maps for each chapter, chapter summaries, word roots, chapter tests, and a variety of interactive questions in various formats.

Practicing Biology: A Student Workbook, Third Edition
(978-0-321-52293-1/0-321-52293-1)
Jean Heitz and Cynthia Giffen, University of Wisconsin, Madison
This workbook offers a variety of activities to suit different learning styles. Activities such as modeling and mapping allow students to visualize and understand biological processes. New activities focus on basic skills, such as reading and developing graphs.

Biological Inquiry: A Workbook of Investigative Cases, Second Edition (978-0-321-51320-5/0-321-51320-7)
Margaret Waterman, Southeast Missouri State University, and Ethel Stanley, Beloit College and BioQUEST Curriculum Consortium
This workbook offers ten investigative cases, including new cases on avian influenza and hedgehog developmental pathways. Students pose questions, analyze data, think critically, examine the relationship between evidence and conclusions, construct hypotheses, investigate options, graph data, interpret results, and communicate scientific arguments. A student website is at **www.masteringbio.com**.

Study Card for *Biology,* Eighth Edition
(978-0-321-49436-8/0-321-49436-9)
This quick reference card provides an overview of the entire field of biology and helps students quickly review before a test.

Kaplan MCAT®/GRE® Biology Test Preparation Guide for *Biology,* Eighth Edition
(978-0-321-53463-7/0-321-53463-8)
This exclusive supplement includes sample questions from the Kaplan test preparation guides with page references to *Biology*, Eighth Edition.

Spanish Glossary (978-0-321-49434-4/0-321-49434-2)
Laura P. Zanello, University of California, Riverside

NEW! **Into the Jungle: Great Adventures in the Search for Evolution**
(978-0-321-55671-4/0-321-55671-2)
Sean B. Carroll, University of Wisconsin, Madison
This book of nine short tales vividly depicts key discoveries in evolutionary biology and the excitement of the scientific process.

NEW! **Get Ready for Biology**
(978-0-321-50057-1/0-321-50057-1)
This engaging workbook helps students brush up on important math and study skills and get up to speed on biological terminology and the basics of chemistry and cell biology.

A Short Guide to Writing About Biology, Sixth Edition
(978-0-321-51716-6/0-321-51716-4)
Jan A. Pechenik, Tufts University
This best-selling writing guide teaches students to think as biologists and to express ideas clearly and concisely through their writing.

An Introduction to Chemistry for Biology Students, Ninth Edition (978-0-8053-9571-6/0-8053-9571-7)
George I. Sackheim, University of Illinois, Chicago
This text/workbook helps students review and master all the basic facts, concepts, and terminology of chemistry that they need for their life science course.

NEW! **Pearson Tutor Services**
www.masteringbio.com
Access to MasteringBiology™ includes complimentary access to highly interactive one-on-one biology coaching by qualified instructors seven nights per week during peak study hours. Students can "drop-in" for live online help, submit questions to an e-structor anytime, or pre-schedule a tutoring session with an e-structor. (For college students only.)

Special Topics in Biology Series (booklets)
- Alzheimer's Disease (978-0-1318-3834-5/0-1318-3834-2)
- Biological Terrorism (978-0-8053-4868-2/0-8053-4868-9)
- Biology of Cancer (978-0-8053-4867-5/0-8053-4867-0)
- Emerging Infectious Diseases (978-0-8053-3955-0/0-8053-3955-8)
- Gene Therapy (978-0-8053-3819-5/0-8053-3819-5)
- Genetic Testimony: A Guide to Forensic DNA Profiling (978-0-1314-2338-1/0-1314-2338-X)
- HIV and AIDS (978-0-8053-3956-7/0-8053-3956-6)
- Mad Cows and Cannibals: A Guide to the Transmissible Spongiform Encephalopathies (978-0-1314-2339-8/0-1314-2339-8)
- Stem Cells and Cloning (978-0-8053-4864-4/0-8053-4864-6)
- Understanding the Human Genome Project, Second Edition (978-0-8053-4877-4/0-8053-4877-8)

Supplements for the Instructor

Instructor Resource CD/DVD-ROM Set

(978-0-321-52292-4/0-321-52292-3)

The instructor media for Campbell/Reece *Biology*, Eighth Edition, is combined into one chapter-by-chapter resource along with a **Quick Reference Guide**. **DVDs** provide convenient one-stop access to all the visual media for each chapter. Assets are now organized by chapter folders, making it easy to access files. The **Test Bank CD-ROM** includes test bank questions in Word and TestGen. Assets on the DVDs include:

NEW! *BioFlix*™

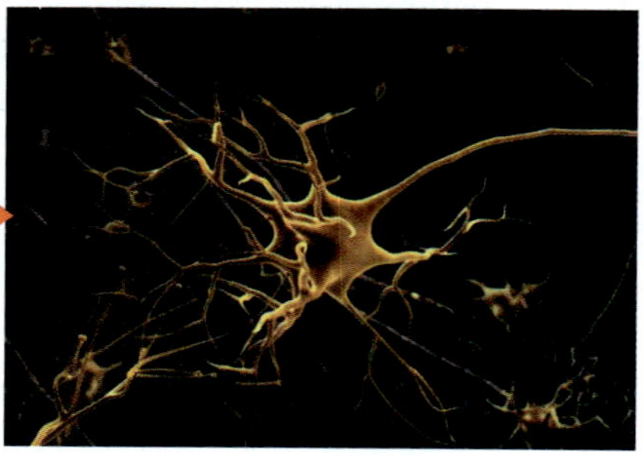

BioFlix™ 3-D Animations invigorate lectures with movie-quality 3-D animations. Topics include: Tour of an Animal Cell, Tour of a Plant Cell, Membrane Transport, Cellular Respiration, Photosynthesis, Mitosis, Meiosis, DNA Replication, Protein Synthesis, Water Transport in Plants, How Neurons Work, and Muscle Contraction. An additional 123 Animations based on student Activities are also included.

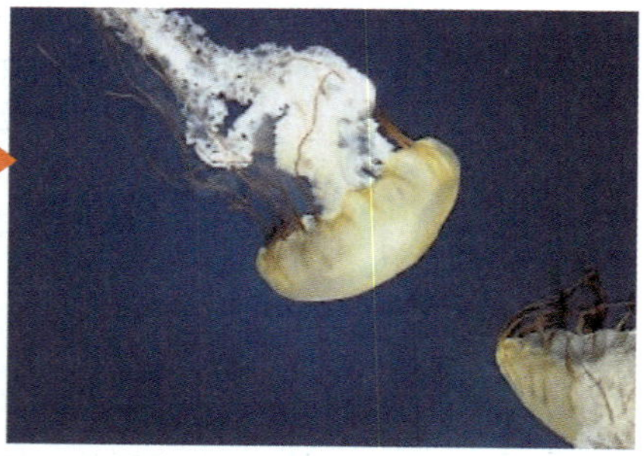

NEW! 29 Discovery Channel™ Videos bring biology to life and show students the process of science. Plus, 65 new **Cell Biology Videos** have been added to the video collection, for a total of 182 videos.

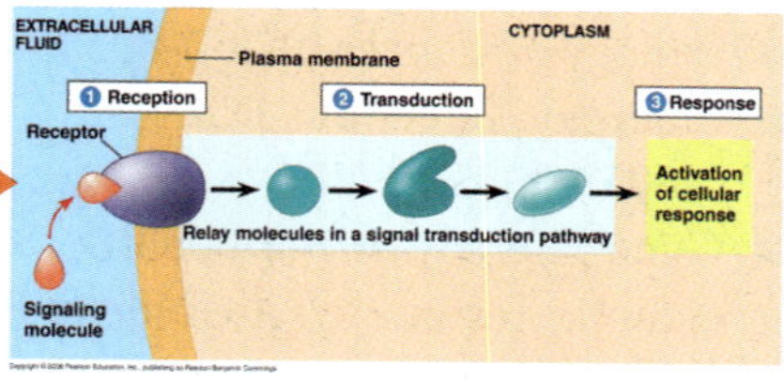

PowerPoint® Label Edit Images include all the art, photos, and tables from the book embedded in PowerPoint® plus **Step Edit Art** broken down into steps. These PowerPoint® files are on the DVDs plus the **Quick Start CD-ROM** for easy access. **PowerPoint® Lecture Presentations** include all of the above plus lecture outlines and links to animations and selected videos. In all PowerPoint files, the text and labels can be edited directly in PowerPoint® and have been enlarged for optimal viewing in large lecture halls. Multiple versions of figures are provided to choose from.

JPEG Images include all the art, photos, and tables from the book with and without labels, selected art layered for step-by-step presentation, and hundreds of extra photos in JPEG format, for a total of more than 1600 photos.

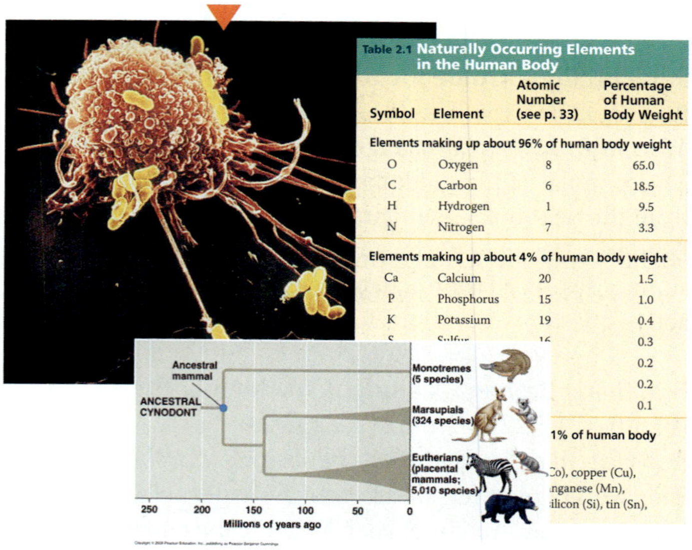

MasteringBIOLOGY™

www.masteringbio.com

NEW! The **MasteringBiology™ For Instructors** area includes all the media from the Instructor Resource DVDs:

- PowerPoint® Label Edit Images
- PowerPoint® Lecture Presentations
- Extra Photos (jpegs) plus captions
- Labeled Images (jpegs)
- Unlabeled Images (jpegs)
- BioFlix™ Animations
- Animations
- Discovery Channel™ Videos plus scripts
- Videos plus descriptions

Additional Online Resources:

- PowerPoint® Active Lecture Questions
- Lecture Outlines
- Student Misconceptions with Pre-Tests and Post-Tests
- Quick Reference Guide
- *Practicing Biology* Instructor Guide
- *Inquiry in Action* Instructor Guide
- Answers to selected textbook essay questions
- Lab Media for Instructors

Many instructor assets are also available at the Instructor Resource Center for *Biology*, Eighth Edition, at www.pearsonhighered.com.

Transparency Acetates
(978-0-321-52328-0/0-321-52328-8)
This transparency package provides all the illustrations and tables from the text, many of which incorporate photos. In addition, key figures are broken down into steps.

Electronic Test Bank

Printed Test Bank (978-0-321-49431-3/0-321-49431-8)
More than 4,500 test questions are available in print, TestGen®, Microsoft® Word, and from within CourseCompass, Blackboard, and WebCT Course Management Systems. Test Bank questions can also be assigned through MasteringBiology™. The questions have been refined through class testing, and over 30% of the questions are new for the Eighth Edition, including three new question types that encourage critical thinking: interpreting art questions, interpreting graphs and data questions focused on quantitative skills, and scenario-based questions.

Instructor Guide for Biological Inquiry: A Workbook of Investigative Cases, Second Edition
(978-0-321-49435-1/0-321-49435-0)
Margaret Waterman, Southeast Missouri State University, and Ethel Stanley, Beloit College and BioQUEST Curriculum Consortium
This Instructor Guide provides insightful coaching for instructors on how to teach using a case-based problem-solving approach, as well as suggested answers.

Course Management Systems
Test Bank questions, quizzes, and selected content from the Study Area of MasteringBiology™ are available in these popular course management systems:
- **CourseCompass™** (www.aw-bc.com/coursecompass)
- **Blackboard** (www.aw-bc.com/blackboard)
- **WebCT** (www.aw-bc.com/webct)

Supplements for the Lab

NEW! Lab Media at www.masteringbio.com
A new section in the Study Area of MasteringBiology™ brings together all the media assets that can be used to teach scientific inquiry, including Investigating Biology Lab Data Tables in Excel®, Biology Labs On-Line, Investigations, GraphIt!, and LabBench. In the For Instructors area, the Lab Media section includes Investigating Biology Lab Preparation Guide, Answers to Investigations, GraphIt! Instructor Versions, Answers to the LabBench Quizzes, and the *Bio-Explorations* Instructor Guide.

Investigating Biology, Sixth Edition
(978-0-321-53660-0/0-321-53660-6)

Annotated Instructor Edition for Investigating Biology
(978-0-321-54194-9/0-321-54194-4)

Preparation Guide for Investigating Biology
(978-0-321-54166-6/0-321-54166-9)
Judith Giles Morgan, Emory University, and M. Eloise Brown Carter, Oxford College of Emory University
This best-selling laboratory manual encourages students to participate in the process of science and develop creative and critical reasoning skills by posing hypotheses, making predictions, conducting open-ended experiments, collecting data, and applying results to new problems. The Sixth Edition includes a new Bioinformatics Lab and features references to online resources available at www.masteringbio.com, including Lab Data Tables in Excel® for recording data. The Annotated Instructor Edition provides teaching information and marginal notes. A Preparation Guide is also available for instructors in print and online.

New Designs for Bio-Explorations
(978-0-8053-7229-8/0-8053-7229-6)

Instructor Guide for Bio-Explorations
(978-0-8053-7228-1/0-8053-7228-8)
Janet Lanza, University of Arkansas at Little Rock
Eight inquiry-based laboratory exercises offer students creative control over the projects they undertake. Each lab exercise provides students with background information and materials that can be used in the lab projects. The Instructor Guide is provided in the For Instructors area of www.masteringbio.com under Lab Media.

Symbiosis: The Benjamin Cummings Custom Laboratory Program for Biological Sciences
www.pearsoncustom.com/database/symbiosis/bc.html
With Symbiosis, instructors can build a customized lab manual that includes selections from the Benjamin Cummings database along with their own original material.

Biology Labs On-Line
www.biologylabsonline.com
Twelve on-line labs enable students to expand their scientific horizons beyond the traditional wet lab setting and perform potentially dangerous, lengthy, or expensive experiments in an electronic environment. Each experiment can be repeated as often as necessary, employing a unique set of variables each time. The labs are available for purchase individually or in a 12-pack with the printed Student Lab Manual (978-0-8053-7017-1/0-8053-7017-X). An Instructor Lab Manual is also available (978-0-8053-7018-8/0-8053-7018-8).

Reviewers

Eighth Edition Reviewers

Dominique Adriaens, Ghent University
George R. Aliaga, Tarrant County College
J. David Archibald, San Diego State University
David M. Armstrong, University of Colorado-Boulder
Angela S. Aspbury, Texas State University
Ellen Baker, Santa Monica College
Rebecca A. Bartow, Western Kentucky University
Tim Beagley, Salt Lake Community College
Kenneth Birnbaum, New York University
Michael W. Black, California Polytechnic State University, San Luis Obispo
Edward Blumenthal, Marquette University
Jason E. Bond, East Carolina University
Cornelius Bondzi, Hampton University
Oliver Bossdorf, State University of New York, Stony Book
Edward Braun, Iowa State University
Chad Brommer, Emory University
Judith L. Bronstein, University of Arizona
Robb T. Brumfield, Louisiana State University
Richard C. Brusca, University of Arizona, Arizona-Sonora Desert Museum
Jorge Busciglio, University of California, Irvine
Guy A. Caldwell, University of Alabama
Jane Caldwell, West Virginia University
Kim A. Caldwell, University of Alabama
R. Andrew Cameron, California Institute of Technology
W. Zacheus Cande, University of California, Berkeley
Frank R. Cantelmo, St. John's University
Jeffrey Carmichael, University of North Dakota
Laura L. Carruth, Georgia State University
J. Aaron Cassill, University of Texas at San Antonio
P. Bryant Chase, Florida State University
Jung H. Choi, Georgia Institute of Technology
Geoffrey Church, Fairfield University
Patricia J. Clark, Indiana University-Purdue University, Indianapolis
Janice J. Clymer, San Diego Mesa College
Jan Colpaert, Hasselt University
Jay Comeaux, McNeese State University
Gregory Copenhaver, University of North Carolina, Chapel Hill
Karen Curto, University of Pittsburgh
Marymegan Daly, The Ohio State University
Cynthia Dassler, The Ohio State University
Michael A. Davis, Central Connecticut State University
Maria E. de Bellard, California State University, Northridge
Patricia A. DeLeon, University of Delaware
Charles F. Delwiche, University of Maryland
William L. Dentler, University of Kansas
Jean DeSaix, University of North Carolina
Michael Dini, Texas Tech University
Douglas J. Eernisse, California State University, Fullerton
Brad Elder, Doane College
Michelle Elekonich, University of Nevada, Las Vegas
Mary Ellard-Ivey, Pacific Lutheran University
Johnny El-Rady, University of South Florida
John A. Endler, University of California, Santa Barbara
Frederick B. Essig, University of South Florida
Olukemi Fadayomi, Ferris State University
Ellen H. Fanning, Vanderbilt University
Lewis Feldman, University of California, Berkeley
Rebecca Ferrell, Metropolitan State College of Denver
Jonathan S. Fisher, St. Louis University
Kirk Fitzhugh, Natural History Museum of Los Angeles County
Norma Fowler, University of Texas, Austin
Robert Gilbert Fowler, San Jose State University
Jed Fuhrman, University of Southern California
Zofia E. Gagnon, Marist College
Michael Gaines, University of Miami
Stephen Gammie, University of Wisconsin, Madison
Andrea Gargas, University of Wisconsin, Madison
Lauren Garner, California Polytechnic State University, San Luis Obispo
Simon Gilroy, Pennsylvania State University
Alan D. Gishlick, Gustavus Adolphus College
Jessica Gleffe, University of California, Irvine
Tricia Glidewell, Marist School
Elizabeth Godrick, Boston University
Ken Halanych, Auburn University
E. William Hamilton, Washington and Lee University
William F. Hanna, Massasoit Community College
Laszlo Hanzely, Northern Illinois University
Lisa Harper, University of California, Berkeley
Bernard A. Hauser, University of Florida
Evan B. Hazard, Bemidji State University (Emeritus)
S. Blair Hedges, Pennsylvania State University
Brian Hedlund, University of Nevada, Las Vegas
Jean Heitz, University of Wisconsin, Madison
Susan Hengeveld, Indiana University
Albert Herrera, University of Southern California
Kenneth Hillers, California Polytechnic State University, San Luis Obispo
A. Scott Holaday, Texas Tech University
N. Michele Holbrook, Harvard University
Alan R. Holyoak, Brigham Young University, Idaho
Sandra M. Horikami, Daytona Beach Community College
Becky Houck, University of Portland
Daniel J. Howard, New Mexico State University
Cristin Hulslander, University of Oregon
Linda L. Hyde, Gordon College
Jeffrey Ihara, Mira Costa College
Lee Johnson, The Ohio State University
Chad Jordan, North Carolina State University
Walter S. Judd, University of Florida
Thomas W. Jurik, Iowa State University
Caroline M. Kane, University of California, Berkeley
Jennifer Katcher, Pima Community College
Laura A. Katz, Smith College
Maureen Kearney, Field Museum of Natural History
Patrick Keeling, University of British Columbia
Elizabeth A. Kellogg, University of Missouri-St. Louis
Chris Kennedy, Simon Fraser University
Rebecca T. Kimball, University of Florida
Jennifer Knight, University of Colorado
Margareta Krabbe, Uppsala University
Anselm Kratochwil, Universität Osnabrück
Deborah M. Kristan, California State University at San Marcos
William Kroll, Loyola University, Chicago
Justin P. Kumar, Indiana University
Marc-André Lachance, University of Western Ontario
Mohamed Lakrim, Kingsborough Community College
John Latto, University of California, Santa Barbara
Daewoo Lee, Ohio University

Michael R. Leonardo, Coe College
John J. Lepri, University of North Carolina at Greensboro
Graeme Lindbeck, Valencia Community College
Diana Lipscomb, George Washington University
Christopher Little, The University of Texas-Pan American
Kevin D. Livingstone, Trinity University
Andrea Lloyd, Middlebury College
Christopher A. Loretz, State University of New York at Buffalo
Douglas B. Luckie, Michigan State University
Christine R. Maher, University of Southern Maine
Keith Malmos, Valencia Community College – East Campus
Cindy Malone, California State University, Northridge
Carol Mapes, Kutztown University of Pennsylvania
Kathleen A. Marrs, Indiana University-Purdue University, Indianapolis
Diane L. Marshall, University of New Mexico
Andrew McCubbin, Washington State University
Lisa Marie Meffert, Rice University
Scott Meissner, Cornell University
John Merrill, Michigan State University
Michael J. Misamore, Texas Christian University
Alan Molumby, University of Illinois, Chicago
Joseph P. Montoya, Georgia Institute of Technology
Janice Moore, Colorado State University
Jeanette Mowery, Madison Area Technical College
Tom Neils, Grand Rapids Community College
Ray Neubauer, University of Texas, Austin
James Newcomb, New England College
Anders Nilsson, University of Umeå
Mohamed A. F. Noor, Duke University
Shawn Nordell, St. Louis University
Richard S. Norman, University of Michigan, Dearborn (Emeritus)
Gretchen North, Occidental College
Mark P. Oemke, Alma College
Nathan O. Okia, Auburn University, Montgomery
John Oross, University of California, Riverside
Charissa Osborne, Butler University
Thomas G. Owens, Cornell University
Kevin Padian, University of California, Berkeley
Anthony T. Paganini, Michigan State University
Michael A. Palladino, Monmouth University
Imara Y. Perera, North Carolina State University
David S. Pilliod, California Polytechnic State University, San Luis Obispo
J. Chris Pires, University of Missouri-Columbia
Angela R. Porta, Kean University
Daniel Potter, University of California, Davis
Mary V. Price, University of California, Riverside
Mitch Price, Pennsylvania State University
Peter Quinby, University of Pittsburgh
Robert H. Reaves, Glendale Community College
Erin Rempala, San Diego Mesa College
Eric Ribbens, Western Illinois University
Christina Richards, New York University
Loren Rieseberg, University of British Columbia
Bruce B. Riley, Texas A&M University
Laurel Roberts, University of Pittsburgh
Mike Rosenzweig, Virginia Polytechnic Institute and State University
Tyson Sacco, Cornell University
Rowan F. Sage, University of Toronto
Tammy Lynn Sage, University of Toronto
Thomas R. Sawicki, Spartanburg Community College
Inder Saxena, University of Texas, Austin
Maynard H. Schaus, Virginia Wesleyan College
Renate Scheibe, University of Osnabrück
Mark Schlissel, University of California, Berkeley
Christopher J. Schneider, Boston University
Thomas W. Schoener, University of California, Davis
Patricia M. Schulte, University of British Columbia
Karen S. Schumaker, University of Arizona
David J. Schwartz, Houston Community College
Robert W. Seagull, Hofstra University
Duane Sears, University of California, Santa Barbara
Joan Sharp, Simon Fraser University
Timothy E. Shannon, Francis Marion University
Richard M. Showman, University of South Carolina
Rebecca Simmons, University of North Dakota
Anne Simon, University of Maryland, College Park
Robert Simons, University of California, Los Angeles
Julio G. Soto, San Jose State University
John Stachowicz, University of California, Davis
Gail A. Stewart, Camden County College
Michael A. Sypes, Pennsylvania State University
Emily Taylor, California Polytechnic State University, San Luis Obispo
John W. Taylor, University of California, Berkeley
William Thwaites, Tillamook Bay Community College
Eric Toolson, University of New Mexico
Paul Q. Trombley, Florida State University
Nancy J. Trun, Duquesne University
Claudia Uhde-Stone, California State University, East Bay
Saba Valadkhan, Case Western Reserve University School of Medicine
Steven D. Verhey, Central Washington University
Kathleen Verville, Washington College
Sara Via, University of Maryland
Leif Asbjørn Vøllestad, University of Oslo
Linda Walters, University of Central Florida
Nickolas M. Waser, University of California, Riverside
Andrea Weeks, George Mason University
Richard Wetts, University of California, Irvine
Susan Whittemore, Keene State College
Ernest H. Williams, Hamilton College
Kathy Williams, San Diego State University
Paul Wilson, California State University, Northridge
Peter Wimberger, University of Puget Sound
Robert Winning, Eastern Michigan University
E. William Wischusen, Louisiana State University
Vickie L. Wolfe, Marshall University
Denise Woodward, Pennsylvania State University
Sarah E. Wyatt, Ohio University
Ramin Yadegari, University of Arizona
Paul Yancey, Whitman College
Gina M. Zainelli, Loyola University, Chicago
Miriam Zolan, Indiana University

BioFlix Reviewers

Mitch Albers, Minneapolis Community and Technical College
Kirk Bartholomew, Sacred Heart University
Gretchen Bernard, Moraine Valley Community College
Peggy Brickman, University of Georgia
Uriel Buitrago-Suarez, Harper College
Nancy Butler, Kutztown University of Pennsylvania
Guy A. Caldwell, University of Alabama
Kim A. Caldwell, University of Alabama
Jose L. Egremy, Northwest Vista College
Kurt J. Elliott, Northwest Vista College
Gerald G. Farr, Texas State University
Lewis Feldman, University of California, Berkeley
Sandra Gibbons, Moraine Valley Community College
Douglas A. Hamilton, Hartwick College
W. Wyatt Hoback, University of Nebraska at Kearney
Elizabeth Hodgson, York College of Pennsylvania
Kelly Hogan, University of North Carolina at Chapel Hill
Mary Rose Lamb, University of Puget Sound
Cody Locke, University of Alabama
Marvin Brandon Lowery, Sam Houston State University
David Mirman, Mt. San Antonio College
James Newcomb, New England College

Thomas G. Owens, Cornell University
Deb Pires, University of California, Los Angeles
Mitch Price, Pennsylvania State University
David A. Rintoul, Kansas State University
Renee Rivas, University of Alabama
Laurel Roberts, University of Pittsburgh
Chris Romero, Front Range Community College
Juliet Spencer, University of San Francisco
Linda Brooke Stabler, University of Central Oklahoma
Beth Stall, El Centro College
Brian Stout, Northwest Vista College
Diane Sweeney, Crystal Springs Uplands School
Jamey Thompson, Hudson Valley Community College
Paul Q. Trombley, Florida State University
Robert S. Wallace, Iowa State University
Susan Whittemore, Keene State College
Miriam Zolan, Indiana University
Michelle Zurawski, Moraine Valley Community College

MasteringBiology Class-Testers and Reviewers

Peter B. Berget, Carnegie Mellon University
Michael W. Black, California Polytechnic State University, San Luis Obispo
Scott Bowling, Auburn University
Suzanne Butler, Miami Dade College
Alejandro Calderon-Urrea, California State University, Fresno
Kim A. Caldwell, University of Alabama
Jeffrey Carmichael, University of North Dakota
Jung H. Choi, Georgia Institute of Technology
Karen Curto, University of Pittsburgh
Lydia Daniels, University of Pittsburgh
Jill Feinstein, Richland Community College
Donald Glassman, Des Moines Area Community College
Joyce Gordon, University of British Columbia
David Grise, Texas A&M University, Corpus Christi
Douglas A. Hamilton, Hartwick College
Mark Hens, University of North Carolina at Greensboro
John C. Kay, Iolani School
Tracy Kickox, University of Illinois at Urbana-Champaign
Mary Rose Lamb, University of Puget Sound
Deb Maddalena, University of Vermont
C. Smoot Major, University of South Alabama
Nilo Marin, Broward Community College
John Merrill, Michigan State University
Melissa Michael, University of Illinois at Urbana-Champaign
Nancy Rice, Western Kentucky University
Chris Romero, Front Range Community College, Larimer
John Salerno, Kennesaw State University
Brian Stout, Northwest Vista College
Sukanya Subramanian, Collin County Community College
Elizabeth Willott, University of Arizona
Lauren Yaich, University of Pittsburgh at Bradford

Reviewers of Previous Editions

Kenneth Able (State University of New York, Albany), Thomas Adams (Michigan State University), Martin Adamson (University of British Columbia), Shylaja Akkaraju (Bronx Community College of CUNY), John Alcock (Arizona State University), Richard Almon (State University of New York, Buffalo), Bonnie Amos (Angelo State University), Katherine Anderson (University of California, Berkeley), Richard J. Andren (Montgomery County Community College), Estry Ang (University of Pittsburgh at Greensburg), Jeff Appling (Clemson University), J. David Archibald (San Diego State University), David Armstrong (University of Colorado at Boulder), Howard J. Arnott (University of Texas at Arlington), Mary Ashley (University of Illinois at Chicago), Robert Atherton (University of Wyoming), Karl Aufderheide (Texas A&M University), Leigh Auleb (San Francisco State University), P. Stephen Baenziger (University of Nebraska), Ellen Baker (Santa Monica College), Katherine Baker (Millersville University), William Barklow (Framingham State College), Susan Barman (Michigan State University), Steven Barnhart (Santa Rosa Junior College), Andrew Barton (University of Maine Farmington), Ron Basmajian (Merced College), David Bass (University of Central Oklahoma), Bonnie Baxter (Hobart & William Smith), Tim Beagley (Salt Lake Community College), Margaret E. Beard (College of the Holy Cross), Tom Beatty (University of British Columbia), Chris Beck (Emory University), Wayne Becker (University of Wisconsin, Madison), Patricia Bedinger (Colorado State University), Jane Beiswenger (University of Wyoming), Anne Bekoff (University of Colorado, Boulder), Marc Bekoff (University of Colorado, Boulder), Tania Beliz (College of San Mateo), Adrianne Bendich (Hoffman-La Roche, Inc.), Barbara Bentley (State University of New York, Stony Brook), Darwin Berg (University of California, San Diego), Werner Bergen (Michigan State University), Gerald Bergstrom (University of Wisconsin, Milwaukee), Anna W. Berkovitz (Purdue University), Dorothy Berner (Temple University), Annalisa Berta (San Diego State University), Paulette Bierzychudek (Pomona College), Charles Biggers (Memphis State University), Robert Blanchard (University of New Hampshire), Andrew R. Blaustein (Oregon State University), Judy Bluemer (Morton College), Robert Blystone (Trinity University), Robert Boley (University of Texas, Arlington), Eric Bonde (University of Colorado, Boulder), Richard Boohar (University of Nebraska, Omaha), Carey L. Booth (Reed College), Allan Bornstein (Southeast Missouri State University), James L. Botsford (New Mexico State University), Lisa Boucher (University of Nebraska-Omaha), J. Michael Bowes (Humboldt State University), Richard Bowker (Alma College), Robert Bowker (Glendale Community College - Arizona), Barbara Bowman (Mills College), Barry Bowman (University of California, Santa Cruz), Deric Bownds (University of Wisconsin, Madison), Robert Boyd (Auburn University), Sunny Boyd (University of Notre Dame), Jerry Brand (University of Texas, Austin), Theodore A. Bremner (Howard University), James Brenneman (University of Evansville), Charles H. Brenner (Berkeley, California), Lawrence Brewer (University of Kentucky), Donald P. Briskin (University of Illinois, Urbana), Paul Broady (University of Canterbury), Danny Brower (University of Arizona), Carole Browne (Wake Forest University), Mark Browning (Purdue University), David Bruck (San Jose State University), Herbert Bruneau (Oklahoma State University), Gary Brusca (Humboldt State University), Richard C. Brusca (University of Arizona, Arizona-Sonora Desert Museum), Alan H. Brush (University of Connecticut, Storrs), Howard Buhse (University of Illinois at Chicago), Arthur Buikema (Virginia Tech), Al Burchsted (College of Staten Island), Meg Burke (University of North Dakota), Edwin Burling (De Anza College), William Busa (Johns Hopkins University), John Bushnell (University of Colorado), Linda Butler (University of Texas, Austin), David Byres (Florida Community College, Jacksonville), Alison Campbell (University of Waikato), Iain Campbell (University of Pittsburgh), Robert E. Cannon (University of North Carolina at Greensboro), Deborah Canington (University of California, Davis), Frank Cantelmo (St John's University), John Capeheart (University of Houston-Downtown), Gregory Capelli (College of William and Mary), Richard Cardullo (University of California, Riverside), Nina Caris (Texas A&M University), Robert Carroll (East Carolina University), David Champlin (University of Southern Maine), Bruce Chase (University of Nebraska, Omaha), Doug Cheeseman (De Anza College), Shepley Chen (University of Illinois, Chicago), Giovina Chinchar (Tougaloo College), Joseph P.

Chinnici (Virginia Commonwealth University), Henry Claman (University of Colorado Health Science Center), Anne Clark (Binghamton University), Greg Clark (University of Texas), Ross C. Clark (Eastern Kentucky University), Lynwood Clemens (Michigan State University), William P. Coffman (University of Pittsburgh), Austin Randy Cohen (California State University, Northridge), J. John Cohen (University of Colorado Health Science Center), Jim Colbert (Iowa State University), Robert Colvin (Ohio University), David Cone (Saint Mary's University), Elizabeth Connor (University of Massachusetts), Joanne Conover (University of Connecticut), John Corliss (University of Maryland), James T. Costa (Western Carolina University), Stuart J. Coward (University of Georgia), Charles Creutz (University of Toledo), Bruce Criley (Illinois Wesleyan University), Norma Criley (Illinois Wesleyan University), Joe W. Crim (University of Georgia), Greg Crowther (University of Washington), Karen Curto (University of Pittsburgh), Anne Cusic (University of Alabama at Birmingham), Richard Cyr (Pennsylvania State University), W. Marshall Darley (University of Georgia), Marianne Dauwalder (University of Texas, Austin), Larry Davenport (Samford University), Bonnie J. Davis (San Francisco State University), Jerry Davis (University of Wisconsin, La Crosse), Thomas Davis (University of New Hampshire), John Dearn (University of Canberra), Teresa DeGolier (Bethel College), James Dekloe (University of California, Santa Cruz), Veronique Delesalle (Gettysburg College), T. Delevoryas (University of Texas, Austin), Roger Del Moral (University of Washington), Diane C. DeNagel (Northwestern University), Daniel Dervartanian (University of Georgia), Jean DeSaix (University of North Carolina at Chapel Hill), Michael Dini (Texas Tech University), Biao Ding (Ohio State University), Andrew Dobson (Princeton University), Stanley Dodson (University of Wisconsin-Madison), Mark Drapeau (University of California, Irvine), John Drees (Temple University School of Medicine), Charles Drewes (Iowa State University), Marvin Druger (Syracuse University), Gary Dudley (University of Georgia), Susan Dunford (University of Cincinnati), Betsey Dyer (Wheaton College), Robert Eaton (University of Colorado), Robert S. Edgar (University of California, Santa Cruz), Douglas Eernisse (California State University, Fullerton), Betty J. Eidemiller (Lamar University), Brad Elder (University of Oklahoma), William D. Eldred (Boston University), Norman Ellstrand (University of California, Riverside), Dennis Emery (Iowa State University), John Endler (University of California, Santa Barbara), Margaret T. Erskine (Lansing Community College), Gerald Esch (Wake Forest University), Frederick B. Essig (University of South Florida), Mary Eubanks (Duke University), David Evans (University of Florida), Robert C. Evans (Rutgers University, Camden), Sharon Eversman (Montana State University), Lincoln Fairchild (Ohio State University), Peter Fajer (Florida State University), Bruce Fall (University of Minnesota), Lynn Fancher (College of DuPage), Paul Farnsworth (University of Texas at San Antonio), Larry Farrell (Idaho State University), Jerry F. Feldman (University of California, Santa Cruz), Eugene Fenster (Longview Community College), Russell Fernald (University of Oregon), Kim Finer (Kent State University), Milton Fingerman (Tulane University), Barbara Finney (Regis College), Frank Fish (West Chester University), David Fisher (University of Hawaii, Manoa), Steven Fisher (University of California, Santa Barbara), Lloyd Fitzpatrick (University of North Texas), William Fixsen (Harvard University), Abraham Flexer (Manuscript Consultant, Boulder, Colorado), Kerry Foresman (University of Montana), Norma Fowler (University of Texas, Austin), Robert G. Fowler (San Jose State University), David Fox (University of Tennessee, Knoxville), Carl Frankel (Pennsylvania State University, Hazleton), James Franzen (University of Pittsburgh), Bill Freedman (Dalhousie University), Otto Friesen (University of Virginia), Frank Frisch (Chapman University), Virginia Fry (Monterey Peninsula College), Bernard Frye (University of Texas at Arlington), Alice Fulton (University of Iowa), Chandler Fulton (Brandeis University), Sara Fultz (Stanford University), Berdell Funke (North Dakota State University), Anne Funkhouser (University of the Pacific), Michael Gaines (University of Miami), Arthur W. Galston (Yale University), Carl Gans (University of Michigan), John Gapter (University of Northern Colorado), Reginald Garrett (University of Virginia), Patricia Gensel (University of North Carolina), Chris George (California Polytechnic State University, San Luis Obispo), Robert George (University of Wyoming), J. Whitfield Gibbons (University of Georgia), J. Phil Gibson (Agnes Scott College), Frank Gilliam (Marshall University), Simon Gilroy (Pennsylvania State University), Alan Gishlick (National Center for Science Education), Todd Gleeson (University of Colorado), John Glendinning (Barnard College), David Glenn-Lewin (Wichita State University), William Glider (University of Nebraska), Elizabeth A. Godrick (Boston University), Lynda Goff (University of California, Santa Cruz), Elliott Goldstein (Arizona State University), Paul Goldstein (University of Texas, El Paso), Sandra Gollnick (State University of New York at Buffalo) Anne Good (University of California, Berkeley), Judith Goodenough (University of Massachusetts, Amherst), Wayne Goodey (University of British Columbia), Robert Goodman (University of Wisconsin-Madison), Ester Goudsmit (Oakland University), Linda Graham (University of Wisconsin, Madison), Robert Grammer (Belmont University), Joseph Graves (Arizona State University), Phyllis Griffard (University of Houston-Downtown), A. J. F. Griffiths (University of British Columbia), William Grimes (University of Arizona), Mark Gromko (Bowling Green State University), Serine Gropper (Auburn University), Katherine L. Gross (Ohio State University), Gary Gussin (University of Iowa), Mark Guyer (National Human Genome Research Institute), Ruth Levy Guyer (Bethesda, Maryland), R. Wayne Habermehl (Montgomery County Community College), Mac Hadley (University of Arizona), Joel Hagen (Radford University), Jack P. Hailman (University of Wisconsin), Leah Haimo (University of California, Riverside), Jody Hall (Brown University), Douglas Hallett, (Northern Arizona University), Rebecca Halyard (Clayton State College), Sam Hammer (Boston University), Penny Hanchey-Bauer (Colorado State University), Laszlo Hanzely (Northern Illinois University), Jeff Hardin (University of Wisconsin, Madison), Richard Harrison (Cornell University), Carla Hass (Pennsylvania State University), Chris Haufler (University of Kansas), Chris Haynes (Shelton State Community College), H. D. Heath (California State University, Hayward), George Hechtel (State University of New York, Stony Brook), Blair Hedges (Pennsylvania State), David Heins (Tulane University), Jean Heitz (University of Wisconsin, Madison), John D. Helmann (Cornell University), Colin Henderson (University of Montana), Michelle Henricks (University of California, Los Angeles), Caroll Henry (Chicago State University), Frank Heppner (University of Rhode Island), Scott Herrick (Missouri Western State College), Ira Herskowitz (University of California, San Francisco), Paul E. Hertz (Barnard College), David Hibbett (Clark University), R. James Hickey (Miami University), William Hillenius (College of Charleston), Ralph Hinegardner (University of California, Santa Cruz), William Hines (Foothill College), Robert Hinrichsen (Indiana University of Pennsylvania), Helmut Hirsch (State University of New York, Albany), Tuan-hua David Ho (Washington University), Carl Hoagstrom (Ohio Northern University), James Hoffman (University of Vermont), A. Scott Holaday (Texas Tech), James Holland (Indiana State University, Bloomington), Charles Holliday (Lafayette College), Lubbock Karl Holte (Idaho State University), Laura Hoopes (Occidental College), Nancy Hopkins (Massachusetts Institute of Technology), Sandra Horikami (Daytona Beach Community College), Kathy Hornberger (Widener University), Pius F. Horner (San Bernardino Valley College), Margaret Houk (Ripon College), Ronald R. Hoy (Cornell University), Donald Humphrey (Emory University School of Medicine), Robert J. Huskey (University of Virginia), Steven Hutcheson (University of Maryland, College Park), Sandra Hsu (Skyline College), Bradley Hyman (University of California, Riverside), Mark Iked (San Bernardino Valley College), Cheryl Ingram-Smith (Clemson University), Alice Jacklet (State University of New York, Albany), John Jackson (North Hennepin

Community College), John C. Jahoda (Bridgewater State College), Dan Johnson (East Tennessee State University), Randall Johnson (University of California, San Diego), Stephen Johnson (William Penn University), Wayne Johnson (Ohio State University), Kenneth C. Jones (California State University, Northridge), Russell Jones (University of California, Berkeley), Alan Journet (Southeast Missouri State University), Walter Judd (University of Florida), Thomas C. Kane (University of Cincinnati), Tamos Kapros (University of Missouri), E. L. Karlstrom (University of Puget Sound), Jennifer Katcher (Pima Community College), Norm Kenkel (University of Manitoba), George Khoury (National Cancer Institute), Mark Kirk (University of Missouri-Columbia), Robert Kitchin (University of Wyoming), Attila O. Klein (Brandeis University), Daniel Klionsky (University of Michigan), Ned Knight (Linfield College), David Kohl (University of California, Santa Barbara), Greg Kopf (University of Pennsylvania School of Medicine), Thomas Koppenheffer (Trinity University), Janis Kuby (San Francisco State University), David Kurijaka (Ohio University), J. A. Lackey (State University of New York, Oswego), Elaine Lai (Brandeis University), Lynn Lamoreux (Texas A&M University), William L'Amoreaux (College of Staten Island), Carmine A. Lanciani (University of Florida), Kenneth Lang (Humboldt State University), Dominic Lannutti (El Paso Community College), Allan Larson (Washington University), Diane K. Lavett (State University of New York, Cortland, and Emory University), Charles Leavell (Fullerton College), C. S. Lee (University of Texas), Robert Leonard (University of California, Riverside), John Lepri (University of North Carolina at Greensboro), Donald Levin (University of Texas), Austin Mike Levine (University of California, Berkeley), Joseph Levine (Boston College), Bill Lewis (Shoreline Community College), John Lewis (Loma Linda University), Lorraine Lica (California State University, Hayward), Harvey Liftin (Broward Community College), Harvey Lillywhite (University of Florida, Gainesville), Clark Lindgren (Grinnell College), Sam Loker (University of New Mexico), Jane Lubchenco (Oregon State University), Margaret A. Lynch (Tufts University), Steven Lynch (Louisiana State University at Shreveport), Richard Machemer Jr. (St. John Fisher College), Elizabeth Machunis-Masuoka (University of Virginia), James MacMahon (Utah State University), Linda Maier (University of Alabama in Huntsville), Jose Maldonado (El Paso Community College), Richard Malkin (University of California, Berkeley), Charles Mallery (University of Miami), William Margolin (University of Texas Medical School), Lynn Margulis (Boston University), Edith Marsh (Angelo State University), Diane Marshall (University of New Mexico), Linda Martin Morris (University of Washington), Karl Mattox (Miami University of Ohio), Joyce Maxwell (California State University, Northridge), Jeffrey D. May (Marshall University), Lee McClenaghan (San Diego State University), Richard McCracken (Purdue University), Kerry McDonald (University of Missouri-Columbia), Jacqueline McLaughlin (Pennsylvania State University, Lehigh Valley), Neal McReynolds (Texas A&M International), Lisa Meffert (Rice University), Michael Meighan (University of California, Berkeley), Scott Meissner (Cornell University), Paul Melchior (North Hennepin Community College), Phillip Meneely (Haverford College), John Merrill (Michigan State University), Brian Metscher (University of California, Irvine), Ralph Meyer (University of Cincinnati), James Mickle (North Carolina State University), Roger Milkman (University of Iowa), Helen Miller (Oklahoma State University), John Miller (University of California, Berkeley), Kenneth R. Miller (Brown University), John E. Minnich (University of Wisconsin, Milwaukee), Michael Misamore (Louisiana State University), Kenneth Mitchell (Tulane University School of Medicine), Alan Molumby (University of Illinois at Chicago), Nicholas Money (Miami University), Russell Monson (University of Colorado, Boulder), Frank Moore (Oregon State University), Randy Moore (Wright State University), William Moore (Wayne State University), Carl Moos (Veterans Administration Hospital, Albany, New York), Michael Mote (Temple University), Alex Motten (Duke University), Deborah Mowshowitz (Columbia University), Rita Moyes (Texas A&M College Station), Darrel L. Murray (University of Illinois at Chicago), John Mutchmor (Iowa State University), Elliot Myerowitz (California Institute of Technology), Gavin Naylor (Iowa State University), John Neess (University of Wisconsin, Madison), Raymond Neubauer (University of Texas, Austin), Todd Newbury (University of California, Santa Cruz), Harvey Nichols (University of Colorado, Boulder), Deborah Nickerson (University of South Florida), Bette Nicotri (University of Washington), Caroline Niederman (Tomball College), Maria Nieto (California State University, Hayward), Greg Nishiyama (College of the Canyons), Charles R. Noback (College of Physicians and Surgeons, Columbia University), Jane Noble-Harvey (Delaware University), Mary C. Nolan (Irvine Valley College), Peter Nonacs (University of California, Los Angeles), Richard Norman (University of Michigan-Dearborn), David O. Norris (University of Colorado, Boulder), Steven Norris (California State, Channel Islands), Cynthia Norton (University of Maine, Augusta), Steve Norton (East Carolina University), Steve Nowicki (Duke University), Bette H. Nybakken (Hartnell College), Brian O'Conner (University of Massachusetts, Amherst), Gerard O'Donovan (University of North Texas), Eugene Odum (University of Georgia), Linda Ogren (University of California, Santa Cruz), Patricia O'Hern (Emory University), Jeanette Oliver (St. Louis Community College Florissant Valley), Gary P. Olivetti (University of Vermont), John Olsen (Rhodes College), Laura J. Olsen (University of Michigan), Sharman O'Neill (University of California, Davis), Wan Ooi (Houston Community College), Gay Ostarello (Diablo Valley College), Catherine Ortega (Fort Lewis College), Charissa Osborne (Butler University), Thomas G. Owens (Cornell University), Penny Padgett (University of North Carolina at Chapel Hill), Kevin Padian (University of California, Berkeley), Dianna Padilla (State University of New York, Stony Brook), Barry Palevitz (University of Georgia), Daniel Papaj (University of Arizona), Peter Pappas (County College of Morris), Bulah Parker (North Carolina State University), Stanton Parmeter (Chemeketa Community College), Robert Patterson (San Francisco State University), Ronald Patterson (Michigan State University), Crellin Pauling (San Francisco State University), Kay Pauling (Foothill Community College), Daniel Pavuk (Bowling Green State University), Debra Pearce (Northern Kentucky University), Patricia Pearson (Western Kentucky University), Shelley Penrod (North Harris College), Beverly Perry (Houston Community College), David Pfennig (University of North Carolina at Chapel Hill), Bob Pittman (Michigan State University), James Platt (University of Denver), Martin Poenie (University of Texas, Austin), Scott Poethig (University of Pennsylvania), Jeffrey Pommerville (Texas A&M University), Warren Porter (University of Wisconsin), Daniel Potter (University of California, Davis), Donald Potts (University of California, Santa Cruz), Andy Pratt (University of Canterbury), David Pratt (University of California, Davis), Halina Presley (University of Illinois, Chicago), Mitch Price (Pennsylvania State University), Rong Sun Pu (Kean University), Rebecca Pyles (East Tennessee State University), Scott Quackenbush (Florida International University), Ralph Quatrano (Oregon State University), Val Raghavan (Ohio State University), Deanna Raineri (University of Illinois, Champaign-Urbana), Talitha Rajah (Indiana University Southeast), Charles Ralph (Colorado State University), Thomas Rand (Saint Mary's University), Kurt Redborg (Coe College), Ahnya Redman (Pennsylvania State), Brian Reeder (Morehead State University), Bruce Reid (Kean University), David Reid (Blackburn College), C. Gary Reiness (Lewis & Clark College), Charles Remington (Yale University), David Reznick (University of California, Riverside), Douglas Rhoads (University of Arkansas), Fred Rhoades (Western Washington State University), Christopher Riegle (Irvine Valley College), Donna Ritch (Pennsylvania State University), Carol Rivin (Oregon State University East), Laurel Roberts (University of Pittsburgh), Thomas Rodella (Merced College), Rodney Rogers (Drake University), William Roosenburg

(Ohio University), Wayne Rosing (Middle Tennessee State University), Thomas Rost (University of California, Davis), Stephen I. Rothstein (University of California, Santa Barbara), John Ruben (Oregon State University), Albert Ruesink (Indiana University), Neil Sabine (Indiana University), Tyson Sacco (Cornell University), Rowan Sage (University of Toronto), Don Sakaguchi (Iowa State University), Walter Sakai (Santa Monica College), Mark F. Sanders (University of California, Davis), Ted Sargent (University of Massachusetts, Amherst), K. Sathasivan (University of Texas, Austin), Gary Saunders (University of New Brunswick), Carl Schaefer (University of Connecticut), David Schimpf (University of Minnesota, Duluth), William H. Schlesinger (Duke University), Robert Schorr (Colorado State University), David Schwartz (Houston Community College), Christa Schwintzer (University of Maine), Erik P. Scully (Towson State University), Edna Seaman (Northeastern University), Orono Shukdeb Sen (Bethune-Cookman College), Wendy Sera (Seton Hill University), Timothy Shannon (Francis Marion University), Joan Sharp (Simon Fraser University), Victoria C. Sharpe (Blinn College), Elaine Shea (Loyola College, Maryland), Stephen Sheckler (Virginia Polytechnic Institute and State University), Richard Sherwin (University of Pittsburgh), Lisa Shimeld (Crafton Hills College), James Shinkle (Trinity University), Barbara Shipes (Hampton University), Richard Showman (University of South Carolina), Peter Shugarman (University of Southern California), Alice Shuttey (DeKalb Community College), James Sidie (Ursinus College), Daniel Simberloff (Florida State University), Anne Simon (University of Maryland), Alastair Simpson (Dalhousie University), Susan Singer (Carleton College), Roger Sloboda (Dartmouth University), John Smarrelli (Le Moyne College), Andrew T. Smith (Arizona State University), Kelly Smith (University of North Florida), Nancy Smith-Huerta (Miami Ohio University), John Smol (Queen's University), Andrew J. Snope (Essex Community College), Mitchell Sogin (Woods Hole Marine Biological Laboratory), Susan Sovonick-Dunford (University of Cincinnati), Frederick W. Spiegel (University of Arkansas), Amanda Starnes (Emory University), Karen Steudel (University of Wisconsin), Barbara Stewart (Swarthmore College), Cecil Still (Rutgers University, New Brunswick), Margery Stinson (Southwestern College), James Stockand (University of Texas Health Science Center, San Antonio), John Stolz (California Institute of Technology), Richard D. Storey (Colorado College), Stephen Strand (University of California, Los Angeles), Eric Strauss (University of Massachusetts, Boston), Antony Stretton (University of Wisconsin-Madison), Russell Stullken (Augusta College), Mark Sturtevant (University of Michigan-Flint), John Sullivan (Southern Oregon State University), Gerald Summers (University of Missouri), Judith Sumner (Assumption College), Marshall D. Sundberg (Emporia State University), Lucinda Swatzell (Southeast Missouri State University), Daryl Sweeney (University of Illinois, Champaign-Urbana), Samuel S. Sweet (University of California, Santa Barbara), Janice Swenson (University of North Florida), Lincoln Taiz (University of California, Santa Cruz), Samuel Tarsitano (Southwest Texas State University), David Tauck (Santa Clara University), James Taylor (University of New Hampshire), John Taylor (University of California, Berkeley), Martha R. Taylor (Cornell University), Thomas Terry (University of Connecticut), Roger Thibault (Bowling Green State University), William Thomas (Colby-Sawyer College), Cyril Thong (Simon Fraser University), John Thornton (Oklahoma State University), Robert Thornton (University of California, Davis), Stephen Timme (Pittsburg State University), Leslie Towill (Arizona State University), James Traniello (Boston University), Constantine Tsoukas (San Diego State University), Marsha Turell (Houston Community College), Robert Tuveson (University of Illinois, Urbana), Maura G. Tyrrell (Stonehill College), Catherine Uekert (Northern Arizona University), Gordon Uno (University of Oklahoma), Lisa A. Urry (Mills College), James W. Valentine (University of California, Santa Barbara), Joseph Vanable (Purdue University), Theodore Van Bruggen (University of South Dakota), Kathryn VandenBosch (Texas A&M University), Gerald Van Dyke (North Carolina State University), Brandi Van Roo (Framingham State College), Moira Van Staaden (Bowling Green State), Frank Visco (Orange Coast College), Laurie Vitt (University of California, Los Angeles), Neal Voelz (St. Cloud State University), Thomas J. Volk (University of Wisconsin, La Crosse), Susan D. Waaland (University of Washington), William Wade (Dartmouth Medical College), D. Alexander Wait (Southwest Missouri State University), John Waggoner (Loyola Marymount University), Jyoti Wagle (Houston Community College), Edward Wagner (University of California, Irvine), Dan Walker (San Jose State University), Robert L. Wallace (Ripon College), Jeffrey Walters (North Carolina State University), Margaret Waterman (University of Pittsburgh), Charles Webber (Loyola University of Chicago), Peter Webster (University of Massachusetts, Amherst), Terry Webster (University of Connecticut, Storrs), Beth Wee (Tulane University), Peter Wejksnora (University of Wisconsin, Milwaukee), Kentwood Wells (University of Connecticut), David J. Westenberg, (University of Missouri, Rolla), Matt White (Ohio University), Stephen Williams (Glendale Community College), Elizabeth Willott (University of Arizona), Christopher Wills (University of California, San Diego), Fred Wilt (University of California, Berkeley), E. William Wischusen (Louisiana State University), Clarence Wolfe (Northern Virginia Community College), Robert T. Woodland (University of Massachusetts Medical School), Joseph Woodring (Louisiana State University), Patrick Woolley (East Central College), Philip Yant (University of Michigan), Linda Yasui (Northern Illinois University), Hideo Yonenaka (San Francisco State University), Edward Zalisko (Blackburn College), Zai Ming Zhao (University of Texas, Austin), John Zimmerman (Kansas State University), Uko Zylstra (Calvin College)

Acknowledgments

The authors wish to express their gratitude to the global community of instructors, researchers, students, and publishing professionals who have contributed to this edition.

As authors of this text, we are mindful of the daunting challenge of keeping up to date in all areas of our rapidly expanding subject. We are grateful to the numerous scientists who helped shape this edition by discussing their research fields with us, answering specific questions in their areas of expertise, and, often, sharing their ideas about biology education.

For advice in updating the phylogeny of Galápagos finches in Chapter 1, we are indebted to Kevin Burns and Peter Grant. For assistance with the chapters of Units 1 through 3 (chemistry, cell biology, and genetics), we first wish to thank the members of the Mills College Biology and Chemistry/Physics Department, notably Barbara Bowman and Elisabeth Wade. We are also grateful to Tom Owens and Mimi Zolan, who were each exceptionally generous with their time and knowledge, and to Michael Black, Laurie Heyer, and Ed Blake, for noteworthy contributions to figures. And we thank the individuals who took the time to share their expertise on early atmospheric conditions (Laura Schaefer), cell biology (Pat Zambryski, Steve King, Jeremy Reiter, and Jeff Hardin), gene regulation (Phil Zamore, Dave Bartel, Tom Gingeras, Steve Bell, Saba Valadkhan, Joe Heilig, Lorraine Pillus, and Mike Levine), current cloning approaches (Caroline Kane and Andy Cameron), genomics (Nikos Krypides, Emir Khatipov, and Rebekah Rasooly), and homeobox genes (Bill McGinnis). In addition, we thank Lisa Weasel for her feedback on the new "Tips for Genetics Problems."

For the chapters of Units 4 and 5, on evolution and the diversity of life, researchers who generously shared their expertise with us included Richard Anthony, Nick Barton, Toby Bradshaw, Keith Clay, Kevin de Queiroz, Peter and Rosemary Grant, Daniel J. Howard, Patrick Keeling, Andrew H. Knoll, Jon Mallatt, Amy McCune, Axel Meyer, Kevin J. Peterson, Loren Rieseberg, Ole Seehausen, and Mark Webster.

For Units 6 through 8, the units on plant and animal form and function, and ecology, we benefited greatly from the expertise of Charles Michel, Eric Britt, Don Boyer, and Alan French. In addition, Tom Deerinck, Peter Gillespie, Mark Chappell, and Doug DeSimone provided important assistance with figures. We also thank Eric Simon, a coauthor of the Campbell nonmajors texts, for helping us think through some terminology and presentation dilemmas in Unit 7.

And finally, for her many contributions throughout the book, we sincerely thank Marty Taylor, the author of the *Student Study Guide* and a coauthor of *Biology: Concepts & Connections*.

A total of 228 biologists, listed on pages xxiv–xxix, provided detailed reviews of one or more chapters for this edition, helping us ensure the book's scientific accuracy and improve its pedagogical effectiveness. Special thanks for exceptional contributions go to Johnny El-Rady, Graeme Lindbeck, Bruce Riley, Robert Fowler, Alan Gishlick, Alastair Simpson, Ken Halanych, Kevin Padian, John Taylor, Jay Comeaux, Grace Wyngaard, Lauren Garner, Missy Holbrook, Toby Kellogg, Eduardo Zeiger, Richard Norman, Albert Herrera, and Patricia Schulte. Thanks also to the numerous other professors and students, from all over the world, who offered suggestions directly to the authors. Of course, we alone bear the responsibility for any errors that remain in the text, but the dedication of our consultants, reviewers, and other correspondents makes us especially confident in the accuracy and effectiveness of this edition.

Conducting the unit-opening interviews was again one of the great pleasures of revising *BIOLOGY*. For the Eighth Edition, we are proud to include interviews with Deborah Gordon, Paul Nurse, Terry Orr-Weaver, Scott Edwards, Sean Carroll, Pat Zambryski, Masashi Yanagisawa, and Diana Wall (see p. xi). We thank these busy people for generously sharing their experiences with us.

The value of *BIOLOGY* as a learning tool is greatly enhanced by the supplementary materials that have been created for instructors and students. We recognize that the dedicated authors of these materials are essentially writing mini (and not so mini) books. We much appreciate the hard work and creativity of the following: Ruth Buskirk and Christopher Gillen (authors of the new *Inquiry in Action: Interpreting Scientific Papers*); Judith Morgan and Eloise Brown Carter (*Investigating Biology*, 6th Edition); Jean Heitz and Cynthia Giffen (*Practicing Biology*, 3rd Edition); Margaret Waterman and Ethel Stanley (*Biological Inquiry: A Workbook of Investigative Cases*, 2nd Edition); Bill Barstow, Louise Paquin, Michael Dini, John Lepri, John Zarnetske, C. O. Patterson, and Jean DeSaix (*Test Bank*); Ed Zalisko, Margaret Ricci, Lauren Garner, Jung Choi, and Virginia White (*Media Quizzes*); Joan Sharp (*Lecture Outlines, PowerPoint Lectures,* and *Student Misconceptions*); Erin Barley (*PowerPoint Lectures*); Bill Wischusen, Ruth Buskirk, Jung Choi, John Merrill, Melissa Michael, Randy Phillis, Mark Lyford, and Chris Gregg (*Active Learning Questions*); and Laura Zanello (*Spanish Glossary*). Once again, we thank our long-time colleague Marty Taylor for her excellent work on the *Student Study Guide*; she has now completed eight editions of this popular student aid. Special thanks go to Tom Owens for his visionary work on the *BioFlix* animations and his creative, collaborative work as lead author on our new *MasteringBiology* tutorials. We also thank Brad Williamson, Jennifer Yeh, Dawn Keller, and Scott Bowling for their excellent work on the *BioFlix* animations and the accompanying student tools. In addition, we are grateful to the many other people—biology instructors, editors, artists, production experts, and narrators—who are listed in the credits for these and other elements of the electronic media that accompany the book. Finally, we thank the class testers and reviewers of *BioFlix* and *MasteringBiology* who are listed on pages xxv–xxvi.

BIOLOGY, Eighth Edition, results from an unusually strong synergy between a team of scientists and a team of publishing professionals. The expansion of the author team and major revision of many chapters, the creation of new pedagogical features and the improvement of old ones, and the exceptionally rich package of supplements created unprecedented challenges for the publishing team.

The members of our core editorial team at Benjamin Cummings—our Fab Five—brought unmatched talents, commitment, and pedagogical insights to this revision, and working with this team over the past three years has been a great pleasure. Our Editor-in-Chief, Beth Wilbur, continues to be a full colleague in the book's ongoing evolution and a respected advocate for biology education in the academic community. Our extraordinary Supervising Editors, Pat Burner and Beth Winickoff, once again had the awesome responsibility of overseeing in detail the work of all the authors, developmental editors, and developmental artists. Together, Beth and Pat ensured that every page of every chapter has the text, figures, and pedagogy to make this edition the most effective biology textbook ever. Deborah Gale, Executive Director of Development, and our Senior Editorial Manager, the incomparable Ginnie Simione Jutson, oversaw the entire project on a day-by-day basis, a feat equivalent to running a three-ring circus. Ginnie's patience and resourcefulness and Deborah's oversight of the project as a whole have enabled the entire book team to operate at a level of sanity that would not have been possible without their guidance.

We were fortunate to have on our team some of college publishing's top developmental editors. In addition to Beth Winickoff and Pat Burner (who did important hands-on editing themselves, as well as their many other tasks), the primary developmental editors for this edition were John Burner and Matt Lee, joined as the project progressed by Alice Fugate and Suzanne Olivier. We are deeply grateful to all our editors for making us better writers, teachers, and biologists.

Biology is a visual subject, and we are indebted to our developmental artists Hilair Chism, Carla Simmons, Andrew Recher, Connie Balek, and Kelly Murphy for helping us make all our figures better tools for teaching and learning—as well as visually appealing. In addition, the support of our bright, efficient, and good-natured Editorial Assistants—Julia Khait, Ben Pearson, and Logan Triglia—is much appreciated. We couldn't have finished the book without them! We also want to thank Robin Heyden for organizing the annual Benjamin Cummings Biology Leadership Conferences, which always bring us closer to the teaching community and offer a fresh supply of creative teaching ideas from outstanding biology educators.

You would not have a book in your hands today if not for the herculean efforts of the book production team, which has the crucial responsibility of converting the text manuscript and illustrations to pages ready for the printer. For the Eighth Edition, these efforts were headed up by Managing Editor Mike Early. We thank him, as well as our longtime copyeditor Janet Greenblatt, proofreaders Joanna Dinsmore and Marie Dartman, Permissions Editors Sue Ewing and Marcy Lunetta, and indexers Lynn Armstrong and Charlotte Shane. Handling the illustrations were Art Editors Laura Murray and Kelly Murphy; the final rendering of the new and revised illustrations was carried out by the artists of Precision Graphics, working under Kristina Seymour. Senior Photo Editor Donna Kalal and photo researcher Maureen Spuhler obtained a large number of handsome and informative photos for this edition. We are indebted to the entire art and photo team. For the beautiful design of the book's interior, we want to thank Art and Design Director Mark Ong and Design Manager Marilyn Perry for their design of text and art styles that show off the words and pictures in a way that will appeal to readers and help them learn. (And thanks to both of them for their endless patience with all our concerns!) For the user-friendly page layouts, we are grateful to Jennifer Dunn and Jana Anderson. And many thanks to Yvo Riezebos for designing the striking cover. Putting together all the pieces of this complicated book were the staff at S4Carlisle Publishing Services, led by Production Manager Lori Dalberg and Composition Supervisor Holly Paige. Thank you, Lori and Holly!

We are pleased to thank the topnotch publishing professionals who worked on the book's printed supplements: Senior Supplements Project Editor Susan Berge, who coordinated the entire print supplements package; Production Supervisor Jane Brundage; Developmental Editor Susan Weisberg; and Project Editors Mary Douglas, Kim Wimpsett, Elizabeth Campbell, and Melanie Field.

With regard to the wonderful package of electronic media that accompanies the book, we offer special thanks to Senior Media Producer Jon Ballard, Project Editors Nora Lally-Graves and Brienn Buchanan, media correlators Nina Lewallen Hufford and Sarah Kaminker, and proofreader Pete Shanks for their work on the website. We also are grateful to Associate Web Developer Linda Young and Web Technologies Manager Steve Wright. For their work on the *Instructor Resource CD/DVD-ROM*, we thank Project Manager James Bruce; Developmental Artist Hilair Chism; Developmental Editors John Burner, Pat Burner, Matt Lee, and Beth Winickoff; copyeditor John Hammett (*PPT Lectures*); proofreader Pete Shanks (*PPT Lectures*); and Donna King, Production Project Manager at Progressive Information Technologies.

We are especially grateful to the team of publishing professionals whose combined talents are evident in the *BioFlix* animations and student tools: Pat Burner, Russell Chun (storyboard artist extraordinaire), Ginnie Simione Jutson, Jon Ballard, Karen Gulliver (tutorial developmental editor), Animated Biomedical Productions (animation production); and Groove 11 (tutorial production). For their work on *MasteringBiology*, we thank the aforementioned Pat Burner, Ginnie Simione Jutson, and Jon Ballard, plus Tania Mlawer, Director of Content Development and Project Management; Mary Catherine Hager, Developmental Editor; Deb Greco, Media Producer; Kristen Sutton, Content Lead; and Developmental Artist Jay McElroy and the artists at Pearson Production Solutions. For their hard work and support, our appreciation goes to the MasteringX Team (in alphabetical order): Ruth Berry, Lewis Costas, Katherine Foley, Julia Henderson, Joseph Ignazi, Jeff King, David Kokorowski, Mary Lee, Claire Masson, Nissi Mathews, Adam Morton, Fred Mueller, Ian Nordby, Maria Panos, Andrea Pascarella, Mary Ann Perry, Caroline Power, Sarah Smith, Margaret Trombley, and Rasil Warnakulasooriya. Last, but not least, we thank Lauren Fogel, Director of Media Development at Benjamin Cummings, for her continued leadership on all things media.

For their important roles in marketing the book, we are very grateful to Director of Marketing Christy Lawrence, Executive Marketing Manager Lauren Harp, and Market Development Manager Josh Frost. For the creation of visually stunning print and electronic promotional materials, we thank Creative Director Lillian Carr; Marketing Communication Specialists Jane Campbell, Kristi Hlaing, and Jessica Perry; Designer Laurie Campbell; Web Designer Mansour Bethoney, who led the creation of the e-brochure; and Webmaster Anna Molodtsova.

Linda Davis, President of Pearson Math, Economics, and Science, has shared our commitment to excellence and provided strong support for four editions now, and we are happy to thank her once again. We also want to thank Paul Corey, now President of Pearson Science, for his enthusiasm, encouragement, and support.

The Pearson Science sales team, which represents *BIOLOGY* on campus, is our living link to the students and professors who use the text. The field representatives tell us what you like and don't like about the book, and they provide prompt service to biology departments. They are strong allies in biology education, and we thank them for their professionalism in communicating the features of our book. For representing our book and its teaching values to our wider international audience, we thank the sustained work of our sales and marketing partners throughout the world, including (but by no means only) Marlene Olsavsky, Ann Oravetz, and Pablo Rendina.

Finally, we wish to thank our families and friends for their encouragement and patience throughout this long project. Our special thanks to: Paul, Dan, Maria, Armelle, and Sean (J.R.); Lily, Grant, Ross, Lilytoo, and Alex (L.U.); Debra and Hannah (M.C.); Harry, Elga, Aaron, Sophie, Noah, and Gabriele (S.W.); Natalie (P.M.); and Sally, Robert, David, and Will (R.J.). And as always, Rochelle and Allison.

Jane Reece, Lisa Urry, Michael Cain,
Steve Wasserman, Peter Minorsky, and Rob Jackson

Detailed Contents

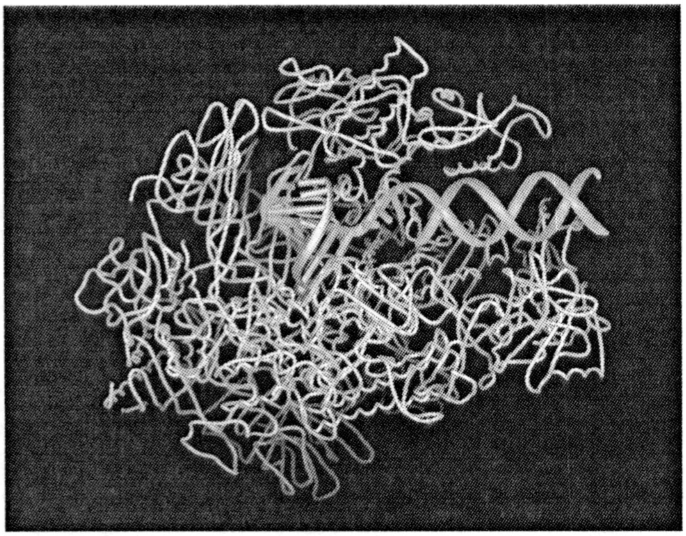

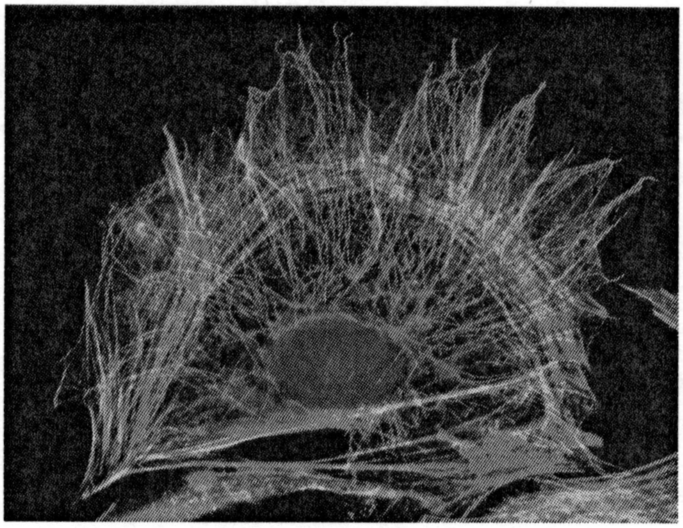

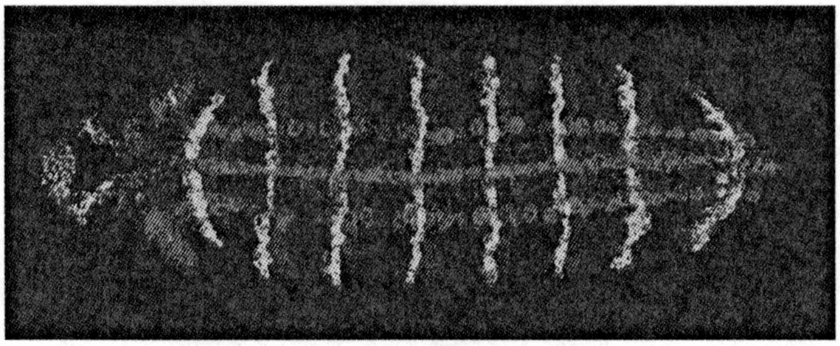

1 Introduction: Themes in the Study of Life

▲ **Figure 1.1 What properties of life are demonstrated by this flower?**

KEY CONCEPTS

1.1 **Themes connect the concepts of biology**

1.2 ***The Core Theme:* Evolution accounts for the unity and diversity of life**

1.3 **Scientists use two main forms of inquiry in their study of nature**

OVERVIEW

Inquiring About the World of Life

The flower featured on the cover of this book and in **Figure 1.1** is from a magnolia, a tree of ancient lineage that is native to Asian and American forests. The magnolia blossom is a sign of the plant's status as a living organism, for flowers contain organs of sexual reproduction, and reproduction is a key property of life, as you will learn later.

Like all organisms, the magnolia tree in **Figure 1.2** is living in close association with other organisms, though it is a lone specimen far from its ancestral forest. For example, it depends on beetles to carry pollen from one flower to another, and the beetles, in turn, eat from its flowers. The flowers are adapted to the beetles in several ways: Their bowl shape allows easy access, and their multiple reproductive organs and tough petals (see Figure 1.1) help ensure that some survive the voracious beetles. Such adaptations are the result of **evolution**, the process of change that has transformed life on Earth from its earliest beginnings to the diversity of organisms living today. As discussed later in this chapter, evolution is the fundamental organizing principle of biology and the main theme of this book.

Although biologists know a great deal about magnolias and other plants, many mysteries remain. For instance, what exactly led to the origin of flowering plants? Posing questions about the living world and seeking science-based answers—scientific inquiry—are the central activities of **biology**, the scientific study of life. Biologists' questions can be ambitious. They may ask how a single tiny cell becomes a tree or a dog, how the human mind works, or how the different forms of life in a forest interact. Can you think of some questions about living organisms that interest you? When you do, you are already starting to think like a biologist. More than anything else, biology is a quest, an ongoing inquiry about the nature of life.

Perhaps some of your questions relate to health or to societal or environmental issues. Biology is woven into the fabric of our culture more than ever before and can help answer many questions that affect our lives. Research breakthroughs in genetics and cell biology are transforming medicine and agriculture. Neuroscience and evolutionary biology are reshaping psychology and sociology. New models in ecology are helping societies evaluate environmental issues, such as global warming. There has never been a more important time to embark on a study of life.

▲ **Figure 1.2 A magnolia tree in early spring.**

▼ **Order.** This close-up of a sunflower illustrates the highly ordered structure that characterizes life.

▲ **Evolutionary adaptation.** The appearance of this pygmy sea horse camouflages the animal in its environment. Such adaptations evolve over many generations by the reproductive success of those individuals with heritable traits that are best suited to their environments.

▲ **Response to the environment.** This Venus flytrap closed its trap rapidly in response to the environmental stimulus of a damselfly landing on the open trap.

▲ **Regulation.** The regulation of blood flow through the blood vessels of this jackrabbit's ears helps maintain a constant body temperature by adjusting heat exchange with the surrounding air.

▲ **Energy processing.** This hummingbird obtains fuel in the form of nectar from flowers. The hummingbird will use chemical energy stored in its food to power flight and other work.

▼ **Growth and development.** Inherited information carried by genes controls the pattern of growth and development of organisms, such as this Nile crocodile.

▲ **Reproduction.** Organisms (living things) reproduce their own kind. Here an emperor penguin protects its baby.

▲ **Figure 1.3 Some properties of life.**

? *Is a gasoline-powered lawn mower alive? Which of these properties does it have? Which properties does it lack?*

But what is life? Even a small child realizes that a dog or a plant is alive, while a rock is not. Yet the phenomenon we call life defies a simple, one-sentence definition. We recognize life by what living things do. **Figure 1.3** highlights some of the properties and processes we associate with life.

While limited to a handful of images, Figure 1.3 reminds us that the living world is wondrously varied. How do biologists make sense of this diversity and complexity? This opening chapter sets up a framework for answering this question. The first part of the chapter provides a panoramic view of the biological "landscape," organized around some unifying themes. We then focus on biology's overarching theme, evolution, with an introduction to the reasoning that led Charles Darwin to his explanatory theory. Finally, we look at scientific inquiry—how scientists raise and attempt to answer questions about the natural world.

CONCEPT 1.1

Themes connect the concepts of biology

Biology is a subject of enormous scope, and anyone who follows the news knows that biological knowledge is expanding at an ever-increasing rate. Simply memorizing the factual details of this huge subject is not a reasonable option. How, then, can you, as a student, go beyond the facts to develop a coherent view of life? One approach is to fit the many things you learn into a set of themes that pervade all of biology—ways of thinking about life that will still apply decades from now. Focusing on a few big ideas will help you organize and make sense of all the information you'll encounter as you study biology. To help you, we have selected seven unifying themes to serve as touchstones as you proceed through this book.

Evolution, the Overarching Theme of Biology

Evolution is biology's core theme—the one idea that makes sense of everything we know about living organisms. Life has been evolving on Earth for billions of years, resulting in a vast diversity of past and present organisms. But along with the diversity we find many shared features. For example, while the sea horse, jackrabbit, hummingbird, crocodile, and penguins in Figure 1.3 look very different, their skeletons are basically similar. The scientific explanation for this unity and diversity—and for the suitability of organisms to their environments—is evolution: the idea that the organisms living on Earth today are the modified descendants of common ancestors. In other words, we can explain traits shared by two organisms with the idea that they have descended from a common ancestor, and we can account for differences with the idea that heritable changes have occurred along the way. Many kinds of evidence support the occurrence of evolution and the theory that describes how it takes place. We'll return to evolution later in the chapter, after surveying some other themes and painting a fuller picture of the scope of biology.

Theme: New properties emerge at each level in the biological hierarchy

The study of life extends from the microscopic scale of the molecules and cells that make up organisms to the global scale of the entire living planet. We can divide this enormous range into different levels of biological organization.

Imagine zooming in from space to take a closer and closer look at life on Earth. It is spring, and our destination is a forest in Ontario, Canada, where we will eventually explore a maple leaf right down to the molecular level. **Figure 1.4** (on the next two pages) narrates this journey into life, with the circled numbers leading you through the levels of biological organization illustrated by the photographs.

Emergent Properties

If we now zoom back out from the molecular level in Figure 1.4, we can see that novel properties emerge at each step, properties that are not present at the preceding level. These **emergent properties** are due to the arrangement and interactions of parts as complexity increases. For example, if you make a test-tube mixture of chlorophyll and all the other kinds of molecules found in a chloroplast, photosynthesis will not occur. Photosynthesis can take place only when the molecules are arranged in a specific way in an intact chloroplast. To take another example, if a serious head injury disrupts the intricate architecture of a human brain, the mind may cease to function properly even though all of the brain parts are still present. Our thoughts and memories are emergent properties of a complex network of nerve cells. At a much higher level of biological organization—at the ecosystem level—the recycling of chemical elements essential to life, such as carbon, depends on a network of diverse organisms interacting with each other and with the soil, water, and air.

Emergent properties are not unique to life. We can see the importance of arrangement in the distinction between a box of bicycle parts and a working bicycle. And while graphite and diamonds are both pure carbon, they have very different properties because their carbon atoms are arranged differently. But compared to such nonliving examples, the unrivaled complexity of biological systems makes the emergent properties of life especially challenging to study.

The Power and Limitations of Reductionism

Because the properties of life emerge from complex organization, scientists seeking to understand biological systems confront a dilemma. On the one hand, we cannot fully explain a higher level of order by breaking it down into its parts. A dissected animal no longer functions; a cell reduced to its chemical ingredients is no longer a cell. Disrupting a living system interferes with its functioning. On the other hand, something as complex as an organism or a cell cannot be analyzed without taking it apart.

Reductionism—the reduction of complex systems to simpler components that are more manageable to study—is a powerful strategy in biology. For example, by studying the molecular structure of DNA that had been extracted from cells, James Watson and Francis Crick inferred, in 1953, how this molecule could serve as the chemical basis of inheritance. The central role of DNA in cells and organisms became better understood, however, when scientists were able to study the interactions of DNA with other molecules. Biologists must balance the reductionist strategy with the larger-scale, holistic objective of understanding emergent properties—how the parts of cells, organisms, and higher levels of order, such as ecosystems, work together. At the cutting edge of research today is the approach called systems biology.

▼ Figure 1.4

Exploring Levels of Biological Organization

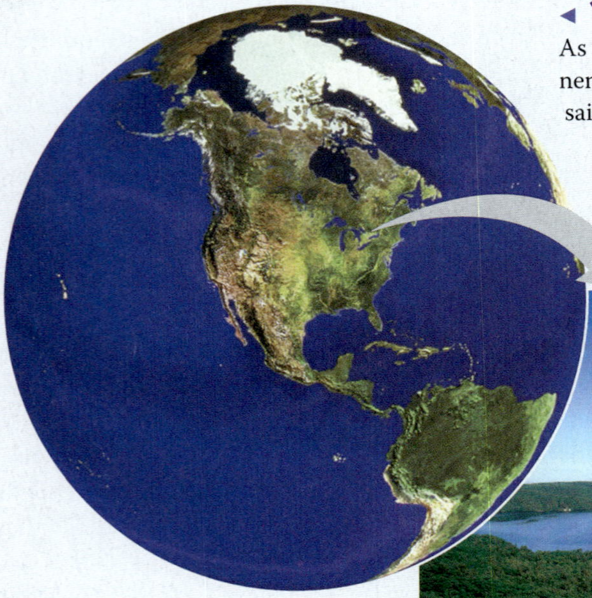

◀ 1 The Biosphere

As soon as we are near enough to Earth to make out its continents and oceans, we begin to see signs of life—in the green mosaic of the planet's forests, for example. This is our first view of the biosphere, which consists of all the environments on Earth that are inhabited by life. The biosphere includes most regions of land, most bodies of water, and the atmosphere to an altitude of several kilometers.

◀ 2 Ecosystems

As we approach Earth's surface for an imaginary landing in Ontario, we can begin to make out a forest with an abundance of deciduous trees (trees that lose their leaves in one season and grow new ones in another). Such a deciduous forest is an example of an ecosystem. Grasslands, deserts, and the ocean's coral reefs are other types of ecosystems. An ecosystem consists of all the living things in a particular area, along with all the nonliving components of the environment with which life interacts, such as soil, water, atmospheric gases, and light. All of Earth's ecosystems combined make up the biosphere.

▶ 3 Communities

The entire array of organisms inhabiting a particular ecosystem is called a biological community. The community in our forest ecosystem includes many kinds of trees and other plants, a diversity of animals, various mushrooms and other fungi, and enormous numbers of diverse microorganisms, which are living forms, such as bacteria, that are too small to see without a microscope. Each of these forms of life is called a *species.*

▶ 4 Populations

A population consists of all the individuals of a species living within the bounds of a specified area. For example, our Ontario forest includes a population of sugar maple trees and a population of white-tailed deer. We can now refine our definition of a community as the set of populations that inhabit a particular area.

▲ 5 Organisms

Individual living things are called organisms. Each of the maple trees and other plants in the forest is an organism, and so is each forest animal, such as a frog, squirrel, deer, and beetle. The soil teems with microorganisms such as bacteria.

▼ 6 Organs and Organ Systems

The structural hierarchy of life continues to unfold as we explore the architecture of the more complex organisms. A maple leaf is an example of an organ, a body part consisting of two or more tissues (which we'll see upon our next scale change). An organ carries out a particular function in the body. Stems and roots are the other major organs of plants. Examples of human organs are the brain, heart, and kidney. The organs of humans, other complex animals, and plants are organized into organ systems, each a team of organs that cooperate in a specific function. For example, the human digestive system includes such organs as the tongue, stomach, and intestines.

▶ 7 Tissues

Our next scale change—to see a leaf's tissues—requires a microscope. The leaf shown here has been cut on an angle. The honeycombed tissue in the interior of the leaf (left portion of photo) is the main location of photosynthesis, the process that converts light energy to the chemical energy of sugar and other food. We are viewing the sliced leaf from a perspective that also enables us to see the jigsaw puzzle–like tissue called epidermis, the "skin" on the surface of the leaf (right part of photo). The pores through the epidermis allow the gas carbon dioxide, a raw material for sugar production, to reach the photosynthetic tissue inside the leaf. At this scale, we can also see that each tissue has a cellular structure. In fact, each kind of tissue is a group of similar cells.

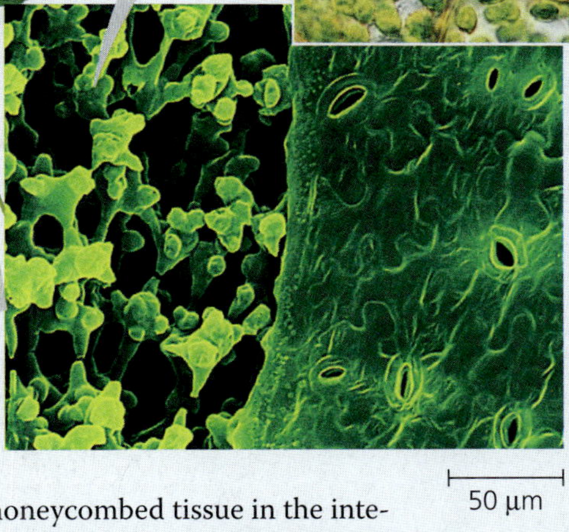

▼ 8 Cells

The cell is life's fundamental unit of structure and function. Some organisms, such as amoebas and most bacteria, are single cells. Other organisms, including plants and animals, are multicellular. Instead of a single cell performing all the functions of life, a multicellular organism has a division of labor among specialized cells. A human body consists of trillions of microscopic cells of many different kinds, such as muscle cells and nerve cells, which are organized into the various specialized tissues. For example, muscle tissue consists of bundles of muscle cells. In the photo below, we see a more highly magnified view of some of the cells in a leaf tissue. Each of the cells is only about 25 micrometers (µm) across. It would take more than 700 of these cells to reach across a penny. As small as these cells are, you can see that each contains numerous green structures called chloroplasts, which are responsible for photosynthesis.

▼ 9 Organelles

Chloroplasts are examples of organelles, the various functional components that make up cells. In this image, a very powerful tool called an electron microscope brings a single chloroplast into sharp focus.

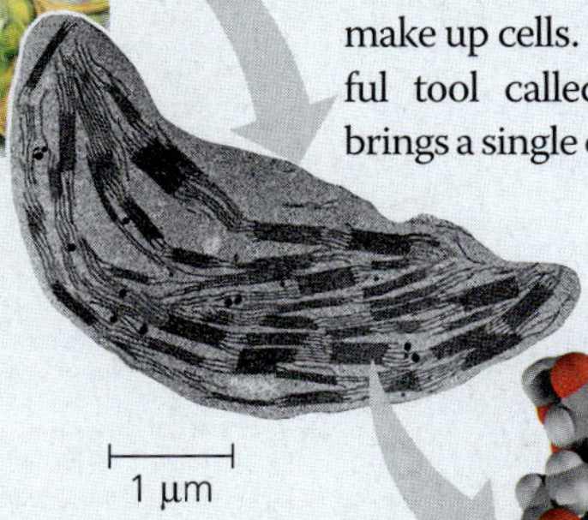

▶ 10 Molecules

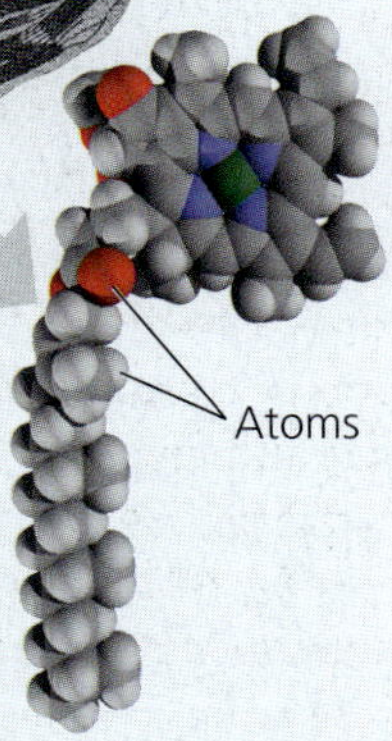

Our last scale change vaults us into a chloroplast for a view of life at the molecular level. A molecule is a chemical structure consisting of two or more small chemical units called *atoms*, which are represented as balls in this computer graphic of a chlorophyll molecule. Chlorophyll is the pigment molecule that makes a maple leaf green. One of the most important molecules on Earth, chlorophyll absorbs sunlight during the first step of photosynthesis. Within each chloroplast, millions of chlorophylls and other molecules are organized into the equipment that converts light energy to the chemical energy of food.

Systems Biology

A system is simply a combination of components that function together. A biologist can study a system at any level of organization. A single leaf cell can be considered a system, as can a frog, an ant colony, or a desert ecosytem. To understand how such systems work, it is not enough to have a "parts list," even a complete one. Realizing this, many researchers are now complementing the reductionist approach with new strategies for studying whole systems. This changing perspective is analogous to moving from ground level on a street corner to a helicopter high above a city, from which you can see how variables such as time of day, construction projects, accidents, and traffic-signal malfunctions affect traffic throughout the city.

The goal of **systems biology** is to construct models for the dynamic behavior of whole biological systems. Successful models enable biologists to predict how a change in one or more variables will affect other components and the whole system. Thus, the systems approach enables us to pose new kinds of questions. How might a drug that lowers blood pressure affect the functions of organs throughout the human body? How might increasing a crop's water supply affect processes in the plants, such as the storage of molecules essential for human nutrition? How might a gradual increase in atmospheric carbon dioxide alter ecosystems and the entire biosphere? The ultimate aim of systems biology is to answer big questions like the last one.

Systems biology is relevant to the study of life at all levels. During the early years of the 20th century, biologists studying animal physiology (functioning) began integrating data on how multiple organs coordinate processes such as the regulation of sugar concentration in the blood. And in the 1960s, scientists investigating ecosystems pioneered a more mathematically sophisticated systems approach with elaborate models diagramming the network of interactions between organisms and nonliving components of ecosystems such as salt marshes. Such models have already been useful for predicting the responses of these systems to changing variables. More recently, systems biology has taken hold at the cellular and molecular levels, as we'll describe later when we discuss DNA.

Theme: Organisms interact with their environments, exchanging matter and energy

Turn back again to Figure 1.4, this time focusing on the forest. In this or any other ecosystem, each organism interacts continuously with its environment, which includes both nonliving factors and other organisms. A tree, for example, absorbs water and minerals from the soil, through its roots. At the same time, its leaves take in carbon dioxide from the air and use sunlight absorbed by chlorophyll to drive photosynthesis, converting water and carbon dioxide to sugar and oxygen. The tree releases oxygen to the air, and its roots help form soil by breaking up rocks. Both organism and environment are affected by the interactions between them. The tree also interacts with other organisms, such as soil microorganisms associated with its roots and animals that eat its leaves and fruit.

Ecosystem Dynamics

The operation of any ecosystem involves two major processes. One process is the cycling of nutrients. For example, minerals acquired by a tree will eventually be returned to the soil by organisms that decompose leaf litter, dead roots, and other organic debris. The second major process in an ecosystem is the one-way flow of energy from sunlight to producers to consumers. Producers are plants and other photosynthetic organisms, which use light energy to make sugar. Consumers are organisms, such as animals, that feed on producers and other consumers. The diagram in **Figure 1.5** outlines the two processes acting in an African ecosystem.

Energy Conversion

Moving, growing, reproducing, and the other activities of life are work, and work requires energy. The exchange of energy between an organism and its surroundings often involves the transformation of one form of energy to another. For example, the leaves of a plant absorb light energy and convert it to chemical energy stored in sugar molecules. When an animal's muscle cells use sugar as fuel to power movements, they convert chemical energy to kinetic energy, the energy of motion. And

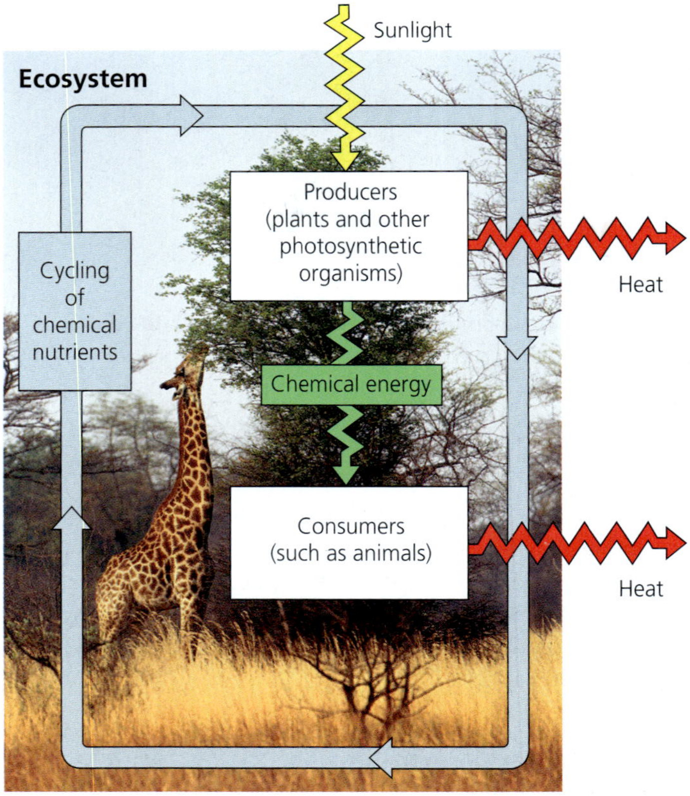

▲ **Figure 1.5 Nutrient cycling and energy flow in an ecosystem.**

(a) A bird's wings have an aerodynamically efficient shape.

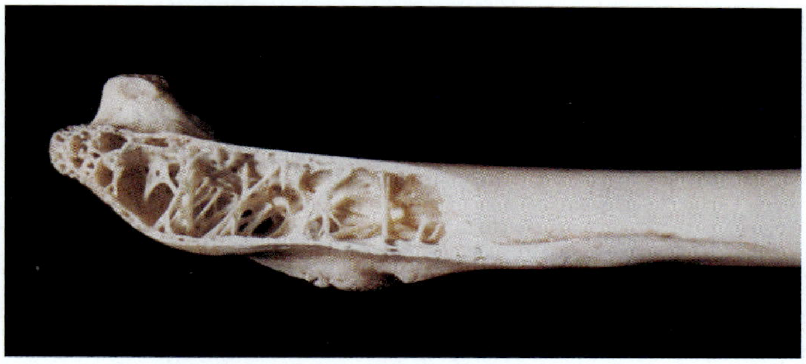

(b) Wing bones have a honeycombed internal structure that is strong but lightweight.

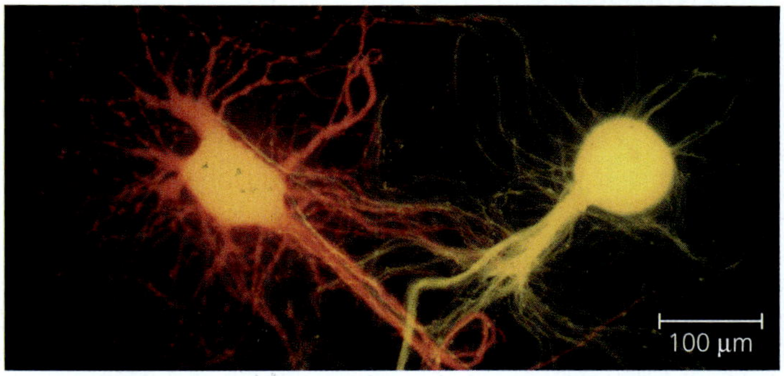

(c) The flight muscles are controlled by neurons (nerve cells), which transmit signals. With long extensions, neurons are especially well structured for communication within the body.

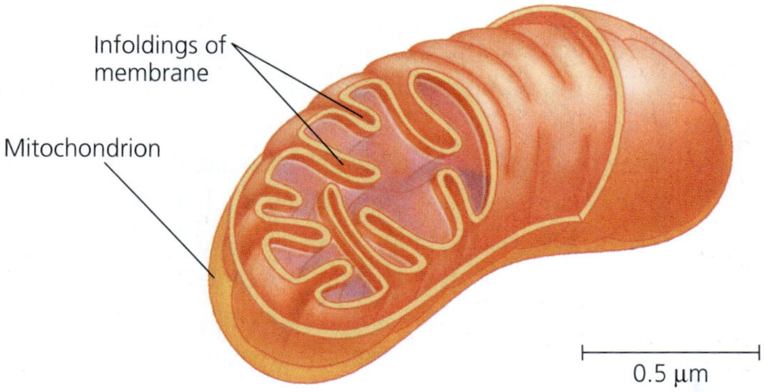

(d) The flight muscles obtain energy in a usable form from organelles called mitochondria. A mitochondrion has an inner membrane with many infoldings. Molecules embedded in the inner membrane carry out many of the steps in energy production, and the infoldings pack a large amount of this membrane into a small container.

▲ **Figure 1.6 Form fits function in a gull's wing.** A bird's build and the structures of its components make flight possible.

? *How does form fit function in a human hand?*

in all these energy conversions, some of the energy is converted to thermal energy, which dissipates to the surroundings as heat. In contrast to chemical nutrients, which recycle within an ecosystem, energy flows through an ecosystem, usually entering as light and exiting as heat (see Figure 1.5).

Theme: Structure and function are correlated at all levels of biological organization

Another theme evident in Figure 1.4 is the idea that form fits function, which you'll recognize from everyday life. For example, a screwdriver is suited to tighten or loosen screws, a hammer to pound nails. How a device works is correlated with its structure. Applied to biology, this theme is a guide to the anatomy of life at all its structural levels. An example from Figure 1.4 is seen in the leaf: Its thin, flat shape maximizes the amount of sunlight that can be captured by its chloroplasts. Analyzing a biological structure gives us clues about what it does and how it works. Conversely, knowing the function of something provides insight into its construction. An example from the animal kingdom, the wing of a bird, provides additional instances of the structure-function theme **(Figure 1.6)**. In exploring life on its different structural levels, we discover functional beauty at every turn.

Theme: Cells are an organism's basic units of structure and function

In life's structural hierarchy, the cell has a special place as the lowest level of organization that can perform all activities required for life. Moreover, the activities of organisms are all based on the activities of cells. For instance, the division of cells to form new cells is the basis for all reproduction and for the growth and repair of multicellular organisms **(Figure 1.7)**. To

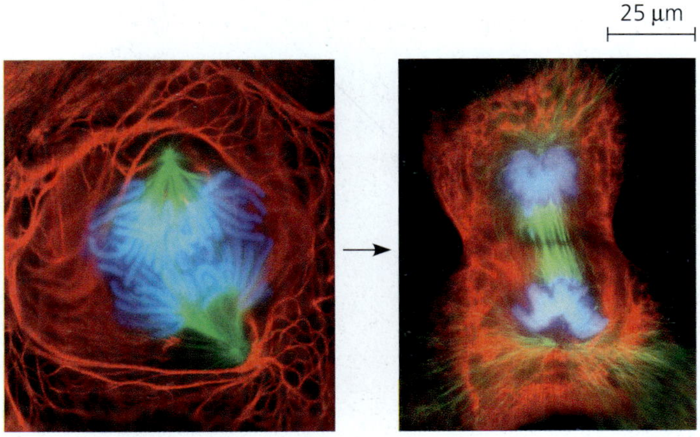

▲ **Figure 1.7 A lung cell from a newt divides into two smaller cells that will grow and divide again.**

cite another example, the movement of your eyes as you read this line is based on activities of muscle and nerve cells. Even a global process such as the recycling of carbon is the cumulative product of cellular activities, including the photosynthesis that occurs in the chloroplasts of leaf cells. Understanding how cells work is a major focus of biological research.

All cells share certain characteristics. For example, every cell is enclosed by a membrane that regulates the passage of materials between the cell and its surroundings. And every cell uses DNA as its genetic information. However, we can distinguish between two main forms of cells: prokaryotic cells and eukaryotic cells. The cells of two groups of microorganisms called bacteria and archaea are prokaryotic. All other forms of life, including plants and animals, are composed of eukaryotic cells.

A **eukaryotic cell** is subdivided by internal membranes into various membrane-enclosed organelles, such as the ones you see in **Figure 1.8** and the chloroplast you saw in Figure 1.4. In most eukaryotic cells, the largest organelle is the nucleus, which contains the cell's DNA. The other organelles are located in the cytoplasm, the entire region between the nucleus and outer membrane of the cell. As Figure 1.8 also shows, prokaryotic cells are much simpler and generally smaller than eukaryotic cells. In a **prokaryotic cell**, the DNA is not separated from the rest of the cell by enclosure in a membrane-bounded nucleus. Prokaryotic cells also lack the other kinds of membrane-enclosed organelles that characterize eukaryotic cells. But whether an organism has prokaryotic or eukaryotic cells, its structure and function depend on cells.

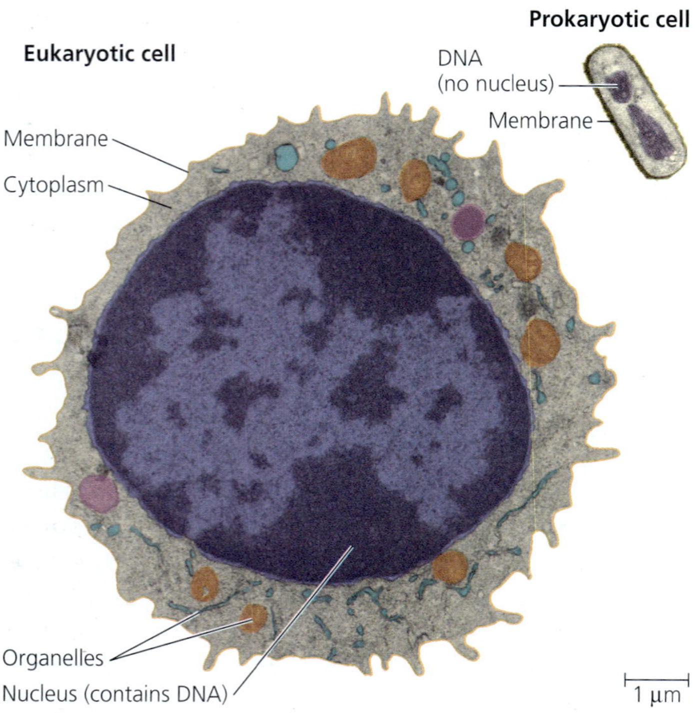

▲ **Figure 1.8 Contrasting eukaryotic and prokaryotic cells in size and complexity.**

Theme: The continuity of life is based on heritable information in the form of DNA

Inside the dividing cell in Figure 1.7 (on the previous page), you can see structures called chromosomes, which are stained with a blue-glowing dye. The chromosomes have almost all of the cell's genetic material, its **DNA** (short for deoxyribonucleic acid). DNA is the substance of **genes**, the units of inheritance that transmit information from parents to offspring. Your blood group (A, B, AB, or O), for example, is the result of certain genes that you inherited from your parents.

DNA Structure and Function

Each chromosome has one very long DNA molecule, with hundreds or thousands of genes arranged along its length. The DNA of chromosomes replicates as a cell prepares to divide, and each of the two cellular offspring inherits a complete set of genes.

Each of us began life as a single cell stocked with DNA inherited from our parents. Replication of that DNA with each round of cell division transmitted copies of it to our trillions of cells. In each cell, the genes along the length of the DNA molecules encode the information for building the cell's other molecules. In this way, DNA controls the development and maintenance of the entire organism and, indirectly, everything it does **(Figure 1.9)**. The DNA serves as a central database.

The molecular structure of DNA accounts for its ability to store information. Each DNA molecule is made up of two long chains arranged in a double helix. Each chain link is one of four kinds of chemical building blocks called nucleotides **(Figure 1.10)**. The way DNA encodes information is analogous to the way we arrange the letters of the alphabet into precise sequences with specific meanings. The word *rat,* for example, evokes a rodent; the words *tar* and *art,* which contain the same letters, mean very different things. Libraries are filled with books containing information encoded in varying sequences of only 26 letters. We can think of nucleotides as the alphabet of inheritance. Specific sequential arrangements of these four chemical letters encode the precise information in genes, which are typically hundreds or thousands of nucleotides long. One gene in a bacterial cell may be translated as "Build a certain component of the cell membrane." A particular human gene may mean "Make growth hormone."

More generally, genes like those just mentioned program the cell's production of large molecules called proteins. Other human proteins include a muscle cell's contraction proteins and the defensive proteins called antibodies. A class of proteins crucial to all cells are enzymes, which catalyze (speed up) specific chemical reactions. Thus, DNA provides the blueprints, and proteins serve as the tools that actually build and maintain the cell and carry out its activities.

The DNA of genes controls protein production indirectly, using a related kind of molecule called RNA as an intermediary.

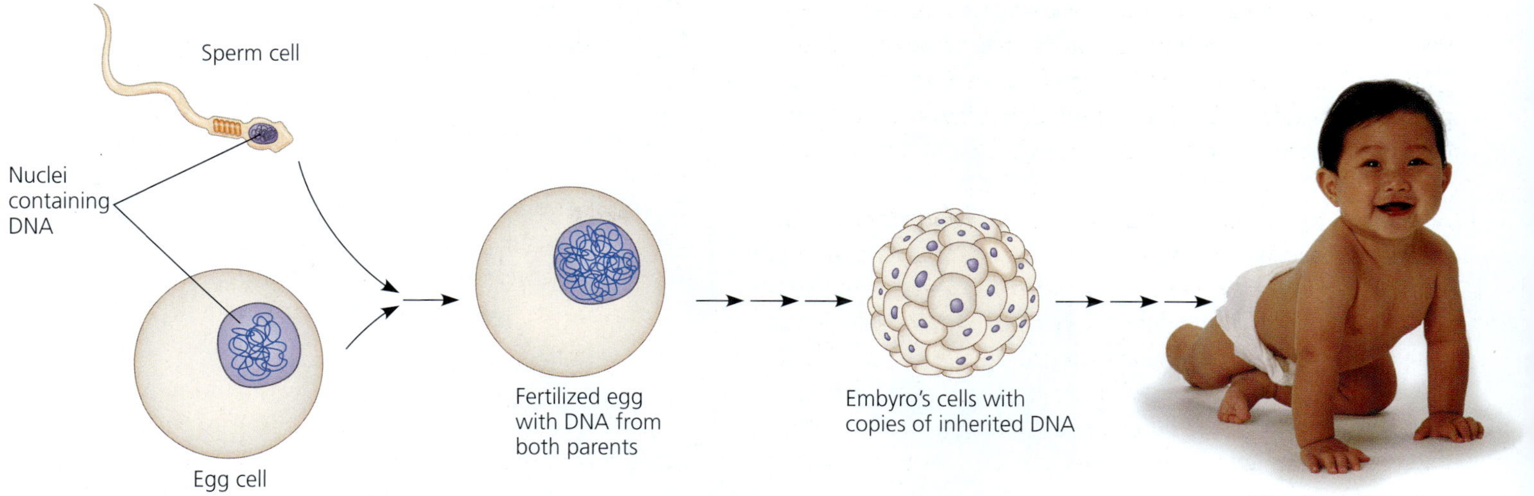

▲ **Figure 1.9 Inherited DNA directs development of an organism.**

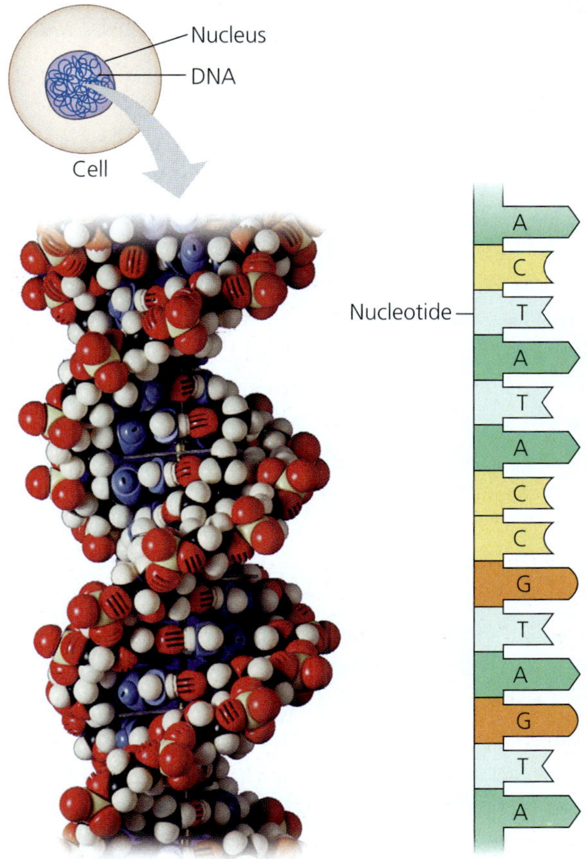

(a) DNA double helix. This model shows each atom in a segment of DNA. Made up of two long chains of building blocks called nucleotides, a DNA molecule takes the three-dimensional form of a double helix.

(b) Single strand of DNA. These geometric shapes and letters are simple symbols for the nucleotides in a small section of one chain of a DNA molecule. Genetic information is encoded in specific sequences of the four types of nucleotides. (Their names are abbreviated here as A, T, C, and G.)

▲ **Figure 1.10 DNA: The genetic material.**

The sequence of nucleotides along a gene is transcribed into RNA, which is then translated into a specific protein with a unique shape and function. In the translation process, all forms of life employ essentially the same genetic code. A particular sequence of nucleotides says the same thing to one organism as it does to another. Differences between organisms reflect differences between their nucleotide sequences.

Not all RNA in the cell is translated into protein. We have known for decades that some types of RNA molecules are actually components of the cellular machinery that manufactures proteins. Recently, scientists have discovered whole new classes of RNA that play other roles in the cell, such as regulating the functioning of protein-coding genes.

The entire "library" of genetic instructions that an organism inherits is called its **genome**. A typical human cell has two similar sets of chromosomes, and each set has DNA totaling about 3 billion nucleotides. If the one-letter symbols for these nucleotides were written in letters the size of those you are now reading, the genetic text would fill about 600 books the size of this one. Within this genomic library of nucleotide sequences are genes for about 75,000 kinds of proteins and an as yet unknown number of RNA molecules.

Systems Biology at the Levels of Cells and Molecules

The entire sequence of nucleotides in the human genome is now known, along with the genome sequences of many other organisms, including bacteria, archaea, fungi, plants, and animals. These accomplishments have been made possible by the development of new methods and DNA-sequencing machines, such as those shown in **Figure 1.11**, on the next page.

The sequencing of the human genome is a scientific and technological achievement comparable to landing the *Apollo* astronauts on the moon in 1969. But it is only the beginning of

▶ **Figure 1.11 Modern biology as an information science.** Automatic DNA-sequencing machines and abundant computing power made the sequencing of the human genome possible. This facility in Walnut Creek, California, was one of many labs that collaborated in the international Human Genome Project.

an even bigger research endeavor, an effort to learn how the activities of the myriad proteins encoded by the DNA are coordinated in cells and whole organisms.

The best way to make sense of the deluge of data from genome-sequencing projects and the growing catalog of known protein functions is to apply a systems approach at the cellular and molecular levels. **Figure 1.12** illustrates the results of a large study that mapped a network of protein interactions within a cell of a fruit fly, a popular research organism. The model is based on a database of thousands of known proteins and their known interactions with other proteins. For example, protein A may attach to and alter the activities of proteins B, C, and D, which then go on to interact with still other proteins. The figure maps these protein partnerships to their cellular locales.

The basics of the systems strategy are straightforward. First, it is necessary to inventory as many parts of the system as possible, such as all the known genes and proteins in a cell (an application of reductionism). Then it is necessary to investigate how each part behaves in relation to others in the working system—all the protein-protein interactions, in the case of our fly cell example. Finally, with the help of computers and specialized software, it is possible to pool all the data into the kind of system network pictured in Figure 1.12.

Though the basic idea of systems biology is simple, the practice is not, as you would expect from the complexity of biological systems. It has taken three key research developments to bring systems biology within reach. One is "high-throughput" technology, tools that can analyze biological materials very rapidly and produce enormous amounts of data. The automatic DNA-sequencing machines that made the sequencing of

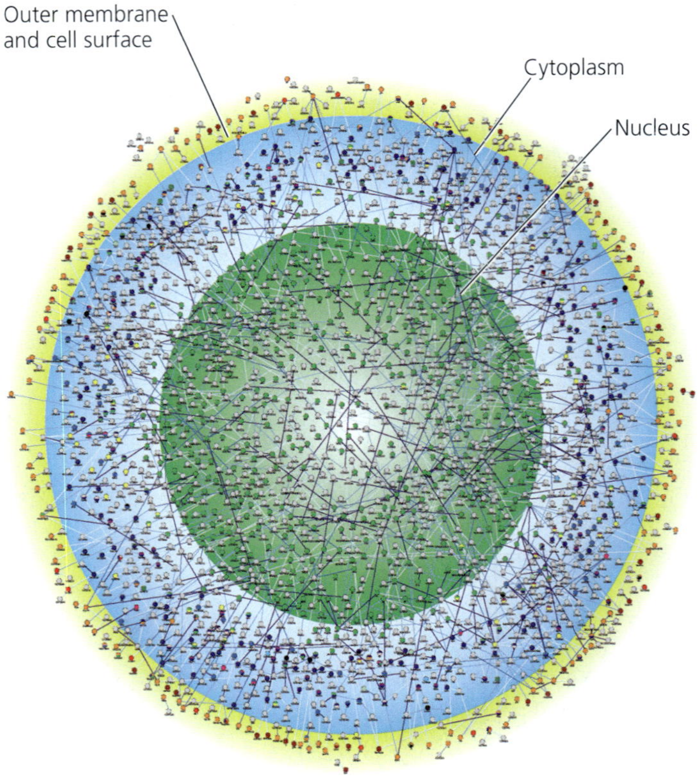

▲ **Figure 1.12 A systems map of interactions among proteins in a cell.** This diagram maps 2,346 proteins (dots) and their network of interactions (lines connecting the proteins) in a fruit fly cell. Systems biologists develop such models from huge databases of information about molecules and their interactions in the cell. A major goal of this systems approach is to use the models to predict how one change, such as an increase in the activity of a particular protein, can ripple through the cell's molecular circuitry to cause other changes. The total number of proteins in this type of cell is probably in the range of 4,000 to 7,000.

the human genome possible are examples of high-throughput devices (see Figure 1.11). The second is **bioinformatics**, which is the use of computational tools to store, organize, and analyze the huge volume of data that result from high-throughput methods. The third key development is the formation of interdisciplinary research teams—melting pots of diverse specialists that may include computer scientists, mathematicians, engineers, chemists, physicists, and, of course, biologists from a variety of fields.

Theme: Feedback mechanisms regulate biological systems

A kind of supply-and-demand economy applies to many biological systems. Consider your muscles, for instance. When your muscle cells require more energy during exercise, they increase their consumption of the sugar molecules that provide fuel. In contrast, when you rest, a different set of chemical reactions converts surplus sugar to storage molecules.

Like most of the cell's chemical processes, those that decompose or store sugar are accelerated, or catalyzed, by the specialized proteins called enzymes. Each type of enzyme catalyzes a specific chemical reaction. In many cases, these reactions are linked into chemical pathways, each step with its own enzyme. How does the cell coordinate its various chemical pathways? In our example of sugar management, how does the cell match fuel supply to demand, regulating its opposing pathways of sugar consumption and storage? The key is the ability of many biological processes to self-regulate by a mechanism called feedback.

In feedback regulation, the output, or product, of a process regulates that very process. In life, the most common form of regulation is **negative feedback**, in which accumulation of an end product of a process slows that process. For example, the cell's breakdown of sugar generates chemical energy in the form of a substance called ATP. When a cell makes more ATP than it can use, the excess ATP "feeds back" and inhibits an enzyme near the beginning of the pathway **(Figure 1.13a)**.

Though less common than processes regulated by negative feedback, there are also many biological processes regulated by **positive feedback**, in which an end product *speeds up* its production **(Figure 1.13b)**. The clotting of your blood in response to injury is an example. When a blood vessel is damaged, structures in the blood called platelets begin to aggregate at the site. Positive feedback occurs as chemicals released by the platelets attract *more* platelets. The platelet pile then initiates a complex process that seals the wound with a clot.

Feedback is a regulatory motif common to life at all levels, from the molecular level to ecosystems and the biosphere. Such regulation is an example of the integration that makes living systems much greater than the sum of their parts.

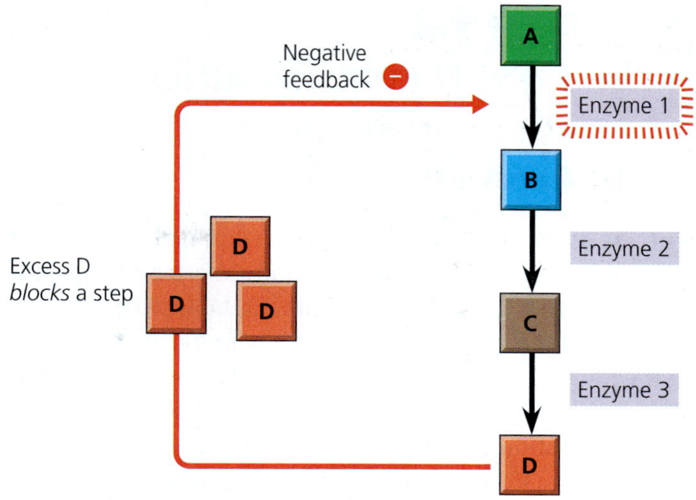

(a) Negative feedback. This three-step chemical pathway converts substance A to substance D. A specific enzyme catalyzes each chemical reaction. Accumulation of the final product (D) inhibits the first enzyme in the sequence, thus slowing down production of more D.

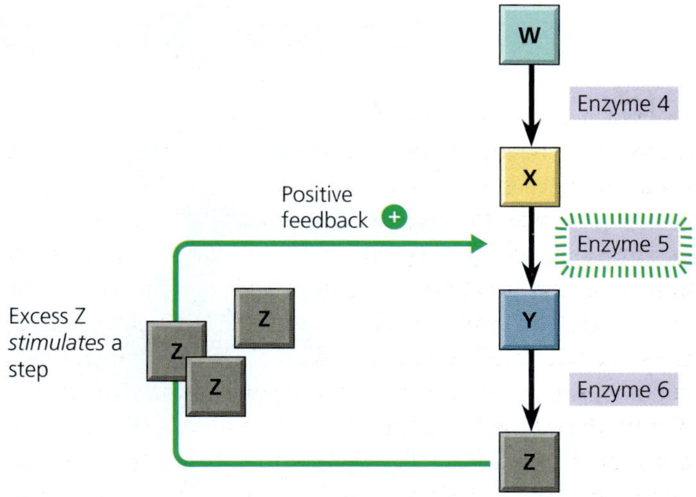

(b) Positive feedback. In a biochemical pathway regulated by positive feedback, a product stimulates an enzyme in the reaction sequence, increasing the rate of production of the product.

▲ **Figure 1.13 Regulation by feedback mechanisms.**

? *What would happen if enzyme 2 were missing?*

CONCEPT CHECK 1.1

1. For each biological level in Figure 1.4, write a sentence that includes the next "lower" level. Example: "A community consists of *populations* of the various species inhabiting a specific area."
2. What theme or themes are exemplified by (a) the sharp spines of a porcupine, (b) the cloning of a plant from a single cell, and (c) a hummingbird using sugar to power its flight?
3. **WHAT IF?** For each theme discussed in this section, give an example not mentioned in the book.

For suggested answers, see Appendix A.

CONCEPT 1.2

The Core Theme: Evolution accounts for the unity and diversity of life

The list of biological themes discussed under Concept 1.1 is not absolute; some people might find a shorter or longer list more useful. There is consensus among biologists, however, as to the core theme of biology: It is evolution. To quote one of the founders of modern evolutionary theory, Theodosius Dobzhansky, "Nothing in biology makes sense except in the light of evolution."

In addition to encompassing a hierarchy of size scales from molecules to the biosphere, biology extends across the great diversity of species that have ever lived on Earth. To understand Dobzhansky's statement, we need to discuss how biologists think about this vast diversity.

Organizing the Diversity of Life

Diversity is a hallmark of life. Biologists have so far identified and named about 1.8 million species. To date, this diversity of life is known to include at least 6,300 species of prokaryotes (organisms with prokaryotic cells), 100,000 fungi, 290,000 plants, 52,000 vertebrates (animals with backbones), and 1 million insects (more than half of all known forms of life). Researchers identify thousands of additional species each year. Estimates of the total number of species range from about 10 million to over 100 million. Whatever the actual number, the enormous variety of life gives biology a very broad scope. Biologists face a major challenge in attempting to make sense of this variety **(Figure 1.14)**.

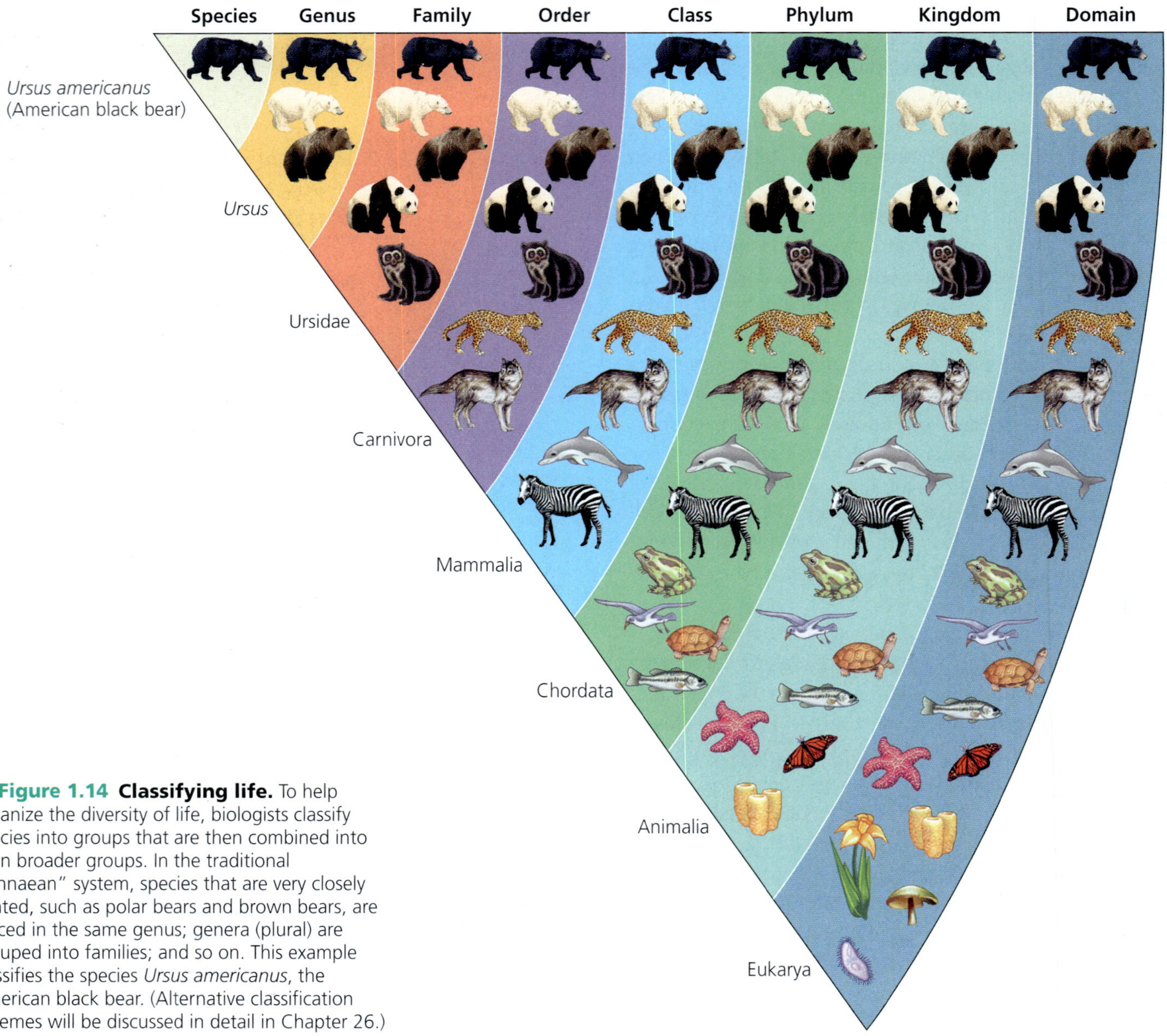

▲ **Figure 1.14 Classifying life.** To help organize the diversity of life, biologists classify species into groups that are then combined into even broader groups. In the traditional "Linnaean" system, species that are very closely related, such as polar bears and brown bears, are placed in the same genus; genera (plural) are grouped into families; and so on. This example classifies the species *Ursus americanus*, the American black bear. (Alternative classification schemes will be discussed in detail in Chapter 26.)

Grouping Species: The Basic Idea

There is a human tendency to group diverse items according to similarities. For instance, perhaps you organize your music collection by artist. And then maybe you group the various artists into broader categories, such as rock, jazz, and classical. In the same way, grouping species that are similar is natural for us. We may speak of squirrels and butterflies, though we recognize that many different species belong to each group. We may even sort groups into broader categories, such as rodents (which include squirrels) and insects (which include butterflies). Taxonomy, the branch of biology that names and classifies species, formalizes this ordering of species into groups of increasing breadth (see Figure 1.14). You will learn more about this taxonomic scheme in Chapter 26. For now, we will focus on kingdoms and domains, the broadest units of classification.

The Three Domains of Life

Until a few decades ago, most biologists adopted a taxonomic scheme that divided the diversity of life into five kingdoms: plants, animals, fungi, single-celled eukaryotic organisms, and prokaryotes. Since then, new methods, such as comparisons of DNA sequences from different species, have led to an ongoing reevaluation of the number and boundaries of kingdoms. Researchers have proposed anywhere from six kingdoms to dozens of kingdoms. But as debate continues at the kingdom level, there is a consensus that the kingdoms of life can now be grouped into three even higher levels of classification called domains. The three domains are named Bacteria, Archaea, and Eukarya **(Figure 1.15)**.

The organisms making up **domain Bacteria** and **domain Archaea** are all prokaryotic. Most prokaryotes are single-celled and microscopic. In the five-kingdom system, bacteria

(a) DOMAIN BACTERIA

▼ **Bacteria** are the most diverse and widespread prokaryotes and are now divided among multiple kingdoms. Each of the rod-shaped structures in this photo is a bacterial cell.

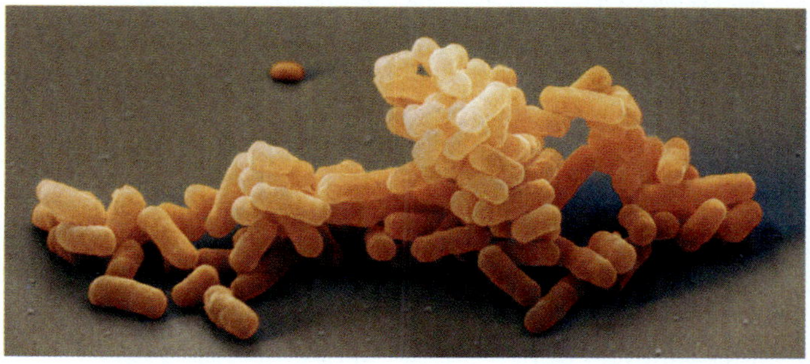

(b) DOMAIN ARCHAEA

▼ Many of the prokaryotes known as **archaea** live in Earth's extreme environments, such as salty lakes and boiling hot springs. Domain Archaea includes multiple kingdoms. The photo shows a colony composed of many cells.

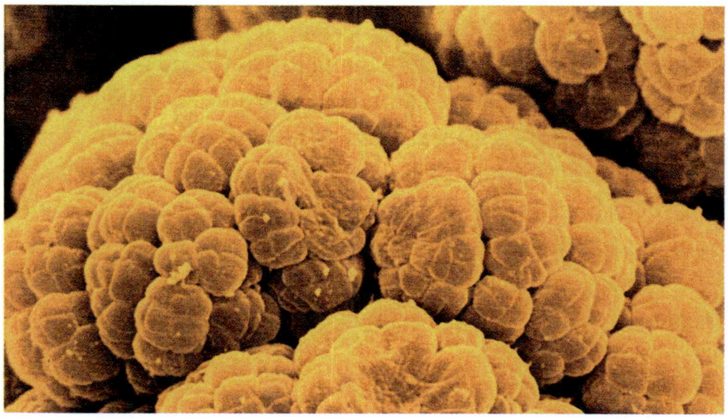

(c) DOMAIN EUKARYA

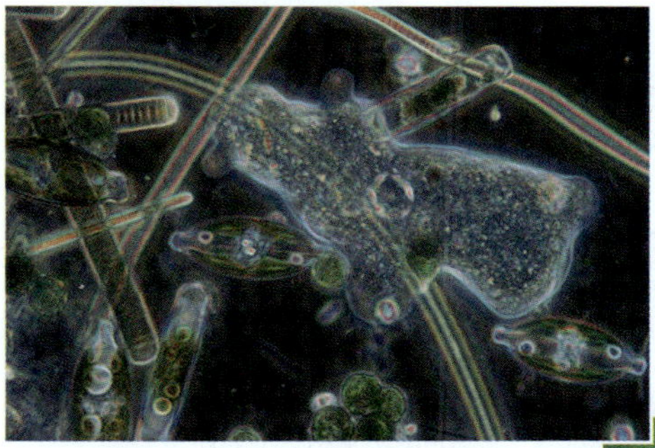

▲ **Protists** (multiple kingdoms) are unicellular eukaryotes and their relatively simple multicellular relatives. Pictured here is an assortment of protists inhabiting pond water. Scientists are currently debating how to assign the protists to kingdoms that accurately reflect their evolutionary relationships.

► **Kingdom Fungi** is defined in part by the nutritional mode of its members, such as this mushroom, which absorb nutrients from their surroundings.

▲ **Kingdom Plantae** consists of multicellular eukaryotes that carry out photosynthesis, the conversion of light energy to the chemical energy in food.

▼ **Kingdom Animalia** consists of multicellular eukaryotes that ingest other organisms.

▲ **Figure 1.15 The three domains of life.**

and archaea were combined in a single kingdom because they shared the prokaryotic form of cell structure. But much evidence now supports the view that bacteria and archaea represent two very distinct branches of prokaryotic life, different in key ways that you'll learn about in Chapter 27. There is also evidence that archaea are at least as closely related to eukaryotic organisms as they are to bacteria.

All the eukaryotes (organisms with eukaryotic cells) are now grouped in **domain Eukarya**. In the era of the five-kingdom scheme, most single-celled eukaryotes, such as the microorganisms known as protozoans, were placed in a single kingdom, "Protista." Many biologists extended the boundaries of kingdom Protista to include some multicellular forms, such as seaweeds, that are closely related to certain unicellular protists. The recent taxonomic trend has been to split the protists into several groups at the kingdom level. In addition to these protistan groups, domain Eukarya includes three kingdoms of multicellular eukaryotes: kingdoms Plantae, Fungi, and Animalia. These three kingdoms are distinguished partly by their modes of nutrition. Plants produce their own sugars and other foods by photosynthesis. Fungi absorb dissolved nutrients from their surroundings; many decompose dead organisms and organic wastes (such as leaf litter and animal feces) and absorb nutrients from these sources. Animals obtain food by ingestion, which is the eating and digesting of other organisms. Animalia is, of course, the kingdom to which we belong.

Unity in the Diversity of Life

As diverse as life is, it also displays remarkable unity. Earlier we mentioned the similar skeletons of different vertebrate animals, but similarities are even more striking at the molecular and cellular levels. For example, the universal genetic language of DNA is common to organisms as different as bacteria and animals. Unity is also evident in many features of cell structure **(Figure 1.16)**.

How can we account for life's dual nature of unity and diversity? The process of evolution, explained next, illuminates both the similarities and differences in the world of life and introduces another dimension of biology: historical time.

Charles Darwin and the Theory of Natural Selection

The history of life, as documented by fossils and other evidence, is a saga of a changing Earth billions of years old, inhabited by an evolving cast of living forms **(Figure 1.17)**. This evolutionary view of life came into sharp focus in November 1859, when Charles Robert Darwin published one of the most important and influential books ever written. Entitled *On the Origin of Species by Means of Natural Selection*, Darwin's book was an immediate bestseller and soon made "Darwinism" almost synonymous with the concept of evolution **(Figure 1.18)**.

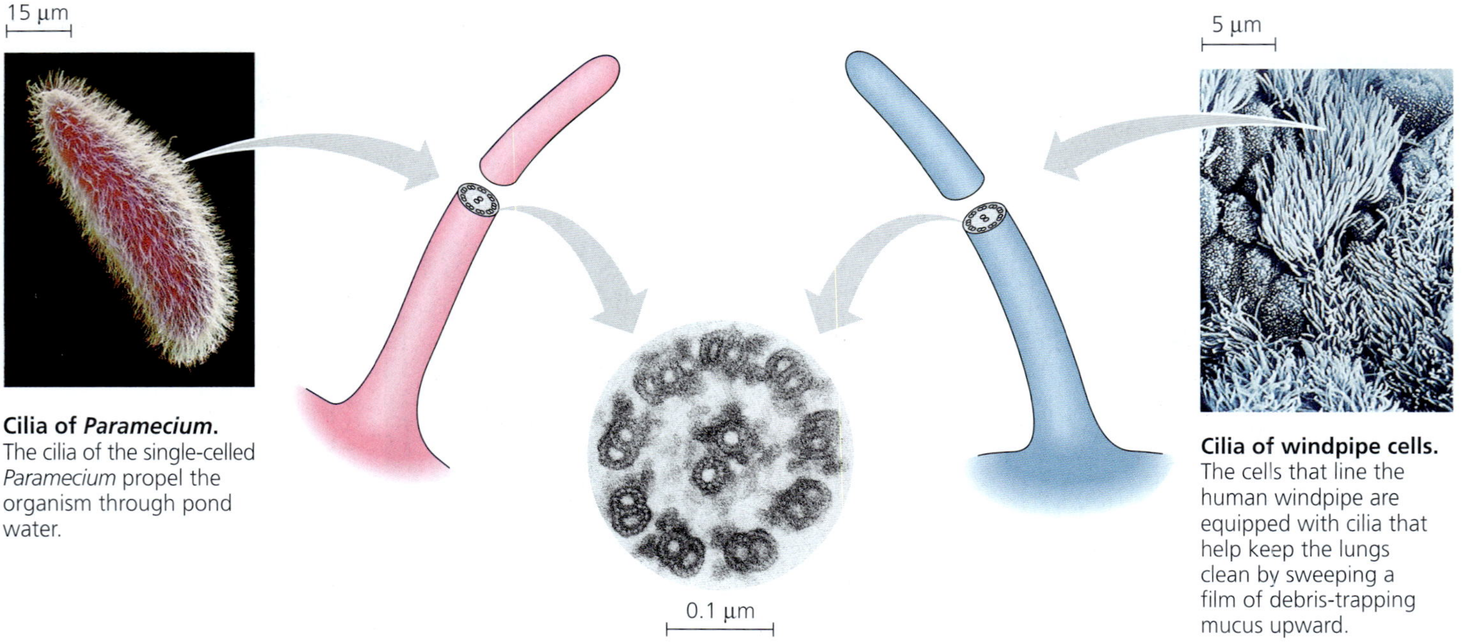

Cilia of *Paramecium*. The cilia of the single-celled *Paramecium* propel the organism through pond water.

Cilia of windpipe cells. The cells that line the human windpipe are equipped with cilia that help keep the lungs clean by sweeping a film of debris-trapping mucus upward.

Cross section of a cilium, as viewed with an electron microscope

▲ **Figure 1.16 An example of unity underlying the diversity of life: the architecture of cilia in eukaryotes.** Cilia (singular, *cilium*) are extensions of cells that function in locomotion. They occur in eukaryotes as diverse as paramecia and humans. Even organisms so different share a common architecture for their cilia, which have an elaborate system of tubules that is striking in cross-sectional views.

▲ **Figure 1.17 Digging into the past.** Paleontologist Paul Sereno, of the University of Chicago, gingerly excavates the leg bones of a dinosaur fossil in Niger.

The Origin of Species articulated two main points. First, Darwin presented evidence to support his view that contemporary species arose from a succession of ancestors. (We will discuss the evidence for evolution in detail in Chapter 22.) Darwin called this evolutionary history of species "descent with modification." It was an insightful phrase, as it captured the duality of life's unity and diversity—unity in the kinship among species that descended from common ancestors; diversity in the modifications that evolved as species branched from their common ancestors **(Figure 1.19)**. Darwin's second main point was to propose a mechanism for descent with modification. He called this evolutionary mechanism natural selection.

▲ **Figure 1.18 Charles Darwin as a young man.**

Darwin synthesized his theory of natural selection from observations that by themselves were neither new nor profound. Others had the pieces of the puzzle, but Darwin saw how they fit together. He started with the following observations from nature: Individuals in a population vary in their traits, many of which seem to be heritable (passed on from parents to offspring). Also, a population can produce far more offspring than can survive to produce offspring of their own. With more individuals than the environment can support, competition is inevitable. Lastly, species generally suit their environments. For instance, birds living where tough seeds are a good food source may have especially strong beaks.

Darwin made inferences from these observations to arrive at his theory of evolution. He reasoned that individuals with inherited traits that are best suited to the local environment are more likely to survive and reproduce than less fit individuals. Over many generations, a higher and higher proportion

▲ **Figure 1.19 Unity and diversity in the orchid family.** These three rain forest orchids are variations on a common floral theme. For example, each of these flowers has a liplike petal that helps attract pollinating insects and provides a landing platform for the pollinators.

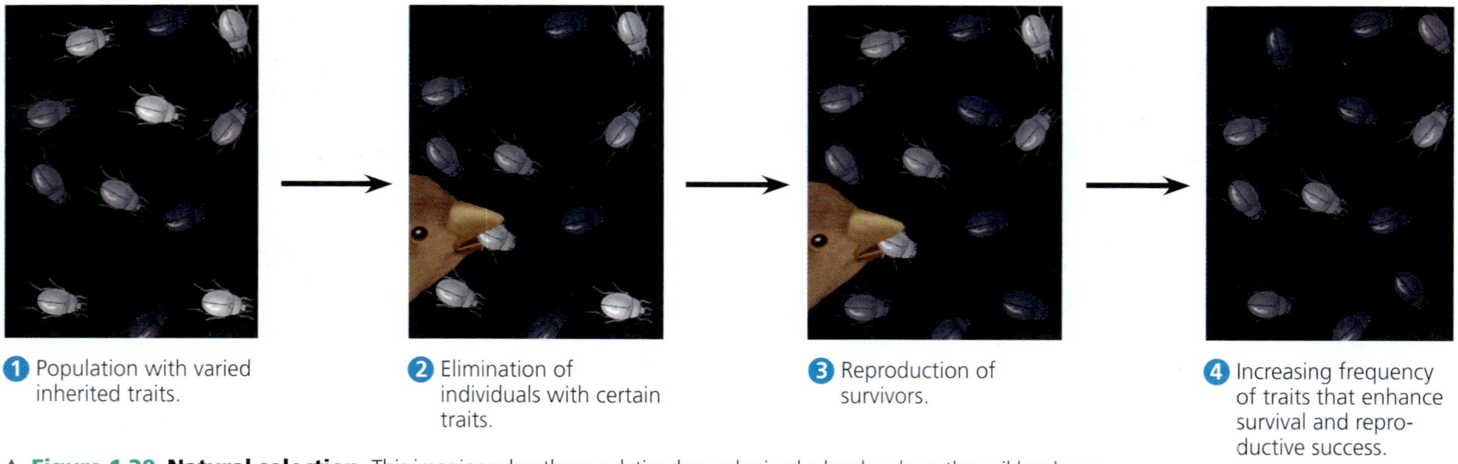

▲ **Figure 1.20 Natural selection.** This imaginary beetle population has colonized a locale where the soil has been blackened by a recent brush fire. Initially, the population varies extensively in the inherited coloration of the individuals, from very light gray to charcoal. For hungry birds that prey on the beetles, it is easiest to spot the beetles that are lightest in color.

of individuals in a population will have the advantageous traits. Evolution occurs as the unequal reproductive success of individuals adapts the population to its environment.

Darwin called this mechanism of evolutionary adaptation "natural selection" because the natural environment "selects" for the propagation of certain traits. The example in **Figure 1.20** illustrates the ability of natural selection to "edit" a population's heritable variations in color. We see the products of natural selection in the exquisite adaptations of various organisms to the special circumstances of their way of life and their environment **(Figure 1.21)**.

▲ **Figure 1.21 Evolutionary adaptation.** Bats, the only mammals capable of active flight, have wings with webbing between extended "fingers." In the Darwinian view of life, such adaptations are refined by natural selection.

The Tree of Life

Take another look at the skeletal architecture of the bat's wings in Figure 1.21. These forelimbs, though adapted for flight, actually have all the same bones, joints, nerves, and blood vessels found in other limbs as diverse as the human arm, the horse's foreleg, and the whale's flipper. Indeed, all mammalian forelimbs are anatomical variations of a common architecture, much as the flowers in Figure 1.19 are variations on an underlying "orchid" theme. Such examples of kinship connect life's unity in diversity to the Darwinian concept of descent with modification. In this view, the unity of mammalian limb anatomy reflects inheritance of that structure from a common ancestor—the "prototype" mammal from which all other mammals descended. The diversity of mammalian forelimbs results from modification by natural selection operating over millions of generations in different environmental contexts. Fossils and other evidence corroborate anatomical unity in supporting this view of mammalian descent from a common ancestor.

Darwin proposed that natural selection, by its cumulative effects over long periods of time, could cause an ancestral species to give rise to two or more descendant species. This could occur, for example, if one population fragmented into several subpopulations isolated in different environments. In these separate arenas of natural selection, one species could gradually radiate into multiple species as the geographically isolated populations adapted over many generations to different sets of environmental factors.

The "family tree" of 14 finches in **Figure 1.22** illustrates a famous example of adaptive radiation of new species from a common ancestor. Darwin collected specimens of these birds during his 1835 visit to the remote Galápagos Islands, 900 kilometers (km) off the Pacific coast of South America. These relatively young, volcanic islands are home to many species of plants and animals found nowhere else in the world, though most Galápagos organisms are clearly related to species on the South American mainland. After volcanism built the Galápagos several million years ago, finches probably diversified on the various islands from an ancestral finch species that by chance reached the archipelago from elsewhere. (Once

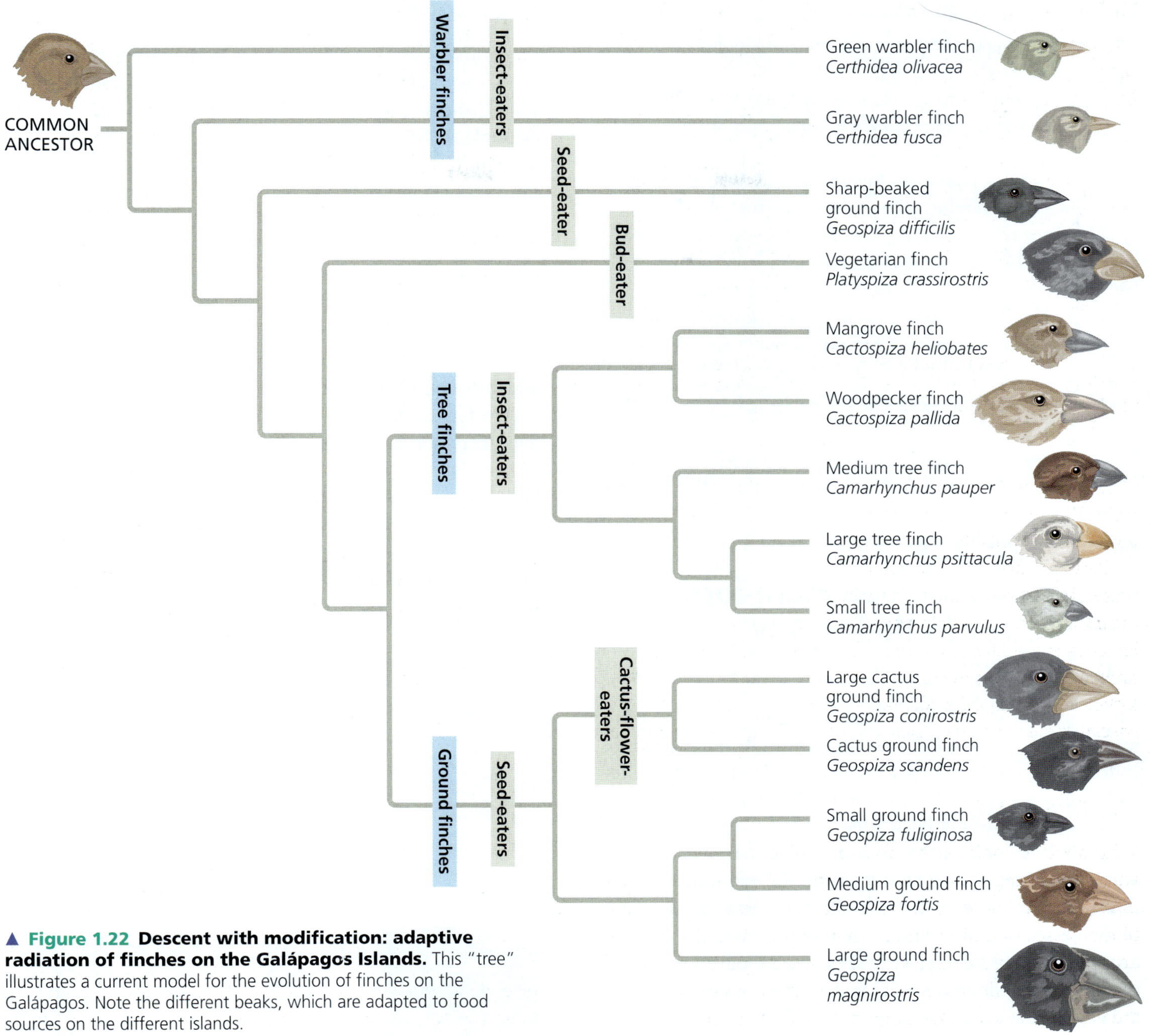

▲ **Figure 1.22 Descent with modification: adaptive radiation of finches on the Galápagos Islands.** This "tree" illustrates a current model for the evolution of finches on the Galápagos. Note the different beaks, which are adapted to food sources on the different islands.

thought to have originated on the mainland of South America like many Galápagos organisms, the ancestral finches are now thought to have come from islands of the Carribean.) Years after Darwin's collection of Galápagos finches, researchers began to sort out the relationships among the finch species, first from anatomical and geographic data and more recently with the help of DNA sequence comparisons.

Biologists' diagrams of evolutionary relationships generally take treelike forms, though today biologists usually turn the trees sideways as in Figure 1.22. Tree diagrams make sense: Just as an individual has a genealogy that can be diagrammed as a family tree, each species of organism is one twig of a branching tree of life extending back in time through ancestral species more and more remote. Species that are very similar, such as the Galápagos finches, share a common ancestor at a relatively recent branch point on the tree of life. But through an ancestor that lived much farther back in time, finches are related to sparrows, hawks, penguins, and all other birds. And birds, mammals, and all other vertebrates share a common ancestor even more ancient. We find evidence of still broader relationships in such similarities as the identical construction of all eukaryotic cilia (see Figure 1.16). Trace life back far enough, and there are only fossils of the primeval prokaryotes that inhabited Earth over 3.5 billion years ago. We can recognize their vestiges in our own cells—in the universal genetic code, for example. All of life is connected through its long evolutionary history.

CONCEPT CHECK 1.2

1. How is a mailing address analogous to biology's hierarchical taxonomic system?
2. Explain why "editing" is an appropriate metaphor for how natural selection acts on a population's heritable variation.
3. **WHAT IF?** The three domains you learned about in Concept 1.2 can be represented in the tree of life as the three main branches. On the eukaryotic branch, three of the subbranches are the kingdoms Plantae, Fungi, and Animalia. What if fungi and animals are more closely related to each other than either of these kingdoms is to plants—as recent evidence strongly suggests? Draw a simple branching pattern that symbolizes the proposed relationship between these three eukaryotic kingdoms.

For suggested answers, see Appendix A.

CONCEPT 1.3

Scientists use two main forms of inquiry in their study of nature

The word *science* is derived from a Latin verb meaning "to know." Science is a way of knowing about the natural world. It developed out of our curiosity about ourselves, other life-forms, our planet, and the universe. Striving to understand seems to be one of our basic urges.

At the heart of science is **inquiry**, a search for information and explanation, often focusing on specific questions. Inquiry drove Darwin to seek answers in nature for how species adapt to their environments. And today inquiry drives the genome analyses that are helping us understand biological unity and diversity at the molecular level. In fact, the inquisitive mind is the engine that drives all progress in biology.

There is no formula for successful scientific inquiry, no single scientific method with a rule book that researchers must rigidly follow. As in all quests, science includes elements of challenge, adventure, and luck, along with careful planning, reasoning, creativity, cooperation, competition, patience, and the persistence to overcome setbacks. Such diverse elements of inquiry make science far less structured than most people realize. That said, it is possible to distill certain characteristics that help to distinguish science from other ways of describing and explaining nature.

Biologists use two main types of scientific inquiry: discovery science and hypothesis-based science. Discovery science is mostly about *describing* nature. Hypothesis-based science is mostly about *explaining* nature. Most scientific inquiries combine these two research approaches.

Discovery Science

Sometimes called descriptive science, **discovery science** describes natural structures and processes as accurately as possible through careful observation and analysis of data. For example, it is discovery science that has built our understanding of cell structure, and it is discovery science that is expanding our databases of genomes of diverse species.

Types of Data

Observation is the use of the senses to gather information, either directly or indirectly with the help of tools such as microscopes that extend our senses. Recorded observations are called **data**. Put another way, data are items of information on which scientific inquiry is based.

The term *data* implies numbers to many people. But some data are *qualitative*, often in the form of recorded descriptions rather than numerical measurements. For example, Jane Goodall spent decades recording her observations of chimpanzee behavior during field research in a jungle in Tanzania **(Figure 1.23)**. She also documented her observations with photographs and movies. Along with these qualitative data, Goodall also enriched the field of animal behavior with volumes of *quantitative* data, which are generally recorded as measurements. Skim through any of the scientific journals in your college library, and you'll see many examples of quantitative data organized into tables and graphs.

▲ **Figure 1.23 Jane Goodall collecting qualitative data on chimpanzee behavior.** Goodall recorded her observations in field notebooks, often with sketches of the animals' behavior.

Induction in Discovery Science

Discovery science can lead to important conclusions based on a type of logic called induction, or **inductive reasoning**. Through induction, we derive generalizations from a large number of specific observations. "The sun always rises in the east" is an example. And so is "All organisms are made of cells." The latter generalization, part of the so-called cell theory, was based on two centuries of biologists discovering cells in the diverse biological specimens they observed with microscopes. The careful observations and data analyses of discovery science, along with the generalizations reached by induction, are fundamental to our understanding of nature.

Hypothesis-Based Science

The observations and inductions of discovery science stimulate us to seek natural causes and explanations for those observations. What *caused* the diversification of finches on the Galápagos Islands? What *causes* the roots of a plant seedling to grow downward and the leaf-bearing shoot to grow upward? What *explains* the generalization that the sun always rises in the east? In science, such inquiry usually involves the proposing and testing of hypothetical explanations—that is, hypotheses.

The Role of Hypotheses in Inquiry

In science, a **hypothesis** is a tentative answer to a well-framed question—an explanation on trial. It is usually an educated guess, based on experience and on the data available from discovery science. A scientific hypothesis leads to predictions that can be tested by making additional observations or by performing experiments.

We all use hypotheses in solving everyday problems. Let's say, for example, that your flashlight fails during a camp-out. That's an observation. The question is obvious: Why doesn't the flashlight work? Two reasonable hypotheses based on your experience are that (1) the batteries in the flashlight are dead or (2) the bulb is burnt out. Each of these alternative hypotheses leads to predictions you can test with experiments. For example, the dead-battery hypothesis predicts that replacing the batteries will fix the problem. **Figure 1.24** diagrams this campground inquiry. Of course, we rarely dissect our thought processes this way when we are solving a problem using hypotheses, predictions, and experiments. But hypothesis-based science clearly has its origins in the human tendency to figure things out by trial and error.

Deduction: The "If . . . Then" Logic of Hypothesis-Based Science

A type of logic called deduction is built into hypothesis-based science. Deduction contrasts with induction, which, remember, is reasoning from a set of specific observations to reach a general conclusion. In **deductive reasoning**, the logic flows in the opposite direction, from the general to the specific.

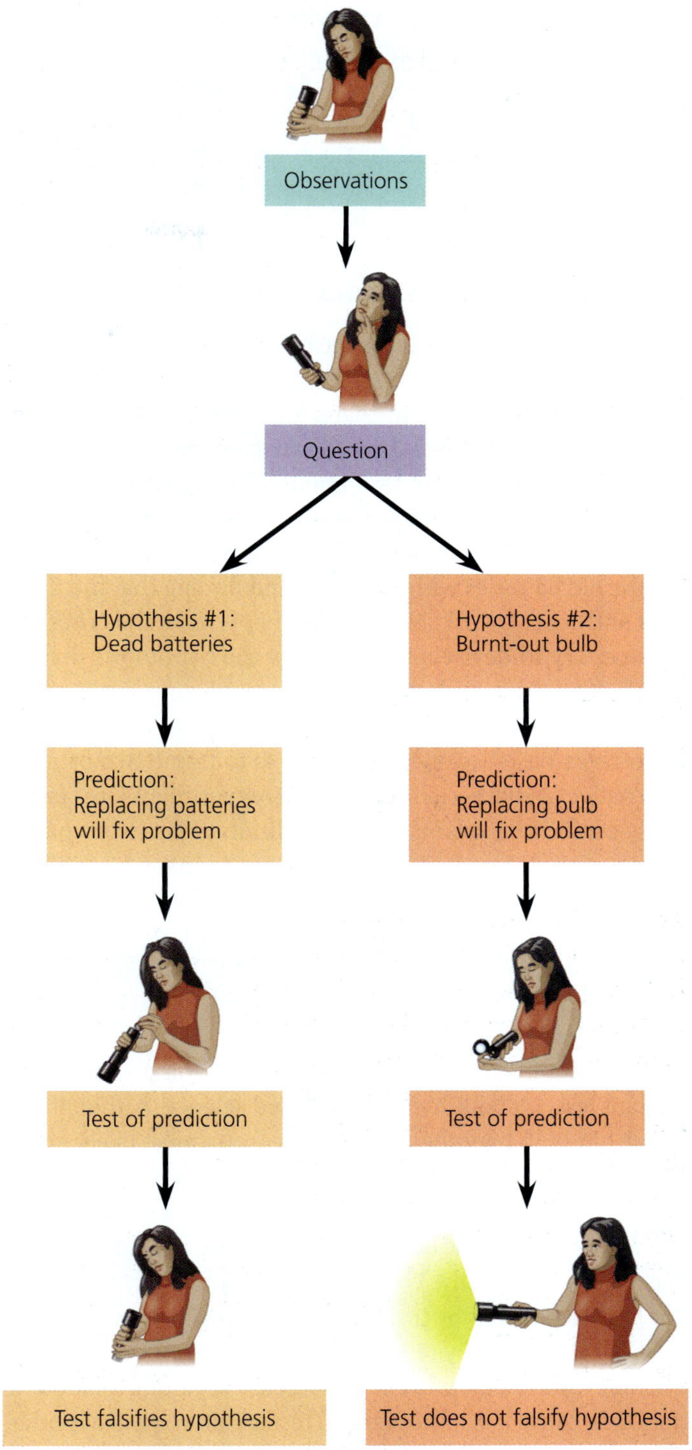

▲ **Figure 1.24 A campground example of hypothesis-based inquiry.**

From general premises, we extrapolate to the specific results we should expect if the premises are true. If all organisms are made of cells (premise 1), and humans are organisms (premise 2), then humans are composed of cells (deductive prediction about a specific case).

In hypothesis-based science, deductions usually take the form of predictions of experimental or observational results that will be found if a particular hypothesis (premise) is correct. We then test the hypothesis by carrying out the experiments or observations to see whether or not the results are as predicted. This deductive testing takes the form of "*If... then*" logic. In the case of the flashlight example: *If* the dead-battery hypothesis is correct, and you replace the batteries with new ones, *then* the flashlight should work.

A Closer Look at Hypotheses in Scientific Inquiry

The flashlight example illustrates two important qualities of scientific hypotheses. First, a hypothesis must be *testable*; there must be some way to check the validity of the idea. Second, a hypothesis must be *falsifiable*; there must be some observation or experiment that could reveal if such an idea is actually *not* true. The hypothesis that dead batteries are the sole cause of the broken flashlight could be falsified by replacing the old batteries with new ones and finding that the flashlight still doesn't work. Not all hypotheses meet the criteria of science: Try to devise a test to falsify the hypothesis that invisible campground ghosts are fooling with your flashlight!

The flashlight inquiry illustrates another key point about hypothesis-based science. The ideal is to frame two or more alternative hypotheses and design experiments to falsify those candidate explanations. In addition to the two explanations tested in Figure 1.24, another of the many possible alternative hypotheses is that *both* the batteries *and* the bulb are bad. What does this hypothesis predict about the outcome of the experiments in Figure 1.24? What additional experiment would you design to test this hypothesis of multiple malfunction?

We can mine the flashlight scenario for still one more important lesson about hypothesis-based science. Although the burnt-out bulb hypothesis stands up as the most likely explanation, notice that the testing supports that hypothesis *not* by proving that it is correct, but by not eliminating it through falsification. Perhaps the bulb was simply loose and the new bulb was inserted correctly. We could attempt to falsify the burnt-out bulb hypothesis by trying another experiment—removing the bulb and carefully reinstalling it. But no amount of experimental testing can *prove* a hypothesis beyond a shadow of doubt, because it is impossible to test *all* alternative hypotheses. A hypothesis gains credibility by surviving attempts to falsify it while testing eliminates (falsifies) alternative hypotheses.

The Myth of the Scientific Method

The flashlight example of Figure 1.24 traces an idealized process of inquiry called *the scientific method.* We can recognize the elements of this process in most of the research articles published by scientists, but rarely in such structured form. Very few scientific inquiries adhere rigidly to the sequence of steps prescribed by the "textbook" scientific method. For example, a scientist may start to design an experiment, but then backtrack upon realizing that more observations are necessary. In other cases, puzzling observations simply don't prompt well-defined questions until other research places those observations in a new context. For example, Darwin collected specimens of the Galápagos finches, but it wasn't until years later, as the idea of natural selection began to gel, that biologists began asking key questions about the history of those birds.

Moreover, scientists sometimes redirect their research when they realize they have been asking the wrong question. For example, in the early 20th century, much research on schizophrenia and manic-depressive disorder (now called bipolar disorder) got sidetracked by focusing too much on the question of how life experiences might cause these serious maladies. Research on the causes and potential treatments became more productive when it was refocused on questions of how certain chemical imbalances in the brain contribute to mental illness. To be fair, we acknowledge that such twists and turns in scientific inquiry become more evident with the advantage of historical perspective.

There is still another reason that good science need not conform exactly to any one method of inquiry: Discovery science has contributed much to our understanding of nature without most of the steps of the so-called scientific method.

It is important for you to get some experience with the power of the scientific method—by using it for some of the laboratory inquiries in your biology course, for example. But it is also important to avoid stereotyping science as lock-step adherence to this method.

A Case Study in Scientific Inquiry: Investigating Mimicry in Snake Populations

Now that we have highlighted the key features of discovery science and hypothesis-based science, you should be able to recognize these forms of inquiry in a case study of actual scientific research.

The story begins with a set of observations and generalizations from discovery science. Many poisonous animals are brightly colored, often with distinctive patterns that stand out against the background. This is called warning coloration because it apparently signals "dangerous species" to potential predators. But there are also mimics. These imposters look like poisonous species but are actually harmless. A question that follows from these observations is: What is the function of such mimicry? A reasonable hypothesis is that such "deception" is an evolutionary adaptation that reduces the harmless animal's risk of being eaten because predators mistake it for the poisonous species. This hypothesis was first formulated by British scientist Henry Bates in 1862.

As obvious as this hypothesis may seem, it has been relatively difficult to test, especially with field experiments. But in

2001, biologists David and Karin Pfennig, of the University of North Carolina, along with William Harcombe, an undergraduate, designed a simple but elegant set of field experiments to test Bates's mimicry hypothesis.

The team investigated a case of mimicry among snakes that live in North and South Carolina **(Figure 1.25)**. A poisonous snake called the eastern coral snake has warning coloration: bold, alternating rings of red, yellow (or white), and black. Predators rarely attack these snakes. It is unlikely that the predators learn this avoidance behavior by trial and error, as a first encounter with a coral snake is usually deadly. In areas where coral snakes live, natural selection has apparently increased the frequency of predators that have inherited an instinctive avoidance of the coral snake's coloration. A nonpoisonous snake named the scarlet kingsnake mimics the ringed coloration of the coral snake.

Both types of snakes live in the Carolinas, but the kingsnakes' geographic range also extends into regions where no coral snakes are found (see Figure 1.25). The geographic distribution of the snakes made it possible to test the key prediction of the mimicry hypothesis. Avoiding snakes with warning coloration is an adaptation we expect to be present only in predator populations that evolved in areas where the poisonous coral snakes are present. Therefore, mimicry should help protect kingsnakes from predators, but *only* in regions where coral snakes also live. The mimicry hypothesis predicts that predators adapted to the warning coloration of coral snakes will attack kingsnakes less frequently than will predators in areas where coral snakes are absent.

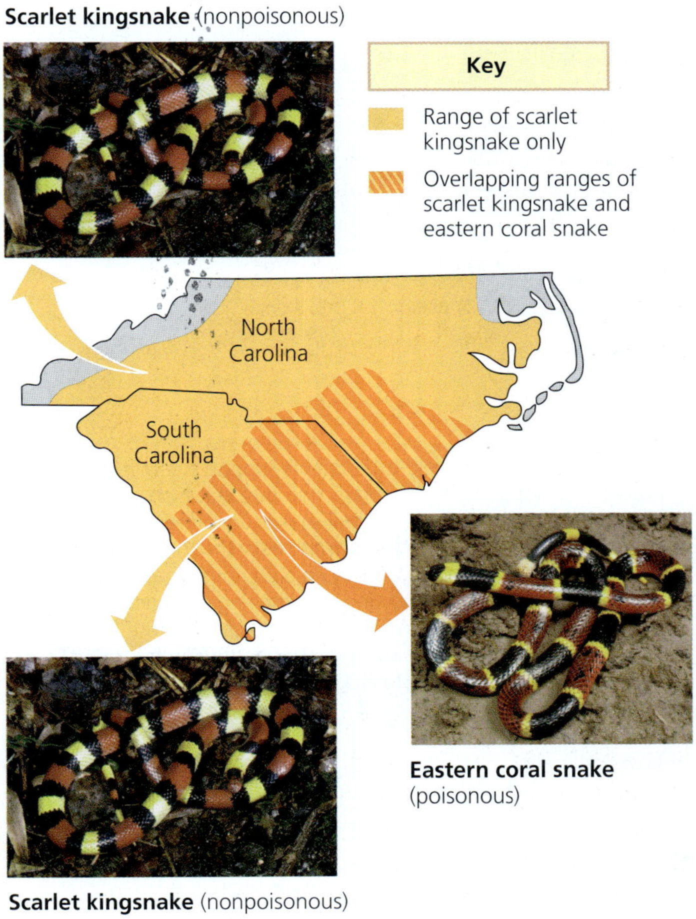

▲ **Figure 1.25 The geographic ranges of a poisonous snake and its mimic.** The scarlet kingsnake (*Lampropeltis triangulum*) mimics the warning coloration of the poisonous eastern coral snake (*Micrurus fulvius*).

Field Experiments with Artificial Snakes

To test the prediction, Harcombe made hundreds of artificial snakes out of wire covered with plasticine. He fashioned two versions of fake snakes: an *experimental group* with the red, black, and white ring pattern of kingsnakes; and a *control group* of plain brown artificial snakes as a basis of comparison **(Figure 1.26)**.

The researchers placed equal numbers of the two types of artificial snakes in field sites throughout North and South Carolina, including the region where coral snakes are absent. After four weeks, the scientists retrieved the fake snakes and recorded how many had been attacked by looking for bite or claw marks. The most common predators were foxes, coyotes, and raccoons, but black bears also attacked some of the artificial snakes (see Figure 1.26b).

The data fit the key prediction of the mimicry hypothesis. Compared to the brown artificial snakes, the ringed artificial snakes were attacked by predators less frequently *only* in field sites within the geographic range of the poisonous coral

(a) Artificial kingsnake

(b) Brown artificial snake that has been attacked

▲ **Figure 1.26 Artificial snakes used in field experiments to test the mimicry hypothesis.** You can see where a bear chomped on the brown artificial snake in (b).

snakes. **Figure 1.27** summarizes the field experiments that the researchers carried out. This figure also introduces a format we will use throughout the book for other examples of biological inquiry.

Designing Controlled Experiments

The snake mimicry experiment is an example of a **controlled experiment**, one that is designed to compare an experimental group (the artificial kingsnakes, in this case) with a control group (the brown artificial snakes). Ideally, the experimental and control groups differ only in the one factor the experiment is designed to test—in our example, the effect of the snakes' coloration on the behavior of predators. Without the control group, the researchers would not have been able to rule out other factors as causes of the more frequent attacks on the artificial kingsnakes—such as different numbers of predators or different temperatures in the different test areas. The clever experimental design left coloration as the only factor that could account for the low predation rate on the artificial kingsnakes placed within the range of coral snakes. It was not the absolute number of attacks on the artificial kingsnakes that counted, but the difference between that number and the number of attacks on the brown snakes.

A common misconception is that the term *controlled experiment* means that scientists control the experimental environment to keep everything constant except the one variable being tested. But that's impossible in field research and not realistic even in highly regulated laboratory environments. Researchers usually "control" unwanted variables not by *eliminating* them through environmental regulation, but by *canceling* their effects by using control groups.

Limitations of Science

Scientific inquiry is a powerful way to learn about nature, but there are limitations to the kinds of questions it can answer. These limits are set by science's requirements that hypotheses be testable and falsifiable and that observations and experimental results be repeatable.

Observations that can't be verified may be interesting or even entertaining, but they cannot count as evidence in scientific inquiry. The headlines of supermarket tabloids would have you believe that humans are occasionally born with the head of a dog and that some of your classmates are extraterrestrials. The unconfirmed eyewitness accounts and the computer-rigged photos are amusing but unconvincing. In science, evidence from observations and experiments is only convincing if it stands up to the criterion of repeatability. The scientists who investigated snake mimicry in the Carolinas obtained similar data when they repeated their experiments with different species of coral snakes and kingsnakes in Arizona. And *you* should be able to obtain similar results if you were to repeat the snake experiments.

Figure 1.27 Inquiry

Does the presence of poisonous coral snakes affect predation rates on their mimics, kingsnakes?

EXPERIMENT David Pfennig and his colleagues made artificial snakes to test a prediction of the mimicry hypothesis: that kingsnakes benefit from mimicking the warning coloration of poisonous coral snakes *only* in regions where coral snakes are present. The researchers placed equal numbers of artificial kingsnakes (experimental group) and brown artificial snakes (control group) at 14 field sites, half in the area the two snakes cohabit and half in the area where coral snakes were absent. The researchers recovered the artificial snakes after four weeks and tabulated predation data based on teeth and claw marks on the snakes.

RESULTS In field sites where coral snakes were present, most attacks were on brown artificial snakes. Where coral snakes were absent, most attacks were on artificial kingsnakes.

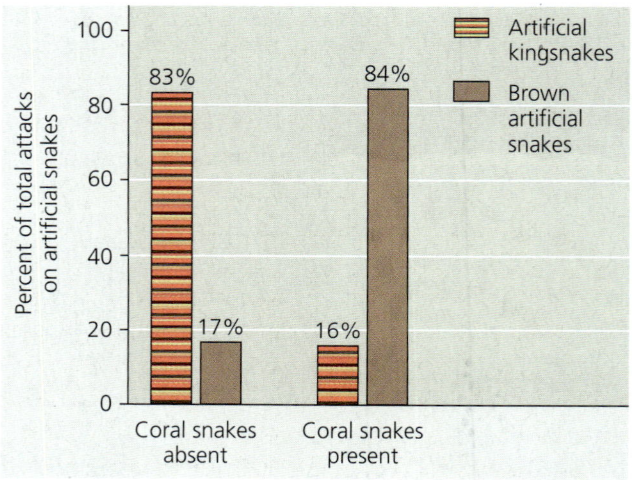

CONCLUSION The field experiments support the mimicry hypothesis by not falsifying the prediction that imitation of coral snakes is only effective where coral snakes are present. The experiments also tested an alternative hypothesis: that predators generally avoid all snakes with brightly colored rings. That hypothesis was falsified by the data showing that the ringed coloration failed to repel predators where coral snakes were absent. (The fake kingsnakes may have been attacked more often in those areas because their bright pattern made them easier to spot than the brown fakes.)

SOURCE D. W. Pfennig, W. R. Harcombe, and K. S. Pfennig, Frequency-dependent Batesian mimicry, *Nature* 410:323 (2001).

Inquiry in Action Read and analyze the original paper in *Inquiry in Action: Interpreting Scientific Papers*.

WHAT IF? What experimental results would you predict if predators throughout the Carolinas avoid all snakes with brightly colored ring patterns?

Because science requires natural explanations for natural phenomena, it can neither support nor falsify hypotheses that angels, ghosts, or spirits, whether benevolent or evil, cause storms, rainbows, illnesses, and cures. Such supernatural explanations are simply outside the bounds of science.

Theories in Science

"It's just a theory!" Our everyday use of the term *theory* often implies an untested speculation. But the term *theory* has a different meaning in science. What is a scientific theory, and how is it different from a hypothesis or from mere speculation?

First, a scientific **theory** is much broader in scope than a hypothesis. *This* is a hypothesis: "Mimicking the coloration of poisonous snakes is an adaptation that protects nonpoisonous snakes from predators." But *this* is a theory: "Evolutionary adaptations arise by natural selection." Darwin's theory of natural selection accounts for an enormous diversity of adaptations, including mimicry.

Second, a theory is general enough to spin off many new, specific hypotheses that can be tested. For example, two researchers at Princeton University, Peter and Rosemary Grant, were motivated by the theory of natural selection to test the specific hypothesis that the beaks of Galápagos finches evolve in response to changes in the types of available food. (Their results supported their hypothesis; see p. 468.)

And third, compared to any one hypothesis, a theory is generally supported by a much greater body of evidence. Those theories that become widely adopted in science (such as the theory of natural selection) explain a great diversity of observations and are supported by a vast accumulation of evidence. In fact, scrutiny of theories continues through testing of the specific, falsifiable hypotheses they spawn.

In spite of the body of evidence supporting a widely accepted theory, scientists must sometimes modify or even reject theories when new research methods produce results that don't fit. For example, the five-kingdom theory of biological diversity began to erode when new methods for comparing cells and molecules made it possible to test some of the hypothetical relationships between organisms that were based on the theory. If there is "truth" in science, it is conditional, based on the preponderance of available evidence.

Model Building in Science

You may work with many models in your biology course this year. Perhaps you'll model cell division by using pipe cleaners as chromosomes. Or maybe you'll practice using mathematical models to predict the growth of a bacterial population. Scientists often construct models as representations of natural phenomena. Scientific **models** can take many forms, including diagrams (such as the evolutionary tree in Figure 1.22), graphs, three-dimensional objects, computer programs, or mathematical equations.

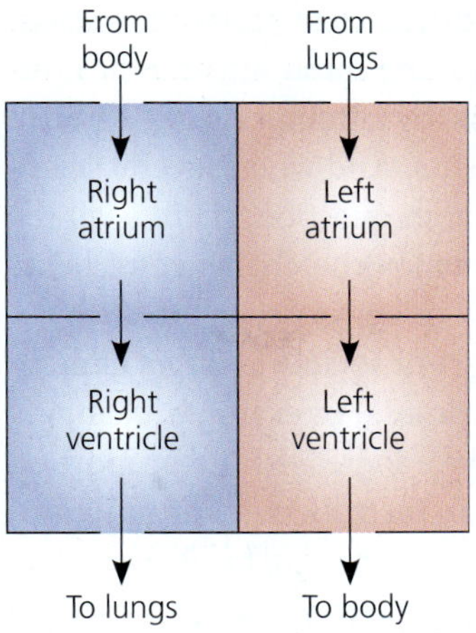

◄ **Figure 1.28 A model of the blood flow through the four chambers of a human heart.**

? *The blood picks up oxygen (and releases carbon dioxide) in the lungs and releases oxygen (and picks up carbon dioxide) in the rest of the body. From the model shown here, predict what would happen if there were a small hole between the two ventricles.*

Choosing the most appropriate type of model depends on what needs to be communicated and explained about the object, idea, or process the model is to represent. Some models need to be as lifelike as possible. Other models are more useful if they are simple schematics. For example, the simple diagram in **Figure 1.28** does a good job of modeling blood flow through the chambers of a human heart without looking anything like a real heart. A heart model designed to help train a physician to perform heart surgery would look very different. Whatever the design of a model, the test of its success is how well it fits the available data, how comfortably it accommodates new observations, how accurately it predicts the outcomes of new experiments or observations, and how effectively it communicates.

The Culture of Science

Movies and cartoons sometimes portray scientists as loners working in isolated labs. In reality, science is an intensely social activity. Most scientists work in teams, which often include both graduate and undergraduate students **(Figure 1.29)**. And to succeed in science, it helps to be a good communicator. Research results have no impact until shared with a community of peers through seminars, publications, and websites.

◄ **Figure 1.29 Science as a social process.** In her New York University laboratory, plant biologist Gloria Coruzzi mentors one of her students in the methods of molecular biology.

Both cooperation and competition characterize the scientific culture. Scientists working in the same research field often check one another's claims by attempting to confirm observations or repeat experiments. And when several scientists converge on the same research question, there is all the excitement of a race. Scientists enjoy the challenge of being first with an important discovery or key experiment.

The biology community is part of society at large, embedded in the cultural milieu of the times. For example, changing attitudes about career choices have increased the proportion of women in biology, which has in turn affected the emphasis in certain research fields. A few decades ago, for instance, biologists who studied the mating behavior of animals focused mostly on competition among males for access to females. More recent research, however, emphasizes the important role that females play in choosing mates. For example, in many bird species, females prefer the bright coloration that "advertises" a male's vigorous health, a behavior that enhances the female's probability of having healthy offspring.

Some philosophers of science argue that scientists are so influenced by cultural and political values that science is no more objective than other ways of understanding nature. At the other extreme are people who speak of scientific theories as though they were natural laws instead of human interpretations of nature. The reality of science is probably somewhere in between—rarely perfectly objective, but continuously vetted through the expectation that observations and experiments be repeatable and hypotheses be testable and falsifiable.

Science, Technology, and Society

The relationship of science to society becomes clearer when we add technology to the picture. Though science and technology sometimes employ similar inquiry patterns, their basic goals differ. The goal of science is to understand natural phenomena. In contrast, **technology** generally *applies* scientific knowledge for some specific purpose. Biologists and other scientists often speak of "discoveries," while engineers and other technologists more often speak of "inventions." And the beneficiaries of those inventions include scientists, who put new technology to work in their research; the impact of information technology on systems biology is just one example. Thus, science and technology are interdependent.

The potent combination of science and technology has dramatic effects on society. For example, discovery of the structure of DNA by Watson and Crick half a century ago and subsequent achievements in DNA science led to the technologies of DNA engineering that are transforming many applied fields, including medicine, agriculture, and forensics **(Figure 1.30)**. Perhaps Watson and Crick envisioned that their discovery would someday produce important applications, but it is unlikely that they could have predicted exactly what all those applications would be.

▲ **Figure 1.30 DNA technology and crime scene investigation.** Forensic technicians can use traces of DNA extracted from a blood sample or other body tissue collected at a crime scene to produce molecular "fingerprints." The stained bands you see in this photograph represent fragments of DNA, and the pattern of bands varies from person to person.

The directions that technology takes depend less on the curiosity that drives basic science than on the current needs and wants of people and on the social environment of the times. Debates about technology center more on "*should* we do it" than "*can* we do it." With advances in technology come difficult choices. For example, under what circumstances is it acceptable to use DNA technology to find out if particular people have genes for hereditary diseases? Should such tests always be voluntary, or are there circumstances when genetic testing should be mandatory? Should insurance companies or employers have access to the information, as they do for many other types of personal health data?

Such ethical issues have as much to do with politics, economics, and cultural values as with science and technology. All citizens—not only professional scientists—have a responsibility to be informed about how science works and about the potential benefits and risks of technology. The relationship between science, technology, and society increases the significance and value of any biology course.

CONCEPT CHECK 1.3

1. Contrast inductive reasoning with deductive reasoning.
2. Why is natural selection called a theory?
3. **WHAT IF?** Suppose you extended the snake mimicry experiment to an area of Virginia where neither type of snake is known to live. What results would you predict at your field site?

For suggested answers, see Appendix A.

Chapter 1 Review

MEDIA Go to the Study Area at **www.masteringbio.com** for BioFlix 3-D Animations, MP3 Tutors, Videos, Practice Tests, an eBook, and more.

SUMMARY OF KEY CONCEPTS

CONCEPT 1.1
Themes connect the concepts of biology (pp. 3–11)

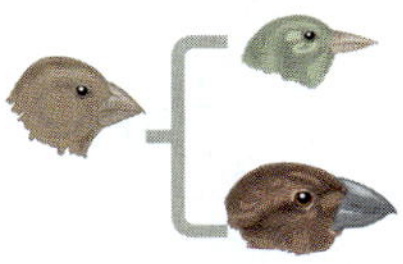

▶ **Evolution, the Overarching Theme of Biology** Evolution accounts for the unity and diversity of life, and also for the match of organisms to their environments.

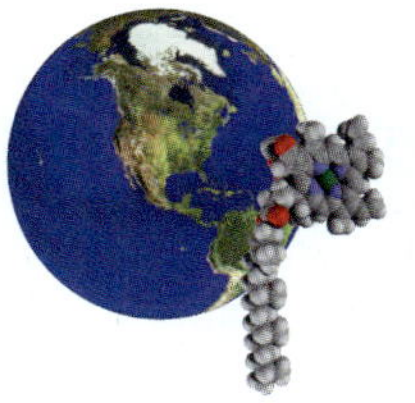

▶ ***Theme:* New properties emerge at each level in the biological hierarchy** The hierarchy of life unfolds as follows: biosphere > ecosystem > community > population > organism > organ system > organ > tissue > cell > organelle > molecule > atom. With each step "upward" from atoms, new properties emerge as a result of interactions among components at the lower levels. In an approach called reductionism, complex systems are broken down to simpler components that are more manageable to study. In systems biology, scientists make models of complex biological systems.

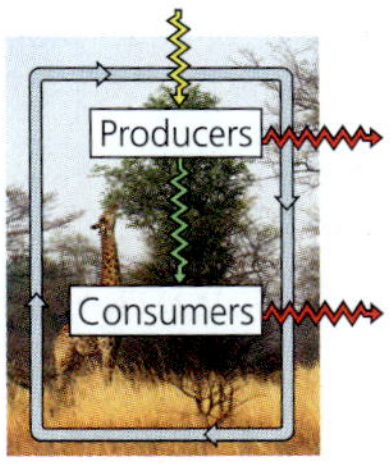

▶ ***Theme:* Organisms interact with their environments, exchanging matter and energy** An organism's environment includes other organisms as well as nonliving factors. Whereas chemical nutrients recycle within an ecosystem, energy flows through an ecosystem. All organisms must perform work, which requires energy. Energy flows from sunlight to producers to consumers.

▶ ***Theme:* Structure and function are correlated at all levels of biological organization** The form of a biological structure suits its function and vice versa.

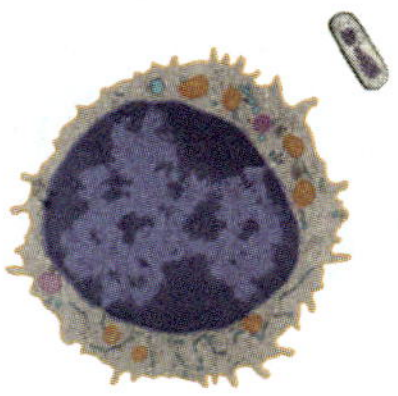

▶ ***Theme:* Cells are an organism's basic units of structure and function** The cell is the lowest level of organization that can perform all activities required for life. Cells are either prokaryotic or eukaryotic. Eukaryotic cells contain membrane-enclosed organelles, including a DNA-containing nucleus. Prokaryotic cells lack such organelles.

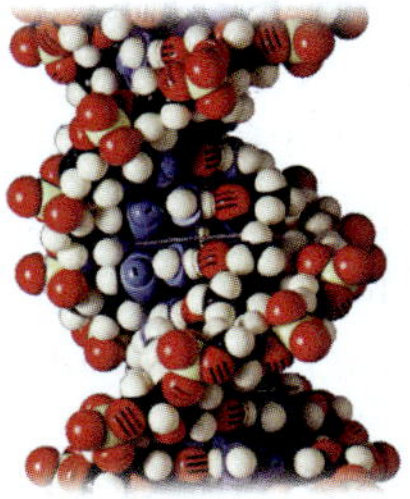

▶ ***Theme:* The continuity of life is based on heritable information in the form of DNA** Genetic information is encoded in the nucleotide sequences of DNA. It is DNA that transmits heritable information from parents to offspring. DNA sequences program a cell's protein production by being transcribed into RNA and then translated into specific proteins. RNA that is not translated into protein serves other important functions.

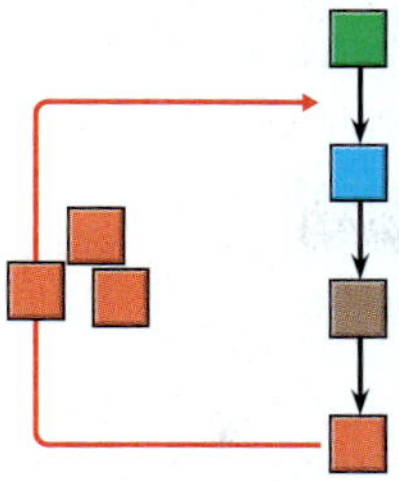

▶ ***Theme:* Feedback mechanisms regulate biological systems** In negative feedback, accumulation of an end product slows the process that makes that product. In positive feedback, the end product stimulates the production of more product. Feedback is a type of regulation common to life at all levels, from molecules to ecosystems.

MEDIA

Activity The Levels of Life Card Game
Activity Energy Flow and Chemical Cycling
Activity Form Fits Function: Cells
Activity Comparing Prokaryotic and Eukaryotic Cells
Activity Heritable Information: DNA
Activity Regulation: Negative and Positive Feedback

CONCEPT 1.2
The Core Theme: Evolution accounts for the unity and diversity of life (pp. 12–18)

▶ **Organizing the Diversity of Life** Biologists classify species according to a system of broader and broader groups. Domain Bacteria and domain Archaea consist of prokaryotes. Domain Eukarya, the eukaryotes, includes various groups of protists and the kingdoms Plantae, Fungi, and Animalia. As diverse as life is, there is also evidence of remarkable unity, which is revealed in the similarities between different kinds of organisms.

▶ **Charles Darwin and the Theory of Natural Selection** Darwin proposed natural selection as the mechanism for evolutionary adaptation of populations to their environments.

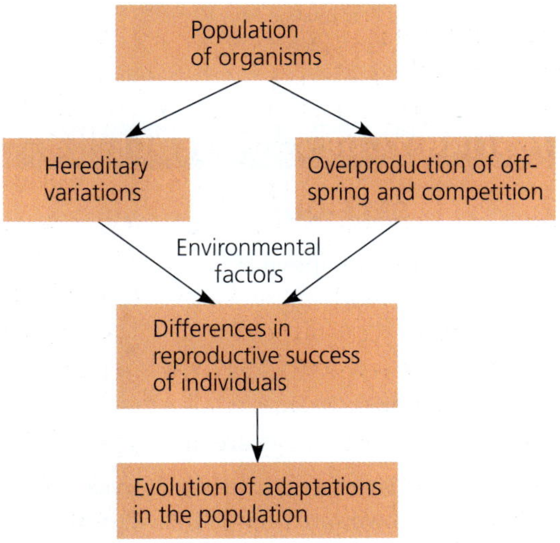

▶ **The Tree of Life** Each species is one twig of a branching tree of life extending back in time through ancestral species more and more remote. All of life is connected through its long evolutionary history.

MEDIA

Activity Classification Schemes
Investigation How Do Environmental Changes Affect a Population?
Biology Labs On-Line EvolutionLab

CONCEPT 1.3

Scientists use two main forms of inquiry in their study of nature (pp. 18–24)

- **Discovery Science** In discovery science, scientists observe and describe some aspect of the world and use inductive reasoning to draw general conclusions.
- **Hypothesis-Based Science** Based on observations, scientists propose hypotheses that lead to predictions and then test the hypotheses by seeing if the predictions come true. Deductive reasoning is used in testing hypotheses: If a hypothesis is correct, and we test it, then we can expect a particular outcome. Hypotheses must be testable and falsifiable.
- **A Case Study in Scientific Inquiry: Investigating Mimicry in Snake Populations** Experiments must be designed to demonstrate the effect of one variable by testing control groups and experimental groups that differ in only that one variable.
- **Limitations of Science** Science cannot address the possibility of supernatural phenomena because hypotheses must be testable and falsifiable, and observations and experimental results must be repeatable.
- **Theories in Science** A scientific theory is broad in scope, generates new hypotheses, and is supported by a large body of evidence.
- **Model Building in Science** Models of ideas, structures, and processes help us understand scientific phenomena and make predictions.
- **The Culture of Science** Science is a social activity characterized by both cooperation and competition.
- **Science, Technology, and Society** Technology is a method or device that applies scientific knowledge for some specific purpose.

MEDIA

GraphIt! An Introduction to Graphing
Investigation How Does Acid Precipitation Affect Trees?
Activity Science, Technology, and Society: DDT

TESTING YOUR KNOWLEDGE

SELF-QUIZ

1. All the organisms on your campus make up
   a. an ecosystem.
   b. a community.
   c. a population.
   d. an experimental group.
   e. a taxonomic domain.

2. Which of the following is a correct sequence of levels in life's hierarchy, proceeding downward from an individual animal?
   a. brain, organ system, nerve cell, nervous tissue
   b. organ system, nervous tissue, brain
   c. organism, organ system, tissue, cell, organ
   d. nervous system, brain, nervous tissue, nerve cell
   e. organ system, tissue, molecule, cell

3. Which of the following is *not* an observation or inference on which Darwin's theory of natural selection is based?
   a. Poorly adapted individuals never produce offspring.
   b. There is heritable variation among individuals.
   c. Because of overproduction of offspring, there is competition for limited resources.
   d. Individuals whose inherited characteristics best fit them to the environment will generally produce more offspring.
   e. A population can become adapted to its environment over time.

4. Systems biology is mainly an attempt to
   a. understand the integration of all levels of biological organization from molecules to the biosphere.
   b. simplify complex problems by reducing the system into smaller, less complex units.
   c. construct models of the behavior of entire biological systems.
   d. build high-throughput machines for the rapid acquisition of biological data.
   e. speed up the technological application of scientific knowledge.

5. Protists and bacteria are grouped into different domains because
   a. protists eat bacteria.
   b. bacteria are not made of cells.
   c. protists have a membrane-bounded nucleus, which bacterial cells lack.
   d. bacteria decompose protists.
   e. protists are photosynthetic.

6. Which of the following best demonstrates the unity among all organisms?
   a. matching DNA nucleotide sequences
   b. descent with modification
   c. the structure and function of DNA
   d. natural selection
   e. emergent properties

7. Which of the following is an example of qualitative data?
   a. The temperature decreased from 20°C to 15°C.
   b. The plant's height is 25 centimeters (cm).
   c. The fish swam in a zig-zag motion.
   d. The six pairs of robins hatched an average of three chicks.
   e. The contents of the stomach are mixed every 20 seconds.

8. Which of the following best describes the logic of hypothesis-based science?
   a. If I generate a testable hypothesis, tests and observations will support it.
   b. If my prediction is correct, it will lead to a testable hypothesis.
   c. If my observations are accurate, they will support my hypothesis.
   d. If my hypothesis is correct, I can expect certain test results.
   e. If my experiments are set up right, they will lead to a testable hypothesis.

9. A controlled experiment is one that
   a. proceeds slowly enough that a scientist can make careful records of the results.
   b. may include experimental groups and control groups tested in parallel.

c. is repeated many times to make sure the results are accurate.
d. keeps all environmental variables constant.
e. is supervised by an experienced scientist.

10. Which of the following statements best distinguishes hypotheses from theories in science?
a. Theories are hypotheses that have been proved.
b. Hypotheses are guesses; theories are correct answers.
c. Hypotheses usually are relatively narrow in scope; theories have broad explanatory power.
d. Hypotheses and theories are essentially the same thing.
e. Theories are proved true in all cases; hypotheses are usually falsified by tests.

11. **DRAW IT** With rough sketches, draw a biological hierarchy similar to the one in Figure 1.4 but using a coral reef as the ecosystem, a fish as the organism, its stomach as the organ, and DNA as the molecule. Include all levels in the hierarchy.

For Self-Quiz answers, see Appendix A.

MEDIA Visit the Study Area at **www.masteringbio.com** for a Practice Test.

EVOLUTION CONNECTION

12. A typical prokaryotic cell has about 3,000 genes in its DNA, while a human cell has about 20,500 genes. About 1,000 of these genes are present in both types of cells. Based on your understanding of evolution, explain how such different organisms could have this same subset of genes. What sorts of functions might these shared genes have?

SCIENTIFIC INQUIRY

13. Based on the results of the snake mimicry case study, suggest another hypothesis researchers might use to extend the investigation.

SCIENCE, TECHNOLOGY, AND SOCIETY

14. The fruits of wild species of tomato are tiny compared to the giant beefsteak tomatoes available today. This difference in fruit size is almost entirely due to the larger number of cells in the domesticated fruits. Plant molecular biologists have recently discovered genes that are responsible for controlling cell division in tomatoes. Why would such a discovery be important to producers of other kinds of fruits and vegetables? To the study of human development and disease? To our basic understanding of biology?

UNIT 1

The Chemistry of Life

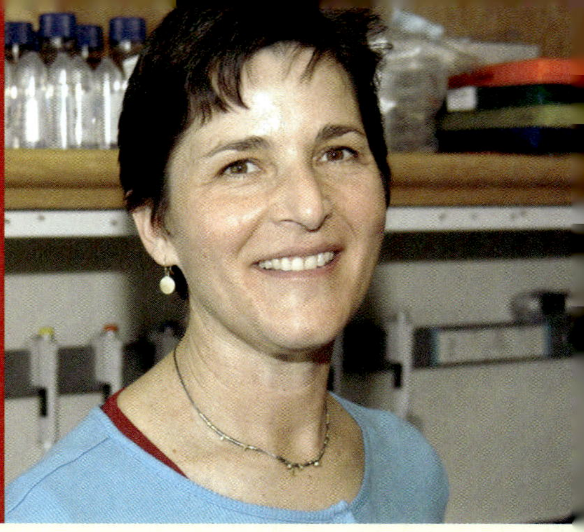

AN INTERVIEW WITH

Deborah M. Gordon

What does an ant sense as it goes about its daily chores? Mainly chemicals—because these are the cues ants use to navigate their environment. The interactions of ants with each other and their surroundings are the research focus of Deborah M. Gordon, a professor of biological sciences at Stanford University. While at Stanford, Dr. Gordon has won several awards for excellence in teaching, as well as recognition for her research in the Arizona desert and the tropics of South America. And through appearances on radio, TV nature shows, and her book *Ants at Work: How an Insect Society Is Organized* (Free Press, 1999), Dr. Gordon has shared her fascination with ant society with people around the world.

How did you get interested in biology?

In my first year in college, thinking of a career in medicine as a possibility, I took introductory chemistry and biology. But those courses just gave me a lot of information that I couldn't really put together. I ended up majoring in French. But I was also very interested in math and in music theory, because I like looking at patterns and understanding how they change over time. Then, in my senior year, I took a course in comparative anatomy, which completely changed my view of biology. That course showed me that evolution is a process that changes patterns in interesting ways.

After graduation, curious about the human body and health, I came to Stanford to take the medical school course in human anatomy. I stayed to complete a master's degree in biology. Then, although I still wasn't sure what I wanted to do with my life, I entered a Ph.D. program at Duke. It was there that I began research on ant behavior—and I loved it.

Students of biology have to study chemistry as well. How is chemistry relevant to ant behavior?

Ants don't see very well; they operate mostly by chemical communication. If you work on ants, you have to think about chemistry because chemicals are critically important in the ant's world. For example, the ants I study in Arizona use long-lasting chemical cues to identify themselves and to mark their nest area. Ants also use many short-term chemical cues called pheromones, which they secrete in certain situations. The best known are alarm pheromones, which are what make ants run around in circles when they're disturbed. Some ants secrete a pheromone from the tip of the abdomen that marks where they walk and creates a trail that other ants can follow. Ants have 12 or 14 different glands that secrete different substances. We really don't know what they're all for, or how many chemical combinations an ant can respond to. In addition to chemicals used in communication, some ants produce antibiotics or chemical defenses against predators. Other ants use chemicals to kill certain plants [see Chapter 2, pp. 30–31].

Why do you study ants?

What interests me about ants is that ants live in societies without any central control. Yet individual ants are very limited in what they can do; each can take in only local information. No ant can figure out what needs to be done for the good of the colony. The big question for me is: How can an ant colony function when nobody's in charge and each ant can only perceive what's right around it?

What about the queens? Please tell us more about how ants live.

There are 10,000 to 12,000 species of ants. They all live in colonies, each with one or a few reproductive females, called queens. The queens lay eggs, using sperm stored from a mating that preceded her establishment of the colony. The rest of the colony—all the ants you see walking around—are her daughters, sterile female workers. These workers do all the work, and they do it without any direction from the queen. Males, born from unfertilized eggs, are produced only once a year, just in time to join virgin queens in a mating flight. Soon after mating, they die.

What exactly do worker ants do?

The ants I study in Arizona, called harvester ants, perform four kinds of tasks. Some workers forage for food. Some patrol; that is, they go out early in the morning and decide where the foragers should go that day. Others do nest-maintenance work, building chambers underground and then carrying out the excess sand. And still others work on the refuse pile, or midden, which they mark with the colony's specific odor. Different groups of ants perform each of the four tasks.

Tell us about your research in Arizona.

At my research site in Arizona, where I've been working for more than 20 years, I study a population of about 300 colonies of harvester ants. Each year my students and I map the locations of the colonies that make up this population. We identify all the colonies that were there the year before, figure out which ones have died, and map the new ones. In this way I can follow the same colonies year after year. I get to know them quite well. I have found that colonies last 15 to 20 years.

In addition to observing the ant colonies, we do simple experiments where we change the ants' environment in some way and observe how the ants respond. For example, in the last few years we've been studying how a colony regulates the number of ants that go out to forage. We've learned that each ant uses its recent history of interactions with other ants to decide what to do next. An ant uses odors to identify the task of the ants it meets. Each ant is coated with a layer of grease, made of chemicals called hydrocarbons. Each task group—the foragers, the patrollers, the nest maintenance workers, and the midden workers—has a unique mixture of hydrocarbons, which the ants secrete and spread on each other by grooming. We've found that, as an ant spends time outside, the proportions of different hydrocarbons on its body changes.

What causes this change?
The heat of the sun. We learned this from an experiment in which we took ants that had been working inside the nest and exposed them to different conditions. After exposure to high temperatures and low humidity for long enough times, these ants came to smell like foragers. So it's not that foragers secrete something different as they do their task but that doing the task changes them. Just as a carpenter gets calluses from holding tools, an ant comes to smell different from the work it does. The ants of a colony all secrete the same mixture of hydrocarbons, but each ant's chemical profile changes depending on what it does.

What tells an ant what to do? For example, what tells an ant to forage?
We've learned that ants use their recent experience of quick antennal contacts with each other to decide what to do. The antennae are their organs of chemical perception. Anyone who has watched ants has seen them meet and touch antennae, and when they do that, they smell each other, detecting the chemicals on each other's body. But a single interaction is not an instruction; a forager meeting another ant is not saying, "Go forage." It's the *pattern* of encounters that conveys the message.

How did you figure that out?
Working with Michael Greene, who's now at the University of Colorado at Denver, we've been able to extract the hydrocarbons from the ants' bodies and put them on little glass beads, and we've found that the ants respond to a bead coated with ant hydrocarbons as if it were an ant. So by dropping these coated beads into the nest, we can figure out how the ants react to encounters with an ant of a particular task group.

Recently we've been working on how foragers use the rate at which other foragers come in with food to decide whether to go out again. This rate gives feedback about how much food is out there. A forager looking for food won't come back until it finds something; if it has to stay out for 45 minutes, it will. But when there's an abundance of food near the nest, the foragers return quickly. The returning ants provide positive feedback: The faster ants come back, the more ants go out.

We've found that harvester ants respond surprisingly quickly to changes in the frequency of encounters with other ants. This rapid response is probably driven by the ants' short memories—only about 10 seconds.

So the answer to the question of what controls a colony is the aggregate of "decisions" made in the simple interactions between individual ants. Are there analogies in other areas of biology?
A system without central control, built from simple, interacting components, is called a "complex system," and scientists in many areas of biology are interested in such systems. One obvious analog to an ant colony is a brain. Your brain is composed of neurons (nerve cells), but no single neuron knows how to think about, for example, the subject of ants—although your brain as a whole can think about ants. The brain operates without central control, in the sense that there isn't a master neuron in there that says, "OK, you guys, you do ants." Yet somehow all of the simple interactions among the neurons add up to the brain's very complex functioning. Another analogy is the growth of an embryo: The cells of an embryo all have the same DNA, but as the embryo grows, its cells take on different forms and functions. Nobody says, "OK, you become liver, you become bone." Instead, as a result of molecular interactions among cells, the embryo develops tissues of different types.

In your book, the colony is spoken of as if it were an organism. Does evolution operate on the level of the colony?
Yes, because the whole colony cooperates to make more queens and males that go out and start new colonies. A colony's behavior can determine how many offspring colonies it makes. In the long term, I'd like to understand how natural selection is acting on ant behavior (if it is). Why does it matter to the colony that it behaves in a certain way?

Now for a practical question: How can we deal with ant invasions in our homes?
It depends on where you live. In Northern California, the invading ants are usually Argentine ants, whose activity is clearly connected to the weather. The ants come into everybody's houses at the same time—when it rains or is very hot and dry—and they go out at the same time. The most important thing to remember is that putting out pesticide, especially when you don't have ants, sends pesticide into the groundwater but doesn't have much effect on the numbers of ants. I don't like having ants in my kitchen—I always take it personally—but I know that when I do, everybody else does, too. And covering or washing away a trail works only briefly—about 20 minutes for Argentine ants, as I found out in an experiment. Blocking off the places where ants are coming in is the best approach.

Ant-bait devices, from which foraging ants are supposed to carry poison to their nest, can work for ants that have one queen in a central nest. Such ants include carpenter ants, which enter houses in many parts of the United States. These ants nest in decaying wood, though contrary to their reputation, they don't eat wood. Ant baits don't work at all for the Argentine ants because these ants have many queens and many nests and you are unlikely to reach all the queens with the poison.

What advice do you have for undergraduates interested in a research career?
Students should try to experience several kinds of research, involving different kinds of activities. The best way to find out if you like doing research—whether it's working in the field or in the lab—is to try it.

How can an ant colony function when nobody's in charge and each ant can only perceive what's right around it?

Inquiry in Action

Learn about an experiment by Deborah Gordon and her graduate student Megan Frederickson in Inquiry Figure 2.2 on page 31. Read and analyze the original paper in *Inquiry in Action: Interpreting Scientific Papers.*

Left to right: Deborah Gordon, Megan Frederickson, Lisa Urry, and Jane Reece

2 The Chemical Context of Life

▲ **Figure 2.1 Who tends this garden?**

KEY CONCEPTS

2.1 **Matter consists of chemical elements in pure form and in combinations called compounds**

2.2 **An element's properties depend on the structure of its atoms**

2.3 **The formation and function of molecules depend on chemical bonding between atoms**

2.4 **Chemical reactions make and break chemical bonds**

OVERVIEW

A Chemical Connection to Biology

The Amazon rain forest in South America is a showcase for the diversity of life on Earth. Colorful birds, insects, and other animals live among a myriad of trees, shrubs, vines, and wildflowers, and an excursion along a waterway or a forest path typically reveals a lush variety of plant life. Visitors traveling near the Amazon's headwaters in Peru are therefore surprised to come across tracts of forest like that seen in the foreground of the photo in **Figure 2.1**. This patch is almost completely dominated by a single plant species—a willowy flowering tree called *Duroia hirsuta*. Travelers may wonder if the garden is planted and maintained by local people, but the indigenous people are as mystified as the visitors. They call these stands of *Duroia* trees "devil's gardens," from a legend attributing them to an evil forest spirit.

Seeking a scientific explanation, a research team working under Deborah Gordon, who is interviewed on pages 28–29, recently solved the "devil's garden" mystery. **Figure 2.2** describes their main experiment. The researchers showed that the "farmers" who create and maintain these gardens are actually ants that live in the hollow stems of the *Duroia* trees. The ants do not plant the *Duroia* trees, but they prevent other plant species from growing in the garden by injecting intruders with a poisonous chemical. In this way, the ants create space for the growth of the *Duroia* trees that serve as their home. With the ability to maintain and expand its habitat, a single colony of devil's garden ants can live for hundreds of years.

The chemical the ants use to weed their garden turns out to be formic acid. This substance is produced by many species of ants and in fact got its name from the Latin word for ant, *formica*. In many cases, the formic acid probably serves as a disinfectant that protects the ants against microbial parasites. The devil's garden ant is the first ant species found to use formic acid as a herbicide. This use of a chemical is an important addition to the list of functions mediated by chemicals in the insect world. Scientists already know that chemicals play an important role in insect communication, attraction of mates, and defense against predators.

Research on devil's gardens is only one example of the relevance of chemistry to the study of life. Unlike a list of college courses, nature is not neatly packaged into the individual natural sciences—biology, chemistry, physics, and so forth. Biologists specialize in the study of life, but organisms and their environments are natural systems to which the concepts of chemistry and physics apply. Biology is a multidisciplinary science.

This unit of chapters introduces basic concepts of chemistry that will apply throughout our study of life. We will make many connections to the themes introduced in Chapter 1. One of these themes is the organization of life into a hierarchy of structural levels, with additional properties emerging at each successive level. In this unit, we will see how emergent properties are apparent at the lowest levels of biological organization—such as the ordering of atoms into molecules and the interactions of those molecules within cells. Somewhere in the transition from molecules to cells, we will cross the blurry boundary between nonlife and life. This chapter focuses on the chemical components that make up all matter.

▼ **Figure 2.2** **Inquiry**

What creates "devil's gardens" in the rain forest?

EXPERIMENT Working under Deborah Gordon and with Michael Greene, graduate student Megan Frederickson sought the cause of "devil's gardens," stands of a single species of tree, *Duroia hirsuta*. One hypothesis was that ants living in these trees, *Myrmelachista schumanni*, produce a poisonous chemical that kills trees of other species; another was that the *Duroia* trees themselves kill competing trees, perhaps by means of a chemical.

To test these hypotheses, Frederickson did field experiments in Peru. Two saplings of a local nonhost tree species, *Cedrela odorata,* were planted inside each of ten devil's gardens. At the base of one, a sticky insect barrier was applied; the other was unprotected. Two more *Cedrela* saplings, with and without barriers, were planted about 50 meters outside each garden.

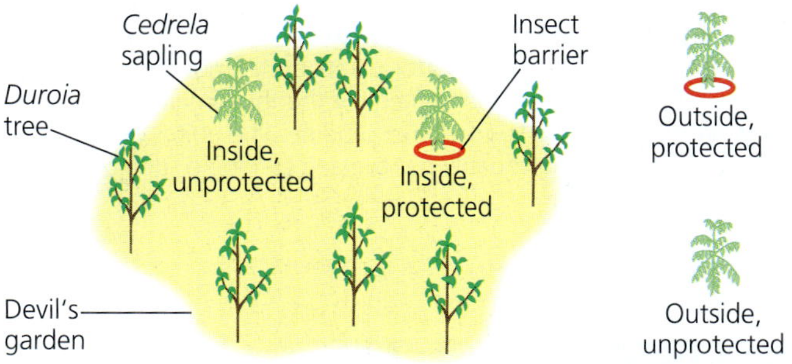

The researchers observed ant activity on the *Cedrela* leaves and measured areas of dead leaf tissue after one day. They also chemically analyzed contents of the ants' poison glands.

RESULTS The ants made injections from the tips of their abdomens into leaves of unprotected saplings in their gardens (see photo). Within one day, these leaves developed dead areas (see graph). The protected saplings were uninjured, as were the saplings planted outside the gardens. Formic acid was the only chemical detected in the poison glands of the ants.

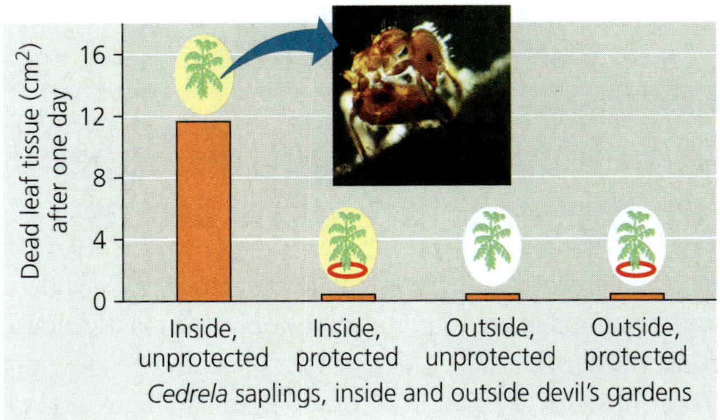

CONCLUSION Ants of the species *Myrmelachista schumanni* kill nonhost trees by injecting the leaves with formic acid, thus creating hospitable habitats (devil's gardens) for the ant colony.

SOURCE M. E. Frederickson, M. J. Greene, and D. M. Gordon, "Devil's gardens" bedevilled by ants, *Nature* 437:495–496 (2005).

Inquiry in Action Read and analyze the original paper in *Inquiry in Action: Interpreting Scientific Papers*.

WHAT IF? What would be the results if the unprotected saplings' inability to grow in the devil's gardens was caused by a chemical released by the *Duroia* trees rather than by the ants?

CONCEPT **2.1**

Matter consists of chemical elements in pure form and in combinations called compounds

Organisms are composed of **matter**, which is anything that takes up space and has mass.* Matter exists in many diverse forms. Rocks, metals, oils, gases, and humans are just a few examples of what seems an endless assortment of matter.

Elements and Compounds

Matter is made up of elements. An **element** is a substance that cannot be broken down to other substances by chemical reactions. Today, chemists recognize 92 elements occurring in nature; gold, copper, carbon, and oxygen are examples. Each element has a symbol, usually the first letter or two of its name. Some symbols are derived from Latin or German; for instance, the symbol for sodium is Na, from the Latin word *natrium.*

A **compound** is a substance consisting of two or more different elements combined in a fixed ratio. Table salt, for example, is sodium chloride (NaCl), a compound composed of the elements sodium (Na) and chlorine (Cl) in a 1:1 ratio. Pure sodium is a metal, and pure chlorine is a poisonous gas. When chemically combined, however, sodium and chlorine form an edible compound. Water (H_2O), another compound, consists of the elements hydrogen (H) and oxygen (O) in a 2:1 ratio. These are simple examples of organized matter having emergent properties: A compound has characteristics different from those of its elements **(Figure 2.3)**.

▲ **Figure 2.3 The emergent properties of a compound.** The metal sodium combines with the poisonous gas chlorine, forming the edible compound sodium chloride, or table salt.

* Sometimes we substitute the term *weight* for mass, although the two are not identical. Mass is the amount of matter in an object, whereas the weight of an object is how strongly that mass is pulled by gravity. The weight of an astronaut walking on the moon is approximately ⅙ that of the astronaut's weight on Earth, but his or her mass is the same. However, as long as we are earthbound, the weight of an object is a measure of its mass; in everyday language, therefore, we tend to use the terms interchangeably.

Essential Elements of Life

About 25 of the 92 natural elements are known to be essential to life. Just four of these—carbon (C), oxygen (O), hydrogen (H), and nitrogen (N)—make up 96% of living matter. Phosphorus (P), sulfur (S), calcium (Ca), potassium (K), and a few other elements account for most of the remaining 4% of an organism's weight. **Table 2.1** lists by percentage the elements that make up the human body; the percentages for other organisms are similar. **Figure 2.4a** illustrates the effect of a deficiency of nitrogen, an essential element, in a plant.

Trace elements are those required by an organism in only minute quantities. Some trace elements, such as iron (Fe), are needed by all forms of life; others are required only by certain species. For example, in vertebrates (animals with backbones), the element iodine (I) is an essential ingredient of a hormone produced by the thyroid gland. A daily intake of only 0.15 milligram (mg) of iodine is adequate for normal activity of the human thyroid. An iodine deficiency in the diet causes the thyroid gland to grow to abnormal size, a condition called goiter **(Figure 2.4b)**. Where it is available, iodized salt has reduced the incidence of goiter.

Table 2.1 Naturally Occurring Elements in the Human Body

Symbol	Element	Atomic Number (see p. 33)	Percentage of Human Body Weight
Elements making up about 96% of human body weight			
O	Oxygen	8	65.0
C	Carbon	6	18.5
H	Hydrogen	1	9.5
N	Nitrogen	7	3.3
Elements making up about 4% of human body weight			
Ca	Calcium	20	1.5
P	Phosphorus	15	1.0
K	Potassium	19	0.4
S	Sulfur	16	0.3
Na	Sodium	11	0.2
Cl	Chlorine	17	0.2
Mg	Magnesium	12	0.1
Elements making up less than 0.01% of human body weight (trace elements)			
Boron (B), chromium (Cr), cobalt (Co), copper (Cu), fluorine (F), iodine (I), iron (Fe), manganese (Mn), molybdenum (Mo), selenium (Se), silicon (Si), tin (Sn), vanadium (V), zinc (Zn)			

(a) Nitrogen deficiency (b) Iodine deficiency

▲ **Figure 2.4 The effects of essential-element deficiencies.** **(a)** This photo shows the effect of nitrogen deficiency in corn (maize). In this controlled experiment, the taller plants on the left are growing in nitrogen-rich soil, and the shorter plants on the right in nitrogen-poor soil. **(b)** Goiter is an enlargement of the thyroid gland, resulting from a deficiency of the trace element iodine. The goiter of this Malaysian woman can probably be reversed by iodine supplements.

CONCEPT CHECK 2.1

1. Explain how table salt has emergent properties.
2. Is a trace element an essential element? Explain.
3. **WHAT IF?** Iron (Fe) is a trace element required for the proper functioning of hemoglobin, the molecule that carries oxygen in red blood cells. What might be the effects of an iron deficiency?

For suggested answers, see Appendix A.

CONCEPT **2.2**

An element's properties depend on the structure of its atoms

Each element consists of a certain kind of atom that is different from the atoms of any other element. An **atom** is the smallest unit of matter that still retains the properties of an element. Atoms are so small that it would take about a million of them to stretch across the period printed at the end of this sentence. We symbolize atoms with the same abbreviation used for the element that is made up of those atoms. For example, the symbol C stands for both the element carbon and a single carbon atom.

Subatomic Particles

Although the atom is the smallest unit having the properties of its element, these tiny bits of matter are composed of even smaller parts, called *subatomic particles*. Physicists have split the atom into more than a hundred types of particles, but only

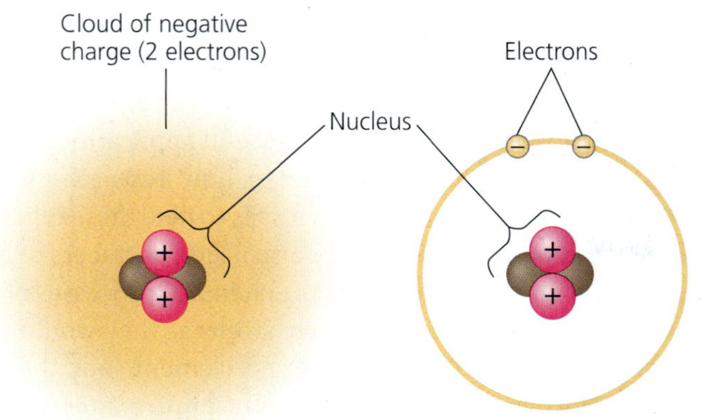

(a) This model represents the electrons as a cloud of negative charge.

(b) In this even more simplified model, the electrons are shown as two small yellow spheres on a circle around the nucleus.

▲ **Figure 2.5 Simplified models of a helium (He) atom.** The helium nucleus consists of 2 neutrons (brown) and 2 protons (pink). Two electrons (yellow) exist outside the nucleus. These models are not to scale; they greatly overestimate the size of the nucleus in relation to the electron cloud.

three kinds of particles are relevant here: **neutrons**, **protons**, and **electrons**. Protons and electrons are electrically charged. Each proton has one unit of positive charge, and each electron has one unit of negative charge. A neutron, as its name implies, is electrically neutral.

Protons and neutrons are packed together tightly in a dense core, or **atomic nucleus**, at the center of an atom; protons give the nucleus a positive charge. The electrons form a sort of cloud of negative charge around the nucleus, and it is the attraction between opposite charges that keeps the electrons in the vicinity of the nucleus. **Figure 2.5** shows two models of the structure of the helium atom as an example.

The neutron and proton are almost identical in mass, each about 1.7×10^{-24} gram (g). Grams and other conventional units are not very useful for describing the mass of objects so minuscule. Thus, for atoms and subatomic particles (and for molecules, too), we use a unit of measurement called the **dalton**, in honor of John Dalton, the British scientist who helped develop atomic theory around 1800. (The dalton is the same as the *atomic mass unit*, or *amu*, a unit you may have encountered elsewhere.) Neutrons and protons have masses close to 1 dalton. Because the mass of an electron is only about $\frac{1}{2{,}000}$ that of a neutron or proton, we can ignore electrons when computing the total mass of an atom.

Atomic Number and Atomic Mass

Atoms of the various elements differ in their number of subatomic particles. All atoms of a particular element have the same number of protons in their nuclei. This number of protons, which is unique to that element, is called the **atomic number** and is written as a subscript to the left of the symbol for the element. The abbreviation $_2$He, for example, tells us that an atom of the element helium has 2 protons in its nucleus. Unless otherwise indicated, an atom is neutral in electrical charge, which means that its protons must be balanced by an equal number of electrons. Therefore, the atomic number tells us the number of protons and also the number of electrons in an electrically neutral atom.

We can deduce the number of neutrons from a second quantity, the **mass number**, which is the sum of protons plus neutrons in the nucleus of an atom. The mass number is written as a superscript to the left of an element's symbol. For example, we can use this shorthand to write an atom of helium as $^4_2$He. Because the atomic number indicates how many protons there are, we can determine the number of neutrons by subtracting the atomic number from the mass number: The helium atom, $^4_2$He, has 2 neutrons. An atom of sodium, $^{23}_{11}$Na, has 11 protons, 11 electrons, and 12 neutrons. The simplest atom is hydrogen, $^1_1$H, which has no neutrons; it consists of a single proton with a single electron.

As mentioned earlier, the contribution of electrons to mass is negligible. Therefore, almost all of an atom's mass is concentrated in its nucleus. Because neutrons and protons each have a mass very close to 1 dalton, the mass number is an approximation of the total mass of an atom, called its **atomic mass**. So we might say that the atomic mass of sodium ($^{23}_{11}$Na) is 23 daltons, although more precisely it is 22.9898 daltons.

Isotopes

All atoms of a given element have the same number of protons, but some atoms have more neutrons than other atoms of the same element and therefore have greater mass. These different atomic forms are called **isotopes** of the element. In nature, an element occurs as a mixture of its isotopes. For example, consider the three isotopes of the element carbon, which has the atomic number 6. The most common isotope is carbon-12, $^{12}_6$C, which accounts for about 99% of the carbon in nature. The isotope $^{12}_6$C has 6 neutrons. Most of the remaining 1% of carbon consists of atoms of the isotope $^{13}_6$C, with 7 neutrons. A third, even rarer isotope, $^{14}_6$C, has 8 neutrons. Notice that all three isotopes of carbon have 6 protons; otherwise, they would not be carbon. Although the isotopes of an element have slightly different masses, they behave identically in chemical reactions. (The number usually given as the atomic mass of an element, such as 22.9898 daltons for sodium, is actually an average of the atomic masses of all the element's naturally occurring isotopes.)

Both ^{12}C and ^{13}C are stable isotopes, meaning that their nuclei do not have a tendency to lose particles. The isotope ^{14}C, however, is unstable, or radioactive. A **radioactive isotope** is one in which the nucleus decays spontaneously, giving off particles and energy. When the decay leads to a change in the

▼ Figure 2.6 **Research Method**

Radioactive Tracers

APPLICATION Scientists use radioactive isotopes to label certain chemical compounds, creating tracers that can be used to follow a metabolic process or locate the compound within an organism. In this example, radioactive tracers are being used to determine the effect of temperature on the rate at which cells make copies of their DNA.

TECHNIQUE

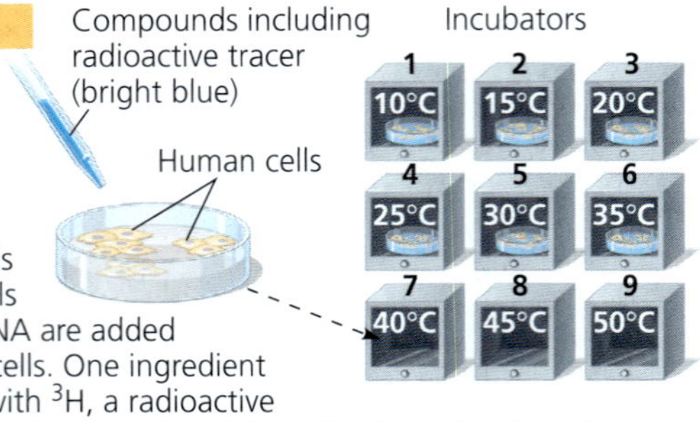

1 Compounds used by cells to make DNA are added to human cells. One ingredient is labeled with ^3H, a radioactive isotope of hydrogen. Nine dishes of cells are incubated at different temperatures. The cells make new DNA, incorporating the radioactive tracer with ^3H.

2 The cells are placed in test tubes; their DNA is isolated; and unused labeled compounds are removed.

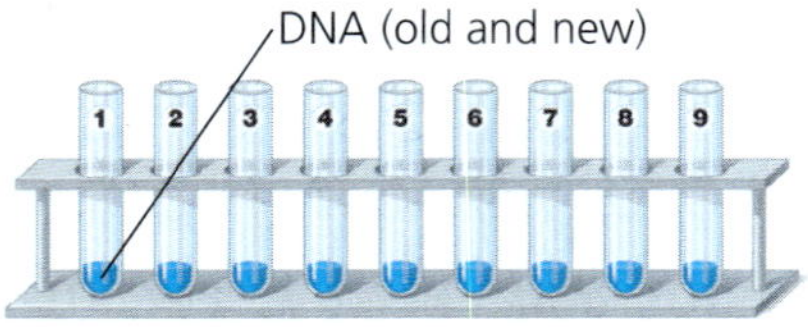

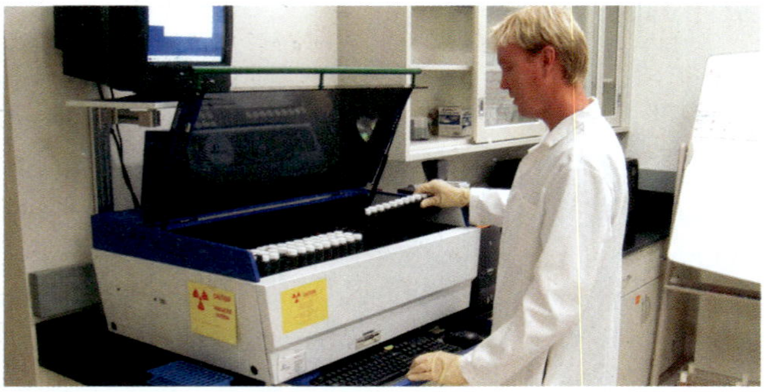

3 A solution called scintillation fluid is added to the test tubes and they are placed in a scintillation counter. As the ^3H in the newly made DNA decays, it emits radiation that excites chemicals in the scintillation fluid, causing them to give off light. Flashes of light are recorded by the scintillation counter.

RESULTS The frequency of flashes, which is recorded as counts per minute, is proportional to the amount of the radioactive tracer present, indicating the amount of new DNA. In this experiment, when the counts per minute are plotted against temperature, it is clear that temperature affects the rate of DNA synthesis; the most DNA was made at 35°C.

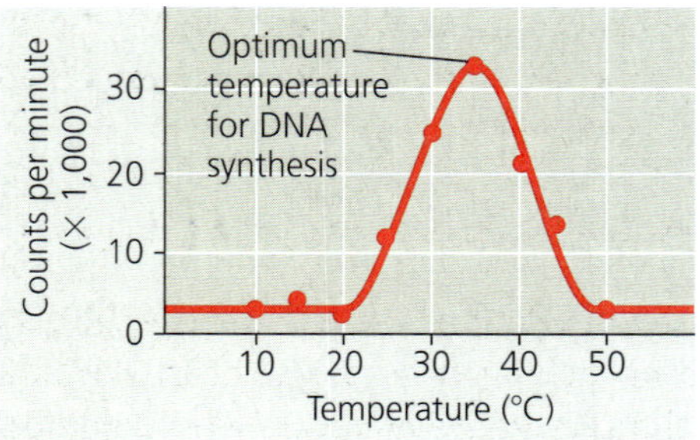

number of protons, it transforms the atom to an atom of a different element. For example, radioactive carbon decays to form nitrogen.

Radioactive isotopes have many useful applications in biology. In Chapter 25, you will learn how researchers use measurements of radioactivity in fossils to date these relics of past life. Radioactive isotopes are also useful as tracers to follow atoms through metabolism, the chemical processes of an organism. Cells use the radioactive atoms as they would use nonradioactive isotopes of the same element, but the radioactive tracers can be readily detected. **Figure 2.6** presents an example of how biologists use radioactive tracers to monitor biological processes, in this case the synthesis of DNA by human cells.

Radioactive tracers are important diagnostic tools in medicine. For example, certain kidney disorders can be diagnosed by injecting small doses of substances containing radioactive isotopes into the blood and then measuring the amount of tracer excreted in the urine. Radioactive tracers are also used in combination with sophisticated imaging instruments, such as PET scanners, which can monitor chemical processes, such as those involved in cancerous growth, as they actually occur in the body **(Figure 2.7)**.

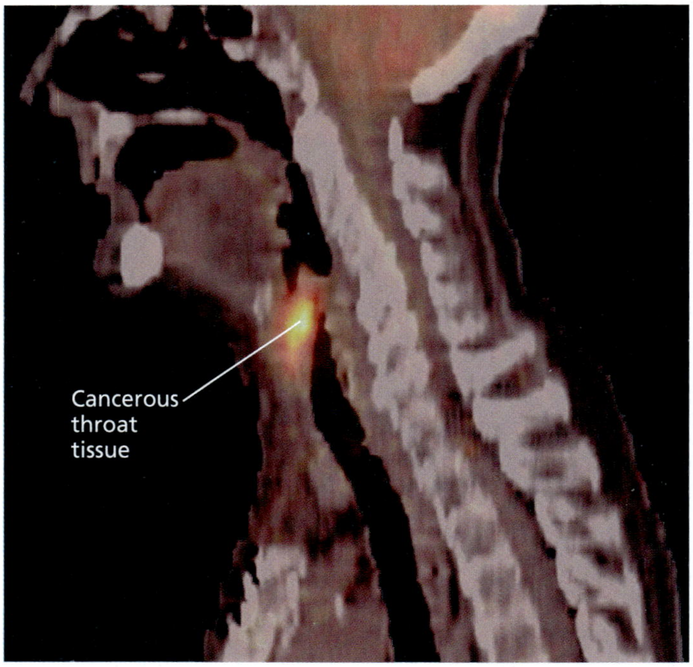

▲ **Figure 2.7 A PET scan, a medical use for radioactive isotopes.** PET, an acronym for positron-emission tomography, detects locations of intense chemical activity in the body. The patient is first injected with a nutrient such as glucose labeled with a radioactive isotope that emits subatomic particles. These particles collide with electrons made available by chemical reactions in the body. A PET scanner detects the energy released in these collisions and maps "hot spots," the regions of an organ that are most chemically active at the time. The color of the image varies with the amount of the isotope present, with the bright yellow color here identifying a region of cancerous throat tissue.

Although radioactive isotopes are very useful in biological research and medicine, radiation from decaying isotopes also poses a hazard to life by damaging cellular molecules. The severity of this damage depends on the type and amount of radiation an organism absorbs. One of the most serious environmental threats is radioactive fallout from nuclear accidents. The doses of most isotopes used in medical diagnosis, however, are relatively safe.

The Energy Levels of Electrons

The simplified models of the atom in Figure 2.5 greatly exaggerate the size of the nucleus relative to the volume of the whole atom. If an atom of helium were the size of Yankee Stadium, the nucleus would be only the size of a pencil eraser in the center of the field. Moreover, the electrons would be like two tiny gnats buzzing around the stadium. Atoms are mostly empty space.

When two atoms approach each other during a chemical reaction, their nuclei do not come close enough to interact. Of the three kinds of subatomic particles we have discussed, only electrons are directly involved in the chemical reactions between atoms.

An atom's electrons vary in the amount of energy they possess. **Energy** is defined as the capacity to cause change—for instance, by doing work. **Potential energy** is the energy that matter possesses because of its location or structure. For example, water in a reservoir on a hill has potential energy because of its altitude. When the gates of the reservoir's dam are opened and the water runs downhill, the energy can be used to do work, such as turning generators. Because energy has been expended, the water has less energy at the bottom of the hill than it did in the reservoir. Matter has a natural tendency to move to the lowest possible state of potential energy; in this example, the water runs downhill. To restore the potential energy of a reservoir, work must be done to elevate the water against gravity.

The electrons of an atom have potential energy because of how they are arranged in relation to the nucleus. The negatively charged electrons are attracted to the positively charged nucleus. It takes work to move a given electron farther away from the nucleus, so the more distant an electron is from the nucleus, the greater its potential energy. Unlike the continuous flow of water downhill, changes in the potential energy of electrons can occur only in steps of fixed amounts. An electron having a certain amount of energy is something like a ball on a staircase **(Figure 2.8a)**. The ball can have different amounts of potential energy, depending on which step it is on, but it cannot spend much time between the steps. Similarly, an electron's potential energy is determined by its energy level. An electron cannot exist in between energy levels.

An electron's energy level is correlated with its average distance from the nucleus. Electrons are found in different

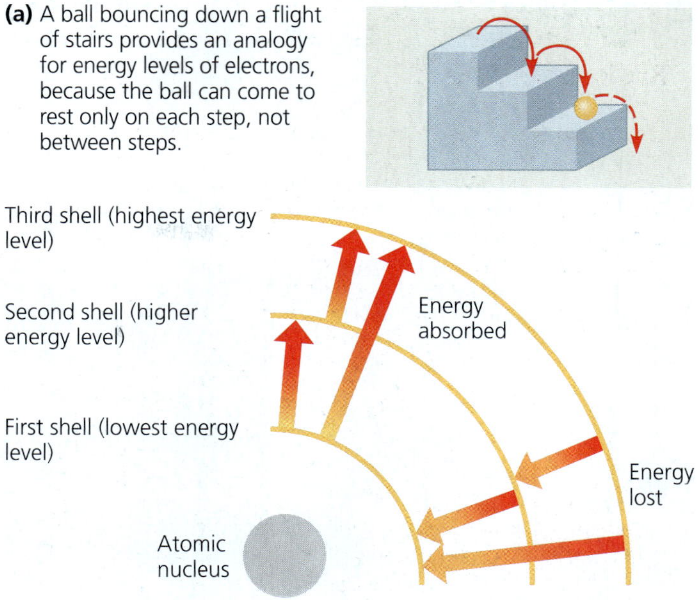

(a) A ball bouncing down a flight of stairs provides an analogy for energy levels of electrons, because the ball can come to rest only on each step, not between steps.

(b) An electron can move from one shell to another only if the energy it gains or loses is exactly equal to the difference in energy between the energy levels of the two shells. Arrows indicate some of the stepwise changes in potential energy that are possible.

▲ **Figure 2.8 Energy levels of an atom's electrons.** Electrons exist only at fixed levels of potential energy called electron shells.

electron shells, each with a characteristic average distance and energy level. In diagrams, shells can be represented by concentric circles **(Figure 2.8b)**. The first shell is closest to the nucleus, and electrons in this shell have the lowest potential energy. Electrons in the second shell have more energy, and electrons in the third shell even more energy. An electron can change the shell it occupies, but only by absorbing or losing an amount of energy equal to the difference in potential energy between its position in the old shell and that in the new shell. When an electron absorbs energy, it moves to a shell farther out from the nucleus. For example, light energy can excite an electron to a higher energy level. (Indeed, this is the first step taken when plants harness the energy of sunlight for photosynthesis, the process that produces food from carbon dioxide and water.) When an electron loses energy, it "falls back" to a shell closer to the nucleus, and the lost energy is usually released to the environment as heat. For example, sunlight excites electrons in the surface of a car to higher energy levels. When the electrons fall back to their original levels, the car's surface heats up. This thermal energy can be transferred to the air or to your hand if you touch the car.

Electron Distribution and Chemical Properties

The chemical behavior of an atom is determined by the distribution of electrons in the atom's electron shells. Beginning with hydrogen, the simplest atom, we can imagine building the atoms of the other elements by adding 1 proton and 1 electron

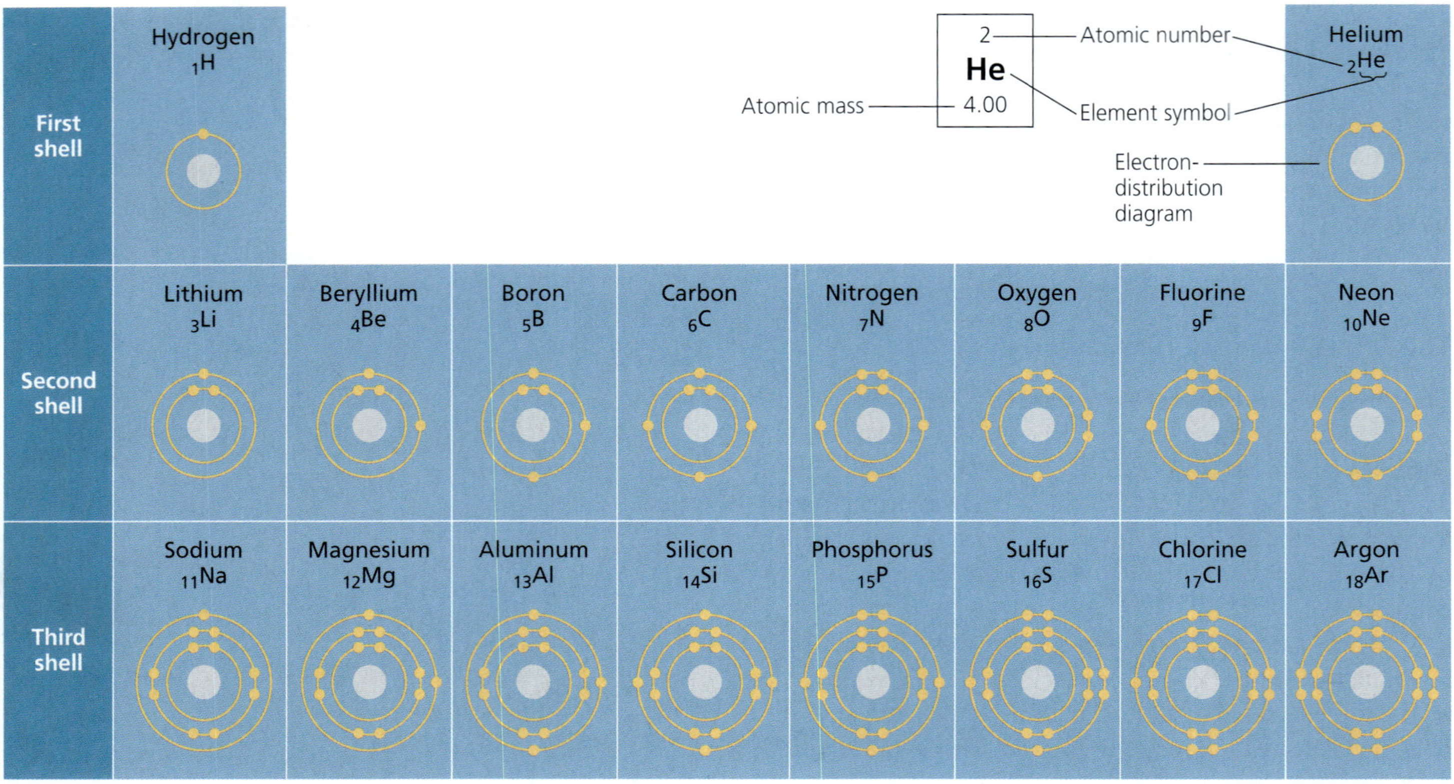

▲ **Figure 2.9 Electron-distribution diagrams for the first 18 elements in the periodic table.** In a standard periodic table (see Appendix B), information for each element is presented as shown for helium in the inset. In the diagrams in this table, electrons are represented as yellow dots and electron shells as concentric circles. These diagrams are a convenient way to picture the distribution of an atom's electrons among its electron shells, but these simplified models do not accurately represent the shape of the atom or the location of its electrons. The elements are arranged in rows, each representing the filling of an electron shell. As electrons are added, they occupy the lowest available shell.

? *What is the atomic number of magnesium? How many protons and electrons does it have? How many electron shells? How many valence electrons?*

at a time (along with an appropriate number of neutrons). **Figure 2.9**, an abbreviated version of what is called the *periodic table of the elements*, shows this distribution of electrons for the first 18 elements, from hydrogen ($_1H$) to argon ($_{18}Ar$). The elements are arranged in three rows, or periods, corresponding to the number of electron shells in their atoms. The left-to-right sequence of elements in each row corresponds to the sequential addition of electrons and protons. (See Appendix B for the complete periodic table.)

Hydrogen's 1 electron and helium's 2 electrons are located in the first shell. Electrons, like all matter, tend to exist in the lowest available state of potential energy. In an atom, this state is in the first shell. However, the first shell can hold no more than 2 electrons; thus, hydrogen and helium are the only elements in the first row of the table. An atom with more than 2 electrons must use higher shells because the first shell is full. The next element, lithium, has 3 electrons. Two of these electrons fill the first shell, while the third electron occupies the second shell. The second shell holds a maximum of 8 electrons. Neon, at the end of the second row, has 8 electrons in the second shell, giving it a total of 10 electrons.

The chemical behavior of an atom depends mostly on the number of electrons in its *outermost* shell. We call those outer electrons **valence electrons** and the outermost electron shell the **valence shell**. In the case of lithium, there is only 1 valence electron, and the second shell is the valence shell. Atoms with the same number of electrons in their valence shells exhibit similar chemical behavior. For example, fluorine (F) and chlorine (Cl) both have 7 valence electrons, and both form compounds when combined with the element sodium (see Figure 2.3). An atom with a completed valence shell is unreactive; that is, it will not interact readily with other atoms. At the far right of the periodic table are helium, neon, and argon, the only three elements shown in Figure 2.9 that have full valence shells. These elements are said to be *inert*, meaning chemically unreactive. All the other atoms in Figure 2.9 are chemically reactive because they have incomplete valence shells.

Electron Orbitals

In the early 1900s, the electron shells of an atom were visualized as concentric paths of electrons orbiting the nucleus, somewhat like planets orbiting the sun. It is still convenient to use two-dimensional concentric-circle diagrams to symbolize electron shells, as in Figure 2.9. However, you need to remember that each concentric circle represents only the

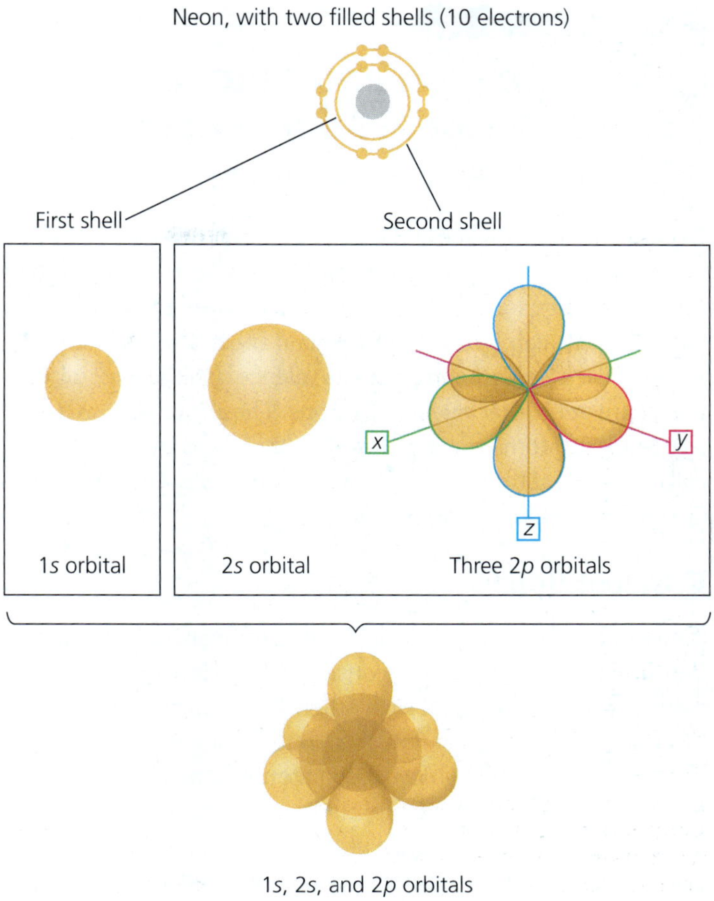

(a) Electron-distribution diagram. An electron-distribution diagram is shown here for a neon atom, which has a total of 10 electrons. Each concentric circle represents an electron shell, which can be subdivided into electron orbitals.

(b) Separate electron orbitals. The three-dimensional shapes represent electron orbitals—the volumes of space where the electrons of an atom are most likely to be found. Each orbital holds a maximum of 2 electrons. The first electron shell, on the left, has one spherical (*s*) orbital, designated 1*s*. The second shell, on the right, has one larger *s* orbital (designated 2*s* for the second shell) plus three dumbbell-shaped orbitals called *p* orbitals (2*p* for the second shell). The three 2*p* orbitals lie at right angles to one another along imaginary *x*-, *y*-, and *z*-axes of the atom. Each 2*p* orbital is outlined here in a different color.

(c) Superimposed electron orbitals. To show the complete picture of the electron orbitals of neon, we superimpose the 1*s* orbital of the first shell and the 2*s* and three 2*p* orbitals of the second shell.

▲ **Figure 2.10 Electron orbitals.**

average distance between an electron in that shell and the nucleus. Accordingly, the concentric-circle diagrams do not give a real picture of an atom. In reality, we can never know the exact location of an electron. What we can do instead is describe the space in which an electron spends most of its time. The three-dimensional space where an electron is found 90% of the time is called an **orbital**.

Each electron shell contains electrons at a particular energy level, distributed among a specific number of orbitals of distinctive shapes and orientations. **Figure 2.10** shows the orbitals of neon as an example. You can think of an orbital as a component of an electron shell. The first electron shell has only one spherical *s* orbital (called 1*s*), but the second shell has four orbitals: one large spherical *s* orbital (called 2*s*) and three dumbbell-shaped *p* orbitals (called 2*p* orbitals). (The third shell and other higher electron shells also have *s* and *p* orbitals, as well as orbitals of more complex shapes.)

No more than 2 electrons can occupy a single orbital. The first electron shell can therefore accommodate up to 2 electrons in its *s* orbital. The lone electron of a hydrogen atom occupies the 1*s* orbital, as do the 2 electrons of a helium atom. The four orbitals of the second electron shell can hold up to 8 electrons. Electrons in each of the four orbitals have nearly the same energy, but they move in different volumes of space.

The reactivity of atoms arises from the presence of unpaired electrons in one or more orbitals of their valence shells. As you will see in the next section, atoms interact in a way that completes their valence shells. When they do so, it is the *unpaired* electrons that are involved.

CONCEPT CHECK 2.2

1. A lithium atom has 3 protons and 4 neutrons. What is its atomic mass in daltons?
2. A nitrogen atom has 7 protons, and the most common isotope of nitrogen has 7 neutrons. A radioactive isotope of nitrogen has 8 neutrons. Write the atomic number and mass number of this radioactive nitrogen as a chemical symbol with a subscript and superscript.
3. How many electrons does fluorine have? How many electron shells? Name the orbitals that are occupied. How many electrons are needed to fill the valence shell?
4. **WHAT IF?** In Figure 2.9, if two or more elements are in the same row, what do they have in common? If two or more elements are in the same column, what do they have in common?

For suggested answers, see Appendix A.

CONCEPT 2.3

The formation and function of molecules depend on chemical bonding between atoms

Now that we have looked at the structure of atoms, we can move up the hierarchy of organization and see how atoms combine to form molecules and ionic compounds. Atoms with incomplete valence shells can interact with certain other atoms in such a way that each partner completes its valence shell: The atoms either share or transfer valence electrons. These interactions usually result in atoms staying close together, held by attractions called **chemical bonds**. The strongest kinds of chemical bonds are covalent bonds and ionic bonds.

Covalent Bonds

A **covalent bond** is the sharing of a pair of valence electrons by two atoms. For example, let's consider what happens when two hydrogen atoms approach each other. Recall that hydrogen has 1 valence electron in the first shell, but the shell's capacity is 2 electrons. When the two hydrogen atoms come close enough for their 1*s* orbitals to overlap, they can share their electrons **(Figure 2.11)**. Each hydrogen atom now has 2 electrons associated with it in what amounts to a completed valence shell. Two or more atoms held together by covalent bonds constitute a **molecule**. In this case, the example is a hydrogen molecule.

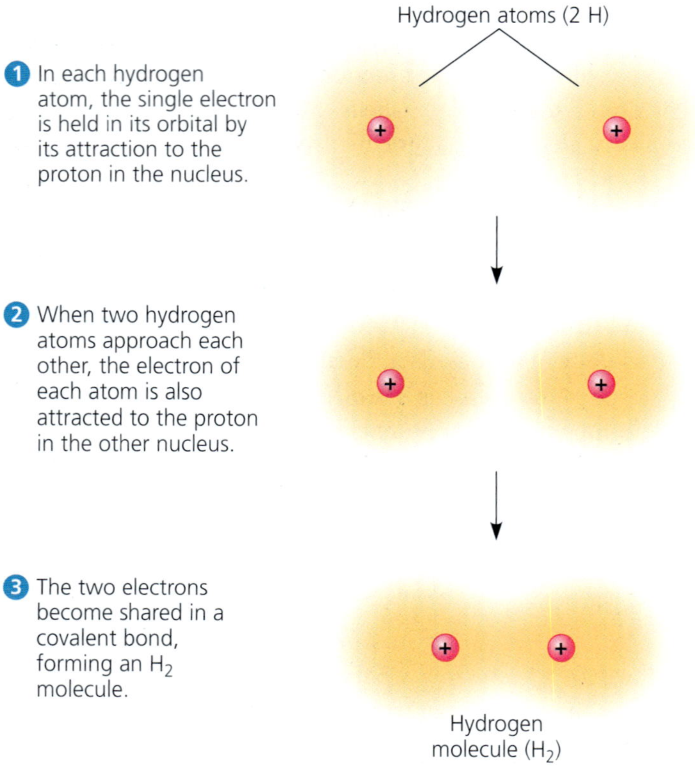

▲ **Figure 2.11 Formation of a covalent bond.**

Electron sharing can be depicted using element symbols, with dots representing the outermost electrons. You have probably seen such diagrams, called Lewis dot structures, in your chemistry book. The Lewis dot structure for a hydrogen molecule, H:H, is shown in **Figure 2.12a**. We can abbreviate the structure of this molecule as H—H, where the line represents a single covalent bond, or simply a **single bond**—that is, a pair of shared electrons. This notation, which represents both atoms and bonding, is called a **structural formula**. We can abbreviate even further by writing H_2, a **molecular formula** indicating simply that the molecule consists of two atoms of hydrogen.

Oxygen has 6 electrons in its second electron shell and therefore needs 2 more electrons to complete its valence shell. Two oxygen atoms form a molecule by sharing *two* pairs of valence electrons **(Figure 2.12b)**. The atoms are thus joined by what is called a double covalent bond, or **double bond**.

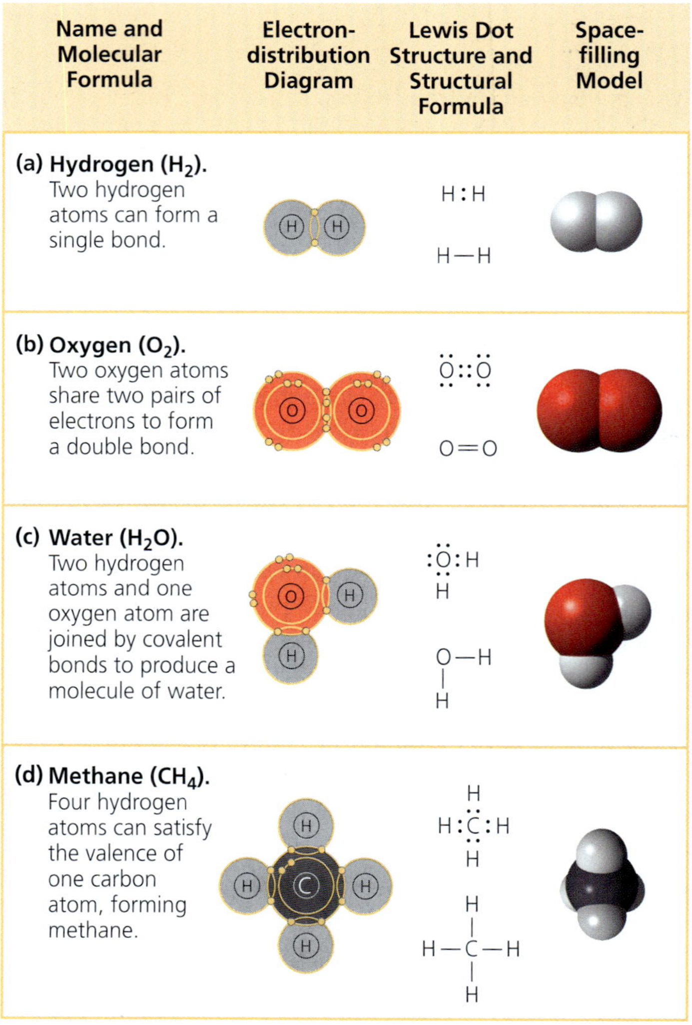

▲ **Figure 2.12 Covalent bonding in four molecules.** A single covalent bond consists of a pair of shared electrons. The number of electrons required to complete an atom's valence shell generally determines how many bonds that atom will form. Four ways of indicating bonds are shown; the space-filling model comes closest to representing the actual shape of the molecule (see also Figure 2.17).

Each atom that can share valence electrons has a bonding capacity corresponding to the number of covalent bonds the atom can form. When the bonds form, they give the atom a full complement of electrons in the valence shell. The bonding capacity of oxygen, for example, is 2. This bonding capacity is called the atom's **valence** and usually equals the number of unpaired electrons required to complete the atom's outermost (valence) shell. See if you can determine the valences of hydrogen, oxygen, nitrogen, and carbon by studying the electron distribution diagrams in Figure 2.9. You can see that the valence of hydrogen is 1; oxygen, 2; nitrogen, 3; and carbon, 4. However, there are more complicated cases, such as phosphorus (P), another element important to life. Phosphorus can have a valence of 3, as we would predict from the presence of 3 unpaired electrons in its valence shell. In biologically important molecules, however, phosphorus can form three single bonds and one double bond. Therefore, it can also have a valence of 5.

The molecules H_2 and O_2 are pure elements rather than compounds because a compound is a combination of two or more *different* elements. Water, with the molecular formula H_2O, is a compound. Two atoms of hydrogen are needed to satisfy the valence of one oxygen atom. **Figure 2.12c** shows the structure of a water molecule. Water is so important to life that Chapter 3 is devoted entirely to its structure and behavior.

Another molecule that is a compound is methane, the main component of natural gas, with the molecular formula CH_4 **(Figure 2.12d)**. It takes four hydrogen atoms, each with a valence of 1, to complement one atom of carbon, with its valence of 4. We will look at many other compounds of carbon in Chapter 4.

The attraction of a particular kind of atom for the electrons of a covalent bond is called its **electronegativity**. The more electronegative an atom, the more strongly it pulls shared electrons toward itself. In a covalent bond between two atoms of the same element, the outcome of the tug-of-war for common electrons is a standoff; the two atoms are equally electronegative. Such a bond, in which the electrons are shared equally, is a **nonpolar covalent bond**. For example, the covalent bond of H_2 is nonpolar, as is the double bond of O_2. In other compounds, however, where one atom is bonded to a more electronegative atom, the electrons of the bond are not shared equally. This type of bond is called a **polar covalent bond**. Such bonds vary in their polarity, depending on the relative electronegativity of the two atoms. For example, the bonds between the oxygen and hydrogen atoms of a water molecule are quite polar **(Figure 2.13)**. Oxygen is one of the most electronegative of all the elements, attracting shared electrons much more strongly than hydrogen does. In a covalent bond between oxygen and hydrogen, the electrons spend more time near the oxygen nucleus than they do near the hydrogen nucleus. Because electrons have a negative charge, the unequal sharing of electrons in water causes the oxygen atom to have a partial negative charge (indicated by the Greek letter δ with a minus sign, δ−, or "delta minus") and each hydrogen atom to have a partial positive charge (δ+, or "delta plus"). In contrast, the individual bonds of methane (CH_4) are much less polar because carbon and hydrogen differ much less in electronegativity than do oxygen and hydrogen.

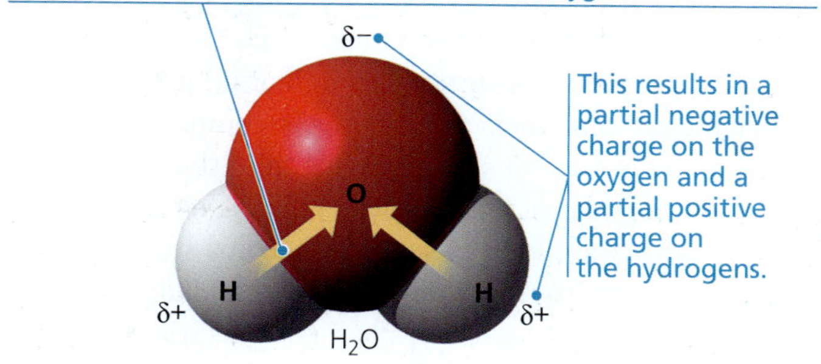

▲ **Figure 2.13 Polar covalent bonds in a water molecule.**

Ionic Bonds

In some cases, two atoms are so unequal in their attraction for valence electrons that the more electronegative atom strips an electron completely away from its partner. This is what happens when an atom of sodium ($_{11}Na$) encounters an atom of chlorine ($_{17}Cl$) **(Figure 2.14)**. A sodium atom has a total of 11 electrons,

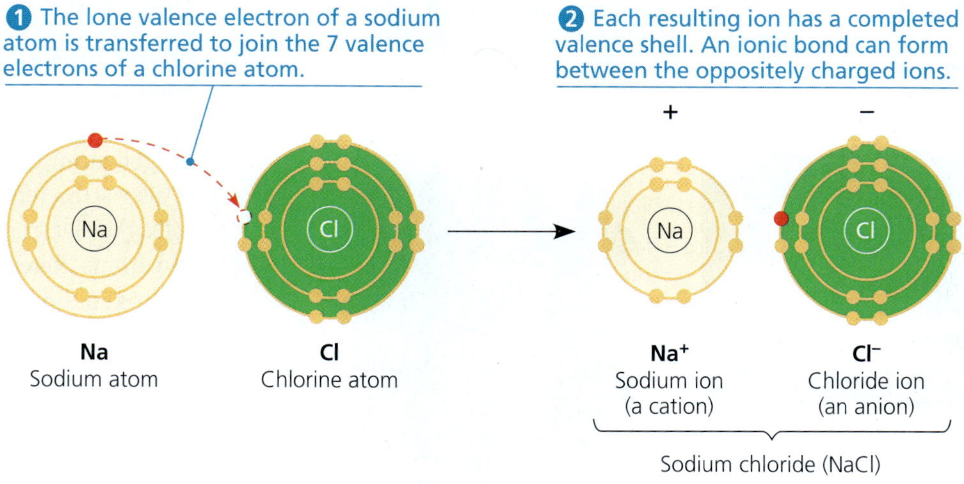

▲ **Figure 2.14 Electron transfer and ionic bonding.** The attraction between oppositely charged atoms, or ions, is an ionic bond. An ionic bond can form between any two oppositely charged ions, even if they have not been formed by transfer of an electron from one to the other.

with its single valence electron in the third electron shell. A chlorine atom has a total of 17 electrons, with 7 electrons in its valence shell. When these two atoms meet, the lone valence electron of sodium is transferred to the chlorine atom, and both atoms end up with their valence shells complete. (Because sodium no longer has an electron in the third shell, the second shell is now the valence shell.)

The electron transfer between the two atoms moves one unit of negative charge from sodium to chlorine. Sodium, now with 11 protons but only 10 electrons, has a net electrical charge of 1+. A charged atom (or molecule) is called an **ion**. When the charge is positive, the ion is specifically called a **cation**; the sodium atom has become a cation. Conversely, the chlorine atom, having gained an extra electron, now has 17 protons and 18 electrons, giving it a net electrical charge of 1−. It has become a chloride ion—an **anion**, or negatively charged ion. Because of their opposite charges, cations and anions attract each other; this attraction is called an **ionic bond**. The transfer of an electron is not the formation of a bond; rather, it allows a bond to form because it results in two ions. Any two ions of opposite charge can form an ionic bond. The ions do not need to have acquired their charge by an electron transfer with each other.

Compounds formed by ionic bonds are called **ionic compounds**, or **salts**. We know the ionic compound sodium chloride (NaCl) as table salt **(Figure 2.15)**. Salts are often found in nature as crystals of various sizes and shapes. Each salt crystal is an aggregate of vast numbers of cations and anions bonded by their electrical attraction and arranged in a three-dimensional lattice. Unlike a covalent compound, which consists of molecules having a definite size and number of atoms, an ionic compound does not consist of molecules in the same sense. The formula for an ionic compound, such as NaCl, indicates only the ratio of elements in a crystal of the salt. "NaCl" by itself is not a molecule.

Not all salts have equal numbers of cations and anions. For example, the ionic compound magnesium chloride ($MgCl_2$) has two chloride ions for each magnesium ion. Magnesium ($_{12}Mg$) must lose 2 outer electrons if the atom is to have a complete valence shell, so it tends to become a cation with a net charge of 2+ (Mg^{2+}). One magnesium cation can therefore form ionic bonds with two chloride anions.

▲ **Figure 2.15 A sodium chloride crystal.** The sodium ions (Na^+) and chloride ions (Cl^-) are held together by ionic bonds. The formula NaCl tells us that the ratio of Na^+ to Cl^- is 1:1.

The term *ion* also applies to entire molecules that are electrically charged. In the salt ammonium chloride (NH_4Cl), for instance, the anion is a single chloride ion (Cl^-), but the cation is ammonium (NH_4^+), a nitrogen atom with four covalently bonded hydrogen atoms. The whole ammonium ion has an electrical charge of 1+ because it is 1 electron short.

Environment affects the strength of ionic bonds. In a dry salt crystal, the bonds are so strong that it takes a hammer and chisel to break enough of them to crack the crystal in two. If the same salt crystal is dissolved in water, however, the ionic bonds are much weaker because each ion is partially shielded by its interactions with water molecules. Most drugs are manufactured as salts because they are quite stable when dry but can dissociate easily in water. In the next chapter, you will learn how water dissolves salts.

Weak Chemical Bonds

In organisms, most of the strongest chemical bonds are covalent bonds, which link atoms to form a cell's molecules. But weaker bonding within and between molecules is also indispensable in the cell, contributing greatly to the emergent properties of life. Most important large biological molecules are held in their functional form by weak bonds. In addition, when two molecules in the cell make contact, they may adhere temporarily by weak bonds. The reversibility of weak bonding can be an advantage: Two molecules can come together, respond to one another in some way, and then separate.

Certain types of weak chemical bonds are important in organisms. One is the ionic bond as it exists between ions dissociated in water, which we just discussed. Another type of weak bond, known as a hydrogen bond, is also crucial to life.

Hydrogen Bonds

Among the various kinds of weak chemical bonds, hydrogen bonds are so important in the chemistry of life that they deserve special attention. A **hydrogen bond** forms when a hydrogen atom covalently bonded to one electronegative atom is also attracted to another electronegative atom. In living cells, the electronegative partners are usually oxygen or nitrogen atoms. Refer to **Figure 2.16** to examine the simple case of hydrogen bonding between water (H_2O) and ammonia (NH_3). In the next chapter, we'll see how the hydrogen bonds between water molecules allow some insects to walk on water.

Van der Waals Interactions

Even a molecule with nonpolar covalent bonds may have positively and negatively charged regions. Electrons are not always symmetrically distributed in such a molecule; at any

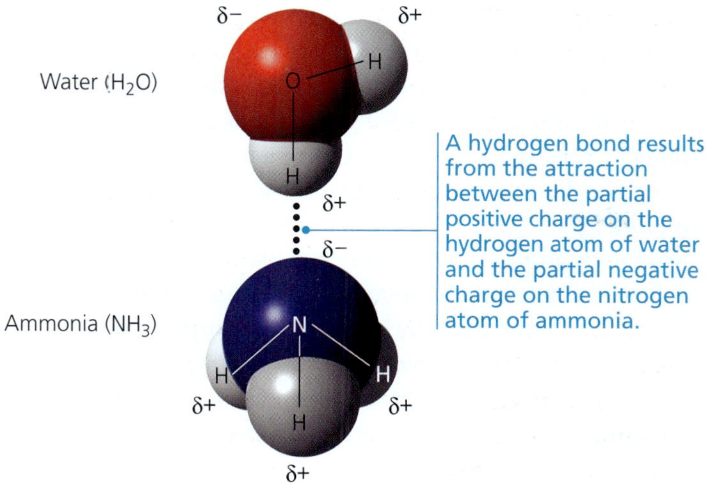

▲ **Figure 2.16 A hydrogen bond.**

DRAW IT *Draw five water molecules using structural formulas and indicating partial charges, and show how they can make hydrogen bonds with each other.*

instant, they may accumulate by chance in one part of the molecule or another. The results are ever-changing regions of positive and negative charge that enable all atoms and molecules to stick to one another. These **van der Waals interactions** are weak and occur only when atoms and molecules are very close together. In spite of their weakness, van der Waals interactions were recently shown to be responsible for the ability of a gecko lizard (left) to walk up a wall. Each gecko toe has hundreds of thousands of tiny hairs, with multiple projections at the hair's tip that increase surface area. Apparently, the van der Waals interactions between the hair tip molecules and the molecules of the wall's surface are so numerous that despite their individual weakness, together they can support the gecko's body weight.

Van der Waals interactions, hydrogen bonds, ionic bonds in water, and other weak bonds may form not only between molecules but also between different regions of a single large molecule, such as a protein. Although these bonds are individually weak, their cumulative effect is to reinforce the three-dimensional shape of a large molecule. You will learn more about the very important biological roles of weak bonds in Chapter 5.

Molecular Shape and Function

A molecule has a characteristic size and shape. The precise shape of a molecule is usually very important to its function in the living cell.

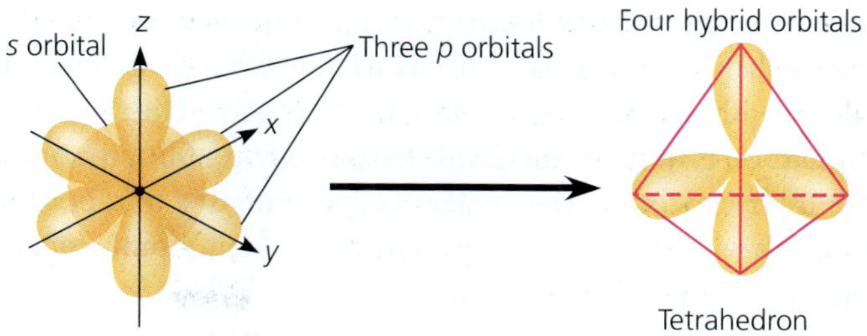

(a) Hybridization of orbitals. The single *s* and three *p* orbitals of a valence shell involved in covalent bonding combine to form four teardrop-shaped hybrid orbitals. These orbitals extend to the four corners of an imaginary tetrahedron (outlined in red).

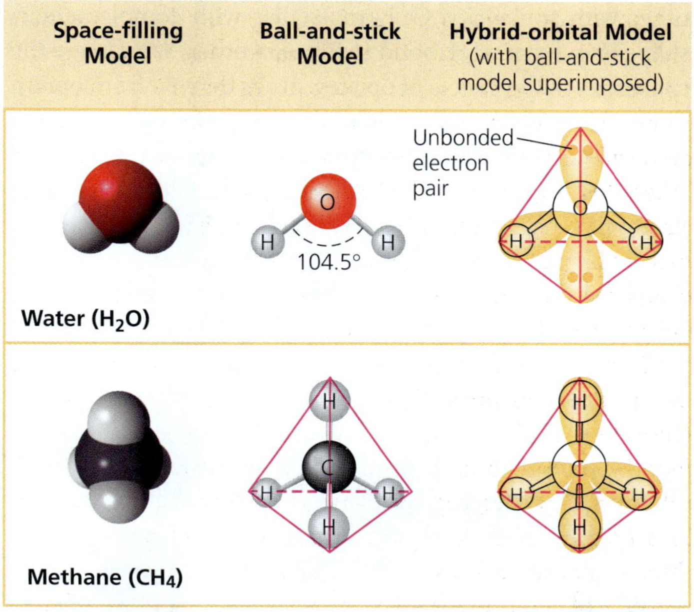

(b) Molecular-shape models. Three models representing molecular shape are shown for water and methane. The positions of the hybrid orbitals determine the shapes of the molecules.

▲ **Figure 2.17 Molecular shapes due to hybrid orbitals.**

A molecule consisting of two atoms, such as H_2 or O_2, is always linear, but molecules with more than two atoms have more complicated shapes. These shapes are determined by the positions of the atoms' orbitals. When an atom forms covalent bonds, the orbitals in its valence shell rearrange. For atoms with valence electrons in both *s* and *p* orbitals (review Figure 2.10), the single *s* and three *p* orbitals hybridize to form four new hybrid orbitals shaped like identical teardrops extending from the region of the atomic nucleus **(Figure 2.17a)**. If we connect the larger ends of the teardrops with lines, we have the outline of a geometric shape called a tetrahedron, similar to a pyramid.

For the water molecule (H_2O), two of the hybrid orbitals in the oxygen atom's valence shell are shared with hydrogen atoms **(Figure 2.17b)**. The result is a molecule shaped roughly like a V, with its two covalent bonds spread apart at an angle of 104.5°.

The methane molecule (CH_4) has the shape of a completed tetrahedron because all four hybrid orbitals of carbon are shared with hydrogen atoms (see Figure 2.17b). The nucleus of the carbon is at the center, with its four covalent bonds radiating to hydrogen nuclei at the corners of the tetrahedron. Larger molecules containing multiple carbon atoms, including many of the molecules that make up living matter, have more complex overall shapes. However, the tetrahedral shape of a carbon atom bonded to four other atoms is often a repeating motif within such molecules.

Molecular shape is crucial in biology because it determines how biological molecules recognize and respond to one another with specificity. Only molecules with complementary shapes can form weak bonds with each other. We can see this specificity in the effects of opiates, drugs derived from opium. Opium's narcotic effects have been known since ancient times. During the 1800s, morphine was isolated from opium, and heroin was synthesized from morphine. These opiates relieve pain and alter mood by binding to specific receptor molecules on the surface of brain cells. Why would brain cells carry receptors for opiates, compounds not made by our bodies? The discovery of endorphins in 1975 answered this question. Endorphins are signaling molecules made by the pituitary gland that bind to the receptors, relieving pain and producing euphoria during times of stress, such as intense exercise. It turns out that opiates have shapes similar to endorphins and mimic them by binding to endorphin receptors in the brain. That is why opiates and endorphins have similar effects **(Figure 2.18)**. The role of molecular shape in brain chemistry illustrates the relationship between structure and function, one of biology's unifying themes.

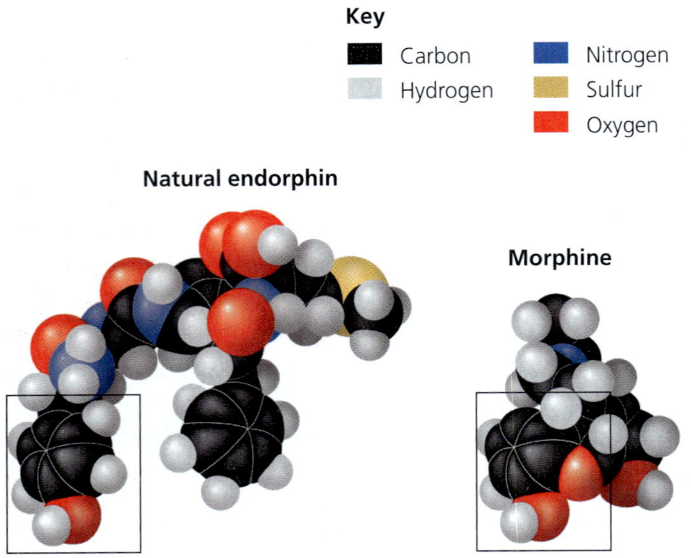

(a) Structures of endorphin and morphine. The boxed portion of the endorphin molecule (left) binds to receptor molecules on target cells in the brain. The boxed portion of the morphine molecule (right) is a close match.

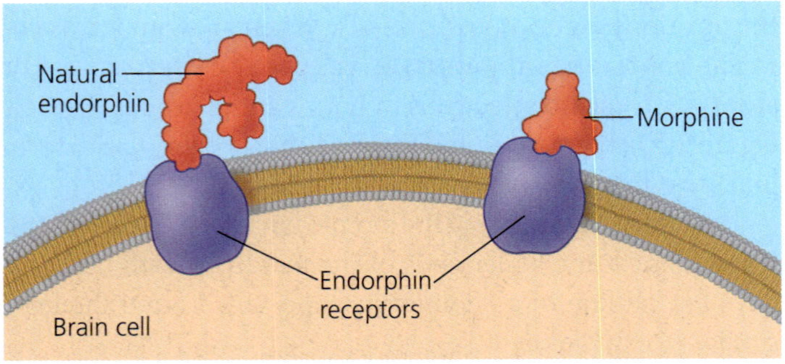

(b) Binding to endorphin receptors. Both endorphin and morphine can bind to endorphin receptors on the surface of a brain cell.

▲ **Figure 2.18 A molecular mimic.** Morphine affects pain perception and emotional state by mimicking the brain's natural endorphins.

CONCEPT CHECK 2.3

1. Why does the following structure fail to make sense chemically?

   H—C=C—H

2. Explain what holds together the atoms in a crystal of magnesium chloride ($MgCl_2$).
3. **WHAT IF?** If you were a pharmaceutical researcher, why would you want to learn the three-dimensional shapes of naturally occurring signaling molecules?

For suggested answers, see Appendix A.

CONCEPT 2.4

Chemical reactions make and break chemical bonds

The making and breaking of chemical bonds, leading to changes in the composition of matter, are called **chemical reactions**. An example is the reaction between hydrogen and oxygen that forms water:

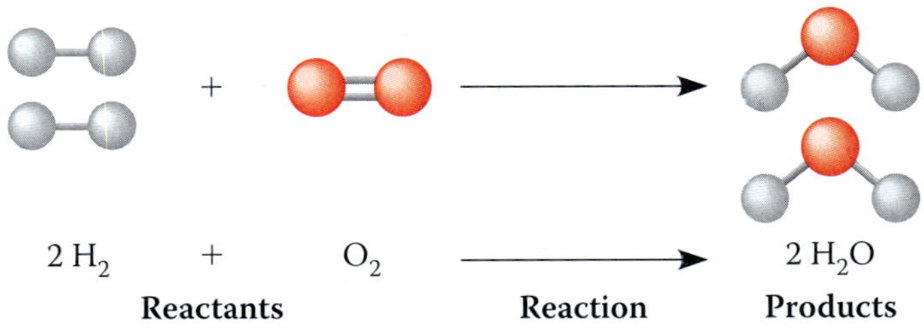

This reaction breaks the covalent bonds of H_2 and O_2 and forms the new bonds of H_2O. When we write a chemical reaction, we use an arrow to indicate the conversion of the starting materials, called the **reactants**, to the **products**. The coefficients indicate the number of molecules involved; for example, the coefficient 2 in front of the H_2 means that the reaction starts with two molecules of hydrogen. Notice that all atoms of the reactants must be accounted for in the products. Matter is conserved in a chemical reaction: Reactions cannot create or destroy matter but can only rearrange it.

Photosynthesis, which takes place within the cells of green plant tissues, is a particularly important example of how chemical reactions rearrange matter. Humans and other animals ultimately depend on photosynthesis for food and oxygen, and this process is at the foundation of almost all ecosystems. The following chemical shorthand summarizes the process of photosynthesis:

$$6\,CO_2 + 6\,H_2O \longrightarrow C_6H_{12}O_6 + 6\,O_2$$

The raw materials of photosynthesis are carbon dioxide (CO_2), which is taken from the air, and water (H_2O), which is absorbed from the soil. Within the plant cells, sunlight powers the conversion of these ingredients to a sugar called glucose ($C_6H_{12}O_6$) and oxygen molecules (O_2), a by-product that the plant releases into the surroundings **(Figure 2.19)**. Although photosynthesis is actually a sequence of many chemical reactions, we still end up with the same number and kinds of atoms we had when we started. Matter has simply been rearranged, with an input of energy provided by sunlight.

▲ **Figure 2.19 Photosynthesis: a solar-powered rearrangement of matter.** *Elodea*, a freshwater plant, produces sugar by rearranging the atoms of carbon dioxide and water in the chemical process known as photosynthesis, which is powered by sunlight. Much of the sugar is then converted to other food molecules. Oxygen gas (O_2) is a by-product of photosynthesis; notice the bubbles of oxygen escaping from the leaves in the photo.

? *Explain how this photo relates to the reactants and products in the equation for photosynthesis given in the above text. (You will learn more about photosynthesis in Chapter 10.)*

All chemical reactions are reversible, with the products of the forward reaction becoming the reactants for the reverse reaction. For example, hydrogen and nitrogen molecules can combine to form ammonia, but ammonia can also decompose to regenerate hydrogen and nitrogen:

$$3\,H_2 + N_2 \rightleftharpoons 2\,NH_3$$

The two opposite-headed arrows indicate that the reaction is reversible.

One of the factors affecting the rate of a reaction is the concentration of reactants. The greater the concentration of reactant molecules, the more frequently they collide with one another and have an opportunity to react and form products. The same holds true for products. As products accumulate, collisions resulting in the reverse reaction become more frequent. Eventually, the forward and reverse reactions occur at the same rate, and the relative concentrations of products and reactants stop changing. The point at which the reactions offset one another exactly is called **chemical equilibrium**. This is a dynamic equilibrium; reactions are still going on, but with no net effect on the concentrations of reactants and products. Equilibrium does *not* mean that the reactants and products are equal in concentration, but only that their concentrations have stabilized at a particular ratio. The reaction involving ammonia reaches equilibrium when ammonia decomposes as rapidly as it forms. In some chemical reactions, the equilibrium point may lie so far to the right that these reactions go essentially to completion; that is, virtually all the reactants are converted to products.

We will return to the subject of chemical reactions after more detailed study of the various types of molecules that are important to life. In the next chapter, we focus on water, the substance in which all the chemical processes of organisms occur.

CONCEPT CHECK 2.4

1. Refer to the reaction between hydrogen and oxygen that forms water, shown with ball-and-stick models on page 42. Draw the Lewis dot structures representing this reaction.
2. Which types of chemical reactions occur faster at equilibrium, the formation of products from reactants, or reactants from products?
3. WHAT IF? Write an equation that uses the products of photosynthesis as reactants and uses the reactants as products. Add energy as another product. This new equation describes a process that occurs in your cells. Describe this equation in words. How does this equation relate to breathing?

For suggested answers, see Appendix A.

Chapter 2 Review

MEDIA Go to the Study Area at **www.masteringbio.com** for BioFlix 3-D Animations, MP3 Tutors, Videos, Practice Tests, an eBook, and more.

SUMMARY OF KEY CONCEPTS

CONCEPT 2.1

Matter consists of chemical elements in pure form and in combinations called compounds (pp. 31–32)

- **Elements and Compounds** Elements cannot be broken down chemically to other substances. A compound contains two or more different elements in a fixed ratio.
- **Essential Elements of Life** Carbon, oxygen, hydrogen, and nitrogen make up approximately 96% of living matter.

MEDIA

Investigation How Are Space Rocks Analyzed for Signs of Life?

CONCEPT 2.2

An element's properties depend on the structure of its atoms (pp. 32–37)

- **Subatomic Particles** An atom, the smallest unit of an element, has the following components:

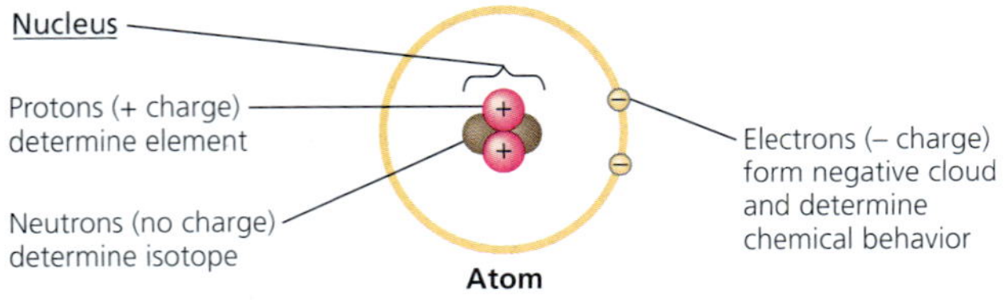

- **Atomic Number and Atomic Mass** An electrically neutral atom has equal numbers of electrons and protons; the number of protons determines the atomic number. The atomic mass is measured in daltons and is roughly equal to the sum of protons plus neutrons.
- **Isotopes** Isotopes of an element differ from each other in neutron number and therefore mass. Unstable isotopes give off particles and energy as radioactivity.
- **The Energy Levels of Electrons** In an atom, electrons occupy specific energy shells; the electrons in a shell have a characteristic energy level.
- **Electron Distribution and Chemical Properties** Electron distribution in shells determines the chemical behavior of an atom. An atom that has an incomplete valence shell is reactive.
- **Electron Orbitals** Electrons exist in orbitals, three-dimensional spaces with specific shapes that are components of electron shells.

MEDIA

Activity Structure of the Atomic Nucleus
Activity Electron Arrangement
Activity Build an Atom

CONCEPT 2.3

The formation and function of molecules depend on chemical bonding between atoms (pp. 38–42)

- **Covalent Bonds** Chemical bonds form when atoms interact and complete their valence shells. Covalent bonds form when pairs of electrons are shared.

Molecules consist of two or more covalently bonded atoms. Electrons of a polar covalent bond are pulled closer to the more electronegative atom. If both atoms are the same, they have the same electronegativity, and the covalent bond is nonpolar.

- **Ionic Bonds**

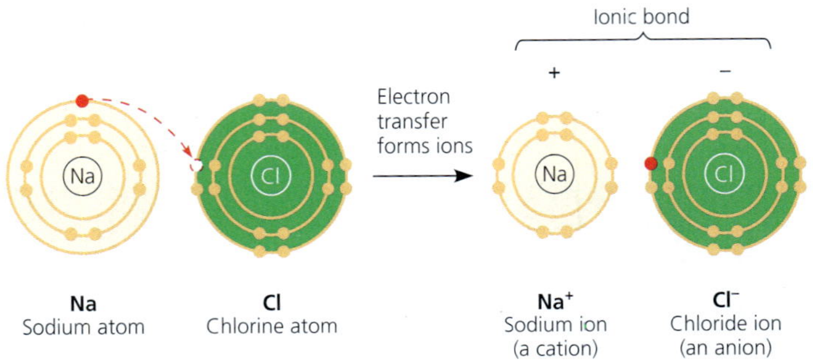

- **Weak Chemical Bonds** A hydrogen bond is an attraction between a hydrogen atom carrying a partial positive charge (δ+) and an electronegative atom (δ−). Van der Waals interactions occur between transiently positive and negative regions of molecules. Weak bonds reinforce the shapes of large molecules and help molecules adhere to each other.
- **Molecular Shape and Function** A molecule's shape is determined by the positions of its atoms' valence orbitals. Covalent bonds result in hybrid orbitals, which are responsible for the shapes of H_2O, CH_4, and many more complex biological molecules. Shape is usually the basis for the recognition of one biological molecule by another.

MEDIA

Activity Covalent Bonds
Activity Nonpolar and Polar Molecules
Activity Ionic Bonds
Activity Hydrogen Bonds

CONCEPT 2.4

Chemical reactions make and break chemical bonds (pp. 42–43)

- Chemical reactions change reactants into products while conserving matter. All chemical reactions are theoretically reversible. Chemical equilibrium is reached when the forward and reverse reaction rates are equal.

TESTING YOUR KNOWLEDGE

SELF-QUIZ

1. In the term *trace element,* the modifier *trace* means
   a. the element is required in very small amounts.
   b. the element can be used as a label to trace atoms through an organism's metabolism.
   c. the element is very rare on Earth.
   d. the element enhances health but is not essential for the organism's long-term survival.
   e. the element passes rapidly through the organism.

2. Compared with ^{31}P, the radioactive isotope ^{32}P has
   a. a different atomic number.
   b. one more neutron.
   c. one more proton.
   d. one more electron.
   e. a different charge.

3. Atoms can be represented by simply listing the number of protons, neutrons, and electrons—for example, $2p^+$; $2n^0$; $2e^-$ for helium. Which one of the following lists represents the ^{18}O isotope of oxygen?
   a. $6p^+$; $8n^0$; $6e^-$
   b. $8p^+$; $10n^0$; $8e^-$
   c. $9p^+$; $9n^0$; $9e^-$
   d. $7p^+$; $2n^0$; $9e^-$
   e. $10p^+$; $8n^0$; $9e^-$

4. The atomic number of sulfur is 16. Sulfur combines with hydrogen by covalent bonding to form a compound, hydrogen sulfide. Based on the number of valence electrons in a sulfur atom, predict the molecular formula of the compound:
   a. HS b. HS_2 c. H_2S d. H_3S_2 e. H_4S

5. The reactivity of an atom arises from
   a. the average distance of the outermost electron shell from the nucleus.
   b. the existence of unpaired electrons in the valence shell.
   c. the sum of the potential energies of all the electron shells.
   d. the potential energy of the valence shell.
   e. the energy difference between the *s* and *p* orbitals.

6. Which statement is true of all atoms that are anions?
   a. The atom has more electrons than protons.
   b. The atom has more protons than electrons.
   c. The atom has fewer protons than does a neutral atom of the same element.
   d. The atom has more neutrons than protons.
   e. The net charge is 1−.

7. What coefficients must be placed in the following blanks so that all atoms are accounted for in the products?

$$C_6H_{12}O_6 \longrightarrow \_\_\_\_ C_2H_6O + \_\_\_\_ CO_2$$

   a. 1; 2 b. 2; 2 c. 1; 3 d. 1; 1 e. 3; 1

8. Which of the following statements correctly describes any chemical reaction that has reached equilibrium?
   a. The concentrations of products and reactants are equal.
   b. The rates of the forward and reverse reactions are equal.
   c. Both forward and reverse reactions have halted.
   d. The reaction is now irreversible.
   e. No reactants remain.

9. **DRAW IT** Draw Lewis structures for each hypothetical molecule shown below, using the correct number of valence electrons for each atom. Determine which molecule makes sense because each atom has a complete valence shell and each bond has the correct number of electrons. Explain what makes the other molecules nonsensical, considering the number of bonds each type of atom can make.

a. 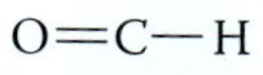

c. 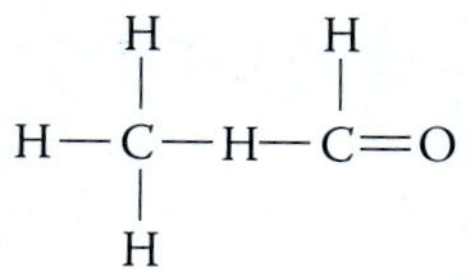

b. 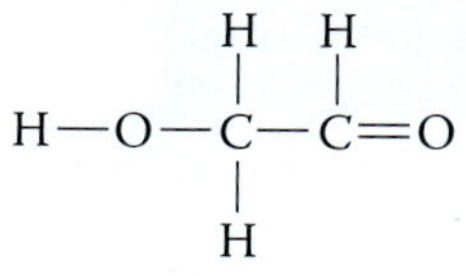

d. 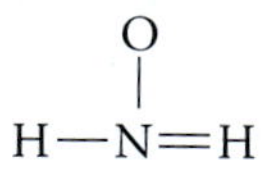

For Self-Quiz answers, see Appendix A.

MEDIA Visit the Study Area at **www.masteringbio.com** for a Practice Test.

EVOLUTION CONNECTION

10. The percentages of naturally occurring elements making up the human body (see Table 2.1) are similar to the percentages of these elements found in other organisms. How could you account for this similarity among organisms?

SCIENTIFIC INQUIRY

11. Female silkworm moths (*Bombyx mori*) attract males by emitting chemical signals that spread through the air. A male hundreds of meters away can detect these molecules and fly toward their source. The sensory organs responsible for this behavior are the comblike antennae visible in the photograph here. Each filament of an antenna is equipped with thousands of receptor cells that detect the sex attractant. Based on what you learned in this chapter, propose a hypothesis to account for the ability of the male moth to detect a specific molecule in the presence of many other molecules in the air. What predictions does your hypothesis make? Design an experiment to test one of these predictions.

SCIENCE, TECHNOLOGY, AND SOCIETY

12. While waiting at an airport, Neil Campbell once overheard this claim: "It's paranoid and ignorant to worry about industry or agriculture contaminating the environment with their chemical wastes. After all, this stuff is just made of the same atoms that were already present in our environment." How would you counter this argument?

3 Water and the Fitness of the Environment

▲ **Figure 3.1 Why does the abundance of water allow life to exist on the planet Earth?**

KEY CONCEPTS

3.1 The polarity of water molecules results in hydrogen bonding

3.2 Four emergent properties of water contribute to Earth's fitness for life

3.3 Acidic and basic conditions affect living organisms

OVERVIEW

The Molecule That Supports All of Life

As astronomers study newly discovered planets orbiting distant stars, they hope to find evidence of water on these far-off celestial bodies, for water is the substance that makes possible life as we know it here on Earth. All organisms familiar to us are made mostly of water and live in an environment dominated by water. Water is the biological medium here on Earth, and possibly on other planets as well.

Three-quarters of Earth's surface is submerged in water **(Figure 3.1)**. Although most of this water is in liquid form, water is also present on Earth as ice and vapor. Water is the only common substance to exist in the natural environment in all three physical states of matter: solid, liquid, and gas. The abundance of water is a major reason Earth is habitable. In a classic book called *The Fitness of the Environment,* ecologist Lawrence Henderson highlights the importance of water to life. While acknowledging that life adapts to its environment through natural selection, Henderson emphasizes that for life to exist at all, the environment must first be a suitable abode.

Life on Earth began in water and evolved there for 3 billion years before spreading onto land. Modern life, even terrestrial (land-dwelling) life, remains tied to water. All living organisms require water more than any other substance. Human beings, for example, can survive for quite a few weeks without food, but only a week or so without water. Molecules of water participate in many chemical reactions necessary to sustain life. Most cells are surrounded by water, and cells themselves are about 70–95% water.

What properties of the simple water molecule allow it to function as a support to all living organisms? In this chapter, you will learn how the structure of a water molecule allows it to interact with other molecules, including other water molecules. This ability leads to unique emergent properties that support and maintain living systems on our planet. Your objective in this chapter is to develop a conceptual understanding of how water contributes to the fitness of Earth for life.

CONCEPT **3.1**

The polarity of water molecules results in hydrogen bonding

Water is so common that it is easy to overlook the fact that it is an exceptional substance with many extraordinary qualities. Following the theme of emergent properties, we can trace water's unique behavior to the structure and interactions of its molecules.

Studied in isolation, the water molecule is deceptively simple. It is shaped something like a wide V, with its two hydrogen atoms joined to the oxygen atom by single covalent bonds. Because oxygen is more electronegative than hydrogen, the electrons of the covalent bonds spend more time closer to oxygen than to hydrogen; in other words, they are polar covalent bonds (see Figure 2.13). This unequal distribution of electrons makes water a **polar molecule**, meaning that the two ends of the molecule have opposite charges: The oxygen region of the molecule has a partial negative charge ($\delta-$), and the hydrogens have a partial positive charge ($\delta+$).

The anomalous properties of water arise from attractions between its polar molecules: The slightly positive hydrogen of

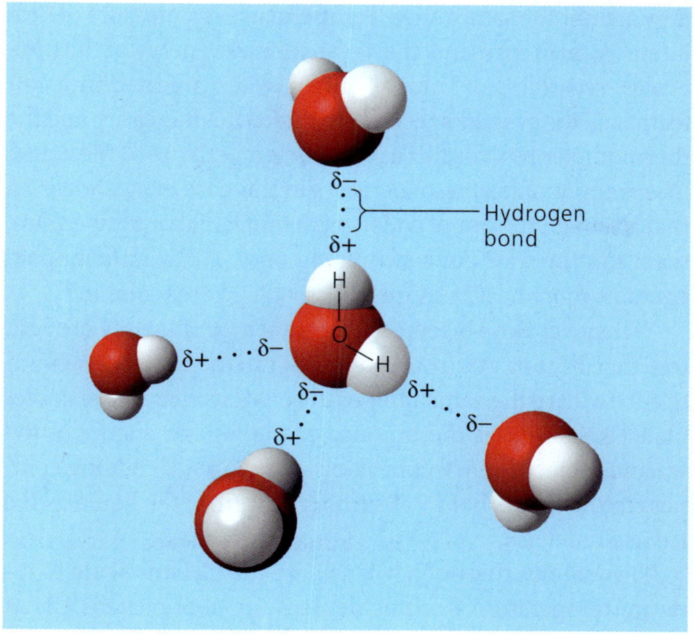

▲ **Figure 3.2 Hydrogen bonds between water molecules.** The charged regions of a polar water molecule are attracted to oppositely charged parts of neighboring molecules. Each molecule can hydrogen-bond to multiple partners, and these associations are constantly changing.

one molecule is attracted to the slightly negative oxygen of a nearby molecule. The two molecules are thus held together by a hydrogen bond **(Figure 3.2)**. When water is in its liquid form, its hydrogen bonds are very fragile, each about 1/20 as strong as a covalent bond. The hydrogen bonds form, break, and re-form with great frequency. Each lasts only a few trillionths of a second, but the molecules are constantly forming new hydrogen bonds with a succession of partners. Therefore, at any instant, a substantial percentage of all the water molecules are hydrogen-bonded to their neighbors. The extraordinary qualities of water are emergent properties resulting from the hydrogen bonding that orders molecules into a higher level of structural organization.

CONCEPT CHECK 3.1

1. What is electronegativity, and how does it affect interactions between water molecules?
2. Why is it unlikely that two neighboring water molecules would be arranged like this?

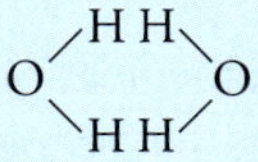

3. **WHAT IF?** What would be the effect on the properties of the water molecule if oxygen and hydrogen had equal electronegativity?

For suggested answers, see Appendix A.

CONCEPT **3.2**

Four emergent properties of water contribute to Earth's fitness for life

We will examine four emergent properties of water that contribute to Earth's suitability as an environment for life: cohesive behavior, ability to moderate temperature, expansion upon freezing, and versatility as a solvent.

Cohesion

Water molecules stay close to each other as a result of hydrogen bonding. Although the arrangement of molecules in a sample of liquid water is constantly changing, at any given moment many of the molecules are linked by multiple hydrogen bonds. These linkages make water more structured than most other liquids. Collectively, the hydrogen bonds hold the substance together, a phenomenon called **cohesion**.

Cohesion due to hydrogen bonding contributes to the transport of water and dissolved nutrients against gravity in plants **(Figure 3.3)**. Water from the roots reaches the leaves through a network of water-conducting cells. As water evaporates from a leaf, hydrogen bonds cause water molecules leaving the veins to tug on molecules farther down, and the upward pull is transmitted through the water-conducting cells all the way to the roots. **Adhesion**, the clinging of one substance to

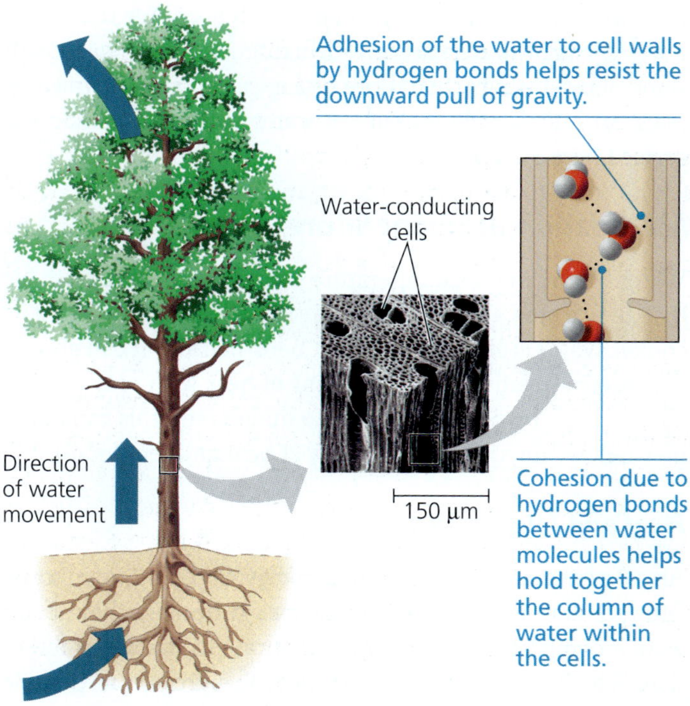

▲ **Figure 3.3 Water transport in plants.** Evaporation from leaves pulls water upward from the roots through water-conducting cells. Because of the properties of cohesion and adhesion, the tallest trees can transport water more than 100 m upward—approximately one-quarter the height of the Empire State Building in New York City.

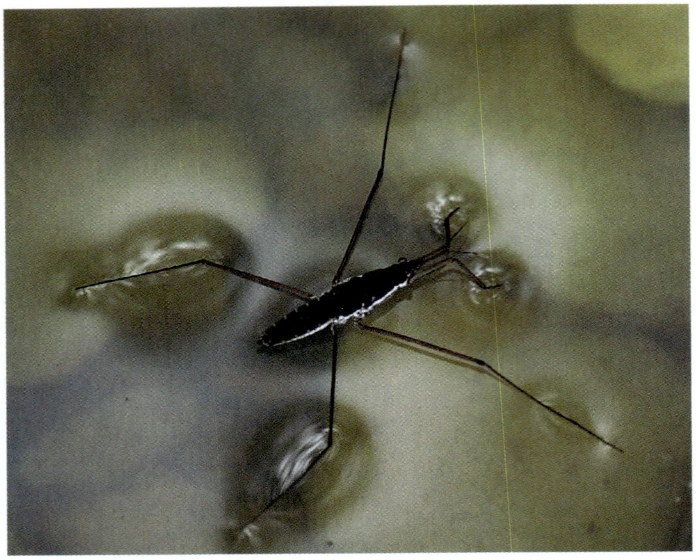

▲ **Figure 3.4 Walking on water.** The high surface tension of water, resulting from the collective strength of its hydrogen bonds, allows the water strider to walk on the surface of a pond.

another, also plays a role. Adhesion of water to cell walls by hydrogen bonds helps counter the downward pull of gravity (see Figure 3.3).

Related to cohesion is **surface tension**, a measure of how difficult it is to stretch or break the surface of a liquid. Water has a greater surface tension than most other liquids. At the interface between water and air is an ordered arrangement of water molecules, hydrogen-bonded to one another and to the water below. This makes the water behave as though coated with an invisible film. You can observe the surface tension of water by slightly overfilling a drinking glass; the water will stand above the rim. In a more biological example, some animals can stand, walk, or run on water without breaking the surface **(Figure 3.4)**.

Moderation of Temperature

Water moderates air temperature by absorbing heat from air that is warmer and releasing the stored heat to air that is cooler. Water is effective as a heat bank because it can absorb or release a relatively large amount of heat with only a slight change in its own temperature. To understand this capability of water, we must first look briefly at heat and temperature.

Heat and Temperature

Anything that moves has **kinetic energy**, the energy of motion. Atoms and molecules have kinetic energy because they are always moving, although not necessarily in any particular direction. The faster a molecule moves, the greater its kinetic energy. **Heat** is a form of energy. For a given body of matter, the amount of heat is a measure of the matter's *total* kinetic energy due to motion of its molecules; thus, heat depends in part on the matter's volume. Although heat is related to temperature, they are not the same thing. **Temperature** is a measure of heat intensity that represents the *average* kinetic energy of the molecules, regardless of volume. When water is heated in a coffeemaker, the average speed of the molecules increases, and the thermometer records this as a rise in temperature of the liquid. The amount of heat also increases in this case. Note, however, that although the pot of coffee has a much higher temperature than, say, the water in a swimming pool, the swimming pool contains more heat because of its much greater volume.

Whenever two objects of different temperature are brought together, heat passes from the warmer to the cooler object until the two are the same temperature. Molecules in the cooler object speed up at the expense of the kinetic energy of the warmer object. An ice cube cools a drink not by adding coldness to the liquid, but by absorbing heat from the liquid as the ice itself melts.

Throughout this book, we will use the **Celsius scale** to indicate temperature (Celsius degrees are abbreviated °C). At sea level, water freezes at 0°C and boils at 100°C. The temperature of the human body averages 37°C, and comfortable room temperature is about 20–25°C.

One convenient unit of heat used in this book is the **calorie (cal)**. A calorie is the amount of heat it takes to raise the temperature of 1 g of water by 1°C. Conversely, a calorie is also the amount of heat that 1 g of water releases when it cools by 1°C. A **kilocalorie (kcal)**, 1,000 cal, is the quantity of heat required to raise the temperature of 1 kilogram (kg) of water by 1°C. (The "calories" on food packages are actually kilocalories.) Another energy unit used in this book is the **joule (J)**. One joule equals 0.239 cal; one calorie equals 4.184 J.

Water's High Specific Heat

The ability of water to stabilize temperature stems from its relatively high specific heat. The **specific heat** of a substance is defined as the amount of heat that must be absorbed or lost for 1 g of that substance to change its temperature by 1°C. We already know water's specific heat because we have defined a calorie as the amount of heat that causes 1 g of water to change its temperature by 1°C. Therefore, the specific heat of water is 1 calorie per gram per degree Celsius, abbreviated as 1 cal/g/°C. Compared with most other substances, water has an unusually high specific heat. For example, ethyl alcohol, the type of alcohol in alcoholic beverages, has a specific heat of 0.6 cal/g/°C; that is, only 0.6 cal is required to raise the temperature of 1 g of ethyl alcohol 1°C.

Because of the high specific heat of water relative to other materials, water will change its temperature less when it absorbs or loses a given amount of heat. The reason you can burn your fingers by touching the side of a metal pot on the stove when the water in the pot is still lukewarm is that the specific heat of water is ten times greater than that of iron. In other words, the same amount of heat will raise the temperature of

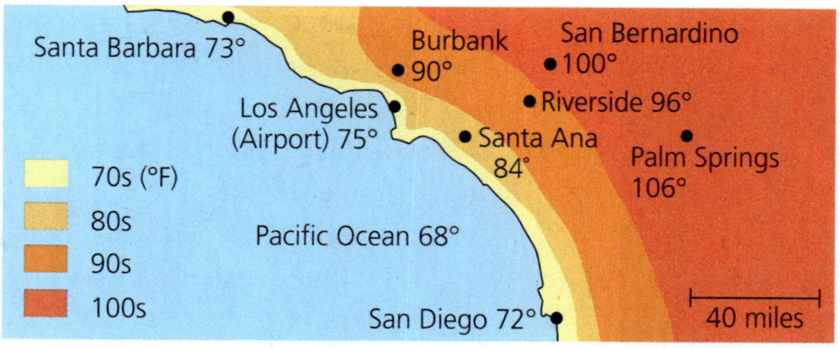

▲ **Figure 3.5 Effect of a large body of water on climate.** By absorbing or releasing heat, oceans moderate coastal climates. In this example from an August day in Southern California, the relatively cool ocean reduces coastal air temperatures by absorbing heat.

1 g of the iron much faster than the temperature of 1 g of the water. Specific heat can be thought of as a measure of how well a substance resists changing its temperature when it absorbs or releases heat. Water resists changing its temperature; when it does change its temperature, it absorbs or loses a relatively large quantity of heat for each degree of change.

We can trace water's high specific heat, like many of its other properties, to hydrogen bonding. Heat must be absorbed in order to break hydrogen bonds, and heat is released when hydrogen bonds form. A calorie of heat causes a relatively small change in the temperature of water because much of the heat is used to disrupt hydrogen bonds before the water molecules can begin moving faster. And when the temperature of water drops slightly, many additional hydrogen bonds form, releasing a considerable amount of energy in the form of heat.

What is the relevance of water's high specific heat to life on Earth? A large body of water can absorb and store a huge amount of heat from the sun in the daytime and during summer while warming up only a few degrees. And at night and during winter, the gradually cooling water can warm the air. This is the reason coastal areas generally have milder climates than inland regions **(Figure 3.5)**. The high specific heat of water also tends to stabilize ocean temperatures, creating a favorable environment for marine life. Thus, because of its high specific heat, the water that covers most of Earth keeps temperature fluctuations on land and in water within limits that permit life. Also, because organisms are made primarily of water, they are more able to resist changes in their own temperature than if they were made of a liquid with a lower specific heat.

Evaporative Cooling

Molecules of any liquid stay close together because they are attracted to one another. Molecules moving fast enough to overcome these attractions can depart the liquid and enter the air as gas. This transformation from a liquid to a gas is called vaporization, or *evaporation*. Recall that the speed of molecular movement varies and that temperature is the *average* kinetic energy of molecules. Even at low temperatures, the speediest molecules can escape into the air. Some evaporation occurs at any temperature; a glass of water at room temperature, for example, will eventually evaporate. If a liquid is heated, the average kinetic energy of molecules increases and the liquid evaporates more rapidly.

Heat of vaporization is the quantity of heat a liquid must absorb for 1 g of it to be converted from the liquid to the gaseous state. For the same reason that water has a high specific heat, it also has a high heat of vaporization relative to most other liquids. To evaporate 1 g of water at 25°C, about 580 cal of heat is needed—nearly double the amount needed to vaporize a gram of alcohol or ammonia. Water's high heat of vaporization is another emergent property caused by hydrogen bonds, which must be broken before the molecules can make their exodus from the liquid.

The high amount of energy required to vaporize water has a wide range of effects. On a global scale, for example, it helps moderate Earth's climate. A considerable amount of solar heat absorbed by tropical seas is consumed during the evaporation of surface water. Then, as moist tropical air circulates poleward, it releases heat as it condenses and forms rain. On an organismal level, water's high heat of vaporization accounts for the severity of steam burns. These burns are caused by the heat energy released when steam condenses into liquid on the skin.

As a liquid evaporates, the surface of the liquid that remains behind cools down. This **evaporative cooling** occurs because the "hottest" molecules, those with the greatest kinetic energy, are the most likely to leave as gas. It is as if the hundred fastest runners at a college transferred to another school; the average speed of the remaining students would decline.

Evaporative cooling of water contributes to the stability of temperature in lakes and ponds and also provides a mechanism that prevents terrestrial organisms from overheating. For example, evaporation of water from the leaves of a plant helps keep the tissues in the leaves from becoming too warm in the sunlight. Evaporation of sweat from human skin dissipates body heat and helps prevent overheating on a hot day or when excess heat is generated by strenuous activity. High humidity on a hot day increases discomfort because the high concentration of water vapor in the air inhibits the evaporation of sweat from the body.

Insulation of Bodies of Water by Floating Ice

Water is one of the few substances that are less dense as a solid than as a liquid. In other words, ice floats in liquid water. While other materials contract when they solidify, water expands. The cause of this exotic behavior is, once again, hydrogen bonding. At temperatures above 4°C, water behaves like other liquids, expanding as it warms and contracting as it cools. Water begins to freeze when its molecules are no longer moving vigorously enough to break their hydrogen bonds. As the temperature falls to 0°C, the water becomes locked into a crystalline lattice, each water molecule hydrogen-bonded to

▲ **Figure 3.6 Ice: crystalline structure and floating barrier.** In ice, each molecule is hydrogen-bonded to four neighbors in a three-dimensional crystal. Because the crystal is spacious, ice has fewer molecules than an equal volume of liquid water. In other words, ice is less dense than liquid water. Floating ice becomes a barrier that protects the liquid water below from the colder air. The marine organism shown here is a type of shrimp called krill; it was photographed beneath floating ice in the Antarctic Ocean.

? *If water did not form hydrogen bonds, what would happen to the shrimp's environment?*

four partners **(Figure 3.6)**. The hydrogen bonds keep the molecules at "arm's length," far enough apart to make ice about 10% less dense (10% fewer molecules for the same volume) than liquid water at 4°C. When ice absorbs enough heat for its temperature to rise above 0°C, hydrogen bonds between molecules are disrupted. As the crystal collapses, the ice melts, and molecules are free to slip closer together. Water reaches its greatest density at 4°C and then begins to expand as the molecules move faster. Keep in mind, however, that even in liquid water, many of the molecules are connected by hydrogen bonds, though only transiently: The hydrogen bonds are constantly breaking and re-forming.

The ability of ice to float because of the expansion of water as it solidifies is an important factor in the fitness of the environment. If ice sank, then eventually all ponds, lakes, and even oceans would freeze solid, making life as we know it impossible on Earth. During summer, only the upper few inches of the ocean would thaw. Instead, when a deep body of water cools, the floating ice insulates the liquid water below, preventing it from freezing and allowing life to exist under the frozen surface, as shown in the photo in Figure 3.6.

The Solvent of Life

A sugar cube placed in a glass of water will dissolve. The glass will then contain a uniform mixture of sugar and water; the concentration of dissolved sugar will be the same everywhere in the mixture. A liquid that is a completely homogeneous mixture of two or more substances is called a **solution**. The dissolving agent of a solution is the **solvent**, and the substance that is dissolved is the **solute**. In this case, water is the solvent and sugar is the solute. An **aqueous solution** is one in which water is the solvent.

The medieval alchemists tried to find a universal solvent, one that would dissolve anything. They learned that nothing works better than water. However, water is not a universal solvent; if it were, it would dissolve any container in which it was stored, including our cells. But water is a very versatile solvent, a quality we can trace to the polarity of the water molecule.

Suppose, for example, that a spoonful of table salt, the ionic compound sodium chloride (NaCl), is placed in water **(Figure 3.7)**. At the surface of each grain, or crystal, of salt, the sodium and chloride ions are exposed to the solvent. These ions and the water molecules have a mutual affinity owing to the attraction between opposite charges. The oxygen

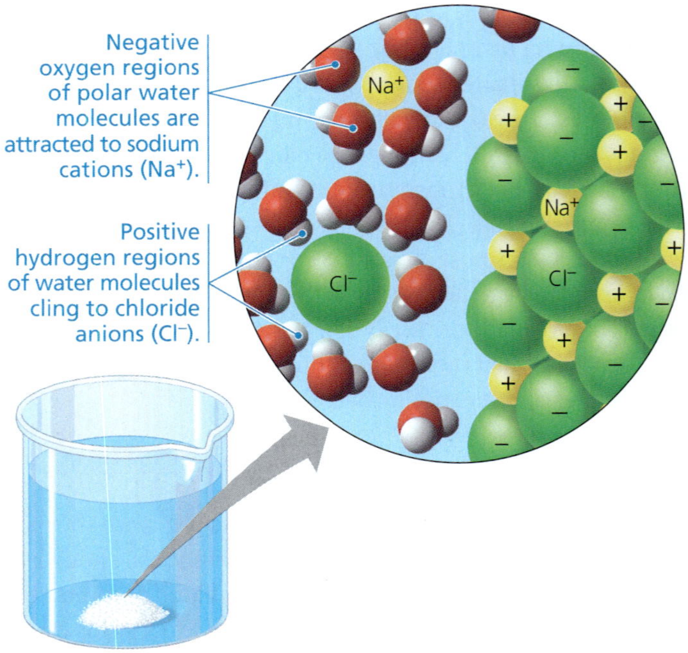

▲ **Figure 3.7 Table salt dissolving in water.** A sphere of water molecules, called a hydration shell, surrounds each solute ion.

? *What would happen if you heated this solution for a long time?*

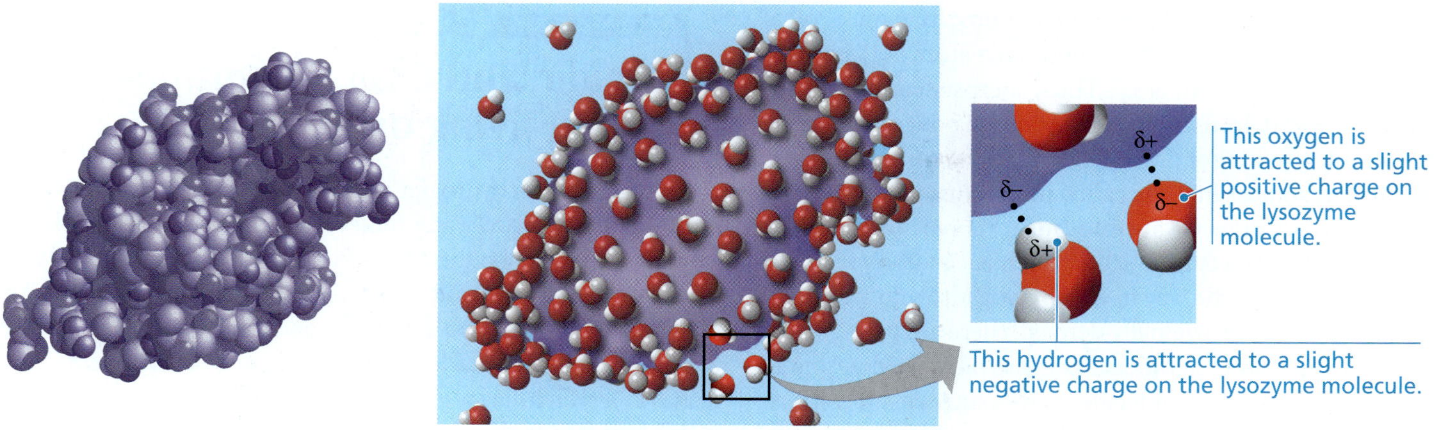

(a) Lysozyme molecule in a nonaqueous environment

(b) Lysozyme molecule (purple) in an aqueous environment such as tears or saliva

(c) Ionic and polar regions on the protein's surface attract water molecules.

▲ **Figure 3.8 A water-soluble protein.** This figure shows human lysozyme, a protein found in tears and saliva that has antibacterial action.

regions of the water molecules are negatively charged and cling to sodium cations. The hydrogen regions are positively charged and are attracted to chloride anions. As a result, water molecules surround the individual sodium and chloride ions, separating and shielding them from one another. The sphere of water molecules around each dissolved ion is called a **hydration shell**. Working inward from the surface of each salt crystal, water eventually dissolves all the ions. The result is a solution of two solutes, sodium cations and chloride anions, homogeneously mixed with water, the solvent. Other ionic compounds also dissolve in water. Seawater, for instance, contains a great variety of dissolved ions, as do living cells.

A compound does not need to be ionic to dissolve in water; many compounds made up of nonionic polar molecules, such as sugars, are also water-soluble. Such compounds dissolve when water molecules surround each of the solute molecules, forming hydrogen bonds with them. Even molecules as large as proteins can dissolve in water if they have ionic and polar regions on their surface **(Figure 3.8)**. Many different kinds of polar compounds are dissolved (along with ions) in the water of such biological fluids as blood, the sap of plants, and the liquid within all cells. Water is the solvent of life.

Hydrophilic and Hydrophobic Substances

Any substance that has an affinity for water is said to be **hydrophilic** (from the Greek *hydro*, water, and *philios*, loving). In some cases, substances can be hydrophilic without actually dissolving. For example, some molecules in cells are so large that they do not dissolve. Instead, they remain suspended in the aqueous liquid of the cell. Such a mixture is an example of a **colloid**, a stable suspension of fine particles in a liquid. Another example of a hydrophilic substance that does not dissolve is cotton, a plant product. Cotton consists of giant molecules of cellulose, a compound with numerous regions of partial positive and partial negative charges that can form hydrogen bonds with water. Water adheres to the cellulose fibers. Thus, a cotton towel does a great job of drying the body, yet does not dissolve in the washing machine. Cellulose is also present in the walls of water-conducting cells in a plant; you read earlier how the adhesion of water to these hydrophilic walls allows water transport to occur.

There are, of course, substances that do not have an affinity for water. Substances that are nonionic and nonpolar (or for some other reason cannot form hydrogen bonds) actually seem to repel water; these substances are said to be **hydrophobic** (from the Greek *phobos*, fearing). An example from the kitchen is vegetable oil, which, as you know, does not mix stably with water-based substances such as vinegar. The hydrophobic behavior of the oil molecules results from a prevalence of relatively nonpolar bonds, in this case bonds between carbon and hydrogen, which share electrons almost equally. Hydrophobic molecules related to oils are major ingredients of cell membranes. (Imagine what would happen to a cell if its membrane dissolved!)

Solute Concentration in Aqueous Solutions

Biological chemistry is "wet" chemistry. Most of the chemical reactions in organisms involve solutes dissolved in water. To understand such reactions, we must know how many atoms and molecules are involved and be able to calculate the concentration of solutes in an aqueous solution (the number of solute molecules in a volume of solution).

When carrying out experiments, we use mass to calculate the number of molecules. We know the mass of each atom in a given molecule, so we can calculate its **molecular mass**, which is simply the sum of the masses of all the atoms in a molecule. As an example, let's calculate the molecular mass of table sugar (sucrose), which has the molecular formula

$C_{12}H_{22}O_{11}$. In round numbers of daltons, the mass of a carbon atom is 12, the mass of a hydrogen atom is 1, and the mass of an oxygen atom is 16. Thus, sucrose has a molecular mass of 342 daltons. Of course, weighing out small numbers of molecules is not practical. For this reason, we usually measure substances in units called moles. Just as a dozen always means 12 objects, a **mole (mol)** represents an exact number of objects—6.02×10^{23}, which is called Avogadro's number. Because of the way in which Avogadro's number and the unit *dalton* were originally defined, there are 6.02×10^{23} daltons in 1 g. This is significant because once we determine the molecular mass of a molecule such as sucrose, we can use the same number (342), but with the unit *gram*, to represent the mass of 6.02×10^{23} molecules of sucrose, or 1 mol of sucrose (this is sometimes called the *molar mass*). To obtain 1 mol of sucrose in the lab, therefore, we weigh out 342 g.

The practical advantage of measuring a quantity of chemicals in moles is that a mole of one substance has exactly the same number of molecules as a mole of any other substance. If the molecular mass of substance A is 342 daltons and that of substance B is 10 daltons, then 342 g of A will have the same number of molecules as 10 g of B. A mole of ethyl alcohol (C_2H_6O) also contains 6.02×10^{23} molecules, but its mass is only 46 g because the mass of a molecule of ethyl alcohol is less than that of a molecule of sucrose. Measuring in moles makes it convenient for scientists working in the laboratory to combine substances in fixed ratios of molecules.

How would we make a liter (L) of solution consisting of 1 mol of sucrose dissolved in water? We would measure out 342 g of sucrose and then gradually add water, while stirring, until the sugar was completely dissolved. We would then add enough water to bring the total volume of the solution up to 1 L. At that point, we would have a 1-molar (1 *M*) solution of sucrose. **Molarity**—the number of moles of solute per liter of solution—is the unit of concentration most often used by biologists for aqueous solutions.

CONCEPT CHECK 3.2

1. Describe how properties of water contribute to the upward movement of water in a tree.
2. Explain the saying "It's not the heat; it's the humidity."
3. How can the freezing of water crack boulders?
4. If you were a pharmacist, how would you make a 0.5-molar (0.5 *M*) solution of sodium chloride (NaCl)? (The atomic mass of Na is 23 daltons and that of Cl is 35.5 daltons.)
5. **WHAT IF?** A water strider's legs (see Figure 3.4) are coated with a hydrophobic substance. What might be the benefit? What would happen if the substance were hydrophilic?

For suggested answers, see Appendix A.

CONCEPT 3.3

Acidic and basic conditions affect living organisms

Occasionally, a hydrogen atom participating in a hydrogen bond between two water molecules shifts from one molecule to the other. When this happens, the hydrogen atom leaves its electron behind, and what is actually transferred is a **hydrogen ion** (H^+), a single proton with a charge of 1+. The water molecule that lost a proton is now a **hydroxide ion** (OH^-), which has a charge of 1−. The proton binds to the other water molecule, making that molecule a **hydronium ion** (H_3O^+). We can picture the chemical reaction this way:

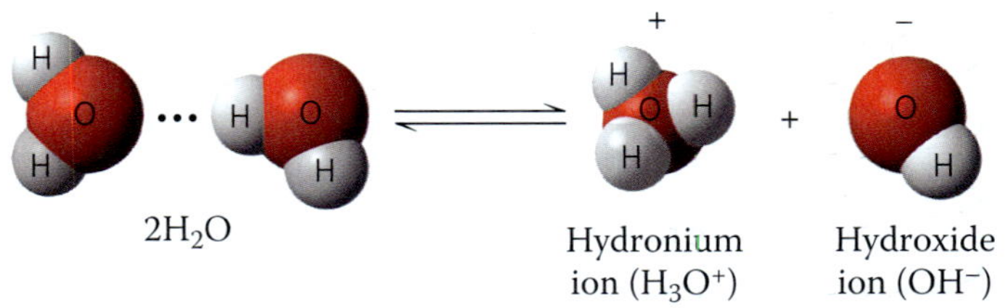

By convention, H^+ (the hydrogen ion) is used to represent H_3O^+ (the hydronium ion), and we follow that practice here. Keep in mind, though, that H^+ does not exist on its own in an aqueous solution. It is always associated with another water molecule in the form of H_3O^+.

As indicated by the double arrows, this is a reversible reaction that reaches a state of dynamic equilibrium when water molecules dissociate at the same rate that they are being re-formed from H^+ and OH^-. At this equilibrium point, the concentration of water molecules greatly exceeds the concentrations of H^+ and OH^-. In pure water, only one water molecule in every 554 million is dissociated. The concentration of each ion in pure water is 10^{-7} *M* (at 25°C). This means there is only one ten-millionth of a mole of hydrogen ions per liter of pure water and an equal number of hydroxide ions.

Although the dissociation of water is reversible and statistically rare, it is exceedingly important in the chemistry of life. H^+ and OH^- are very reactive. Changes in their concentrations can drastically affect a cell's proteins and other complex molecules. As we have seen, the concentrations of H^+ and OH^- are equal in pure water, but adding certain kinds of solutes, called acids and bases, disrupts this balance. Biologists use something called the pH scale to describe how acidic or basic (the opposite of acidic) a solution is. In the remainder of this chapter, you will learn about acids, bases, and pH and why changes in pH can adversely affect organisms.

Effects of Changes in pH

Before discussing the pH scale, let's see what acids and bases are and how they interact with water.

Acids and Bases

What would cause an aqueous solution to have an imbalance in H^+ and OH^- concentrations? When acids dissolve in water, they donate additional H^+ to the solution. An **acid** is a substance that increases the hydrogen ion concentration of a solution. For example, when hydrochloric acid (HCl) is added to water, hydrogen ions dissociate from chloride ions:

$$HCl \longrightarrow H^+ + Cl^-$$

This source of H^+ (dissociation of water is the other source) results in an acidic solution—one having more H^+ than OH^-.

A substance that reduces the hydrogen ion concentration of a solution is called a **base**. Some bases reduce the H^+ concentration directly by accepting hydrogen ions. Ammonia (NH_3), for instance, acts as a base when the unshared electron pair in nitrogen's valence shell attracts a hydrogen ion from the solution, resulting in an ammonium ion (NH_4^+):

$$NH_3 + H^+ \rightleftharpoons NH_4^+$$

Other bases reduce the H^+ concentration indirectly by dissociating to form hydroxide ions, which combine with hydrogen ions and form water. One such base is sodium hydroxide (NaOH), which in water dissociates into its ions:

$$NaOH \longrightarrow Na^+ + OH^-$$

In either case, the base reduces the H^+ concentration. Solutions with a higher concentration of OH^- than H^+ are known as basic solutions. A solution in which the H^+ and OH^- concentrations are equal is said to be neutral.

Notice that single arrows were used in the reactions for HCl and NaOH. These compounds dissociate completely when mixed with water, and so hydrochloric acid is called a strong acid and sodium hydroxide a strong base. In contrast, ammonia is a relatively weak base. The double arrows in the reaction for ammonia indicate that the binding and release of hydrogen ions are reversible reactions, although at equilibrium there will be a fixed ratio of NH_4^+ to NH_3.

There are also weak acids, which reversibly release and accept back hydrogen ions. An example is carbonic acid:

$$\underset{\text{Carbonic acid}}{H_2CO_3} \rightleftharpoons \underset{\text{Bicarbonate ion}}{HCO_3^-} + \underset{\text{Hydrogen ion}}{H^+}$$

Here the equilibrium so favors the reaction in the left direction that when carbonic acid is added to water, only 1% of the molecules are dissociated at any particular time. Still, that is enough to shift the balance of H^+ and OH^- from neutrality.

The pH Scale

In any aqueous solution at 25°C, the *product* of the H^+ and OH^- concentrations is constant at 10^{-14}. This can be written

$$[H^+][OH^-] = 10^{-14}$$

In such an equation, brackets indicate molar concentration. In a neutral solution at room temperature (25°C), $[H^+] = 10^{-7}$ and $[OH^-] = 10^{-7}$, so in this case, 10^{-14} is the product of $10^{-7} \times 10^{-7}$. If enough acid is added to a solution to increase $[H^+]$ to 10^{-5} *M*, then $[OH^-]$ will decline by an equivalent amount to 10^{-9} *M* (note that $10^{-5} \times 10^{-9} = 10^{-14}$). This constant relationship expresses the behavior of acids and bases in an aqueous solution. An acid not only adds hydrogen ions to a solution, but also removes hydroxide ions because of the tendency for H^+ to combine with OH^-, forming water. A base has the opposite effect, increasing OH^- concentration but also reducing H^+ concentration by the formation of water. If enough of a base is added to raise the OH^- concentration to 10^{-4} *M*, it will cause the H^+ concentration to drop to 10^{-10} *M*. Whenever we know the concentration of either H^+ or OH^- in an aqueous solution, we can deduce the concentration of the other ion.

Because the H^+ and OH^- concentrations of solutions can vary by a factor of 100 trillion or more, scientists have developed a way to express this variation more conveniently than in moles per liter. The pH scale **(Figure 3.9)** compresses the

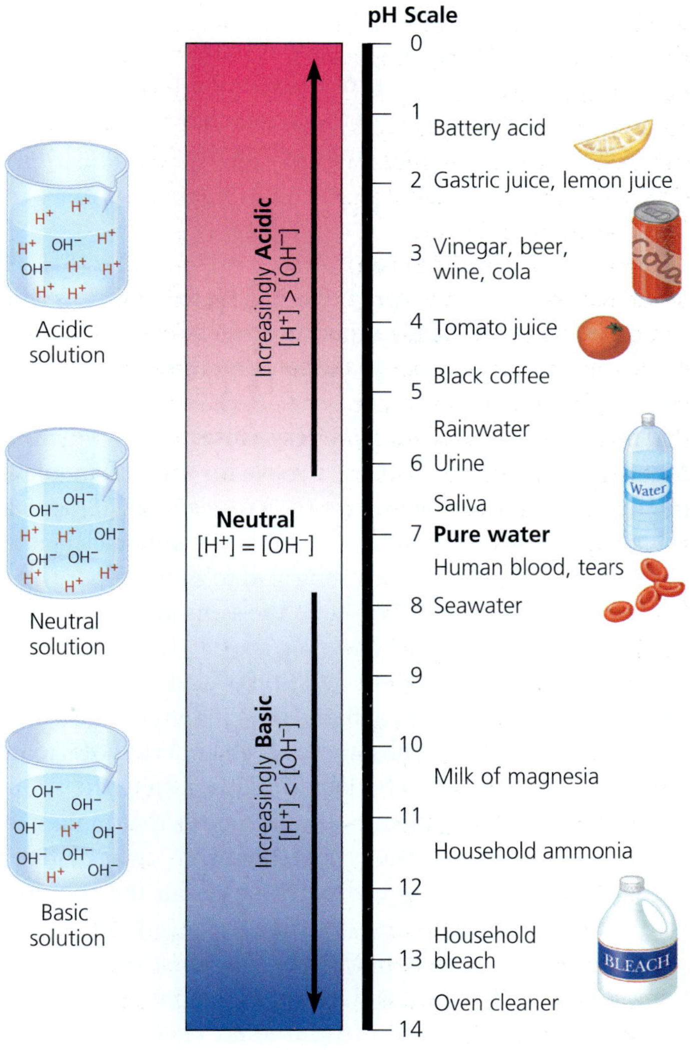

▲ **Figure 3.9 The pH scale and pH values of some aqueous solutions.**

range of H^+ and OH^- concentrations by employing logarithms. The **pH** of a solution is defined as the negative logarithm (base 10) of the hydrogen ion concentration:

$$\text{pH} = -\log\,[H^+]$$

For a neutral aqueous solution, $[H^+]$ is 10^{-7} *M*, giving us

$$-\log 10^{-7} = -(-7) = 7$$

Notice that pH *declines* as H^+ concentration *increases.* Notice, too, that although the pH scale is based on H^+ concentration, it also implies OH^- concentration. A solution of pH 10 has a hydrogen ion concentration of 10^{-10} *M* and a hydroxide ion concentration of 10^{-4} *M.*

The pH of a neutral aqueous solution at 25°C is 7, the midpoint of the scale. A pH value less than 7 denotes an acidic solution; the lower the number, the more acidic the solution. The pH for basic solutions is above 7. Most biological fluids are within the range pH 6–8. There are a few exceptions, however, including the strongly acidic digestive juice of the human stomach, which has a pH of about 2.

Remember that each pH unit represents a tenfold difference in H^+ and OH^- concentrations. It is this mathematical feature that makes the pH scale so compact. A solution of pH 3 is not twice as acidic as a solution of pH 6, but a thousand times more acidic. When the pH of a solution changes slightly, the actual concentrations of H^+ and OH^- in the solution change substantially.

Buffers

The internal pH of most living cells is close to 7. Even a slight change in pH can be harmful, because the chemical processes of the cell are very sensitive to the concentrations of hydrogen and hydroxide ions.

The pH of human blood is very close to 7.4, which is slightly basic. A person cannot survive for more than a few minutes if the blood pH drops to 7 or rises to 7.8, and a chemical system exists in the blood that maintains a stable pH. If you add 0.01 mol of a strong acid to a liter of pure water, the pH drops from 7.0 to 2.0. If the same amount of acid is added to a liter of blood, however, the pH decrease is only from 7.4 to 7.3. Why does the addition of acid have so much less of an effect on the pH of blood than it does on the pH of water? The presence of substances called buffers allows for a relatively constant pH in biological fluids despite the addition of acids or bases. **Buffers** are substances that minimize changes in the concentrations of H^+ and OH^- in a solution. They do so by accepting hydrogen ions from the solution when they are in excess and donating hydrogen ions to the solution when they have been depleted. Most buffer solutions contain a weak acid and its corresponding base, which combine reversibly with hydrogen ions. There are several buffers that contribute to pH stability in human blood and many other biological solutions. One of these is carbonic acid (H_2CO_3), formed when CO_2 reacts with water in blood plasma. As mentioned earlier, carbonic acid dissociates to yield a bicarbonate ion (HCO_3^-) and a hydrogen ion (H^+):

$$\underset{\substack{H^+\text{ donor}\\ \text{(acid)}}}{H_2CO_3} \underset{\text{Response to a drop in pH}}{\overset{\text{Response to a rise in pH}}{\rightleftharpoons}} \underset{\substack{H^+\text{ acceptor}\\ \text{(base)}}}{HCO_3^-} + \underset{\substack{\text{Hydrogen}\\ \text{ion}}}{H^+}$$

The chemical equilibrium between carbonic acid and bicarbonate acts as a pH regulator, the reaction shifting left or right as other processes in the solution add or remove hydrogen ions. If the H^+ concentration in blood begins to fall (that is, if pH rises), the reaction proceeds to the right and more carbonic acid dissociates, replenishing hydrogen ions. But when H^+ concentration in blood begins to rise (when pH drops), the reaction proceeds to the left, with HCO_3^- (the base) removing the hydrogen ions from the solution and forming H_2CO_3. Thus, the carbonic acid–bicarbonate buffering system consists of an acid and a base in equilibrium with each other. Most other buffers are also acid-base pairs.

Threats to Water Quality on Earth

Considering the dependence of all life on water, contamination of rivers, lakes, seas, and rain is a dire environmental problem. Many threats to water quality have been posed by human activities. Consider, for example, the burning of fossil fuels (coal, oil, and gas). This practice, which has been increasing since the Industrial Revolution in the 1800s, releases gaseous compounds into the atmosphere, including prodigious amounts of CO_2. The chemical reactions of these compounds with water alter the delicate balance of conditions for life on Earth by affecting water pH and temperature.

The burning of fossil fuels is a major source of sulfur oxides and nitrous oxides. These react with water in the air to form strong acids, which fall to Earth with rain or snow. **Acid precipitation** refers to rain, snow, or fog with a pH lower (more acidic) than pH 5.2. (Uncontaminated rain has a pH of about 5.6, slightly acidic, owing to the formation of carbonic acid from carbon dioxide and water.) Electric power plants that burn coal produce more of these oxides than any other single source. Winds carry the pollutants away, and acid rain may fall hundreds of kilometers away from industrial centers. In certain sites in Pennsylvania and New York, the pH of rainfall in December 2001 averaged 4.3, about 20 times more acidic than normal rain. Acid precipitation falls on many other regions, including eastern Canada, the Cascade Mountains of the Pacific Northwest, and certain parts of Europe and Asia **(Figure 3.10)**.

Acid precipitation can damage life in lakes and streams. Also, acid precipitation falling on land adversely affects soil chemistry and has taken a toll on some North American and European forests (see Figure 3.10). Nevertheless, studies indicate that the majority of North American forests are not

▲ **Figure 3.10 Acid precipitation and its effects on a forest.** Acid rain is thought to be responsible for killing trees in many forests, including the fir forest shown here in the Czech Republic.

currently suffering substantially from acid precipitation, in large part due to amendments made in 1990 to the Clean Air Act.

Carbon dioxide, the main product of fossil fuel combustion, causes other problems. Its release into the atmosphere has been increasing steadily and is expected to double by the year 2065, relative to 1880 levels. About half of the CO_2 stays in the atmosphere, acting like a reflective blanket over the planet that prevents heat from radiating into outer space. This "greenhouse" effect and the problems associated with it will be discussed in Chapter 55. A portion of the CO_2 is taken up by trees and other organisms during photosynthesis, as mentioned in Chapter 2. The remainder—about 30% or so—is absorbed by the oceans. In spite of the huge volume of water in the oceans, scientists worry that this absorption of so much CO_2 will harm marine life and ecosystems.

When CO_2 dissolves in seawater, it reacts with water (H_2O) to form carbonic acid (H_2CO_3). Almost all of the carbonic acid in turn dissociates, producing protons and a balance between two ions, bicarbonate (HCO_3^-) and carbonate (CO_3^{2-}). As seawater acidifies due to the extra protons, the balance shifts toward HCO_3^-, lowering the concentration of CO_3^{2-}. Many studies have shown that calcification, the production of calcium carbonate ($CaCO_3$) by corals and other organisms, is directly affected by the concentration of CO_3^{2-}. Any decrease in CO_3^{2-} is therefore of great concern because calcification accounts for the formation of coral reefs in our tropical seas. These sensitive ecosystems act as havens for a great diversity of organisms.

Perhaps one of the best-known and longest studies on coral reef calcification was carried out by scientists at the ecosystem center in Arizona known as Biosphere-2. The center includes an artificial coral reef system in which the temperature and chemistry of the seawater can be controlled and manipulated. Chris Langdon and colleagues used this system to test the effects of varying the concentration of CO_3^{2-} on the rate of calcification in the coral reef **(Figure 3.11)**. Together with

▼ **Figure 3.11 Inquiry**

What is the effect of carbonate ion concentration on coral reef calcification?

EXPERIMENT Chris Langdon and colleagues at Columbia University wondered how the increase in CO_2 emissions due to burning of fossil fuels and the resulting decrease in CO_3^{2-} concentration in the oceans affect coral reefs. They took advantage of the artificial coral reef system at Biosphere-2. This 2,650-cubic-meter (m^3) aquarium behaves like a natural coral reef community. For almost four years, these researchers varied the carbonate concentration in the seawater under controlled conditions and measured the rate of calcification by reef organisms.

RESULTS The calcification rate was observed to be lower at lower concentrations of carbonate ion ($[CO_3^{2-}]$).

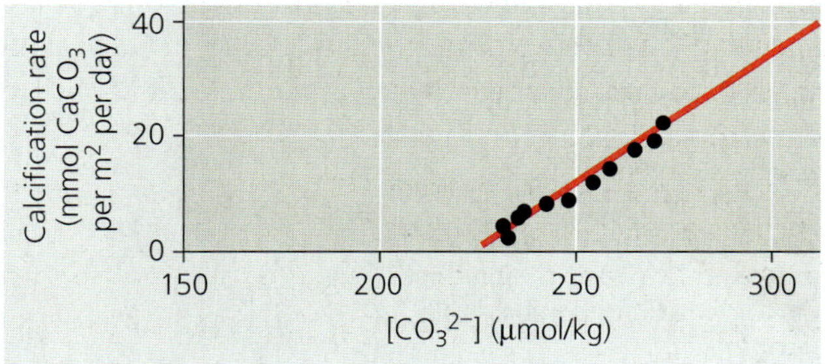

CONCLUSION Coral reefs could be endangered by reduced $[CO_3^{2-}]$. Other studies have predicted a doubling of CO_2 emissions from 1880 to 2065 and a resulting decrease in carbonate ion concentration. Combining those predictions with the results of this study, the authors predict that by 2065 the rate of coral reef calcification may decrease by 40% relative to preindustrial levels.

SOURCE C. Langdon et al., Effect of calcium carbonate saturation state on the calcification rate of an experimental coral reef, *Global Biogeochemical Cycles* 14:639–654 (2000).

WHAT IF? The data above were for a particular concentration of calcium (Ca^{2+}). The series of experiments included measurements at two higher concentrations of Ca^{2+}. Given that the following reaction represents calcification, predict how the $[Ca^{2+}]$ would affect the results and explain why.

$$Ca^{2+} + CO_3^{2-} \longrightarrow CaCO_3$$

conclusions from other studies, their results led them to predict that the expected doubling of CO_2 emissions by the year 2065 could lead to a 40% decrease in coral reef calcification. Although scientists may not agree on the exact percentage, most concur that this and other studies provide cause for grave concern.

If there is reason for optimism about the future quality of water resources on our planet, it is that we have made progress in learning about the delicate chemical balances in oceans and other bodies of water. Continued progress can come only from the actions of people who are concerned about environmental quality. This requires understanding the crucial role that water plays in the environment's fitness for continued life on Earth.

CONCEPT CHECK 3.3

1. Compared with a basic solution at pH 9, the same volume of an acidic solution at pH 4 has ___ times as many hydrogen ions (H^+).
2. HCl is a strong acid that dissociates in water: $HCl \longrightarrow H^+ + Cl^-$. What is the pH of 0.01 *M* HCl?
3. Acetic acid (CH_3COOH) can be a buffer, similar to carbonic acid. Write the dissociation reaction, identifying the acid, base, H^+ acceptor, and H^+ donor.
4. **WHAT IF?** Given a liter of pure water and a liter solution of acetic acid, what would happen to the pH if you added 0.01 mol of a strong acid to each? Use the reaction equation to explain the result.

For suggested answers, see Appendix A.

Chapter 3 Review

MEDIA Go to the Study Area at **www.masteringbio.com** for BioFlix 3-D Animations, MP3 Tutors, Videos, Practice Tests, an eBook, and more.

SUMMARY OF KEY CONCEPTS

CONCEPT 3.1
The polarity of water molecules results in hydrogen bonding (pp. 46–47)

- A hydrogen bond forms when the slightly negatively charged oxygen of one water molecule is attracted to the slightly positively charged hydrogen of a nearby molecule. Hydrogen bonding between water molecules is the basis for water's unusual properties.

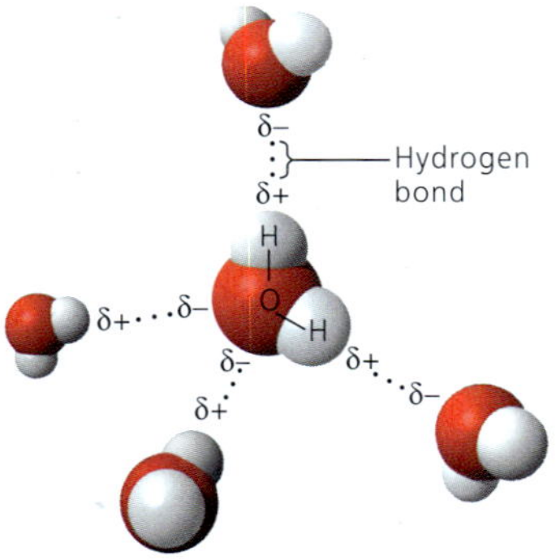

MEDIA

Activity The Polarity of Water

CONCEPT 3.2
Four emergent properties of water contribute to Earth's fitness for life (pp. 47–52)

- **Cohesion** Hydrogen bonding keeps water molecules close to each other, and this cohesion helps pull water upward in the microscopic water-conducting cells of plants. Hydrogen bonding is also responsible for water's surface tension.
- **Moderation of Temperature** Water has a high specific heat: Heat is absorbed when hydrogen bonds break and is released when hydrogen bonds form. This helps keep temperatures relatively steady, within limits that permit life. Evaporative cooling is based on water's high heat of vaporization. The evaporative loss of the most energetic water molecules cools a surface.
- **Insulation of Bodies of Water by Floating Ice** Ice floats because it is less dense than liquid water. This allows life to exist under the frozen surfaces of lakes and polar seas.

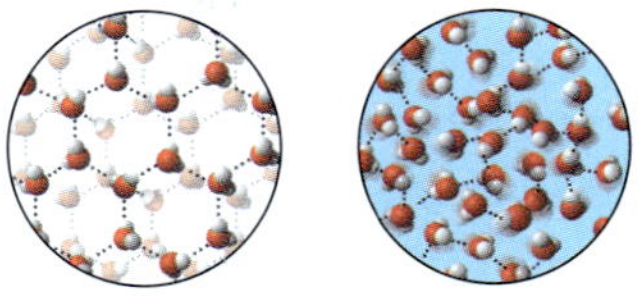

Ice: stable hydrogen bonds

Liquid water: transient hydrogen bonds

- **The Solvent of Life** Water is an unusually versatile solvent because its polar molecules are attracted to charged and polar substances capable of forming hydrogen bonds. Hydrophilic substances have an affinity for water; hydrophobic substances do not. Molarity, the number of moles of solute per liter of solution, is used as a measure of solute concentration in solutions. A mole is a certain number of molecules of a substance. The mass of a mole of the substance in grams is the same as the molecular mass in daltons.

MEDIA

MP3 Tutor The Properties of Water

Activity Cohesion of Water

CONCEPT 3.3
Acidic and basic conditions affect living organisms (pp. 52–56)

- A water molecule can transfer an H^+ to another water molecule to form H_3O^+ (represented simply by H^+) and OH^-.
- **Effects of Changes in pH** The concentration of H^+ is expressed as pH, where $pH = -\log [H^+]$. Buffers in biological fluids resist changes in pH. A buffer consists of an acid-base pair that combines reversibly with hydrogen ions.

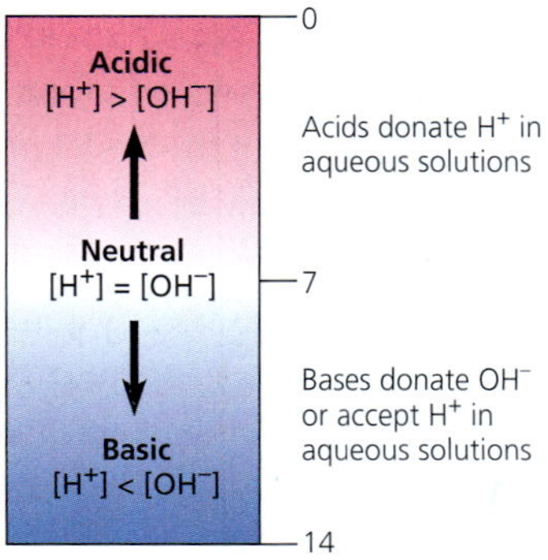

- **Threats to Water Quality on Earth** The burning of fossil fuels results in emission of oxides (leading to acid precipitation) and increasing amounts of CO_2. Some CO_2 becomes dissolved in the oceans, lowering pH and potentially affecting the rate of calcification on coral reefs.

MEDIA

Activity Dissociation of Water Molecules
Activity Acids, Bases, and pH
Investigation How Does Acid Precipitation Affect Trees?

TESTING YOUR KNOWLEDGE

SELF-QUIZ

1. Many mammals control their body temperature by sweating. Which property of water is most directly responsible for the ability of sweat to lower body temperature?
   a. water's change in density when it condenses
   b. water's ability to dissolve molecules in the air
   c. the release of heat by the formation of hydrogen bonds
   d. the absorption of heat by the breaking of hydrogen bonds
   e. water's high surface tension

2. A slice of pizza has 500 kcal. If we could burn the pizza and use all the heat to warm a 50-L container of cold water, what would be the approximate increase in the temperature of the water? (*Note:* A liter of cold water weighs about 1 kg.)
   a. 50°C
   b. 5°C
   c. 10°C
   d. 100°C
   e. 1°C

3. The bonds that are broken when water vaporizes are
   a. ionic bonds.
   b. hydrogen bonds between water molecules.
   c. covalent bonds between atoms within water molecules.
   d. polar covalent bonds.
   e. nonpolar covalent bonds.

4. Which of the following is a hydrophobic material?
   a. paper
   b. table salt
   c. wax
   d. sugar
   e. pasta

5. We can be sure that a mole of table sugar and a mole of vitamin C are equal in their
   a. mass in daltons.
   b. mass in grams.
   c. number of molecules.
   d. number of atoms.
   e. volume.

6. How many grams of acetic acid ($C_2H_4O_2$) would you use to make 10 L of a 0.1 M aqueous solution of acetic acid? (*Note:* The atomic masses, in daltons, are approximately 12 for carbon, 1 for hydrogen, and 16 for oxygen.)
   a. 10.0 g
   b. 0.1 g
   c. 6.0 g
   d. 60.0 g
   e. 0.6 g

7. Measurements show that the pH of a particular lake is 4.0. What is the hydrogen ion concentration of the lake?
   a. 4.0 M
   b. 10^{-10} M
   c. 10^{-4} M
   d. 10^{4} M
   e. 4%

8. What is the *hydroxide* ion concentration of the lake described in question 7?
   a. 10^{-7} M
   b. 10^{-4} M
   c. 10^{-10} M
   d. 10^{-14} M
   e. 10 M

9. **DRAW IT** Draw three water molecules, using space-filling models, and label the atoms. Draw solid lines to indicate covalent bonds and dotted lines for hydrogen bonds. Add partial-charge labels as appropriate.

For Self-Quiz answers, see Appendix A.

MEDIA Visit the Study Area at **www.masteringbio.com** for a Practice Test.

EVOLUTION CONNECTION

10. The surface of the planet Mars has many landscape features reminiscent of those formed by flowing water on Earth, including what appear to be meandering channels and outwash areas. In 2004, images of Meridiani Planum on Mars taken by NASA's *Opportunity* rover suggested that liquid water was once present on its surface. For example, one image (left, above) shows polygonal fractures in the rock. Similar fracture patterns in rocks on Earth's surface (above, right) correlate with the earlier presence of water. Ice exists at the Martian poles today, and some scientists suspect a great deal more water may be present beneath the Martian surface. Why has there been so much interest in the presence of water on Mars? Does the presence of water make it more likely that life arose there? What other physical factors might also be important?

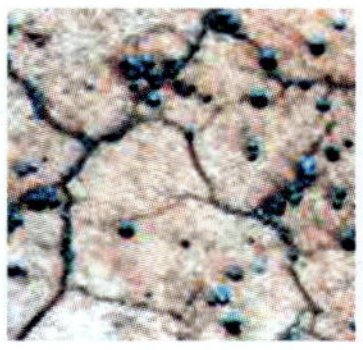
Surface of Mars

Surface of Earth

SCIENTIFIC INQUIRY

11. Design a controlled experiment to test the hypothesis that acid precipitation inhibits the growth of *Elodea,* a common freshwater plant (see Figure 2.19).

12. In agricultural areas, farmers pay close attention to the weather forecast. Right before a predicted overnight freeze, farmers spray water on crops to protect the plants. Use the properties of water to explain how this method works. Be sure to mention why hydrogen bonds are responsible for this phenomenon.

SCIENCE, TECHNOLOGY, AND SOCIETY

13. Agriculture, industry, and the growing populations of cities all compete, through political influence, for water. If you were in charge of water resources in an arid region, what would your priorities be for allocating the limited water supply for various uses? How would you try to build consensus among the different special-interest groups?

Carbon and the Molecular Diversity of Life

▲ **Figure 4.1 What properties of carbon underlie its role as the molecular basis of life?**

KEY CONCEPTS

4.1 Organic chemistry is the study of carbon compounds

4.2 Carbon atoms can form diverse molecules by bonding to four other atoms

4.3 A small number of chemical groups are key to the functioning of biological molecules

OVERVIEW

Carbon: The Backbone of Life

Although water is the universal medium for life on Earth, living organisms, such as the plants and trilobite beetle in **Figure 4.1**, are made up of chemicals that are based mostly on the element carbon. Carbon enters the biosphere through the action of plants, which use solar energy to transform atmospheric CO_2 into the molecules of life. These molecules are passed along to animals that feed on plants.

Of all chemical elements, carbon is unparalleled in its ability to form molecules that are large, complex, and diverse, and this molecular diversity has made possible the diversity of organisms that have evolved on Earth. Proteins, DNA, carbohydrates, and other molecules that distinguish living matter from inanimate material are all composed of carbon atoms bonded to one another and to atoms of other elements. Hydrogen (H), oxygen (O), nitrogen (N), sulfur (S), and phosphorus (P) are other common ingredients of these compounds, but it is the element carbon (C) that accounts for the large diversity of biological molecules.

Proteins and other very large molecules are the main focus of Chapter 5. Here we investigate the properties of smaller molecules. We will use these molecules to illustrate concepts of molecular architecture that will help explain why carbon is so important to life, at the same time highlighting the theme that emergent properties arise from the organization of matter in living organisms.

CONCEPT 4.1

Organic chemistry is the study of carbon compounds

For historical reasons, compounds containing carbon are said to be organic, and the branch of chemistry that specializes in the study of carbon compounds is called **organic chemistry**. Organic compounds range from simple molecules, such as methane (CH_4), to colossal ones, such as proteins, with thousands of atoms. Most organic compounds contain hydrogen atoms in addition to carbon atoms.

The overall percentages of the major elements of life—C, H, O, N, S, and P—are quite uniform from one organism to another. Because of carbon's versatility, however, this limited assortment of atomic building blocks, taken in roughly the same proportions, can be used to build an inexhaustible variety of organic molecules. Different species of organisms, and different individuals within a species, are distinguished by variations in their organic molecules.

Since the dawn of human history, people have used other organisms as sources of valued substances—from foods and medicines to fabrics. The science of organic chemistry originated in attempts to purify and improve the yield of such products. By the early 1800s, chemists had learned to make many simple compounds in the laboratory by combining elements under the right conditions. Artificial synthesis of the complex molecules extracted from living matter seemed impossible, however. At that time, the Swedish chemist Jöns Jakob Berzelius made the distinction between organic compounds, those thought to arise only in living organisms, and inorganic compounds, those found only in the nonliving world. *Vitalism*, the belief in a life force outside the jurisdiction of physical and chemical laws, provided the foundation for the new discipline of organic chemistry.

Chemists began to chip away at the foundation of vitalism when they finally learned to synthesize organic compounds in laboratories. In 1828, Friedrich Wöhler, a German chemist who had studied with Berzelius, attempted to make an "inorganic" salt, ammonium cyanate, by mixing solutions of ammonium ions (NH_4^+) and cyanate ions (CNO^-). Wöhler was astonished to find that instead he had made urea, an organic compound present in the urine of animals. Wöhler challenged the vitalists when he wrote, "I must tell you that I can prepare urea without requiring a kidney or an animal, either man or dog." However, one of the ingredients used in the synthesis, the cyanate, had been extracted from animal blood, and the vitalists were not swayed by Wöhler's discovery. A few years later, however, Hermann Kolbe, a student of Wöhler's, made the organic compound acetic acid from inorganic substances that could be prepared directly from pure elements.

Vitalism crumbled completely after several decades of laboratory synthesis of some increasingly complex organic compounds. In 1953, Stanley Miller, a graduate student of Harold Urey at the University of Chicago, helped bring this abiotic (nonliving) synthesis of organic compounds into the context of evolution in a classic experiment described in **Figure 4.2**. Miller's experiment, testing whether complex organic molecules could arise spontaneously under conditions thought to have existed on the early Earth, stimulated interest and further research on the origin of organic compounds. Some scientists have questioned whether the gases Miller used as starting materials were really present in the primitive Earth's atmosphere. Recent work supports a slightly different recipe for early Earth's conditions; when used in the experiment, it led to the compounds Miller found. Although the jury is still out, these experiments support the idea that abiotic synthesis of organic compounds could have been an early stage in the origin of life.

The pioneers of organic chemistry helped shift the mainstream of biological thought from vitalism to *mechanism*, the view that physical and chemical laws govern all natural phenomena, including the processes of life. Organic chemistry was redefined as the study of carbon compounds, regardless of origin. Organisms produce most of the naturally occurring organic compounds, and these molecules represent a diversity and range of complexity unrivaled by inorganic compounds. However, the rules of chemistry apply to all molecules. The foundation of organic chemistry is not some intangible life force, but the unique chemical versatility of the element carbon.

CONCEPT CHECK 4.1

1. What conclusion did Stanley Miller draw when he found amino acids in the products of his experiment?
2. **WHAT IF?** When Miller tried the experiment in Figure 4.2 without the electrical discharge, no organic compounds were found. What might explain this result?

For suggested answers, see Appendix A.

▼ **Figure 4.2 Inquiry**

Can organic molecules form under conditions believed to simulate those on the early Earth?

EXPERIMENT In 1953, Stanley Miller set up a closed system to simulate conditions thought to have existed on the early Earth. A flask of water simulated the primeval sea. The water was heated so that some vaporized and moved into a second, higher flask containing the "atmosphere"—a mixture of gases. Sparks were discharged in the synthetic atmosphere to mimic lightning.

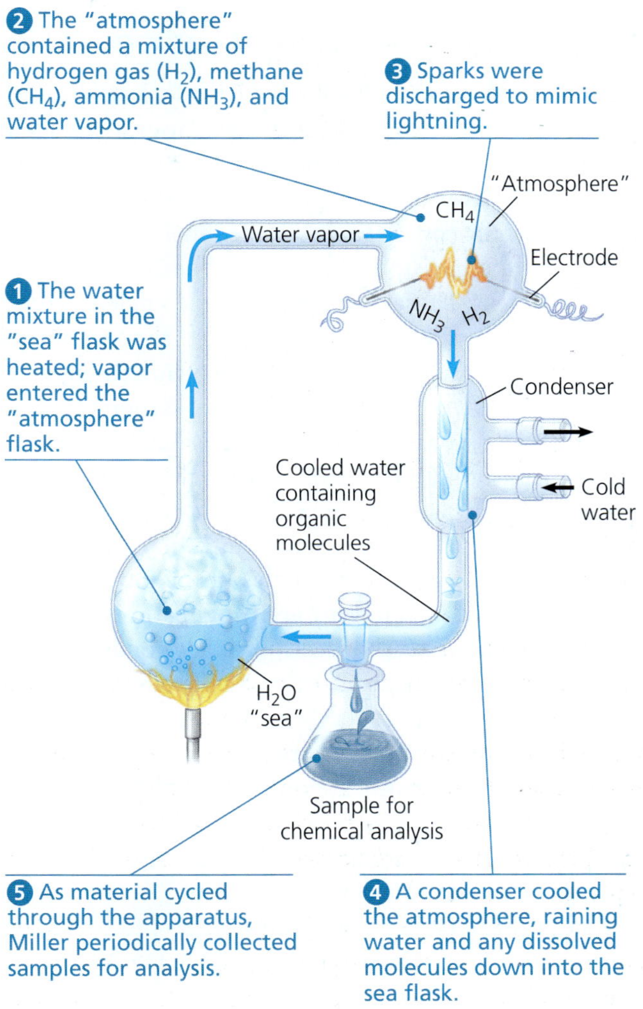

RESULTS Miller identified a variety of organic molecules that are common in organisms. These included simple compounds such as formaldehyde (CH_2O) and hydrogen cyanide (HCN) and more complex molecules such as amino acids and long chains of carbon and hydrogen known as hydrocarbons.

CONCLUSION Organic molecules, a first step in the origin of life, may have been synthesized abiotically on the early Earth. (We will explore this hypothesis in more detail in Chapter 25.)

SOURCE S. Miller, A production of amino acids under possible primitive Earth conditions, *Science* 117:528–529 (1953).

WHAT IF? If Miller had increased the concentration of NH_3 in his experiment, how might the relative amounts of the products HCN and CH_2O have differed?

CONCEPT **4.2**

Carbon atoms can form diverse molecules by bonding to four other atoms

The key to an atom's chemical characteristics is its electron configuration. This configuration determines the kinds and number of bonds an atom will form with other atoms.

The Formation of Bonds with Carbon

Carbon has 6 electrons, with 2 in the first electron shell and 4 in the second shell. Having 4 valence electrons in a shell that holds 8, carbon would have to donate or accept 4 electrons to complete its valence shell and become an ion. Instead, a carbon atom usually completes its valence shell by sharing its 4 electrons with other atoms in covalent bonds so that 8 electrons are present. These bonds may include single and double covalent bonds. Each carbon atom thus acts as an intersection point from which a molecule can branch off in as many as four directions. This *tetravalence* is one facet of carbon's versatility that makes large, complex molecules possible.

When a carbon atom forms four single covalent bonds, the arrangement of its four hybrid orbitals causes the bonds to angle toward the corners of an imaginary tetrahedron (see Figure 2.17b). The bond angles in methane (CH_4) are 109.5° **(Figure 4.3a)**, and they are roughly the same in any group of atoms where carbon has four single bonds. For example, ethane (C_2H_6) is shaped like two overlapping tetrahedrons **(Figure 4.3b)**. In molecules with more carbons, every grouping of a carbon bonded to four other atoms has a tetrahedral shape. But when two carbon atoms are joined by a double bond, all bonds around those carbons are in the same plane. For example, ethene (C_2H_4) is a flat molecule; its atoms all lie in the same plane **(Figure 4.3c)**. We find it convenient to write all structural formulas as though the molecules represented were flat, but keep in mind that molecules are three-dimensional and that the shape of a molecule often determines its function.

The electron configuration of carbon gives it covalent compatibility with many different elements. **Figure 4.4** shows the valences of carbon and its most frequent partners—oxygen, hydrogen, and nitrogen. These are the four major atomic components of organic molecules. These valences are the basis for the rules of covalent bonding in organic chemistry—the building code for the architecture of organic molecules.

Let's consider how the rules of covalent bonding apply to carbon atoms with partners other than hydrogen. We'll look at two examples, the simple molecules carbon dioxide and urea.

In the carbon dioxide molecule (CO_2), a single carbon atom is joined to two atoms of oxygen by double covalent bonds. The structural formula for CO_2 is shown here:

$$O{=}C{=}O$$

Each line in a structural formula represents a pair of shared electrons. The two double bonds formed by the carbon atom are the equivalent of four single covalent bonds. The arrangement completes the valence shells of all atoms in the molecule. Because CO_2 is a very simple molecule and lacks hydrogen, it is often considered inorganic, even though it contains carbon.

Name and Comment	Molecular Formula	Structural Formula	Ball-and-Stick Model	Space-Filling Model
(a) Methane. When a carbon atom has four single bonds to other atoms, the molecule is tetrahedral.	CH_4	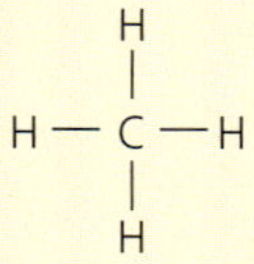	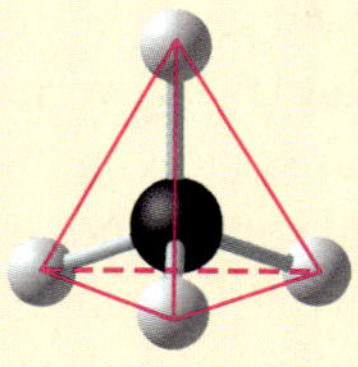	
(b) Ethane. A molecule may have more than one tetrahedral group of single-bonded atoms. (Ethane consists of two such groups.)	C_2H_6	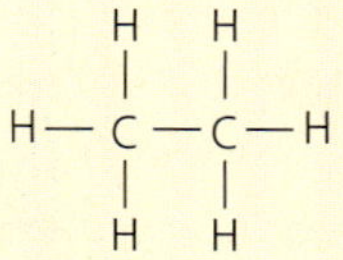	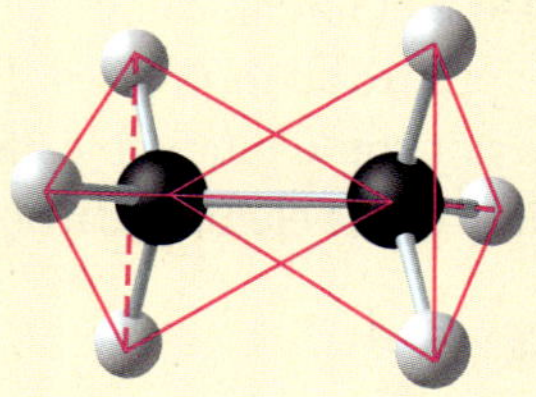	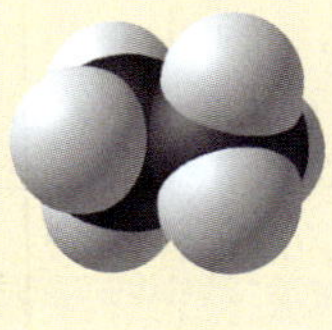
(c) Ethene (ethylene). When two carbon atoms are joined by a double bond, all atoms attached to those carbons are in the same plane; the molecule is flat.	C_2H_4	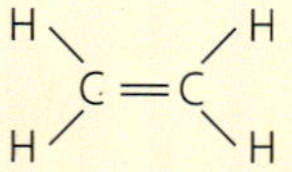	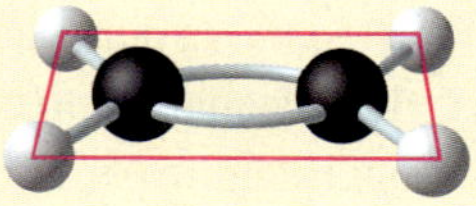	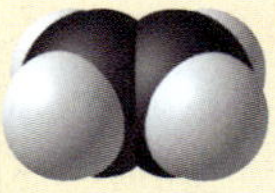

▲ **Figure 4.3 The shapes of three simple organic molecules.**

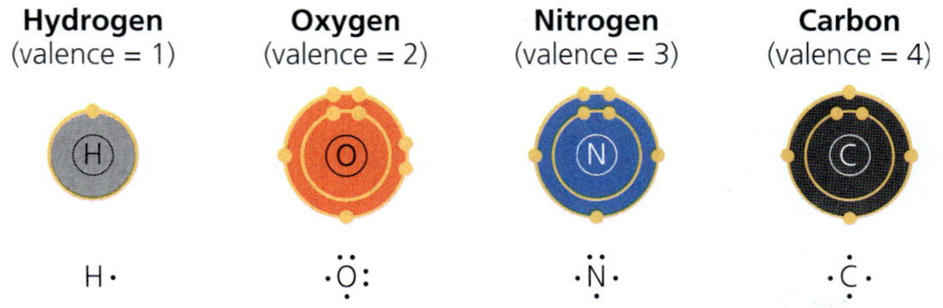

▲ **Figure 4.4 Valences of the major elements of organic molecules.** Valence is the number of covalent bonds an atom can form. It is generally equal to the number of electrons required to complete the valence (outermost) shell (see Figure 2.9). All the electrons are shown for each atom in the electron distribution diagrams (top). Only the electrons in the valence shell are presented in the Lewis dot structures.

DRAW IT *Refer to Figure 2.9 and draw the Lewis dot structures for sodium, phosphorus, sulfur, and chlorine.*

Whether we call CO_2 organic or inorganic, however, it is clearly important to the living world as the source of carbon for all organic molecules in organisms.

Urea, $CO(NH_2)_2$, is the organic compound found in urine that Wöhler synthesized in the early 1800s. The structural formula for urea is shown at the right. Again, each atom has the required number of covalent bonds. In this case, one carbon atom is involved in both single and double bonds.

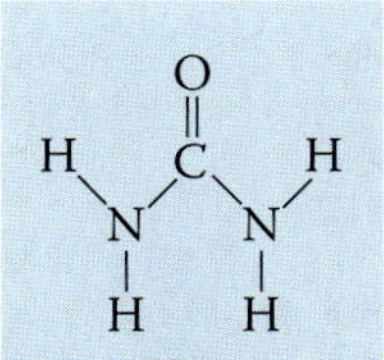

Urea

Urea and carbon dioxide are molecules with one carbon atom. But as Figure 4.3 shows, a carbon atom can also use one or more valence electrons to form covalent bonds to other carbon atoms, linking the atoms into chains of seemingly infinite variety.

Molecular Diversity Arising from Carbon Skeleton Variation

Carbon chains form the skeletons of most organic molecules **(Figure 4.5)**. The skeletons vary in length and may be straight, branched, or arranged in closed rings. Some carbon skeletons have double bonds, which vary in number and location. Such variation in carbon skeletons is one important source of the molecular complexity and diversity that characterize living matter. In addition, atoms of other elements can be bonded to the skeletons at available sites.

Hydrocarbons

All of the molecules that are shown in Figures 4.3 and 4.5 are **hydrocarbons**, organic molecules consisting of only carbon and hydrogen. Atoms of hydrogen are attached to the carbon skeleton wherever electrons are available for covalent bonding. Hydrocarbons are the major components of petroleum, which is called a fossil fuel because it consists of the partially decomposed remains of organisms that lived millions of years ago.

Although hydrocarbons are not prevalent in living organisms, many of a cell's organic molecules have regions consisting of only carbon and hydrogen. For example, the molecules known as fats have long hydrocarbon tails attached to a nonhydrocarbon component (**Figure 4.6**, on next page). Neither petroleum nor fat dissolves in water; both are hydrophobic compounds because the great majority of their bonds are relatively nonpolar carbon-to-hydrogen linkages. Another characteristic of hydrocarbons is that they can undergo reactions that release a relatively large amount of energy. The gasoline that fuels a car consists of hydrocarbons, and the hydrocarbon tails of fat molecules serve as stored fuel for animal bodies.

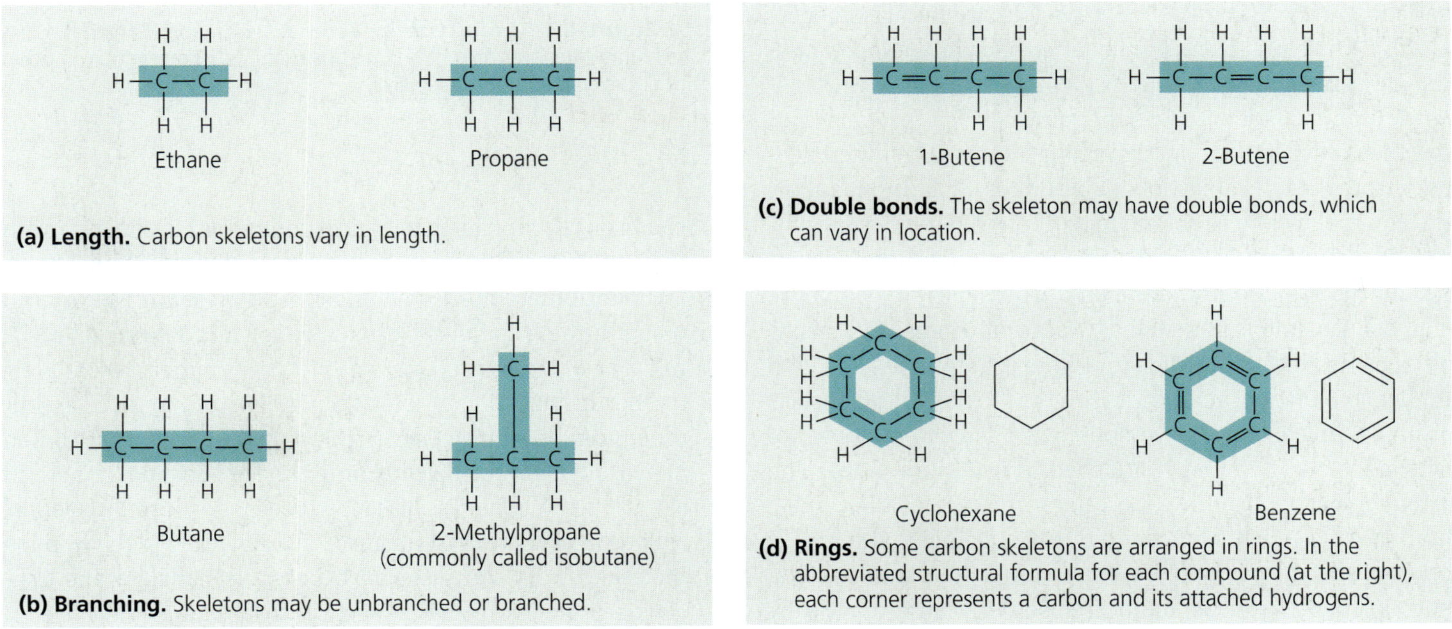

(a) Length. Carbon skeletons vary in length.

(b) Branching. Skeletons may be unbranched or branched.

(c) Double bonds. The skeleton may have double bonds, which can vary in location.

(d) Rings. Some carbon skeletons are arranged in rings. In the abbreviated structural formula for each compound (at the right), each corner represents a carbon and its attached hydrogens.

▲ **Figure 4.5 Variations in carbon skeletons.** Hydrocarbons, organic molecules consisting only of carbon and hydrogen, illustrate the diversity of the carbon skeletons of organic molecules.

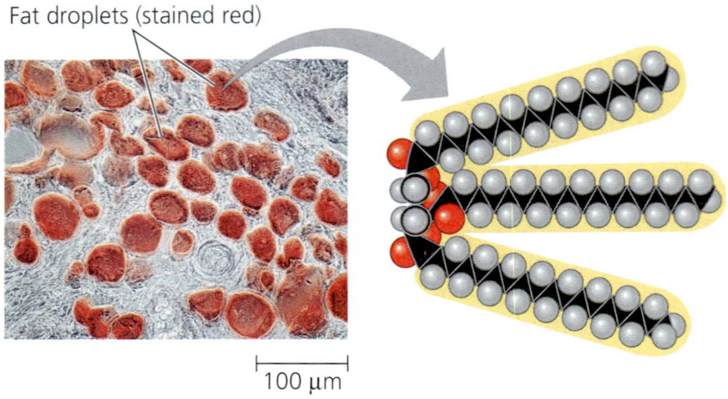

(a) Mammalian adipose cells **(b) A fat molecule**

▲ **Figure 4.6 The role of hydrocarbons in fats. (a)** Mammalian adipose cells stockpile fat molecules as a fuel reserve. Each adipose cell in this micrograph is almost filled by a large fat droplet, which contains a huge number of fat molecules. **(b)** A fat molecule consists of a small, nonhydrocarbon component joined to three hydrocarbon tails. The tails can be broken down to provide energy. They also account for the hydrophobic behavior of fats. (Black = carbon; gray = hydrogen; red = oxygen.)

Isomers

Variation in the architecture of organic molecules can be seen in **isomers**, compounds that have the same numbers of atoms of the same elements but different structures and hence different properties. Compare, for example, the two five-carbon compounds in **Figure 4.7a**. Both have the molecular formula C_5H_{12}, but they differ in the covalent arrangement of their carbon skeletons. The skeleton is straight in one compound but branched in the other. We will examine three types of isomers: structural isomers, geometric isomers, and enantiomers.

Structural isomers differ in the covalent arrangements of their atoms. The number of possible isomers increases tremendously as carbon skeletons increase in size. There are only 3 forms of C_5H_{12} (2 are shown in Figure 4.7a), but there are 18 variations of C_8H_{18} and 366,319 possible structural isomers of $C_{20}H_{42}$. Structural isomers may also differ in the location of double bonds.

Geometric isomers have the same covalent partnerships, but they differ in their spatial arrangements. The differences arise from the inflexibility of double bonds. Single bonds allow the atoms they join to rotate freely about the bond axis without changing the compound. In contrast, double bonds do not permit such rotation, resulting in the possibility of geometric isomers. If a double bond joins two carbon atoms, and each C also has two different atoms (or groups of atoms) attached to it, then two distinct geometric isomers are possible. Consider a simple molecule with two double-bonded carbons, each of which has an H and an X attached to it **(Figure 4.7b)**. The arrangement with both Xs on the same side of the double bond is called a *cis* isomer, and the arrangement with the Xs on opposite sides is called a *trans* isomer. The subtle difference in shape between geometric isomers can dramatically affect the biological activities of organic molecules. For example, the biochemistry of vision involves a light-induced

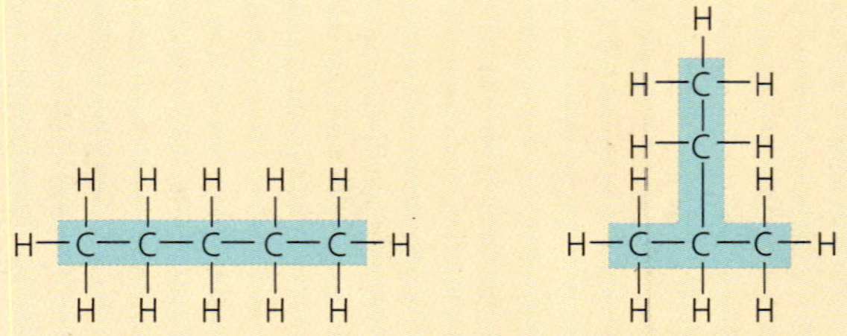

(a) Structural isomers differ in covalent partners, as shown in this example of two isomers of C_5H_{12}: pentane (left) and 2-methyl butane (right).

cis isomer: The two Xs are on the same side.

trans isomer: The two Xs are on opposite sides.

(b) Geometric isomers differ in arrangement about a double bond. In these diagrams, X represents an atom or group of atoms attached to a double-bonded carbon.

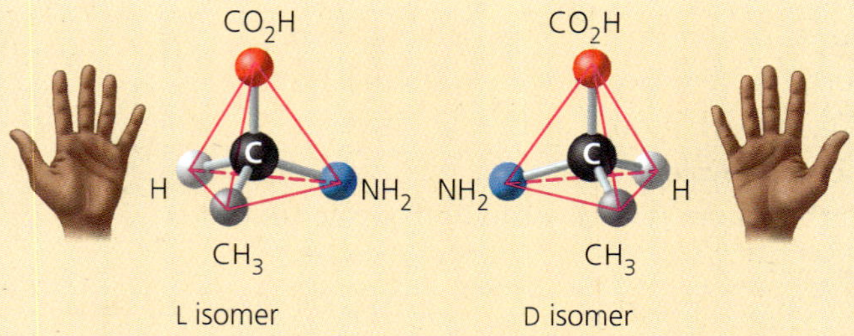

(c) Enantiomers differ in spatial arrangement around an asymmetric carbon, resulting in molecules that are mirror images, like left and right hands. The two isomers are designated the L and D isomers from the Latin for left and right (*levo* and *dextro*). Enantiomers cannot be superimposed on each other.

▲ **Figure 4.7 Three types of isomers.** Compounds with the same molecular formula but different structures, isomers are a source of diversity in organic molecules.

DRAW IT *There are three structural isomers of C_5H_{12}; draw the one not shown in (a).*

change of rhodopsin, a chemical compound in the eye, from the *cis* isomer to the *trans* isomer (see Chapter 50).

Enantiomers are isomers that are mirror images of each other. In the ball-and-stick models shown in **Figure 4.7c**, the middle carbon is called an *asymmetric carbon* because it is attached to four different atoms or groups of atoms. The four groups can be arranged in space around the asymmetric carbon in two different ways that are mirror images. Enantiomers are, in a way, left-handed and right-handed versions of the molecule. Just as your right hand won't fit into a left-handed glove, the working molecules in a cell can distinguish the two versions by shape. Usually, one isomer is biologically active, and the other is inactive.

The concept of enantiomers is important in the pharmaceutical industry because the two enantiomers of a drug may not be equally

Drug	Condition	Effective Enantiomer	Ineffective Enantiomer
Ibuprofen	Pain; inflammation	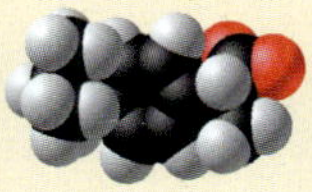 *S*-Ibuprofen	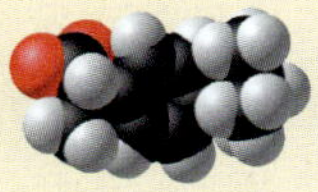 *R*-Ibuprofen
Albuterol	Asthma	 *R*-Albuterol	 *S*-Albuterol

▲ **Figure 4.8 The pharmacological importance of enantiomers.** Ibuprofen and albuterol are examples of drugs whose enantiomers have different effects. (*S* and *R* are letters used in one system to distinguish two enantiomers.) Ibuprofen reduces inflammation and pain. It is commonly sold as a mixture of the two enantiomers. The *S* enantiomer is 100 times more effective than the other. Albuterol is used to relax bronchial muscles, improving airflow in asthma patients. Only *R*-albuterol is synthesized and sold as a drug; the *S* form counteracts the active *R* form.

effective **(Figure 4.8)**. In some cases, one of the isomers may even produce harmful effects. This was the case with thalidomide, a drug prescribed for thousands of pregnant women in the late 1950s and early 1960s. The drug was a mixture of two enantiomers. One enantiomer reduced morning sickness, the desired effect, but the other caused severe birth defects. (Unfortunately, even if the "good" thalidomide enantiomer is used in purified form, some of it soon converts to the "bad" enantiomer in the patient's body.) The differing effects of enantiomers in the body demonstrate that organisms are sensitive to even the most subtle variations in molecular architecture. Once again, we see that molecules have emergent properties that depend on the specific arrangement of their atoms.

CONCEPT CHECK 4.2

1. Draw a structural formula for C_2H_4.
2. Which molecules in Figure 4.5 are isomers? For each pair, identify the type of isomer.
3. How are gasoline and fat chemically similar?
4. **WHAT IF?** Can propane (C_3H_8) form isomers?

For suggested answers, see Appendix A.

CONCEPT 4.3

A small number of chemical groups are key to the functioning of biological molecules

The distinctive properties of an organic molecule depend not only on the arrangement of its carbon skeleton but also on the molecular components attached to that skeleton. We can think of hydrocarbons, the simplest organic molecules, as the underlying framework for more complex organic molecules. A number of chemical groups can replace one or more of the hydrogens bonded to the carbon skeleton of the hydrocarbon. (Some groups include atoms of the carbon skeleton, as we will see.) These groups may participate in chemical reactions or may contribute to function indirectly by their effects on molecular shape. The number and arrangement of the groups help give each molecule its unique properties.

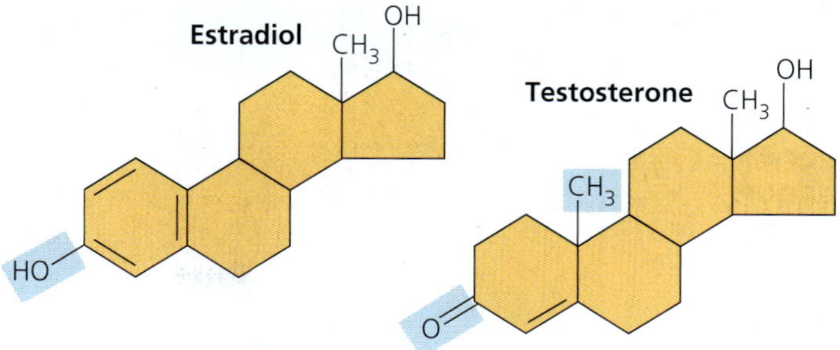

▲ **Figure 4.9 A comparison of chemical groups of female (estradiol) and male (testosterone) sex hormones.** The two molecules differ only in the chemical groups attached to a common carbon skeleton of four fused rings, shown here in abbreviated form. These subtle variations in molecular architecture (shaded in blue) influence the development of the anatomical and physiological differences between female and male vertebrates.

The Chemical Groups Most Important in the Processes of Life

Consider the differences between testosterone and estradiol (a type of estrogen). These compounds are male and female sex hormones, respectively, in humans and other vertebrates **(Figure 4.9)**. Both are steroids, organic molecules with a common carbon skeleton in the form of four fused rings. These sex hormones differ only in the chemical groups attached to the rings. The different actions of these two molecules on many targets throughout the body help produce the contrasting features of males and females. Thus, even our sexuality has its biological basis in variations of molecular architecture.

In the example of sex hormones, different chemical groups contribute to function by affecting the molecule's shape. In other cases, the chemical groups affect molecular function by being directly involved in chemical reactions; these important chemical groups are known as **functional groups**. Each functional group participates in chemical reactions in a characteristic way, from one organic molecule to another.

The seven chemical groups most important in biological processes are the hydroxyl, carbonyl, carboxyl, amino, sulfhydryl, phosphate, and methyl groups. The first six groups can act as functional groups; they are also hydrophilic and thus increase the solubility of organic compounds in water. The methyl group is not reactive, but instead often acts as a recognizable tag on biological molecules. Before reading further, study **Figure 4.10** on the next two pages to familiarize yourself with these biologically important chemical groups.

▼ Figure 4.10

Exploring Some Biologically Important Chemical Groups

Hydroxyl

CHEMICAL GROUP: Hydroxyl

STRUCTURE

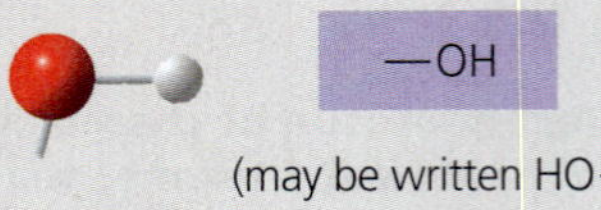

(may be written HO—)

In a **hydroxyl group** (—OH), a hydrogen atom is bonded to an oxygen atom, which in turn is bonded to the carbon skeleton of the organic molecule. (Do not confuse this functional group with the hydroxide ion, OH^-.)

NAME OF COMPOUND

Alcohols (their specific names usually end in *-ol*)

EXAMPLE

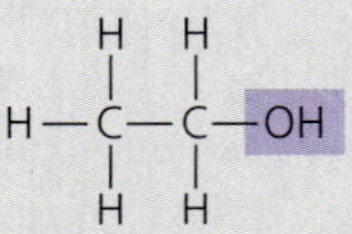

Ethanol, the alcohol present in alcoholic beverages

FUNCTIONAL PROPERTIES

- Is polar as a result of the electrons spending more time near the electronegative oxygen atom.
- Can form hydrogen bonds with water molecules, helping dissolve organic compounds such as sugars (see Figure 5.3).

Carbonyl

CHEMICAL GROUP: Carbonyl

STRUCTURE

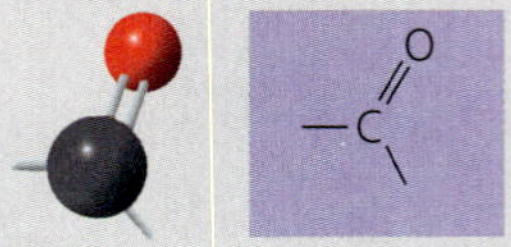

The **carbonyl group** (>CO) consists of a carbon atom joined to an oxygen atom by a double bond.

NAME OF COMPOUND

Ketones if the carbonyl group is within a carbon skeleton

Aldehydes if the carbonyl group is at the end of the carbon skeleton

EXAMPLE

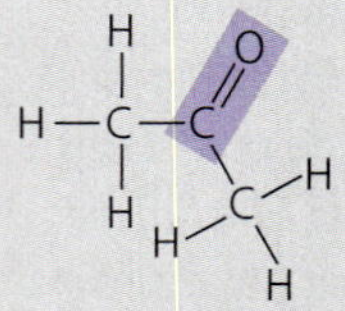

Acetone, the simplest ketone

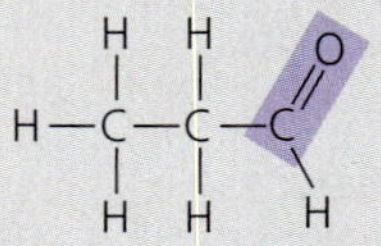

Propanal, an aldehyde

FUNCTIONAL PROPERTIES

- A ketone and an aldehyde may be structural isomers with different properties, as is the case for acetone and propanal.
- These two groups are also found in sugars, giving rise to two major groups of sugars: aldoses (containing an aldehyde) and ketoses (containing a ketone).

Carboxyl

CHEMICAL GROUP: Carboxyl

STRUCTURE

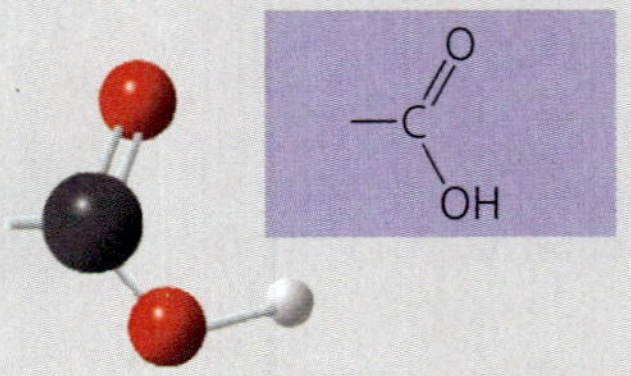

When an oxygen atom is double-bonded to a carbon atom that is also bonded to an —OH group, the entire assembly of atoms is called a **carboxyl group** (—COOH).

NAME OF COMPOUND

Carboxylic acids, or organic acids

EXAMPLE

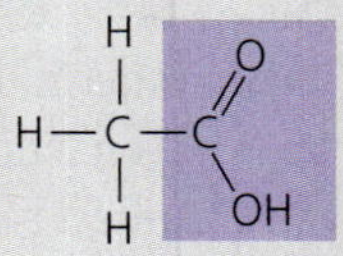

Acetic acid, which gives vinegar its sour taste

FUNCTIONAL PROPERTIES

- Has acidic properties (is a source of hydrogen ions) because the covalent bond between oxygen and hydrogen is so polar; for example,

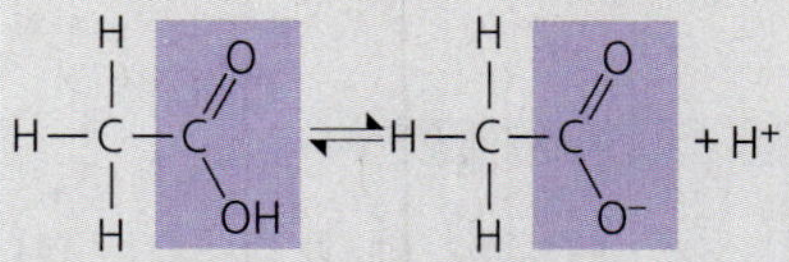

Acetic acid | Acetate ion

- Found in cells in the ionized form with a charge of 1− and called a carboxylate ion (here, specifically, the acetate ion).

Amino

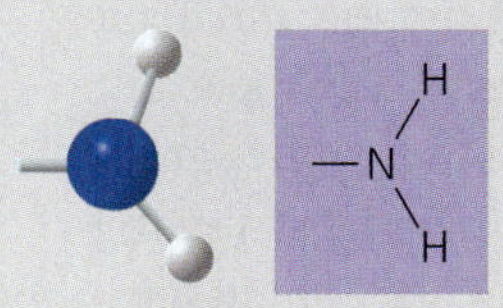

The **amino group** (—NH_2) consists of a nitrogen atom bonded to two hydrogen atoms and to the carbon skeleton.

Amines

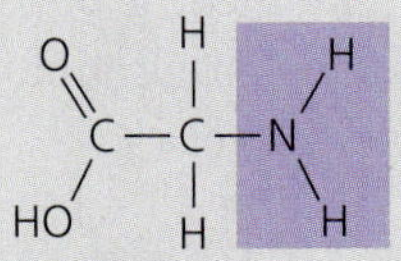

Glycine

Because it also has a carboxyl group, glycine is both an amine and a carboxylic acid; compounds with both groups are called **amino acids**.

- Acts as a base; can pick up an H^+ from the surrounding solution (water, in living organisms).

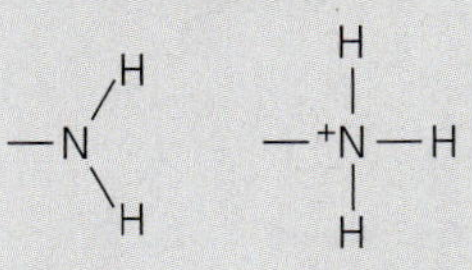

(nonionized) (ionized)

- Ionized, with a charge of 1+, under cellular conditions.

Sulfhydryl

(may be written HS—)

The **sulfhydryl group** consists of a sulfur atom bonded to an atom of hydrogen; resembles a hydroxyl group in shape.

Thiols

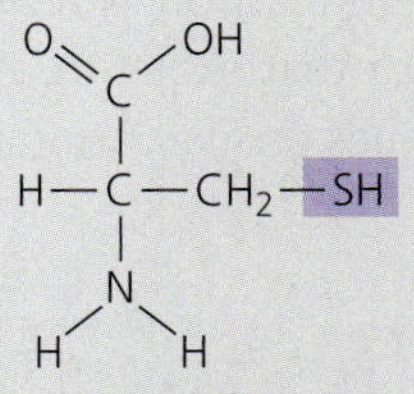

Cysteine

Cysteine is an important sulfur-containing amino acid.

- Two sulfhydryl groups can react, forming a covalent bond. This "cross-linking" helps stabilize protein structure (see Figure 5.21).
- Cross-linking of cysteines in hair proteins maintains the curliness or straightness of hair. Straight hair can be "permanently" curled by shaping it around curlers, then breaking and re-forming the cross-linking bonds.

Phosphate

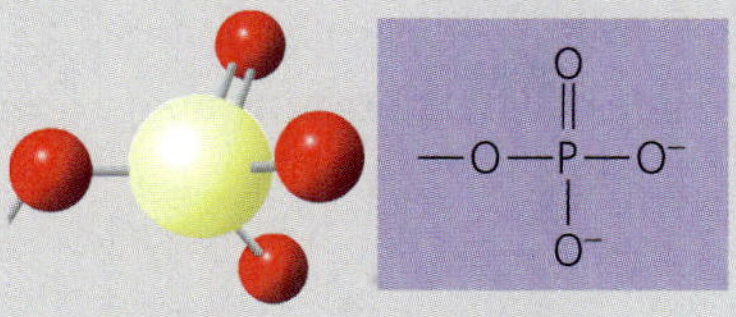

In a **phosphate group**, a phosphorus atom is bonded to four oxygen atoms; one oxygen is bonded to the carbon skeleton; two oxygens carry negative charges. The phosphate group (—OPO_3^{2-}, abbreviated Ⓟ) is an ionized form of a phosphoric acid group (—OPO_3H_2; note the two hydrogens).

Organic phosphates

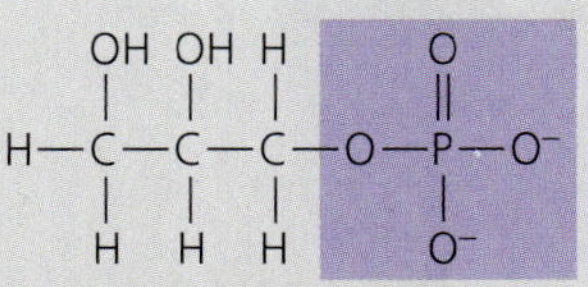

Glycerol phosphate

In addition to taking part in many important chemical reactions in cells, glycerol phosphate provides the backbone for phospholipids, the most prevalent molecules in cell membranes.

- Contributes negative charge to the molecule of which it is a part (2− when at the end of a molecule, as above; 1− when located internally in a chain of phosphates).
- Has the potential to react with water, releasing energy.

Methyl

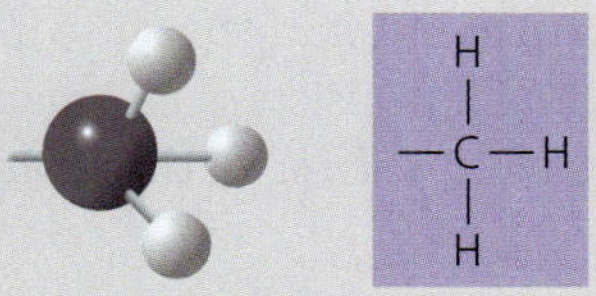

A **methyl group** consists of a carbon bonded to three hydrogen atoms. The methyl group may be attached to a carbon or to a different atom.

Methylated compounds

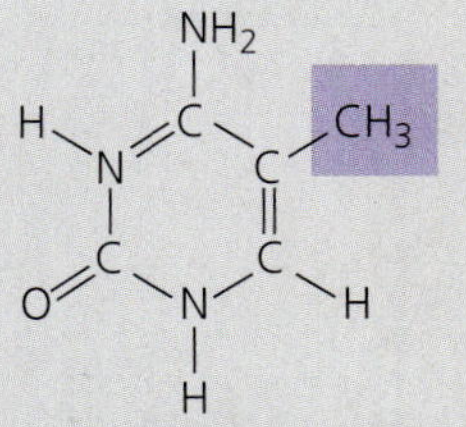

5-Methyl cytidine

5-Methyl cytidine is a component of DNA that has been modified by addition of the methyl group.

- Addition of a methyl group to DNA, or to molecules bound to DNA, affects expression of genes.
- Arrangement of methyl groups in male and female sex hormones affects their shape and function (see Figure 4.9).

? *Given the information in this figure and what you know about the electronegativity of oxygen, predict which of the following molecules would be the stronger acid. Explain your answer.*

a. 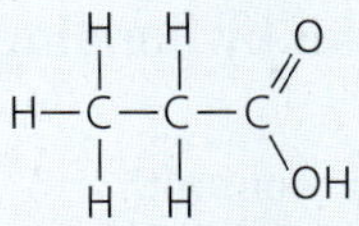

b. 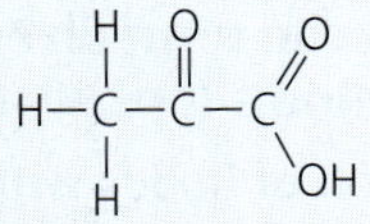

ATP: An Important Source of Energy for Cellular Processes

The "Phosphate" column in Figure 4.10 shows a simple example of an organic phosphate molecule. A more complicated organic phosphate, **adenosine triphosphate**, or **ATP**, is worth mentioning because its function in the cell is so important. ATP consists of an organic molecule called adenosine attached to a string of three phosphate groups:

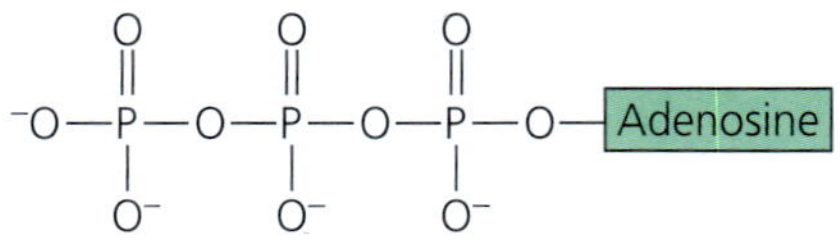

Where three phosphates are present in series, as in ATP, one phosphate may be split off as a result of a reaction with water. This inorganic phosphate ion, $HOPO_3^{2-}$, is often abbreviated Ⓟ$_i$ in this book. Having lost one phosphate, ATP becomes adenosine *di*phosphate, or ADP. Although ATP is sometimes said to "store" energy, it is more accurate to think of it as "storing" the potential to react with water. This reaction releases energy that can be used by the cell. You will learn about this in more detail in Chapter 8.

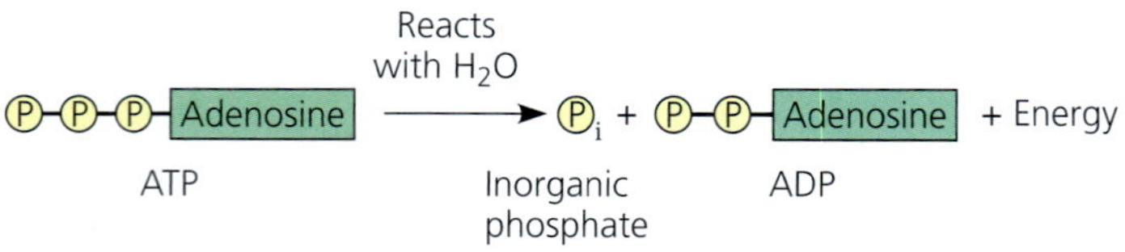

CONCEPT CHECK 4.3

1. What does the term *amino acid* signify about the structure of such a molecule?
2. What chemical change occurs when ATP reacts with water and releases energy?
3. **WHAT IF?** Suppose you had an organic molecule such as glycine (see Figure 4.10, amino group example), and you chemically removed the —NH_2 group and replaced it with —COOH. Draw the structural formula for this molecule and speculate about its chemical properties.

For suggested answers, see Appendix A.

The Chemical Elements of Life: *A Review*

Living matter, as you have learned, consists mainly of carbon, oxygen, hydrogen, and nitrogen, with smaller amounts of sulfur and phosphorus. These elements all form strong covalent bonds, an essential characteristic in the architecture of complex organic molecules. Of all these elements, carbon is the virtuoso of the covalent bond. The versatility of carbon makes possible the great diversity of organic molecules, each with particular properties that emerge from the unique arrangement of its carbon skeleton and the chemical groups appended to that skeleton. At the foundation of all biological diversity lies this variation at the molecular level.

Chapter 4 Review

MB **MEDIA** Go to the Study Area at **www.masteringbio.com** for BioFlix 3-D Animations, MP3 Tutors, Videos, Practice Tests, an eBook, and more.

SUMMARY OF KEY CONCEPTS

CONCEPT 4.1

Organic chemistry is the study of carbon compounds (pp. 58–59)

- Organic compounds were once thought to arise only within living organisms, but this idea (vitalism) was disproved when chemists were able to synthesize organic compounds in the laboratory.

CONCEPT 4.2

Carbon atoms can form diverse molecules by bonding to four other atoms (pp. 60–63)

- **The Formation of Bonds with Carbon** Carbon, with a valence of 4, can bond to various other atoms, including O, H, and N. Carbon can also bond to other carbon atoms, forming the carbon skeletons of organic compounds.
- **Molecular Diversity Arising from Carbon Skeleton Variation** The carbon skeletons of organic molecules vary in length and shape and have bonding sites for atoms of other elements. Hydrocarbons consist only of carbon and hydrogen. Isomers are compounds with the same molecular formula but different structures and properties. Three types of isomers are structural isomers, geometric isomers, and enantiomers.

MEDIA

Activity Diversity of Carbon-Based Molecules
Activity Isomers
Investigation What Factors Determine the Effectiveness of Drugs?

CONCEPT 4.3

A small number of chemical groups are key to the functioning of biological molecules (pp. 63–66)

- **The Chemical Groups Most Important in the Processes of Life** Chemical groups attached to the carbon skeletons of organic molecules participate in chemical reactions (functional groups) or contribute to function by affecting molecular shape.

▶ ATP: An Important Source of Energy for Cellular Processes

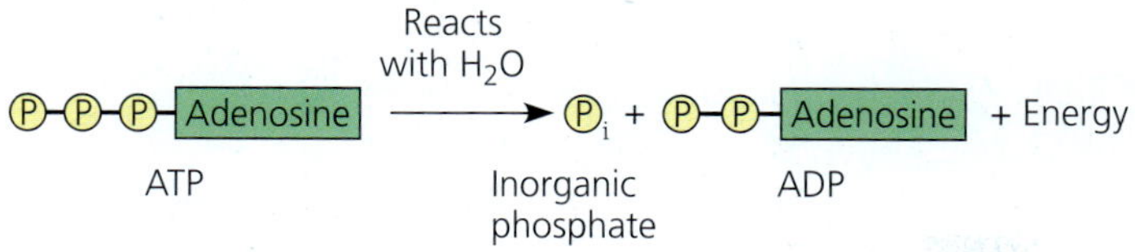

MEDIA

Activity Functional Groups

▶ The Chemical Elements of Life: *A Review* Living matter is made mostly of carbon, oxygen, hydrogen, and nitrogen, with some sulfur and phosphorus. Biological diversity has its molecular basis in carbon's ability to form a huge number of molecules with particular shapes and chemical properties.

TESTING YOUR KNOWLEDGE

SELF-QUIZ

1. Organic chemistry is currently defined as
   a. the study of compounds made only by living cells.
   b. the study of carbon compounds.
   c. the study of vital forces.
   d. the study of natural (as opposed to synthetic) compounds.
   e. the study of hydrocarbons.

2. Which of the following hydrocarbons has a double bond in its carbon skeleton?
   a. C_3H_8
   b. C_2H_6
   c. CH_4
   d. C_2H_4
   e. C_2H_2

3. Choose the term that correctly describes the relationship between these two sugar molecules:

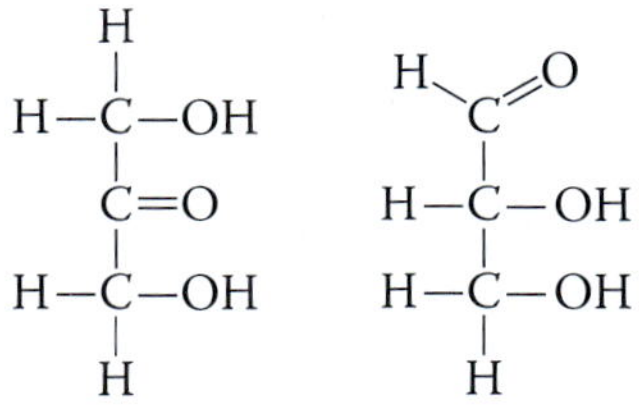

   a. structural isomers
   b. geometric isomers
   c. enantiomers
   d. isotopes

4. Identify the asymmetric carbon in this molecule:

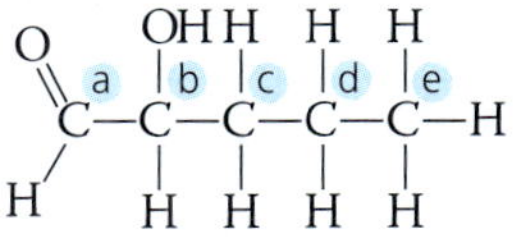

5. Which functional group is *not* present in this molecule?

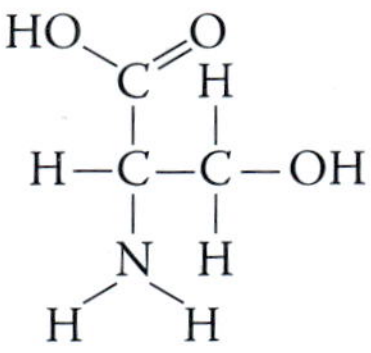

   a. carboxyl   b. sulfhydryl   c. hydroxyl   d. amino

6. Which action could produce a carbonyl group?
   a. the replacement of the —OH of a carboxyl group with hydrogen
   b. the addition of a thiol to a hydroxyl
   c. the addition of a hydroxyl to a phosphate
   d. the replacement of the nitrogen of an amine with oxygen
   e. the addition of a sulfhydryl to a carboxyl

7. Which chemical group is most likely to be responsible for an organic molecule behaving as a base?
   a. hydroxyl
   b. carbonyl
   c. carboxyl
   d. amino
   e. phosphate

For Self-Quiz answers, see Appendix A.

MEDIA Visit the Study Area at **www.masteringbio.com** for a Practice Test.

EVOLUTION CONNECTION

8. **DRAW IT** Some scientists believe that life elsewhere in the universe might be based on the element silicon, rather than on carbon, as on Earth. Look at the electron distribution diagram for silicon in Figure 2.9 and draw the Lewis dot structure for silicon. What properties does silicon share with carbon that would make silicon-based life more likely than, say, neon-based life or aluminum-based life?

SCIENTIFIC INQUIRY

9. In 1918, an epidemic of sleeping sickness caused an unusual rigid paralysis in some survivors, similar to symptoms of advanced Parkinson's disease. Years later, L-dopa (below, left), a chemical used to treat Parkinson's disease, was given to some of these patients, as dramatized in the movie *Awakenings*. L-dopa was remarkably effective at eliminating the paralysis, at least temporarily. However, its enantiomer, D-dopa (right), was subsequently shown to have no effect at all, as is the case for Parkinson's disease. Suggest a hypothesis to explain why, for both diseases, one enantiomer is effective and the other is not.

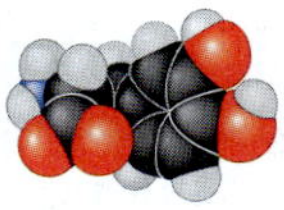

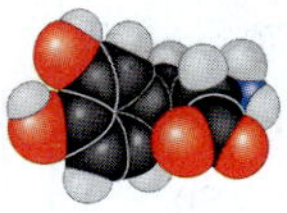

L-dopa **D-dopa**

SCIENCE, TECHNOLOGY, AND SOCIETY

10. Thalidomide achieved notoriety 50 years ago because of a wave of birth defects among children born to women who took thalidomide during pregnancy as a treatment for morning sickness. However, in 1998 the U.S. Food and Drug Administration (FDA) approved this drug for the treatment of certain conditions associated with Hansen's disease (leprosy). In clinical trials, thalidomide also shows promise for use in treating patients suffering from AIDS, tuberculosis, and some types of cancer. Do you think approval of this drug is appropriate? If so, under what conditions? What criteria do you think the FDA should use in weighing a drug's benefits against its dangers?

The Structure and Function of Large Biological Molecules

5

▲ **Figure 5.1 Why do scientists study the structures of macromolecules?**

KEY CONCEPTS

- **5.1** Macromolecules are polymers, built from monomers
- **5.2** Carbohydrates serve as fuel and building material
- **5.3** Lipids are a diverse group of hydrophobic molecules
- **5.4** Proteins have many structures, resulting in a wide range of functions
- **5.5** Nucleic acids store and transmit hereditary information

OVERVIEW

The Molecules of Life

Given the rich complexity of life on Earth, we might expect organisms to have an enormous diversity of molecules. Remarkably, however, the critically important large molecules of all living things—from bacteria to elephants—fall into just four main classes: carbohydrates, lipids, proteins, and nucleic acids. On the molecular scale, members of three of these classes—carbohydrates, proteins, and nucleic acids—are huge and are thus called **macromolecules**. For example, a protein may consist of thousands of atoms that form a molecular colossus with a mass well over 100,000 daltons. Considering the size and complexity of macromolecules, it is noteworthy that biochemists have determined the detailed structures of so many of them **(Figure 5.1)**.

The architecture of a large biological molecule helps explain how that molecule works. Like water and simple organic molecules, large biological molecules exhibit unique emergent properties arising from the orderly arrangement of their atoms. In this chapter, we'll first consider how macromolecules are built. Then we'll examine the structure and function of all four classes of large biological molecules: carbohydrates, lipids, proteins, and nucleic acids.

CONCEPT **5.1**

Macromolecules are polymers, built from monomers

The macromolecules in three of the four classes of life's organic compounds—carbohydrates, proteins, and nucleic acids—are chain-like molecules called polymers (from the Greek *polys*, many, and *meris*, part). A **polymer** is a long molecule consisting of many similar or identical building blocks linked by covalent bonds, much as a train consists of a chain of cars. The repeating units that serve as the building blocks of a polymer are smaller molecules called **monomers**. Some of the molecules that serve as monomers also have other functions of their own.

The Synthesis and Breakdown of Polymers

The classes of polymers differ in the nature of their monomers, but the chemical mechanisms by which cells make and break down polymers are basically the same in all cases **(Figure 5.2)**. Monomers are connected by a reaction in which two molecules are covalently bonded to each other through loss of a water molecule; this is known as a **condensation reaction**, specifically a **dehydration reaction**, because water is the molecule that is lost **(Figure 5.2a)**. When a bond forms between two monomers, each monomer contributes part of the water molecule that is lost: One molecule provides a hydroxyl group (—OH), while the other provides a hydrogen (—H). This reaction can be repeated as monomers are added to the chain one by one, making a polymer. The dehydration process is facilitated by **enzymes**, specialized macromolecules that speed up chemical reactions in cells.

Polymers are disassembled to monomers by **hydrolysis**, a process that is essentially the reverse of the dehydration

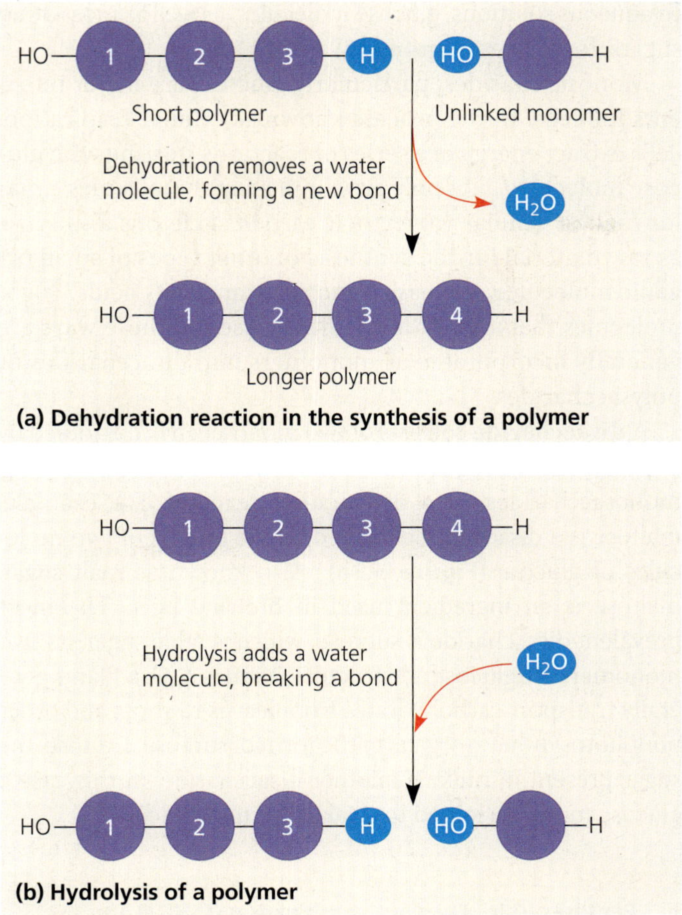

▲ **Figure 5.2 The synthesis and breakdown of polymers.**

reaction **(Figure 5.2b)**. Hydrolysis means to break using water (from the Greek *hydro*, water, and *lysis*, break). Bonds between the monomers are broken by the addition of water molecules, with a hydrogen from the water attaching to one monomer and a hydroxyl group attaching to the adjacent monomer. An example of hydrolysis working within our bodies is the process of digestion. The bulk of the organic material in our food is in the form of polymers that are much too large to enter our cells. Within the digestive tract, various enzymes attack the polymers, speeding up hydrolysis. The released monomers are then absorbed into the bloodstream for distribution to all body cells. Those cells can then use dehydration reactions to assemble the monomers into new, different polymers that can perform specific functions required by the cell.

The Diversity of Polymers

Each cell has thousands of different kinds of macromolecules; the collection varies from one type of cell to another even in the same organism. The inherent differences between human siblings reflect variations in polymers, particularly DNA and proteins. Molecular differences between unrelated individuals are more extensive and those between species greater still. The diversity of macromolecules in the living world is vast, and the possible variety is effectively limitless.

What is the basis for such diversity in life's polymers? These molecules are constructed from only 40 to 50 common monomers and some others that occur rarely. Building a huge variety of polymers from such a limited number of monomers is analogous to constructing hundreds of thousands of words from only 26 letters of the alphabet. The key is arrangement—the particular linear sequence that the units follow. However, this analogy falls far short of describing the great diversity of macromolecules because most biological polymers have many more monomers than the number of letters in the longest word. Proteins, for example, are built from 20 kinds of amino acids arranged in chains that are typically hundreds of amino acids long. The molecular logic of life is simple but elegant: Small molecules common to all organisms are ordered into unique macromolecules.

Despite this immense diversity, molecular structure and function can still be grouped roughly by class. Let's look at each of the four major classes of large biological molecules. For each class, the large molecules have emergent properties not found in their individual building blocks.

CONCEPT CHECK 5.1

1. What are the four main classes of large biological molecules?
2. How many molecules of water are needed to completely hydrolyze a polymer that is ten monomers long?
3. **WHAT IF?** Suppose you eat a serving of green beans. What reactions must occur for the amino acid monomers in the protein of the beans to be converted to proteins in your body?

For suggested answers, see Appendix A.

CONCEPT **5.2**

Carbohydrates serve as fuel and building material

Carbohydrates include both sugars and polymers of sugars. The simplest carbohydrates are the monosaccharides, also known as simple sugars. Disaccharides are double sugars, consisting of two monosaccharides joined by a covalent bond. Carbohydrates also include macromolecules called polysaccharides, polymers composed of many sugar building blocks.

Sugars

Monosaccharides (from the Greek *monos*, single, and *sacchar*, sugar) generally have molecular formulas that are some multiple of the unit CH_2O **(Figure 5.3)**. Glucose ($C_6H_{12}O_6$), the most common monosaccharide, is of central importance in the chemistry of life. In the structure of glucose, we can see the trademarks of a sugar: The molecule has a carbonyl group (>C=O) and multiple hydroxyl groups (—OH). Depending on the location of the carbonyl group, a sugar is either an aldose (aldehyde sugar) or a ketose (ketone sugar). Glucose, for example, is an aldose; fructose, a structural isomer of glucose, is a ketose. (Most names for sugars end in *-ose*.) Another criterion for classifying sugars is the size of the carbon skeleton, which ranges from three to seven carbons long. Glucose, fructose, and other sugars that have six carbons are called hexoses. Trioses (three-carbon sugars) and pentoses (five-carbon sugars) are also common.

Still another source of diversity for simple sugars is in the spatial arrangement of their parts around asymmetric carbons. (Recall that an asymmetric carbon is a carbon attached to four different atoms or groups of atoms.) Glucose and galactose, for example, differ only in the placement of parts around one asymmetric carbon (see the purple boxes in Figure 5.3). What seems like a small difference is significant enough to give the two sugars distinctive shapes and behaviors.

Although it is convenient to draw glucose with a linear carbon skeleton, this representation is not completely accurate. In aqueous solutions, glucose molecules, as well as most other sugars, form rings **(Figure 5.4)**.

Monosaccharides, particularly glucose, are major nutrients for cells. In the process known as cellular respiration, cells extract energy in a series of reactions starting with glucose molecules. Not only are simple-sugar molecules a major fuel for cellular work; their carbon skeletons also serve as raw material for the synthesis of other types of small organic molecules, such as amino acids and fatty acids. Sugar molecules that are not immediately used in these ways are generally incorporated as monomers into disaccharides or polysaccharides.

A **disaccharide** consists of two monosaccharides joined by a **glycosidic linkage**, a covalent bond formed between two monosaccharides by a dehydration reaction. For example, maltose is a disaccharide formed by the linking of two molecules of glucose **(Figure 5.5a)**. Also known as malt sugar, maltose is an ingredient used in brewing beer. The most prevalent disaccharide is sucrose, which is table sugar. Its two monomers are glucose and fructose **(Figure 5.5b)**. Plants generally transport carbohydrates from leaves to roots and other nonphotosynthetic organs in the form of sucrose. Lactose, the sugar present in milk, is another disaccharide, in this case a glucose molecule joined to a galactose molecule.

▶ **Figure 5.3 The structure and classification of some monosaccharides.** Sugars may be aldoses (aldehyde sugars, top row) or ketoses (ketone sugars, bottom row), depending on the location of the carbonyl group (dark orange). Sugars are also classified according to the length of their carbon skeletons. A third point of variation is the spatial arrangement around asymmetric carbons (compare, for example, the purple portions of glucose and galactose).

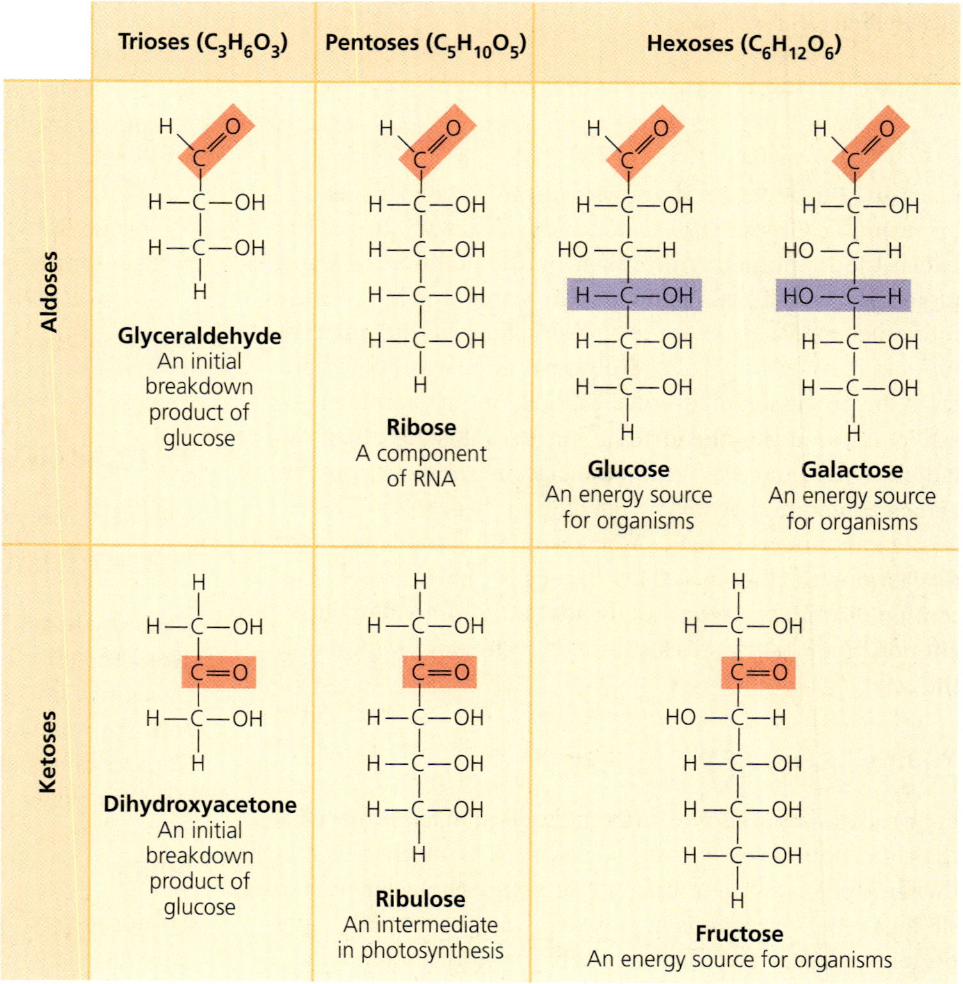

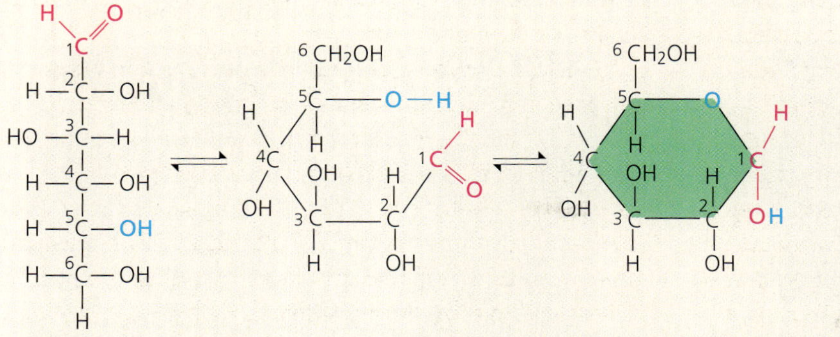

(a) Linear and ring forms. Chemical equilibrium between the linear and ring structures greatly favors the formation of rings. The carbons of the sugar are numbered 1 to 6, as shown. To form the glucose ring, carbon 1 bonds to the oxygen attached to carbon 5.

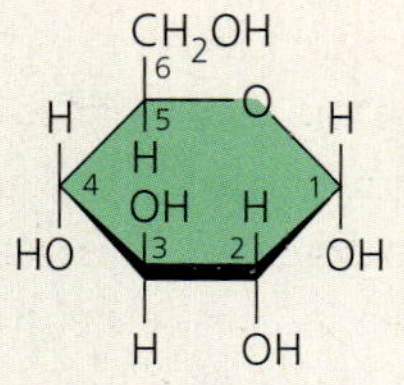

(b) Abbreviated ring structure. Each corner represents a carbon. The ring's thicker edge indicates that you are looking at the ring edge-on; the components attached to the ring lie above or below the plane of the ring.

▲ **Figure 5.4 Linear and ring forms of glucose.**

DRAW IT *Start with the linear form of fructose (see Figure 5.3) and draw the formation of the fructose ring in two steps. Number the carbons. Attach carbon 5 via oxygen to carbon 2. Compare the number of carbons in the fructose and glucose rings.*

(a) Dehydration reaction in the synthesis of maltose. The bonding of two glucose units forms maltose. The glycosidic linkage joins the number 1 carbon of one glucose to the number 4 carbon of the second glucose. Joining the glucose monomers in a different way would result in a different disaccharide.

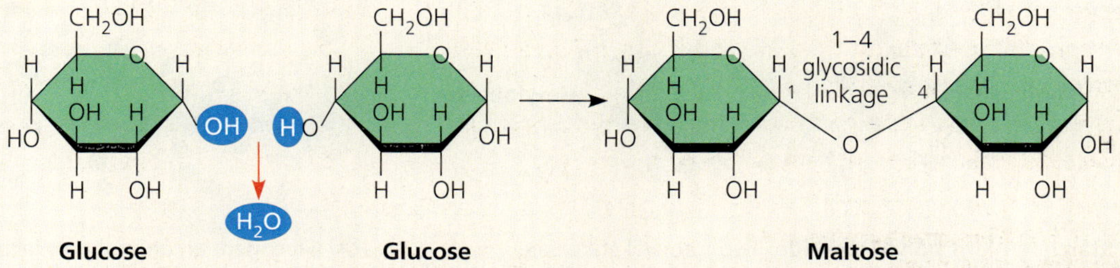

(b) Dehydration reaction in the synthesis of sucrose. Sucrose is a disaccharide formed from glucose and fructose. Notice that fructose, though a hexose like glucose, forms a five-sided ring.

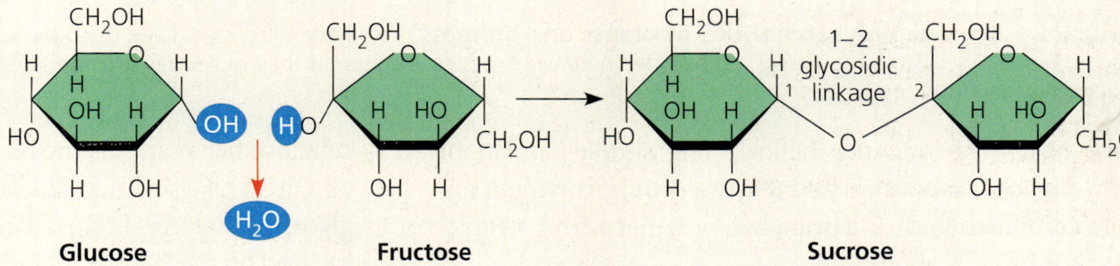

▲ **Figure 5.5 Examples of disaccharide synthesis.**

Polysaccharides

Polysaccharides are macromolecules, polymers with a few hundred to a few thousand monosaccharides joined by glycosidic linkages. Some polysaccharides serve as storage material, hydrolyzed as needed to provide sugar for cells. Other polysaccharides serve as building material for structures that protect the cell or the whole organism. The architecture and function of a polysaccharide are determined by its sugar monomers and by the positions of its glycosidic linkages.

Storage Polysaccharides

Both plants and animals store sugars for later use in the form of storage polysaccharides. Plants store **starch**, a polymer of glucose monomers, as granules within cellular structures known as plastids, which include chloroplasts. Synthesizing starch enables the plant to stockpile surplus glucose. Because glucose is a major cellular fuel, starch represents stored energy. The sugar can later be withdrawn from this carbohydrate "bank" by hydrolysis, which breaks the bonds between the glucose monomers. Most animals, including humans, also have enzymes that can hydrolyze plant starch, making glucose available as a nutrient for cells. Potato tubers and grains—the fruits of wheat, maize (corn), rice, and other grasses—are the major sources of starch in the human diet.

Most of the glucose monomers in starch are joined by 1–4 linkages (number 1 carbon to number 4 carbon), like the glucose units in maltose (see Figure 5.5a). The angle of these

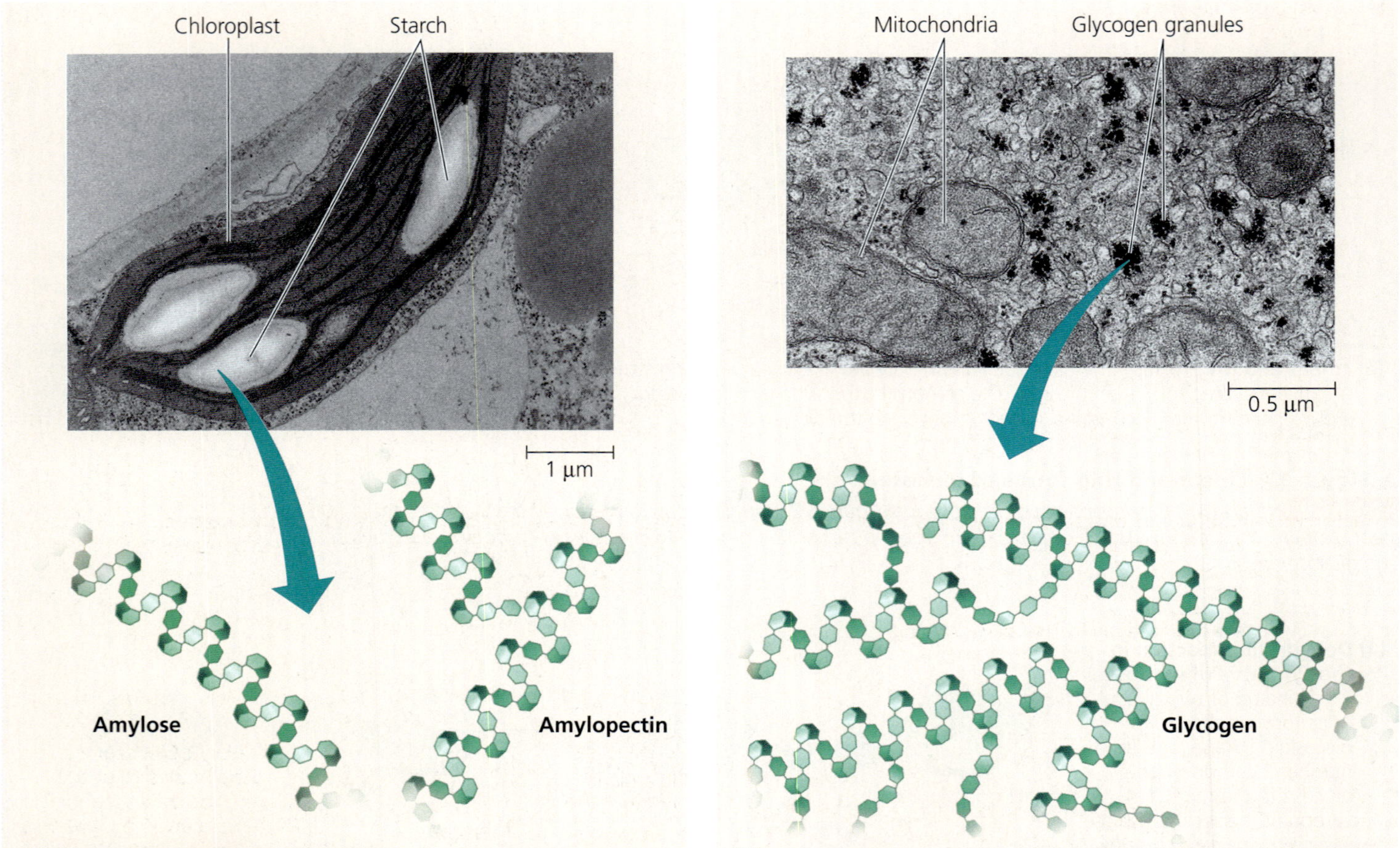

(a) Starch: a plant polysaccharide. Two forms of starch are amylose (unbranched) and amylopectin (branched). The light ovals in the micrograph are granules of starch within a chloroplast of a plant cell.

(b) Glycogen: an animal polysaccharide. Glycogen is more branched than amylopectin. Animal cells stockpile glycogen as dense clusters of granules within liver and muscle cells. (The micrograph shows part of a liver cell; mitochondria are organelles that help break down sugars.)

▲ **Figure 5.6 Storage polysaccharides of plants and animals.** These examples, starch and glycogen, are composed entirely of glucose monomers, represented here by hexagons. Because of their molecular structure, the polymer chains tend to form helices.

bonds makes the polymer helical. The simplest form of starch, amylose, is unbranched **(Figure 5.6a)**. Amylopectin, a more complex starch, is a branched polymer with 1–6 linkages at the branch points.

Animals store a polysaccharide called **glycogen**, a polymer of glucose that is like amylopectin but more extensively branched **(Figure 5.6b)**. Humans and other vertebrates store glycogen mainly in liver and muscle cells. Hydrolysis of glycogen in these cells releases glucose when the demand for sugar increases. This stored fuel cannot sustain an animal for long, however. In humans, for example, glycogen stores are depleted in about a day unless they are replenished by consumption of food. This is an issue of concern in low-carbohydrate diets.

Structural Polysaccharides

Organisms build strong materials from structural polysaccharides. For example, the polysaccharide called **cellulose** is a major component of the tough walls that enclose plant cells. On a global scale, plants produce almost 10^{14} kg (100 billion tons) of cellulose per year; it is the most abundant organic compound on Earth. Like starch, cellulose is a polymer of glucose, but the glycosidic linkages in these two polymers differ. The difference is based on the fact that there are actually two slightly different ring structures for glucose **(Figure 5.7a)**. When glucose forms a ring, the hydroxyl group attached to the number 1 carbon is positioned either below or above the plane of the ring. These two ring forms for glucose are called alpha (α) and beta (β), respectively. In starch, all the glucose monomers are in the α configuration **(Figure 5.7b)**, the arrangement we saw in Figures 5.4 and 5.5. In contrast, the glucose monomers of cellulose are all in the β configuration, making every other glucose monomer upside down with respect to its neighbors **(Figure 5.7c)**.

The differing glycosidic linkages in starch and cellulose give the two molecules distinct three-dimensional shapes. Whereas a starch molecule is mostly helical, a cellulose molecule is straight. Cellulose is never branched, and some hydroxyl groups on its glucose monomers are free to hydrogen-bond with the hydroxyls of other cellulose molecules lying parallel to it. In plant cell walls, parallel cellulose molecules held together in this way are grouped into units called microfibrils **(Figure 5.8)**. These cable-like

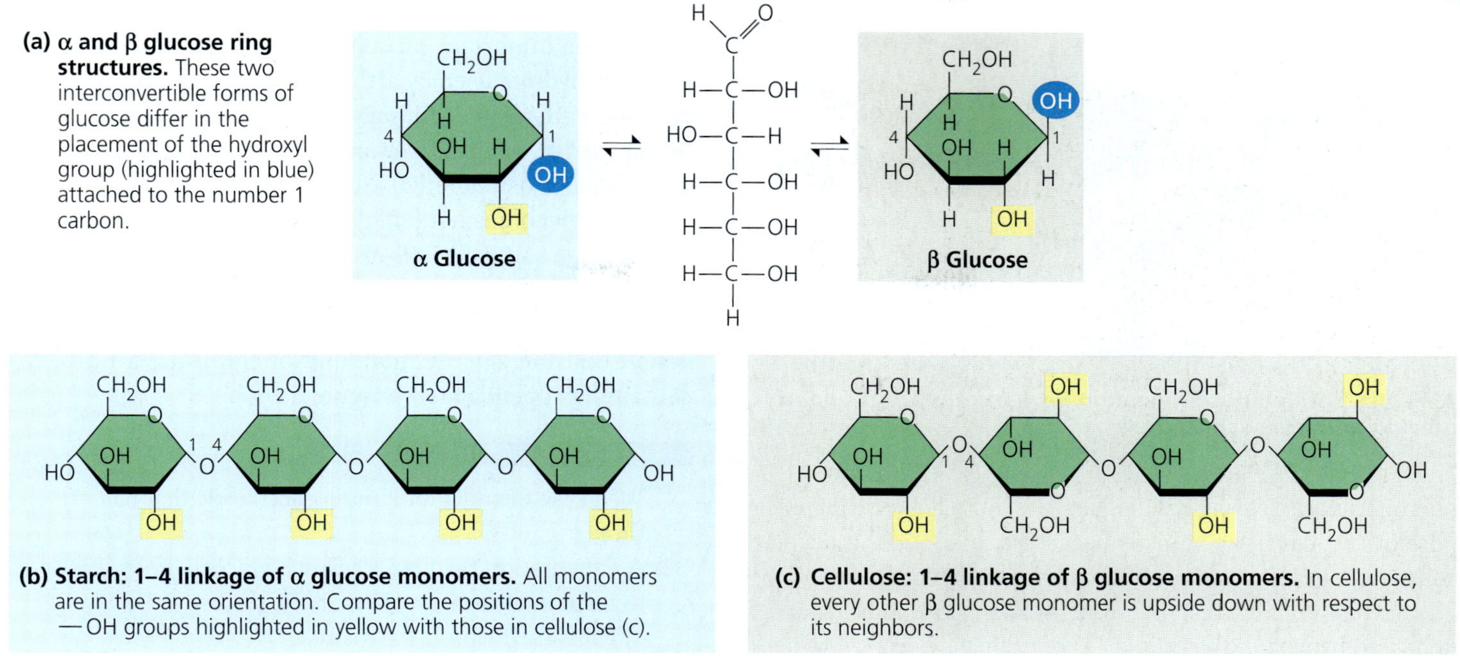

(a) α and β glucose ring structures. These two interconvertible forms of glucose differ in the placement of the hydroxyl group (highlighted in blue) attached to the number 1 carbon.

(b) Starch: 1–4 linkage of α glucose monomers. All monomers are in the same orientation. Compare the positions of the —OH groups highlighted in yellow with those in cellulose (c).

(c) Cellulose: 1–4 linkage of β glucose monomers. In cellulose, every other β glucose monomer is upside down with respect to its neighbors.

▲ **Figure 5.7 Starch and cellulose structures.**

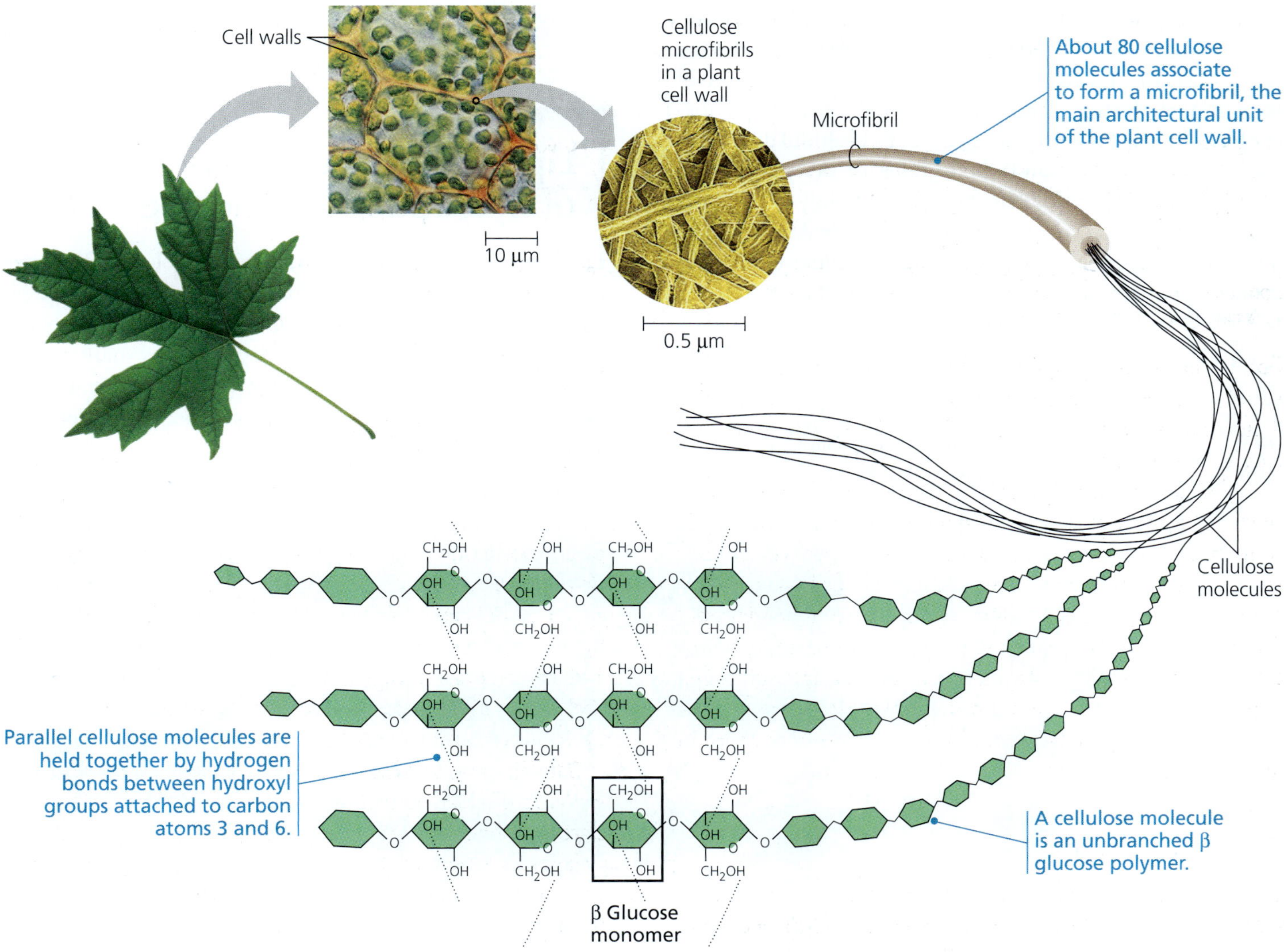

▲ **Figure 5.8 The arrangement of cellulose in plant cell walls.**

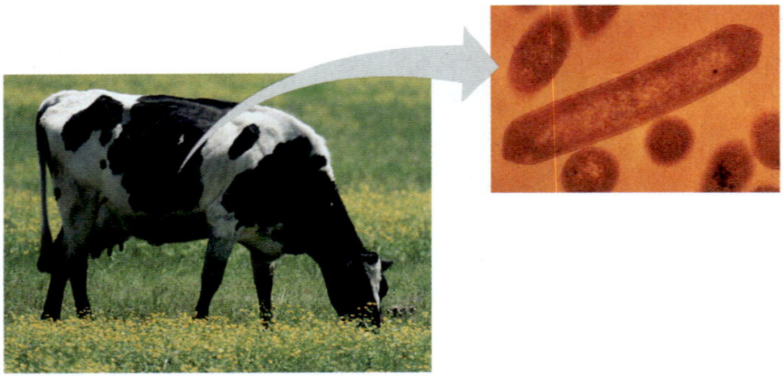

▲ **Figure 5.9 Cellulose-digesting prokaryotes are found in grazing animals such as this cow.**

microfibrils are a strong building material for plants and an important substance for humans because cellulose is the major constituent of paper and the only component of cotton.

Enzymes that digest starch by hydrolyzing its α linkages are unable to hydrolyze the β linkages of cellulose because of the distinctly different shapes of these two molecules. In fact, few organisms possess enzymes that can digest cellulose. Humans do not; the cellulose in our food passes through the digestive tract and is eliminated with the feces. Along the way, the cellulose abrades the wall of the digestive tract and stimulates the lining to secrete mucus, which aids in the smooth passage of food through the tract. Thus, although cellulose is not a nutrient for humans, it is an important part of a healthful diet. Most fresh fruits, vegetables, and whole grains are rich in cellulose. On food packages, "insoluble fiber" refers mainly to cellulose.

Some prokaryotes can digest cellulose, breaking it down into glucose monomers. A cow harbors cellulose-digesting prokaryotes in its rumen, the first compartment in its stomach (**Figure 5.9**). The prokaryotes hydrolyze the cellulose of hay and grass and convert the glucose to other nutrients that nourish the cow. Similarly, a termite, which is unable to digest cellulose by itself, has prokaryotes living in its gut that can make a meal of wood. Some fungi can also digest cellulose, thereby helping recycle chemical elements within Earth's ecosystems.

Another important structural polysaccharide is **chitin**, the carbohydrate used by arthropods (insects, spiders, crustaceans, and related animals) to build their exoskeletons (**Figure 5.10**). An exoskeleton is a hard case that surrounds the soft parts of an animal. Pure chitin is leathery and flexible, but it becomes hardened when encrusted with calcium carbonate, a salt. Chitin is also found in many fungi, which use this polysaccharide rather than cellulose as the building material for their cell walls. Chitin is similar to cellulose, except that the glucose monomer of chitin has a nitrogen-containing appendage (see Figure 5.10a).

CONCEPT CHECK 5.2

1. Write the formula for a monosaccharide that has three carbons.
2. A dehydration reaction joins two glucose molecules to form maltose. The formula for glucose is $C_6H_{12}O_6$. What is the formula for maltose?
3. **WHAT IF?** What would happen if a cow were given antibiotics that killed all the prokaryotes in its stomach?

For suggested answers, see Appendix A.

CONCEPT **5.3**

Lipids are a diverse group of hydrophobic molecules

Lipids are the one class of large biological molecules that does not include true polymers, and they are generally not big enough to be considered macromolecules. The compounds called **lipids** are grouped together because they share one important trait: They mix poorly, if at all, with water. The hydrophobic behavior of lipids is based on their molecular structure. Although they may have some polar bonds associated with oxygen, lipids consist mostly of hydrocarbon regions.

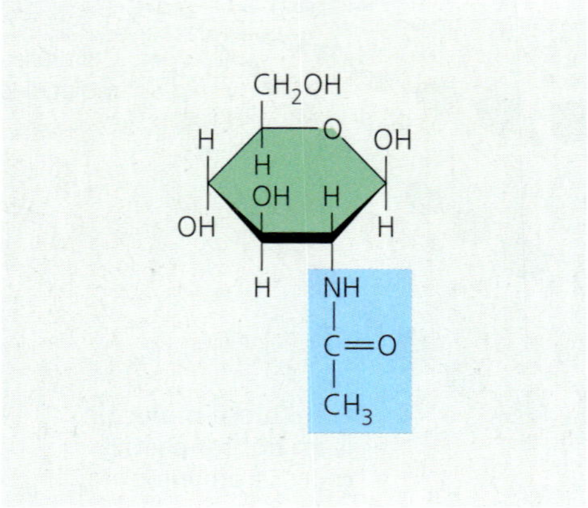

(a) The structure of the chitin monomer.

(b) Chitin forms the exoskeleton of arthropods. This cicada is molting, shedding its old exoskeleton and emerging in adult form.

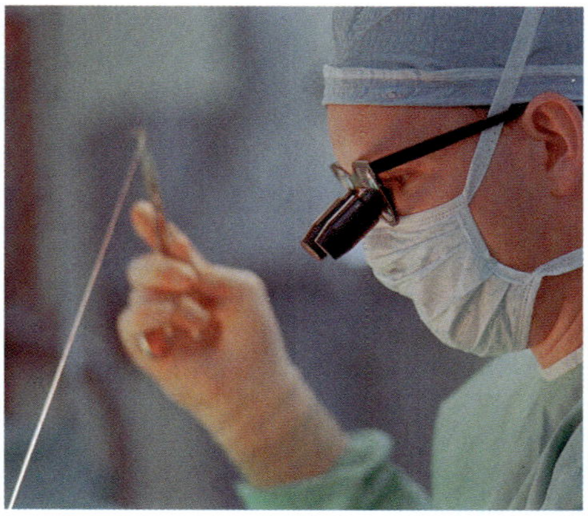

(c) Chitin is used to make a strong and flexible surgical thread that decomposes after the wound or incision heals.

▲ **Figure 5.10 Chitin, a structural polysaccharide.**

Lipids are varied in form and function. They include waxes and certain pigments, but we will focus on the most biologically important types of lipids: fats, phospholipids, and steroids.

Fats

Although fats are not polymers, they are large molecules assembled from a few smaller molecules by dehydration reactions. A **fat** is constructed from two kinds of smaller molecules: glycerol and fatty acids **(Figure 5.11a)**. Glycerol is an alcohol with three carbons, each bearing a hydroxyl group. A **fatty acid** has a long carbon skeleton, usually 16 or 18 carbon atoms in length. The carbon at one end of the fatty acid is part of a carboxyl group, the functional group that gives these molecules the name fatty *acid.* Attached to the carboxyl group is a long hydrocarbon chain. The relatively nonpolar C—H bonds in the hydrocarbon chains of fatty acids are the reason fats are hydrophobic. Fats separate from water because the water molecules hydrogen-bond to one another and exclude the fats. This is the reason that vegetable oil (a liquid fat) separates from the aqueous vinegar solution in a bottle of salad dressing.

In making a fat, three fatty acid molecules each join to glycerol by an ester linkage, a bond between a hydroxyl group and a carboxyl group. The resulting fat, also called a **triacylglycerol**, thus consists of three fatty acids linked to one glycerol molecule. (Still another name for a fat is *triglyceride*, a word often found in the list of ingredients on packaged foods.) The fatty acids in a fat can be the same, as in **Figure 5.11b**, or they can be of two or three different kinds.

Fatty acids vary in length and in the number and locations of double bonds. The terms *saturated fats* and *unsaturated fats* are commonly used in the context of nutrition **(Figure 5.12)**.

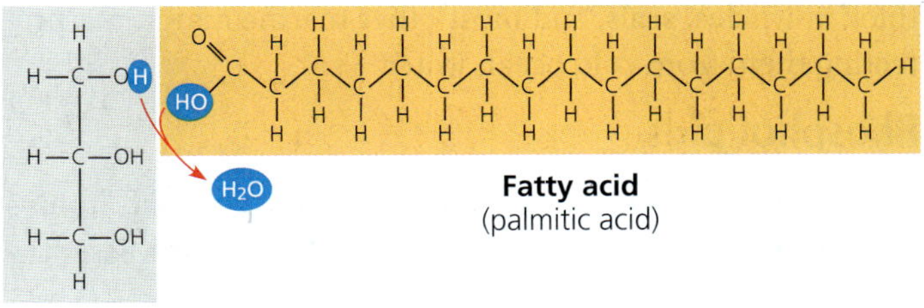

(a) Dehydration reaction in the synthesis of a fat

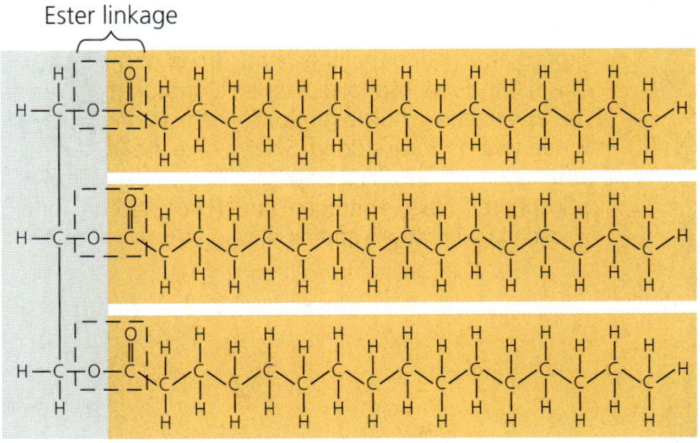

(b) Fat molecule (triacylglycerol)

▲ **Figure 5.11 The synthesis and structure of a fat, or triacylglycerol.** The molecular building blocks of a fat are one molecule of glycerol and three molecules of fatty acids. **(a)** One water molecule is removed for each fatty acid joined to the glycerol. **(b)** A fat molecule with three identical fatty acid units. The carbons of the fatty acids are arranged zig-zag to suggest the actual orientations of the four single bonds extending from each carbon (see Figure 4.3a).

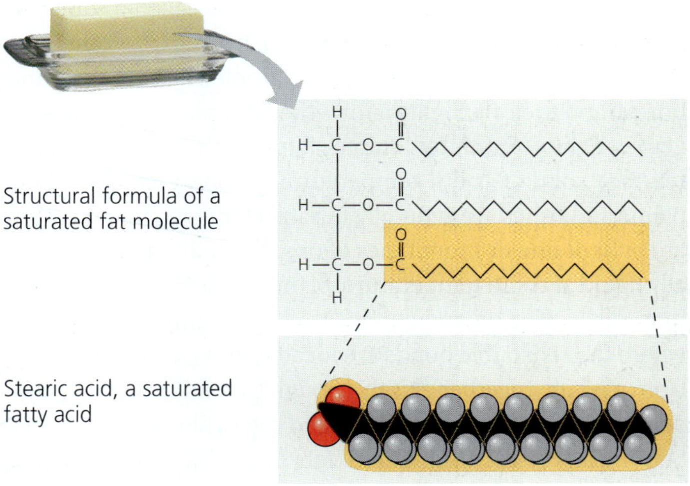

(a) Saturated fat. At room temperature, the molecules of a saturated fat such as this butter are packed closely together, forming a solid.

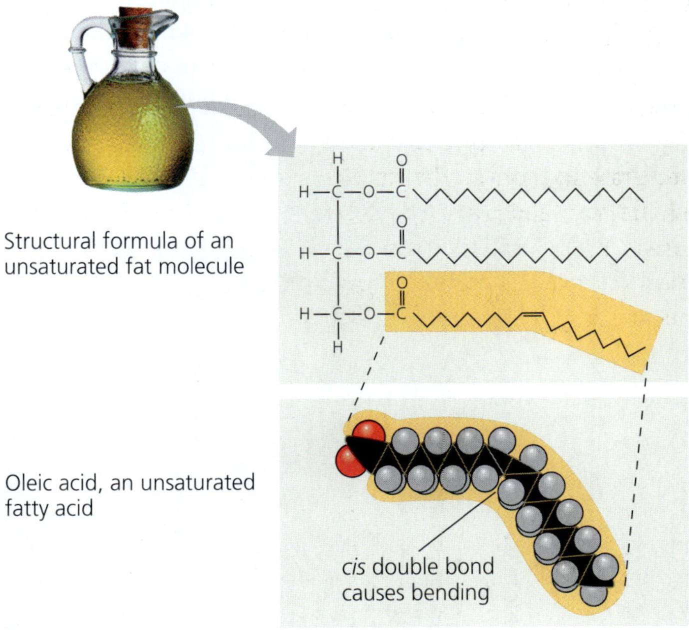

(b) Unsaturated fat. At room temperature, the molecules of an unsaturated fat such as this olive oil cannot pack together closely enough to solidify because of the kinks in some of their fatty acid hydrocarbon chains.

▲ **Figure 5.12 Examples of saturated and unsaturated fats and fatty acids.** The structural formula for each fat follows a common chemical convention of omitting the carbons and attached hydrogens of the hydrocarbon regions. In the space-filling models of the fatty acids, black = carbon, gray = hydrogen, and red = oxygen.

These terms refer to the structure of the hydrocarbon chains of the fatty acids. If there are no double bonds between carbon atoms composing the chain, then as many hydrogen atoms as possible are bonded to the carbon skeleton. Such a structure is described as being *saturated* with hydrogen, so the resulting fatty acid is called a **saturated fatty acid** **(Figure 5.12a)**. An **unsaturated fatty acid** has one or more double bonds, formed by the removal of hydrogen atoms from the carbon skeleton. The fatty acid will have a kink in its hydrocarbon chain wherever a *cis* double bond occurs **(Figure 5.12b)**.

A fat made from saturated fatty acids is called a saturated fat. Most animal fats are saturated: The hydrocarbon chains of their fatty acids—the "tails" of the fat molecules—lack double bonds, and their flexibility allows the fat molecules to pack together tightly. Saturated animal fats—such as lard and butter—are solid at room temperature. In contrast, the fats of plants and fishes are generally unsaturated, meaning that they are built of one or more types of unsaturated fatty acids. Usually liquid at room temperature, plant and fish fats are referred to as oils—olive oil and cod liver oil are examples. The kinks where the *cis* double bonds are located prevent the molecules from packing together closely enough to solidify at room temperature. The phrase "hydrogenated vegetable oils" on food labels means that unsaturated fats have been synthetically converted to saturated fats by adding hydrogen. Peanut butter, margarine, and many other products are hydrogenated to prevent lipids from separating out in liquid (oil) form.

A diet rich in saturated fats is one of several factors that may contribute to the cardiovascular disease known as atherosclerosis. In this condition, deposits called plaques develop within the walls of blood vessels, causing inward bulges that impede blood flow and reduce the resilience of the vessels. Recent studies have shown that the process of hydrogenating vegetable oils produces not only saturated fats but also unsaturated fats with *trans* double bonds. These **trans fats** may contribute more than saturated fats to atherosclerosis (see Chapter 42) and other problems. Because trans fats are especially common in baked goods and processed foods, the USDA requires trans fat content information on nutritional labels.

Fat has come to have such a negative connotation in our culture that you might wonder what useful purpose fats serve. The major function of fats is energy storage. The hydrocarbon chains of fats are similar to gasoline molecules and just as rich in energy. A gram of fat stores more than twice as much energy as a gram of a polysaccharide, such as starch. Because plants are relatively immobile, they can function with bulky energy storage in the form of starch. (Vegetable oils are generally obtained from seeds, where more compact storage is an asset to the plant.) Animals, however, must carry their energy stores with them, so there is an advantage to having a more compact reservoir of fuel—fat. Humans and other mammals stock their long-term food reserves in adipose cells (see Figure 4.6a), which swell and shrink as fat is deposited and withdrawn from storage. In addition to storing energy, adipose tissue also cushions such vital organs as the kidneys, and a layer of fat beneath the skin insulates the body. This subcutaneous layer is especially thick in whales, seals, and most other marine mammals, protecting them from cold ocean water.

Phospholipids

Cells could not exist without another type of lipid—**phospholipids** **(Figure 5.13)**. Phospholipids are essential

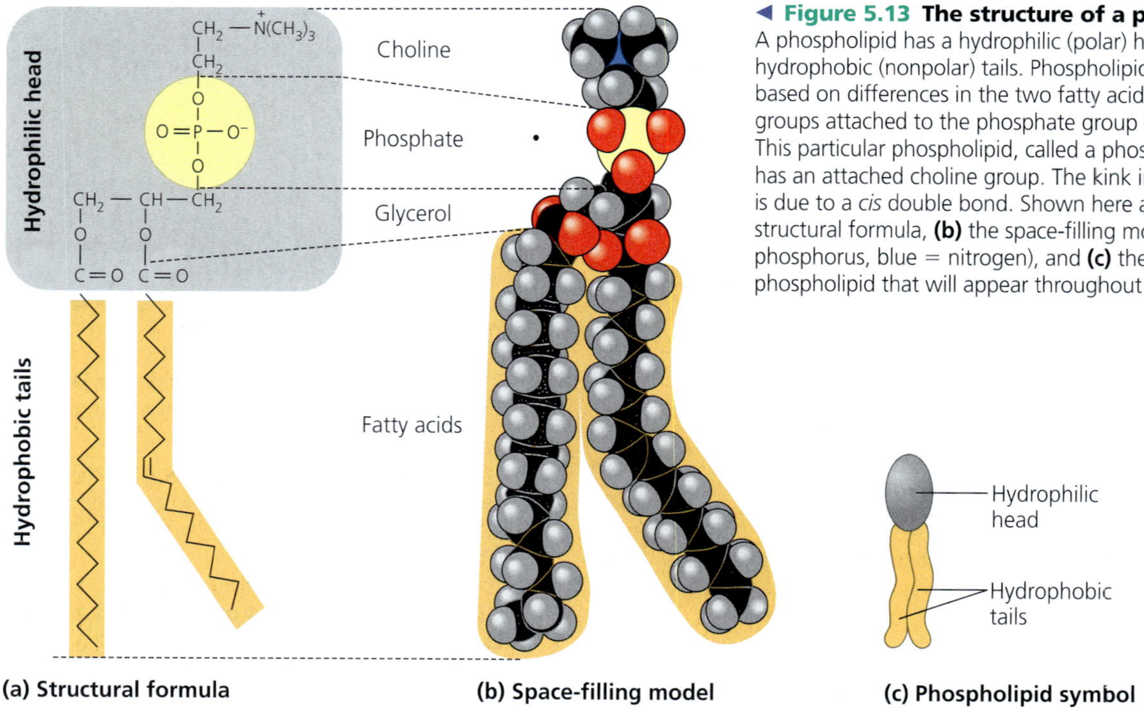

◄ **Figure 5.13 The structure of a phospholipid.** A phospholipid has a hydrophilic (polar) head and two hydrophobic (nonpolar) tails. Phospholipid diversity is based on differences in the two fatty acids and in the groups attached to the phosphate group of the head. This particular phospholipid, called a phosphatidylcholine, has an attached choline group. The kink in one of its tails is due to a *cis* double bond. Shown here are **(a)** the structural formula, **(b)** the space-filling model (yellow = phosphorus, blue = nitrogen), and **(c)** the symbol for a phospholipid that will appear throughout this book.

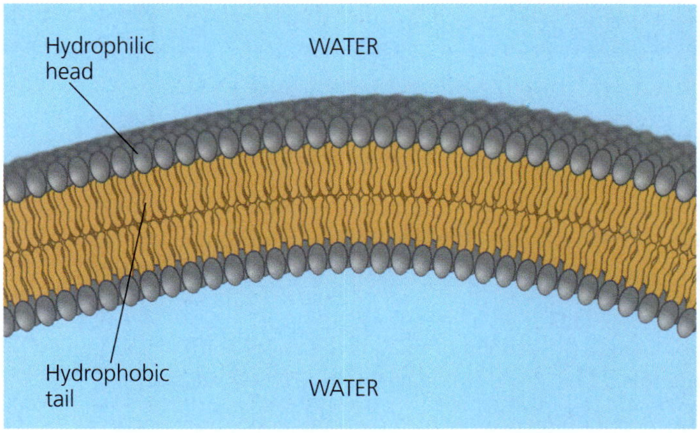

▲ **Figure 5.14 Bilayer structure formed by self-assembly of phospholipids in an aqueous environment.** The phospholipid bilayer shown here is the main fabric of biological membranes. Note that the hydrophilic heads of the phospholipids are in contact with water in this structure, whereas the hydrophobic tails are in contact with each other and remote from water.

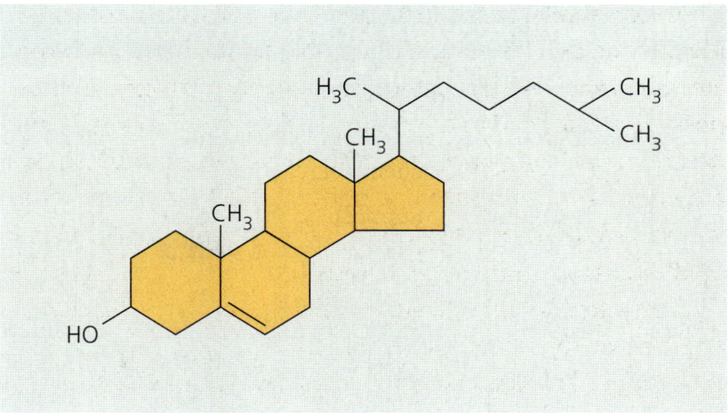

▲ **Figure 5.15 Cholesterol, a steroid.** Cholesterol is the molecule from which other steroids, including the sex hormones, are synthesized. Steroids vary in the chemical groups attached to their four interconnected rings (shown in gold).

for cells because they make up cell membranes. Their structure provides a classic example of how form fits function at the molecular level. As shown in Figure 5.13, a phospholipid is similar to a fat molecule but has only two fatty acids attached to glycerol rather than three. The third hydroxyl group of glycerol is joined to a phosphate group, which has a negative electrical charge. Additional small molecules, which are usually charged or polar, can be linked to the phosphate group to form a variety of phospholipids.

The two ends of phospholipids show different behavior toward water. The hydrocarbon tails are hydrophobic and are excluded from water. However, the phosphate group and its attachments form a hydrophilic head that has an affinity for water. When phospholipids are added to water, they self-assemble into double-layered aggregates—bilayers—that shield their hydrophobic portions from water **(Figure 5.14)**.

At the surface of a cell, phospholipids are arranged in a similar bilayer. The hydrophilic heads of the molecules are on the outside of the bilayer, in contact with the aqueous solutions inside and outside of the cell. The hydrophobic tails point toward the interior of the bilayer, away from the water. The phospholipid bilayer forms a boundary between the cell and its external environment; in fact, cells could not exist without phospholipids.

Steroids

Many hormones, as well as cholesterol, are **steroids**, which are lipids characterized by a carbon skeleton consisting of four fused rings **(Figure 5.15)**. Different steroids vary in the chemical groups attached to this ensemble of rings. **Cholesterol** is a common component of animal cell membranes and is also the precursor from which other steroids are synthesized. In vertebrates, cholesterol is synthesized in the liver. Many hormones, including vertebrate sex hormones, are steroids produced from cholesterol (see Figure 4.9). Thus, cholesterol is a crucial molecule in animals, although a high level of it in the blood may contribute to atherosclerosis. Both saturated fats and trans fats exert their negative impact on health by affecting cholesterol levels.

CONCEPT CHECK 5.3

1. Compare the structure of a fat (triglyceride) with that of a phospholipid.
2. Why are human sex hormones considered lipids?
3. **WHAT IF?** Suppose a membrane surrounded an oil droplet, as it does in the cells of plant seeds. Describe and explain the form it might take.

For suggested answers, see Appendix A.

CONCEPT **5.4**

Proteins have many structures, resulting in a wide range of functions

Nearly every dynamic function of a living being depends on proteins. In fact, the importance of proteins is underscored by their name, which comes from the Greek word *proteios*, meaning "first place." Proteins account for more than 50% of the dry mass of most cells, and they are instrumental in almost everything organisms do. Some proteins speed up chemical reactions, while others play a role in structural support,

Table 5.1 An Overview of Protein Functions

Type of Protein	Function	Examples
Enzymatic proteins	Selective acceleration of chemical reactions	Digestive enzymes catalyze the hydrolysis of the polymers in food.
Structural proteins	Support	Insects and spiders use silk fibers to make their cocoons and webs, respectively. Collagen and elastin provide a fibrous framework in animal connective tissues. Keratin is the protein of hair, horns, feathers, and other skin appendages.
Storage proteins	Storage of amino acids	Ovalbumin is the protein of egg white, used as an amino acid source for the developing embryo. Casein, the protein of milk, is the major source of amino acids for baby mammals. Plants have storage proteins in their seeds.
Transport proteins	Transport of other substances	Hemoglobin, the iron-containing protein of vertebrate blood, transports oxygen from the lungs to other parts of the body. Other proteins transport molecules across cell membranes.
Hormonal proteins	Coordination of an organism's activities	Insulin, a hormone secreted by the pancreas, helps regulate the concentration of sugar in the blood of vertebrates.
Receptor proteins	Response of cell to chemical stimuli	Receptors built into the membrane of a nerve cell detect chemical signals released by other nerve cells.
Contractile and motor proteins	Movement	Actin and myosin are responsible for the contraction of muscles. Other proteins are responsible for the undulations of the organelles called cilia and flagella.
Defensive proteins	Protection against disease	Antibodies combat bacteria and viruses.

storage, transport, cellular communication, movement, and defense against foreign substances **(Table 5.1)**.

Life would not be possible without **enzymes**, most of which are proteins. Enzymatic proteins regulate metabolism by acting as **catalysts**, chemical agents that selectively speed up chemical reactions without being consumed by the reaction **(Figure 5.16)**.

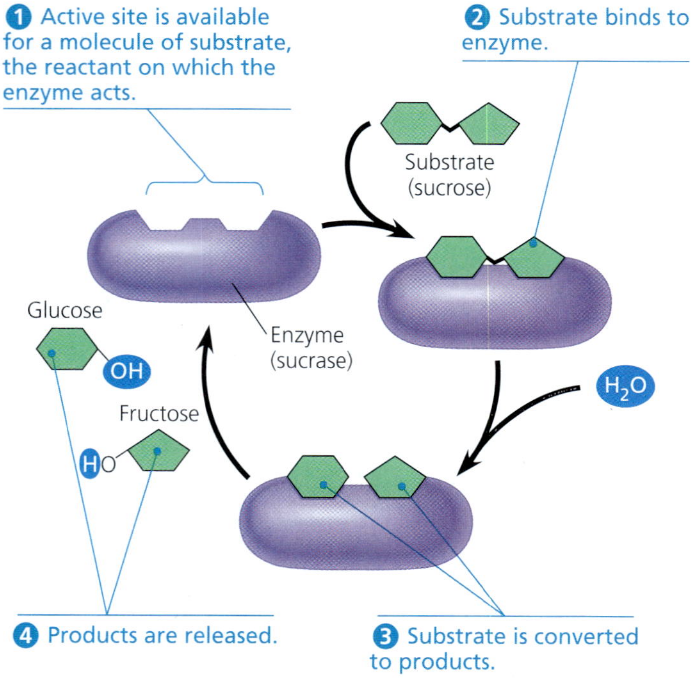

▲ **Figure 5.16 The catalytic cycle of an enzyme.** The enzyme sucrase accelerates hydrolysis of sucrose into glucose and fructose. Acting as a catalyst, the sucrase protein is not consumed during the cycle, but remains available for further catalysis.

Because an enzyme can perform its function over and over again, these molecules can be thought of as workhorses that keep cells running by carrying out the processes of life.

A human has tens of thousands of different proteins, each with a specific structure and function; proteins, in fact, are the most structurally sophisticated molecules known. Consistent with their diverse functions, they vary extensively in structure, each type of protein having a unique three-dimensional shape.

Polypeptides

Diverse as proteins are, they are all polymers constructed from the same set of 20 amino acids. Polymers of amino acids are called **polypeptides**. A **protein** consists of one or more polypeptides, each folded and coiled into a specific three-dimensional structure.

Amino Acid Monomers

All amino acids share a common structure. **Amino acids** are organic molecules possessing both carboxyl and amino groups (see Chapter 4). The illustration at the right shows the general formula for an amino acid. At the center of the amino acid is an asymmetric carbon atom called the *alpha (α) carbon.* Its four different partners are an amino group, a carboxyl group, a hydrogen atom, and a variable group symbolized by R. The R group, also called the side chain, differs with each amino acid.

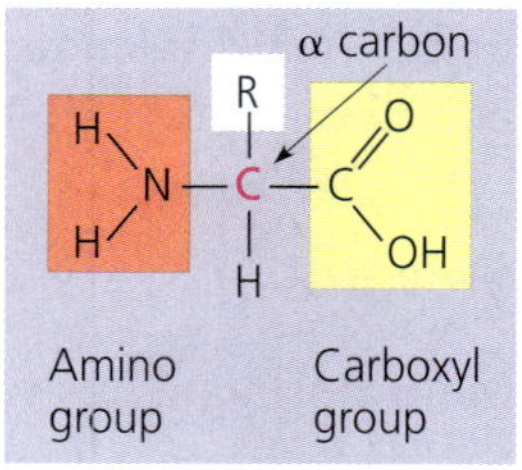

Figure 5.17 shows the 20 amino acids that cells use to build their thousands of proteins. Here the amino and carboxyl

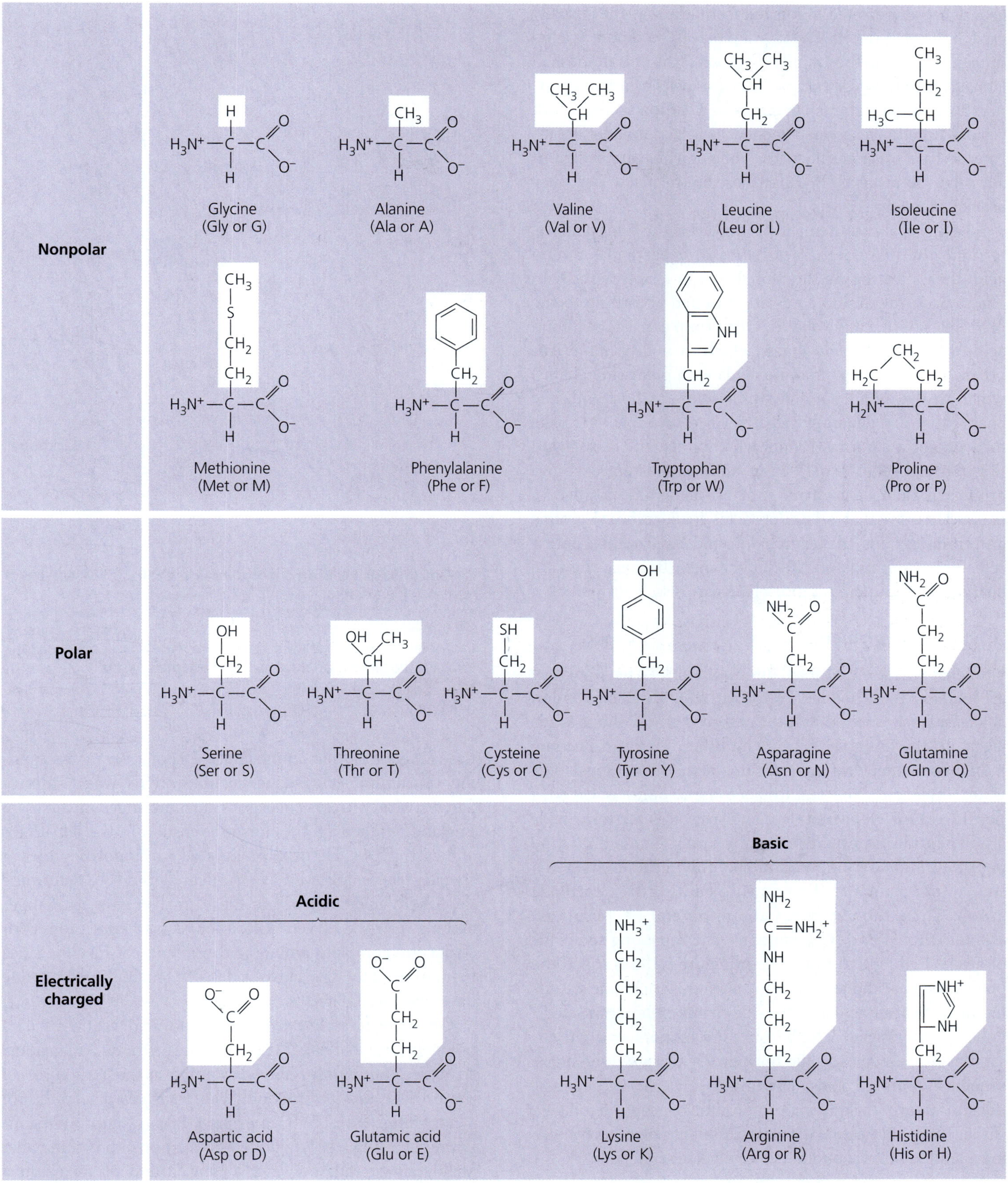

▲ **Figure 5.17 The 20 amino acids of proteins.** The amino acids are grouped here according to the properties of their side chains (R groups), highlighted in white. The amino acids are shown in their prevailing ionic forms at pH 7.2, the pH within a cell. The three-letter and more commonly used one-letter abbreviations for the amino acids are in parentheses. All the amino acids used in proteins are the same enantiomer, called the L form, as shown here (see Figure 4.7).

groups are all depicted in ionized form, the way they usually exist at the pH in a cell. The R group may be as simple as a hydrogen atom, as in the amino acid glycine (the one amino acid lacking an asymmetric carbon, since two of its α carbon's partners are hydrogen atoms), or it may be a carbon skeleton with various functional groups attached, as in glutamine. (Organisms do have other amino acids, some of which are occasionally found in proteins. Because these are relatively rare, they are not shown in Figure 5.17.)

The physical and chemical properties of the side chain determine the unique characteristics of a particular amino acid, thus affecting its functional role in a polypeptide. In Figure 5.17, the amino acids are grouped according to the properties of their side chains. One group consists of amino acids with nonpolar side chains, which are hydrophobic. Another group consists of amino acids with polar side chains, which are hydrophilic. Acidic amino acids are those with side chains that are generally negative in charge owing to the presence of a carboxyl group, which is usually dissociated (ionized) at cellular pH. Basic amino acids have amino groups in their side chains that are generally positive in charge. (Notice that *all* amino acids have carboxyl groups and amino groups; the terms *acidic* and *basic* in this context refer only to groups on the side chains.) Because they are charged, acidic and basic side chains are also hydrophilic.

Amino Acid Polymers

Now that we have examined amino acids, let's see how they are linked to form polymers **(Figure 5.18)**. When two amino acids are positioned so that the carboxyl group of one is adjacent to the amino group of the other, they can become joined by a dehydration reaction, with the removal of a water molecule. The resulting covalent bond is called a **peptide bond**. Repeated over and over, this process yields a polypeptide, a polymer of many amino acids linked by peptide bonds. At one end of the polypeptide chain is a free amino group; at the opposite end is a free carboxyl group. Thus, the chain has an amino end (N-terminus) and a carboxyl end (C-terminus). The repeating sequence of atoms highlighted in purple in Figure 5.18b is called the polypeptide backbone. Extending from this backbone are different kinds of appendages, the side chains of the amino acids. Polypeptides range in length from a few monomers to a thousand or more. Each specific polypeptide has a unique linear sequence of amino acids. The immense variety of polypeptides in nature illustrates an important concept introduced earlier—that cells can make many different polymers by linking a limited set of monomers into diverse sequences.

Protein Structure and Function

The specific activities of proteins result from their intricate three-dimensional architecture, the simplest level of which is the sequence of their amino acids. The pioneer in determining the amino acid sequence of proteins was Frederick Sanger, who, with his colleagues at Cambridge University in England, worked on the hormone insulin in the late 1940s and early 1950s. He used agents that break polypeptides at specific places, followed by chemical methods to determine the amino acid sequence in these small fragments. Sanger and his co-workers were able, after years of effort, to reconstruct the complete amino acid sequence of insulin. Since then, most of the steps involved in sequencing a polypeptide have been automated.

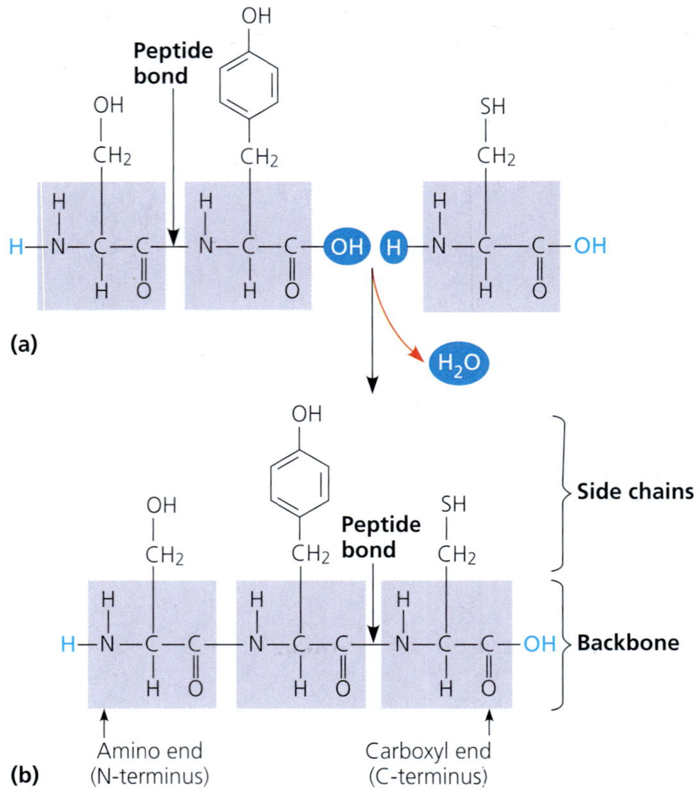

▲ **Figure 5.18 Making a polypeptide chain. (a)** Peptide bonds formed by dehydration reactions link the carboxyl group of one amino acid to the amino group of the next. **(b)** The peptide bonds are formed one at a time, starting with the amino acid at the amino end (N-terminus). The polypeptide has a repetitive backbone (purple) to which the amino acid side chains are attached.

DRAW IT *In (a), circle and label the carboxyl and amino groups that will form the peptide bond shown in (b).*

Once we have learned the amino acid sequence of a polypeptide, what can it tell us about the three-dimensional structure (commonly referred to simply as the "structure") of the protein and its function? The term *polypeptide* is not synonymous with the term *protein*. Even for a protein consisting of a single polypeptide, the relationship is somewhat analogous to that between a long strand of yarn and a sweater of particular size and shape that can be knit from the yarn. A functional protein is not *just* a polypeptide chain, but one or more polypeptides precisely twisted, folded, and coiled into a molecule of unique shape **(Figure 5.19)**. And it

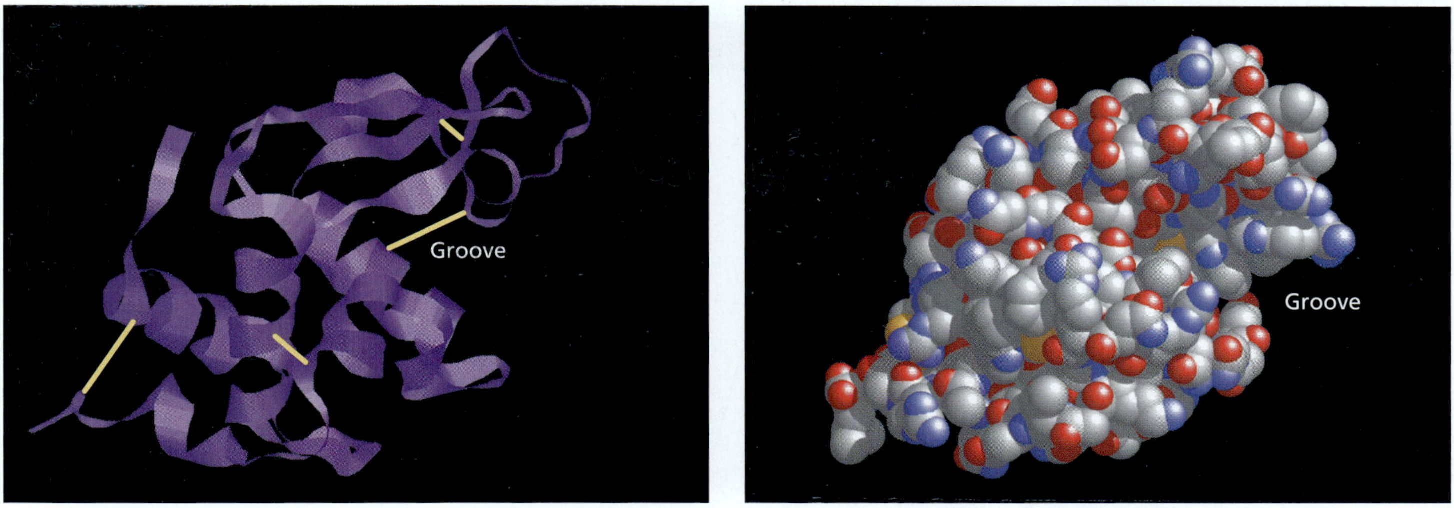

(a) A **ribbon model** shows how the single polypeptide chain folds and coils to form the functional protein. (The yellow lines represent cross-linking bonds between cysteines that stabilize the protein's shape.)

(b) A **space-filling model** shows more clearly the globular shape seen in many proteins, as well as the specific three-dimensional structure unique to lysozyme.

▲ **Figure 5.19 Structure of a protein, the enzyme lysozyme.** Present in our sweat, tears, and saliva, lysozyme is an enzyme that helps prevent infection by binding to and destroying specific molecules on the surface of many kinds of bacteria. The groove is the part of the protein that recognizes and binds to the target molecules on bacterial walls.

is the amino acid sequence of each polypeptide that determines what three-dimensional structure the protein will have.

When a cell synthesizes a polypeptide, the chain generally folds spontaneously, assuming the functional structure for that protein. This folding is driven and reinforced by the formation of a variety of bonds between parts of the chain, which in turn depends on the sequence of amino acids. Many proteins are roughly spherical (*globular proteins*), while others are shaped like long fibers (*fibrous proteins*). Even within these broad categories, countless variations exist.

A protein's specific structure determines how it works. In almost every case, the function of a protein depends on its ability to recognize and bind to some other molecule. In an especially striking example of the marriage of form and function, **Figure 5.20** shows the exact match of shape between an antibody (a protein in the body) and the particular foreign substance on a flu virus that the antibody binds to and marks for destruction. A second example is an enzyme, which must recognize and bind closely to its substrate, the substance the enzyme works on (see Figure 5.16). Also, you learned in Chapter 2 that natural signaling molecules called endorphins bind to specific receptor proteins on the surface of brain cells in humans, producing euphoria and relieving pain. Morphine, heroin, and other opiate drugs are able to mimic endorphins because they all share a similar shape with endorphins and can thus fit into and bind to endorphin receptors in the brain. This fit is very specific, something like a lock and key (see Figure 2.18). Thus, the function of a protein—for instance, the ability of a receptor protein to bind to a particular pain-relieving signaling molecule—is an emergent property resulting from exquisite molecular order.

Four Levels of Protein Structure

With the goal of understanding the function of a protein, learning about its structure is often productive. In spite of their great diversity, all proteins share three superimposed levels of structure, known as primary, secondary, and tertiary structure. A fourth level, quaternary structure, arises when a protein consists of two or more polypeptide chains. **Figure 5.21**, on the following two pages, describes these four levels of protein structure. Be sure to study this figure thoroughly before going on to the next section.

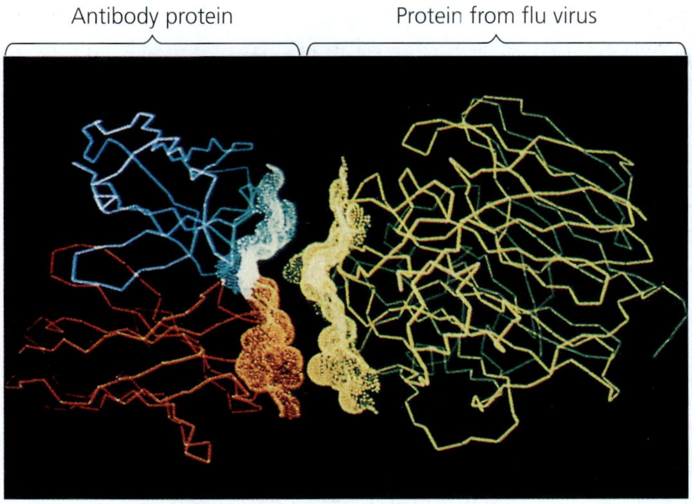

▲ **Figure 5.20 An antibody binding to a protein from a flu virus.** A technique called X-ray crystallography was used to generate a computer model of an antibody protein (blue and orange, left) bound to a flu virus protein (green and yellow, right). Computer software was then used to back the images away from each other, revealing the exact complementarity of shape between the two protein surfaces.

▼ **Figure 5.21**

Exploring Levels of Protein Structure

Primary Structure

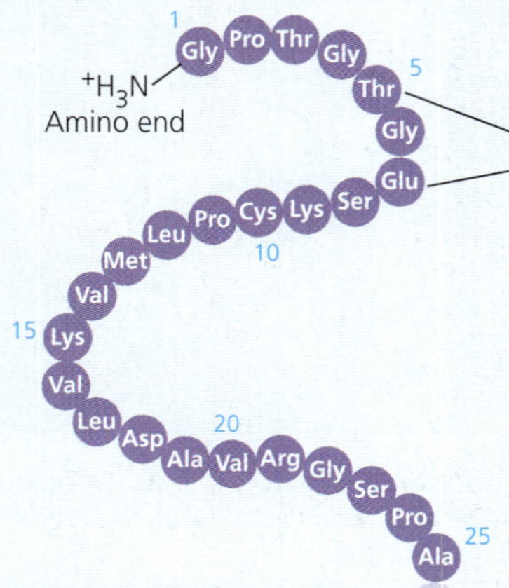

The **primary structure** of a protein is its unique sequence of amino acids. As an example, let's consider transthyretin, a globular protein found in the blood that transports vitamin A and one of the thyroid hormones throughout the body. Each of the four identical polypeptide chains that together make up transthyretin is composed of 127 amino acids. Shown here is one of these chains unraveled for a closer look at its primary structure. Each of the 127 positions along the chain is occupied by one of the 20 amino acids, indicated here by its three-letter abbreviation. The primary structure is like the order of letters in a very long word. If left to chance, there would be 20^{127} different ways of making a polypeptide chain 127 amino acids long. However, the precise primary structure of a protein is determined not by the random linking of amino acids, but by inherited genetic information.

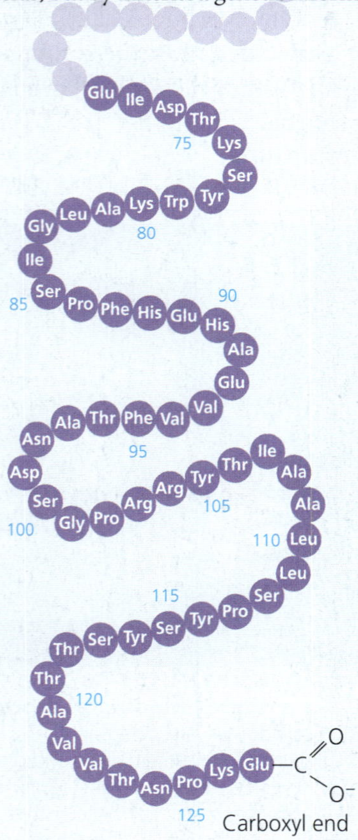

Secondary Structure

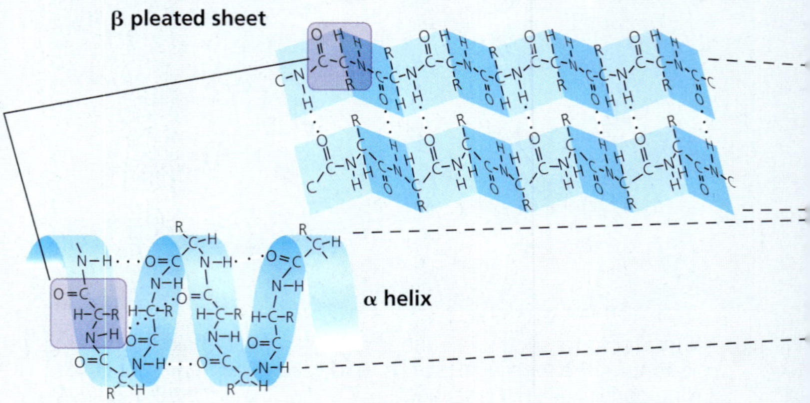

Most proteins have segments of their polypeptide chains repeatedly coiled or folded in patterns that contribute to the protein's overall shape. These coils and folds, collectively referred to as **secondary structure**, are the result of hydrogen bonds between the repeating constituents of the polypeptide backbone (not the amino acid side chains). Both the oxygen and the nitrogen atoms of the backbone are electronegative, with partial negative charges (see Figure 2.16). The weakly positive hydrogen atom attached to the nitrogen atom has an affinity for the oxygen atom of a nearby peptide bond. Individually, these hydrogen bonds are weak, but because they are repeated many times over a relatively long region of the polypeptide chain, they can support a particular shape for that part of the protein.

One such secondary structure is the **α helix**, a delicate coil held together by hydrogen bonding between every fourth amino acid, shown above. Although transthyretin has only one α helix region (see tertiary structure), other globular proteins have multiple stretches of α helix separated by nonhelical regions. Some fibrous proteins, such as α-keratin, the structural protein of hair, have the α helix formation over most of their length.

The other main type of secondary structure is the **β pleated sheet**. As shown above, in this structure two or more regions of the polypeptide chain lying side by side are connected by hydrogen bonds between parts of the two parallel polypeptide backbones. Pleated sheets make up the core of many globular proteins, as is the case for transthyretin, and dominate some fibrous proteins, including the silk protein of a spider's web. The teamwork of so many hydrogen bonds makes each spider silk fiber stronger than a steel strand of the same weight.

Tertiary Structure

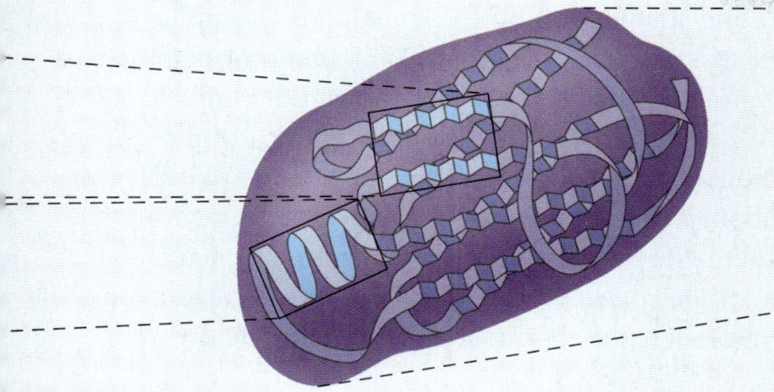

Superimposed on the patterns of secondary structure is a protein's **tertiary structure**, shown above for the transthyretin polypeptide. While secondary structure involves interactions between backbone constituents, tertiary structure is the overall shape of a polypeptide resulting from interactions between the side chains (R groups) of the various amino acids. One type of interaction that contributes to tertiary structure is—somewhat misleadingly—called a **hydrophobic interaction**. As a polypeptide folds into its functional shape, amino acids with hydrophobic (nonpolar) side chains usually end up in clusters at the core of the protein, out of contact with water. Thus, what we call a hydrophobic interaction is actually caused by the action of water molecules, which exclude nonpolar substances as they form hydrogen bonds with each other and with hydrophilic parts of the protein. Once nonpolar amino acid side chains are close together, van der Waals interactions help hold them together. Meanwhile, hydrogen bonds between polar side chains and ionic bonds between positively and negatively charged side chains also help stabilize tertiary structure. These are all weak interactions, but their cumulative effect helps give the protein a unique shape.

The shape of a protein may be reinforced further by covalent bonds called **disulfide bridges**. Disulfide bridges form where two cysteine monomers, amino acids with sulfhydryl groups (—SH) on their side chains (see Figure 4.10), are brought close together by the folding of the protein. The sulfur of one cysteine bonds to the sulfur of the second, and the disulfide bridge (—S—S—) rivets parts of the protein together (see yellow lines in Figure 5.19a). All of these different kinds of bonds can occur in one protein, as shown here in a small part of a hypothetical protein.

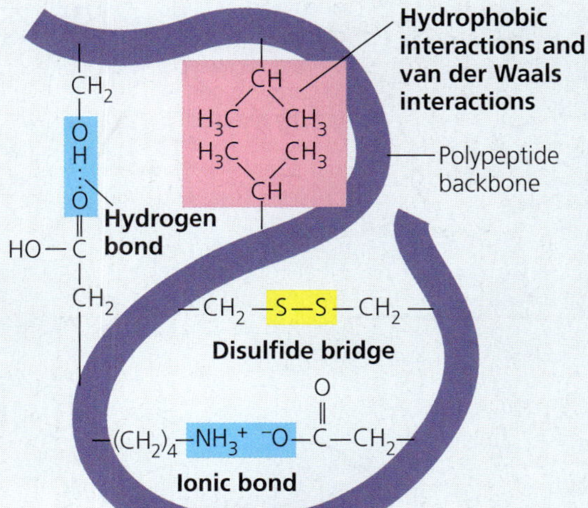

Quaternary Structure

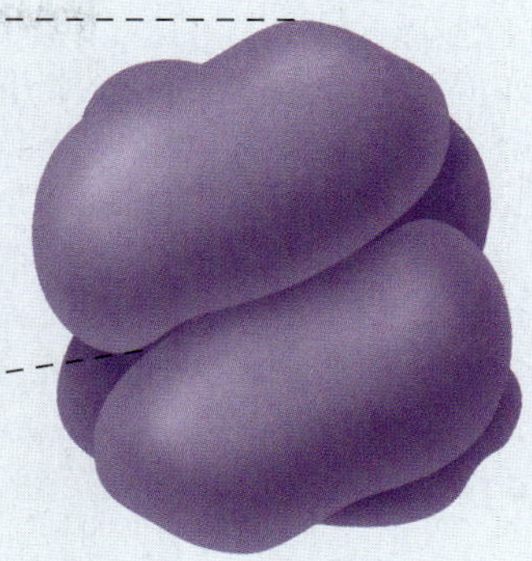

Some proteins consist of two or more polypeptide chains aggregated into one functional macromolecule. **Quaternary structure** is the overall protein structure that results from the aggregation of these polypeptide subunits. For example, shown above is the complete, globular transthyretin protein, made up of its four polypeptides. Another example is collagen, shown below left, which is a fibrous protein that has helical subunits intertwined into a larger triple helix, giving the long fibers great strength. This suits collagen fibers to their function as the girders of connective tissue in skin, bone, tendons, ligaments, and other body parts (collagen accounts for 40% of the protein in a human body). Hemoglobin, the oxygen-binding protein of red blood cells shown below right, is another example of a globular protein with quaternary structure. It consists of four polypeptide subunits, two of one kind ("α chains") and two of another kind ("β chains"). Both α and β subunits consist primarily of α-helical secondary structure. Each subunit has a nonpolypeptide component, called heme, with an iron atom that binds oxygen.

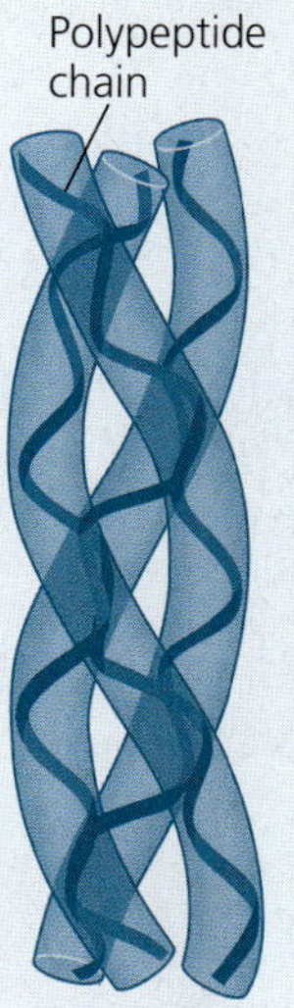

Collagen

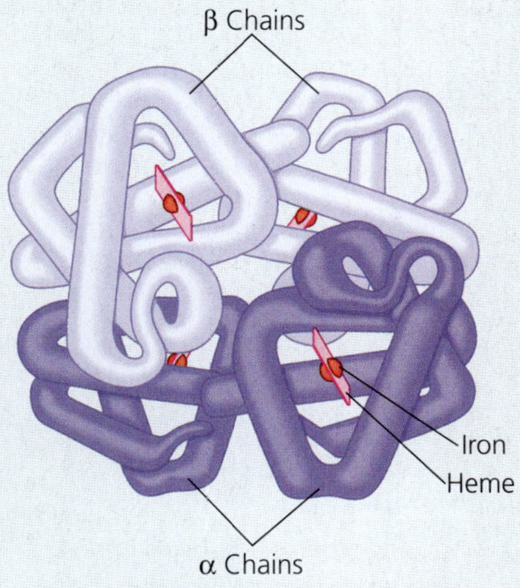

Hemoglobin

Sickle-Cell Disease: A Change in Primary Structure

Even a slight change in primary structure can affect a protein's shape and ability to function. For instance, *sickle-cell disease*, an inherited blood disorder, is caused by the substitution of one amino acid (valine) for the normal one (glutamic acid) at a particular position in the primary structure of hemoglobin, the protein that carries oxygen in red blood cells. Normal red blood cells are disk-shaped, but in sickle-cell disease, the abnormal hemoglobin molecules tend to crystallize, deforming some of the cells into a sickle shape **(Figure 5.22)**. The life of someone with the disease is punctuated by "sickle-cell crises," which occur when the angular cells clog tiny blood vessels, impeding blood flow. The toll taken on such patients is a dramatic example of how a simple change in protein structure can have devastating effects on protein function.

What Determines Protein Structure?

You've learned that a unique shape endows each protein with a specific function. But what are the key factors determining protein structure? You already know most of the answer: A polypeptide chain of a given amino acid sequence can spontaneously arrange itself into a three-dimensional shape determined and maintained by the interactions responsible for secondary and tertiary structure. This folding normally occurs as the protein is being synthesized within the cell. However, protein structure also depends on the physical and chemical conditions of the protein's environment. If the pH, salt concentration, temperature, or other aspects of its environment are altered, the protein may unravel and lose its native shape, a change called **denaturation** **(Figure 5.23)**. Because it is misshapen, the denatured protein is biologically inactive.

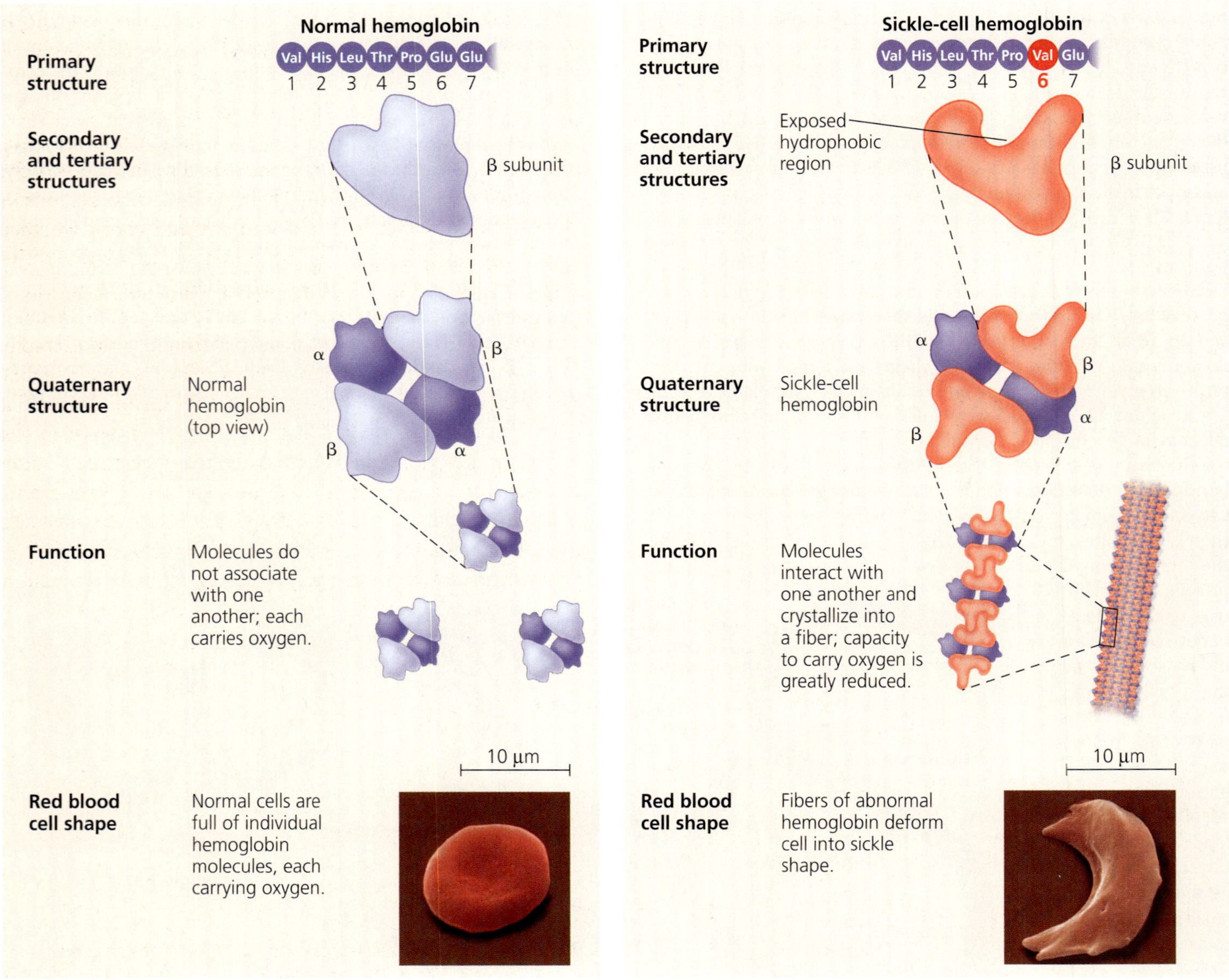

▲ **Figure 5.22 A single amino acid substitution in a protein causes sickle-cell disease.** To show fiber formation clearly, the orientation of the hemoglobin molecule here is different from that in Figure 5.21.

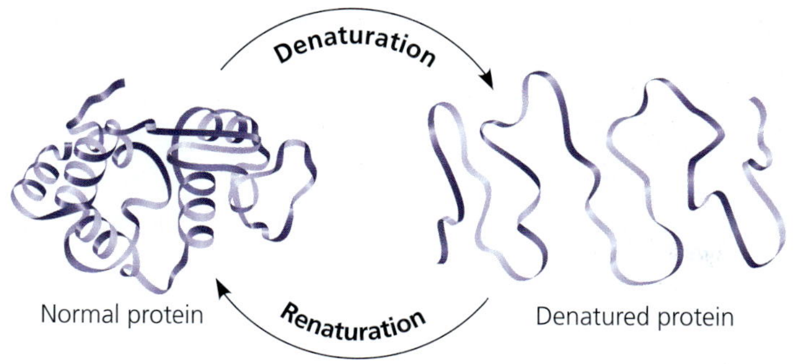

▲ **Figure 5.23 Denaturation and renaturation of a protein.** High temperatures or various chemical treatments will denature a protein, causing it to lose its shape and hence its ability to function. If the denatured protein remains dissolved, it can often renature when the chemical and physical aspects of its environment are restored to normal.

Most proteins become denatured if they are transferred from an aqueous environment to an organic solvent, such as ether or chloroform; the polypeptide chain refolds so that its hydrophobic regions face outward toward the solvent. Other denaturation agents include chemicals that disrupt the hydrogen bonds, ionic bonds, and disulfide bridges that maintain a protein's shape. Denaturation can also result from excessive heat, which agitates the polypeptide chain enough to overpower the weak interactions that stabilize the structure. The white of an egg becomes opaque during cooking because the denatured proteins are insoluble and solidify. This also explains why excessively high fevers can be fatal: Proteins in the blood can denature at very high body temperatures.

When a protein in a test-tube solution has been denatured by heat or chemicals, it can sometimes return to its functional shape when the denaturing agent is removed. We can conclude that the information for building specific shape is intrinsic to the protein's primary structure. The sequence of amino acids determines the protein's shape—where an α helix can form, where β pleated sheets can occur, where disulfide bridges are located, where ionic bonds can form, and so on. In the crowded environment inside a cell, there are also specific proteins that aid in the folding of other proteins.

Protein Folding in the Cell

Biochemists now know the amino acid sequences of more than 1.2 million proteins and the three-dimensional shapes of about 8,500. One would think that by correlating the primary structures of many proteins with their three-dimensional structures, it would be relatively easy to discover the rules of protein folding. Unfortunately, the protein-folding process is not that simple. Most proteins probably go through several intermediate structures on their way to a stable shape, and looking at the mature structure does not reveal the stages of folding required to achieve that form. However, biochemists have developed methods for tracking a protein through such stages. Researchers have also discovered **chaperonins** (also called chaperone proteins), protein molecules that assist in the proper folding of other proteins (**Figure 5.24**). Chaperonins do not specify the final structure of a polypeptide. Instead, they keep the new polypeptide segregated from "bad influences" in the cytoplasmic environment while it folds spontaneously. The chaperonin shown in Figure 5.24, from the bacterium *E. coli*, is a giant multiprotein complex shaped like a hollow cylinder. The cavity provides a shelter for folding polypeptides.

Misfolding of polypeptides is a serious problem in cells. Many diseases, such as Alzheimer's and Parkinson's, are associated with an accumulation of misfolded proteins. Recently, researchers have begun to shed light on molecular systems in the cell that interact with chaperonins and check whether proper folding has occurred. Such systems either refold the misfolded proteins correctly or mark them for destruction.

Even when scientists have a correctly folded protein in hand, determining its exact three-dimensional structure is not simple, for a single protein molecule has thousands of atoms. The first 3-D structures were worked out in 1959, for hemoglobin and a related protein. The method that made these feats possible was **X-ray crystallography**, which has since been used to determine the 3-D structures of many other proteins. In a recent example, Roger Kornberg and his colleagues at Stanford University used this method in order to elucidate the structure of RNA polymerase, an enzyme that plays a crucial role in the expression

▶ **Figure 5.24 A chaperonin in action.** The computer graphic (left) shows a large chaperonin protein complex. It has an interior space that provides a shelter for the proper folding of newly made polypeptides. The complex consists of two proteins: One protein is a hollow cylinder; the other is a cap that can fit on either end.

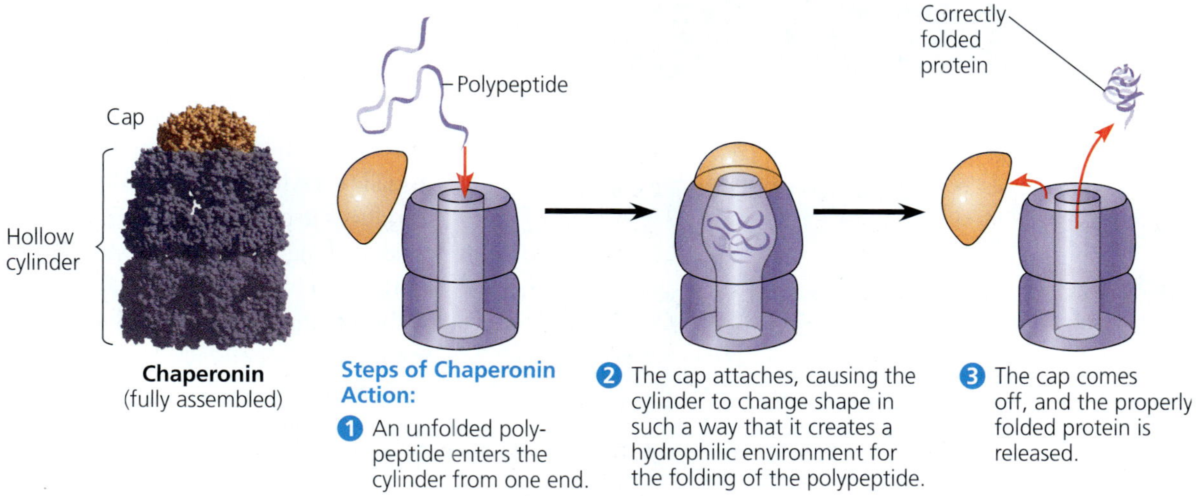

of genes **(Figure 5.25)**. Another method now in use is nuclear magnetic resonance (NMR) spectroscopy, which does not require protein crystallization.

A still newer approach uses bioinformatics (see Chapter 1) to predict the 3-D structures of polypeptides from their amino acid sequences. In 2005, researchers in Austria used computers to analyze the sequences and structures of 129 common plant protein allergens. They were able to classify all of these allergens into 20 families of proteins out of 3,849 families and 65% into just 4 families. This result suggests that the shared structures in these protein families may play a role in generating allergic reactions. These structures may provide targets for new allergy medications.

X-ray crystallography, NMR spectroscopy, and bioinformatics are complementary approaches to understanding protein structure. Together they have also given us valuable hints about protein function.

▼ **Figure 5.25** **Inquiry**

What can the 3-D shape of the enzyme RNA polymerase II tell us about its function?

EXPERIMENT In 2006, Roger Kornberg was awarded the Nobel Prize in Chemistry for using X-ray crystallography to determine the 3-D shape of RNA polymerase II, which binds to the DNA double helix and synthesizes RNA. After crystallizing a complex of all three components, Kornberg and his colleagues aimed an X-ray beam through the crystal. The atoms of the crystal diffracted (bent) the X-rays into an orderly array that a digital detector recorded as a pattern of spots called an X-ray diffraction pattern.

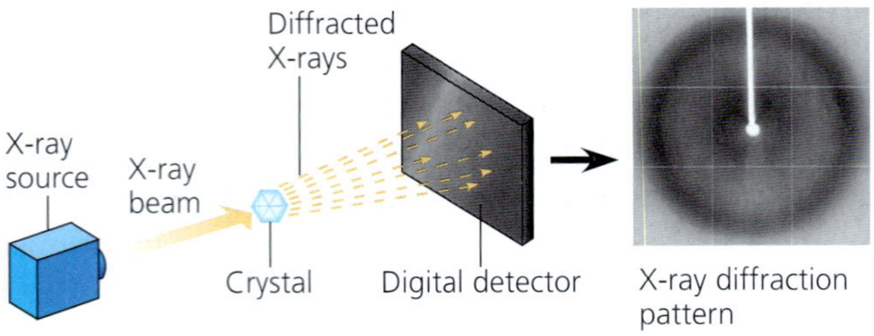

RESULTS Using data from X-ray diffraction patterns, as well as the amino acid sequence determined by chemical methods, Kornberg and colleagues built a 3-D model of the complex with the help of computer software.

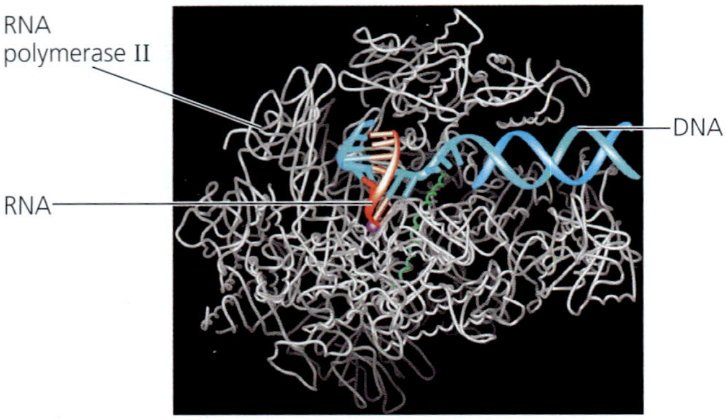

CONCLUSION By analyzing their model, the researchers developed a hypothesis about the functions of different regions of RNA polymerase II. For example, the region above the DNA may act as a clamp that holds the nucleic acids in place. (You'll learn more about this enzyme in Chapter 17.)

SOURCE A. L. Gnatt et al., Structural basis of transcription: an RNA polymerase II elongation complex at 3.3Å, *Science* 292:1876–1882 (2001).

WHAT IF? If you were an author of the paper and were describing the model, what type of protein structure would you call the small green polypeptide spiral in the center?

CONCEPT CHECK 5.4

1. Why does a denatured protein no longer function normally?
2. What parts of a polypeptide chain participate in the bonds that hold together secondary structure? What parts participate in tertiary structure?
3. **WHAT IF?** If a genetic mutation changes primary structure, how might it destroy the protein's function?

For suggested answers, see Appendix A.

CONCEPT **5.5**

Nucleic acids store and transmit hereditary information

If the primary structure of polypeptides determines a protein's shape, what determines primary structure? The amino acid sequence of a polypeptide is programmed by a unit of inheritance known as a **gene**. Genes consist of DNA, a polymer belonging to the class of compounds known as **nucleic acids**.

The Roles of Nucleic Acids

The two types of nucleic acids, **deoxyribonucleic acid (DNA)** and **ribonucleic acid (RNA)**, enable living organisms to reproduce their complex components from one generation to the next. Unique among molecules, DNA provides directions for its own replication. DNA also directs RNA synthesis and, through RNA, controls protein synthesis **(Figure 5.26)**.

DNA is the genetic material that organisms inherit from their parents. Each chromosome contains one long DNA molecule, usually carrying several hundred or more genes. When a cell reproduces itself by dividing, its DNA molecules are copied and passed along from one generation of cells to the next. Encoded in the structure of DNA is the information that programs all the cell's activities. The DNA, however, is not directly involved in running the operations of the cell, any more than computer software by itself can print a bank statement or read the bar code on a box of cereal. Just as a printer is needed to print out a statement and a scanner is needed to read a bar code, proteins are required

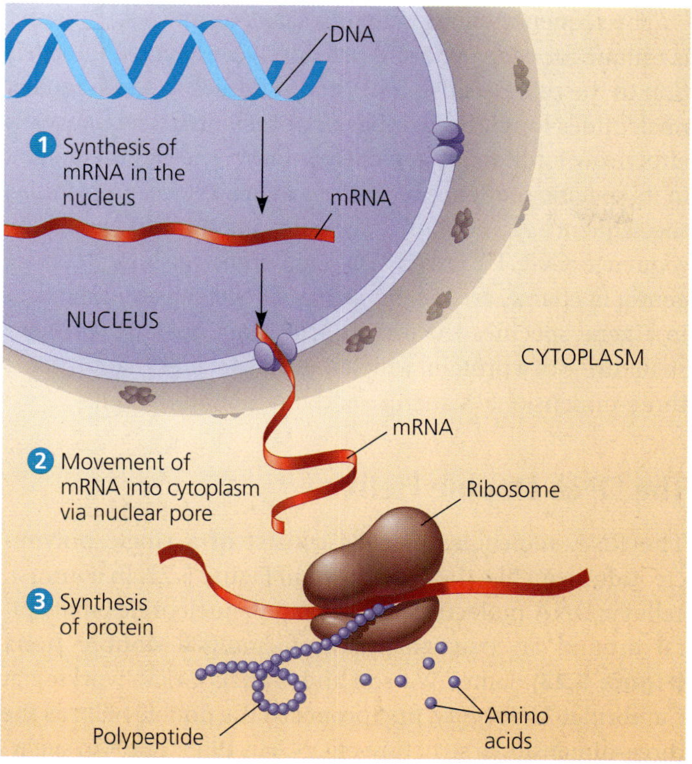

▲ **Figure 5.26 DNA → RNA → protein.** In a eukaryotic cell, DNA in the nucleus programs protein production in the cytoplasm by dictating synthesis of messenger RNA (mRNA). (The cell nucleus is actually much larger relative to the other elements of this figure.)

to implement genetic programs. The molecular hardware of the cell—the tools for biological functions—consists mostly of proteins. For example, the oxygen carrier in red blood cells is the protein hemoglobin, not the DNA that specifies its structure.

How does RNA, the other type of nucleic acid, fit into the flow of genetic information from DNA to proteins? Each gene along a DNA molecule directs synthesis of a type of RNA called *messenger RNA* (*mRNA*). The mRNA molecule interacts with the cell's protein-synthesizing machinery to direct production of a polypeptide, which folds into all or part of a protein. We can summarize the flow of genetic information as DNA → RNA → protein (see Figure 5.26). The sites of protein synthesis are tiny structures called ribosomes. In a eukaryotic cell, ribosomes are in the cytoplasm, but DNA resides in the nucleus. Messenger RNA conveys genetic instructions for building proteins from the nucleus to the cytoplasm. Prokaryotic cells lack nuclei but still use RNA to convey a message from the DNA to ribosomes and other cellular equipment that translate the coded information into amino acid sequences. RNA also plays many other roles in the cell.

The Structure of Nucleic Acids

Nucleic acids are macromolecules that exist as polymers called **polynucleotides** (**Figure 5.27a**). As indicated by the

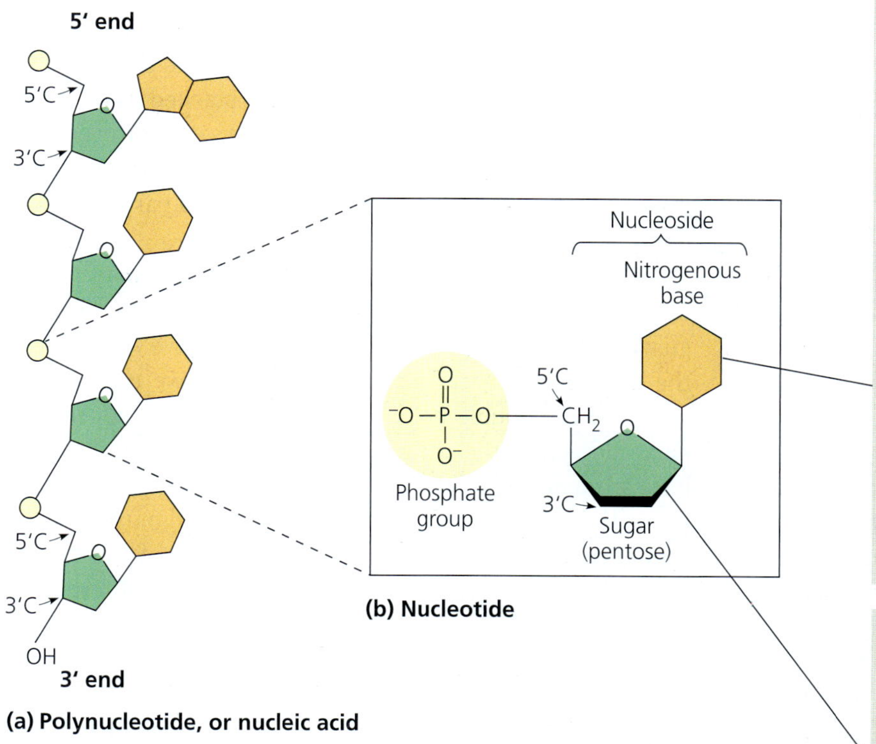

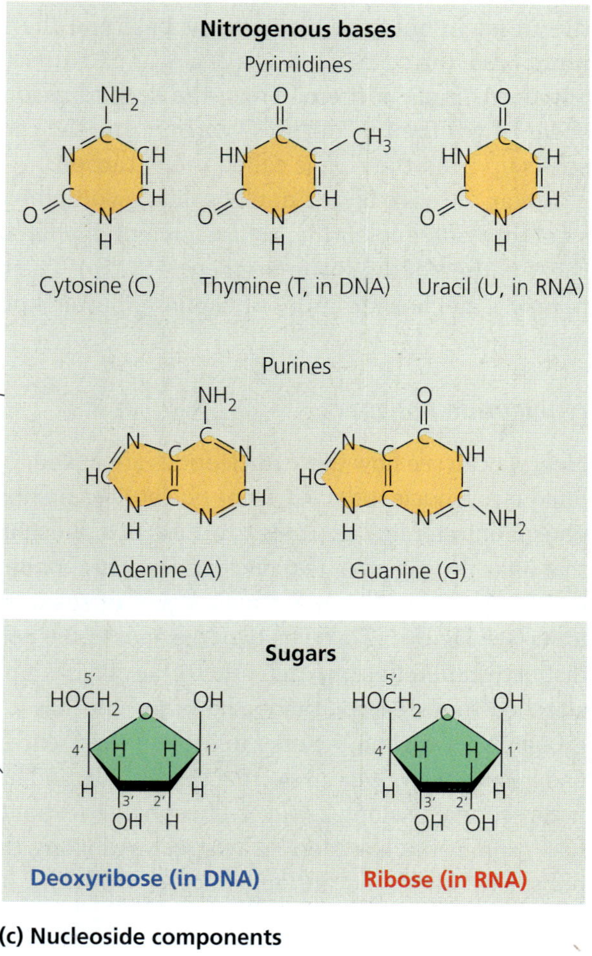

▲ **Figure 5.27 Components of nucleic acids. (a)** A polynucleotide has a sugar-phosphate backbone with variable appendages, the nitrogenous bases. **(b)** A nucleotide monomer includes a nitrogenous base, a sugar, and a phosphate group. Without the phosphate group, the structure is called a nucleoside. **(c)** A nucleoside includes a nitrogenous base (purine or pyrimidine) and a five-carbon sugar (deoxyribose or ribose).

name, each polynucleotide consists of monomers called **nucleotides**. A nucleotide is itself composed of three parts: a nitrogenous base, a five-carbon sugar (a pentose), and a phosphate group **(Figure 5.27b)**. The portion of this unit without the phosphate group is called a *nucleoside*.

Nucleotide Monomers

To build a nucleotide, let's first consider the two components of the nucleoside: the nitrogenous base and the sugar **(Figure 5.27c)**. There are two families of nitrogenous bases: pyrimidines and purines. A **pyrimidine** has a six-membered ring of carbon and nitrogen atoms. (The nitrogen atoms tend to take up H^+ from solution, which explains why it's called a nitrogenous *base*.) The members of the pyrimidine family are cytosine (C), thymine (T), and uracil (U). **Purines** are larger, with a six-membered ring fused to a five-membered ring. The purines are adenine (A) and guanine (G). The specific pyrimidines and purines differ in the chemical groups attached to the rings. Adenine, guanine, and cytosine are found in both types of nucleic acid; thymine is found only in DNA and uracil only in RNA.

The sugar connected to the nitrogenous base is **ribose** in the nucleotides of RNA and **deoxyribose** in DNA (see Figure 5.27c). The only difference between these two sugars is that deoxyribose lacks an oxygen atom on the second carbon in the ring; hence the name *deoxy*ribose. Because the atoms in both the nitrogenous base and the sugar are numbered, the sugar atoms have a prime (′) after the number to distinguish them. Thus, the second carbon in the sugar ring is the 2′ ("2 prime") carbon, and the carbon that sticks up from the ring is called the 5′ carbon.

So far, we have built a nucleoside. To complete the construction of a nucleotide, we attach a phosphate group to the 5′ carbon of the sugar (see Figure 5.27b). The molecule is now a nucleoside monophosphate, better known as a nucleotide.

Nucleotide Polymers

Now we can see how these nucleotides are linked together to build a polynucleotide. Adjacent nucleotides are joined by a phosphodiester linkage, which consists of a phosphate group that links the sugars of two nucleotides. This bonding results in a backbone with a repeating pattern of sugar-phosphate units (see Figure 5.27a). The two free ends of the polymer are distinctly different from each other. One end has a phosphate attached to a 5′ carbon, and the other end has a hydroxyl group on a 3′ carbon; we refer to these as the 5′ end and the 3′ end, respectively. We can say that the DNA strand has a built-in directionality along its sugar-phosphate backbone, from 5′ to 3′, somewhat like a one-way street. All along this sugar-phosphate backbone are appendages consisting of the nitrogenous bases.

The sequence of bases along a DNA (or mRNA) polymer is unique for each gene and provides very specific information to the cell. Because genes are hundreds to thousands of nucleotides long, the number of possible base sequences is effectively limitless. A gene's meaning to the cell is encoded in its specific sequence of the four DNA bases. For example, the sequence AGGTAACTT means one thing, whereas the sequence CGCTTTAAC has a different meaning. (Entire genes, of course, are much longer.) The linear order of bases in a gene specifies the amino acid sequence—the primary structure—of a protein, which in turn specifies that protein's three-dimensional structure and function in the cell.

The DNA Double Helix

The RNA molecules of cells consist of a single polynucleotide chain like the one shown in Figure 5.27. In contrast, cellular DNA molecules have two polynucleotides that spiral around an imaginary axis, forming a **double helix** **(Figure 5.28)**. James Watson and Francis Crick, working at Cambridge University, first proposed the double helix as the three-dimensional structure of DNA in 1953. The two sugar-phosphate backbones run in opposite 5′ → 3′ directions from each other, an arrangement referred to as **antiparallel**, somewhat like a divided highway. The sugar-phosphate backbones are on the outside of the helix, and the nitrogenous bases are paired in the interior of the helix. The two polynucleotides, or strands, as they are called, are held together by hydrogen bonds between the paired bases and by van der Waals interactions between the stacked bases. Most DNA molecules are very long, with thousands or even millions of base pairs connecting the two chains. One long DNA double helix includes many genes, each one a particular segment of the molecule.

Only certain bases in the double helix are compatible with each other. Adenine (A) always pairs with thymine (T), and guanine (G) always pairs with cytosine (C). If we were to read the sequence of bases along one strand as we traveled the length of the double helix, we would know the sequence of bases along the other strand. If a stretch of one strand has the base sequence 5′-AGGTCCG-3′, then the base-pairing rules tell us that the same stretch of the other strand must have the sequence 3′-TCCAGGC-5′. The two strands of the double helix are *complementary*, each the predictable counterpart of the other. It is this feature of DNA that makes possible the precise copying of genes that is responsible for inheritance (see Figure 5.28). In preparation for cell division, each of the two strands of a DNA molecule serves as a template to order nucleotides into a new complementary strand. The result is two identical copies of the original double-stranded DNA molecule, which are then distributed to the two daughter cells. Thus, the structure of DNA accounts for its function in transmitting genetic information whenever a cell reproduces.

DNA and Proteins as Tape Measures of Evolution

We are accustomed to thinking of shared traits, such as hair and milk production in mammals, as evidence of shared ancestors. Because we now understand that DNA carries heritable information in the form of genes, we can see that genes and their products (proteins) document the hereditary background of an organism. The linear sequences of nucleotides in DNA molecules are passed from parents to offspring; these sequences determine the amino acid sequences of proteins. Siblings have greater similarity in their DNA and proteins than do unrelated individuals of the same species. If the evolutionary view of life is valid, we should be able to extend this concept of "molecular genealogy" to relationships *between* species: We should expect two species that appear to be closely related based on fossil and anatomical evidence to also share a greater proportion of their DNA and protein sequences than do more distantly related species. In fact, that is the case. An example is the comparison of a polypeptide chain of human hemoglobin with the corresponding hemoglobin polypeptide in five other vertebrates. In this chain of 146 amino acids, humans and gorillas differ in just 1 amino acid, while humans and frogs differ in 67 amino acids. (Clearly, these changes do not make the protein nonfunctional.) Molecular biology has added a new tape measure to the toolkit biologists use to assess evolutionary kinship.

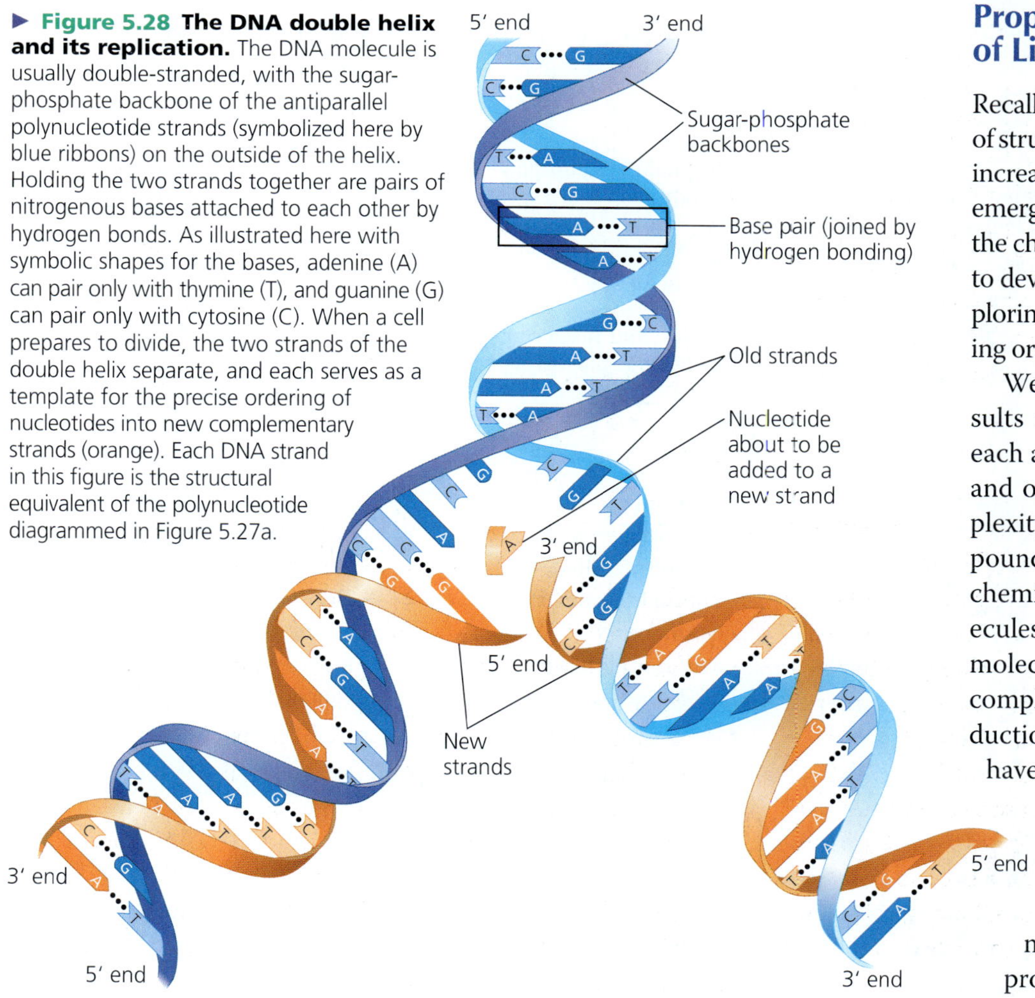

▶ **Figure 5.28 The DNA double helix and its replication.** The DNA molecule is usually double-stranded, with the sugar-phosphate backbone of the antiparallel polynucleotide strands (symbolized here by blue ribbons) on the outside of the helix. Holding the two strands together are pairs of nitrogenous bases attached to each other by hydrogen bonds. As illustrated here with symbolic shapes for the bases, adenine (A) can pair only with thymine (T), and guanine (G) can pair only with cytosine (C). When a cell prepares to divide, the two strands of the double helix separate, and each serves as a template for the precise ordering of nucleotides into new complementary strands (orange). Each DNA strand in this figure is the structural equivalent of the polynucleotide diagrammed in Figure 5.27a.

CONCEPT CHECK 5.5

1. Go to Figure 5.27a and number all the carbons in the sugars for the top three nucleotides; circle the nitrogenous bases and star the phosphates.
2. In a DNA double helix, a region along one DNA strand has this sequence of nitrogenous bases: 5′-TAGGCCT-3′. Write down this strand and its complementary strand, clearly indicating the 5′ and 3′ ends of the complementary strand.
3. **WHAT IF?** (a) Suppose a substitution occurred in one DNA strand of the double helix in question 2, resulting in:

   5′-TAAGCCT-3′
   3′-ATCCGGA-5′

   Draw the two strands; circle and label the mismatched bases. (b) If the modified top strand is replicated, what would its matching strand be?

For suggested answers, see Appendix A.

The Theme of Emergent Properties in the Chemistry of Life: *A Review*

Recall that life is organized along a hierarchy of structural levels (see Figure 1.4). With each increasing level of order, new properties emerge. In Chapters 2–5, we have dissected the chemistry of life. But we have also begun to develop a more integrated view of life, exploring how properties emerge with increasing order.

We have seen that water's behavior results from interactions of its molecules, each an ordered arrangement of hydrogen and oxygen atoms. We reduced the complexity and diversity of organic compounds to carbon skeletons and appended chemical groups. We saw that macromolecules are assembled from small organic molecules, taking on new properties. By completing our overview with an introduction to macromolecules and lipids, we have built a bridge to Unit Two, where we will study cell structure and function. We will keep a balance between the need to reduce life to simpler processes and the ultimate satisfaction of viewing those processes in their integrated context.

Chapter 5 Review

MEDIA Go to the Study Area at **www.masteringbio.com** for BioFlix 3-D Animations, MP3 Tutors, Videos, Practice Tests, an eBook, and more.

SUMMARY OF KEY CONCEPTS

CONCEPT 5.1
Macromolecules are polymers, built from monomers (pp. 68–69)

▶ **The Synthesis and Breakdown of Polymers** Carbohydrates, proteins, and nucleic acids are polymers, chains of monomers. The components of lipids vary. Monomers form larger molecules by dehydration reactions, in which water molecules are released. Polymers can disassemble by the reverse process, hydrolysis.

▶ **The Diversity of Polymers** An immense variety of polymers can be built from a small set of monomers.

MEDIA
Activity Making and Breaking Polymers

Large Biological Molecules	Components	Examples	Functions
Concept 5.2 **Carbohydrates serve as fuel and building material (pp. 69–74)** **MEDIA** Activity Models of Glucose Activity Carbohydrates	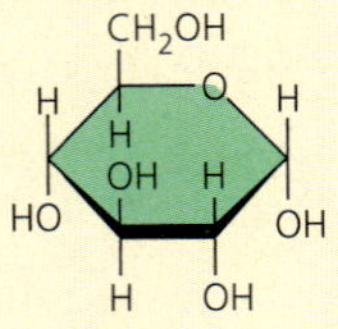 Monosaccharide monomer	Monosaccharides: glucose, fructose Disaccharides: lactose, sucrose	Fuel; carbon sources that can be converted to other molecules or combined into polymers
		Polysaccharides: • Cellulose (plants) • Starch (plants) • Glycogen (animals) • Chitin (animals and fungi)	• Strengthens plant cell walls • Stores glucose for energy • Stores glucose for energy • Strengthens exoskeletons and fungal cell walls
Concept 5.3 **Lipids are a diverse group of hydrophobic molecules (pp. 74–77)** **MEDIA** Activity Lipids	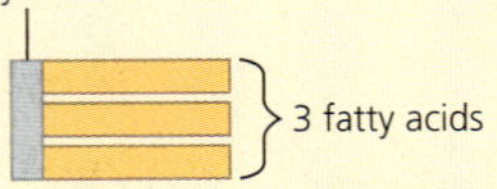	Triacylglycerols (fats or oils): glycerol + 3 fatty acids	Important energy source 
	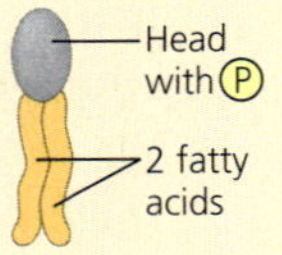	Phospholipids: phosphate group + 2 fatty acids	Lipid bilayers of membranes 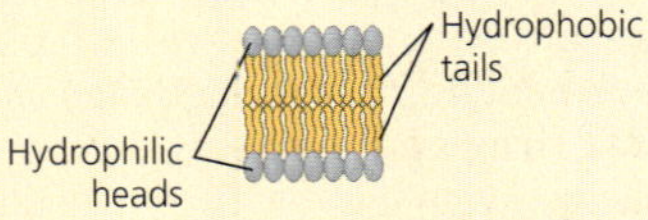
	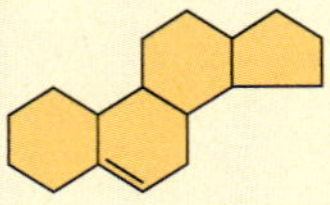 Steroid backbone	Steroids: four fused rings with attached chemical groups	• Component of cell membranes (cholesterol) • Signaling molecules that travel through the body (hormones)
Concept 5.4 **Proteins have many structures, resulting in a wide range of functions (pp. 77–86)** **MEDIA** MP3 Tutor Protein Structure and Function Activity Protein Functions Activity Protein Structure Biology Labs On-Line HemoglobinLab	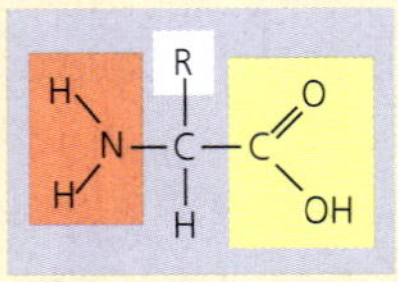 Amino acid monomer (20 types)	• Enzymes • Structural proteins • Storage proteins • Transport proteins • Hormones • Receptor proteins • Motor proteins • Defensive proteins	• Catalyze chemical reactions • Provide structural support • Store amino acids • Transport substances • Coordinate organismal responses • Receive signals from outside cell • Function in cell movement • Protect against disease
Concept 5.5 **Nucleic acids store and transmit hereditary information (pp. 86–89)** **MEDIA** Activity Nucleic Acid Functions Activity Nucleic Acid Structure	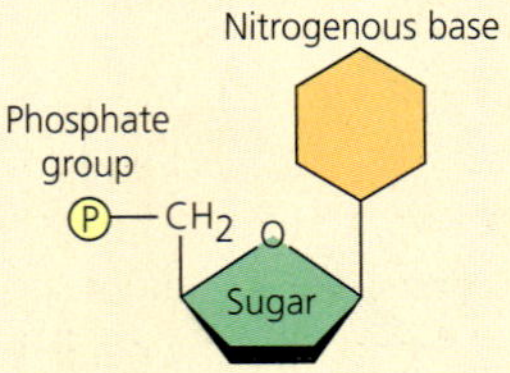 Nucleotide monomer	DNA:  • Sugar = deoxyribose • Nitrogenous bases = C, G, A, T • Usually double-stranded	Stores all hereditary information
		RNA:  • Sugar = ribose • Nitrogenous bases = C, G, A, U • Usually single-stranded	Carries protein-coding instructions from DNA to protein-synthesizing machinery

TESTING YOUR KNOWLEDGE

SELF-QUIZ

1. Which term includes all others in the list?
   a. monosaccharide
   b. disaccharide
   c. starch
   d. carbohydrate
   e. polysaccharide

2. The molecular formula for glucose is $C_6H_{12}O_6$. What would be the molecular formula for a polymer made by linking ten glucose molecules together by dehydration reactions?
   a. $C_{60}H_{120}O_{60}$
   b. $C_6H_{12}O_6$
   c. $C_{60}H_{102}O_{51}$
   d. $C_{60}H_{100}O_{50}$
   e. $C_{60}H_{111}O_{51}$

3. The enzyme amylase can break glycosidic linkages between glucose monomers only if the monomers are the α form. Which of the following could amylase break down?
   a. glycogen, starch, and amylopectin
   b. glycogen and cellulose
   c. cellulose and chitin
   d. starch and chitin
   e. starch, amylopectin, and cellulose

4. Which of the following statements concerning *unsaturated* fats is true?
   a. They are more common in animals than in plants.
   b. They have double bonds in the carbon chains of their fatty acids.
   c. They generally solidify at room temperature.
   d. They contain more hydrogen than saturated fats having the same number of carbon atoms.
   e. They have fewer fatty acid molecules per fat molecule.

5. The structural level of a protein least affected by a disruption in hydrogen bonding is the
   a. primary level.
   b. secondary level.
   c. tertiary level.
   d. quaternary level.
   e. All structural levels are equally affected.

6. Which of the following pairs of base sequences could form a short stretch of a normal double helix of DNA?
   a. 5′-purine-pyrimidine-purine-pyrimidine-3′ with 3′-purine-pyrimidine-purine-pyrimidine-5′
   b. 5′-AGCT-3′ with 5′-TCGA-3′
   c. 5′-GCGC-3′ with 5′-TATA-3′
   d. 5′-ATGC-3′ with 5′-GCAT-3′
   e. All of these pairs are correct.

7. Enzymes that break down DNA catalyze the hydrolysis of the covalent bonds that join nucleotides together. What would happen to DNA molecules treated with these enzymes?
   a. The two strands of the double helix would separate.
   b. The phosphodiester linkages between deoxyribose sugars would be broken.
   c. The purines would be separated from the deoxyribose sugars.
   d. The pyrimidines would be separated from the deoxyribose sugars.
   e. All bases would be separated from the deoxyribose sugars.

8. Construct a table that organizes the following terms, and label the columns and rows.

phosphodiester linkages	polypeptides	monosaccharides
peptide bonds	triacylglycerols	nucleotides
glycosidic linkages	polynucleotides	amino acids
ester linkages	polysaccharides	fatty acids

9. **DRAW IT** Draw the polynucleotide strand in Figure 5.27a and label the bases G, T, C, and T, starting from the 5′ end. Now draw the complementary strand of the double helix, using the same symbols for phosphates (circles), sugars (pentagons), and bases. Label the bases. Draw arrows showing the 5′ → 3′ direction of each strand. Use the arrows to make sure the second strand is antiparallel to the first. *Hint:* After you draw the first strand vertically, turn the paper upside down; it is easier to draw the second strand from the 5′ toward the 3′ direction as you go from top to bottom.

For Self-Quiz answers, see Appendix A.

MEDIA Visit the Study Area at **www.masteringbio.com** for a Practice Test.

EVOLUTION CONNECTION

10. Comparisons of amino acid sequences can shed light on the evolutionary divergence of related species. Would you expect all the proteins of a given set of living species to show the same degree of divergence? Why or why not?

SCIENTIFIC INQUIRY

11. During the Napoleonic Wars in the early 1800s, there was a sugar shortage in Europe because supply ships could not enter blockaded harbors. To create artificial sweeteners, German scientists hydrolyzed wheat starch. They did this by adding hydrochloric acid to heated starch solutions, breaking some of the glycosidic linkages between the glucose monomers. The graph here shows the percentage of glycosidic linkages broken over time. Why do you think consumers found the sweetener to be less sweet than sugar? Sketch a glycosidic linkage in starch using Figures 5.5a and 5.7b for reference. Show how the acid was able to break this bond. Why do you think the acid broke only 50% of the linkages in the wheat starch?

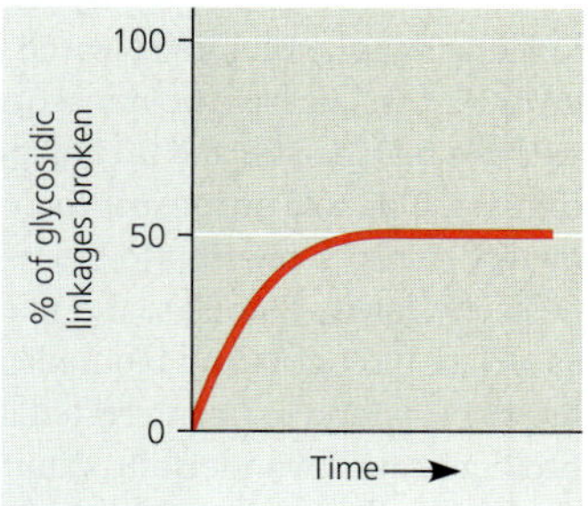

Biological Inquiry: A Workbook of Investigative Cases Explore large biological molecules further with the case "Picture Perfect."

SCIENCE, TECHNOLOGY, AND SOCIETY

12. Some amateur and professional athletes take anabolic steroids to help them "bulk up" or build strength. The health risks of this practice are extensively documented. Apart from health considerations, how do you feel about the use of chemicals to enhance athletic performance? Is an athlete who takes anabolic steroids cheating, or is such use part of the preparation that is required to succeed in competition? Explain.

UNIT 2

The Cell

AN INTERVIEW WITH

Paul Nurse

British biologist Sir Paul Nurse shared a Nobel Prize in 2001 with Leland H. Hartwell and R. Timothy Hunt for groundbreaking discoveries about how the eukaryotic cell controls its reproduction. Educated at the Universities of Birmingham and East Anglia, Dr. Nurse was a professor at the University of Oxford and a researcher at the Imperial Cancer Research Fund (now called Cancer Research UK), eventually heading up the latter organization. In 2003, he left the United Kingdom for Rockefeller University in New York City, where he serves as president while also running an active research laboratory.

Tell us about Rockefeller University.

Rockefeller is basically a research institute that gives Ph.D.'s. We have no departments; instead, we have between 70 and 80 independent laboratories. Our sole purpose is to do the highest-quality research relevant to biomedicine. We have physicists, chemists, and mathematicians, as well as biologists and biomedical scientists, but they all tend to be interested in biological problems. It's a wonderfully stimulating environment— and rather anarchic, but I'm happy with that. In my role as president, I focus very much on having the best people around, and then just letting them loose to investigate what they want. I'm very privileged to be here.

How did you first become interested in science, and biology in particular?

When I was a child, I had a long walk to my school through a couple of parks, and I liked to look at the plants, the birds, and the insects. I remember wondering why certain plants seemed to have big leaves when they grew in the shade and small leaves when they grew in sunlight. I also had an interest in astronomy, which I still have. So the origin of my interest in science was really through natural history.

I wasn't the greatest of students at school. My parents were working class, and we didn't have many books at home. It took me quite a while to get up to speed. But I had a tremendous curiosity to learn about the world around me. Then, as a teenager, I had a science teacher who really encouraged me, named Keith Neil. I did a number of biology projects with him—a fruit fly project, censuses of butterflies and beetles, and a trout egg-laying project in the lab. After I got the Nobel Prize, I nominated him for a mentoring award and did a TV program with him. I think my interest in biology was also influenced by a sense that there were so many unanswered questions in biology that even an ordinary mortal could contribute to the field in some way!

At university, I was at first going to concentrate on ecology and evolution, but I ended up going for subjects that seemed easier to study in a laboratory—cell biology and genetics.

What led you to study the cell cycle?

As a graduate student, I asked myself, "What's really important in biology?" I thought it would be important to focus on the features that distinguish life from nonlife, and a key one of those is the ability of an organism to reproduce itself. You see reproduction in its simplest form in the division of a cell because the cell is the simplest unit of life.

What did you want to find out about cell division?

The cell cycle is a series of events from the "birth" of a cell to its later division into two cells. I wanted to understand how the different events in the cell cycle were coordinated—what controlled the progression of cell-cycle events. In fact, I have ended up spending most of my life trying to answer that question!

There were two alternative ideas about how the cell cycle works: One was that the cell simply proceeds automatically through the events in the pathway, from A to B to C to D, and so forth. Another was that a smaller number of rate-limiting steps within the pathway serve as control points. These control points would determine how rapidly the cell cycle progressed.

How did you proceed to figure out which hypothesis was correct?

My inspiration came from Lee Hartwell, who had used genetics in budding yeast (the yeast used by bakers and brewers) to find mutants with defects in the cell cycle. This is the classical genetic approach. You look for what stops a system from working to tell you how it should be working. I used Lee's work as a template, but I used a different sort of yeast, called fission yeast, which is actually not closely related to budding yeast. Like Lee, I isolated mutants with defects in genes that defined particular steps in the cell-cycle process. We called them *cdc* genes—*cdc* for *c*ell *d*ivision *c*ycle.

In reproducing, a cell of fission yeast first grows to twice its starting size and then divides in two. I looked for mutants that couldn't divide, where the cells just got bigger and bigger. Finding such mutants told us that the genes defective in these cells were necessary for the cells to divide. And when you find a number of different kinds of mutants that are defective in a process, the obvious interpretation is that the process is a pathway of events, each of which is controlled by one or more genes. So the cell cycle could have worked like a straightforward pathway, with each step simply leading to the next.

But one day I happened to notice a different sort of mutant under the microscope: yeast cells that were dividing but at an unusually small size. I realized that if a cell went through the cell cycle faster than normal, it would reach the end of the cell cycle before it had doubled in size. The mutant I'd spotted had only a single mutation, in a single gene, but it was sufficient to make the whole cell cycle go faster. That meant there had to be at least some major rate-limiting steps in the cell cycle, because if they didn't exist you couldn't make the cell cycle go faster. So the second hypothesis seemed to be correct.

All of that came out of just looking at those small mutant cells for five minutes. I could say that most of the next twenty years of my career was based on those five minutes!

It's essential to add that many other scientists have contributed to the field of cell cycle regulation, working in many different places. In addition to Lee Hartwell, who is in Seattle, these people included Yoshio Masui in Toronto and Jim Maller in Colorado, who both worked with frog eggs, and my longtime friend Tim Hunt, in England, who studied the eggs of sea urchins.

You focused in on a mutant with a defect in a gene you called *cdc2*. What does this gene do?

It turned out that the *cdc2* gene codes for an enzyme called a protein kinase. Protein kinases are heavily used by cells as a means of regulating what other proteins do. There are hundreds of different kinds of these enzymes—over 500 in human cells, for example. What they do is phosphorylate other proteins: They take phosphates from ATP molecules and transfer them to proteins. Phosphate groups are big lumps of negative charge, and they can change the shape of a protein and therefore its properties. So showing that *cdc2* coded for a protein kinase was important in identifying protein phosphorylation as a key regulatory mechanism in the cell cycle. Eventually we showed that, in fission yeast, the *cdc2* protein kinase is used fairly early in the cell cycle, where it controls the replication of DNA, and then again later in the cycle, when the replicated chromosomes are ready to separate from each other in mitosis. This is soon before the cell divides in two.

Do similar enzymes control the cell cycle in other kinds of organisms?

A postdoc in my lab, Melanie Lee, was able to track down the human equivalent of the yeast *cdc2* gene, by showing that it was able to substitute for a defective *cdc2* gene in a yeast cell. After we put the human gene in a yeast cell that had a defective *cdc2* gene, the yeast cell was able to divide normally!

It turns out that the Cdc2 protein and many others involved in cell-cycle regulation are very similar in all eukaryotic organisms. What this has to mean is that this system of controlling cell division must have evolved very soon after eukaryotic cells first appeared on the planet, and that it was so crucial to cell survival that it remained unchanged. We have found the *cdc2* gene, for example, in every eukaryotic organism we've looked at.

What is the medical relevance of research on the cell cycle?

The main medical relevance is to cancer, because cancer occurs when cells grow and divide out of control. Now, growth and division is a good thing in the right place at the right time; it's how a fertilized human egg develops into a baby and how wounds in the body are repaired. But if you start getting growth and division in the wrong place at the wrong time, then you can get tumors that destroy the function of the organs or tissue in which they are located.

But maybe, in the end, the more important connection to cancer has to do with what's called genome stability. You have to precisely replicate all your genes in every cell cycle and then separate them precisely into the two progeny cells—that's what the cell cycle is all about. If there are mistakes, if the DNA does not replicate properly or the chromosomes don't separate properly, you end up generating genome instability, in which the number of chromosomes may be altered and parts of chromosomes rearranged. Such changes can lead to cancer. So understanding how genome stability is maintained is crucial for understanding how cancer arises.

What is your approach to mentoring young scientists in your lab? And what about collaboration with other labs?

I've always run a pretty disorganized lab. With students and postdocs, I look upon my job as not so much directing them as helping them follow their own interests, although I do try to keep them from falling into too many elephant traps. The lab is a bit inefficient, to be honest, because we're constantly starting new projects. But since people take these projects away with them when they leave, this practice helps the field expand very fast.

My collaboration with people outside my own lab is mostly in the form of talking. I find I think much better when I can bounce around ideas with other people. This sort of conversation is especially useful when there is a very honest and open relationship—it's good to be sufficiently comfortable with someone to be able to say, "What you just said was stupid." I've benefited from such frank discussions for decades with Tim Hunt, for example, but we've never published a paper together. Tim discovered another kind of cell-cycle control protein, called cyclin, which works in partnership with protein kinases like Cdc2. The protein kinases we've been talking about in this interview are therefore called cyclin-dependent kinases, or CDKs.

What responsibilities do scientists have toward society?

I always say that scientists need to have a "license to operate." We have to earn that license; we cannot assume society will be pro-science. When I was younger, I used to think: Well, I'm doing this because I'm curious, and science should be supported because it's an important cultural endeavor, something like art or music. But we have to realize that if we can justify science only in cultural terms, science budgets will plummet. The public and their governmental representatives want to use scientific discoveries to benefit humankind, and that's completely reasonable. I think it's critical that scientists communicate effectively with the public so that we can influence policymakers in government. Most importantly, we scientists have to listen to the public and understand how they see the issues; we need to have a dialogue with the public. Without that dialogue, we just don't know what misunderstandings are out there. I think we need more grassroots involvement by scientists.

. . . this system of controlling cell division must have evolved very soon after eukaryotic cells first appeared on the planet.

Paul Nurse and Jane Reece

Inquiry in Action

Learn about an experiment by Paul Nurse and colleagues in Inquiry Figure 12.16 on page 240.

A Tour of the Cell

6

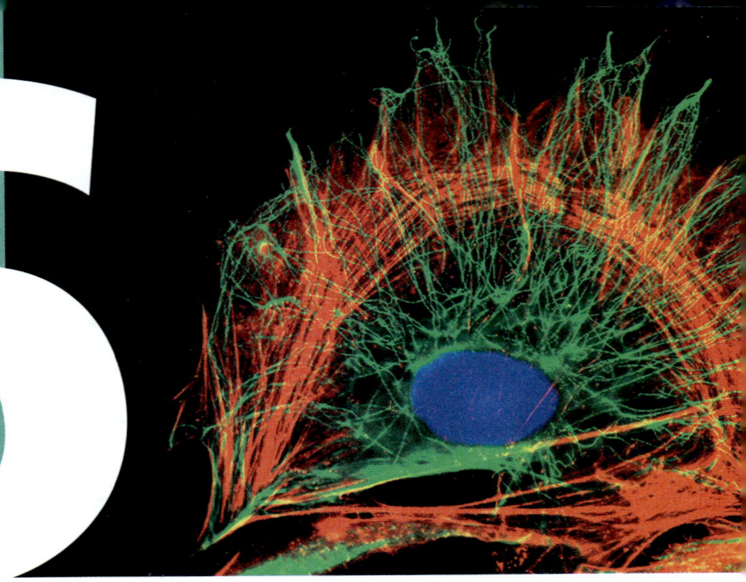

▲ **Figure 6.1 How do cellular components cooperate to help the cell function?**

KEY CONCEPTS

6.1 **To study cells, biologists use microscopes and the tools of biochemistry**

6.2 **Eukaryotic cells have internal membranes that compartmentalize their functions**

6.3 **The eukaryotic cell's genetic instructions are housed in the nucleus and carried out by the ribosomes**

6.4 **The endomembrane system regulates protein traffic and performs metabolic functions in the cell**

6.5 **Mitochondria and chloroplasts change energy from one form to another**

6.6 **The cytoskeleton is a network of fibers that organizes structures and activities in the cell**

6.7 **Extracellular components and connections between cells help coordinate cellular activities**

OVERVIEW

The Fundamental Units of Life

What do a small compartment in a honeycomb, a prison room, and the area covered by a mobile phone tower have in common with a microscopic part of your body? Each is the simplest unit of function in a larger system, and each is described by the word *cell.* The cell is as fundamental to the living systems of biology as the atom is to chemistry: All organisms are made of cells.

In the hierarchy of biological organization, the cell is the simplest collection of matter that can live. Indeed, there are diverse forms of life existing as single-celled organisms. More complex organisms, including plants and animals, are multicellular; their bodies are cooperatives of many kinds of specialized cells that could not survive for long on their own. However, even when they are arranged into higher levels of organization, such as tissues and organs, cells are an organism's basic units of structure and function. The contraction of muscle cells moves your eyes as you read this sentence; when you decide to turn the next page, nerve cells will transmit that decision from your brain to the muscle cells of your hand. Each action of an organism begins at the cellular level.

The cell is a microcosm that demonstrates most of the themes introduced in Chapter 1. Life at the cellular level arises from structural order, reflecting emergent properties and the correlation between structure and function. For example, the movement of an animal cell depends on an intricate interplay of the structures that make up a cellular skeleton (the colored fibers in the micrograph in **Figure 6.1**). Another recurring theme in biology is the interaction of organisms with their environment. Cells sense and respond to environmental fluctuations. And keep in mind the one biological theme that unifies all others: evolution. All cells are related by their descent from earlier cells. However, they have been modified in many different ways during the long evolutionary history of life on Earth.

Although cells can differ substantially from one another, they share certain common characteristics. In this chapter, we'll first examine the tools and experimental approaches that allow us to understand subcellular details; then we'll tour the cell and become acquainted with its components.

CONCEPT 6.1

To study cells, biologists use microscopes and the tools of biochemistry

It can be difficult to understand how a cell, usually too small to be seen by the unaided eye, can be so complex. How can cell biologists possibly investigate the inner workings of such tiny entities? Before we tour the cell, it will be helpful to learn how cells are studied.

Microscopy

The development of instruments that extend the human senses has gone hand in hand with the advance of science. The discovery and early study of cells progressed with the invention of microscopes in 1590 and their refinement during the 1600s. Microscopes are still indispensable for the study of cells.

The microscopes first used by Renaissance scientists, as well as the microscopes you are likely to use in the laboratory, are all light microscopes. In a **light microscope (LM)**, visible light is passed through the specimen and then through glass lenses. The lenses refract (bend) the light in such a way that the image of the specimen is magnified as it is projected into the eye, onto photographic film or a digital sensor, or onto a video screen. (See the diagram of microscope structure in Appendix D.)

Two important parameters in microscopy are magnification and resolving power, or resolution. *Magnification* is the ratio of an object's image size to its real size. *Resolution* is a measure of the clarity of the image; it is the minimum distance two points can be separated and still be distinguished as two points. For example, what appears to the unaided eye as one star in the sky may be resolved as twin stars with a telescope.

Just as the resolving power of the human eye is limited, the light microscope cannot resolve detail finer than about 0.2 micrometer (μm), or 200 nanometers (nm), the size of a small bacterium, regardless of the magnification factor **(Figure 6.2)**. This resolution is limited by the shortest wavelength of light used to illuminate the specimen. Light microscopes can magnify effectively to about 1,000 times the actual size of the specimen; at greater magnifications, additional details cannot be seen clearly. A third important parameter in microscopy is *contrast*, which accentuates differences in parts of the sample. In fact, most improvements in light microscopy in the last hundred years have involved new methods for enhancing contrast, such as staining or labeling cell components to stand out visually (**Figure 6.3**, on the next page).

Cell walls were first seen by Robert Hooke in 1665 as he looked through a microscope at dead cells from the bark of an oak tree. But it took the wonderfully crafted lenses of Antoni van Leeuwenhoek to visualize living cells. Imagine Hooke's awe when he visited van Leeuwenhoek in 1674 and the world of microorganisms—what his host called "very little animalcules"—was revealed to him. In spite of these early observations, the cell's geography remained largely uncharted for some time. Most subcellular structures—including **organelles**, which are membrane-enclosed compartments—are simply too small to be resolved by the light microscope.

Cell biology advanced rapidly in the 1950s with the introduction of the electron microscope. Instead of using light, the **electron microscope (EM)** focuses a beam of electrons through the specimen or onto its surface (see Appendix D). Resolution is inversely related to the wavelength of the radiation a microscope uses for imaging, and electron beams have much shorter wavelengths than visible light. Modern electron microscopes can theoretically achieve a resolution of about 0.002 nm, although for practical purposes they usually cannot resolve biological structures smaller than about 2 nm. Still, this resolution is a hundredfold improvement over the light

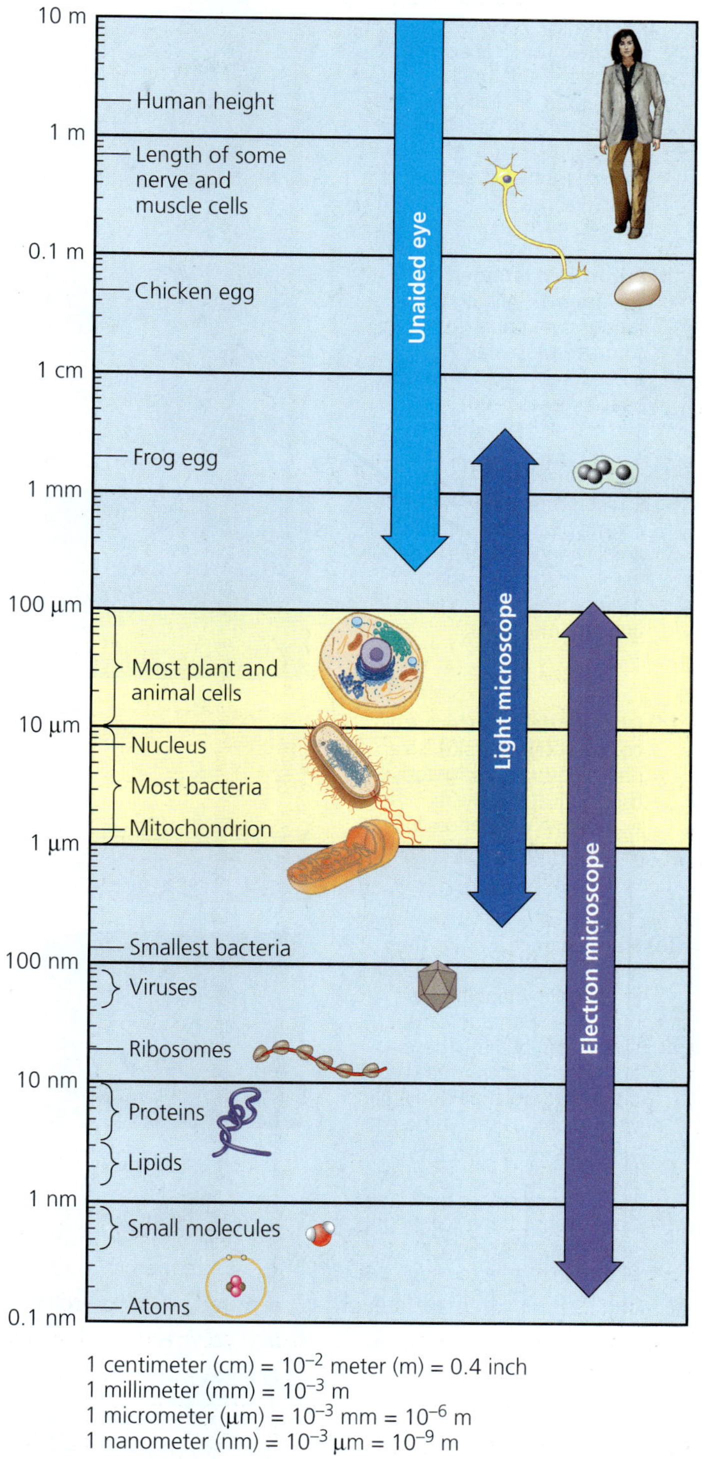

▲ **Figure 6.2 The size range of cells.** Most cells are between 1 and 100 μm in diameter (yellow region of chart) and are therefore visible only under a microscope. Notice that the scale along the left side is logarithmic to accommodate the range of sizes shown. Starting at the top of the scale with 10 m and going down, each reference measurement marks a tenfold decrease in diameter or length. For a complete table of the metric system, see Appendix C.

microscope. The term *cell ultrastructure* refers to the cellular anatomy revealed by an electron microscope.

The **scanning electron microscope (SEM)** is especially useful for detailed study of the surface of a specimen **(Figure 6.4a)**. The electron beam scans the surface of the sample, which is usually coated with a thin film of gold. The beam excites electrons on the surface, and these secondary electrons are detected by a device that translates the pattern of electrons into an electronic signal to a video screen. The result is an image of the specimen's topography. The SEM has great depth of field, resulting in an image that appears three-dimensional.

The **transmission electron microscope (TEM)** is used to study the internal ultrastructure of cells **(Figure 6.4b)**. The TEM aims an electron beam through a very thin section of the specimen, similar to the way a light microscope transmits light through a slide. The specimen has been stained with atoms of heavy metals, which attach to certain cellular structures, thus enhancing the electron density of some parts of the cell more than others. The electrons passing through the specimen are scattered more in the denser regions, so fewer are transmitted. The image displays the pattern of transmitted electrons. Instead of using glass lenses, the TEM uses electromagnets as lenses to bend the paths of the electrons, ultimately focusing the image onto a screen for viewing or onto photographic film. Some microscopes are equipped with a digital camera to photograph the image on the screen; others have a digital detector in place of both screen and camera.

▼ Figure 6.3 Research Method

Light Microscopy

TECHNIQUE | **RESULTS**

(a) Brightfield (unstained specimen). Passes light directly through specimen. Unless cell is naturally pigmented or artificially stained, image has little contrast. [Parts (a)–(d) show a human cheek epithelial cell.]

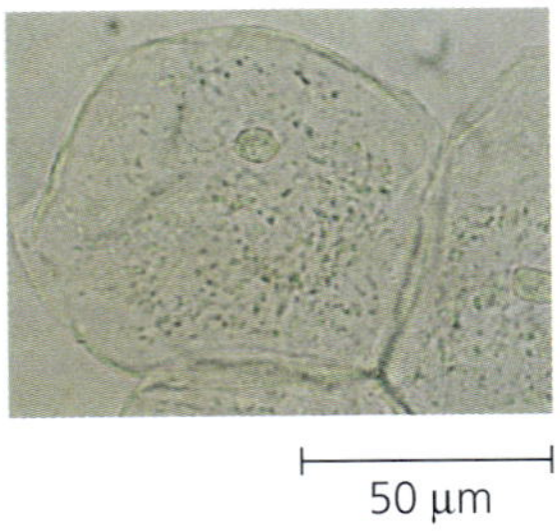

(b) Brightfield (stained specimen). Staining with various dyes enhances contrast. Most staining procedures require that cells be fixed (preserved).

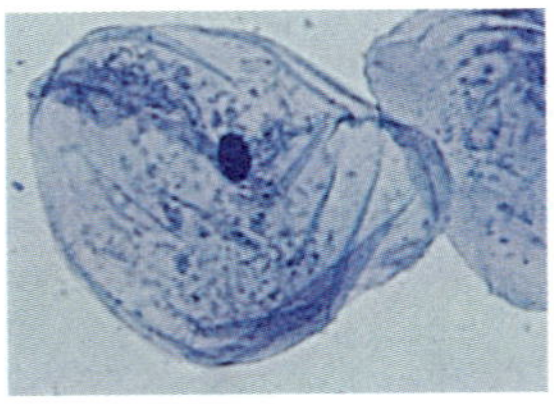

(c) Phase-contrast. Enhances contrast in unstained cells by amplifying variations in density within specimen; especially useful for examining living, unpigmented cells.

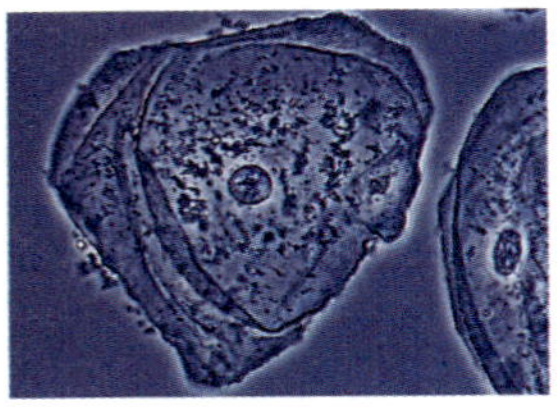

(d) Differential-interference-contrast (Nomarski). Like phase-contrast microscopy, uses optical modifications to exaggerate differences in density, making the image appear almost 3-D.

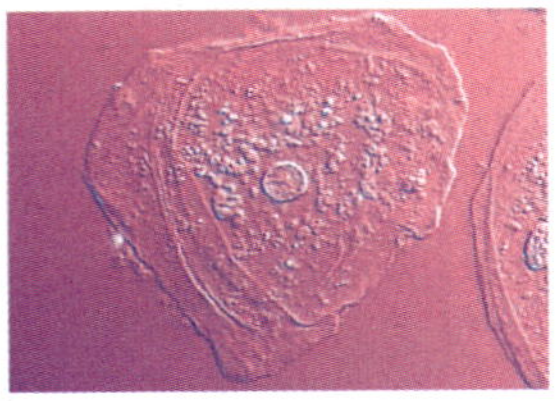

(e) Fluorescence. Shows the locations of specific molecules in the cell by tagging the molecules with fluorescent dyes or antibodies. These fluorescent substances absorb ultraviolet radiation and emit visible light, as shown here in a cell from an artery.

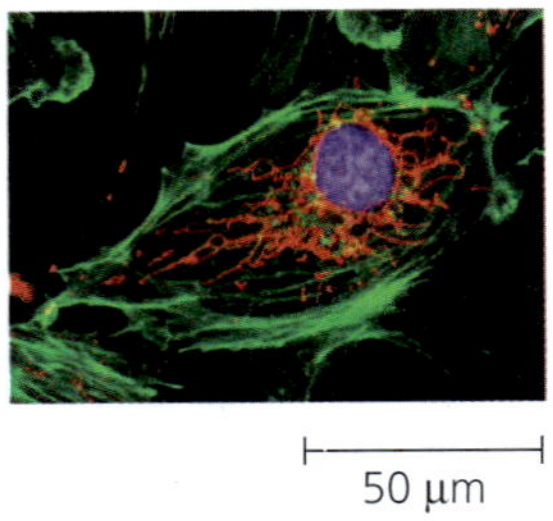

(f) Confocal. A fluorescent "optical sectioning" technique that uses a pinhole aperture to eliminate out-of-focus light from a thick sample, creating a single plane of fluorescence in the image. By capturing sharp images at many different planes, a 3-D reconstruction can be created. At the right are confocal (top) and standard fluorescent micrographs of stained nervous tissue, where nerve cells are green, support cells are red, and regions of overlap are yellow. The standard image is blurry because the out-of-focus light is not excluded.

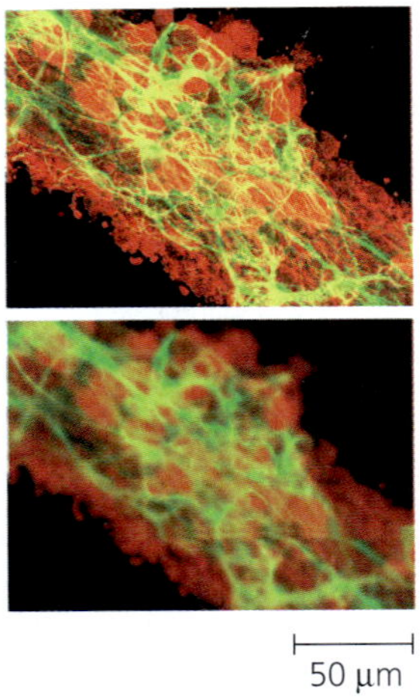

▼ Figure 6.4 Research Method

Electron Microscopy

TECHNIQUE | **RESULTS**

(a) Scanning electron microscopy (SEM). Micrographs taken with a scanning electron microscope show a 3-D image of the surface of a specimen. This SEM shows the surface of a cell from a rabbit trachea (windpipe) covered with motile organelles called cilia. Beating of the cilia helps move inhaled debris upward toward the throat.

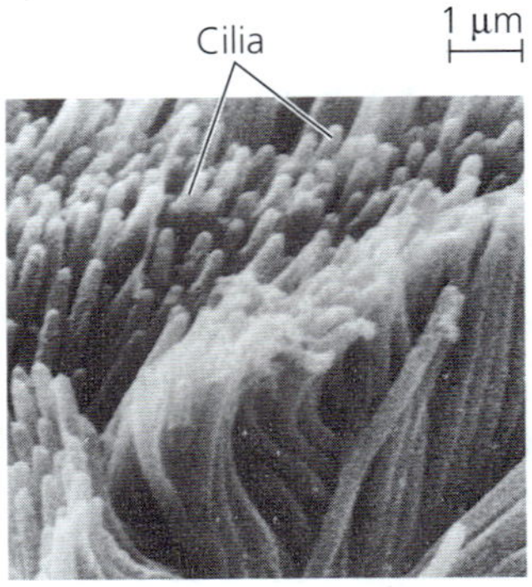

(b) Transmission electron microscopy (TEM). A transmission electron microscope profiles a thin section of a specimen. Here we see a section through a tracheal cell, revealing its ultrastructure. In preparing the TEM, some cilia were cut along their lengths, creating longitudinal sections, while other cilia were cut straight across, creating cross sections.

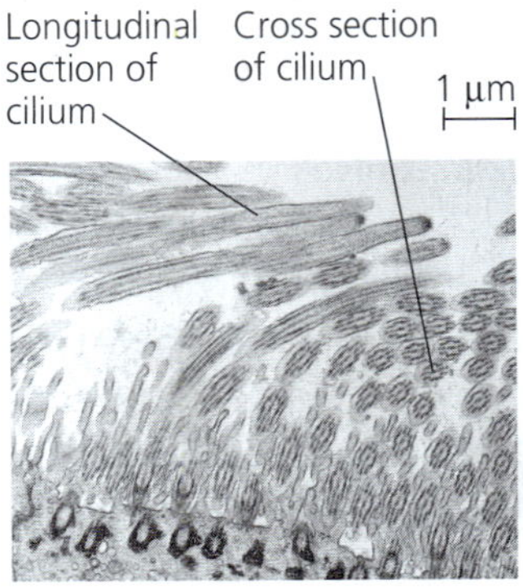

Electron microscopes reveal many organelles and other subcellular structures that are impossible to resolve with the light microscope. But the light microscope offers advantages, especially in studying living cells. A disadvantage of electron microscopy is that the methods used to prepare the specimen kill the cells. Also, specimen preparation can introduce artifacts, structural features seen in micrographs that do not exist in the living cell (as is true for all microscopy techniques). From this point on in the book, micrographs are identified by the type of microscopy: LM for a light micrograph, SEM for a scanning electron micrograph, and TEM for a transmission electron micrograph. Also, micrograph images may be artificially "colorized" to highlight particular structures.

Microscopes are the most important tools of *cytology*, the study of cell structure. But simply describing the diverse organelles and other structures within the cell reveals little about their function. Modern cell biology developed from an integration of cytology with *biochemistry*, the study of the molecules and chemical processes (metabolism) of cells.

Cell Fractionation

A useful technique for studying cell structure and function is **cell fractionation**, which takes cells apart and separates the major organelles and other subcellular structures from one another **(Figure 6.5)**. The instrument used is the centrifuge, which spins test tubes holding mixtures of disrupted cells at various speeds. The resulting forces cause a fraction of the cell components to settle to the bottom of the tube, forming a pellet. At lower speeds, the pellet consists of larger components, and higher speeds yield a pellet with smaller components. The most powerful machines, called *ultracentrifuges*, spin up to 130,000 revolutions per minute (rpm) and apply forces on particles of more than 1 million times the force of gravity (1,000,000 *g*).

Cell fractionation enables researchers to prepare specific cell components in bulk and identify their functions, a task that would be far more difficult with intact cells. For example, biochemical tests showed that one of the cell fractions produced by centrifugation included enzymes involved in cellular respiration. Electron microscopy revealed that this fraction contained large numbers of the organelles called mitochondria. Together, these data helped biologists determine that mitochondria are the sites of cellular respiration. Biochemistry and cytology thus complement each other in correlating cell function with structure.

CONCEPT CHECK 6.1

1. How do stains used for light microscopy compare with those used for electron microscopy?
2. **WHAT IF?** Which type of microscope would you use to study (a) the changes in shape of a living white blood cell, (b) the details of surface texture of a hair, and (c) the detailed structure of an organelle?

For suggested answers, see Appendix A.

▼ **Figure 6.5 Research Method**

Cell Fractionation

APPLICATION Cell fractionation is used to isolate (fractionate) cell components based on size and density.

TECHNIQUE First, cells are homogenized in a blender to break them up. The resulting mixture (cell homogenate) is then centrifuged at various speeds and durations to fractionate the cell components, forming a series of pellets, overlaid by the remaining homogenate (supernatant).

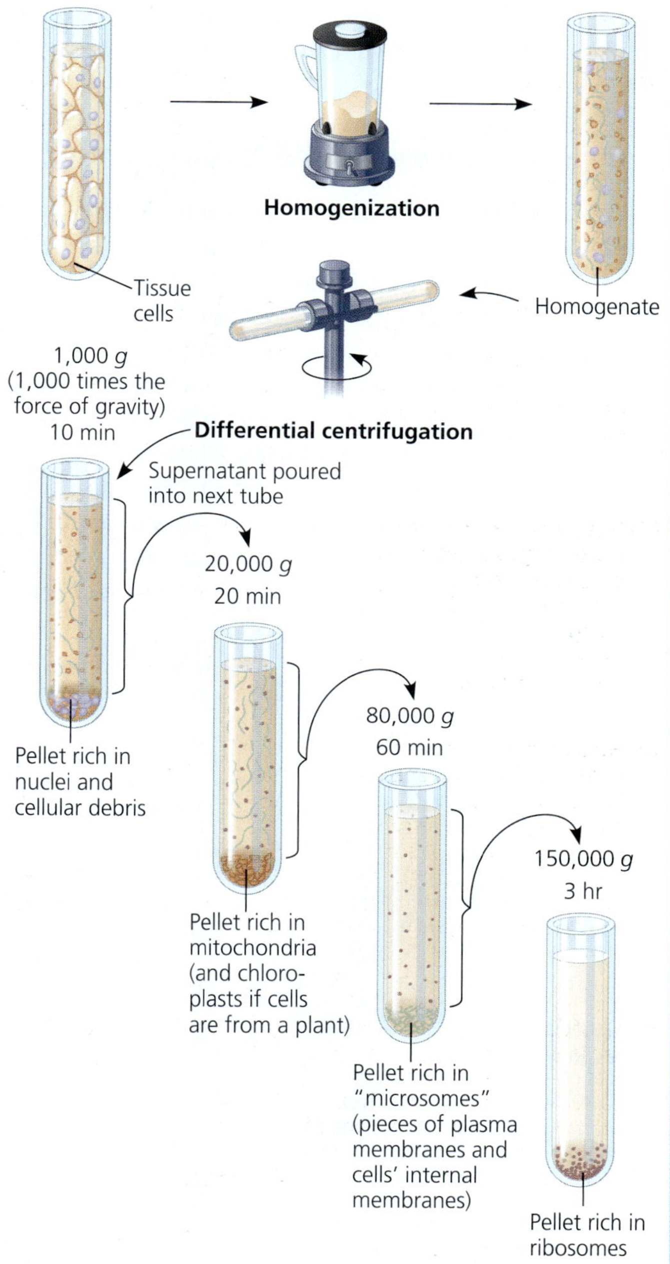

RESULTS In early experiments, researchers used microscopy to identify the organelles in each pellet and biochemical methods to determine their metabolic functions. These identifications established a baseline for this method, enabling today's researchers to know which cell fraction they should collect in order to isolate and study particular organelles.

CONCEPT **6.2**

Eukaryotic cells have internal membranes that compartmentalize their functions

The basic structural and functional unit of every organism is one of two types of cells—prokaryotic or eukaryotic. Only organisms of the domains Bacteria and Archaea consist of prokaryotic cells. Protists, fungi, animals, and plants all consist of eukaryotic cells. This chapter focuses on generalized animal and plant cells after first comparing them with prokaryotic cells.

Comparing Prokaryotic and Eukaryotic Cells

All cells have several basic features in common: They are all bounded by a selective barrier, called the *plasma membrane*. Enclosed by the membrane is a semifluid, jellylike substance called **cytosol**, in which organelles and other components are found. All cells contain *chromosomes*, which carry genes in the form of DNA. And all cells have *ribosomes*, tiny complexes that make proteins according to instructions from the genes.

A major difference between prokaryotic and eukaryotic cells is the location of their DNA, as reflected in their names. In a **eukaryotic cell**, most of the DNA is in an organelle called the *nucleus*, which is bounded by a double membrane (see Figure 6.9, on pp. 100–101). (The word *eukaryotic* is from the Greek *eu*, true, and *karyon*, kernel, here referring to the nucleus.) In a **prokaryotic cell** (from the Greek *pro*, before, and *karyon*), the DNA is concentrated in a region that is not membrane-enclosed, called the **nucleoid** (**Figure 6.6**). The interior of a prokaryotic cell is called the **cytoplasm**; this term is also used for the region between the nucleus and the plasma membrane of a eukaryotic cell. Within the cytoplasm of a eukaryotic cell, suspended in cytosol, are a variety of organelles of specialized form and function. These membrane-bounded structures are absent in prokaryotic cells. Thus, the presence or absence of a true nucleus is just one example of the disparity in structural complexity between the two types of cells.

Eukaryotic cells are generally much larger than prokaryotic cells (see Figure 6.2). Size is a general aspect of cell structure that relates to function. The logistics of carrying out cellular metabolism sets limits on cell size. At the lower limit, the smallest cells known are bacteria called mycoplasmas, which have diameters between 0.1 and 1.0 μm. These are perhaps the smallest packages with enough DNA to program metabolism and enough enzymes and other cellular equipment to carry out the activities necessary for a cell to sustain itself and reproduce. Typical bacteria are 1–5 μm in diameter, a dimension about ten times greater than that of mycoplasmas. Eukaryotic cells are typically 10–100 μm in diameter.

Metabolic requirements also impose theoretical upper limits on the size that is practical for a single cell. At the boundary of every cell, the **plasma membrane** functions as a selective barrier that allows sufficient passage of oxygen, nutrients, and wastes to

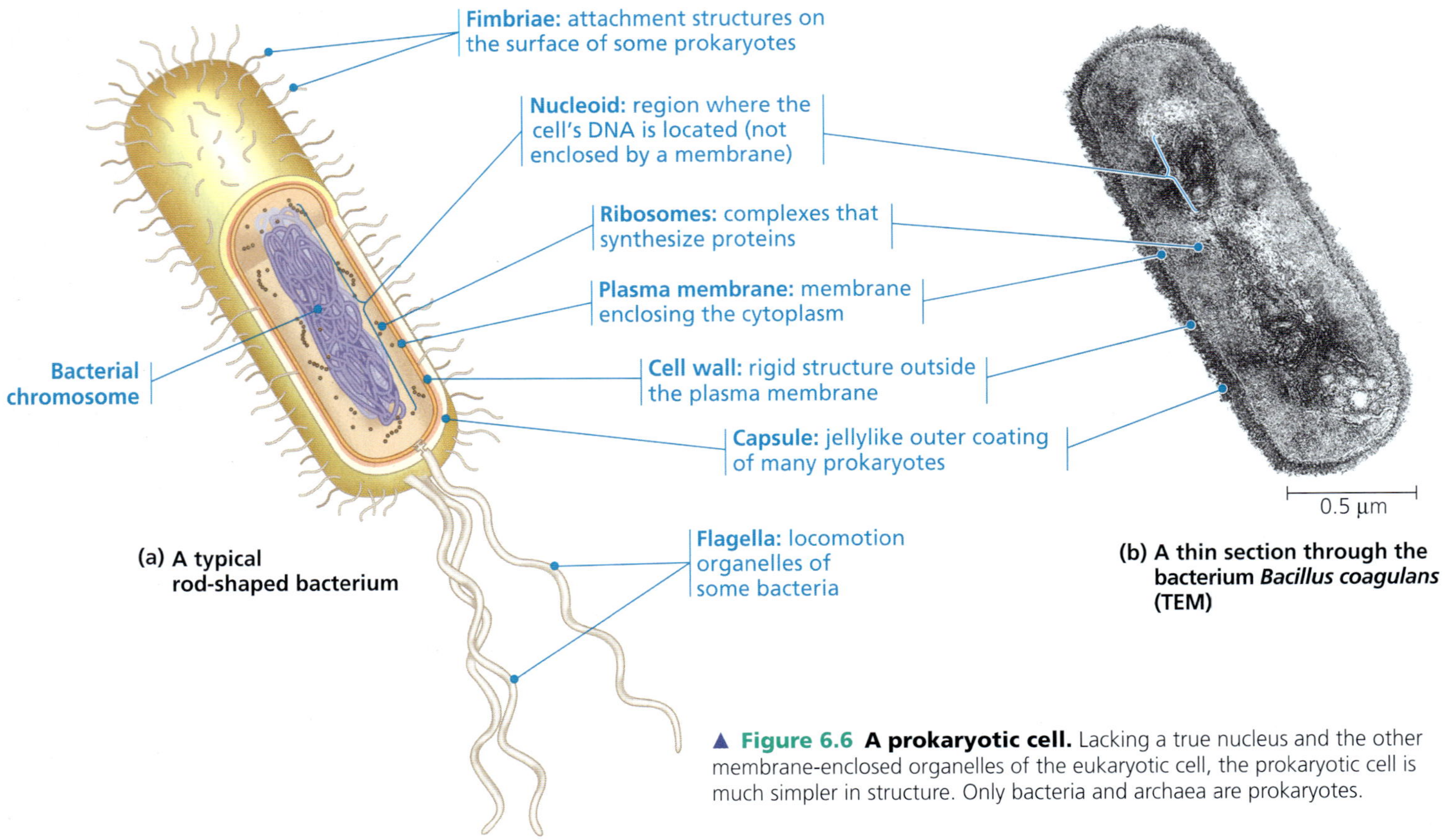

▲ **Figure 6.6 A prokaryotic cell.** Lacking a true nucleus and the other membrane-enclosed organelles of the eukaryotic cell, the prokaryotic cell is much simpler in structure. Only bacteria and archaea are prokaryotes.

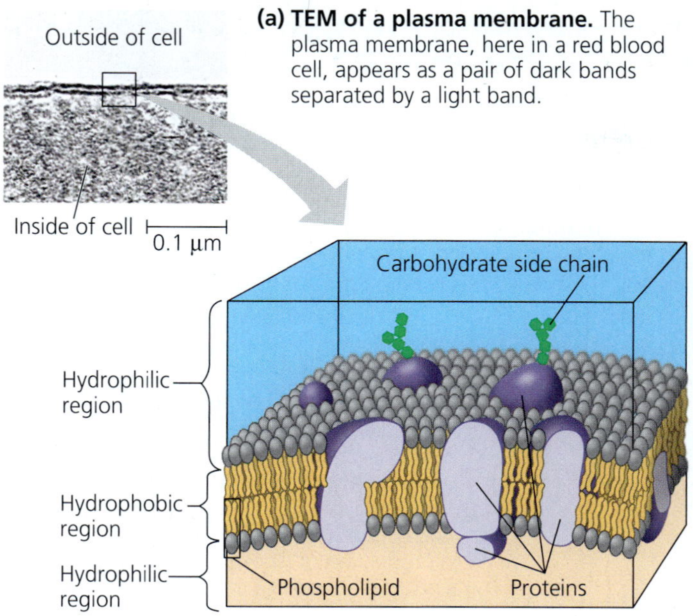

(b) Structure of the plasma membrane

▲ **Figure 6.7 The plasma membrane.** The plasma membrane and the membranes of organelles consist of a double layer (bilayer) of phospholipids with various proteins attached to or embedded in it. In the interior of a membrane, the phospholipid tails are hydrophobic, as are the interior portions of membrane proteins in contact with them. The phospholipid heads are hydrophilic, as are proteins or parts of proteins in contact with the aqueous solution on either side of the membrane. (Channels through certain proteins are also hydrophilic.) Carbohydrate side chains are found only attached to proteins or lipids on the outer surface of the plasma membrane.

? *Describe the components of a phospholipid (see Figure 5.13) that allow it to function as the major element in the plasma membrane.*

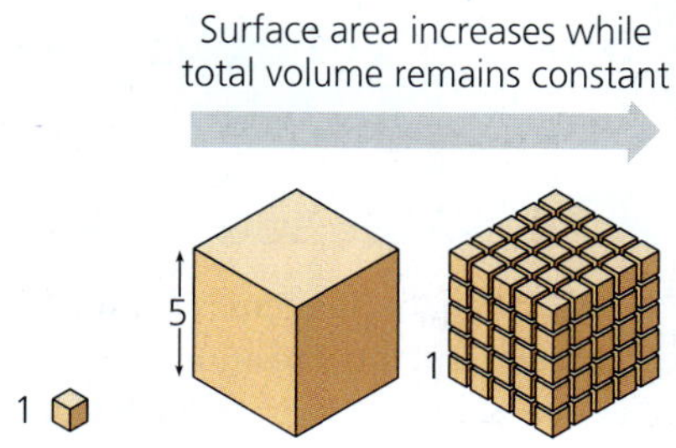

Total surface area [Sum of the surface areas (height × width) of all box sides × number of boxes]	6	150	750
Total volume [height × width × length × number of boxes]	1	125	125
Surface-to-volume (S-to-V) ratio [surface area ÷ volume]	6	1.2	6

▲ **Figure 6.8 Geometric relationships between surface area and volume.** In this diagram, cells are represented as boxes. Using arbitrary units of length, we can calculate the cell's surface area (in square units, or units^2), volume (in cubic units, or units^3), and ratio of surface area to volume. A high surface-to-volume ratio facilitates the exchange of materials between a cell and its environment.

service the entire cell **(Figure 6.7)**. For each square micrometer of membrane, only a limited amount of a particular substance can cross per second, so the ratio of surface area to volume is critical. As a cell (or any other object) increases in size, its volume grows proportionately more than its surface area. (Area is proportional to a linear dimension squared, whereas volume is proportional to the linear dimension cubed.) Thus, a smaller object has a greater ratio of surface area to volume **(Figure 6.8)**.

The need for a surface area sufficiently large to accommodate the volume helps explain the microscopic size of most cells and the narrow, elongated shapes of others, such as nerve cells. Larger organisms do not generally have *larger* cells than smaller organisms—simply *more* cells (see Figure 6.8). A sufficiently high ratio of surface area to volume is especially important in cells that exchange a lot of material with their surroundings, such as intestinal cells. Such cells may have many long, thin projections from their surface called microvilli, which increase surface area without an appreciable increase in volume.

The possible evolutionary relationships between prokaryotic and eukaryotic cells will be discussed in Chapter 25, and prokaryotic cells will be described in detail in Chapter 27. Most of the discussion of cell structure that follows in this chapter applies to eukaryotic cells.

A Panoramic View of the Eukaryotic Cell

In addition to the plasma membrane at its outer surface, a eukaryotic cell has extensive and elaborately arranged internal membranes, which divide the cell into compartments—the organelles mentioned earlier. The cell's compartments provide different local environments that facilitate specific metabolic functions, so incompatible processes can go on simultaneously inside a single cell. The plasma and organelle membranes also participate directly in the cell's metabolism, because many enzymes are built right into the membranes.

Because membranes are fundamental to the organization of the cell, Chapter 7 will discuss them in detail. In general, biological membranes consist of a double layer of phospholipids and other lipids. Embedded in this lipid bilayer or attached to its surfaces are diverse proteins (see Figure 6.7). However, each type of membrane has a unique composition of lipids and proteins suited to that membrane's specific functions. For example, enzymes embedded in the membranes of the organelles called mitochondria function in cellular respiration.

Before continuing with this chapter, examine the overviews of eukaryotic cells in **Figure 6.9**, on the next two pages. These generalized cell diagrams introduce the various organelles and provide a map of the cell for the detailed tour upon which we will now embark. Figure 6.9 also contrasts animal and plant cells. As eukaryotic cells, they have much more in common than either has with any prokaryotic cell. As you will see, however, there are important differences between animal and plant cells.

▼ Figure 6.9

Exploring Animal and Plant Cells

Animal Cell

This drawing of a generalized animal cell incorporates the most common structures of animal cells (no cell actually looks just like this). As shown by this cutaway view, the cell has a variety of components, including organelles ("little organs"), which are bounded by membranes. The most prominent organelle in an animal cell is usually the nucleus. Most of the cell's metabolic activities occur in the cytoplasm, the entire region between the nucleus and the plasma membrane. The cytoplasm contains many organelles and other cell components suspended in a semifluid medium, the cytosol. Pervading much of the cytoplasm is a labyrinth of membranes called the endoplasmic reticulum (ER).

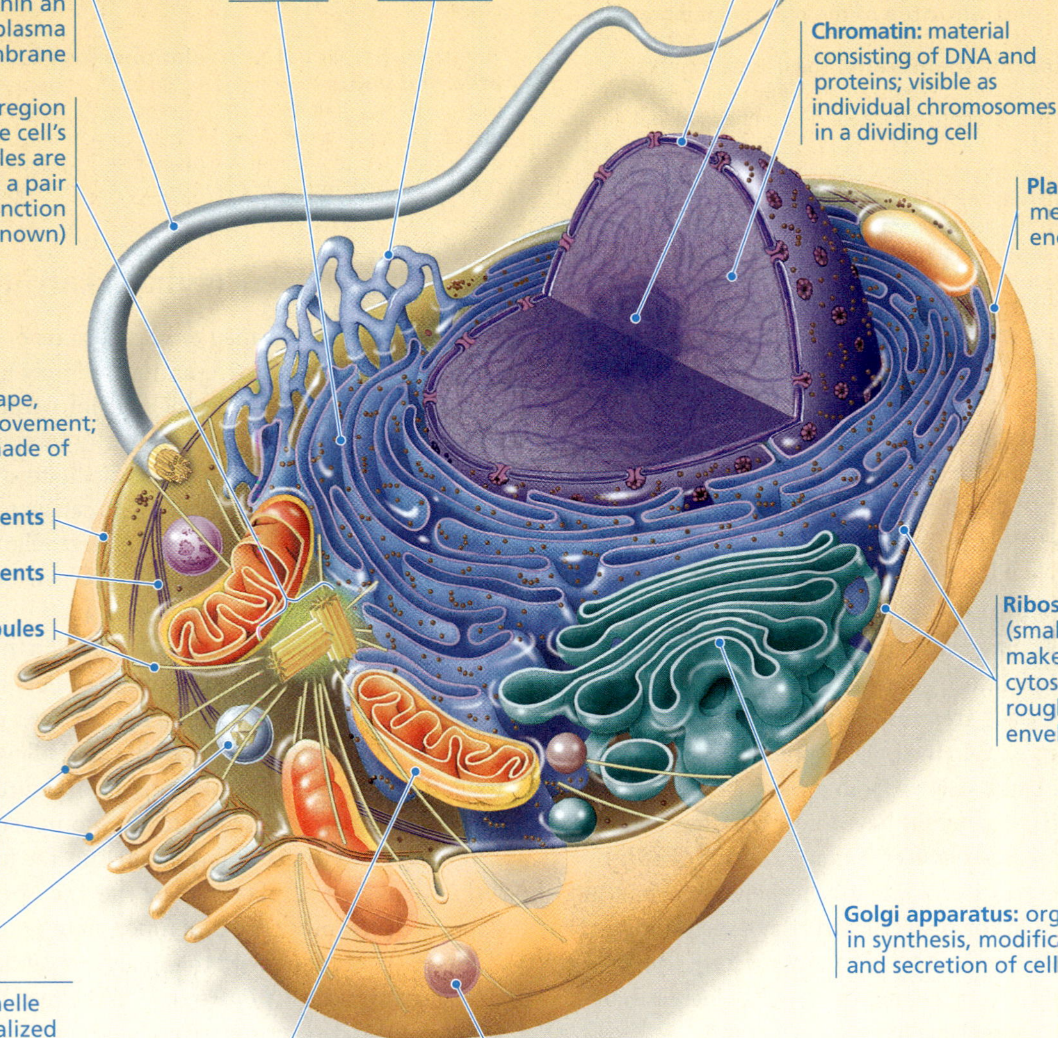

In animal cells but not plant cells:
Lysosomes
Centrosomes, with centrioles
Flagella (but present in some plant sperm)

Plant Cell

This drawing of a generalized plant cell reveals the similarities and differences between an animal cell and a plant cell. In addition to most of the features seen in an animal cell, a plant cell has organelles called plastids. The most important type of plastid is the chloroplast, which carries out photosynthesis. Many plant cells have a large central vacuole; some may have one or more smaller vacuoles. Among other tasks, vacuoles carry out functions performed by lysosomes in animal cells. Outside a plant cell's plasma membrane is a thick cell wall, perforated by channels called plasmodesmata.

MEDIA **BioFlix** Visit the Study Area at **www.masteringbio.com** for the BioFlix 3-D Animations called Tour of an Animal Cell and Tour of a Plant Cell.

If you preview the rest of the chapter now, you'll see Figure 6.9 repeated in miniature as orientation diagrams. In each case, a particular organelle is highlighted, color-coded to its appearance in Figure 6.9. As we take a closer look at individual organelles, the orientation diagrams will help you place those structures in the context of the whole cell.

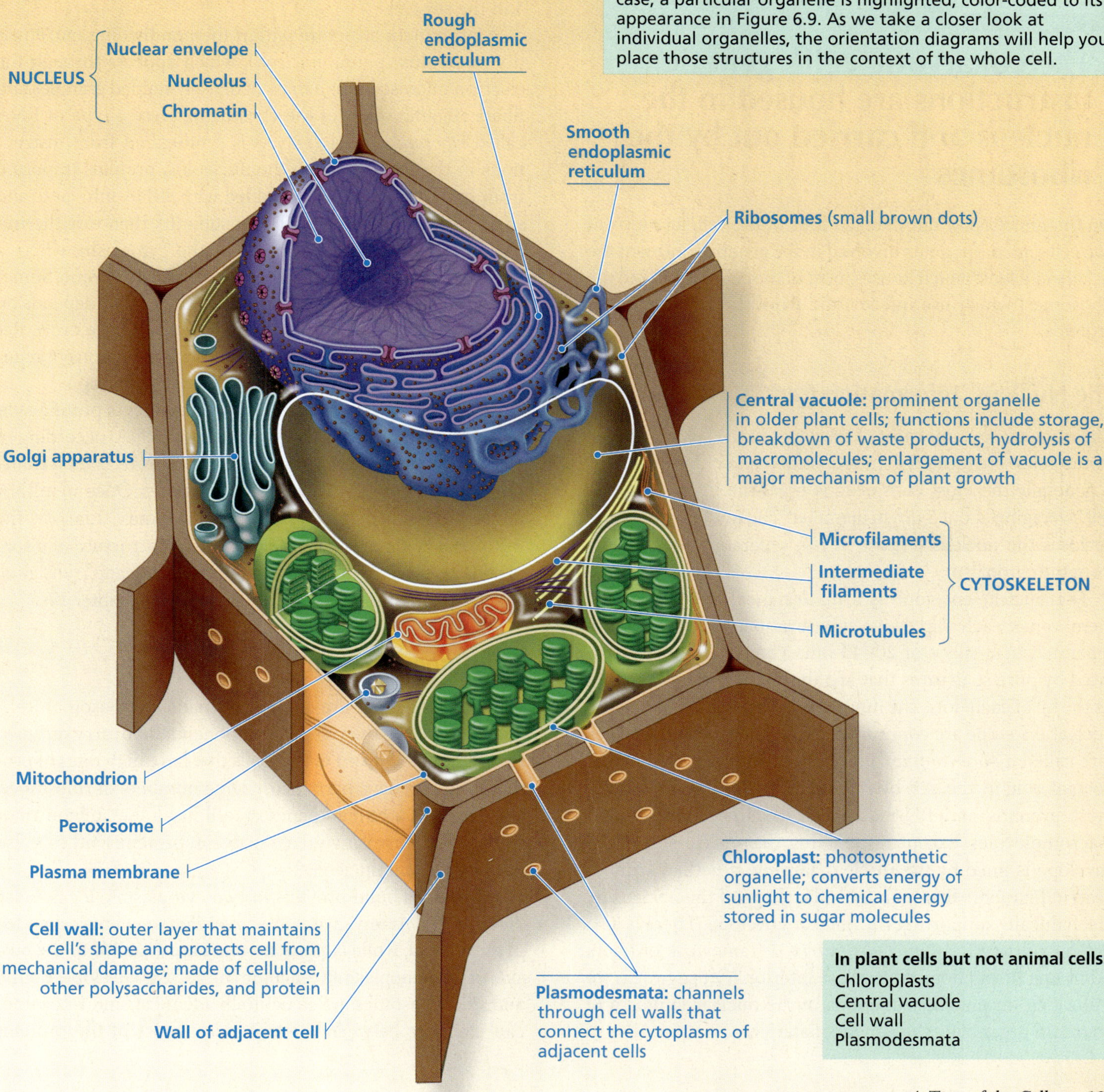

In plant cells but not animal cells:
Chloroplasts
Central vacuole
Cell wall
Plasmodesmata

CONCEPT CHECK **6.2**

1. After carefully reviewing Figure 6.9, briefly describe the structure and function of the nucleus, the mitochondrion, the chloroplast, and the endoplasmic reticulum.
2. WHAT IF? Imagine an elongated cell (such as a nerve cell) that is $125 \times 1 \times 1$, using arbitrary units similar to the ones in Figure 6.8. Predict where its surface-to-volume ratio would lie in Figure 6.8. Then calculate and check your prediction.

For suggested answers, see Appendix A.

CONCEPT **6.3**

The eukaryotic cell's genetic instructions are housed in the nucleus and carried out by the ribosomes

On the first stop of our detailed tour of the cell, let's look at two cellular components involved in the genetic control of the cell: the nucleus, which houses most of the cell's DNA, and the ribosomes, which use information from the DNA to make proteins.

The Nucleus: Information Central

The **nucleus** contains most of the genes in the eukaryotic cell (some genes are located in mitochondria and chloroplasts). It is generally the most conspicuous organelle in a eukaryotic cell, averaging about 5 μm in diameter. The **nuclear envelope** encloses the nucleus **(Figure 6.10)**, separating its contents from the cytoplasm.

The nuclear envelope is a *double* membrane. The two membranes, each a lipid bilayer with associated proteins, are separated by a space of 20–40 nm. The envelope is perforated by pore structures that are about 100 nm in diameter. At the lip of each pore, the inner and outer membranes of the nuclear envelope are continuous. An intricate protein structure called a *pore complex* lines each pore and plays an important role in the cell by regulating the entry and exit of most proteins and RNAs, as well as large complexes of macromolecules. Except at the pores, the nuclear side of the envelope is lined by the **nuclear lamina**, a netlike array of protein filaments that maintains the shape of the nucleus by mechanically supporting the nuclear envelope. There is also much evidence for a *nuclear matrix*, a framework of fibers extending throughout the nuclear interior. (On page 322, we will touch on possible functions of the nuclear lamina and matrix in organizing the genetic material.)

Within the nucleus, the DNA is organized into discrete units called **chromosomes**, structures that carry the genetic information. Each chromosome is made up of a material called **chromatin**, a complex of proteins and DNA. Stained chromatin usually appears as a diffuse mass through both light microscopes and electron microscopes. As a cell prepares to divide, however, the thin chromatin fibers coil up (condense), becoming thick enough to be distinguished as the familiar separate structures we know as chromosomes. Each eukaryotic species has a characteristic number of chromosomes. A typical human cell, for example, has 46 chromosomes in its nucleus; the exceptions are the sex cells (eggs and sperm), which have only 23 chromosomes in humans. A fruit fly cell has 8 chromosomes in most cells and 4 in the sex cells.

A prominent structure within the nondividing nucleus is the **nucleolus** (plural, *nucleoli*), which appears through the electron microscope as a mass of densely stained granules and fibers adjoining part of the chromatin. Here a type of RNA called *ribosomal RNA* (rRNA) is synthesized from instructions in the DNA. Also in the nucleolus, proteins imported from the cytoplasm are assembled with rRNA into large and small ribosomal subunits. These subunits then exit the nucleus through the nuclear pores to the cytoplasm, where a large and a small subunit can assemble into a ribosome. Sometimes there are two or more nucleoli; the number depends on the species and the stage in the cell's reproductive cycle. Recent studies suggest that the nucleolus also functions in regulation of some cellular processes, such as cell division.

As we saw in Figure 5.26, the nucleus directs protein synthesis by synthesizing messenger RNA (mRNA) according to instructions provided by the DNA. The mRNA is then transported to the cytoplasm via the nuclear pores. Once an mRNA molecule reaches the cytoplasm, ribosomes translate the mRNA's genetic message into the primary structure of a specific polypeptide. This process of transcribing and translating genetic information is described in detail in Chapter 17.

Ribosomes: Protein Factories

Ribosomes, which are complexes made of ribosomal RNA and protein, are the cellular components that carry out protein synthesis **(Figure 6.11)**. Cells that have high rates of protein synthesis have particularly large numbers of ribosomes. For example, a human pancreas cell has a few million ribosomes. Not surprisingly, cells active in protein synthesis also have prominent nucleoli.

Ribosomes build proteins in two cytoplasmic locales (see Figure 6.11). At any given time, *free ribosomes* are suspended in the cytosol, while *bound ribosomes* are attached to the outside of the endoplasmic reticulum or nuclear envelope. Bound and free ribosomes are structurally identical, and ribosomes can alternate between the two roles. Most of the proteins

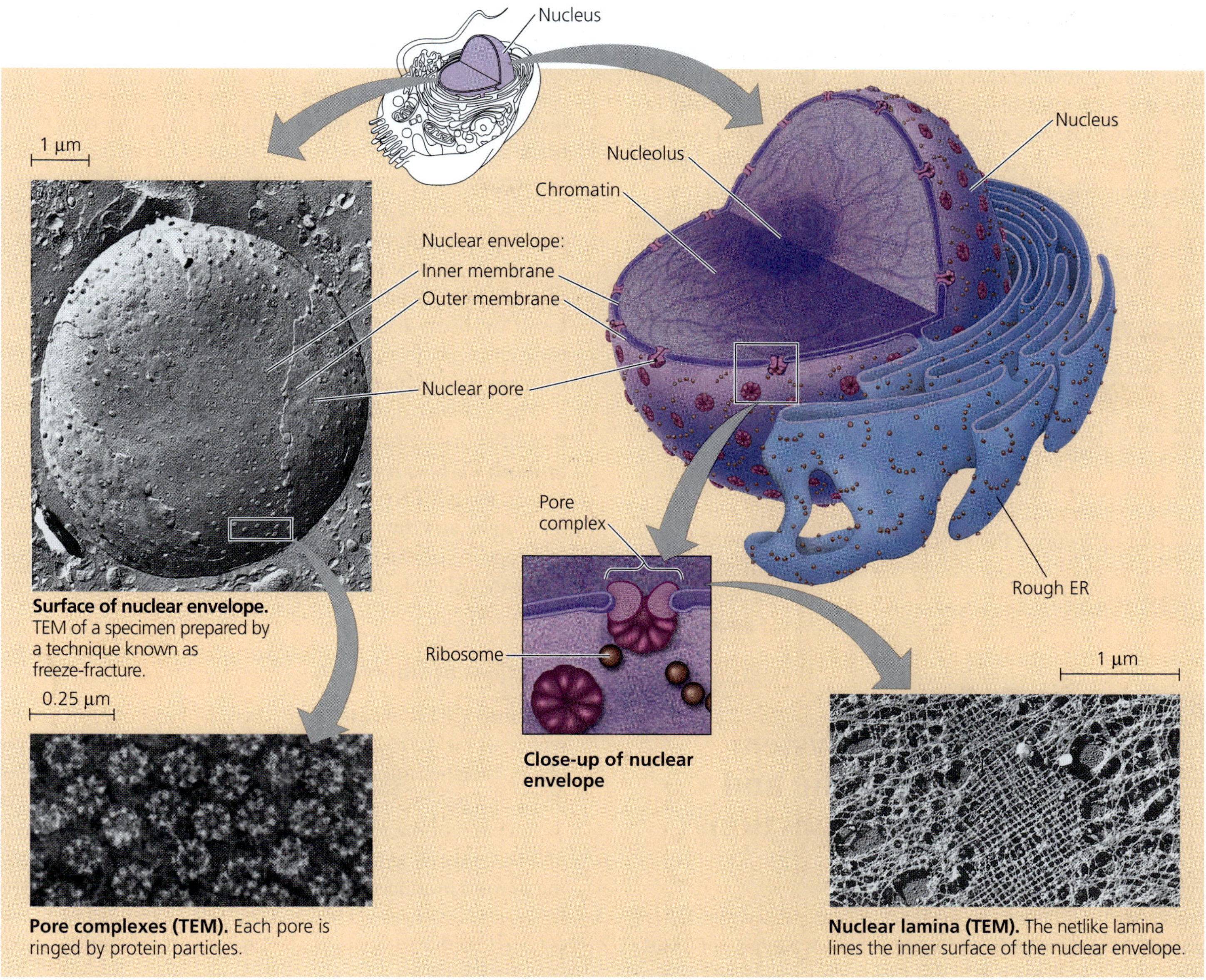

▲ **Figure 6.10 The nucleus and its envelope.** Within the nucleus are the chromosomes, which appear as a mass of chromatin (DNA and associated proteins), and one or more nucleoli (singular, *nucleolus*), which function in ribosome synthesis. The nuclear envelope, which consists of two membranes separated by a narrow space, is perforated with pores and lined by the nuclear lamina.

► **Figure 6.11 Ribosomes.** This electron micrograph of part of a pancreas cell shows many ribosomes, both free (in the cytosol) and bound (to the endoplasmic reticulum). The simplified diagram of a ribosome shows its two subunits.

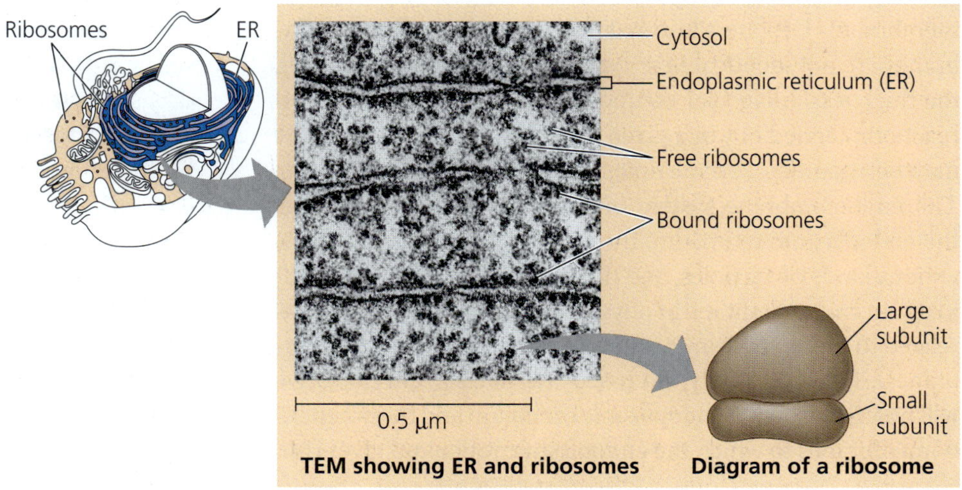

made on free ribosomes function within the cytosol; examples are enzymes that catalyze the first steps of sugar breakdown. Bound ribosomes generally make proteins that are destined for insertion into membranes, for packaging within certain organelles such as lysosomes (see Figure 6.9), or for export from the cell (secretion). Cells that specialize in protein secretion—for instance, the cells of the pancreas that secrete digestive enzymes—frequently have a high proportion of bound ribosomes. You will learn more about ribosome structure and function in Chapter 17.

CONCEPT CHECK 6.3

1. What role do the ribosomes play in carrying out genetic instructions?
2. Describe the molecular composition of nucleoli and explain their function.
3. **WHAT IF?** If the function of a particular protein in a eukaryotic cell is to make up part of the chromatin, describe the process of its synthesis. Include the cellular locations of all relevant molecules.

For suggested answers, see Appendix A.

CONCEPT 6.4

The endomembrane system regulates protein traffic and performs metabolic functions in the cell

Many of the different membranes of the eukaryotic cell are part of an **endomembrane system**, which carries out a variety of tasks in the cell. These tasks include synthesis of proteins and their transport into membranes and organelles or out of the cell, metabolism and movement of lipids, and detoxification of poisons. The membranes of this system are related either through direct physical continuity or by the transfer of membrane segments as tiny **vesicles** (sacs made of membrane). Despite these relationships, the various membranes are not identical in structure and function. Moreover, the thickness, molecular composition, and types of chemical reactions carried out in a given membrane are not fixed, but may be modified several times during the membrane's life. The endomembrane system includes the nuclear envelope, the endoplasmic reticulum, the Golgi apparatus, lysosomes, various kinds of vacuoles, and the plasma membrane (not actually an *endo*membrane in physical location, but nevertheless related to the endoplasmic reticulum and other internal membranes). Having already discussed the nuclear envelope, we will now focus on the endoplasmic reticulum and the other endomembranes to which the endoplasmic reticulum gives rise.

The Endoplasmic Reticulum: Biosynthetic Factory

The **endoplasmic reticulum (ER)** is such an extensive network of membranes that it accounts for more than half the total membrane in many eukaryotic cells. (The word *endoplasmic* means "within the cytoplasm," and *reticulum* is Latin for "little net.") The ER consists of a network of membranous tubules and sacs called cisternae (from the Latin *cisterna*, a reservoir for a liquid). The ER membrane separates the internal compartment of the ER, called the ER lumen (cavity) or cisternal space, from the cytosol. And because the ER membrane is continuous with the nuclear envelope, the space between the two membranes of the envelope is continuous with the lumen of the ER **(Figure 6.12)**.

There are two distinct, though connected, regions of the ER that differ in structure and function: smooth ER and rough ER. **Smooth ER** is so named because its outer surface lacks ribosomes. **Rough ER** has ribosomes on the outer surface of the membrane and thus appears rough through the electron microscope. As already mentioned, ribosomes are also attached to the cytoplasmic side of the nuclear envelope's outer membrane, which is continuous with rough ER.

Functions of Smooth ER

The smooth ER functions in diverse metabolic processes, which vary with cell type. These processes include synthesis of lipids, metabolism of carbohydrates, and detoxification of drugs and poisons.

Enzymes of the smooth ER are important in the synthesis of lipids, including oils, phospholipids, and steroids. Among the steroids produced by the smooth ER in animal cells are the sex hormones of vertebrates and the various steroid hormones secreted by the adrenal glands. The cells that synthesize and secrete these hormones—in the testes and ovaries, for example—are rich in smooth ER, a structural feature that fits the function of these cells.

Other enzymes of the smooth ER help detoxify drugs and poisons, especially in liver cells. Detoxification usually involves adding hydroxyl groups to drug molecules, making them more soluble and easier to flush from the body. The sedative phenobarbital and other barbiturates are examples of drugs metabolized in this manner by smooth ER in liver cells. In fact, barbiturates, alcohol, and many other drugs induce the proliferation of smooth ER and its associated detoxification enzymes, thus increasing the rate of detoxification. This, in turn, increases tolerance to the drugs, meaning that higher doses are required to achieve a particular effect, such as sedation. Also, because some of the detoxification enzymes have relatively broad action, the proliferation of smooth ER in response to one drug can increase tolerance to other drugs as well. Barbiturate abuse, for example, can decrease the effectiveness of certain antibiotics and other useful drugs.

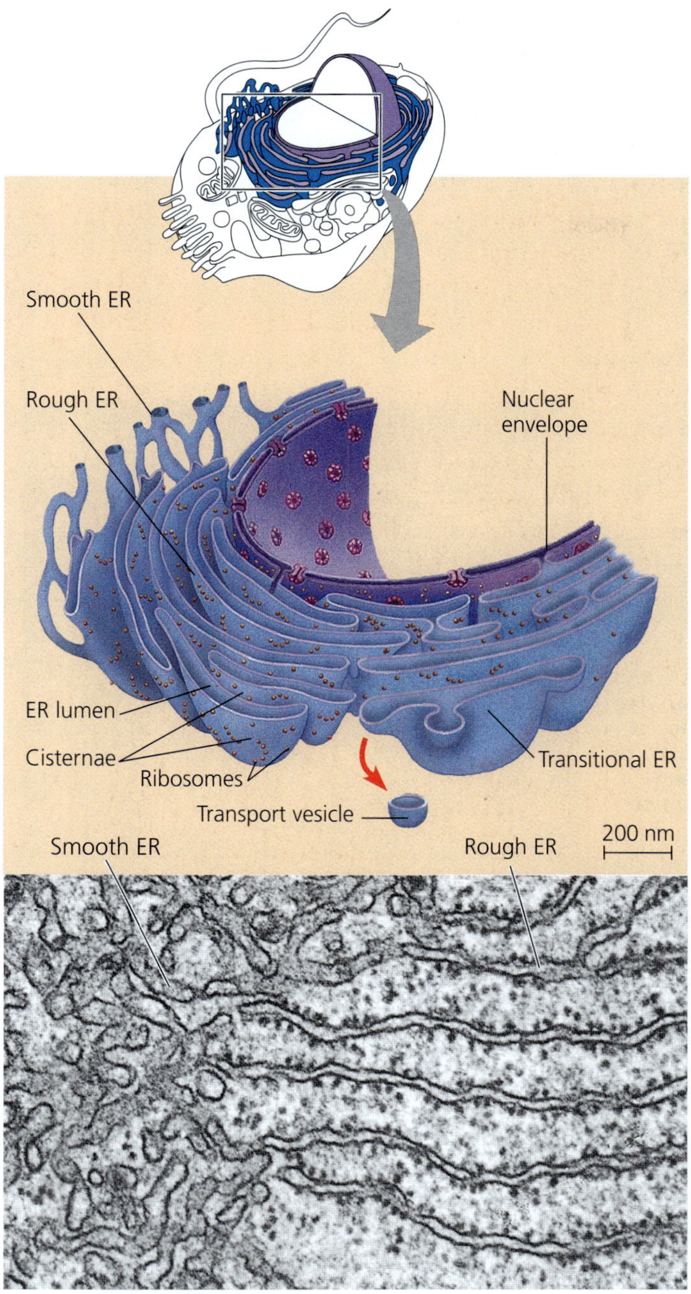

▲ **Figure 6.12 Endoplasmic reticulum (ER).** A membranous system of interconnected tubules and flattened sacs called cisternae, the ER is also continuous with the nuclear envelope. (The drawing is a cutaway view.) The membrane of the ER encloses a continuous compartment called the ER lumen (or cisternal space). Rough ER, which is studded on its outer surface with ribosomes, can be distinguished from smooth ER in the electron micrograph (TEM). Transport vesicles bud off from a region of the rough ER called transitional ER and travel to the Golgi apparatus and other destinations.

The smooth ER also stores calcium ions. In muscle cells, for example, a specialized smooth ER membrane pumps calcium ions from the cytosol into the ER lumen. When a muscle cell is stimulated by a nerve impulse, calcium ions rush back across the ER membrane into the cytosol and trigger contraction of the muscle cell. In other cell types, calcium ion release from the smooth ER triggers different responses.

Functions of Rough ER

Many types of cells secrete proteins produced by ribosomes attached to rough ER. For example, certain pancreatic cells synthesize the protein insulin on the ER and secrete this hormone into the bloodstream. As a polypeptide chain grows from a bound ribosome, it is threaded into the ER lumen through a pore formed by a protein complex in the ER membrane. As the new protein enters the ER lumen, it folds into its native shape. Most secretory proteins are **glycoproteins**, proteins that have carbohydrates covalently bonded to them. The carbohydrates are attached to the proteins in the ER by specialized molecules built into the ER membrane.

After secretory proteins are formed, the ER membrane keeps them separate from proteins that are produced by free ribosomes and will remain in the cytosol. Secretory proteins depart from the ER wrapped in the membranes of vesicles that bud like bubbles from a specialized region called transitional ER (see Figure 6.12). Vesicles in transit from one part of the cell to another are called **transport vesicles**; we will discuss their fate shortly.

In addition to making secretory proteins, rough ER is a membrane factory for the cell; it grows in place by adding membrane proteins and phospholipids to its own membrane. As polypeptides destined to be membrane proteins grow from the ribosomes, they are inserted into the ER membrane itself and are anchored there by their hydrophobic portions. The rough ER also makes its own membrane phospholipids; enzymes built into the ER membrane assemble phospholipids from precursors in the cytosol. The ER membrane expands and is transferred in the form of transport vesicles to other components of the endomembrane system.

The Golgi Apparatus: Shipping and Receiving Center

After leaving the ER, many transport vesicles travel to the **Golgi apparatus**. We can think of the Golgi as a center of manufacturing, warehousing, sorting, and shipping. Here, products of the ER, such as proteins, are modified and stored and then sent to other destinations. Not surprisingly, the Golgi apparatus is especially extensive in cells specialized for secretion.

The Golgi apparatus consists of flattened membranous sacs—cisternae—looking like a stack of pita bread (**Figure 6.13**, on the next page). A cell may have many, even hundreds, of these stacks. The membrane of each cisterna in a stack separates its internal space from the cytosol. Vesicles concentrated in the vicinity of the Golgi apparatus are engaged in the transfer of material between parts of the Golgi and other structures.

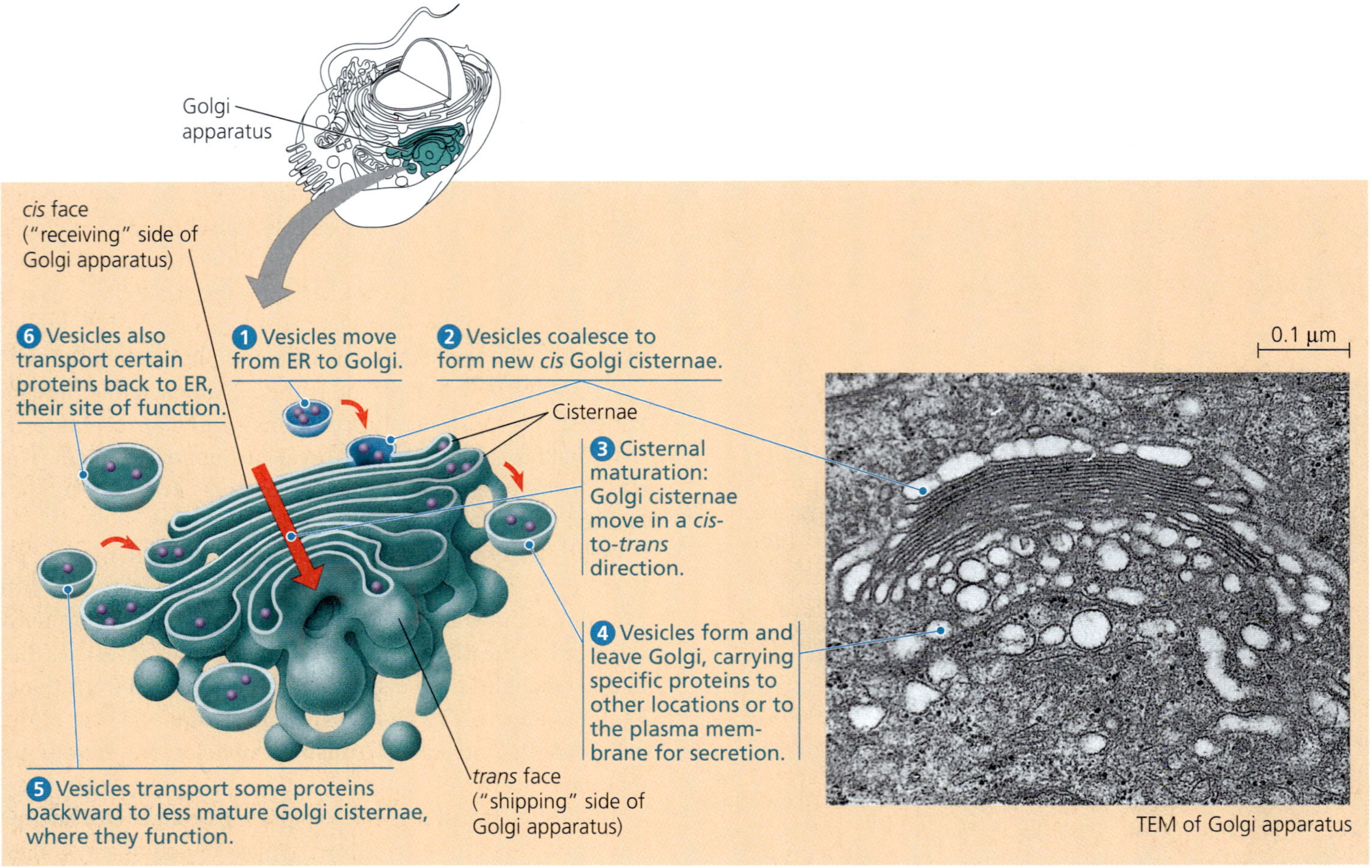

▲ **Figure 6.13 The Golgi apparatus.** The Golgi apparatus consists of stacks of flattened sacs, or cisternae, which, unlike ER cisternae, are not physically connected. (The drawing is a cutaway view.) A Golgi stack receives and dispatches transport vesicles and the products they contain. A Golgi stack has a structural and functional polarity, with a *cis* face that receives vesicles containing ER products and a *trans* face that dispatches vesicles. The cisternal maturation model proposes that the Golgi cisternae themselves "mature," moving from the *cis* to the *trans* face while carrying some proteins along. In addition, some vesicles recycle enzymes that had been carried forward in moving cisternae, transporting them "backward" to a less mature region where their functions are needed.

A Golgi stack has a distinct structural polarity, with the membranes of cisternae on opposite sides of the stack differing in thickness and molecular composition. The two poles of a Golgi stack are referred to as the *cis* face and the *trans* face; these act, respectively, as the receiving and shipping departments of the Golgi apparatus. The *cis* face is usually located near the ER. Transport vesicles move material from the ER to the Golgi apparatus. A vesicle that buds from the ER can add its membrane and the contents of its lumen to the *cis* face by fusing with a Golgi membrane. The *trans* face gives rise to vesicles, which pinch off and travel to other sites.

Products of the ER are usually modified during their transit from the *cis* region to the *trans* region of the Golgi. For example, various Golgi enzymes modify the carbohydrate portions of glycoproteins. Carbohydrates are first added to proteins in the rough ER, often during the process of polypeptide synthesis. The carbohydrate on the resulting glycoprotein is then modified as it passes through the rest of the ER and the Golgi. The Golgi removes some sugar monomers and substitutes others, producing a large variety of carbohydrates. Membrane phospholipids may also be altered in the Golgi.

In addition to its finishing work, the Golgi apparatus manufactures certain macromolecules by itself. Many polysaccharides secreted by cells are Golgi products, including pectins and certain other noncellulose polysaccharides made by plant cells and incorporated along with cellulose into their cell walls. (Cellulose is made by enzymes located within the plasma membrane, which directly deposit this polysaccharide on the outside surface.) Like secretory proteins, non-protein Golgi products that will be secreted depart from the *trans* face of the Golgi inside transport vesicles that eventually fuse with the plasma membrane.

The Golgi manufactures and refines its products in stages, with different cisternae containing unique teams of enzymes. Until recently, biologists viewed the Golgi as a static structure, with products in various stages of processing transferred from one cisterna to the next by vesicles. While this may occur, recent research has given rise to a new model of the Golgi as a more dynamic structure. According to the model called the *cisternal maturation model*, the cisternae of the Golgi actually progress forward from the *cis* to the *trans* face of the Golgi, carrying and modifying their cargo as they move. Figure 6.13 shows the details of this model.

Before a Golgi stack dispatches its products by budding vesicles from the *trans* face, it sorts these products and targets them for various parts of the cell. Molecular identification tags, such as phosphate groups added to the Golgi products, aid in sorting by acting like ZIP codes on mailing labels. Finally, transport vesicles budded from the Golgi may have external molecules on their membranes that recognize "docking sites" on the surface of specific organelles or on the plasma membrane, thus targeting the vesicles appropriately.

Lysosomes: Digestive Compartments

A **lysosome** is a membranous sac of hydrolytic enzymes that an animal cell uses to digest macromolecules. Lysosomal enzymes work best in the acidic environment found in lysosomes. If a lysosome breaks open or leaks its contents, the released enzymes are not very active because the cytosol has a neutral pH. However, excessive leakage from a large number of lysosomes can destroy a cell by autodigestion.

Hydrolytic enzymes and lysosomal membrane are made by rough ER and then transferred to the Golgi apparatus for further processing. At least some lysosomes probably arise by budding from the *trans* face of the Golgi apparatus (see Figure 6.13). Proteins of the inner surface of the lysosomal membrane and the digestive enzymes themselves are thought to be spared from destruction by having three-dimensional shapes that protect vulnerable bonds from enzymatic attack.

Lysosomes carry out intracellular digestion in a variety of circumstances. Amoebas and many other protists eat by engulfing smaller organisms or other food particles, a process called **phagocytosis** (from the Greek *phagein*, to eat, and *kytos*, vessel, referring here to the cell). The *food vacuole* formed in this way then fuses with a lysosome, whose enzymes digest the food (**Figure 6.14a**, bottom). Digestion products, including simple sugars, amino acids, and other monomers, pass into the cytosol and become nutrients for the cell. Some human cells also carry out phagocytosis. Among them are macrophages, a type of white blood cell that helps defend the body by engulfing and destroying bacteria and other invaders (see Figure 6.14a, top, and Figure 6.33).

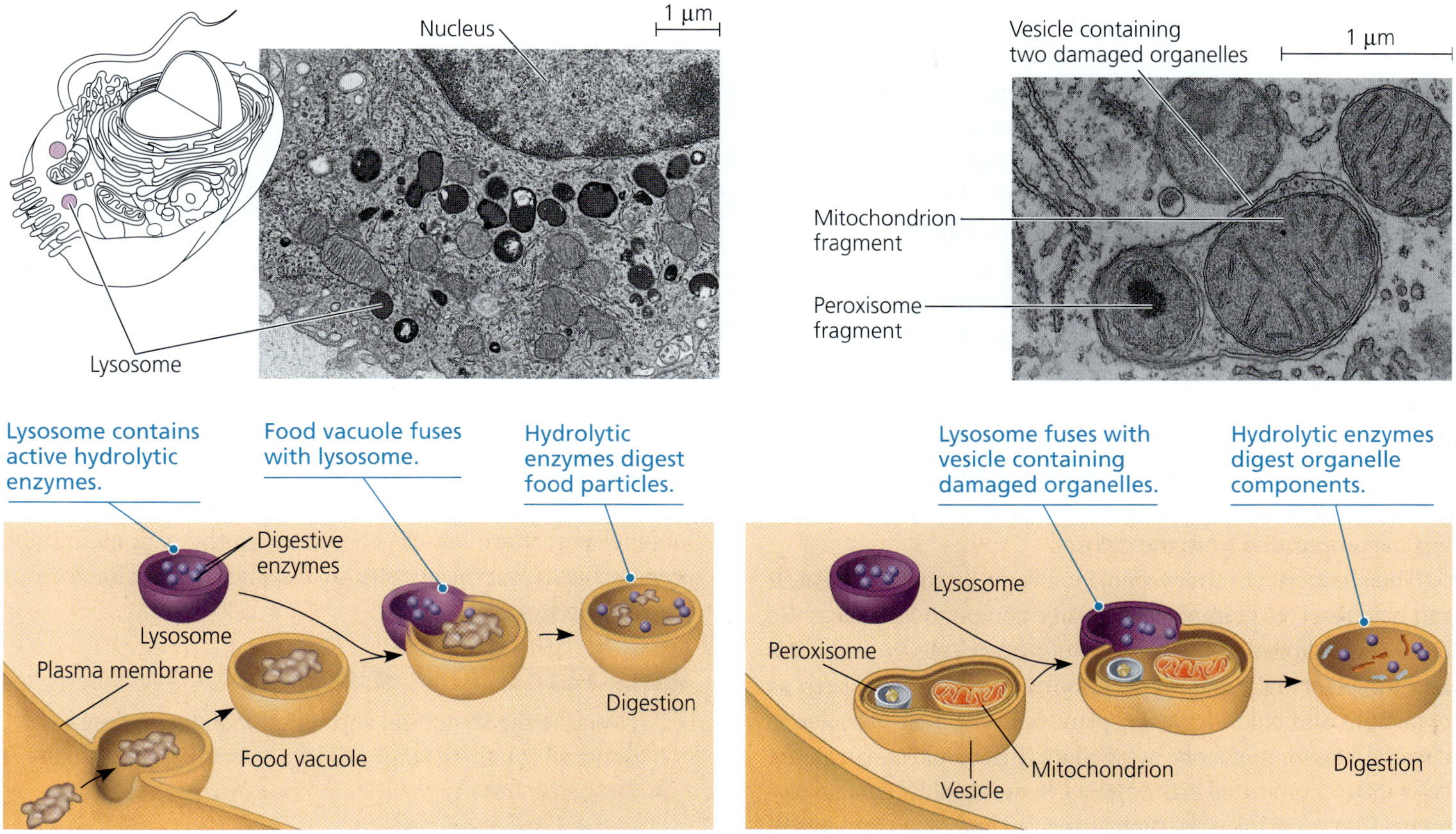

(a) Phagocytosis: lysosome digesting food

(b) Autophagy: lysosome breaking down damaged organelles

▲ **Figure 6.14 Lysosomes.** Lysosomes digest (hydrolyze) materials taken into the cell and recycle intracellular materials. **(a)** *Top:* In this macrophage (a type of white blood cell) from a rat, the lysosomes are very dark because of a stain that reacts with one of the products of digestion within the lysosome (TEM). Macrophages ingest bacteria and viruses and destroy them using lysosomes. *Bottom:* This diagram shows one lysosome fusing with a food vacuole during the process of phagocytosis by a protist. **(b)** *Top:* In the cytoplasm of this rat liver cell is a vesicle containing two disabled organelles; the vesicle will fuse with a lysosome in the process of autophagy (TEM). *Bottom:* This diagram shows fusion of such a vesicle with a lysosome. This type of vesicle has a double membrane of unknown origin. The outer membrane fuses with the lysosome, and the inner membrane is degraded along with the damaged organelles.

Lysosomes also use their hydrolytic enzymes to recycle the cell's own organic material, a process called *autophagy.* During autophagy, a damaged organelle or small amount of cytosol becomes surrounded by a double membrane, which is of unknown origin, and a lysosome fuses with the outer membrane of this vesicle **(Figure 6.14b)**. The lysosomal enzymes dismantle the enclosed material, and the organic monomers are returned to the cytosol for reuse. With the help of lysosomes, the cell continually renews itself. A human liver cell, for example, recycles half of its macromolecules each week.

The cells of people with inherited lysosomal storage diseases lack a functioning hydrolytic enzyme normally present in lysosomes. The lysosomes become engorged with indigestible substrates, which begin to interfere with other cellular activities. In Tay-Sachs disease, for example, a lipid-digesting enzyme is missing or inactive, and the brain becomes impaired by an accumulation of lipids in the cells. Fortunately, lysosomal storage diseases are rare in the general population.

Vacuoles: Diverse Maintenance Compartments

Vacuoles are membrane-bounded vesicles whose functions vary in different kinds of cells. **Food vacuoles**, formed by phagocytosis, have already been mentioned (see Figure 6.14a). Many freshwater protists have **contractile vacuoles** that pump excess water out of the cell, thereby maintaining a suitable concentration of ions and molecules inside the cell (see Figure 7.14). In plants and fungi, which lack lysosomes, vacuoles carry out hydrolysis; however, they play other roles as well. Mature plant cells generally contain a large **central vacuole** **(Figure 6.15)**. The central vacuole develops by the coalescence of smaller vacuoles, themselves derived from the endoplasmic reticulum and Golgi apparatus. The vacuole is thus an integral part of a plant cell's endomembrane system. Like all cellular membranes, the vacuolar membrane is selective in transporting solutes; as a result, the solution inside the central vacuole, called cell sap, differs in composition from the cytosol.

The plant cell's central vacuole is a versatile compartment. It can hold reserves of important organic compounds, such as the proteins stockpiled in the vacuoles of storage cells in seeds. It is also the plant cell's main repository of inorganic ions, such as potassium and chloride. Many plant cells use their vacuoles as disposal sites for metabolic by-products that would endanger the cell if they accumulated in the cytosol. Some vacuoles contain pigments that color the cells, such as the red and blue pigments of petals that help attract pollinating insects to flowers. Vacuoles may also help protect the plant against predators by containing compounds that are poisonous or unpalatable to animals. The vacuole has a major role in the growth of plant cells, which enlarge as their vacuoles absorb water, enabling the cell to become larger with a minimal investment in new cytoplasm. The cytosol often occupies only a thin layer between the central vacuole and the plasma membrane, so the ratio of plasma membrane surface to cytosolic volume is great, even for a large plant cell.

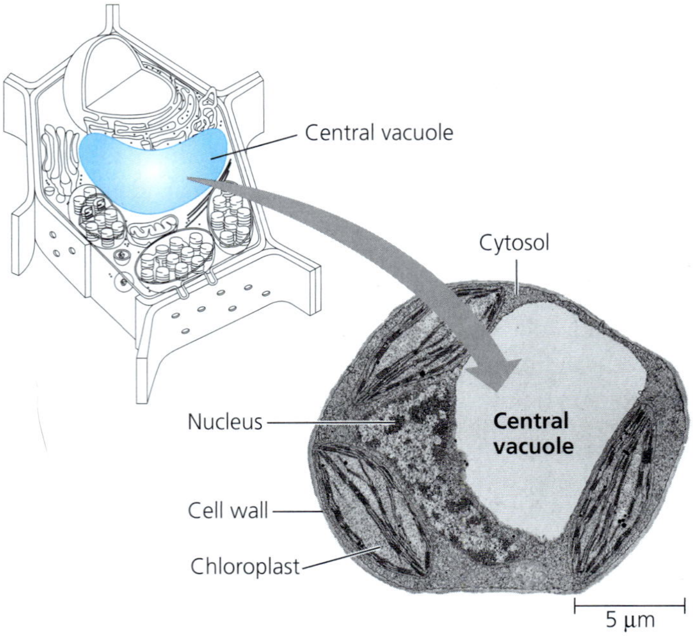

▲ **Figure 6.15 The plant cell vacuole.** The central vacuole is usually the largest compartment in a plant cell; the rest of the cytoplasm is generally confined to a narrow zone between the vacuolar membrane and the plasma membrane (TEM).

The Endomembrane System: *A Review*

Figure 6.16 reviews the endomembrane system, which shows the flow of membrane lipids and proteins through the various organelles. As the membrane moves from the ER to the Golgi and then elsewhere, its molecular composition and metabolic functions are modified, along with those of its contents. The endomembrane system is a complex and dynamic player in the cell's compartmental organization.

We'll continue our tour of the cell with some membranous organelles that are *not* closely related to the endomembrane system but play crucial roles in the energy transformations carried out by cells.

CONCEPT CHECK 6.4

1. Describe the structural and functional distinctions between rough and smooth ER.
2. Describe how transport vesicles integrate the endomembrane system.
3. **WHAT IF?** Imagine a protein that functions in the ER but requires modification in the Golgi apparatus before it can achieve that function. Describe the protein's path through the cell, starting with the mRNA molecule that specifies the protein.

For suggested answers, see Appendix A.

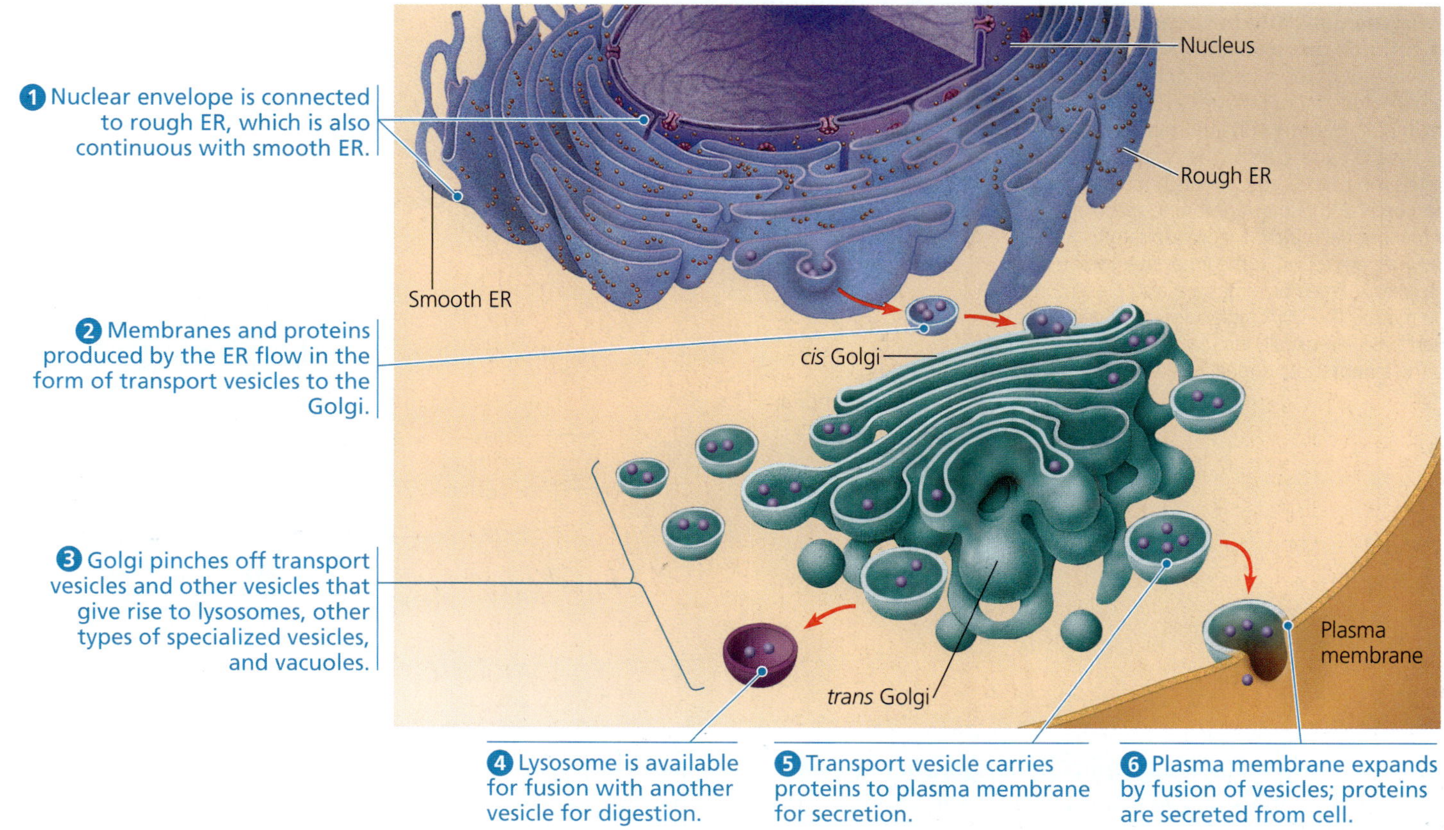

▲ **Figure 6.16 Review: relationships among organelles of the endomembrane system.** The red arrows show some of the migration pathways for membranes and the materials they enclose.

CONCEPT **6.5**

Mitochondria and chloroplasts change energy from one form to another

Organisms transform the energy they acquire from their surroundings. In eukaryotic cells, mitochondria and chloroplasts are the organelles that convert energy to forms that cells can use for work. **Mitochondria** (singular, *mitochondrion*) are the sites of cellular respiration, the metabolic process that generates ATP by extracting energy from sugars, fats, and other fuels with the help of oxygen. **Chloroplasts**, found in plants and algae, are the sites of photosynthesis. They convert solar energy to chemical energy by absorbing sunlight and using it to drive the synthesis of organic compounds such as sugars from carbon dioxide and water.

Although mitochondria and chloroplasts are enclosed by membranes, they are not part of the endomembrane system. In contrast to organelles of the endomembrane system, mitochondria have two membranes separating their innermost space from the cytosol, and chloroplasts typically have three. (Chloroplasts and related organelles in some algae have *four* membranes.) The membrane proteins of mitochondria and chloroplasts are made not by ribosomes bound to the ER, but by free ribosomes in the cytosol and by ribosomes contained within these organelles themselves. These organelles also contain a small amount of DNA. It is this DNA that programs the synthesis of the proteins made on the organelle's ribosomes. (Proteins imported from the cytosol—most of the organelle's proteins—are programmed by nuclear DNA.) Mitochondria and chloroplasts are semiautonomous organelles that grow and reproduce within the cell. In Chapters 9 and 10, we will focus on how mitochondria and chloroplasts function. We will consider the evolution of these organelles in Chapter 25. Here we are concerned mainly with the structure of these energy transformers.

In this section, we will also consider the **peroxisome**, an oxidative organelle that is not part of the endomembrane system. Like mitochondria and chloroplasts, the peroxisome imports its proteins primarily from the cytosol.

Mitochondria: Chemical Energy Conversion

Mitochondria are found in nearly all eukaryotic cells, including those of plants, animals, fungi, and most protists. Even in exceptions, such as the human intestinal parasite *Giardia* and some other protists, recent studies have identified closely related organelles that probably evolved from mitochondria.

▶ **Figure 6.17 The mitochondrion, site of cellular respiration.** The inner and outer membranes of the mitochondrion are evident in the drawing and micrograph (TEM). The cristae are infoldings of the inner membrane, which increase its surface area. The cutaway drawing shows the two compartments bounded by the membranes: the intermembrane space and the mitochondrial matrix. Many respiratory enzymes are found in the inner membrane and the matrix. Free ribosomes are also present in the matrix. The DNA molecules, too small to be seen here, are usually circular and are attached to the inner mitochondrial membrane.

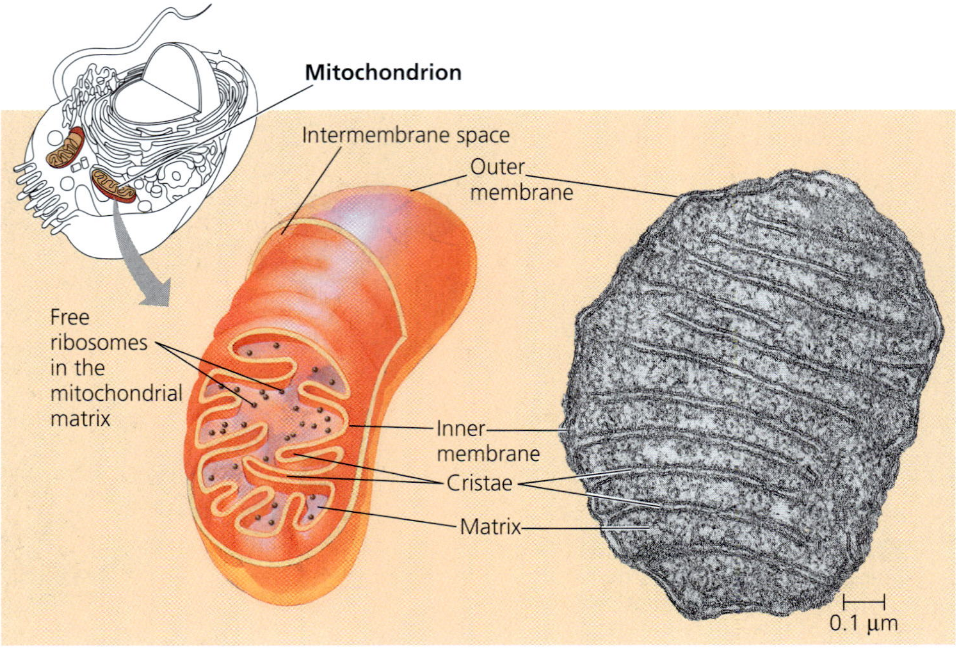

Some cells have a single large mitochondrion, but more often a cell has hundreds or even thousands of mitochondria; the number correlates with the cell's level of metabolic activity. For example, motile or contractile cells have proportionally more mitochondria per volume than less active cells. Mitochondria are about 1–10 μm long. Time-lapse films of living cells reveal mitochondria moving around, changing their shapes, and fusing or dividing in two, unlike the static cylinders seen in electron micrographs of dead cells.

The mitochondrion is enclosed by two membranes, each a phospholipid bilayer with a unique collection of embedded proteins **(Figure 6.17)**. The outer membrane is smooth, but the inner membrane is convoluted, with infoldings called **cristae**. The inner membrane divides the mitochondrion into two internal compartments. The first is the intermembrane space, the narrow region between the inner and outer membranes. The second compartment, the **mitochondrial matrix**, is enclosed by the inner membrane. The matrix contains many different enzymes as well as the mitochondrial DNA and ribosomes. Enzymes in the matrix catalyze some steps of cellular respiration. Other proteins that function in respiration, including the enzyme that makes ATP, are built into the inner membrane. As highly folded surfaces, the cristae give the inner mitochondrial membrane a large surface area, thus enhancing the productivity of cellular respiration. This is another example of structure fitting function.

Chloroplasts: Capture of Light Energy

The chloroplast is a specialized member of a family of closely related plant organelles called **plastids**. Some others are amyloplasts, colorless plastids that store starch (amylose), particularly in roots and tubers, and chromoplasts, which have pigments that give fruits and flowers their orange and yellow hues. Chloroplasts contain the green pigment chlorophyll, along with enzymes and other molecules that function in the photosynthetic production of sugar. These lens-shaped organelles, measuring about 2 μm by 5 μm, are found in leaves and other green organs of plants and in algae **(Figure 6.18)**.

The contents of a chloroplast are partitioned from the cytosol by an envelope consisting of two membranes separated by a very narrow intermembrane space. Inside the chloroplast is another membranous system in the form of flattened, interconnected sacs called **thylakoids**. In some regions, thylakoids are stacked like poker chips; each stack is called a **granum** (plural, *grana*). The fluid outside the thylakoids is the **stroma**, which contains the chloroplast DNA and ribosomes as well as many enzymes. The membranes of the chloroplast divide the chloroplast space into three compartments: the intermembrane space, the stroma, and the thylakoid space. In Chapter 10, you will learn how this compartmental organization enables the chloroplast to convert light energy to chemical energy during photosynthesis.

As with mitochondria, the static and rigid appearance of chloroplasts in micrographs or schematic diagrams is not true to their dynamic behavior in the living cell. Their shapes are changeable, and they grow and occasionally pinch in two, reproducing themselves. They are mobile and, with mitochondria and other organelles, move around the cell along tracks of the cytoskeleton, a structural network we will consider later in this chapter.

Peroxisomes: Oxidation

The peroxisome is a specialized metabolic compartment that is bounded by a single membrane **(Figure 6.19)**. Peroxisomes

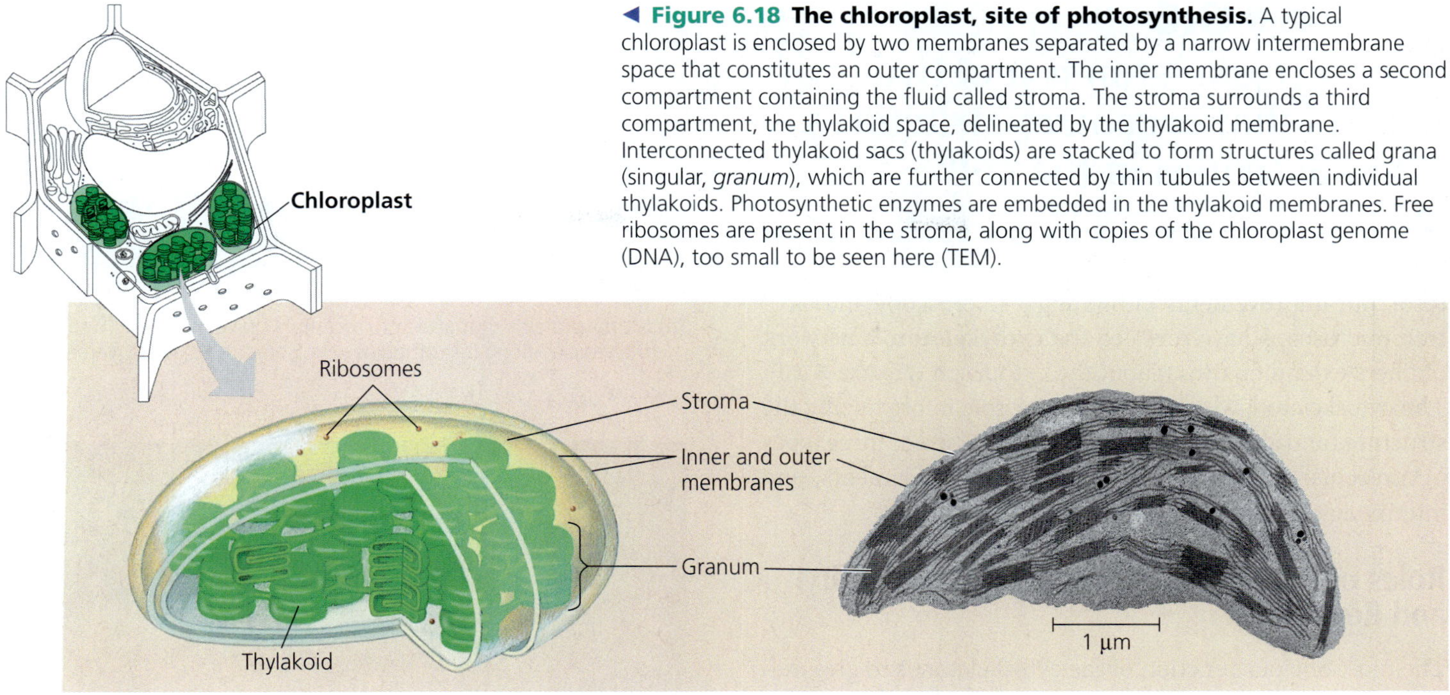

◀ **Figure 6.18 The chloroplast, site of photosynthesis.** A typical chloroplast is enclosed by two membranes separated by a narrow intermembrane space that constitutes an outer compartment. The inner membrane encloses a second compartment containing the fluid called stroma. The stroma surrounds a third compartment, the thylakoid space, delineated by the thylakoid membrane. Interconnected thylakoid sacs (thylakoids) are stacked to form structures called grana (singular, *granum*), which are further connected by thin tubules between individual thylakoids. Photosynthetic enzymes are embedded in the thylakoid membranes. Free ribosomes are present in the stroma, along with copies of the chloroplast genome (DNA), too small to be seen here (TEM).

contain enzymes that transfer hydrogen from various substrates to oxygen (O_2), producing hydrogen peroxide (H_2O_2) as a by-product, from which the organelle derives its name. These reactions may have many different functions. Some peroxisomes use oxygen to break fatty acids down into smaller molecules that can then be transported to mitochondria, where they are used as fuel for cellular respiration. Peroxisomes in the liver detoxify alcohol and other harmful compounds by transferring hydrogen from the poisons to oxygen. The H_2O_2 formed by peroxisomes is itself toxic, but the organelle also contains an enzyme that converts H_2O_2 to water. This is an excellent example of how the cell's compartmental structure is crucial to its functions: The enzymes that produce hydrogen peroxide and those that dispose of this toxic compound are sequestered in the same space, away from other cellular components that could otherwise be damaged.

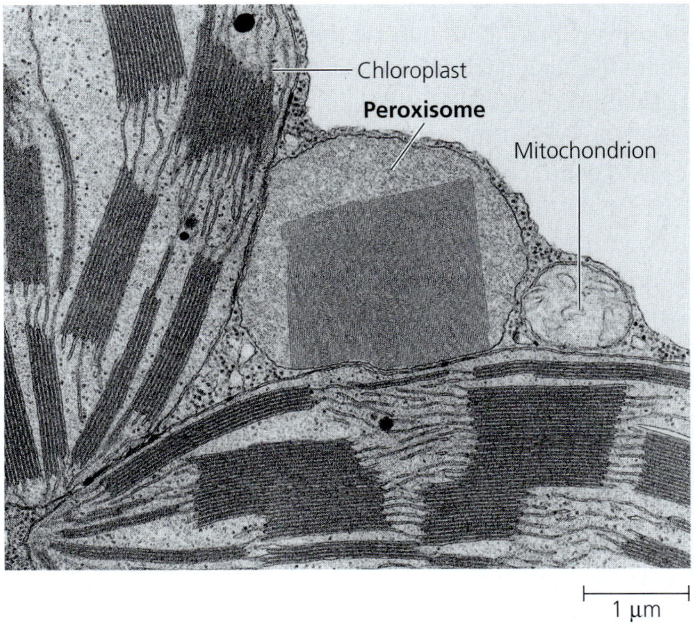

▲ **Figure 6.19 A peroxisome.** Peroxisomes are roughly spherical and often have a granular or crystalline core that is thought to be a dense collection of enzyme molecules. This peroxisome is in a leaf cell (TEM). Notice its proximity to two chloroplasts and a mitochondrion. These organelles cooperate with peroxisomes in certain metabolic functions.

Specialized peroxisomes called *glyoxysomes* are found in the fat-storing tissues of plant seeds. These organelles contain enzymes that initiate the conversion of fatty acids to sugar, which the emerging seedling uses as a source of energy and carbon until it can produce its own sugar by photosynthesis.

Unlike lysosomes, peroxisomes do not bud from the endomembrane system. They grow larger by incorporating proteins made primarily in the cytosol, lipids made in the ER, and lipids synthesized within the peroxisome itself. Peroxisomes may increase in number by splitting in two when they reach a certain size.

CONCEPT CHECK 6.5

1. Describe two common characteristics of chloroplasts and mitochondria. Consider both function and membrane structure.
2. **WHAT IF?** A classmate proposes that mitochondria, chloroplasts, and peroxisomes should be classified in the endomembrane system. Argue against the proposal.

For suggested answers, see Appendix A.

CONCEPT 6.6

The cytoskeleton is a network of fibers that organizes structures and activities in the cell

In the early days of electron microscopy, biologists thought that the organelles of a eukaryotic cell floated freely in the cytosol. But improvements in both light microscopy and electron microscopy have revealed the **cytoskeleton**, a network of fibers extending throughout the cytoplasm **(Figure 6.20)**. The cytoskeleton, which plays a major role in organizing the structures and activities of the cell, is composed of three types of molecular structures: microtubules, microfilaments, and intermediate filaments.

Roles of the Cytoskeleton: Support, Motility, and Regulation

The most obvious function of the cytoskeleton is to give mechanical support to the cell and maintain its shape. This is especially important for animal cells, which lack walls. The remarkable strength and resilience of the cytoskeleton as a whole is based on its architecture. Like a geodesic dome, the cytoskeleton is stabilized by a balance between opposing forces exerted by its elements. And just as the skeleton of an animal helps fix the positions of other body parts, the cytoskeleton provides anchorage for many organelles and even cytosolic enzyme molecules. The cytoskeleton is more dynamic than an animal skeleton, however. It can be quickly dismantled in one part of the cell and reassembled in a new location, changing the shape of the cell.

Several types of cell motility (movement) also involve the cytoskeleton. The term *cell motility* encompasses both changes in cell location and more limited movements of parts of the cell. Cell motility generally requires the interaction of the cytoskeleton with **motor proteins**. Examples of such cell motility abound. Cytoskeletal elements and motor proteins work together with plasma membrane molecules to allow whole cells to move along fibers outside the cell. Motor proteins bring about the bending of cilia and flagella by gripping microtubules within those organelles and sliding them against each other. A similar mechanism involving microfilaments causes muscle cells to contract. Inside the cell, vesicles and other organelles often travel to their destinations along "monorails" provided by the cytoskeleton. For example, this is how vesicles containing neurotransmitter molecules migrate to the tips of axons, the long extensions of nerve cells that release these molecules as chemical signals to adjacent nerve cells **(Figure 6.21)**. The vesicles that bud off from the ER travel to the Golgi along cytoskeletal tracks. The cytoskeleton also manipulates the plasma membrane in a way that forms food vacuoles or other phagocytic vesicles. And the streaming of cytoplasm that circulates materials within many large plant cells is yet another kind of cellular movement brought about by the cytoskeleton.

The cytoskeleton is also involved in regulating biochemical activities in the cell in response to mechanical stimulation. Forces exerted by extracellular molecules via cell-surface proteins are apparently transmitted into the cell by cytoskeletal elements, and the forces may even reach the nucleus. In one experiment, investigators used a micromanipulation device to

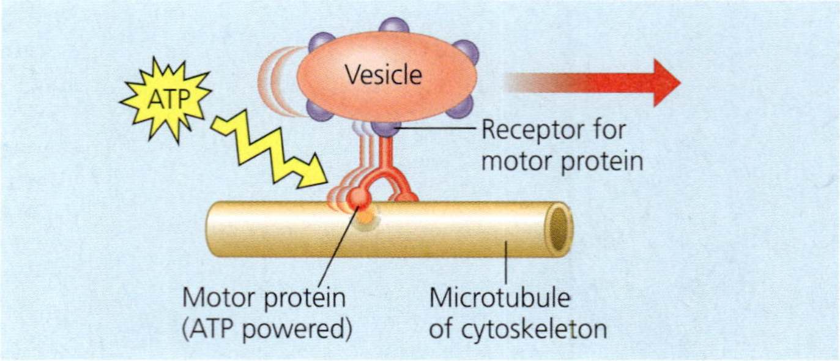

(a) Motor proteins that attach to receptors on vesicles can "walk" the vesicles along microtubules or, in some cases, microfilaments.

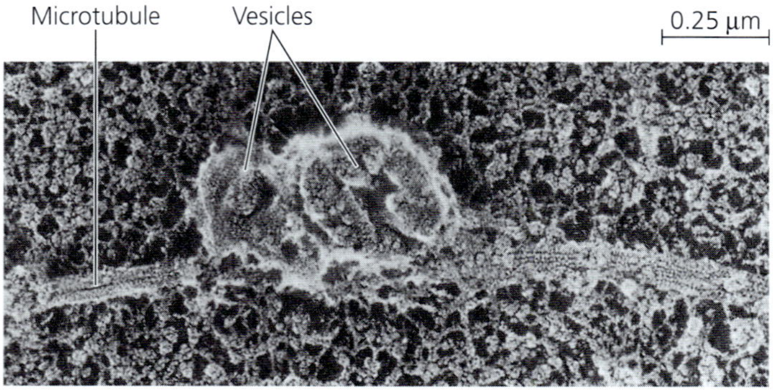

(b) Vesicles containing neurotransmitters migrate to the tips of nerve cell axons via the mechanism in (a). In this SEM of a squid giant axon, two vesicles can be seen moving along a microtubule. (A separate part of the experiment provided the evidence that they were in fact moving.)

▲ **Figure 6.21 Motor proteins and the cytoskeleton.**

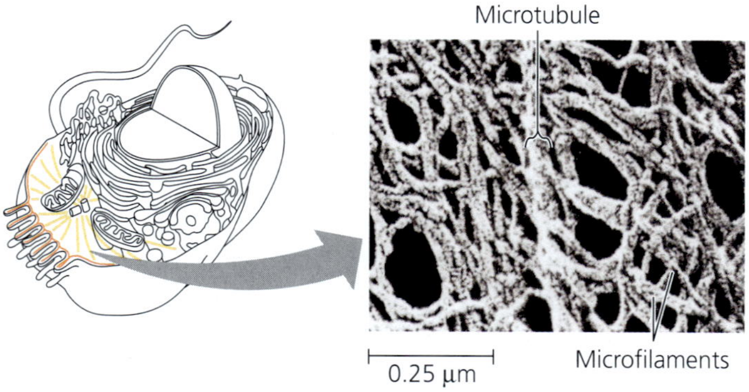

▲ **Figure 6.20 The cytoskeleton.** In this TEM, prepared by a method known as deep-etching, the thicker, hollow microtubules and the thinner, solid microfilaments are visible. A third component of the cytoskeleton, intermediate filaments, is not evident here.

Table 6.1 The Structure and Function of the Cytoskeleton

Property	Microtubules (Tubulin Polymers)	Microfilaments (Actin Filaments)	Intermediate Filaments
Structure	Hollow tubes; wall consists of 13 columns of tubulin molecules	Two intertwined strands of actin, each a polymer of actin subunits	Fibrous proteins supercoiled into thicker cables
Diameter	25 nm with 15-nm lumen	7 nm	8–12 nm
Protein subunits	Tubulin, a dimer consisting of α-tubulin and β-tubulin	Actin	One of several different proteins of the keratin family, depending on cell type
Main functions	Maintenance of cell shape (compression-resisting "girders") Cell motility (as in cilia or flagella) Chromosome movements in cell division Organelle movements	Maintenance of cell shape (tension-bearing elements) Changes in cell shape Muscle contraction Cytoplasmic streaming Cell motility (as in pseudopodia) Cell division (cleavage furrow formation)	Maintenance of cell shape (tension-bearing elements) Anchorage of nucleus and certain other organelles Formation of nuclear lamina

Micrographs of fibroblasts, a favorite cell type for cell biology studies. Each has been experimentally treated to fluorescently tag the structure of interest.

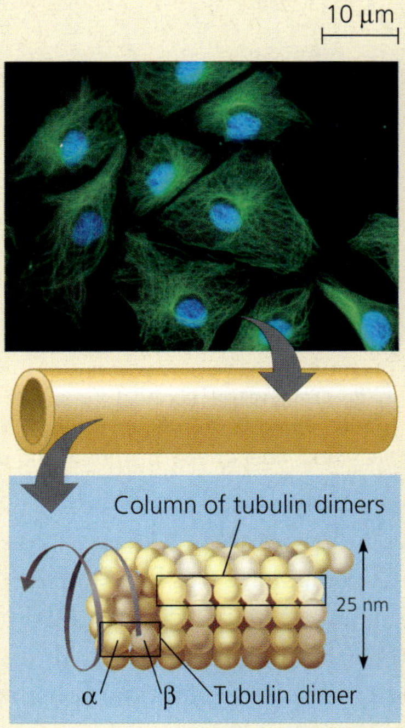

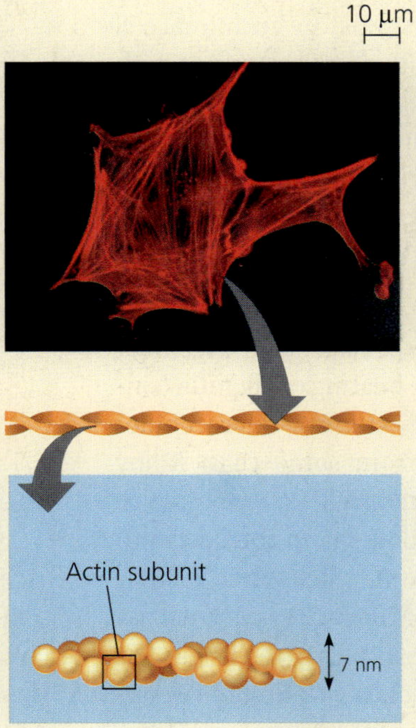

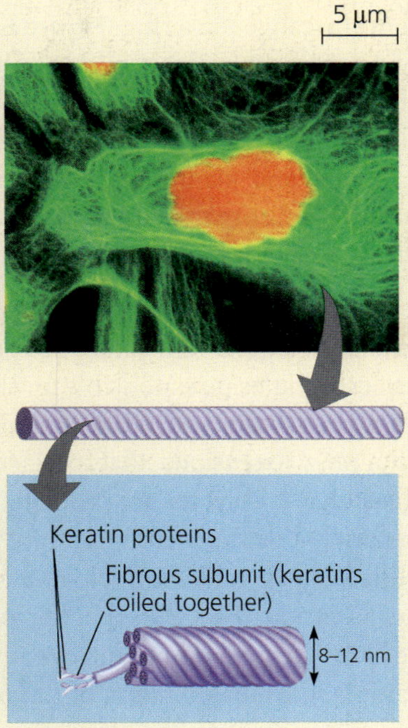

pull on certain plasma membrane proteins attached to the cytoskeleton. A video microscope captured the almost instantaneous rearrangements of nucleoli and other structures in the nucleus. In this way, cytoskeletal transmission of naturally occurring mechanical signals may help regulate and coordinate the cell's response.

Components of the Cytoskeleton

Now let's look more closely at the three main types of fibers that make up the cytoskeleton (**Table 6.1**). *Microtubules* are the thickest of the three types; *microfilaments* (also called actin filaments) are the thinnest; and *intermediate filaments* are fibers with diameters in a middle range.

Microtubules

All eukaryotic cells have **microtubules**, hollow rods measuring about 25 nm in diameter and from 200 nm to 25 µm in length. The wall of the hollow tube is constructed from a globular protein called tubulin. Each tubulin protein is a *dimer*, a molecule made up of two subunits. A tubulin dimer consists of two slightly different polypeptides, α-tubulin and β-tubulin. Microtubules grow in length by adding tubulin dimers; they can also be disassembled and their tubulin used to build microtubules elsewhere in the cell. Because of the architecture of a microtubule, its two ends are slightly different. One end can accumulate or release tubulin dimers at a much higher rate than the other, thus growing and shrinking significantly during cellular activities. (This is called the "plus end," not because it can only add tubulin proteins but because it's the end where both "on" and "off" rates are much higher.)

Microtubules shape and support the cell and also serve as tracks along which organelles equipped with motor proteins can move. To mention an example different from the one in Figure 6.21, microtubules guide secretory vesicles from the Golgi apparatus to the plasma membrane. Microtubules also separate chromosomes during cell division (see Chapter 12).

Centrosomes and Centrioles In animal cells, microtubules grow out from a **centrosome**, a region that is often located near the nucleus and is considered a "microtubule-organizing center." These microtubules function as compression-resisting girders of the cytoskeleton. Within the centrosome are a pair of **centrioles**, each composed of nine sets of triplet microtubules arranged in a ring **(Figure 6.22)**. Before an animal cell divides, the centrioles replicate. Although centrosomes with centrioles may help organize microtubule assembly in animal cells, they are not essential for this function in all eukaryotes; yeast cells and plant cells lack centrosomes with centrioles but have well-organized microtubules. Clearly, other microtubule-organizing centers must play the role of centrosomes in these cells.

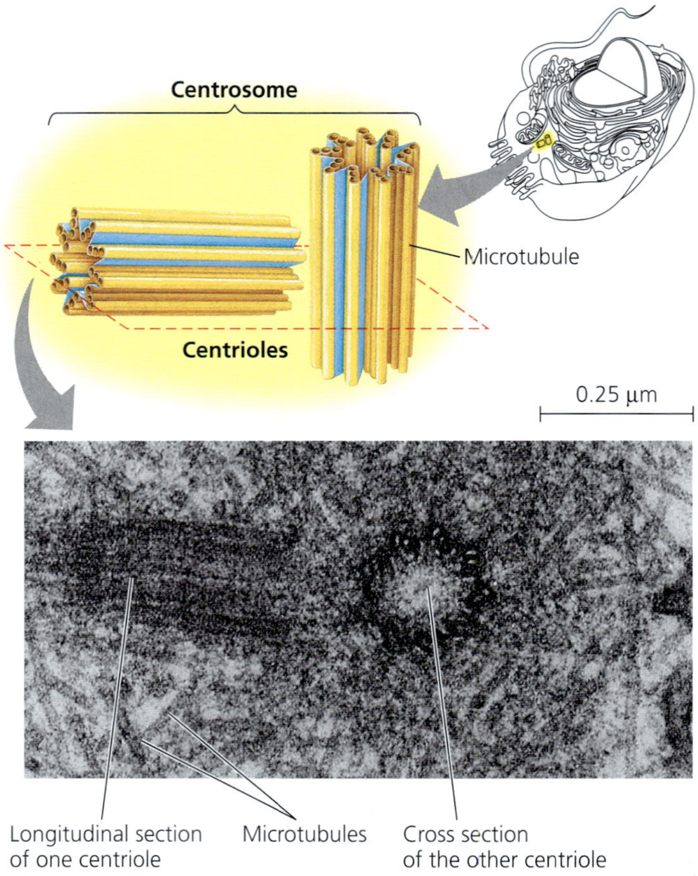

▲ **Figure 6.22 Centrosome containing a pair of centrioles.** Most animal cells have a centrosome, a region near the nucleus where the cell's microtubules are initiated. Within the centrosome is a pair of centrioles, each about 250 nm (0.25 μm) in diameter. The two centrioles are at right angles to each other, and each is made up of nine sets of three microtubules. The blue portions of the drawing represent nontubulin proteins that connect the microtubule triplets (TEM).

? *How many microtubules are in a centrosome? In the drawing, circle and label one microtubule and describe its structure.*

Cilia and Flagella In eukaryotes, a specialized arrangement of microtubules is responsible for the beating of **flagella** (singular, *flagellum*) and **cilia** (singular, *cilium*), microtubule-containing extensions that project from some cells. Many unicellular eukaryotes are propelled through water by cilia or flagella that act as locomotor appendages, and the sperm of animals, algae, and some plants have flagella. When cilia or flagella extend from cells that are held in place as part of a tissue layer, they can move fluid over the surface of the tissue. For example, the ciliated lining of the trachea (windpipe) sweeps mucus containing trapped debris out of the lungs (see Figure 6.4). In a woman's reproductive tract, the cilia lining the oviducts help move an egg toward the uterus.

Motile cilia usually occur in large numbers on the cell surface. They are about 0.25 μm in diameter and about 2–20 μm long. Flagella are the same diameter but longer, 10–200 μm. Also, flagella are usually limited to just one or a few per cell.

Flagella and cilia differ in their beating patterns **(Figure 6.23)**. A flagellum has an undulating motion that generates force in the same direction as the flagellum's axis. In contrast, cilia work more like oars, with alternating power and recovery strokes generating force in a direction perpendicular to the cilium's axis, much as the oars of a crew boat extend outward at right angles to the boat's forward movement.

A cilium may also act as a signal-receiving "antenna" for the cell. Cilia that have this function are generally nonmotile, and there is only one per cell. (In fact, in vertebrate animals, almost all cells seem to have such a cilium, which is called a *primary cilium*.) Membrane proteins on this kind of cilium transmit molecular signals from the cell's environment to its interior, triggering signaling pathways that may lead to changes in the cell's activities. Cilia-based signaling appears to be crucial to brain function and to embryonic development.

Though different in length, number per cell, and beating pattern, motile cilia and flagella share a common ultrastructure. Each has a core of microtubules sheathed in an extension of the plasma membrane **(Figure 6.24)**. Nine doublets of microtubules, the members of each pair sharing part of their walls, are arranged in a ring. In the center of the ring are two single microtubules. This arrangement, referred to as the "9 + 2" pattern, is found in nearly all eukaryotic flagella and motile cilia. (Nonmotile primary cilia have a "9 + 0" pattern, lacking the central pair of microtubules.) The microtubule assembly of a cilium or

▶ Figure 6.23 A comparison of the beating of flagella and cilia.

(a) Motion of flagella. A flagellum usually undulates, its snakelike motion driving a cell in the same direction as the axis of the flagellum. Propulsion of a human sperm cell is an example of flagellate locomotion (LM).

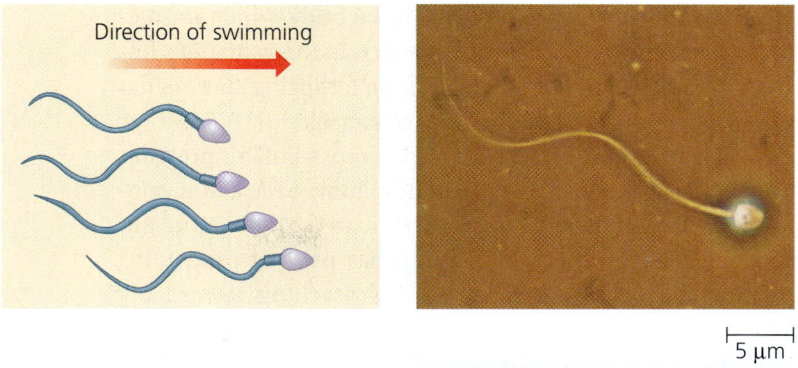

(b) Motion of cilia. Cilia have a back-and-forth motion. The rapid power stroke moves the cell in a direction perpendicular to the axis of the cilium. Then, during the slower recovery stroke, the cilium bends and sweeps sideways, closer to the surface. A dense nap of cilia, beating at a rate of about 40 to 60 strokes a second, covers this *Colpidium*, a freshwater protozoan (colorized SEM).

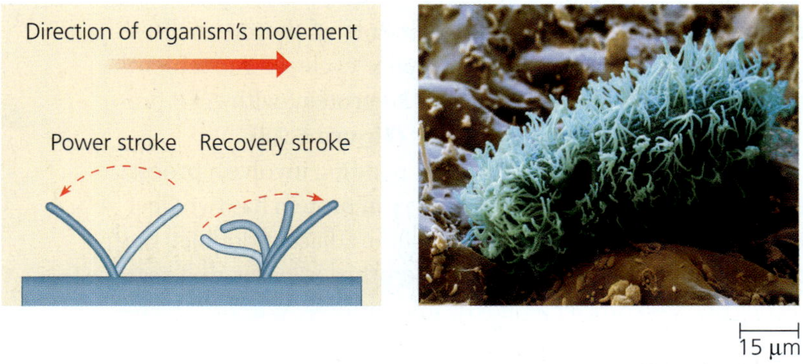

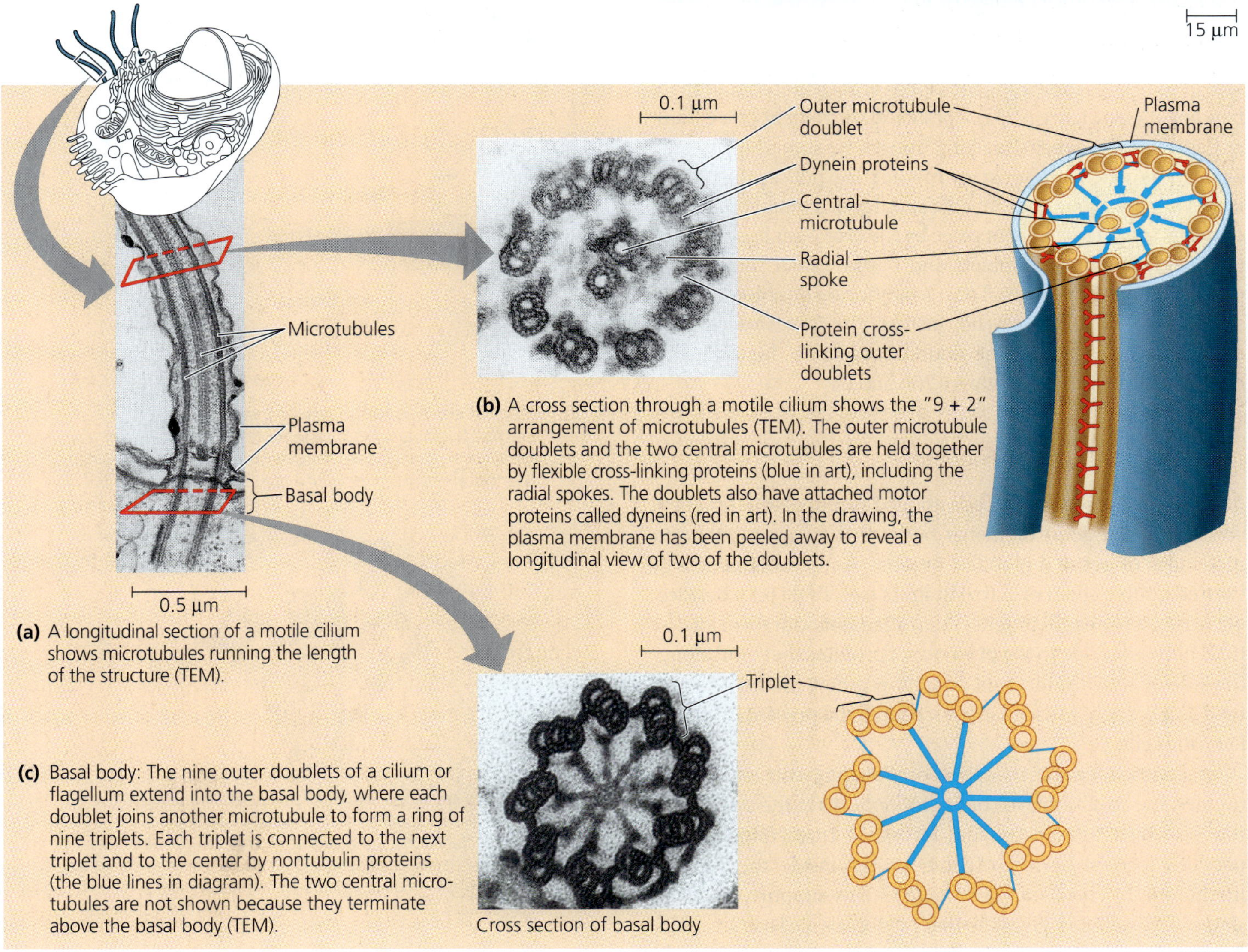

(a) A longitudinal section of a motile cilium shows microtubules running the length of the structure (TEM).

(b) A cross section through a motile cilium shows the "9 + 2" arrangement of microtubules (TEM). The outer microtubule doublets and the two central microtubules are held together by flexible cross-linking proteins (blue in art), including the radial spokes. The doublets also have attached motor proteins called dyneins (red in art). In the drawing, the plasma membrane has been peeled away to reveal a longitudinal view of two of the doublets.

(c) Basal body: The nine outer doublets of a cilium or flagellum extend into the basal body, where each doublet joins another microtubule to form a ring of nine triplets. Each triplet is connected to the next triplet and to the center by nontubulin proteins (the blue lines in diagram). The two central microtubules are not shown because they terminate above the basal body (TEM).

Cross section of basal body

▲ Figure 6.24 Ultrastructure of a eukaryotic flagellum or motile cilium.

flagellum is anchored in the cell by a **basal body**, which is structurally very similar to a centriole. In fact, in many animals (including humans), the basal body of the fertilizing sperm's flagellum enters the egg and becomes a centriole.

In flagella and motile cilia, flexible cross-linking proteins, evenly spaced along the length of the cilium or flagellum, connect the outer doublets to each other and to the two central microtubules. Each outer doublet also has pairs of protruding proteins spaced along its length and reaching toward the neighboring doublet; these are large motor proteins called **dyneins**, each composed of several polypeptides. Dyneins are responsible for the bending movements of the organelle. A dynein molcule performs a complex cycle of movements caused by changes in the shape of the protein, with ATP providing the energy for these changes **(Figure 6.25)**.

The mechanics of dynein-based bending involve a process that resembles walking. A typical dynein protein has two "feet" that "walk" along the microtubule of the adjacent doublet, one foot maintaining contact while the other releases and reattaches one step further along the microtubule. Without any restraints on the movement of the microtubule doublets, one doublet would continue to "walk" along and slide past the surface of the other, elongating the cilium or flagellum rather than bending it (see Figure 6.25a). For lateral movement of a cilium or flagellum, the dynein "walking" must have something to pull against, as when the muscles in your leg pull against your bones to move your knee. In cilia and flagella, the microtubule doublets seem to be held in place by the cross-linking proteins just inside the outer doublets and by the radial spokes and other structural elements. Thus, neighboring doublets cannot slide past each other very far. Instead, the forces exerted by dynein "walking" cause the doublets to curve, bending the cilium or flagellum (see Figure 6.25b and c).

Microfilaments (Actin Filaments)

Microfilaments are solid rods about 7 nm in diameter. They are also called actin filaments because they are built from molecules of **actin**, a globular protein. A microfilament is a twisted double chain of actin subunits (see Table 6.1). Besides occurring as linear filaments, microfilaments can form structural networks, due to the presence of proteins that bind along the side of an actin filament and allow a new filament to extend as a branch. Microfilaments seem to be present in all eukaryotic cells.

In contrast to the compression-resisting role of microtubules, the structural role of microfilaments in the cytoskeleton is to bear tension (pulling forces). A three-dimensional network formed by microfilaments just inside the plasma membrane (*cortical microfilaments*) helps support the cell's shape. This network gives the outer cytoplasmic layer of a cell, called the **cortex**, the semisolid consistency of a gel, in contrast with the more fluid (sol) state of the interior cytoplasm.

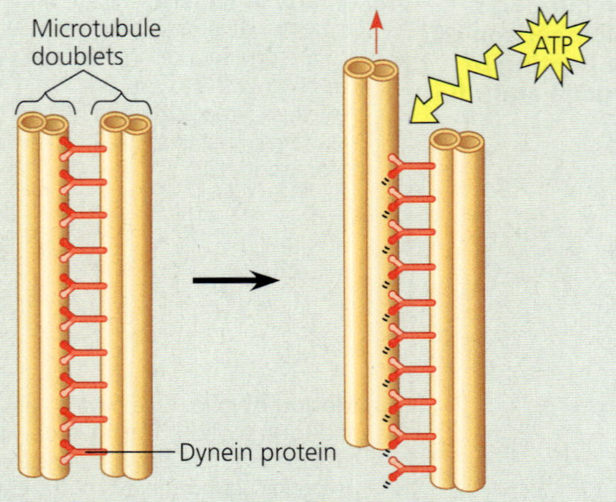

(a) Effect of unrestrained dynein movement. If a cilium or flagellum had no cross-linking proteins, the two feet of each dynein along one doublet (powered by ATP) would alternately grip and release the adjacent doublet. This "walking" motion would push the adjacent doublet up. Instead of bending, the doublets would slide past each other.

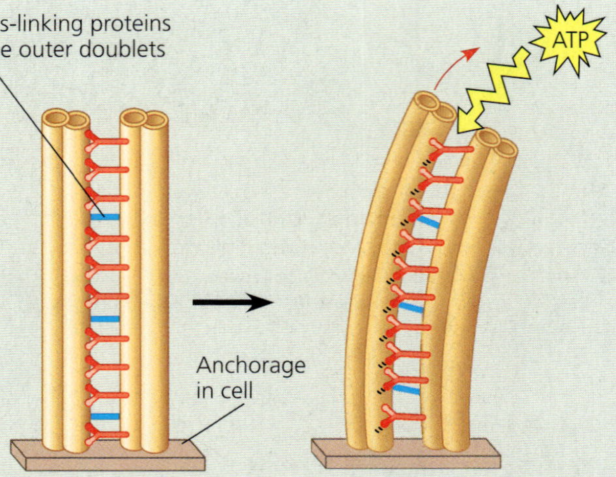

(b) Effect of cross-linking proteins. In a cilium or flagellum, two adjacent doublets cannot slide far because they are physically restrained by proteins, so they bend. (Only two of the nine outer doublets in Figure 6.24b are shown here.)

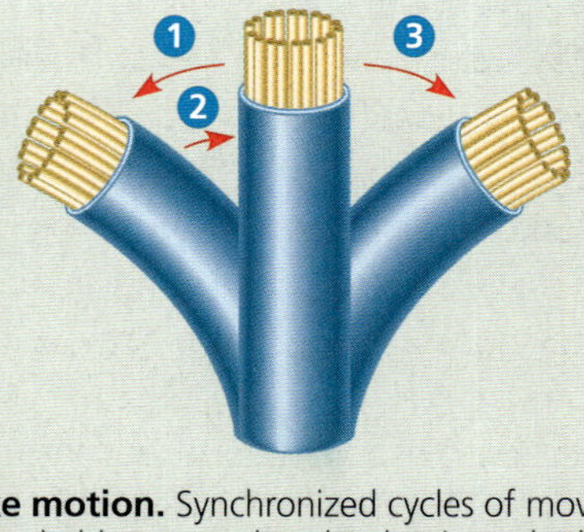

(c) Wavelike motion. Synchronized cycles of movement of many dyneins probably cause a bend to begin at the base of the cilium or flagellum and move outward toward the tip. Many successive bends, such as the ones shown here to the left and right, result in a wavelike motion. In this diagram, the two central microtubules and the cross-linking proteins are not shown.

▲ **Figure 6.25 How dynein "walking" moves flagella and cilia.**

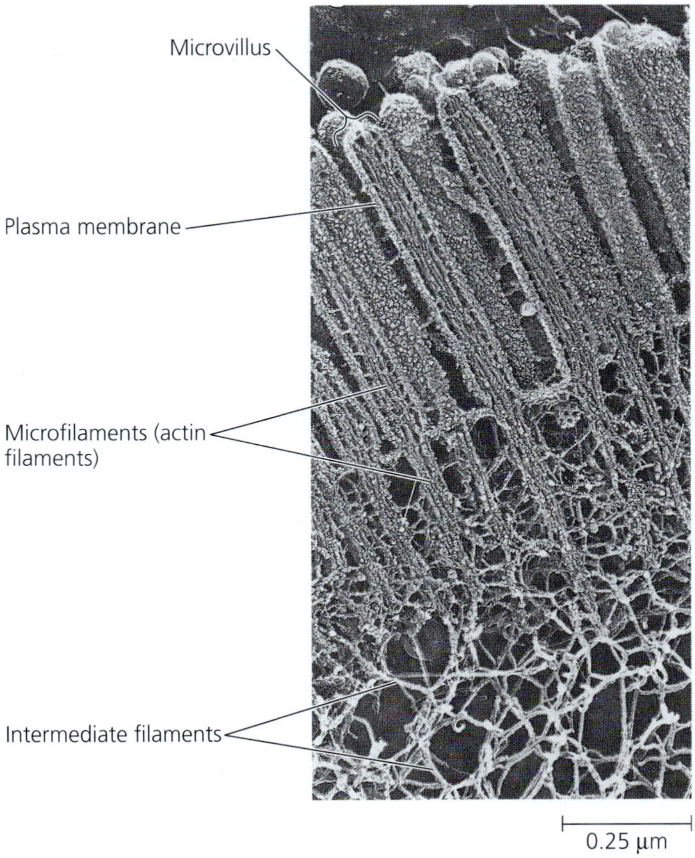

▲ **Figure 6.26 A structural role of microfilaments.** The surface area of this nutrient-absorbing intestinal cell is increased by its many microvilli (singular, *microvillus*), cellular extensions reinforced by bundles of microfilaments. These actin filaments are anchored to a network of intermediate filaments (TEM).

In animal cells specialized for transporting materials across the plasma membrane, such as intestinal cells, bundles of microfilaments make up the core of microvilli, the previously mentioned delicate projections that increase the cell surface area there **(Figure 6.26)**.

Microfilaments are well known for their role in cell motility, particularly as part of the contractile apparatus of muscle cells. Thousands of actin filaments are arranged parallel to one another along the length of a muscle cell, interdigitated with thicker filaments made of a protein called **myosin** **(Figure 6.27a)**. Like dynein when it interacts with microtubules, myosin acts as a microfilament-based motor protein by means of projections that "walk" along the actin filaments. Contraction of the muscle cell results from the actin and myosin filaments sliding past one another in this way, shortening the cell. In other kinds of cells, actin filaments are associated with myosin in miniature and less elaborate versions of the arrangement in muscle cells. These actin-myosin aggregates are responsible for localized contractions of cells. For example, a contracting belt of microfilaments forms a cleavage furrow that pinches a dividing animal cell into two daughter cells.

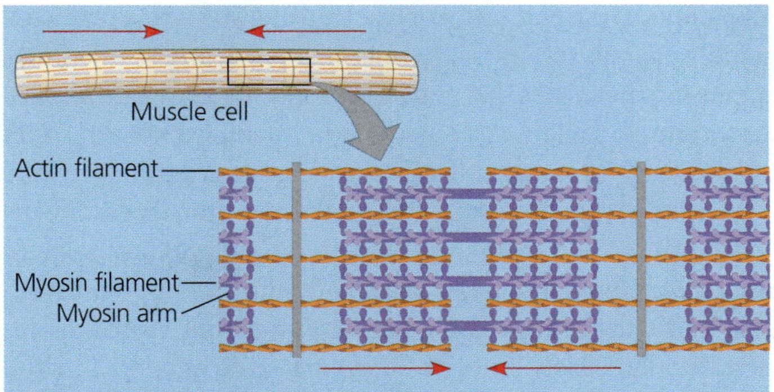

(a) Myosin motors in muscle cell contraction. The "walking" of myosin arms drives the parallel myosin and actin filaments past each other so that the actin filaments approach each other in the middle (red arrows). This shortens the muscle cell. Muscle contraction involves the shortening of many muscle cells at the same time.

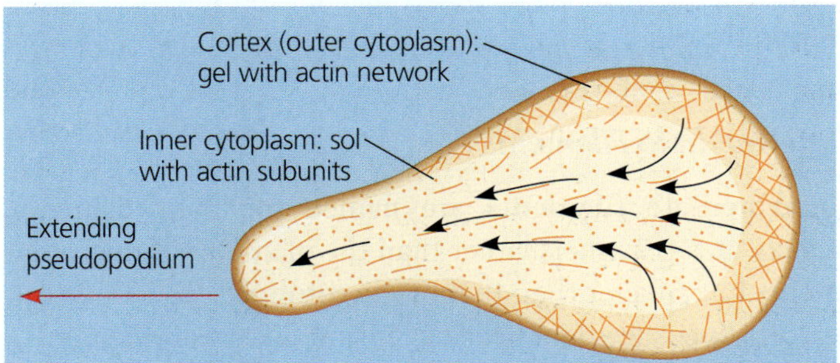

(b) Amoeboid movement. Interaction of actin filaments with myosin near the cell's trailing end (at right) squeezes the interior, more fluid cytoplasm forward (to the left) into the pseudopodium.

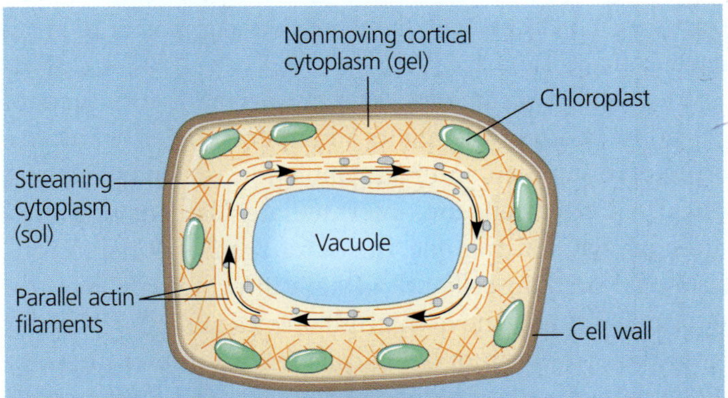

(c) Cytoplasmic streaming in plant cells. A layer of cytoplasm cycles around the cell, moving over a carpet of parallel actin filaments. Myosin motors attached to organelles in the fluid cytosol may drive the streaming by interacting with the actin.

▲ **Figure 6.27 Microfilaments and motility.** In the three examples shown in this figure, cell nuclei and most other organelles have been omitted for clarity.

Localized contraction brought about by actin and myosin also plays a role in amoeboid movement **(Figure 6.27b)**, in which a cell such as an amoeba crawls along a surface by extending and flowing into cellular extensions called **pseudopodia** (from the Greek *pseudes*, false, and *pod*, foot).

Pseudopodia extend and contract through the reversible assembly of actin subunits into microfilaments and of microfilaments into networks that convert cytoplasm from a sol to a gel. According to a widely accepted model, filaments near the cell's trailing end interact with myosin, causing contraction. Like squeezing on a toothpaste tube, this contraction forces the interior, more fluid cytoplasm into the pseudopodium, where the actin network has been weakened. The pseudopodium extends until the actin reassembles into a network. Amoebas are not the only cells that move by crawling; so do many cells in the animal body, including some white blood cells.

In plant cells, both actin-myosin interactions and sol-gel transformations brought about by actin may be involved in **cytoplasmic streaming**, a circular flow of cytoplasm within cells **(Figure 6.27c)**. This movement, which is especially common in large plant cells, speeds the distribution of materials within the cell.

Intermediate Filaments

Intermediate filaments are named for their diameter, which, at 8–12 nm, is larger than the diameter of microfilaments but smaller than that of microtubules (see Table 6.1, p. 113). Specialized for bearing tension (like microfilaments), intermediate filaments are a diverse class of cytoskeletal elements. Each type is constructed from a different molecular subunit belonging to a family of proteins whose members include the keratins. Microtubules and microfilaments, in contrast, are consistent in diameter and composition in all eukaryotic cells.

Intermediate filaments are more permanent fixtures of cells than are microfilaments and microtubules, which are often disassembled and reassembled in various parts of a cell. Even after cells die, intermediate filament networks often persist; for example, the outer layer of our skin consists of dead skin cells full of keratin proteins. Chemical treatments that remove microfilaments and microtubules from the cytoplasm of living cells leave a web of intermediate filaments that retains its original shape. Such experiments suggest that intermediate filaments are especially important in reinforcing the shape of a cell and fixing the position of certain organelles. For instance, the nucleus commonly sits within a cage made of intermediate filaments, fixed in location by branches of the filaments that extend into the cytoplasm. Other intermediate filaments make up the nuclear lamina that lines the interior of the nuclear envelope (see Figure 6.10). In cases where the shape of the entire cell is correlated with function, intermediate filaments support that shape. A case in point is the long extensions (axons) of nerve cells that transmit impulses, which are strengthened by one class of intermediate filament. Thus, the various kinds of intermediate filaments may function as the framework of the entire cytoskeleton.

CONCEPT CHECK 6.6

1. Describe shared features of microtubule-based motion of flagella and microfilament-based muscle contraction.
2. How do cilia and flagella bend?
3. **WHAT IF?** Males afflicted with Kartagener's syndrome are sterile because of immotile sperm, tend to suffer lung infections, and frequently have internal organs, such as the heart, on the wrong side of the body. This disorder has a genetic basis. Suggest what the underlying defect might be.

For suggested answers, see Appendix A.

CONCEPT 6.7

Extracellular components and connections between cells help coordinate cellular activities

Having crisscrossed the interior of the cell to explore its interior components, we complete our tour of the cell by returning to the surface of this microscopic world, where there are additional structures with important functions. The plasma membrane is usually regarded as the boundary of the living cell, but most cells synthesize and secrete materials that are external to the plasma membrane. Although these materials and the structures they form are outside the cell, their study is central to cell biology because they are involved in a great many cellular functions.

Cell Walls of Plants

The **cell wall** is an extracellular structure of plant cells that distinguishes them from animal cells. The wall protects the plant cell, maintains its shape, and prevents excessive uptake of water. On the level of the whole plant, the strong walls of specialized cells hold the plant up against the force of gravity. Prokaryotes, fungi, and some protists also have cell walls, but we will postpone discussion of them until Unit Five.

Plant cell walls are much thicker than the plasma membrane, ranging from 0.1 μm to several micrometers. The exact chemical composition of the wall varies from species to species and even from one cell type to another in the same plant, but the basic design of the wall is consistent. Microfibrils made of the polysaccharide cellulose (see Figure 5.8) are synthesized by an enzyme called cellulose synthase and secreted to the extracellular space, where they become embedded in a matrix of other polysaccharides and proteins. This combination of materials, strong fibers in a "ground substance" (matrix), is the same basic architectural design found in steel-reinforced concrete and in fiberglass.

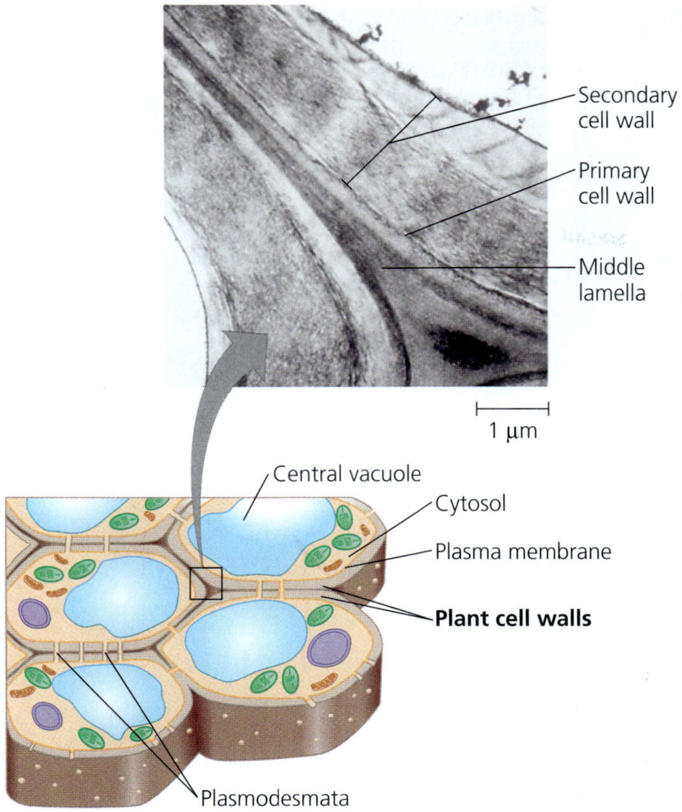

▲ **Figure 6.28 Plant cell walls.** The drawing shows several cells, each with a large vacuole, a nucleus, and several chloroplasts and mitochondria. The transmission electron micrograph (TEM) shows the cell walls where two cells come together. The multilayered partition between plant cells consists of adjoining walls individually secreted by the cells.

▼ **Figure 6.29 Inquiry**

What role do microtubules play in orienting deposition of cellulose in cell walls?

EXPERIMENT Previous experiments on preserved plant tissues had shown alignment of microtubules in the cell cortex with cellulose fibrils in the cell wall. Also, drugs that disrupted microtubules were observed to cause disoriented cellulose fibrils. To further investigate the possible role of cortical microtubules in guiding fibril deposition, David Ehrhardt and colleagues at Stanford University used a type of confocal microscopy to study cell wall deposition in living cells. In these cells, they labeled both cellulose synthase and microtubules with fluorescent markers and observed them over time.

RESULTS The path of cellulose synthase movement and the positions of existing microtubules coincided highly over time. The fluorescent micrographs below represent an average of five images, taken 10 seconds apart. The labeling molecules caused cellulose synthase to fluoresce green and the microtubules to fluoresce red. The arrowheads indicate prominent areas where the two are seen to align.

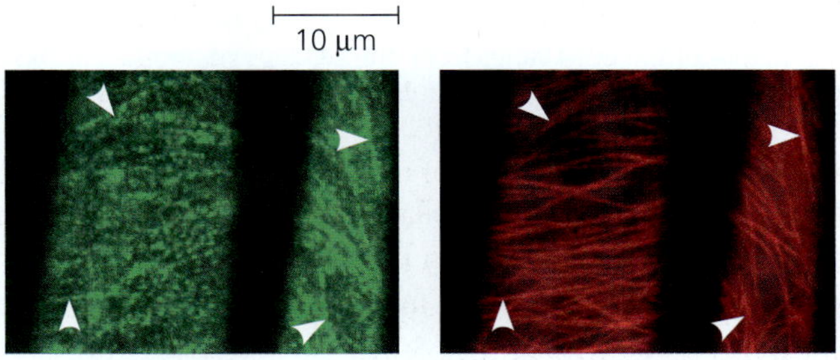

Distribution of cellulose synthase over time

Distribution of microtubules over time

CONCLUSION The organization of microtubules appears to directly guide the path of cellulose synthase as it lays down cellulose, thus determining the orientation of cellulose fibrils.

SOURCE A. R. Paradez et al., Visualization of cellulose synthase demonstrates functional association with microtubules, *Science* 312:1491–1495 (2006).

WHAT IF? In a second experiment, the researchers exposed the plant cells to blue light, previously shown to cause reorientation of microtubules. What events would you predict would follow blue light exposure?

A young plant cell first secretes a relatively thin and flexible wall called the **primary cell wall** (**Figure 6.28**). In actively growing cells, the cellulose fibrils are oriented at right angles to the direction of cell expansion, possibly affecting the growth pattern. David Ehrhardt and colleagues investigated the role of microtubules in orienting these fibrils (**Figure 6.29**). Their observations strongly supported the idea that microtubules in the cell cortex guide cellulose synthase as it synthesizes and deposits the fibrils. By orienting cellulose deposition, microtubules thus affect the growth pattern of the cells.

Between primary walls of adjacent cells is the **middle lamella**, a thin layer rich in sticky polysaccharides called pectins. The middle lamella glues adjacent cells together (pectin is used as a thickening agent in jams and jellies). When the cell matures and stops growing, it strengthens its wall. Some plant cells do this simply by secreting hardening substances into the primary wall. Other cells add a **secondary cell wall** between the plasma membrane and the primary wall. The secondary wall, often deposited in several laminated layers, has a strong and durable matrix that affords the cell protection and support. Wood, for example, consists mainly of secondary walls. Plant cell walls are commonly perforated by channels between adjacent cells called plasmodesmata (see Figure 6.28), which will be discussed shortly.

The Extracellular Matrix (ECM) of Animal Cells

Although animal cells lack walls akin to those of plant cells, they do have an elaborate **extracellular matrix (ECM)** (**Figure 6.30**, on the next page). The main ingredients of the ECM are glycoproteins secreted by the cells. (Recall that glycoproteins are proteins with covalently bonded carbohydrate, usually short chains of sugars.) The most abundant glycoprotein in the ECM of most animal cells is **collagen**, which forms strong fibers outside the

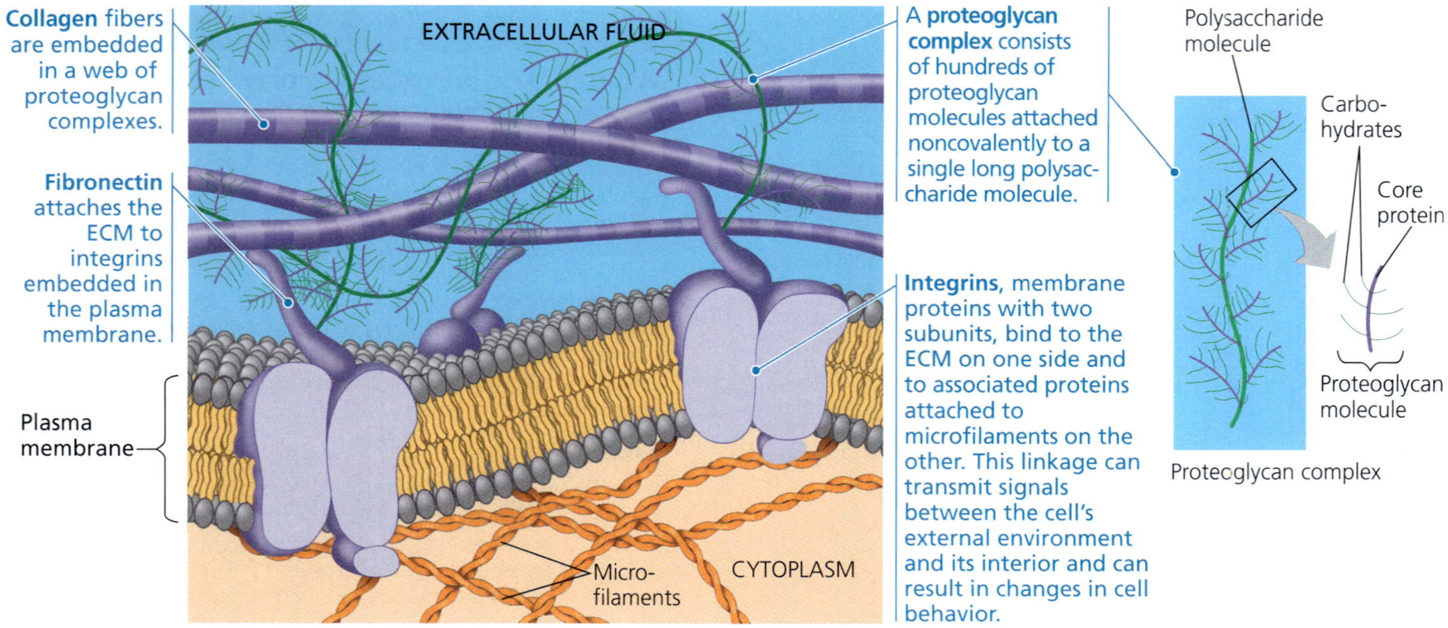

▲ **Figure 6.30 Extracellular matrix (ECM) of an animal cell.** The molecular composition and structure of the ECM varies from one cell type to another. In this example, three different types of glycoproteins are present: proteoglycans, collagen, and fibronectin.

cells (see Figure 5.21). In fact, collagen accounts for about 40% of the total protein in the human body. The collagen fibers are embedded in a network woven from **proteoglycans**. A proteoglycan molecule consists of a small core protein with many carbohydrate chains covalently attached, so that it may be up to 95% carbohydrate. Large proteoglycan complexes can form when hundreds of proteoglycans become noncovalently attached to a single long polysaccharide molecule, as shown in Figure 6.30. Some cells are attached to the ECM by still other ECM glycoproteins, such as **fibronectin**. Fibronectin and other ECM proteins bind to cell surface receptor proteins called **integrins** that are built into the plasma membrane. Integrins span the membrane and bind on their cytoplasmic side to associated proteins attached to microfilaments of the cytoskeleton. The name *integrin* is based on the word *integrate*: Integrins are in a position to transmit signals between the ECM and the cytoskeleton and thus to integrate changes occurring outside and inside the cell.

Current research on fibronectin, other ECM molecules, and integrins is revealing the influential role of the extracellular matrix in the lives of cells. By communicating with a cell through integrins, the ECM can regulate a cell's behavior. For example, some cells in a developing embryo migrate along specific pathways by matching the orientation of their microfilaments to the "grain" of fibers in the extracellular matrix. Researchers are also learning that the extracellular matrix around a cell can influence the activity of genes in the nucleus. Information about the ECM probably reaches the nucleus by a combination of mechanical and chemical signaling pathways. Mechanical signaling involves fibronectin, integrins, and microfilaments of the cytoskeleton. Changes in the cytoskeleton may in turn trigger chemical signaling pathways inside the cell, leading to changes in the set of proteins being made by the cell and therefore changes in the cell's function. In this way, the extracellular matrix of a particular tissue may help coordinate the behavior of all the cells within that tissue. Direct connections between cells also function in this coordination, as we discuss next.

Intercellular Junctions

Cells in an animal or plant are organized into tissues, organs, and organ systems. Cells often adhere, interact, and communicate through direct physical contact.

Plasmodesmata in Plant Cells

It might seem that the nonliving cell walls of plants would isolate cells from one another. But in fact, as shown in **Figure 6.31**, cell walls are perforated with channels called **plasmodesmata**

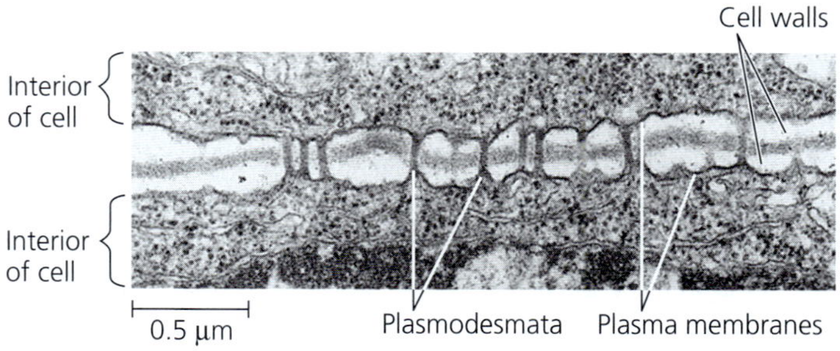

▲ **Figure 6.31 Plasmodesmata between plant cells.** The cytoplasm of one plant cell is continuous with the cytoplasm of its neighbors via plasmodesmata, channels through the cell walls (TEM).

(singular, *plasmodesma*; from the Greek *desmos*, to bind). Cytosol passes through the plasmodesmata and connects the chemical environments of adjacent cells. These connections unify most of the plant into one living continuum. The plasma membranes of adjacent cells line the channel of each plasmodesma and thus are continuous. Water and small solutes can pass freely from cell to cell, and recent experiments have shown that in some circumstances, certain proteins and RNA molecules can also do this (see Concept 36.6). The macromolecules transported to neighboring cells seem to reach the plasmodesmata by moving along fibers of the cytoskeleton.

Tight Junctions, Desmosomes, and Gap Junctions in Animal Cells

In animals, there are three main types of intercellular junctions: *tight junctions*, *desmosomes*, and *gap junctions* (the latter of which are most like the plasmodesmata of plants). All three types of intercellular junctions are especially common in epithelial tissue, which lines the external and internal surfaces of the body. **Figure 6.32** uses epithelial cells of the intestinal lining to illustrate these junctions; you should study this figure before moving on.

▼ **Figure 6.32**

Exploring Intercellular Junctions in Animal Tissues

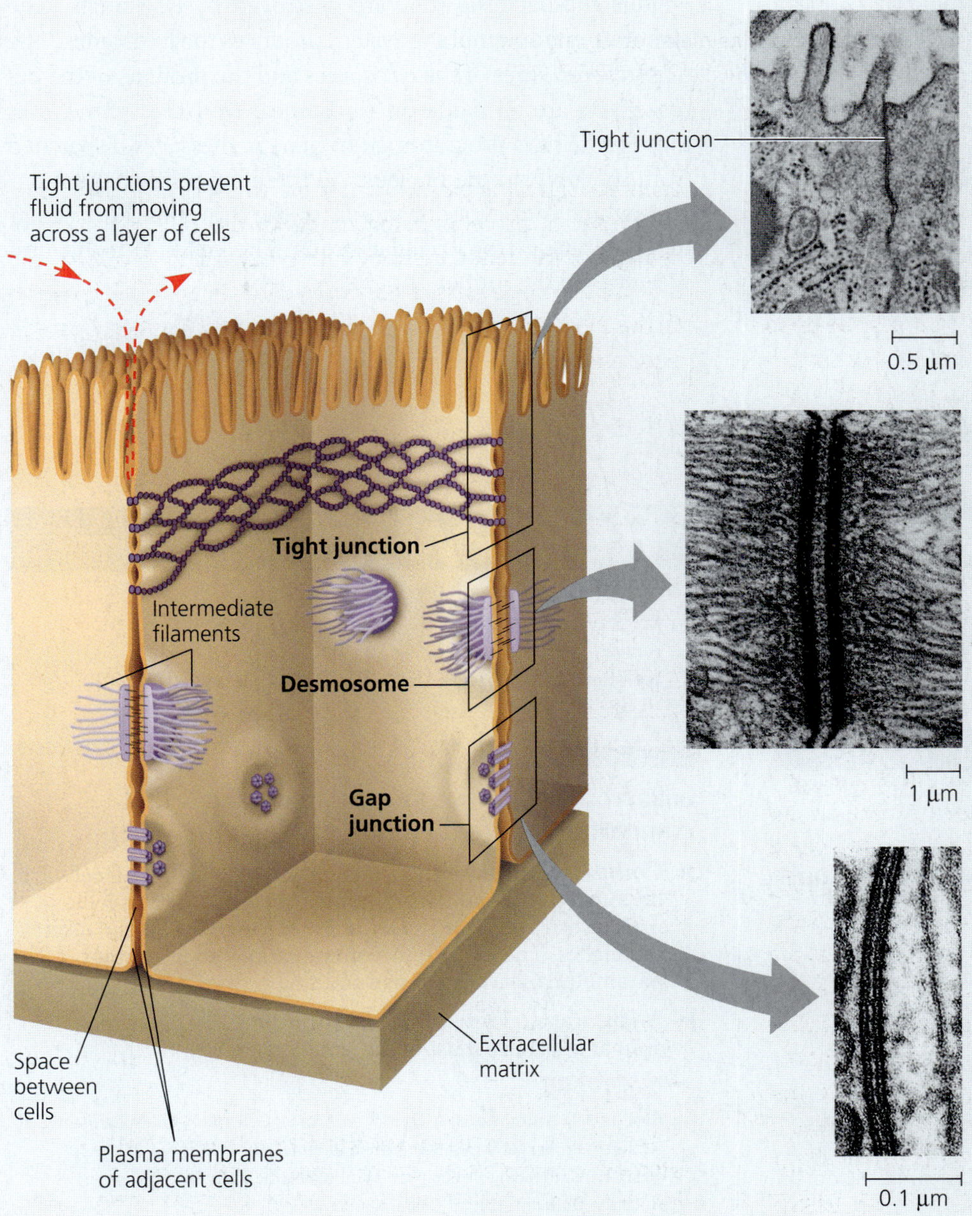

Tight Junctions

At **tight junctions**, the plasma membranes of neighboring cells are very tightly pressed against each other, bound together by specific proteins (purple). Forming continuous seals around the cells, tight junctions prevent leakage of extracellular fluid across a layer of epithelial cells. For example, tight junctions between skin cells make us watertight by preventing leakage between cells in our sweat glands.

Desmosomes

Desmosomes (also called *anchoring junctions*) function like rivets, fastening cells together into strong sheets. Intermediate filaments made of sturdy keratin proteins anchor desmosomes in the cytoplasm. Desmosomes attach muscle cells to each other in a muscle. Some "muscle tears" involve the rupture of desmosomes.

Gap Junctions

Gap junctions (also called *communicating junctions*) provide cytoplasmic channels from one cell to an adjacent cell and in this way are similar in their function to the plasmodesmata in plants. Gap junctions consist of membrane proteins that surround a pore through which ions, sugars, amino acids, and other small molecules may pass. Gap junctions are necessary for communication between cells in many types of tissues, including heart muscle, and in animal embryos.

CONCEPT CHECK 6.7

1. In what way are the cells of plants and animals structurally different from single-celled eukaryotes?
2. **WHAT IF?** If the plant cell wall or the animal extracellular matrix were impermeable, what effect would this have on cell function?

For suggested answers, see Appendix A.

The Cell: A Living Unit Greater Than the Sum of Its Parts

From our panoramic view of the cell's overall compartmental organization to our close-up inspection of each organelle's architecture, this tour of the cell has provided many opportunities to correlate structure with function. (This would be a good time to review cell structure by returning to Figure 6.9, on pp. 100 and 101.) But even as we dissect the cell, remember that none of its components works alone. As an example of cellular integration, consider the microscopic scene in **Figure 6.33**. The large cell is a macrophage (see Figure 6.14a). It helps defend the mammalian body against infections by ingesting bacteria (the smaller cells) into phagocytic vesicles. The macrophage crawls along a surface and reaches out to the bacteria with thin pseudopodia (called filopodia). Actin filaments interact with other elements of the cytoskeleton in these movements. After the macrophage engulfs the bacteria, they are destroyed by lysosomes. The elaborate endomembrane system produces the lysosomes. The digestive enzymes of the lysosomes and the proteins of the cytoskeleton are all made on ribosomes. And the synthesis of these proteins is programmed by genetic messages dispatched from the DNA in the nucleus. All these processes require energy, which mitochondria supply in the form of ATP. Cellular functions arise from cellular order: The cell is a living unit greater than the sum of its parts.

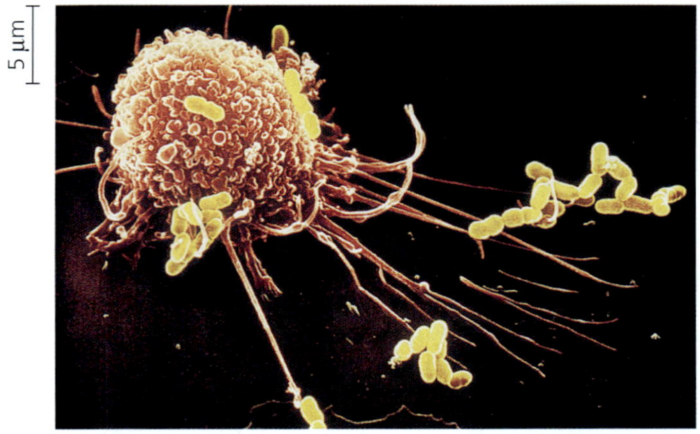

▲ **Figure 6.33 The emergence of cellular functions.** The ability of this macrophage (brown) to recognize, apprehend, and destroy bacteria (yellow) is a coordinated activity of the whole cell. Its cytoskeleton, lysosomes, and plasma membrane are among the components that function in phagocytosis (colorized SEM).

Chapter 6 Review

MEDIA Go to the Study Area at **www.masteringbio.com** for BioFlix 3-D Animations, MP3 Tutors, Videos, Practice Tests, an eBook, and more.

SUMMARY OF KEY CONCEPTS

CONCEPT 6.1

To study cells, biologists use microscopes and the tools of biochemistry (pp. 94–97)

- **Microscopy** Improvements in microscopy that affect the parameters of magnification, resolution, and contrast have catalyzed progress in the study of cell structure. Light and electron microscopy (LM and EM) remain important tools.
- **Cell Fractionation** Cell biologists can obtain pellets enriched in particular cellular components by centrifuging disrupted cells at sequential speeds. Larger components are in the pellet after lower speed centrifugation, and smaller components after higher speed centrifugation.

MEDIA

Activity Metric System Review
Investigation What Is the Size and Scale of Our World?

CONCEPT 6.2

Eukaryotic cells have internal membranes that compartmentalize their functions (pp. 98–102)

- **Comparing Prokaryotic and Eukaryotic Cells** All cells are bounded by a plasma membrane. Unlike eukaryotic cells, prokaryotic cells lack nuclei and other membrane-enclosed organelles. The surface-to-volume ratio is an important parameter affecting cell size and shape.
- **A Panoramic View of the Eukaryotic Cell** Plant and animal cells have most of the same organelles.

MEDIA

BioFlix 3-D Animations A Tour of an Animal Cell and A Tour of a Plant Cell
Activity Prokaryotic Cell Structure and Function
Activity Comparing Prokaryotic and Eukaryotic Cells
Activity Build an Animal Cell and a Plant Cell

	Cell Component	Structure	Function
Concept 6.3 **The eukaryotic cell's genetic instructions are housed in the nucleus and carried out by the ribosomes** **(pp. 102–104)** MEDIA Activity Role of the Nucleus and Ribosomes in Protein Synthesis	Nucleus 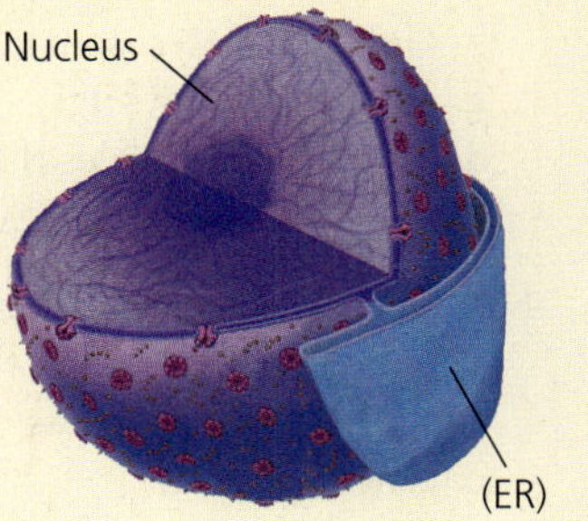	Surrounded by nuclear envelope (double membrane) perforated by nuclear pores. The nuclear envelope is continuous with the endoplasmic reticulum (ER).	Houses chromosomes, made of chromatin (DNA, the genetic material, and proteins); contains nucleoli, where ribosomal subunits are made. Pores regulate entry and exit of materials.
	Ribosome 	Two subunits made of ribosomal RNA and proteins; can be free in cytosol or bound to ER	Protein synthesis
Concept 6.4 **The endomembrane system regulates protein traffic and performs metabolic functions in the cell** **(pp. 104–108)** MEDIA Activity The Endomembrane System	Endoplasmic reticulum 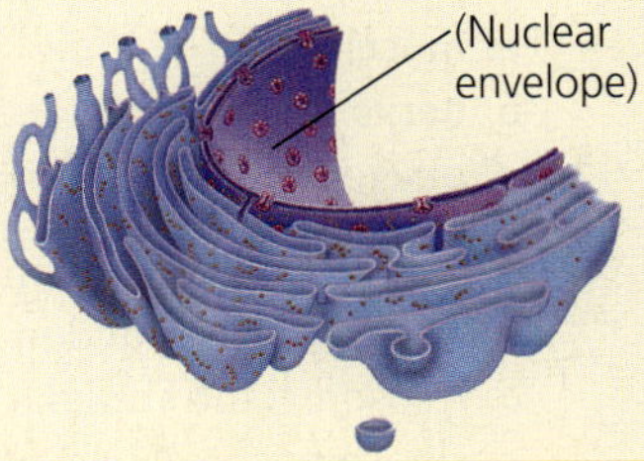	Extensive network of membrane-bounded tubules and sacs; membrane separates lumen from cytosol; continuous with the nuclear envelope	Smooth ER: synthesis of lipids, metabolism of carbohydrates, Ca^{2+} storage, detoxification of drugs and poisons Rough ER: Aids in synthesis of secretory and other proteins from bound ribosomes; adds carbohydrates to glycoproteins; produces new membrane
	Golgi apparatus 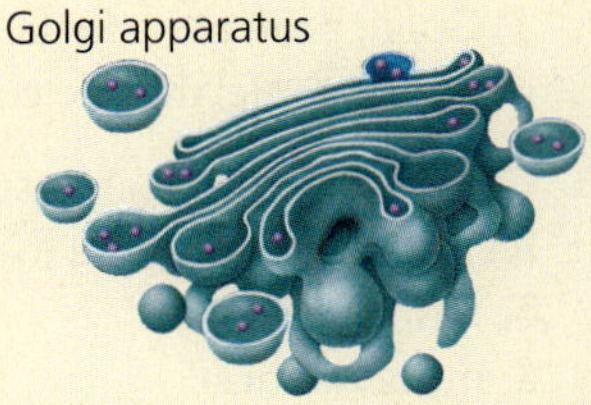	Stacks of flattened membranous sacs; has polarity (*cis* and *trans* faces)	Modification of proteins, carbohydrates on proteins, and phospholipids; synthesis of many polysaccharides; sorting of Golgi products, which are then released in vesicles
	Lysosome 	Membranous sac of hydrolytic enzymes (in animal cells)	Breakdown of ingested substances, cell macromolecules, and damaged organelles for recycling
	Vacuole 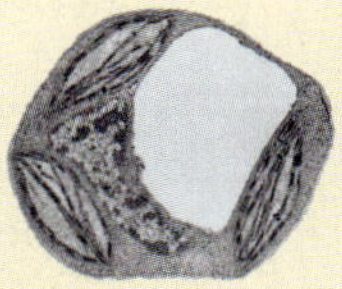	Large membrane-bounded vesicle in plants	Digestion, storage, waste disposal, water balance, cell growth, and protection
Concept 6.5 **Mitochondria and chloroplasts change energy from one form to another** **(pp. 109–111)** MEDIA Activity Build a Chloroplast and a Mitochondrion	Mitochondrion 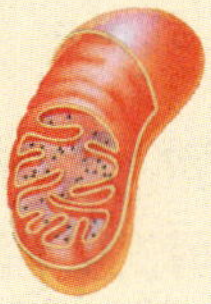	Bounded by double membrane; inner membrane has infoldings (cristae)	Cellular respiration
	Chloroplast 	Typically two membranes around fluid stroma, which contains membranous thylakoids stacked into grana (in plants)	Photosynthesis
	Peroxisome 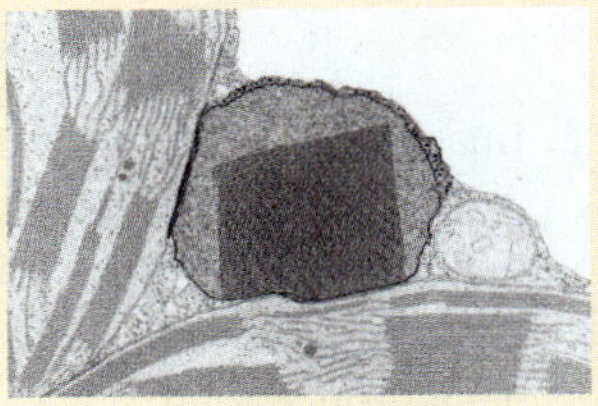	Specialized metabolic compartment bounded by a single membrane	Contains enzymes that transfer hydrogen to water, producing hydrogen peroxide (H_2O_2) as a by-product, which is converted to water by other enzymes in the peroxisome

CONCEPT **6.6**

The cytoskeleton is a network of fibers that organizes structures and activities in the cell (pp. 112–118)

- **Roles of the Cytoskeleton: Support, Motility, and Regulation** The cytoskeleton functions in structural support for the cell and in motility and signal transmission.
- **Components of the Cytoskeleton** Microtubules shape the cell, guide organelle movement, and separate chromosomes in dividing cells. Cilia and flagella are motile appendages containing microtubules. Primary cilia also play sensory and signaling roles. Microfilaments are thin rods functioning in muscle contraction, amoeboid movement, cytoplasmic streaming, and microvillus support. Intermediate filaments support cell shape and fix organelles in place.

MEDIA

Activity Cilia and Flagella

CONCEPT **6.7**

Extracellular components and connections between cells help coordinate cellular activities (pp. 118–122)

- **Cell Walls of Plants** Plant cell walls are made of cellulose fibers embedded in other polysaccharides and proteins. Cellulose deposition is oriented along microtubules.
- **The Extracellular Matrix (ECM) of Animal Cells** Animal cells secrete glycoproteins that form the ECM, which functions in support, adhesion, movement, and regulation.
- **Intercellular Junctions** Plants have plasmodesmata that pass through adjoining cell walls. Animal cells have tight junctions, desmosomes, and gap junctions.
- **The Cell: A Living Unit Greater Than the Sum of Its Parts**

MEDIA

Activity Cell Junctions
Activity Review: Animal Cell Structure and Function
Activity Review: Plant Cell Structure and Function

TESTING YOUR KNOWLEDGE

SELF-QUIZ

1. Which statement correctly characterizes bound ribosomes?
a. Bound ribosomes are enclosed in their own membrane.
b. Bound and free ribosomes are structurally different.
c. Bound ribosomes generally synthesize membrane proteins and secretory proteins.
d. The most common location for bound ribosomes is the cytoplasmic surface of the plasma membrane.
e. All of the above.

2. Which structure is *not* part of the endomembrane system?
a. nuclear envelope
b. chloroplast
c. Golgi apparatus
d. plasma membrane
e. ER

3. Cells of the pancreas will incorporate radioactively labeled amino acids into proteins. This "tagging" of newly synthesized proteins enables a researcher to track their location. In this case, we are tracking an enzyme secreted by pancreatic cells. What is its most likely pathway?
a. ER→Golgi→nucleus
b. Golgi→ER→lysosome
c. nucleus→ER→Golgi
d. ER→Golgi→vesicles that fuse with plasma membrane
e. ER→lysosomes→vesicles that fuse with plasma membrane

4. Which structure is common to plant *and* animal cells?
a. chloroplast
b. wall made of cellulose
c. central vacuole
d. mitochondrion
e. centriole

5. Which of the following is present in a prokaryotic cell?
a. mitochondrion
b. ribosome
c. nuclear envelope
d. chloroplast
e. ER

6. Which cell would be best for studying lysosomes?
a. muscle cell
b. nerve cell
c. phagocytic white blood cell
d. leaf cell of a plant
e. bacterial cell

7. Which structure-function pair is *mismatched*?
a. nucleolus; production of ribosomal subunits
b. lysosome; intracellular digestion
c. ribosome; protein synthesis
d. Golgi; protein trafficking
e. microtubule; muscle contraction

8. Cyanide binds with at least one molecule involved in producing ATP. If a cell is exposed to cyanide, most of the cyanide would be found within the
a. mitochondria.
b. ribosomes.
c. peroxisomes.
d. lysosomes.
e. endoplasmic reticulum.

9. **DRAW IT** From memory, draw two cells, showing the structures below and any connections between them.

nucleus, rough ER, smooth ER, mitochondrion, centrosome, chloroplast, vacuole, lysosome, microtubule, cell wall, ECM, microfilament, Golgi apparatus, intermediate filament, plasma membrane, peroxisome, ribosome, nucleolus, nuclear pore, vesicle, flagellum, microvilli, plasmodesma

For Self-Quiz Answers, see Appendix A.

MEDIA Visit the Study Area at **www.masteringbio.com** for a Practice Test.

EVOLUTION CONNECTION

10. Which aspects of cell structure best reveal evolutionary unity? What are some examples of specialized modifications?

SCIENTIFIC INQUIRY

11. Imagine protein X, destined to go to the plasma membrane. Assume that the mRNA carrying the genetic message for protein X has already been translated by ribosomes in a cell culture. If you fractionate the cell (see Figure 6.5), in which fraction would you find protein X? Explain by describing its transit.

7 Membrane Structure and Function

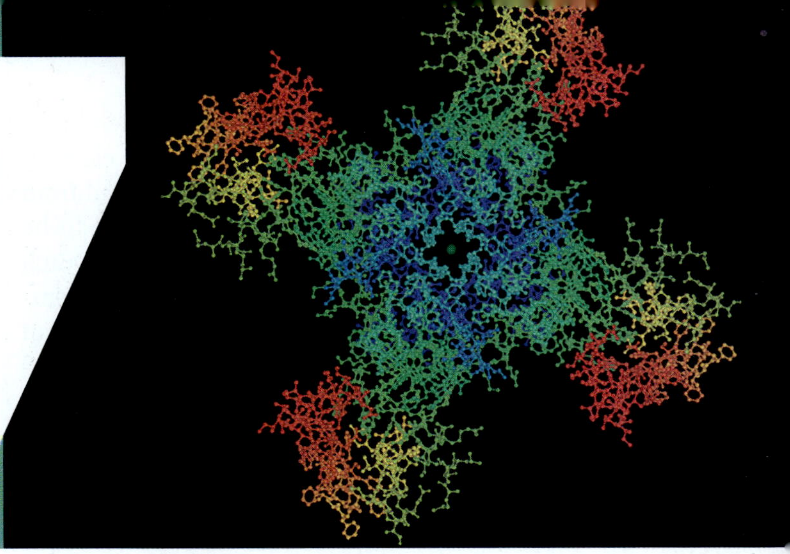

▲ **Figure 7.1 How do cell membrane proteins help regulate chemical traffic?**

KEY CONCEPTS

- **7.1 Cellular membranes are fluid mosaics of lipids and proteins**
- **7.2 Membrane structure results in selective permeability**
- **7.3 Passive transport is diffusion of a substance across a membrane with no energy investment**
- **7.4 Active transport uses energy to move solutes against their gradients**
- **7.5 Bulk transport across the plasma membrane occurs by exocytosis and endocytosis**

OVERVIEW

Life at the Edge

The plasma membrane is the edge of life, the boundary that separates the living cell from its surroundings. A remarkable film only about 8 nm thick—it would take over 8,000 to equal the thickness of this page—the plasma membrane controls traffic into and out of the cell it surrounds. Like all biological membranes, the plasma membrane exhibits **selective permeability**; that is, it allows some substances to cross it more easily than others. One of the earliest episodes in the evolution of life may have been the formation of a membrane that enclosed a solution different from the surrounding solution while still permitting the uptake of nutrients and elimination of waste products. The ability of the cell to discriminate in its chemical exchanges with its environment is fundamental to life, and it is the plasma membrane and its component molecules that make this selectivity possible.

In this chapter, you will learn how cellular membranes control the passage of substances. The image in **Figure 7.1** shows the elegant structure of a eukaryotic plasma membrane protein that plays a crucial role in nerve cell signaling. This protein restores the ability of the nerve cell to fire again by providing a channel for a stream of potassium ions (K^+) to exit the cell at a precise moment after nerve stimulation. (The green ball in the center represents one K^+ moving through the channel.) In this case, the plasma membrane and its proteins not only act as an outer boundary but also enable the cell to carry out its functions. The same applies to the many varieties of internal membranes that partition the eukaryotic cell: The molecular makeup of each membrane allows compartmentalized specialization in cells. To understand how membranes work, we'll begin by examining their architecture.

CONCEPT **7.1**

Cellular membranes are fluid mosaics of lipids and proteins

Lipids and proteins are the staple ingredients of membranes, although carbohydrates are also important. The most abundant lipids in most membranes are phospholipids. The ability of phospholipids to form membranes is inherent in their molecular structure. A phospholipid is an **amphipathic** molecule, meaning it has both a hydrophilic region and a hydrophobic region (see Figure 5.13). Other types of membrane lipids are also amphipathic. Furthermore, most of the proteins within membranes have both hydrophobic and hydrophilic regions.

How are phospholipids and proteins arranged in the membranes of cells? You encountered the currently accepted model for the arrangement of these molecules in Chapter 6 (see Figure 6.7). In this **fluid mosaic model**, the membrane is a fluid structure with a "mosaic" of various proteins embedded in or attached to a double layer (bilayer) of phospholipids. Scientists propose models as hypotheses, ways of organizing and explaining existing information. We'll discuss the fluid mosaic model in detail, starting with the story of how it was developed.

Membrane Models: *Scientific Inquiry*

Scientists began building molecular models of the membrane decades before membranes were first seen with the electron microscope in the 1950s. In 1915, membranes isolated from red blood cells were chemically analyzed and found to be composed of lipids and proteins. Ten years later, two Dutch scientists, E. Gorter and F. Grendel, reasoned that cell membranes must be phospholipid bilayers. Such a double layer of molecules could exist as a stable boundary between two aqueous compartments because the molecular arrangement shelters the hydrophobic tails of the phospholipids from water while exposing the hydrophilic heads to water **(Figure 7.2)**.

Building on the idea that a phospholipid bilayer was the main fabric of a membrane, the next question was where the proteins were located. Although the heads of phospholipids are hydrophilic, the surface of a membrane consisting of a pure phospholipid bilayer adheres less strongly to water than does the surface of a biological membrane. Given these data, Hugh Davson and James Danielli suggested in 1935 that this difference could be accounted for if the membrane were coated on both sides with hydrophilic proteins. They proposed a sandwich model: a phospholipid bilayer between two layers of proteins.

When researchers first used electron microscopes to study cells in the 1950s, the pictures seemed to support the Davson-Danielli model. By the 1960s, the Davson-Danielli sandwich had become widely accepted as the structure not only of the plasma membrane but also of all the cell's internal membranes. By the end of that decade, however, many cell biologists recognized two problems with the model. The first problem was the generalization that all membranes of the cell are identical. Whereas the plasma membrane is 7–8 nm thick and has a three-layered structure in electron micrographs, the inner membrane of the mitochondrion is only 6 nm thick and looks like a row of beads. Mitochondrial membranes also have a higher percentage of proteins and different kinds of phospholipids and other lipids. In short, membranes with different functions differ in chemical composition and structure.

A second, more serious problem with the sandwich model was the protein placement. Unlike proteins dissolved in the cytosol, membrane proteins are not very soluble in water, because they are amphipathic; that is, they have hydrophobic regions as well as hydrophilic regions. If such proteins were layered on the surface of the membrane, their hydrophobic parts would be in aqueous surroundings.

In 1972, S. J. Singer and G. Nicolson proposed that membrane proteins are dispersed, individually inserted into the phospholipid bilayer with their hydrophilic regions protruding **(Figure 7.3)**. This molecular arrangement would maximize contact of hydrophilic regions of proteins and phospholipids with water in the cytosol and extracellular fluid, while providing their hydrophobic parts with a nonaqueous environment. In this fluid mosaic model, the membrane is a mosaic of protein molecules bobbing in a fluid bilayer of phospholipids.

A method of preparing cells for electron microscopy called freeze-fracture has demonstrated visually that proteins are indeed embedded in the phospholipid bilayer of the membrane. Freeze-fracture splits a membrane along the middle of the phospholipid bilayer, somewhat like pulling apart a chunky peanut butter sandwich. When the membrane layers are viewed in the electron microscope, the interior of the bilayer appears cobblestoned, with protein particles interspersed in a smooth matrix, as in the fluid mosaic model **(Figure 7.4)**. Some proteins travel with one layer or the other, like the peanut chunks in the sandwich.

Because models are hypotheses, replacing one model of membrane structure with another does not imply that the original model was worthless. The acceptance or rejection of a model depends on how well it fits observations and explains experimental results. A good model also makes predictions that shape future research. Models inspire experiments, and few models survive these tests without modification. New findings may make a model obsolete; even then, it may not be totally scrapped, but revised to incorporate the new observations. The fluid mosaic model is continually being refined. For example, recent research suggests that membranes may be "more mosaic than fluid." Often, multiple proteins semipermanently associate in specialized patches, where they carry out common functions. Also, the membrane may be much more packed with proteins than imagined in the classic fluid mosaic model. Let's now take a closer look at membrane structure.

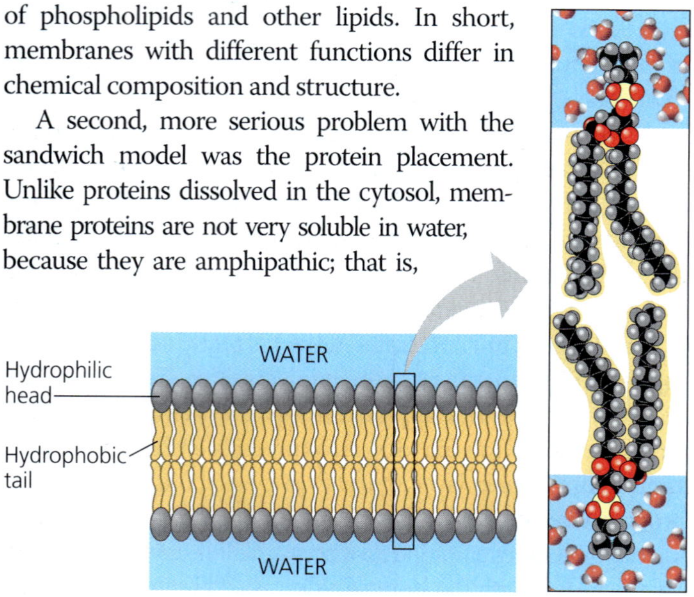

▲ **Figure 7.2 Phospholipid bilayer (cross section).**

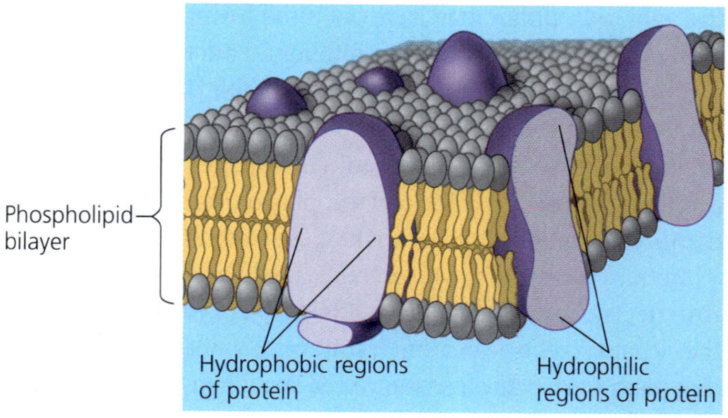

▲ **Figure 7.3 The fluid mosaic model for membranes.**

▼ **Figure 7.4 Research Method**

Freeze-Fracture

APPLICATION A cell membrane can be split into its two layers, revealing the ultrastructure of the membrane's interior.

TECHNIQUE A cell is frozen and fractured with a knife. The fracture plane often follows the hydrophobic interior of a membrane, splitting the phospholipid bilayer into two separated layers. The membrane proteins go wholly with one of the layers.

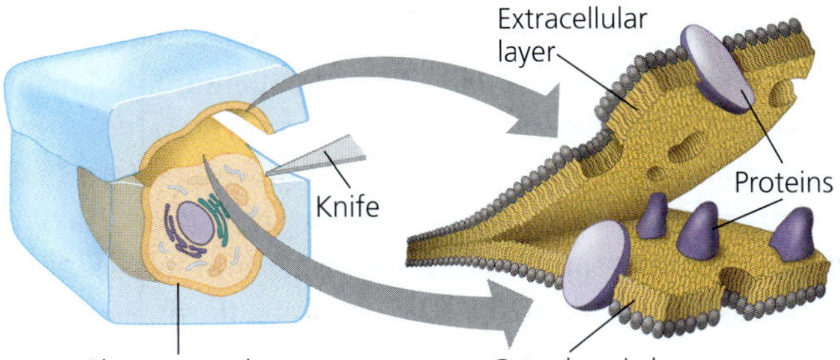

RESULTS These SEMs show membrane proteins (the "bumps") in the two layers, demonstrating that proteins are embedded in the phospholipid bilayer.

Inside of extracellular layer

Inside of cytoplasmic layer

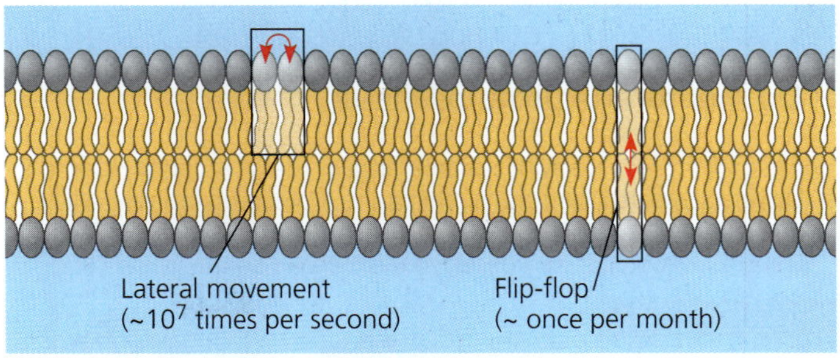

(a) Movement of phospholipids. Lipids move laterally in a membrane, but flip-flopping across the membrane is quite rare.

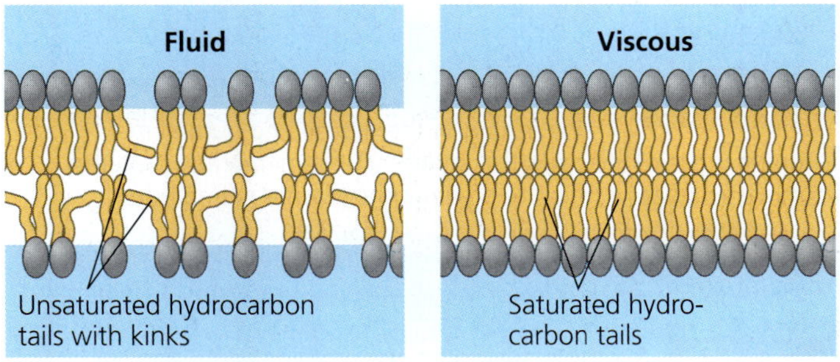

(b) Membrane fluidity. Unsaturated hydrocarbon tails of phospholipids have kinks that keep the molecules from packing together, enhancing membrane fluidity.

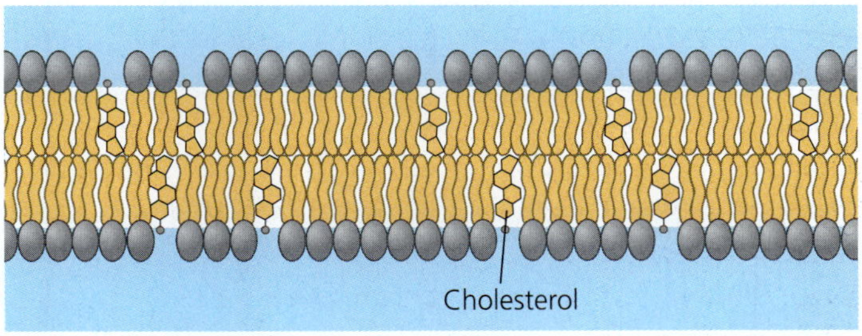

(c) Cholesterol within the animal cell membrane. Cholesterol reduces membrane fluidity at moderate temperatures by reducing phospholipid movement, but at low temperatures it hinders solidification by disrupting the regular packing of phospholipids.

▲ **Figure 7.5 The fluidity of membranes.**

The Fluidity of Membranes

Membranes are not static sheets of molecules locked rigidly in place. A membrane is held together primarily by hydrophobic interactions, which are much weaker than covalent bonds (see Figure 5.21). Most of the lipids and some of the proteins can shift about laterally—that is, in the plane of the membrane, like partygoers elbowing their way through a crowded room **(Figure 7.5a)**. It is quite rare, however, for a molecule to flip-flop transversely across the membrane, switching from one phospholipid layer to the other; to do so, the hydrophilic part of the molecule must cross the hydrophobic core of the membrane.

The lateral movement of phospholipids within the membrane is rapid. Adjacent phospholipids switch positions about 10^7 times per second, which means that a phospholipid can travel about 2 μm—the length of many bacterial cells—in 1 second. Proteins are much larger than lipids and move more slowly, but some membrane proteins do drift, as shown in a classic experiment by David Frye and Michael Edidin (**Figure 7.6**, on the next page). And some membrane proteins seem to move in a highly directed manner, perhaps driven along cytoskeletal fibers by motor proteins connected to the membrane proteins' cytoplasmic regions. However, many other membrane proteins seem to be held virtually immobile by their attachment to the cytoskeleton.

A membrane remains fluid as temperature decreases until finally the phospholipids settle into a closely packed arrangement and the membrane solidifies, much as bacon grease forms lard when it cools. The temperature at which a membrane solidifies depends on the types of lipids it is made of. The membrane remains fluid to a lower temperature if it is rich in phospholipids with unsaturated hydrocarbon tails (see Figures 5.12 and 5.13). Because of kinks in the tails where double bonds are located, unsaturated hydrocarbon tails cannot pack together as closely as saturated hydrocarbon tails, and this makes the membrane more fluid **(Figure 7.5b)**.

The steroid cholesterol, which is wedged between phospholipid molecules in the plasma membranes of animal cells, has different effects on membrane fluidity at different temperatures **(Figure 7.5c)**. At relatively higher temperatures—at 37°C, the

▼ Figure 7.6 Inquiry

Do membrane proteins move?

EXPERIMENT David Frye and Michael Edidin, at Johns Hopkins University, labeled the plasma membrane proteins of a mouse cell and a human cell with two different markers and fused the cells. Using a microscope, they observed the markers on the hybrid cell.

RESULTS

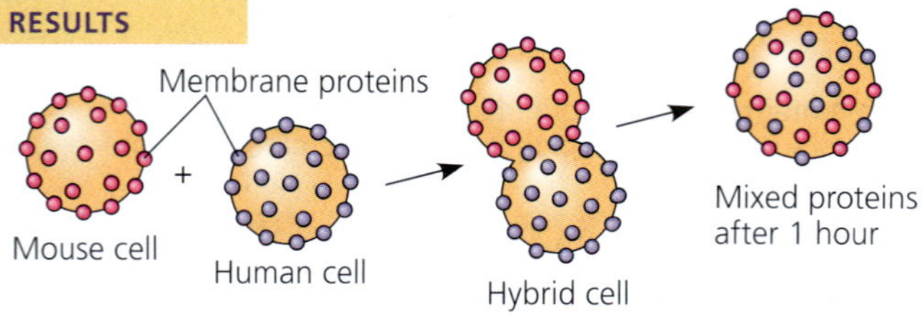

CONCLUSION The mixing of the mouse and human membrane proteins indicates that at least some membrane proteins move sideways within the plane of the plasma membrane.

SOURCE L. D. Frye and M. Edidin, The rapid intermixing of cell surface antigens after formation of mouse-human heterokaryons, *J. Cell Sci.* 7:319 (1970).

WHAT IF? If, after many hours, the protein distribution still looked like that in the third image above, would you be able to conclude that proteins don't move within the membrane? What other explanation could there be?

body temperature of humans, for example—cholesterol makes the membrane less fluid by restraining phospholipid movement. However, because cholesterol also hinders the close packing of phospholipids, it lowers the temperature required for the membrane to solidify. Thus, cholesterol can be thought of as a "temperature buffer" for the membrane, resisting changes in membrane fluidity that can be caused by changes in temperature.

Membranes must be fluid to work properly; they are usually about as fluid as salad oil. When a membrane solidifies, its permeability changes, and enzymatic proteins in the membrane may become inactive—for example, if their activity requires them to be able to move laterally in the membrane. The lipid composition of cell membranes can change as an adjustment to changing temperature. For instance, in many plants that tolerate extreme cold, such as winter wheat, the percentage of unsaturated phospholipids increases in autumn, an adaptation that keeps the membranes from solidifying during winter.

Membrane Proteins and Their Functions

Now we come to the *mosaic* aspect of the fluid mosaic model. A membrane is a collage of different proteins embedded in the fluid matrix of the lipid bilayer **(Figure 7.7)**. More than 50 kinds of proteins have been found so far in the plasma mem-

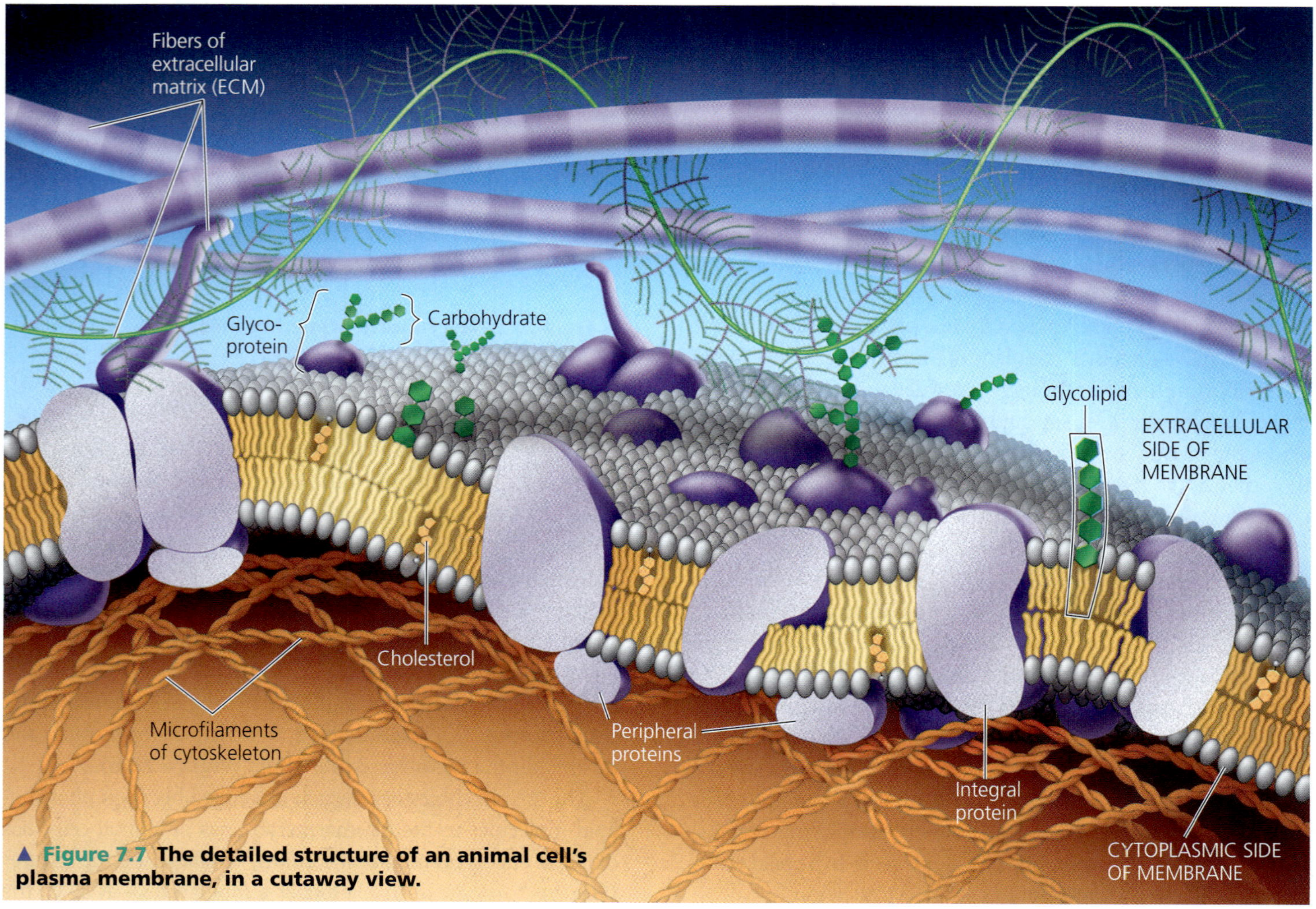

▲ Figure 7.7 The detailed structure of an animal cell's plasma membrane, in a cutaway view.

brane of red blood cells, for example. Phospholipids form the main fabric of the membrane, but proteins determine most of the membrane's functions. Different types of cells contain different sets of membrane proteins, and the various membranes within a cell each have a unique collection of proteins.

Notice in Figure 7.7 that there are two major populations of membrane proteins: integral proteins and peripheral proteins. **Integral proteins** penetrate the hydrophobic core of the lipid bilayer. Many are *transmembrane proteins*, which span the membrane; other integral proteins extend only partway into the hydrophobic core. The hydrophobic regions of an integral protein consist of one or more stretches of nonpolar amino acids (see Figure 5.17), usually coiled into α helices **(Figure 7.8)**. The hydrophilic parts of the molecule are exposed to the aqueous solutions on either side of the membrane. Some proteins also have a hydrophilic channel through their center that allows passage of hydrophilic substances (see Figure 7.1). **Peripheral proteins** are not embedded in the lipid bilayer at all; they are appendages loosely bound to the surface of the membrane, often to exposed parts of integral proteins (see Figure 7.7).

On the cytoplasmic side of the plasma membrane, some membrane proteins are held in place by attachment to the cytoskeleton. And on the extracellular side, certain membrane proteins are attached to fibers of the extracellular matrix (see Figure 6.30; *integrins* are one type of integral protein). These attachments combine to give animal cells a stronger framework than the plasma membrane alone could provide.

Figure 7.9 gives an overview of six major functions performed by proteins of the plasma membrane. A single cell may

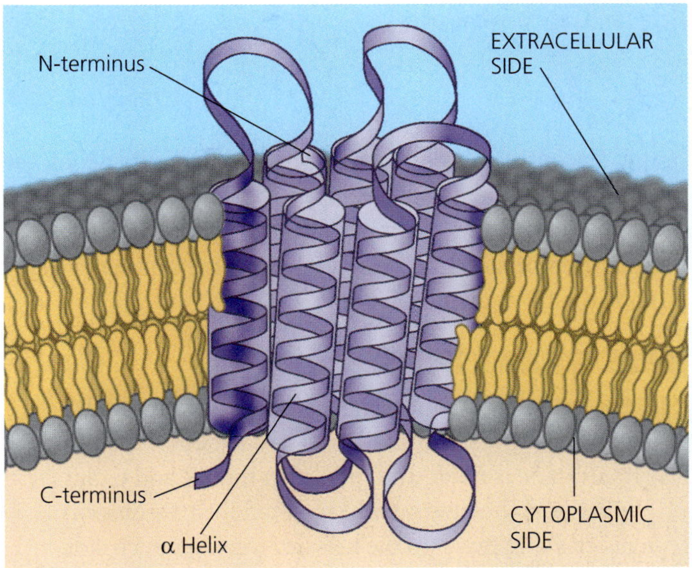

▲ **Figure 7.8 The structure of a transmembrane protein.** This protein, bacteriorhodopsin (a bacterial transport protein), has a distinct orientation in the membrane, with the N-terminus outside the cell and the C-terminus inside. This ribbon model highlights the α-helical secondary structure of the hydrophobic parts, which lie mostly within the hydrophobic core of the membrane. The protein includes seven transmembrane helices (outlined with cylinders for emphasis). The nonhelical hydrophilic segments are in contact with the aqueous solutions on the extracellular and cytoplasmic sides of the membrane.

(a) Transport. *Left:* A protein that spans the membrane may provide a hydrophilic channel across the membrane that is selective for a particular solute. *Right:* Other transport proteins shuttle a substance from one side to the other by changing shape. Some of these proteins hydrolyze ATP as an energy source to actively pump substances across the membrane.

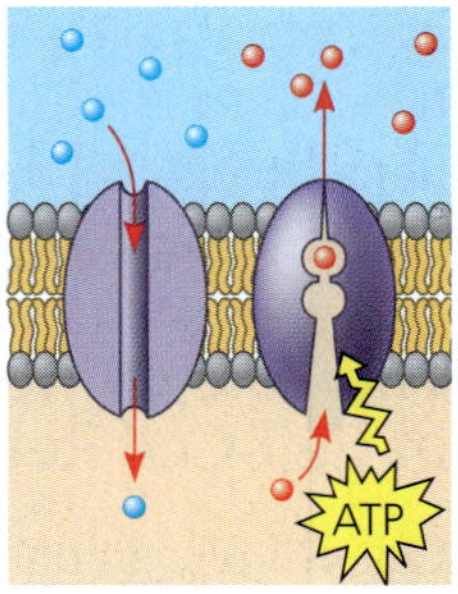

(b) Enzymatic activity. A protein built into the membrane may be an enzyme with its active site exposed to substances in the adjacent solution. In some cases, several enzymes in a membrane are organized as a team that carries out sequential steps of a metabolic pathway.

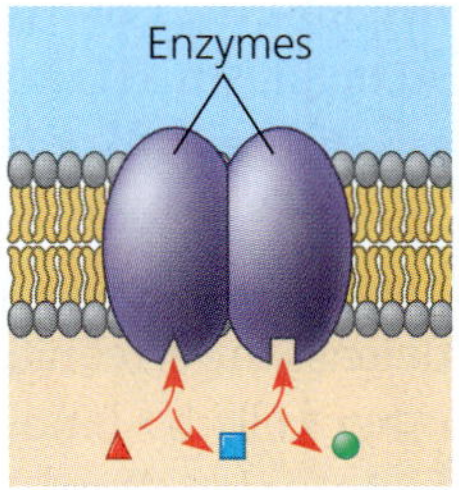

(c) Signal transduction. A membrane protein (receptor) may have a binding site with a specific shape that fits the shape of a chemical messenger, such as a hormone. The external messenger (signaling molecule) may cause a shape change in the protein that relays the message to the inside of the cell, usually by binding to a cytoplasmic protein. (See Figure 11.6.)

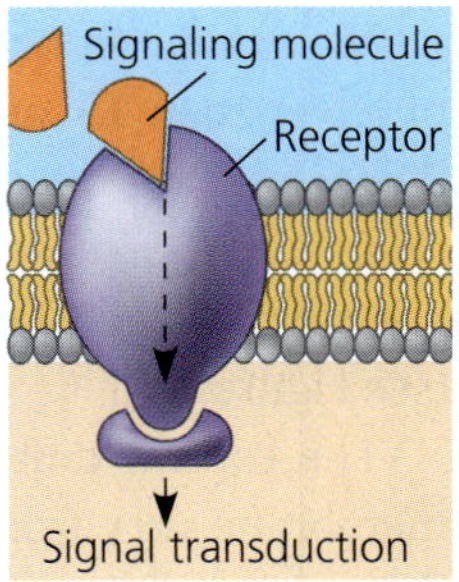

(d) Cell-cell recognition. Some glycoproteins serve as identification tags that are specifically recognized by membrane proteins of other cells.

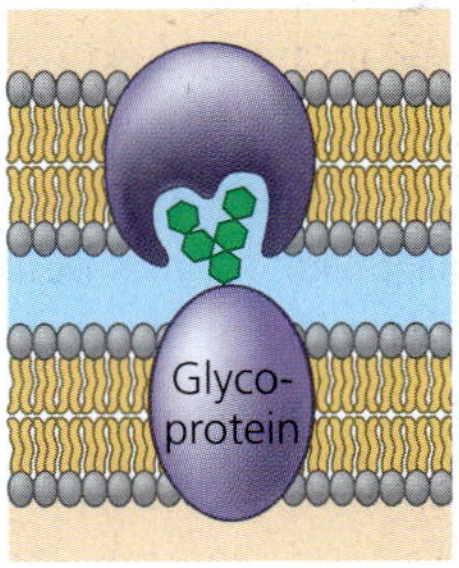

(e) Intercellular joining. Membrane proteins of adjacent cells may hook together in various kinds of junctions, such as gap junctions or tight junctions (see Figure 6.32).

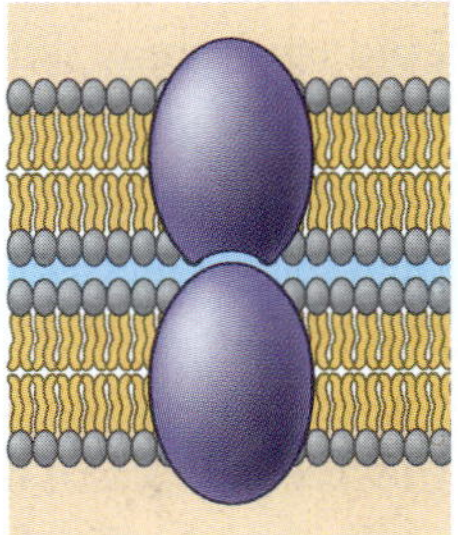

(f) Attachment to the cytoskeleton and extracellular matrix (ECM). Microfilaments or other elements of the cytoskeleton may be noncovalently bound to membrane proteins, a function that helps maintain cell shape and stabilizes the location of certain membrane proteins. Proteins that can bind to ECM molecules can coordinate extracellular and intracellular changes (see Figure 6.30).

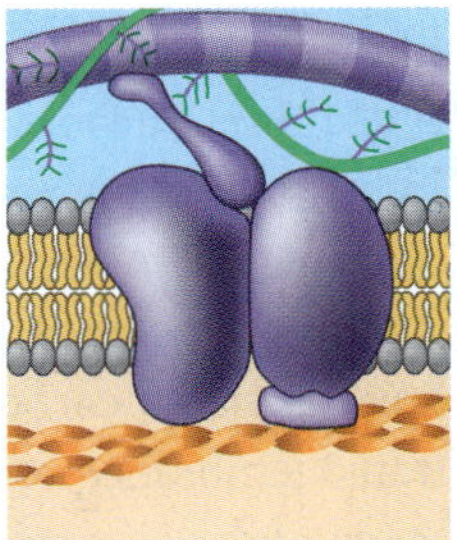

▲ **Figure 7.9 Some functions of membrane proteins.** In many cases, a single protein performs multiple tasks.

? *Some transmembrane proteins can bind to a particular ECM molecule and, when bound, transmit a signal into the cell. Use the proteins shown here to explain how this might occur.*

have membrane proteins carrying out several of these functions, and a single membrane protein may have multiple functions. In this way, the membrane is a functional mosaic as well as a structural one.

The Role of Membrane Carbohydrates in Cell-Cell Recognition

Cell-cell recognition, a cell's ability to distinguish one type of neighboring cell from another, is crucial to the functioning of an organism. It is important, for example, in the sorting of cells into tissues and organs in an animal embryo. It is also the basis for the rejection of foreign cells (including those of transplanted organs) by the immune system, an important line of defense in vertebrate animals (see Chapter 43). Cells recognize other cells by binding to surface molecules, often to carbohydrates, on the plasma membrane (see Figure 7.9d).

Membrane carbohydrates are usually short, branched chains of fewer than 15 sugar units. Some are covalently bonded to lipids, forming molecules called **glycolipids**. (Recall that *glyco* refers to the presence of carbohydrate.) However, most are covalently bonded to proteins, which are thereby **glycoproteins** (see Figure 7.7).

The carbohydrates on the extracellular side of the plasma membrane vary from species to species, among individuals of the same species, and even from one cell type to another in a single individual. The diversity of the molecules and their location on the cell's surface enable membrane carbohydrates to function as markers that distinguish one cell from another. For example, the four human blood types designated A, B, AB, and O reflect variation in the carbohydrates on the surface of red blood cells.

Synthesis and Sidedness of Membranes

Membranes have distinct inside and outside faces. The two lipid layers may differ in specific lipid composition, and each protein has directional orientation in the membrane (see Figure 7.8). When a vesicle fuses with the plasma membrane, the outside layer of the vesicle becomes continuous with the cytoplasmic (inner) layer of the plasma membrane. Therefore, molecules that start out on the *inside* face of the ER end up on the *outside* face of the plasma membrane.

The process, shown in **Figure 7.10**, starts with ❶ the synthesis of membrane proteins and lipids in the endoplasmic reticulum. Carbohydrates (green) are added to the proteins (purple), making them glycoproteins. The carbohydrate portions may then be modified. ❷ Inside the Golgi apparatus, the glycoproteins undergo further carbohydrate modification, and lipids acquire carbohydrates, becoming glycolipids. ❸ The transmembrane proteins (purple dumbbells), membrane glycolipids, and secretory proteins (purple spheres) are transported in vesicles to the plasma membrane. ❹ There the vesicles fuse with the membrane, releasing secretory proteins from the cell. Vesicle fusion positions the carbohydrates of membrane glycoproteins and glycolipids on the outside of the plasma membrane. Thus, the asymmetrical arrangement of proteins, lipids, and their associated carbohydrates in the plasma membrane is determined as the membrane is being built by the ER and Golgi apparatus.

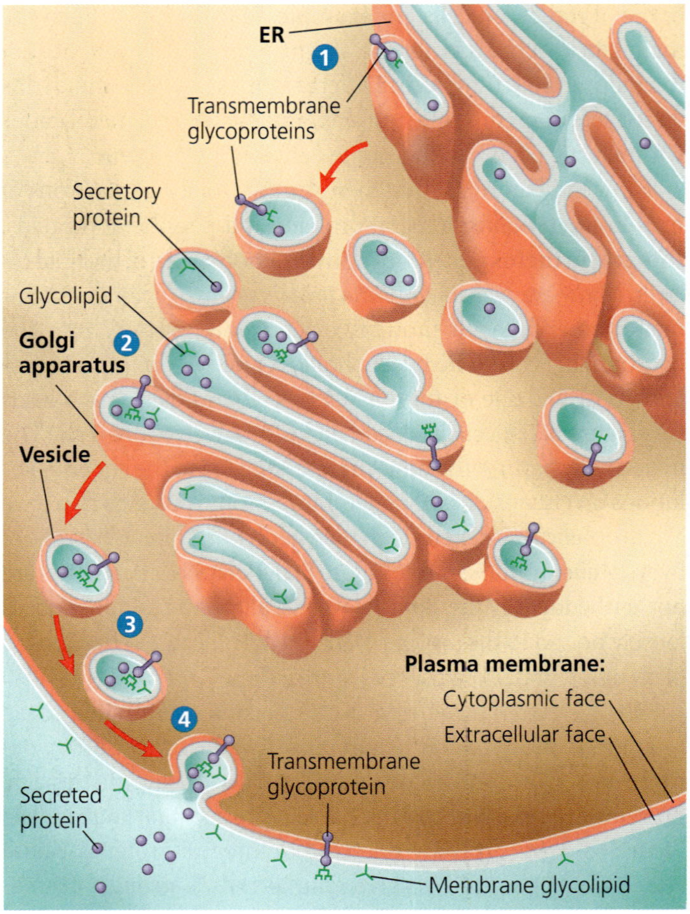

▲ **Figure 7.10 Synthesis of membrane components and their orientation on the resulting membrane.** The plasma membrane has distinct cytoplasmic (orange) and extracellular (aqua) faces, with the extracellular face arising from the inside face of ER, Golgi, and vesicle membranes.

CONCEPT CHECK 7.1

1. The carbohydrates attached to some proteins and lipids of the plasma membrane are added as the membrane is made and refined in the ER and Golgi apparatus; the new membrane then forms transport vesicles that travel to the cell surface. On which side of the vesicle membrane are the carbohydrates?
2. WHAT IF? How would you expect the saturation levels of membrane phospholipid fatty acids to differ in plants adapted to cold environments and plants adapted to hot environments?

For suggested answers, see Appendix A.

CONCEPT 7.2

Membrane structure results in selective permeability

The biological membrane is an exquisite example of a supramolecular structure—many molecules ordered into a higher level of organization—with emergent properties beyond those of the individual molecules. The remainder of this chapter focuses on one of the most important of those properties: the ability to regulate transport across cellular boundaries, a function essential to the cell's existence. We will see once again that form fits function: The fluid mosaic model helps explain how membranes regulate the cell's molecular traffic.

A steady traffic of small molecules and ions moves across the plasma membrane in both directions. Consider the chemical exchanges between a muscle cell and the extracellular fluid that bathes it. Sugars, amino acids, and other nutrients enter the cell, and metabolic waste products leave it. The cell takes in oxygen for use in cellular respiration and expels carbon dioxide. Also, the cell regulates its concentrations of inorganic ions, such as Na^+, K^+, Ca^{2+}, and Cl^-, by shuttling them one way or the other across the plasma membrane. Although traffic through the membrane is extensive, cell membranes are selectively permeable, and substances do not cross the barrier indiscriminately. The cell is able to take up many varieties of small molecules and ions and exclude others. Moreover, substances that move through the membrane do so at different rates.

The Permeability of the Lipid Bilayer

Nonpolar molecules, such as hydrocarbons, carbon dioxide, and oxygen, are hydrophobic and can therefore dissolve in the lipid bilayer of the membrane and cross it easily, without the aid of membrane proteins. However, the hydrophobic core of the membrane impedes the direct passage of ions and polar molecules, which are hydrophilic, through the membrane. Polar molecules such as glucose and other sugars pass only slowly through a lipid bilayer, and even water, an extremely small polar molecule, does not cross very rapidly. A charged atom or molecule and its surrounding shell of water (see Figure 3.7) find the hydrophobic layer of the membrane even more difficult to penetrate. Furthermore, the lipid bilayer is only one aspect of the gatekeeper system responsible for the selective permeability of a cell. Proteins built into the membrane play key roles in regulating transport.

Transport Proteins

Cell membranes *are* permeable to specific ions and a variety of polar molecules. These hydrophilic substances can avoid contact with the lipid bilayer by passing through **transport proteins** that span the membrane.

Some transport proteins, called *channel proteins*, function by having a hydrophilic channel that certain molecules or atomic ions use as a tunnel through the membrane (see Figure 7.9a, left). For example, the passage of water molecules through the membrane in certain cells is greatly facilitated by channel proteins known as **aquaporins**. Each aquaporin allows entry of up to 3 *billion* (3×10^9) water molecules per second, passing single file through its central channel, which fits ten at a time. Without aquaporins, only a tiny fraction of these water molecules would diffuse through the same area of the cell membrane in a second, so the channel protein brings about a tremendous increase in rate. Other transport proteins, called *carrier proteins*, hold onto their passengers and change shape in a way that shuttles them across the membrane (see Figure 7.9a, right). A transport protein is specific for the substance it translocates (moves), allowing only a certain substance (or substances) to cross the membrane. For example, glucose, carried in the blood and needed by red blood cells for cellular activities, enters the red blood cells rapidly via specific carrier proteins in the plasma membrane. The glucose passes through the membrane 50,000 times faster than if diffusing through on its own. This "glucose transporter" is so selective as a carrier protein that it even rejects fructose, a structural isomer of glucose.

Thus, the selective permeability of a membrane depends on both the discriminating barrier of the lipid bilayer and the specific transport proteins built into the membrane. But what establishes the *direction* of traffic across a membrane? At a given time, what determines whether a particular substance will enter the cell or leave the cell? And what mechanisms actually drive molecules across membranes? We will address these questions next as we explore two modes of membrane traffic: passive transport and active transport.

CONCEPT CHECK 7.2

1. Two molecules that can cross a lipid bilayer without help from membrane proteins are O_2 and CO_2. What properties allow this to occur?
2. Why would water molecules need a transport protein to move rapidly and in large quantities across a membrane?
3. WHAT IF? Aquaporins exclude passage of hydronium ions (H_3O^+). But recent research has revealed a role for some aquaporins in fat metabolism, in which they allow passage of glycerol, a three-carbon alcohol (see Figure 5.11), as well as H_2O. Since H_3O^+ is much closer in size to water than is glycerol, what do you suppose is the basis of this selectivity?

For suggested answers, see Appendix A.

CONCEPT **7.3**

Passive transport is diffusion of a substance across a membrane with no energy investment

Molecules have a type of energy called thermal motion (heat). One result of thermal motion is **diffusion**, the movement of molecules of any substance so that they spread out evenly into the available space. Each molecule moves randomly, yet diffusion of a *population* of molecules may be directional. To understand this process, let's imagine a synthetic membrane separating pure water from a solution of a dye in water. Assume that this membrane has microscopic pores and is permeable to the dye molecules **(Figure 7.11a)**. Each dye molecule wanders randomly, but there will be a *net* movement of the dye molecules across the membrane to the side that began as pure water. The dye molecules will continue to spread across the membrane until both solutions have equal concentrations of the dye. Once that point is reached, there will be a dynamic equilibrium, with as many dye molecules crossing the membrane each second in one direction as in the other.

We can now state a simple rule of diffusion: In the absence of other forces, a substance will diffuse from where it is more concentrated to where it is less concentrated. Put another way, any substance will diffuse down its **concentration gradient**, the region along which the density of a chemical substance decreases. No work must be done in order to make this happen; diffusion is a spontaneous process, needing no input of energy. Note that each substance diffuses down its *own* concentration gradient, unaffected by the concentration differences of other substances **(Figure 7.11b)**.

Much of the traffic across cell membranes occurs by diffusion. When a substance is more concentrated on one side of a membrane than on the other, there is a tendency for the substance to diffuse across the membrane down its concentration gradient (assuming that the membrane is permeable to that substance). One important example is the uptake of oxygen by a cell performing cellular respiration. Dissolved oxygen diffuses into the cell across the plasma membrane. As long as cellular respiration consumes the O_2 as it enters, diffusion into the cell will continue because the concentration gradient favors movement in that direction.

The diffusion of a substance across a biological membrane is called **passive transport** because the cell does not have to expend energy to make it happen. The concentration gradient itself represents potential energy (see Chapter 2, p. 35) and drives diffusion. Remember, however, that membranes are selectively permeable and therefore have different effects on the rates of diffusion of various molecules. In the case of water, aquaporins allow water to diffuse very rapidly across the membranes of certain cells. As we'll see next, the movement of water across the plasma membrane has important consequences for cells.

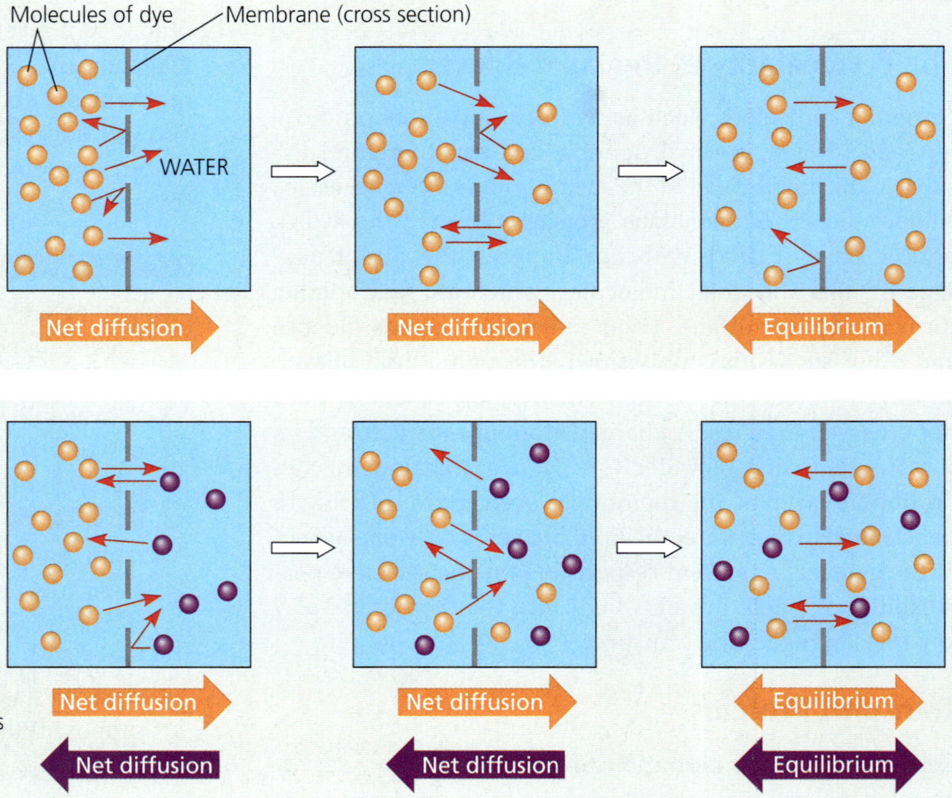

(a) Diffusion of one solute. The membrane has pores large enough for molecules of dye to pass through. Random movement of dye molecules will cause some to pass through the pores; this will happen more often on the side with more molecules. The dye diffuses from where it is more concentrated to where it is less concentrated (called diffusing down a concentration gradient). This leads to a dynamic equilibrium: The solute molecules continue to cross the membrane, but at equal rates in both directions.

(b) Diffusion of two solutes. Solutions of two different dyes are separated by a membrane that is permeable to both. Each dye diffuses down its own concentration gradient. There will be a net diffusion of the purple dye toward the left, even though the *total* solute concentration was initially greater on the left side.

▲ **Figure 7.11 The diffusion of solutes across a membrane.** Each of the large arrows under the diagrams shows the net diffusion of the dye molecules of that color.

Effects of Osmosis on Water Balance

To see how two solutions with different solute concentrations interact, picture a U-shaped glass tube with a selectively permeable membrane separating two sugar solutions **(Figure 7.12)**. Pores in this synthetic membrane are too small for sugar molecules to pass through but large enough for water molecules. How does this affect the *water* concentration? It seems logical that the solution with the higher concentration of solute would have the lower concentration of water and that water would diffuse into it from the other side for that reason. However, for a dilute solution like most biological fluids, solutes do not affect the water concentration significantly. Instead, tight clustering of water molecules around the hydrophilic solute molecules makes some of the water unavailable to cross the membrane. It is the difference in *free* water concentration that is important. In the end, the effect is the same: Water diffuses across the membrane from the region of lower solute concentration to that of higher solute concentration until the solute concentrations on both sides of the membrane are equal. The diffusion of water across a selectively permeable membrane is called **osmosis**. The movement of water across cell membranes and the balance of water between the cell and its environment are crucial to organisms. Let's now apply to living cells what we have learned about osmosis in artificial systems.

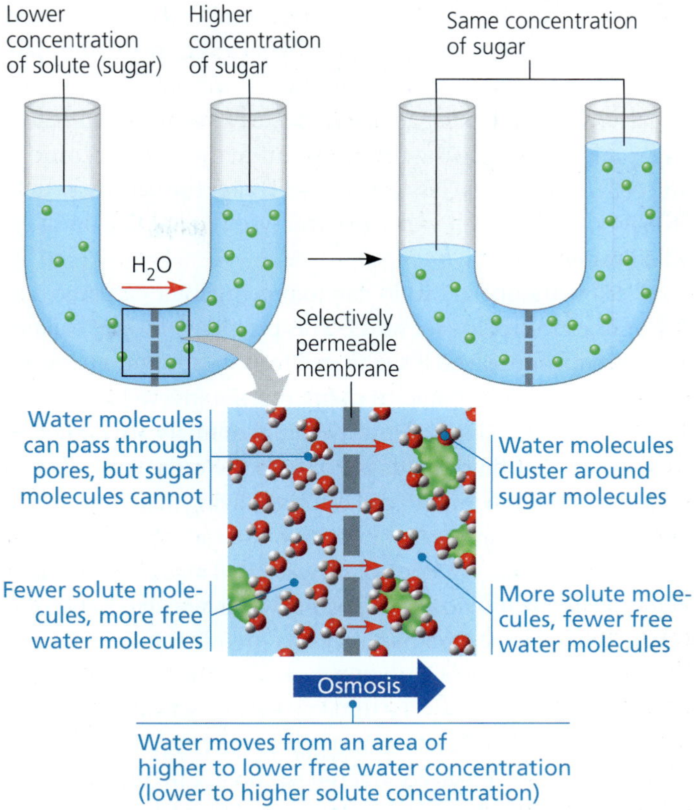

▲ **Figure 7.12 Osmosis.** Two sugar solutions of different concentrations are separated by a membrane, which the solvent (water) can pass through but the solute (sugar) cannot. Water molecules move randomly and may cross in either direction, but overall, water diffuses from the solution with less concentrated solute to that with more concentrated solute. This transport of water, or osmosis, equalizes the sugar concentrations on both sides.

WHAT IF? *If an orange dye capable of passing through the membrane was added to the left side of the tube above, how would it be distributed at the end of the process? (See Figure 7.11.) Would the solution levels in the tube on the right be affected?*

Water Balance of Cells Without Walls

When considering the behavior of a cell in a solution, both solute concentration and membrane permeability must be considered. Both factors are taken into account in the concept of **tonicity**, the ability of a solution to cause a cell to gain or lose water. The tonicity of a solution depends in part on its concentration of solutes that cannot cross the membrane (nonpenetrating solutes), relative to that inside the cell. If there is a higher concentration of nonpenetrating solutes in the surrounding solution, water will tend to leave the cell, and vice versa.

If a cell without a wall, such as an animal cell, is immersed in an environment that is **isotonic** to the cell (*iso* means "same"), there will be no *net* movement of water across the plasma membrane. Water flows across the membrane, but at the same rate in both directions. In an isotonic environment, the volume of an animal cell is stable **(Figure 7.13a)**.

Now let's transfer the cell to a solution that is **hypertonic** to the cell (*hyper* means "more," in this case referring to nonpenetrating solutes). The cell will lose water to its environment, shrivel, and probably die. This is one way an

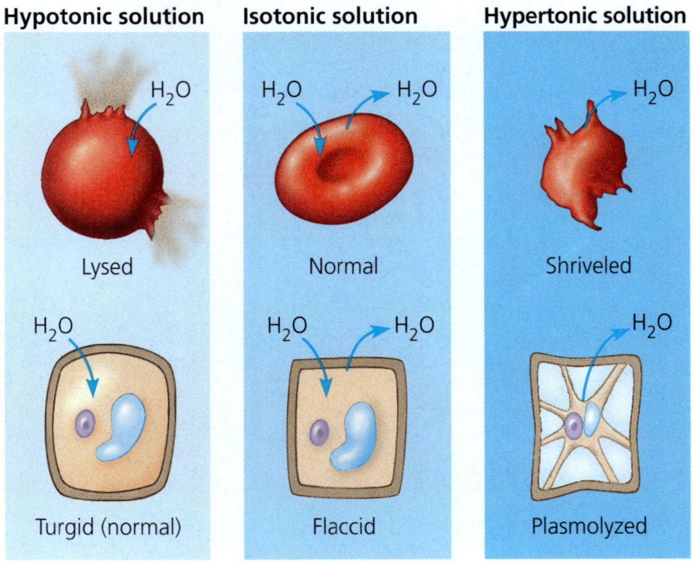

(a) Animal cell. An animal cell fares best in an isotonic environment unless it has special adaptations that offset the osmotic uptake or loss of water.

(b) Plant cell. Plant cells are turgid (firm) and generally healthiest in a hypotonic environment, where the uptake of water is eventually balanced by the wall pushing back on the cell.

▲ **Figure 7.13 The water balance of living cells.** How living cells react to changes in the solute concentration of their environment depends on whether or not they have cell walls. **(a)** Animal cells, such as this red blood cell, do not have cell walls. **(b)** Plant cells do. (Arrows indicate net water movement after the cells were first placed in these solutions.)

increase in the salinity (saltiness) of a lake can kill animals there; if the lake water becomes hypertonic to the animals' cells, the cells might shrivel and die. However, taking up too much water can be just as hazardous to an animal cell as losing water. If we place the cell in a solution that is **hypotonic** to the cell (*hypo* means "less"), water will enter the cell faster than it leaves, and the cell will swell and lyse (burst) like an overfilled water balloon.

A cell without rigid walls can tolerate neither excessive uptake nor excessive loss of water. This problem of water balance is automatically solved if such a cell lives in isotonic surroundings. Seawater is isotonic to many marine invertebrates. The cells of most terrestrial (land-dwelling) animals are bathed in an extracellular fluid that is isotonic to the cells. Animals and other organisms without rigid cell walls living in hypertonic or hypotonic environments must have special adaptations for **osmoregulation**, the control of water balance. For example, the protist *Paramecium* lives in pond water, which is hypotonic to the cell. *Paramecium* has a plasma membrane that is much less permeable to water than the membranes of most other cells, but this only slows the uptake of water, which continually enters the cell. The *Paramecium* cell doesn't burst because it is also equipped with a contractile vacuole, an organelle that functions as a bilge pump to force water out of the cell as fast as it enters by osmosis **(Figure 7.14)**. We will examine other evolutionary adaptations for osmoregulation in Chapter 44.

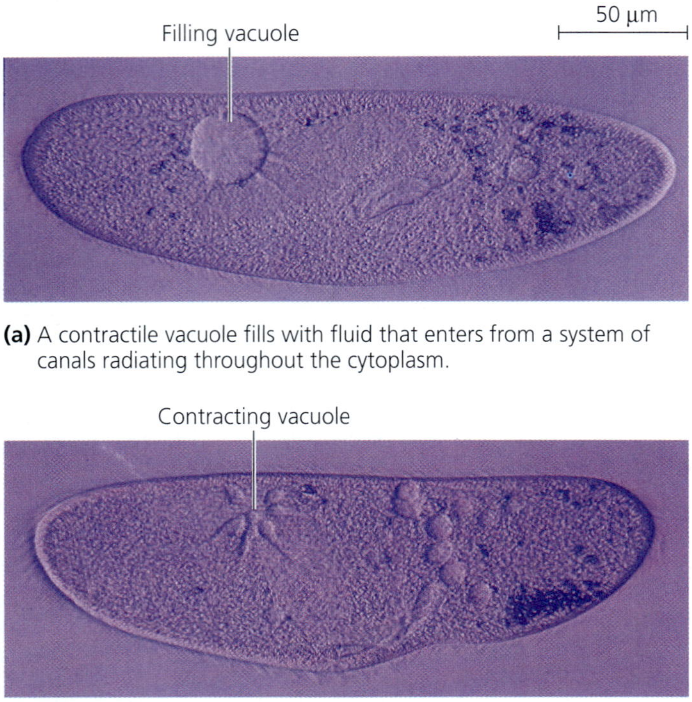

(a) A contractile vacuole fills with fluid that enters from a system of canals radiating throughout the cytoplasm.

(b) When full, the vacuole and canals contract, expelling fluid from the cell.

▲ **Figure 7.14 The contractile vacuole of *Paramecium*: an evolutionary adaptation for osmoregulation.** The contractile vacuole of this freshwater protist offsets osmosis by pumping water out of the cell (LM).

Water Balance of Cells with Walls

The cells of plants, prokaryotes, fungi, and some protists have walls (see Figure 6.28). When such a cell is immersed in a hypotonic solution—bathed in rainwater, for example—the wall helps maintain the cell's water balance. Consider a plant cell. Like an animal cell, the plant cell swells as water enters by osmosis **(Figure 7.13b)**. However, the relatively inelastic wall will expand only so much before it exerts a back pressure on the cell that opposes further water uptake. At this point, the cell is **turgid** (very firm), which is the healthy state for most plant cells. Plants that are not woody, such as most houseplants, depend for mechanical support on cells kept turgid by a surrounding hypotonic solution. If a plant's cells and their surroundings are isotonic, there is no net tendency for water to enter, and the cells become **flaccid** (limp).

However, a wall is of no advantage if the cell is immersed in a hypertonic environment. In this case, a plant cell, like an animal cell, will lose water to its surroundings and shrink. As the plant cell shrivels, its plasma membrane pulls away from the wall. This phenomenon, called **plasmolysis**, causes the plant to wilt and can lead to plant death. The walled cells of bacteria and fungi also plasmolyze in hypertonic environments.

Facilitated Diffusion: Passive Transport Aided by Proteins

Let's look more closely at how water and certain hydrophilic solutes cross a membrane. As mentioned earlier, many polar molecules and ions impeded by the lipid bilayer of the membrane diffuse passively with the help of transport proteins that span the membrane. This phenomenon is called **facilitated diffusion**. Cell biologists are still trying to learn exactly how various transport proteins facilitate diffusion. Most transport proteins are very specific: They transport some substances but not others.

As described earlier, the two types of transport proteins are channel proteins and carrier proteins. Channel proteins simply provide corridors that allow a specific molecule or ion to cross the membrane **(Figure 7.15a)**. The hydrophilic passageways provided by these proteins can allow water molecules or small ions to flow very quickly from one side of the membrane to the other. Although water molecules are small enough to cross through the phospholipid bilayer, the rate of water movement by this route is relatively slow because of the polarity of the water molecules. Aquaporins, the water channel proteins, facilitate the massive amounts of diffusion that occur in plant cells and in animal cells such as red blood cells (see Figure 7.13). Kidney cells also have a high number of aquaporins, allowing them to reclaim water from urine before it is excreted. It has been estimated that a person would have to drink 50 gallons of water a day and excrete the same volume if the kidneys did not perform this function.

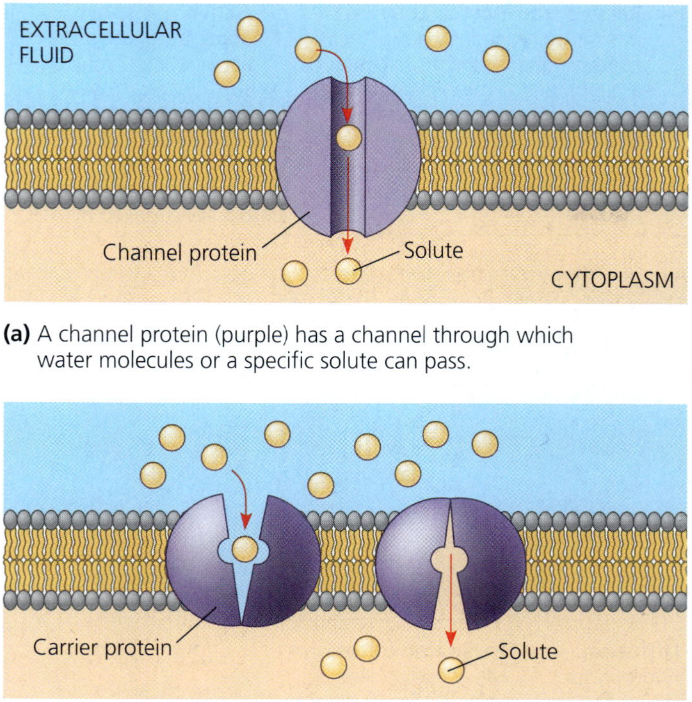

(a) A channel protein (purple) has a channel through which water molecules or a specific solute can pass.

(b) A carrier protein alternates between two shapes, moving a solute across the membrane during the shape change.

▲ **Figure 7.15 Two types of transport proteins that carry out facilitated diffusion.** In both cases, the protein can transport the solute in either direction, but the net movement is down the concentration gradient of the solute.

Another group of channel proteins are **ion channels**, many of which function as **gated channels**, which open or close in response to a stimulus. The stimulus may be electrical or chemical; if chemical, the stimulus is a substance other than the one to be transported. For example, stimulation of a nerve cell by certain neurotransmitter molecules opens gated channels that allow sodium ions into the cell. Later, an electrical stimulus activates the ion channel protein shown in Figure 7.1, and potassium ions rush out of the cell.

Carrier proteins, such as the glucose transporter mentioned earlier, seem to undergo a subtle change in shape that somehow translocates the solute-binding site across the membrane **(Figure 7.15b)**. These changes in shape may be triggered by the binding and release of the transported molecule.

In certain inherited diseases, specific transport systems are either defective or missing altogether. An example is cystinuria, a human disease characterized by the absence of a carrier protein that transports cysteine and some other amino acids across the membranes of kidney cells. Kidney cells normally reabsorb these amino acids from the urine and return them to the blood, but an individual afflicted with cystinuria develops painful stones from amino acids that accumulate and crystallize in the kidneys.

CONCEPT CHECK **7.3**

1. How do you think a cell performing cellular respiration rids itself of the resulting CO_2?
2. In the supermarket, produce is often sprayed with water. Explain why this makes vegetables look crisp.
3. WHAT IF? If a *Paramecium* swims from a hypotonic environment to an isotonic one, will its contractile vacuole become more active or less? Why?

For suggested answers, see Appendix A.

CONCEPT **7.4**

Active transport uses energy to move solutes against their gradients

Despite the help of transport proteins, facilitated diffusion is considered passive transport because the solute is moving down its concentration gradient. Facilitated diffusion speeds transport of a solute by providing efficient passage through the membrane, but it does not alter the direction of transport. Some transport proteins, however, can move solutes against their concentration gradients, across the plasma membrane from the side where they are less concentrated (whether inside or outside) to the side where they are more concentrated.

The Need for Energy in Active Transport

To pump a solute across a membrane against its gradient requires work; the cell must expend energy. Therefore, this type of membrane traffic is called **active transport**. The transport proteins that move solutes against a concentration gradient are all carrier proteins, rather than channel proteins. This makes sense because when channel proteins are open, they merely allow solutes to flow down their concentration gradient, rather than picking them up and transporting them against their gradient.

Active transport enables a cell to maintain internal concentrations of small solutes that differ from concentrations in its environment. For example, compared with its surroundings, an animal cell has a much higher concentration of potassium ions and a much lower concentration of sodium ions. The plasma membrane helps maintain these steep gradients by pumping sodium out of the cell and potassium into the cell.

As in other types of cellular work, ATP supplies the energy for most active transport. One way ATP can power active transport is by transferring its terminal phosphate group directly to the transport protein. This can induce the protein to change its shape in a manner that translocates a solute bound to the protein across the membrane. One transport system that works this way is the **sodium-potassium pump**, which exchanges sodium (Na^+) for

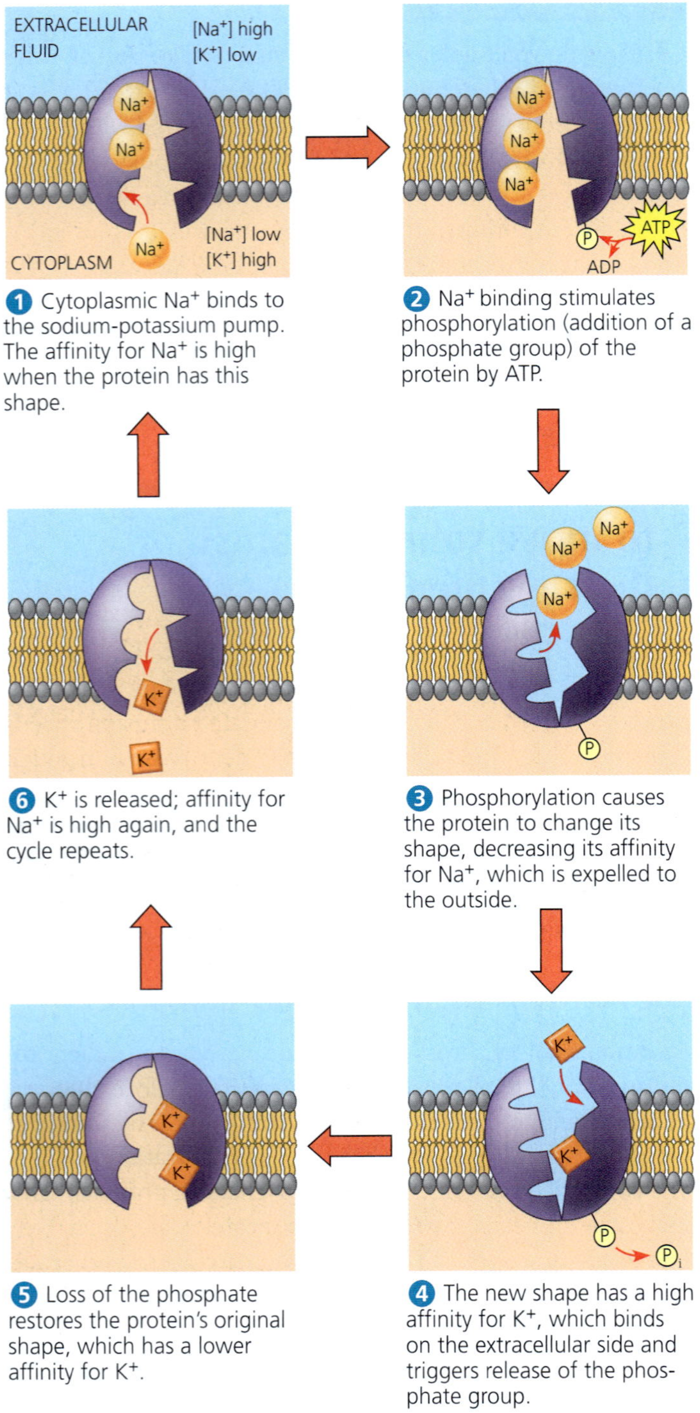

▲ **Figure 7.16 The sodium-potassium pump: a specific case of active transport.** This transport system pumps ions against steep concentration gradients: Sodium ion concentration (represented as [Na^+]) is high outside the cell and low inside, while potassium ion concentration ([K^+]) is low outside the cell and high inside. The pump oscillates between two shapes in a pumping cycle that translocates three sodium ions out of the cell for every two potassium ions pumped into the cell. The two shapes have different affinities for the two types of ions. ATP powers the shape change by phosphorylating the transport protein (that is, by transferring a phosphate group to the protein).

potassium (K^+) across the plasma membrane of animal cells **(Figure 7.16)**. The distinction between passive transport and active transport is reviewed in **Figure 7.17**.

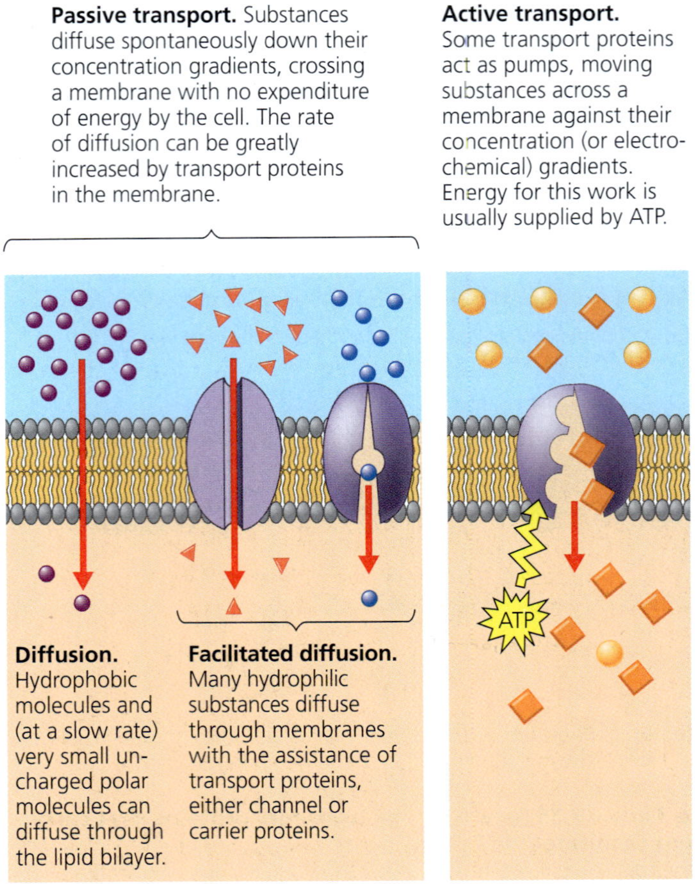

▲ **Figure 7.17 Review: passive and active transport.**

How Ion Pumps Maintain Membrane Potential

All cells have voltages across their plasma membranes. Voltage is electrical potential energy—a separation of opposite charges. The cytoplasm is negative in charge relative to the extracellular fluid because of an unequal distribution of anions and cations on opposite sides of the membrane. The voltage across a membrane, called a **membrane potential**, ranges from about -50 to -200 millivolts (mV). (The minus sign indicates that the inside of the cell is negative relative to the outside.)

The membrane potential acts like a battery, an energy source that affects the traffic of all charged substances across the membrane. Because the inside of the cell is negative compared with the outside, the membrane potential favors the passive transport of cations into the cell and anions out of the cell. Thus, *two* forces drive the diffusion of ions across a membrane: a chemical force (the ion's concentration gradient) and an electrical force (the effect of the membrane potential on the ion's movement). This combination of forces acting on an ion is called the **electrochemical gradient**.

In the case of ions, then, we must refine our concept of passive transport: An ion diffuses not simply down its *concentration* gradient but, more exactly, down its *electrochemical* gradient. For example, the concentration of sodium ions (Na^+) inside a resting nerve cell is much lower than outside it. When the cell is stimulated, gated channels open that facilitate Na^+ diffusion. Sodium ions then "fall" down their electrochemical gradient, driven by the concentration gradient of Na^+ and by the attraction of these cations to the negative side of the membrane. In this example, both electrical and chemical contributions to the electrochemical gradient act in the same direction across the membrane, but this is not always so. In cases where electrical forces due to the membrane potential oppose the simple diffusion of an ion down its concentration gradient, active transport may be necessary. In Chapter 48, you'll learn about the importance of electrochemical gradients and membrane potentials in the transmission of nerve impulses.

Some membrane proteins that actively transport ions contribute to the membrane potential. An example is the sodium-potassium pump. Notice in Figure 7.16 that the pump does not translocate Na^+ and K^+ one for one, but pumps three sodium ions out of the cell for every two potassium ions it pumps into the cell. With each "crank" of the pump, there is a net transfer of one positive charge from the cytoplasm to the extracellular fluid, a process that stores energy as voltage. A transport protein that generates voltage across a membrane is called an **electrogenic pump**. The sodium-potassium pump seems to be the major electrogenic pump of animal cells. The main electrogenic pump of plants, fungi, and bacteria is a **proton pump**, which actively transports hydrogen ions (protons) out of the cell. The pumping of H^+ transfers positive charge from the cytoplasm to the extracellular solution **(Figure 7.18)**. By generating voltage across membranes, electrogenic pumps store energy that can be tapped for cellular work. One important use of proton gradients in the cell is for ATP synthesis during cellular respiration, as you will see in Chapter 9. Another is a type of membrane traffic called cotransport.

Cotransport: Coupled Transport by a Membrane Protein

A single ATP-powered pump that transports a specific solute can indirectly drive the active transport of several other solutes in a mechanism called **cotransport**. A substance that has been pumped across a membrane can do work as it moves back across the membrane by diffusion, analogous to water that has been pumped uphill and performs work as it flows back down. Another transport protein, a cotransporter separate from the pump, can couple the "downhill" diffusion of this substance to the "uphill" transport of a second substance against its own concentration gradient. For example, a plant cell uses the gradient of hydrogen ions generated by its proton pumps to drive the active transport of amino acids, sugars, and several other nutrients into the cell. One transport protein couples the return of hydrogen ions to the transport of sucrose into the cell **(Figure 7.19)**. This protein can translocate sucrose into the cell against a concentration gradient, but only if the sucrose molecule travels in the company of a hydrogen ion. The hydrogen ion uses the transport protein as an avenue to diffuse down the electrochemical gradient maintained by the proton pump. Plants use sucrose-H^+ cotransport to load sucrose produced by photosynthesis into cells in the veins of leaves. The vascular tissue of the plant can then distribute the sugar to nonphotosynthetic organs, such as roots.

What we know about cotransport proteins, osmosis, and water balance in animal cells has helped us find more effective treat-

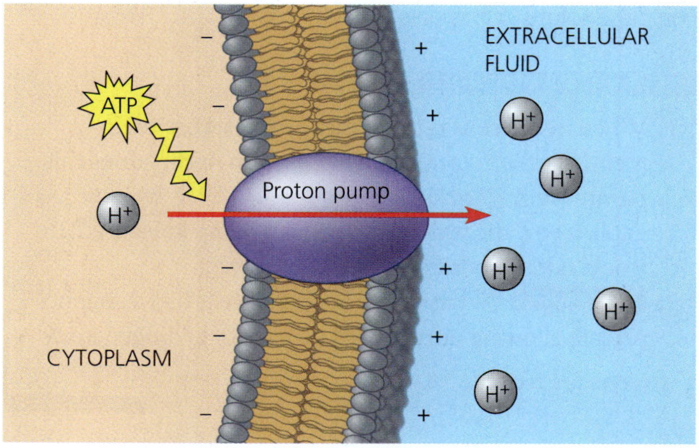

▲ **Figure 7.18 An electrogenic pump.** Proton pumps, the main electrogenic pumps of plants, fungi, and bacteria, are membrane proteins that store energy by generating voltage (charge separation) across membranes. Using ATP for power, a proton pump translocates positive charge in the form of hydrogen ions. The voltage and H^+ concentration gradient represent a dual energy source that can drive other processes, such as the uptake of nutrients.

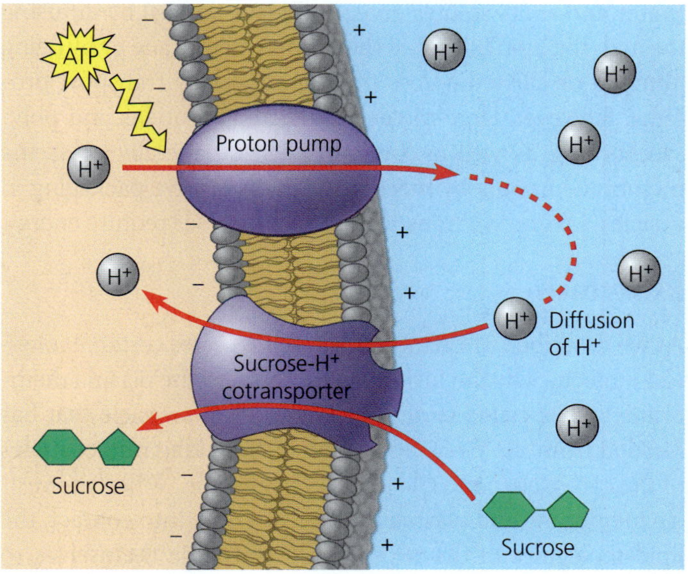

▲ **Figure 7.19 Cotransport: active transport driven by a concentration gradient.** A carrier protein such as this sucrose-H^+ cotransporter is able to use the diffusion of H^+ down its electrochemical gradient into the cell to drive the uptake of sucrose. The H^+ gradient is maintained by an ATP-driven proton pump that concentrates H^+ outside the cell, thus storing potential energy that can be used for active transport, in this case of sucrose. Thus, ATP is indirectly providing the energy necessary for cotransport.

ments for the dehydration resulting from diarrhea, a serious problem in developing countries where intestinal parasites are prevalent. Patients are given a solution to drink containing a high concentration of glucose and salt. The solutes are taken up by transport proteins on the surface of intestinal cells and passed through the cells into the blood. The increase in the blood's solute concentration causes a flow of water from the intestine through the intestinal cells into the blood, rehydrating the patient. Because of the transport proteins involved, both glucose *and* sodium ions from salt must be present. This is why athletes consume solute-rich sports drinks.

CONCEPT CHECK 7.4

1. When nerve cells establish a voltage across their membrane with a sodium-potassium pump, does this pump use ATP or does it produce ATP? Why?
2. Explain why the sodium-potassium pump in Figure 7.16 would not be considered a cotransporter.
3. WHAT IF? What would happen if cells had a channel protein allowing unregulated passage of hydrogen ions?

For suggested answers, see Appendix A.

CONCEPT 7.5

Bulk transport across the plasma membrane occurs by exocytosis and endocytosis

Water and small solutes enter and leave the cell by diffusing through the lipid bilayer of the plasma membrane or by being pumped or carried across the membrane by transport proteins. However, large molecules, such as proteins and polysaccharides, as well as larger particles, generally cross the membrane in bulk by mechanisms that involve packaging in vesicles. Like active transport, these processes require energy.

Exocytosis

As we described in Chapter 6, the cell secretes certain biological molecules by the fusion of vesicles with the plasma membrane; this is called **exocytosis**. A transport vesicle that has budded from the Golgi apparatus moves along microtubules of the cytoskeleton to the plasma membrane. When the vesicle membrane and plasma membrane come into contact, the lipid molecules of the two bilayers rearrange themselves so that the two membranes fuse. The contents of the vesicle then spill to the outside of the cell, and the vesicle membrane becomes part of the plasma membrane (see Figure 7.10).

Many secretory cells use exocytosis to export products. For example, some cells in the pancreas make insulin and secrete it into the extracellular fluid by exocytosis. Another example is the neuron (nerve cell), which uses exocytosis to release neurotransmitters that signal other neurons or muscle cells. When plant cells are making walls, exocytosis delivers proteins and carbohydrates from Golgi vesicles to the outside of the cell.

Endocytosis

In **endocytosis**, the cell takes in biological molecules and particulate matter by forming new vesicles from the plasma membrane. Although the proteins involved in the processes are different, the events of endocytosis look like the reverse of exocytosis. A small area of the plasma membrane sinks inward to form a pocket. As the pocket deepens, it pinches in, forming a vesicle containing material that had been outside the cell. There are three types of endocytosis: **phagocytosis** ("cellular eating"), **pinocytosis** ("cellular drinking"), and **receptor-mediated endocytosis**. (Study **Figure 7.20**.)

Human cells use receptor-mediated endocytosis to take in cholesterol for use in the synthesis of membranes and other steroids. Cholesterol travels in the blood in particles called low-density lipoproteins (LDLs), complexes of lipids and proteins. LDLs act as **ligands** (a term for any molecule that binds specifically to a receptor site of another molecule) by binding to LDL receptors on plasma membranes and then entering the cells by endocytosis. In humans with familial hypercholesterolemia, an inherited disease characterized by a very high level of cholesterol in the blood, the LDL receptor proteins are defective or missing, and the LDL particles cannot enter cells. Instead, cholesterol accumulates in the blood, where it contributes to early atherosclerosis, the buildup of lipid deposits within the walls of blood vessels. This buildup causes the walls to bulge inward, thereby narrowing the vessel and impeding blood flow.

Vesicles not only transport substances between the cell and its surroundings but also provide a mechanism for rejuvenating or remodeling the plasma membrane. Endocytosis and exocytosis occur continually in most eukaryotic cells, yet the amount of plasma membrane in a nongrowing cell remains fairly constant. Apparently, the addition of membrane by one process offsets the loss of membrane by the other.

Energy and cellular work have figured prominently in our study of membranes. We have seen, for example, that active transport is powered by ATP. In the next three chapters, you will learn more about how cells acquire chemical energy to do the work of life.

CONCEPT CHECK 7.5

1. As a cell grows, its plasma membrane expands. Does this involve endocytosis or exocytosis? Explain.
2. WHAT IF? To send a signal, a neuron may carry out exocytosis of signaling molecules that are recognized by a second neuron. In some cases, the first neuron ends the signal by taking up the molecules by endocytosis. Would you expect this to occur by pinocytosis or by receptor-mediated endocytosis? Explain.

For suggested answers, see Appendix A.

▼ Figure 7.20

Exploring Endocytosis in Animal Cells

Phagocytosis

In **phagocytosis**, a cell engulfs a particle by wrapping pseudopodia (singular, *pseudopodium*) around it and packaging it within a membrane-enclosed sac that can be large enough to be classified as a vacuole. The particle is digested after the vacuole fuses with a lysosome containing hydrolytic enzymes.

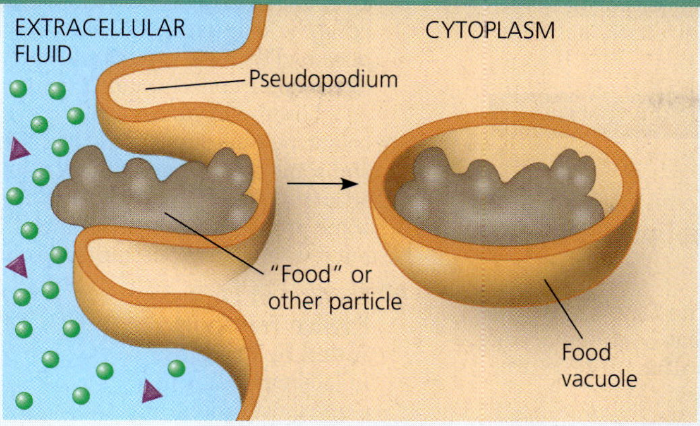

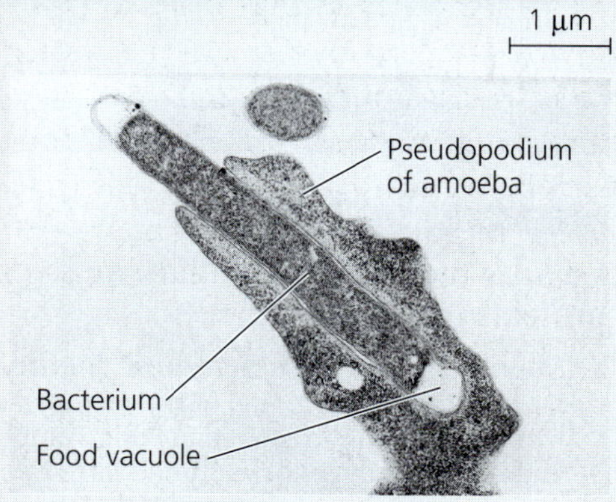

An amoeba engulfing a bacterium via phagocytosis (TEM)

Pinocytosis

In **pinocytosis**, the cell "gulps" droplets of extracellular fluid into tiny vesicles. It is not the fluid itself that is needed by the cell, but the molecules dissolved in the droplets. Because any and all included solutes are taken into the cell, pinocytosis is nonspecific in the substances it transports.

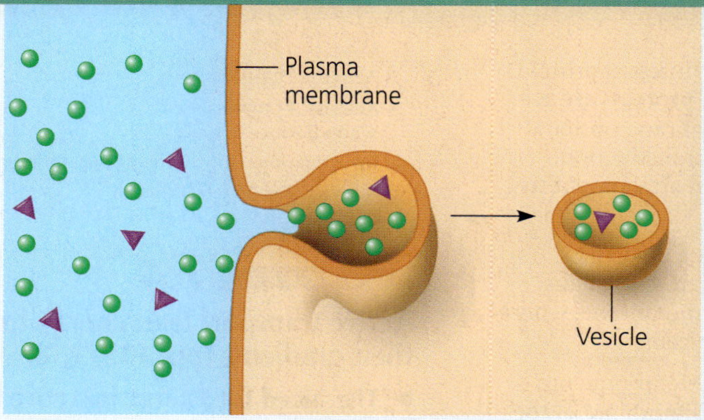

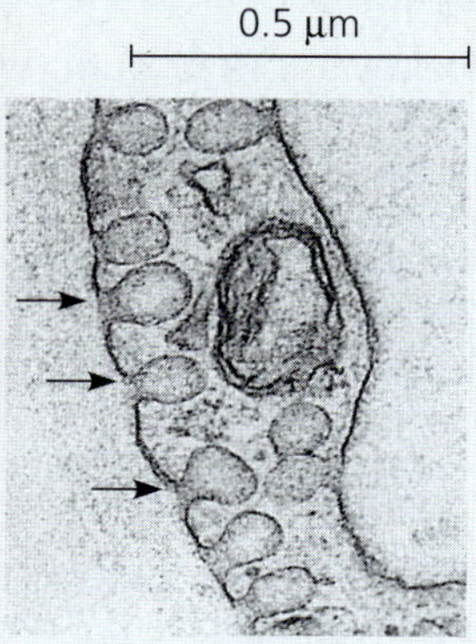

Pinocytosis vesicles forming (arrows) in a cell lining a small blood vessel (TEM)

Receptor-Mediated Endocytosis

Receptor-mediated endocytosis enables the cell to acquire bulk quantities of specific substances, even though those substances may not be very concentrated in the extracellular fluid. Embedded in the membrane are proteins with specific receptor sites exposed to the extracellular fluid. The receptor proteins are usually already clustered in regions of the membrane called coated pits, which are lined on their cytoplasmic side by a fuzzy layer of coat proteins. The specific substances (ligands) bind to these receptors. When binding occurs, the coated pit forms a vesicle containing the ligand molecules. Notice that there are relatively more bound molecules (purple) inside the vesicle, but other molecules (green) are also present. After this ingested material is liberated from the vesicle, the receptors are recycled to the plasma membrane by the same vesicle.

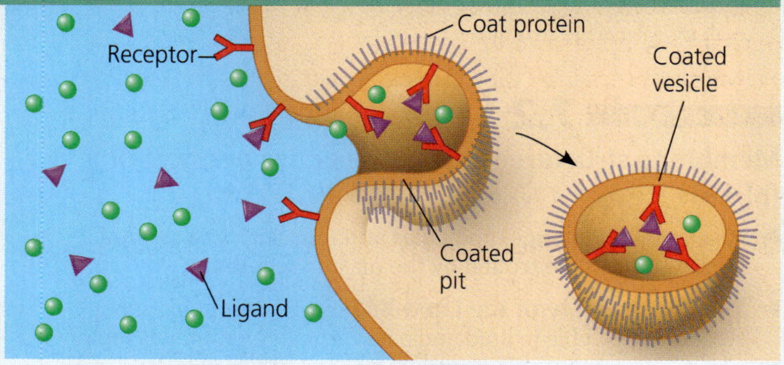

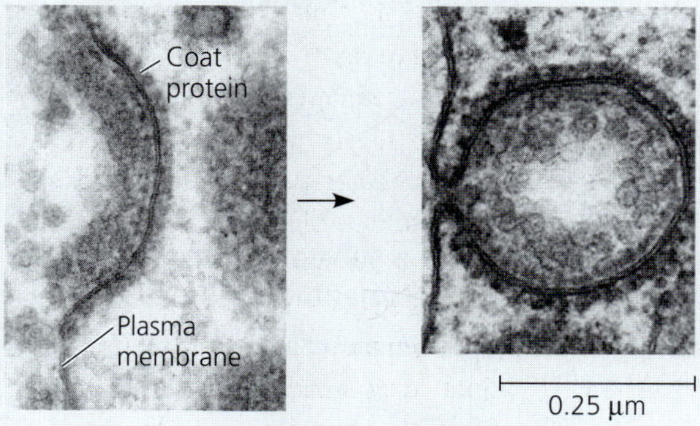

A coated pit and a coated vesicle formed during receptor-mediated endocytosis (TEMs)

Chapter 7 Review

MEDIA Go to the Study Area at **www.masteringbio.com** for BioFlix 3-D Animations, MP3 Tutors, Videos, Practice Tests, an eBook, and more.

SUMMARY OF KEY CONCEPTS

CONCEPT 7.1

Cellular membranes are fluid mosaics of lipids and proteins (pp. 125–130)

- **Membrane Models: *Scientific Inquiry*** The Davson-Danielli sandwich model of the membrane has been replaced by the fluid mosaic model, in which amphipathic proteins are embedded in the phospholipid bilayer.
- **The Fluidity of Membranes** Phospholipids and, to a lesser extent, proteins move laterally within the membrane. The unsaturated hydrocarbon tails of some phospholipids keep membranes fluid at lower temperatures, while cholesterol acts as a temperature buffer, resisting changes in fluidity caused by temperature changes.
- **Membrane Proteins and Their Functions** Integral proteins are embedded in the lipid bilayer; peripheral proteins are attached to the surfaces. The functions of membrane proteins include transport, enzymatic activity, signal transduction, cell-cell recognition, intercellular joining, and attachment to the cytoskeleton and extracellular matrix.
- **The Role of Membrane Carbohydrates in Cell-Cell Recognition** Short chains of sugars are linked to proteins and lipids on the exterior side of the plasma membrane, where they interact with surface molecules of other cells.
- **Synthesis and Sidedness of Membranes** Membrane proteins and lipids are synthesized in the ER and modified in the ER and Golgi apparatus. The inside and outside faces of the membrane differ in molecular composition.

MEDIA
Activity Membrane Structure

CONCEPT 7.2

Membrane structure results in selective permeability (p. 131)

- A cell must exchange molecules and ions with its surroundings, a process controlled by the plasma membrane.
- **The Permeability of the Lipid Bilayer** Hydrophobic substances are soluble in lipid and pass through membranes rapidly.
- **Transport Proteins** To cross the membrane, polar molecules and ions generally require specific transport proteins.

MEDIA
Activity Selective Permeability of Membranes

CONCEPT 7.3

Passive transport is diffusion of a substance across a membrane with no energy investment (pp. 132–135)

- Diffusion is the spontaneous movement of a substance down its concentration gradient.
- **Effects of Osmosis on Water Balance** Water diffuses out of a cell if the solution outside has a higher solute concentration (hypertonic) than the cytosol and enters the cell if the solution has a lower solute concentration (hypotonic). If the concentrations are equal (isotonic), no net osmosis occurs. Cell survival depends on balancing water uptake and loss. Cells lacking walls (as in animals and some protists) are isotonic with their environments or have adaptations for osmoregulation. Plants, prokaryotes, fungi, and some protists have relatively inelastic cell walls, so the cells don't burst when in a hypotonic environment.
- **Facilitated Diffusion: Passive Transport Aided by Proteins** In facilitated diffusion, a transport protein speeds the movement of water or a solute across a membrane down its concentration gradient.

Passive transport:
Facilitated diffusion

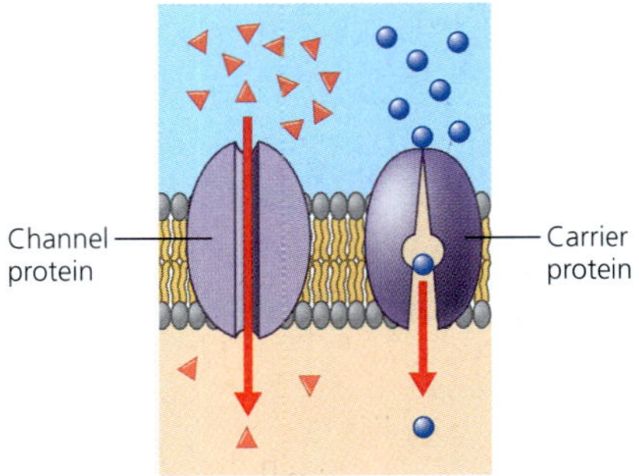

MEDIA
Activity Diffusion
Activity Osmosis and Water Balance in Cells
Investigation How Do Salt Concentrations Affect Cells?
Activity Facilitated Diffusion

CONCEPT 7.4

Active transport uses energy to move solutes against their gradients (pp. 135–138)

Active transport:

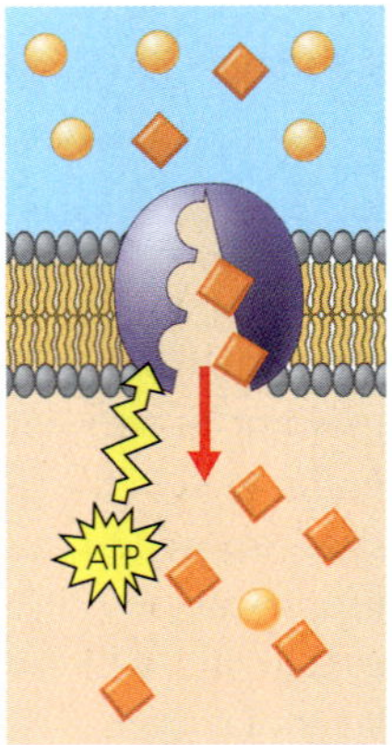

- **The Need for Energy in Active Transport** Specific membrane proteins use energy, usually in the form of ATP, to do the work of active transport.
- **How Ion Pumps Maintain Membrane Potential** Ions can have both a concentration (chemical) gradient and an electrical gradient (voltage). These forces combine in the electrochemical gradient, which determines the net direction of ionic diffusion. Electrogenic pumps, such as sodium-potassium pumps and proton pumps, are transport proteins that contribute to electrochemical gradients.
- **Cotransport: Coupled Transport by a Membrane Protein** One solute's "downhill" diffusion drives the other's "uphill" transport.

MEDIA
Activity Active Transport

CONCEPT 7.5

Bulk transport across the plasma membrane occurs by exocytosis and endocytosis (pp. 138–139)

- **Exocytosis** In exocytosis, transport vesicles migrate to the plasma membrane, fuse with it, and release their contents.

▶ **Endocytosis** In endocytosis, molecules enter cells within vesicles that pinch inward from the plasma membrane. The three types of endocytosis are phagocytosis, pinocytosis, and receptor-mediated endocytosis.

MEDIA

Activity Exocytosis and Endocytosis

TESTING YOUR KNOWLEDGE

SELF-QUIZ

1. In what way do the membranes of a eukaryotic cell vary?
a. Phospholipids are found only in certain membranes.
b. Certain proteins are unique to each membrane.
c. Only certain membranes of the cell are selectively permeable.
d. Only certain membranes are constructed from amphipathic molecules.
e. Some membranes have hydrophobic surfaces exposed to the cytoplasm, while others have hydrophilic surfaces facing the cytoplasm.

2. According to the fluid mosaic model of membrane structure, proteins of the membrane are mostly
a. spread in a continuous layer over the inner and outer surfaces of the membrane.
b. confined to the hydrophobic core of the membrane.
c. embedded in a lipid bilayer.
d. randomly oriented in the membrane, with no fixed inside-outside polarity.
e. free to depart from the fluid membrane and dissolve in the surrounding solution.

3. Which of the following factors would tend to increase membrane fluidity?
a. a greater proportion of unsaturated phospholipids
b. a greater proportion of saturated phospholipids
c. a lower temperature
d. a relatively high protein content in the membrane
e. a greater proportion of relatively large glycolipids compared with lipids having smaller molecular masses

4. Which of the following processes includes all others?
a. osmosis
b. diffusion of a solute across a membrane
c. facilitated diffusion
d. passive transport
e. transport of an ion down its electrochemical gradient

5. Based on Figure 7.19, which of these experimental treatments would increase the rate of sucrose transport into the cell?
a. decreasing extracellular sucrose concentration
b. decreasing extracellular pH
c. decreasing cytoplasmic pH
d. adding an inhibitor that blocks the regeneration of ATP
e. adding a substance that makes the membrane more permeable to hydrogen ions

6. **DRAW IT** An artificial cell consisting of an aqueous solution enclosed in a selectively permeable membrane is immersed in a beaker containing a different solution. The membrane is permeable to water and to the simple sugars glucose and fructose but impermeable to the disaccharide sucrose.
a. Draw solid arrows to indicate the net movement of solutes into and/or out of the cell.

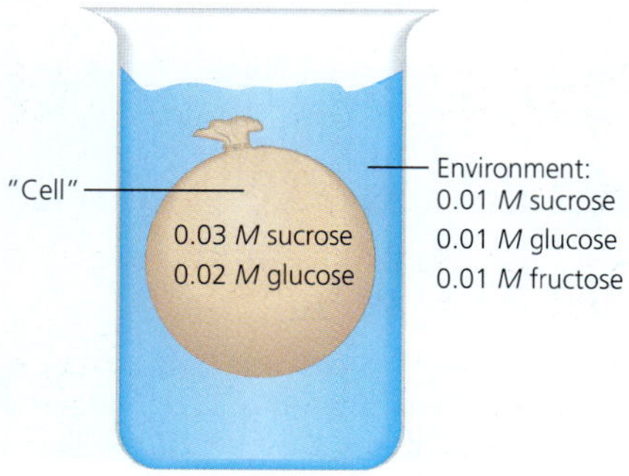

b. Is the solution outside the cell isotonic, hypotonic, or hypertonic?
c. Draw a dashed arrow to show the net osmotic movement of water, if any.
d. Will the artificial cell become more flaccid, more turgid, or stay the same?
e. Eventually, will the two solutions have the same or different solute concentrations?

For Self-Quiz answers, see Appendix A.

MEDIA Visit the Study Area at **www.masteringbio.com** for a Practice Test.

EVOLUTION CONNECTION

7. *Paramecium* and other protists that live in hypotonic environments have cell membranes that slow osmotic water uptake, while those living in isotonic environments have more permeable cell membranes. What water regulation adaptations might have evolved in protists in hypertonic habitats such as Great Salt Lake? In habitats with changing salt concentration?

SCIENTIFIC INQUIRY

8. An experiment is designed to study the mechanism of sucrose uptake by plant cells. Cells are immersed in a sucrose solution, and the pH of the solution is monitored. Samples of the cells are taken at intervals, and their sucrose concentration is measured. Their sucrose uptake correlates with a rise in the solution's pH. This rise is proportional to the starting concentration of sucrose in the solution. A metabolic poison that blocks the ability of cells to regenerate ATP is found to inhibit the pH changes in the solution. Propose a hypothesis accounting for these results. Suggest an experiment to test it.

SCIENCE, TECHNOLOGY, AND SOCIETY

9. Extensive irrigation in arid regions causes salts to accumulate in the soil. (When water evaporates, salts are left behind to concentrate in the soil.) Based on what you learned about water balance in plant cells, why might increased soil salinity (saltiness) be harmful to crops? Suggest ways to minimize damage. What costs are attached to your solutions?

8 An Introduction to Metabolism

▲ **Figure 8.1 What causes the bioluminescence in these fungi?**

KEY CONCEPTS

- 8.1 **An organism's metabolism transforms matter and energy, subject to the laws of thermodynamics**
- 8.2 **The free-energy change of a reaction tells us whether or not the reaction occurs spontaneously**
- 8.3 **ATP powers cellular work by coupling exergonic reactions to endergonic reactions**
- 8.4 **Enzymes speed up metabolic reactions by lowering energy barriers**
- 8.5 **Regulation of enzyme activity helps control metabolism**

OVERVIEW

The Energy of Life

The living cell is a chemical factory in miniature, where thousands of reactions occur within a microscopic space. Sugars can be converted to amino acids that are linked together into proteins when needed, and proteins are dismantled into amino acids that can be converted to sugars when food is digested. Small molecules are assembled into polymers, which may be hydrolyzed later as the needs of the cell change. In multicellular organisms, many cells export chemical products that are used in other parts of the organism. The process known as cellular respiration drives the cellular economy by extracting the energy stored in sugars and other fuels. Cells apply this energy to perform various types of work, such as the transport of solutes across the plasma membrane, which we discussed in Chapter 7. In a more exotic example, cells of the fungus in **Figure 8.1** convert the energy stored in certain organic molecules to light, a process called bioluminescence. (The glow may attract insects that benefit the fungus by dispersing its spores.) Bioluminescence and all other metabolic activities carried out by a cell are precisely coordinated and controlled. In its complexity, its efficiency, its integration, and its responsiveness to subtle changes, the cell is peerless as a chemical factory. The concepts of metabolism that you learn in this chapter will help you understand how matter and energy flow during life's processes and how that flow is regulated.

CONCEPT **8.1**

An organism's metabolism transforms matter and energy, subject to the laws of thermodynamics

The totality of an organism's chemical reactions is called **metabolism** (from the Greek *metabole*, change). Metabolism is an emergent property of life that arises from interactions between molecules within the orderly environment of the cell.

Organization of the Chemistry of Life into Metabolic Pathways

We can picture a cell's metabolism as an elaborate road map of the thousands of chemical reactions that occur in a cell, arranged as intersecting metabolic pathways. A **metabolic pathway** begins with a specific molecule, which is then altered in a series of defined steps, resulting in a certain product. Each step of the pathway is catalyzed by a specific enzyme:

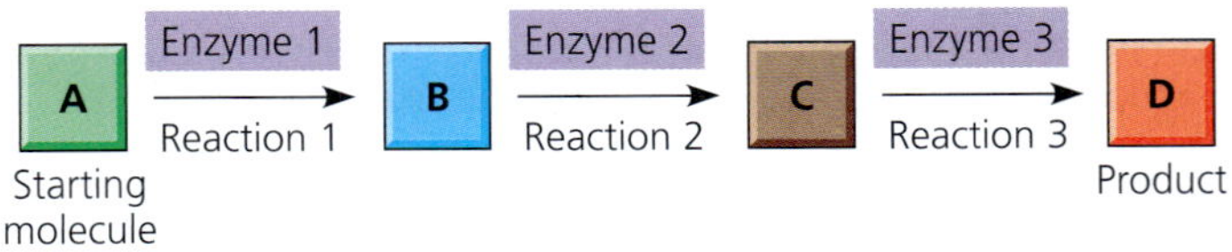

Analogous to the red, yellow, and green stoplights that control the flow of automobile traffic, mechanisms that regulate enzymes balance metabolic supply and demand, averting deficits or surpluses of important cellular molecules.

Metabolism as a whole manages the material and energy resources of the cell. Some metabolic pathways release energy by breaking down complex molecules to simpler compounds. These degradative processes are called **catabolic pathways**, or breakdown pathways. A major pathway of catabolism is cellular respiration, in which the sugar glucose and other organic fuels are broken down in the presence of oxygen to carbon dioxide and water. (Pathways can have more than one starting molecule and/or product.) Energy that was stored in the organic molecules becomes available to do the work of the cell, such as ciliary beating or membrane transport. **Anabolic pathways**, in contrast, consume energy to build complicated molecules from simpler ones; they are sometimes called biosynthetic pathways. An example of anabolism is the synthesis of a protein from amino acids. Catabolic and anabolic pathways are the "downhill" and "uphill" avenues of the metabolic map. Energy released from the downhill reactions of catabolic pathways can be stored and then used to drive the uphill reactions of anabolic pathways.

In this chapter, we will focus on mechanisms common to metabolic pathways. Because energy is fundamental to all metabolic processes, a basic knowledge of energy is necessary to understand how the living cell works. Although we will use some nonliving examples to study energy, the concepts demonstrated by these examples also apply to **bioenergetics**, the study of how energy flows through living organisms.

Forms of Energy

Energy is the capacity to cause change. In everyday life, energy is important because some forms of energy can be used to do work—that is, to move matter against opposing forces, such as gravity and friction. Put another way, energy is the ability to rearrange a collection of matter. For example, you expend energy to turn the pages of this book, and your cells expend energy in transporting certain substances across membranes. Energy exists in various forms, and the work of life depends on the ability of cells to transform energy from one form into another.

Energy can be associated with the relative motion of objects; this energy is called **kinetic energy**. Moving objects can perform work by imparting motion to other matter: A pool player uses the motion of the cue stick to push the cue ball, which in turn moves the other balls; water gushing through a dam turns turbines; and the contraction of leg muscles pushes bicycle pedals. **Heat**, or **thermal energy**, is kinetic energy associated with the random movement of atoms or molecules. Light is also a type of energy that can be harnessed to perform work, such as powering photosynthesis in green plants.

An object not presently moving may still possess energy. Energy that is not kinetic is called **potential energy**; it is energy that matter possesses because of its location or structure. Water behind a dam, for instance, possesses energy because of its altitude above sea level. Molecules possess energy because of the arrangement of their atoms. **Chemical energy** is a term used by biologists to refer to the potential energy available for release in a chemical reaction. Recall that catabolic pathways release energy by breaking down complex molecules. Biologists say that these complex molecules, such as glucose, are high in chemical energy. During a catabolic reaction, atoms are rearranged and energy is released, resulting in lower-energy breakdown products. This transformation also occurs, for example, in the engine of a car when the hydrocarbons of gasoline react explosively with oxygen, releasing the energy that pushes the pistons and producing exhaust. Although less explosive, a similar reaction of food molecules with oxygen provides chemical energy in biological systems, producing carbon dioxide and water as waste products. It is the structures and biochemical pathways of cells that enable them to release chemical energy from food molecules, powering life processes.

How is energy converted from one form to another? Consider the divers in **Figure 8.2**. The young man climbing the steps to the diving platform is releasing chemical energy from the food he ate for lunch and using some of that energy to perform the work

▲ **Figure 8.2 Transformations between potential and kinetic energy.**

of climbing. The kinetic energy of muscle movement is thus being transformed into potential energy due to his increasing height above the water. The young man diving is converting his potential energy to kinetic energy, which is then transferred to the water as he enters it. A small amount of energy is lost as heat due to friction.

Now let's go back one step and consider the original source of the organic food molecules that provided the necessary chemical energy for the diver to climb the steps. This chemical energy was itself derived from light energy by plants during photosynthesis. Organisms are energy transformers.

The Laws of Energy Transformation

The study of the energy transformations that occur in a collection of matter is called **thermodynamics**. Scientists use the word *system* to denote the matter under study; they refer to the rest of the universe—everything outside the system—as the *surroundings*. An *isolated system*, such as that approximated by liquid in a thermos bottle, is unable to exchange either energy or matter with its surroundings. In an *open system*, energy and matter can be transferred between the system and its surroundings. Organisms are open systems. They absorb energy—for instance, light energy or chemical energy in the form of organic molecules—and release heat and metabolic waste products, such as carbon dioxide, to the surroundings. Two laws of thermodynamics govern energy transformations in organisms and all other collections of matter.

The First Law of Thermodynamics

According to the **first law of thermodynamics**, the energy of the universe is constant. *Energy can be transferred and transformed, but it cannot be created or destroyed.* The first law is also known as the *principle of conservation of energy.* The electric company does not make energy, but merely converts it to a form that is convenient for us to use. By converting sunlight to chemical energy, a plant acts as an energy transformer, not an energy producer.

The cheetah in **Figure 8.3a** will convert the chemical energy of the organic molecules in its food to kinetic and other forms of energy as it carries out biological processes. What happens to this energy after it has performed work? The second law helps to answer this question.

The Second Law of Thermodynamics

If energy cannot be destroyed, why can't organisms simply recycle their energy over and over again? It turns out that during every energy transfer or transformation, some energy becomes unusable energy, unavailable to do work. In most energy transformations, more usable forms of energy are at least partly converted to heat, which is the energy associated with the random motion of atoms or molecules. Only a small fraction of the chemical energy from the food in Figure 8.3a is transformed into the motion of the cheetah shown in **Figure 8.3b**; most is lost as heat, which dissipates rapidly through the surroundings.

In the process of carrying out chemical reactions that perform various kinds of work, living cells unavoidably convert other forms of energy to heat. A system can put heat to work only when there is a temperature difference that results in the heat flowing from a warmer location to a cooler one. If temperature is uniform, as it is in a living cell, then the only use for heat energy generated during a chemical reaction is to warm a body of matter, such as the organism. (This can make a room crowded with people uncomfortably warm, as each person is carrying out a multitude of chemical reactions!)

(a) First law of thermodynamics: Energy can be transferred or transformed but neither created nor destroyed. For example, the chemical (potential) energy in food will be converted to the kinetic energy of the cheetah's movement in (b).

(b) Second law of thermodynamics: Every energy transfer or transformation increases the disorder (entropy) of the universe. For example, disorder is added to the cheetah's surroundings in the form of heat and the small molecules that are the by-products of metabolism.

▲ **Figure 8.3 The two laws of thermodynamics.**

A logical consequence of the loss of usable energy during energy transfer or transformation is that each such event makes the universe more disordered. Scientists use a quantity called **entropy** as a measure of disorder, or randomness. The more randomly arranged a collection of matter is, the greater its entropy. We can now state the **second law of thermodynamics** as follows: *Every energy transfer or transformation increases the entropy of the universe.* Although order can increase locally, there is an unstoppable trend toward randomization of the universe as a whole.

In many cases, increased entropy is evident in the physical disintegration of a system's organized structure. For example, you can observe increasing entropy in the gradual decay of an unmaintained building. Much of the increasing entropy of the universe is less apparent, however, because it appears as increasing amounts of heat and less ordered forms of matter. As the cheetah in Figure 8.3b converts chemical energy to kinetic energy, it is also increasing the disorder of its surroundings by producing heat and the small molecules, such as the CO_2 it exhales, that are the breakdown products of food.

The concept of entropy helps us understand why certain processes occur. It turns out that for a process to occur on its own, without outside help (an input of energy), it must increase the entropy of the universe. Let's first agree to use the word *spontaneous* for a process that can occur without an input of energy. Note that as we're using it here, the word *spontaneous* does not imply that such a process would occur quickly. Some spontaneous processes may be virtually instantaneous, such as an explosion, while others may be much slower, such as the rusting of an old car over time. A process that cannot occur on its own is said to be nonspontaneous; it will happen only if energy is added to the system. We know from experience that certain events occur spontaneously and others do not. For instance, we know that water flows downhill spontaneously, but moves uphill only with an input of energy, such as when a machine pumps the water against gravity. In fact, another way to state the second law is: *For a process to occur spontaneously, it must increase the entropy of the universe.*

Biological Order and Disorder

Living systems increase the entropy of their surroundings, as predicted by thermodynamic law. It is true that cells create ordered structures from less organized starting materials. For example, amino acids are ordered into the specific sequences of polypeptide chains. At the organismal level, **Figure 8.4** shows the extremely symmetrical anatomy of a plant's root, formed by biological processes from simpler starting materials. However, an organism also takes in organized forms of matter and energy from the surroundings and replaces them with less ordered forms. For example, an animal obtains starch, proteins, and other complex molecules from the food it eats. As catabolic pathways break these molecules down, the animal releases carbon dioxide and water—small molecules that possess less chemical energy than the food did. The depletion of chemical energy is accounted for by heat generated during metabolism. On a larger scale, energy flows into an ecosystem in the form of light and exits in the form of heat (see Figure 1.5).

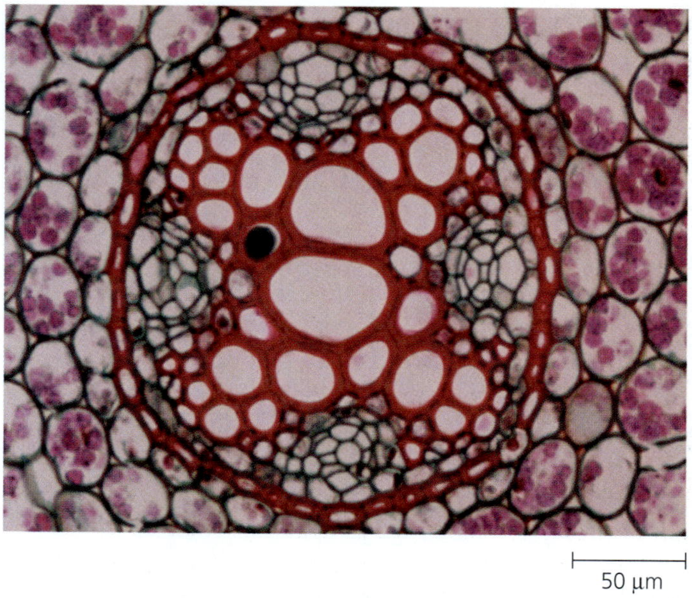

▲ **Figure 8.4 Order as a characteristic of life.** Order is evident in the detailed anatomy of this root tissue from a buttercup plant (LM, cross section). As open systems, organisms can increase their order as long as the order of their surroundings decreases.

During the early history of life, complex organisms evolved from simpler ancestors. For example, we can trace the ancestry of the plant kingdom from much simpler organisms called green algae to more complex flowering plants. However, this increase in organization over time in no way violates the second law. The entropy of a particular system, such as an organism, may actually decrease as long as the total entropy of the *universe*—the system plus its surroundings—increases. Thus, organisms are islands of low entropy in an increasingly random universe. The evolution of biological order is perfectly consistent with the laws of thermodynamics.

CONCEPT CHECK **8.1**

1. How does the second law of thermodynamics help explain the diffusion of a substance across a membrane?
2. Describe the forms of energy found in an apple as it grows on a tree, then falls and is digested by someone who eats it.
3. WHAT IF? If you place a teaspoon of sugar in the bottom of a glass of water, it will dissolve completely over time. Left longer, eventually the water will disappear and the sugar crystals will reappear. Explain these observations in terms of entropy.

For suggested answers, see Appendix A.

CONCEPT 8.2

The free-energy change of a reaction tells us whether or not the reaction occurs spontaneously

The laws of thermodynamics that we've just discussed apply to the universe as a whole. As biologists, we want to understand the chemical reactions of life—for example, which reactions occur spontaneously and which ones require some input of energy from outside. But how can we know this without assessing the energy and entropy changes in the entire universe for each separate reaction?

Free-Energy Change, ΔG

Recall that the universe is really equivalent to "the system" plus "the surroundings." In 1878, J. Willard Gibbs, a professor at Yale, defined a very useful function called the Gibbs free energy of a system (without considering its surroundings), symbolized by the letter G. We'll refer to the Gibbs free energy simply as free energy. **Free energy** is the portion of a system's energy that can perform work when temperature and pressure are uniform throughout the system, as in a living cell. Let's consider how we determine the free-energy change that occurs when a system changes—for example, during a chemical reaction.

The change in free energy, ΔG, can be calculated for a chemical reaction with the following formula:

$$\Delta G = \Delta H - T\Delta S$$

This formula uses only properties of the system (the reaction) itself: ΔH symbolizes the change in the system's *enthalpy* (in biological systems, equivalent to total energy); ΔS is the change in the system's entropy; and T is the absolute temperature in Kelvin (K) units (K = °C + 273; see Appendix C).

Once we know the value of ΔG for a process, we can use it to predict whether the process will be spontaneous (that is, whether it will occur without an input of energy from outside). More than a century of experiments has shown that only processes with a negative ΔG are spontaneous. For a process to occur spontaneously, therefore, the system must either give up enthalpy (H must decrease), give up order (TS must increase), or both: When the changes in H and TS are tallied, ΔG must have a negative value ($\Delta G < 0$) for a process to be spontaneous. This means that every spontaneous process decreases the system's free energy. Processes that have a positive or zero ΔG are never spontaneous.

This information is immensely interesting to biologists, for it gives us the power to predict which kinds of change can happen without help. Such spontaneous changes can be harnessed to perform work. This principle is very important in the study of metabolism, where a major goal is to determine which reactions can supply energy for cellular work.

Free Energy, Stability, and Equilibrium

As we saw in the previous section, when a process occurs spontaneously in a system, we can be sure that ΔG is negative. Another way to think of ΔG is to realize that it represents the difference between the free energy of the final state and the free energy of the initial state:

$$\Delta G = G_{\text{final state}} - G_{\text{initial state}}$$

Thus, ΔG can be negative only when the process involves a loss of free energy during the change from initial state to final state. Because it has less free energy, the system in its final state is less likely to change and is therefore more stable than it was previously.

We can think of free energy as a measure of a system's instability—its tendency to change to a more stable state. Unstable systems (higher G) tend to change in such a way that they become more stable (lower G). For example, a diver on top of a platform is less stable (more likely to fall) than when floating in the water, a drop of concentrated dye is less stable (more likely to disperse) than when the dye is spread randomly through the liquid, and a sugar molecule is less stable (more likely to break down) than the simpler molecules into which it can be split **(Figure 8.5)**. Unless something prevents it, each of these systems will move toward greater stability: The diver falls, the solution becomes uniformly colored, and the sugar molecule is broken down.

Another term that describes a state of maximum stability is *equilibrium*, which you learned about in Chapter 2 in connection with chemical reactions. There is an important relationship between free energy and equilibrium, including chemical equilibrium. Recall that most chemical reactions are reversible and proceed to a point at which the forward and backward reactions occur at the same rate. The reaction is then said to be at chemical equilibrium, and there is no further net change in the relative concentration of products and reactants.

As a reaction proceeds toward equilibrium, the free energy of the mixture of reactants and products decreases. Free energy increases when a reaction is somehow pushed away from equilibrium, perhaps by removing some of the products (and thus changing their concentration relative to that of the reactants). For a system at equilibrium, G is at its lowest possible value in that system. We can think of the equilibrium state as a free-energy valley. Any change from the equilibrium position will have a positive ΔG and will not be spontaneous. For this reason, systems never spontaneously move away from equilibrium. Because a system at equilibrium cannot spontaneously change, it can do no work. A process is spontaneous and can perform work only when it is moving toward equilibrium.

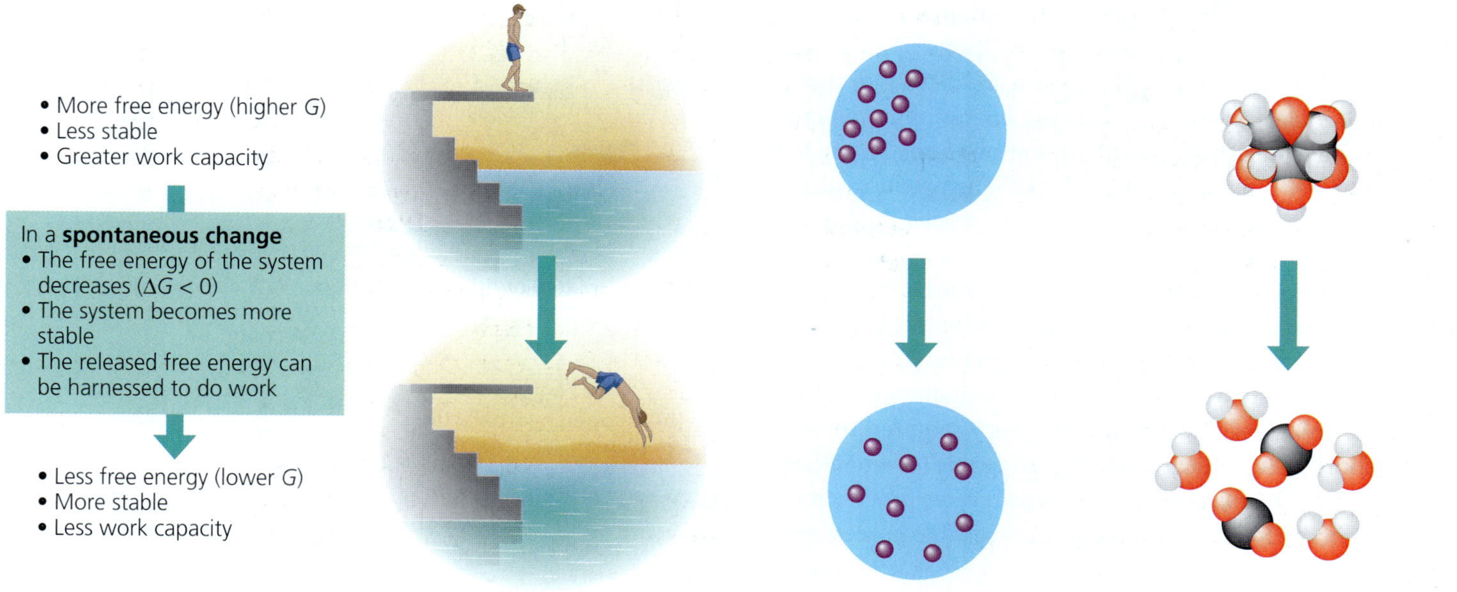

(a) Gravitational motion. Objects move spontaneously from a higher altitude to a lower one.

(b) Diffusion. Molecules in a drop of dye diffuse until they are randomly dispersed.

(c) Chemical reaction. In a cell, a sugar molecule is broken down into simpler molecules.

▲ **Figure 8.5 The relationship of free energy to stability, work capacity, and spontaneous change.** Unstable systems (top diagrams) are rich in free energy, *G*. They have a tendency to change spontaneously to a more stable state (bottom), and it is possible to harness this "downhill" change to perform work.

Free Energy and Metabolism

We can now apply the free-energy concept more specifically to the chemistry of life's processes.

Exergonic and Endergonic Reactions in Metabolism

Based on their free-energy changes, chemical reactions can be classified as either exergonic ("energy outward") or endergonic ("energy inward"). An **exergonic reaction** proceeds with a net release of free energy **(Figure 8.6a)**. Because the chemical mixture loses free energy (*G* decreases), ΔG is negative for an exergonic reaction. Using ΔG as a standard for spontaneity, exergonic reactions are those that occur spontaneously. (Remember, the word *spontaneous* does not imply that a reaction will occur instantaneously or even rapidly.) The magnitude of ΔG for an exergonic reaction represents the maximum amount of work the reaction can perform.* The greater the decrease in free energy, the greater the amount of work that can be done.

We can use the overall reaction for cellular respiration as an example:

$$C_6H_{12}O_6 + 6\ O_2 \rightarrow 6\ CO_2 + 6\ H_2O$$
$$\Delta G = -686\ \text{kcal/mol}\ (-2{,}870\ \text{kJ/mol})$$

* The word *maximum* qualifies this statement, because some of the free energy is released as heat and cannot do work. Therefore, ΔG represents a theoretical upper limit of available energy.

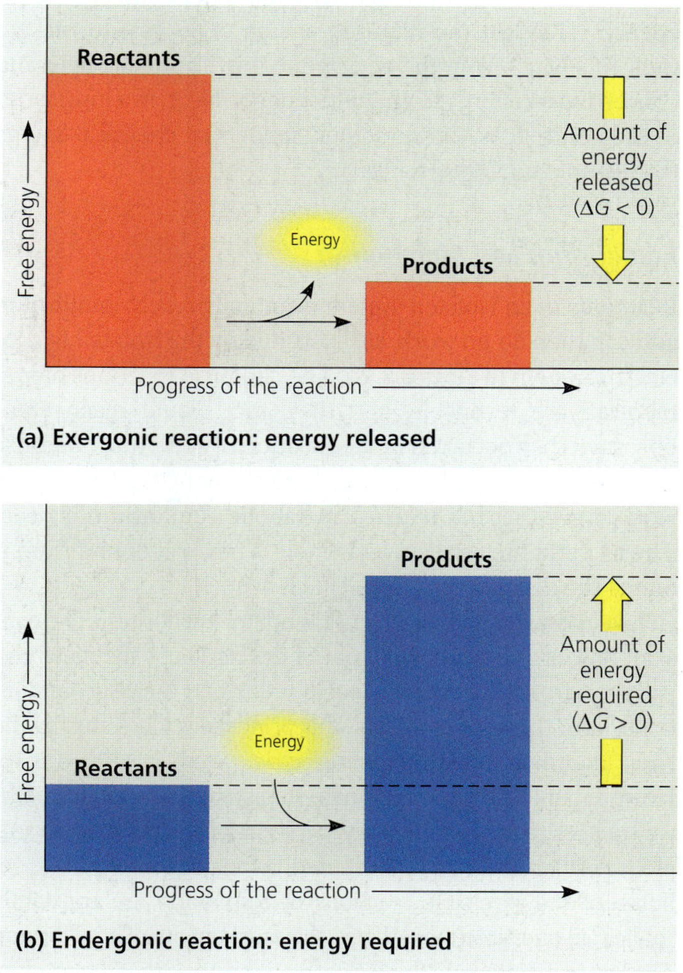

▲ **Figure 8.6 Free energy changes (ΔG) in exergonic and endergonic reactions.**

For each mole (180 g) of glucose broken down by respiration under what are called "standard conditions" (1 *M* of each reactant and product, 25°C, pH 7), 686 kcal (2,870 kJ) of energy are made available for work. Because energy must be conserved, the chemical products of respiration store 686 kcal less free energy per mole than the reactants. The products are, in a sense, the spent exhaust of a process that tapped the free energy stored in the sugar molecules.

An **endergonic reaction** is one that absorbs free energy from its surroundings **(Figure 8.6b)**. Because this kind of reaction essentially *stores* free energy in molecules (*G* increases), ΔG is positive. Such reactions are nonspontaneous, and the magnitude of ΔG is the quantity of energy required to drive the reaction. If a chemical process is exergonic (downhill), releasing energy in one direction, then the reverse process must be endergonic (uphill), using energy. A reversible process cannot be downhill in both directions. If $\Delta G = -686$ kcal/mol for respiration, which converts sugar and oxygen to carbon dioxide and water, then the reverse process—the conversion of carbon dioxide and water to sugar and oxygen—must be strongly endergonic, with $\Delta G = +686$ kcal/mol. Such a reaction would never happen by itself.

How, then, do plants make the sugar that organisms use for energy? They get the required energy—686 kcal to make a mole of sugar—from the environment by capturing light and converting its energy to chemical energy. Next, in a long series of exergonic steps, they gradually spend that chemical energy to assemble sugar molecules.

Equilibrium and Metabolism

Reactions in an isolated system eventually reach equilibrium and can then do no work, as illustrated by the isolated hydroelectric system in **Figure 8.7a**. The chemical reactions of metabolism are reversible, and they, too, would reach equilibrium if they occurred in the isolation of a test tube. Because systems at equilibrium are at a minimum of *G* and can do no work, a cell that has reached metabolic equilibrium is dead! The fact that metabolism as a whole is never at equilibrium is one of the defining features of life.

Like most systems, a living cell is not in equilibrium. The constant flow of materials in and out of the cell keeps the metabolic pathways from ever reaching equilibrium, and the cell continues to do work throughout its life. This principle is illustrated by the open (and more realistic) hydroelectric system in **Figure 8.7b**. However, unlike this simple single-step system, a catabolic pathway in a cell releases free energy in a series of reactions. An example is cellular respiration, illustrated by analogy in **Figure 8.7c**. Some of the reversible reactions of respiration are constantly "pulled" in one direction—that is, they are kept out of equilibrium. The key to maintaining this lack of equilibrium is that the product of a reaction does not accumulate, but instead becomes a reactant in the next step; finally, waste products are expelled from the cell.

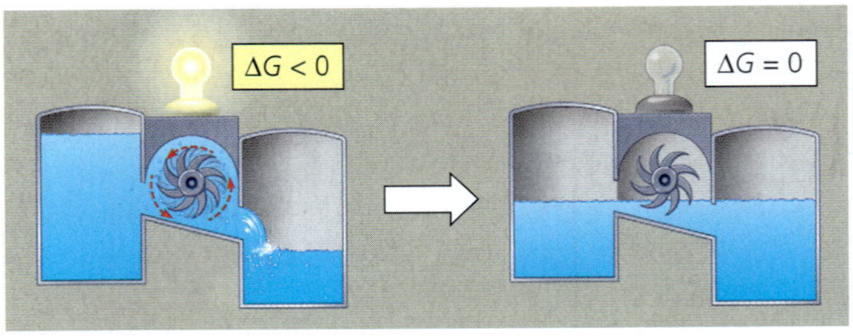

(a) An isolated hydroelectric system. Water flowing downhill turns a turbine that drives a generator providing electricity to a light bulb, but only until the system reaches equilibrium.

(b) An open hydroelectric system. Flowing water keeps driving the generator because intake and outflow of water keep the system from reaching equilibrium.

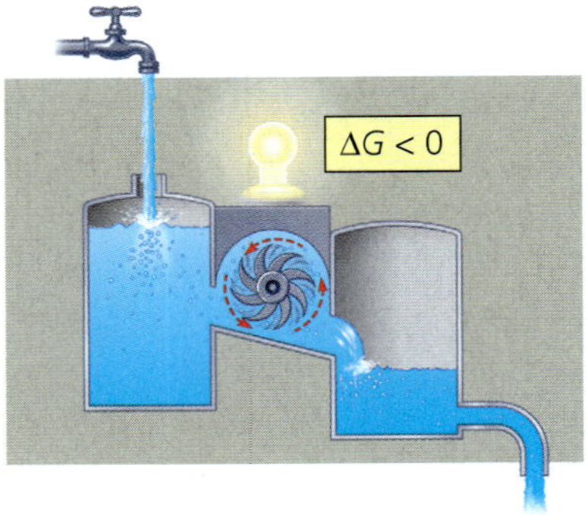

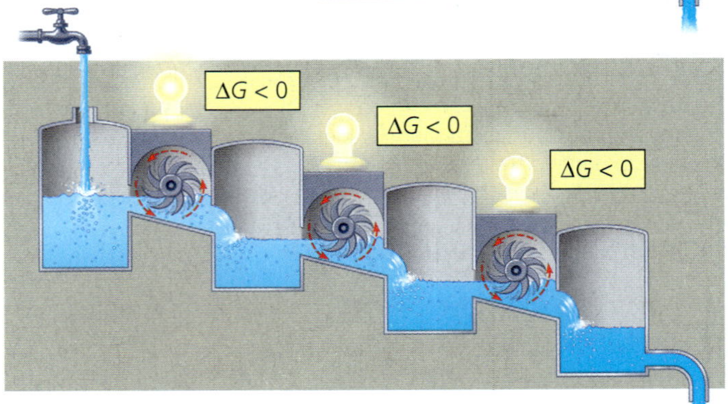

(c) A multistep open hydroelectric system. Cellular respiration is analogous to this system: Glucose is broken down in a series of exergonic reactions that power the work of the cell. The product of each reaction becomes the reactant for the next, so no reaction reaches equilibrium.

▲ **Figure 8.7 Equilibrium and work in isolated and open systems.**

The overall sequence of reactions is kept going by the huge free-energy difference between glucose and oxygen at the top of the energy "hill" and carbon dioxide and water at the "downhill" end. As long as our cells have a steady supply of glucose or other fuels and oxygen and are able to expel waste products to the surroundings, their metabolic pathways never reach equilibrium and can continue to do the work of life.

We see once again how important it is to think of organisms as open systems. Sunlight provides a daily source of free energy for an ecosystem's plants and other photosynthetic organisms. Animals and other nonphotosynthetic organisms in an ecosystem must have a source of free energy in the form of the organic products of photosynthesis. Now that we have applied the free-energy concept to metabolism, we are ready to see how a cell actually performs the work of life.

CONCEPT CHECK 8.2

1. Cellular respiration uses glucose and oxygen, which have high levels of free energy, and releases CO_2 and water, which have low levels of free energy. Is respiration spontaneous or not? Is it exergonic or endergonic? What happens to the energy released from glucose?
2. A key process in metabolism is the transport of hydrogen ions (H^+) across a membrane to create a concentration gradient. Other processes can result in an equal concentration of hydrogen ions on each side. Which arrangement of hydrogen ions allows the H^+ to perform work in this system?
3. **WHAT IF?** At nighttime celebrations, revelers can sometimes be seen wearing glow-in-the-dark necklaces. The necklaces start glowing once they are "activated," which usually involves snapping the necklace in a way that allows two chemicals to react and emit light in the form of "chemiluminescence." Is the chemical reaction exergonic or endergonic? Explain your answer.

For suggested answers, see Appendix A.

CONCEPT 8.3

ATP powers cellular work by coupling exergonic reactions to endergonic reactions

A cell does three main kinds of work:

- *Chemical work*, the pushing of endergonic reactions, which would not occur spontaneously, such as the synthesis of polymers from monomers (chemical work will be discussed further here and will come up again in Chapters 9 and 10)
- *Transport work*, the pumping of substances across membranes against the direction of spontaneous movement (see Chapter 7)
- *Mechanical work*, such as the beating of cilia (see Chapter 6), the contraction of muscle cells, and the movement of chromosomes during cellular reproduction

A key feature in the way cells manage their energy resources to do this work is **energy coupling**, the use of an exergonic process to drive an endergonic one. ATP is responsible for mediating most energy coupling in cells, and in most cases it acts as the immediate source of energy that powers cellular work.

The Structure and Hydrolysis of ATP

ATP (adenosine triphosphate) was introduced in Chapter 4 when we discussed the phosphate group as a functional group. ATP contains the sugar ribose, with the nitrogenous base adenine and a chain of three phosphate groups bonded to it **(Figure 8.8)**. In addition to its role in energy coupling, ATP is also one of the nucleoside triphosphates used to make RNA (see Figure 5.27).

The bonds between the phosphate groups of ATP can be broken by hydrolysis. When the terminal phosphate bond is broken, a molecule of inorganic phosphate ($HOPO_3^{2-}$, abbreviated Ⓟ$_i$ throughout this book) leaves the ATP, which becomes adenosine diphosphate, or ADP **(Figure 8.9)**. The reaction is exergonic and releases 7.3 kcal of energy per mole of ATP hydrolyzed:

$$\text{ATP} + H_2O \rightarrow \text{ADP} + \text{Ⓟ}_i$$
$$\Delta G = -7.3 \text{ kcal/mol } (-30.5 \text{ kJ/mol})$$

This is the free-energy change measured under standard conditions. In the cell, conditions do not conform to standard conditions, primarily because reactant and product concentrations differ from 1 *M*. For example, when ATP hydrolysis occurs under cellular conditions, the actual ΔG is about -13 kcal/mol, 78% greater than the energy released by ATP hydrolysis under standard conditions.

Because their hydrolysis releases energy, the phosphate bonds of ATP are sometimes referred to as high-energy phosphate bonds, but the term is misleading. The phosphate bonds of ATP

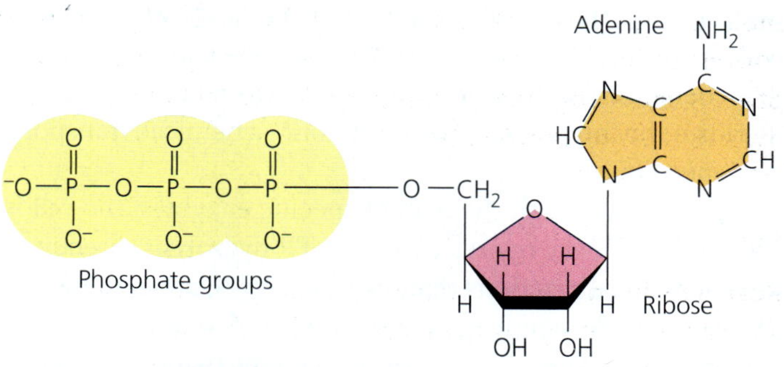

▲ **Figure 8.8 The structure of adenosine triphosphate (ATP).** In the cell, most hydroxyl groups of phosphates are ionized (—O⁻).

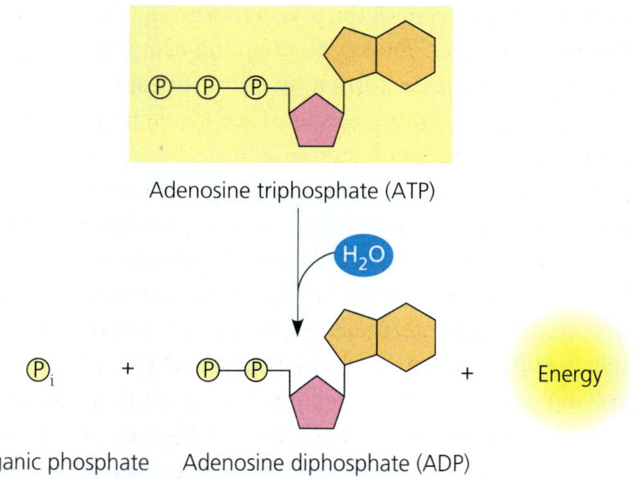

▲ **Figure 8.9 The hydrolysis of ATP.** The reaction of ATP and water yields inorganic phosphate (Ⓟ$_i$) and ADP and releases energy.

are not unusually strong bonds, as "high-energy" may imply; rather, the reactants (ATP and water) themselves have high energy relative to the energy of the products (ADP and $Ⓟ_i$). The release of energy during the hydrolysis of ATP comes from the chemical change to a state of lower free energy, not from the phosphate bonds themselves.

ATP is useful to the cell because the energy it releases on losing a phosphate group is somewhat greater than the energy most other molecules could deliver. But why does this hydrolysis release so much energy? If we reexamine the ATP molecule in Figure 8.8, we can see that all three phosphate groups are negatively charged. These like charges are crowded together, and their mutual repulsion contributes to the instability of this region of the ATP molecule. The triphosphate tail of ATP is the chemical equivalent of a compressed spring.

How ATP Performs Work

When ATP is hydrolyzed in a test tube, the release of free energy merely heats the surrounding water. In an organism, this same generation of heat can sometimes be beneficial. For instance, the process of shivering uses ATP hydrolysis during muscle contraction to generate heat and warm the body. In most cases in the cell, however, the generation of heat alone would be an inefficient (and potentially dangerous) use of a valuable energy resource. Instead, the cell's proteins harness the energy released during ATP hydrolysis in several ways to perform the three types of cellular work—chemical, transport, and mechanical.

For example, with the help of specific enzymes, the cell is able to use the energy released by ATP hydrolysis directly to drive chemical reactions that, by themselves, are endergonic. If the ΔG of an endergonic reaction is less than the amount of energy released by ATP hydrolysis, then the two reactions can be coupled so that, overall, the coupled reactions are exergonic **(Figure 8.10)**. This usually involves the transfer of a phosphate group from ATP to some other molecule, such as the reactant. The recipient of the phosphate group is then said to be **phosphorylated**. The key to coupling exergonic and endergonic reactions is the formation of this phosphorylated intermediate, which is more reactive (less stable) than the original unphosphorylated molecule.

Transport and mechanical work in the cell are also nearly always powered by the hydrolysis of ATP. In these cases, ATP hydrolysis leads to a change in a protein's shape and often its ability to bind another molecule. Sometimes this occurs via a phosphorylated intermediate, as seen for the transport protein in **Figure 8.11a**. In most instances of mechanical work involving motor proteins "walking" along cytoskeletal elements **(Figure 8.11b)**, a cycle occurs in which ATP is first bound noncovalently to the motor protein. Next, ATP is hydrolyzed, releasing ADP and $Ⓟ_i$; another ATP molecule can then bind. At each stage, the motor protein changes its shape and ability to bind the cytoskeleton, resulting in movement of the protein along the cytoskeletal track.

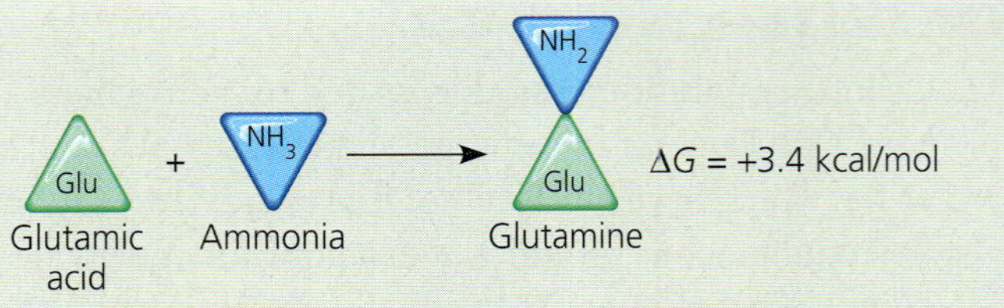

(a) Endergonic reaction. Amino acid conversion by itself is endergonic (ΔG is positive), so it is not spontaneous.

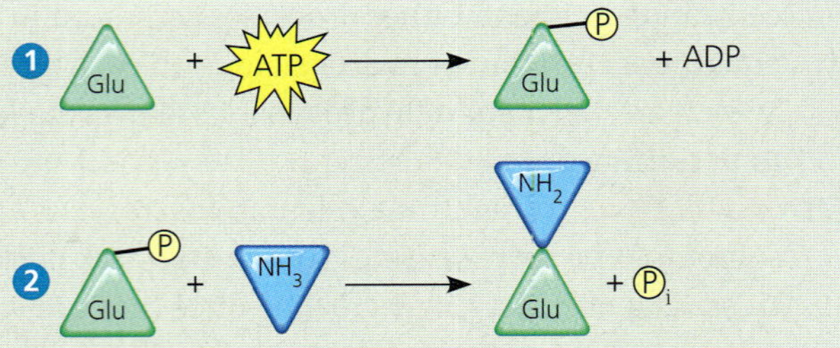

(b) Coupled with ATP hydrolysis, an exergonic reaction. In the cell, glutamine synthesis occurs in two steps, coupled by a phosphorylated intermediate. ❶ ATP phosphorylates glutamic acid, making the amino acid less stable. ❷ Ammonia displaces the phosphate group, forming glutamine.

$$\text{Glu} + NH_3 \longrightarrow \text{Glu—}NH_2 \qquad \Delta G = +3.4 \text{ kcal/mol}$$
$$\text{ATP} \longrightarrow \text{ADP} + Ⓟ_i \qquad \Delta G = -7.3 \text{ kcal/mol}$$
$$\text{Net } \Delta G = -3.9 \text{ kcal/mol}$$

(c) Overall free-energy change. Adding the ΔG (under standard conditions) for the amino acid conversion to the ΔG for ATP hydrolysis gives the free-energy change for the overall reaction. Because the overall process is exergonic (ΔG is negative), it occurs spontaneously.

▲ **Figure 8.10 How ATP drives chemical work: Energy coupling using ATP hydrolysis.** In this example, the exergonic process of ATP hydrolysis is used to drive an endergonic process—the cellular synthesis of the amino acid glutamine from glutamic acid and ammonia.

The Regeneration of ATP

An organism at work uses ATP continuously, but ATP is a renewable resource that can be regenerated by the addition of phosphate to ADP **(Figure 8.12)**. The free energy required to phosphorylate ADP comes from exergonic breakdown reactions (catabolism) in the cell. This shuttling of inorganic phosphate and energy is called the ATP cycle, and it couples the cell's energy-yielding (exergonic) processes to the energy-consuming (endergonic) ones. The ATP cycle moves at an astonishing pace. For example, a working muscle cell recycles its entire pool of ATP in less than a minute. That turnover represents 10 million molecules of ATP consumed and regenerated per second per cell. If ATP could not be regenerated by the phosphorylation of ADP, humans would use up nearly their body weight in ATP each day.

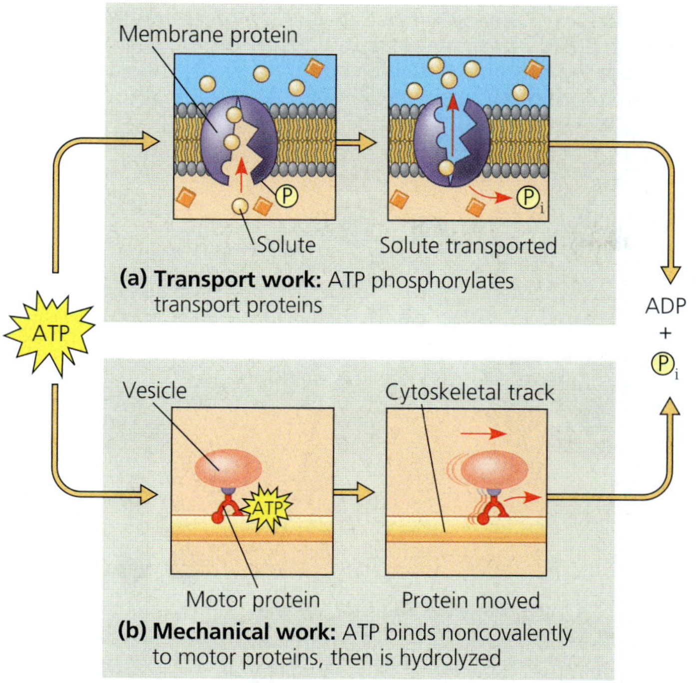

▲ **Figure 8.11 How ATP drives transport and mechanical work.** ATP hydrolysis causes changes in the shapes and binding affinities of proteins. This can occur either **(a)** directly, by phosphorylation, as shown for membrane proteins involved in active transport of solutes, or **(b)** indirectly, via noncovalent binding of ATP and its hydrolytic products, as is the case for motor proteins that move vesicles (and organelles) along cytoskeletal "tracks" in the cell.

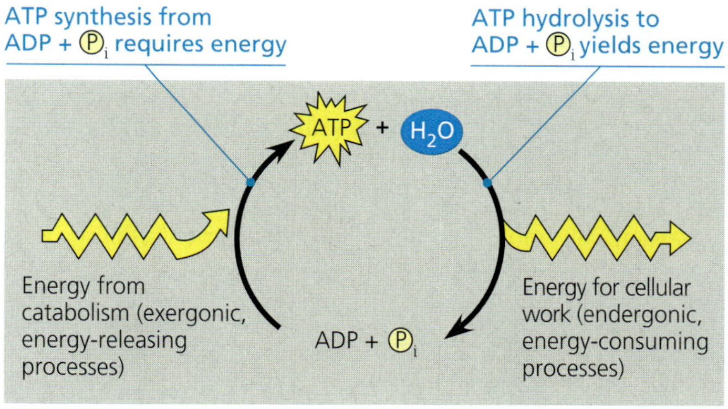

▲ **Figure 8.12 The ATP cycle.** Energy released by breakdown reactions (catabolism) in the cell is used to phosphorylate ADP, regenerating ATP. Chemical potential energy stored in ATP drives most cellular work.

Because both directions of a reversible process cannot go downhill, the regeneration of ATP from ADP and Ⓟ$_i$ is necessarily endergonic:

$$\text{ADP} + \text{Ⓟ}_i \rightarrow \text{ATP} + H_2O$$
$$\Delta G = +7.3 \text{ kcal/mol } (+30.5 \text{ kJ/mol}) \text{ (standard conditions)}$$

Because ATP formation from ADP and Ⓟ$_i$ is not spontaneous, free energy must be spent to make it occur. Catabolic (exergonic) pathways, especially cellular respiration, provide the energy for the endergonic process of making ATP. Plants also use light energy to produce ATP.

Thus, the ATP cycle is a turnstile through which energy passes during its transfer from catabolic to anabolic pathways. In fact, the chemical potential energy temporarily stored in ATP drives most cellular work.

CONCEPT CHECK 8.3

1. In most cases, how does ATP transfer energy from exergonic to endergonic reactions in the cell?
2. **WHAT IF?** Which of the following combinations has more free energy: glutamic acid + ammonia + ATP, or glutamine + ADP + Ⓟ$_i$? Explain your answer.

For suggested answers, see Appendix A.

CONCEPT **8.4**

Enzymes speed up metabolic reactions by lowering energy barriers

The laws of thermodynamics tell us what will and will not happen under given conditions but say nothing about the rate of these processes. A spontaneous chemical reaction occurs without any requirement for outside energy, but it may occur so slowly that it is imperceptible. For example, even though the hydrolysis of sucrose (table sugar) to glucose and fructose is exergonic, occurring spontaneously with a release of free energy ($\Delta G = -7$ kcal/mol), a solution of sucrose dissolved in sterile water will sit for years at room temperature with no appreciable hydrolysis. However, if we add a small amount of the enzyme sucrase to the solution, then all the sucrose may be hydrolyzed within seconds **(Figure 8.13)**. How does the enzyme do this?

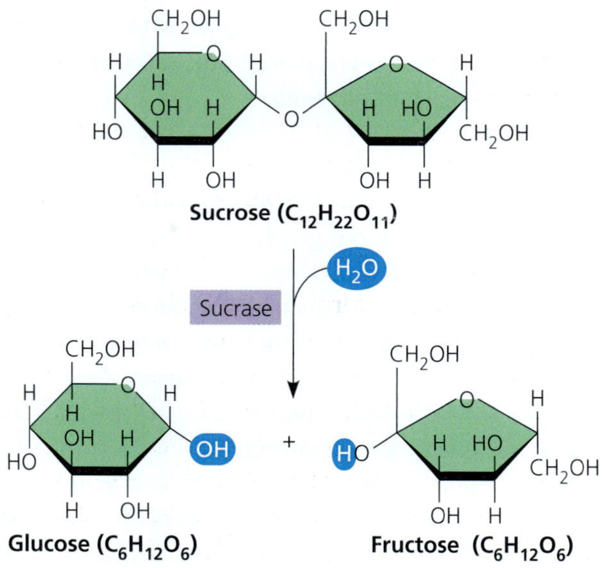

▲ **Figure 8.13 Example of an enzyme-catalyzed reaction: hydrolysis of sucrose by sucrase.**

An **enzyme** is a macromolecule that acts as a **catalyst**, a chemical agent that speeds up a reaction without being consumed by the reaction. In this chapter, we are focusing on enzymes that are proteins. (RNA enzymes, also called ribozymes, are discussed in Chapters 17 and 25.) In the absence of regulation by enzymes, chemical traffic through the pathways of metabolism would become terribly congested because many chemical reactions would take such a long time. In the next two sections, we will see what impedes a spontaneous reaction from occurring faster and how an enzyme changes the situation.

The Activation Energy Barrier

Every chemical reaction between molecules involves both bond breaking and bond forming. For example, the hydrolysis of sucrose involves breaking the bond between glucose and fructose and one of the bonds of a water molecule and then forming two new bonds, as shown in Figure 8.13. Changing one molecule into another generally involves contorting the starting molecule into a highly unstable state before the reaction can proceed. This contortion can be compared to the bending of a metal key ring when you pry it open to add a new key. The key ring is highly unstable in its opened form but returns to a stable state once the key is threaded all the way onto the ring. To reach the contorted state where bonds can change, reactant molecules must absorb energy from their surroundings. When the new bonds of the product molecules form, energy is released as heat, and the molecules return to stable shapes with lower energy than the contorted state.

The initial investment of energy for starting a reaction—the energy required to contort the reactant molecules so the bonds can break—is known as the *free energy of activation*, or **activation energy**, abbreviated E_A in this book. We can think of activation energy as the amount of energy needed to push the reactants over an energy barrier, or hill, so that the "downhill" part of the reaction can begin. **Figure 8.14** graphs the energy changes for a hypothetical exergonic reaction that swaps portions of two reactant molecules:

$$AB + CD \rightarrow AC + BD$$

The energizing, or activation, of the reactants is represented by the uphill portion of the graph, in which the free-energy content of the reactant molecules is increasing. At the summit, the reactants are in an unstable condition known as the *transition state*: They are activated, and their bonds can be broken. The subsequent bond-forming phase of the reaction corresponds to the downhill part of the curve, which shows the loss of free energy by the molecules.

Activation energy is often supplied in the form of heat that the reactant molecules absorb from the surroundings. The bonds of the reactants break only when the molecules have absorbed enough energy to become unstable—to enter the transition state. The absorption of thermal energy increases the speed of

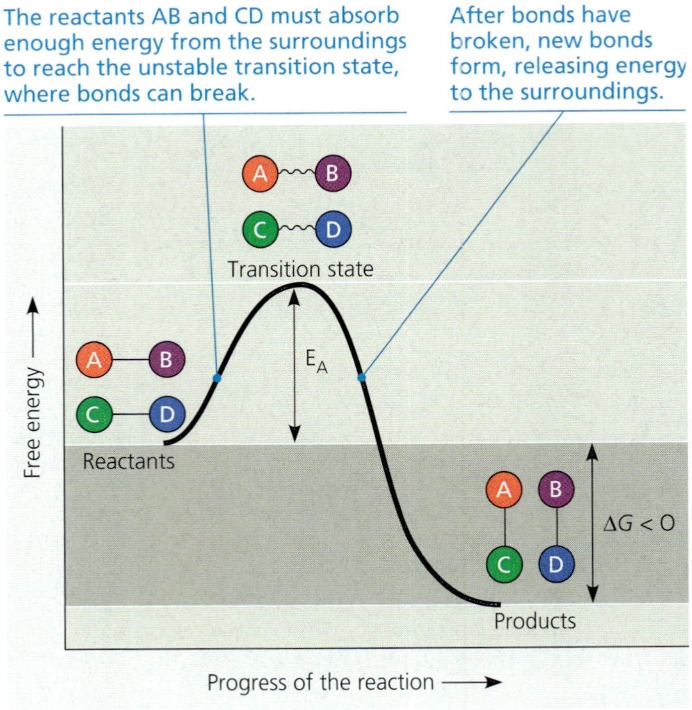

▲ **Figure 8.14 Energy profile of an exergonic reaction.** The "molecules" are hypothetical, with A, B, C, and D representing portions of the molecules. Thermodynamically, this is an exergonic reaction, with a negative ΔG, and the reaction occurs spontaneously. However, the activation energy (E_A) provides a barrier that determines the rate of the reaction.

DRAW IT *Graph the progress of an endergonic reaction in which EF and GH form products EG and FH, assuming that the reactants must pass through a transition state.*

the reactant molecules, so they collide more often and more forcefully. Also, thermal agitation of the atoms within the molecules makes the bonds more likely to break. As the atoms settle into their new, more stable bonding arrangements, energy is released to the surroundings. If the reaction is exergonic, E_A will be repaid with interest, as the formation of new bonds releases more energy than was invested in the breaking of old bonds.

The reaction shown in Figure 8.14 is exergonic and occurs spontaneously. However, the activation energy provides a barrier that determines the rate of the reaction. The reactants must absorb enough energy to reach the top of the activation energy barrier before the reaction can occur. For some reactions, E_A is modest enough that even at room temperature there is sufficient thermal energy for many of the reactants to reach the transition state in a short time. In most cases, however, E_A is so high and the transition state is reached so rarely that the reaction will hardly proceed at all. In these cases, the reaction will occur at a noticeable rate only if the reactants are heated. For example, the reaction of gasoline and oxygen is exergonic and will occur spontaneously, but energy is required for the molecules to reach the transition state and react. Only when the spark plugs fire in an automobile engine can there be the explosive release of energy that pushes the pistons. Without a spark, a mixture of gasoline

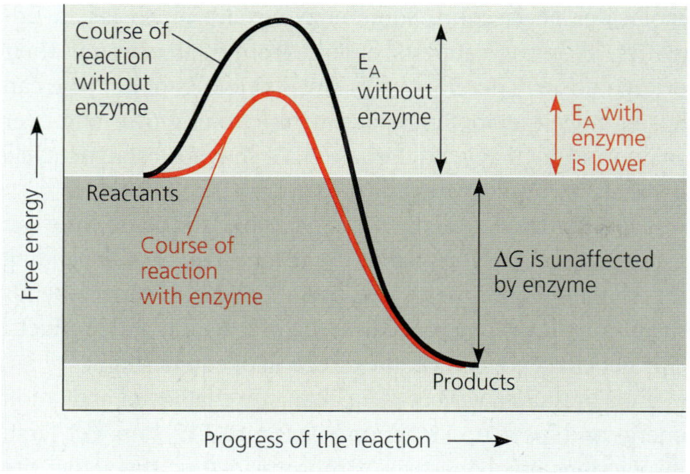

▲ **Figure 8.15 The effect of an enzyme on activation energy.** Without affecting the free-energy change (ΔG) for a reaction, an enzyme speeds the reaction by reducing its activation energy (E_A).

hydrocarbons and oxygen will not react because the E_A barrier is too high.

How Enzymes Lower the E_A Barrier

Proteins, DNA, and other complex molecules of the cell are rich in free energy and have the potential to decompose spontaneously; that is, the laws of thermodynamics favor their breakdown. These molecules persist only because at temperatures typical for cells, few molecules can make it over the hump of activation energy. However, the barriers for selected reactions must occasionally be surmounted for cells to carry out the processes needed for life. Heat speeds a reaction by allowing reactants to attain the transition state more often, but this solution would be inappropriate for biological systems. First, high temperature denatures proteins and kills cells. Second, heat would speed up *all* reactions, not just those that are needed. Organisms therefore use an alternative: catalysis.

An enzyme catalyzes a reaction by lowering the E_A barrier **(Figure 8.15)**, enabling the reactant molecules to absorb enough energy to reach the transition state even at moderate temperatures. An enzyme cannot change the ΔG for a reaction; it cannot make an endergonic reaction exergonic. Enzymes can only hasten reactions that would occur eventually anyway, but this function makes it possible for the cell to have a dynamic metabolism, routing chemicals smoothly through the cell's metabolic pathways. And because enzymes are very specific for the reactions they catalyze, they determine which chemical processes will be going on in the cell at any particular time.

Substrate Specificity of Enzymes

The reactant an enzyme acts on is referred to as the enzyme's **substrate**. The enzyme binds to its substrate (or substrates, when there are two or more reactants), forming an **enzyme-substrate complex**. While enzyme and substrate are joined, the catalytic action of the enzyme converts the substrate to the product (or products) of the reaction. The overall process can be summarized as follows:

$$\begin{array}{ccccc} \text{Enzyme +} & & \text{Enzyme-} & & \text{Enzyme +} \\ \text{Substrate(s)} & \rightleftharpoons & \text{substrate} & \rightleftharpoons & \text{Product(s)} \\ & & \text{complex} & & \end{array}$$

For example, the enzyme sucrase (most enzyme names end in *-ase*) catalyzes the hydrolysis of the disaccharide sucrose into its two monosaccharides, glucose and fructose (see Figure 8.13):

$$\begin{array}{ccccc} \text{Sucrase +} & & \text{Sucrase-} & & \text{Sucrase +} \\ \text{Sucrose +} & \rightleftharpoons & \text{sucrose-H}_2\text{O} & \rightleftharpoons & \text{Glucose +} \\ \text{H}_2\text{O} & & \text{complex} & & \text{Fructose} \end{array}$$

The reaction catalyzed by each enzyme is very specific; an enzyme can recognize its specific substrate even among closely related compounds, such as isomers. For instance, sucrase will act only on sucrose and will not bind to other disaccharides, such as maltose. What accounts for this molecular recognition? Recall that most enzymes are proteins, and proteins are macromolecules with unique three-dimensional configurations. The specificity of an enzyme results from its shape, which is a consequence of its amino acid sequence.

Only a restricted region of the enzyme molecule actually binds to the substrate. This region, called the **active site**, is typically a pocket or groove on the surface of the protein where catalysis occurs **(Figure 8.16a)**. Usually, the active site is formed by

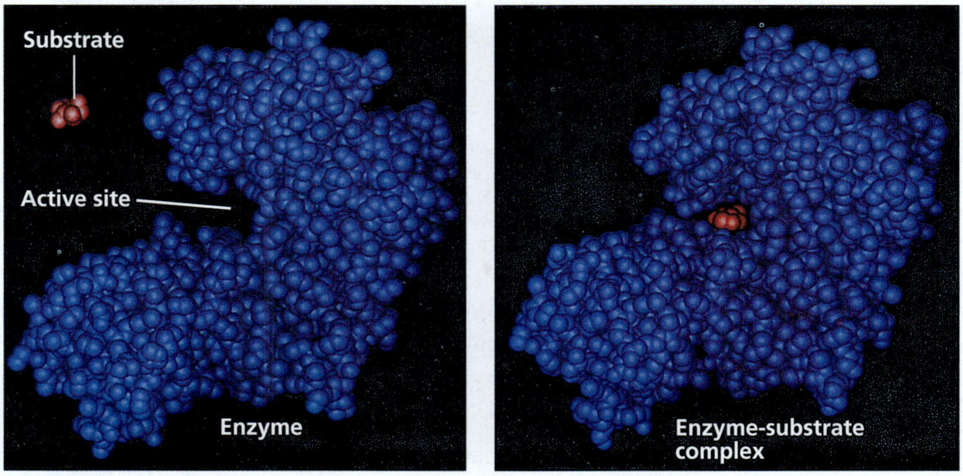

(a) In this computer graphic model, the active site of this enzyme (hexokinase, shown in blue) forms a groove on its surface. Its substrate is glucose (red).

(b) When the substrate enters the active site, it induces a change in the shape of the protein. This change allows more weak bonds to form, causing the active site to enfold the substrate and hold it in place.

▲ **Figure 8.16 Induced fit between an enzyme and its substrate.**

only a few of the enzyme's amino acids, with the rest of the protein molecule providing a framework that determines the configuration of the active site. The specificity of an enzyme is attributed to a compatible fit between the shape of its active site and the shape of the substrate. The active site, however, is not a rigid receptacle for the substrate. As the substrate enters the active site, interactions between its chemical groups and those on the R groups (side chains) of the amino acids that form the active site of the protein cause the enzyme to change its shape slightly so that the active site fits even more snugly around the substrate **(Figure 8.16b)**. This **induced fit** is like a clasping handshake. Induced fit brings chemical groups of the active site into positions that enhance their ability to catalyze the chemical reaction.

Catalysis in the Enzyme's Active Site

In most enzymatic reactions, the substrate is held in the active site by so-called weak interactions, such as hydrogen bonds and ionic bonds. R groups of a few of the amino acids that make up the active site catalyze the conversion of substrate to product, and the product departs from the active site. The enzyme is then free to take another substrate molecule into its active site. The entire cycle happens so fast that a single enzyme molecule typically acts on about a thousand substrate molecules per second. Some enzymes are much faster. Enzymes, like other catalysts, emerge from the reaction in their original form. Therefore, very small amounts of enzyme can have a huge metabolic impact by functioning over and over again in catalytic cycles. **Figure 8.17** shows a catalytic cycle involving two substrates and two products.

Most metabolic reactions are reversible, and an enzyme can catalyze either the forward or the reverse reaction, depending on which direction has a negative ΔG. This in turn depends mainly on the relative concentrations of reactants and products. The net effect is always in the direction of equilibrium.

Enzymes use a variety of mechanisms that lower activation energy and speed up a reaction (see Figure 8.17, step 3). First, in reactions involving two or more reactants, the active site provides a template on which the substrates can come together in the proper orientation for a reaction to occur between them. Second, as the active site of an enzyme clutches the bound substrates, the enzyme may stretch the substrate molecules toward their transition-state form, stressing and bending critical chemical bonds that must be broken during the reaction. Because E_A is proportional to the difficulty of breaking the bonds, distorting the substrate helps it approach the transition state and thus reduces the amount of free energy that must be absorbed to achieve that state.

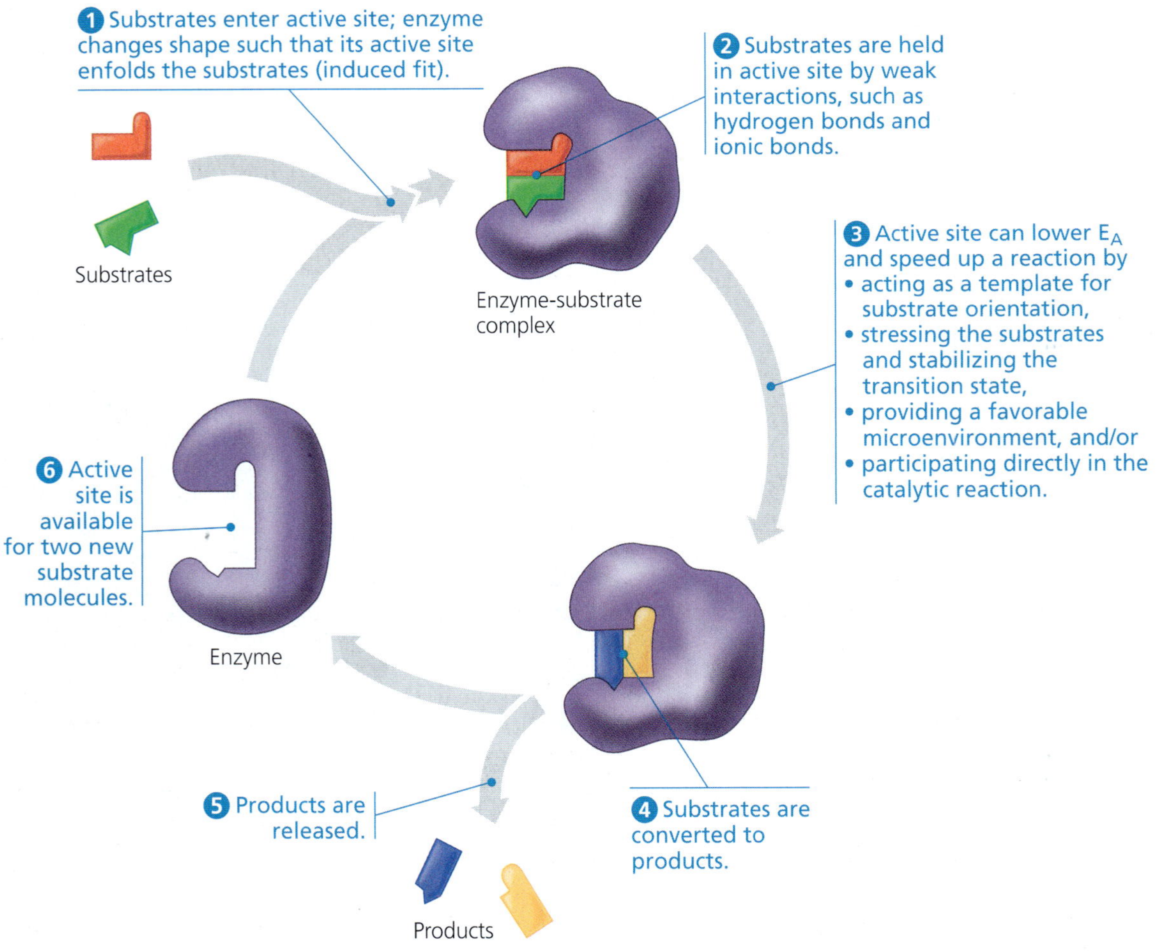

▶ **Figure 8.17 The active site and catalytic cycle of an enzyme.** An enzyme can convert one or more reactant molecules to one or more product molecules. The enzyme shown here converts two substrate molecules to two product molecules.

Third, the active site may also provide a microenvironment that is more conducive to a particular type of reaction than the solution itself would be without the enzyme. For example, if the active site has amino acids with acidic R groups, the active site may be a pocket of low pH in an otherwise neutral cell. In such cases, an acidic amino acid may facilitate H^+ transfer to the substrate as a key step in catalyzing the reaction.

A fourth mechanism of catalysis is the direct participation of the active site in the chemical reaction. Sometimes this process even involves brief covalent bonding between the substrate and an R group of an amino acid of the enzyme. Subsequent steps of the reaction restore the R groups to their original states, so that the active site is the same after the reaction as it was before.

The rate at which a particular amount of enzyme converts substrate to product is partly a function of the initial concentration of the substrate: The more substrate molecules that are available, the more frequently they access the active sites of the enzyme molecules. However, there is a limit to how fast the reaction can be pushed by adding more substrate to a fixed concentration of enzyme. At some point, the concentration of substrate will be high enough that all enzyme molecules have their active sites engaged. As soon as the product exits an active site, another substrate molecule enters. At this substrate concentration, the enzyme is said to be *saturated,* and the rate of the reaction is determined by the speed at which the active site converts substrate to product. When an enzyme population is saturated, the only way to increase the rate of product formation is to add more enzyme. Cells sometimes increase the rate of a reaction by producing more enzyme molecules.

Effects of Local Conditions on Enzyme Activity

The activity of an enzyme—how efficiently the enzyme functions—is affected by general environmental factors, such as temperature and pH. It can also be affected by chemicals that specifically influence that enzyme. In fact, researchers have learned much about enzyme function by employing such chemicals.

Effects of Temperature and pH

Recall from Chapter 5 that the three-dimensional structures of proteins are sensitive to their environment. As a consequence, each enzyme works better under some conditions than under others, because these *optimal conditions* favor the most active shape for the enzyme molecule.

Temperature and pH are environmental factors important in the activity of an enzyme. Up to a point, the rate of an enzymatic reaction increases with increasing temperature, partly because substrates collide with active sites more frequently when the molecules move rapidly. Above that temperature, however, the speed of the enzymatic reaction drops sharply. The thermal agitation of the enzyme molecule disrupts the hydrogen bonds, ionic bonds, and other weak interactions that stabilize the active shape of the enzyme, and the protein molecule eventually denatures. Each enzyme has an optimal temperature at which its reaction rate is greatest. Without denaturing the enzyme, this temperature allows the greatest number of molecular collisions and the fastest conversion of the reactants to product molecules. Most human enzymes have optimal temperatures of about 35–40°C (close to human body temperature). The thermophilic bacteria that live in hot springs contain enzymes with optimal temperatures of 70°C or higher **(Figure 8.18a)**.

Just as each enzyme has an optimal temperature, it also has a pH at which it is most active. The optimal pH values for most enzymes fall in the range of pH 6–8, but there are exceptions. For example, pepsin, a digestive enzyme in the human stomach, works best at pH 2. Such an acidic environment denatures most enzymes, but pepsin is adapted to maintain its functional three-dimensional structure in the acidic environment of the stomach. In contrast, trypsin, a digestive enzyme residing in the alkaline environment of the human intestine, has an optimal pH of 8 and would be denatured in the stomach **(Figure 8.18b)**.

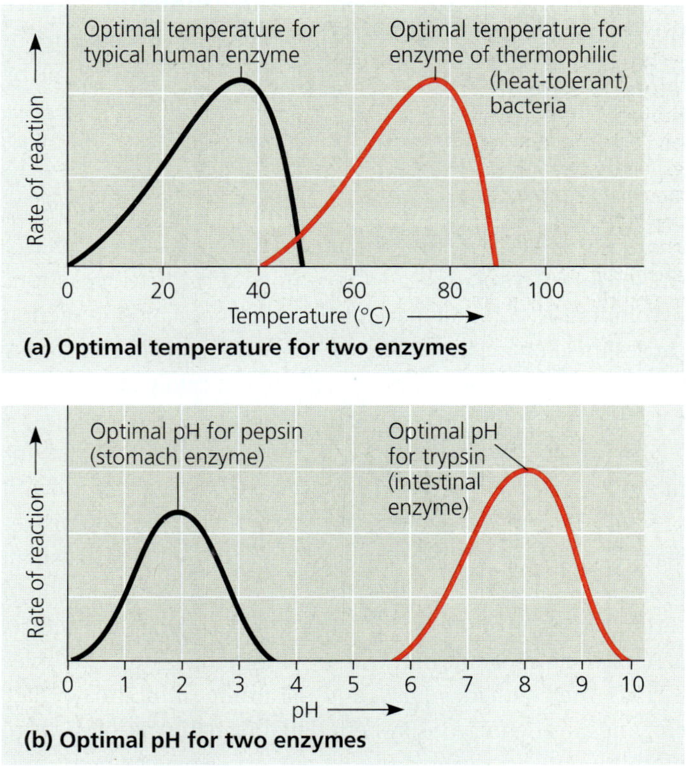

▲ **Figure 8.18 Environmental factors affecting enzyme activity.** Each enzyme has an optimal **(a)** temperature and **(b)** pH that favor the most active shape of the protein molecule.

DRAW IT *Given that a mature lysosome has an internal pH of around 4.5, draw a curve in (b) showing what you would predict for a lysosomal enzyme, labeling its optimal pH.*

Cofactors

Many enzymes require nonprotein helpers for catalytic activity. These adjuncts, called **cofactors**, may be bound tightly to the enzyme as permanent residents, or they may bind loosely and reversibly along with the substrate. The cofactors of some enzymes are inorganic, such as the metal atoms zinc, iron, and copper in ionic form. If the cofactor is an organic molecule, it is more specifically called a **coenzyme**. Most vitamins are important in nutrition because they act as coenzymes or raw materials from which coenzymes are made. Cofactors function in various ways, but in all cases where they are used, they perform a crucial function in catalysis. You'll encounter examples of cofactors later in the book.

Enzyme Inhibitors

Certain chemicals selectively inhibit the action of specific enzymes, and we have learned a lot about enzyme function by studying the effects of these molecules. If the inhibitor attaches to the enzyme by covalent bonds, inhibition is usually irreversible.

Many enzyme inhibitors, however, bind to the enzyme by weak interactions, in which case inhibition is reversible. Some reversible inhibitors resemble the normal substrate molecule and compete for admission into the active site **(Figure 8.19a** and **b)**. These mimics, called **competitive inhibitors**, reduce the productivity of enzymes by blocking substrates from entering active sites. This kind of inhibition can be overcome by increasing the concentration of substrate so that as active sites become available, more substrate molecules than inhibitor molecules are around to gain entry to the sites.

In contrast, **noncompetitive inhibitors** do not directly compete with the substrate to bind to the enzyme at the active site **(Figure 8.19c)**. Instead, they impede enzymatic reactions by binding to another part of the enzyme. This interaction causes the enzyme molecule to change its shape in such a way that the active site becomes less effective at catalyzing the conversion of substrate to product.

Toxins and poisons are often irreversible enzyme inhibitors. An example is sarin, a nerve gas that caused the death of several people and injury to many others when it was released by terrorists in the Tokyo subway in 1995. This small molecule binds covalently to the R group on the amino acid serine, which is found in the active site of acetylcholinesterase, an enzyme important in the nervous system. Other examples include the pesticides DDT and parathion, inhibitors of key enzymes in the nervous system. Finally, many antibiotics are inhibitors of specific enzymes in bacteria. For instance, penicillin blocks the active site of an enzyme that many bacteria use to make their cell walls.

Citing enzyme inhibitors that are metabolic poisons may give the impression that enzyme inhibition is generally abnormal and harmful. In fact, molecules naturally present in the cell often regulate enzyme activity by acting as inhibitors. Such regulation—selective inhibition—is essential to the control of cellular metabolism, as we discuss next.

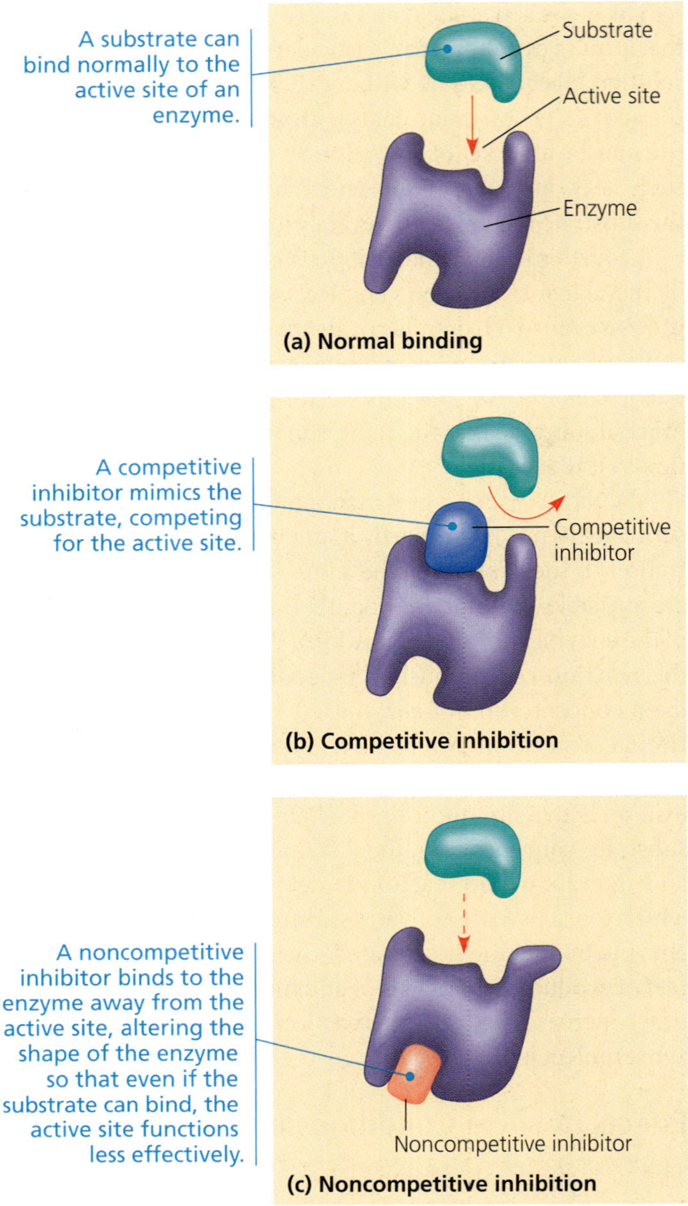

▲ **Figure 8.19 Inhibition of enzyme activity.**

CONCEPT CHECK 8.4

1. Many spontaneous reactions occur very slowly. Why don't all spontaneous reactions occur instantly?
2. Why do enzymes act only on very specific substrates?
3. **WHAT IF?** Malonate is an inhibitor of the enzyme succinate dehydrogenase. How would you determine whether malonate is a competitive or noncompetitive inhibitor?

For suggested answers, see Appendix A.

CONCEPT 8.5

Regulation of enzyme activity helps control metabolism

Chemical chaos would result if all of a cell's metabolic pathways were operating simultaneously. Intrinsic to the process of life is a cell's ability to tightly regulate its metabolic pathways by controlling when and where its various enzymes are active. It does this either by switching on and off the genes that encode specific enzymes (as we will discuss in Unit Three) or, as we discuss here, by regulating the activity of enzymes once they are made.

Allosteric Regulation of Enzymes

In many cases, the molecules that naturally regulate enzyme activity in a cell behave something like reversible noncompetitive inhibitors (see Figure 8.19c): These regulatory molecules change an enzyme's shape and the functioning of its active site by binding to a site elsewhere on the molecule, via noncovalent interactions. **Allosteric regulation** is the term used to describe any case in which a protein's function at one site is affected by the binding of a regulatory molecule to a separate site. It may result in either inhibition or stimulation of an enzyme's activity.

Allosteric Activation and Inhibition

Most enzymes known to be allosterically regulated are constructed from two or more subunits, each composed of a polypeptide chain and having its own active site **(Figure 8.20)**. Each subunit has its own active site. The entire complex oscillates between two different shapes, one catalytically active and the other inactive **(Figure 8.20a)**. In the simplest case of allosteric regulation, an activating or inhibiting regulatory molecule binds to a regulatory site (sometimes called an allosteric site), often located where subunits join. The binding of an *activator* to a regulatory site stabilizes the shape that has functional active sites, whereas the binding of an *inhibitor* stabilizes the inactive form of the enzyme. The subunits of an allosteric enzyme fit together in such a way that a shape change in one subunit is transmitted to all others. Through this interaction of subunits, a single activator or inhibitor molecule that binds to one regulatory site will affect the active sites of all subunits.

Fluctuating concentrations of regulators can cause a sophisticated pattern of response in the activity of cellular enzymes. The products of ATP hydrolysis (ADP and $\textcircled{P}_i$), for example, play a complex role in balancing the flow of traffic between anabolic and catabolic pathways by their effects on key enzymes. ATP binds to several catabolic enzymes allosterically, lowering their affinity for substrate and thus inhibiting their activity. ADP, however, functions as an activator of the same enzymes. This is logical because a major function of catabolism is to regenerate ATP. If ATP production lags behind its use, ADP accumulates and activates the enzymes that speed up catabolism, producing more ATP. If the supply of ATP exceeds demand, then catabolism slows down as ATP molecules accumulate and bind these same enzymes, inhibiting them. (You'll see specific examples of this type of regulation when you learn about cellular respiration in the next chapter.) ATP, ADP, and other related

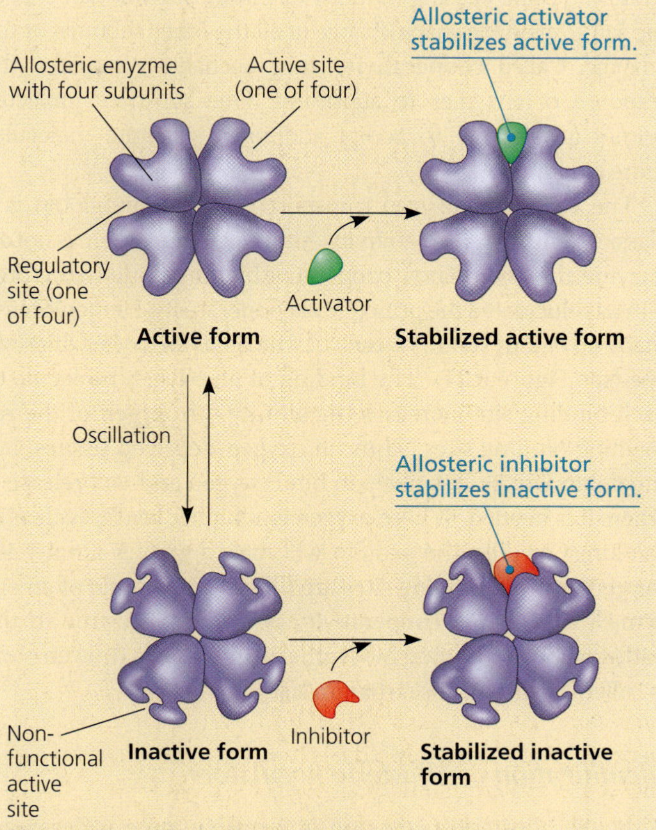

(a) Allosteric activators and inhibitors. In the cell, activators and inhibitors dissociate when at low concentrations. The enzyme can then oscillate again.

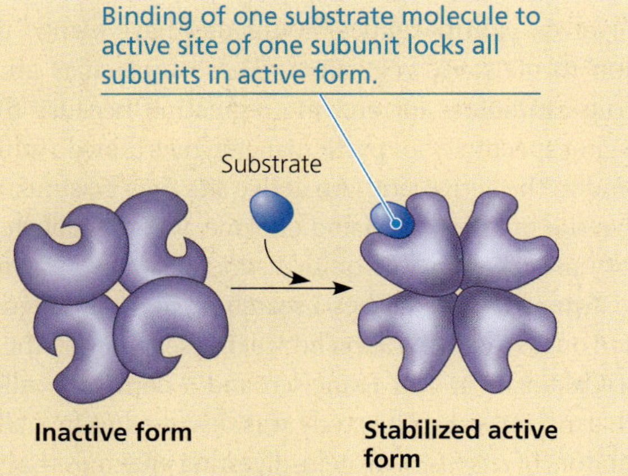

(b) Cooperativity: another type of allosteric activation. The inactive form shown on the left oscillates back and forth with the active form when the active form is not stabilized by substrate.

▲ **Figure 8.20 Allosteric regulation of enzyme activity.**

molecules also affect key enzymes in anabolic pathways. In this way, allosteric enzymes control the rates of key reactions in both sorts of metabolic pathways.

In another kind of allosteric activation, a *substrate* molecule binding to one active site may stimulate the catalytic powers of a multisubunit enzyme by affecting the other active sites **(Figure 8.20b)**. If an enzyme has two or more subunits, a substrate molecule causing induced fit in one subunit can trigger the same favorable shape change in all the other subunits of the enzyme. Called **cooperativity**, this mechanism amplifies the response of enzymes to substrates: One substrate molecule primes an enzyme to accept additional substrate molecules more readily.

The vertebrate oxygen transport protein hemoglobin is a classic example of cooperativity. Although hemoglobin is not an enzyme, the study of how cooperative binding works in this protein has elucidated the principle of cooperativity. Hemoglobin is made up of four subunits, each of which has an oxygen-binding site (see Figure 5.21). The binding of an oxygen molecule to each binding site increases the affinity for oxygen of the remaining binding sites. Thus, in oxygen-deprived tissues, hemoglobin will be less likely to bind oxygen and will release it where it is needed. Where oxygen is at higher levels, such as in the lungs or gills, the protein will have a greater affinity for oxygen as more binding sites are filled. An example of an enzyme that exhibits cooperativity is the first enzyme in the pathway for pyrimidine biosynthesis in bacteria (this enzyme is called aspartyl transcarbamoylase).

Identification of Allosteric Regulators

Although allosteric regulation is probably quite widespread, relatively few of the many known metabolic enzymes have been shown to be regulated in this way. Allosteric regulatory molecules are hard to characterize, in part because they tend to bind the enzyme at low affinity and are thus hard to isolate. Recently, however, pharmaceutical companies have turned their attention to allosteric regulators. These molecules are attractive drug candidates for enzyme regulation because they exhibit higher specificity for particular enzymes than do inhibitors that bind to the active site. (An active site may be similar to the active site in another, related enzyme, whereas allosteric regulatory sites appear to be quite distinct between enzymes.)

Figure 8.21 describes a search for allosteric regulators, carried out as a collaboration between researchers at the University of California at San Francisco and a company called Sunesis Pharmaceuticals. The study was designed to find allosteric inhibitors of *caspases*, protein-digesting enzymes that play an active role in inflammation and cell death. (You'll learn more about caspases and cell death in Chapter 11.) By specifically regulating these enzymes, we may be able to better manage inappropriate inflammatory responses, such as those commonly seen in vascular and neurodegenerative diseases.

▼ Figure 8.21 Inquiry

Are there allosteric inhibitors of caspase enzymes?

EXPERIMENT In an effort to identify allosteric inhibitors of caspases, Justin Scheer and co-workers screened close to 8,000 compounds for their ability to bind to a possible allosteric binding site in caspase 1 and inhibit the enzyme's activity. Each compound was designed to form a disulfide bond with a cysteine near the site in order to stabilize the low-affinity interaction that is expected of an allosteric inhibitor. As the caspases are known to exist in both active and inactive forms, the researchers hypothesized that this linkage might lock the enzyme in the inactive form.

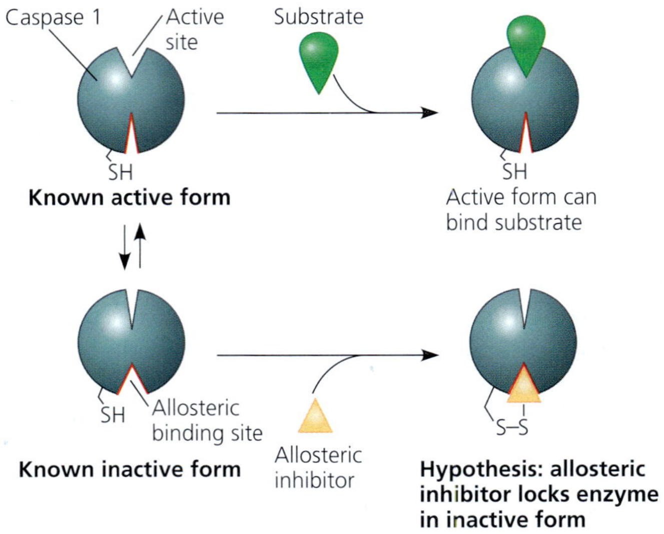

To test this model, X-ray diffraction analysis was used to determine the structure of caspase 1 when bound to one of the inhibitors and to compare it with the active and inactive structures.

RESULTS Fourteen compounds were identified that could bind to the proposed allosteric site (red) of caspase 1 and block enzymatic activity. The enzyme's shape when one such inhibitor was bound resembled the inactive caspase 1 more than the active form.

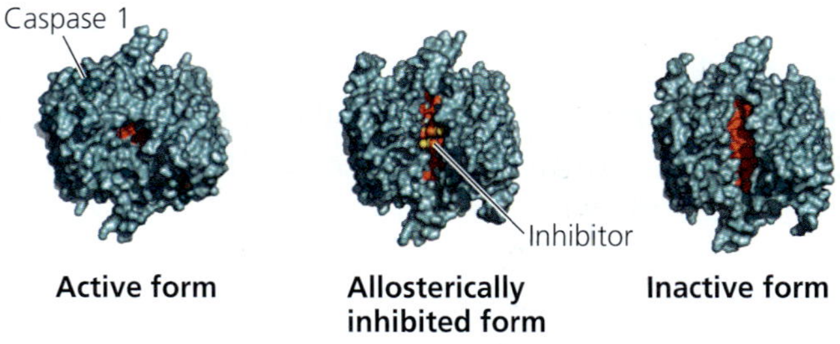

CONCLUSION The inhibitory compound that was studied apparently locks the enzyme in its inactive form, as expected for a true allosteric regulator. The data therefore support the existence of an allosteric inhibitory site on caspase 1, which can be used to control enzymatic activity.

SOURCE J. M. Scheer et al., A common allosteric site and mechanism in caspases, *PNAS* 103:7595–7600 (2006).

WHAT IF? As a control, the researchers broke the disulfide linkage between one of the inhibitors and the caspase. Assuming that the experimental solution contains no other inhibitors, how would you expect the resulting caspase 1 activity to be affected?

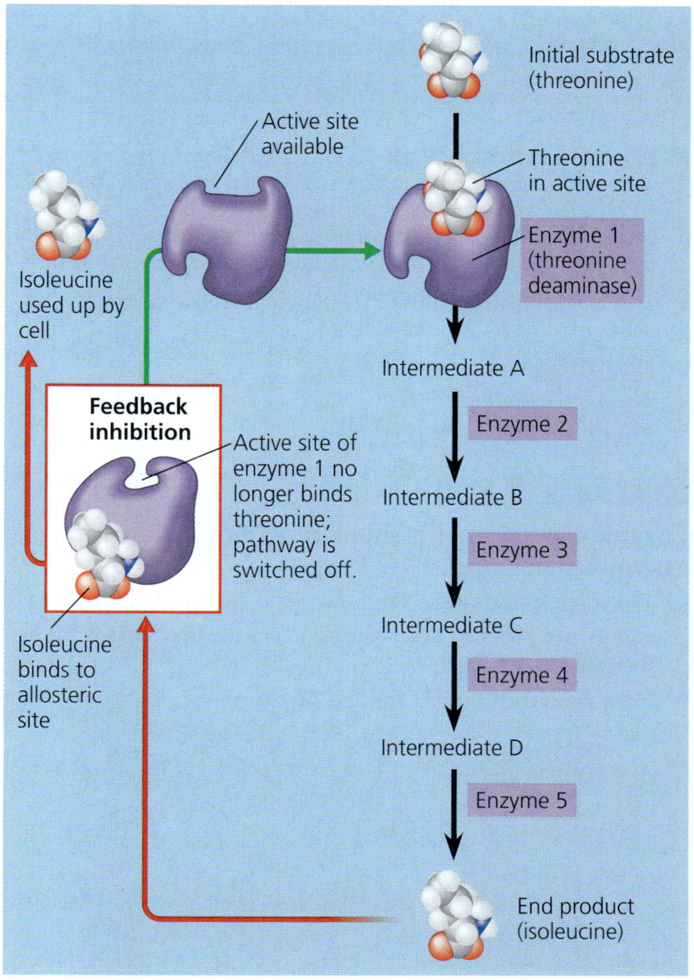

▲ **Figure 8.22 Feedback inhibition in isoleucine synthesis.**

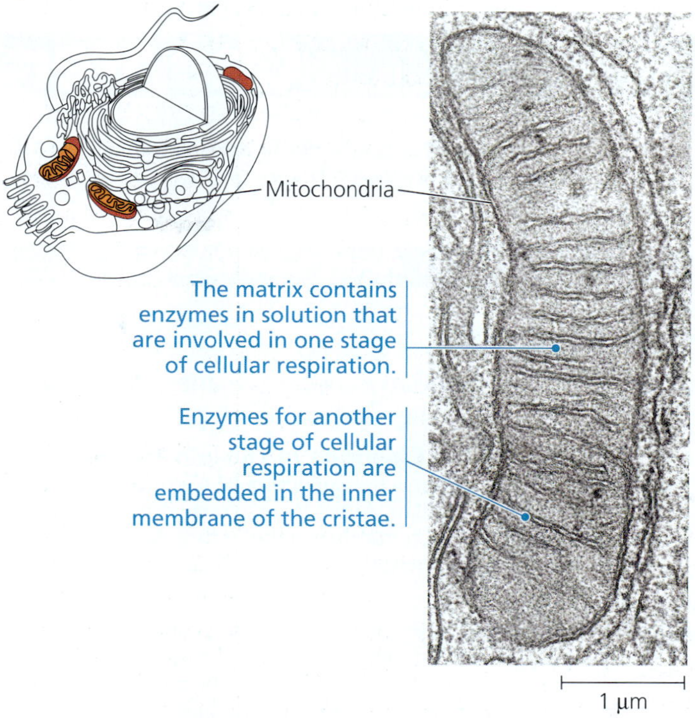

▲ **Figure 8.23 Organelles and structural order in metabolism.** Organelles such as these mitochondria (TEM) contain enzymes that carry out specific functions, in this case cellular respiration.

Feedback Inhibition

When ATP allosterically inhibits an enzyme in an ATP-generating pathway, as we discussed earlier, the result is feedback inhibition, a common method of metabolic control. In **feedback inhibition**, a metabolic pathway is switched off by the inhibitory binding of its end product to an enzyme that acts early in the pathway. **Figure 8.22** shows an example of this control mechanism operating on an anabolic pathway. Some cells use this five-step pathway to synthesize the amino acid isoleucine from threonine, another amino acid. As isoleucine accumulates, it slows down its own synthesis by allosterically inhibiting the enzyme for the first step of the pathway. Feedback inhibition thereby prevents the cell from wasting chemical resources by making more isoleucine than is necessary.

Specific Localization of Enzymes Within the Cell

The cell is not just a bag of chemicals with thousands of different kinds of enzymes and substrates in a random mix. The cell is compartmentalized, and cellular structures help bring order to metabolic pathways. In some cases, a team of enzymes for several steps of a metabolic pathway are assembled into a multienzyme complex. The arrangement facilitates the sequence of reactions, with the product from the first enzyme becoming the substrate for an adjacent enzyme in the complex, and so on, until the end product is released. Some enzymes and enzyme complexes have fixed locations within the cell and act as structural components of particular membranes. Others are in solution within specific membrane-enclosed eukaryotic organelles, each with its own internal chemical environment. For example, in eukaryotic cells, the enzymes for cellular respiration reside in specific locations within mitochondria **(Figure 8.23)**.

In this chapter, you have learned that metabolism, the intersecting set of chemical pathways characteristic of life, is a choreographed interplay of thousands of different kinds of cellular molecules. In the next chapter, we explore cellular respiration, the major catabolic pathway that breaks down organic molecules, releasing energy for the crucial processes of life.

CONCEPT CHECK 8.5

1. How can an activator and an inhibitor have different effects on an allosterically regulated enzyme?
2. **WHAT IF?** Imagine you are a pharmacological researcher who wants to design a drug that inhibits a particular enzyme. Upon reading the scientific literature, you find that the enzyme's active site is similar to that of several other enzymes. What might be the best approach to developing your inhibitor drug?

For suggested answers, see Appendix A.

Chapter 8 Review

MEDIA Go to the Study Area at **www.masteringbio.com** for BioFlix 3-D Animations, MP3 Tutors, Videos, Practice Tests, an eBook, and more.

SUMMARY OF KEY CONCEPTS

CONCEPT 8.1

An organism's metabolism transforms matter and energy, subject to the laws of thermodynamics (pp. 142–145)

- **Organization of the Chemistry of Life into Metabolic Pathways** Metabolism is the collection of chemical reactions that occur in an organism. Aided by enzymes, it follows intersecting pathways, which may be catabolic (breaking down molecules, releasing energy) or anabolic (building molecules, consuming energy).
- **Forms of Energy** Energy is the capacity to cause change; some forms of energy do work by moving matter. Kinetic energy is associated with motion. Potential energy is related to the location or structure of matter and includes chemical energy possessed by a molecule due to its structure.
- **The Laws of Energy Transformation** The first law, conservation of energy, states that energy cannot be created or destroyed, only transferred or transformed. The second law states that spontaneous changes, those requiring no outside input of energy, increase the entropy (disorder) of the universe.

MEDIA

MP3 Tutor Basic Energy Concepts
Activity Energy Transformations

CONCEPT 8.2

The free-energy change of a reaction tells us whether or not the reaction occurs spontaneously (pp. 146–149)

- **Free-Energy Change, ΔG** A living system's free energy is energy that can do work under cellular conditions. The change in free energy (ΔG) during a biological process is related directly to enthalpy change (ΔH) and to the change in entropy (ΔS): $\Delta G = \Delta H - T\Delta S$.
- **Free Energy, Stability, and Equilibrium** Organisms live at the expense of free energy. During a spontaneous change, free energy decreases and the stability of a system increases. At maximum stability, the system is at equilibrium and can do no work.
- **Free Energy and Metabolism** In an exergonic (spontaneous) chemical reaction, the products have less free energy than the reactants ($-\Delta G$). Endergonic (nonspontaneous) reactions require an input of energy ($+\Delta G$). The addition of starting materials and the removal of end products prevent metabolism from reaching equilibrium.

CONCEPT 8.3

ATP powers cellular work by coupling exergonic reactions to endergonic reactions (pp. 149–151)

- **The Structure and Hydrolysis of ATP** ATP is the cell's energy shuttle. Hydrolysis at its terminal phosphate group produces ADP and phosphate and releases free energy.
- **How ATP Performs Work** ATP hydrolysis drives endergonic reactions by phosphorylation, the transfer of a phosphate group to specific reactants, making them more reactive. ATP hydrolysis (sometimes with protein phosphorylation) also causes changes in the shape and binding affinities of transport and motor proteins.
- **The Regeneration of ATP** Catabolic pathways drive the regeneration of ATP from ADP and phosphate.

MEDIA

Activity The Structure of ATP
Activity Chemical Reactions and ATP

CONCEPT 8.4

Enzymes speed up metabolic reactions by lowering energy barriers (pp. 151–156)

- **The Activation Energy Barrier** In a chemical reaction, the energy necessary to break the bonds of the reactants is the activation energy, E_A.
- **How Enzymes Lower the E_A Barrier**

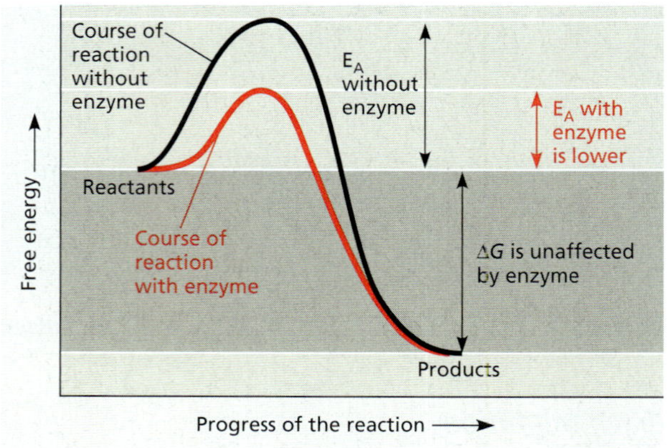

- **Substrate Specificity of Enzymes** Each type of enzyme has a unique active site that combines specifically with its substrate, the reactant molecule on which it acts. The enzyme changes shape slightly when it binds the substrate (induced fit).
- **Catalysis in the Enzyme's Active Site** The active site can lower an E_A barrier by orienting substrates correctly, straining their bonds, providing a favorable microenvironment, and even covalently bonding with the substrate.
- **Effects of Local Conditions on Enzyme Activity** Each enzyme has an optimal temperature and pH. Inhibitors reduce enzyme function. A competitive inhibitor binds to the active site, while a noncompetitive inhibitor binds to a different site on the enzyme.

MEDIA

Activity How Enzymes Work
Investigation How Is the Rate of Enzyme Catalysis Measured?
Biology Labs On-Line EnzymeLab

CONCEPT 8.5

Regulation of enzyme activity helps control metabolism (pp. 157–159)

- **Allosteric Regulation of Enzymes** Many enzymes are allosterically regulated: Regulatory molecules, either activators or inhibitors, bind to specific regulatory sites, affecting the shape and function of the enzyme. In cooperativity, binding of one substrate molecule can stimulate binding or activity at other active sites. In feedback inhibition, the end product of a metabolic pathway allosterically inhibits the enzyme for a previous step in the pathway.

- **Specific Localization of Enzymes Within the Cell** Some enzymes are grouped into complexes, some are incorporated into membranes, and some are contained inside organelles, increasing the efficiency of metabolic processes.

TESTING YOUR KNOWLEDGE

SELF-QUIZ

1. Choose the pair of terms that correctly completes this sentence: Catabolism is to anabolism as __________ is to __________.
   a. exergonic; spontaneous
   b. exergonic; endergonic
   c. free energy; entropy
   d. work; energy
   e. entropy; enthalpy

2. Most cells cannot harness heat to perform work because
   a. heat is not a form of energy.
   b. cells do not have much heat; they are relatively cool.
   c. temperature is usually uniform throughout a cell.
   d. heat can never be used to do work.
   e. heat must remain constant during work.

3. Which of the following metabolic processes can occur without a net influx of energy from some other process?
   a. ADP + Ⓟ$_i$ → ATP + H_2O
   b. $C_6H_{12}O_6 + 6\,O_2 \rightarrow 6\,CO_2 + 6\,H_2O$
   c. $6\,CO_2 + 6\,H_2O \rightarrow C_6H_{12}O_6 + 6\,O_2$
   d. amino acids → protein
   e. glucose + fructose → sucrose

4. If an enzyme in solution is saturated with substrate, the most effective way to obtain a faster yield of products is to
   a. add more of the enzyme.
   b. heat the solution to 90°C.
   c. add more substrate.
   d. add an allosteric inhibitor.
   e. add a noncompetitive inhibitor.

5. If an enzyme is added to a solution where its substrate and product are in equilibrium, what would occur?
   a. Additional product would be formed.
   b. Additional substrate would be formed.
   c. The reaction would change from endergonic to exergonic.
   d. The free energy of the system would change.
   e. Nothing; the reaction would stay at equilibrium.

6. Some bacteria are metabolically active in hot springs because
   a. they are able to maintain a lower internal temperature.
   b. high temperatures make catalysis unnecessary.
   c. their enzymes have high optimal temperatures.
   d. their enzymes are completely insensitive to temperature.
   e. they use molecules other than proteins or RNAs as their main catalysts.

7. **DRAW IT** Using a series of arrows, draw the branched metabolic reaction pathway described by the following statements, and then answer the question at the end. Use red arrows and minus signs to indicate inhibition.
   L can form either M or N.
   M can form O.
   O can form either P or R.
   P can form Q.
   R can form S.
   O inhibits the reaction of L to form M.
   Q inhibits the reaction of O to form P.
   S inhibits the reaction of O to form R.

   Which reaction would prevail if both Q and S were present in the cell in high concentrations?
   a. L → M
   b. M → O
   c. L → N
   d. O → P
   e. R → S

For Self-Quiz answers, see Appendix A.

MEDIA Visit the Study Area at **www.masteringbio.com** for a Practice Test.

EVOLUTION CONNECTION

8. A recent revival of the antievolutionary "intelligent design" argument holds that biochemical pathways are too complex to have evolved, because all intermediate steps in a given pathway must be present to produce the final product. Critique this argument. How could you use the diversity of metabolic pathways that produce the same or similar products to support your case?

SCIENTIFIC INQUIRY

9. **DRAW IT** A researcher has developed an assay to measure the activity of an important enzyme present in liver cells being grown in culture. She adds the enzyme's substrate to a dish of cells and then measures the appearance of reaction products. The results are graphed as the amount of product on the y-axis versus time on the x-axis. The researcher notes four sections of the graph. For a short period of time, no products appear (section A). Then (section B) the reaction rate is quite high (the slope of the line is steep). Next, the reaction gradually slows down (section C). Finally, the graph line becomes flat (section D). Draw and label the graph, and propose a model to explain the molecular events occurring at each stage of this reaction profile.

SCIENCE, TECHNOLOGY, AND SOCIETY

10. The EPA is evaluating the safety of the most commonly used organophosphate insecticides (organic compounds containing phosphate groups). Organophosphates typically interfere with nerve transmission by inhibiting the enzymes that degrade transmitter molecules diffusing from one neuron to another. Noxious insects are not uniquely susceptible; humans and other vertebrates can be affected as well. Thus, the use of organophosphate pesticides creates some health risks. As a consumer, what level of risk are you willing to accept in exchange for an abundant and affordable food supply?

Cellular Respiration
Harvesting Chemical Energy

9

▲ **Figure 9.1 How do these leaves power the work of life for the giant panda?**

KEY CONCEPTS

9.1 Catabolic pathways yield energy by oxidizing organic fuels

9.2 Glycolysis harvests chemical energy by oxidizing glucose to pyruvate

9.3 The citric acid cycle completes the energy-yielding oxidation of organic molecules

9.4 During oxidative phosphorylation, chemiosmosis couples electron transport to ATP synthesis

9.5 Fermentation and anaerobic respiration enable cells to produce ATP without the use of oxygen

9.6 Glycolysis and the citric acid cycle connect to many other metabolic pathways

OVERVIEW

Life Is Work

Living cells require transfusions of energy from outside sources to perform their many tasks—for example, assembling polymers, pumping substances across membranes, moving, and reproducing. The giant panda in **Figure 9.1** obtains energy for its cells by eating plants; some animals feed on other organisms that eat plants. The energy stored in the organic molecules of food ultimately comes from the sun. Energy flows into an ecosystem as sunlight and leaves as heat **(Figure 9.2)**. In contrast, the chemical elements essential to life are recycled. Photosynthesis generates oxygen and organic molecules used by the mitochondria of eukaryotes (including plants and algae) as fuel for cellular respiration. Respiration breaks this fuel down, generating ATP. The waste products of this type of respiration, carbon dioxide and water, are the raw materials for photosynthesis. In this chapter, we consider how cells harvest the chemical energy stored in organic molecules and use it to generate ATP, the molecule that drives most cellular work. After presenting some basics about respiration, we will focus on the three key pathways of respiration: glycolysis, the citric acid cycle, and oxidative phosphorylation.

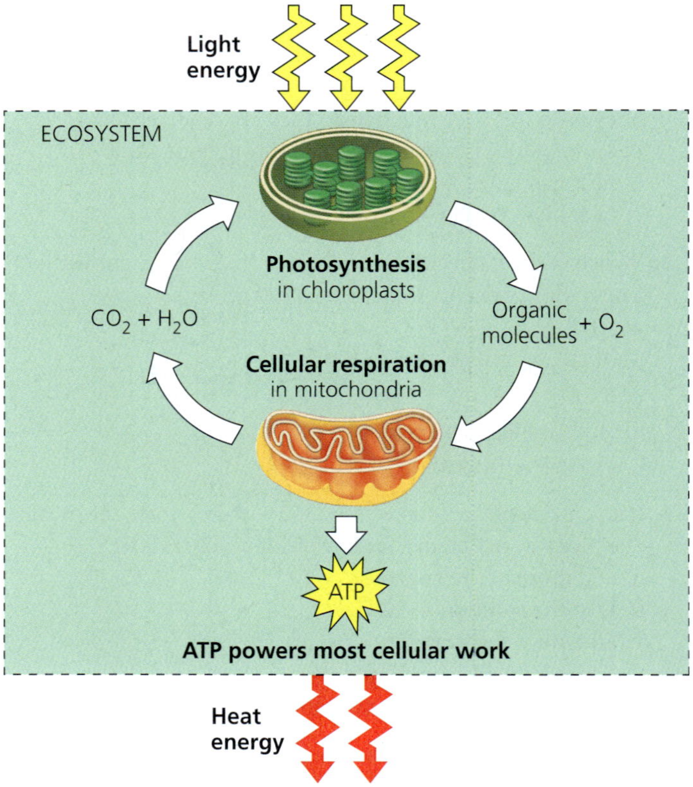

▲ **Figure 9.2 Energy flow and chemical recycling in ecosystems.** Energy flows into an ecosystem as sunlight and ultimately leaves as heat, while the chemical elements essential to life are recycled.

CONCEPT **9.1**

Catabolic pathways yield energy by oxidizing organic fuels

As you learned in Chapter 8, metabolic pathways that release stored energy by breaking down complex molecules are called catabolic pathways. Electron transfer plays a major role in these pathways. In this section, we consider these processes, which are central to cellular respiration.

Catabolic Pathways and Production of ATP

Organic compounds possess potential energy as a result of their arrangement of atoms. Compounds that can participate in exergonic reactions can act as fuels. With the help of enzymes, a cell systematically degrades complex organic molecules that are rich in potential energy to simpler waste products that have less energy. Some of the energy taken out of chemical storage can be used to do work; the rest is dissipated as heat.

One catabolic process, **fermentation**, is a partial degradation of sugars that occurs without the use of oxygen. However, the most prevalent and efficient catabolic pathway is **aerobic respiration**, in which oxygen is consumed as a reactant along with the organic fuel (*aerobic* is from the Greek *aer*, air, and *bios*, life). The cells of most eukaryotic and many prokaryotic organisms can carry out aerobic respiration. Some prokaryotes use substances other than oxygen as reactants in a similar process that harvests chemical energy without using any oxygen at all; this process is called *anaerobic respiration* (the prefix *an-* means "without"). Technically, the term **cellular respiration** includes both aerobic and anaerobic processes. However, it originated as a synonym for aerobic respiration because of the relationship of that process to organismal respiration, in which an animal breathes in oxygen. Thus, *cellular respiration* is often used to refer to the aerobic process, a practice we follow in most of this chapter.

Although very different in mechanism, aerobic respiration is in principle similar to the combustion of gasoline in an automobile engine after oxygen is mixed with the fuel (hydrocarbons). Food provides the fuel for respiration, and the exhaust is carbon dioxide and water. The overall process can be summarized as follows:

$$\text{Organic compounds} + \text{Oxygen} \longrightarrow \text{Carbon dioxide} + \text{Water} + \text{Energy}$$

Although carbohydrates, fats, and proteins can all be processed and consumed as fuel, it is helpful to learn the steps of cellular respiration by tracking the degradation of the sugar glucose ($C_6H_{12}O_6$):

$$C_6H_{12}O_6 + 6\,O_2 \longrightarrow 6\,CO_2 + 6\,H_2O + \text{Energy (ATP + heat)}$$

Glucose is the fuel that cells most often use; we will discuss other organic molecules contained in foods later in the chapter.

This breakdown of glucose is exergonic, having a free-energy change of –686 kcal (2,870 kJ) per mole of glucose decomposed ($\Delta G = -686$ kcal/mol). Recall that a negative ΔG indicates that the products of the chemical process store less energy than the reactants and that the reaction can happen spontaneously—in other words, without an input of energy.

Catabolic pathways do not directly move flagella, pump solutes across membranes, polymerize monomers, or perform other cellular work. Catabolism is linked to work by a chemical drive shaft—ATP, which you learned about in Chapter 8. To keep working, the cell must regenerate its supply of ATP from ADP and Ⓟ$_i$ (see Figure 8.12). To understand how cellular respiration accomplishes this, let's examine the fundamental chemical processes known as oxidation and reduction.

Redox Reactions: Oxidation and Reduction

How do the catabolic pathways that decompose glucose and other organic fuels yield energy? The answer is based on the transfer of electrons during the chemical reactions. The relocation of electrons releases energy stored in organic molecules, and this energy ultimately is used to synthesize ATP.

The Principle of Redox

In many chemical reactions, there is a transfer of one or more electrons (e^-) from one reactant to another. These electron transfers are called oxidation-reduction reactions, or **redox reactions** for short. In a redox reaction, the loss of electrons from one substance is called **oxidation**, and the addition of electrons to another substance is known as **reduction**. (Note that *adding* electrons is called *reduction*; negatively charged electrons added to an atom *reduce* the amount of positive charge of that atom.) To take a simple, nonbiological example, consider the reaction between the elements sodium (Na) and chlorine (Cl) that forms table salt:

becomes oxidized (loses electron): Na → Na^+

$$Na + Cl \longrightarrow Na^+ + Cl^-$$

becomes reduced (gains electron): Cl → Cl^-

We could generalize a redox reaction this way:

becomes oxidized: Xe^- → X

$$Xe^- + Y \longrightarrow X + Ye^-$$

becomes reduced: Y → Ye^-

In the generalized reaction, substance Xe^-, the electron donor, is called the **reducing agent**; it reduces Y, which accepts the donated electron. Substance Y, the electron acceptor, is the **oxidizing agent**; it oxidizes Xe^- by removing its electron. Because an electron transfer requires both a donor and an acceptor, oxidation and reduction always go together.

Not all redox reactions involve the complete transfer of electrons from one substance to another; some change the degree of electron sharing in covalent bonds. The reaction between methane and oxygen, shown in **Figure 9.3** on the next page, is an example. As explained in Chapter 2, the covalent electrons in methane are shared nearly equally between the bonded atoms because carbon and hydrogen have about the same affinity for valence electrons; they are about equally electronegative. But when methane reacts with oxygen, forming carbon dioxide, electrons end up shared less equally between the carbon atom and its new covalent partners, the oxygen atoms, which are very electronegative. In effect, the carbon atom has partially "lost" its shared electrons; thus, methane has been oxidized.

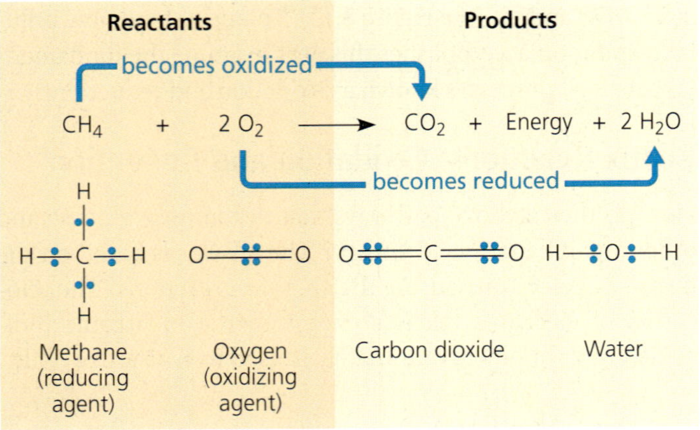

▲ **Figure 9.3 Methane combustion as an energy-yielding redox reaction.** The reaction releases energy to the surroundings because the electrons lose potential energy when they end up being shared unequally, spending more time near electronegative atoms such as oxygen.

Now let's examine the fate of the reactant O_2. The two atoms of the oxygen molecule (O_2) share their electrons equally. But when oxygen reacts with the hydrogen from methane, forming water, the electrons of the covalent bonds spend more time near the oxygen (see Figure 9.3). In effect, each oxygen atom has partially "gained" electrons, so the oxygen molecule has been reduced. Because oxygen is so electronegative, it is one of the most potent of all oxidizing agents.

Energy must be added to pull an electron away from an atom, just as energy is required to push a ball uphill. The more electronegative the atom (the stronger its pull on electrons), the more energy is required to take an electron away from it. An electron loses potential energy when it shifts from a less electronegative atom toward a more electronegative one, just as a ball loses potential energy when it rolls downhill. A redox reaction that moves electrons closer to oxygen, such as the burning (oxidation) of methane, therefore releases chemical energy that can be put to work.

Oxidation of Organic Fuel Molecules During Cellular Respiration

The oxidation of methane by oxygen is the main combustion reaction that occurs at the burner of a gas stove. The combustion of gasoline in an automobile engine is also a redox reaction; the energy released pushes the pistons. But the energy-yielding redox process of greatest interest to biologists is respiration: the oxidation of glucose and other molecules in food. Examine again the summary equation for cellular respiration, but this time think of it as a redox process:

$$\overbrace{C_6H_{12}O_6 \;+\; 6\,O_2 \;\longrightarrow\; 6\,CO_2}^{\text{becomes oxidized}} \;+\; \underbrace{6\,H_2O}_{\text{becomes reduced (from } O_2\text{)}} \;+\; \text{Energy}$$

As in the combustion of methane or gasoline, the fuel (glucose) is oxidized and oxygen is reduced. The electrons lose potential energy along the way, and energy is released.

In general, organic molecules that have an abundance of hydrogen are excellent fuels because their bonds are a source of "hilltop" electrons, whose energy may be released as these electrons "fall" down an energy gradient when they are transferred to oxygen. The summary equation for respiration indicates that hydrogen is transferred from glucose to oxygen. But the important point, not visible in the summary equation, is that the energy state of the electron changes as hydrogen (with its electron) is transferred to oxygen. In respiration, the oxidation of glucose transfers electrons to a lower energy state, liberating energy that becomes available for ATP synthesis.

The main energy foods, carbohydrates and fats, are reservoirs of electrons associated with hydrogen. Only the barrier of activation energy holds back the flood of electrons to a lower energy state (see Figure 8.14). Without this barrier, a food substance like glucose would combine almost instantaneously with O_2. When we supply the activation energy by igniting glucose, it burns in air, releasing 686 kcal (2,870 kJ) of heat per mole of glucose (about 180 g). Body temperature is not high enough to initiate burning, of course. Instead, if you swallow some glucose, enzymes in your cells will lower the barrier of activation energy, allowing the sugar to be oxidized in a series of steps.

Stepwise Energy Harvest via NAD^+ and the Electron Transport Chain

If energy is released from a fuel all at once, it cannot be harnessed efficiently for constructive work. For example, if a gasoline tank explodes, it cannot drive a car very far. Cellular respiration does not oxidize glucose in a single explosive step either. Rather, glucose and other organic fuels are broken down in a series of steps, each one catalyzed by an enzyme. At key steps, electrons are stripped from the glucose. As is often the case in oxidation reactions, each electron travels with a proton—thus, as a hydrogen atom. The hydrogen atoms are not transferred directly to oxygen, but instead are usually passed first to an electron carrier, a coenzyme called **NAD^+** (nicotinamide adenine dinucleotide, a derivative of the vitamin niacin). As an electron acceptor, NAD^+ functions as an oxidizing agent during respiration.

How does NAD^+ trap electrons from glucose and other organic molecules? Enzymes called dehydrogenases remove a pair of hydrogen atoms (2 electrons and 2 protons) from the substrate (glucose, in this example), thereby oxidizing it. The enzyme delivers the 2 electrons along with 1 proton to its coenzyme, NAD^+ **(Figure 9.4)**. The other proton is released as a hydrogen ion (H^+) into the surrounding solution:

$$\text{H–}\overset{|}{\underset{|}{\text{C}}}\text{–OH} + NAD^+ \xrightarrow{\text{Dehydrogenase}} \overset{|}{\underset{|}{\text{C}}}\text{=O} + NADH + H^+$$

By receiving 2 negatively charged electrons but only 1 positively charged proton, NAD^+ has its charge neutralized when it is reduced to NADH. The name NADH shows the hydrogen that has been received in the reaction. NAD^+ is the most versatile electron

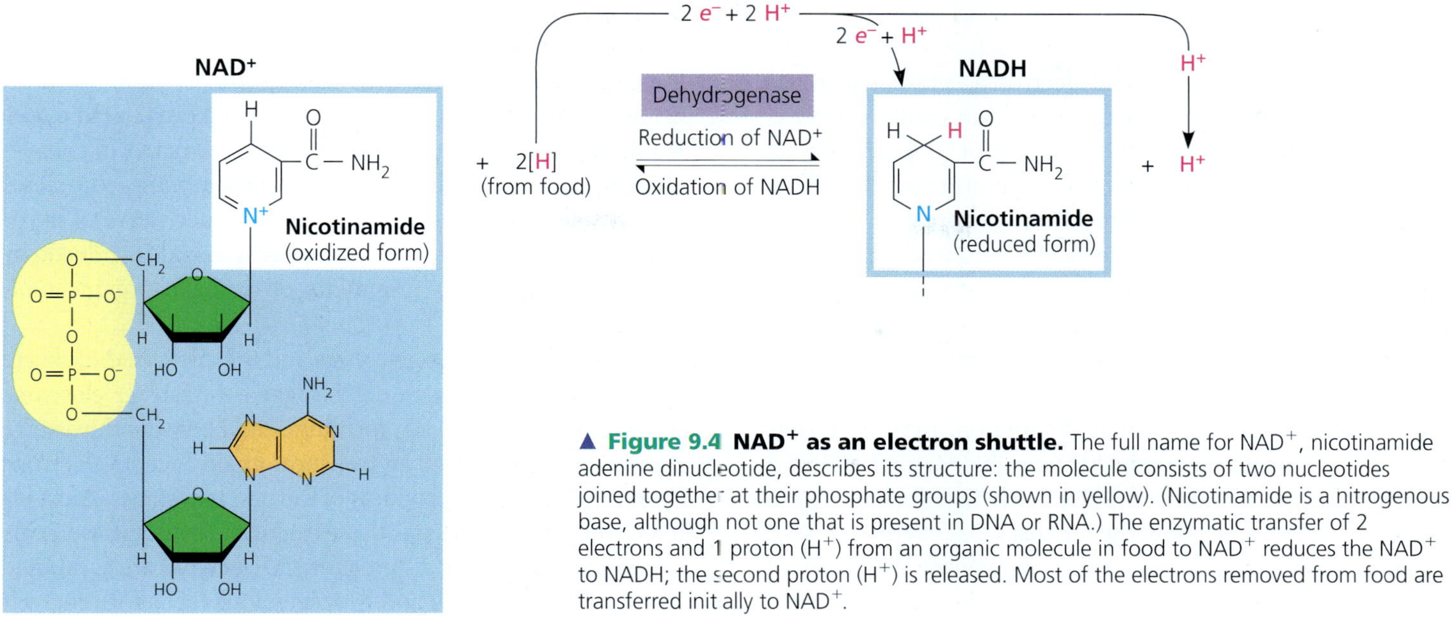

▲ **Figure 9.4 NAD$^+$ as an electron shuttle.** The full name for NAD$^+$, nicotinamide adenine dinucleotide, describes its structure: the molecule consists of two nucleotides joined together at their phosphate groups (shown in yellow). (Nicotinamide is a nitrogenous base, although not one that is present in DNA or RNA.) The enzymatic transfer of 2 electrons and 1 proton (H^+) from an organic molecule in food to NAD$^+$ reduces the NAD$^+$ to NADH; the second proton (H^+) is released. Most of the electrons removed from food are transferred initially to NAD$^+$.

acceptor in cellular respiration and functions in several of the redox steps during the breakdown of glucose.

Electrons lose very little of their potential energy when they are transferred from glucose to NAD$^+$. Each NADH molecule formed during respiration represents stored energy that can be tapped to make ATP when the electrons complete their "fall" down an energy gradient from NADH to oxygen.

How do electrons that are extracted from glucose and stored as potential energy in NADH finally reach oxygen? It will help to compare the redox chemistry of cellular respiration to a much simpler reaction: the reaction between hydrogen and oxygen to form water **(Figure 9.5a)**. Mix H_2 and O_2, provide a spark for activation energy, and the gases combine explosively. In fact, combustion of liquid H_2 and O_2 is harnessed to power the main engines of the space shuttle after it is launched, boosting it into orbit. The explosion represents a release of energy as the electrons of hydrogen "fall" closer to the electronegative oxygen atoms. Cellular respiration also brings hydrogen and oxygen together to form water, but there are two important differences. First, in cellular respiration, the hydrogen that reacts with oxygen is derived from organic molecules rather than H_2. Second, instead of occurring in one explosive reaction, respiration uses an **electron transport chain** to break the fall of electrons to oxygen into several energy-releasing steps **(Figure 9.5b)**. An electron transport chain consists of a number of molecules, mostly proteins, built into the inner membrane of mitochondria of eukaryotic cells and the plasma membrane of aerobically respiring prokaryotes. Electrons removed from glucose are shuttled by NADH to the "top," higher-energy end of the chain. At the "bottom," lower-energy end, O_2 captures these electrons along with hydrogen nuclei (H^+), forming water.

Electron transfer from NADH to oxygen is an exergonic reaction with a free-energy change of −53 kcal/mol (−222 kJ/mol). Instead of this energy being released and wasted in a single explosive step, electrons cascade down the chain from

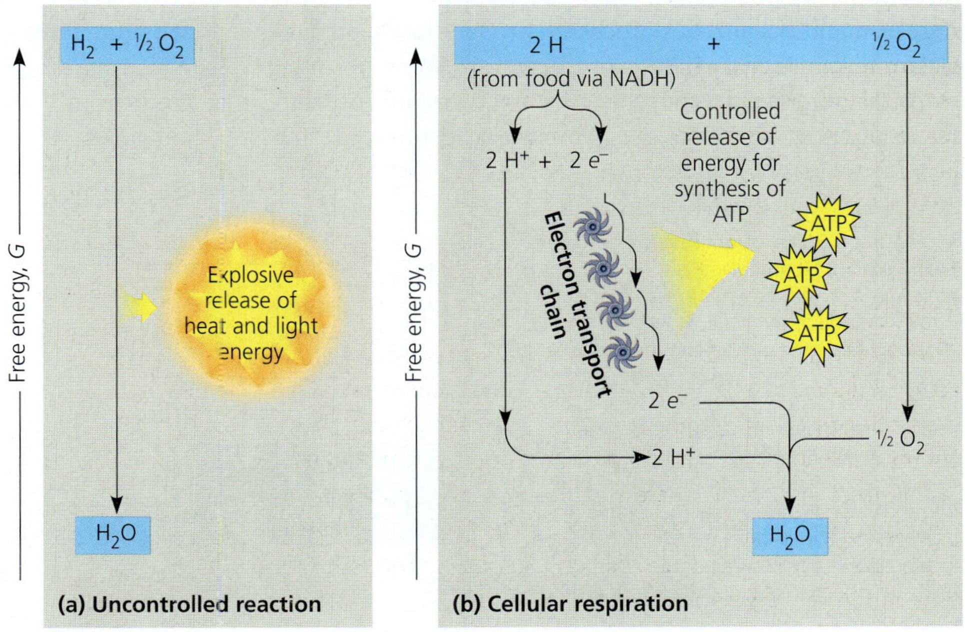

▲ **Figure 9.5 An introduction to electron transport chains. (a)** The one-step exergonic reaction of hydrogen with oxygen to form water releases a large amount of energy in the form of heat and light: an explosion. **(b)** In cellular respiration, the same reaction occurs in stages: An electron transport chain breaks the "fall" of electrons in this reaction into a series of smaller steps and stores some of the released energy in a form that can be used to make ATP. (The rest of the energy is released as heat.)

one carrier molecule to the next in a series of redox reactions, losing a small amount of energy with each step until they finally reach oxygen, the terminal electron acceptor, which has a very great affinity for electrons. Each "downhill" carrier is more electronegative than, and thus capable of oxidizing, its "uphill" neighbor, with oxygen at the bottom of the chain. Therefore, the electrons removed from glucose by NAD^+ fall down an energy gradient in the electron transport chain to a far more stable location in the electronegative oxygen atom. Put another way, oxygen pulls electrons down the chain in an energy-yielding tumble analogous to gravity pulling objects downhill.

In summary, during cellular respiration, most electrons travel the following "downhill" route: glucose → NADH → electron transport chain → oxygen. Later in this chapter, you will learn more about how the cell uses the energy released from this exergonic electron fall to regenerate its supply of ATP. For now, having covered the basic redox mechanisms of cellular respiration, let's look at the entire process.

The Stages of Cellular Respiration: *A Preview*

Respiration is a cumulative function of three metabolic stages:

1. **Glycolysis (color-coded teal throughout the chapter)**
2. **The citric acid cycle (color-coded salmon)**
3. **Oxidative phosphorylation: electron transport and chemiosmosis (color-coded violet)**

Cellular respiration is sometimes defined as including only the citric acid cycle and oxidative phosphorylation. We include glycolysis, however, because most respiring cells deriving energy from glucose use this process to produce starting material for the citric acid cycle.

As diagrammed in **Figure 9.6**, the first two stages of cellular respiration, glycolysis and the citric acid cycle, are the catabolic pathways that break down glucose and other organic fuels. **Glycolysis**, which occurs in the cytosol, begins the degradation process by breaking glucose into two molecules of a compound called pyruvate. The **citric acid cycle**, which takes place within the mitochondrial matrix of eukaryotic cells or simply in the cytosol of prokaryotes, completes the breakdown of glucose by oxidizing a derivative of pyruvate to carbon dioxide. Thus, the carbon dioxide produced by respiration represents fragments of oxidized organic molecules.

Some of the steps of glycolysis and the citric acid cycle are redox reactions in which dehydrogenases transfer electrons from substrates to NAD^+, forming NADH. In the third stage of respiration, the electron transport chain accepts electrons from the breakdown products of the first two stages (most often via NADH) and passes these electrons from one molecule to another. At the end of the chain, the electrons are combined with molecular oxygen and hydrogen ions (H^+), forming water (see Figure 9.5b). The energy released at each step of the chain is stored in a form the mitochondrion (or prokaryotic cell) can use to make ATP. This mode of ATP synthesis is called **oxidative phosphorylation** because it is powered by the redox reactions of the electron transport chain.

In eukaryotic cells, the inner membrane of the mitochondrion is the site of electron transport and chemiosmosis, the processes that together constitute oxidative phosphorylation. In prokaryotes, these processes take place in the plasma membrane. Oxidative phosphorylation accounts for almost 90% of the ATP generated by respiration. A smaller amount of ATP is formed directly in a few reactions of glycolysis and the citric acid cycle by a mechanism called **substrate-level phosphorylation** (**Figure 9.7**). This mode of ATP synthesis occurs when an enzyme transfers a phosphate group from a substrate molecule to ADP, rather than adding an inorganic phosphate to ADP as in oxidative phosphorylation.

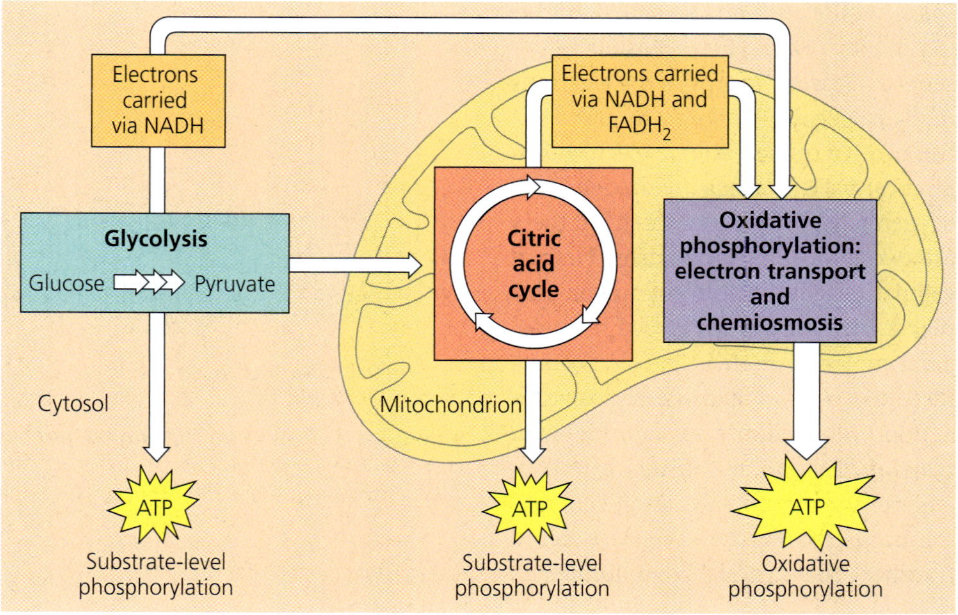

▶ **Figure 9.6 An overview of cellular respiration.** During glycolysis, each glucose molecule is broken down into two molecules of the compound pyruvate. In eukaryotic cells, as shown here, the pyruvate enters the mitochondrion, where the citric acid cycle oxidizes it to carbon dioxide. NADH and a similar electron carrier, a coenzyme called $FADH_2$, transfer electrons derived from glucose to electron transport chains, which are built into the inner mitochondrial membrane. (In prokaryotes, the electron transport chains are located in the plasma membrane.) During oxidative phosphorylation, electron transport chains convert the chemical energy to a form used for ATP synthesis in the process called chemiosmosis.

BioFlix Visit the Study Area at **www.masteringbio.com** for the BioFlix 3-D Animation on Cellular Respiration.

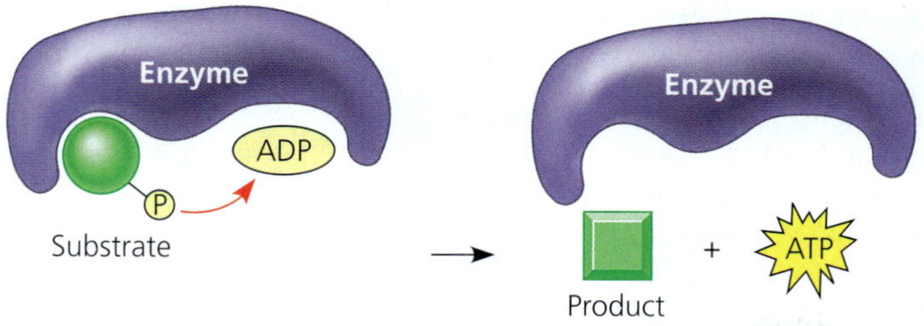

▲ **Figure 9.7 Substrate-level phosphorylation.** Some ATP is made by direct transfer of a phosphate group from an organic substrate to ADP by an enzyme. (For examples in glycolysis, see Figure 9.9, steps 7 and 10.)

? *Do you think the potential energy is higher for the reactants or the products? Explain.*

"Substrate molecule" here refers to an organic molecule generated as an intermediate during the catabolism of glucose.

For each molecule of glucose degraded to carbon dioxide and water by respiration, the cell makes up to about 38 molecules of ATP, each with 7.3 kcal/mol of free energy. Respiration cashes in the large denomination of energy banked in a single molecule of glucose (686 kcal/mol) for the small change of many molecules of ATP, which is more practical for the cell to spend on its work.

This preview has introduced you to how glycolysis, the citric acid cycle, and oxidative phosphorylation fit into the process of cellular respiration. We are now ready to take a closer look at each of these three stages of respiration.

CONCEPT CHECK 9.1

1. Compare and contrast aerobic and anaerobic respiration.
2. **WHAT IF?** If the following redox reaction occurred, which compound would be oxidized and which reduced?

$$C_4H_6O_5 + NAD^+ \longrightarrow C_4H_4O_5 + NADH + H^+$$

For suggested answers, see Appendix A.

CONCEPT **9.2**

Glycolysis harvests chemical energy by oxidizing glucose to pyruvate

The word *glycolysis* means "sugar splitting," and that is exactly what happens during this pathway. Glucose, a six-carbon sugar, is split into two three-carbon sugars. These smaller sugars are then oxidized and their remaining atoms rearranged to form two molecules of pyruvate. (Pyruvate is the ionized form of pyruvic acid.)

As summarized in **Figure 9.8**, glycolysis can be divided into two phases: energy investment and energy payoff. During the energy investment phase, the cell actually spends ATP. This investment is repaid with interest during the energy payoff phase, when ATP is produced by substrate-level phosphorylation and NAD^+ is reduced to NADH by electrons released

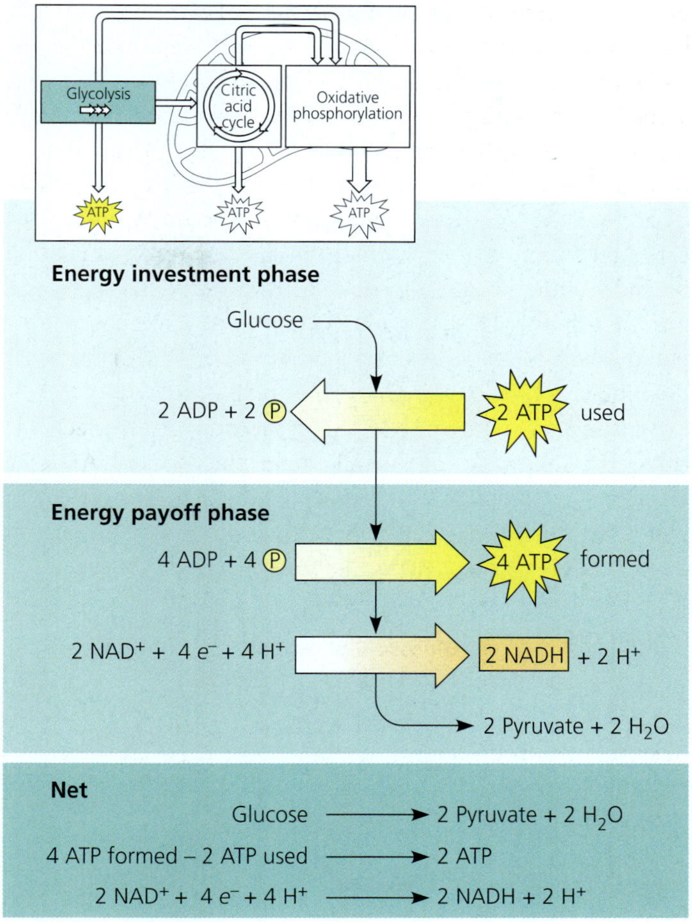

▲ **Figure 9.8 The energy input and output of glycolysis.**

from the oxidation of glucose. The net energy yield from glycolysis, per glucose molecule, is 2 ATP plus 2 NADH. The ten steps of the glycolytic pathway are described in more detail in **Figure 9.9**, on the next two pages, which you should study carefully before continuing.

In the end, all of the carbon originally present in glucose is accounted for in the two molecules of pyruvate; no CO_2 is released during glycolysis. Glycolysis occurs whether or not O_2 is present. However, if O_2 *is* present, the chemical energy stored in pyruvate and NADH can be extracted by the citric acid cycle and oxidative phosphorylation.

CONCEPT CHECK 9.2

1. During the redox reaction in glycolysis (step 6 in Figure 9.9), which molecule acts as the oxidizing agent? The reducing agent?
2. **WHAT IF?** Step 3 in Figure 9.9 is a major point of regulation of glycolysis. The enzyme phosphofructokinase is allosterically regulated by ATP and related molecules. Considering the overall result of glycolysis, would you expect ATP to inhibit or stimulate activity of this enzyme? (*Hint:* Make sure you consider the role of ATP as an allosteric regulator, not as a substrate of the enzyme.)

For suggested answers, see Appendix A.

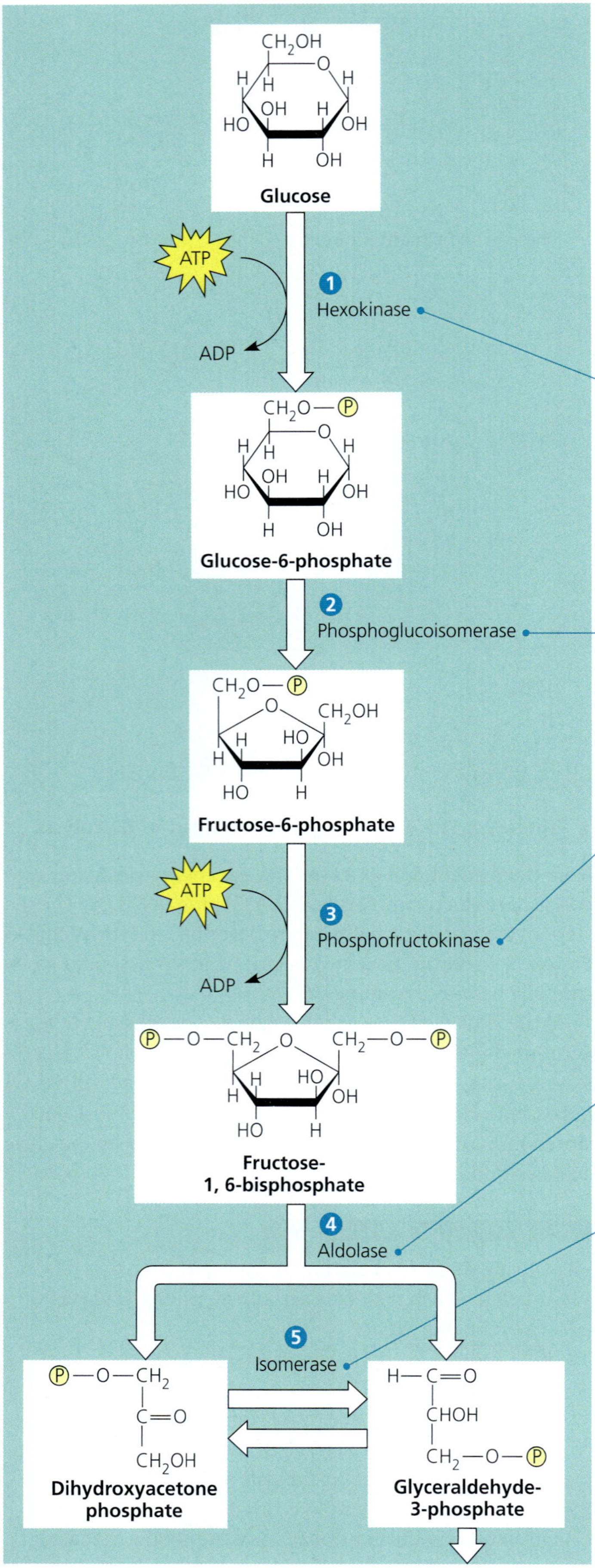

▼ **Figure 9.9 A closer look at glycolysis.** The orientation diagram at the right relates glycolysis to the entire process of respiration. Do not let the chemical detail in the main diagram block your view of glycolysis as a source of ATP and NADH.

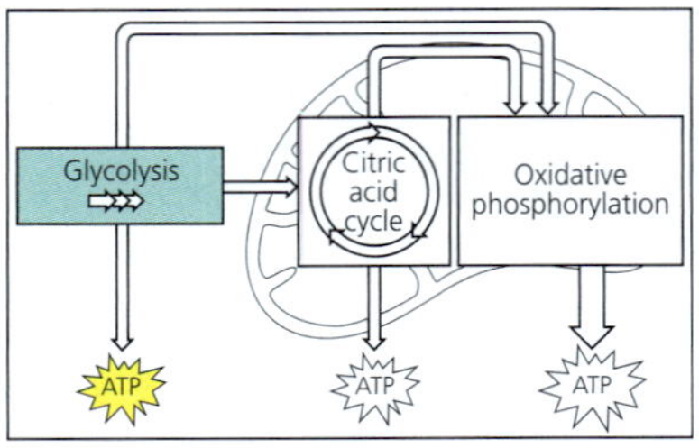

ENERGY INVESTMENT PHASE

1 Glucose enters the cell and is phosphorylated by the enzyme hexokinase, which transfers a phosphate group from ATP to the sugar. The charge of the phosphate group traps the sugar in the cell because the plasma membrane is impermeable to large ions. Phosphorylation also makes glucose more chemically reactive. In this diagram, the transfer of a phosphate group or pair of electrons from one reactant to another is indicated by coupled arrows:

2 Glucose-6-phosphate is converted to its isomer, fructose-6-phosphate.

3 This enzyme transfers a phosphate group from ATP to the sugar, investing another molecule of ATP in glycolysis. So far, 2 ATP have been used. With phosphate groups on its opposite ends, the sugar is now ready to be split in half. This is a key step for regulation of glycolysis; phosphofructokinase is allosterically regulated by ATP and its products.

4 This is the reaction from which glycolysis gets its name. The enzyme cleaves the sugar molecule into two different three-carbon sugars: dihydroxyacetone phosphate and glyceraldehyde-3-phosphate. These two sugars are isomers of each other.

5 Isomerase catalyzes the reversible conversion between the two three-carbon sugars. This reaction never reaches equilibrium in the cell because the next enzyme in glycolysis uses only glyceraldehyde-3-phosphate as its substrate (and not dihydroxyacetone phosphate). This pulls the equilibrium in the direction of glyceraldehyde-3-phosphate, which is removed as fast as it forms. Thus, the net result of steps 4 and 5 is cleavage of a six-carbon sugar into two molecules of glyceraldehyde-3-phosphate; each will progress through the remaining steps of glycolysis.

WHAT IF? *What would happen if you removed dihydroxyacetone phosphate as fast as it was produced?*

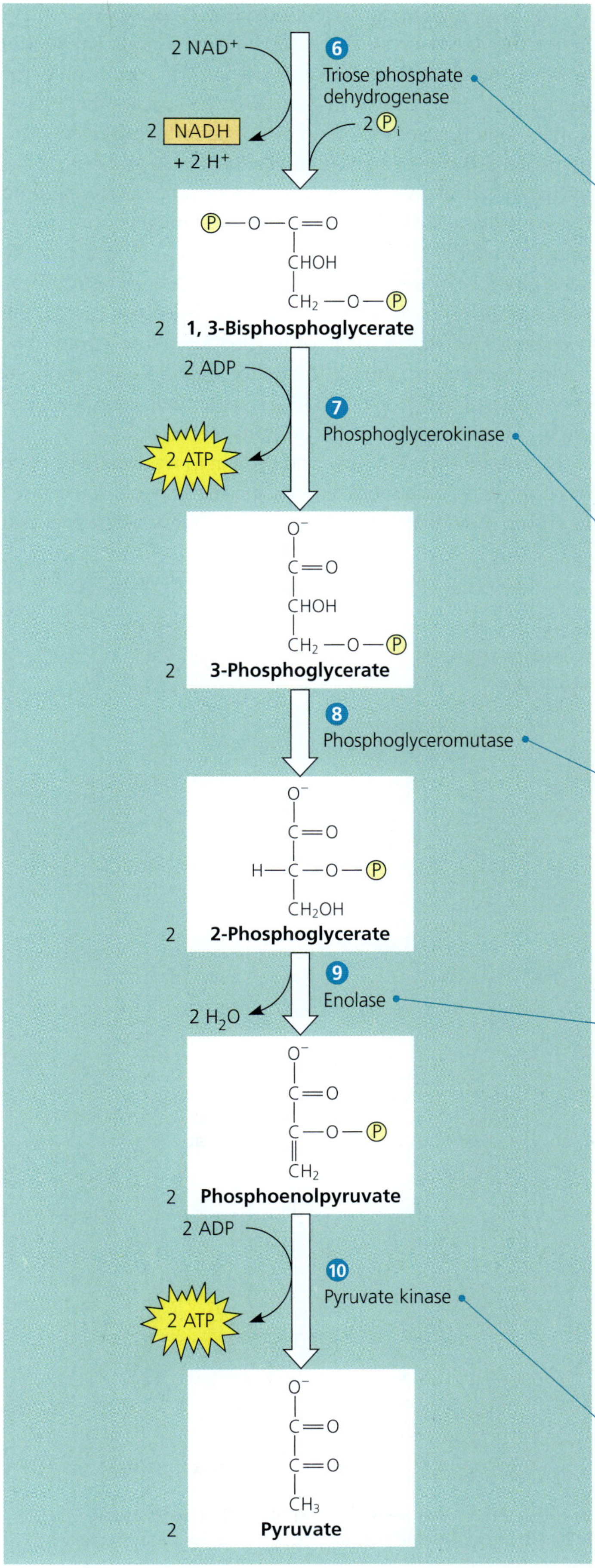

ENERGY PAYOFF PHASE

6 This enzyme catalyzes two sequential reactions while it holds glyceraldehyde-3-phosphate in its active site. First, the sugar is oxidized by the transfer of electrons and H^+ to NAD^+, forming NADH (a redox reaction). This reaction is very exergonic, and the enzyme uses the released energy to attach a phosphate group to the oxidized substrate, making a product of very high potential energy. The source of the phosphates is the pool of inorganic phosphate ions that are always present in the cytosol. Notice that the coefficient 2 precedes all molecules in the energy payoff phase; these steps occur after glucose has been split into two three-carbon sugars (step 4).

7 Glycolysis produces some ATP by substrate-level phosphorylation. The phosphate group added in the previous step is transferred to ADP in an exergonic reaction. For each glucose molecule that began glycolysis, step 7 produces 2 ATP, since every product after the sugar-splitting step (step 4) is doubled. Recall that 2 ATP were invested to get sugar ready for splitting; this ATP debt has now been repaid. Glucose has been converted to two molecules of 3-phosphoglycerate, which is not a sugar. The carbonyl group that characterizes a sugar has been oxidized to a carboxyl group (— COO^-), the hallmark of an organic acid. The sugar was oxidized in step 6, and now the energy made available by that oxidation has been used to make ATP.

8 This enzyme relocates the remaining phosphate group, preparing the substrate for the next reaction.

9 This enzyme causes a double bond to form in the substrate by extracting a water molecule, yielding phosphoenolpyruvate (PEP). The electrons of the substrate are rearranged in such a way that the resulting phosphorylated compound has a very high potential energy, allowing step 10 to occur.

10 The last reaction of glycolysis produces more ATP by transferring the phosphate group from PEP to ADP, a second instance of substrate-level phosphorylation. Since this step occurs twice for each glucose molecule, 2 ATP are produced. Overall, glycolysis has used 2 ATP in the energy investment phase (steps 1 and 3) and produced 4 ATP in the energy payoff phase (steps 7 and 10), for a net gain of 2 ATP. Glycolysis has repaid the ATP investment with 100% interest. Additional energy was stored by step 6 in NADH, which can be used to make ATP by oxidative phosphorylation if oxygen is present. Glucose has been broken down and oxidized to two molecules of pyruvate, the end product of the glycolytic pathway. If oxygen is present, the chemical energy in pyruvate can be extracted by the citric acid cycle. If oxygen is not present, fermentation may occur; this will be described later.

CONCEPT 9.3

The citric acid cycle completes the energy-yielding oxidation of organic molecules

Glycolysis releases less than a quarter of the chemical energy stored in glucose; most of the energy remains stockpiled in the two molecules of pyruvate. If molecular oxygen is present, the pyruvate enters a mitochondrion (in eukaryotic cells), where the enzymes of the citric acid cycle complete the oxidation of glucose. (In prokaryotic cells, this process occurs in the cytosol.)

Upon entering the mitochondrion via active transport, pyruvate is first converted to a compound called acetyl coenzyme A, or **acetyl CoA** (**Figure 9.10**). This step, the junction between glycolysis and the citric acid cycle, is accomplished by a multienzyme complex that catalyzes three reactions: ❶ Pyruvate's carboxyl group ($—COO^-$), which is already fully oxidized and thus has little chemical energy, is removed and given off as a molecule of CO_2. (This is the first step in which CO_2 is released during respiration.) ❷ The remaining two-carbon fragment is oxidized, forming a compound named acetate (the ionized form of acetic acid). An enzyme transfers the extracted electrons to NAD^+, storing energy in the form of NADH. ❸ Finally, coenzyme A (CoA), a sulfur-containing compound derived from a B vitamin, is attached to the acetate by an unstable bond (the wavy line in Figure 9.10) that makes the acetyl group (the attached acetate) very reactive. Because of the chemical nature of the CoA group, the product of this chemical grooming, acetyl CoA, has a high potential energy; in other words, the reaction of acetyl CoA to yield lower-energy products is highly exergonic. This molecule is now ready to feed its acetyl group into the citric acid cycle for further oxidation.

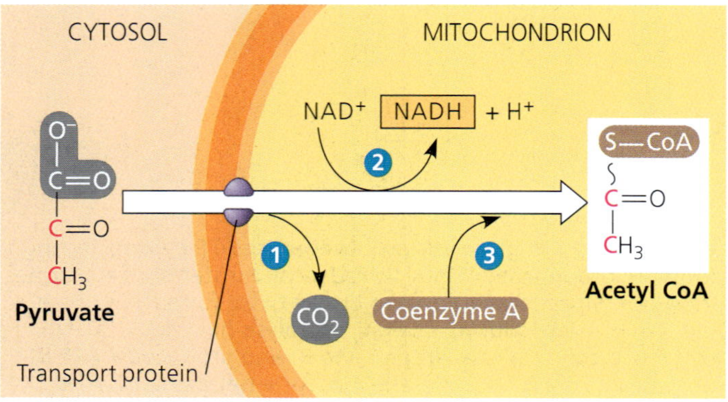

▲ **Figure 9.10 Conversion of pyruvate to acetyl CoA, the junction between glycolysis and the citric acid cycle.** Pyruvate is a charged molecule, so in eukaryotic cells it must enter the mitochondrion via active transport, with the help of a transport protein. Next, a complex of several enzymes (the pyruvate dehydrogenase complex) catalyzes the three numbered steps, which are described in the text. The acetyl group of acetyl CoA will enter the citric acid cycle. The CO_2 molecule will diffuse out of the cell.

The citric acid cycle is also called the tricarboxylic acid cycle or the Krebs cycle, the latter honoring Hans Krebs, the German-British scientist who was largely responsible for working out the pathway in the 1930s. The cycle functions as a metabolic furnace that oxidizes organic fuel derived from pyruvate. **Figure 9.11** summarizes the inputs and outputs as pyruvate is broken down to three CO_2 molecules, including the molecule of CO_2 released during the conversion of pyruvate to acetyl CoA. The cycle generates 1 ATP per turn by substrate-level phosphorylation, but most of the chemical energy is transferred to NAD^+ and a related electron carrier, the coenzyme FAD (flavin adenine dinucleotide, derived from riboflavin, a B vitamin), during the redox reactions. The reduced coenzymes, NADH and $FADH_2$, shuttle their cargo of high-energy electrons to the electron transport chain.

Now let's look at the citric acid cycle in more detail. The cycle has eight steps, each catalyzed by a specific enzyme. You can see in **Figure 9.12** that for each turn of the citric acid cycle, two

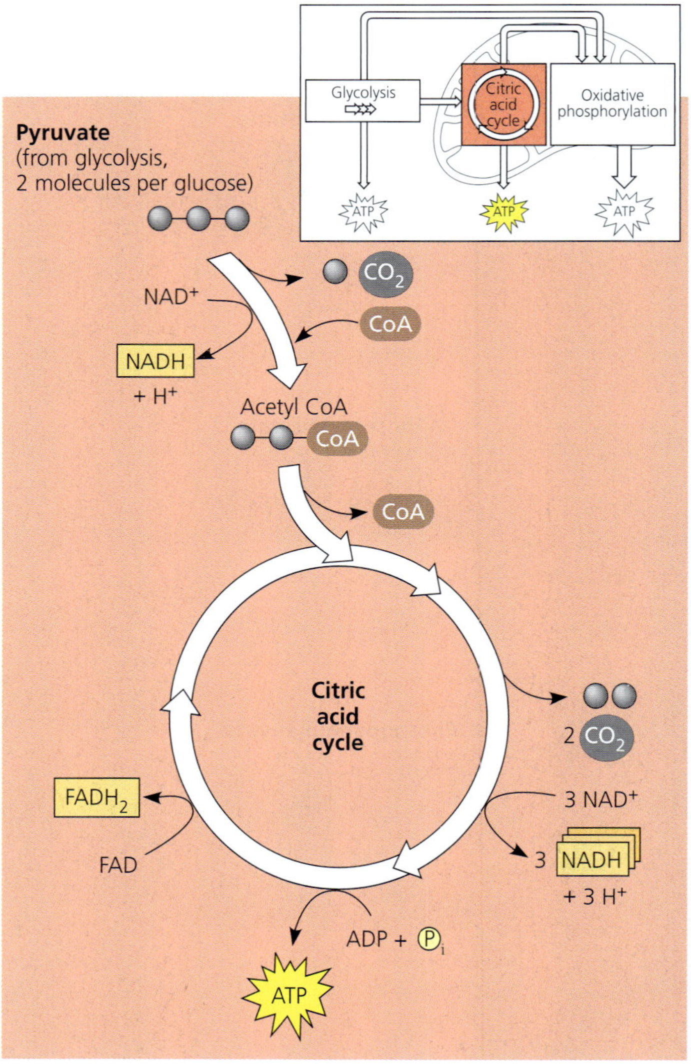

▲ **Figure 9.11 An overview of the citric acid cycle.** To calculate the inputs and outputs on a per-glucose basis, multiply by 2, because each glucose molecule is split during glycolysis into two pyruvate molecules.

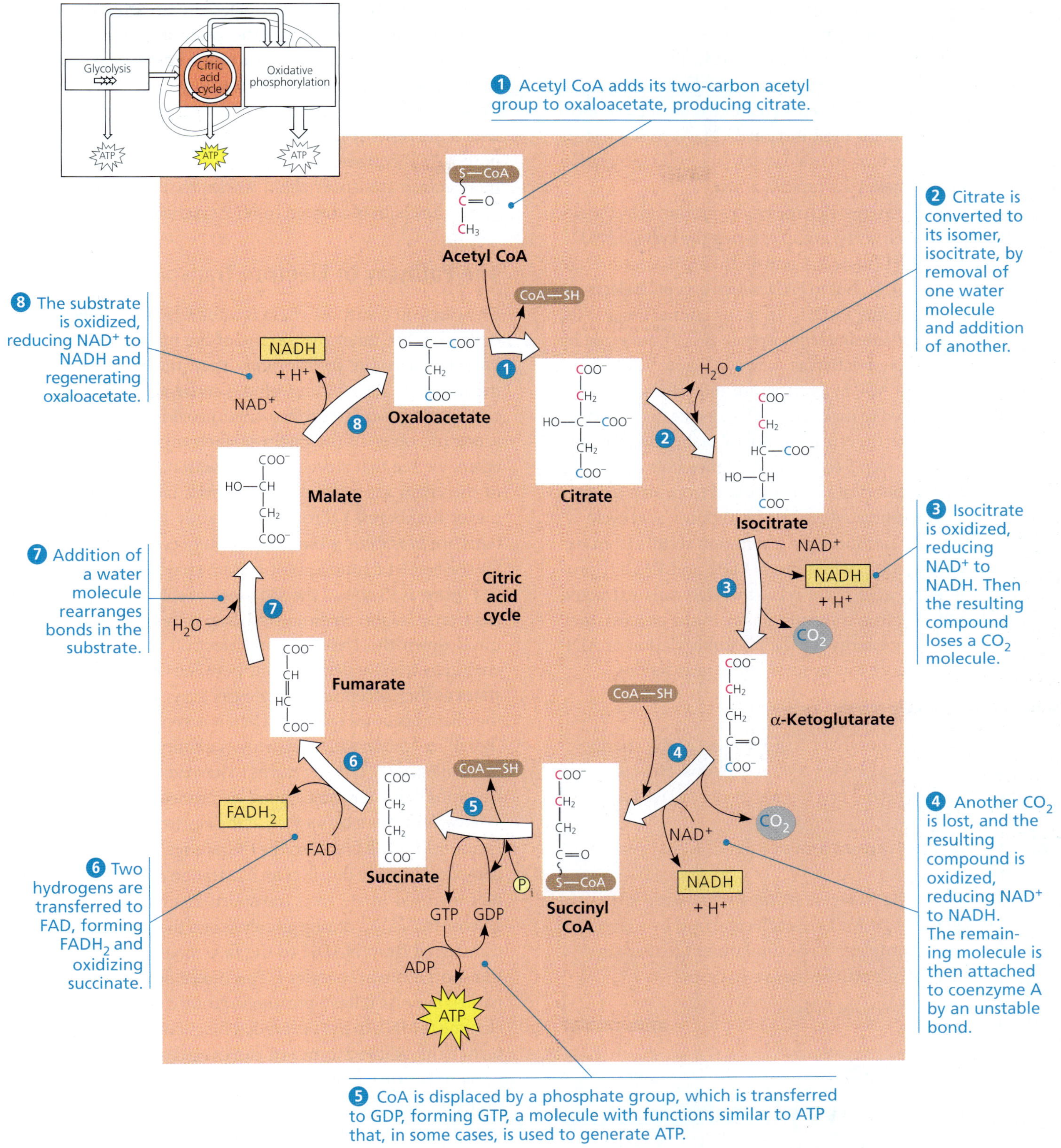

▲ **Figure 9.12 A closer look at the citric acid cycle.** In the chemical structures, red type traces the fate of the two carbon atoms that enter the cycle via acetyl CoA (step 1), and blue type indicates the two carbons that exit the cycle as CO_2 in steps 3 and 4. (The red labeling goes only through step 5 because the succinate molecule is symmetrical; the two ends cannot be distinguished from each other.) Notice that the carbon atoms that enter the cycle from acetyl CoA do not leave the cycle in the same turn. They remain in the cycle, occupying a different location in the molecules on their next turn, after another acetyl group is added. As a consequence, the oxaloacetate that is regenerated at step 8 is composed of different carbon atoms each time around. In eukaryotic cells, all the citric acid cycle enzymes are located in the mitochondrial matrix except for the enzyme that catalyzes step 6, which resides in the inner mitochondrial membrane. Carboxylic acids are represented in their ionized forms, as $—COO^-$, because the ionized forms prevail at the pH within the mitochondrion. For example, citrate is the ionized form of citric acid.

carbons (red) enter in the relatively reduced form of an acetyl group (step 1), and two different carbons (blue) leave in the completely oxidized form of CO_2 molecules (steps 3 and 4). The acetyl group of acetyl CoA joins the cycle by combining with the compound oxaloacetate, forming citrate (step 1). (Citrate is the ionized form of citric acid, for which the cycle is named.) The next seven steps decompose the citrate back to oxaloacetate. It is this regeneration of oxaloacetate that makes this process a *cycle.*

Now let's tally the energy-rich molecules produced by the citric acid cycle. For each acetyl group entering the cycle, 3 NAD^+ are reduced to NADH (steps 3, 4, and 8). In step 6, electrons are transferred not to NAD^+, but to FAD, which accepts 2 electrons and 2 protons to become $FADH_2$. In many animal tissue cells, step 5 produces a guanosine triphosphate (GTP) molecule by substrate-level phosphorylation as shown in Figure 9.12. GTP is a molecule similar to ATP in its structure and cellular function. This GTP may be used to make an ATP molecule (as shown) or directly power work in the cell. In the cells of plants, bacteria, and some animal tissues, step 5 forms an ATP molecule directly by substrate-level phosphorylation. The output from step 5 represents the only ATP generated directly by the citric acid cycle.

Most of the ATP produced by respiration results from oxidative phosphorylation, when the NADH and $FADH_2$ produced by the citric acid cycle relay the electrons extracted from food to the electron transport chain. In the process, they supply the necessary energy for the phosphorylation of ADP to ATP. We will explore this process in the next section.

CONCEPT CHECK 9.3

1. Name the molecules that conserve most of the energy from the citric acid cycle's redox reactions. How is this energy converted to a form that can be used to make ATP?
2. What cellular processes produce the CO_2 that you exhale?
3. **WHAT IF?** The conversions shown in Figure 9.10 and step 4 of Figure 9.12 are each catalyzed by a large multienzyme complex. What similarities are there in the reactions that occur in these two cases?

For suggested answers, see Appendix A.

CONCEPT 9.4

During oxidative phosphorylation, chemiosmosis couples electron transport to ATP synthesis

Our main objective in this chapter is to learn how cells harvest the energy of glucose and other nutrients in food to make ATP. But the metabolic components of respiration we have dissected so far, glycolysis and the citric acid cycle, produce only 4 ATP molecules per glucose molecule, all by substrate-level phosphorylation: 2 net ATP from glycolysis and 2 ATP from the citric acid cycle. At this point, molecules of NADH (and $FADH_2$) account for most of the energy extracted from the glucose. These electron escorts link glycolysis and the citric acid cycle to the machinery of oxidative phosphorylation, which uses energy released by the electron transport chain to power ATP synthesis. In this section, you will learn first how the electron transport chain works, then how electron flow down the chain is coupled to ATP synthesis.

The Pathway of Electron Transport

The electron transport chain is a collection of molecules embedded in the inner membrane of the mitochondrion in eukaryotic cells (in prokaryotes, they reside in the plasma membrane). The folding of the inner membrane to form cristae increases its surface area, providing space for thousands of copies of the chain in each mitochondrion. (Once again, we see that structure fits function.) Most components of the chain are proteins, which exist in multiprotein complexes numbered I through IV. Tightly bound to these proteins are *prosthetic groups*, nonprotein components essential for the catalytic functions of certain enzymes.

Figure 9.13 shows the sequence of electron carriers in the electron transport chain and the drop in free energy as electrons travel down the chain. During electron transport along the chain, electron carriers alternate between reduced and oxidized states as they accept and donate electrons. Each component of the chain becomes reduced when it accepts electrons from its "uphill" neighbor, which has a lower affinity for electrons (is less electronegative). It then returns to its oxidized form as it passes electrons to its "downhill," more electronegative neighbor.

Now let's take a closer look at the electron transport chain in Figure 9.13. We'll first describe the passage of electrons through complex I in some detail, as an illustration of the general principles involved in electron transport. Electrons removed from glucose by NAD^+, during glycolysis and the citric acid cycle, are transferred from NADH to the first molecule of the electron transport chain in complex I. This molecule is a flavoprotein, so named because it has a prosthetic group called flavin mononucleotide (FMN). In the next redox reaction, the flavoprotein returns to its oxidized form as it passes electrons to an iron-sulfur protein (Fe·S in complex I), one of a family of proteins with both iron and sulfur tightly bound. The iron-sulfur protein then passes the electrons to a compound called ubiquinone (Q in Figure 9.13). This electron carrier is a small hydrophobic molecule, the only member of the electron transport chain that is not a protein. Ubiquinone is individually mobile within the membrane rather than residing in a particular complex. (Another name for ubiquinone is coenzyme Q, or CoQ; you may have seen it sold as a nutritional supplement.)

Most of the remaining electron carriers between ubiquinone and oxygen are proteins called **cytochromes**. Their prosthetic

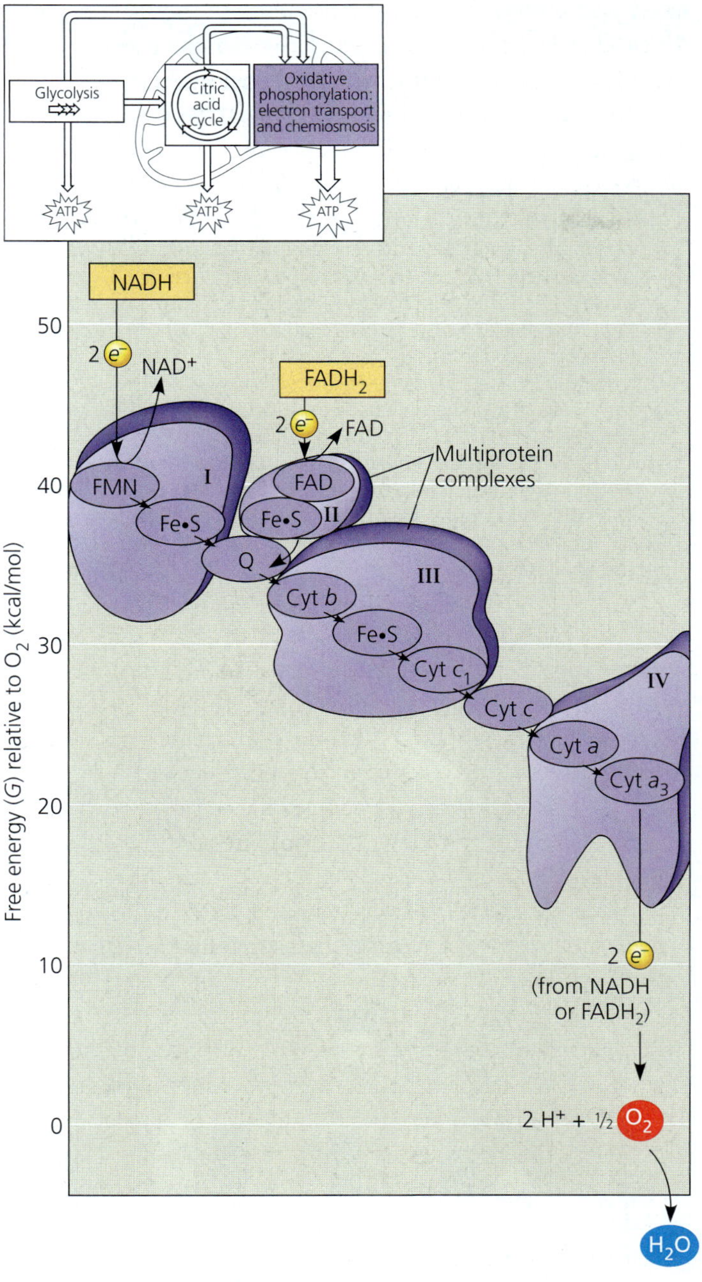

▲ **Figure 9.13 Free-energy change during electron transport.** The overall energy drop (ΔG) for electrons traveling from NADH to oxygen is 53 kcal/mol, but this "fall" is broken up into a series of smaller steps by the electron transport chain. (An oxygen atom is represented here as ½ O_2 to emphasize that the electron transport chain reduces molecular oxygen, O_2, not individual oxygen atoms.)

group, called a heme group, has an iron atom that accepts and donates electrons. (It is similar to the heme group in hemoglobin, the protein of red blood cells, except that the iron in hemoglobin carries oxygen, not electrons.) The electron transport chain has several types of cytochromes, each a different protein with a slightly different electron-carrying heme group. The last cytochrome of the chain, cyt a_3, passes its electrons to oxygen, which is *very* electronegative. Each oxygen atom also picks up a pair of hydrogen ions from the aqueous solution, forming water.

Another source of electrons for the transport chain is $FADH_2$, the other reduced product of the citric acid cycle. Notice in Figure 9.13 that $FADH_2$ adds its electrons to the electron transport chain at complex II, at a lower energy level than NADH does. Consequently, although NADH and $FADH_2$ each donate an equivalent number of electrons (2) for oxygen reduction, the electron transport chain provides about one-third less energy for ATP synthesis when the electron donor is $FADH_2$ rather than NADH. We'll see why in the next section.

The electron transport chain makes no ATP directly. Instead, it eases the fall of electrons from food to oxygen, breaking a large free-energy drop into a series of smaller steps that release energy in manageable amounts. How does the mitochondrion (or the prokaryotic plasma membrane) couple this electron transport and energy release to ATP synthesis? The answer is a mechanism called chemiosmosis.

Chemiosmosis: The Energy-Coupling Mechanism

Populating the inner membrane of the mitochondrion or the prokaryotic plasma membrane are many copies of a protein complex called **ATP synthase**, the enzyme that actually makes ATP from ADP and inorganic phosphate. ATP synthase works like an ion pump running in reverse. Recall from Chapter 7 that ion pumps usually use ATP as an energy source to transport ions against their gradients. In fact, the proton pump shown in Figure 7.19 is an ATP synthase. As we mentioned in Chapter 8, enzymes can catalyze a reaction in either direction, depending on the ΔG for the reaction, which is affected by the local concentrations of reactants and products. Rather than hydrolyzing ATP to pump protons against their concentration gradient, under the conditions of cellular respiration, ATP synthase uses the energy of an existing ion gradient to power ATP synthesis. The power source for the ATP synthase is a difference in the concentration of H^+ on opposite sides of the inner mitochondrial membrane. (We can also think of this gradient as a difference in pH, since pH is a measure of H^+ concentration.) This process, in which energy stored in the form of a hydrogen ion gradient across a membrane is used to drive cellular work such as the synthesis of ATP, is called **chemiosmosis** (from the Greek *osmos*, push). We have previously used the word *osmosis* in discussing water transport, but here it refers to the flow of H^+ across a membrane.

From studying the structure of ATP synthase, scientists have learned how the flow of H^+ through this large enzyme powers ATP generation. ATP synthase is a multisubunit complex with four main parts, each made up of multiple polypeptides. Protons move one by one into binding sites on one of the parts (the rotor), causing it to spin in a way that catalyzes ATP production from ADP and inorganic phosphate. The flow of

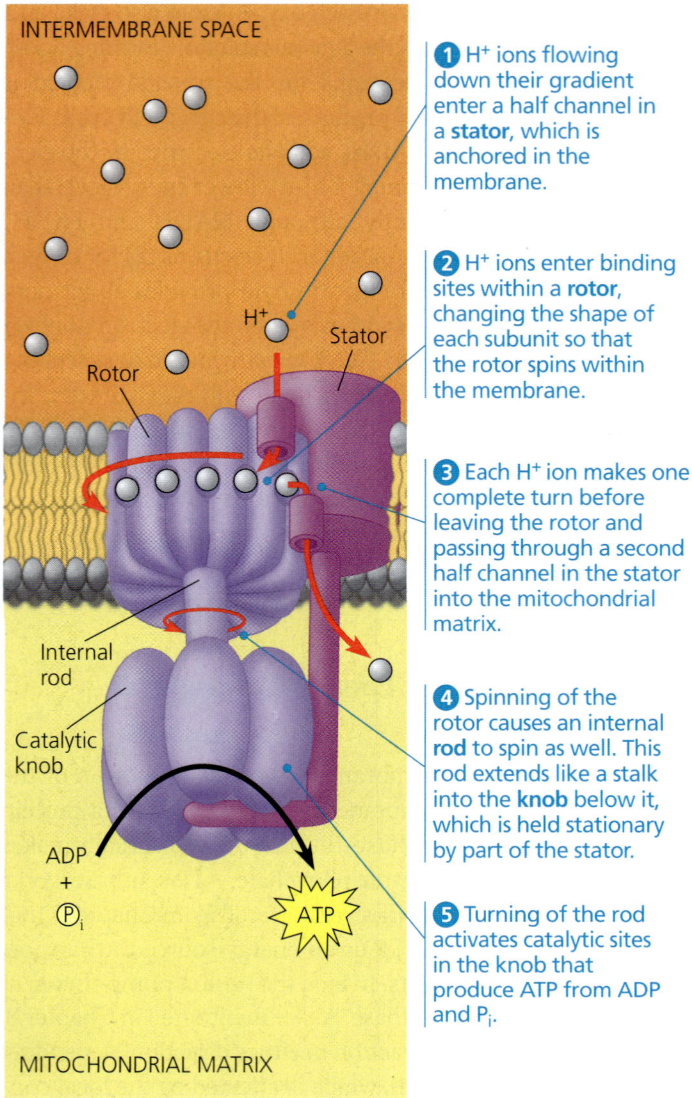

▲ **Figure 9.14 ATP synthase, a molecular mill.** The ATP synthase protein complex functions as a mill, powered by the flow of hydrogen ions. This complex resides in mitochondrial and chloroplast membranes of eukaryotes and in the plasma membranes of prokaryotes. Each of the four parts of ATP synthase consists of a number of polypeptide subunits.

protons thus behaves somewhat like a rushing stream that turns a waterwheel **(Figure 9.14)**.

ATP synthase is the smallest molecular rotary motor known in nature. The research that led to a detailed description of this enzyme's activity first showed that part of the complex actually spun around in the membrane when the reaction proceeded in the direction of ATP *hydrolysis*. Although biochemists assumed that the same rotational mechanism was responsible for ATP *synthesis*, there was no definitive support for this model until 2004, when several research institutions in collaboration with a private company were able to tackle this issue using *nanotechnology* (techniques involving control of matter on the molecular scale; from the Greek *nanos*, meaning "dwarf"). **Figure 9.15** describes the elegant experiment performed by these investigators to demonstrate that the direction of rotation of one part of

▼ **Figure 9.15 Inquiry**

Is the rotation of the internal rod in ATP synthase responsible for ATP synthesis?

EXPERIMENT Previous experiments on ATP synthase had demonstrated that the "internal rod" rotated when ATP was hydrolyzed (see Figure 9.14). Hiroyasu Itoh and colleagues set out to investigate whether simply rotating the rod in the opposite direction would cause ATP synthesis to occur. They isolated the internal rod and catalytic knob, which was then anchored to a nickel plate. A magnetic bead was bound to the rod. This complex was placed in a chamber containing an array of electromagnets, and the bead was manipulated by the sequential activation of the magnets to rotate the internal rod in either direction. The investigators hypothesized that if the bead were rotated in the direction opposite to that observed during hydrolysis, ATP synthesis would occur. ATP levels were monitored by a "reporter enzyme" in the solution that emits a discrete amount of light (a photon) when it cleaves ATP. Their hypothesis was that rotation in one direction would result in more photons than rotation in the other direction or no rotation at all.

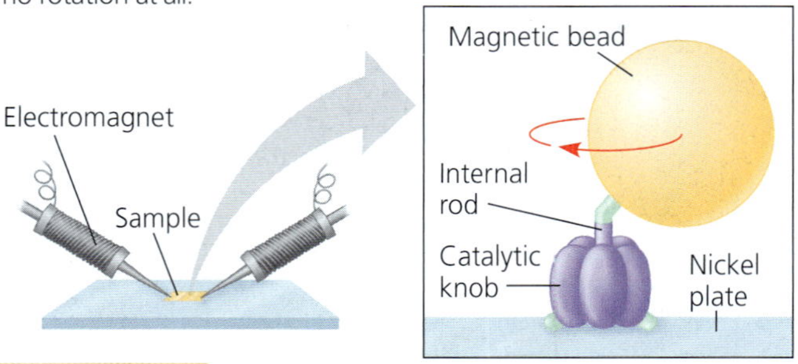

RESULTS More photons were emitted by spinning the rod for 5 minutes in one direction (yellow bars) than by no rotation (gray bars) or rotation in the opposite direction (blue bars).

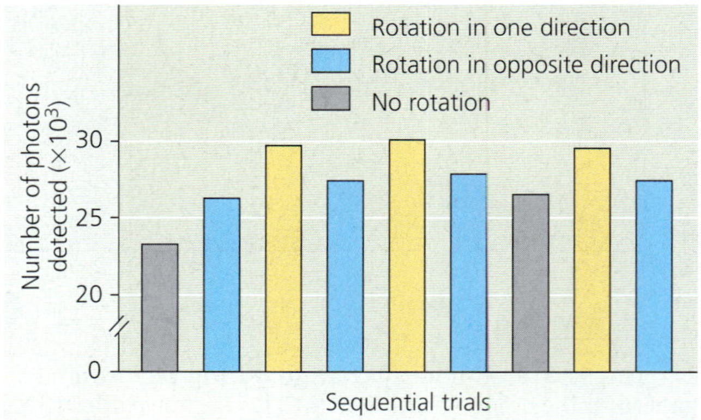

CONCLUSION The researchers concluded that the mechanical rotation of the internal rod in a particular direction within ATP synthase appears to be all that is required for generating ATP. As ATP synthase is the smallest rotary motor known, one of the goals in this type of research is to learn how to use its activity in artificial ways.

SOURCE H. Itoh et al., Mechanically driven ATP synthesis by F_1-ATPase, *Nature* 427:465–468 (2004).

Inquiry in Action Read and analyze the original paper in *Inquiry in Action: Interpreting Scientific Papers.*

WHAT IF? The "no rotation" (gray) bars represent the background level of ATP in the experiment. When the enzyme is rotated one way (yellow bars), the increase in ATP level suggests synthesis is occurring. For enzymes rotating the other way (blue bars), what level of ATP would you expect compared to the gray bars? (Note: this may not be what is observed.)

the protein complex in relation to another is solely responsible for either ATP synthesis or ATP hydrolysis.

How does the inner mitochondrial membrane or the prokaryotic plasma membrane generate and maintain the H^+ gradient that drives ATP synthesis by the ATP synthase protein complex? Establishing the H^+gradient is a major function of the electron transport chain, which is shown in its mitochondrial location in **Figure 9.16**. The chain is an energy converter that uses the exergonic flow of electrons from NADH and $FADH_2$ to pump H^+ across the membrane, from the mitochondrial matrix into the intermembrane space. The H^+ has a tendency to move back across the membrane, diffusing down its gradient. And the ATP synthases are the only sites that provide a route through the membrane for H^+. As we described previously, their passage through ATP synthase uses the exergonic flow of H^+ to drive the phosphorylation of ADP. Thus, the energy stored in an H^+ gradient across a membrane couples the redox reactions of the electron transport chain to ATP synthesis, an example of chemiosmosis.

At this point, you may be wondering how the electron transport chain pumps hydrogen ions. Researchers have found that certain members of the electron transport chain accept and release protons (H^+) along with electrons. (The aqueous solutions inside and surrounding the cell are a ready source of H^+.) At certain steps along the chain, electron transfers cause H^+ to be taken up and released into the

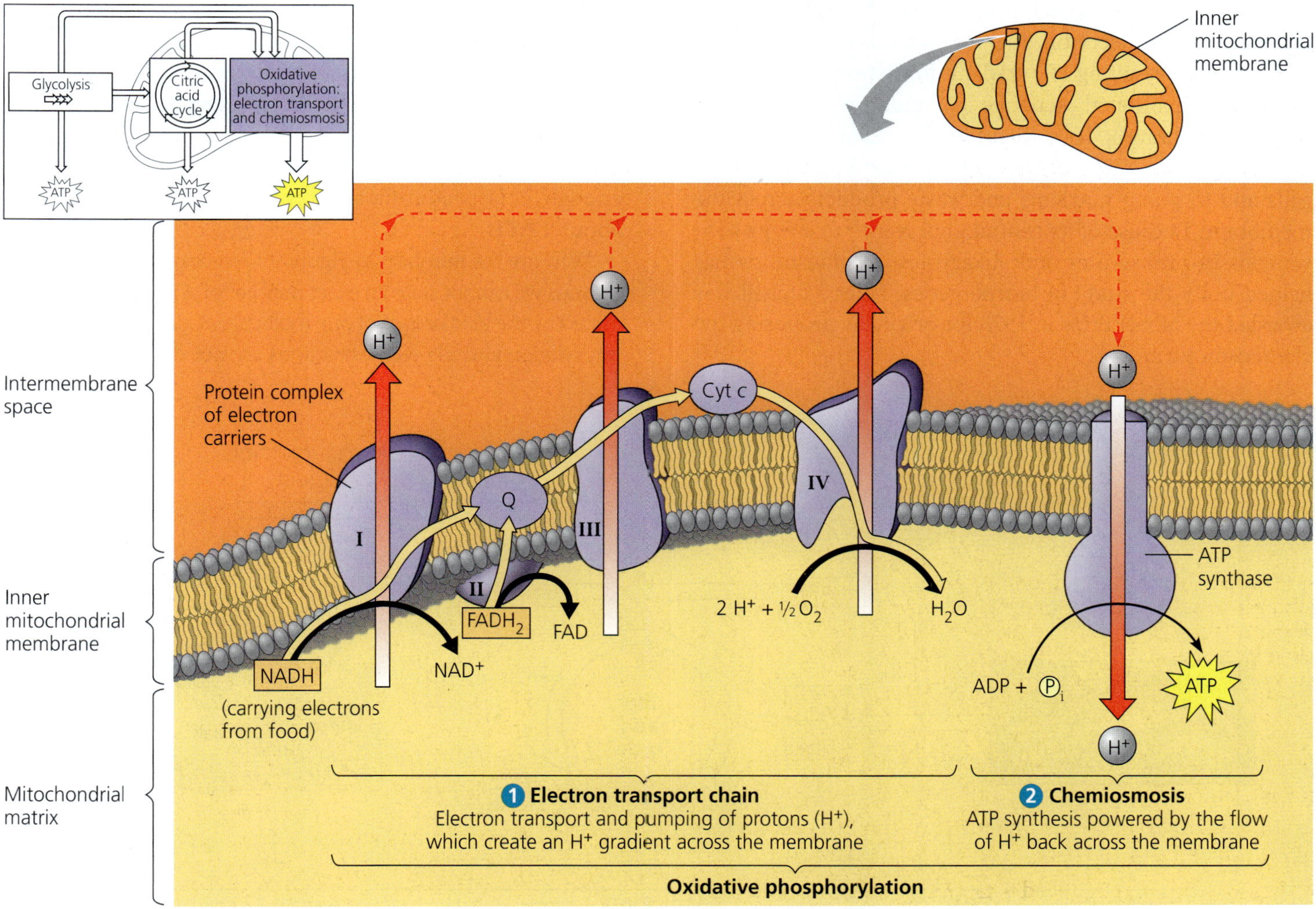

▲ **Figure 9.16 Chemiosmosis couples the electron transport chain to ATP synthesis.** ❶ NADH and $FADH_2$ shuttle high-energy electrons extracted from food during glycolysis and the citric acid cycle to an electron transport chain built into the inner mitochondrial membrane. The gold arrows trace the transport of electrons, which finally pass to oxygen at the "downhill" end of the chain, forming water. As Figure 9.13 showed, most of the electron carriers of the chain are grouped into four complexes. Two mobile carriers, ubiquinone (Q) and cytochrome c (Cyt c), move rapidly, ferrying electrons between the large complexes. As complexes I, III, and IV accept and then donate electrons, they pump protons from the mitochondrial matrix into the intermembrane space. (In prokaryotes, protons are pumped outside the plasma membrane.) Note that $FADH_2$ deposits its electrons via complex II and so results in fewer protons being pumped into the intermembrane space than occurs with NADH. Chemical energy originally harvested from food is transformed into a proton-motive force, a gradient of H^+ across the membrane. ❷ During chemiosmosis, the protons flow back down their gradient via ATP synthase, which is built into the membrane nearby. The ATP synthase harnesses the proton-motive force to phosphorylate ADP, forming ATP. Together, electron transport and chemiosmosis make up oxidative phosphorylation.

WHAT IF? *If complex IV were nonfunctional, could chemiosmosis produce any ATP, and if so, how would the rate of synthesis differ?*

surrounding solution. In eukaryotic cells, the electron carriers are spatially arranged in the membrane in such a way that H^+ is accepted from the mitochondrial matrix and deposited in the intermembrane space (see Figure 9.16). The H^+ gradient that results is referred to as a **proton-motive force**, emphasizing the capacity of the gradient to perform work. The force drives H^+ back across the membrane through the H^+ channels provided by ATP synthases.

In general terms, *chemiosmosis is an energy-coupling mechanism that uses energy stored in the form of an H^+ gradient across a membrane to drive cellular work.* In mitochondria, the energy for gradient formation comes from exergonic redox reactions, and ATP synthesis is the work performed. But chemiosmosis also occurs elsewhere and in other variations. Chloroplasts use chemiosmosis to generate ATP during photosynthesis; in these organelles, light (rather than chemical energy) drives both electron flow down an electron transport chain and the resulting H^+ gradient formation. Prokaryotes, as already mentioned, generate H^+ gradients across their plasma membranes. They then tap the proton-motive force not only to make ATP inside the cell but also to rotate their flagella and to pump nutrients and waste products across the membrane. Because of its central importance to energy conversions in prokaryotes and eukaryotes, chemiosmosis has helped unify the study of bioenergetics. Peter Mitchell was awarded the Nobel Prize in 1978 for originally proposing the chemiosmotic model.

An Accounting of ATP Production by Cellular Respiration

In the last few sections, we have looked more closely at the key processes of cellular respiration. Now, let's take a step back and remind ourselves of its overall function: harvesting the energy of glucose for ATP synthesis.

During respiration, most energy flows in this sequence: glucose → NADH → electron transport chain → proton-motive force → ATP. We can do some bookkeeping to calculate the ATP profit when cellular respiration oxidizes a molecule of glucose to six molecules of carbon dioxide. The three main departments of this metabolic enterprise are glycolysis, the citric acid cycle, and the electron transport chain, which drives oxidative phosphorylation. **Figure 9.17** gives a detailed accounting of the ATP yield per glucose molecule oxidized. The tally adds the 4 ATP produced directly by substrate-level phosphorylation during glycolysis and the citric acid cycle to the many more molecules of ATP generated by oxidative phosphorylation. Each NADH that transfers a pair of electrons from glucose to the electron transport chain contributes enough to the proton-motive force to generate a maximum of about 3 ATP.

Why are the numbers in Figure 9.17 inexact? There are three reasons we cannot state an exact number of ATP molecules generated by the breakdown of one molecule of glucose. First, phosphorylation and the redox reactions are not directly coupled to

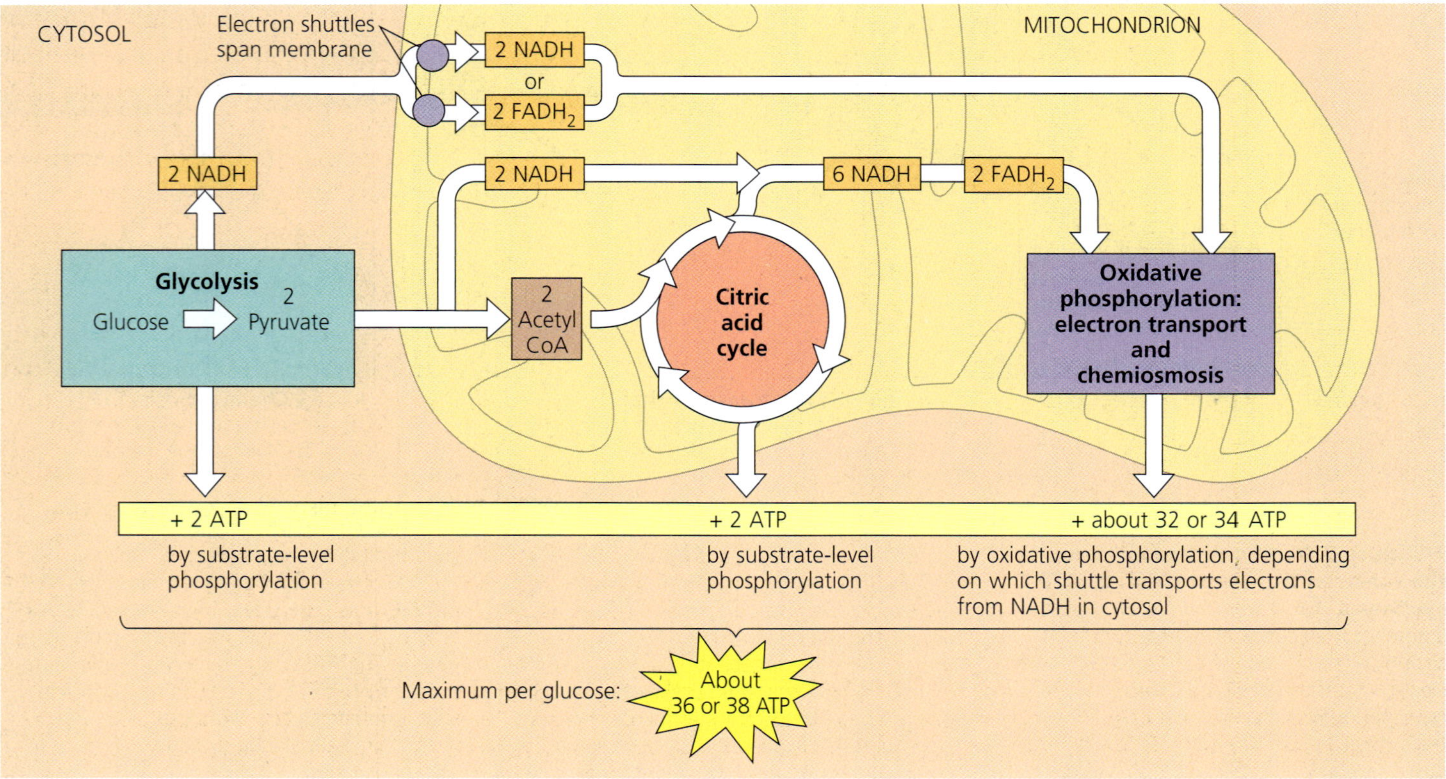

▲ **Figure 9.17 ATP yield per molecule of glucose at each stage of cellular respiration.**

each other, so the ratio of number of NADH molecules to number of ATP molecules is not a whole number. We know that 1 NADH results in 10 H^+ being transported out across the inner mitochondrial membrane, and we also know that somewhere between 3 and 4 H^+ must reenter the mitochondrial matrix via ATP synthase to generate 1 ATP. Therefore, a single molecule of NADH generates enough proton-motive force for synthesis of 2.5 to 3.3 ATP; generally, we round off and say that 1 NADH can generate about 3 ATP. The citric acid cycle also supplies electrons to the electron transport chain via $FADH_2$, but since it enters later in the chain, each molecule of this electron carrier is responsible for transport of only enough H^+ for the synthesis of 1.5 to 2 ATP. These numbers also take into account the slight energetic cost of moving the ATP formed in the mitochondrion out into the rest of the cytoplasm where it will be used.

Second, the ATP yield varies slightly depending on the type of shuttle used to transport electrons from the cytosol into the mitochondrion. The mitochondrial inner membrane is impermeable to NADH, so NADH in the cytosol is segregated from the machinery of oxidative phosphorylation. The two electrons of NADH captured in glycolysis must be conveyed into the mitochondrion by one of several electron shuttle systems. Depending on the type of shuttle in a particular cell type, the electrons are passed either to NAD^+ or to FAD in the mitochondrial matrix (see Figure 9.17). If the electrons are passed to FAD, as in brain cells, only about 2 ATP can result from each cytosolic NADH. If the electrons are passed to mitochondrial NAD^+, as in liver cells and heart cells, the yield is about 3 ATP.

A third variable that reduces the yield of ATP is the use of the proton-motive force generated by the redox reactions of respiration to drive other kinds of work. For example, the proton-motive force powers the mitochondrion's uptake of pyruvate from the cytosol. However, if *all* the proton-motive force generated by the electron transport chain were used to drive ATP synthesis, one glucose molecule could generate a maximum of 34 ATP produced by oxidative phosphorylation plus 4 ATP (net) from substrate-level phosphorylation to give a total yield of about 38 ATP (or only about 36 ATP if the less efficient shuttle were functioning).

We can now make a rough estimate of the efficiency of respiration—that is, the percentage of chemical energy possessed by glucose that has been transferred to ATP. Recall that the complete oxidation of a mole of glucose releases 686 kcal of energy under standard conditions ($\Delta G = -686$ kcal/mol). Phosphorylation of ADP to form ATP stores at least 7.3 kcal per mole of ATP. Therefore, the efficiency of respiration is 7.3 kcal per mole of ATP times 38 moles of ATP per mole of glucose divided by 686 kcal per mole of glucose, which equals 0.4. Thus, about 40% of the potential chemical energy in glucose has been transferred to ATP; the actual percentage is probably higher because ΔG is lower under cellular conditions. The rest of the stored energy is lost as heat. We humans use some of this heat to maintain our relatively high body temperature (37°C), and we dissipate the rest through sweating and other cooling mechanisms. Cellular respiration is remarkably efficient in its energy conversion. By comparison, the most efficient automobile converts only about 25% of the energy stored in gasoline to energy that moves the car.

CONCEPT CHECK 9.4

1. What effect would an absence of O_2 have on the process shown in Figure 9.16?
2. **WHAT IF?** In the absence of O_2, as in question 1, what do you think would happen if you decreased the pH of the intermembrane space of the mitochondrion? Explain your answer.

For suggested answers, see Appendix A.

CONCEPT **9.5**

Fermentation and anaerobic respiration enable cells to produce ATP without the use of oxygen

Because most of the ATP generated by cellular respiration is due to the work of oxidative phosphorylation, our estimate of ATP yield from aerobic respiration is contingent on an adequate supply of oxygen to the cell. Without the electronegative oxygen to pull electrons down the transport chain, oxidative phosphorylation ceases. However, there are two general mechanisms by which certain cells can oxidize organic fuel and generate ATP *without* the use of oxygen: anaerobic respiration and fermentation. The distinction between these two is based on whether an electron transport chain is present. (The electron transport chain is also called the respiratory chain because of its role in cellular respiration.)

We have already mentioned anaerobic respiration, which takes place in certain prokaryotic organisms that live in environments without oxygen. These organisms have an electron transport chain but do not use oxygen as a final electron acceptor at the end of the chain. Oxygen performs this function very well because it is extremely electronegative, but other, less electronegative substances can also serve as final electron acceptors. Some "sulfate-reducing" marine bacteria, for instance, use the sulfate ion (SO_4^{2-}) at the end of their respiratory chain. Operation of the chain builds up a proton-motive force used to produce ATP, but H_2S (hydrogen sulfide) is produced as a by-product rather than water.

Fermentation is a way of harvesting chemical energy without using either oxygen or any electron transport chain—in other words, without cellular respiration. How can food be oxidized without cellular respiration? Remember, oxidation simply refers to the loss of electrons to an electron acceptor, so it does not need to involve oxygen. Glycolysis oxidizes glucose to two molecules of pyruvate. The oxidizing agent of glycolysis is NAD^+, and neither oxygen nor any electron transfer chain is involved. Overall, glycolysis is exergonic, and some of the energy made available is used to produce 2 ATP (net) by substrate-level phosphorylation. If oxygen *is* present, then additional ATP is made by oxidative phosphorylation when NADH passes electrons removed from glucose to the electron transport chain. But glycolysis generates 2 ATP whether oxygen is present or not—that is, whether conditions are aerobic or anaerobic.

As an alternative to respiratory oxidation of organic nutrients, fermentation is an expansion of glycolysis that allows continuous generation of ATP by the substrate-level phosphorylation of glycolysis. For this to occur, there must be a sufficient supply of NAD^+ to accept electrons during the oxidation step of glycolysis. Without some mechanism to recycle NAD^+ from NADH, glycolysis would soon deplete the cell's pool of NAD^+ by reducing it all to NADH and would shut itself down for lack of an oxidizing agent. Under aerobic conditions, NAD^+ is recycled from NADH by the transfer of electrons to the electron transport chain. An anaerobic alternative is to transfer electrons from NADH to pyruvate, the end product of glycolysis.

Types of Fermentation

Fermentation consists of glycolysis plus reactions that regenerate NAD^+ by transferring electrons from NADH to pyruvate or derivatives of pyruvate. The NAD^+ can then be reused to oxidize sugar by glycolysis, which nets two molecules of ATP by substrate-level phosphorylation. There are many types of fermentation, differing in the end products formed from pyruvate. Two common types are alcohol fermentation and lactic acid fermentation.

In **alcohol fermentation (Figure 9.18a)**, pyruvate is converted to ethanol (ethyl alcohol) in two steps. The first step releases carbon dioxide from the pyruvate, which is converted to the two-carbon compound acetaldehyde. In the second step, acetaldehyde is reduced by NADH to ethanol. This regenerates the supply of NAD^+ needed for the continuation of glycolysis. Many bacteria carry out alcohol fermentation under anaerobic conditions. Yeast (a fungus) also carries out alcohol fermentation. For thousands of years, humans have used yeast in brewing, winemaking, and baking. The CO_2 bubbles generated by baker's yeast during alcohol fermentation allow bread to rise.

During **lactic acid fermentation (Figure 9.18b)**, pyruvate is reduced directly by NADH to form lactate as an end product, with no release of CO_2. (Lactate is the ionized form of lactic acid.) Lactic acid fermentation by certain fungi and bacteria is used in the dairy industry to make cheese and yogurt.

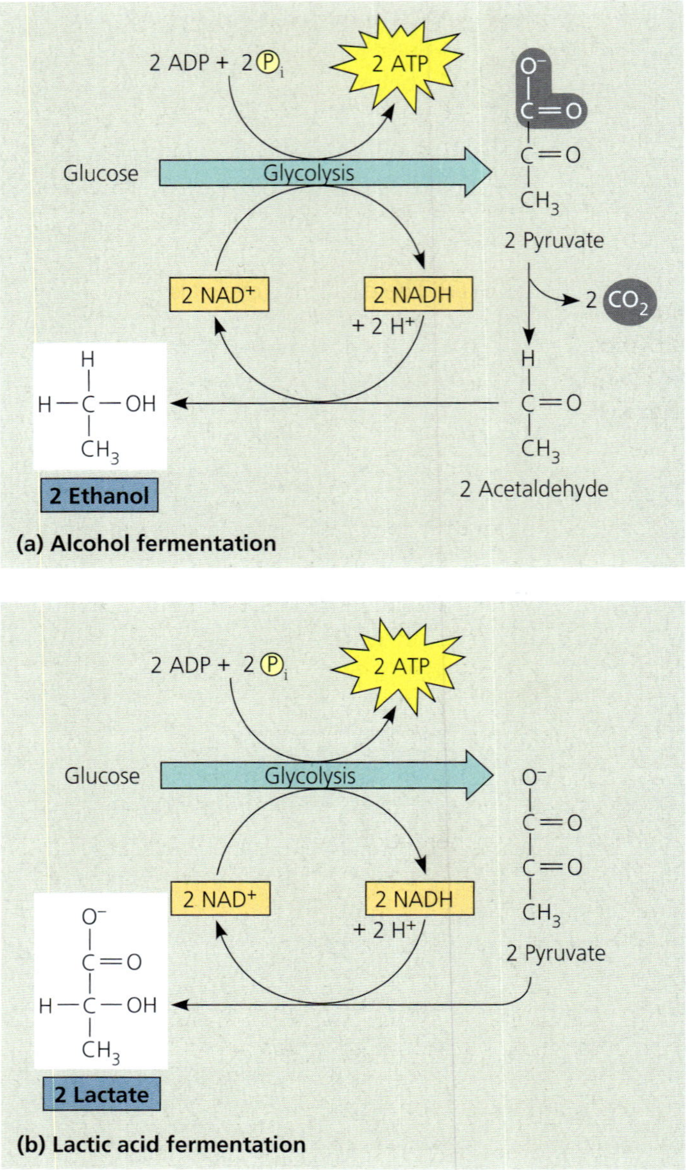

▲ **Figure 9.18 Fermentation.** In the absence of oxygen, many cells use fermentation to produce ATP by substrate-level phosphorylation. Pyruvate, the end product of glycolysis, serves as an electron acceptor for oxidizing NADH back to NAD^+, which can then be reused in glycolysis. Two of the common end products formed from fermentation are **(a)** ethanol and **(b)** lactate, the ionized form of lactic acid.

Human muscle cells make ATP by lactic acid fermentation when oxygen is scarce. This occurs during the early stages of strenuous exercise, when sugar catabolism for ATP production outpaces the muscle's supply of oxygen from the blood. Under these conditions, the cells switch from aerobic respiration to fermentation. The lactate that accumulates was previously thought to cause muscle fatigue and pain, but recent research suggests instead that increased levels of potassium ions (K^+) may be to blame, while lactate appears to enhance muscle performance. In any case, the excess lactate is gradually carried away by the blood to the liver. Lactate is converted back to pyruvate by liver cells.

Fermentation and Aerobic Respiration Compared

Fermentation and aerobic cellular respiration are anaerobic and aerobic alternatives, respectively, for producing ATP by harvesting the chemical energy of food. Both pathways use glycolysis to oxidize glucose and other organic fuels to pyruvate, with a net production of 2 ATP by substrate-level phosphorylation. And in both fermentation and respiration, NAD^+ is the oxidizing agent that accepts electrons from food during glycolysis. A key difference is the contrasting mechanisms for oxidizing NADH back to NAD^+, which is required to sustain glycolysis. In fermentation, the final electron acceptor is an organic molecule such as pyruvate (lactic acid fermentation) or acetaldehyde (alcohol fermentation). In aerobic respiration, by contrast, the final acceptor for electrons from NADH is oxygen. This process not only regenerates the NAD^+ required for glycolysis but pays an ATP bonus when the stepwise electron transport from this NADH to oxygen drives oxidative phosphorylation. An even bigger ATP payoff comes from the oxidation of pyruvate in the citric acid cycle, which is unique to respiration. Without oxygen, the energy still stored in pyruvate is unavailable to the cell. Thus, cellular respiration harvests much more energy from each sugar molecule than fermentation can. In fact, respiration yields up to 19 times as much ATP per glucose molecule as does fermentation—up to 38 molecules of ATP for respiration, compared with 2 molecules of ATP produced by substrate-level phosphorylation in fermentation.

Some organisms, called **obligate anaerobes**, carry out only fermentation or anaerobic respiration and in fact cannot survive in the presence of oxygen. A few cell types, such as cells of the vertebrate brain, can carry out only aerobic oxidation of pyruvate, but not fermentation. Other organisms, including yeasts and many bacteria, can make enough ATP to survive using either fermentation or respiration. Such species are called **facultative anaerobes**. On the cellular level, our muscle cells behave as facultative anaerobes. In such cells, pyruvate is a fork in the metabolic road that leads to two alternative catabolic routes **(Figure 9.19)**. Under aerobic conditions, pyruvate can be converted to acetyl CoA, and oxidation continues in the citric acid cycle. Under anaerobic conditions, pyruvate is diverted from the citric acid cycle, serving instead as an electron acceptor to recycle NAD^+. To make the same amount of ATP, a facultative anaerobe would have to consume sugar at a much faster rate when fermenting than when respiring.

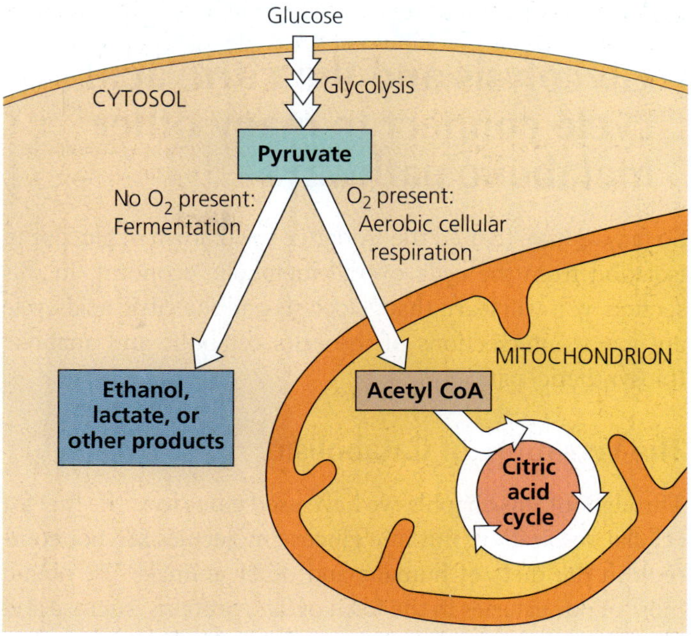

▲ **Figure 9.19 Pyruvate as a key juncture in catabolism.** Glycolysis is common to fermentation and cellular respiration. The end product of glycolysis, pyruvate, represents a fork in the catabolic pathways of glucose oxidation. In a facultative anaerobe, which is capable of both aerobic cellular respiration and fermentation, pyruvate is committed to one of those two pathways, usually depending on whether or not oxygen is present.

The Evolutionary Significance of Glycolysis

The role of glycolysis in both fermentation and respiration has an evolutionary basis. Ancient prokaryotes probably used glycolysis to make ATP long before oxygen was present in Earth's atmosphere. The oldest known fossils of bacteria date back 3.5 billion years, but appreciable quantities of oxygen probably did not begin to accumulate in the atmosphere until about 2.7 billion years ago. Cyanobacteria produced this O_2 as a by-product of photosynthesis. Therefore, early prokaryotes may have generated ATP exclusively from glycolysis. The fact that glycolysis is today the most widespread metabolic pathway among Earth's organisms suggests that it evolved very early in the history of life. The cytosolic location of glycolysis also implies great antiquity; the pathway does not require any of the membrane-bounded organelles of the eukaryotic cell, which evolved approximately 1 billion years after the prokaryotic cell. Glycolysis is a metabolic heirloom from early cells that continues to function in fermentation and as the first stage in the breakdown of organic molecules by respiration.

CONCEPT CHECK 9.5

1. Consider the NADH formed during glycolysis. What is the final acceptor for its electrons during fermentation? What is the final acceptor for its electrons during aerobic respiration?
2. **WHAT IF?** A glucose-fed yeast cell is moved from an aerobic environment to an anaerobic one. For the cell to continue generating ATP at the same rate, how would its rate of glucose consumption need to change?

For suggested answers, see Appendix A.

CONCEPT **9.6**

Glycolysis and the citric acid cycle connect to many other metabolic pathways

So far, we have treated the oxidative breakdown of glucose in isolation from the cell's overall metabolic economy. In this section, you will learn that glycolysis and the citric acid cycle are major intersections of the cell's catabolic and anabolic (biosynthetic) pathways.

The Versatility of Catabolism

Throughout this chapter, we have used glucose as the fuel for cellular respiration. But free glucose molecules are not common in the diets of humans and other animals. We obtain most of our calories in the form of fats, proteins, sucrose and other disaccharides, and starch, a polysaccharide. All these organic molecules in food can be used by cellular respiration to make ATP **(Figure 9.20)**.

Glycolysis can accept a wide range of carbohydrates for catabolism. In the digestive tract, starch is hydrolyzed to glucose, which can then be broken down in the cells by glycolysis and the citric acid cycle. Similarly, glycogen, the polysaccharide that humans and many other animals store in their liver and muscle cells, can be hydrolyzed to glucose between meals as fuel for respiration. The digestion of disaccharides, including sucrose, provides glucose and other monosaccharides as fuel for respiration.

Proteins can also be used for fuel, but first they must be digested to their constituent amino acids. Many of the amino acids, of course, are used by the organism to build new proteins. Amino acids present in excess are converted by enzymes to intermediates of glycolysis and the citric acid cycle. Before amino acids can feed into glycolysis or the citric acid cycle, their amino groups must be removed, a process called deamination. The nitrogenous refuse is excreted from the animal in the form of ammonia, urea, or other waste products.

Catabolism can also harvest energy stored in fats obtained either from food or from storage cells in the body. After fats are digested to glycerol and fatty acids, the glycerol is converted to glyceraldehyde-3-phosphate, an intermediate of glycolysis. Most of the energy of a fat is stored in the fatty acids. A metabolic sequence called **beta oxidation** breaks the fatty acids down to two-carbon fragments, which enter the citric acid cycle as acetyl CoA. NADH and $FADH_2$ are also generated during beta oxidation; they can enter the electron transport chain, leading to further ATP production. Fats make excellent fuel, in large part due to their chemical structure and the high energy level of their electrons compared to those of carbohydrates. A gram of fat oxidized by

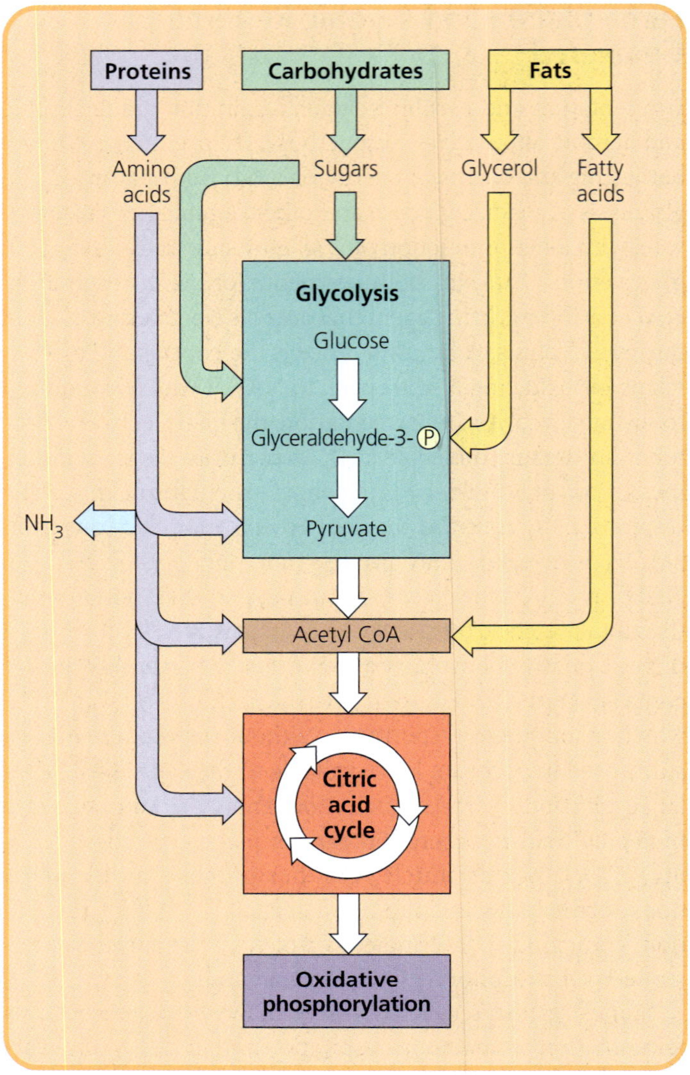

▲ **Figure 9.20 The catabolism of various molecules from food.** Carbohydrates, fats, and proteins can all be used as fuel for cellular respiration. Monomers of these molecules enter glycolysis or the citric acid cycle at various points. Glycolysis and the citric acid cycle are catabolic funnels through which electrons from all kinds of organic molecules flow on their exergonic fall to oxygen.

respiration produces more than twice as much ATP as a gram of carbohydrate. Unfortunately, this also means that a person trying to lose weight must work hard to use up fat stored in the body because so many calories are stockpiled in each gram of fat.

Biosynthesis (Anabolic Pathways)

Cells need substance as well as energy. Not all the organic molecules of food are destined to be oxidized as fuel to make ATP. In addition to calories, food must also provide the carbon skeletons that cells require to make their own molecules. Some organic monomers obtained from digestion can be used directly. For example, as previously mentioned, amino acids from the hydrolysis of proteins in food can be incorporated into the organism's own proteins. Often, however, the body needs specific molecules that are not present as such in food. Compounds formed as intermediates of glycolysis and the citric acid cycle can be diverted into anabolic pathways as precursors from which the cell can synthesize the molecules it requires. For example, humans can make about half of the 20 amino acids in proteins by modifying compounds siphoned away from the citric acid cycle; the rest are "essential amino acids" that must be obtained in the diet. Also, glucose can be made from pyruvate, and fatty acids can be synthesized from acetyl CoA. Of course, these anabolic, or biosynthetic, pathways do not generate ATP, but instead consume it.

In addition, glycolysis and the citric acid cycle function as metabolic interchanges that enable our cells to convert some kinds of molecules to others as we need them. For example, an intermediate compound generated during glycolysis, dihydroxyacetone phosphate (see Figure 9.9, step 5), can be converted to one of the major precursors of fats. If we eat more food than we need, we store fat even if our diet is fat-free. Metabolism is remarkably versatile and adaptable.

Regulation of Cellular Respiration via Feedback Mechanisms

Basic principles of supply and demand regulate the metabolic economy. The cell does not waste energy making more of a particular substance than it needs. If there is a glut of a certain amino acid, for example, the anabolic pathway that synthesizes that amino acid from an intermediate of the citric acid cycle is switched off. The most common mechanism for this control is feedback inhibition: The end product of the anabolic pathway inhibits the enzyme that catalyzes an early step of the pathway (see Figure 8.22). This prevents the needless diversion of key metabolic intermediates from uses that are more urgent.

The cell also controls its catabolism. If the cell is working hard and its ATP concentration begins to drop, respiration speeds up. When there is plenty of ATP to meet demand, respiration slows down, sparing valuable organic molecules for other functions. Again, control is based mainly on regulating the activity of enzymes at strategic points in the catabolic pathway. As shown in **Figure 9.21**, one important switch is phosphofructokinase, the enzyme that catalyzes step 3 of glycolysis (see Figure 9.9). That is the first step that commits substrate irreversibly to the glycolytic pathway. By controlling the rate of this step, the cell can speed up or slow down the entire catabolic process. Phosphofructokinase can thus be considered the pacemaker of respiration.

Phosphofructokinase is an allosteric enzyme with receptor sites for specific inhibitors and activators. It is inhibited by ATP and stimulated by AMP (adenosine monophosphate), which the cell derives from ADP. As

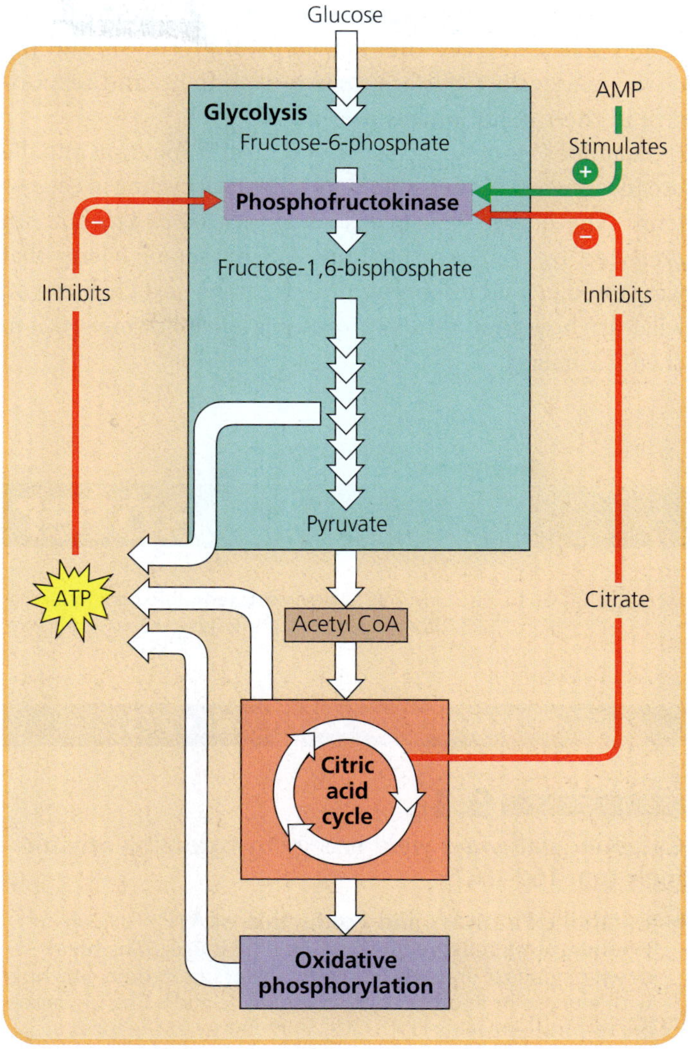

▲ **Figure 9.21 The control of cellular respiration.** Allosteric enzymes at certain points in the respiratory pathway respond to inhibitors and activators that help set the pace of glycolysis and the citric acid cycle. Phosphofructokinase, which catalyzes an early step in glycolysis (see Figure 9.9), is one such enzyme. It is stimulated by AMP (derived from ADP) but is inhibited by ATP and by citrate. This feedback regulation adjusts the rate of respiration as the cell's catabolic and anabolic demands change.

ATP accumulates, inhibition of the enzyme slows down glycolysis. The enzyme becomes active again as cellular work converts ATP to ADP (and AMP) faster than ATP is being regenerated. Phosphofructokinase is also sensitive to citrate, the first product of the citric acid cycle. If citrate accumulates in mitochondria, some of it passes into the cytosol and inhibits phosphofructokinase. This mechanism helps synchronize the rates of glycolysis and the citric acid cycle. As citrate accumulates, glycolysis slows down, and the supply of acetyl groups to the citric acid cycle decreases. If citrate consumption increases, either because of a demand for more ATP or because anabolic pathways are draining off intermediates of the citric acid cycle, glycolysis accelerates and meets the demand.

Metabolic balance is augmented by the control of enzymes that catalyze other key steps of glycolysis and the citric acid cycle. Cells are thrifty, expedient, and responsive in their metabolism.

Examine Figure 9.2 again to put cellular respiration into the broader context of energy flow and chemical cycling in ecosystems. The energy that keeps us alive is *released*, but not *produced*, by cellular respiration. We are tapping energy that was stored in food by photosynthesis. In the next chapter, you will learn how photosynthesis captures light and converts it to chemical energy.

CONCEPT CHECK 9.6

1. Compare the structure of a fat (see Figure 5.11) with that of a carbohydrate (see Figure 5.3). What features of their structures make fat a much better fuel?
2. Under what circumstances might your body synthesize fat molecules?
3. **WHAT IF?** What will happen in a muscle cell that has used up its supply of oxygen and ATP? (See Figures 9.19 and 9.21.)

For suggested answers, see Appendix A.

Chapter 9 Review

MEDIA Go to the Study Area at **www.masteringbio.com** for BioFlix 3-D Animations, MP3 Tutors, Videos, Practice Tests, an eBook, and more.

SUMMARY OF KEY CONCEPTS

CONCEPT 9.1

Catabolic pathways yield energy by oxidizing organic fuels (pp. 162–167)

- **Catabolic Pathways and Production of ATP** To keep working, a cell must regenerate the ATP it uses. The breakdown of glucose and other organic fuels is exergonic. Starting with glucose or another organic molecule and using O_2, aerobic respiration yields H_2O, CO_2, and energy in the form of ATP and heat. Cellular respiration includes both aerobic and anaerobic respiration; the latter uses another electron acceptor at the end of the electron transport chain instead of O_2, but also yields ATP.
- **Redox Reactions: Oxidation and Reduction** The cell taps the energy stored in food molecules through redox reactions, in which one substance partially or totally shifts electrons to another. The substance receiving electrons is reduced; the substance losing electrons is oxidized. During cellular respiration, glucose ($C_6H_{12}O_6$) is oxidized to CO_2, and O_2 is reduced to H_2O. Electrons lose potential energy during their transfer from organic compounds to oxygen. Electrons from organic compounds are usually passed first to NAD^+, reducing it to NADH. NADH passes the electrons to an electron transport chain, which conducts them to O_2 in energy-releasing steps. The energy is used to make ATP.
- **The Stages of Cellular Respiration: *A Preview*** Glycolysis and the citric acid cycle supply electrons (via NADH or $FADH_2$) to the electron transport chain, which drives oxidative phosphorylation. Oxidative phosphorylation generates ATP.

MEDIA
BioFlix 3-D Animation Cellular Respiration
Activity Build a Chemical Cycling System
Activity Overview of Cellular Respiration

CONCEPT 9.2

Glycolysis harvests chemical energy by oxidizing glucose to pyruvate (pp. 167–169)

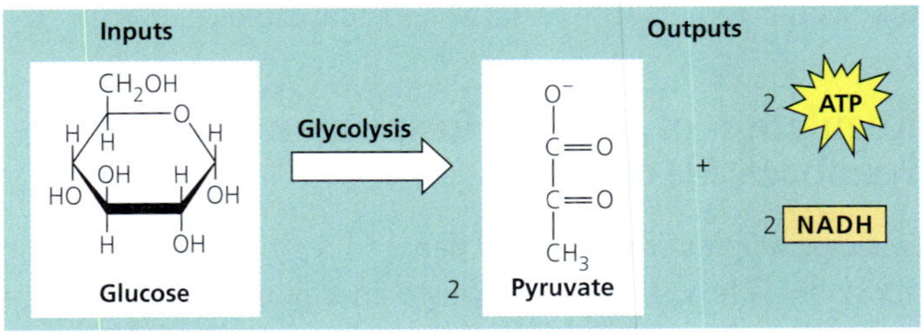

MEDIA
MP3 Tutor Cellular Respiration Part 1—Glycolysis
Activity Glycolysis

CONCEPT 9.3

The citric acid cycle completes the energy-yielding oxidation of organic molecules (pp. 170–172)

- In eukaryotic cells, the import of pyruvate into the mitochondrion and its conversion to acetyl CoA links glycolysis to the citric acid cycle. (In prokaryotic cells, the citric acid cycle occurs in the cytosol.)

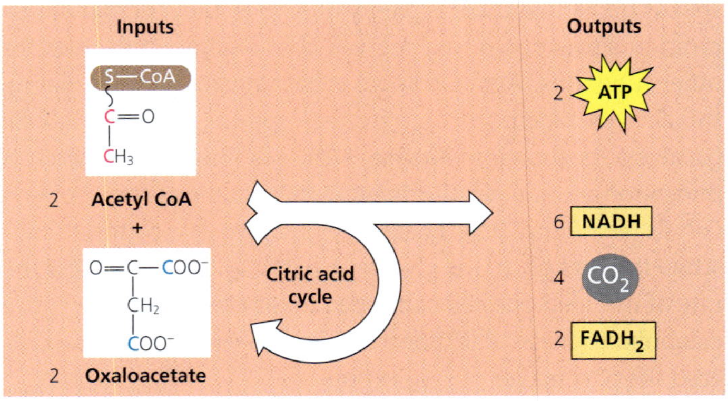

MEDIA
Activity The Citric Acid Cycle

CONCEPT 9.4

During oxidative phosphorylation, chemiosmosis couples electron transport to ATP synthesis (pp. 172–177)

- NADH and $FADH_2$ donate electrons to the electron transport chain, which powers ATP synthesis via oxidative phosphorylation.
- **The Pathway of Electron Transport** In the electron transport chain, electrons from NADH and $FADH_2$ lose energy in several energy-releasing steps. At the end of the chain, electrons are passed to O_2, reducing it to H_2O.
- **Chemiosmosis: The Energy-Coupling Mechanism** At certain steps along the electron transport chain, electron transfer causes protein complexes in eukaryotes to move H^+ from the mitochondrial matrix to the intermembrane space, storing energy as a proton-motive force (H^+ gradient). As H^+ diffuses back into the matrix through ATP synthase, its passage drives the phosphorylation of ADP. Prokaryotes generate an H^+ gradient across their plasma membrane and use this gradient to synthesize ATP in the cell.

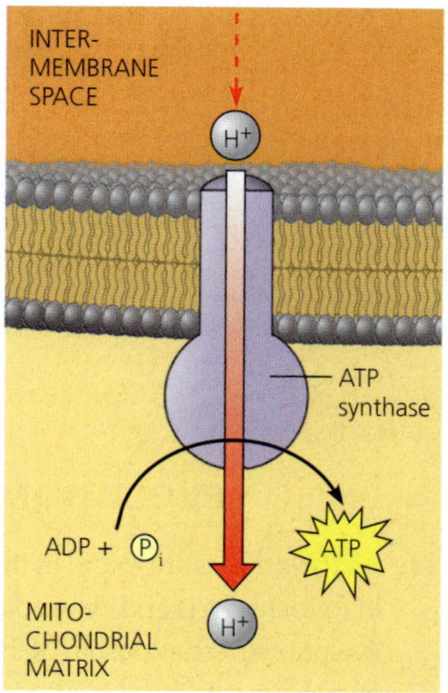

- **An Accounting of ATP Production by Cellular Respiration** About 40% of the energy stored in a glucose molecule is transferred to ATP during cellular respiration, producing a maximum of about 38 ATP.

MEDIA

MP3 Tutor Cellular Respiration Part 2—Citric Acid Cycle and Electron Transport

Activity Electron Transport

Biology Labs On-Line MitochondriaLab

Investigation How Is the Rate of Cellular Respiration Measured?

CONCEPT 9.5

Fermentation and anaerobic respiration enable cells to produce ATP without the use of oxygen (pp. 177–179)

- **Types of Fermentation** Glycolysis nets 2 ATP by substrate-level phosphorylation, whether oxygen is present or not. Under anaerobic conditions, either anaerobic respiration or fermentation can take place. In anaerobic respiration, an electron transport chain is present with a final electron acceptor other than oxygen. In fermentation, the electrons from NADH are passed to pyruvate or a derivative of pyruvate, regenerating the NAD^+ required to oxidize more glucose. Two common types of fermentation are alcohol fermentation and lactic acid fermentation.
- **Fermentation and Aerobic Respiration Compared** Both use glycolysis to oxidize glucose but differ in their final electron acceptor. Respiration yields more ATP.
- **The Evolutionary Significance of Glycolysis** Glycolysis occurs in nearly all organisms and probably evolved in ancient prokaryotes before there was O_2 in the atmosphere.

MEDIA

Activity Fermentation

CONCEPT 9.6

Glycolysis and the citric acid cycle connect to many other metabolic pathways (pp. 180–182)

- **The Versatility of Catabolism** Catabolic pathways funnel electrons from many kinds of organic molecules into cellular respiration.
- **Biosynthesis (Anabolic Pathways)** Cells can use small molecules from food directly or use them to build other substances through glycolysis or the citric acid cycle.
- **Regulation of Cellular Respiration via Feedback Mechanisms** Cellular respiration is controlled by allosteric enzymes at key points in glycolysis and the citric acid cycle.

TESTING YOUR KNOWLEDGE

SELF-QUIZ

1. What is the reducing agent in the following reaction?

$$\text{Pyruvate} + \text{NADH} + H^+ \rightarrow \text{Lactate} + NAD^+$$

a. oxygen
b. NADH
c. NAD^+
d. lactate
e. pyruvate

2. The *immediate* energy source that drives ATP synthesis by ATP synthase during oxidative phosphorylation is the
a. oxidation of glucose and other organic compounds.
b. flow of electrons down the electron transport chain.
c. affinity of oxygen for electrons.
d. H^+ concentration across the membrane holding ATP synthase.
e. transfer of phosphate to ADP.

3. Which metabolic pathway is common to both fermentation and cellular respiration of a glucose molecule?
a. the citric acid cycle
b. the electron transport chain
c. glycolysis
d. synthesis of acetyl CoA from pyruvate
e. reduction of pyruvate to lactate

4. In mitochondria, exergonic redox reactions
a. are the source of energy driving prokaryotic ATP synthesis.
b. are directly coupled to substrate-level phosphorylation.
c. provide the energy that establishes the proton gradient.
d. reduce carbon atoms to carbon dioxide.
e. are coupled via phosphorylated intermediates to endergonic processes.

5. The final electron acceptor of the electron transport chain that functions in aerobic oxidative phosphorylation is
   a. oxygen.
   b. water.
   c. NAD^+.
   d. pyruvate.
   e. ADP.

6. When electrons flow along the electron transport chains of mitochondria, which of the following changes occurs?
   a. The pH of the matrix increases.
   b. ATP synthase pumps protons by active transport.
   c. The electrons gain free energy.
   d. The cytochromes phosphorylate ADP to form ATP.
   e. NAD^+ is oxidized.

7. Cells do not catabolize carbon dioxide because
   a. its double bonds are too stable to be broken.
   b. CO_2 has fewer bonding electrons than other organic compounds.
   c. CO_2 is already completely reduced.
   d. CO_2 is already completely oxidized.
   e. the molecule has too few atoms.

8. Which of the following is a true distinction between fermentation and cellular respiration?
   a. Only respiration oxidizes glucose.
   b. NADH is oxidized by the electron transport chain in respiration only.
   c. Fermentation, but not respiration, is an example of a catabolic pathway.
   d. Substrate-level phosphorylation is unique to fermentation.
   e. NAD^+ functions as an oxidizing agent only in respiration.

9. Most CO_2 from catabolism is released during
   a. glycolysis.
   b. the citric acid cycle.
   c. lactate fermentation.
   d. electron transport.
   e. oxidative phosphorylation.

10. **DRAW IT** The graph here shows the pH difference across the inner mitochondrial membrane over time in an actively respiring cell. At the time indicated by the vertical arrow, a metabolic poison is added that specifically and completely inhibits all function of mitochondrial ATP synthase. Draw what you would expect to see for the rest of the graphed line.

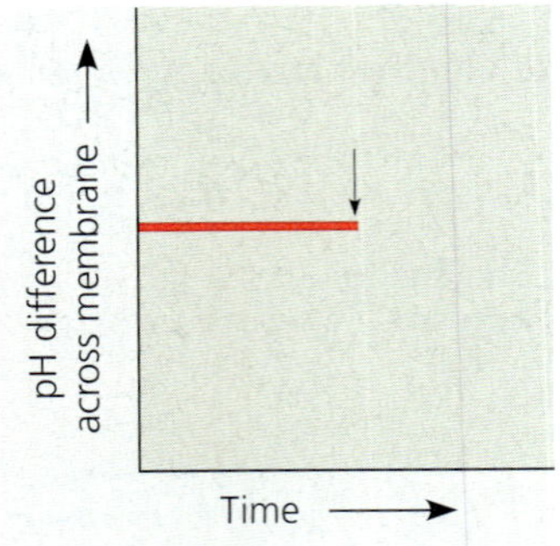

For Self-Quiz answers, see Appendix A.

MEDIA Visit the Study Area at **www.masteringbio.com** for a Practice Test.

EVOLUTION CONNECTION

11. ATP synthases are found in the prokaryotic plasma membrane and in mitochondria and chloroplasts. What does this suggest about the evolutionary relationship of these eukaryotic organelles to prokaryotes? How might the amino acid sequences of the ATP synthases from the different sources support or refute your hypothesis?

SCIENTIFIC INQUIRY

12. In the 1940s, some physicians prescribed low doses of a drug called dinitrophenol (DNP) to help patients lose weight. This unsafe method was abandoned after a few patients died. DNP uncouples the chemiosmotic machinery by making the lipid bilayer of the inner mitochondrial membrane leaky to H^+. Explain how this causes weight loss.

SCIENCE, TECHNOLOGY, AND SOCIETY

13. Nearly all human societies use fermentation to produce alcoholic drinks such as beer and wine. The practice dates back to the earliest days of agriculture. How do you suppose this use of fermentation was first discovered? Why did wine prove to be a more useful beverage, especially to a preindustrial culture, than the grape juice from which it was made?

Biological Inquiry: A Workbook of Investigative Cases Explore fermentation further in the case "Bean Brew."

10 Photosynthesis

▲ **Figure 10.1 How can sunlight, seen here as a spectrum of colors in a rainbow, power the synthesis of organic substances?**

KEY CONCEPTS

10.1 Photosynthesis converts light energy to the chemical energy of food

10.2 The light reactions convert solar energy to the chemical energy of ATP and NADPH

10.3 The Calvin cycle uses ATP and NADPH to convert CO_2 to sugar

10.4 Alternative mechanisms of carbon fixation have evolved in hot, arid climates

OVERVIEW

The Process That Feeds the Biosphere

Life on Earth is solar powered. The chloroplasts of plants capture light energy that has traveled 150 million kilometers from the sun and convert it to chemical energy stored in sugar and other organic molecules. This conversion process is called **photosynthesis**. Let's begin by placing photosynthesis in its ecological context.

Photosynthesis nourishes almost the entire living world directly or indirectly. An organism acquires the organic compounds it uses for energy and carbon skeletons by one of two major modes: autotrophic nutrition or heterotrophic nutrition. **Autotrophs** are "self-feeders" (*auto* means "self," and *trophos* means "feed"); they sustain themselves without eating anything derived from other living beings. Autotrophs produce their organic molecules from CO_2 and other inorganic raw materials obtained from the environment. They are the ultimate sources of organic compounds for all nonautotrophic organisms, and for this reason, biologists refer to autotrophs as the *producers* of the biosphere.

Almost all plants are autotrophs; the only nutrients they require are water and minerals from the soil and carbon dioxide from the air. Specifically, plants are *photo*autotrophs, organisms that use light as a source of energy to synthesize organic substances **(Figure 10.1)**. Photosynthesis also occurs in algae, certain other protists, and some prokaryotes (**Figure 10.2**, on the next page). In this chapter, we will touch on these other groups in passing, but our emphasis will be on plants. Variations in autotrophic nutrition that occur in prokaryotes and algae will be detailed in Chapters 27 and 28.

Heterotrophs obtain their organic material by the second major mode of nutrition. Unable to make their own food, they live on compounds produced by other organisms (*hetero* means "other"). Heterotrophs are the biosphere's *consumers.* The most obvious form of this "other-feeding" occurs when an animal eats plants or other animals. But heterotrophic nutrition may be more subtle. Some heterotrophs consume the remains of dead organisms by decomposing and feeding on organic litter such as carcasses, feces, and fallen leaves; they are known as decomposers. Most fungi and many types of prokaryotes get their nourishment this way. Almost all heterotrophs, including humans, are completely dependent, either directly or indirectly, on photoautotrophs for food—and also for oxygen, a by-product of photosynthesis.

In this chapter, you will learn how photosynthesis works. After a discussion of the general principles of photosynthesis, we will consider the two stages of photosynthesis: the light reactions, in which solar energy is captured and transformed into chemical energy; and the Calvin cycle, in which the chemical energy is used to make organic molecules of food. Finally, we will consider a few aspects of photosynthesis from an evolutionary perspective.

▼ **Figure 10.2 Photoautotrophs.** These organisms use light energy to drive the synthesis of organic molecules from carbon dioxide and (in most cases) water. They feed not only themselves, but the entire living world. **(a)** On land, plants are the predominant producers of food. In aquatic environments, photosynthetic organisms include **(b)** multicellular algae, such as this kelp; **(c)** some unicellular protists, such as *Euglena*; **(d)** the prokaryotes called cyanobacteria; and **(e)** other photosynthetic prokaryotes, such as these purple sulfur bacteria, which produce sulfur (spherical globules) (c, d, e: LMs).

(a) Plants

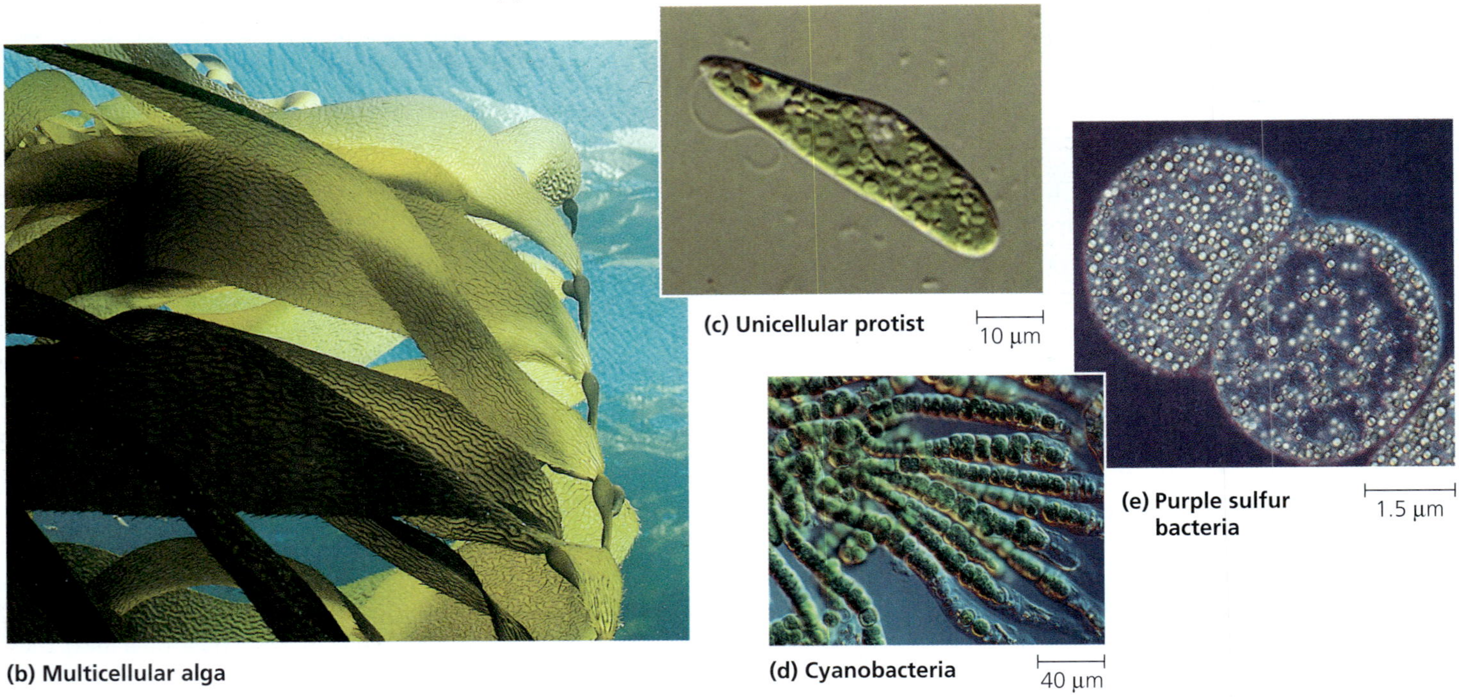

(b) Multicellular alga

(c) Unicellular protist

(d) Cyanobacteria

(e) Purple sulfur bacteria

CONCEPT **10.1**

Photosynthesis converts light energy to the chemical energy of food

The remarkable ability of an organism to harness light energy and use it to drive the synthesis of organic compounds emerges from structural organization in the cell: Photosynthetic enzymes and other molecules are grouped together in a biological membrane, enabling the necessary series of chemical reactions to be carried out efficiently. The process of photosynthesis most likely originated in a group of bacteria that had infolded regions of the plasma membrane containing clusters of such molecules. In existing photosynthetic bacteria, infolded photosynthetic membranes function similarly to the internal membranes of the chloroplast, a eukaryotic organelle you learned about in Chapter 6. In fact, the original chloroplast is believed to have been a photosynthetic prokaryote that lived inside a eukaryotic cell. (You'll learn more about this hypothesis in Chapter 25.) Chloroplasts are present in a variety of photosynthesizing organisms (see Figure 10.2), but here we will focus on plants.

Chloroplasts: The Sites of Photosynthesis in Plants

All green parts of a plant, including green stems and unripened fruit, have chloroplasts, but the leaves are the major sites of photosynthesis in most plants **(Figure 10.3)**. There are about half a million chloroplasts per square millimeter of leaf surface. The color of the leaf is from **chlorophyll**, the green pigment located within chloroplasts. It is the light energy absorbed by chlorophyll that drives the synthesis of organic molecules in the chloroplast. Chloroplasts are found mainly in the cells of the **mesophyll**, the tissue in the interior of the leaf.

Carbon dioxide enters the leaf, and oxygen exits, by way of microscopic pores called **stomata** (singular, *stoma*; from the Greek, meaning "mouth"). Water absorbed by the roots is delivered to the leaves in veins. Leaves also use veins to export sugar to roots and other nonphotosynthetic parts of the plant.

A typical mesophyll cell has about 30 to 40 chloroplasts, each organelle measuring about 2–4 μm by 4–7 μm. An envelope of two membranes encloses the **stroma**, the dense fluid within the chloroplast. An elaborate system of interconnected membranous sacs called **thylakoids** segregates the stroma from another compartment, the interior of the thylakoids, or *thylakoid space.* In some places, thylakoid sacs are stacked in columns called *grana* (singular, *granum*). Chlorophyll resides in the thylakoid membranes. (The infolded photosynthetic membranes of prokaryotes are also called thylakoid membranes; see Figure 27.7b.) Now that we have looked at the sites of photosynthesis in plants, we are ready to look more closely at the process of photosynthesis.

Tracking Atoms Through Photosynthesis: *Scientific Inquiry*

Scientists have tried for centuries to piece together the process by which plants make food. Although some of the steps are still not completely understood, the overall photosynthetic equation has been known since the 1800s: In the presence of light, the green parts of plants produce organic compounds and oxygen from carbon dioxide and water. Using molecular formulas, we can summarize the complex series of chemical reactions in photosynthesis with this chemical equation:

$$6\ CO_2 + 12\ H_2O + \text{Light energy} \rightarrow C_6H_{12}O_6 + 6\ O_2 + 6\ H_2O$$

We use glucose ($C_6H_{12}O_6$) here to simplify the relationship between photosynthesis and respiration, but the direct product of photosynthesis is actually a three-carbon sugar that can be used to make glucose. Water appears on both sides of the equation because 12 molecules are consumed and 6 molecules are newly formed during photosynthesis. We can simplify the equation by indicating only the net consumption of water:

$$6\ CO_2 + 6\ H_2O + \text{Light energy} \rightarrow C_6H_{12}O_6 + 6\ O_2$$

Writing the equation in this form, we can see that the overall chemical change during photosynthesis is the reverse of the one that occurs during cellular respiration. Both of these metabolic processes occur in plant cells. However, as you will soon learn, chloroplasts do not synthesize sugars by simply reversing the steps of respiration.

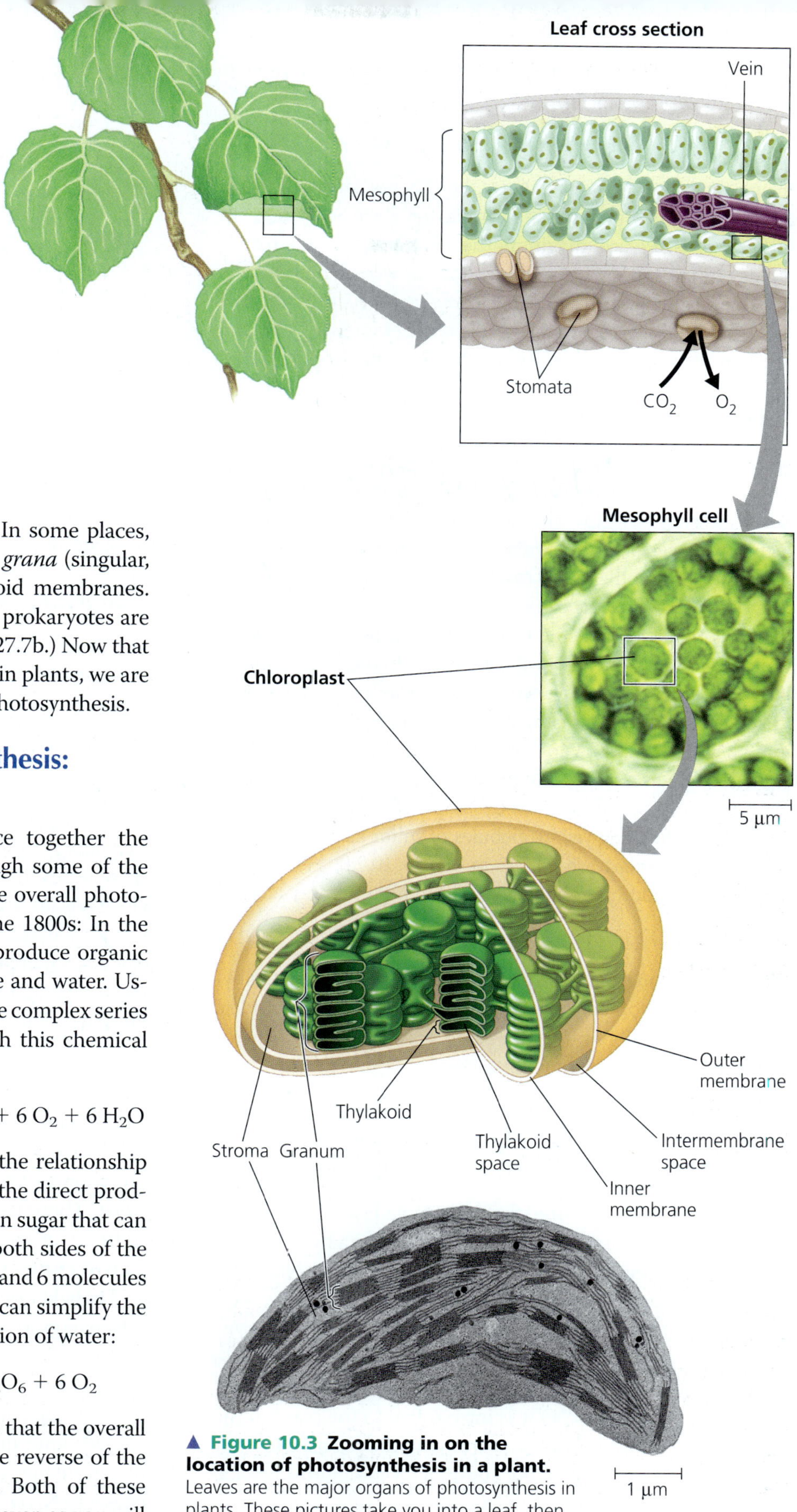

▲ **Figure 10.3 Zooming in on the location of photosynthesis in a plant.** Leaves are the major organs of photosynthesis in plants. These pictures take you into a leaf, then into a cell, and finally into a chloroplast, the organelle where photosynthesis occurs (middle, LM; bottom, TEM).

Now let's divide the photosynthetic equation by 6 to put it in its simplest possible form:

$$CO_2 + H_2O \rightarrow [CH_2O] + O_2$$

Here, the brackets indicate that CH_2O is not an actual sugar but represents the general formula for a carbohydrate. In other words, we are imagining the synthesis of a sugar molecule one carbon at a time. Six repetitions would theoretically produce a glucose molecule. Let's now use this simplified formula to see how researchers tracked the elements C, H, and O from the reactants of photosynthesis to the products.

The Splitting of Water

One of the first clues to the mechanism of photosynthesis came from the discovery that the O_2 given off by plants is derived from H_2O and not from CO_2. The chloroplast splits water into hydrogen and oxygen. Before this discovery, the prevailing hypothesis was that photosynthesis split carbon dioxide ($CO_2 \rightarrow C + O_2$) and then added water to the carbon ($C + H_2O \rightarrow [CH_2O]$). This hypothesis predicted that the O_2 released during photosynthesis came from CO_2. This idea was challenged in the 1930s by C. B. van Niel, of Stanford University. Van Niel was investigating photosynthesis in bacteria that make their carbohydrate from CO_2 but do not release O_2. Van Niel concluded that, at least in these bacteria, CO_2 is not split into carbon and oxygen. One group of bacteria used hydrogen sulfide (H_2S) rather than water for photosynthesis, forming yellow globules of sulfur as a waste product (these globules are visible in Figure 10.2e). Here is the chemical equation for photosynthesis in these sulfur bacteria:

$$CO_2 + 2\,H_2S \rightarrow [CH_2O] + H_2O + 2\,S$$

Van Niel reasoned that the bacteria split H_2S and used the hydrogen atoms to make sugar. He then generalized that idea, proposing that all photosynthetic organisms require a hydrogen source but that the source varies:

$$\begin{aligned} \text{Sulfur bacteria: } CO_2 + 2\,H_2S &\rightarrow [CH_2O] + H_2O + 2\,S \\ \text{Plants: } CO_2 + 2\,H_2O &\rightarrow [CH_2O] + H_2O + O_2 \\ \text{General: } CO_2 + 2\,H_2X &\rightarrow [CH_2O] + H_2O + 2\,X \end{aligned}$$

Thus, van Niel hypothesized that plants split H_2O as a source of electrons from hydrogen atoms, releasing O_2 as a by-product.

Nearly 20 years later, scientists confirmed van Niel's hypothesis by using oxygen-18 (^{18}O), a heavy isotope, as a tracer to follow the fate of oxygen atoms during photosynthesis. The experiments showed that the O_2 from plants was labeled with ^{18}O *only* if water was the source of the tracer (experiment 1). If the ^{18}O was introduced to the plant in the form of CO_2, the label did not turn up in the released O_2 (experiment 2). In the following summary, red denotes labeled atoms of oxygen (^{18}O):

$$\begin{aligned} \text{Experiment 1: } CO_2 + 2\,H_2O &\rightarrow [CH_2O] + H_2O + O_2 \\ \text{Experiment 2: } CO_2 + 2\,H_2O &\rightarrow [CH_2O] + H_2O + O_2 \end{aligned}$$

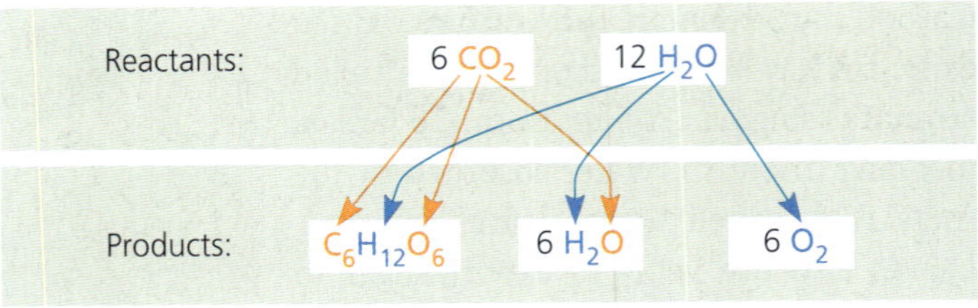

▲ **Figure 10.4 Tracking atoms through photosynthesis.** The atoms from CO_2 are shown in orange, and the atoms from H_2O are shown in blue.

A significant result of the shuffling of atoms during photosynthesis is the extraction of hydrogen from water and its incorporation into sugar. The waste product of photosynthesis, O_2, is released to the atmosphere. **Figure 10.4** shows the fates of all atoms in photosynthesis.

Photosynthesis as a Redox Process

Let's briefly compare photosynthesis with cellular respiration. Both processes involve redox reactions. During cellular respiration, energy is released from sugar when electrons associated with hydrogen are transported by carriers to oxygen, forming water as a by-product. The electrons lose potential energy as they "fall" down the electron transport chain toward electronegative oxygen, and the mitochondrion harnesses that energy to synthesize ATP (see Figure 9.16). Photosynthesis reverses the direction of electron flow. Water is split, and electrons are transferred along with hydrogen ions from the water to carbon dioxide, reducing it to sugar. Because the electrons increase in potential energy as they move from water to sugar, this process requires energy, in other words is endergonic. This energy boost is provided by light.

The Two Stages of Photosynthesis: *A Preview*

The equation for photosynthesis is a deceptively simple summary of a very complex process. Actually, photosynthesis is not a single process, but two processes, each with multiple steps. These two stages of photosynthesis are known as the **light reactions** (the *photo* part of photosynthesis) and the **Calvin cycle** (the *synthesis* part) **(Figure 10.5)**.

The light reactions are the steps of photosynthesis that convert solar energy to chemical energy. Water is split, providing a source of electrons and protons (hydrogen ions, H^+) and giving off O_2 as a by-product. Light absorbed by chlorophyll drives a transfer of the electrons and hydrogen ions from water to an acceptor called **NADP$^+$** (nicotinamide adenine dinucleotide phosphate), where they are temporarily stored. The electron acceptor NADP$^+$ is first cousin to NAD$^+$, which functions as an electron carrier in cellular respiration; the two molecules differ only by the presence of an extra phosphate group in the NADP$^+$ molecule. The light reactions use solar power to reduce NADP$^+$ to NADPH by adding a

▶ **Figure 10.5 An overview of photosynthesis: cooperation of the light reactions and the Calvin cycle.** In the chloroplast, the thylakoid membranes are the sites of the light reactions, whereas the Calvin cycle occurs in the stroma. The light reactions use solar energy to make ATP and NADPH, which supply chemical energy and reducing power, respectively, to the Calvin cycle. The Calvin cycle incorporates CO_2 into organic molecules, which are converted to sugar. (Recall that most simple sugars have formulas that are some multiple of CH_2O.)

A smaller version of this diagram will reappear in several subsequent figures as a reminder of whether the events being described occur in the light reactions or in the Calvin cycle.

BioFlix Visit the Study Area at **www.masteringbio.com** for the BioFlix 3-D Animation on Photosynthesis.

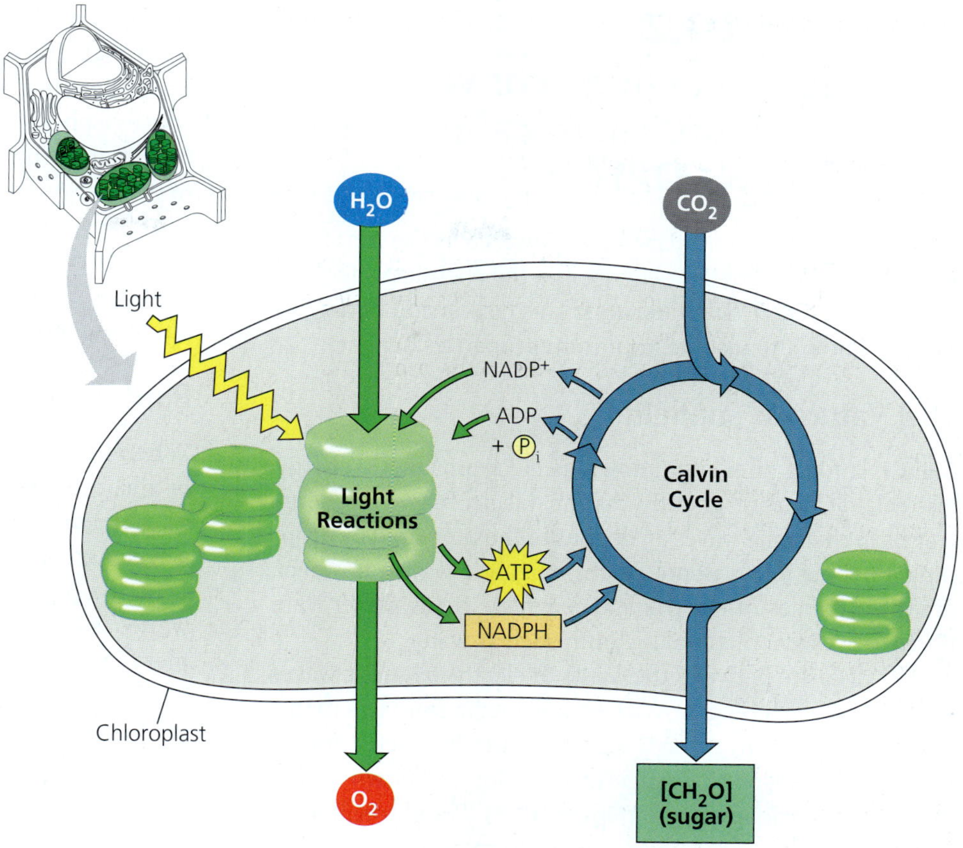

pair of electrons along with an H^+. The light reactions also generate ATP, using chemiosmosis to power the addition of a phosphate group to ADP, a process called **photophosphorylation**. Thus, light energy is initially converted to chemical energy in the form of two compounds: NADPH, a source of electrons as "reducing power" that can be passed along to an electron acceptor, reducing it, and ATP, the versatile energy currency of cells. Notice that the light reactions produce no sugar; that happens in the second stage of photosynthesis, the Calvin cycle.

The Calvin cycle is named for Melvin Calvin, who, along with his colleagues, began to elucidate its steps in the late 1940s. The cycle begins by incorporating CO_2 from the air into organic molecules already present in the chloroplast. This initial incorporation of carbon into organic compounds is known as **carbon fixation**. The Calvin cycle then reduces the fixed carbon to carbohydrate by the addition of electrons. The reducing power is provided by NADPH, which acquired its cargo of electrons in the light reactions. To convert CO_2 to carbohydrate, the Calvin cycle also requires chemical energy in the form of ATP, which is also generated by the light reactions. Thus, it is the Calvin cycle that makes sugar, but it can do so only with the help of the NADPH and ATP produced by the light reactions. The metabolic steps of the Calvin cycle are sometimes referred to as the dark reactions, or light-independent reactions, because none of the steps requires light *directly*. Nevertheless, the Calvin cycle in most plants occurs during daylight, for only then can the light reactions provide the NADPH and ATP that the Calvin cycle requires. In essence, the chloroplast uses light energy to make sugar by coordinating the two stages of photosynthesis.

As Figure 10.5 indicates, the thylakoids of the chloroplast are the sites of the light reactions, while the Calvin cycle occurs in the stroma. In the thylakoids, molecules of $NADP^+$ and ADP pick up electrons and phosphate, respectively, and NADPH and ATP are then released to the stroma, where they play crucial roles in the Calvin cycle. The two stages of photosynthesis are treated in this figure as metabolic modules that take in ingredients and crank out products. Our next step toward understanding photosynthesis is to look more closely at how the two stages work, beginning with the light reactions.

CONCEPT CHECK 10.1

1. How do the reactant molecules of photosynthesis reach the chloroplasts in leaves?
2. How did the use of an oxygen isotope help elucidate the chemistry of photosynthesis?
3. **WHAT IF?** The Calvin cycle clearly requires the products of the light reactions, ATP and NADPH. Suppose a classmate asserts that the converse is not true—that the light reactions don't depend on the Calvin cycle and, with continual light, could just keep on producing ATP and NADPH. Do you agree or disagree? Explain.

For suggested answers, see Appendix A.

CONCEPT **10.2**

The light reactions convert solar energy to the chemical energy of ATP and NADPH

Chloroplasts are chemical factories powered by the sun. Their thylakoids transform light energy into the chemical energy of ATP and NADPH. To understand this conversion better, we need to know about some important properties of light.

The Nature of Sunlight

Light is a form of energy known as electromagnetic energy, also called electromagnetic radiation. Electromagnetic energy travels in rhythmic waves analogous to those created by dropping a pebble into a pond. Electromagnetic waves, however, are disturbances of electric and magnetic fields rather than disturbances of a material medium such as water.

The distance between the crests of electromagnetic waves is called the **wavelength**. Wavelengths range from less than a nanometer (for gamma rays) to more than a kilometer (for radio waves). This entire range of radiation is known as the **electromagnetic spectrum** (**Figure 10.6**). The segment most important to life is the narrow band from about 380 nm to 750 nm in wavelength. This radiation is known as **visible light** because it can be detected as various colors by the human eye.

The model of light as waves explains many of light's properties, but in certain respects light behaves as though it consists of discrete particles, called **photons**. Photons are not tangible objects, but they act like objects in that each of them has a fixed quantity of energy. The amount of energy is inversely related to the wavelength of the light: the shorter the wavelength, the greater the energy of each photon of that light. Thus, a photon of violet light packs nearly twice as much energy as a photon of red light.

Although the sun radiates the full spectrum of electromagnetic energy, the atmosphere acts like a selective window, allowing visible light to pass through while screening out a substantial fraction of other radiation. The part of the spectrum we can see—visible light—is also the radiation that drives photosynthesis.

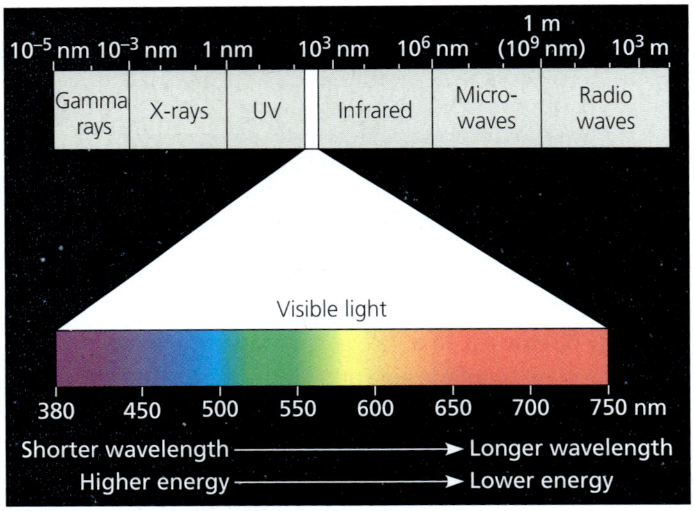

▲ **Figure 10.6 The electromagnetic spectrum.** White light is a mixture of all wavelengths of visible light. A prism can sort white light into its component colors by bending light of different wavelengths at different angles. (Droplets of water in the atmosphere can act as prisms, forming a rainbow; see Figure 10.1.) Visible light drives photosynthesis.

Photosynthetic Pigments: The Light Receptors

When light meets matter, it may be reflected, transmitted, or absorbed. Substances that absorb visible light are known as *pigments.* Different pigments absorb light of different wavelengths, and the wavelengths that are absorbed disappear. If a pigment is illuminated with white light, the color we see is the color most reflected or transmitted by the pigment. (If a pigment absorbs all wavelengths, it appears black.) We see green when we look at a leaf because chlorophyll absorbs violet-blue and red light while transmitting and reflecting green light (**Figure 10.7**). The ability of a pigment to absorb various wavelengths of light can be measured with an instrument called a **spectrophotometer**. This machine directs beams of light of different wavelengths through a solution of the pigment and measures the fraction of the light transmitted at each wavelength. A graph plotting

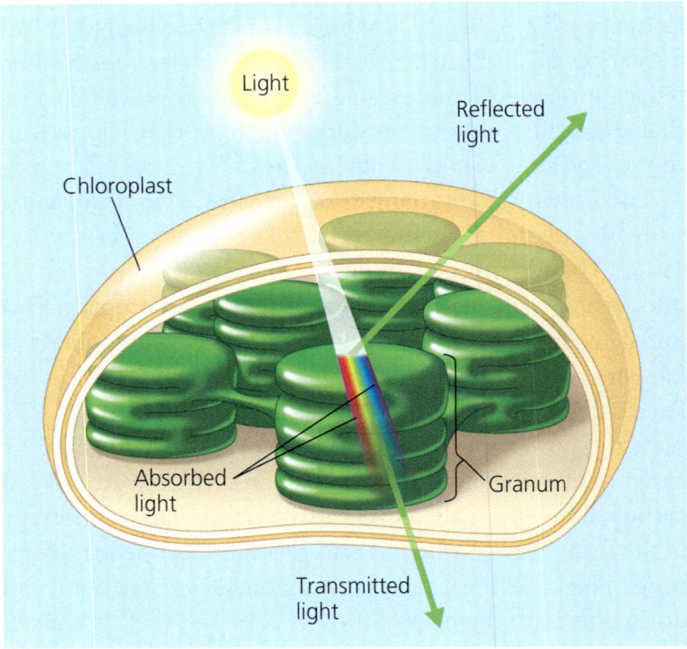

▲ **Figure 10.7 Why leaves are green: interaction of light with chloroplasts.** The chlorophyll molecules of chloroplasts absorb violet-blue and red light (the colors most effective in driving photosynthesis) and reflect or transmit green light. This is why leaves appear green.

a pigment's light absorption versus wavelength is called an **absorption spectrum** (**Figure 10.8**).

The absorption spectra of chloroplast pigments provide clues to the relative effectiveness of different wavelengths for driving photosynthesis, since light can perform work in chloroplasts only if it is absorbed. **Figure 10.9a** shows the absorption spectra of three types of pigments in chloroplasts: **chlorophyll *a***, which participates directly in the light reactions; the accessory pigment *chlorophyll b*; and a group of accessory pigments called carotenoids. The spectrum of chlorophyll *a* suggests that

▼ Figure 10.8 Research Method

Determining an Absorption Spectrum

APPLICATION An absorption spectrum is a visual representation of how well a particular pigment absorbs different wavelengths of visible light. Absorption spectra of various chloroplast pigments help scientists decipher each pigment's role in a plant.

TECHNIQUE A spectrophotometer measures the relative amounts of light of different wavelengths absorbed and transmitted by a pigment solution.

1. White light is separated into colors (wavelengths) by a prism.
2. One by one, the different colors of light are passed through the sample (chlorophyll in this example). Green light and blue light are shown here.
3. The transmitted light strikes a photoelectric tube, which converts the light energy to electricity.
4. The electrical current is measured by a galvanometer. The meter indicates the fraction of light transmitted through the sample, from which we can determine the amount of light absorbed.

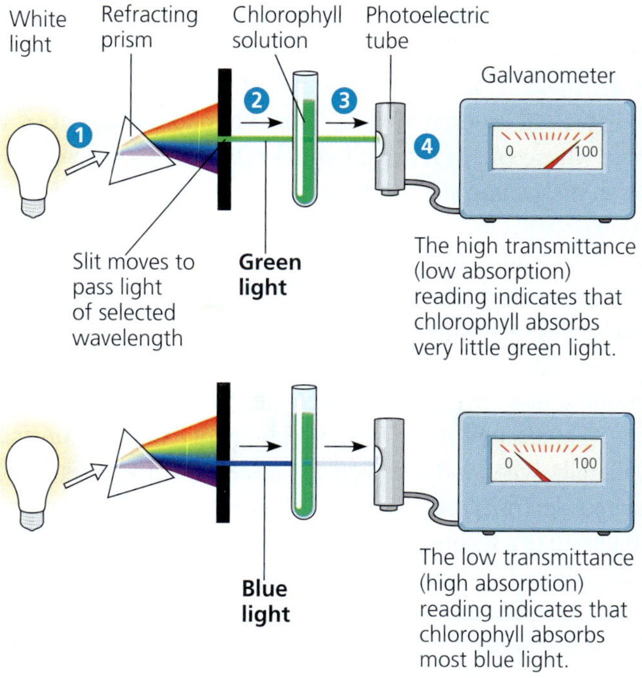

RESULTS See Figure 10.9a for absorption spectra of three types of chloroplast pigments.

▼ Figure 10.9 Inquiry

Which wavelengths of light are most effective in driving photosynthesis?

EXPERIMENT Absorption and action spectra, along with a classic experiment by Theodor W. Engelmann, reveal which wavelengths of light are photosynthetically important.

RESULTS

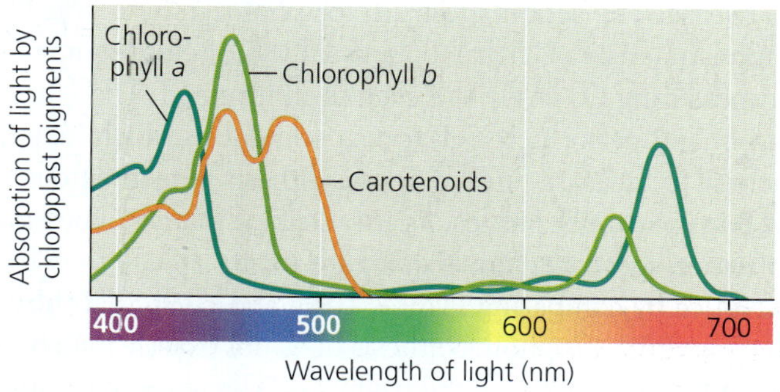

(a) Absorption spectra. The three curves show the wavelengths of light best absorbed by three types of chloroplast pigments.

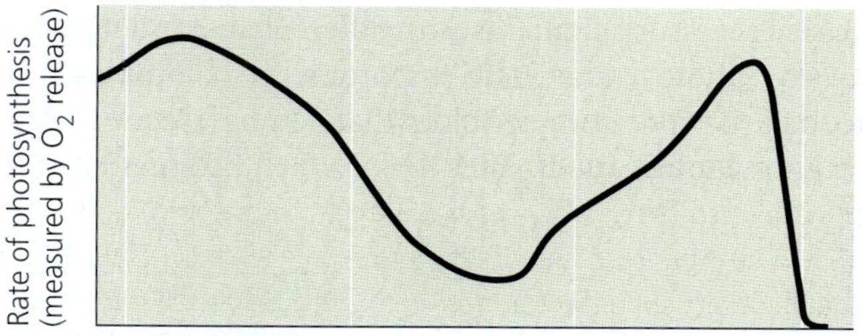

(b) Action spectrum. This graph plots the rate of photosynthesis versus wavelength. The resulting action spectrum resembles the absorption spectrum for chlorophyll *a* but does not match exactly (see part a). This is partly due to the absorption of light by accessory pigments such as chlorophyll *b* and carotenoids.

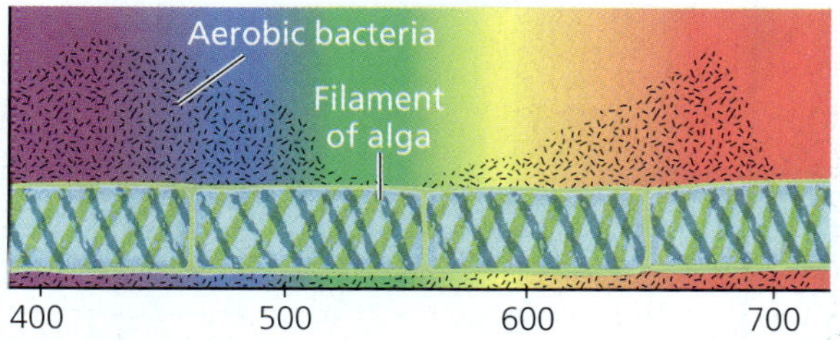

(c) Engelmann's experiment. In 1883, Theodor W. Engelmann illuminated a filamentous alga with light that had been passed through a prism, exposing different segments of the alga to different wavelengths. He used aerobic bacteria, which concentrate near an oxygen source, to determine which segments of the alga were releasing the most O_2 and thus photosynthesizing most. Bacteria congregated in greatest numbers around the parts of the alga illuminated with violet-blue or red light.

CONCLUSION Light in the violet-blue and red portions of the spectrum is most effective in driving photosynthesis.

SOURCE T. W. Engelmann, *Bacterium photometricum*. Ein Betrag zur vergleichenden Physiologie des Licht- und farbensinnes, *Archiv. für Physiologie*. 30:95–124 (1883).

WHAT IF? If Engelmann had placed a red-colored filter between the prism and the alga, how would the results have differed?

violet-blue and red light work best for photosynthesis, since they are absorbed, while green is the least effective color. This is confirmed by an **action spectrum** for photosynthesis **(Figure 10.9b)**, which profiles the relative effectiveness of different wavelengths of radiation in driving the process. An action spectrum is prepared by illuminating chloroplasts with light of different colors and then plotting wavelength against some measure of photosynthetic rate, such as CO_2 consumption or O_2 release. The action spectrum for photosynthesis was first demonstrated by a German botanist in 1883. Before equipment for measuring O_2 levels had even been invented, Theodor W. Engelmann performed a clever experiment in which he used bacteria to measure rates of photosynthesis in filamentous algae **(Figure 10.9c)**. His results are a striking match to the modern action spectrum shown in Figure 10.9b.

Notice by comparing Figures 10.9a and 10.9b that the action spectrum for photosynthesis does not exactly match the absorption spectrum of chlorophyll *a*. The absorption spectrum of chlorophyll *a* alone underestimates the effectiveness of certain wavelengths in driving photosynthesis. This is partly because accessory pigments with different absorption spectra are also photosynthetically important in chloroplasts and broaden the spectrum of colors that can be used for photosynthesis. **Figure 10.10** shows chlorophyll *a* compared to one of these accessory pigments, **chlorophyll *b***. A slight structural difference between them is enough to cause the two pigments to absorb at slightly different wavelengths in the red and blue parts of the spectrum (see Figure 10.9a). As a result, chlorophyll *a* is blue green and chlorophyll *b* is olive green.

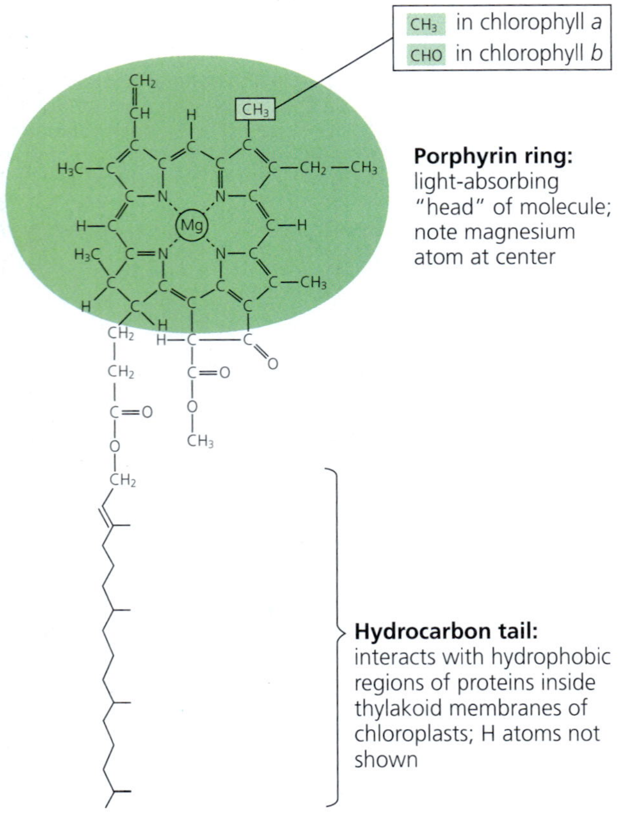

▲ **Figure 10.10 Structure of chlorophyll molecules in chloroplasts of plants.** Chlorophyll *a* and chlorophyll *b* differ only in one of the functional groups bonded to the organic structure called a porphyrin ring.

Other accessory pigments include **carotenoids**, hydrocarbons that are various shades of yellow and orange because they absorb violet and blue-green light (see Figure 10.9a). Carotenoids may broaden the spectrum of colors that can drive photosynthesis. However, a more important function of at least some carotenoids seems to be *photoprotection*: These compounds absorb and dissipate excessive light energy that would otherwise damage chlorophyll or interact with oxygen, forming reactive oxidative molecules that are dangerous to the cell. Interestingly, carotenoids similar to the photoprotective ones in chloroplasts have a photoprotective role in the human eye. These and related molecules, often found in health food products, are valued as "phytochemicals" (from the Greek *phyton*, plant), compounds with antioxidant properties. Plants can synthesize all the antioxidants they require, but humans and other animals must obtain some of them from their diets.

Excitation of Chlorophyll by Light

What exactly happens when chlorophyll and other pigments absorb light? The colors corresponding to the absorbed wavelengths disappear from the spectrum of the transmitted and reflected light, but energy cannot disappear. When a molecule absorbs a photon of light, one of the molecule's electrons is elevated to an orbital where it has more potential energy. When the electron is in its normal orbital, the pigment molecule is said to be in its ground state. Absorption of a photon boosts an electron to an orbital of higher energy, and the pigment molecule is then said to be in an excited state. The only photons absorbed are those whose energy is exactly equal to the energy difference between the ground state and an excited state, and this energy difference varies from one kind of molecule to another. Thus, a particular compound absorbs only photons corresponding to specific wavelengths, which is why each pigment has a unique absorption spectrum.

Once absorption of a photon raises an electron from the ground state to an excited state, the electron cannot remain there long. The excited state, like all high-energy states, is unstable. Generally, when isolated pigment molecules absorb light, their excited electrons drop back down to the ground-state orbital in a billionth of a second, releasing their excess energy as heat. This conversion of light energy to heat is what makes the top of an automobile so hot on a sunny day. (White cars are coolest because their paint reflects all wavelengths of visible light, although it may absorb ultraviolet and other invisible radiation.) In isolation, some pigments, including chlorophyll, emit light as well as heat after absorbing photons. As excited electrons fall back to the ground state, photons are given off. This afterglow is called

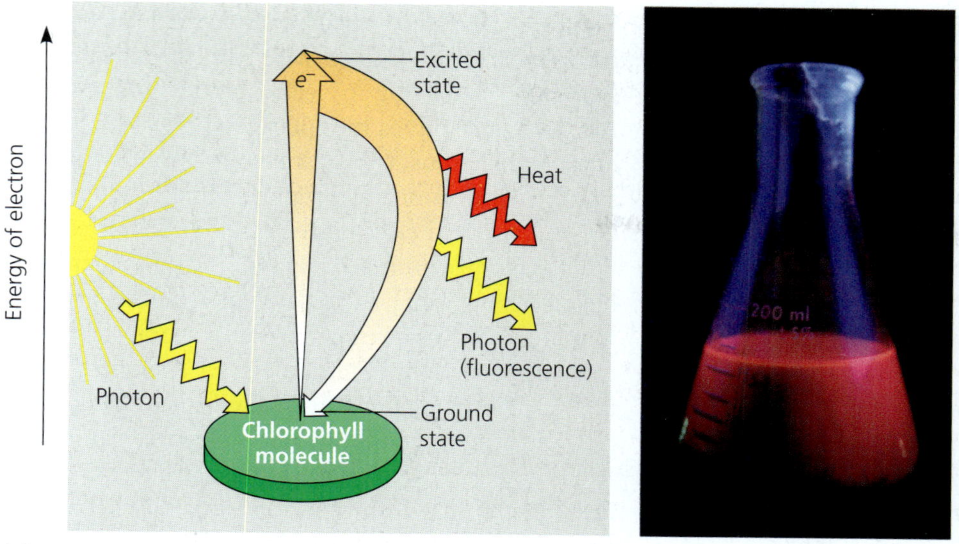

(a) Excitation of isolated chlorophyll molecule

(b) Fluorescence

◀ **Figure 10.11 Excitation of isolated chlorophyll by light. (a)** Absorption of a photon causes a transition of the chlorophyll molecule from its ground state to its excited state. The photon boosts an electron to an orbital where it has more potential energy. If the illuminated molecule exists in isolation, its excited electron immediately drops back down to the ground-state orbital, and its excess energy is given off as heat and fluorescence (light). **(b)** A chlorophyll solution excited with ultraviolet light fluoresces with a red-orange glow.

WHAT IF? *If a leaf containing a similar concentration of chlorophyll as the solution was exposed to the same ultraviolet light, no fluorescence would be seen. Explain the difference in fluorescence emission between the solution and the leaf.*

fluorescence. If a solution of chlorophyll isolated from chloroplasts is illuminated, it will fluoresce in the red-orange part of the spectrum and also give off heat **(Figure 10.11)**.

A Photosystem: A Reaction-Center Complex Associated with Light-Harvesting Complexes

Chlorophyll molecules excited by the absorption of light energy produce very different results in an intact chloroplast than they do in isolation (see Figure 10.11). In their native environment of the thylakoid membrane, chlorophyll molecules are organized along with other small organic molecules and proteins into photosystems.

A **photosystem** is composed of a protein complex called a **reaction-center complex** surrounded by several light-harvesting complexes **(Figure 10.12)**. The reaction-center complex includes a special pair of chlorophyll *a* molecules. Each **light-harvesting complex** consists of various pigment molecules (which may include chlorophyll *a*, chlorophyll *b*, and carotenoids) bound to proteins. The number and variety of pigment molecules enable a photosystem to harvest light over a larger surface and a larger portion of the spectrum than any single pigment molecule alone could. Together, these light-harvesting complexes act as an antenna for the reaction-center complex. When a pigment molecule absorbs a photon, the energy is transferred from pigment molecule to pigment molecule within a light-harvesting complex, somewhat like a human "wave" at a sports arena, until it is passed into the reaction-center complex. The reaction-center complex contains a molecule capable of accepting electrons and becoming reduced; it is called the **primary electron acceptor**. The pair of chlorophyll *a* molecules in the reaction-center complex are special because their molecular environment—their location and the other molecules with which they are associated—enables

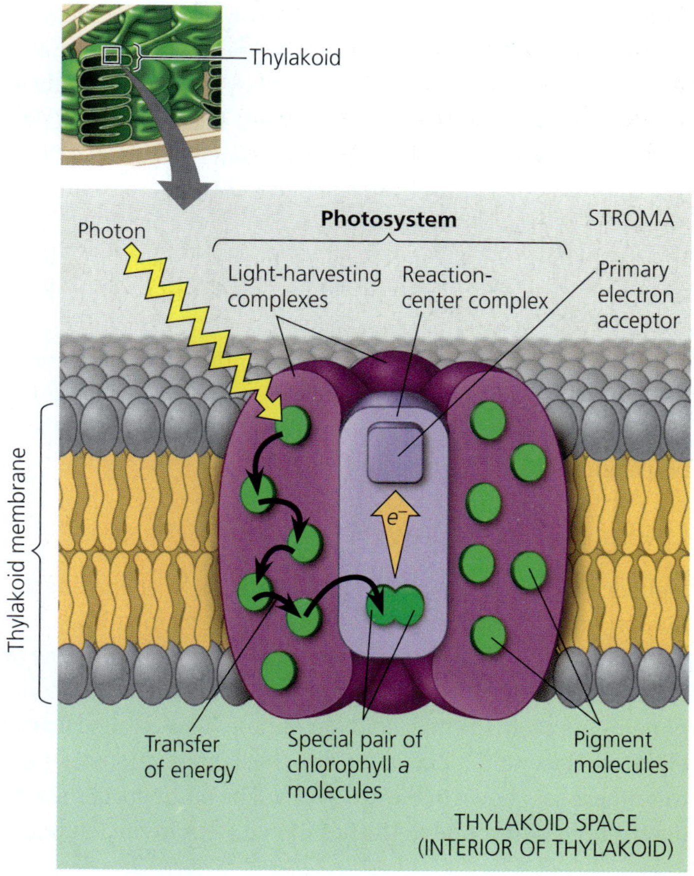

▲ **Figure 10.12 How a photosystem harvests light.** When a photon strikes a pigment molecule in a light-harvesting complex, the energy is passed from molecule to molecule until it reaches the reaction-center complex. Here, an excited electron from the special pair of chlorophyll *a* molecules is transferred to the primary electron acceptor.

them to use the energy from light not only to boost one of their electrons to a higher energy level, but also to transfer it to a different molecule—the primary electron acceptor.

The solar-powered transfer of an electron from the reaction-center chlorophyll *a* pair to the primary electron acceptor is the first step of the light reactions. As soon as the chlorophyll electron is excited to a higher energy level, the primary electron acceptor captures it; this is a redox reaction. Isolated chlorophyll fluoresces because there is no electron acceptor, so electrons of photoexcited chlorophyll drop right back to the ground state. In a chloroplast, the potential energy represented by the excited electron is not lost. Thus, each photosystem—a reaction-center complex surrounded by light-harvesting complexes—functions in the chloroplast as a unit. It converts light energy to chemical energy, which will ultimately be used for the synthesis of sugar.

The thylakoid membrane is populated by two types of photosystems that cooperate in the light reactions of photosynthesis. They are called **photosystem II (PS II)** and **photosystem I (PS I)**. (They were named in order of their discovery, but photosystem II functions first in the light reactions.) Each has a characteristic reaction-center complex—a particular kind of primary electron acceptor next to a special pair of chlorophyll *a* molecules associated with specific proteins. The reaction-center chlorophyll *a* of photosystem II is known as P680 because this pigment is best at absorbing light having a wavelength of 680 nm (in the red part of the spectrum). The chlorophyll *a* at the reaction-center complex of photosystem I is called P700 because it most effectively absorbs light of wavelength 700 nm (in the far-red part of the spectrum). These two pigments, P680 and P700, are nearly identical chlorophyll *a* molecules. However, their association with different proteins in the thylakoid membrane affects the electron distribution in the two pigments and accounts for the slight differences in their light-absorbing properties. Now let's see how the two photosystems work together in using light energy to generate ATP and NADPH, the two main products of the light reactions.

Linear Electron Flow

Light drives the synthesis of ATP and NADPH by energizing the two photosystems embedded in the thylakoid membranes of chloroplasts. The key to this energy transformation is a flow of electrons through the photosystems and other molecular components built into the thylakoid membrane. This is called **linear electron flow**, and it occurs during the light reactions of photosynthesis, as shown in **Figure 10.13**. The numbers in the text description correspond to the numbered steps in the figure.

1. A photon of light strikes a pigment molecule in a light-harvesting complex, boosting one of its electrons to a higher energy level. As this electron falls back to its ground state, an electron in a nearby pigment molecule is simultaneously raised to an excited state. The process continues, with the energy being relayed to other pigment molecules until it reaches the P680 pair of chlorophyll *a* molecules in the PS II reaction-center complex. It excites an electron in this pair of chlorophylls to a higher energy state.
2. This electron is transferred from the excited P680 to the primary electron acceptor. We can refer to the resulting form of P680, missing an electron, as $P680^+$.
3. An enzyme catalyzes the splitting of a water molecule into two electrons, two hydrogen ions, and an oxygen atom. The electrons are supplied one by one to the $P680^+$ pair, each electron replacing one transferred to the primary electron acceptor. ($P680^+$ is the strongest biological oxidizing agent known; its electron "hole" must be filled. This greatly facilitates the transfer of electrons from the split water molecule.) The oxygen atom immediately combines with an oxygen atom generated by the splitting of another water molecule, forming O_2.
4. Each photoexcited electron passes from the primary electron acceptor of PS II to PS I via an electron transport chain, the components of which are similar to those of the electron transport chain that functions in cellular respiration. The electron transport chain between PS II and PS I is made up of the electron carrier plastoquinone (Pq), a cytochrome complex, and a protein called plastocyanin (Pc).
5. The exergonic "fall" of electrons to a lower energy level provides energy for the synthesis of ATP. As electrons pass through the cytochrome complex, the pumping of protons builds a proton gradient that is subsequently used in chemiosmosis.
6. Meanwhile, light energy was transferred via light-harvesting complex pigments to the PS I reaction-center complex, exciting an electron of the P700 pair of chlorophyll *a* molecules located there. The photoexcited electron was then transferred to PS I's primary electron acceptor, creating an electron "hole" in the P700—which we now can call $P700^+$. In other words, $P700^+$ can now act as an electron acceptor, accepting an electron that reaches the bottom of the electron transport chain from PS II.
7. Photoexcited electrons are passed in a series of redox reactions from the primary electron acceptor of PS I down a second electron transport chain through the protein ferredoxin (Fd). (This chain does not create a proton gradient and thus does not produce ATP.)
8. The enzyme $NADP^+$ reductase catalyzes the transfer of electrons from Fd to $NADP^+$. Two electrons are required for its reduction to NADPH. This molecule is at a higher energy level than water, and its electrons are more readily available for the reactions of the Calvin cycle than were those of water.

As complicated as the scheme shown in Figure 10.13 is, do not lose track of its functions. The light reactions use solar power to generate ATP and NADPH, which provide chemical energy and reducing power, respectively, to the carbohydrate-synthesizing reactions of the Calvin cycle. The energy changes of electrons as they flow through the light reactions are shown in a mechanical analogy in **Figure 10.14**.

▼ **Figure 10.13 How linear electron flow during the light reactions generates ATP and NADPH.** The gold arrows trace the current of light-driven electrons from water to NADPH.

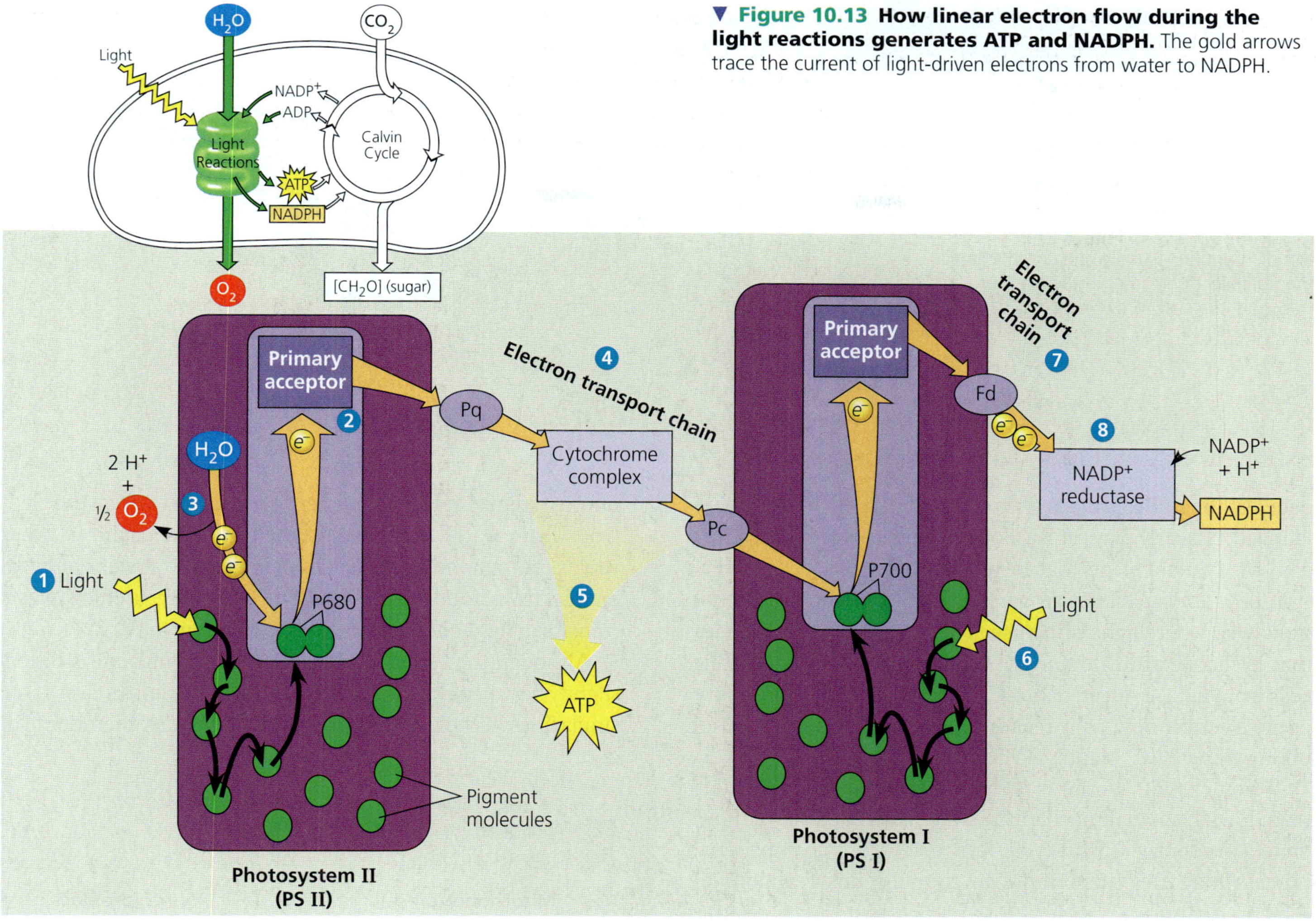

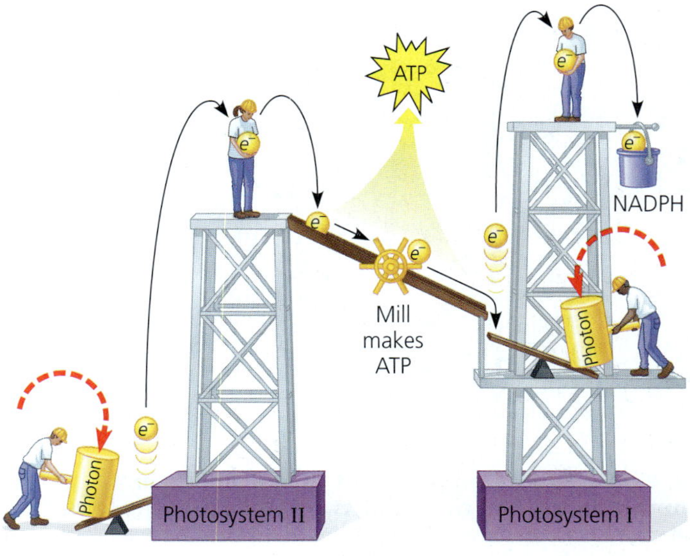

▲ **Figure 10.14 A mechanical analogy for the light reactions.**

Cyclic Electron Flow

In certain cases, photoexcited electrons can take an alternative path called **cyclic electron flow**, which uses photosystem I but not photosystem II. You can see in **Figure 10.15**, on the next page, that cyclic flow is a short circuit: The electrons cycle back from ferredoxin (Fd) to the cytochrome complex and from there continue on to a P700 chlorophyll in the PS I reaction-center complex. There is no production of NADPH and no release of oxygen. Cyclic flow does, however, generate ATP.

Several of the currently existing groups of photosynthetic bacteria are known to have photosystem I but not photosystem II; for these species, which include the purple sulfur bacteria (see Figure 10.2e), cyclic electron flow is the sole means of generating ATP in photosynthesis. Evolutionary biologists believe that these bacterial groups are descendants of the bacteria in which photosynthesis first evolved, in a form similar to cyclic electron flow.

Cyclic electron flow can also occur in photosynthetic species that possess both photosystems; this includes some prokaryotes, such as the cyanobacteria shown in Figure 10.2d, as well as the eukaryotic photosynthetic species that have been tested to

▶ **Figure 10.15 Cyclic electron flow.** Photoexcited electrons from PS I are occasionally shunted back from ferredoxin (Fd) to chlorophyll via the cytochrome complex and plastocyanin (Pc). This electron shunt supplements the supply of ATP (via chemiosmosis) but produces no NADPH. The "shadow" of linear electron flow is included in the diagram for comparison with the cyclic route. The two ferredoxin molecules shown in this diagram are actually one and the same—the final electron carrier in the electron transport chain of PS I.

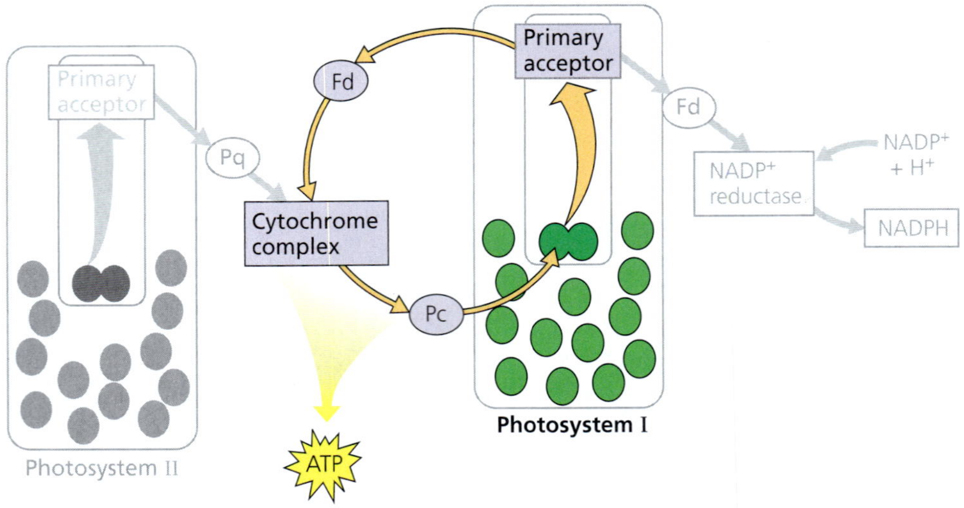

date. Although the process is probably in part an "evolutionary leftover," it clearly plays at least one beneficial role for these organisms. Mutant plants that are not able to carry out cyclic electron flow are capable of growing well in low light, but do not grow well where light is intense. This is evidence for the idea that cyclic electron flow may be photoprotective, protecting cells from light-induced damage. Later you'll learn more about cyclic electron flow as it relates to a particular adaptation of photosynthesis (C_4 plants; see Concept 10.4).

Whether ATP synthesis is driven by linear or cyclic electron flow, the actual mechanism is the same. Before we move on to consider the Calvin cycle, let's review chemiosmosis, the process that uses membranes to couple redox reactions to ATP production.

A Comparison of Chemiosmosis in Chloroplasts and Mitochondria

Chloroplasts and mitochondria generate ATP by the same basic mechanism: chemiosmosis. An electron transport chain assembled in a membrane pumps protons across the membrane as electrons are passed through a series of carriers that are progressively more electronegative. In this way, electron transport chains transform redox energy to a proton-motive force, potential energy stored in the form of an H^+ gradient across a membrane. Built into the same membrane is an ATP synthase complex that couples the diffusion of hydrogen ions down their gradient to the phosphorylation of ADP. Some of the electron carriers, including the iron-containing proteins called cytochromes, are very similar in chloroplasts and mitochondria. The ATP synthase complexes of the two organelles are also very much alike. But there are noteworthy differences between oxidative phosphorylation in mitochondria and photophosphorylation in chloroplasts. In mitochondria, the high-energy electrons dropped down the transport chain are extracted from organic molecules (which are thus oxidized), while in chloroplasts, the source of electrons is water. Chloroplasts do not need molecules from food to make ATP; their photosystems capture light energy and use it to drive the electrons from water to the top of the transport chain. In other words, mitochondria use chemiosmosis to transfer chemical energy from food molecules to ATP, whereas chloroplasts transform light energy into chemical energy in ATP.

Although the spatial organization of chemiosmosis differs slightly between chloroplasts and mitochondria, it is easy to see similarities in the two **(Figure 10.16)**. The inner membrane of

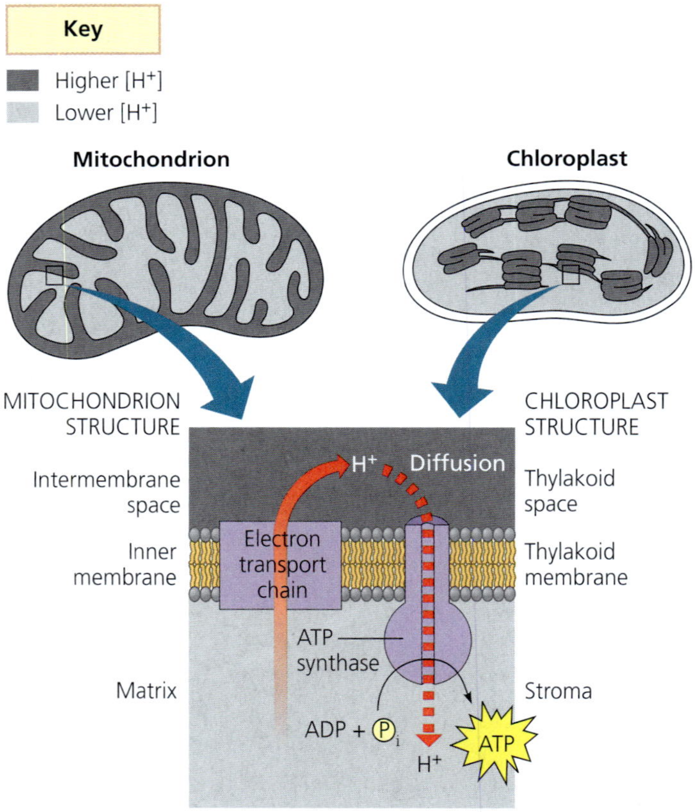

▲ **Figure 10.16 Comparison of chemiosmosis in mitochondria and chloroplasts.** In both kinds of organelles, electron transport chains pump protons (H^+) across a membrane from a region of low H^+ concentration (light gray in this diagram) to one of high H^+ concentration (dark gray). The protons then diffuse back across the membrane through ATP synthase, driving the synthesis of ATP.

the mitochondrion pumps protons from the mitochondrial matrix out to the intermembrane space, which then serves as a reservoir of hydrogen ions. The thylakoid membrane of the chloroplast pumps protons from the stroma into the thylakoid space (interior of the thylakoid), which functions as the H^+ reservoir. If you imagine the cristae of mitochondria pinching off from the inner membrane, this may help you see how the thylakoid space and the intermembrane space are comparable spaces in the two organelles, while the mitochondrial matrix is analogous to the stroma of the chloroplast. In the mitochondrion, protons diffuse down their concentration gradient from the intermembrane space through ATP synthase to the matrix, driving ATP synthesis. In the chloroplast, ATP is synthesized as the hydrogen ions diffuse from the thylakoid space back to the stroma through ATP synthase complexes, whose catalytic knobs are on the stroma side of the membrane. Thus, ATP forms in the stroma, where it is used to help drive sugar synthesis during the Calvin cycle **(Figure 10.17)**.

The proton (H^+) gradient, or pH gradient, across the thylakoid membrane is substantial. When chloroplasts in an experimental setting are illuminated, the pH in the thylakoid space drops to about 5 (the H^+ concentration increases), and the pH in the stroma increases to about 8 (the H^+ concentration decreases). This gradient of three pH units corresponds to a thousandfold

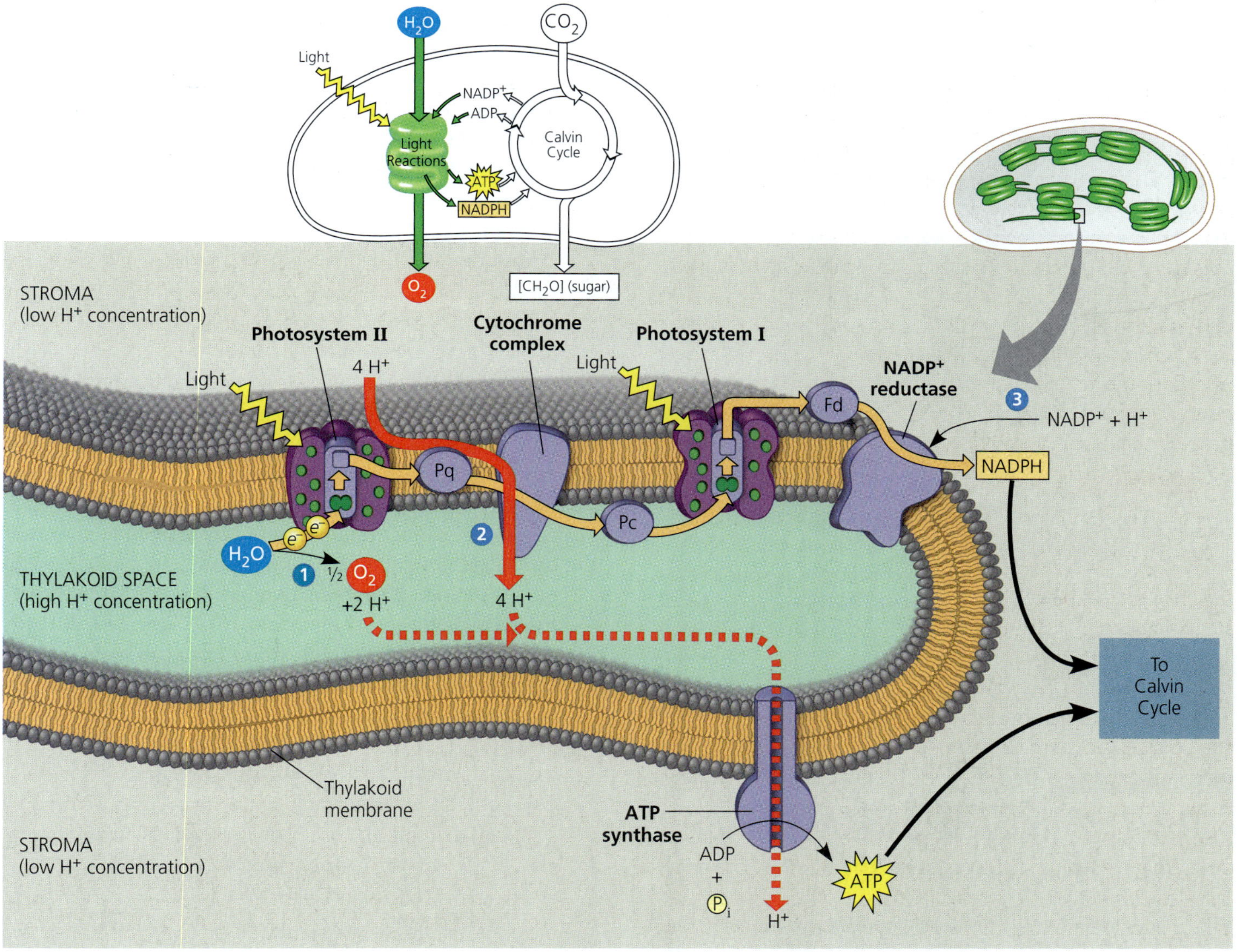

▲ **Figure 10.17 The light reactions and chemiosmosis: the organization of the thylakoid membrane.** This diagram shows a current model for the organization of the thylakoid membrane. The gold arrows track the linear electron flow outlined in Figure 10.13. As electrons pass from carrier to carrier in redox reactions, hydrogen ions removed from the stroma are deposited in the thylakoid space, storing energy as a proton-motive force (H^+ gradient). At least three steps in the light reactions contribute to the proton gradient: ❶ Water is split by photosystem II on the side of the membrane facing the thylakoid space; ❷ as plastoquinone (Pq), a mobile carrier, transfers electrons to the cytochrome complex, four protons are translocated across the membrane into the thylakoid space; and ❸ a hydrogen ion is removed from the stroma when it is taken up by $NADP^+$. Notice how, as in Figure 10.16, hydrogen ions are being pumped from the stroma into the thylakoid space. The diffusion of H^+ from the thylakoid space back to the stroma (along the H^+ concentration gradient) powers the ATP synthase. These light-driven reactions store chemical energy in NADPH and ATP, which shuttle the energy to the carbohydrate-producing Calvin cycle.

difference in H^+ concentration. If in the laboratory the lights are turned off, the pH gradient is abolished, but it can quickly be restored by turning the lights back on. Experiments such as this provided strong evidence in support of the chemiosmotic model.

Based on studies in several laboratories, Figure 10.17 shows a current model for the organization of the light-reaction "machinery" within the thylakoid membrane. Each of the molecules and molecular complexes in the figure is present in numerous copies in each thylakoid. Notice that NADPH, like ATP, is produced on the side of the membrane facing the stroma, where the Calvin cycle reactions take place.

Let's summarize the light reactions. Electron flow pushes electrons from water, where they are at a low state of potential energy, ultimately to NADPH, where they are stored at a high state of potential energy. The light-driven electron current also generates ATP. Thus, the equipment of the thylakoid membrane converts light energy to chemical energy stored in ATP and NADPH. (Oxygen is a by-product.) Let's now see how the Calvin cycle uses the products of the light reactions to synthesize sugar from CO_2.

CONCEPT CHECK 10.2

1. What color of light is *least* effective in driving photosynthesis? Explain.
2. Compared to a solution of isolated chlorophyll, why do intact chloroplasts release less heat and fluorescence when illuminated?
3. In the light reactions, what is the initial electron donor? Where do the electrons end up?
4. **WHAT IF?** In an experiment, isolated chloroplasts placed in a solution with the appropriate components can carry out ATP synthesis. Predict what would happen to the rate of synthesis if a compound is added to the solution that makes membranes freely permeable to hydrogen ions.

For suggested answers, see Appendix A.

CONCEPT 10.3
The Calvin cycle uses ATP and NADPH to convert CO_2 to sugar

The Calvin cycle is similar to the citric acid cycle in that a starting material is regenerated after molecules enter and leave the cycle. However, while the citric acid cycle is catabolic, oxidizing glucose and using the energy to synthesize ATP, the Calvin cycle is anabolic, building carbohydrates from smaller molecules and consuming energy. Carbon enters the Calvin cycle in the form of CO_2 and leaves in the form of sugar. The cycle spends ATP as an energy source and consumes NADPH as reducing power for adding high-energy electrons to make the sugar.

As we mentioned previously, the carbohydrate produced directly from the Calvin cycle is actually not glucose, but a three-carbon sugar; the name of this sugar is **glyceraldehyde-3-phosphate (G3P)**. For the net synthesis of one molecule of G3P, the cycle must take place three times, fixing three molecules of CO_2. (Recall that carbon fixation refers to the initial incorporation of CO_2 into organic material.) As we trace the steps of the cycle, keep in mind that we are following three molecules of CO_2 through the reactions. **Figure 10.18** divides the Calvin cycle into three phases: carbon fixation, reduction, and regeneration of the CO_2 acceptor.

Phase 1: Carbon fixation. The Calvin cycle incorporates each CO_2 molecule, one at a time, by attaching it to a five-carbon sugar named ribulose bisphosphate (abbreviated RuBP). The enzyme that catalyzes this first step is RuBP carboxylase, or **rubisco.** (This is the most abundant protein in chloroplasts and is also said to be the most abundant protein on Earth.) The product of the reaction is a six-carbon intermediate so unstable that it immediately splits in half, forming two molecules of 3-phosphoglycerate (for each CO_2 fixed).

Phase 2: Reduction. Each molecule of 3-phosphoglycerate receives an additional phosphate group from ATP, becoming 1,3-bisphosphoglycerate. Next, a pair of electrons donated from NADPH reduces 1,3-bisphosphoglycerate, which also loses a phosphate group, becoming G3P. Specifically, the electrons from NADPH reduce a carboyxl group on 1,3-bisphosphoglycerate to the aldehyde group of G3P, which stores more potential energy. G3P is a sugar—the same three-carbon sugar formed in glycolysis by the splitting of glucose (see Figure 9.9). Notice in Figure 10.18 that for every *three* molecules of CO_2 that enter the cycle, there are *six* molecules of G3P formed. But only one molecule of this three-carbon sugar can be counted as a net gain of carbohydrate. The cycle began with 15 carbons' worth of carbohydrate in the form of three molecules of the five-carbon sugar RuBP. Now there are 18 carbons' worth of carbohydrate in the form of six molecules of G3P. One molecule exits the cycle to be used by the plant cell, but the other five molecules must be recycled to regenerate the three molecules of RuBP.

Phase 3: Regeneration of the CO_2 acceptor (RuBP). In a complex series of reactions, the carbon skeletons of five molecules of G3P are rearranged by the last steps of the Calvin cycle into three molecules of RuBP. To accomplish this, the cycle spends three more molecules of ATP. The RuBP is now prepared to receive CO_2 again, and the cycle continues.

For the net synthesis of one G3P molecule, the Calvin cycle consumes a total of nine molecules of ATP and six molecules

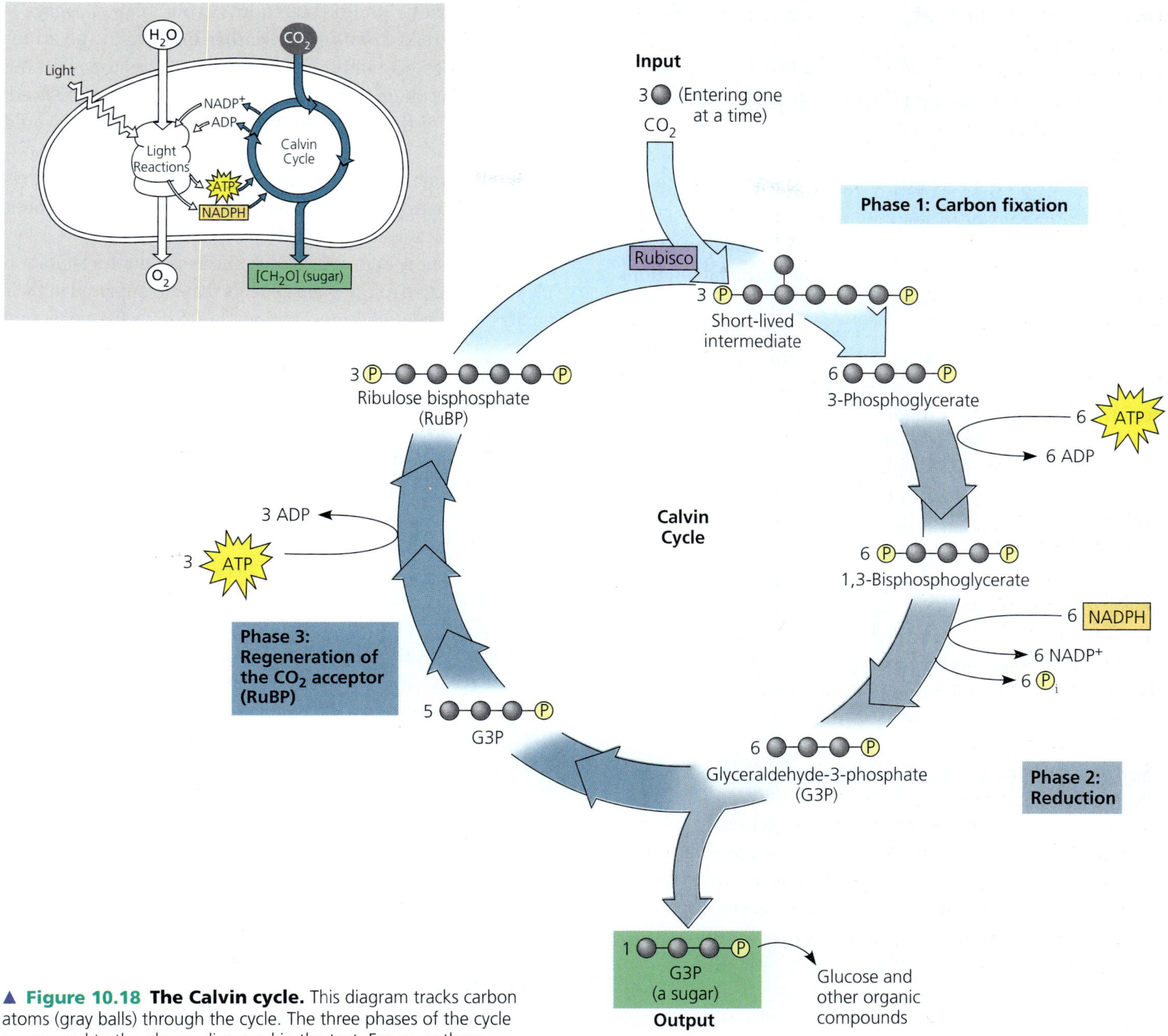

▲ **Figure 10.18 The Calvin cycle.** This diagram tracks carbon atoms (gray balls) through the cycle. The three phases of the cycle correspond to the phases discussed in the text. For every three molecules of CO_2 that enter the cycle, the net output is one molecule of glyceraldehyde-3-phosphate (G3P), a three-carbon sugar. The light reactions sustain the Calvin cycle by regenerating ATP and NADPH.

DRAW IT *Redraw this cycle using numerals to indicate the numbers of carbons instead of gray balls, multiplying at each step to ensure that you have accounted for all carbons. In what forms do the carbon atoms enter and leave the cycle?*

of NADPH. The light reactions regenerate the ATP and NADPH. The G3P spun off from the Calvin cycle becomes the starting material for metabolic pathways that synthesize other organic compounds, including glucose and other carbohydrates. Neither the light reactions nor the Calvin cycle alone can make sugar from CO_2. Photosynthesis is an emergent property of the intact chloroplast, which integrates the two stages of photosynthesis.

CONCEPT CHECK 10.3

1. To synthesize one glucose molecule, the Calvin cycle uses ________ molecules of CO_2, ________ molecules of ATP, and ________ molecules of NADPH.
2. Explain why the large numbers of ATP and NADPH molecules used during the Calvin cycle are consistent with the high value of glucose as an energy source.
3. **WHAT IF?** Explain why a poison that inhibits an enzyme of the Calvin cycle will also inhibit the light reactions.

For suggested answers, see Appendix A.

CONCEPT 10.4

Alternative mechanisms of carbon fixation have evolved in hot, arid climates

Ever since plants first moved onto land about 475 million years ago, they have been adapting to the problems of terrestrial life, particularly the problem of dehydration. In Chapters 29 and 36, we will consider anatomical adaptations that help plants conserve water. Here we are concerned with metabolic adaptations. The solutions often involve trade-offs. An important example is the compromise between photosynthesis and the prevention of excessive water loss from the plant. The CO_2 required for photosynthesis enters a leaf via stomata, the pores through the leaf surface (see Figure 10.3). However, stomata are also the main avenues of transpiration, the evaporative loss of water from leaves. On a hot, dry day, most plants close their stomata, a response that conserves water. This response also reduces photosynthetic yield by limiting access to CO_2. With stomata even partially closed, CO_2 concentrations begin to decrease in the air spaces within the leaf, and the concentration of O_2 released from the light reactions begins to increase. These conditions within the leaf favor an apparently wasteful process called photorespiration.

Photorespiration: An Evolutionary Relic?

In most plants, initial fixation of carbon occurs via rubisco, the Calvin cycle enzyme that adds CO_2 to ribulose bisphosphate. Such plants are called **C_3 plants** because the first organic product of carbon fixation is a three-carbon compound, 3-phosphoglycerate (see Figure 10.18). Rice, wheat, and soybeans are C_3 plants that are important in agriculture. When their stomata partially close on hot, dry days, C_3 plants produce less sugar because the declining level of CO_2 in the leaf starves the Calvin cycle. In addition, rubisco can bind O_2 in place of CO_2. As CO_2 becomes scarce within the air spaces of the leaf, rubisco adds O_2 to the Calvin cycle instead of CO_2. The product splits, and a two-carbon compound leaves the chloroplast. Peroxisomes and mitochondria rearrange and split this compound, releasing CO_2. The process is called **photorespiration** because it occurs in the light (*photo*) and consumes O_2 while producing CO_2 (*respiration*). However, unlike normal cellular respiration, photorespiration generates no ATP; in fact, photorespiration consumes ATP. And unlike photosynthesis, photorespiration produces no sugar. In fact, photorespiration *decreases* photosynthetic output by siphoning organic material from the Calvin cycle and releasing CO_2 that would otherwise be fixed.

How can we explain the existence of a metabolic process that seems to be counterproductive for the plant? According to one hypothesis, photorespiration is evolutionary baggage—a metabolic relic from a much earlier time when the atmosphere had less O_2 and more CO_2 than it does today. In the ancient atmosphere that prevailed when rubisco first evolved, the inability of the enzyme's active site to exclude O_2 would have made little difference. The hypothesis suggests that modern rubisco retains some of its chance affinity for O_2, which is now so concentrated in the atmosphere that a certain amount of photorespiration is inevitable.

We now know that, at least in some cases, photorespiration plays a protective role in plants. Plants that are impaired in their ability to carry out photorespiration (due to defective genes) are more susceptible to damage induced by excess light. Researchers consider this clear evidence that photorespiration acts to neutralize the otherwise damaging products of the light reactions, which build up when a low CO_2 concentration limits the progress of the Calvin cycle. Whether there are other benefits of photorespiration is still unknown. In many types of plants—including a significant number of crop plants—photorespiration drains away as much as 50% of the carbon fixed by the Calvin cycle. As heterotrophs that depend on carbon fixation in chloroplasts for our food, we naturally view photorespiration as wasteful. Indeed, if photorespiration could be reduced in certain plant species without otherwise affecting photosynthetic productivity, crop yields and food supplies might increase.

In some plant species, alternate modes of carbon fixation have evolved that minimize photorespiration and optimize the Calvin cycle—even in hot, arid climates. The two most important of these photosynthetic adaptations are C_4 photosynthesis and CAM.

C_4 Plants

The **C_4 plants** are so named because they preface the Calvin cycle with an alternate mode of carbon fixation that forms a four-carbon compound as its first product. Several thousand species in at least 19 plant families use the C_4 pathway. Among the C_4 plants important to agriculture are sugarcane and corn, members of the grass family.

A unique leaf anatomy is correlated with the mechanism of C_4 photosynthesis (**Figure 10.19**; compare with Figure 10.3). In C_4 plants, there are two distinct types of photosynthetic cells: bundle-sheath cells and mesophyll cells. **Bundle-sheath cells** are arranged into tightly packed sheaths around the veins of the leaf. Between the bundle sheath and the leaf surface are the more loosely arranged **mesophyll cells**. The Calvin cycle is confined to the chloroplasts of the bundle-sheath cells. However, the cycle is preceded by incorporation of CO_2 into organic compounds in the mesophyll cells (see the numbered steps in Figure 10.19). (1) The first step is carried out by an enzyme present only in mesophyll cells called **PEP carboxylase**. This enzyme adds CO_2 to phosphoenolpyruvate (PEP), forming the four-carbon product oxaloacetate. PEP carboxylase

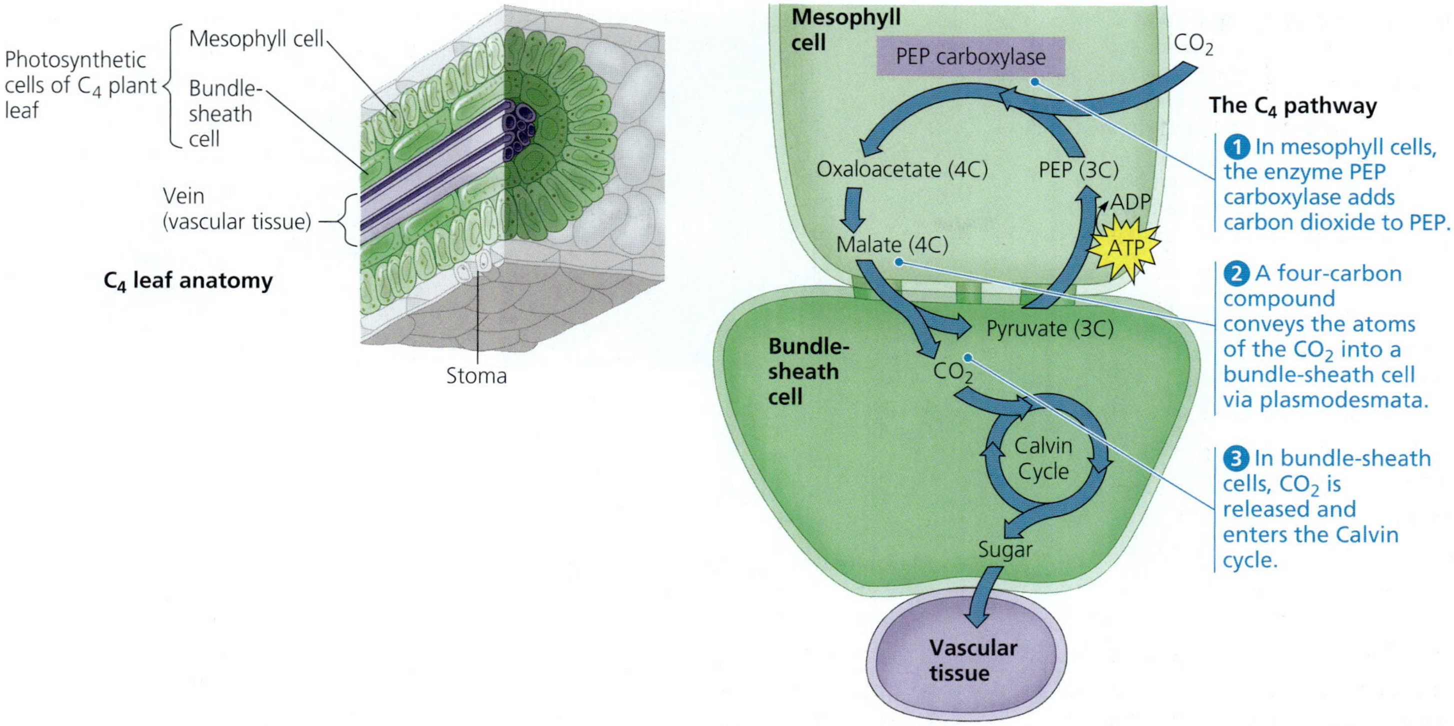

▲ **Figure 10.19 C_4 leaf anatomy and the C_4 pathway.** The structure and biochemical functions of the leaves of C_4 plants are an evolutionary adaptation to hot, dry climates. This adaptation maintains a CO_2 concentration in the bundle sheath that favors photosynthesis over photorespiration.

has a much higher affinity for CO_2 than does rubisco and no affinity for O_2. Therefore, PEP carboxylase can fix carbon efficiently when rubisco cannot—that is, when it is hot and dry and stomata are partially closed, causing CO_2 concentration in the leaf to fall and O_2 concentration to rise. 2 After the C_4 plant fixes carbon from CO_2, the mesophyll cells export their four-carbon products (malate in the example shown in Figure 10.19) to bundle-sheath cells through plasmodesmata (see Figure 6.31). 3 Within the bundle-sheath cells, the four-carbon compounds release CO_2, which is reassimilated into organic material by rubisco and the Calvin cycle. The same reaction regenerates pyruvate, which is transported to mesophyll cells. There, ATP is used to convert pyruvate to PEP, allowing the reaction cycle to continue; this ATP can be thought of as the "price" of concentrating CO_2 in the bundle-sheath cells. To generate this extra ATP, bundle-sheath cells carry out cyclic electron flow, the process described earlier in this chapter (see Figure 10.15). In fact, these cells contain PS I but no PS II, so cyclic electron flow is their only photosynthetic mode of generating ATP.

In effect, the mesophyll cells of a C_4 plant pump CO_2 into the bundle sheath, keeping the CO_2 concentration in the bundle-sheath cells high enough for rubisco to bind carbon dioxide rather than oxygen. The cyclic series of reactions involving PEP carboxylase and the regeneration of PEP can be thought of as a CO_2-concentrating pump that is powered by ATP. In this way, C_4 photosynthesis minimizes photorespiration and enhances sugar production. This adaptation is especially advantageous in hot regions with intense sunlight, where stomata partially close during the day, and it is in such environments that C_4 plants evolved and thrive today.

CAM Plants

A second photosynthetic adaptation to arid conditions has evolved in many succulent (water-storing) plants, numerous cacti, pineapples, and representatives of several other plant families. These plants open their stomata during the night and close them during the day, just the reverse of how other plants behave. Closing stomata during the day helps desert plants conserve water, but it also prevents CO_2 from entering the leaves. During the night, when their stomata are open, these plants take up CO_2 and incorporate it into a variety of organic acids. This mode of carbon fixation is called **crassulacean acid metabolism**, or **CAM**, after the plant family Crassulaceae, the succulents in which the process was first discovered. The mesophyll cells of **CAM plants** store the organic acids they make during the night in their vacuoles until morning, when the stomata close. During the day, when the light reactions can supply ATP and NADPH for the Calvin cycle, CO_2 is released from the organic acids made the night before to become incorporated into sugar in the chloroplasts.

▶ **Figure 10.20 C_4 and CAM photosynthesis compared.** Both adaptations are characterized by ❶ preliminary incorporation of CO_2 into organic acids, followed by ❷ transfer of CO_2 to the Calvin cycle. The C_4 and CAM pathways are two evolutionary solutions to the problem of maintaining photosynthesis with stomata partially or completely closed on hot, dry days.

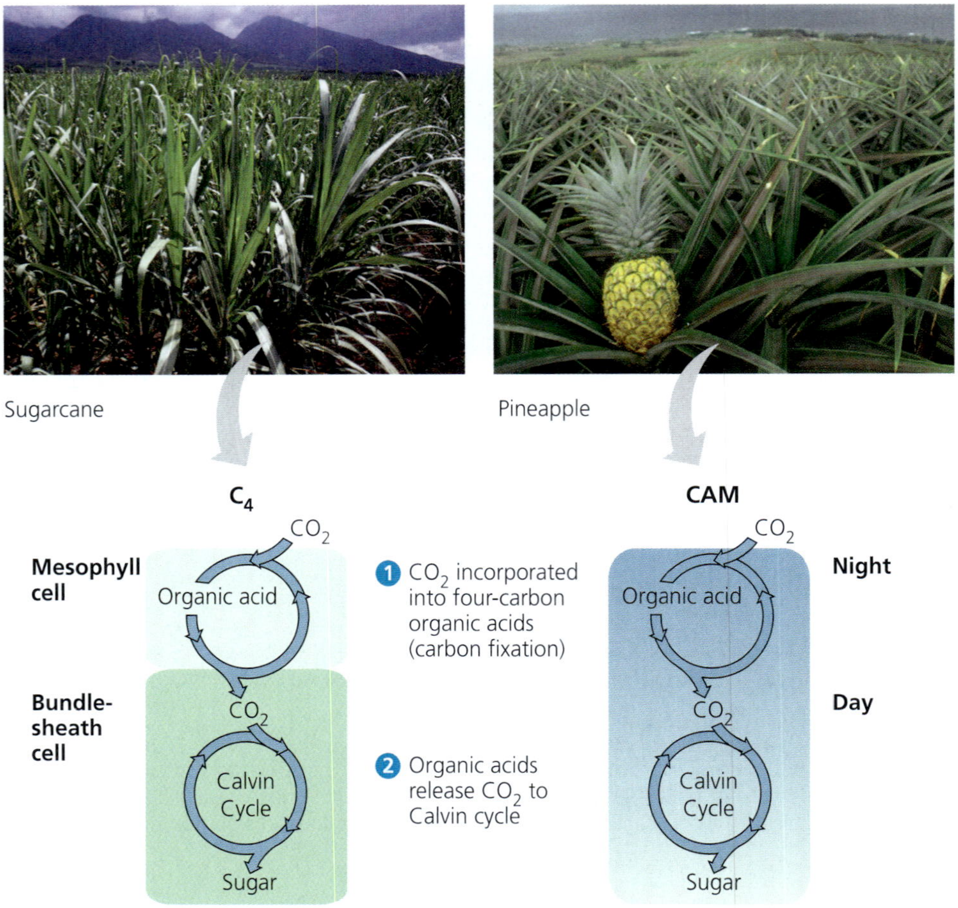

(a) Spatial separation of steps. In C_4 plants, carbon fixation and the Calvin cycle occur in different types of cells.

(b) Temporal separation of steps. In CAM plants, carbon fixation and the Calvin cycle occur in the same cells at different times.

Notice in **Figure 10.20** that the CAM pathway is similar to the C_4 pathway in that carbon dioxide is first incorporated into organic intermediates before it enters the Calvin cycle. The difference is that in C_4 plants, the initial steps of carbon fixation are separated structurally from the Calvin cycle, whereas in CAM plants, the two steps occur at separate times but within the same cell. (Keep in mind that CAM, C_4, and C_3 plants all eventually use the Calvin cycle to make sugar from carbon dioxide.)

CONCEPT CHECK 10.4

1. Explain why photorespiration lowers photosynthetic output for plants.
2. The presence of only PS I, not PS II, in the bundle-sheath cells of C_4 plants has an effect on O_2 concentration. What is that effect, and how might that benefit the plant?
3. **WHAT IF?** How would you expect the relative abundance of C_3 versus C_4 and CAM species to change in a geographic region whose climate becomes much hotter and drier?

For suggested answers, see Appendix A.

The Importance of Photosynthesis: *A Review*

In this chapter, we have followed photosynthesis from photons to food. The light reactions capture solar energy and use it to make ATP and transfer electrons from water to $NADP^+$, forming NADPH. The Calvin cycle uses the ATP and NADPH to produce sugar from carbon dioxide. The energy that enters the chloroplasts as sunlight becomes stored as chemical energy in organic compounds. See **Figure 10.21** for a review of the entire process.

What are the fates of photosynthetic products? The sugar made in the chloroplasts supplies the entire plant with chemical energy and carbon skeletons for the synthesis of all the major organic molecules of plant cells. About 50% of the organic material made by photosynthesis is consumed as fuel for cellular respiration in the mitochondria of the plant cells. Sometimes there is a loss of photosynthetic products to photorespiration.

Technically, green cells are the only autotrophic parts of the plant. The rest of the plant depends on organic molecules exported from leaves via veins. In most plants, carbohydrate is transported out of the leaves in the form of

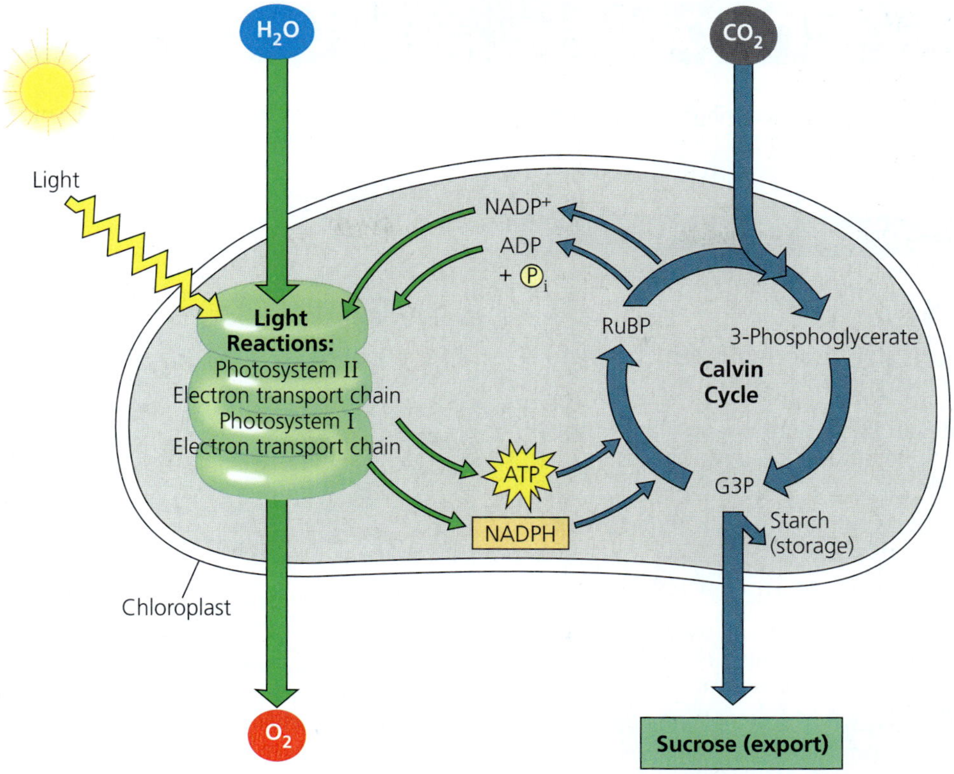

Light Reactions:

- Are carried out by molecules in the thylakoid membranes
- Convert light energy to the chemical energy of ATP and NADPH
- Split H_2O and release O_2 to the atmosphere

Calvin Cycle Reactions:

- Take place in the stroma
- Use ATP and NADPH to convert CO_2 to the sugar G3P
- Return ADP, inorganic phosphate, and $NADP^+$ to the light reactions

▲ **Figure 10.21 A review of photosynthesis.** This diagram outlines the main reactants and products of the light reactions and the Calvin cycle as they occur in the chloroplasts of plant cells. The entire ordered operation depends on the structural integrity of the chloroplast and its membranes. Enzymes in the chloroplast and cytosol convert glyceraldehyde-3-phosphate (G3P), the direct product of the Calvin cycle, to many other organic compounds.

sucrose, a disaccharide. After arriving at nonphotosynthetic cells, the sucrose provides raw material for cellular respiration and a multitude of anabolic pathways that synthesize proteins, lipids, and other products. A considerable amount of sugar in the form of glucose is linked together to make the polysaccharide cellulose, especially in plant cells that are still growing and maturing. Cellulose, the main ingredient of cell walls, is the most abundant organic molecule in the plant—and probably on the surface of the planet.

Most plants manage to make more organic material each day than they need to use as respiratory fuel and precursors for biosynthesis. They stockpile the extra sugar by synthesizing starch, storing some in the chloroplasts themselves and some in storage cells of roots, tubers, seeds, and fruits. In accounting for the consumption of the food molecules produced by photosynthesis, let's not forget that most plants lose leaves, roots, stems, fruits, and sometimes their entire bodies to heterotrophs, including humans.

On a global scale, photosynthesis is the process responsible for the presence of oxygen in our atmosphere. Furthermore, in terms of food production, the collective productivity of the minuscule chloroplasts is prodigious: Photosynthesis makes an estimated 160 billion metric tons of carbohydrate per year (a metric ton is 1,000 kg, about 1.1 tons). That's organic matter equivalent in mass to a stack of about 60 trillion copies of this textbook—17 stacks of books reaching from Earth to the sun! No other chemical process on the planet can match the output of photosynthesis. And no process is more important than photosynthesis to the welfare of life on Earth.

Chapter 10 Review

MEDIA Go to the Study Area at **www.masteringbio.com** for BioFlix 3-D Animations, MP3 Tutors, Videos, Practice Tests, an eBook, and more.

SUMMARY OF KEY CONCEPTS

CONCEPT 10.1

Photosynthesis converts light energy to the chemical energy of food (pp. 186–189)

- **Chloroplasts: The Sites of Photosynthesis in Plants** In autotrophic eukaryotes, photosynthesis occurs in chloroplasts, organelles containing thylakoids. Stacks of thylakoids form grana.
- **Tracking Atoms Through Photosynthesis: *Scientific Inquiry*** Photosynthesis is summarized as

$$6\,CO_2 + 12\,H_2O + \text{Light energy} \rightarrow C_6H_{12}O_6 + 6\,O_2 + 6\,H_2O$$

Chloroplasts split water into hydrogen and oxygen, incorporating the electrons of hydrogen into sugar molecules. Photosynthesis is a redox process: H_2O is oxidized, CO_2 is reduced.

- **The Two Stages of Photosynthesis: *A Preview*** The light reactions in the thylakoid membranes split water, releasing O_2, producing ATP, and forming NADPH. The Calvin cycle in the stroma forms sugar from CO_2, using ATP for energy and NADPH for reducing power.

MEDIA

BioFlix 3-D Animation Photosynthesis
MP3 Tutor Photosynthesis
Activity The Sites of Photosynthesis
Activity Overview of Photosynthesis

CONCEPT 10.2

The light reactions convert solar energy to the chemical energy of ATP and NADPH (pp. 190–198)

- **The Nature of Sunlight** Light is a form of electromagnetic energy. The colors we see as visible light include those wavelengths that drive photosynthesis.
- **Photosynthetic Pigments: The Light Receptors** A pigment absorbs visible light of specific wavelengths. Chlorophyll *a* is the main photosynthetic pigment in plants. Other accessory pigments absorb different wavelengths of light and pass the energy on to chlorophyll *a*.
- **Excitation of Chlorophyll by Light** A pigment goes from a ground state to an excited state when a photon boosts one of its electrons to a higher-energy orbital. This excited state is unstable. Electrons from isolated pigments tend to fall back to the ground state, giving off heat and/or light.
- **A Photosystem: A Reaction-Center Complex Associated with Light-Harvesting Complexes** A photosystem is composed of a reaction-center complex surrounded by light-harvesting complexes that funnel the energy of photons to the reaction-center complex. When a special pair of reaction-center chlorophyll *a* molecules absorbs energy, one of its electrons is boosted to a higher energy level and transferred to the primary electron acceptor. Photosystem II contains P680 chlorophyll *a* molecules in the reaction-center complex; photosystem I contains P700 molecules.
- **Linear Electron Flow** The flow of electrons during the light reactions produces NADPH, ATP, and oxygen:

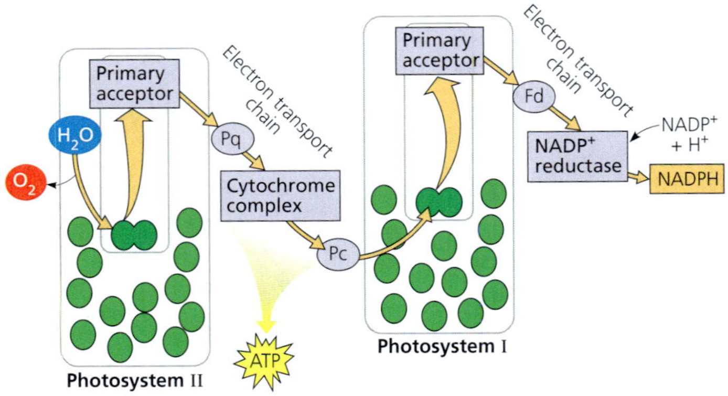

- **Cyclic Electron Flow** Cyclic electron flow employs only photosystem I, producing ATP but no NADPH or O_2.
- **A Comparison of Chemiosmosis in Chloroplasts and Mitochondria** In both organelles, redox reactions of electron transport chains generate an H^+ gradient across a membrane. ATP synthase uses this proton-motive force to make ATP.

MEDIA

Activity Light Energy and Pigments
Investigation How Does Paper Chromatography Separate Plant Pigments?
Activity The Light Reactions

CONCEPT 10.3

The Calvin cycle uses ATP and NADPH to convert CO_2 to sugar (pp. 198–199)

- The Calvin cycle occurs in the stroma, using electrons from NADPH and energy from ATP. One molecule of G3P exits the cycle per three CO_2 molecules fixed and is converted to glucose and other organic molecules.

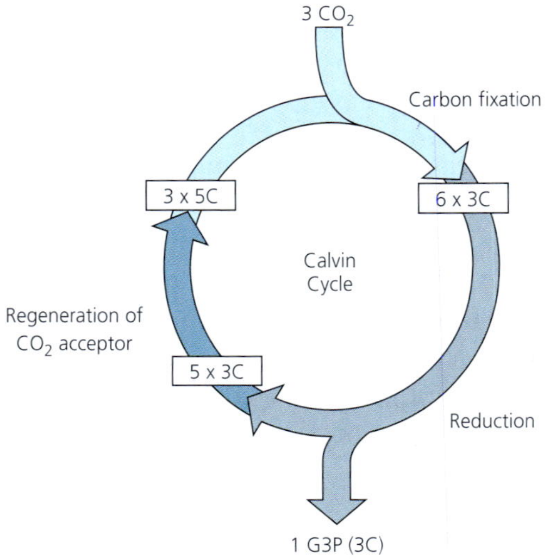

MEDIA

Activity The Calvin Cycle
Investigation How Is the Rate of Photosynthesis Measured?
Biology Labs On-Line LeafLab

CONCEPT 10.4

Alternative mechanisms of carbon fixation have evolved in hot, arid climates (pp. 200–202)

- **Photorespiration: An Evolutionary Relic?** On dry, hot days, C_3 plants close their stomata, conserving water. Oxygen from the light reactions builds up. In photorespiration, O_2 substitutes for CO_2 in the active site of rubisco. This process consumes organic fuel and releases CO_2 without producing ATP or carbohydrate.
- **C_4 Plants** C_4 plants minimize the cost of photorespiration by incorporating CO_2 into four-carbon compounds in mesophyll cells. These compounds are exported to bundle-sheath cells, where they release carbon dioxide for use in the Calvin cycle.
- **CAM Plants** CAM plants open their stomata at night, incorporating CO_2 into organic acids, which are stored in mesophyll cells. During the day, the stomata close, and the CO_2 is released from the organic acids for use in the Calvin cycle.

MEDIA

Activity Photosynthesis in Dry Climates

- **The Importance of Photosynthesis: *A Review*** Organic compounds produced by photosynthesis provide the energy and building material for ecosystems.

TESTING YOUR KNOWLEDGE

SELF-QUIZ

1. The light reactions of photosynthesis supply the Calvin cycle with
   a. light energy.
   b. CO_2 and ATP.
   c. H_2O and NADPH.
   d. ATP and NADPH.
   e. sugar and O_2.

2. Which of the following sequences correctly represents the flow of electrons during photosynthesis?
   a. NADPH → O_2 → CO_2
   b. H_2O → NADPH → Calvin cycle
   c. NADPH → chlorophyll → Calvin cycle
   d. H_2O → photosystem I → photosystem II
   e. NADPH → electron transport chain → O_2

3. In *mechanism*, photophosphorylation is most similar to
   a. substrate-level phosphorylation in glycolysis.
   b. oxidative phosphorylation in cellular respiration.
   c. the Calvin cycle.
   d. carbon fixation.
   e. reduction of $NADP^+$.

4. How is photosynthesis similar in C_4 plants and CAM plants?
   a. In both cases, only photosystem I is used.
   b. Both types of plants make sugar without the Calvin cycle.
   c. In both cases, rubisco is not used to fix carbon initially.
   d. Both types of plants make most of their sugar in the dark.
   e. In both cases, thylakoids are not involved in photosynthesis.

5. Which process is most directly driven by light energy?
   a. creation of a pH gradient by pumping protons across the thylakoid membrane
   b. carbon fixation in the stroma
   c. reduction of $NADP^+$ molecules
   d. removal of electrons from chlorophyll molecules
   e. ATP synthesis

6. Which of the following statements is a correct distinction between autotrophs and heterotrophs?
   a. Only heterotrophs require chemical compounds from the environment.
   b. Cellular respiration is unique to heterotrophs.
   c. Only heterotrophs have mitochondria.
   d. Autotrophs, but not heterotrophs, can nourish themselves beginning with CO_2 and other nutrients that are inorganic.
   e. Only heterotrophs require oxygen.

7. Which of the following does *not* occur during the Calvin cycle?
   a. carbon fixation
   b. oxidation of NADPH
   c. release of oxygen
   d. regeneration of the CO_2 acceptor
   e. consumption of ATP

For Self-Quiz answers, see Appendix A.

MEDIA Visit the Study Area at **www.masteringbio.com** for a Practice Test.

EVOLUTION CONNECTION

8. Photorespiration can decrease soybeans' photosynthetic output by about 50%. Would you expect this figure to be higher or lower in wild relatives of soybeans? Why?

SCIENTIFIC INQUIRY

9. **DRAW IT** The following diagram represents an experiment with isolated chloroplasts. The chloroplasts were first made acidic by soaking them in a solution at pH 4. After the thylakoid space reached pH 4, the chloroplasts were transferred to a basic solution at pH 8. The chloroplasts then made ATP in the dark.

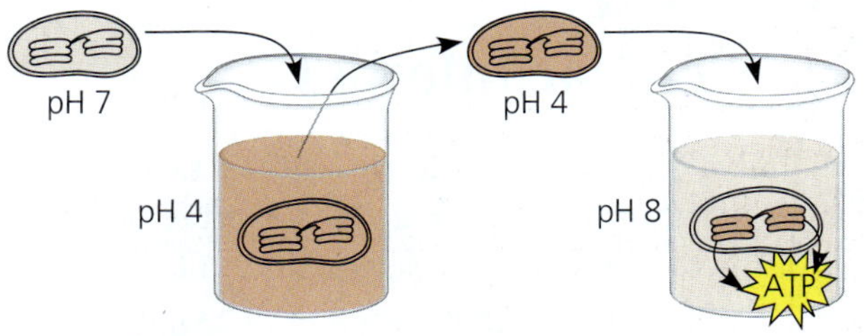

Draw an enlargement of part of the thylakoid membrane in the beaker with the solution at pH 8. Draw ATP synthase. Label the areas of high H^+ concentration and low H^+ concentration. Show the direction protons flow through the enzyme, and show the reaction where ATP is synthesized. Would ATP end up in the thylakoid or outside of it? Explain why the chloroplasts in the experiment were able to make ATP in the dark.

SCIENCE, TECHNOLOGY, AND SOCIETY

10. Scientific evidence indicates that the CO_2 added to the air by the burning of wood and fossil fuels is contributing to "global warming," a rise in global temperature. Tropical rain forests are estimated to be responsible for more than 20% of global photosynthesis, yet their consumption of large amounts of CO_2 is thought to make little or no *net* contribution to reduction of global warming. Why might this be? (*Hint:* What happens to the food produced by a rain forest tree when it is eaten by animals or the tree dies?)

11 Cell Communication

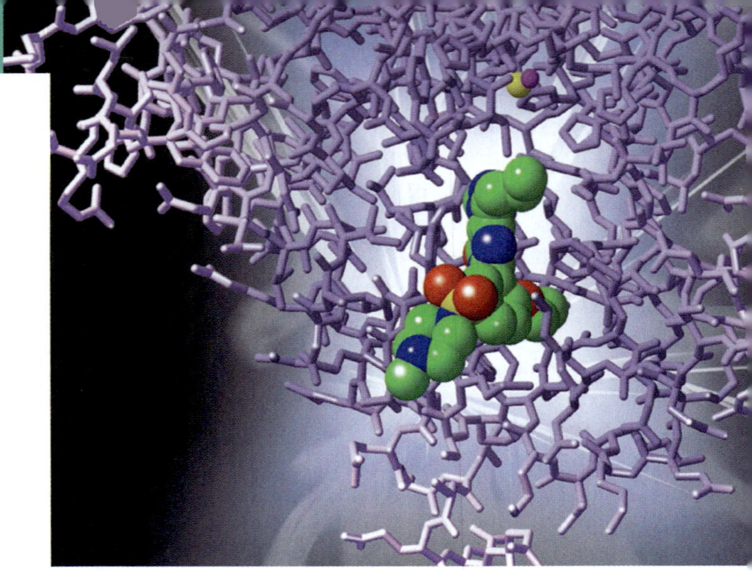

▲ **Figure 11.1 How do the effects of Viagra (multicolored) result from its inhibition of a signaling-pathway enzyme (purple)?**

KEY CONCEPTS

11.1 External signals are converted to responses within the cell

11.2 Reception: A signaling molecule binds to a receptor protein, causing it to change shape

11.3 Transduction: Cascades of molecular interactions relay signals from receptors to target molecules in the cell

11.4 Response: Cell signaling leads to regulation of transcription or cytoplasmic activities

11.5 Apoptosis (programmed cell death) integrates multiple cell-signaling pathways

OVERVIEW

The Cellular Internet

A hiker slips and falls down a steep ravine, injuring her leg in the fall. Tragedy is averted when she is able to pull out a cell phone and call for help. Cell phones, the Internet, e-mail, instant messaging—no one would deny the importance of communication in our lives. The role of communication in life at the cellular level is equally critical. Cell-to-cell communication is absolutely essential for multicellular organisms such as humans and oak trees. The trillions of cells in a multicellular organism must communicate with each other to coordinate their activities in a way that enables the organism to develop from a fertilized egg, then survive and reproduce in turn. Communication between cells is also important for many unicellular organisms. Networks of communication between cells can be even more complicated than the World Wide Web.

In studying how cells signal to each other and how they interpret the signals they receive, biologists have discovered some universal mechanisms of cellular regulation, additional evidence for the evolutionary relatedness of all life. The same small set of cell-signaling mechanisms shows up again and again in many lines of biological research—from embryonic development to hormone action to cancer. In one example, a common cell-to-cell signaling pathway leads to dilation of blood vessels. Once the signal subsides, the response is shut down by the enzyme shown in purple in **Figure 11.1**. Also shown is a multicolored molecule that blocks the action of this enzyme and keeps blood vessels dilated. Enzyme-inhibiting compounds like this one are often prescribed for treatment of medical conditions. The action of the multicolored compound, known as Viagra, will be discussed later in the chapter. The signals received by cells, whether originating from other cells or from changes in the physical environment, take various forms, including light and touch. However, cells most often communicate with each other by chemical signals. In this chapter, we focus on the main mechanisms by which cells receive, process, and respond to chemical signals sent from other cells. At the end, we will take a look at *apoptosis*, a type of programmed cell death that integrates input from multiple signaling pathways.

CONCEPT **11.1**

External signals are converted to responses within the cell

What does a "talking" cell say to a "listening" cell, and how does the latter cell respond to the message? Let's approach these questions by first looking at communication among microorganisms, for modern microbes are a window on the role of cell signaling in the evolution of life on Earth.

Evolution of Cell Signaling

One topic of cell "conversation" is sex—at least for the yeast *Saccharomyces cerevisiae*, which people have used for millennia to make bread, wine, and beer. Researchers have learned that cells of this yeast identify their mates by chemical signaling. There are two sexes, or mating types, called **a** and **α** **(Figure 11.2)**. Cells of mating type **a** secrete a signaling molecule called **a** factor, which can bind to specific receptor proteins on nearby **α** cells. At the same time, **α** cells secrete **α** factor, which binds to receptors on **a** cells. Without actually entering the cells, the two mating factors cause the cells to grow toward each other and also bring about other cellular changes. The result is the fusion, or mating, of two cells of opposite type. The new **a**/**α** cell contains all the genes of both original cells, a combination of genetic resources that provides advantages to the cell's descendants, which arise by subsequent cell divisions.

How is the mating signal at the yeast cell surface changed, or *transduced*, into a form that brings about the cellular response of mating? The process by which a signal on a cell's surface is converted to a specific cellular response is a series of steps called a **signal transduction pathway**. Many such pathways have been extensively studied in both yeast and animal cells. Amazingly, the molecular details of signal transduction in yeast and mammals are strikingly similar, even though the last common ancestor of these two groups of organisms lived over a billion years ago. These similarities—and others more recently uncovered between signaling systems in bacteria and plants—suggest that early versions of the cell-signaling mechanisms used today evolved well before the first multicellular creatures appeared on Earth.

Scientists think that signaling mechanisms first evolved in ancient prokaryotes and single-celled eukaryotes and then were adopted for new uses by their multicellular descendants. Meanwhile, cell signaling has remained important in the microbial world. Cells of many bacterial species secrete small molecules that can be detected by other bacterial cells. The concentration of such signaling molecules allows bacteria to sense the local density of bacterial cells, a phenomenon called *quorum sensing*. Furthermore, signaling among members of a bacterial population can lead to coordination of their activities. In response to the signal, bacterial cells are able to come together and form *biofilms*, aggregations of bacteria that often form recognizable structures containing regions of specialized function. **Figure 11.3** shows an aggregation response characteristic of one type of bacterium.

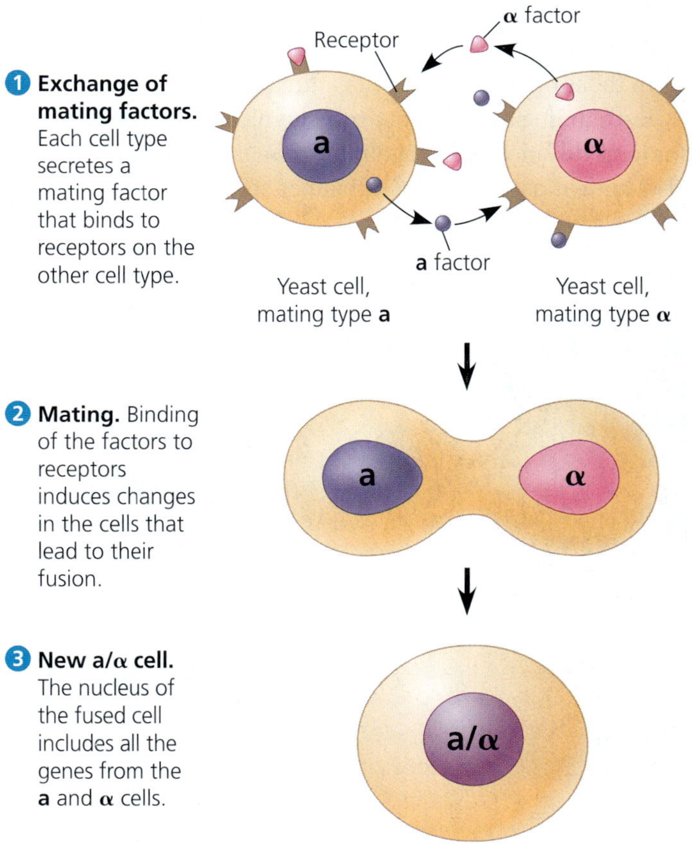

▲ **Figure 11.2 Communication between mating yeast cells.** *Saccharomyces cerevisiae* cells use chemical signaling to identify cells of opposite mating type and initiate the mating process. The two mating types and their corresponding chemical signaling molecules, or mating factors, are called **a** and **α**.

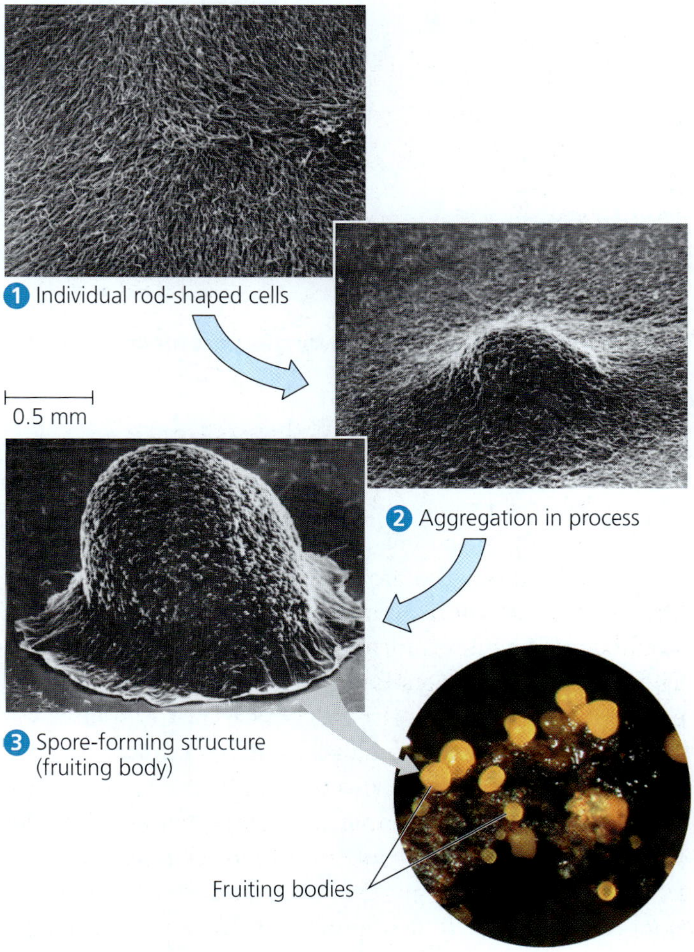

▲ **Figure 11.3 Communication among bacteria.** Soil-dwelling bacteria called myxobacteria ("slime bacteria") use chemical signals to share information about nutrient availability. When food is scarce, starving cells secrete a molecule that reaches neighboring cells and stimulates them to aggregate. The cells form a structure, called a fruiting body, that produces thick-walled spores capable of surviving until the environment improves. The bacteria shown here are *Myxococcus xanthus* (steps 1–3, SEMs; lower photo, LM).

Local and Long-Distance Signaling

Like yeast cells, cells in a multicellular organism usually communicate via chemical messengers targeted for cells that may or may not be immediately adjacent. As we saw in Chapters 6 and 7, cells may communicate by direct contact **(Figure 11.4)**. Both animals and plants have cell junctions that, where present, directly connect the cytoplasms of adjacent cells **(Figure 11.4a)**. In these cases, signaling substances dissolved in the cytosol can pass freely between adjacent cells. Moreover, animal cells may communicate via direct contact between membrane-bound cell-surface molecules, which occurs during a process called cell-cell recognition **(Figure 11.4b)**. This sort of signaling is important in such processes as embryonic development and the immune response.

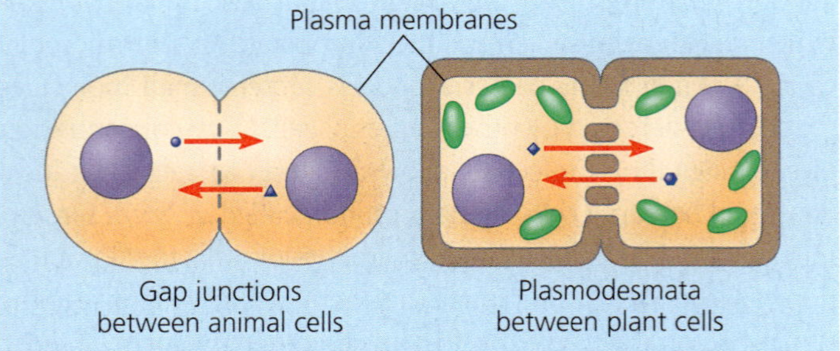

(a) Cell junctions. Both animals and plants have cell junctions that allow molecules to pass readily between adjacent cells without crossing plasma membranes.

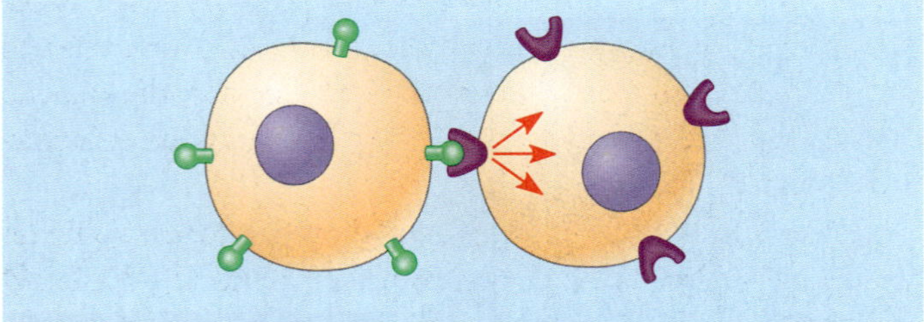

(b) Cell-cell recognition. Two cells in an animal may communicate by interaction between molecules protruding from their surfaces.

▲ **Figure 11.4 Communication by direct contact between cells.**

In many other cases, messenger molecules are secreted by the signaling cell. Some of these travel only short distances; such **local regulators** influence cells in the vicinity. One class of local regulators in animals, *growth factors*, consists of compounds that stimulate nearby target cells to grow and divide. Numerous cells can simultaneously receive and respond to the molecules of growth factor produced by a single cell in their vicinity. This type of local signaling in animals is called *paracrine signaling* **(Figure 11.5a)**.

Another, more specialized type of local signaling called *synaptic signaling* occurs in the animal nervous system **(Figure 11.5b)**. An electrical signal along a nerve cell triggers the secretion of a chemical signal carried by neurotransmitter molecules. These diffuse across the synapse, the narrow space between the nerve cell and its target cell (often another nerve cell). The neurotransmitter stimulates the target cell.

Local signaling in plants is not as well understood. Because of their cell walls, plants use mechanisms somewhat different from those operating locally in animals.

Both animals and plants use chemicals called **hormones** for long-distance signaling. In hormonal signaling in animals, also known as endocrine signaling, specialized cells release

Local signaling

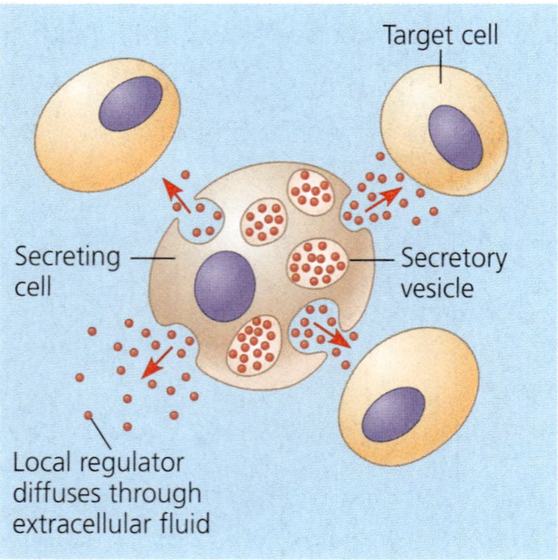

(a) Paracrine signaling. A secreting cell acts on nearby target cells by discharging molecules of a local regulator (a growth factor, for example) into the extracellular fluid.

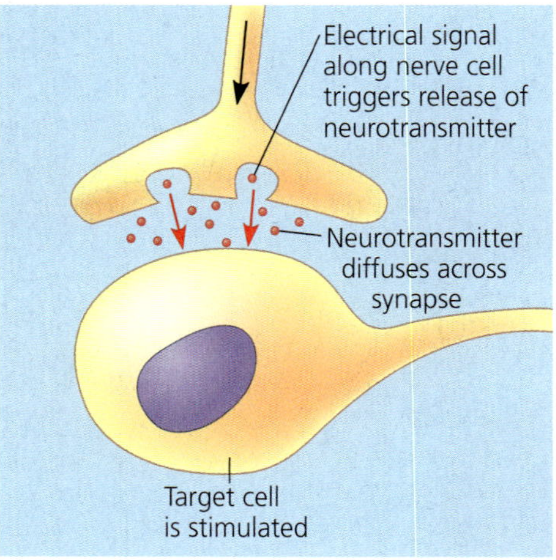

(b) Synaptic signaling. A nerve cell releases neurotransmitter molecules into a synapse, stimulating the target cell.

Long-distance signaling

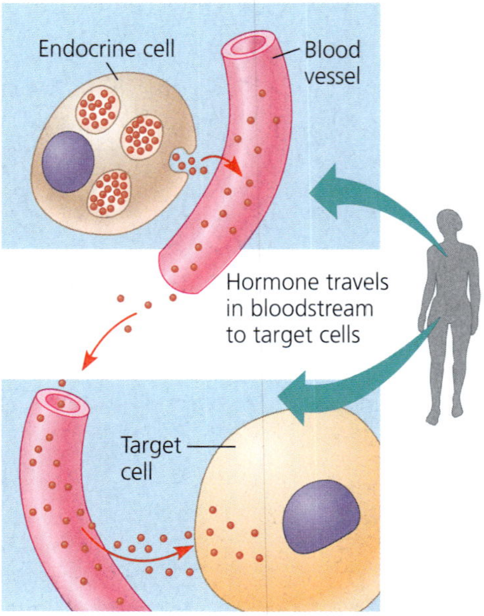

(c) Hormonal signaling. Specialized endocrine cells secrete hormones into body fluids, often the blood. Hormones may reach virtually all body cells.

▲ **Figure 11.5 Local and long-distance cell communication in animals.** In both local and long-distance signaling, only specific target cells recognize and respond to a given signaling molecule.

hormone molecules, which travel via the circulatory system to target cells in other parts of the body **(Figure 11.5c)**. Plant hormones (often called *plant growth regulators*) sometimes travel in vessels but more often reach their targets by moving through cells or by diffusing through the air as a gas (see Chapter 39). Hormones vary widely in molecular size and type, as do local regulators. For instance, the plant hormone ethylene, a gas that promotes fruit ripening and helps regulate growth, is a hydrocarbon of only six atoms (C_2H_4), small enough to pass through cell walls. In contrast, the mammalian hormone insulin, which regulates sugar levels in the blood, is a protein with thousands of atoms.

The transmission of a signal through the nervous system can also be considered an example of long-distance signaling. An electrical signal travels the length of a nerve cell and is then converted back to a chemical signal when a signaling molecule is released and crosses the synapse to another nerve cell. Here it is converted back to an electrical signal. In this way, a nerve signal can travel along a series of nerve cells. Because some nerve cells are quite long, the nerve signal can quickly travel great distances—from your brain to your big toe, for example. This type of long-distance signaling will be covered in detail in Chapter 48.

What happens when a cell encounters a signaling molecule? The molecule must be recognized by a specific receptor molecule, and the information it carries, the signal, must be changed into another form—transduced—inside the cell before the cell can respond. The remainder of the chapter discusses this process, primarily as it occurs in animal cells.

The Three Stages of Cell Signaling: *A Preview*

Our current understanding of how chemical messengers act via signal transduction pathways had its origins in the pioneering work of Earl W. Sutherland, whose research led to a Nobel Prize in 1971. Sutherland and his colleagues at Vanderbilt University were investigating how the animal hormone epinephrine stimulates the breakdown of the storage polysaccharide glycogen within liver cells and skeletal muscle cells. Glycogen breakdown releases the sugar glucose-1-phosphate, which the cell converts to glucose-6-phosphate. The cell (a liver cell, for example) can then use this compound, an early intermediate in glycolysis, for energy production. Alternatively, the compound can be stripped of phosphate and released from the liver cell into the blood as glucose, which can fuel cells throughout the body. Thus, one effect of epinephrine, which is secreted from the adrenal gland during times of physical or mental stress, is the mobilization of fuel reserves.

Sutherland's research team discovered that epinephrine stimulates glycogen breakdown by somehow activating a cytosolic enzyme, glycogen phosphorylase. However, when epinephrine was added to a test-tube mixture containing the enzyme and its substrate, glycogen, no breakdown occurred. Epinephrine could activate glycogen phosphorylase only when the hormone was added to a solution containing *intact* cells. This result told Sutherland two things. First, epinephrine does not interact directly with the enzyme responsible for glycogen breakdown; an intermediate step or series of steps must be occurring inside the cell. Second, the plasma membrane is somehow involved in transmitting the epinephrine signal.

Sutherland's early work suggested that the process going on at the receiving end of a cellular conversation can be dissected into three stages: reception, transduction, and response **(Figure 11.6)**:

1. **Reception.** Reception is the target cell's detection of a signaling molecule coming from outside the cell. A chemical signal is "detected" when the signaling molecule binds to a receptor protein located at the cell's surface or inside the cell.
2. **Transduction.** The binding of the signaling molecule changes the receptor protein in some way, initiating the process of transduction. The transduction stage converts the signal to a form that can bring about a specific

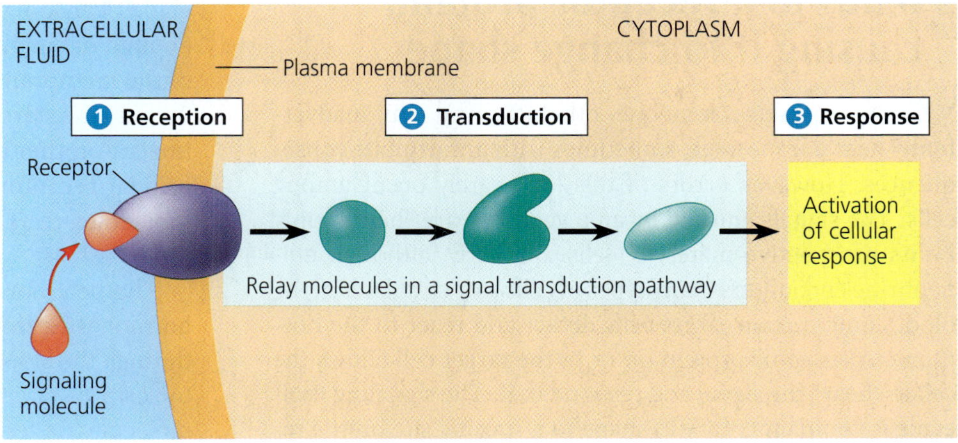

▶ **Figure 11.6 Overview of cell signaling.** From the perspective of the cell receiving the message, cell signaling can be divided into three stages: signal reception, signal transduction, and cellular response. When reception occurs at the plasma membrane, as shown here, the transduction stage is usually a pathway of several steps, with each relay molecule in the pathway bringing about a change in the next molecule. The final molecule in the pathway triggers the cell's response. The three stages are explained in more detail in the text.

? *How does the epinephrine in Sutherland's experiment fit into this diagram of cell signaling?*

cellular response. In Sutherland's system, the binding of epinephrine to a receptor protein in a liver cell's plasma membrane leads to activation of glycogen phosphorylase. Transduction sometimes occurs in a single step but more often requires a sequence of changes in a series of different molecules—a *signal transduction pathway.* The molecules in the pathway are often called relay molecules.

3. **Response.** In the third stage of cell signaling, the transduced signal finally triggers a specific cellular response. The response may be almost any imaginable cellular activity—such as catalysis by an enzyme (for example, glycogen phosphorylase), rearrangement of the cytoskeleton, or activation of specific genes in the nucleus. The cell-signaling process helps ensure that crucial activities like these occur in the right cells, at the right time, and in proper coordination with the other cells of the organism. We'll now explore the mechanisms of cell signaling in more detail.

CONCEPT CHECK **11.1**

1. Explain how signaling is involved in ensuring that yeast cells only fuse with cells of the opposite mating type.
2. Explain how nerve cells provide examples of both local and long-distance signaling.
3. When epinephrine is mixed with glycogen phosphorylase and glycogen in a test tube, is glucose-1-phosphate generated? Why or why not?
4. **WHAT IF?** In liver cells, glycogen phosphorylase acts in which of the three stages of the signaling pathway associated with an epinephrine-initiated signal?

For suggested answers, see Appendix A.

CONCEPT **11.2**

Reception: A signaling molecule binds to a receptor protein, causing it to change shape

When we speak to someone, others nearby may inadvertently hear our message, sometimes with unfortunate consequences. However, errors of this kind rarely occur among cells. The signals emitted by an **a** yeast cell are "heard" only by its prospective mates, **α** cells. Similarly, although epinephrine encounters many types of cells as it circulates in the blood, only certain target cells detect and react to the hormone. A receptor protein on or in the target cell allows the cell to "hear" the signal and respond to it. The signaling molecule is complementary in shape to a specific site on the receptor and attaches there, like a key in a lock or a substrate in the catalytic site of an enzyme. The signaling molecule behaves as a **ligand**, the term for a molecule that specifically binds to another molecule, often a larger one. Ligand binding generally causes a receptor protein to undergo a change in shape. For many receptors, this shape change directly activates the receptor, enabling it to interact with other cellular molecules. For other kinds of receptors, the immediate effect of ligand binding is to cause the aggregation of two or more receptor molecules, which leads to further molecular events inside the cell.

In a general way, ligand binding is similar to the binding of an allosteric regulator to an enzyme, causing a shape change that either promotes or inhibits enzyme activity. In the case of signal transduction, binding of the ligand alters the ability of the receptor to transmit the signal.

Most signal receptors are plasma membrane proteins. Their ligands are water-soluble and generally too large to pass freely through the plasma membrane. Other signal receptors, however, are located inside the cell. We discuss both of these next.

Receptors in the Plasma Membrane

Most water-soluble signaling molecules bind to specific sites on receptor proteins embedded in the cell's plasma membrane. Such a receptor transmits information from the extracellular environment to the inside of the cell by changing shape or aggregating when a specific ligand binds to it. We can see how membrane receptors work by looking at three major types: G protein-coupled receptors, receptor tyrosine kinases, and ion channel receptors. These receptors are discussed and illustrated in **Figure 11.7**, on the next three pages; study this figure before going on.

Intracellular Receptors

Intracellular receptor proteins are found in either the cytoplasm or nucleus of target cells. To reach such a receptor, a chemical messenger passes through the target cell's plasma membrane. A number of important signaling molecules can do this because they are either hydrophobic enough or small enough to cross the phospholipid interior of the membrane. Such hydrophobic chemical messengers include the steroid hormones and thyroid hormones of animals. Another chemical signaling molecule with an intracellular receptor is nitric oxide (NO), a gas; its very small molecules readily pass between the membrane phospholipids.

The behavior of testosterone is representative of steroid hormones. Secreted by cells of the testis, the hormone travels through the blood and enters cells all over the body. In the cytoplasm of target cells, the only cells that contain receptor molecules for testosterone, the hormone binds to the receptor

▼ **Figure 11.7**

Exploring Membrane Receptors

G Protein-Coupled Receptors

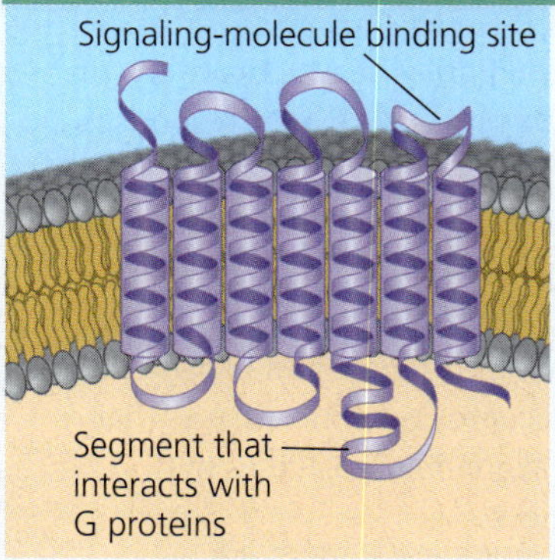

G protein-coupled receptor

A **G protein-coupled receptor** is a plasma membrane receptor that works with the help of a **G protein**, a protein that binds the energy-rich molecule GTP. Many different signaling molecules, including yeast mating factors, epinephrine and many other hormones, and neurotransmitters, use G protein-coupled receptors. These receptors vary in the binding sites for both their signaling molecules (also called their ligands) and for different G proteins inside the cell. Nevertheless, G protein-coupled receptor proteins are all remarkably similar in structure. They each have seven α helices spanning the membrane, as shown above.

A large family of eukaryotic receptor proteins has this secondary structure, where the single polypeptide, represented here as a ribbon, has seven transmembrane α helices, represented as cylinders and depicted in a row for clarity. Specific loops between the helices form binding sites for signaling and G-protein molecules.

G protein-coupled receptor systems are extremely widespread and diverse in their functions, including roles in embryonic development and sensory reception. In humans, for example, both vision and smell depend on such proteins. Similarities in structure among G proteins and G protein-coupled receptors in diverse organisms suggest that G proteins and associated receptors evolved very early.

G-protein systems are involved in many human diseases, including bacterial infections. The bacteria that cause cholera, pertussis (whooping cough), and botulism, among others, make their victims ill by producing toxins that interfere with G-protein function. Pharmacologists now realize that up to 60% of all medicines used today exert their effects by influencing G-protein pathways.

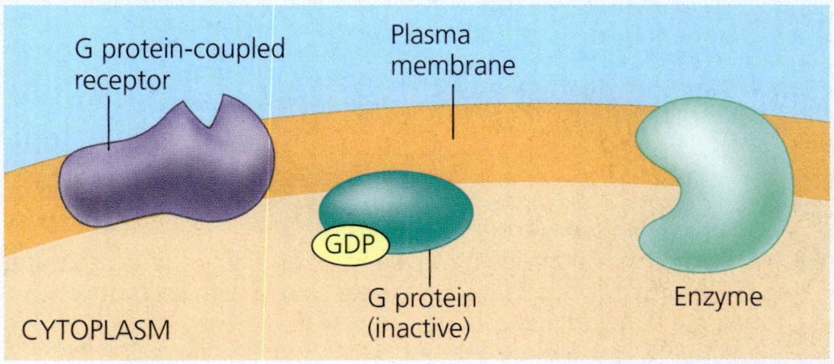

1 Loosely attached to the cytoplasmic side of the membrane, the G protein functions as a molecular switch that is either on or off, depending on which of two guanine nucleotides is attached, GDP or GTP—hence the term *G protein*. (GTP, or guanosine triphosphate, is similar to ATP.) When GDP is bound to the G protein, as shown above, the G protein is inactive. The receptor and G protein work together with another protein, usually an enzyme.

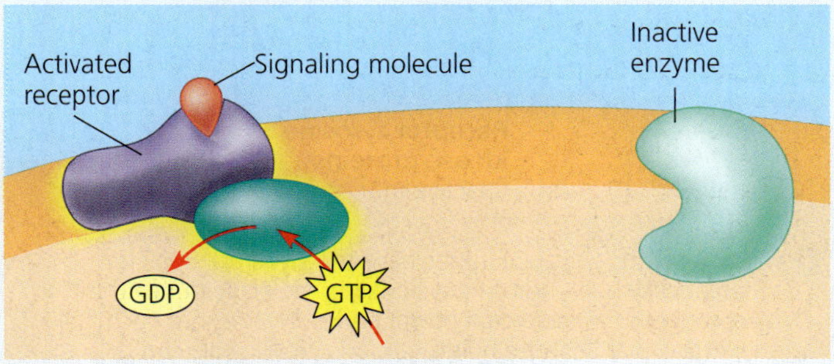

2 When the appropriate signaling molecule binds to the extracellular side of the receptor, the receptor is activated and changes shape. Its cytoplasmic side then binds an inactive G protein, causing a GTP to displace the GDP. This activates the G protein.

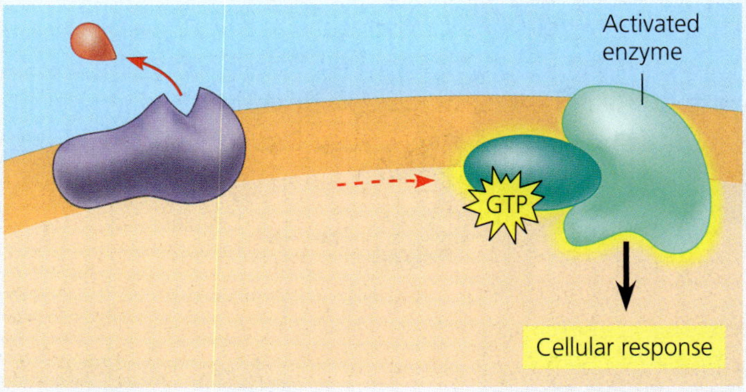

3 The activated G protein dissociates from the receptor, diffuses along the membrane, and then binds to an enzyme, altering the enzymes sh ape and activity. When the enzyme is activated, it can trigger the next step in a pathway leading to a cellular response.

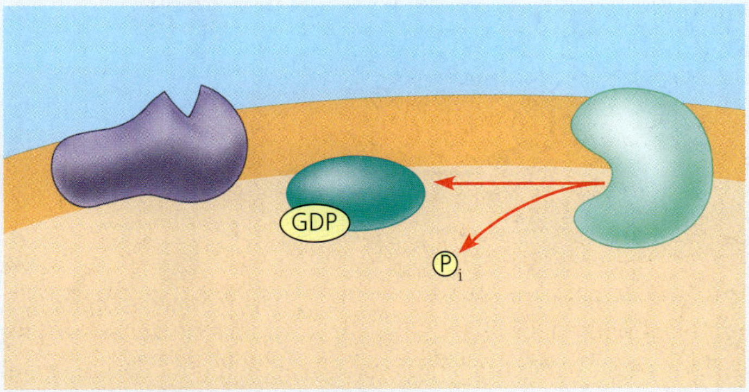

4 The changes in the enzyme and G protein are only temporary because the G protein also functions as a GTPase enzyme—in other words, it then hydrolyzes its bound GTP to GDP. Now inactive again, the G protein leaves the enzyme, which returns to its original state. The G protein is now available for reuse. The GTPase function of the G protein allows the pathway to shut down rapidly when the signali ng molecule is no longer present.

Continued on next page

▼ Figure 11.7 (continued)

Exploring Membrane Receptors

Receptor Tyrosine Kinases

Receptor tyrosine kinases belong to a major class of plasma membrane receptors characterized by having enzymatic activity. A *kinase* is an enzyme that catalyzes the transfer of phosphate groups. The part of the receptor protein extending into the cytoplasm functions as a tyrosine kinase, an enzyme that catalyzes the transfer of a phosphate group from ATP to the amino acid tyrosine on a substrate protein. Thus, receptor tyrosine kinases are membrane receptors that attach phosphates to tyrosines.

One receptor tyrosine kinase complex may activate ten or more different transduction pathways and cellular responses. Often, more than one signal transduction pathway can be triggered at once, helping the cell regulate and coordinate many aspects of cell growth and cell reproduction. The ability of a single ligand-binding event to trigger so many pathways is a key difference between receptor tyrosine kinases and G protein-coupled receptors. Abnormal receptor tyrosine kinases that function even in the absence of signaling molecules may contribute to some kinds of cancer.

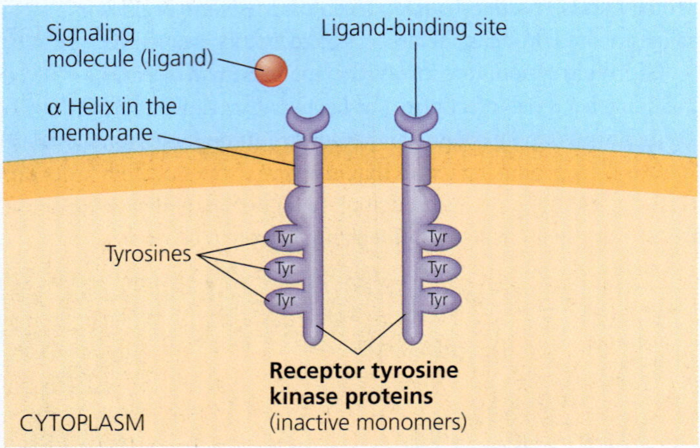

1 Many receptor tyrosine kinases have the structure depicted schematically here. Before the signaling molecule binds, the receptors exist as individual polypeptides. Notice that each has an extracellular ligand-binding site, an α helix spanning the membrane, and an intracellular tail containing multiple tyrosines.

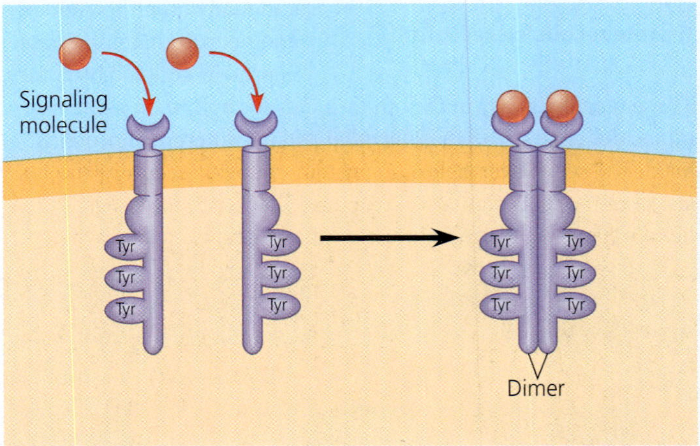

2 The binding of a signaling molecule (such as a growth factor) causes two receptor polypeptides to associate closely with each other, forming a dimer (dimerization).

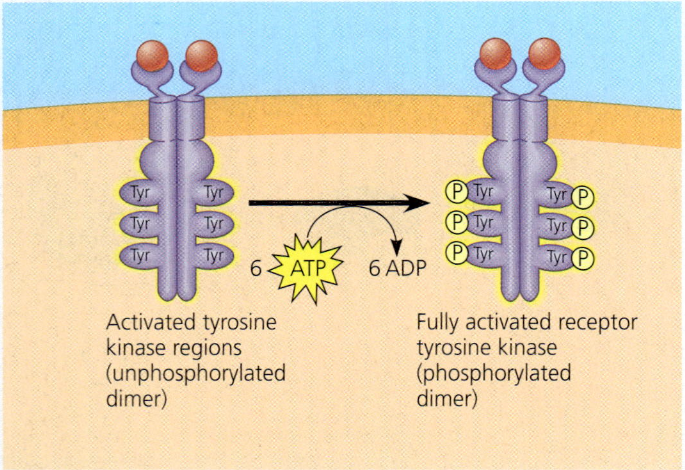

3 Dimerization activates the tyrosine kinase region of each polypeptide; each tyrosine kinase adds a phosphate from an ATP molecule to a tyrosine on the tail of the other polypeptide.

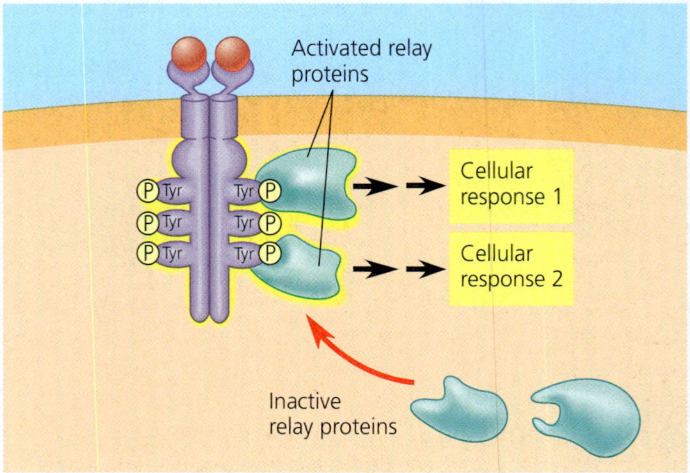

4 Now that the receptor protein is fully activated, it is recognized by specific relay proteins inside the cell. Each such protein binds to a specific phosphorylated tyrosine, undergoing a resulting structural change that activates the bound protein. Each activated protein triggers a transduction pathway, leading to a cellular response.

Ion Channel Receptors

A **ligand-gated ion channel** is a type of membrane receptor containing a region that can act as a "gate" when the receptor changes shape. When a signaling molecule binds as a ligand to the receptor protein, the gate opens or closes, allowing or blocking the flow of specific ions, such as Na^+ or Ca^{2+}, through a channel in the receptor. Like the other receptors we have discussed, these proteins bind the ligand at a specific site on their extracellular sides.

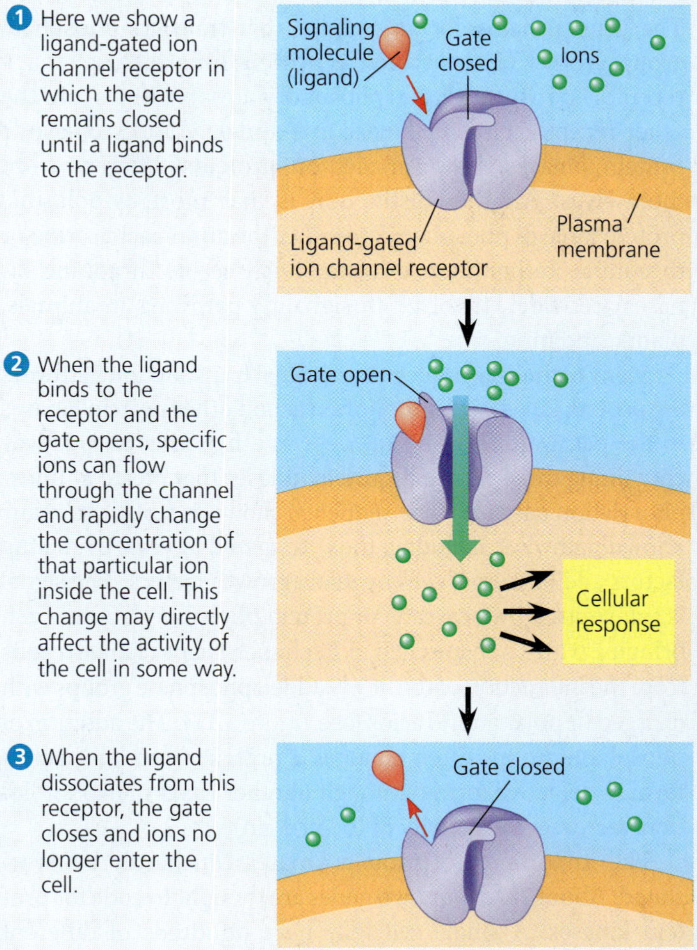

Ligand-gated ion channels are very important in the nervous system. For example, the neurotransmitter molecules released at a synapse between two nerve cells (see Figure 11.5b) bind as ligands to ion channels on the receiving cell, causing the channels to open. Ions flow in (or, in some cases, out), triggering an electrical signal that propagates down the length of the receiving cell. Some gated ion channels are controlled by electrical signals instead of ligands; these *voltage-gated ion channels* are also crucial to the functioning of the nervous system, as we will discuss in Chapter 48.

protein, activating it **(Figure 11.8)**. With the hormone attached, the active form of the receptor protein then enters the nucleus and turns on specific genes that control male sex characteristics.

How does the activated hormone-receptor complex turn on genes? Recall that the genes in a cell's DNA function by being transcribed and processed into messenger RNA (mRNA), which leaves the nucleus and is translated into a specific protein by ribosomes in the cytoplasm (see Figure 5.26). Special proteins called *transcription factors* control which genes are turned on—that is, which genes are transcribed into mRNA—in a particular cell at a particular time. The testosterone receptor, when activated, acts as a transcription factor that turns on specific genes.

By acting as a transcription factor, the testosterone receptor itself carries out the complete transduction of the signal. Most other intracellular receptors function in the same way, although many of them are already in the nucleus before the signaling molecule reaches them (an example is the thyroid hormone receptor). Interestingly, many of these intracellular receptor proteins are structurally similar, suggesting an evolutionary

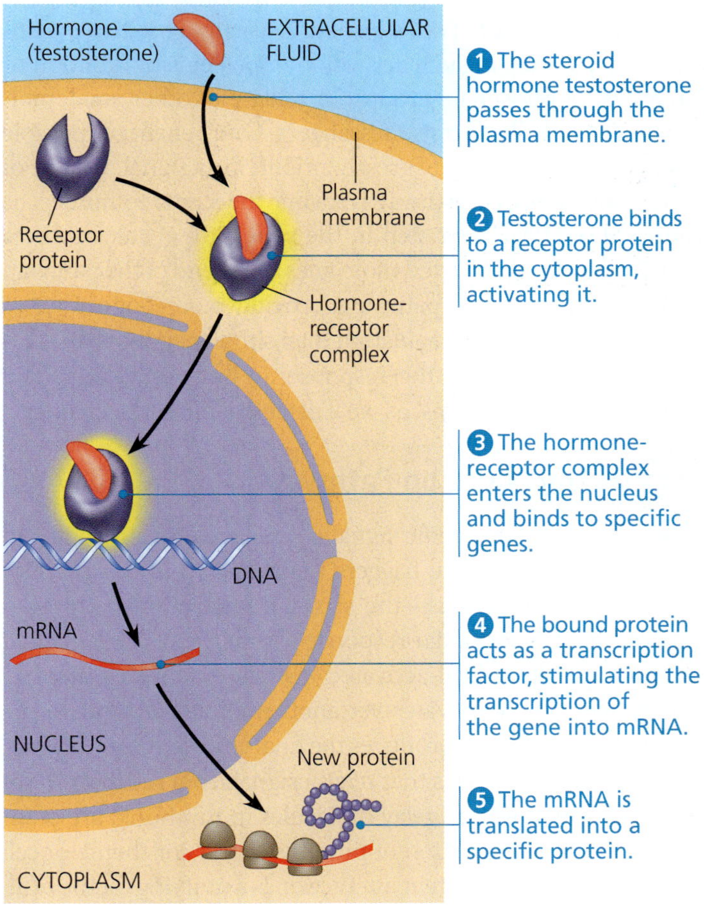

▲ **Figure 11.8 Steroid hormone interacting with an intracellular receptor.**

? *Why is a cell-surface receptor protein not required for this steroid hormone to enter the cell?*

kinship. We will look more closely at hormones with intracellular receptors in Chapter 45.

CONCEPT CHECK 11.2

1. Nerve growth factor (NGF) is a water-soluble signaling molecule. Would you expect the receptor for NGF to be intracellular or in the plasma membrane? Why?
2. **WHAT IF?** What would the effect be if a cell made defective receptor tyrosine kinase proteins that were unable to dimerize?

For suggested answers, see Appendix A.

CONCEPT 11.3
Transduction: Cascades of molecular interactions relay signals from receptors to target molecules in the cell

When receptors for signaling molecules are plasma membrane proteins, like most of those we have discussed, the transduction stage of cell signaling is usually a multistep pathway. Steps often include activation of proteins by addition or removal of phosphate groups, or release of other small molecules or ions that act as messengers. One benefit of multiple steps is the possibility of greatly amplifying a signal. If some of the molecules in a pathway transmit the signal to numerous molecules at the next step in the series, the result can be a large number of activated molecules at the end of the pathway. Moreover, multistep pathways provide more opportunities for coordination and regulation than simpler systems do. This allows fine-tuning of the response, in both unicellular and multicellular organisms, as we'll discuss later in the chapter.

Signal Transduction Pathways

The binding of a specific signaling molecule to a receptor in the plasma membrane triggers the first step in the chain of molecular interactions—the signal transduction pathway—that leads to a particular response within the cell. Like falling dominoes, the signal-activated receptor activates another molecule, which activates yet another molecule, and so on, until the protein that produces the final cellular response is activated. The molecules that relay a signal from receptor to response, which we call relay molecules in this book, are often proteins. The interaction of proteins is a major theme of cell signaling. Indeed, protein interaction is a unifying theme of all regulation at the cellular level.

Keep in mind that the original signaling molecule is not physically passed along a signaling pathway; in most cases, it never even enters the cell. When we say that the signal is relayed along a pathway, we mean that certain information is passed on. At each step, the signal is transduced into a different form, commonly a shape change in a protein. Very often, the shape change is brought about by phosphorylation.

Protein Phosphorylation and Dephosphorylation

Previous chapters introduced the concept of activating a protein by adding one or more phosphate groups to it (see Figure 8.11a). In Figure 11.7, we have already seen how phosphorylation is involved in the activation of receptor tyrosine kinases. In fact, the phosphorylation and dephosphorylation of proteins is a widespread cellular mechanism for regulating protein activity. The general name for an enzyme that transfers phosphate groups from ATP to a protein is **protein kinase**. Recall that a receptor tyrosine kinase phosphorylates tyrosines on the other receptor tyrosine kinase in a dimer. Most cytoplasmic protein kinases, however, act on proteins different from themselves. Another distinction is that most cytoplasmic protein kinases phosphorylate either the amino acid serine or threonine, rather than tyrosine. Such serine/threonine kinases are widely involved in signaling pathways in animals, plants, and fungi.

Many of the relay molecules in signal transduction pathways are protein kinases, and they often act on other protein kinases in the pathway. **Figure 11.9** depicts a hypothetical pathway containing three different protein kinases that create a "phosphorylation cascade." The sequence shown is similar to many known pathways, including those triggered in yeast by mating factors and in animal cells by many growth factors. The signal is transmitted by a cascade of protein phosphorylations, each bringing with it a shape change. Each such shape change results from the interaction of the newly added phosphate groups with charged or polar amino acids (see Figure 5.17). The addition of phosphate groups often changes a protein from an inactive form to an active form (although in other cases phosphorylation *decreases* the activity of the protein).

The importance of protein kinases can hardly be overstated. About 2% of our own genes are thought to code for protein kinases. A single cell may have hundreds of different kinds, each specific for a different substrate protein. Together, they probably regulate a large proportion of the thousands of proteins in a cell. Among these are most of the proteins that, in turn, regulate cell reproduction. Abnormal activity of such a kinase can cause abnormal cell growth and contribute to the development of cancer.

Equally important in the phosphorylation cascade are the **protein phosphatases**, enzymes that can rapidly remove phosphate groups from proteins, a process called dephosphorylation. By dephosphorylating and thus inactivating protein kinases, phosphatases provide the mechanism for turning off the signal transduction pathway when the initial signal is no

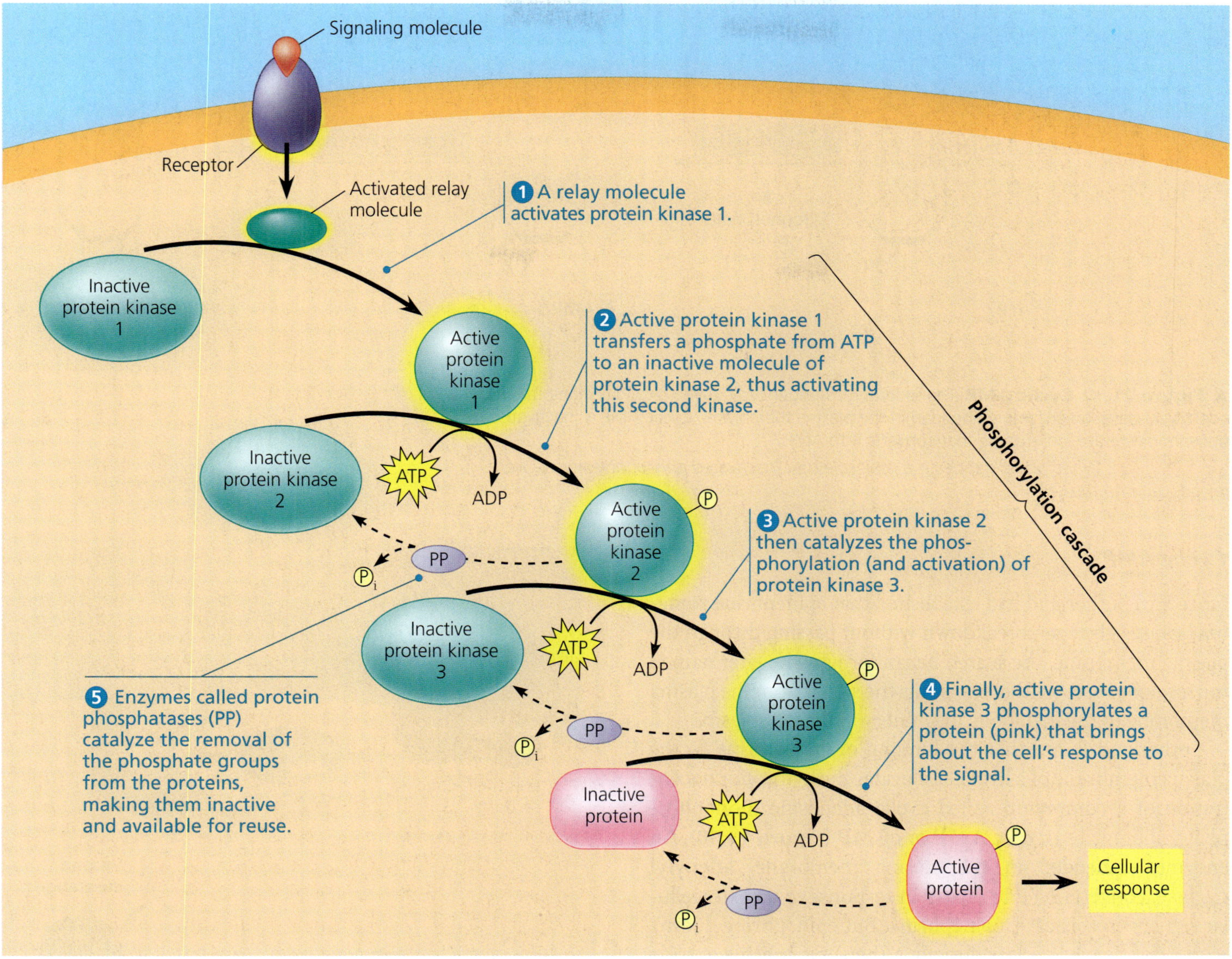

▲ **Figure 11.9 A phosphorylation cascade.** In a phosphorylation cascade, a series of different molecules in a pathway are phosphorylated in turn, each molecule adding a phosphate group to the next one in line. In this example, phosphorylation activates each molecule, and dephosphorylation returns it to its inactive form. The active and inactive forms of each protein are represented by different shapes to remind you that activation is usually associated with a change in molecular shape.

? *Which protein is responsible for activation of protein kinase 3?*

longer present. Phosphatases also make the protein kinases available for reuse, enabling the cell to respond again to an extracellular signal. At any given moment, the activity of a protein regulated by phosphorylation depends on the balance in the cell between active kinase molecules and active phosphatase molecules. The phosphorylation/dephosphorylation system acts as a molecular switch in the cell, turning activities on or off as required.

Small Molecules and Ions as Second Messengers

Not all components of signal transduction pathways are proteins. Many signaling pathways also involve small, nonprotein, water-soluble molecules or ions called **second messengers**. (The extracellular signaling molecule that binds to the membrane receptor is a pathway's "first messenger.") Because second messengers are both small and water-soluble, they can readily spread throughout the cell by diffusion. For example, as we'll see shortly, it is a second messenger called cyclic AMP that carries the signal initiated by epinephrine from the plasma membrane of a liver or muscle cell into the cell's interior, where it brings about glycogen breakdown. Second messengers participate in pathways initiated by both G protein-coupled receptors and receptor tyrosine kinases. The two most widely used second messengers are cyclic AMP and calcium ions, Ca^{2+}. A large variety of relay proteins are sensitive to the cytosolic concentration of one or the other of these second messengers.

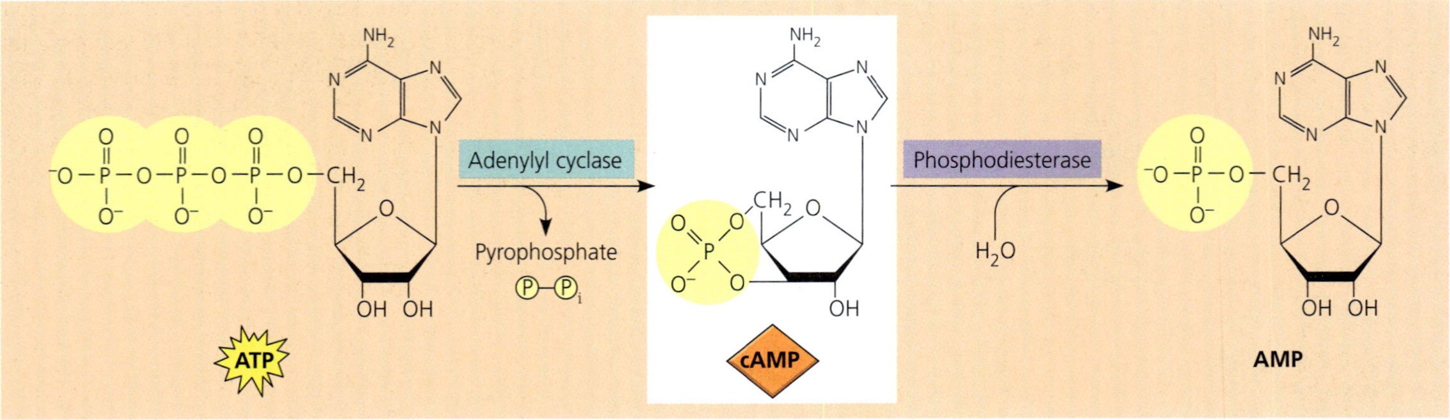

▲ **Figure 11.10 Cyclic AMP.** The second messenger cyclic AMP (cAMP) is made from ATP by adenylyl cyclase, an enzyme embedded in the plasma membrane. Cyclic AMP is inactivated by phosphodiesterase, an enzyme that converts it to AMP.

WHAT IF? *What would happen if a molecule that inactivated phosphodiesterase were introduced into the cell?*

Cyclic AMP

Once Earl Sutherland had established that epinephrine somehow causes glycogen breakdown without passing through the plasma membrane, the search began for what he later named the *second messenger* that transmits the signal from the plasma membrane to the metabolic machinery in the cytoplasm.

Sutherland found that the binding of epinephrine to the plasma membrane of a liver cell elevates the cytosolic concentration of a compound called cyclic adenosine monophosphate, abbreviated **cyclic AMP** or **cAMP (Figure 11.10)**. An enzyme embedded in the plasma membrane, **adenylyl cyclase**, converts ATP to cAMP in response to an extracellular signal—in this case, epinephrine. But epinephrine doesn't stimulate adenylyl cyclase directly. When epinephrine outside the cell binds to a specific receptor protein, the protein activates adenylyl cyclase, which in turn can catalyze the synthesis of many molecules of cAMP. In this way, the normal cellular concentration of cAMP can be boosted 20-fold in a matter of seconds. The cAMP broadcasts the signal to the cytoplasm. It does not persist for long in the absence of the hormone because another enzyme, called phosphodiesterase, converts cAMP to AMP. Another surge of epinephrine is needed to boost the cytosolic concentration of cAMP again.

Subsequent research has revealed that epinephrine is only one of many hormones and other signaling molecules that trigger the formation of cAMP. It has also brought to light the other components of cAMP pathways, including G proteins, G protein-coupled receptors, and protein kinases **(Figure 11.11)**. The immediate effect of cAMP is usually the activation of a serine/threonine kinase called *protein kinase A*. The activated kinase then phosphorylates various other proteins, depending on the cell type. (The complete pathway for epinephrine's stimulation of glycogen breakdown is shown later, in Figure 11.15.)

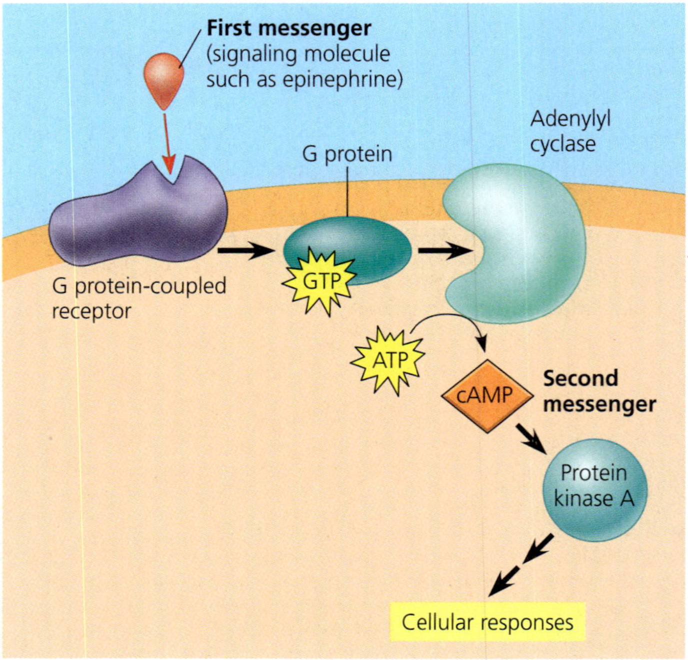

▲ **Figure 11.11 cAMP as a second messenger in a G-protein-signaling pathway.** The first messenger activates a G protein-coupled receptor, which activates a specific G protein. In turn, the G protein activates adenylyl cyclase, which catalyzes the conversion of ATP to cAMP. The cAMP then acts as a second messenger and activates another protein, usually protein kinase A, leading to cellular responses.

Further regulation of cell metabolism is provided by other G-protein systems that *inhibit* adenylyl cyclase. In these systems, a different signaling molecule activates a different receptor, which activates an *inhibitory* G protein.

Now that we know about the role of cAMP in G-protein-signaling pathways, we can explain in molecular detail how certain microbes cause disease. Consider cholera, a disease that is frequently epidemic in places where the water supply is contaminated with human feces. People acquire the cholera bacterium, *Vibrio cholerae,* by drinking contaminated water. The bacteria colonize the lining of the small intestine and produce a toxin. The cholera toxin is an enzyme that chemically modifies a G protein involved in regulating salt and water secretion. Because the modified G protein is unable to hydrolyze GTP to GDP, it remains stuck in its active form, continuously stimulating adenylyl cyclase to make cAMP. The resulting high concentration of cAMP causes the intestinal cells to secrete large amounts of salts, with water following by osmosis, into the intestines. An infected person quickly develops profuse diarrhea and if left untreated can soon die from the loss of water and salts.

Our understanding of signaling pathways involving cyclic AMP or related messengers has allowed us to develop treatments for certain conditions in humans. In one pathway *cyclic GMP*, or *cGMP*, acts as a signaling molecule whose effects include relaxation of smooth muscle cells in artery walls. A compound that inhibits the hydrolysis of cGMP to GMP, thus prolonging the signal, was originally prescribed for chest pains because it increased blood flow to the heart muscle. Under the trade name Viagra (see Figure 11.1), this compound is now widely used as a treatment for erectile dysfunction in human males. Because Viagra leads to dilation of blood vessels, it also allows increased blood flow to the penis, optimizing physiological conditions for penile erections. The similarities between external reproductive structures in males and females (see Chapter 46) have motivated medical researchers to initiate clinical studies exploring whether Viagra might also be used to treat sexual dysfunction in females; these studies are currently under way.

Calcium Ions and Inositol Trisphosphate (IP_3)

Many signaling molecules in animals, including neurotransmitters, growth factors, and some hormones, induce responses in their target cells via signal transduction pathways that increase the cytosolic concentration of calcium ions (Ca^{2+}). Calcium is even more widely used than cAMP as a second messenger. Increasing the cytosolic concentration of Ca^{2+} causes many responses in animal cells, including muscle cell contraction, secretion of certain substances, and cell division. In plant cells, a wide range of hormonal and environmental stimuli can cause brief increases in cytosolic Ca^{2+} concentration, triggering various signaling pathways, such as the pathway for greening in response to light (see Figure 39.4). Cells use Ca^{2+} as a second messenger in both G-protein and receptor tyrosine kinase pathways.

Although cells always contain some Ca^{2+}, this ion can function as a second messenger because its concentration in the cytosol is normally much lower than the concentration outside the cell **(Figure 11.12)**. In fact, the level of Ca^{2+} in the blood and extracellular fluid of an animal often exceeds that in the cytosol by more than 10,000 times. Calcium ions are actively transported out of the cell and are actively imported from the cytosol into the endoplasmic reticulum (and, under some conditions, into mitochondria and chloroplasts) by various protein pumps (see Figure 11.12). As a result, the calcium concentration in the ER is usually much higher than that in the cytosol. Because the cytosolic calcium level is low, a small change in absolute numbers of ions represents a relatively large percentage change in calcium concentration.

In response to a signal relayed by a signal transduction pathway, the cytosolic calcium level may rise, usually by a mechanism that releases Ca^{2+} from the cell's ER. The pathways leading to calcium release involve still other second messengers, **inositol trisphosphate (IP_3)** and **diacylglycerol (DAG)**. These two messengers are produced by cleavage of a certain kind of phospholipid in the plasma membrane.

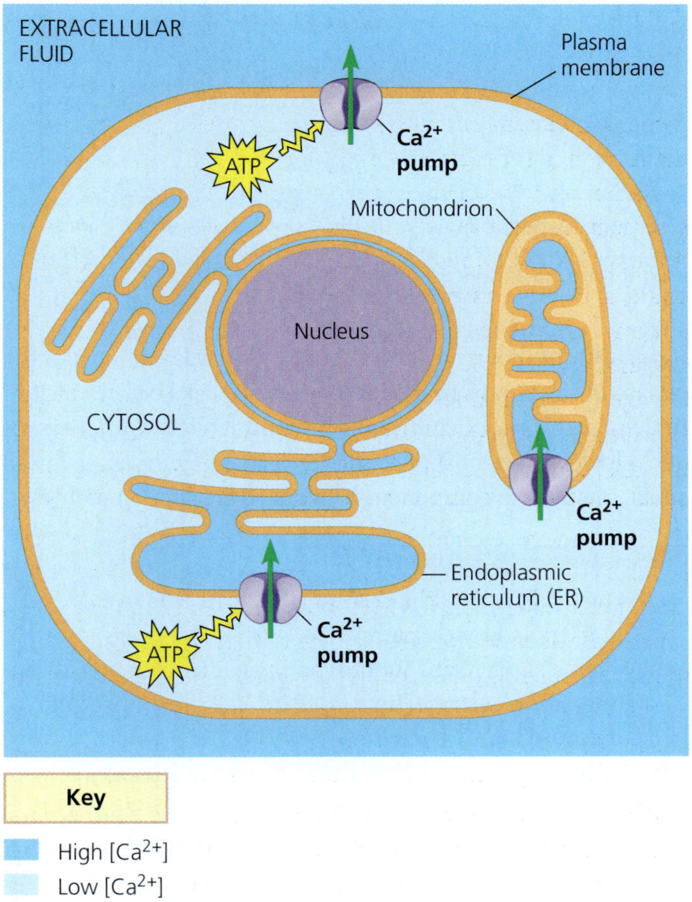

▲ **Figure 11.12 The maintenance of calcium ion concentrations in an animal cell.** The Ca^{2+} concentration in the cytosol is usually much lower (light blue) than that in the extracellular fluid and ER (darker blue). Protein pumps in the plasma membrane and the ER membrane, driven by ATP, move Ca^{2+} from the cytosol into the extracellular fluid and into the lumen of the ER. Mitochondrial pumps, driven by chemiosmosis (see Chapter 9), move Ca^{2+} into mitochondria when the calcium level in the cytosol rises significantly.

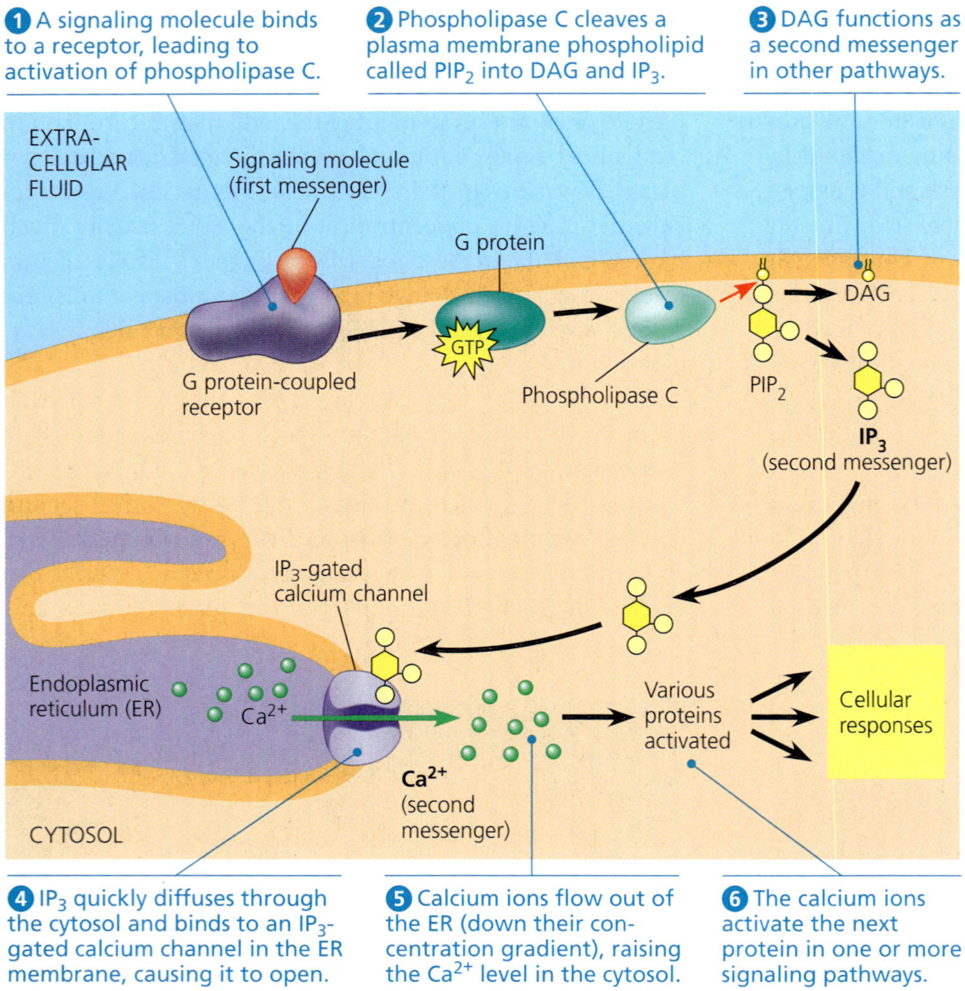

◀ **Figure 11.13 Calcium and IP_3 in signaling pathways.** Calcium ions (Ca^{2+}) and inositol trisphosphate (IP_3) function as second messengers in many signal transduction pathways. In this figure, the process is initiated by the binding of a signaling molecule to a G protein-coupled receptor. A receptor tyrosine kinase could also initiate this pathway by activating phospholipase C.

Figure 11.13 shows how this occurs and how IP_3 stimulates the release of calcium from the ER. Because IP_3 acts before calcium in these pathways, calcium could be considered a "*third* messenger." However, scientists use the term *second messenger* for all small, nonprotein components of signal transduction pathways.

CONCEPT CHECK 11.3

1. What is a protein kinase, and what is its role in a signal transduction pathway?
2. When a signal transduction pathway involves a phosphorylation cascade, how does the cell's response get turned off?
3. What is the actual "signal" that is being transduced in any signal transduction pathway, such as those shown in Figures 11.6 and 11.9? In other words, in what way is information being passed from the exterior to the interior of the cell?
4. **WHAT IF?** Upon activation of phospholipase C by ligand binding to a receptor, what effect does the IP_3-gated calcium channel have on Ca^{2+} concentration in the cytosol?

For suggested answers, see Appendix A.

CONCEPT **11.4**

Response: Cell signaling leads to regulation of transcription or cytoplasmic activities

We now take a closer look at the cell's subsequent response to an extracellular signal—what some researchers call the "output response." What is the nature of the final step in a signaling pathway?

Nuclear and Cytoplasmic Responses

Ultimately, a signal transduction pathway leads to the regulation of one or more cellular activities. The response at the end of the pathway may occur in the nucleus of the cell or in the cytoplasm.

Many signaling pathways ultimately regulate protein synthesis, usually by turning specific genes on or off in the nucleus. Like an activated steroid receptor (see Figure 11.8), the final activated molecule in a signaling pathway may function as a transcription factor. **Figure 11.14** shows an example in which a signaling pathway activates a transcription factor

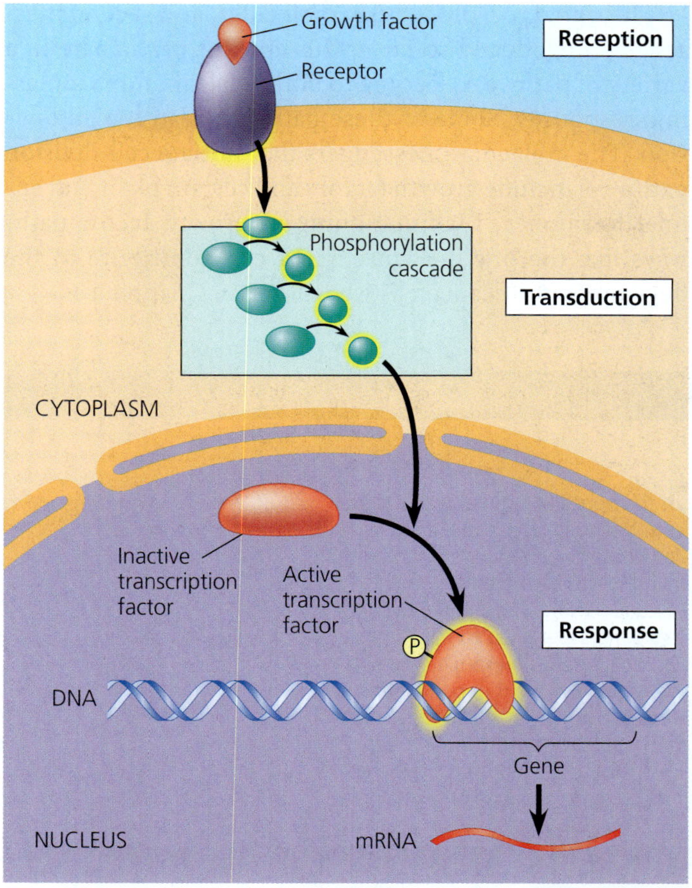

▲ **Figure 11.14 Nuclear responses to a signal: the activation of a specific gene by a growth factor.** This diagram is a simplified representation of a typical signaling pathway that leads to the regulation of gene activity in the cell nucleus. The initial signaling molecule, a local regulator called a growth factor, triggers a phosphorylation cascade. (The ATP molecules that serve as sources of phosphate are not shown.) Once phosphorylated, the last kinase in the sequence enters the nucleus and there activates a gene-regulating protein, a transcription factor. This protein stimulates a specific gene so that an mRNA is synthesized, which then directs the synthesis of a particular protein in the cytoplasm.

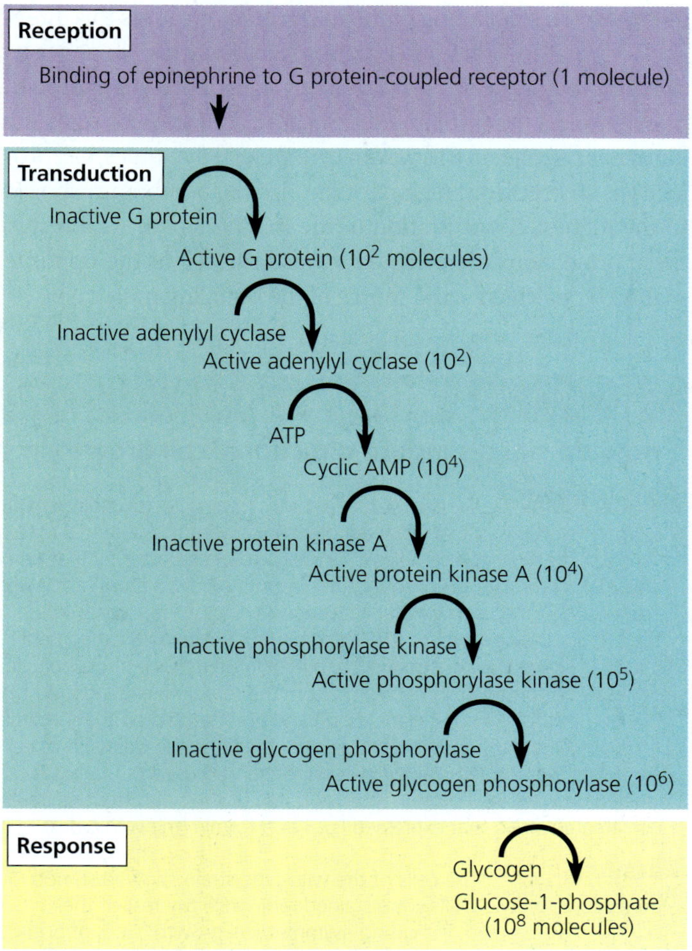

▲ **Figure 11.15 Cytoplasmic response to a signal: the stimulation of glycogen breakdown by epinephrine.** In this signaling system, the hormone epinephrine acts through a G protein-coupled receptor to activate a succession of relay molecules, including cAMP and two protein kinases (see also Figure 11.11). The final protein to be activated is the enzyme glycogen phosphorylase, which uses inorganic phosphate to release glucose monomers from glycogen in the form of glucose-1-phosphate molecules. This pathway amplifies the hormonal signal, because one receptor protein can activate about 100 molecules of G protein, and each enzyme in the pathway, once activated, can act on many molecules of its substrate, the next molecule in the cascade. The number of activated molecules given for each step is approximate.

that turns a gene on: The response to the growth factor signal is the synthesis of mRNA, which will be translated in the cytoplasm into a specific protein. In other cases, the transcription factor might regulate a gene by turning it off. Often a transcription factor regulates several different genes.

Sometimes a signaling pathway may regulate the *activity* of proteins rather than their *synthesis*, directly affecting proteins that function outside the nucleus. For example, a signal may cause the opening or closing of an ion channel in the plasma membrane or a change in cell metabolism. As we have discussed already, the response of liver cells to signaling by the hormone epinephrine helps regulate cellular energy metabolism by affecting the activity of an enzyme. The final step in the signaling pathway that begins with epinephrine binding activates the enzyme that catalyzes the breakdown of glycogen. **Figure 11.15** shows the complete pathway leading to the release of glucose-1-phosphate molecules from glycogen. Note that as each molecule is activated, the response is amplified, as we will discuss later.

In addition to the regulation of enzymes, signaling events may also affect other cellular attributes, such as overall cell shape. An example of this regulation can be found in the activities leading to the mating of yeast cells (see Figure 11.2). Yeast cells are not motile; their mating process depends on the growth of localized projections in one cell toward a cell of the opposite mating type. As shown in **Figure 11.16**, binding of the mating factor causes this directional growth. When the mating factor binds, it activates signaling-pathway kinases that affect the orientation of growth of cytoskeletal microfilaments. Because activation of signaling kinases is coupled in this way to cytoskeletal dynamics, cell projections emerge from regions of the plasma membrane exposed to the highest concentration of the mating factor. As a result, these projections are oriented toward the cell of the opposite mating type, which is the source of the signaling molecule.

The signal receptors, relay molecules, and second messengers introduced so far in this chapter participate in a variety of pathways, leading to both nuclear and cytoplasmic responses. Some of these pathways lead to cell division. The molecular messengers that initiate cell-division pathways include growth factors and certain plant and animal hormones. Malfunctioning of growth factor pathways like the one in Figure 11.14 can contribute to the development of cancer, as we will see in Chapter 18.

▼ Figure 11.16 Inquiry

How do signals induce directional cell growth in yeast?

EXPERIMENT When a yeast cell binds mating factor molecules from a cell of the opposite mating type, a signaling pathway causes it to grow a projection toward the potential mate. The cell with the projection is called a "shmoo" because it resembles a 1950s cartoon character by that name. Dina Matheos and colleagues in Mark Rose's lab at Princeton University sought to determine how mating factor signaling is linked to this asymmetrical growth. Previous work had shown that activation of one of the kinases in the signaling cascade (Fus3) caused it to move to the membrane near where the factor bound. Preliminary experiments by these researchers identified formin, a protein that directs the construction of microfilaments, as a phosphorylation target of Fus3 kinase. To examine the role of Fus3 and formin in shmoo formation, the researchers generated two mutant yeast strains: one that no longer had the kinase (this strain is called ΔFus3) and one that lacked the formin (Δformin). To observe the effects of these mutations on cell growth induced by the mating factor, the cell walls of each strain were first stained with a green fluorescent dye. These green-stained cells were then exposed to mating factor and stained with a red fluorescent dye that labeled new cell wall growth. Images taken of the cells after the staining procedure were then compared with a similarly treated strain that expressed Fus3 and formin (the wild type).

RESULTS The cells of the wild-type strain showed shmoo projections, whose walls were stained red, while the rest of their cell walls were green, indicating asymmetrical growth. Cells of both the ΔFus3 and Δformin strains showed no shmoo formation, and their cell walls were stained almost uniformly yellow. This color resulted from merged green and red stains, indicating symmetrical growth, characteristic of cells not exposed to mating factor.

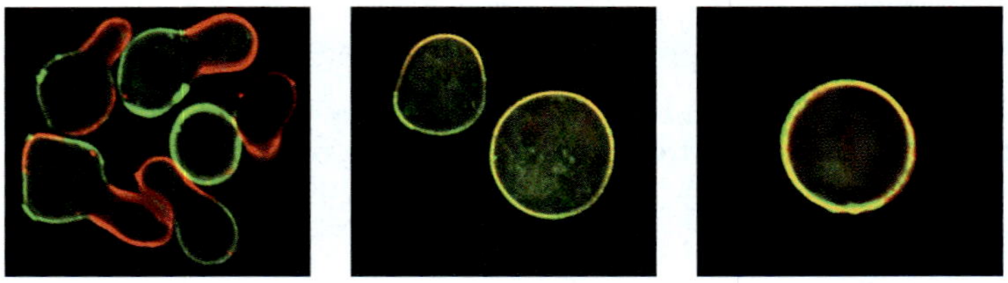

Wild-type (shmoos) | ΔFus3 | Δformin

CONCLUSION The similar defect (lack of ability to form shmoos) in strains lacking either Fus3 or formin suggests that both proteins are required for shmoo formation. These results led the investigators to propose the model shown here for the induction of directed asymmetrical growth in the receiving cell toward the cell of the opposite mating type.

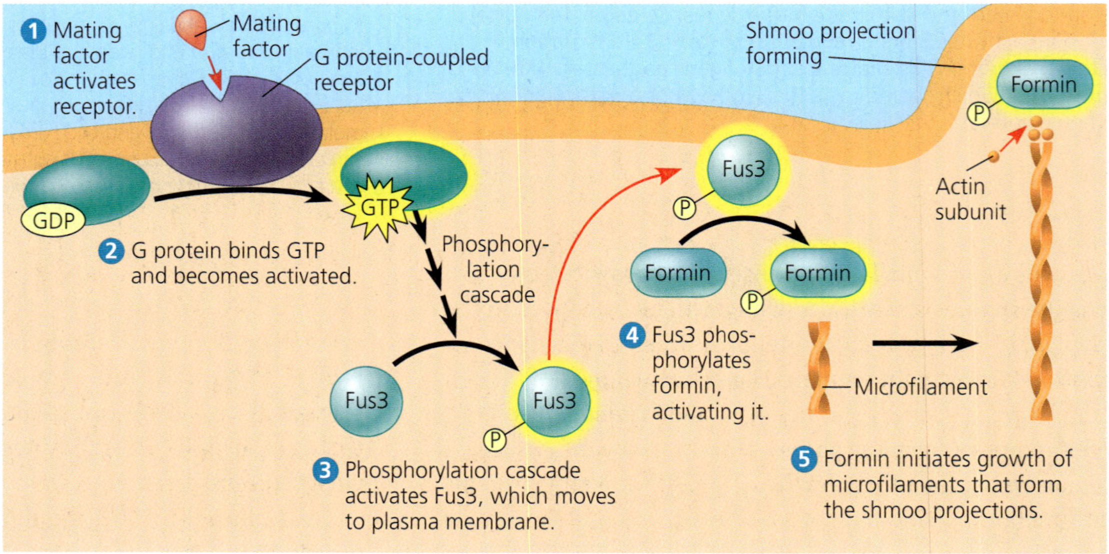

SOURCE D. Matheos et al., Pheromone-induced polarization is dependent on the Fus3p MAPK acting through the formin Bni1p, *Journal of Cell Biology* 165:99–109 (2004).

WHAT IF? Based on these results and the proposed model from this work, what would happen to a cell if its Fus3 kinase were not able to associate with the membrane upon activation?

Fine-Tuning of the Response

Regardless of whether the response occurs in the nucleus or in the cytoplasm, it is fine-tuned at multiple points. As mentioned earlier, signaling pathways with numerous steps between a signaling event at the cell surface and the cell's response have two important benefits: They amplify the signal (and thus the response), and they provide different points at which a cell's response can be regulated. This allows coordination of signaling pathways and also contributes to the specificity of the response. The overall efficiency of the response is also enhanced by scaffolding proteins. Finally, a crucial point in fine-tuning the response is the termination of the signal.

Signal Amplification

Elaborate enzyme cascades amplify the cell's response to a signal. At each catalytic step in the cascade, the number of activated products is much greater than in the preceding step. For example, in the epinephrine-triggered pathway in Figure 11.15, each adenylyl cyclase molecule catalyzes the formation of many cAMP molecules, each molecule of protein kinase A phosphorylates many molecules of the next kinase in the pathway, and so on. The amplification effect stems from the fact that these proteins persist in the active form long enough to process numerous molecules of substrate before they become inactive again. As a result of the signal's amplification, a small number of epinephrine molecules binding to receptors on the surface of a liver cell or muscle cell can lead to the release of hundreds of millions of glucose molecules from glycogen.

The Specificity of Cell Signaling and Coordination of the Response

Consider two different cells in your body—a liver cell and a heart muscle cell, for example. Both are in contact with your bloodstream and are therefore constantly exposed to many different hormone molecules, as well as to local regulators secreted by nearby cells. Yet the liver cell responds to some signals but ignores others, and the same is true for the heart cell. And some kinds of signals trigger responses in both cells—but different responses. For instance, epinephrine stimulates the liver cell to break down glycogen, but the main response of the heart cell to epinephrine is contraction, leading to a more rapid heartbeat. How do we account for this difference?

The explanation for the specificity exhibited in cellular responses to signals is the same as the basic explanation for virtually all differences between cells: *Different kinds of cells have different collections of proteins* **(Figure 11.17)**. (This is because different kinds of cells turn on different sets of genes.) The response of a particular cell to a signal depends on its particular collection of signal receptor proteins, relay proteins, and proteins needed to carry out the response. A liver cell, for example, is poised to respond appropriately to epinephrine by having the proteins listed in Figure 11.15 as well as those needed to manufacture glycogen.

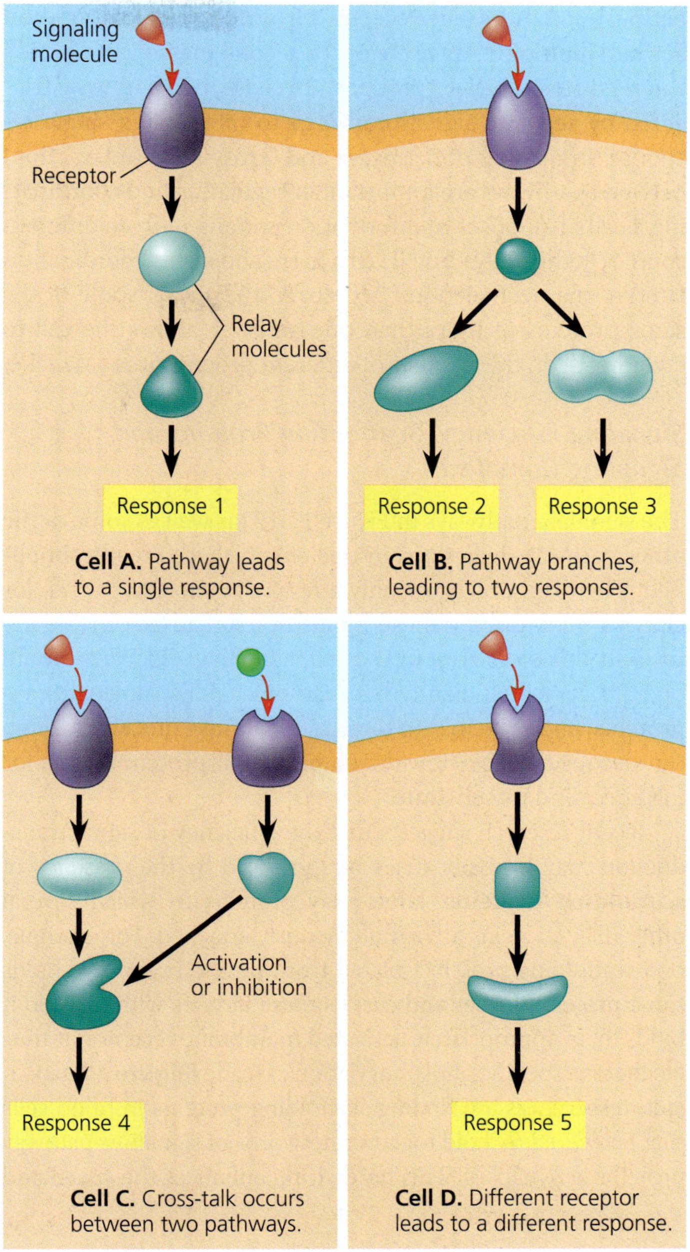

▲ **Figure 11.17 The specificity of cell signaling.** The particular proteins a cell possesses determine what signaling molecules it responds to and the nature of the response. The four cells in these diagrams respond to the same signaling molecule (red) in different ways because each has a different set of proteins (purple and teal shapes). Note, however, that the same kinds of molecules can participate in more than one pathway.

Thus, two cells that respond differently to the same signal differ in one or more of the proteins that handle and respond to the signal. Notice in Figure 11.17 that different pathways may have some molecules in common. For example, cells A, B, and C all use the same receptor protein for the red signaling molecule; differences in other proteins account for their differing responses. In cell D, a different receptor protein is used for the same signaling molecule, leading to yet another response. In cell B, a pathway that is triggered by a single kind of signal diverges to produce two responses; such branched pathways often involve receptor tyrosine kinases (which can activate multiple relay proteins) or second messengers (which can regulate numerous proteins). In cell C, two pathways triggered by separate signals converge to modulate a single response. Branching of pathways and "cross-talk" (interaction) between pathways are important in regulating and coordinating a cell's responses to information coming in from different sources in the body. (You'll learn more about this coordination later, in the next section.) Moreover, the use of some of the same proteins in more than one pathway allows the cell to economize on the number of different proteins it must make.

Signaling Efficiency: Scaffolding Proteins and Signaling Complexes

The signaling pathways in Figure 11.17 (as well as some of the other pathway depictions in this chapter) are greatly simplified. The diagrams show only a few relay molecules and, for clarity's sake, display these molecules spread out in the cytosol. If this were true in the cell, signaling pathways would operate very inefficiently because most relay molecules are proteins, and proteins are too large to diffuse quickly through the viscous cytosol. How does a particular protein kinase, for instance, find its substrate?

Recent research suggests that the efficiency of signal transduction may in many cases be increased by the presence of **scaffolding proteins**, large relay proteins to which several other relay proteins are simultaneously attached. For example, one scaffolding protein isolated from mouse brain cells holds three protein kinases and carries these kinases with it when it binds to an appropriately activated membrane receptor; it thus facilitates a specific phosphorylation cascade **(Figure 11.18)**. In fact, researchers are finding scaffolding proteins in brain cells that *permanently* hold together networks of signaling-pathway proteins at synapses. This hardwiring enhances the speed and accuracy of signal transfer between cells, because the rate of protein-protein interaction is not limited by diffusion.

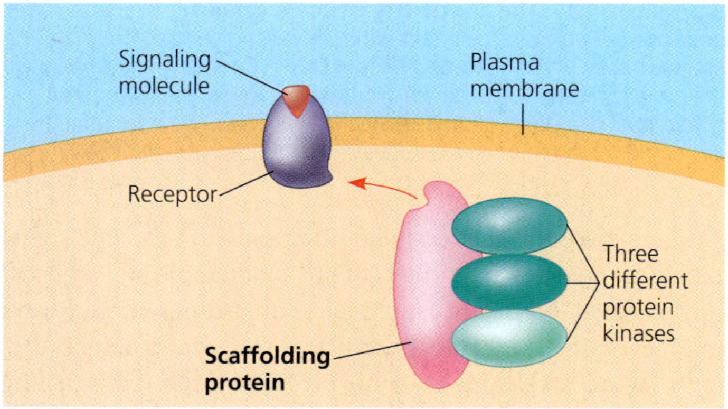

▲ **Figure 11.18 A scaffolding protein.** The scaffolding protein shown here (pink) simultaneously binds to a specific activated membrane receptor and three different protein kinases. This physical arrangement facilitates signal transduction by these molecules.

When signaling pathways were first discovered, they were thought to be linear, independent pathways. Our understanding of the processes of cellular communication has benefited from the realization that things are not that simple. In fact, as seen in Figure 11.17, some proteins may participate in more than one pathway, either in different cell types or in the same cell at different times or under different conditions. This view underscores the importance of permanent or transient protein complexes in the functioning of a cell.

The importance of the relay proteins that serve as points of branching or intersection in signaling pathways is highlighted by the problems arising when these proteins are defective or missing. For instance, in an inherited disorder called Wiskott-Aldrich syndrome (WAS), the absence of a single relay protein leads to such diverse effects as abnormal bleeding, eczema, and a predisposition to infections and leukemia. These symptoms are thought to arise primarily from the absence of the protein in cells of the immune system. By studying normal cells, scientists found that the WAS protein is located just beneath the cell surface. The protein interacts both with microfilaments of the cytoskeleton and with several different components of signaling pathways that relay information from the cell surface, including pathways regulating immune cell proliferation. This multifunctional relay protein is thus both a branch point and an important intersection point in a complex signal transduction network that controls immune cell behavior. When the WAS protein is absent, the cytoskeleton is not properly organized and signaling pathways are disrupted, leading to the WAS symptoms.

Termination of the Signal

To keep Figure 11.17 simple, we did not indicate the *inactivation* mechanisms that are an essential aspect of cell signaling. For a cell of a multicellular organism to remain alert and capable of responding to incoming signals, each molecular change in its signaling pathways must last only a short time. As we saw in the cholera example, if a signaling pathway component becomes locked into one state, whether active or inactive, the consequences for the organism can be dire.

Thus, a key to a cell's continuing receptiveness to regulation by signaling is the reversibility of the changes that signals produce. The binding of signaling molecules to receptors is reversible; the lower the concentration of signaling molecules is, the fewer will be bound at any given moment. When signaling molecules leave the receptor, the receptor reverts to its inactive form. Then, by a variety of means, the relay molecules return to their inactive forms:

The GTPase activity intrinsic to a G protein hydrolyzes its bound GTP; the enzyme phosphodiesterase converts cAMP to AMP; protein phosphatases inactivate phosphorylated kinases and other proteins; and so forth. As a result, the cell is soon ready to respond to a fresh signal.

In this section, we explored the complexity of signaling initiation and termination in a single pathway, and we saw the potential for pathways to intersect with each other. In the next section, we'll consider an important network of interacting pathways in the cell.

CONCEPT CHECK 11.4

1. How can a target cell's response to a hormone be amplified more than a millionfold?
2. **WHAT IF?** If two cells have different scaffolding proteins, explain how they could behave differently in response to the same signaling molecule.

For suggested answers, see Appendix A.

CONCEPT 11.5

Apoptosis (programmed cell death) integrates multiple cell-signaling pathways

One of the most elaborate networks of signaling pathways in the cell seems to ask and answer the basic question posed by Hamlet: To be or not to be? Cells that are infected or damaged or that have simply reached the end of their functional life span often enter a program of controlled cell suicide called **apoptosis** (from the Greek, meaning "falling off," and used in a classic Greek poem to refer to leaves falling from a tree). During this process, cellular agents chop up the DNA and fragment the organelles and other cytoplasmic components. The cell shrinks and becomes lobed (called "blebbing") **(Figure 11.19)**, and the cell's parts are packaged up in vesicles that are engulfed and digested by specialized scavenger cells, leaving no trace. Apoptosis protects neighboring cells from damage that they would otherwise suffer if a dying cell merely leaked out all its contents, including its many digestive and other enzymes.

Apoptosis in the Soil Worm *Caenorhabditis elegans*

Embryonic development is a period during which apoptosis is widespread and plays a crucial role. The molecular mechanisms underlying apoptosis were worked out in detail by researchers studying embryonic development of a small soil worm, a nematode called *Caenorhabditis elegans*. Because the adult worm has only about a thousand cells, the researchers were able to work out the entire ancestry of each cell. The timely suicide of cells occurs exactly 131 times during normal development of *C. elegans*, at precisely the same points in the cell lineage of each worm. In worms and other species, apoptosis is triggered by signals that activate a cascade of "suicide" proteins in the cells destined to die.

Genetic research on *C. elegans* has revealed two key apoptosis genes, called *ced-3* and *ced-4* (*ced* stands for "cell death"), which encode proteins essential for apoptosis. (The proteins are called Ced-3 and Ced-4, respectively.) These and most other proteins involved in apoptosis are continually present in cells, but in inactive form; thus, protein activity is regulated rather than protein synthesis (by way of gene activity).

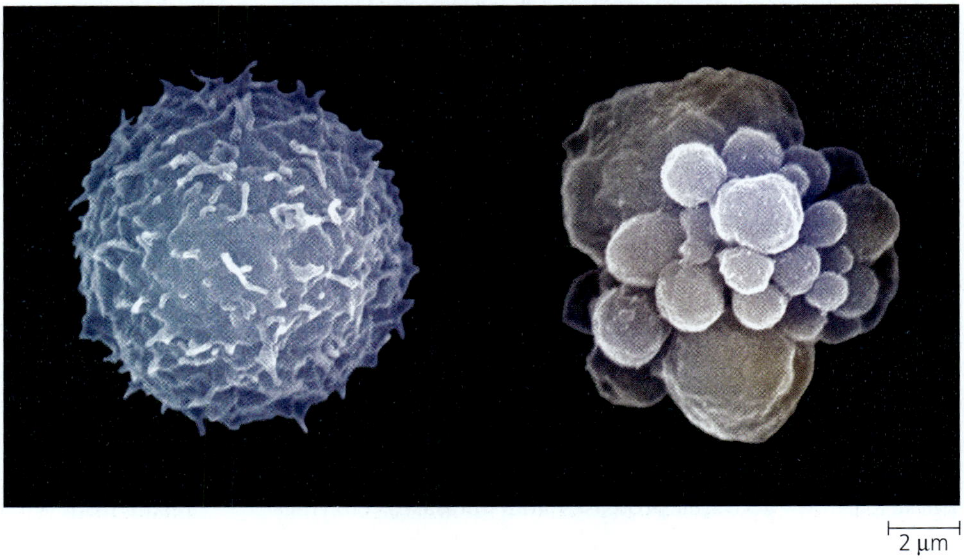

▶ **Figure 11.19 Apoptosis of a human white blood cell.** We can compare a normal white blood cell (left) with a white blood cell undergoing apoptosis (right). The apoptotic cell is shrinking and forming lobes ("blebs"), which eventually are shed as membrane-bounded cell fragments (colorized SEMs).

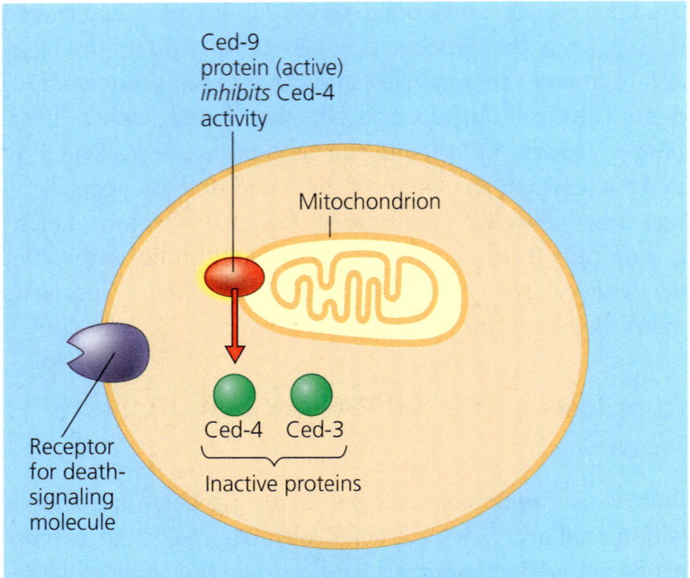

(a) No death signal. As long as Ced-9, located in the outer mitochondrial membrane, is active, apoptosis is inhibited, and the cell remains alive.

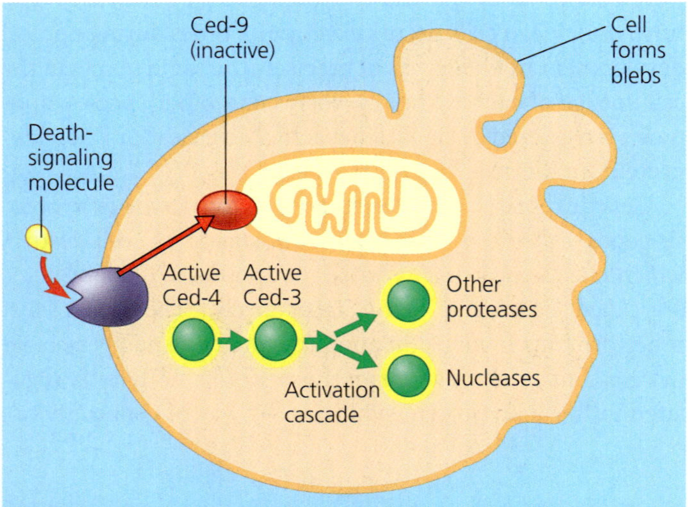

(b) Death signal. When a cell receives a death signal, Ced-9 is inactivated, relieving its inhibition of Ced-3 and Ced-4. Active Ced-3, a protease, triggers a cascade of reactions leading to activation of nucleases and other proteases. The action of these enzymes causes the changes seen in apoptotic cells and eventual cell death.

▲ **Figure 11.20 Molecular basis of apoptosis in *C. elegans.*** Three proteins, Ced-3, Ced-4, and Ced-9, are critical to apoptosis and its regulation in the nematode. Apoptosis is more complicated in mammals but involves proteins similar to those in the nematode.

In *C. elegans*, a protein in the outer mitochondrial membrane, called Ced-9 (the product of the *ced-9* gene), serves as a master regulator of apoptosis, acting as a brake in the absence of a signal promoting apoptosis **(Figure 11.20)**. When a death signal is received by the cell, it overrides the brake, and the apoptotic pathway activates proteases and nucleases, enzymes that cut up the proteins and DNA of the cell. The main proteases of apoptosis are called *caspases*; in the nematode, the chief caspase is Ced-3.

Apoptotic Pathways and the Signals That Trigger Them

In humans and other mammals, several different pathways, involving about 15 different caspases, can carry out apoptosis. The pathway that is used depends on the type of cell and on the particular signal that triggers apoptosis. One major pathway involves mitochondrial proteins. Apoptotic proteins can form molecular pores in the mitochondrial outer membrane, causing it to leak and release proteins that promote apoptosis. Surprisingly, these include cytochrome *c*, which functions in mitochondrial electron transport in healthy cells (see Figure 9.16) but acts as a cell death factor when released from mitochondria. The process of mitochondrial apoptosis in mammals uses proteins similar to the nematode proteins Ced-3, Ced-4, and Ced-9.

At key points in the apoptotic program, proteins integrate signals from several different sources and can send a cell down an apoptotic pathway. Often, the signal originates outside the cell, like the death-signaling molecule depicted in Figure 11.20, which presumably was released by a neighboring cell. When a death-signaling ligand occupies a cell-surface receptor, this binding leads to activation of caspases and other enzymes that carry out apoptosis, without involving the mitochondrial pathway. Two other types of alarm signals originate from *inside* the cell. One comes from the nucleus, generated when the DNA has suffered irreparable damage, and a second comes from the endoplasmic reticulum when excessive protein misfolding occurs. Mammalian cells make life-or-death "decisions" by somehow integrating the death signals and life signals they receive from these external and internal sources.

A built-in cell suicide mechanism is essential to development and maintenance in all animals. The similarities between apoptosis genes in nematodes and mammals, as well as the observation that apoptosis occurs in multicellular fungi and even in single-celled yeasts, indicate that the basic mechanism evolved early in animal evolution. In vertebrates, apoptosis is essential for normal development of the nervous system, for normal operation of the immune system, and for normal morphogenesis of hands and feet in humans and paws in other mammals **(Figure 11.21)**. A lower level of apoptosis in developing limbs accounts for the webbed feet of ducks and other water birds, in contrast to chickens and other land birds with nonwebbed feet. In the case of humans, the failure of appropriate apoptosis can result in webbed fingers and toes.

Significant evidence points to the involvement of apoptosis in certain degenerative diseases of the nervous system, such as Parkinson's disease and Alzheimer's disease. Also, cancer can result from a failure of cell suicide; some cases of human melanoma, for example, have been linked to faulty forms of

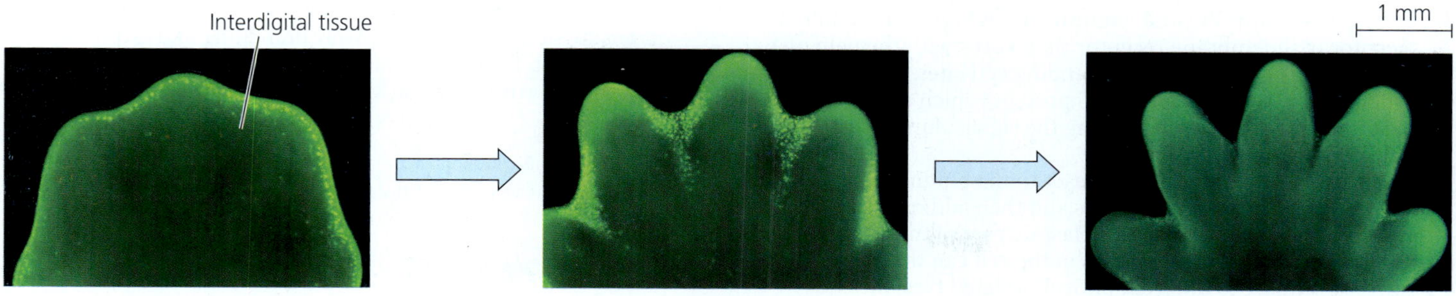

▲ **Figure 11.21 Effect of apoptosis during paw development in the mouse.** In mice, humans, and other mammals, as well as in land birds, the embryonic region that develops into feet or hands initially has a solid, platelike structure. Apoptosis eliminates the cells in the interdigital regions, thus forming the digits. The embryonic mouse paws shown in these fluorescence light micrographs are stained so that cells undergoing apoptosis appear bright yellow. Apoptosis of cells begins at the margin of each interdigital region (left), peaks as the tissue in these regions is reduced (middle), and is no longer visible when the interdigital tissue has been eliminated (right).

the human version of the *C. elegans* Ced-4 protein. It is not surprising, therefore, that the signaling pathways feeding into apoptosis are quite elaborate. After all, the life-or-death question is the most fundamental one imaginable for a cell.

This chapter has introduced you to many of the general mechanisms of cell communication, such as ligand binding, protein-protein interactions and shape changes, cascades of interactions, and protein phosphorylation. As you continue through the text, you will encounter numerous examples of cell signaling.

CONCEPT CHECK 11.5

1. Give an example of apoptosis during embryonic development, and explain its function in the developing embryo.
2. **WHAT IF?** What type of protein defects could result in apoptosis occurring when it should not? What type could result in apoptosis not occurring when it should?

For suggested answers, see Appendix A.

Chapter 11 Review

MEDIA Go to the Study Area at **www.masteringbio.com** for BioFlix 3-D Animations, MP3 Tutors, Videos, Practice Tests, an eBook, and more.

SUMMARY OF KEY CONCEPTS

CONCEPT 11.1

External signals are converted to responses within the cell (pp. 206–210)

▶ **Evolution of Cell Signaling** Signaling in microbes has much in common with processes in multicellular organisms, suggesting an early origin of signaling mechanisms. Bacterial cells can sense the local density of bacterial cells (quorum sensing) by binding molecules secreted by other cells. In some cases, such signals lead to aggregation of these cells into biofilms.

▶ **Local and Long-Distance Signaling** In local signaling, animal cells may communicate by direct contact or by secreting local regulators, such as growth factors or neurotransmitters. For signaling over long distances, both animals and plants use hormones; animals also signal along nerve cells.

▶ **The Three Stages of Cell Signaling: *A Preview*** Earl Sutherland discovered how the hormone epinephrine acts on cells, shown here as an example of a cell-signaling pathway:

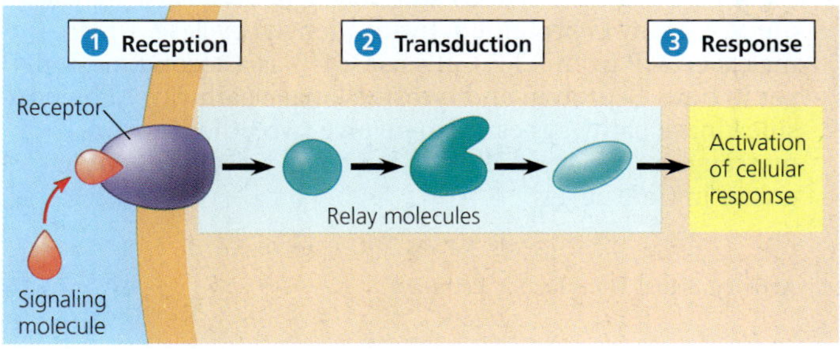

As discussed in Sections 11.2 and 11.3, the signal is transmitted by successive shape changes in the receptor and relay molecules.

MEDIA
Investigation How Do Cells Communicate with Each Other?
Activity Overview of Cell Signaling

CONCEPT 11.2

Reception: A signaling molecule binds to a receptor protein, causing it to change shape (pp. 210–214)

▶ The binding between signaling molecule (ligand) and receptor is highly specific. A shape change in a receptor is often the initial transduction of the signal.

▶ **Receptors in the Plasma Membrane** A G protein-coupled receptor is a membrane receptor that works with the help of a cytoplasmic G protein. Ligand binding activates the receptor, which then activates a specific G protein, which activates yet another protein, thus propagating the signal along a signal transduction pathway.

Receptor tyrosine kinases react to the binding of signaling molecules by forming dimers and then adding phosphate groups to tyrosines on the cytoplasmic part of the other subunit of the dimer. Relay proteins in the cell can then be activated by binding to different phosphorylated tyrosines, allowing this receptor to trigger several pathways at once.

Specific signaling molecules cause ligand-gated ion channels in a membrane to open or close, regulating the flow of specific ions.

▶ **Intracellular Receptors** Intracellular receptors are cytoplasmic or nuclear proteins. Signaling molecules that are small or hydrophobic and can readily cross the plasma membrane use these receptors.

MEDIA

Activity Reception

CONCEPT **11.3**

Transduction: Cascades of molecular interactions relay signals from receptors to target molecules in the cell (pp. 214–218)

▶ **Signal Transduction Pathways** At each step in a pathway, the signal is transduced into a different form, commonly a shape change in a protein.

▶ **Protein Phosphorylation and Dephosphorylation** Many signal transduction pathways include phosphorylation cascades, in which a series of protein kinases each add a phosphate group to the next one in line, activating it. Phosphatase enzymes soon remove the phosphates.

▶ **Small Molecules and Ions as Second Messengers** Second messengers, such as cyclic AMP (cAMP) and Ca^{2+}, diffuse readily through the cytosol and thus help broadcast signals quickly. Many G proteins activate adenylyl cyclase, which makes cAMP from ATP. Cells use Ca^{2+} as a second messenger in both G-protein and tyrosine kinase pathways. The tyrosine kinase pathways can also involve two other second messengers, DAG and IP_3. IP_3 can trigger a subsequent increase in Ca^{2+} levels.

MEDIA

Activity Signal Transduction Pathways

CONCEPT **11.4**

Response: Cell signaling leads to regulation of transcription or cytoplasmic activities (pp. 218–223)

▶ **Nuclear and Cytoplasmic Responses** Some pathways regulate genes by activating transcription factors, proteins that turn specific genes on or off. In the cytoplasm, signaling pathways regulate, for example, enzyme activity and cytoskeleton rearrangement, which can lead to cell shape changes.

▶ **Fine-Tuning of the Response** Each catalytic protein in a signaling pathway amplifies the signal by activating multiple copies of the next component of the pathway; for long pathways, the total amplification may be a millionfold or more. The particular combination of proteins in a cell gives the cell great specificity in both the signals it detects and the responses it carries out. Scaffolding proteins can increase signal transduction efficiency. Pathway branching and cross-talk further help the cell coordinate incoming signals. Signal response is terminated quickly by the reversal of ligand binding.

MEDIA

Activity Cellular Responses
Activity Build a Signaling Pathway

CONCEPT **11.5**

Apoptosis (programmed cell death) integrates multiple cell-signaling pathways (pp. 223–225)

▶ Apoptosis is a type of programmed cell death in which cell components are disposed of in an orderly fashion, without damage to neighboring cells.

▶ **Apoptosis in the Soil Worm *Caenorhabditis elegans*** Apoptosis occurs at defined times during embryonic development of *C. elegans.* A protein (Ced-9) in the mitochondrial membrane acts as a brake; when released by a death signal, it allows activation of caspases that carry out apoptosis.

▶ **Apoptotic Pathways and the Signals That Trigger Them** Several apoptotic pathways exist in the cells of humans and other mammals, and these pathways may be triggered in different ways. A major pathway involves pore formation in the outer mitochondrial membrane, which leads to release of factors that activate caspases. Signals can originate from outside or inside the cell.

TESTING YOUR KNOWLEDGE

SELF-QUIZ

1. Phosphorylation cascades involving a series of protein kinases are useful for cellular signal transduction because
a. they are species specific.
b. they always lead to the same cellular response.
c. they amplify the original signal manyfold.
d. they counter the harmful effects of phosphatases.
e. the number of molecules used is small and fixed.

2. Binding of a signaling molecule to which type of receptor leads directly to a change in the distribution of ions on opposite sides of the membrane?
a. receptor tyrosine kinase
b. G protein-coupled receptor
c. phosphorylated receptor tyrosine kinase dimer
d. ligand-gated ion channel
e. intracellular receptor

3. The activation of receptor tyrosine kinases is characterized by
a. dimerization and phosphorylation.
b. IP_3 binding.
c. a phosphorylation cascade.
d. GTP hydrolysis.
e. channel protein shape change.

4. Which observation suggested to Sutherland the involvement of a second messenger in epinephrine's effect on liver cells?
a. Enzymatic activity was proportional to the amount of calcium added to a cell-free extract.
b. Receptor studies indicated that epinephrine was a ligand.

c. Glycogen breakdown was observed only when epinephrine was administered to intact cells.
d. Glycogen breakdown was observed when epinephrine and glycogen phosphorylase were combined.
e. Epinephrine was known to have different effects on different types of cells.

5. Protein phosphorylation is commonly involved with all of the following *except*
a. regulation of transcription by extracellular signaling molecules.
b. enzyme activation.
c. activation of G protein-coupled receptors.
d. activation of receptor tyrosine kinases.
e. activation of protein kinase molecules.

6. Lipid-soluble signaling molecules, such as testosterone, cross the membranes of all cells but affect only target cells because
a. only target cells retain the appropriate DNA segments.
b. intracellular receptors are present only in target cells.
c. most cells lack the Y chromosome required.
d. only target cells possess the cytosolic enzymes that transduce the testosterone.
e. only in target cells is testosterone able to initiate the phosphorylation cascade leading to activated transcription factor.

7. Consider this pathway: epinephrine → G protein-coupled receptor → G protein → adenylyl cyclase → cAMP. Identify the second messenger.
a. cAMP
b. G protein
c. GTP
d. adenylyl cyclase
e. G protein-coupled receptor

8. Apoptosis involves all but the following:
a. fragmentation of the DNA
b. cell-signaling pathways
c. activation of cellular enzymes
d. lysis of the cell
e. digestion of cellular contents by scavenger cells

9. **DRAW IT** Draw the following apoptotic pathway, which operates in human immune cells. A death signal is received when a molecule called Fas binds its cell-surface receptor. The binding of many Fas molecules to receptors causes receptor clustering. The intracellular regions of the receptors, when together, bind adapter proteins. These in turn bind to inactive forms of caspase-8, which become activated and activate caspase-3, in turn. Once activated, caspase-3 initiates apoptosis.

For Self-Quiz answers, see Appendix A.

MEDIA Visit the Study Area at **www.masteringbio.com** for a Practice Test.

EVOLUTION CONNECTION

10. What evolutionary mechanisms might account for the origin and persistence of cell-to-cell signaling systems in unicellular prokaryotes?

SCIENTIFIC INQUIRY

11. Epinephrine initiates a signal transduction pathway that involves production of cyclic AMP (cAMP) and leads to the breakdown of glycogen to glucose, a major energy source for cells. But glycogen breakdown is actually only part of a "fight-or-flight response" that epinephrine brings about; the overall effect on the body includes increased heart rate and alertness, as well as a burst of energy. Given that caffeine blocks the activity of cAMP phosphodiesterase, propose a mechanism by which caffeine ingestion leads to heightened alertness and sleeplessness.

Biological Inquiry: A Workbook of Investigative Cases Explore cell signaling processes in the hedgehog signaling pathway with the case "Shh: Silencing the Hedgehog Pathway."

SCIENCE, TECHNOLOGY, AND SOCIETY

12. The aging process is thought to be initiated at the cellular level. Among the changes that can occur after a certain number of cell divisions is the loss of a cell's ability to respond to growth factors and other chemical signals. Much research into aging is aimed at understanding such losses, with the ultimate goal of significantly extending the human life span. Not everyone, however, agrees that this is a desirable goal. If life expectancy were greatly increased, what might be the social and ecological consequences? How might we cope with them?

12

The Cell Cycle

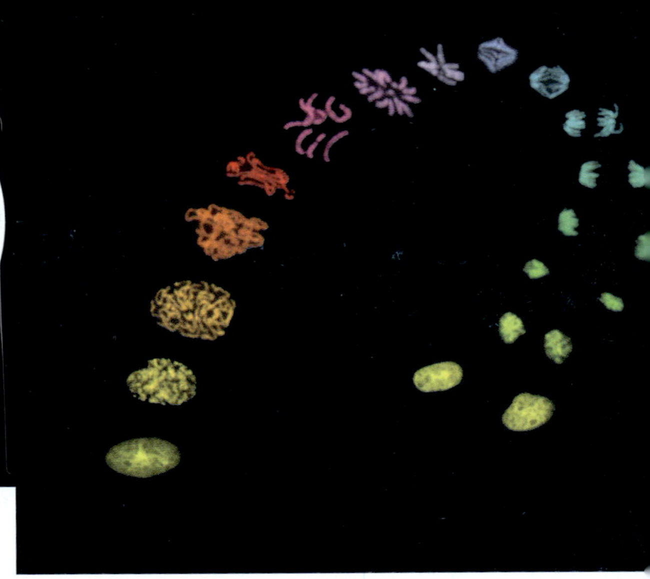

▲ **Figure 12.1 How do a cell's chromosomes change during cell division?**

KEY CONCEPTS

12.1 Cell division results in genetically identical daughter cells

12.2 The mitotic phase alternates with interphase in the cell cycle

12.3 The eukaryotic cell cycle is regulated by a molecular control system

OVERVIEW

The Key Roles of Cell Division

The ability of organisms to reproduce their own kind is the one characteristic that best distinguishes living things from nonliving matter. This unique capacity to procreate, like all biological functions, has a cellular basis. Rudolf Virchow, a German physician, put it this way in 1855: "Where a cell exists, there must have been a preexisting cell, just as the animal arises only from an animal and the plant only from a plant." He summarized this concept with the Latin axiom "*Omnis cellula e cellula*," meaning "Every cell from a cell." The continuity of life is based on the reproduction of cells, or **cell division**. The series of fluorescence micrographs in **Figure 12.1** follows an animal cell's chromosomes, from lower left to lower right, as one cell divides into two.

Cell division plays several important roles in the life of an organism. When a unicellular organism, such as an amoeba, divides and forms duplicate offspring, the division of one cell reproduces an entire organism **(Figure 12.2a)**. Cell division on a larger scale can produce progeny from some multicellular organisms (such as plants that grow from cuttings). Cell division also enables sexually reproducing organisms to develop from a single cell—the fertilized egg, or zygote **(Figure 12.2b)**. And after an organism is fully grown, cell division continues to function in renewal and repair, replacing cells that die from normal wear and tear or accidents. For example, dividing cells in your bone marrow continuously make new blood cells **(Figure 12.2c)**.

The cell division process is an integral part of the **cell cycle**, the life of a cell from the time it is first formed from a dividing parent cell until its own division into two cells. Passing identical genetic material to cellular offspring is a crucial function of cell

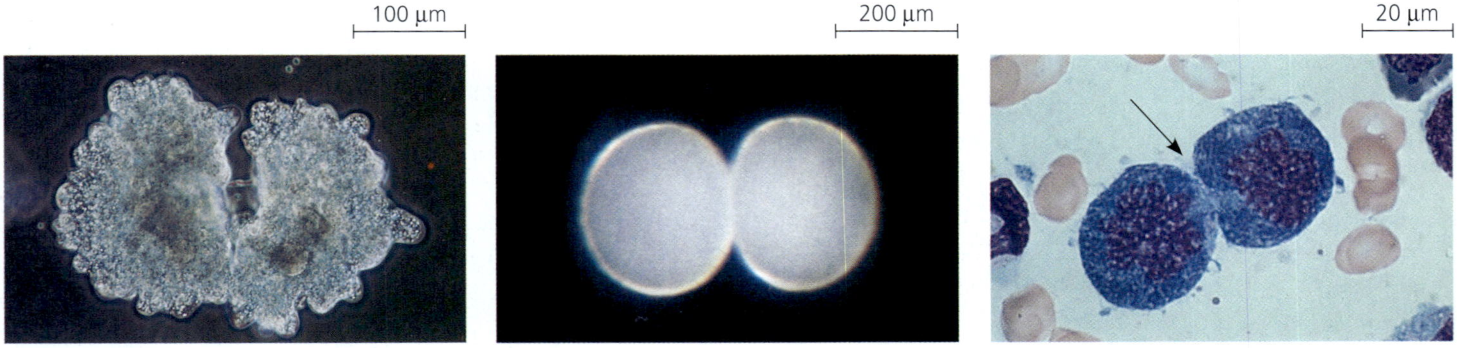

(a) Reproduction. An amoeba, a single-celled eukaryote, is dividing into two cells. Each new cell will be an individual organism (LM).

(b) Growth and development. This micrograph shows a sand dollar embryo shortly after the fertilized egg divided, forming two cells (LM).

(c) Tissue renewal. These dividing bone marrow cells (arrow) will give rise to new blood cells (LM).

▲ **Figure 12.2 The functions of cell division.**

division. In this chapter, you will learn how cell division distributes identical genetic material to daughter cells.* After studying the cellular mechanics of cell division in eukaryotes and bacteria, you will learn about the molecular control system that regulates progress through the eukaryotic cell cycle and what happens when the control system malfunctions. Because cell cycle regulation, or a lack thereof, plays a major role in cancer development, this aspect of cell biology is an active area of research.

CONCEPT 12.1

Cell division results in genetically identical daughter cells

The reproduction of an ensemble as complex as a cell cannot occur by a mere pinching in half; a cell is not like a soap bubble that simply enlarges and splits in two. Most cell division involves the distribution of identical genetic material—DNA—to two daughter cells. (The special type of cell division that produces sperm and eggs results in daughter cells that are *not* genetically identical.) What is most remarkable about cell division is the fidelity with which the DNA is passed along from one generation of cells to the next. A dividing cell duplicates its DNA, allocates the two copies to opposite ends of the cell, and only then splits into daughter cells.

Cellular Organization of the Genetic Material

A cell's endowment of DNA, its genetic information, is called its **genome**. Although a prokaryotic genome is often a single long DNA molecule, eukaryotic genomes usually consist of a number of DNA molecules. The overall length of DNA in a eukaryotic cell is enormous. A typical human cell, for example, has about 2 m of DNA—a length about 250,000 times greater than the cell's diameter. Yet before the cell can divide to form genetically identical daughter cells, all of this DNA must be copied and then the two copies separated so that each daughter cell ends up with a complete genome.

The replication and distribution of so much DNA is manageable because the DNA molecules are packaged into **chromosomes**, so named because they take up certain dyes used in microscopy (from the Greek *chroma*, color, and *soma*, body) **(Figure 12.3)**. Every eukaryotic species has a characteristic number of chromosomes in each cell nucleus. For example, the nuclei of human **somatic cells** (all body cells except the reproductive cells) each contain 46 chromosomes made up of two sets of 23, one set inherited from each parent. Reproductive cells, or **gametes**—sperm and eggs—have half as many chromosomes as somatic cells, or one set of 23 chromosomes in humans. The number of chromosomes in somatic cells varies widely among species: 18 in cabbage plants, 56 in elephants, 90 in hedgehogs, and 148 in one species of alga.

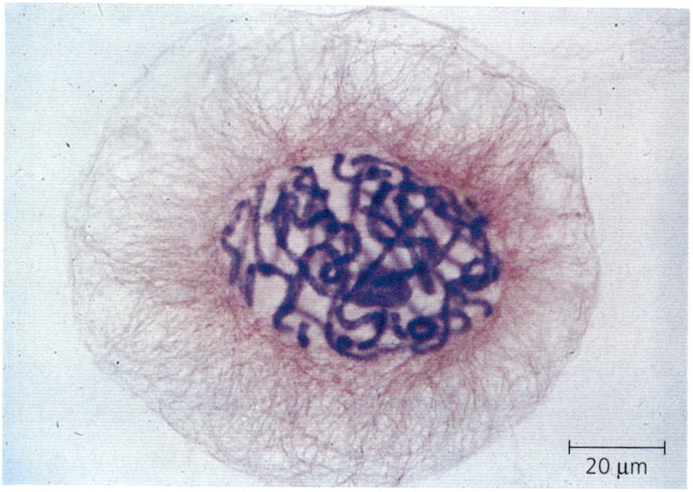

▲ **Figure 12.3 Eukaryotic chromosomes.** Chromosomes (stained purple) are visible within the nucleus of this cell from an African blood lily. The thinner red threads in the surrounding cytoplasm are the cytoskeleton. The cell is preparing to divide (LM).

Eukaryotic chromosomes are made of **chromatin**, a complex of DNA and associated protein molecules. Each single chromosome contains one very long, linear DNA molecule that carries several hundred to a few thousand genes, the units that specify an organism's inherited traits. The associated proteins maintain the structure of the chromosome and help control the activity of the genes.

Distribution of Chromosomes During Eukaryotic Cell Division

When a cell is not dividing, and even as it duplicates its DNA in preparation for cell division, each chromosome is in the form of a long, thin chromatin fiber. After DNA duplication, however, the chromosomes condense: Each chromatin fiber becomes densely coiled and folded, making the chromosomes much shorter and so thick that we can see them with a light microscope.

Each duplicated chromosome has two **sister chromatids**. The two chromatids, each containing an identical DNA molecule, are initially attached all along their lengths by adhesive protein complexes called *cohesins*; this attachment is known as *sister chromatid cohesion*. In its condensed form, the duplicated chromosome has a narrow "waist" at the **centromere**, a specialized region where the two chromatids are most closely attached. The part of a chromatid on either side of the centromere is referred to as an *arm* of the chromatid. Later in the cell division process, the two sister chromatids of each duplicated chromosome separate and move into two new nuclei, one forming at each end of the cell. Once the sister chromatids separate, they are considered individual chromosomes. Thus, each new nucleus receives a collection of chromosomes identical to that of

* Although the terms *daughter cells* and *sister chromatids* (a term you will encounter later in the chapter) are traditional and will be used throughout this book, the structures they refer to have no gender.

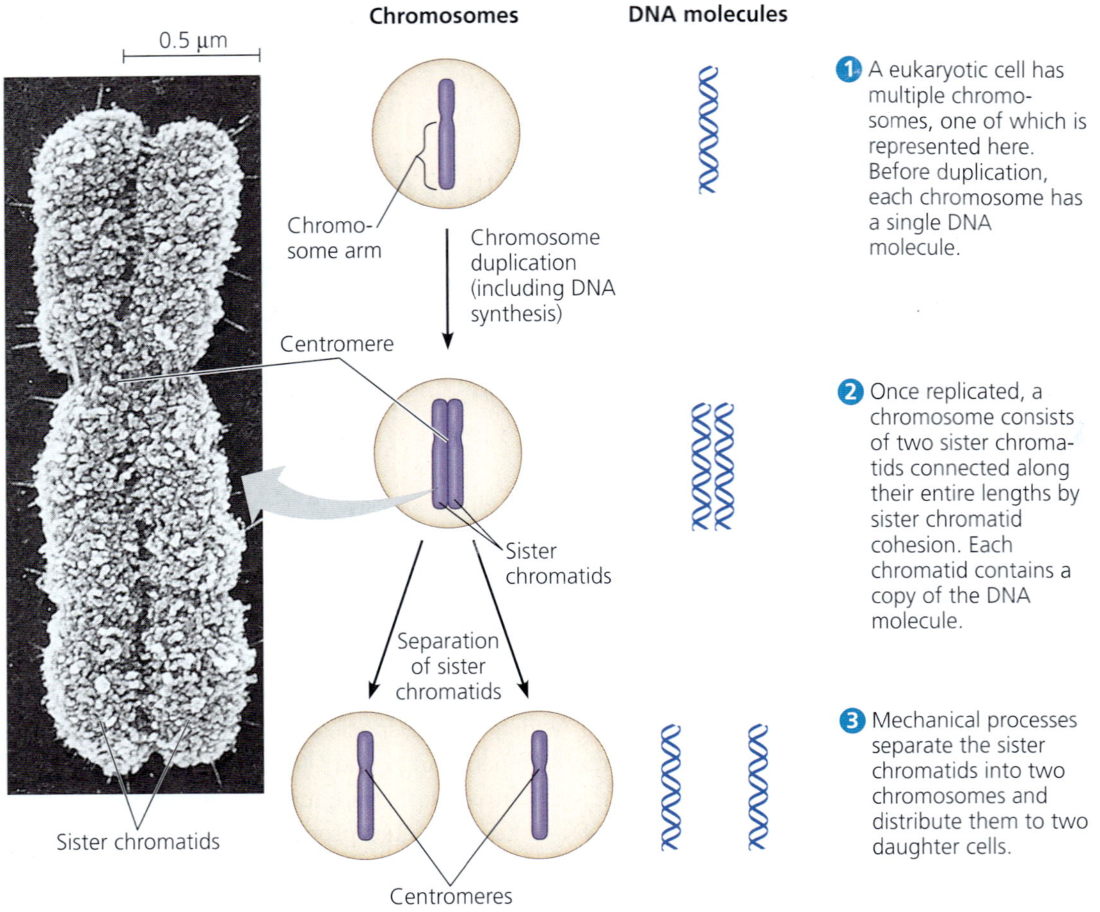

◀ **Figure 12.4 Chromosome duplication and distribution during cell division.** A eukaryotic cell preparing to divide duplicates each of its chromosomes. Next to each chromosome drawing is a simplified double helix representing each DNA molecule. (In an actual chromosome, each DNA molecule would be tightly folded and coiled, complexed with proteins.) The micrograph shows a highly condensed duplicated human chromosome (SEM). The sister chromatids of each duplicated chromosome are distributed to two daughter cells during cell division. (Chromosomes normally exist in the highly condensed state shown here only during the process of cell division; the chromosomes in the top and bottom cells are shown in condensed form for illustration purposes only.)

? *Circle one chromatid in the chromosome in the micrograph. How many arms does the chromosome have?*

the parent cell **(Figure 12.4)**. **Mitosis**, the division of the nucleus, is usually followed immediately by **cytokinesis**, the division of the cytoplasm. Where there was one cell, there are now two, each the genetic equivalent of the parent cell.

What happens to the chromosome number as we follow the human life cycle through the generations? You inherited 46 chromosomes, one set of 23 from each parent. They were combined in the nucleus of a single cell when a sperm from your father united with an egg from your mother, forming a fertilized egg, or zygote. Mitosis and cytokinesis produced the 200 trillion somatic cells that now make up your body, and the same processes continue to generate new cells to replace dead and damaged ones. In contrast, you produce gametes—eggs or sperm—by a variation of cell division called **meiosis**, which yields nonidentical daughter cells that have only one set of chromosomes, thus half as many chromosomes as the parent cell. Meiosis occurs only in your gonads (ovaries or testes). In each generation of humans, meiosis reduces the chromosome number from 46 (two sets of chromosomes) to 23 (one set). Fertilization fuses two gametes together and returns the chromosome number to 46, and mitosis conserves that number in every somatic cell nucleus of the new individual. In Chapter 13, we will examine the role of meiosis in reproduction and inheritance in more detail. In the remainder of this chapter, we focus on mitosis and the rest of the cell cycle in eukaryotes.

CONCEPT CHECK 12.1

1. Starting with a fertilized egg (zygote), a series of five cell divisions would produce an early embryo with how many cells?
2. How many chromatids are in a duplicated chromosome?
3. **WHAT IF?** A chicken has 78 chromosomes in its somatic cells. How many chromosomes did the chicken inherit from each parent? How many chromosomes are in each of the chicken's gametes? How many chromosomes will be in each somatic cell of the chicken's offspring?

For suggested answers, see Appendix A.

CONCEPT **12.2**

The mitotic phase alternates with interphase in the cell cycle

In 1882, a German anatomist named Walther Flemming developed dyes that allowed him to observe, for the first time, the behavior of chromosomes during mitosis and cytokinesis. (In fact, Flemming coined the terms *mitosis* and *chromatin*.) During the period between one cell division and the next, it appeared to Flemming that the cell was simply growing larger. But we now know that many critical events occur during this stage in the life of a cell.

Phases of the Cell Cycle

Mitosis is just one part of the cell cycle **(Figure 12.5)**. In fact, the **mitotic (M) phase**, which includes both mitosis and cytokinesis, is usually the shortest part of the cell cycle. Mitotic cell division alternates with a much longer stage called **interphase**, which often accounts for about 90% of the cycle. It is during interphase that the cell grows and copies its chromosomes in preparation for cell division. Interphase can be divided into subphases: the **G_1 phase** ("first gap"), the **S phase** ("synthesis"), and the **G_2 phase** ("second gap"). During all three subphases, the cell grows by producing proteins and cytoplasmic organelles such as mitochondria and endoplasmic reticulum. However, chromosomes are duplicated only during the S phase (we will discuss synthesis of DNA in Chapter 16). Thus, a cell grows (G_1), continues to grow as it copies its chromosomes (S), grows more as it completes preparations for cell division (G_2), and divides (M). The daughter cells may then repeat the cycle.

A particular human cell might undergo one division in 24 hours. Of this time, the M phase would occupy less than 1 hour, while the S phase might occupy about 10–12 hours, or about half the cycle. The rest of the time would be apportioned between the G_1 and G_2 phases. The G_2 phase usually takes 4–6 hours; in our example, G_1 would occupy about 5–6 hours. G_1 is the most variable in length in different types of cells.

Mitosis is conventionally broken down into five stages: **prophase, prometaphase, metaphase, anaphase,** and **telophase.** Overlapping with the latter stages of mitosis, cytokinesis completes the mitotic phase. **Figure 12.6**, on the next two pages, describes these stages in an animal cell. Be sure to study this figure thoroughly before progressing to the next two sections, which examine mitosis and cytokinesis more closely.

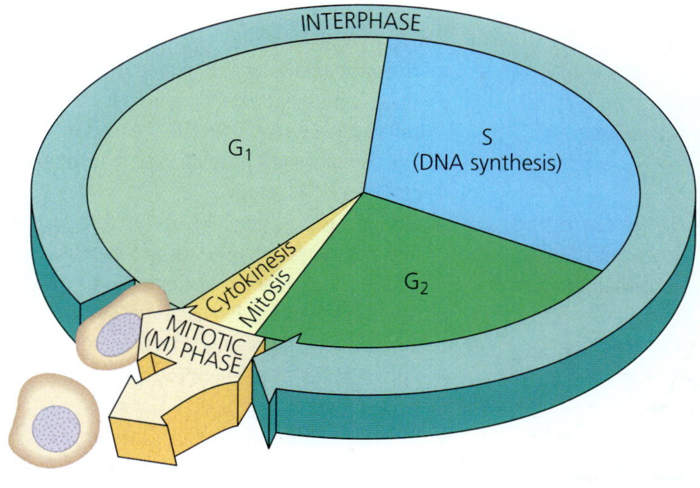

▲ **Figure 12.5 The cell cycle.** In a dividing cell, the mitotic (M) phase alternates with interphase, a growth period. The first part of interphase (G_1) is followed by the S phase, when the chromosomes replicate; G_2 is the last part of interphase. In the M phase, mitosis divides the nucleus and distributes its chromosomes to the daughter nuclei, and cytokinesis divides the cytoplasm, producing two daughter cells. The relative durations of G_1, S, and G_2 may vary.

The Mitotic Spindle: *A Closer Look*

Many of the events of mitosis depend on the **mitotic spindle**, which begins to form in the cytoplasm during prophase. This structure consists of fibers made of microtubules and associated proteins. While the mitotic spindle assembles, the other microtubules of the cytoskeleton partially disassemble, probably providing the material used to construct the spindle. The spindle microtubules elongate (polymerize) by incorporating more subunits of the protein tubulin and shorten (depolymerize) by losing subunits (see Table 6.1).

In animal cells, the assembly of spindle microtubules starts at the **centrosome**, a subcellular region containing material that functions throughout the cell cycle to organize the cell's microtubules (it is also called the *microtubule-organizing center*). A pair of centrioles is located at the center of the centrosome, but they are not essential for cell division: If the centrioles are destroyed with a laser microbeam, a spindle nevertheless forms during mitosis. In fact, centrioles are not even present in plant cells, which do form mitotic spindles.

During interphase in animal cells, the single centrosome replicates, forming two centrosomes, which remain together near the nucleus. The two centrosomes move apart during prophase and prometaphase of mitosis as spindle microtubules grow out from them. By the end of prometaphase, the two centrosomes, one at each pole of the spindle, are at opposite ends of the cell. An **aster**, a radial array of short microtubules, extends from each centrosome. The spindle includes the centrosomes, the spindle microtubules, and the asters.

Each of the two sister chromatids of a replicated chromosome has a **kinetochore**, a structure of proteins associated with specific sections of chromosomal DNA at the centromere. The chromosome's two kinetochores face in opposite directions. During prometaphase, some of the spindle microtubules attach to the kinetochores; these are called kinetochore microtubules. (The number of microtubules attached to a kinetochore varies among species, from one microtubule in yeast cells to 40 or so in some mammalian cells.) When one of a chromosome's kinetochores is "captured" by microtubules, the chromosome begins to move toward the pole from which those microtubules extend. However, this movement is checked as soon as microtubules from the opposite pole attach to the other kinetochore. What happens next is like a tug-of-war that ends in a draw. The chromosome moves first in one direction, then the other, back and forth, finally settling midway between the two ends of the cell. At metaphase, the centromeres of all the duplicated chromosomes are on a plane midway between the spindle's two poles. This imaginary plane is called the **metaphase plate** of the cell

▼ Figure 12.6

Exploring The Mitotic Division of an Animal Cell

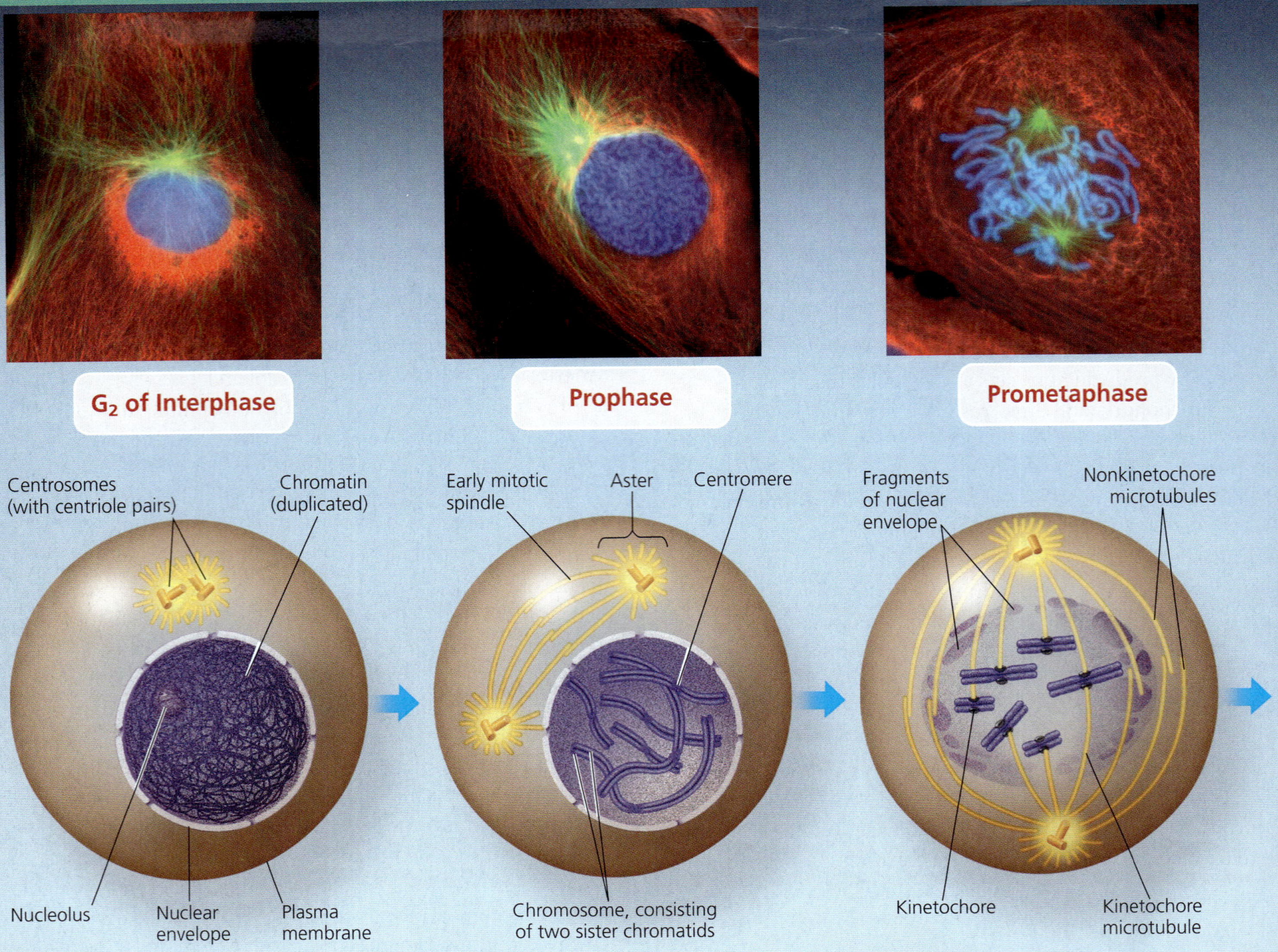

G_2 of Interphase

- A nuclear envelope bounds the nucleus.
- The nucleus contains one or more nucleoli (singular, *nucleolus*).
- Two centrosomes have formed by replication of a single centrosome.
- In animal cells, each centrosome features two centrioles.
- Chromosomes, duplicated during S phase, cannot be seen individually because they have not yet condensed.

The light micrographs show dividing lung cells from a newt, which has 22 chromosomes in its somatic cells (chromosomes appear blue, microtubules green, and intermediate filaments red). For simplicity, the drawings show only six chromosomes.

Prophase

- The chromatin fibers become more tightly coiled, condensing into discrete chromosomes observable with a light microscope.
- The nucleoli disappear.
- Each duplicated chromosome appears as two identical sister chromatids joined together at their centromeres and all along their arms by cohesins (sister chromatid cohesion).
- The mitotic spindle (named for its shape) begins to form. It is composed of the centrosomes and the microtubules that extend from them. The radial arrays of shorter microtubules that extend from the centrosomes are called asters ("stars").
- The centrosomes move away from each other, apparently propelled by the lengthening microtubules between them.

Prometaphase

- The nuclear envelope fragments.
- The microtubules extending from each centrosome can now invade the nuclear area.
- The chromosomes have become even more condensed.
- Each of the two chromatids of each chromosome now has a kinetochore, a specialized protein structure located at the centromere.
- Some of the microtubules attach to the kinetochores, becoming "kinetochore microtubules"; these jerk the chromosomes back and forth.
- Nonkinetochore microtubules interact with those from the opposite pole of the spindle.

? *How many molecules of DNA are in the prometaphase drawing? How many molecules per chromosome? How many double helices are there per chromosome? Per chromatid?*

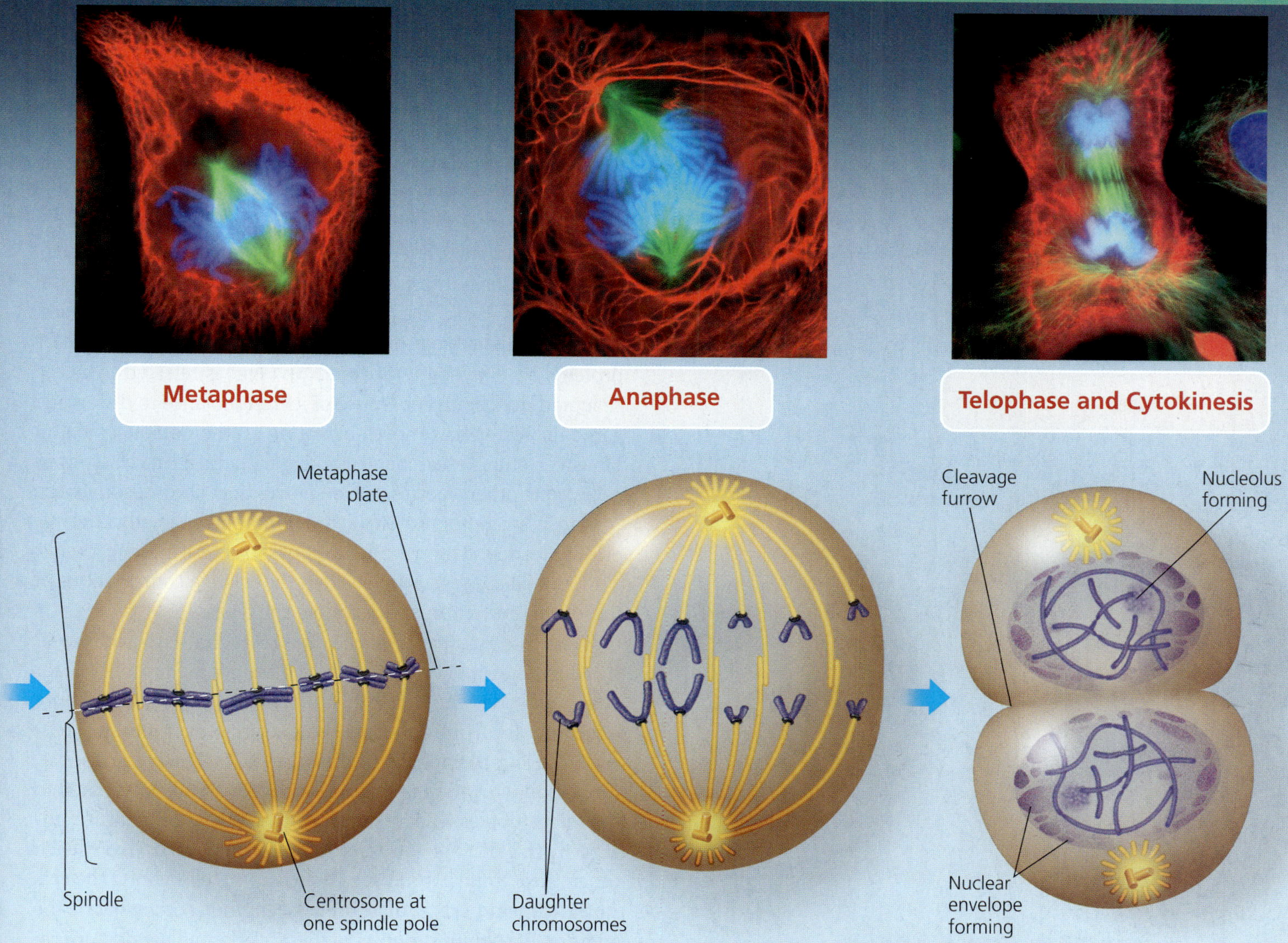

Metaphase

- Metaphase is the longest stage of mitosis, often lasting about 20 minutes.
- The centrosomes are now at opposite poles of the cell.
- The chromosomes convene on the metaphase plate, an imaginary plane that is equidistant between the spindle's two poles. The chromosomes' centromeres lie on the metaphase plate.
- For each chromosome, the kinetochores of the sister chromatids are attached to kinetochore microtubules coming from opposite poles.

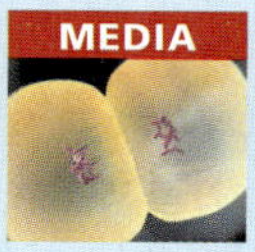

BioFlix Visit the Study Area at **www.masteringbio.com** for the BioFlix 3-D Animation on Mitosis.

Anaphase

- Anaphase is the shortest stage of mitosis, often lasting only a few minutes.
- Anaphase begins when the cohesin proteins are cleaved. This allows the two sister chromatids of each pair to part suddenly. Each chromatid thus becomes a full-fledged chromosome.
- The two liberated daughter chromosomes begin moving toward opposite ends of the cell as their kinetochore microtubules shorten. Because these microtubules are attached at the centromere region, the chromosomes move centromere first (at about 1 μm/min).
- The cell elongates as the nonkinetochore microtubules lengthen.
- By the end of anaphase, the two ends of the cell have equivalent—and complete—collections of chromosomes.

Telophase

- Two daughter nuclei form in the cell.
- Nuclear envelopes arise from the fragments of the parent cell's nuclear envelope and other portions of the endomembrane system.
- Nucleoli reappear.
- The chromosomes become less condensed.
- Mitosis, the division of one nucleus into two genetically identical nuclei, is now complete.

Cytokinesis

- The division of the cytoplasm is usually well under way by late telophase, so the two daughter cells appear shortly after the end of mitosis.
- In animal cells, cytokinesis involves the formation of a cleavage furrow, which pinches the cell in two.

(Figure 12.7). Meanwhile, microtubules that do not attach to kinetochores have been elongating, and by metaphase they overlap and interact with other nonkinetochore microtubules from the opposite pole of the spindle. (These are sometimes called "polar" microtubules.) By metaphase, the microtubules of the asters have also grown and are in contact with the plasma membrane. The spindle is now complete.

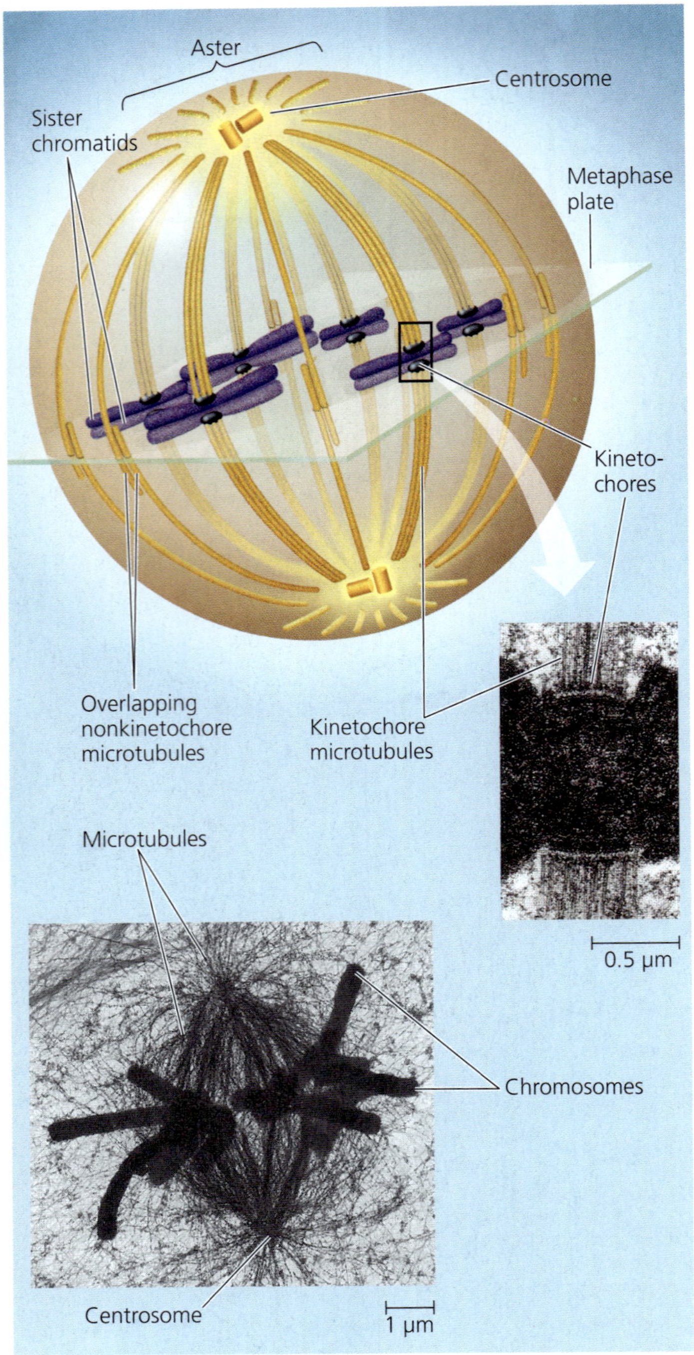

▲ **Figure 12.7 The mitotic spindle at metaphase.** The kinetochores of each chromosome's two sister chromatids face in opposite directions. Here, each kinetochore is attached to a *cluster* of kinetochore microtubules extending from the nearest centrosome. Nonkinetochore microtubules overlap at the metaphase plate (TEMs).

DRAW IT *On the lower micrograph, draw a line indicating the metaphase plate. Circle an aster. Draw arrows indicating the directions of chromosome movement once anaphase begins.*

Let's now see how the structure of the completed spindle correlates with its function during anaphase. Anaphase commences suddenly when the cohesins holding together the sister chromatids of each chromosome are cleaved by enzymes. Once the chromatids become separate, full-fledged chromosomes, they move toward opposite ends of the cell.

How do the kinetochore microtubules function in this poleward movement of chromosomes? Apparently, two mechanisms are in play, both involving motor proteins. (To review how motor proteins move an object along a microtubule, see Figure 6.21.) A clever experiment carried out in Gary Borisy's lab at the University of Wisconsin in 1987 suggested that motor proteins on the kinetochores "walk" the chromosomes along the microtubules, which depolymerize at their kinetochore ends after the motor proteins have passed (Figure 12.8). (This is referred to as the "Pacman" mechanism because of its resemblance to the arcade game character that moves by eating all the dots in its path.) However, other researchers, working with different cell types or cells from other species, have shown that chromosomes are "reeled in" by motor proteins at the spindle poles and that the microtubules depolymerize after they pass by these motor proteins. The general consensus now is that the relative contributions of these two mechanisms vary among cell types.

What is the function of the *non*kinetochore microtubules? In a dividing animal cell, these microtubules are responsible for elongating the whole cell during anaphase. Nonkinetochore microtubules from opposite poles overlap each other extensively during metaphase (see Figure 12.7). During anaphase, the region of overlap is reduced as motor proteins attached to the microtubules walk them away from one another, using energy from ATP. As the microtubules push apart from each other, their spindle poles are pushed apart, elongating the cell. At the same time, the microtubules lengthen somewhat by the addition of tubulin subunits to their overlapping ends. As a result, the microtubules continue to overlap.

At the end of anaphase, duplicate groups of chromosomes have arrived at opposite ends of the elongated parent cell. Nuclei re-form during telophase. Cytokinesis generally begins during anaphase or telophase, and the spindle eventually disassembles.

Cytokinesis: *A Closer Look*

In animal cells, cytokinesis occurs by a process known as **cleavage**. The first sign of cleavage is the appearance of a **cleavage furrow**, a shallow groove in the cell surface near the old metaphase plate (Figure 12.9a). On the cytoplasmic side of the furrow is a contractile ring of actin microfilaments associated with molecules of the protein myosin. (Actin and myosin are also responsible for muscle contraction and many other kinds of cell movement.) The actin microfilaments interact with the myosin molecules, causing the ring to contract. The contraction of the dividing cell's ring of microfilaments is like the pulling of drawstrings. The cleavage furrow deepens

▼ Figure 12.8 Inquiry

At which end do kinetochore microtubules shorten during anaphase?

EXPERIMENT Gary Borisy and colleagues wanted to determine whether kinetochore microtubules depolymerize at the kinetochore end or the pole end as chromosomes move toward the poles during mitosis. First, they labeled the microtubules of a pig kidney cell in early anaphase with a yellow fluorescent dye.

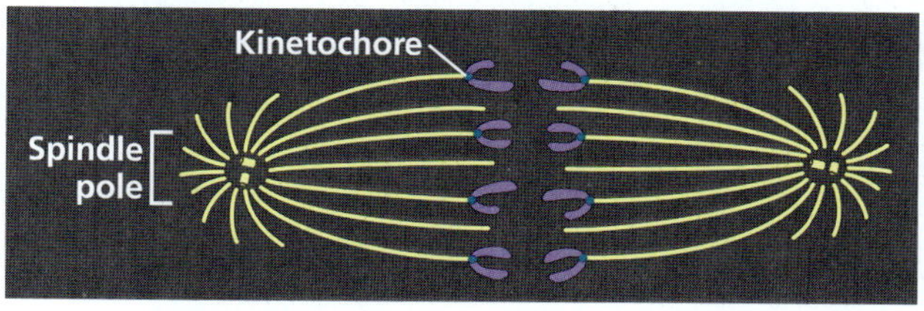

Then they marked a region of the kinetochore microtubules between one spindle pole and the chromosomes by using a laser to eliminate the fluorescence from that region. (The microtubules remained intact.) As anaphase proceeded, they monitored the changes in microtubule length on either side of the mark.

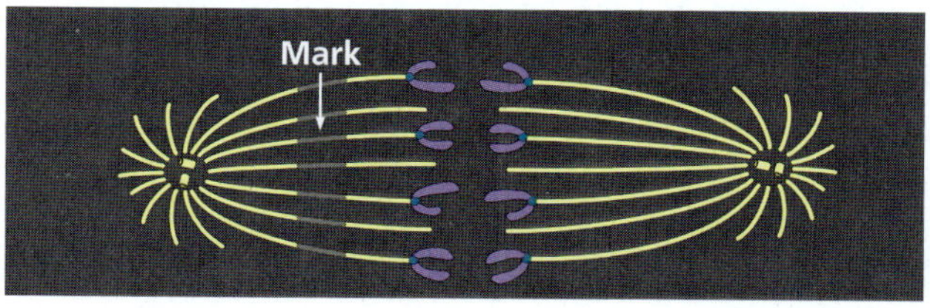

RESULTS As the chromosomes moved poleward, the microtubule segments on the kinetochore side of the mark shortened, while those on the spindle pole side stayed the same length.

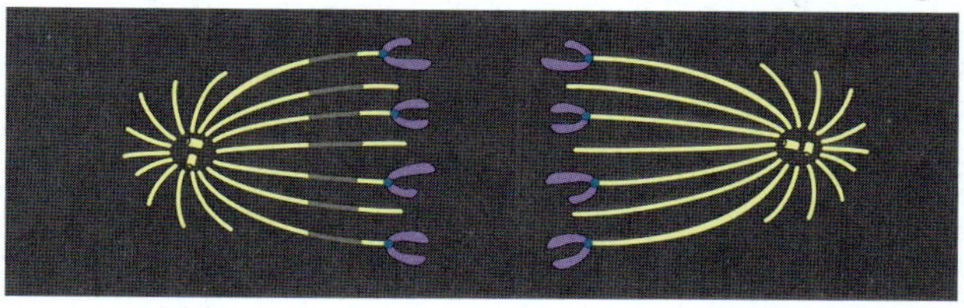

CONCLUSION During anaphase in this cell type, chromosome movement is correlated with kinetochore microtubules shortening at their kinetochore ends and not at their spindle pole ends. This experiment supports the hypothesis that during anaphase, a chromosome is walked along a microtubule as the microtubule depolymerizes at its kinetochore end, releasing tubulin subunits.

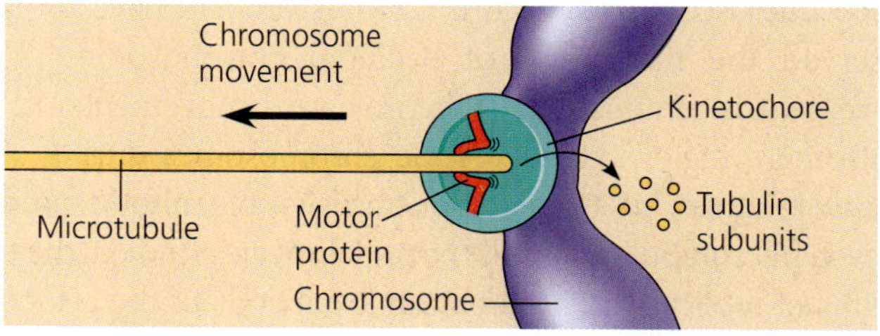

SOURCE G. J. Gorbsky, P. J. Sammak, and G. G. Borisy, Chromosomes move poleward in anaphase along stationary microtubules that coordinately disassemble from their kinetochore ends, *Journal of Cell Biology* 104:9–18 (1987).

WHAT IF? If this experiment had been done on a cell type in which "reeling in" at the poles was the main cause of chromosome movement, how would the mark have moved relative to the poles? How would the microtubule lengths have changed?

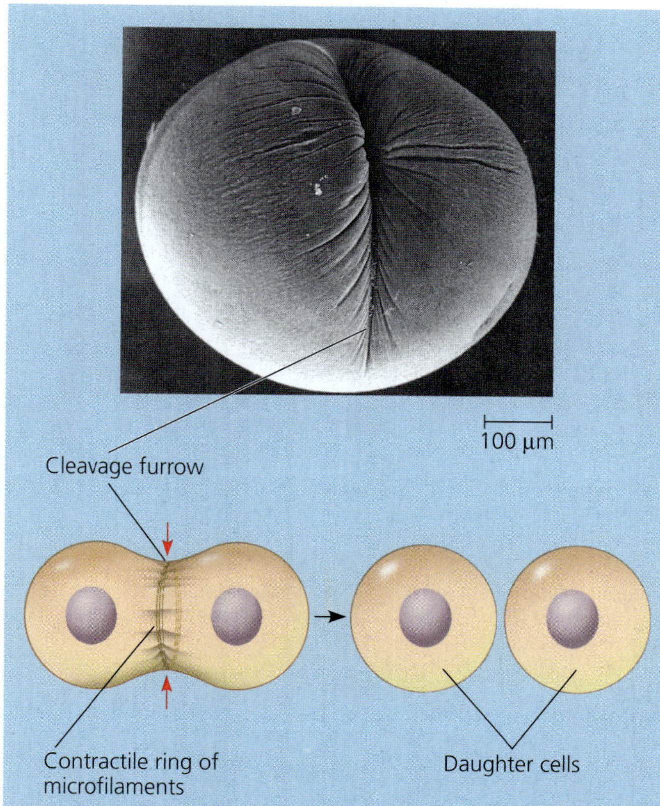

(a) Cleavage of an animal cell (SEM)

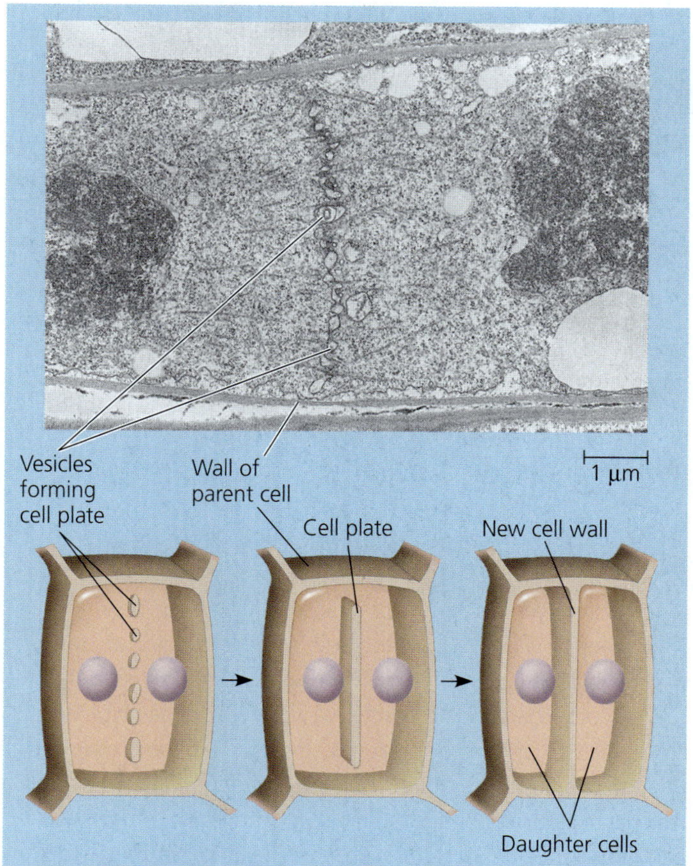

(b) Cell plate formation in a plant cell (TEM)

▲ **Figure 12.9 Cytokinesis in animal and plant cells.**

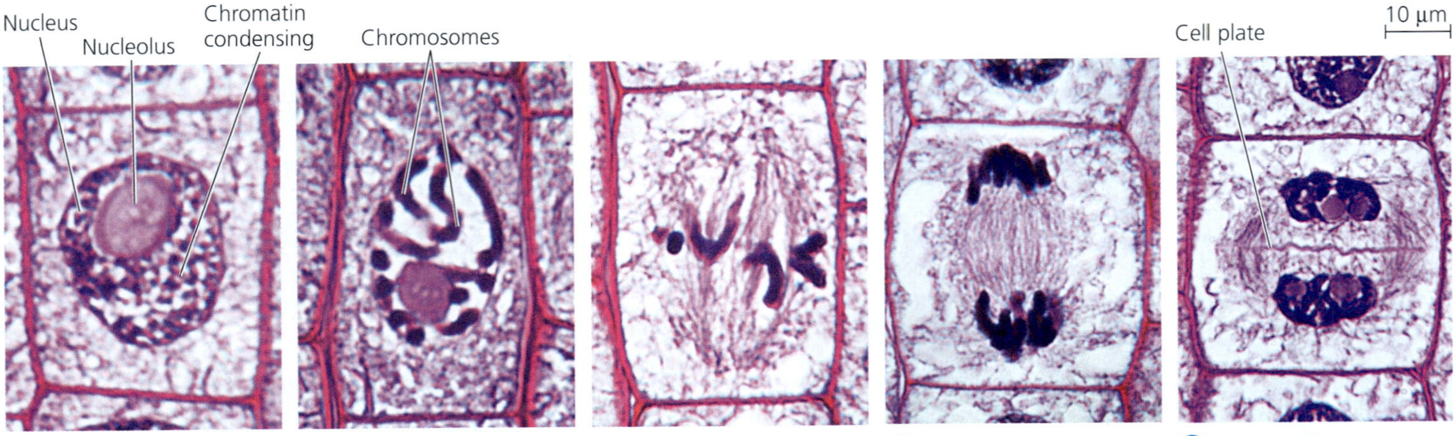

1 **Prophase.** The chromatin is condensing and the nucleolus is beginning to disappear. Although not yet visible in the micrograph, the mitotic spindle is starting to form.

2 **Prometaphase.** Discrete chromosomes are now visible; each consists of two aligned, identical sister chromatids. Later in prometaphase, the nuclear envelope will fragment.

3 **Metaphase.** The spindle is complete, and the chromosomes, attached to microtubules at their kinetochores, are all at the metaphase plate.

4 **Anaphase.** The chromatids of each chromosome have separated, and the daughter chromosomes are moving to the ends of the cell as their kinetochore microtubules shorten.

5 **Telophase.** Daughter nuclei are forming. Meanwhile, cytokinesis has started: The cell plate, which will divide the cytoplasm in two, is growing toward the perimeter of the parent cell.

▲ **Figure 12.10 Mitosis in a plant cell.** These light micrographs show mitosis in cells of an onion root.

until the parent cell is pinched in two, producing two completely separated cells, each with its own nucleus and share of cytosol, organelles, and other subcellular structures.

Cytokinesis in plant cells, which have cell walls, is markedly different. There is no cleavage furrow. Instead, during telophase, vesicles derived from the Golgi apparatus move along microtubules to the middle of the cell, where they coalesce, producing a **cell plate** **(Figure 12.9b)**. Cell wall materials carried in the vesicles collect in the cell plate as it grows. The cell plate enlarges until its surrounding membrane fuses with the plasma membrane along the perimeter of the cell. Two daughter cells result, each with its own plasma membrane. Meanwhile, a new cell wall arising from the contents of the cell plate has formed between the daughter cells.

Figure 12.10 is a series of micrographs of a dividing plant cell. Examining this figure will help you review mitosis and cytokinesis.

Binary Fission

The asexual reproduction of single-celled eukaryotes, such as the amoeba in Figure 12.2a, includes mitosis and occurs by a type of cell division called **binary fission**, meaning "division in half." Prokaryotes (bacteria and archaea) also reproduce by binary fission, but the prokaryotic process does not involve mitosis. In bacteria, most genes are carried on a single *bacterial chromosome* that consists of a circular DNA molecule and associated proteins. Although bacteria are smaller and simpler than eukaryotic cells, the challenge of replicating their genomes in an orderly fashion and distributing the copies equally to two daughter cells is still formidable. The chromosome of the bacterium *Escherichia coli*, for example, when it is fully stretched out, is about 500 times as long as the cell. For such a long chromosome to fit within the cell requires that it be highly coiled and folded.

In *E. coli*, the process of cell division is initiated when the DNA of the bacterial chromosome begins to replicate at a specific place on the chromosome called the **origin of replication**, producing two origins. As the chromosome continues to replicate, one origin moves rapidly toward the opposite end of the cell **(Figure 12.11)**. While the chromosome is replicating, the cell elongates. When replication is complete and the bacterium has reached about twice its initial size, its plasma membrane grows inward, dividing the parent *E. coli* cell into two daughter cells. Each cell inherits a complete genome.

Using the techniques of modern DNA technology to tag the origins of replication with molecules that glow green in fluorescence microscopy (see Figure 6.3), researchers have directly observed the movement of bacterial chromosomes. This movement is reminiscent of the poleward movements of the centromere regions of eukaryotic chromosomes during anaphase of mitosis, but bacteria don't have visible mitotic spindles or even microtubules. In most bacterial species studied, the two origins of replication end up at opposite ends of the cell or in some other very specific location, possibly anchored there by one or more proteins. How bacterial chromosomes move and how their specific location is established and maintained are still not fully understood. However, several proteins have been identified that play important roles: One resembling eukaryotic actin may function in bacterial chromosome movement during cell division, and another that is related to tubulin may help separate the two bacterial daughter cells.

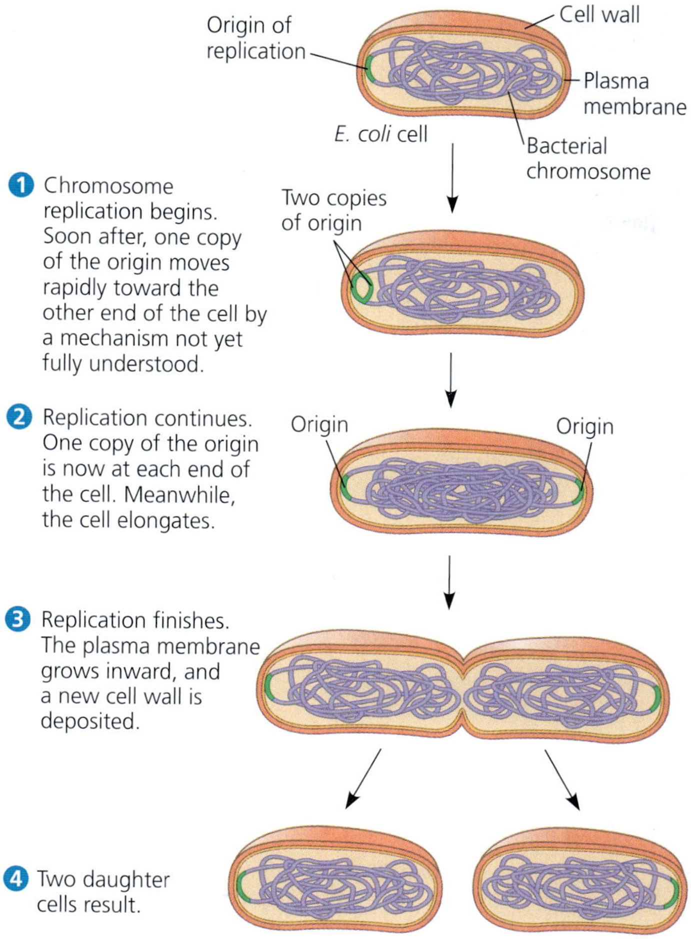

▲ **Figure 12.11 Bacterial cell division by binary fission.** The example shown here is the bacterium *E. coli*, which has a single, circular chromosome.

The Evolution of Mitosis

How did mitosis evolve? Given that prokaryotes preceded eukaryotes on Earth by more than a billion years, we might hypothesize that mitosis had its origins in simpler prokaryotic mechanisms of cell reproduction. The fact that some of the proteins involved in bacterial binary fission are related to eukaryotic proteins that function in mitosis supports that hypothesis.

As eukaryotes evolved, along with their larger genomes and nuclear envelopes, the ancestral process of binary fission, seen today in bacteria, somehow gave rise to mitosis. **Figure 12.12** traces a hypothesis for the stepwise evolution of mitosis. Possible intermediate stages are represented by two unusual types of nuclear division found today in certain unicellular eukaryotes. These two examples of nuclear division are thought to be cases where ancestral mechanisms have remained relatively unchanged over evolutionary time. In both types, the nuclear envelope remains intact. In dinoflagellates, replicated chromosomes are attached to the nuclear envelope and separate as the nucleus elongates prior to dividing. In diatoms and yeasts, a spindle *within* the nucleus separates the chromosomes. In most eukaryotic cells, the nuclear envelope breaks down and a spindle separates the chromosomes.

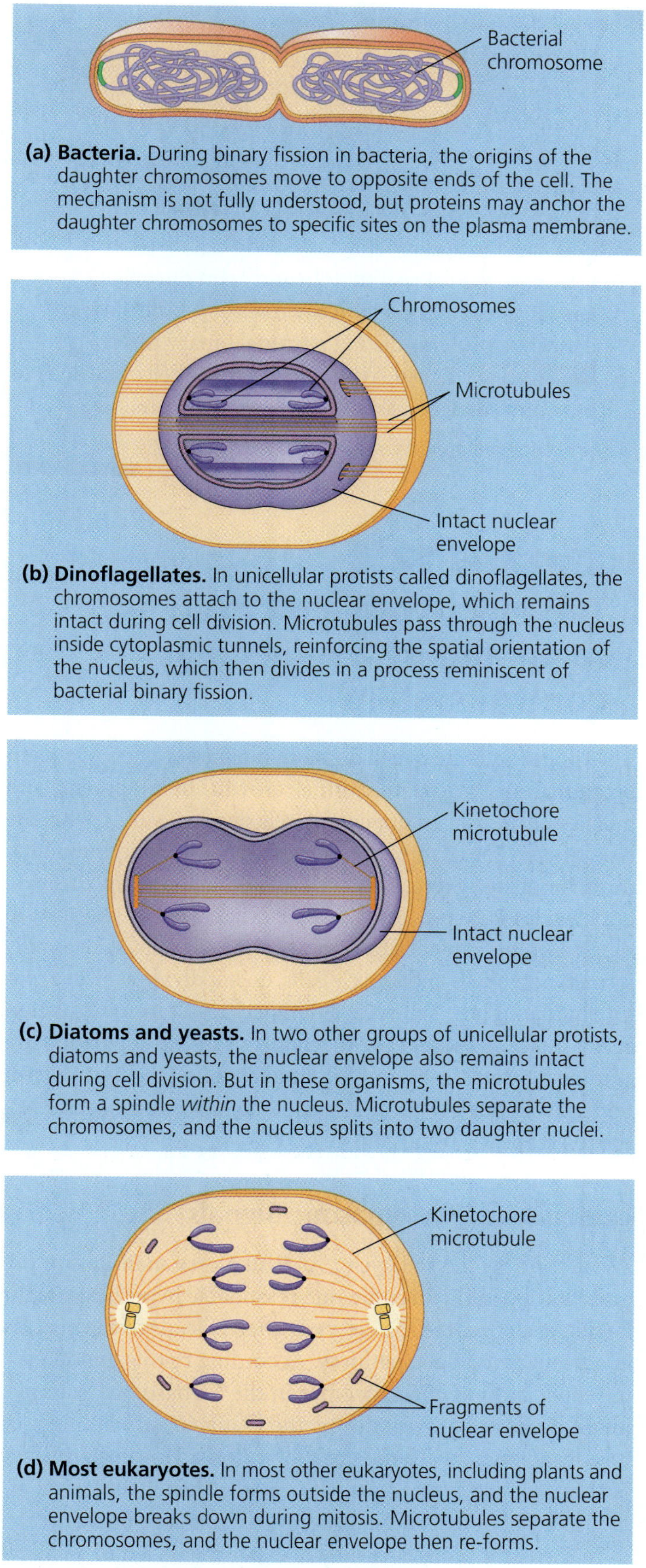

(a) Bacteria. During binary fission in bacteria, the origins of the daughter chromosomes move to opposite ends of the cell. The mechanism is not fully understood, but proteins may anchor the daughter chromosomes to specific sites on the plasma membrane.

(b) Dinoflagellates. In unicellular protists called dinoflagellates, the chromosomes attach to the nuclear envelope, which remains intact during cell division. Microtubules pass through the nucleus inside cytoplasmic tunnels, reinforcing the spatial orientation of the nucleus, which then divides in a process reminiscent of bacterial binary fission.

(c) Diatoms and yeasts. In two other groups of unicellular protists, diatoms and yeasts, the nuclear envelope also remains intact during cell division. But in these organisms, the microtubules form a spindle *within* the nucleus. Microtubules separate the chromosomes, and the nucleus splits into two daughter nuclei.

(d) Most eukaryotes. In most other eukaryotes, including plants and animals, the spindle forms outside the nucleus, and the nuclear envelope breaks down during mitosis. Microtubules separate the chromosomes, and the nuclear envelope then re-forms.

▲ **Figure 12.12 A hypothetical sequence for the evolution of mitosis.** Some unicellular eukaryotes existing today have mechanisms of cell division that appear to be intermediate between the binary fission of bacteria (a) and mitosis as it occurs in most other eukaryotes (d). Except for (a), these schematic diagrams do not show cell walls.

CONCEPT CHECK 12.2

1. How many chromosomes are shown in the diagram in Figure 12.7? How many chromatids are shown?
2. Compare cytokinesis in animal cells and plant cells.
3. What is a function of nonkinetochore microtubules?
4. Identify three similarities between bacterial chromosomes and eukaryotic chromosomes, considering both structure and behavior during cell division.
5. Compare the roles of tubulin and actin during eukaryotic cell division with the roles of tubulin-like and actin-like proteins during bacterial binary fission.
6. **WHAT IF?** During which stages of the cell cycle does a chromosome consist of two identical chromatids?

For suggested answers, see Appendix A.

CONCEPT **12.3**

The eukaryotic cell cycle is regulated by a molecular control system

The timing and rate of cell division in different parts of a plant or animal are crucial to normal growth, development, and maintenance. The frequency of cell division varies with the type of cell. For example, human skin cells divide frequently throughout life, whereas liver cells maintain the ability to divide but keep it in reserve until an appropriate need arises—say, to repair a wound. Some of the most specialized cells, such as fully formed nerve cells and muscle cells, do not divide at all in a mature human. These cell cycle differences result from regulation at the molecular level. The mechanisms of this regulation are of intense interest, not only for understanding the life cycles of normal cells but also for understanding how cancer cells manage to escape the usual controls.

▼ **Figure 12.13 Inquiry**

Do molecular signals in the cytoplasm regulate the cell cycle?

EXPERIMENT Researchers at the University of Colorado wondered whether a cell's progression through the cell cycle is controlled by cytoplasmic molecules. To investigate this, they induced cultured mammalian cells at different phases of the cell cycle to fuse. Two such experiments are shown here.

RESULTS

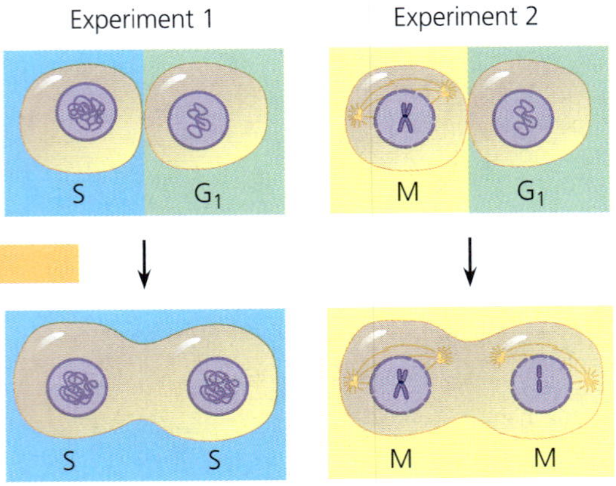

When a cell in the S phase was fused with a cell in G_1, the G_1 nucleus immediately entered the S phase—DNA was synthesized.

When a cell in the M phase was fused with a cell in G_1, the G_1 nucleus immediately began mitosis—a spindle formed and chromatin condensed, even though the chromosome had not been duplicated.

CONCLUSION The results of fusing a G_1 cell with a cell in the S or M phase of the cell cycle suggest that molecules present in the cytoplasm during the S or M phase control the progression to those phases.

SOURCE R. T. Johnson and P. N. Rao, Mammalian cell fusion: Induction of premature chromosome condensation in interphase nuclei, *Nature* 226:717–722 (1970).

WHAT IF? If the progression of phases did not depend on cytoplasmic molecules and each phase began when the previous one was complete, how would the results have differed?

Evidence for Cytoplasmic Signals

What controls the cell cycle? As Paul Nurse mentions in the interview opening this unit, one reasonable hypothesis might be that each event in the cell cycle merely leads to the next, as in a simple metabolic pathway. According to this hypothesis, the replication of chromosomes in the S phase, for example, might cause cell growth during the G_2 phase, which might in turn lead inevitably to the onset of mitosis. However, this hypothesis, which proposes a pathway that is not subject to either internal or external regulation, turns out to be incorrect.

In the early 1970s, a variety of experiments led to an alternative hypothesis: that the cell cycle is driven by specific signaling molecules present in the cytoplasm. Some of the first strong evidence for this hypothesis came from experiments with mammalian cells grown in culture. In these experiments, two cells in different phases of the cell cycle were fused to form a single cell with two nuclei. If one of the original cells was in the S phase and the other was in G_1, the G_1 nucleus immediately entered the S phase, as though stimulated by chemicals present in the cytoplasm of the first cell. Similarly, if a cell undergoing mitosis (M phase) was fused with another cell in any stage of its cell cycle, even G_1, the second nucleus immediately entered mitosis, with condensation of the chromatin and formation of a mitotic spindle **(Figure 12.13)**.

The Cell Cycle Control System

The experiment shown in Figure 12.13 and other experiments on animal cells and yeasts demonstrated that the sequential events of the cell cycle are directed by a distinct **cell cycle control system**, a cyclically operating set of molecules in the cell that both triggers and coordinates key events in the cell cycle. The cell cycle control system has been compared to the control device of an automatic washing machine **(Figure 12.14)**. Like the washer's timing device, the cell cycle control system proceeds on its own, according to a built-in clock. However, just as a washer's cycle is subject to both internal control (such as the sensor that detects when the tub is filled with water) and external adjustment (such as activation of the start mechanism), the cell cycle is regulated at certain checkpoints by both internal and external signals.

A **checkpoint** in the cell cycle is a control point where stop and go-ahead signals can regulate the cycle. (The signals are transmitted within the cell by the kinds of signal transduction pathways discussed in Chapter 11.) Animal cells generally have built-in stop signals that halt the cell cycle at checkpoints until overridden by go-ahead signals. Many signals registered at checkpoints come from cellular surveillance mechanisms inside the cell; the signals report whether crucial cellular processes that should have occurred by that point have in fact been completed correctly and thus whether or not the cell cycle should proceed. Checkpoints also register signals from outside the cell, as we will discuss later. Three major checkpoints are found in the G_1, G_2, and M phases (see Figure 12.14).

For many cells, the G_1 checkpoint—dubbed the "restriction point" in mammalian cells—seems to be the most important. If a cell receives a go-ahead signal at the G_1 checkpoint, it will usually complete the G_1, S, G_2, and M phases and divide. If it does not receive a go-ahead signal at that point, it will exit the cycle, switching into a nondividing state called the **G_0 phase**

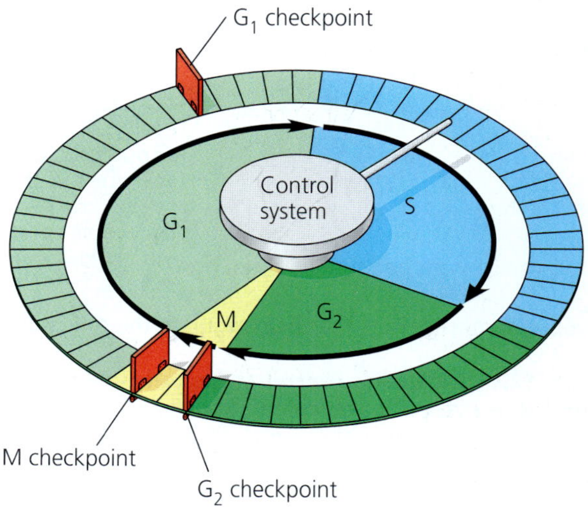

▲ **Figure 12.14 Mechanical analogy for the cell cycle control system.** In this diagram of the cell cycle, the flat "stepping stones" around the perimeter represent sequential events. Like the control device of an automatic washer, the cell cycle control system proceeds on its own, driven by a built-in clock. However, the system is subject to internal and external regulation at various checkpoints, of which three are shown (red).

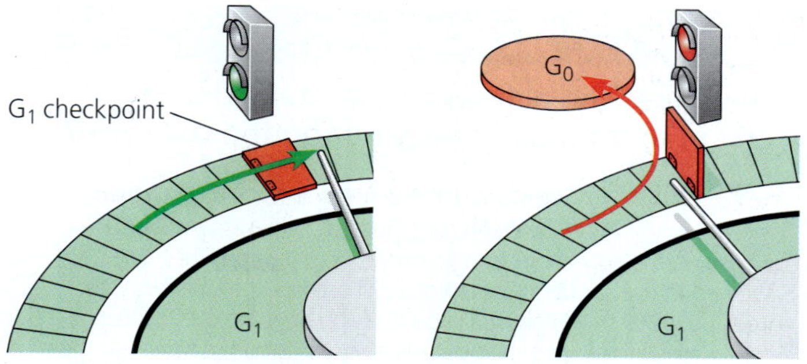

(a) If a cell receives a go-ahead signal at the G_1 checkpoint, the cell continues on in the cell cycle.

(b) If a cell does not receive a go-ahead signal at the G_1 checkpoint, the cell exits the cell cycle and goes into G_0, a nondividing state.

▲ **Figure 12.15 The G_1 checkpoint.**
WHAT IF? *What might be the result if the cell ignored the checkpoint and progressed through the cell cycle?*

(Figure 12.15). Most cells of the human body are actually in the G_0 phase. As mentioned earlier, mature nerve cells and muscle cells never divide. Other cells, such as liver cells, can be "called back" from the G_0 phase to the cell cycle by external cues, such as growth factors released during injury.

To understand how cell cycle checkpoints work, we first need to see what kinds of molecules make up the cell cycle control system (the molecular basis for the cell cycle clock) and how a cell progresses through the cycle. Then we will consider the internal and external checkpoint signals that can make the clock pause or continue.

The Cell Cycle Clock: Cyclins and Cyclin-Dependent Kinases

Rhythmic fluctuations in the abundance and activity of cell cycle control molecules pace the sequential events of the cell cycle. These regulatory molecules are mainly proteins of two types: protein kinases and cyclins. Protein kinases are enzymes that activate or inactivate other proteins by phosphorylating them (see Chapter 11). Particular protein kinases give the go-ahead signals at the G_1 and G_2 checkpoints. **Figure 12.16**, on the next page, describes an experiment from Paul Nurse's laboratory that demonstrates the crucial function of the protein kinase Cdc2 in triggering mitosis at the G_2 checkpoint in one type of yeast. Studies by other researchers have shown that this enzyme plays the same role in sea star eggs and cultured human cells, suggesting that the function of this protein has been conserved during the evolution of eukaryotes and is likely the same in many species.

Many of the kinases that drive the cell cycle are actually present at a constant concentration in the growing cell, but much of the time they are in an inactive form. To be active, such a kinase must be attached to a **cyclin**, a protein that gets its name from its cyclically fluctuating concentration in the cell. Because of this requirement, these kinases are called **cyclin-dependent kinases**, or **Cdks**. The activity of a Cdk

▼ **Figure 12.16 Inquiry**

How does the activity of a protein kinase essential for mitosis vary during the cell cycle?

EXPERIMENT Working with the fission yeast *Schizosaccharomyces pombe*, Paul Nurse and colleagues identified a gene, *cdc2*, whose normal functioning is necessary for cell division. They first showed that its product was a protein kinase (see the Unit Two interview, pp. 92–93). As part of a large study on how the *cdc2* protein kinase is regulated during the cell cycle, they measured its activity as the cell cycle progressed. In a culture of yeast cells synchronized so that they divided simultaneously, the researchers removed samples at intervals over a period of time sufficient for two cycles of cell division.

They submitted each sample to two kinds of analysis: (1) microscopic examination to determine the percentage of cells dividing (as shown by the presence of the cell plate formed during yeast cytokinesis) and (2) measurement of kinase activity in an extract of the cells (as indicated by phosphorylation of a standard protein). Control experiments established that the kinase activity they measured was due primarily to the *cdc2* protein kinase. In this way, they were able to see if an increase in enzymatic activity correlated with cell division.

RESULTS The activity of the *cdc2* protein kinase varied during the cell cycle in a periodic way, rising to a peak just before mitosis and then falling.

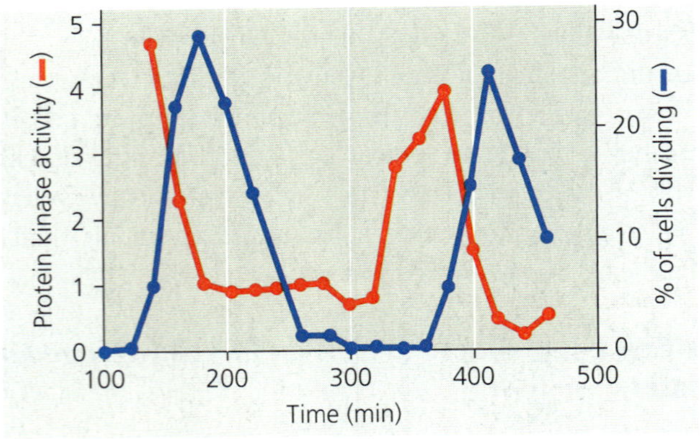

CONCLUSION The correlation between enzyme activity and the onset of mitosis, combined with evidence from other experiments that mitosis does not occur in the absence of *cdc2* kinase activity, supports the hypothesis that the *cdc2* kinase plays an essential role in triggering mitosis.

SOURCE S. Moreno, J. Hayles, and P. Nurse, Regulation of p34^{cdc2} protein kinase during mitosis, *Cell* 58:361–372 (1989).

WHAT IF? What results would you expect—for both kinase activity and percentage of cells dividing—if the cells tested were mutants completely deficient in the *cdc2* protein kinase?

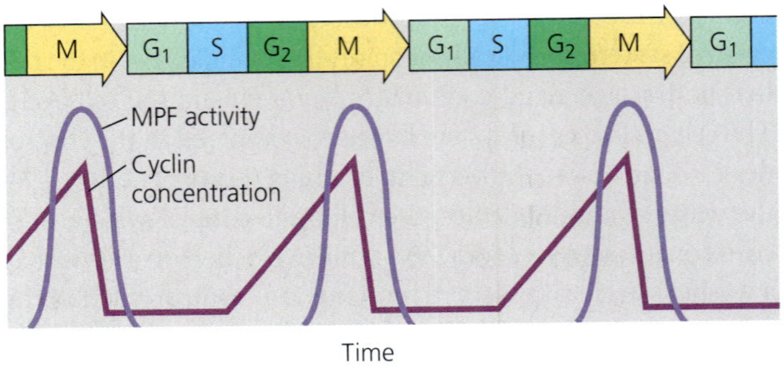

(a) Fluctuation of MPF activity and cyclin concentration during the cell cycle

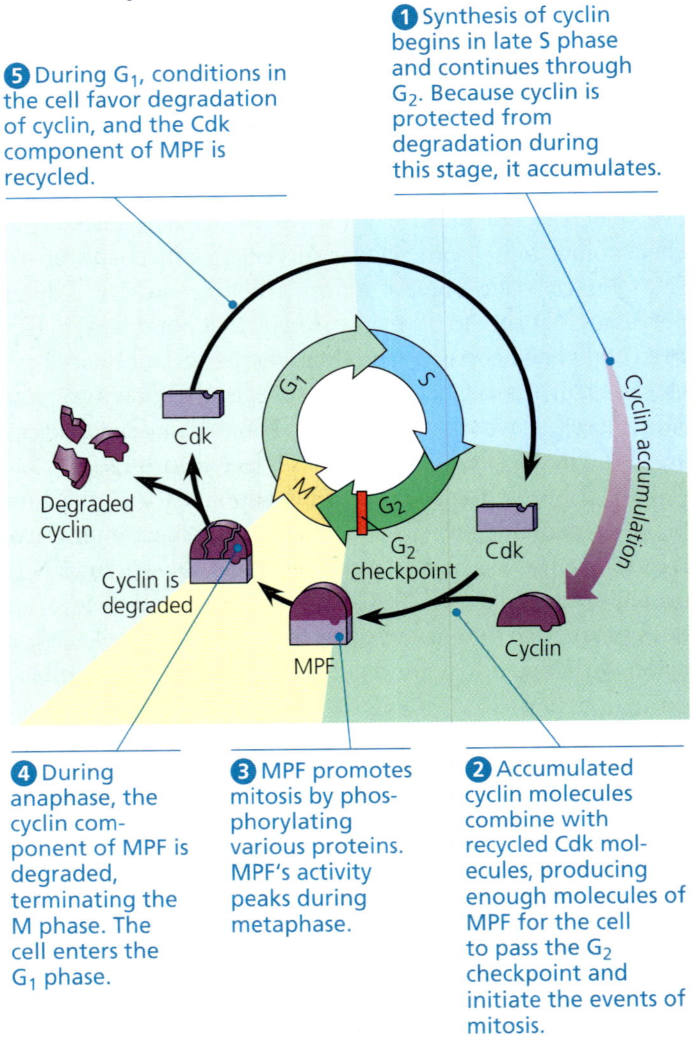

(b) Molecular mechanisms that help regulate the cell cycle

▲ **Figure 12.17 Molecular control of the cell cycle at the G_2 checkpoint.** The steps of the cell cycle are timed by rhythmic fluctuations in the activity of cyclin-dependent kinases (Cdks). Here we focus on a cyclin-Cdk complex in animal cells called MPF, which acts at the G_2 checkpoint as a go-ahead signal, triggering the events of mitosis. (The Cdk of MPF is the same as the *cdc2* protein kinase of fission yeast featured in Figure 12.16.)

rises and falls with changes in the concentration of its cyclin partner. **Figure 12.17a** shows the fluctuating activity of **MPF**, the cyclin-Cdk complex that was discovered first (in frog eggs). Note that the peaks of MPF activity correspond to the peaks of cyclin concentration. The cyclin level rises during the S and G_2 phases and then falls abruptly during M phase. (The red curve in Figure 12.16 shows the cyclic activity of the MPF in fission yeast.)

The initials MPF stand for "maturation-promoting factor," but we can think of MPF as "M-phase-promoting factor" because it triggers the cell's passage past the G_2 checkpoint into M phase **(Figure 12.17b)**. When cyclins that accumulate during G_2 associate with Cdk molecules, the resulting MPF complex phosphorylates a variety of proteins, initiating mitosis.

MPF acts both directly as a kinase and indirectly by activating other kinases. For example, MPF causes phosphorylation of various proteins of the nuclear lamina (see Figure 6.10), which promotes fragmentation of the nuclear envelope during prometaphase of mitosis. There is also evidence that MPF contributes to molecular events required for chromosome condensation and spindle formation during prophase.

During anaphase, MPF helps switch itself off by initiating a process that leads to the destruction of its own cyclin. The noncyclin part of MPF, the Cdk, persists in the cell in inactive form until it associates with new cyclin molecules synthesized during the S and G_2 phases of the next round of the cycle.

What controls cell behavior at the G_1 checkpoint? Animal cells appear to have at least three Cdk proteins and several different cyclins that operate at this checkpoint. The fluctuating activities of different cyclin-Cdk complexes are of major importance in controlling all the stages of the cell cycle.

Stop and Go Signs: Internal and External Signals at the Checkpoints

Research scientists are currently working out the pathways that link signals originating inside and outside the cell with the responses by cyclin-dependent kinases and other proteins. An example of an internal signal occurs at the M phase checkpoint. Anaphase, the separation of sister chromatids, does not begin until all the chromosomes are properly attached to the spindle at the metaphase plate. Researchers have learned that as long as some kinetochores are unattached to spindle microtubules, the sister chromatids remain together, delaying anaphase. Only when the kinetochores of all the chromosomes are attached to the spindle does the appropriate regulatory protein become activated. (In this case, the regulatory protein is not a Cdk.) Once activated, the protein sets off a chain of molecular events that ultimately results in the enzymatic cleavage of cohesins, allowing the sister chromatids to separate. This mechanism ensures that daughter cells do not end up with missing or extra chromosomes.

Studies using animal cells in culture have led to the identification of many external factors, both chemical and physical, that can influence cell division. For example, cells fail to divide if an essential nutrient is lacking in the culture medium. (This is analogous to trying to run an automatic washing machine without the water supply hooked up.) And even if all other conditions are favorable, most types of mammalian cells divide in culture only if the growth medium includes specific growth factors. As mentioned in Chapter 11, a **growth factor** is a protein released by certain cells that stimulates other cells to divide. Researchers have discovered more than 50 growth factors. Different cell types respond specifically to different growth factors or combinations of growth factors.

Consider, for example, *platelet-derived growth factor (PDGF)*, which is made by blood cell fragments called platelets.

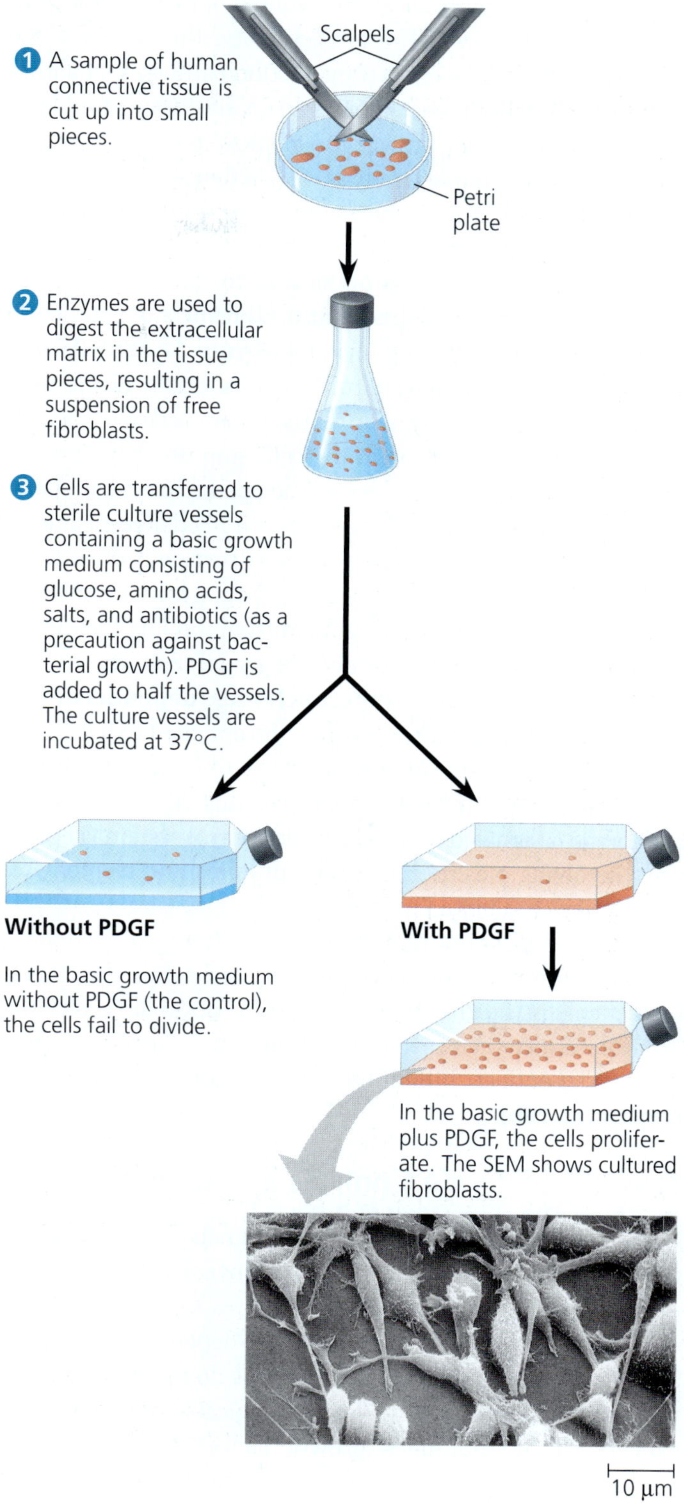

▲ **Figure 12.18 The effect of a growth factor on cell division.** As this experiment shows, adding platelet-derived growth factor (PDGF) to human fibroblasts in culture causes the cells to proliferate.

? *PDGF is known to signal cells by binding to a cell-surface receptor that is a receptor tyrosine kinase. If you added a chemical that prevented phosphorylation of this receptor, how would the results differ?*

The experiment illustrated in **Figure 12.18** demonstrates that PDGF is required for the division of fibroblasts in culture. Fibroblasts, a type of connective tissue cell, have PDGF receptors on their plasma membranes. The binding of PDGF molecules to these receptors (which are receptor tyrosine kinases; see Chapter 11) triggers a signal transduction pathway that allows the cells to pass the G_1 checkpoint and divide. PDGF stimulates fibroblast division not only in the artificial conditions of cell culture, but in an animal's body as well. When an injury occurs, platelets release PDGF in the vicinity. The resulting proliferation of fibroblasts helps heal the wound.

The effect of an external physical factor on cell division is clearly seen in **density-dependent inhibition**, a phenomenon in which crowded cells stop dividing **(Figure 12.19a)**. As first observed many years ago, cultured cells normally divide until they form a single layer of cells on the inner surface of the culture container, at which point the cells stop dividing. If some cells are removed, those bordering the open space begin dividing again and continue until the vacancy is filled. Recent studies have revealed that the binding of a cell-surface protein to its counterpart on an adjoining cell sends a growth-inhibiting signal to both cells, preventing them from moving forward in the cell cycle, even in the presence of growth factors.

Most animal cells also exhibit **anchorage dependence** (see Figure 12.19a). To divide, they must be attached to a substratum, such as the inside of a culture jar or the extracellular matrix of a tissue. Experiments suggest that like cell density, anchorage is signaled to the cell cycle control system via pathways involving plasma membrane proteins and elements of the cytoskeleton linked to them.

Density-dependent inhibition and anchorage dependence appear to function in the body's tissues as well as in cell culture, checking the growth of cells at some optimal density and location. Cancer cells, which we discuss next, exhibit neither density-dependent inhibition nor anchorage dependence **(Figure 12.19b)**.

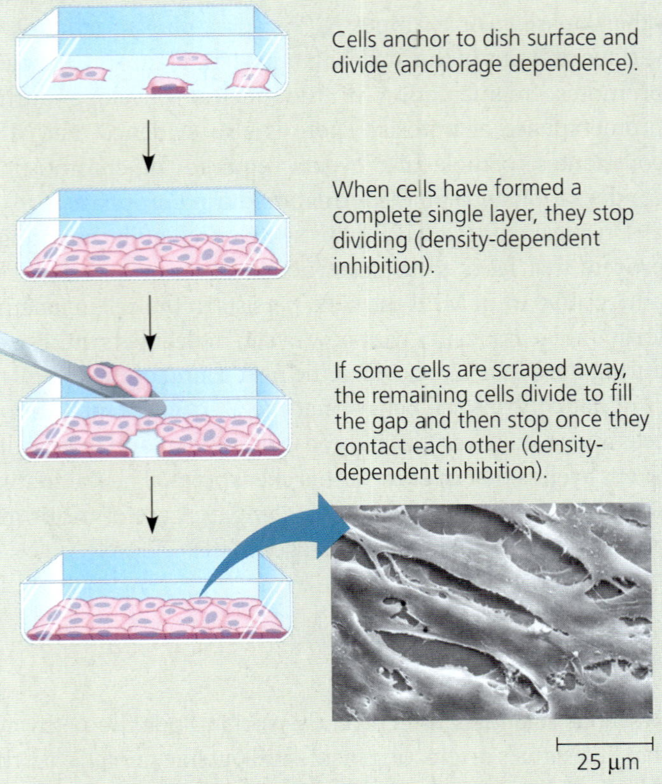

(a) Normal mammalian cells. Contact with neighboring cells and the availability of nutrients, growth factors, and a substratum for attachment limit cell density to a single layer.

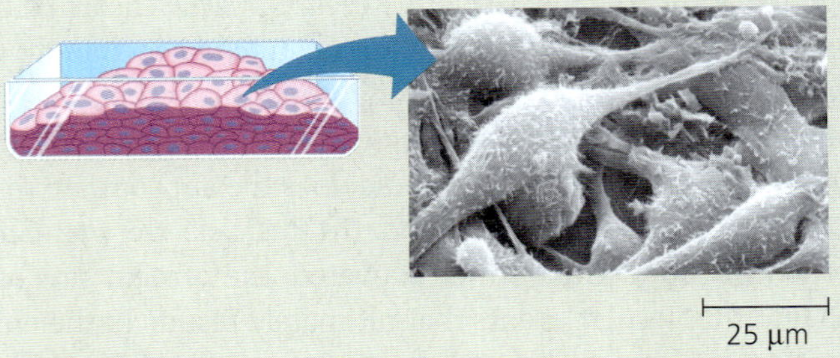

(b) Cancer cells. Cancer cells usually continue to divide well beyond a single layer, forming a clump of overlapping cells. They do not exhibit anchorage dependence or density-dependent inhibition.

▲ **Figure 12.19 Density-dependent inhibition and anchorage dependence of cell division.** Individual cells are shown disproportionately large in the drawings.

Loss of Cell Cycle Controls in Cancer Cells

Cancer cells do not heed the normal signals that regulate the cell cycle. They divide excessively and invade other tissues. If unchecked, they can kill the organism.

In addition to their lack of density-dependent inhibition and anchorage dependence, cancer cells do not stop dividing when growth factors are depleted. A logical hypothesis is that cancer cells do not need growth factors in their culture medium to grow and divide. They may make a required growth factor themselves, or they may have an abnormality in the signaling pathway that conveys the growth factor's signal to the cell cycle control system even in the absence of that factor. Another possibility is an abnormal cell cycle control system. In fact, as you will learn in Chapter 18, these are all conditions that may lead to cancer.

There are other important differences between normal cells and cancer cells that reflect derangements of the cell cycle. If and when they stop dividing, cancer cells do so at random points in the cycle, rather than at the normal checkpoints. Moreover, cancer cells can go on dividing indefinitely in culture if they are given a continual supply of nutrients; in essence, they are "immortal." A striking example is a cell line that has been reproducing in culture since 1951. Cells of this line are called HeLa cells because their original source was a tumor removed from a woman named Henrietta Lacks. By contrast, nearly all normal mammalian cells growing in culture divide only about 20 to 50 times before they stop dividing, age, and die. (We'll see a possible reason for this phenomenon when we discuss chromosome replication in Chapter 16.)

The abnormal behavior of cancer cells can be catastrophic when it occurs in the body. The problem begins when a single cell in a tissue undergoes **transformation**, the process that converts a normal cell to a cancer cell. The body's immune system normally recognizes a transformed cell as an insurgent and destroys it. However, if the cell evades destruction, it may proliferate and form a tumor, a mass of abnormal cells within otherwise normal tissue. If the abnormal cells remain at the original site, the lump is called a **benign tumor**. Most benign tumors do not cause serious problems and can be completely removed by surgery. In contrast, a **malignant tumor** becomes invasive enough to impair the functions of one or more organs **(Figure 12.20)**. An individual with a malignant tumor is said to have cancer.

The cells of malignant tumors are abnormal in many ways besides their excessive proliferation. They may have unusual numbers of chromosomes (whether this is a cause or an effect of transformation is a current topic of debate). Their metabolism may be disabled, and they may cease to function in any constructive way. Abnormal changes on the cell surface cause cancer cells to lose attachments to neighboring cells and the extracellular matrix, which allows them to spread into nearby tissues. Cancer cells may also secrete signal molecules that cause blood vessels to grow toward the tumor. A few tumor cells may separate from the original tumor, enter blood vessels and lymph vessels, and travel to other parts of the body. There, they may proliferate and form a new tumor. This spread of cancer cells to locations distant from their original site is called **metastasis** (see Figure 12.20).

A tumor that appears to be localized may be treated with high-energy radiation, which damages DNA in cancer cells much more than it does in normal cells, apparently because the majority of cancer cells have lost the ability to repair such damage. To treat known or suspected metastatic tumors, chemotherapy is used, in which drugs that are toxic to actively dividing cells are administered through the circulatory system. As you might expect, chemotherapeutic drugs interfere with specific steps in the cell cycle. For example, the drug Taxol freezes the mitotic spindle by preventing microtubule depolymerization, which stops actively dividing cells from proceeding past metaphase. The side effects of chemotherapy are due to the drugs' effects on normal cells that divide often. For example, nausea results from chemotherapy's effects on intestinal cells, hair loss from effects on hair follicle cells, and susceptibility to infection from effects on immune system cells.

Researchers are beginning to understand how a normal cell is transformed into a cancer cell. You will learn more about the molecular biology of cancer in Chapter 18. Though the causes of cancer are diverse, cellular transformation always involves the alteration of genes that somehow influence the cell cycle control system. Our knowledge of how changes in the genome lead to the various abnormalities of cancer cells remains rudimentary, however.

Perhaps the reason we have so many unanswered questions about cancer cells is that there is still so much to learn about how normal cells function. The cell, life's basic unit of structure and function, holds enough secrets to engage researchers well into the future.

CONCEPT CHECK 12.3

1. In Figure 12.13, why do the nuclei resulting from experiment 2 contain different amounts of DNA?
2. What is the go-ahead signal for a cell to pass the G_2 phase checkpoint and enter mitosis? (See Figure 12.17.)
3. What phase are most of your body cells in?
4. Compare and contrast a benign tumor and a malignant tumor.
5. **WHAT IF?** What would happen if you performed the experiment in Figure 12.18 with cancer cells?

For suggested answers, see Appendix A.

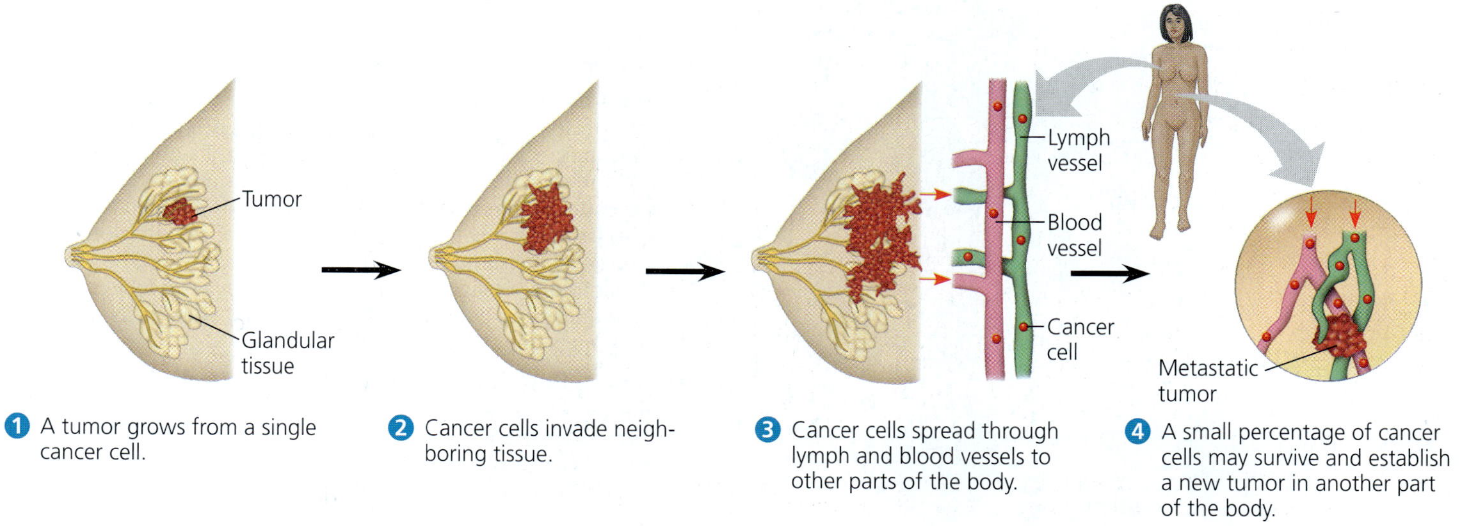

▲ **Figure 12.20 The growth and metastasis of a malignant breast tumor.** The cells of malignant (cancerous) tumors grow in an uncontrolled way and can spread to neighboring tissues and, via lymph and blood vessels, to other parts of the body. The spread of cancer cells beyond their original site is called metastasis.

Chapter 12 Review

MEDIA Go to the Study Area at **www.masteringbio.com** for BioFlix 3-D Animations, MP3 Tutors, Videos, Practice Tests, an eBook, and more.

SUMMARY OF KEY CONCEPTS

- Unicellular organisms reproduce by cell division; multicellular organisms depend on cell division for their development from a fertilized egg and for growth and repair.

MEDIA

Activity Roles of Cell Division

CONCEPT 12.1

Cell division results in genetically identical daughter cells (pp. 229–230)

- Cells duplicate their genetic material before they divide, ensuring that each daughter cell receives an exact copy of the genetic material, DNA.
- **Cellular Organization of the Genetic Material** DNA is partitioned among chromosomes. Eukaryotic chromosomes consist of chromatin, a complex of DNA and protein that condenses during mitosis. In animals, gametes have one set of chromosomes and somatic cells have two sets.
- **Distribution of Chromosomes During Eukaryotic Cell Division** In preparation for cell division, chromosomes replicate, each one then consisting of two identical sister chromatids joined along their lengths by sister chromatid cohesion. When this cohesion is broken, the chromatids separate during cell division, becoming the chromosomes of the new daughter cells. Eukaryotic cell division consists of mitosis (division of the nucleus) and cytokinesis (division of the cytoplasm).

CONCEPT 12.2

The mitotic phase alternates with interphase in the cell cycle (pp. 230–238)

- **Phases of the Cell Cycle** Between divisions, cells are in interphase: the G_1, S, and G_2 phases. The cell grows throughout interphase, but DNA is replicated only during the synthesis (S) phase. Mitosis and cytokinesis make up the mitotic (M) phase of the cell cycle.

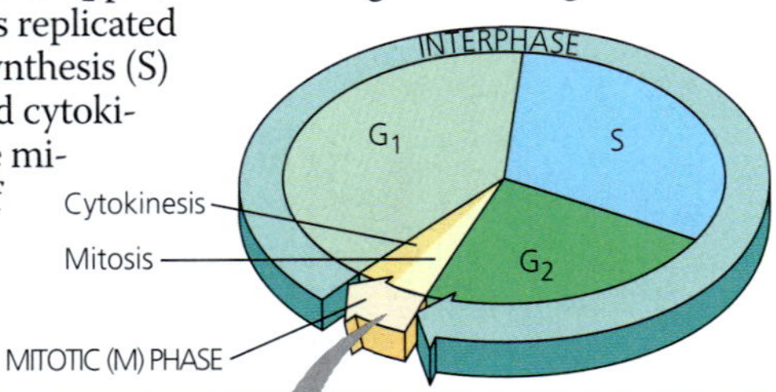

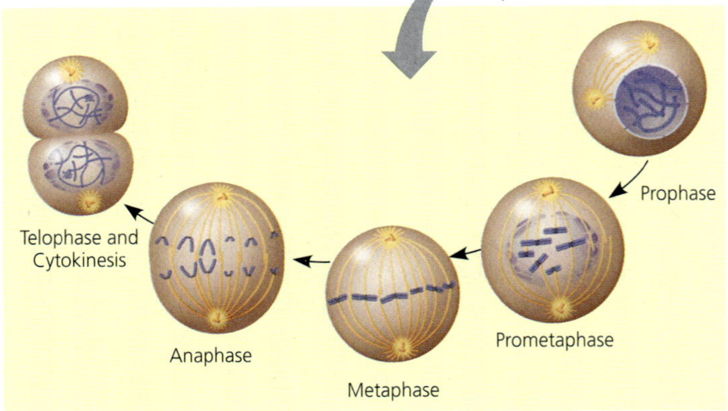

- ***The Mitotic Spindle: A Closer Look*** The mitotic spindle is an apparatus of microtubules that controls chromosome movement during mitosis. In animal cells, the spindle arises from the centrosomes and includes spindle microtubules and asters. Some spindle microtubules attach to the kinetochores of chromosomes and move the chromosomes to the metaphase plate. In anaphase, sister chromatids separate, and motor proteins move them along the kinetochore microtubules toward opposite ends of the cell. Meanwhile, motor proteins push nonkinetochore microtubules from opposite poles away from each other, elongating the cell. In telophase, genetically identical daughter nuclei form at opposite ends of the cell.
- ***Cytokinesis: A Closer Look*** Mitosis is usually followed by cytokinesis. Animal cells carry out cytokinesis by cleavage, and plant cells form a cell plate.
- **Binary Fission** During binary fission in bacteria, the chromosome replicates and the two daughter chromosomes actively move apart. The specific proteins involved in this movement are a subject of current research.
- **The Evolution of Mitosis** Since prokaryotes preceded eukaryotes by more than a billion years, it is likely that mitosis evolved from prokaryotic cell division. Certain protists exhibit types of cell division that seem intermediate between bacterial binary fission and the process of mitosis carried out by most eukaryotic cells.

MEDIA

BioFlix 3-D Animation Mitosis
MP3 Tutor Mitosis
Activity The Cell Cycle
Activity Mitosis and Cytokinesis Animation
Activity Mitosis and Cytokinesis Video
Investigation How Much Time Do Cells Spend in Each Phase of Mitosis?

CONCEPT 12.3

The eukaryotic cell cycle is regulated by a molecular control system (pp. 238–243)

- **Evidence for Cytoplasmic Signals** Molecules present in the cytoplasm regulate progress through the cell cycle.
- **The Cell Cycle Control System** Cyclic changes in regulatory proteins work as a cell cycle clock. The clock has specific checkpoints where the cell cycle stops until a go-ahead signal is received. The key molecules are cyclins and cyclin-dependent kinases (Cdks). Cell culture has enabled researchers to study the molecular details of cell division. Both internal signals and external signals control the cell cycle checkpoints via signal transduction pathways. Most cells exhibit density-dependent inhibition of cell division as well as anchorage dependence.
- **Loss of Cell Cycle Controls in Cancer Cells** Cancer cells elude normal regulation and divide out of control, forming tumors. Malignant tumors invade surrounding tissues and can metastasize, exporting cancer cells to other parts of the body, where they may form secondary tumors.

MEDIA

Activity Causes of Cancer

TESTING YOUR KNOWLEDGE

SELF-QUIZ

1. Through a microscope, you can see a cell plate beginning to develop across the middle of a cell and nuclei re-forming on either side of the cell plate. This cell is most likely
   a. an animal cell in the process of cytokinesis.
   b. a plant cell in the process of cytokinesis.
   c. an animal cell in the S phase of the cell cycle.
   d. a bacterial cell dividing.
   e. a plant cell in metaphase.

2. Vinblastine is a standard chemotherapeutic drug used to treat cancer. Because it interferes with the assembly of microtubules, its effectiveness must be related to
   a. disruption of mitotic spindle formation.
   b. inhibition of regulatory protein phosphorylation.
   c. suppression of cyclin production.
   d. myosin denaturation and inhibition of cleavage furrow formation.
   e. inhibition of DNA synthesis.

3. A particular cell has half as much DNA as some other cells in a mitotically active tissue. The cell in question is most likely in
   a. G_1.
   b. G_2.
   c. prophase.
   d. metaphase.
   e. anaphase.

4. One difference between cancer cells and normal cells is that cancer cells
   a. are unable to synthesize DNA.
   b. are arrested at the S phase of the cell cycle.
   c. continue to divide even when they are tightly packed together.
   d. cannot function properly because they are affected by density-dependent inhibition.
   e. are always in the M phase of the cell cycle.

5. The decline of MPF activity at the end of mitosis is due to
   a. the destruction of the protein kinase Cdk.
   b. decreased synthesis of cyclin.
   c. the degradation of cyclin.
   d. synthesis of DNA.
   e. an increase in the cell's volume-to-genome ratio.

6. The drug cytochalasin B blocks the function of actin. Which of the following aspects of the cell cycle would be most disrupted by cytochalasin B?
   a. spindle formation
   b. spindle attachment to kinetochores
   c. DNA synthesis
   d. cell elongation during anaphase
   e. cleavage furrow formation

7. In the cells of some organisms, mitosis occurs without cytokinesis. This will result in
   a. cells with more than one nucleus.
   b. cells that are unusually small.
   c. cells lacking nuclei.
   d. destruction of chromosomes.
   e. cell cycles lacking an S phase.

8. Which of the following does *not* occur during mitosis?
   a. condensation of the chromosomes
   b. replication of the DNA
   c. separation of sister chromatids
   d. spindle formation
   e. separation of the spindle poles

9. In the light micrograph below of dividing cells near the tip of an onion root, identify a cell in each of the following stages: prophase, prometaphase, metaphase, anaphase, and telophase. Describe the major events occurring at each stage.

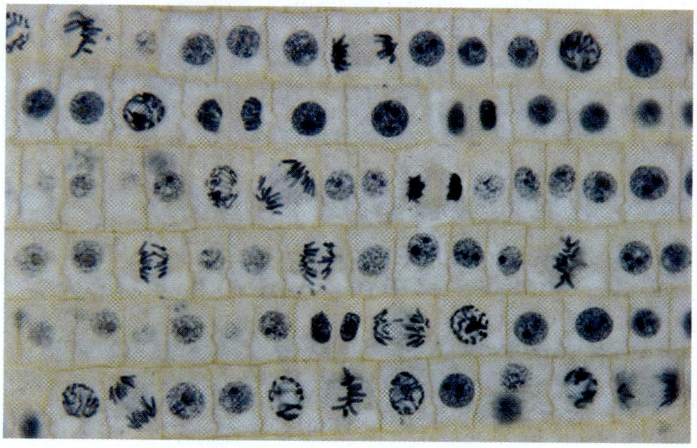

10. **DRAW IT** Draw one eukaryotic chromosome as it would appear during interphase, during each of the stages of mitosis, and during cytokinesis. Also draw and label the nuclear envelope and any microtubules attached to the chromosome(s).

For Self-Quiz answers, see Appendix A.

MEDIA Visit the Study Area at **www.masteringbio.com** for a Practice Test.

EVOLUTION CONNECTION

11. The result of mitosis is that the daughter cells end up with the same number of chromosomes that the parent cell had. Another way to maintain the number of chromosomes would be to carry out cell division first and then duplicate the chromosomes in each daughter cell. Do you think this would be an equally good way of organizing the cell cycle? Why do you suppose that evolution has not led to this alternative?

SCIENTIFIC INQUIRY

12. Although both ends of a microtubule can gain or lose subunits, one end (called the plus end) polymerizes and depolymerizes at a higher rate than the other end (the minus end). For spindle microtubules, the plus ends are in the center of the spindle, and the minus ends are at the poles. Motor proteins that move on microtubules specialize in walking either toward the plus end or toward the minus end; the two types are called plus end–directed and minus end–directed motor proteins, respectively. Given what you know about chromosome movement and spindle changes during anaphase, predict which type of motor proteins would be present on (a) kinetochore microtubules and (b) nonkinetochore microtubules.

UNIT 3

Genetics

AN INTERVIEW WITH

Terry L. Orr-Weaver

How the daughter cells resulting from cell division end up with equal numbers of chromosomes is the focus of much of Terry Orr-Weaver's research; she also studies how cells control the DNA replication that precedes cell division. Her research group has identified a number of proteins involved in these processes. A professor of biology at MIT and the first woman to become a member of the Whitehead Institute for Biomedical Research, Dr. Orr-Weaver has an undergraduate degree in chemistry from UC San Diego and a Ph.D. in biological chemistry from Harvard. She is a member of the U.S. National Academy of Sciences and a past president of the Genetics Society of America.

After receiving your Ph.D., you switched from using single-celled yeast as your model organism to the multicelled fruit fly *Drosophila melanogaster*. Why?

I was excited by the possibility of working at the interface of two fields. At that time, the cell cycle field was exploding from work with single cells—mainly yeasts and mammalian cells grown in culture—and biochemical experiments using extracts. Meanwhile, developmental biology was entering a new era with the discovery of pattern formation genes. I realized there was a key question at the interface of these two fields that was being ignored: If an organism starts as a single cell and ends up as a multicellular entity, how are pattern and cell divisions coordinated? And I realized there had to be intrinsic regulation from the cell cycle components but also extrinsic developmental control feeding into that. This was clearly going to be an important area of study.

In choosing a multicellular model organism, why use *Drosophila*?

A lot of fundamental discoveries in biology have been made with fruit flies. For example, it was research with *Drosophila* that established in 1916 that chromosomes were the physical basis of inheritance, the structures that carried the genes. Decades later, the master regulatory genes that set up the animal body plan in embryonic development were discovered in *Drosophila*. These genes determine, for example, where the fly's head is and where its legs are. Then, to the amazement of biologists, it turned out that exactly the same genes control how the human body plan is set up! So *Drosophila* has a really rich heritage. When I started working on *Drosophila*, it had a 70-year-long history of use in genetics, providing an incredible array of knowledge and methods.

How do geneticists approach biological questions?

Geneticists want to discover the genes that are involved in a biological process and then figure out what the genes do. First, you decide what process you're interested in—in our case, how chromosomes get partitioned during cell division. Next, you figure out what you would see if you had a mutant in which that process was perturbed—how would you recognize it? Then you generate mutations, usually with the help of high-energy radiation or a chemical, hoping that one or more of the resulting mutants are affected in the process of interest. What you're doing by making a mutation is generating a disease state in your organism. When you've found a mutant you're interested in, you can find out what gene has been made defective, and you know that that gene has to play an important role in the process you're interested in.

Now that we have the genome sequences of many organisms, including the fruit fly and the human, you might think that all their genes are known—so what's to discover? But it turned out that we didn't know the functions of many of the genes that showed up in genome sequences. Even in an organism as simple and well-studied as yeast, at the publication of the genome sequence we didn't know what 70% of the genes did. And we still don't have a clue about half of the human genes. So having genome sequences provides a foundation, but we must still find out what all the genes do, and we geneticists feel the best way to go about this is to use genetics.

A big and useful surprise in this gene-discovery enterprise has been how many genes have been "conserved" through evolutionary history and are still very similar in organisms as distantly related as fruit flies and humans. It's very hard to figure out directly what human genes do. But when we discover a new gene in *Drosophila*, invariably it turns out that there's a similar gene in humans that does the same thing. So *Drosophila* turns out to be an even better model organism than we had guessed.

Besides genetics, what other approaches and methods do you use?

A combination of genetics, biochemistry, and cell biology turns out to be incredibly powerful for our research. For instance, with the microscope, a tool of cell biology, we can literally watch the chromosomes as they undergo mitosis or meiosis. If we have a mutation causing a defect in one of those processes, we can look directly at how the chromosomes behave in mutant cells. And in our research on DNA replication, we can look directly at proteins that attach to the DNA during DNA replication. In this lab, we try to do things as directly as possible!

Here's an example from our work on DNA replication in *Drosophila*. Using a genetic approach, we discovered a mutant that couldn't carry out DNA replication—although such a mutant can live and grow for a while by using proteins its mother stockpiled in the egg. It looked like the gene affected might code for a really important protein. But we wanted to find out if it was a protein that played a *direct* role in DNA replication, and we couldn't tell that from the genetics alone. However, by labeling the normal version of the protein with a fluorescent tag, we could use the microscope to see where the protein was located in cells, and we saw that the protein got on the DNA right at a place where DNA synthesis starts. That established that this protein was directly involved.

How is meiosis different from mitosis?

Let me first review mitosis. After a cell duplicates its DNA, mitosis ensures that each duplicated chromosome gets partitioned to the daughter cells so that you end up with two cells that have exactly the same DNA content and chromosome number. That's what happens in normal cell division.

But what about making a sperm or an egg? In fertilization, a sperm and egg are going to fuse and give rise to a new progeny. If the organism is diploid—that is, has two similar copies of each chromosome, as flies and humans do—and the progeny is going to be diploid, then you've got to make sperm and eggs that have only one copy of each chromosome, so that when those sperm and egg come together you restore the right chromosome number. And to produce sperm and eggs with only half the diploid chromosome number, you need a special kind of cell division—meiosis. There has to be a way to bring together the two similar copies of each chromosome and then pull them apart, with each going to a different daughter cell. In meiosis, there is an extra round of chromosome partitioning where what we call the homologous pairs of chromosomes—the chromosome that initially came from dad and the similar chromosome that came from mom—get separated from each other. To sum up, in mitosis you're separating identical copies of each chromosome, whereas in meiosis you have an extra round of division stuck in there, where the copy of each chromosome from dad and the copy from mom get separated from each other.

What important questions about meiosis still need to be answered?

What we don't understand at all is how the chromosomes of a homologous pair find each other. That pairing is unique to meiosis; it doesn't happen in mitosis. The two strands of the DNA don't come apart, so although the homologous chromosomes have very similar DNA sequences, it's not base-pairing that brings the homologous chromosomes together. Given the relatively gigantic volume of the nucleus and the huge mass of chromatin in a eukaryotic cell, how do the right chromosomes find each other? That's the number one mystery about meiosis. And I would say the second big mystery is why humans are so unbelievably bad at carrying out meiosis.

What makes you say that? Are we worse than fruit flies?

We're about a thousand times worse. Here are some amazing statistics: Twenty percent of recognized pregnancies end in spontaneous miscarriage, and out of those at least half are due to a mistake during meiosis. And for every pregnancy that proceeds far enough to be recognized, there have been many that ended without being recognized. So about 10–20% of the time meiosis doesn't work properly in humans. In fruit flies, meiosis seems to occur inaccurately only 0.01–0.05% of the time. Even mice, which are mammals like ourselves, do much better than humans. So I would call this the second big mystery regarding meiosis: Why are humans so bad at it?

Tell us more about what happens when meiosis doesn't work correctly.

Most of the time the result is early death of the embryo. In humans, if any chromosome other than X or Y is present in only one copy or more than two copies, the embryo dies very early. The only exceptions are for chromosomes number 13, 18, and 21: When one of these is present in three copies, the embyo usually dies at an early stage, but it may survive. Even in the case of Down syndrome, a relatively common condition where a person has three copies of chromosome 21, only a minority of fetuses survive to term. And when the individual does survive, there are serious effects. In fact, errors in meiosis are the leading cause of mental retardation in the United States. Some scientists argue that, in evolutionary terms, the human species is able to cope with such a high rate of meiotic errors because most of the pregnancies are lost very early.

What do you like best about research?

What's so great about research is that you get to unwrap presents all the time! When you find an interesting mutant, it's like a beautifully wrapped present. And then you unwrap it, and it can be a big surprise to learn which gene is affected and what its protein product is. Sometimes you unwrap it and it's beautiful and satisfying and completely makes sense; other times you don't at first understand what the gift is, and then you have to figure that out. I want to tell students how wonderful it is getting these presents to unwrap. But they also need to realize that it might take three years to get the ribbon and all the tape off. It can take some real patience to unwrap the present!

In your view, how important is it for scientists to reach out beyond the scientific community?

I think scientists do have an obligation to educate the public. It's especially critical for people to understand the importance of basic research and the study of model organisms. Because if we shortchange basic research, in the long run we won't be able to benefit from medical applications. The reason we've done so well in this country is our willingness to invest in basic research science.

The challenge is figuring out how to convey a message like this in an era of soundbites. The media and the public want to hear that a piece of research is directly going to cure a disease. So how can we get a message across that requires an explanation more complex than a catchphrase? There are many stories we can tell where discoveries in basic research have led to very important applications, but these stories take at least a couple of paragraphs to communicate. Also, there's a real need for scientists to try to build a bridge between science and the public.

It's especially important to educate our representatives in government. A coalition of several biology societies, including the Genetics Society of America, has a Joint Steering Committee for Public Policy, which has been terrific in educating members of Congress and their staffs. They bring in scientists to explain their research and why investing in research is important. Maybe that's a good starting point.

Inquiry in Action

Learn about an experiment by Terry Orr-Weaver in Inquiry Figure 16.22 on page 322.

When you find an interesting mutant, it's like a beautifully wrapped present . . . and it can be a big surprise . . .

Jane Reece and Terry Orr-Weaver

Meiosis and Sexual Life Cycles

13

▲ **Figure 13.1 What accounts for family resemblance?**

KEY CONCEPTS

13.1 Offspring acquire genes from parents by inheriting chromosomes
13.2 Fertilization and meiosis alternate in sexual life cycles
13.3 Meiosis reduces the number of chromosome sets from diploid to haploid
13.4 Genetic variation produced in sexual life cycles contributes to evolution

OVERVIEW

Variations on a Theme

Most people who send out birth announcements mention the sex of the baby, but they don't feel the need to specify that their offspring is a human being! One of the characteristics of life is the ability of organisms to reproduce their own kind—elephants produce little elephants, and oak trees generate oak saplings. Exceptions to this rule show up only as sensational but highly suspect stories in tabloid newspapers.

Another rule often taken for granted is that offspring resemble their parents more than they do unrelated individuals. If you examine the family members shown in **Figure 13.1**—Sissy Spacek and Jack Fisk with daughters Madison and Schuyler Fisk—you can pick out some similar features among them. The transmission of traits from one generation to the next is called inheritance, or **heredity** (from the Latin *heres*, heir). However, sons and daughters are not identical copies of either parent or of their siblings. Along with inherited similarity, there is also **variation**. Farmers have exploited the principles of heredity and variation for thousands of years, breeding plants and animals for desired traits. But what are the biological mechanisms leading to the hereditary similarity and variation that we call a "family resemblance"? This question eluded biologists until the development of genetics in the 20th century.

Genetics is the scientific study of heredity and hereditary variation. In this unit, you will learn about genetics at multiple levels, from organisms to cells to molecules. On the practical side, you will see how genetics continues to revolutionize medicine and agriculture, and you will be asked to consider some social and ethical questions raised by our ability to manipulate DNA, the genetic material. At the end of the unit, you will be able to stand back and consider the whole genome, an organism's entire complement of DNA. The rapid accumulation of the genome sequences of many species, including our own, has taught us a great deal about evolution on the molecular level—in other words, evolution of the genome itself. In fact, genetic methods and discoveries are catalyzing progress in all areas of biology, from cell biology to physiology, developmental biology, behavior, and even ecology.

We begin our study of genetics in this chapter by examining how chromosomes pass from parents to offspring in sexually reproducing organisms. The processes of meiosis (a special type of cell division) and fertilization (the fusion of sperm and egg) maintain a species' chromosome count during the sexual life cycle. We will describe the cellular mechanics of meiosis and how this process differs from mitosis. Finally, we will consider how both meiosis and fertilization contribute to genetic variation, such as the variation obvious in the family shown in Figure 13.1.

CONCEPT 13.1

Offspring acquire genes from parents by inheriting chromosomes

Family friends may tell you that you have your mother's freckles or your father's eyes. However, parents do not, in any literal sense, give their children freckles, eyes, hair, or any other traits. What, then, *is* actually inherited?

Inheritance of Genes

Parents endow their offspring with coded information in the form of hereditary units called **genes**. The genes we inherit from our mothers and fathers are our genetic link to our parents, and they account for family resemblances such as shared eye color or freckles. Our genes program the specific traits that emerge as we develop from fertilized eggs into adults.

The genetic program is written in the language of DNA, the polymer of four different nucleotides you learned about in Chapters 1 and 5. Inherited information is passed on in the form of each gene's specific sequence of DNA nucleotides, much as printed information is communicated in the form of meaningful sequences of letters. In both cases, the language is symbolic. Just as your brain translates the word *apple* into a mental image of the fruit, cells translate genes into freckles and other features. Most genes program cells to synthesize specific enzymes and other proteins, whose cumulative action produces an organism's inherited traits. The programming of these traits in the form of DNA is one of the unifying themes of biology.

The transmission of hereditary traits has its molecular basis in the precise replication of DNA, which produces copies of genes that can be passed along from parents to offspring. In animals and plants, reproductive cells called **gametes** are the vehicles that transmit genes from one generation to the next. During fertilization, male and female gametes (sperm and eggs) unite, thereby passing on genes of both parents to their offspring.

Except for small amounts of DNA in mitochondria and chloroplasts, the DNA of a eukaryotic cell is packaged into chromosomes within the nucleus. Every living species has a characteristic number of chromosomes. For example, humans have 46 chromosomes in almost all of their cells. Each chromosome consists of a single long DNA molecule elaborately coiled in association with various proteins. One chromosome includes several hundred to a few thousand genes, each of which is a specific sequence of nucleotides within the DNA molecule. A gene's specific location along the length of a chromosome is called the gene's **locus** (from the Latin, meaning "place"; plural, *loci*). Our genetic endowment consists of the genes carried on the chromosomes we inherited from our parents.

Comparison of Asexual and Sexual Reproduction

Only organisms that reproduce asexually produce offspring that are exact copies of themselves. In **asexual reproduction**, a single individual is the sole parent and passes copies of all its genes to its offspring. For example, single-celled eukaryotic organisms can reproduce asexually by mitotic cell division, in which DNA is copied and allocated equally to two daughter cells. The genomes of the offspring are virtually exact copies of the parent's genome. Some multicellular organisms are also capable of reproducing asexually **(Figure 13.2)**. Because the cells of the offspring are derived by mitosis in the parent, the "chip off the old block" is usually genetically identical to its parent. An individual that reproduces asexually gives rise to a **clone**, a group of genetically identical individuals. Genetic differences occasionally arise in asexually reproducing organisms as a result of changes in the DNA called mutations, which we will discuss in Chapter 17.

(a) Hydra

(b) Redwoods

▲ **Figure 13.2 Asexual reproduction in two multicellular organisms. (a)** This relatively simple animal, a hydra, reproduces by budding. The bud, a localized mass of mitotically dividing cells, develops into a small hydra, which detaches from the parent (LM). **(b)** Each tree in this circle of redwoods grew from a single parent tree, whose stump is in the center of the circle.

In **sexual reproduction**, two parents give rise to offspring that have unique combinations of genes inherited from the two parents. In contrast to a clone, offspring of sexual reproduction vary genetically from their siblings and both parents: They are variations on a common theme of family resemblance, not exact replicas. Genetic variation like that shown in Figure 13.1 is an important consequence of sexual reproduction. What mechanisms generate this genetic variation? The key is the behavior of chromosomes during the sexual life cycle.

CONCEPT CHECK 13.1

1. How are the traits of parents (such as hair color) transmitted to their offspring?
2. Explain how asexually reproducing organisms produce offspring that are genetically identical to each other and to their parents.
3. **WHAT IF?** A horticulturalist breeds orchids, trying to obtain a plant with a unique combination of desirable traits. After many years, she finally succeeds. To produce more plants like this one, should she breed it or clone it? Why?

For suggested answers, see Appendix A.

CONCEPT 13.2
Fertilization and meiosis alternate in sexual life cycles

A **life cycle** is the generation-to-generation sequence of stages in the reproductive history of an organism, from conception to production of its own offspring. In this section, we use humans as an example to track the behavior of chromosomes through sexual life cycles. We begin by considering the chromosome count in human somatic cells and gametes; we will then explore how the behavior of chromosomes relates to the human life cycle and other types of sexual life cycles.

Sets of Chromosomes in Human Cells

In humans, each **somatic cell**—any cell other than those involved in gamete formation—has 46 chromosomes. During mitosis, the chromosomes become condensed enough to be visible in a light microscope. Because chromosomes differ in size, in the positions of their centromeres, and in the pattern of colored bands produced by certain stains, they can be distinguished from one another by microscopic examination when sufficiently condensed.

Careful examination of a micrograph of the 46 human chromosomes from a single cell in mitosis reveals that there are two chromosomes of each of 23 types. This becomes clear when images of the chromosomes are arranged in pairs, starting with the longest chromosomes. The resulting ordered display is called a **karyotype** (**Figure 13.3**). The two chromosomes composing a pair have the same length, centromere position, and staining pattern: These are called **homologous chromosomes**, or homologs. Both chromosomes of each pair carry genes controlling the same inherited characters. For example, if a gene for eye color is situated at a particular locus on a certain chromosome, then the homolog of that chromosome will also have a gene specifying eye color at the equivalent locus.

The two distinct chromosomes referred to as X and Y are an important exception to the general pattern of homologous chromosomes in human somatic cells. Human females have a homologous pair of X chromosomes (XX), but males have one X and one Y chromosome (XY). Only small parts of the X and Y are homologous. Most of the genes carried on the X chromosome do not have counterparts on the tiny Y, and the Y chromosome has genes lacking on the X. Because they determine an individual's sex, the X and Y chromosomes are called **sex chromosomes**. The other chromosomes are called **autosomes**.

The occurrence of homologous pairs of chromosomes in each human somatic cell is a consequence of our sexual origins. We inherit one chromosome of each pair from each parent. Thus, the 46 chromosomes in our somatic cells are actually two sets of 23 chromosomes—a maternal set (from our mother)

▼ Figure 13.3 Research Method

Preparing a karyotype

APPLICATION A karyotype is a display of condensed chromosomes arranged in pairs. Karyotyping can be used to screen for abnormal numbers of chromosomes or defective chromosomes associated with certain congenital disorders, such as Down syndrome.

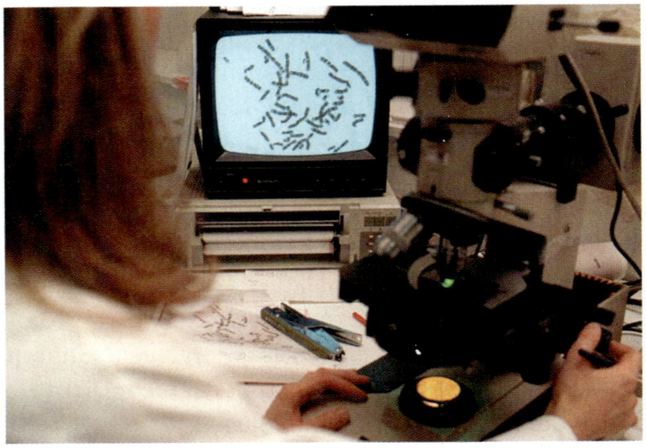

TECHNIQUE Karyotypes are prepared from isolated somatic cells, which are treated with a drug to stimulate mitosis and then grown in culture for several days. Cells arrested in metaphase are stained and then viewed with a microscope equipped with a digital camera. A photograph of the chromosomes is displayed on a computer monitor, and the images of the chromosomes are arranged into pairs according to size and shape.

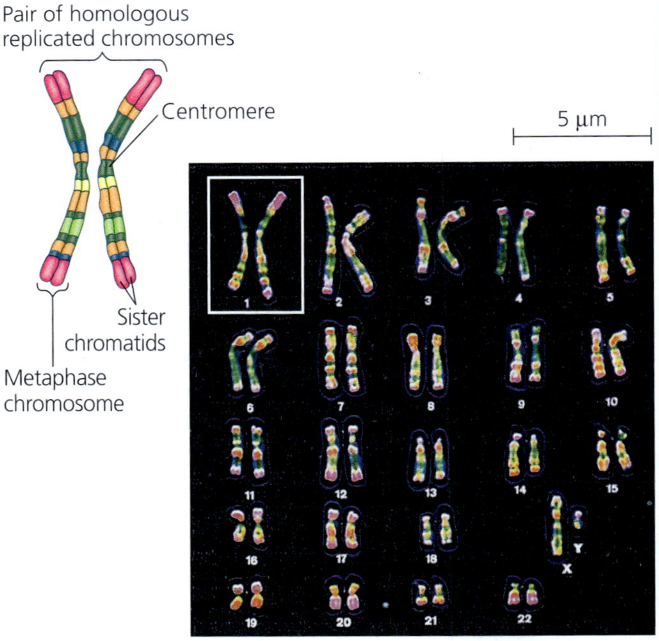

RESULTS This karyotype shows the chromosomes from a normal human male. The size of the chromosome, position of the centromere, and pattern of stained bands help identify specific chromosomes. Although difficult to discern in the karyotype, each metaphase chromosome consists of two closely attached sister chromatids (see the diagram of a pair of homologous replicated chromosomes).

and a paternal set (from our father). The number of chromosomes in a single set is represented by n. Any cell with two chromosome sets is called a **diploid cell** and has a diploid number of chromosomes, abbreviated $2n$. For humans, the diploid number is 46 ($2n = 46$), the number of chromosomes in our somatic cells. In a cell in which DNA synthesis has occurred, all the chromosomes are replicated, and therefore each consists of two identical sister chromatids, associated closely at the centromere and along the arms. **Figure 13.4** helps clarify the various terms that we use in describing replicated chromosomes in a diploid cell. Study this figure so that you understand the differences between homologous chromosomes, sister chromatids, nonsister chromatids, and chromosome sets.

Unlike somatic cells, gametes (sperm and eggs) contain a single chromosome set. Such cells are called **haploid cells**, and each has a haploid number of chromosomes (n). For humans, the haploid number is 23 ($n = 23$). The set of 23 consists of the 22 autosomes plus a single sex chromosome. An unfertilized egg contains an X chromosome, but a sperm may contain an X or a Y chromosome.

Note that each sexually reproducing species has a characteristic diploid number and haploid number. For example, the fruit fly, *Drosophila melanogaster*, has a diploid number of 8 and a haploid number of 4, while dogs have a diploid number of 78 and a haploid number of 39.

Now that you have learned the concepts of diploid and haploid numbers of chromosomes, let's consider chromosome behavior during sexual life cycles. We'll use the human life cycle as an example.

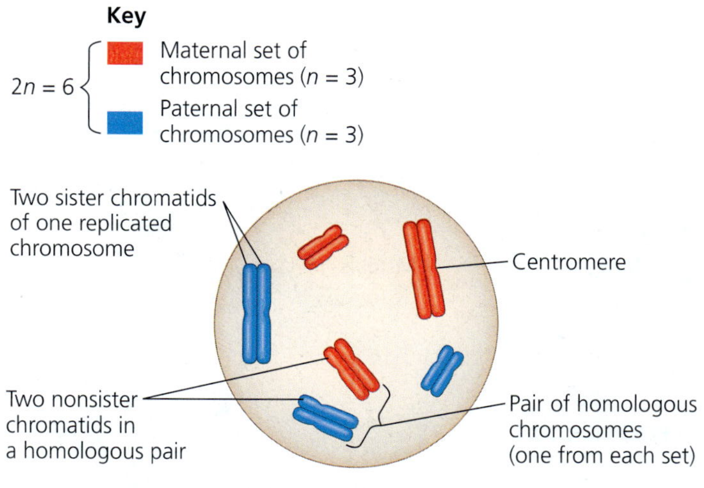

▲ **Figure 13.4 Describing chromosomes.** A cell with a diploid number of 6 ($2n = 6$) is depicted here following chromosome replication and condensation. Each of the six replicated chromosomes consists of two sister chromatids associated closely along their lengths. Each homologous pair is composed of one chromosome from the maternal set (red) and one from the paternal set (blue). Each set is made up of three chromosomes in this example. Nonsister chromatids are any two chromatids in a pair of homologous chromosomes that are not sister chromatids.

? *What is the haploid number of this cell? Is a "set" of chromosomes haploid or diploid?*

Behavior of Chromosome Sets in the Human Life Cycle

The human life cycle begins when a haploid sperm from the father fuses with a haploid egg from the mother. This union of gametes, culminating in fusion of their nuclei, is called **fertilization**. The resulting fertilized egg, or **zygote**, is diploid because it contains two haploid sets of chromosomes bearing genes representing the maternal and paternal family lines. As a human develops into a sexually mature adult, mitosis of the zygote and its descendants generates all the somatic cells of the body. Both chromosome sets in the zygote and all the genes they carry are passed with precision to the somatic cells.

The only cells of the human body not produced by mitosis are the gametes, which develop from specialized cells called *germ cells* in the gonads—ovaries in females and testes in males **(Figure 13.5)**. Imagine what would happen if human gametes were made by mitosis: They would be diploid like the somatic cells. At the next round of fertilization, when two gametes fused, the normal chromosome number of 46 would double to 92, and each subsequent generation would double the

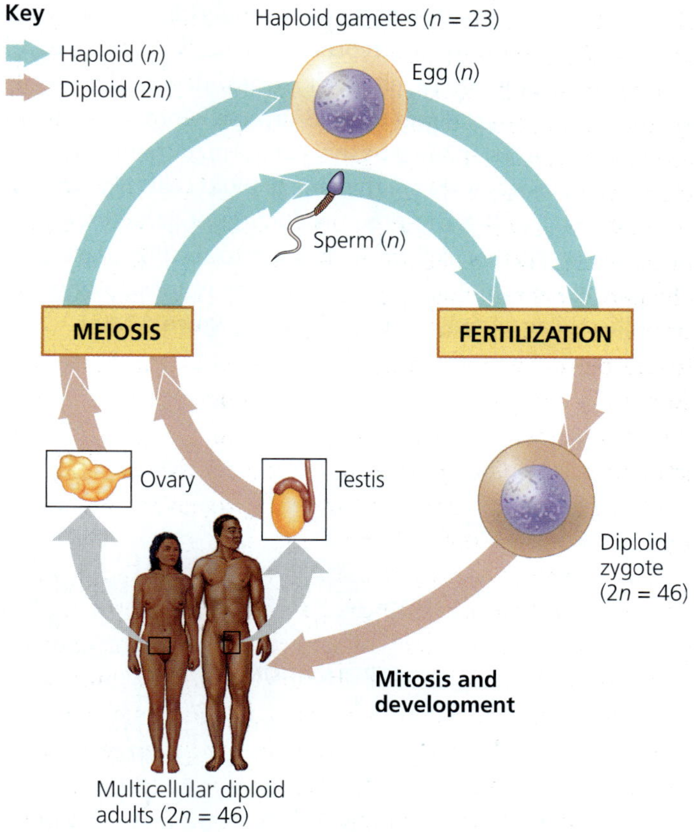

▲ **Figure 13.5 The human life cycle.** In each generation, the number of chromosome sets doubles at fertilization, but is halved during meiosis. For humans, the number of chromosomes in a haploid cell is 23, consisting of one set ($n = 23$); the number of chromosomes in the diploid zygote and all somatic cells arising from it is 46, consisting of two sets ($2n = 46$).

This figure introduces a color code that will be used for other life cycles later in this book. The teal arrows highlight haploid stages of a life cycle, and the beige arrows highlight diploid stages.

number of chromosomes yet again. This does not happen, however, because in sexually reproducing organisms, the gametes are formed by a modified type of cell division called **meiosis**. This type of cell division reduces the number of sets of chromosomes from two to one in the gametes, counterbalancing the doubling that occurs at fertilization. In animals, meiosis occurs only in the ovaries or testes. As a result of meiosis, each human sperm and egg is haploid ($n = 23$). Fertilization restores the diploid condition by combining two haploid sets of chromosomes, and the human life cycle is repeated, generation after generation (see Figure 13.5). You will learn more about the production of sperm and eggs in Chapter 46.

In general, the steps of the human life cycle are typical of many sexually reproducing animals. Indeed, the processes of fertilization and meiosis are the unique trademarks of sexual reproduction, in plants as well as animals. Fertilization and meiosis alternate in sexual life cycles, maintaining a constant number of chromosomes in each species from one generation to the next.

The Variety of Sexual Life Cycles

Although the alternation of meiosis and fertilization is common to all organisms that reproduce sexually, the timing of these two events in the life cycle varies, depending on the species. These variations can be grouped into three main types of life cycles. In the type that occurs in humans and most other animals, gametes are the only haploid cells. Meiosis occurs in germ cells during the production of gametes, which undergo no further cell division prior to fertilization. After fertilization, the diploid zygote divides by mitosis, producing a multicellular organism that is diploid **(Figure 13.6a)**.

Plants and some species of algae exhibit a second type of life cycle called **alternation of generations**. This type includes both diploid and haploid stages that are multicellular. The multicellular diploid stage is called the **sporophyte**. Meiosis in the sporophyte produces haploid cells called **spores**. Unlike a gamete, a haploid spore doesn't fuse with another cell but divides mitotically, generating a multicellular haploid stage called the **gametophyte**. Cells of the gametophyte give rise to gametes by mitosis. Fusion of two haploid gametes at fertilization results in a diploid zygote, which develops into the next sporophyte generation. Therefore, in this type of life cycle, the sporophyte generation produces a gametophyte as its offspring, and the gametophyte generation produces the next sporophyte generation **(Figure 13.6b)**. Clearly, the term *alternation of generations* is a fitting name for this type of life cycle.

A third type of life cycle occurs in most fungi and some protists, including some algae. After gametes fuse and form a diploid zygote, meiosis occurs without a multicellular diploid offspring developing. Meiosis produces not gametes but haploid cells that then divide by mitosis and give rise to either unicellular descendants or a haploid multicellular adult organism. Subsequently, the haploid organism carries out further mitoses, producing the cells that develop into gametes. The only diploid stage found in these species is the single-celled zygote **(Figure 13.6c)**.

Note that *either* haploid or diploid cells can divide by mitosis, depending on the type of life cycle. Only diploid cells, however, can undergo meiosis because haploid cells have a single set of chromosomes that cannot be further reduced. Though the three types of sexual life cycles differ in the timing of meiosis and fertilization, they share a fundamental result: genetic

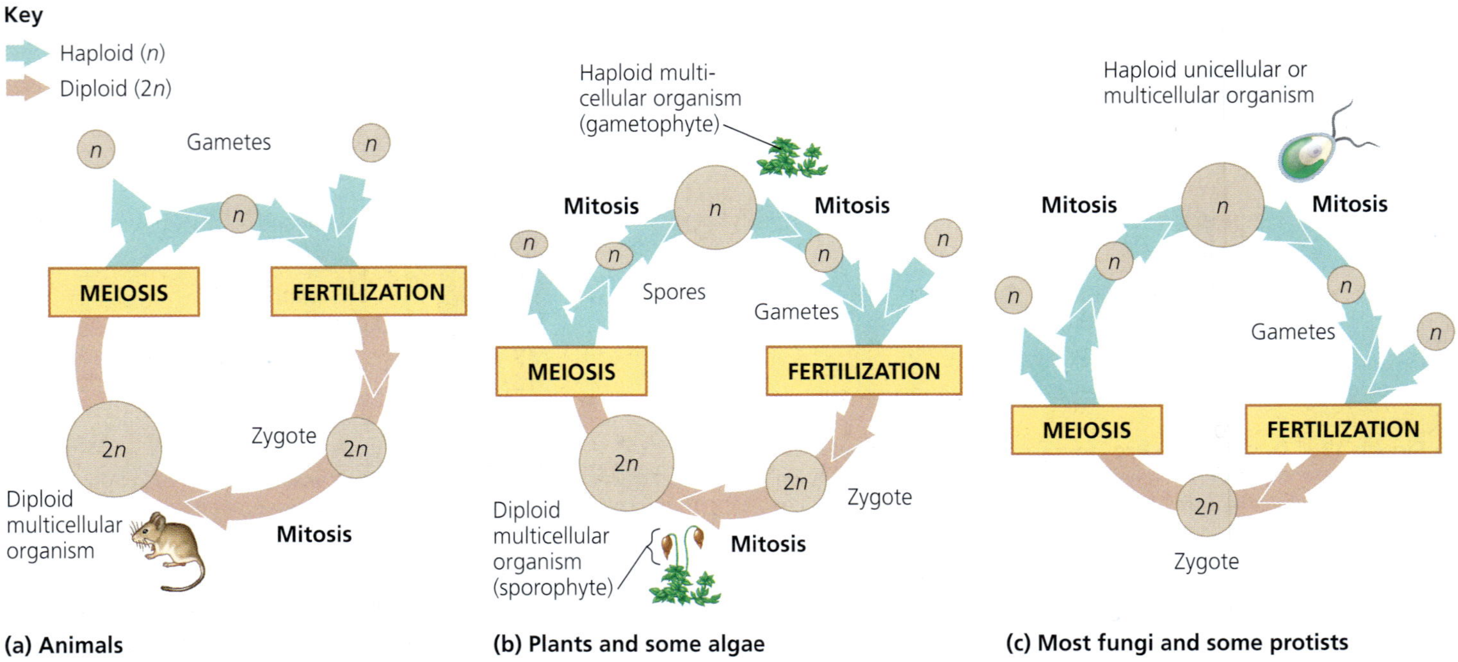

▲ **Figure 13.6 Three types of sexual life cycles.** The common feature of all three cycles is the alternation of meiosis and fertilization, key events that contribute to genetic variation among offspring. The cycles differ in the timing of these two key events.

variation among offspring. A closer look at meiosis will reveal the sources of this variation.

CONCEPT CHECK 13.2

1. How does the karyotype of a human female differ from that of a human male?
2. How does the alternation of meiosis and fertilization in the life cycles of sexually reproducing organisms maintain the normal chromosome count for each species?
3. Each sperm of a pea plant contains seven chromosomes. What are the haploid and diploid numbers for peas?
4. **WHAT IF?** A certain eukaryote lives as a unicellular organism, but during environmental stress, its cells produce gametes. The gametes fuse, and the resulting zygote undergoes meiosis, generating new single cells. What type of organism could this be?

For suggested answers, see Appendix A.

CONCEPT **13.3**

Meiosis reduces the number of chromosome sets from diploid to haploid

Many of the steps of meiosis closely resemble corresponding steps in mitosis. Meiosis, like mitosis, is preceded by the replication of chromosomes. However, this single replication is followed by not one but two consecutive cell divisions, called **meiosis I** and **meiosis II**. These two divisions result in four daughter cells (rather than the two daughter cells of mitosis), each with only half as many chromosomes as the parent cell.

The Stages of Meiosis

The overview of meiosis in **Figure 13.7** shows that both members of a single homologous pair of chromosomes in a diploid cell are replicated and that the copies are then sorted into four haploid daughter cells. Recall that sister chromatids are two copies of *one* chromosome, closely associated all along their lengths; this association is called *sister chromatid cohesion*. Together, the sister chromatids make up one replicated chromosome (see Figure 13.4). In contrast, the two chromosomes of a homologous pair are individual chromosomes that were inherited from different parents. Homologs appear alike in the microscope, but they may have different versions of genes, called *alleles*, at corresponding loci (for example, an allele for freckles on one chromosome and an allele for the absence of freckles at the same locus on the homolog). Homologs are not associated with each other except during meiosis, as you will soon see.

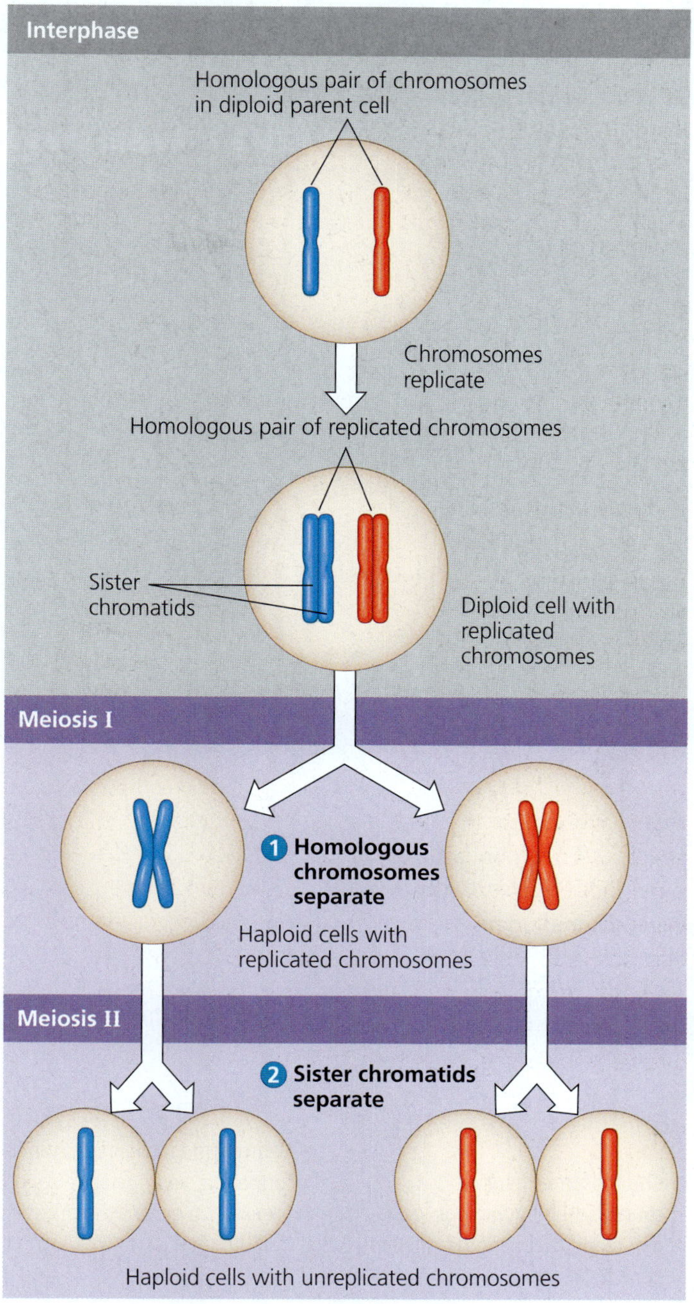

▲ **Figure 13.7 Overview of meiosis: how meiosis reduces chromosome number.** After the chromosomes replicate in interphase, the diploid cell divides *twice*, yielding four haploid daughter cells. This overview tracks just one pair of homologous chromosomes, which for the sake of simplicity are drawn in the condensed state throughout (they would not normally be condensed during interphase). The red chromosome was inherited from the female parent, the blue chromosome from the male parent.

DRAW IT *Redraw the cells in this figure using a simple DNA double helix to represent each DNA molecule.*

Figure 13.8, on the next two pages, describes in detail the stages of the two divisions of meiosis for an animal cell whose diploid number is 6. Meiosis halves the total number of chromosomes in a very specific way, reducing the number of sets from two to one, with each daughter cell receiving one set of chromosomes. Study Figure 13.8 thoroughly before going on.

Figure 13.8

Exploring The Meiotic Division of an Animal Cell

MEIOSIS I: Separates homologous chromosomes

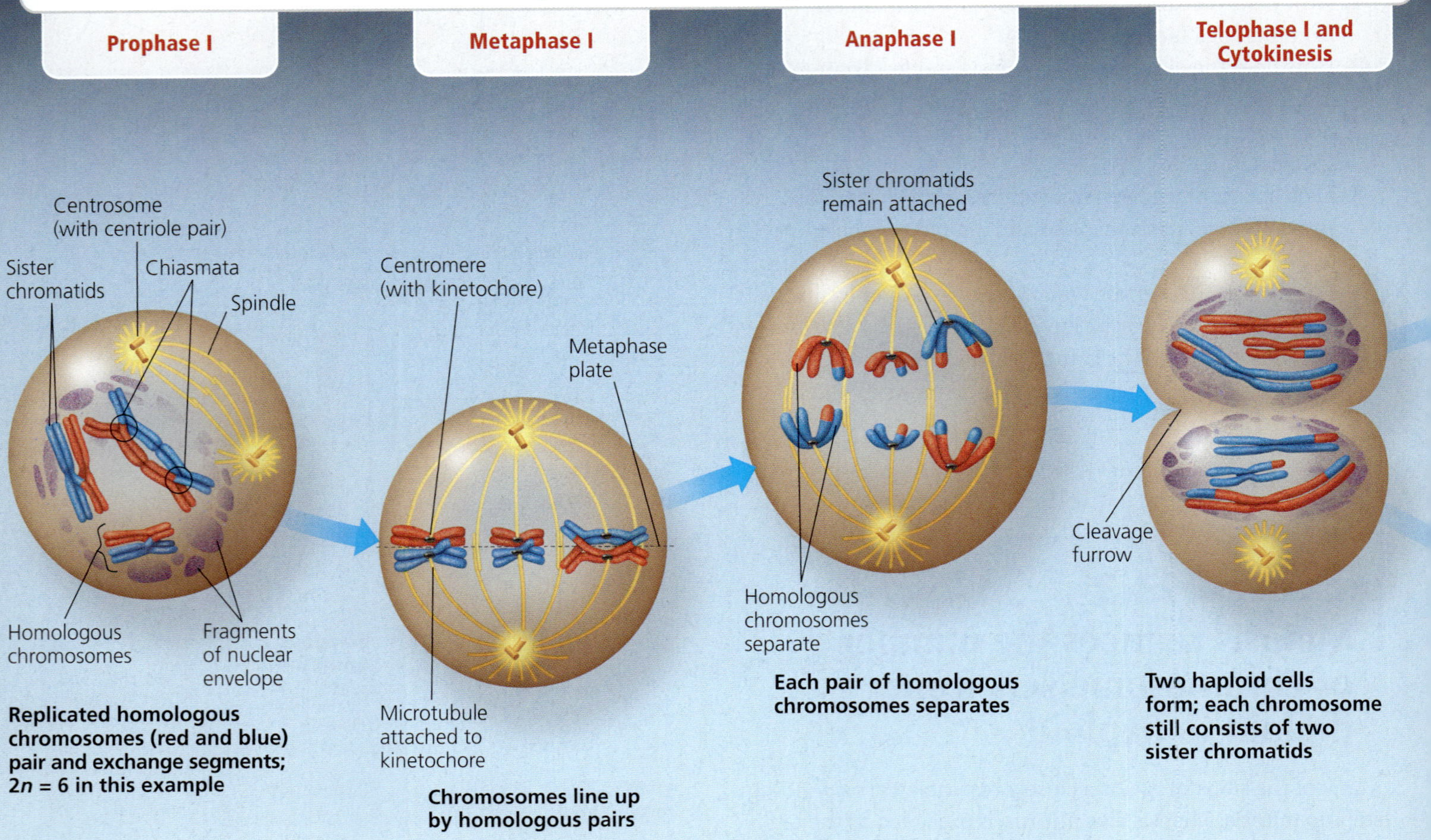

Prophase I

- Chromosomes begin to condense, and homologs loosely pair along their lengths, aligned gene by gene.
- Crossing over (the exchange of corresponding segments of DNA molecules by nonsister chromatids) is completed while homologs are in *synapsis*, held tightly together by proteins along their lengths (before the stage shown).
- Synapsis ends in mid-prophase, and the chromosomes in each pair move apart slightly, as shown above.
- Each homologous pair has one or more chiasmata, points where crossing over has occurred and the homologs are still associated due to cohesion between sister chromatids (*sister chromatid cohesion*).
- Centrosome movement, spindle formation, and nuclear envelope breakdown occur as in mitosis.
- In late prophase I (after the stage shown), microtubules from one pole or the other attach to the two kinetochores, protein structures at the centromeres of the two homologs. The homologous pairs then move toward the metaphase plate.

Metaphase I

- Pairs of homologous chromosomes are now arranged on the metaphase plate, with one chromosome in each pair facing each pole.
- Both chromatids of one homolog are attached to kinetochore microtubules from one pole; those of the other homolog are attached to microtubules from the opposite pole.

Anaphase I

- Breakdown of proteins responsible for sister chromatid cohesion along chromatid arms allows homologs to separate.
- The homologs move toward opposite poles, guided by the spindle apparatus.
- Sister chromatid cohesion persists at the centromere, causing chromatids to move as a unit toward the same pole.

Telophase I and Cytokinesis

- At the beginning of telophase I, each half of the cell has a complete haploid set of replicated chromosomes. Each chromosome is composed of two sister chromatids; one or both chromatids include regions of nonsister chromatid DNA.
- Cytokinesis (division of the cytoplasm) usually occurs simultaneously with telophase I, forming two haploid daughter cells.
- In animal cells, a cleavage furrow forms. (In plant cells, a cell plate forms.)
- In some species, chromosomes decondense and the nuclear envelope re-forms.
- No replication occurs between meiosis I and meiosis II.

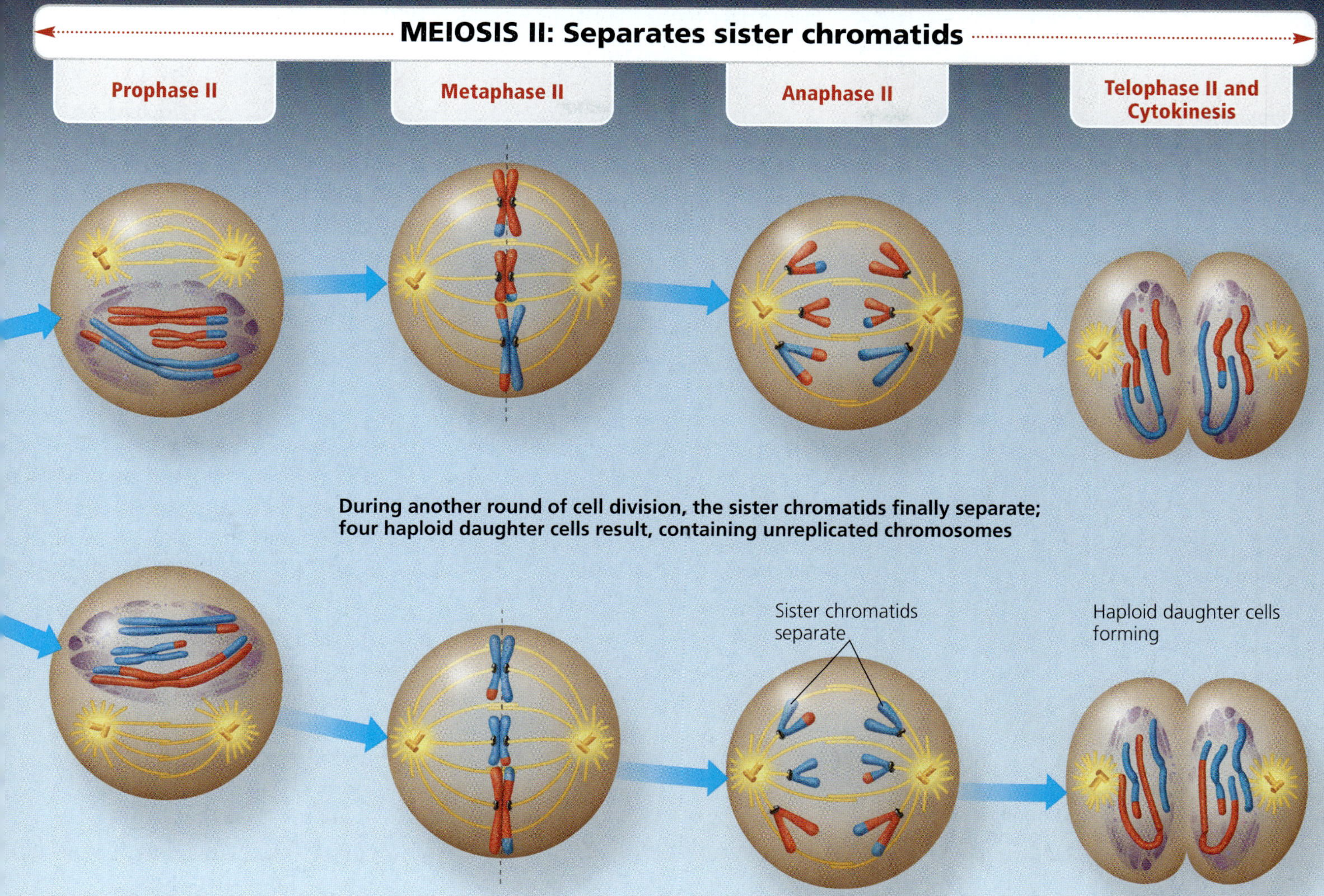

Prophase II

- A spindle apparatus forms.
- In late prophase II (not shown here), chromosomes, each still composed of two chromatids associated at the centromere, move toward the metaphase II plate.

Metaphase II

- The chromosomes are positioned on the metaphase plate as in mitosis.
- Because of crossing over in meiosis I, the two sister chromatids of each chromosome are *not* genetically identical.
- The kinetochores of sister chromatids are attached to microtubules extending from opposite poles.

Anaphase II

- Breakdown of proteins holding the sister chromatids together at the centromere allows the chromatids to separate. The chromatids move toward opposite poles as individual chromosomes.

Telophase II and Cytokinesis

- Nuclei form, the chromosomes begin decondensing, and cytokinesis occurs.
- The meiotic division of one parent cell produces four daughter cells, each with a haploid set of (unreplicated) chromosomes.
- Each of the four daughter cells is genetically distinct from the other daughter cells and from the parent cell.

BioFlix Visit the Study Area at **www.masteringbio.com** for the BioFlix 3-D Animation on Meiosis.

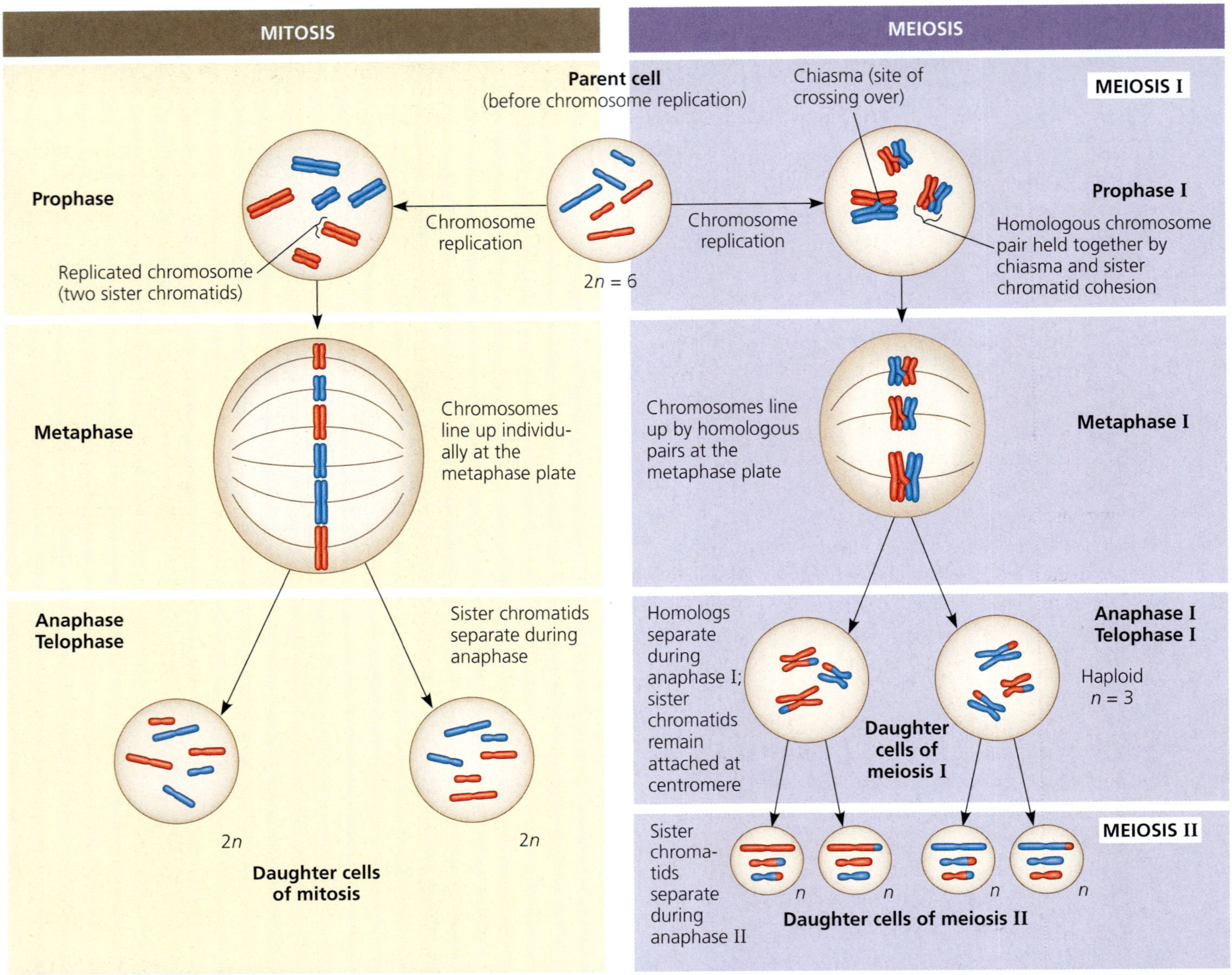

SUMMARY

Property	Mitosis	Meiosis
DNA replication	Occurs during interphase before mitosis begins	Occurs during interphase before meiosis I begins
Number of divisions	One, including prophase, metaphase, anaphase, and telophase	Two, each including prophase, metaphase, anaphase, and telophase
Synapsis of homologous chromosomes	Does not occur	Occurs during prophase I along with crossing over between nonsister chromatids; resulting chiasmata hold pairs together due to sister chromatid cohesion
Number of daughter cells and genetic composition	Two, each diploid (2*n*) and genetically identical to the parent cell	Four, each haploid (*n*), containing half as many chromosomes as the parent cell; genetically different from the parent cell and from each other
Role in the animal body	Enables multicellular adult to arise from zygote; produces cells for growth, repair, and, in some species, asexual reproduction	Produces gametes; reduces number of chromosomes by half and introduces genetic variability among the gametes

▲ **Figure 13.9 A comparison of mitosis and meiosis in diploid cells.**

DRAW IT *Could any other combinations of chromosomes be generated during meiosis II from the specific cells shown in telophase I? Explain.* (Hint: *Draw the cells as they would appear in metaphase II.*)

A Comparison of Mitosis and Meiosis

Figure 13.9 summarizes the key differences between meiosis and mitosis in diploid cells. Basically, meiosis reduces the number of chromosome sets from two (diploid) to one (haploid), whereas mitosis conserves the number of chromosome sets. Therefore, meiosis produces cells that differ genetically from their parent cell and from each other, whereas mitosis produces daughter cells that are genetically identical to their parent cell and to each other.

Three events unique to meiosis occur during meiosis I:

1. **Synapsis and crossing over.** During prophase I, replicated homologs pair up and become physically connected along their lengths by a zipper-like protein structure, the *synaptonemal complex*; this process is called **synapsis.** Genetic rearrangement between nonsister chromatids, known as **crossing over**, is completed during this stage. Following disassembly of the synaptonemal complex in late prophase, the two homologs pull apart slightly but remain connected by at least one X-shaped region called a **chiasma** (plural, *chiasmata*). A chiasma is the physical manifestation of crossing over; it appears as a cross because sister chromatid cohesion still holds the two original sister chromatids together, even in regions where one of them is now part of the other homolog. Synapsis and crossing over normally do not occur during mitosis.
2. **Homologs on the metaphase plate.** At metaphase I of meiosis, chromosomes are positioned on the metaphase plate as pairs of homologs, rather than individual chromosomes, as in metaphase of mitosis.
3. **Separation of homologs.** At anaphase I of meiosis, the replicated chromosomes of each homologous pair move toward opposite poles, but the sister chromatids of each replicated chromosome remain attached. In anaphase of mitosis, by contrast, sister chromatids separate.

How do sister chromatids stay together through meiosis I but separate from each other in meiosis II and mitosis? Sister chromatids are attached along their lengths by protein complexes called *cohesins.* In mitosis, this attachment lasts until the end of metaphase, when enzymes cleave the cohesins, freeing the sister chromatids to move to opposite poles of the cell. In meiosis, sister chromatid cohesion is released in two steps. In metaphase I, homologs are held together by cohesion between sister chromatid arms in regions where DNA has been exchanged. At anaphase I, cohesins are cleaved along the arms, allowing homologs to separate. At anaphase II, cohesins are cleaved at the centromeres, allowing chromatids to separate.

Figure 13.10 shows one of a series of experiments carried out by Yoshinori Watanabe and colleagues at the University of Tokyo. They knew that similar proteins were present in cohesin complexes during mitosis and meiosis, and they wondered what was responsible for preventing cohesin cleavage at the centromere while it was occurring along sister chromatid

▼ Figure 13.10 Inquiry

What prevents the separation of sister chromatids at anaphase I of meiosis?

EXPERIMENT Yoshinori Watanabe and colleagues knew that during anaphase I, the protein shugoshin is present only around centromeres. They wondered whether it protects cohesins there from degradation in meiosis I, ensuring that chromatids stay together while homologs separate. To test this hypothesis, they used a species of yeast in which meiosis produces haploid spores lined up in a specific order inside a spore case. To follow the movement of chromosomes, they fluorescently labeled a region near the centromere of both chromatids in one homolog, leaving the other homolog unlabeled. They then disabled the gene coding for shugoshin and compared this yeast strain (shugoshin$^-$) with normal yeast cells (shugoshin$^+$). The researchers expected that the two labeled chromosomes arising from the labeled chromatids in normal cells would end up in separate spores at one end of the spore case. They further predicted that if shugoshin does protect cohesins from cleavage at the centromere at anaphase I, then the labeled chromosomes in shugoshin$^-$ cells would separate randomly in meiosis II, sometimes ending up in the same spore.

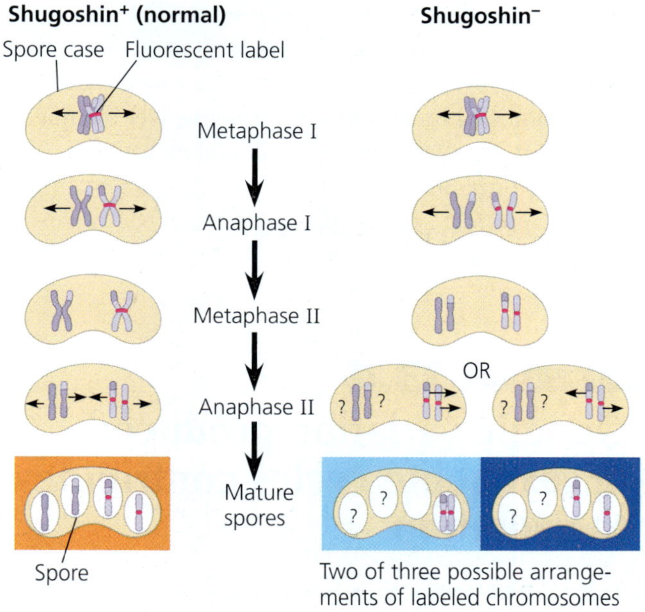

RESULTS In shugoshin$^+$ cells, the two labeled chromosomes ended up in different spores in almost all cases. In shugoshin$^-$ cells, they were in the same spore in about half the cases.

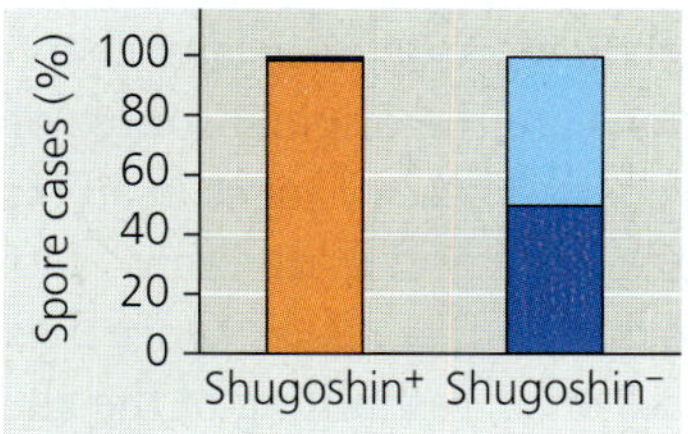

CONCLUSION The researchers concluded that shugoshin protects cohesins at the centromere at anaphase I, thus maintaining the attachment between sister chromatids and ensuring that they separate properly during meiosis II.

SOURCE T. S. Kitajima, S. A. Kawashima, and Y. Watanabe, The conserved kinetochore protein shugoshin protects centromeric cohesion during meiosis, *Nature* 427:510–517 (2004).

WHAT IF? Draw a graph showing what you expect happened to the chromatids of the *unlabeled* chromosome in both strains of cells.

arms at the end of metaphase I. They found a protein they named shugoshin (Japanese for "guardian spirit") that protects cohesins from cleavage at the centromere during meiosis I. Shugoshin is similar to a fruit fly protein identified 10 years earlier by Terry Orr-Weaver, this unit's interviewee.

Meiosis I is called the *reductional division* because it halves the number of chromosome sets per cell—a reduction from two sets (the diploid state) to one set (the haploid state). During the second meiotic division, meiosis II (sometimes called the *equational division*), the sister chromatids separate, producing haploid daughter cells. The mechanism for separating sister chromatids is virtually identical in meiosis II and mitosis. The molecular basis of chromosome behavior during meiosis continues to be a focus of intense research interest.

CONCEPT CHECK 13.3

1. How are the chromosomes in a cell at metaphase of mitosis similar to and different from the chromosomes in a cell at metaphase of meiosis II?
2. **WHAT IF?** Given that the synaptonemal complex disappears by the end of prophase, how would the two homologs be associated if crossing over did not occur? What effect might this ultimately have on gamete formation?

For suggested answers, see Appendix A.

CONCEPT 13.4

Genetic variation produced in sexual life cycles contributes to evolution

How do we account for the genetic variation illustrated in Figure 13.1? As you will learn in more detail in later chapters, mutations are the original source of genetic diversity. These changes in an organism's DNA create the different versions of genes known as *alleles*. Once these differences arise, reshuffling of the alleles during sexual reproduction produces the variation that results in each member of a species having its own unique combination of traits.

Origins of Genetic Variation Among Offspring

In species that reproduce sexually, the behavior of chromosomes during meiosis and fertilization is responsible for most of the variation that arises each generation. Let's examine three mechanisms that contribute to the genetic variation arising from sexual reproduction: independent assortment of chromosomes, crossing over, and random fertilization.

Independent Assortment of Chromosomes

One aspect of sexual reproduction that generates genetic variation is the random orientation of homologous pairs of chromosomes at metaphase of meiosis I. At metaphase I, the homologous pairs, each consisting of one maternal and one paternal chromosome, are situated on the metaphase plate. (Note that the terms *maternal* and *paternal* refer, respectively, to the mother and father of the individual whose cells are undergoing meiosis.) Each pair may orient with either its maternal or paternal homolog closer to a given pole—its orientation is as random as the flip of a coin. Thus, there is a 50% chance that a particular daughter cell of meiosis I will get the maternal chromosome of a certain homologous pair and a 50% chance that it will get the paternal chromosome.

Because each homologous pair of chromosomes is positioned independently of the other pairs at metaphase I, the first meiotic division results in each pair sorting its maternal and paternal homologs into daughter cells independently of every other pair. This is called *independent assortment.* Each daughter cell represents one outcome of all possible combinations of maternal and paternal chromosomes. As shown in **Figure 13.11**, the number of combinations possible for daughter cells formed by meiosis of a diploid cell with two homologous pairs of chromosomes is four (two possible arrangements for the first pair times two possible arrangements for the second pair). Note that only two of the four

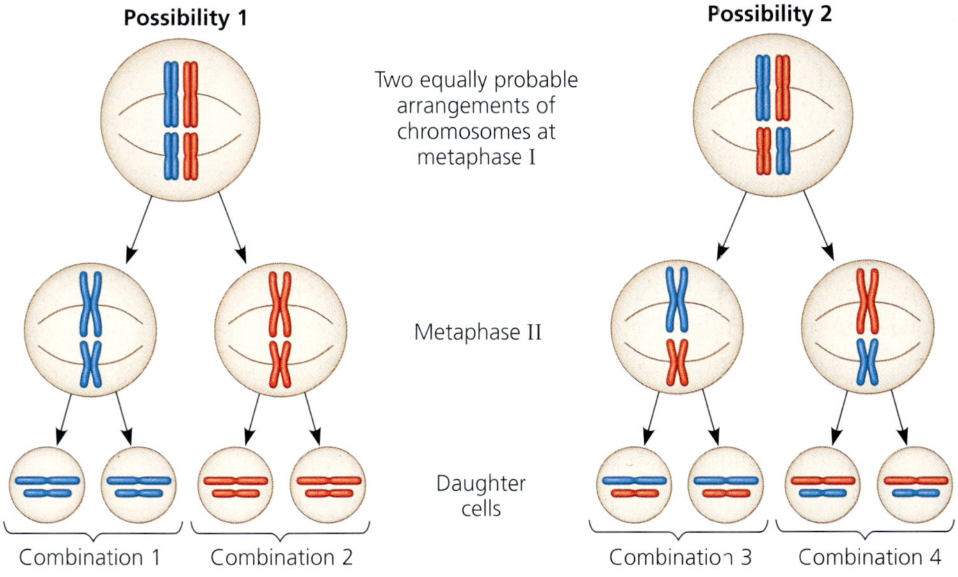

▶ **Figure 13.11 The independent assortment of homologous chromosomes in meiosis.**

combinations of daughter cells shown in the figure would result from meiosis of a *single* diploid cell, because a single parent cell would have one or the other possible chromosomal arrangement at metaphase I, but not both. However, the population of daughter cells resulting from meiosis of a large number of diploid cells contains all four types in approximately equal numbers. In the case of $n = 3$, eight combinations of chromosomes are possible for daughter cells. More generally, the number of possible combinations when chromosomes sort independently during meiosis is 2^n, where n is the haploid number of the organism.

In the case of humans ($n = 23$), the number of possible combinations of maternal and paternal chromosomes in the resulting gametes is 2^{23}, or about 8.4 million. Each gamete that you produce in your lifetime contains one of roughly 8.4 million possible combinations of chromosomes.

Crossing Over

As a consequence of the independent assortment of chromosomes during meiosis, each of us produces a collection of gametes differing greatly in their combinations of the chromosomes we inherited from our two parents. Figure 13.11 suggests that each individual chromosome in a gamete is exclusively maternal or paternal in origin. In fact, this is *not* the case, because crossing over produces **recombinant chromosomes**, individual chromosomes that carry genes (DNA) derived from two different parents **(Figure 13.12)**. In meiosis in humans, an average of one to three crossover events occur per chromosome pair, depending on the size of the chromosomes and the position of their centromeres.

Crossing over begins very early in prophase I, as homologous chromosomes pair loosely along their lengths. Each gene on one homolog is aligned precisely with the corresponding gene on the other homolog. In a single crossover event, specific proteins orchestrate an exchange of corresponding segments of two *nonsister* chromatids—one maternal and one paternal chromatid of a homologous pair. In this way, crossing over produces chromosomes with new combinations of maternal and paternal alleles (see Figure 13.12).

In humans and most other organisms studied so far, crossing over also plays an essential role in the lining up of homologous chromosomes during metaphase I. As seen in Figure 13.8, a chiasma forms as the result of a crossover occurring while sister chromatid cohesion is present along the arms. Chiasmata hold homologs together as the spindle forms for the first meiotic division. During anaphase I, the release of cohesion along sister chromatid arms allows homologs to separate. During anaphase II, the release of sister chromatid cohesion at the centromeres allows the sister chromatids to separate.

At metaphase II, chromosomes that contain one or more recombinant chromatids can be oriented in two alternative, nonequivalent ways with respect to other chromosomes, because their sister chromatids are no longer identical. The different possible arrangements of nonidentical sister chromatids during meiosis II further increases the number of genetic types of daughter cells that can result from meiosis.

You will learn more about crossing over in Chapter 15. The important point for now is that crossing over, by combining DNA inherited from two parents into a single chromosome, is an important source of genetic variation in sexual life cycles.

Random Fertilization

The random nature of fertilization adds to the genetic variation arising from meiosis. In humans, each male and female gamete represents one of about 8.4 million (2^{23}) possible chromosome combinations due to independent assortment. The fusion of a male gamete with a female gamete during fertilization will produce a zygote with any of about 70 trillion ($2^{23} \times 2^{23}$) diploid combinations. If we factor in the variation brought about by crossing over, the number of possibilities is truly astronomical. It may sound trite, but you really *are* unique.

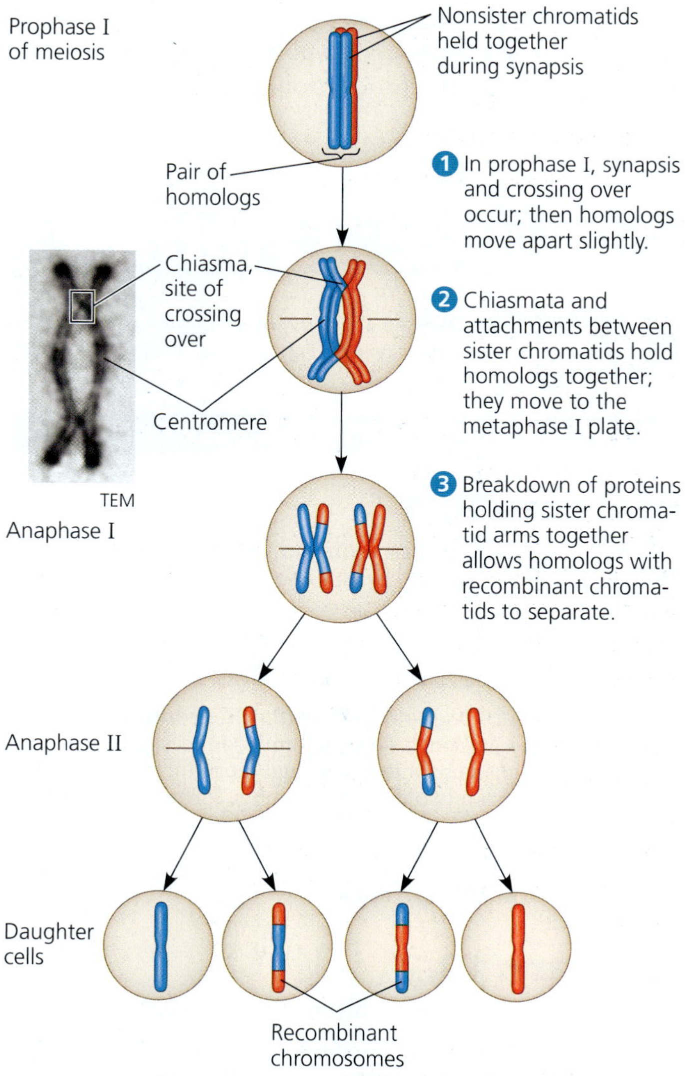

▲ **Figure 13.12 The results of crossing over during meiosis.**

The Evolutionary Significance of Genetic Variation Within Populations

Now that you've learned how new combinations of genes arise among offspring in a sexually reproducing population, let's see how the genetic variation in a population relates to evolution. Darwin recognized that a population evolves through the differential reproductive success of its variant members. On average, those individuals best suited to the local environment leave the most offspring, thus transmitting their genes. This natural selection results in the accumulation of those genetic variations favored by the environment. As the environment changes, the population may survive if, in each generation, at least some of its members can cope effectively with the new conditions. Different combinations of alleles may work better than those that previously prevailed. Mutations are the original source of different alleles, which are then mixed and matched during meiosis. In this chapter, we have seen how sexual reproduction greatly increases the genetic variation present in a population. In fact, the ability of sexual reproduction to generate genetic variation is one of the most commonly proposed explanations for the persistence of sexual reproduction.

Although Darwin realized that heritable variation is what makes evolution possible, he could not explain why offspring resemble—but are not identical to—their parents. Ironically, Gregor Mendel, a contemporary of Darwin, published a theory of inheritance that helps explain genetic variation, but his discoveries had no impact on biologists until 1900, more than 15 years after Darwin (1809–1882) and Mendel (1822–1884) had died. In the next chapter, you will learn how Mendel discovered the basic rules governing the inheritance of specific traits.

CONCEPT CHECK 13.4

1. What is the original source of all the different alleles of a gene?
2. The diploid number for fruit flies is 8, while that for grasshoppers is 46. If no crossing over took place, would the genetic variation among offspring from a given pair of parents be greater in fruit flies or grasshoppers? Explain.
3. **WHAT IF?** Under what circumstances would crossing over during meiosis *not* contribute to genetic variation among daughter cells?

For suggested answers, see Appendix A.

Chapter 13 Review

MEDIA Go to the Study Area at **www.masteringbio.com** for BioFlix 3-D Animations, MP3 Tutors, Videos, Practice Tests, an eBook, and more.

SUMMARY OF KEY CONCEPTS

CONCEPT 13.1

Offspring acquire genes from parents by inheriting chromosomes (pp. 248–249)

- **Inheritance of Genes** Each gene in an organism's DNA exists at a specific locus on a certain chromosome. We inherit one set of chromosomes from our mother and one set from our father.
- **Comparison of Asexual and Sexual Reproduction** In asexual reproduction, a single parent produces genetically identical offspring by mitosis. Sexual reproduction combines sets of genes from two different parents, forming genetically diverse offspring.

MEDIA

Activity Asexual and Sexual Life Cycles

CONCEPT 13.2

Fertilization and meiosis alternate in sexual life cycles (pp. 250–253)

- **Sets of Chromosomes in Human Cells** Normal human somatic cells are diploid. They have 46 chromosomes made up of two sets of 23—one set from each parent. In human diploid cells, there are 22 homologous pairs of autosomes, each with a maternal and a paternal homolog. The 23rd pair, the sex chromosomes, determines whether the person is female (XX) or male (XY).
- **Behavior of Chromosome Sets in the Human Life Cycle** At sexual maturity, ovaries and testes (the gonads) produce haploid gametes by meiosis, each gamete containing a single set of 23 chromosomes ($n = 23$). During fertilization, an egg and sperm unite, forming a diploid ($2n = 46$) single-celled zygote, which develops into a multicellular organism by mitosis.
- **The Variety of Sexual Life Cycles** Sexual life cycles differ in the timing of meiosis relative to fertilization and in the point(s) of the cycle at which a multicellular organism is produced by mitosis.

CONCEPT 13.3

Meiosis reduces the number of chromosome sets from diploid to haploid (pp. 253–258)

- **The Stages of Meiosis** The two cell divisions of meiosis produce four haploid daughter cells. The number of chromosome sets is reduced from two (diploid) to one (haploid) during meiosis I, the reductional division.
- **A Comparison of Mitosis and Meiosis** Meiosis is distinguished from mitosis by three events of meiosis I, as shown on the next page:

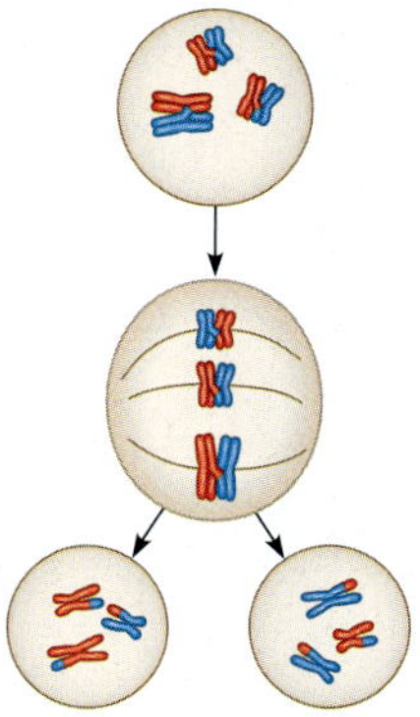

Prophase I: Each homologous pair undergoes synapsis and crossing over between nonsister chromatids.

Metaphase I: Chromosomes line up as homologous pairs on the metaphase plate.

Anaphase I: Homologs separate from each other; sister chromatids remain joined at the centromere.

Meiosis II separates the sister chromatids.

MEDIA

BioFlix 3-D Animation Meiosis
MP3 Tutor Meiosis
MP3 Tutor Mitosis-Meiosis Comparison
Activity Meiosis Animation

CONCEPT **13.4**

Genetic variation produced in sexual life cycles contributes to evolution (pp. 258–260)

- **Origins of Genetic Variation Among Offspring** Three events in sexual reproduction contribute to genetic variation in a population: independent assortment of chromosomes during meiosis, crossing over during meiosis I, and random fertilization of egg cells by sperm. Due to sister chromatid cohesion, crossing over leads to chiasmata, which hold homologs together until anaphase I.
- **The Evolutionary Significance of Genetic Variation Within Populations** Genetic variation is the raw material for evolution by natural selection. Mutations are the original source of this variation; the production of new combinations of variant genes in sexual reproduction generates additional genetic diversity.

MEDIA

Activity Origins of Genetic Variation
Investigation How Can the Frequency of Crossing Over Be Estimated?

TESTING YOUR KNOWLEDGE

SELF-QUIZ

1. A human cell containing 22 autosomes and a Y chromosome is
a. a sperm.
b. an egg.
c. a zygote.
d. a somatic cell of a male.
e. a somatic cell of a female.

2. Which life cycle stage is found in plants but not animals?
a. gamete
b. zygote
c. multicellular diploid
d. multicellular haploid
e. unicellular diploid

3. Homologous chromosomes move toward opposite poles of a dividing cell during
a. mitosis.
b. meiosis I.
c. meiosis II.
d. fertilization.
e. binary fission.

4. Meiosis II is similar to mitosis in that
a. sister chromatids separate during anaphase.
b. DNA replicates before the division.
c. the daughter cells are diploid.
d. homologous chromosomes synapse.
e. the chromosome number is reduced.

5. If the DNA content of a diploid cell in the G_1 phase of the cell cycle is x, then the DNA content of the same cell at metaphase of meiosis I would be
a. $0.25x$. b. $0.5x$. c. x. d. $2x$. e. $4x$.

6. If we continued to follow the cell lineage from question 5, then the DNA content of a single cell at metaphase of meiosis II would be
a. $0.25x$. b. $0.5x$. c. x. d. $2x$. e. $4x$.

7. How many different combinations of maternal and paternal chromosomes can be packaged in gametes made by an organism with a diploid number of 8 ($2n = 8$)?
a. 2 b. 4 c. 8 d. 16 e. 32

Use the diagram of a cell below to answer questions 8–10.

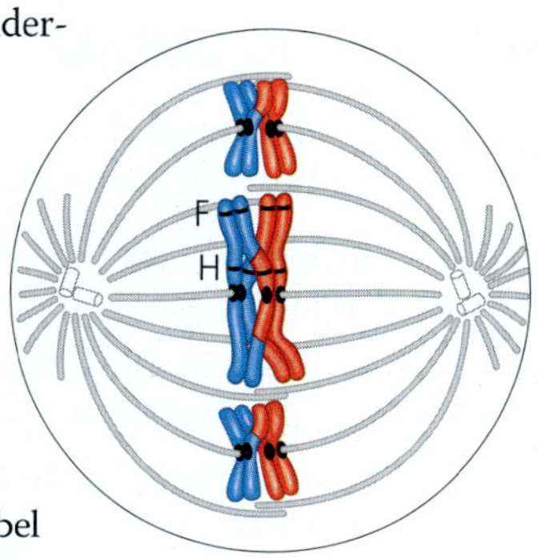

8. How can you tell this cell is undergoing meiosis, not mitosis?

9. Identify the stage of meiosis shown.

10. **DRAW IT** Copy the drawing to a separate sheet of paper and label appropriate structures with these terms, drawing lines or brackets as needed: chromosome (label as replicated or unreplicated), centromere, kinetochore, sister chromatids, nonsister chromatids, homologous pair, homologs, chiasma, sister chromatid cohesion. Describe the makeup of a haploid set and a diploid set.

For Self-Quiz answers, see Appendix A.

MEDIA Visit the Study Area at **www.masteringbio.com** for a Practice Test.

EVOLUTION CONNECTION

11. Many species can reproduce either asexually or sexually. What might be the evolutionary significance of the switch from asexual to sexual reproduction that occurs in some organisms when the environment becomes unfavorable?

SCIENTIFIC INQUIRY

12. The diagram accompanying questions 8–10 represents a meiotic cell in a certain individual. A previous study has shown that the freckles gene is located at the locus marked F, and the hair color gene is located at the locus marked H, both on the long chromosome. The individual from whom this cell was taken has inherited different alleles for each gene ("freckles" and "black hair" from one parent, and "no freckles" and "blond hair" from the other). Predict allele combinations in the gametes resulting from this meiotic event. (It will help if you draw out the rest of meiosis, labeling alleles by name.) List other possible combinations of these alleles in this individual's gametes.

14 Mendel and the Gene Idea

▲ **Figure 14.1 What principles of inheritance did Gregor Mendel discover by breeding garden pea plants?**

KEY CONCEPTS

14.1 Mendel used the scientific approach to identify two laws of inheritance

14.2 The laws of probability govern Mendelian inheritance

14.3 Inheritance patterns are often more complex than predicted by simple Mendelian genetics

14.4 Many human traits follow Mendelian patterns of inheritance

OVERVIEW

Drawing from the Deck of Genes

If you happened to see a woman with bright purple hair walking down the street, you would probably conclude that she hadn't inherited her striking hair color from either parent. Consciously or not, you have transformed a lifetime of observations of hair color and other features into a list of possible variations that occur naturally among people. Eyes of brown, blue, green, or gray; hair of black, brown, blond, or red—these are just a few examples of heritable variations that we may observe among individuals in a population. What are the genetic principles that account for the transmission of such traits from parents to offspring in humans and other organisms?

The explanation of heredity most widely in favor during the 1800s was the "blending" hypothesis, the idea that genetic material contributed by the two parents mixes in a manner analogous to the way blue and yellow paints blend to make green. This hypothesis predicts that over many generations, a freely mating population will give rise to a uniform population of individuals. However, our everyday observations and the results of breeding experiments with animals and plants contradict that prediction. The blending hypothesis also fails to explain other phenomena of inheritance, such as traits reappearing after skipping a generation.

An alternative to the blending model is a "particulate" hypothesis of inheritance: the gene idea. According to this model, parents pass on discrete heritable units—genes—that retain their separate identities in offspring. An organism's collection of genes is more like a deck of cards than a pail of paint. Like playing cards, genes can be shuffled and passed along, generation after generation, in undiluted form.

Modern genetics had its genesis in an abbey garden, where a monk named Gregor Mendel documented a particulate mechanism for inheritance. **Figure 14.1** shows Mendel (back row, holding a sprig of fuchsia) with his fellow monks. Mendel developed his theory of inheritance several decades before chromosomes were observed in the microscope and the significance of their behavior was understood. In this chapter, we will step into Mendel's garden to re-create his experiments and explain how he arrived at his theory of inheritance. We will also explore inheritance patterns more complex than those observed by Mendel in garden peas. Finally, we will see how the Mendelian model applies to the inheritance of human variations, including hereditary disorders such as sickle-cell disease.

CONCEPT 14.1

Mendel used the scientific approach to identify two laws of inheritance

Mendel discovered the basic principles of heredity by breeding garden peas in carefully planned experiments. As we retrace his work, you will recognize the key elements of the scientific process that were introduced in Chapter 1.

Mendel's Experimental, Quantitative Approach

Mendel grew up on his parents' small farm in a region of Austria that is now part of the Czech Republic. In this agricultural area, Mendel and the other children received agricultural training in school along with their basic education. As an adolescent, Mendel overcame financial hardship and illness to excel in high school and, later, at the Olmutz Philosophical Institute.

In 1843, at the age of 21, Mendel entered an Augustinian monastery, a reasonable choice at that time for someone who valued the life of the mind. He considered becoming a teacher but failed the necessary examination. In 1851, he left the monastery to pursue two years of study in physics and chemistry at the University of Vienna. These were very important years for Mendel's development as a scientist, in large part due to the strong influence of two professors. One was the physicist Christian Doppler, who encouraged his students to learn science through experimentation and trained Mendel to use mathematics to help explain natural phenomena. The other was a botanist named Franz Unger, who aroused Mendel's interest in the causes of variation in plants. The instruction Mendel received from these two mentors later played a critical role in his experiments with garden peas.

After attending the university, Mendel returned to the monastery and was assigned to teach at a local school, where several other instructors were enthusiastic about scientific research. In addition, his fellow monks shared a long-standing fascination with the breeding of plants. The monastery therefore provided fertile soil in more ways than one for Mendel's scientific endeavors. Around 1857, Mendel began breeding garden peas in the abbey garden to study inheritance. Although the question of heredity had long been a focus of curiosity at the monastery, Mendel's fresh approach allowed him to deduce principles that had remained elusive to others.

One reason Mendel probably chose to work with peas is that they are available in many varieties. For example, one variety has purple flowers, while another variety has white flowers. A heritable feature that varies among individuals, such as flower color, is called a **character**. Each variant for a character, such as purple or white color for flowers, is termed a **trait**.*

Other advantages of using peas are their short generation time and the large number of offspring from each mating. Furthermore, Mendel could strictly control mating between plants. The reproductive organs of a pea plant are in its flowers, and each pea flower has both pollen-producing organs (stamens) and an egg-bearing organ (carpel). In nature, pea plants usually self-fertilize: Pollen grains from the stamens land on the carpel of the same flower, and sperm released from the pollen grains fertilize eggs present in the carpel. To achieve cross-pollination (fertilization between different plants), Mendel removed the immature stamens of a plant before they produced pollen and then dusted pollen from another plant onto the altered flowers (**Figure 14.2**). Each resulting zygote then developed into a plant embryo encased in a seed (pea). Whether forcing self-pollination or executing artificial cross-pollination, Mendel could always be sure of the parentage of new seeds.

Mendel chose to track only those characters that varied between two distinct alternatives. For example, his plants had either purple flowers or white flowers; there was nothing intermediate between these two varieties. Had Mendel focused instead on characters that varied in a continuum among individuals—seed weight, for example—he would not have discovered the particulate nature of inheritance. (You'll learn why later.)

Mendel also made sure that he started his experiments with varieties that, over many generations of self-pollination,

* Some geneticists use the terms *character* and *trait* synonymously, but in this book we distinguish between them.

▼ **Figure 14.2 Research Method**

Crossing Pea Plants

APPLICATION By crossing (mating) two true-breeding varieties of an organism, scientists can study patterns of inheritance. In this example, Mendel crossed pea plants that varied in flower color.

TECHNIQUE

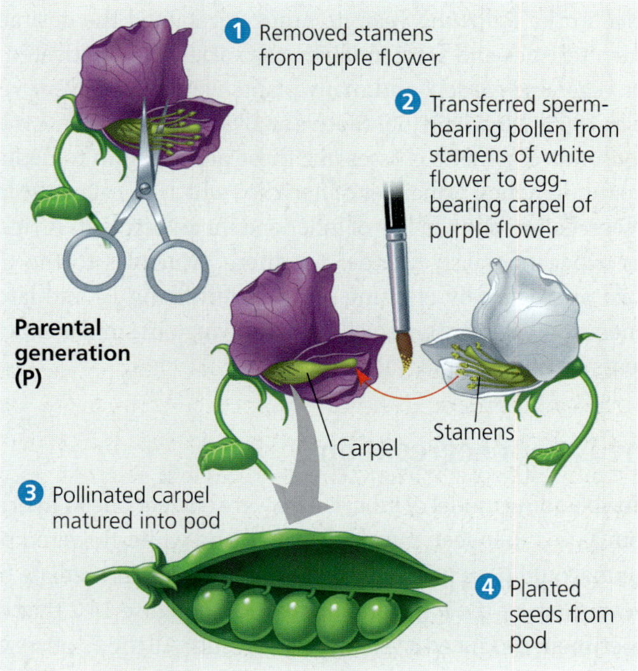

RESULTS When pollen from a white flower was transferred to a purple flower, the first-generation hybrids all had purple flowers. The result was the same for the reciprocal cross, which involved the transfer of pollen from purple flowers to white flowers.

had produced only the same variety as the parent plant. Such plants are said to be **true-breeding**. For example, a plant with purple flowers is true-breeding if the seeds produced by self-pollination in successive generations all give rise to plants that also have purple flowers.

In a typical breeding experiment, Mendel cross-pollinated two contrasting, true-breeding pea varieties—for example, purple-flowered plants and white-flowered plants (see Figure 14.2). This mating, or *crossing*, of two true-breeding varieties is called **hybridization**. The true-breeding parents are referred to as the **P generation** (parental generation), and their hybrid offspring are the **F_1 generation** (first filial generation, the word *filial* from the Latin word for "son"). Allowing these F_1 hybrids to self-pollinate produces an **F_2 generation** (second filial generation). Mendel usually followed traits for at least the P, F_1, and F_2 generations. Had Mendel stopped his experiments with the F_1 generation, the basic patterns of inheritance would have escaped him.

Mendel was a thorough and enthusiastic researcher. In a letter dated 1867, he wrote, "In 1859 I obtained a very fertile descendant with large, tasty seeds from a first generation hybrid. Since in the following year, its progeny retained the desirable characteristics and were uniform, the variety was cultivated in our vegetable garden, and many plants were raised every year up to 1865." Mendel's quantitative analysis of the F_2 plants from thousands of genetic crosses like these allowed him to deduce two fundamental principles of heredity, which he called the law of segregation and the law of independent assortment. A rigorous experimentalist, Mendel put these principles to the test again and again by crossing and self-fertilizing F_2 and later-generation pea plants, as well as by carrying out similar experiments on other plants, such as beans.

The Law of Segregation

If the blending model of inheritance were correct, the F_1 hybrids from a cross between purple-flowered and white-flowered pea plants would have pale purple flowers, a trait intermediate between those of the P generation. Notice in Figure 14.2 that the experiment produced a very different result: All the F_1 offspring had flowers just as purple as the purple-flowered parents. What happened to the white-flowered plants' genetic contribution to the hybrids? If it were lost, then the F_1 plants could produce only purple-flowered offspring in the F_2 generation. But when Mendel allowed the F_1 plants to self-pollinate and planted their seeds, the white-flower trait reappeared in the F_2 generation.

Mendel used very large sample sizes and kept accurate records of his results: 705 of the F_2 plants had purple flowers, and 224 had white flowers. These data fit a ratio of approximately three purple to one white **(Figure 14.3)**. Mendel reasoned that the heritable factor for white flowers did not disappear in the F_1 plants, but was somehow hidden or masked when the purple-flower factor was present. In Mendel's terminology, purple flower color is a *dominant* trait and white flower color is a *recessive* trait. The reappearance of white-flowered plants in the F_2 generation was evidence that the heritable factor causing white flowers had not been diluted or destroyed by coexisting with the purple-flower factor in the F_1 hybrids.

Mendel observed the same pattern of inheritance in six other characters, each represented by two distinctly different

▼ Figure 14.3 Inquiry

When F_1 hybrid pea plants are allowed to self-pollinate, which traits appear in the F_2 generation?

EXPERIMENT Around 1860, in a monastery garden in Brünn, Austria, Gregor Mendel used the character of flower color in pea plants to follow traits through two generations. He crossed true-breeding purple-flowered plants and white-flowered plants (crosses are symbolized by ×). The resulting F_1 hybrids were allowed to self-pollinate or were cross-pollinated with other F_1 hybrids. The F_2 generation plants were then observed for flower color.

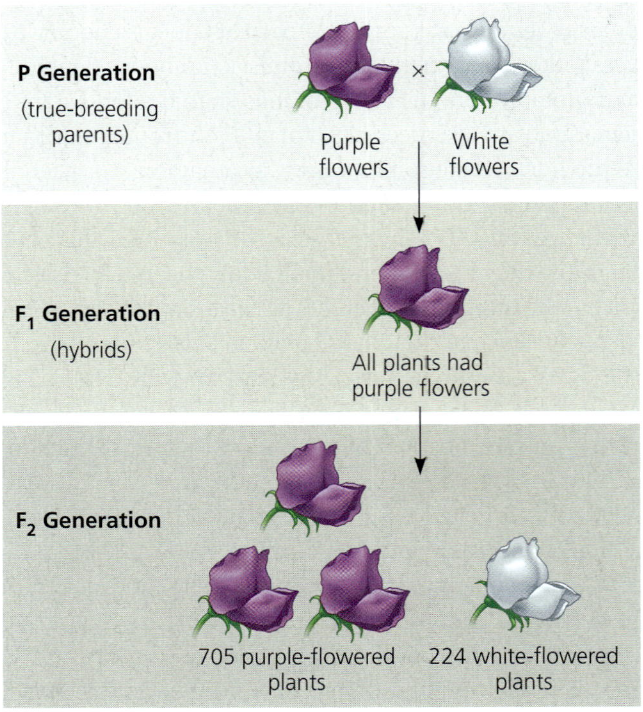

RESULTS Both purple-flowered and white-flowered plants appeared in the F_2 generation, in a ratio of approximately 3:1.

CONCLUSION The "heritable factor" for the recessive trait (white flowers) had not been destroyed or deleted in the F_1 generation, but merely masked by the presence of the factor for purple flowers, which is the dominant trait.

SOURCE G. Mendel, Experiments in plant hybridization, *Proceedings of the Natural History Society of Brünn* 4:3–47 (1866).

WHAT IF? If you mated two purple-flowered plants from the P generation, what ratio of traits would you expect to observe in the offspring? Explain.

traits **(Table 14.1)**. For example, when Mendel crossed a true-breeding variety that produced smooth, round pea seeds with one that produced wrinkled seeds, all the F_1 hybrids produced round seeds; this is the dominant trait for seed shape. In the F_2 generation, 75% of the seeds were round and 25% were wrinkled—a 3:1 ratio, as in Figure 14.3. Now let's see how Mendel deduced the law of segregation from his experimental results. In the discussion that follows, we will use modern terms instead of some of the terms used by Mendel. (For example, we'll use "gene" instead of Mendel's "heritable factor.")

Table 14.1 The Results of Mendel's F_1 Crosses for Seven Characters in Pea Plants

Character	Dominant Trait	x	Recessive Trait	F_2 Generation Dominant:Recessive	Ratio
Flower color	Purple	×	White	705:224	3.15:1
Flower position	Axial	×	Terminal	651:207	3.14:1
Seed color	Yellow	×	Green	6,022:2,001	3.01:1
Seed shape	Round	×	Wrinkled	5,474:1,850	2.96:1
Pod shape	Inflated	×	Constricted	882:299	2.95:1
Pod color	Green	×	Yellow	428:152	2.82:1
Stem length	Tall	×	Dwarf	787:277	2.84:1

Mendel's Model

Mendel developed a model to explain the 3:1 inheritance pattern that he consistently observed among the F_2 offspring in his pea experiments. We describe four related concepts making up this model, the fourth of which is the law of segregation.

First, *alternative versions of genes account for variations in inherited characters.* The gene for flower color in pea plants, for example, exists in two versions, one for purple flowers and the other for white flowers. These alternative versions of a gene are called **alleles** **(Figure 14.4)**. Today, we can relate this concept to chromosomes and DNA. As noted in Chapter 13, each gene is a sequence of nucleotides at a specific place, or locus, along a particular chromosome. The DNA at that locus, however, can vary slightly in its nucleotide sequence and hence in its information content. The purple-flower allele and the white-flower allele are two DNA variations possible at the flower-color locus on one of a pea plant's chromosomes.

Second, *for each character, an organism inherits two alleles, one from each parent.* Remarkably, Mendel made this deduction without knowing about the role of chromosomes. Recall from Chapter 13 that each somatic cell in a diploid organism has two sets of chromosomes, one set inherited from each parent. Thus, a genetic locus is actually represented twice in a diploid cell, once on each homolog of a specific pair of chromosomes. The two alleles at a particular locus may be identical, as in the true-breeding plants of Mendel's P generation. Or the alleles may differ, as in the F_1 hybrids (see Figure 14.4).

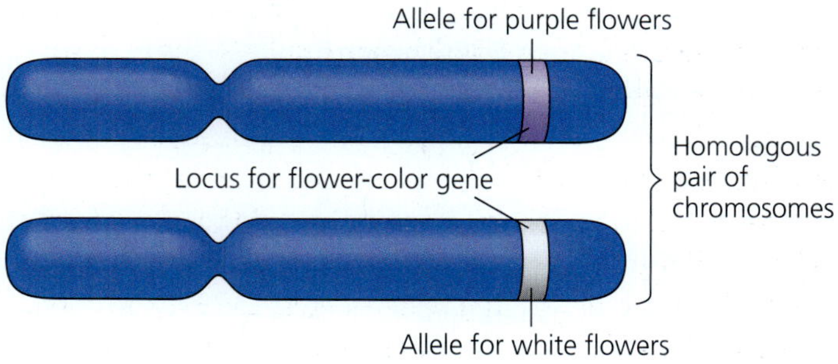

▲ **Figure 14.4 Alleles, alternative versions of a gene.** A somatic cell has two copies of each chromosome (forming a homologous pair) and thus two alleles of each gene, which may be identical or different. This figure depicts a homologous pair of chromosomes in an F_1 hybrid pea plant. The chromosome with an allele for purple flowers was inherited from one parent, and that with an allele for white flowers from the other parent.

Third, *if the two alleles at a locus differ, then one, the* **dominant allele**, *determines the organism's appearance; the other, the* **recessive allele**, *has no noticeable effect on the organism's appearance.* Accordingly, Mendel's F_1 plants had purple flowers because the allele for that trait is dominant and the allele for white flowers is recessive.

The fourth and final part of Mendel's model, the **law of segregation**, states that *the two alleles for a heritable character segregate (separate) during gamete formation and end up in different gametes.* Thus, an egg or a sperm gets only one of the two alleles that are present in the somatic cells of the organism making the gamete. In terms of chromosomes, this segregation corresponds to the distribution of the two members of a homologous pair of chromosomes to different gametes in meiosis (see Figure 13.7). Note that if an organism has identical alleles for a particular character—that is, the organism is true-breeding for that character—then that allele is present in all gametes. But if different alleles are present, as in the F_1 hybrids, then 50% of the gametes receive the dominant allele and 50% receive the recessive allele.

Does Mendel's segregation model account for the 3:1 ratio he observed in the F_2 generation of his numerous crosses? For the flower-color character, the model predicts that the two different alleles present in an F_1 individual will segregate into gametes such that half the gametes will have the purple-flower allele and half will have the white-flower allele. During self-pollination, gametes of each class unite randomly. An egg with a purple-flower allele has an equal chance of being fertilized by a sperm with a purple-flower allele or one with a white-flower allele. Since the same is true for an egg with a white-flower allele, there are four equally likely combinations of sperm and egg. **Figure 14.5** illustrates these combinations using a **Punnett square**, a handy diagrammatic device for predicting the allele composition of offspring from a cross between individuals of known genetic makeup. Notice that we use a capital letter to symbolize a dominant allele and a lowercase letter for a recessive allele. In our example, *P* is the purple-flower allele, and *p* is the white-flower allele; the gene itself may be referred to as the *P/p* gene.

In the F_2 offspring, what color will the flowers be? One-fourth of the plants have inherited two purple-flower alleles; clearly, these plants will have purple flowers. One-half of the F_2 offspring have inherited one purple-flower allele and one white-flower allele; these plants will also have purple flowers, the dominant trait. Finally, one-fourth of the F_2 plants have inherited two white-flower alleles and will express the recessive trait. Thus, Mendel's model accounts for the 3:1 ratio of traits that he observed in the F_2 generation.

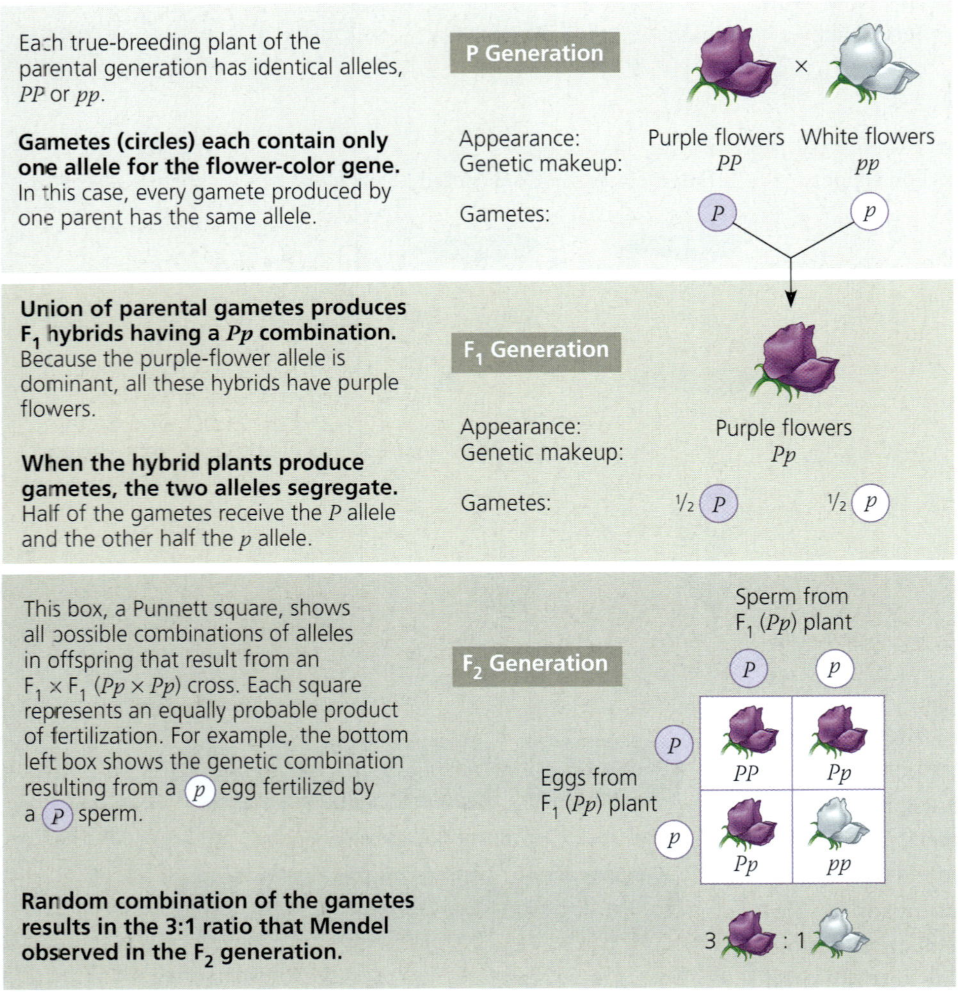

▲ **Figure 14.5 Mendel's law of segregation.** This diagram shows the genetic makeup of the generations in Figure 14.3. It illustrates Mendel's model for inheritance of the alleles of a single gene. Each plant has two alleles for the gene controlling flower color, one allele inherited from each parent. To construct a Punnett square that predicts the F_2 generation offspring, we list all the possible gametes from one parent (here, the F_1 female) along the left side of the square and all the possible gametes from the other parent (here, the F_1 male) along the top. The boxes represent the offspring resulting from all the possible unions of male and female gametes.

Useful Genetic Vocabulary

An organism that has a pair of identical alleles for a character is said to be **homozygous** for the gene controlling that character. In the parental generation in Figure 14.5, the purple pea plant is homozygous for the dominant allele (*PP*), while the white plant is homozygous for the recessive allele (*pp*). Homozygous plants "breed true" because all of their gametes contain the same allele—either *P* or *p* in this example. If we

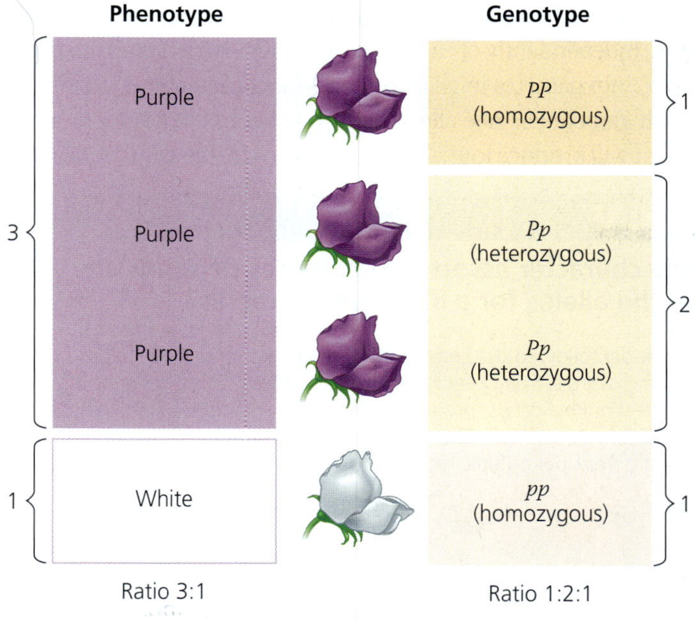

▲ **Figure 14.6 Phenotype versus genotype.** Grouping F_2 offspring from a cross for flower color according to phenotype results in the typical 3:1 phenotypic ratio. In terms of genotype, however, there are actually two categories of purple-flowered plants, *PP* (homozygous) and *Pp* (heterozygous), giving a 1:2:1 genotypic ratio.

cross dominant homozygotes with recessive homozygotes, every offspring will have two different alleles—*Pp* in the case of the F_1 hybrids of our flower-color experiment (see Figure 14.5). An organism that has two different alleles for a gene is said to be **heterozygous** for that gene. Unlike homozygotes, heterozygotes are not true-breeding because they produce gametes with different alleles; for example, *P*- and *p*-containing gametes are both produced by the F_1 hybrids of Figure 14.5. As a result, self-pollination of those F_1 hybrids produces both purple-flowered and white-flowered offspring.

Because of the different effects of dominant and recessive alleles, an organism's traits do not always reveal its genetic composition. Therefore, we distinguish between an organism's appearance or observable traits, called its **phenotype**, and its genetic makeup, its **genotype**. In the case of flower color in pea plants, *PP* and *Pp* plants have the same phenotype (purple) but different genotypes. **Figure 14.6** reviews these terms. Note that "phenotype" refers to physiological traits as well as traits that relate directly to appearance. For example, there is a pea variety that lacks the normal ability to self-pollinate. This physiological variation (non-self-pollination) is a phenotypic trait.

The Testcross

Suppose we have a "mystery" pea plant that has purple flowers. We cannot tell from its flower color if this plant is homozygous (*PP*) or heterozygous (*Pp*) because both genotypes result in the same purple phenotype. To determine the genotype, we can cross this plant with a white-flowered plant (*pp*), which will make only gametes with the recessive allele (*p*). The allele in the gamete contributed by the mystery plant will therefore determine the appearance of the offspring **(Figure 14.7)**. If all the offspring of the cross have purple flowers, then the purple-flowered mystery plant must be homozygous for the dominant allele, because a *PP* × *pp* cross produces all *Pp* offspring. But if both the purple and the white phenotypes appear among the offspring, then the purple-flowered parent must be heterozygous. The offspring of a *Pp* × *pp* cross will have a 1:1 phenotypic ratio. Breeding an organism of unknown genotype with a recessive homozygote is called a **testcross** because it can reveal the genotype of that organism. The testcross was devised by Mendel and continues to be an important tool of geneticists.

▼ **Figure 14.7 Research Method**

The Testcross

APPLICATION An organism that exhibits a dominant trait, such as purple flowers in pea plants, can be either homozygous for the dominant allele or heterozygous. To determine the organism's genotype, geneticists can perform a testcross.

TECHNIQUE In a testcross, the individual with the unknown genotype is crossed with a homozygous individual expressing the recessive trait (white flowers in this example), and Punnett squares are used to predict the possible outcomes.

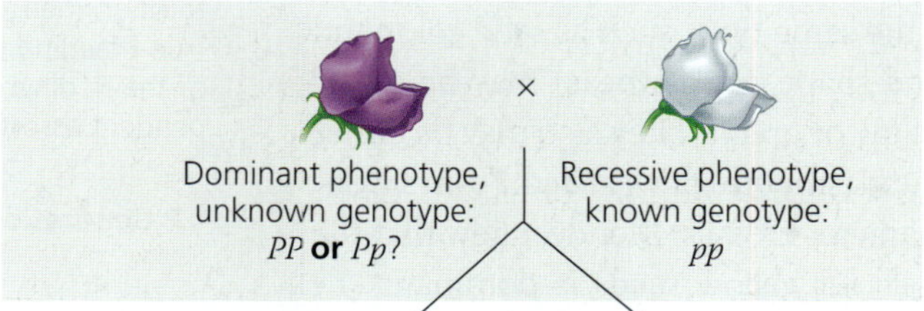

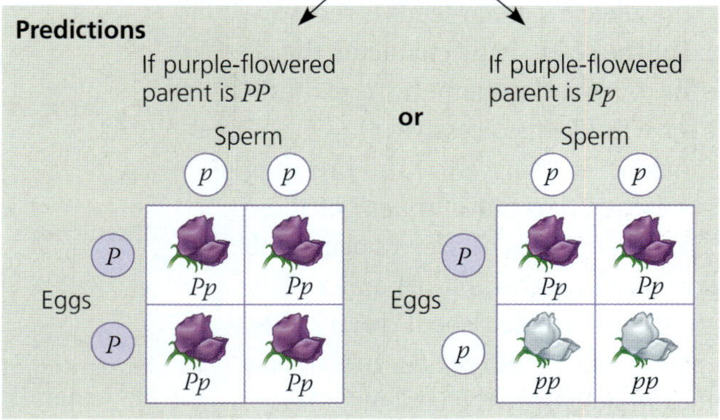

RESULTS Matching the results to either prediction identifies the unknown parental genotype (either *PP* or *Pp* in this example). In this testcross, we transferred pollen from a white-flowered plant to the carpels of a purple-flowered plant; the opposite (reciprocal) cross would have led to the same results.

The Law of Independent Assortment

Mendel derived the law of segregation from experiments in which he followed only a *single* character, such as flower color. All the F_1 progeny produced in his crosses of true-breeding parents were **monohybrids**, meaning that they were heterozygous for one character. We refer to a cross between such heterozygotes as a *monohybrid cross.*

Mendel identified his second law of inheritance by following *two* characters at the same time, such as seed color and seed shape. Seeds (peas) may be either yellow or green. They also may be either round (smooth) or wrinkled. From single-character crosses, Mendel knew that the allele for yellow seeds is dominant (*Y*) and the allele for green seeds is recessive (*y*). For the seed-shape character, the allele for round is dominant (*R*), and the allele for wrinkled is recessive (*r*).

Imagine crossing two true-breeding pea varieties that differ in *both* of these characters—a cross between a plant with yellow-round seeds (*YYRR*) and a plant with green-wrinkled seeds (*yyrr*). The F_1 plants will be **dihybrids**, individuals heterozygous for two characters (*YyRr*). But are these two characters transmitted from parents to offspring as a package? That is, will the *Y* and *R* alleles always stay together, generation after generation? Or are seed color and seed shape inherited independently? **Figure 14.8** illustrates how a *dihybrid cross,* a cross between F_1 dihybrids, can determine which of these two hypotheses is correct.

The F_1 plants, of genotype *YyRr,* exhibit both dominant phenotypes, yellow seeds with round shapes, no matter which hypothesis is correct. The key step in the experiment is to see what happens when F_1 plants self-pollinate and produce F_2 offspring. If the hybrids must transmit their alleles in the same combinations in which the alleles were inherited from the P generation, then the F_1 hybrids will produce only two classes of gametes: *YR* and *yr*. This "dependent assortment" hypothesis predicts that the phenotypic ratio of the F_2 generation will be 3:1, just as in a monohybrid cross (Figure 14.8, left side).

The alternative hypothesis is that the two pairs of alleles segregate independently of each other. In other words, genes are packaged into gametes in all possible allelic combinations, as long as each gamete has one allele for each gene. In our example, an F_1 plant will produce four classes of gametes in equal quantities: *YR,*

▼ **Figure 14.8 Inquiry**

Do the alleles for one character assort into gametes dependently or independently of the alleles for a different character?

EXPERIMENT Gregor Mendel followed the characters of seed color and seed shape through the F_2 generation. He crossed a true-breeding plant with yellow-round seeds with a true-breeding plant with green-wrinkled seeds, producing dihybrid F_1 plants. Self-pollination of the F_1 dihybrids produced the F_2 generation. The two hypotheses (dependent and independent assortment) predict different phenotypic ratios.

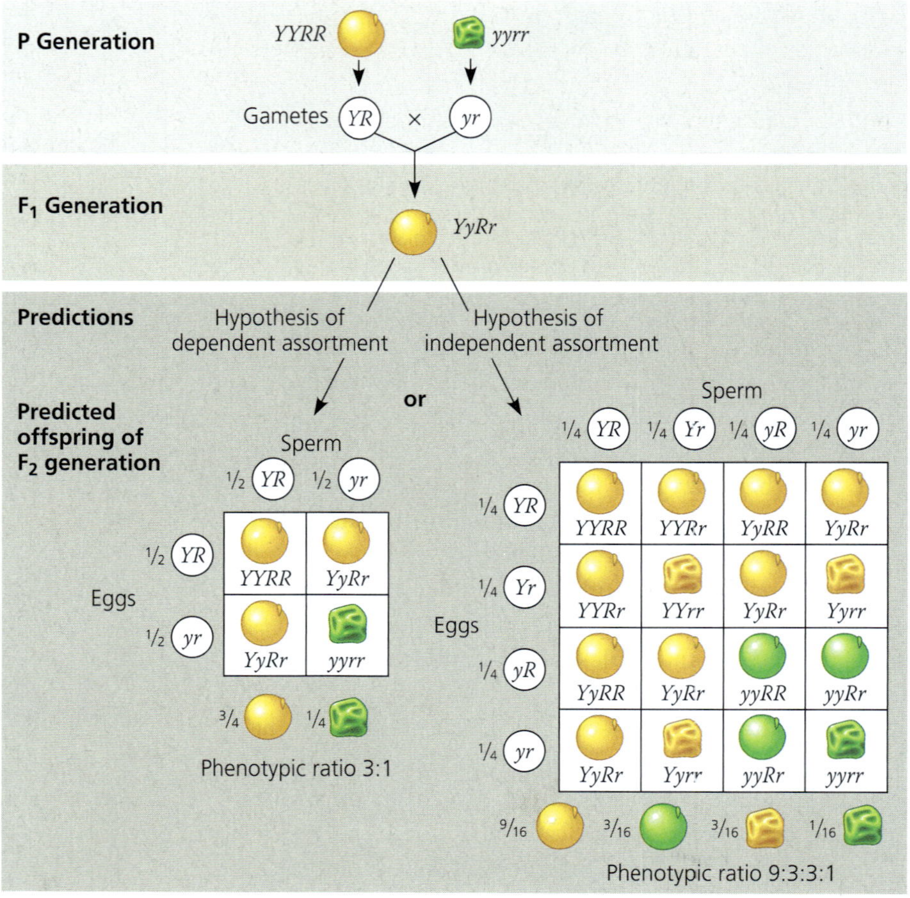

RESULTS

315 108 101 32 Phenotypic ratio approximately 9:3:3:1

CONCLUSION Only the hypothesis of independent assortment predicts the appearance of two of the observed phenotypes: green-round seeds and yellow-wrinkled seeds (see the right-hand Punnett square). The alleles for seed color and seed shape sort into gametes independently of each other.

SOURCE G. Mendel, Experiments in plant hybridization, *Proceedings of the Natural History Society of Brünn* 4:3–47 (1866).

WHAT IF? Suppose Mendel had transferred pollen from an F_1 plant to the carpel of a plant that was homozygous recessive for both genes. Set up the cross and draw Punnett squares that predict the offspring for both hypotheses. Would this cross have supported the hypothesis of independent assortment equally well?

Yr, yR, and *yr.* If sperm of the four classes fertilize eggs of the four classes, there will be 16 (4×4) equally probable ways in which the alleles can combine in the F_2 generation, as shown in Figure 14.8, right side. These combinations make up four phenotypic categories with a ratio of 9:3:3:1 (nine yellow-round to three green-round to three yellow-wrinkled to one green-wrinkled). When Mendel did the experiment and classified the F_2 offspring, his results were close to the predicted 9:3:3:1 phenotypic ratio, supporting the hypothesis that the alleles for one gene—controlling seed color or seed shape, in this example—are sorted into gametes independently of the alleles of other genes.

Mendel tested his seven pea characters in various dihybrid combinations and always observed a 9:3:3:1 phenotypic ratio in the F_2 generation. However, notice in Figure 14.8 that there is a 3:1 phenotypic ratio for each of the two characters if you consider them separately: three yellow to one green, and three round to one wrinkled. As far as a single character is concerned, the alleles segregate as if this were a monohybrid cross. The results of Mendel's dihybrid experiments are the basis for what we now call the **law of independent assortment**, which states that *each pair of alleles segregates independently of each other pair of alleles during gamete formation.*

Strictly speaking, this law applies only to genes (allele pairs) located on different chromosomes—that is, on chromosomes that are not homologous. Genes located near each other on the same chromosome tend to be inherited together and have more complex inheritance patterns than predicted by the law of independent assortment (see Chapter 15). All the pea characters Mendel chose for analysis were controlled by genes on different chromosomes (or behaved as though they were); this situation greatly simplified interpretation of his multicharacter pea crosses. All the examples we consider in the rest of this chapter involve genes located on different chromosomes.

CONCEPT CHECK 14.1

1. **DRAW IT** A pea plant heterozygous for inflated pods (*Ii*) is crossed with a plant homozygous for constricted pods (*ii*). Draw a Punnett square for this cross. Assume pollen comes from the *ii* plant.
2. **DRAW IT** Pea plants heterozygous for flower position and stem length (*AaTt*) are allowed to self-pollinate, and 400 of the resulting seeds are planted. Draw a Punnett square for this cross. How many offspring would be predicted to have terminal flowers and be dwarf? (See Table 14.1.)
3. **WHAT IF?** List the different gametes that could be made by a pea plant heterozygous for seed color, seed shape, and pod shape (*YyRrIi*; see Table 14.1). How large a Punnett square would you need to predict the offspring of a self-pollination of this "trihybrid"?

For suggested answers, see Appendix A.

CONCEPT 14.2

The laws of probability govern Mendelian inheritance

Mendel's laws of segregation and independent assortment reflect the same rules of probability that apply to tossing coins, rolling dice, and drawing cards from a deck. The probability scale ranges from 0 to 1. An event that is certain to occur has a probability of 1, while an event that is certain *not* to occur has a probability of 0. With a coin that has heads on both sides, the probability of tossing heads is 1, and the probability of tossing tails is 0. With a normal coin, the chance of tossing heads is $\frac{1}{2}$, and the chance of tossing tails is $\frac{1}{2}$. The probability of drawing the ace of spades from a 52-card deck is $\frac{1}{52}$. The probabilities of all possible outcomes for an event must add up to 1. With a deck of cards, the chance of picking a card other than the ace of spades is $\frac{51}{52}$.

Tossing a coin illustrates an important lesson about probability. For every toss, the probability of heads is $\frac{1}{2}$. The outcome of any particular toss is unaffected by what has happened on previous trials. We refer to phenomena such as coin tosses as independent events. Each toss of a coin, whether done sequentially with one coin or simultaneously with many, is independent of every other toss. And like two separate coin tosses, the alleles of one gene segregate into gametes independently of another gene's alleles (the law of independent assortment). Two basic rules of probability can help us predict the outcome of the fusion of such gametes in simple monohybrid crosses and more complicated crosses.

The Multiplication and Addition Rules Applied to Monohybrid Crosses

How do we determine the probability that two or more independent events will occur together in some specific combination? For example, what is the chance that two coins tossed simultaneously will both land heads up? The *multiplication rule* states that to determine this probability, we multiply the probability of one event (one coin coming up heads) by the probability of the other event (the other coin coming up heads). By the multiplication rule, then, the probability that both coins will land heads up is $\frac{1}{2} \times \frac{1}{2} = \frac{1}{4}$.

We can apply the same reasoning to an F_1 monohybrid cross. With seed shape in pea plants as the heritable character, the genotype of F_1 plants is *Rr.* Segregation in a heterozygous plant is like flipping a coin: Each egg produced has a $\frac{1}{2}$ chance of carrying the dominant allele (*R*) and a $\frac{1}{2}$ chance of carrying the recessive allele (*r*). The same odds apply to each sperm cell produced. For a particular F_2 plant to have wrinkled seeds, the recessive trait, both the egg and the sperm that come together must carry the *r* allele. The probability that two *r* alleles will be present in gametes at fertilization is found by multiplying $\frac{1}{2}$ (the

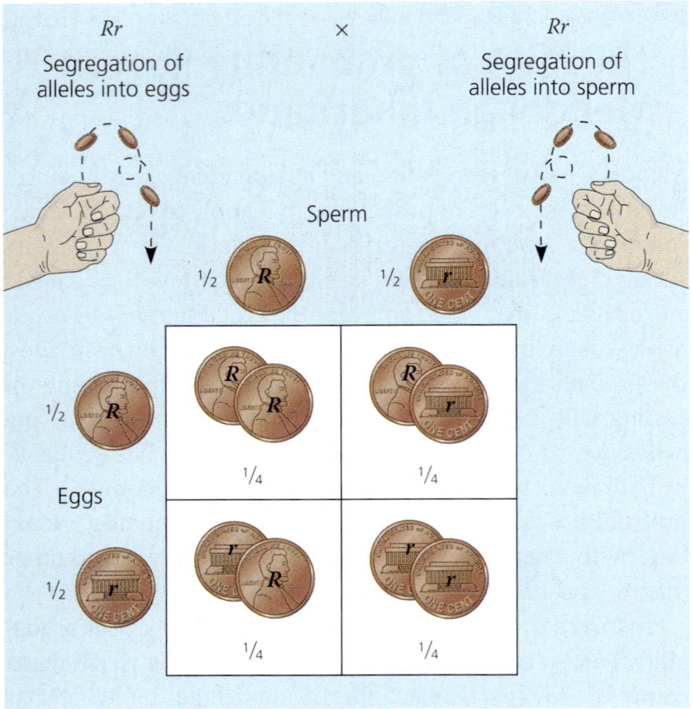

▲ **Figure 14.9 Segregation of alleles and fertilization as chance events.** When a heterozygote (*Rr*) forms gametes, whether a particular gamete ends up with an *R* or an *r* is like the toss of a coin. We can determine the probability for any genotype among the offspring of two heterozygotes by multiplying together the individual probabilities of an egg and sperm having a particular allele (*R* or *r* in this example).

probability that the egg will have an *r*) × ½ (the probability that the sperm will have an *r*). Thus, the multiplication rule tells us that the probability of an F_2 plant with wrinkled seeds (*rr*) is ¼ **(Figure 14.9)**. Likewise, the probability of an F_2 plant carrying both dominant alleles for seed shape (*RR*) is ¼.

To figure out the probability that an F_2 plant from a monohybrid cross will be heterozygous rather than homozygous, we need to invoke a second rule. Notice in Figure 14.9 that the dominant allele can come from the egg and the recessive allele from the sperm, or vice versa. That is, F_1 gametes can combine to produce *Rr* offspring in two independent and mutually exclusive ways: For any particular heterozygous F_2 plant, the dominant allele can come from the egg *or* the sperm, but not from both. According to the *addition rule*, the probability that any one of two or more mutually exclusive events will occur is calculated by adding their individual probabilities. As we have just seen, the multiplication rule gives us the individual probabilities that we will now add together. The probability for one possible way of obtaining an F_2 heterozygote—the dominant allele from the egg and the recessive allele from the sperm—is ¼. The probability for the other possible way—the recessive allele from the egg and the dominant allele from the sperm—is also ¼ (see Figure 14.9). Using the rule of addition, then, we can calculate the probability of an F_2 heterozygote as ¼ + ¼ = ½.

Solving Complex Genetics Problems with the Rules of Probability

We can also apply the rules of probability to predict the outcome of crosses involving multiple characters. Recall that each allelic pair segregates independently during gamete formation (the law of independent assortment). Thus, a dihybrid or other multicharacter cross is equivalent to two or more independent monohybrid crosses occurring simultaneously. By applying what we have learned about monohybrid crosses, we can determine the probability of specific genotypes occurring in the F_2 generation without having to construct unwieldy Punnett squares.

Consider the dihybrid cross between *YyRr* heterozygotes shown in Figure 14.8. We will focus first on the seed-color character. For a monohybrid cross of *Yy* plants, the probabilities of the offspring genotypes are ¼ for *YY*, ½ for *Yy*, and ¼ for *yy*. The same probabilities apply to the offspring genotypes for seed shape: ¼ *RR*, ½ *Rr*, and ¼ *rr*. Knowing these probabilities, we can simply use the multiplication rule to determine the probability of each of the genotypes in the F_2 generation. For example, the probability of an F_2 plant having the *YYRR* genotype is ¼ (*YY*) × ¼ (*RR*) = 1/16. This corresponds to the upper left box in the larger Punnett square in Figure 14.8. To give another example, the probability of an F_2 plant with the *YyRR* genotype is ½ (*Yy*) × ¼ (*RR*) = ⅛. If you look closely at the larger Punnett square in Figure 14.8, you will see that 2 of the 16 boxes (⅛) correspond to the *YyRR* genotype.

Now let's see how we can combine the multiplication and addition rules to solve even more complex problems in Mendelian genetics. For instance, imagine a cross of two pea varieties in which we track the inheritance of three characters. Suppose we cross a trihybrid with purple flowers and yellow, round seeds (heterozygous for all three genes) with a plant with purple flowers and green, wrinkled seeds (heterozygous for flower color but homozygous recessive for the other two characters). Using Mendelian symbols, our cross is *PpYyRr* × *Ppyyrr*. What fraction of offspring from this cross would be predicted to exhibit the recessive phenotypes for *at least two* of the three characters?

To answer this question, we can start by listing all genotypes that fulfill this condition: *ppyyRr*, *ppYyrr*, *Ppyyrr*, *PPyyrr*, and *ppyyrr*. (Because the condition is *at least two* recessive traits, it includes the last genotype, which produces all three recessive traits.) Next, we calculate the probability for each of these genotypes resulting from our *PpYyRr* × *Ppyyrr* cross by multiplying together the individual probabilities for the allele pairs, just as we did in our dihybrid example. Note that in a cross involving heterozygous and homozygous allele pairs (for example, *Yy* × *yy*), the probability of heterozygous offspring is ½ and the probability of homozygous offspring is ½. Finally, we use the addition rule to add the probabilities for all the different genotypes that fulfill the condition of at least two recessive traits, as shown below.

ppyyRr	$\frac{1}{4}$ (probability of *pp*) $\times$ $\frac{1}{2}$ (*yy*) $\times$ $\frac{1}{2}$ (*Rr*)	$= \frac{1}{16}$
ppYyrr	$\frac{1}{4} \times \frac{1}{2} \times \frac{1}{2}$	$= \frac{1}{16}$
Ppyyrr	$\frac{1}{2} \times \frac{1}{2} \times \frac{1}{2}$	$= \frac{2}{16}$
PPyyrr	$\frac{1}{4} \times \frac{1}{2} \times \frac{1}{2}$	$= \frac{1}{16}$
ppyyrr	$\frac{1}{4} \times \frac{1}{2} \times \frac{1}{2}$	$= \frac{1}{16}$
Chance of *at least two* recessive traits		$= \frac{6}{16}$ or $\frac{3}{8}$

With practice, you'll be able to solve genetics problems faster by using the rules of probability than by filling in Punnett squares.

We cannot predict with certainty the exact numbers of progeny of different genotypes resulting from a genetic cross. But the rules of probability give us the *chance* of various outcomes. Usually, the larger the sample size, the closer the results will conform to our predictions. The reason Mendel counted so many offspring from his crosses is that he understood this statistical feature of inheritance and had a keen sense of the rules of chance.

CONCEPT CHECK 14.2

1. For any gene with a dominant allele *C* and recessive allele *c*, what proportions of the offspring from a $CC \times Cc$ cross are expected to be homozygous dominant, homozygous recessive, and heterozygous?
2. An organism with the genotype *BbDD* is mated to one with the genotype *BBDd.* Assuming independent assortment of these two genes, write the genotypes of all possible offspring from this cross and use the rules of probability to calculate the chance of each genotype occurring.
3. **WHAT IF?** Three characters (flower color, seed color, and pod shape) are considered in a cross between two pea plants (*PpYyIi* $\times$ *ppYyii*). What fraction of offspring would be predicted to be homozygous recessive for at least two of the three characters?

For suggested answers, see Appendix A.

CONCEPT **14.3**

Inheritance patterns are often more complex than predicted by simple Mendelian genetics

In the 20th century, geneticists extended Mendelian principles not only to diverse organisms, but also to patterns of inheritance more complex than those described by Mendel. For the work that led to his two laws of inheritance, Mendel chose pea plant characters that turn out to have a relatively simple genetic basis: Each character is determined by one gene, for which there are only two alleles, one completely dominant and the other completely recessive.* But these conditions are not met by all heritable characters, and the relationship between genotype and phenotype is rarely so simple. Mendel himself realized that he could not explain the more complicated patterns he observed in crosses involving other pea characters or other plant species. This does not diminish the utility of Mendelian genetics (also called Mendelism), however, because the basic principles of segregation and independent assortment apply even to more complex patterns of inheritance. In this section, we will extend Mendelian genetics to hereditary patterns that were not reported by Mendel.

Extending Mendelian Genetics for a Single Gene

The inheritance of characters determined by a single gene deviates from simple Mendelian patterns when alleles are not completely dominant or recessive, when a particular gene has more than two alleles, or when a single gene produces multiple phenotypes. We will describe examples of each of these situations in this section.

Degrees of Dominance

Alleles can show different degrees of dominance and recessiveness in relation to each other. In Mendel's classic pea crosses, the F_1 offspring always looked like one of the two parental varieties because one allele in a pair showed **complete dominance** over the other. In such situations, the phenotypes of the heterozygote and the dominant homozygote are indistinguishable.

For some genes, however, neither allele is completely dominant, and the F_1 hybrids have a phenotype somewhere between those of the two parental varieties. This phenomenon, called **incomplete dominance**, is seen when red snapdragons are crossed with white snapdragons: All the F_1 hybrids have pink flowers, as shown in **Figure 14.10** on the next page. This third phenotype results from flowers of the heterozygotes having less red pigment than the red homozygotes (unlike the situation in Mendel's pea plants, where the *Pp* heterozygotes make enough pigment for the flowers to be a purple color indistinguishable from that of *PP* plants).

At first glance, incomplete dominance of either allele seems to provide evidence for the blending hypothesis of inheritance, which would predict that the red or white trait could never be retrieved from the pink hybrids. In fact, interbreeding F_1 hybrids produces F_2 offspring with a phenotypic ratio of one red to two pink to one white. (Because heterozygotes have a separate phenotype, the genotypic and phenotypic ratios for the F_2 generation are the same, 1:2:1.) The segregation of the red-flower and white-flower alleles in the

* There is one exception: Geneticists have found that Mendel's pod-shape character is actually determined by two genes.

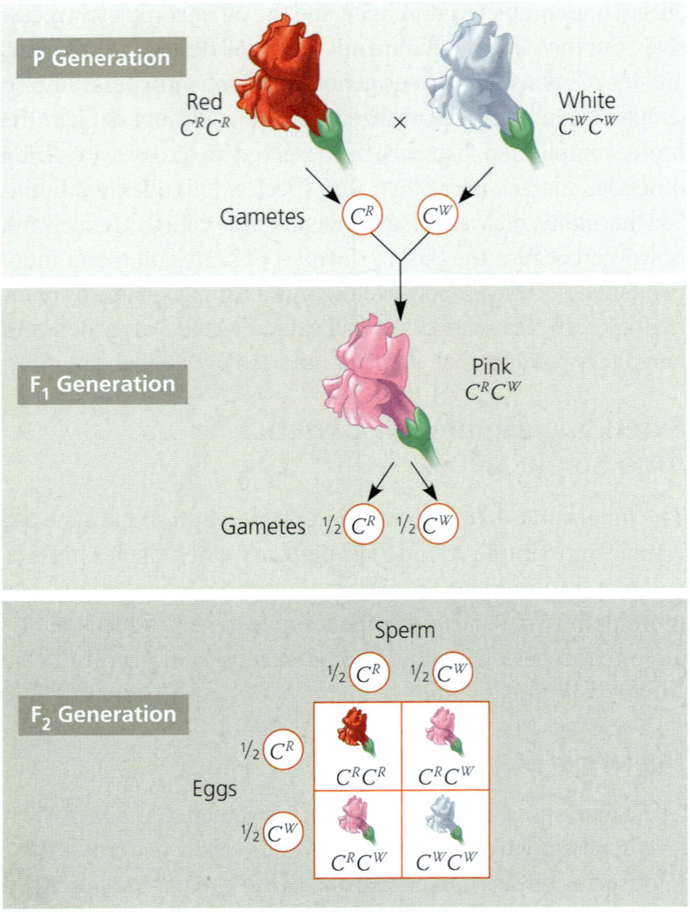

▲ **Figure 14.10 Incomplete dominance in snapdragon color.** When red snapdragons are crossed with white ones, the F_1 hybrids have pink flowers. Segregation of alleles into gametes of the F_1 plants results in an F_2 generation with a 1:2:1 ratio for both genotype and phenotype. The letter *C* with a superscript indicates an allele for flower color: C^R for red and C^W for white.

? *Suppose a classmate argues that this figure supports the blending hypothesis for inheritance. What might your classmate say, and how would you respond?*

gametes produced by the pink-flowered plants confirms that the alleles for flower color are heritable factors that maintain their identity in the hybrids; that is, inheritance is particulate.

Another variation on dominance relationships between alleles is called **codominance**; in this variation, the two alleles both affect the phenotype in separate, distinguishable ways. For example, the human MN blood group is determined by codominant alleles for two specific molecules located on the surface of red blood cells, the M and N molecules. A single gene locus, at which two allelic variations are possible, determines the phenotype of this blood group. Individuals homozygous for the *M* allele (*MM*) have red blood cells with only M molecules; individuals homozygous for the *N* allele (*NN*) have red blood cells with only N molecules. But *both* M and N molecules are present on the red blood cells of individuals heterozygous for the *M* and *N* alleles (*MN*). Note that the MN phenotype is *not* intermediate between the M and N phenotypes, which distinguishes codominance from incomplete dominance. Rather, *both* M and N phenotypes are exhibited by heterozygotes, since both molecules are present.

The Relationship Between Dominance and Phenotype

We've now seen that the relative effects of two alleles range from complete dominance of one allele, through incomplete dominance of either allele, to codominance of both alleles. It is important to understand that an allele is not termed *dominant* because it somehow subdues a recessive allele. Recall that alleles are simply variations in a gene's nucleotide sequence. When a dominant allele coexists with a recessive allele in a heterozygote, they do not actually interact at all. It is in the pathway from genotype to phenotype that dominance and recessiveness come into play.

To illustrate the relationship between dominance and phenotype, we can use one of the characters Mendel studied—round versus wrinkled pea seed shape. The dominant allele (round) codes for an enzyme that helps convert an unbranched form of starch to a branched form in the seed. The recessive allele (wrinkled) codes for a defective form of this enzyme, leading to an accumulation of unbranched starch, which causes excess water to enter the seed by osmosis. Later, when the seed dries, it wrinkles. If a dominant allele is present, no excess water enters the seed and it does not wrinkle when it dries. One dominant allele results in enough of the enzyme to synthesize adequate amounts of branched starch, which means dominant homozygotes and heterozygotes have the same phenotype: round seeds.

A closer look at the relationship between dominance and phenotype reveals an intriguing fact: For any character, the observed dominant/recessive relationship of alleles depends on the level at which we examine phenotype. **Tay-Sachs disease**, an inherited disorder in humans, provides an example. The brain cells of a child with Tay-Sachs disease cannot metabolize certain lipids because a crucial enzyme does not work properly. As these lipids accumulate in brain cells, the child begins to suffer seizures, blindness, and degeneration of motor and mental performance and dies within a few years.

Only children who inherit two copies of the Tay-Sachs allele (homozygotes) have the disease. Thus, at the *organismal* level, the Tay-Sachs allele qualifies as recessive. However, the activity level of the lipid-metabolizing enzyme in heterozygotes is intermediate between that in individuals homozygous for the normal allele and that in individuals with Tay-Sachs disease. The intermediate phenotype observed at the *biochemical* level is characteristic of incomplete dominance of either allele. Fortunately, the heterozygote condition does not lead to disease symptoms, apparently because half the normal enzyme activity is sufficient to prevent lipid accumulation in the brain. Extending our analysis to yet another level, we find that heterozygous individuals produce equal numbers of normal and dysfunctional enzyme molecules. Thus, at the *molecular* level, the normal allele and the Tay-Sachs allele are codominant. As you can see, whether alleles appear to be completely dominant, incompletely dominant, or codominant depends on the level at which the phenotype is analyzed.

Frequency of Dominant Alleles Although you might assume that the dominant allele for a particular character would be more common in a population than the recessive allele for that character, this is not necessarily the case. For example, about one baby out of 400 in the United States is born with extra fingers or toes, a condition known as polydactyly. Some cases of polydactyly are caused by the presence of a dominant allele. The low frequency of polydactyly indicates that the recessive allele, which results in five digits per appendage, is far more prevalent than the dominant allele in the population. In Chapter 23, you will learn how the relative frequencies of alleles in a population are affected by natural selection.

Multiple Alleles

Only two alleles exist for the pea characters that Mendel studied, but most genes exist in more than two allelic forms. The ABO blood groups in humans, for instance, are determined by three alleles of a single gene: I^A, I^B, and *i*. A person's blood group (phenotype) may be one of four types: A, B, AB, or O. These letters refer to two carbohydrates—A and B—that may be found on the surface of red blood cells. A person's blood cells may have carbohydrate A (type A blood), carbohydrate B (type B), both (type AB), or neither (type O), as shown schematically in **Figure 14.11**. Matching compatible blood groups is critical for safe blood transfusions (see Chapter 43).

Allele	Carbohydrate
I^A	A △
I^B	B ○
i	none

(a) The three alleles for the ABO blood groups and their associated carbohydrates. Each allele codes for an enzyme that may add a specific carbohydrate (designated by the superscript on the allele and shown as a triangle or circle) to the red blood cell.

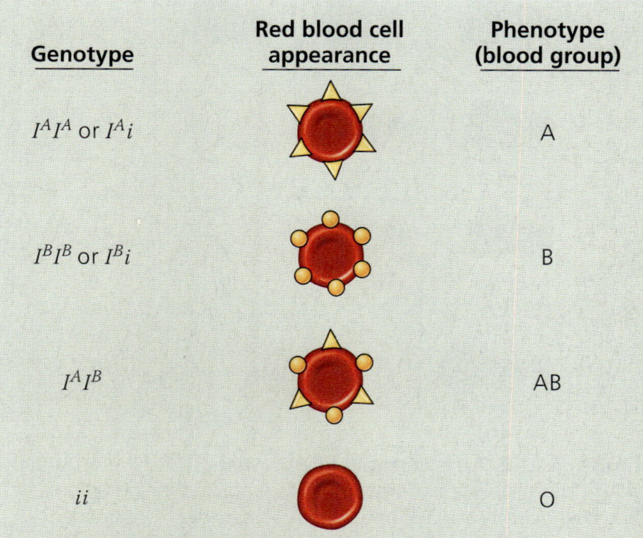

(b) Blood group genotypes and phenotypes. There are six possible genotypes, resulting in four different phenotypes.

▲ **Figure 14.11 Multiple alleles for the ABO blood groups.** The four blood groups result from different combinations of three alleles.

? *Based on the surface carbohydrate phenotype in (b), what are the dominance relationships among the alleles?*

Pleiotropy

So far, we have treated Mendelian inheritance as though each gene affects only one phenotypic character. Most genes, however, have multiple phenotypic effects, a property called **pleiotropy** (from the Greek *pleion*, more). In humans, for example, pleiotropic alleles are responsible for the multiple symptoms associated with certain hereditary diseases, such as cystic fibrosis and sickle-cell disease, discussed later in this chapter. In the garden pea, the gene that determines flower color also affects the color of the coating on the outer surface of the seed, which can be gray or white. Given the intricate molecular and cellular interactions responsible for an organism's development and physiology, it isn't surprising that a single gene can affect a number of characteristics in an organism.

Extending Mendelian Genetics for Two or More Genes

Dominance relationships, multiple alleles, and pleiotropy all have to do with the effects of the alleles of a single gene. We now consider two situations in which two or more genes are involved in determining a particular phenotype.

Epistasis

In **epistasis** (from the Greek for "standing upon"), a gene at one locus alters the phenotypic expression of a gene at a second locus. An example will help clarify this concept. In mice and many other mammals, black coat color is dominant to brown. Let's designate *B* and *b* as the two alleles for this character. For a mouse to have brown fur, its genotype must be *bb*. But there is more to the story. A second gene determines whether or not pigment will be deposited in the hair. The dominant allele, symbolized by *C* (for color), results in the deposition of either black or brown pigment, depending on the genotype at the first locus. But if the mouse is homozygous recessive for the second locus (*cc*), then the coat is white (albino), regardless of the genotype at the black/brown locus. In this case, the gene for pigment deposition (*C/c*) is said to be epistatic to the gene that codes for black or brown pigment (*B/b*).

What happens if we mate black mice that are heterozygous for both genes (*BbCc*)? Although the two genes affect the same phenotypic character (coat color), they follow the law of independent assortment. Thus, our breeding experiment represents an F_1 dihybrid cross, like those that produced a 9:3:3:1

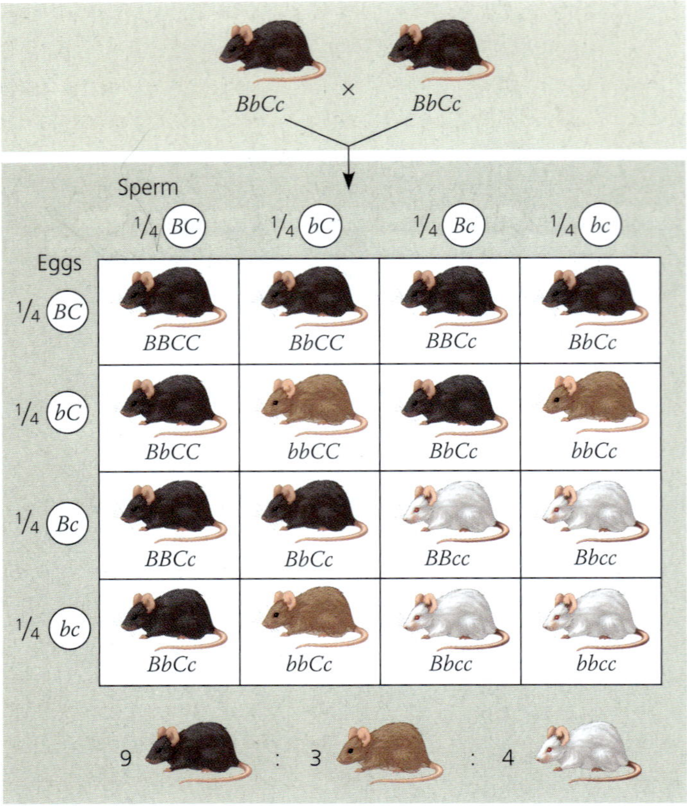

▲ **Figure 14.12 An example of epistasis.** This Punnett square illustrates the genotypes and phenotypes predicted for offspring of matings between two black mice of genotype *BbCc*. The *C*/*c* gene, which is epistatic to the *B*/*b* gene coding for hair pigment, controls whether or not pigment of any color will be deposited in the hair.

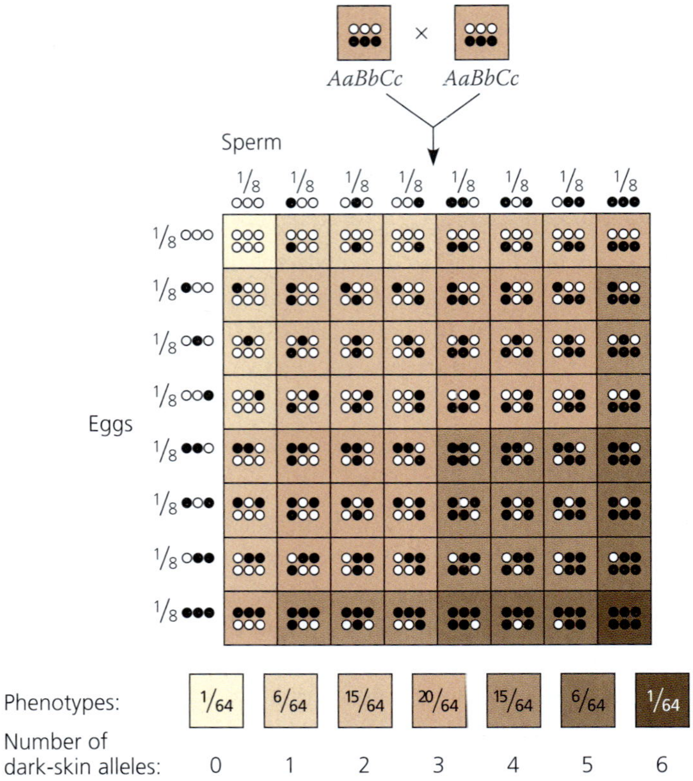

▲ **Figure 14.13 A simplified model for polygenic inheritance of skin color.** According to this model, three separately inherited genes affect the darkness of skin. The heterozygous individuals (*AaBbCc*) represented by the two rectangles at the top of this figure each carry three dark-skin alleles (black circles, which represent *A*, *B*, or *C*) and three light-skin alleles (white circles, which represent *a*, *b*, or *c*). The Punnett square shows all the possible genetic combinations in gametes and in offspring of a large number of hypothetical matings between these heterozygotes. The results are summarized by the phenotypic ratios under the Punnett square.

DRAW IT *Make a bar graph of the results, with skin color (number of dark-skin alleles) along the x-axis and fraction of offspring along the y-axis. Draw a rough curve corresponding to the results and discuss what it shows about the relative proportions of different phenotypes among the offspring.*

ratio in Mendel's experiments. We can use a Punnett square to represent the genotypes of the F_2 offspring **(Figure 14.12)**. As a result of epistasis, the phenotypic ratio among the F_2 offspring is 9 black to 3 brown to 4 white. Other types of epistatic interactions produce different ratios, but all are modified versions of 9:3:3:1.

Polygenic Inheritance

Mendel studied characters that could be classified on an either-or basis, such as purple versus white flower color. But for many characters, such as human skin color and height, an either-or classification is impossible because the characters vary in the population along a continuum (in gradations). These are called **quantitative characters**. Quantitative variation usually indicates **polygenic inheritance**, an additive effect of two or more genes on a single phenotypic character (the converse of pleiotropy, where a single gene affects several phenotypic characters).

There is evidence, for instance, that skin pigmentation in humans is controlled by at least three separately inherited genes (probably more, but we will simplify). Let's consider three genes, with a dark-skin allele for each gene (*A*, *B*, or *C*) contributing one "unit" of darkness to the phenotype and being incompletely dominant to the other allele (*a*, *b*, or *c*). An *AABBCC* person would be very dark, while an *aabbcc* individual would be very light. An *AaBbCc* person would have skin of an intermediate shade. Because the alleles have a cumulative effect, the genotypes *AaBbCc* and *AABbcc* would make the same genetic contribution (three units) to skin darkness. The Punnett square in **Figure 14.13** shows all possible genotypes of offspring from a mating between individuals heterozygous for all three genes. As indicated by the row of squares below the Punnett square, there are seven skin-color phenotypes that could result from this mating. Environmental factors, such as exposure to the sun, also affect the skin-color phenotype.

Nature and Nurture: The Environmental Impact on Phenotype

Another departure from simple Mendelian genetics arises when the phenotype for a character depends on environment

▲ **Figure 14.14 The effect of environment on phenotype.** The outcome of a genotype lies within its norm of reaction, a phenotypic range that depends on the environment in which the genotype is expressed. For example, hydrangea flowers of the same genetic variety range in color from blue-violet to pink, with the shade and intensity of color depending on the acidity and aluminum content of the soil.

as well as genotype. A single tree, locked into its inherited genotype, has leaves that vary in size, shape, and greenness, depending on exposure to wind and sun. For humans, nutrition influences height, exercise alters build, sun-tanning darkens the skin, and experience improves performance on intelligence tests. Even identical twins, who are genetic equals, accumulate phenotypic differences as a result of their unique experiences.

Whether human characteristics are more influenced by genes or the environment—nature or nurture—is a very old and hotly contested debate that we will not attempt to settle here. We can say, however, that a genotype generally is not associated with a rigidly defined phenotype, but rather with a range of phenotypic possibilities due to environmental influences. This phenotypic range is called the **norm of reaction** for a genotype **(Figure 14.14)**. For some characters, such as the ABO blood group system, the norm of reaction has no breadth whatsoever; that is, a given genotype mandates a very specific phenotype. Other characteristics, such as a person's blood count of red and white cells, vary quite a bit, depending on such factors as the altitude, the customary level of physical activity, and the presence of infectious agents.

Generally, norms of reaction are broadest for polygenic characters. Environment contributes to the quantitative nature of these characters, as we have seen in the continuous variation of skin color. Geneticists refer to such characters as **multifactorial**, meaning that many factors, both genetic and environmental, collectively influence phenotype.

Integrating a Mendelian View of Heredity and Variation

Over the past several pages, we have broadened our view of Mendelian inheritance by exploring degrees of dominance as well as multiple alleles, pleiotropy, epistasis, polygenic inheritance, and the phenotypic impact of the environment. How can we integrate these refinements into a comprehensive theory of Mendelian genetics? The key is to make the transition from the reductionist emphasis on single genes and phenotypic characters to the emergent properties of the organism as a whole, one of the themes of this book.

The term *phenotype* can refer not only to specific characters, such as flower color and blood group, but also to an organism in its entirety—*all* aspects of its physical appearance, internal anatomy, physiology, and behavior. Similarly, the term *genotype* can refer to an organism's entire genetic makeup, not just its alleles for a single genetic locus. In most cases, a gene's impact on phenotype is affected by other genes and by the environment. In this integrated view of heredity and variation, an organism's phenotype reflects its overall genotype and unique environmental history.

Considering all that can occur in the pathway from genotype to phenotype, it is indeed impressive that Mendel could uncover the fundamental principles governing the transmission of individual genes from parents to offspring. Mendel's two laws, segregation and independent assortment, explain heritable variations in terms of alternative forms of genes (hereditary "particles," now known as the alleles of genes) that are passed along, generation after generation, according to simple rules of probability. This theory of inheritance is equally valid for peas, flies, fishes, birds, and human beings—indeed for any organism with a sexual life cycle. Furthermore, by extending the principles of segregation and independent assortment to help explain such hereditary patterns as epistasis and quantitative characters, we begin to see how broadly Mendelism applies. From Mendel's abbey garden came a theory of particulate inheritance that anchors modern genetics. In the last section of this chapter, we will apply Mendelian genetics to human inheritance, with emphasis on the transmission of hereditary diseases.

CONCEPT CHECK 14.3

1. *Incomplete dominance* and *epistasis* are both terms that define genetic relationships. What is the most basic distinction between these terms?
2. If a man with type AB blood marries a woman with type O blood, what blood types would you expect in their children?
3. **WHAT IF?** A rooster with gray feathers is mated with a hen of the same phenotype. Among their offspring, 15 chicks are gray, 6 are black, and 8 are white. What is the simplest explanation for the inheritance of these colors in chickens? What phenotypes would you expect in the offspring of a cross between a gray rooster and a black hen?

For suggested answers, see Appendix A.

CONCEPT **14.4**

Many human traits follow Mendelian patterns of inheritance

Whereas peas are convenient subjects for genetic research, humans are not. The human generation span is about 20 years, and human parents produce relatively few offspring compared to peas and most other species. Even more important, no one would consider it ethical to ask pairs of humans to breed so that the phenotypes of their offspring could be analyzed! In spite of these constraints, the study of human genetics continues to advance, spurred on by the desire to understand our own inheritance. New techniques in molecular biology have led to many breakthrough discoveries, as we will see in Chapter 20, but basic Mendelism endures as the foundation of human genetics.

Pedigree Analysis

Unable to manipulate the mating patterns of people, geneticists must analyze the results of matings that have already occurred. They do so by collecting information about a family's history for a particular trait and assembling this information into a family tree describing the traits of parents and children across the generations—the family **pedigree**.

Figure 14.15a shows a three-generation pedigree that traces the occurrence of a pointed contour of the hairline on the forehead. This trait, called a widow's peak, is due to a dominant allele, *W*. Because the widow's-peak allele is dominant, all individuals who lack a widow's peak must be homozygous recessive (*ww*). The two grandparents with widow's peaks must have the *Ww* genotype, since some of their offspring are homozygous recessive. The offspring in the second generation who *do* have widow's peaks must also be heterozygous, because they are the products of *Ww* × *ww* matings. The third generation in this pedigree consists of two sisters. The one who has a widow's peak could be either homozygous (*WW*) or heterozygous (*Ww*), given what we know about the genotypes of her parents (both *Ww*).

Figure 14.15b is a pedigree of the same family, but this time we focus on a recessive trait, attached earlobes. We'll use *f* for the recessive allele and *F* for the dominant allele, which results in free earlobes. As you work your way through the pedigree, notice once again that you can apply what you have learned about Mendelian inheritance to understand the genotypes shown for the family members.

An important application of a pedigree is to help us calculate the probability that a child will have a particular genotype and phenotype. Suppose that the couple represented in the second generation of Figure 14.15 decides to have one more child. What is the probability that the child will have a widow's

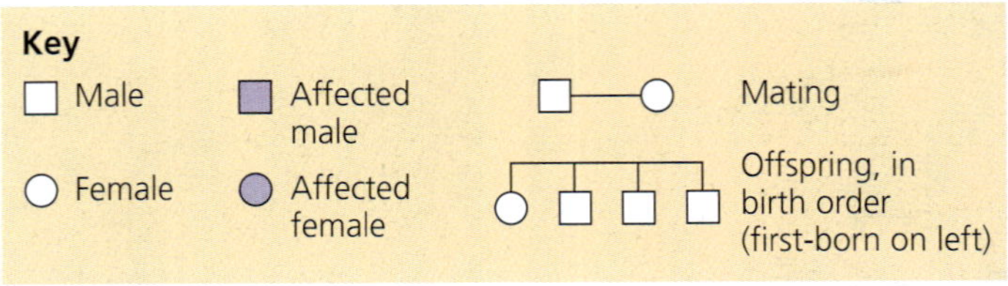

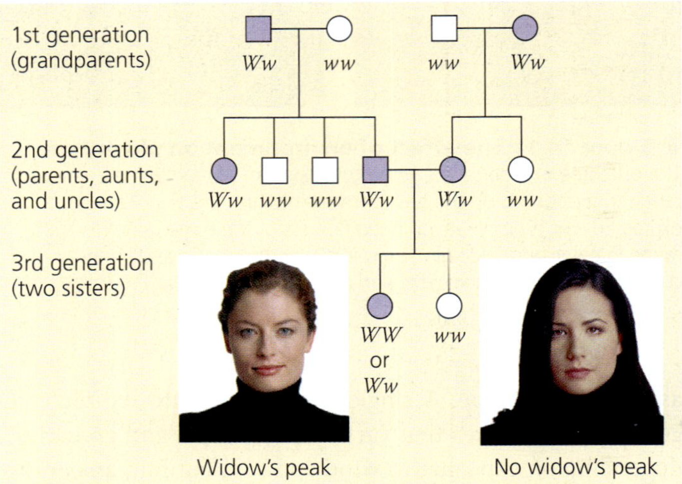

(a) Is a widow's peak a dominant or recessive trait?
Tips for pedigree analysis: Notice in the third generation that the second-born daughter lacks a widow's peak, although both of her parents had the trait. Such a pattern of inheritance supports the hypothesis that the trait is due to a dominant allele. If the trait were due to a *recessive* allele, and both parents had the recessive phenotype, then *all* of their offspring would also have the recessive phenotype.

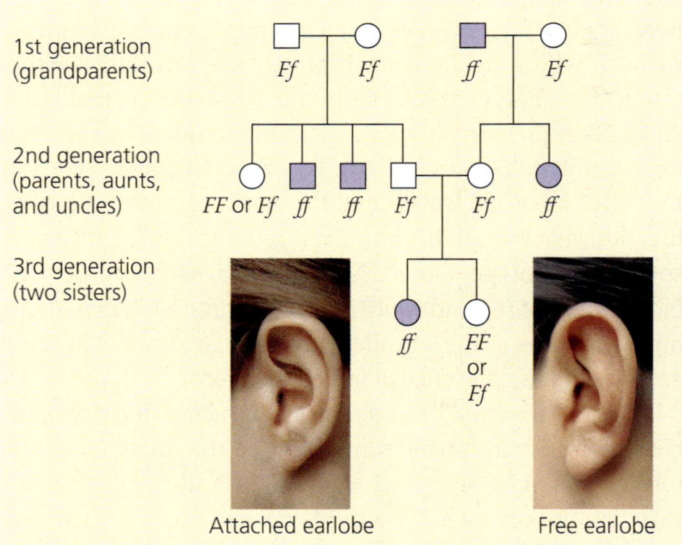

(b) Is an attached earlobe a dominant or recessive trait?
Tips for pedigree analysis: Notice that the first-born daughter in the third generation has attached earlobes, although both of her parents lack that trait (they have free earlobes). Such a pattern is easily explained if the attached-lobe phenotype is due to a recessive allele. If it were due to a *dominant* allele, then at least one parent would also have had the trait.

▲ **Figure 14.15 Pedigree analysis.** Each of these pedigrees traces a trait through three generations of the same family. The two traits have different inheritance patterns, as seen by analysis of the pedigrees.

peak? This is equivalent to a Mendelian F_1 monohybrid cross ($Ww \times Ww$), and thus the probability that a child will inherit a dominant allele and have a widow's peak is $\frac{3}{4}$ ($\frac{1}{4}$ WW + $\frac{1}{2}$ Ww). What is the probability that the child will have attached earlobes? Again, we can treat this as a monohybrid cross ($Ff \times Ff$), but this time we want to know the chance that the offspring will be homozygous recessive (ff). That probability is $\frac{1}{4}$. Finally, what is the chance that the child will have a widow's peak *and* attached earlobes? Assuming that the genes for these two characters are on different chromosomes, the two pairs of alleles will assort independently in this dihybrid cross ($WwFf \times WwFf$). Thus, we can use the multiplication rule: $\frac{3}{4}$ (chance of widow's peak) $\times$ $\frac{1}{4}$ (chance of attached earlobes) = $\frac{3}{16}$ (chance of widow's peak and attached earlobes).

Pedigrees are a more serious matter when the alleles in question cause disabling or deadly diseases instead of innocuous human variations such as hairline or earlobe configuration. However, for disorders inherited as simple Mendelian traits, the same techniques of pedigree analysis apply.

Recessively Inherited Disorders

Thousands of genetic disorders are known to be inherited as simple recessive traits. These disorders range in severity from relatively mild, such as albinism (lack of pigmentation, which results in susceptibility to skin cancers and vision problems), to life-threatening, such as cystic fibrosis.

The Behavior of Recessive Alleles

How can we account for the behavior of alleles that cause recessively inherited disorders? Recall that genes code for proteins of specific function. An allele that causes a genetic disorder (let's call it allele *a*) codes either for a malfunctioning protein or for no protein at all. In the case of disorders classified as recessive, heterozygotes (*Aa*) are normal in phenotype because one copy of the normal allele (*A*) produces a sufficient amount of the specific protein. Thus, a recessively inherited disorder shows up only in the homozygous individuals (*aa*) who inherit one recessive allele from each parent. Although phenotypically normal with regard to the disorder, heterozygotes may transmit the recessive allele to their offspring and thus are called **carriers**. **Figure 14.16** illustrates these ideas using albinism as an example.

Most people who have recessive disorders are born to parents who are carriers of the disorder but themselves have a normal phenotype, as is the case shown in the Punnett square in Figure 14.16. A mating between two carriers corresponds to a Mendelian F_1 monohybrid cross, so the predicted genotypic ratio for the offspring is 1 *AA* : 2 *Aa* : 1 *aa*. Thus, each child has a $\frac{1}{4}$ chance of inheriting a double dose of the recessive allele; in the case of albinism, such a child will be albino. From the genotypic ratio, we also can see that out of three offspring with the *normal* phenotype (one *AA* plus two *Aa*), two are predicted to be heterozygous carriers, a $\frac{2}{3}$ chance. Recessive homozygotes could also result from *Aa* $\times$ *aa* and *aa* $\times$ *aa* matings, but if the disorder is lethal before reproductive age or results in sterility (neither of which is true for albinism), no *aa* individuals will reproduce. Even if recessive homozygotes are able to reproduce, such individuals will still account for a much smaller percentage of the population than heterozygous carriers (for reasons we will examine in Chapter 23).

In general, genetic disorders are not evenly distributed among all groups of people. For example, the incidence of Tay-Sachs disease, which we described earlier in this chapter, is disproportionately high among Ashkenazic Jews, Jewish people whose ancestors lived in central Europe. In that population, Tay-Sachs disease occurs in one out of 3,600 births, an incidence about 100 times greater than that among non-Jews or Mediterranean (Sephardic) Jews. This uneven distribution results from the different genetic histories of the world's peoples during less technological times, when populations were more geographically (and hence genetically) isolated.

When a disease-causing recessive allele is rare, it is relatively unlikely that two carriers of the same harmful allele will meet and mate. However, if the man and woman are close relatives (for example, siblings or first cousins), the probability of passing on recessive traits increases greatly. These are called consanguineous ("same blood") matings, and they are indicated in pedigrees by double lines. Because people with recent common ancestors are more likely to carry the same recessive alleles than are unrelated people, it is more likely that a mating of close relatives will produce offspring homozygous for recessive traits—including harmful ones. Such effects can be observed in many types of domesticated and zoo animals that have become inbred.

There is debate among geneticists about the extent to which human consanguinity increases the risk of inherited

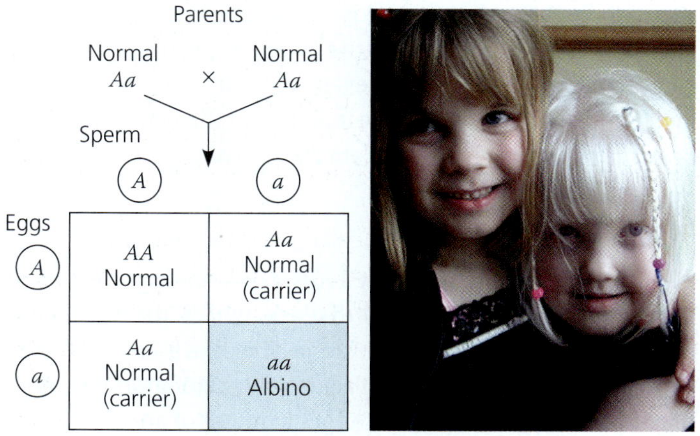

▲ **Figure 14.16 Albinism: a recessive trait.** One of the two sisters shown here has normal coloration; the other is albino. Most recessive homozygotes are born to parents who are carriers of the disorder but themselves have a normal phenotype, the case shown in the Punnett square.

? *What is the probability that the sister with normal coloration is a carrier of the albinism allele?*

diseases. Many deleterious alleles have such severe effects that a homozygous embryo spontaneously aborts long before birth. Still, most societies and cultures have laws or taboos forbidding marriages between close relatives. These rules may have evolved out of empirical observation that in most populations, stillbirths and birth defects are more common when parents are closely related. Social and economic factors have also influenced the development of customs and laws against consanguineous marriages.

Cystic Fibrosis

The most common lethal genetic disease in the United States is **cystic fibrosis**, which strikes one out of every 2,500 people of European descent, though it is much rarer in other groups. Among people of European descent, one out of 25 (4%) are carriers of the cystic fibrosis allele. The normal allele for this gene codes for a membrane protein that functions in the transport of chloride ions between certain cells and the extracellular fluid. These chloride transport channels are defective or absent in the plasma membranes of children who inherit two recessive alleles for cystic fibrosis. The result is an abnormally high concentration of extracellular chloride, which causes the mucus that coats certain cells to become thicker and stickier than normal. The mucus builds up in the pancreas, lungs, digestive tract, and other organs, leading to multiple (pleiotropic) effects, including poor absorption of nutrients from the intestines, chronic bronchitis, and recurrent bacterial infections. Recent research indicates that the high concentration of extracellular chloride also contributes to infection by disabling a natural antibiotic made by some body cells. When immune cells come to the rescue, their remains add to the mucus, creating a vicious cycle.

If untreated, most children with cystic fibrosis die before their 5th birthday. But daily doses of antibiotics to prevent infection, gentle pounding on the chest to clear mucus from clogged airways, and other preventive treatments can prolong life. In the United States, more than half of those with cystic fibrosis now survive into their late 20s or even 30s and beyond.

Sickle-Cell Disease

The most common inherited disorder among people of African descent is **sickle-cell disease**, which affects one out of 400 African-Americans. Sickle-cell disease is caused by the substitution of a single amino acid in the hemoglobin protein of red blood cells. When the oxygen content of an affected individual's blood is low (at high altitudes or under physical stress, for instance), the sickle-cell hemoglobin molecules aggregate into long rods that deform the red cells into a sickle shape (see Figure 5.22). Sickled cells may clump and clog small blood vessels, often leading to other symptoms throughout the body, including physical weakness, pain, organ damage, and even paralysis. The multiple effects of a double dose of the sickle-cell allele are another example of pleiotropy. Regular blood transfusions can ward off brain damage in children with sickle-cell disease, and new drugs can help prevent or treat other problems, but there is no cure.

Although two sickle-cell alleles are necessary for an individual to manifest full-blown sickle-cell disease, the presence of one sickle-cell allele can affect the phenotype. Thus, at the organismal level, the normal allele is incompletely dominant to the sickle-cell allele. Heterozygotes, said to have *sickle-cell trait*, are usually healthy, but they may suffer some sickle-cell symptoms during prolonged periods of reduced blood oxygen content. At the molecular level, the two alleles are codominant; both normal and abnormal (sickle-cell) hemoglobins are made in heterozygotes.

About one out of ten African-Americans have sickle-cell trait, an unusually high frequency of heterozygotes for an allele with severe detrimental effects in homozygotes. One explanation for this is that a single copy of the sickle-cell allele reduces the frequency and severity of malaria attacks, especially among young children. The malaria parasite spends part of its life cycle in red blood cells (see Figure 28.10), and the presence of even heterozygous amounts of sickle-cell hemoglobin results in lower parasite densities and hence reduced malaria symptoms. Thus, in tropical Africa, where infection with the malaria parasite is common, the sickle-cell allele is both boon and bane. The relatively high frequency of African-Americans with sickle-cell trait is a vestige of their African roots.

Dominantly Inherited Disorders

Although many harmful alleles are recessive, a number of human disorders are due to dominant alleles. One example is *achondroplasia*, a form of dwarfism that occurs in one of every 25,000 people. Heterozygous individuals have the dwarf phenotype **(Figure 14.17)**. Therefore, all people who are not achondroplastic dwarfs—99.99% of the population—are homozygous for the recessive allele. Like the presence of extra fingers or toes mentioned earlier, achondroplasia is a trait for which the recessive allele is much more prevalent than the corresponding dominant allele.

Dominant alleles that cause a lethal disease are much less common than recessive alleles that do so. All lethal alleles arise by mutations (changes to the DNA) in cells that produce sperm or eggs; presumably, such mutations occur equally often whether the mutant allele is dominant or recessive. However, if a lethal dominant allele causes the death of offspring before they mature and can reproduce, the allele will not be passed on to future generations. (In contrast, a lethal recessive allele can be perpetuated from generation to generation by heterozygous carriers who have normal phenotypes, since only homozygous recessive offspring will have the lethal disease.)

▲ **Figure 14.17 Achondroplasia: a dominant trait.** The late actor David Rappaport had achondroplasia, a form of dwarfism caused by a dominant allele (*D*). The allele might have arisen as a mutation in the egg or sperm of a parent or could have been inherited from an affected parent, as shown for an affected father in the Punnett square.

Huntington's Disease

A lethal dominant allele can escape elimination if it causes death only after an individual who carries the allele has reached a relatively advanced age. By the time the symptoms become evident, the individual may have already transmitted the lethal allele to his or her children. For example, **Huntington's disease**, a degenerative disease of the nervous system, is caused by a lethal dominant allele that has no obvious phenotypic effect until the individual is about 35 to 45 years old. Once the deterioration of the nervous system begins, it is irreversible and inevitably fatal. Any child born to a parent who has the allele for Huntington's disease has a 50% chance of inheriting the allele and the disorder (see the Punnett square in Figure 14.17). In the United States, this devastating disease afflicts about one in 10,000 people.

Until relatively recently, the onset of symptoms was the only way to know if a person had inherited the Huntington's allele. This is no longer the case. By analyzing DNA samples from a large family with a high incidence of the disorder, geneticists tracked the Huntington's allele to a locus near the tip of chromosome 4, and the gene has now been sequenced. This information led to development of a test that can detect the presence of the Huntington's allele in an individual's genome. (The methods that make such tests possible are discussed in Chapter 20.) The availability of this test poses an agonizing dilemma for those with a family history of Huntington's disease, such as relatives of the folk singer Woody Guthrie, who died of the disease. Does a person currently in good health benefit by finding out whether he or she has inherited a fatal and not yet curable disease? Some individuals may want to be tested for the disease before planning a family, whereas others may decide it would be too stressful to find out. Clearly, this is a personal decision.

Multifactorial Disorders

The hereditary diseases we have discussed so far are sometimes described as simple Mendelian disorders because they result from abnormality of one or both alleles at a single genetic locus. Many more people are susceptible to diseases that have a multifactorial basis—a genetic component plus a significant environmental influence. Heart disease, diabetes, cancer, alcoholism, certain mental illnesses such as schizophrenia and bipolar disorder, and many other diseases are multifactorial. In many cases, the hereditary component is polygenic. For example, many genes affect cardiovascular health, making some of us more prone than others to heart attacks and strokes. No matter what our genotype, however, our lifestyle has a tremendous effect on phenotype for cardiovascular health and other multifactorial characters. Exercise, a healthful diet, abstinence from smoking, and an ability to handle stressful situations all reduce our risk of heart disease and some types of cancer.

At present, so little is understood about the genetic contributions to most multifactorial diseases that the best public health strategy is to educate people about the importance of environmental factors and to promote healthful behavior.

Genetic Testing and Counseling

A preventive approach to simple Mendelian disorders is possible when the risk of a particular genetic disorder can be assessed before a child is conceived or during the early stages of the pregnancy. Many hospitals have genetic counselors who can provide information to prospective parents concerned about a family history for a specific disease.

Counseling Based on Mendelian Genetics and Probability Rules

Consider the case of a hypothetical couple, John and Carol. Both had a brother who died from the same recessively inherited lethal disease. Before conceiving their first child, John and Carol seek genetic counseling to determine the risk of having a child with the disease. From the information about their brothers, we know that both parents of John and both parents of Carol must have been carriers of the recessive allele. Thus, John and Carol are both products of $Aa \times Aa$ crosses, where a symbolizes the allele that causes this particular disease. We also know that John and Carol are not homozygous recessive (aa), because they do not have the disease. Therefore, their genotypes are either AA or Aa.

Given a genotypic ratio of $1\ AA : 2\ Aa : 1\ aa$ for offspring of an $Aa \times Aa$ cross, John and Carol each have a $\frac{2}{3}$ chance of being carriers (Aa). According to the rule of multiplication, the overall probability of their firstborn having the disorder is ⅔ (the chance that John is a carrier) times ⅔ (the chance that Carol is a carrier) times ¼ (the chance of two carriers having a child with the disease), which equals ⅑. Suppose that Carol and John decide to have a child—after all, there is an ⁸⁄₉ chance that their baby will not have the disorder. If, despite these odds, their child is born with the disease, then we would know that *both* John and Carol are, in fact, carriers (*Aa* genotype). If both John and Carol are carriers, there is a ¼ chance that any subsequent child this couple has will have the disease.

When we use Mendel's laws to predict possible outcomes of matings, it is important to remember that each child represents an independent event in the sense that its genotype is unaffected by the genotypes of older siblings. Suppose that John and Carol have three more children, and *all three* have the hypothetical hereditary disease. There is only one chance in 64 (¼ × ¼ × ¼) that such an outcome will occur. Despite this run of misfortune, the chance that still another child of this couple will have the disease remains ¼.

Tests for Identifying Carriers

Because most children with recessive disorders are born to parents with normal phenotypes, the key to accurately assessing the genetic risk for a particular disease is determining whether the prospective parents are heterozygous carriers of the recessive allele. For an increasing number of heritable disorders, tests are available that can distinguish individuals of normal phenotype who are dominant homozygotes from those who are heterozygotes. There are now tests that can identify carriers of the alleles for Tay-Sachs disease, sickle-cell disease, and the most common form of cystic fibrosis.

These tests for identifying carriers enable people with family histories of genetic disorders to make informed decisions about having children. But these new methods for genetic screening pose potential problems. If confidentiality is breached, will carriers be stigmatized? Will they be denied health or life insurance, even though they themselves are healthy? Will misinformed employers equate "carrier" with disease? And will sufficient genetic counseling be available to help a large number of individuals understand their test results? Advances in biotechnology offer possibilities for reducing human suffering, but not before key ethical issues are resolved.

Fetal Testing

Suppose a couple learns that they are both carriers of the Tay-Sachs allele, but they decide to have a child anyway. Tests performed in conjunction with a technique known as **amniocentesis** can determine, beginning at the 14th–16th week of pregnancy, whether the developing fetus has Tay-Sachs disease **(Figure 14.18a)**. To perform this procedure, a physician inserts a needle into the uterus and extracts about 10 mL of amniotic fluid, the liquid that bathes the fetus. Some genetic disorders can be detected from the presence of certain chemicals in the amniotic fluid itself. Tests for other disorders, including Tay-Sachs disease, are performed on cells cultured in the laboratory, descendants of the fetal cells sloughed off into the amniotic fluid. These cultured cells can also be used for karyotyping to identify certain chromosomal defects (see Figure 13.3).

In an alternative technique called **chorionic villus sampling (CVS)**, a physician inserts a narrow tube through the cervix into the uterus and suctions out a tiny sample of tissue from the placenta, the organ that transmits nutrients and fetal wastes between the fetus and the mother **(Figure 14.18b)**. The cells of the chorionic villi of the placenta, the portion sampled, are derived from the fetus and have the same genotype as the new individual. These cells are proliferating rapidly enough to allow karyotyping to be carried out immediately. This rapid analysis is an advantage over amniocentesis, in which the cells must be cultured for several weeks before karyotyping. Another advantage of CVS is that it can be performed as early as the 8th–10th week of pregnancy. However, CVS is not suitable for tests requiring amniotic fluid. Recently, medical scientists have developed methods for isolating fetal cells that have escaped into the mother's blood. Although very few in number, these cells can be cultured and then tested.

Imaging techniques allow a physician to examine a fetus directly for major anatomical abnormalities. In the *ultrasound* technique, sound waves are used to produce an image of the fetus by a simple noninvasive procedure. In *fetoscopy*, a needle-thin tube containing a viewing scope and fiber optics (to transmit light) is inserted into the uterus.

Ultrasound has no known risk to either mother or fetus, while the other procedures can cause complications in a small percentage of cases. Previously, amniocentesis or CVS for diagnostic testing was generally offered only to women over age 35, due to their increased risk of bearing a child with Down syndrome. In 2007, however, reassessment of the risks and possible benefits led to a change in the recommended practice, and now such testing is offered to all pregnant women. If the fetal tests reveal a serious disorder, the parents face the difficult choice of terminating the pregnancy or preparing to care for a child with a genetic disorder.

Newborn Screening

Some genetic disorders can be detected at birth by simple tests that are now routinely performed in most hospitals in the United States. One common screening program is for phenylketonuria (PKU), a recessively inherited disorder that occurs in about one out of every 10,000 to 15,000 births in the United States. Children with this disease cannot properly metabolize the amino acid phenylalanine. This compound and its by-product, phenylpyruvate, can accumulate to toxic levels in the blood, causing mental retardation. However, if the deficiency is detected in the newborn, a special diet low in phenylalanine will usually allow normal development and prevent retardation. Unfortunately, very few other genetic disorders are treatable at the present time.

Screening of newborns and fetuses for serious inherited diseases, tests for identifying carriers, and genetic counseling all rely on the Mendelian model of inheritance. We owe the "gene idea"—the concept of particulate heritable factors transmitted according to simple rules of chance—to the elegant quantitative experiments of Gregor Mendel. The importance of his discoveries was overlooked by most biologists until early in the 20th century, several decades after his findings were reported. In the next chapter, you will learn how Mendel's laws have their physical basis in the behavior of chromosomes during sexual life cycles and how the synthesis of Mendelism and a chromosome theory of inheritance catalyzed progress in genetics.

CONCEPT CHECK 14.4

1. Beth and Tom each have a sibling with cystic fibrosis, but neither Beth nor Tom nor any of their parents have the disease. Calculate the probability that if this couple has a child, the child will have cystic fibrosis. What would be the probability if a test revealed that Tom is a carrier but Beth is not?
2. Joan was born with six toes on each foot, a dominant trait called polydactyly. Two of her five siblings and her mother, but not her father, also have extra digits. What is Joan's genotype for the number-of-digits character? Explain your answer. Use *D* and *d* to symbolize the alleles for this character.
3. **WHAT IF?** What would you suspect if Peter was born with polydactyly, but neither of his biological parents had extra digits?

For suggested answers, see Appendix A.

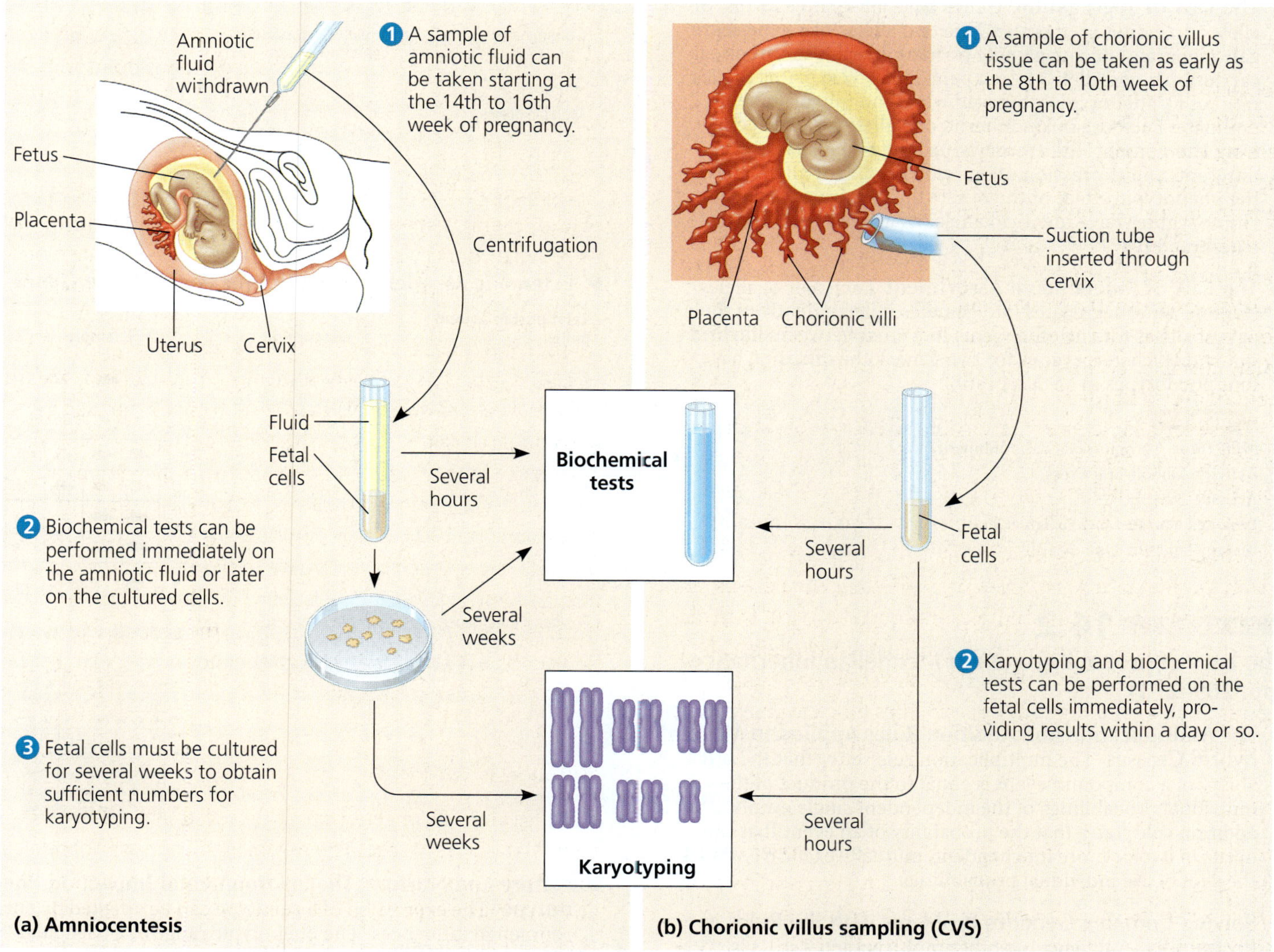

▲ **Figure 14.18 Testing a fetus for genetic disorders.** Biochemical tests may detect substances associated with particular disorders. Karyotyping shows whether the chromosomes of the fetus are normal in number and appearance.

Chapter 14 Review

MEDIA Go to the Study Area at **www.masteringbio.com** for BioFlix 3-D animations, MP3 Tutors, Videos, Practice Tests, an eBook, and more.

SUMMARY OF KEY CONCEPTS

CONCEPT 14.1

Mendel used the scientific approach to identify two laws of inheritance (pp. 262–269)

▶ **Mendel's Experimental, Quantitative Approach** In the 1860s, Gregor Mendel formulated a theory of inheritance based on experiments with garden peas, proposing that parents pass on to their offspring discrete genes that retain their identity through generations.

▶ **The Law of Segregation** Genes have alternative forms, or alleles. In a diploid organism, the two alleles of a gene segregate (separate) during gamete formation; each sperm or egg carries only one allele of each pair. This law explains the 3:1 ratio of F_2 phenotypes observed when monohybrids self-pollinate. Each organism inherits one allele for each gene from each parent. In heterozygotes, the two alleles are different, and expression of one (the dominant allele) masks the phenotypic effect of the other (the recessive allele). Homozygotes have identical alleles of a given gene and are true-breeding.

▶ **The Law of Independent Assortment** Each pair of alleles (for one gene) segregates into gametes independently of the pair of alleles for any other gene. In a cross between dihybrids (individuals heterozygous for two genes), the offspring have four phenotypes in a 9:3:3:1 ratio.

MEDIA
MP3 Tutor Chromosomal Basis of Inheritance
Activity Monohybrid Cross
Activity Dihybrid Cross
Biology Labs On-Line PedigreeLab
Biology Labs On-Line FlyLab

CONCEPT 14.2

The laws of probability govern Mendelian inheritance (pp. 269–271)

▶ **The Multiplication and Addition Rules Applied to Monohybrid Crosses** The multiplication rule states that the probability of a compound event is equal to the product of the individual probabilities of the independent single events. The addition rule states that the probability of an event that can occur in two or more independent, mutually exclusive ways is the sum of the individual probabilities.

▶ **Solving Complex Genetics Problems with the Rules of Probability** A dihybrid or other multicharacter cross is equivalent to two or more independent monohybrid crosses occurring simultaneously. In calculating the chances of the various offspring genotypes from such crosses, each character first is considered separately and then the individual probabilities are multiplied.

MEDIA
Activity Gregor's Garden

CONCEPT 14.3

Inheritance patterns are often more complex than predicted by simple Mendelian genetics (pp. 271–275)

▶ **Extending Mendelian Genetics for a Single Gene**

Degree of dominance	Description	Example
Complete dominance of one allele	Heterozygous phenotype same as that of homozygous dominant	*PP*  *Pp* 
Incomplete dominance of either allele	Heterozygous phenotype intermediate between the two homozygous phenotypes	 C^RC^R C^RC^W C^WC^W
Codominance	Heterozygotes: Both phenotypes expressed	I^AI^B 
Multiple alleles	In the whole population, some genes have more than two alleles	ABO blood group alleles I^A, I^B, i
Pleiotropy	One gene is able to affect multiple phenotypic characters	Sickle-cell disease

▶ **Extending Mendelian Genetics for Two or More Genes**

Relationship among genes	Description	Example
Epistasis	One gene affects the expression of another	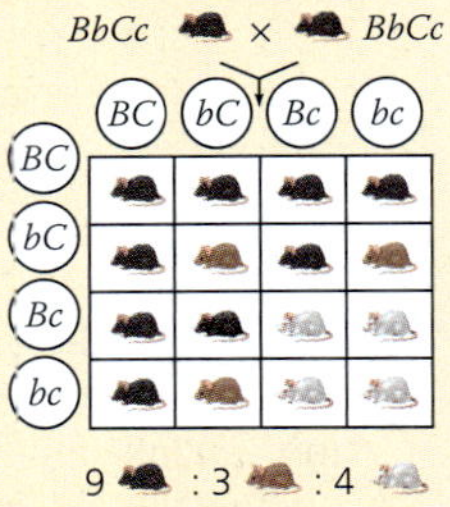
Polygenic inheritance	A single phenotypic character is affected by two or more genes	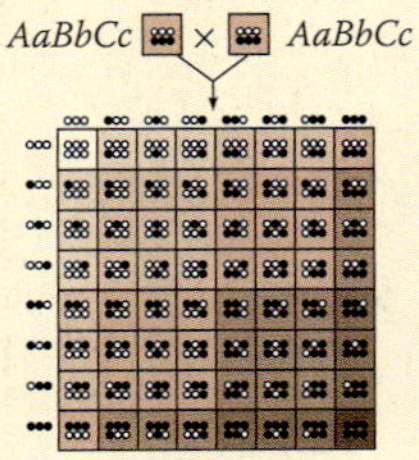

▶ **Nature and Nurture: The Environmental Impact on Phenotype** The expression of a genotype can be affected by environmental influences. The phenotypic range of a particular genotype is called its norm of reaction. Polygenic characters that are also influenced by the environment are called multifactorial characters.

▶ **Integrating a Mendelian View of Heredity and Variation** An organism's overall phenotype, including its physical appearance, internal anatomy, physiology, and behavior, reflects its overall genotype and unique environmental history. Even in more complex inheritance patterns, Mendel's fundamental laws of segregation and independent assortment still apply.

MEDIA

Activity Incomplete Dominance

CONCEPT **14.4**

Many human traits follow Mendelian patterns of inheritance (pp. 276–281)

- **Pedigree Analysis** Family pedigrees can be used to deduce the possible genotypes of individuals and make predictions about future offspring. Predictions are usually statistical probabilities rather than certainties.
- **Recessively Inherited Disorders** Many genetic disorders are inherited as simple recessive traits. Most affected (homozygous recessive) individuals are children of phenotypically normal, heterozygous carriers.
- **Dominantly Inherited Disorders** Lethal dominant alleles are eliminated from the population if affected people die before reproducing. Nonlethal dominant alleles and lethal ones that strike relatively late in life are inherited in a Mendelian way.
- **Multifactorial Disorders** Many human diseases have both genetic and environmental components and do not follow simple Mendelian inheritance patterns.
- **Genetic Testing and Counseling** Using family histories, genetic counselors help couples determine the odds that their children will have genetic disorders. Amniocentesis and chorionic villus sampling can indicate whether a suspected genetic disorder is present in a fetus. Other genetic tests can be performed after birth.

MEDIA

Investigation How Do You Diagnose a Genetic Disorder?

TESTING YOUR KNOWLEDGE

TIPS FOR GENETICS PROBLEMS

1. Write down symbols for the alleles. (These may be given in the problem.) When represented by single letters, the dominant allele is uppercase and the recessive is lowercase.

2. Write down the possible genotypes, as determined by the phenotype.
   a. If the phenotype is that of the dominant trait (for example, purple flowers), then the genotype is either homozygous dominant or heterozygous (*PP* or *Pp*, in this example).
   b. If the phenotype is that of the recessive trait, the genotype must be homozygous recessive (for example, *pp*).
   c. If the problem says "true-breeding," the genotype is homozygous.

3. Determine what the problem is asking for. If asked to do a cross, write it out in the form [Genotype] × [Genotype], using the alleles you've decided on.

4. To figure out the outcome of a cross, set up a Punnett square.
   a. Put the gametes of one parent at the top and those of the other on the left. To determine the allele(s) in each gamete for a given genotype, set up a systematic way to list all the possibilities. (Remember, each gamete has one allele of each gene.) Note that there are 2^n possible types of gametes, where n is the number of gene loci that are heterozygous. For example, an individual with genotype *AaBbCc* would produce $2^3 = 8$ types of gametes. Write the genotypes of the gametes in circles above the columns and to the left of the rows.
   b. Fill in the Punnett square as if each possible sperm were fertilizing each possible egg, making all of the possible offspring. In a cross of *AaBbCc* × *AaBbCc*, for example, the Punnett square would have 8 columns and 8 rows, so there are 64 different offspring; you would know the genotype of each and thus the phenotype. Count genotypes and phenotypes to obtain the genotypic and phenotypic ratios.

5. You can use the rules of probability if the Punnett square would be too big. (For example, see Genetics Problem 4.) You can consider each gene separately (see pp. 270–271).

6. If, instead, the problem gives you the phenotypic ratios of offspring, but not the genotypes of the parents in a given cross, the phenotypes can help you deduce the parents' unknown genotypes.
   a. For example, if ½ the offspring have the recessive phenotype and ½ the dominant, you know that the cross was between a heterozygote and a homozygous recessive.
   b. If the ratio is 3:1, the cross was between two heterozygotes.
   c. If two genes are involved and you see a 9:3:3:1 ratio in the offspring, you know that each parent is heterozygous for both genes. Caution: Don't assume that the reported numbers will exactly equal the predicted ratios. For example, if there are 13 offspring with the dominant trait and 11 with the recessive, assume that the ratio is one dominant to one recessive.

7. For pedigree problems, use the tips in Figure 14.15 and below to determine what kind of trait is involved.
   a. If parents without the trait have offspring with the trait, the trait must be recessive and the parents both carriers.
   b. If the trait is seen in every generation, it is most likely dominant (see the next possibility, though).
   c. If both parents have the trait, then in order for it to be recessive, all offspring must show the trait.
   d. To determine the likely genotype of a certain individual in a pedigree, first label the genotypes of all the family members you can. Even if some of the genotypes are incomplete, label what you do know. For example, if an individual has the dominant phenotype, the genotype must be *AA* or *Aa*; you can write this as *A_*. Try different possibilities to see which fits the results. Use the rules of probability to calculate the probability of each possible genotype being the correct one.

GENETICS PROBLEMS

1. Match each term on the left with a statement on the right.

Term	*Statement*
__ Gene	a. Has no effect on phenotype in a heterozygote
__ Allele	b. A variant for a character
__ Character	c. Having two identical alleles for a gene
__ Trait	d. A cross between individuals heterozygous for a single character
__ Dominant allele	e. An alternative version of a gene
__ Recessive allele	f. Having two different alleles for a gene
__ Genotype	g. A heritable feature that varies among individuals
__ Phenotype	h. An organism's appearance or observable traits
__ Homozygous	i. A cross between an individual with an unknown genotype and a homozygous recessive individual
__ Heterozygous	j. Determines phenotype in a heterozygote
__ Testcross	k. The genetic makeup of an individual
__ Monohybrid cross	l. A heritable unit that determines a character; can exist in different forms

2. **DRAW IT** Two pea plants heterozygous for the characters of pod color and pod shape are crossed. Draw a Punnett square to determine the phenotypic ratios of the offspring.

3. In some plants, a true-breeding, red-flowered strain gives all pink flowers when crossed with a white-flowered strain: C^RC^R (red) × C^WC^W (white) → C^RC^W (pink). If flower position (axial or terminal) is inherited as it is in peas (see Table 14.1), what will be the ratios of genotypes and phenotypes of the F_1 generation resulting from the following cross: axial-red (true-breeding) × terminal-white? What will be the ratios in the F_2 generation?

4. Flower position, stem length, and seed shape were three characters that Mendel studied. Each is controlled by an independently assorting gene and has dominant and recessive expression as follows:

Character	*Dominant*	*Recessive*
Flower position	Axial (*A*)	Terminal (*a*)
Stem length	Tall (*T*)	Dwarf (*t*)
Seed shape	Round (*R*)	Wrinkled (*r*)

If a plant that is heterozygous for all three characters is allowed to self-fertilize, what proportion of the offspring would you expect to be as follows? (*Note:* Use the rules of probability instead of a huge Punnett square.)
a. homozygous for the three dominant traits
b. homozygous for the three recessive traits
c. heterozygous for all three characters
d. homozygous for axial and tall, heterozygous for seed shape

5. A black guinea pig crossed with an albino guinea pig produces 12 black offspring. When the albino is crossed with a second black one, 7 blacks and 5 albinos are obtained. What is the best explanation for this genetic situation? Write genotypes for the parents, gametes, and offspring.

6. In sesame plants, the one-pod condition (*P*) is dominant to the three-pod condition (*p*), and normal leaf (*L*) is dominant to wrinkled leaf (*l*). Pod type and leaf type are inherited independently. Determine the genotypes for the two parents for all possible matings producing the following offspring:
a. 318 one-pod, normal leaf and 98 one-pod, wrinkled leaf
b. 323 three-pod, normal leaf and 106 three-pod, wrinkled leaf
c. 401 one-pod, normal leaf
d. 150 one-pod, normal leaf, 147 one-pod, wrinkled leaf, 51 three-pod, normal leaf, and 48 three-pod, wrinkled leaf
e. 223 one-pod, normal leaf, 72 one-pod, wrinkled leaf, 76 three-pod, normal leaf, and 27 three-pod, wrinkled leaf

7. A man with type A blood marries a woman with type B blood. Their child has type O blood. What are the genotypes of these individuals? What other genotypes, and in what frequencies, would you expect in offspring from this marriage?

8. Phenylketonuria (PKU) is an inherited disease caused by a recessive allele. If a woman and her husband, who are both carriers, have three children, what is the probability of each of the following?
a. All three children are of normal phenotype.
b. One or more of the three children have the disease.
c. All three children have the disease.
d. At least one child is phenotypically normal.
(*Note:* Remember that the probabilities of all possible outcomes always add up to 1.)

9. The genotype of F_1 individuals in a tetrahybrid cross is *AaBbCcDd.* Assuming independent assortment of these four genes, what are the probabilities that F_2 offspring will have the following genotypes?
a. *aabbccdd*
b. *AaBbCcDd*
c. *AABBCCDD*
d. *AaBBccDd*
e. *AaBBCCdd*

10. What is the probability that each of the following pairs of parents will produce the indicated offspring? (Assume independent assortment of all gene pairs.)
a. *AABBCC* × *aabbcc* → *AaBbCc*
b. *AABbCc* × *AaBbCc* → *AAbbCC*
c. *AaBbCc* × *AaBbCc* → *AaBbCc*
d. *aaBbCC* × *AABbcc* → *AaBbCc*

11. Karen and Steve each have a sibling with sickle-cell disease. Neither Karen nor Steve nor any of their parents have the disease, and none of them have been tested to see if they have the sickle-cell trait. Based on this incomplete information, calculate the probability that if this couple has a child, the child will have sickle-cell disease.

12. In 1981, a stray black cat with unusual rounded, curled-back ears was adopted by a family in California. Hundreds of descendants of the cat have since been born, and cat fanciers hope to develop the curl cat into a show breed. Suppose you owned the first curl cat and wanted to develop a true-breeding variety. How would you determine whether the curl allele is dominant or recessive? How would you obtain true-breeding curl cats? How could you be sure they are true-breeding?

13. Imagine that a newly discovered, recessively inherited disease is expressed only in individuals with type O blood, although the disease and blood group are independently inherited. A normal man with type A blood and a normal woman with type B blood have already had one child with the disease. The woman is now pregnant for a second time. What is the probability that the second child will also have the disease? Assume that both parents are heterozygous for the gene that causes the disease.

14. In tigers, a recessive allele causes an absence of fur pigmentation (a white tiger) and a cross-eyed condition. If two phenotypically normal tigers that are heterozygous at this locus are mated, what percentage of their offspring will be cross-eyed? What percentage of cross-eyed tigers will be white?

15. In maize (corn) plants, a dominant allele *I* inhibits kernel color, while the recessive allele *i* permits color when homozygous. At a different locus, the dominant allele *P* causes purple kernel color, while the homozygous recessive genotype *pp* causes red kernels. If plants heterozygous at both loci are crossed, what will be the phenotypic ratio of the offspring?

16. The pedigree below traces the inheritance of alkaptonuria, a biochemical disorder. Affected individuals, indicated here by the colored circles and squares, are unable to metabolize a substance called alkapton, which colors the urine and stains body tissues. Does alkaptonuria appear to be caused by a dominant allele or by a recessive allele? Fill in the genotypes of the individuals whose genotypes can be deduced. What genotypes are possible for each of the other individuals?

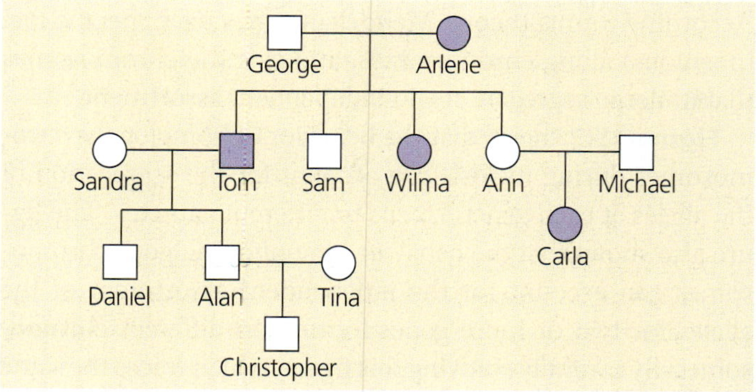

17. A man has six fingers on each hand and six toes on each foot. His wife and their daughter have the normal number of digits. Extra digits is a dominant trait. What fraction of this couple's children would be expected to have extra digits?

18. Imagine that you are a genetic counselor, and a couple planning to start a family comes to you for information. Charles was married once before, and he and his first wife had a child with cystic fibrosis. The brother of his current wife, Elaine, died of cystic fibrosis. What is the probability that Charles and Elaine will have a baby with cystic fibrosis? (Neither Charles nor Elaine has cystic fibrosis.)

19. In mice, black color (*B*) is dominant to white (*b*). At a different locus, a dominant allele (*A*) produces a band of yellow just below the tip of each hair in mice with black fur. This gives a frosted appearance known as agouti. Expression of the recessive allele (*a*) results in a solid coat color. If mice that are heterozygous at both loci are crossed, what is the expected phenotypic ratio of their offspring?

For Genetics Problems answers, see Appendix A.

MEDIA Visit the Study Area at **www.masteringbio.com** for a Practice Test.

EVOLUTION CONNECTION

20. Over the past half century, there has been a trend in the United States and other developed countries for people to marry and start families later in life than did their parents and grandparents. What effects might this trend have on the incidence (frequency) of late-acting dominant lethal alleles in the population?

SCIENTIFIC INQUIRY

21. You are handed a mystery pea plant with tall stems and axial flowers and asked to determine its genotype as quickly as possible. You know that the allele for tall stems (*T*) is dominant to that for dwarf stems (*t*) and that the allele for axial flowers (*A*) is dominant to that for terminal flowers (*a*).
   a. What are *all* the possible genotypes for your mystery plant?
   b. Describe the *one* cross you would do, out in your garden, to determine the exact genotype of your mystery plant.
   c. While waiting for the results of your cross, you predict the results for each possible genotype listed in part a. How do you do this? Why is this not called "performing a cross"?
   d. Explain how the results of your cross and your predictions will help you learn the genotype of your mystery plant.

SCIENCE, TECHNOLOGY, AND SOCIETY

22. Imagine that one of your parents had Huntington's disease. What is the probability that you, too, will someday manifest the disease? There is no cure for Huntington's. Would you want to be tested for the Huntington's allele? Why or why not?

15 The Chromosomal Basis of Inheritance

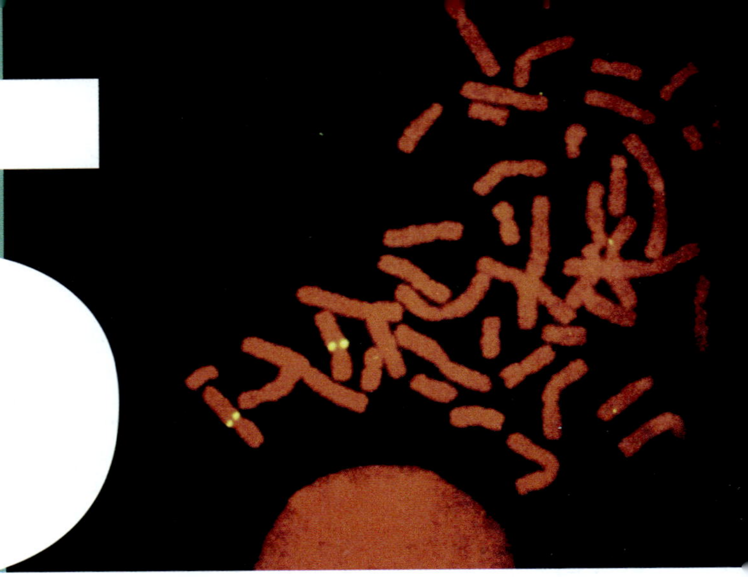

▲ **Figure 15.1 Where are Mendel's hereditary factors located in the cell?**

KEY CONCEPTS

15.1 Mendelian inheritance has its physical basis in the behavior of chromosomes

15.2 Sex-linked genes exhibit unique patterns of inheritance

15.3 Linked genes tend to be inherited together because they are located near each other on the same chromosome

15.4 Alterations of chromosome number or structure cause some genetic disorders

15.5 Some inheritance patterns are exceptions to the standard chromosome theory

OVERVIEW

Locating Genes Along Chromosomes

Gregor Mendel's "hereditary factors" were purely an abstract concept when he proposed their existence in 1860. At that time, no cellular structures were known that could house these imaginary units. Even after chromosomes were first observed, many biologists remained skeptical about Mendel's laws of segregation and independent assortment until there was sufficient evidence that these principles of heredity had a physical basis in chromosomal behavior.

Today, we can show that genes—Mendel's "factors"—are located along chromosomes. We can see the location of a particular gene by tagging chromosomes with a fluorescent dye that highlights that gene. For example, the yellow dots in **Figure 15.1** mark the locus of a specific gene on a homologous pair of human chromosomes. (Because the chromosomes in this light micrograph have already replicated, we see two dots per chromosome, one on each sister chromatid.) In this chapter, which integrates and extends what you learned in the past two chapters, we describe the chromosomal basis for the transmission of genes from parents to offspring, along with some important exceptions to the standard mode of inheritance.

CONCEPT 15.1

Mendelian inheritance has its physical basis in the behavior of chromosomes

Using improved techniques of microscopy, cytologists worked out the process of mitosis in 1875 and meiosis in the 1890s. Cytology and genetics converged when biologists began to see parallels between the behavior of chromosomes and the behavior of Mendel's proposed hereditary factors during sexual life cycles: Chromosomes and genes are both present in pairs in diploid cells; homologous chromosomes separate and alleles segregate during the process of meiosis; and fertilization restores the paired condition for both chromosomes and genes. Around 1902, Walter S. Sutton, Theodor Boveri, and others independently noted these parallels, and the **chromosome theory of inheritance** began to take form. According to this theory, Mendelian genes have specific loci (positions) along chromosomes, and it is the chromosomes that undergo segregation and independent assortment.

Figure 15.2 shows that the behavior of homologous chromosomes during meiosis can account for the segregation of the alleles at each genetic locus to different gametes. The figure also shows that the behavior of nonhomologous chromosomes can account for the independent assortment of the alleles for two or more genes located on different chromosomes. By carefully studying this figure, which traces the same dihybrid pea cross you learned about in Figure 14.8, you can see how the behavior of chromosomes during meiosis in the F_1 generation and subsequent random fertilization give rise to the F_2 phenotypic ratio observed by Mendel.

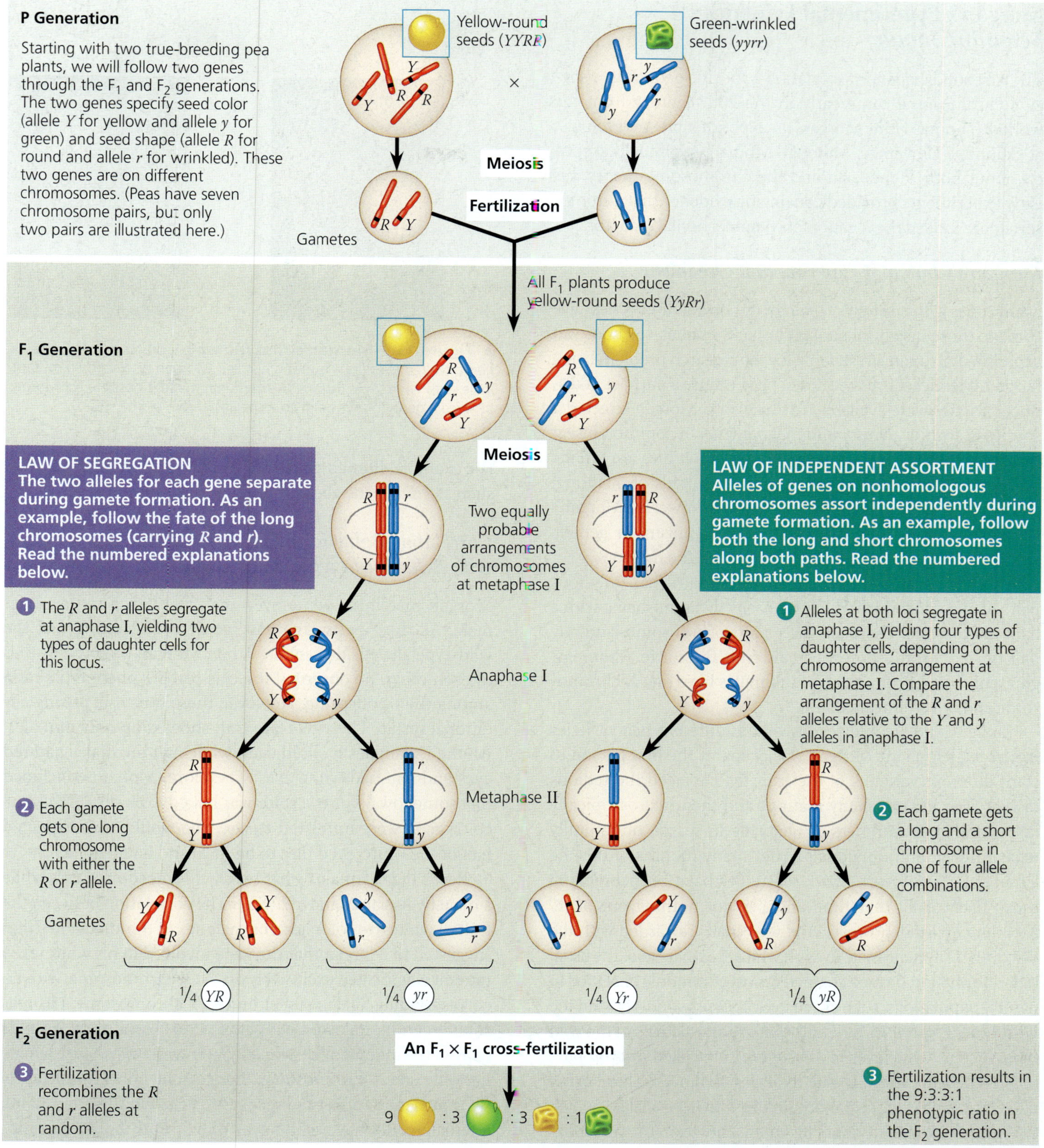

▲ **Figure 15.2 The chromosomal basis of Mendel's laws.** Here we correlate the results of one of Mendel's dihybrid crosses (see Figure 14.8) with the behavior of chromosomes during meiosis (see Figure 13.8). The arrangement of chromosomes at metaphase I of meiosis and their movement during anaphase I account for the segregation and independent assortment of the alleles for seed color and shape. Each cell that undergoes meiosis in an F_1 plant produces two kinds of gametes. If we count the results for all cells, however, each F_1 plant produces equal numbers of all four kinds of gametes because the alternative chromosome arrangements at metaphase I are equally likely.

? *If you crossed an F_1 plant with a plant that was homozygous recessive for both genes* (yyrr), *how would the phenotypic ratio of the offspring compare with the 9:3:3:1 ratio seen here?*

Morgan's Experimental Evidence: *Scientific Inquiry*

The first solid evidence associating a specific gene with a specific chromosome came early in the 20th century from the work of Thomas Hunt Morgan, an experimental embryologist at Columbia University. Although Morgan was initially skeptical about both Mendelism and the chromosome theory, his early experiments provided convincing evidence that chromosomes are indeed the location of Mendel's heritable factors.

Morgan's Choice of Experimental Organism

Many times in the history of biology, important discoveries have come to those insightful enough or lucky enough to choose an experimental organism suitable for the research problem being tackled. Mendel chose the garden pea because a number of distinct varieties were available. For his work, Morgan selected a species of fruit fly, *Drosophila melanogaster*, a common insect that feeds on the fungi growing on fruit. Fruit flies are prolific breeders; a single mating will produce hundreds of offspring, and a new generation can be bred every two weeks. Morgan's laboratory began using this convenient organism for genetic studies in 1907 and soon became known as "the fly room."

Another advantage of the fruit fly is that it has only four pairs of chromosomes, which are easily distinguishable with a light microscope. There are three pairs of autosomes and one pair of sex chromosomes. Female fruit flies have a homologous pair of X chromosomes, and males have one X chromosome and one Y chromosome.

While Mendel could readily obtain different pea varieties from seed suppliers, Morgan was probably the first person to want different varieties of the fruit fly. He faced the tedious task of carrying out many matings and then microscopically inspecting large numbers of offspring in search of naturally occurring variant individuals. After many months of this, he lamented, "Two years' work wasted. I have been breeding those flies for all that time and I've got nothing out of it." Morgan persisted, however, and was finally rewarded with the discovery of a single male fly with white eyes instead of the usual red. The phenotype for a character most commonly observed in natural populations, such as red eyes in *Drosophila*, is called the **wild type** (**Figure 15.3**). Traits that are alternatives to the wild type, such as white eyes in *Drosophila*, are called *mutant phenotypes* because they are due to alleles assumed to have originated as changes, or mutations, in the wild-type allele.

Morgan and his students invented a notation for symbolizing alleles in *Drosophila* that is still widely used for fruit flies. For a given character in flies, the gene takes its symbol from the first mutant (non–wild type) discovered. Thus, the allele for white eyes in *Drosophila* is symbolized by *w*. A superscript + identifies the allele for the wild-type trait—w^+ for the allele for red eyes, for example. Over the years, a variety of gene notation systems have been developed for different organisms.

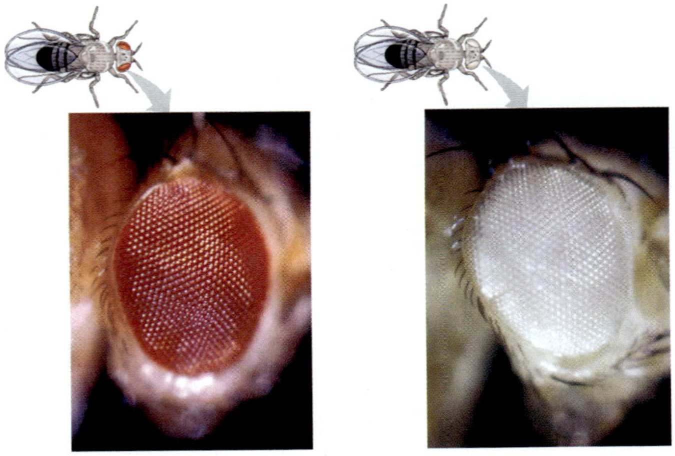

▲ **Figure 15.3 Morgan's first mutant.** Wild-type *Drosophila* flies have red eyes (left). Among his flies, Morgan discovered a mutant male with white eyes (right). This variation made it possible for Morgan to trace a gene for eye color to a specific chromosome (LMs).

For example, human genes are usually written in all capitals, such as *HD* for the allele for Huntington's disease.

Correlating Behavior of a Gene's Alleles with Behavior of a Chromosome Pair

Morgan mated his white-eyed male fly with a red-eyed female. All the F_1 offspring had red eyes, suggesting that the wild-type allele is dominant. When Morgan bred the F_1 flies to each other, he observed the classical 3:1 phenotypic ratio among the F_2 offspring. However, there was a surprising additional result: The white-eye trait showed up only in males. All the F_2 females had red eyes, while half the males had red eyes and half had white eyes. Therefore, Morgan concluded that somehow a fly's eye color was linked to its sex. (If the eye-color gene were unrelated to sex, one would have expected half of the white-eyed flies to be male and half female.)

Recall that a female fly has two X chromosomes (XX), while a male fly has an X and a Y (XY). The correlation between the trait of white eye color and the male sex of the affected F_2 flies suggested to Morgan that the gene involved in his white-eyed mutant was located exclusively on the X chromosome, with no corresponding allele present on the Y chromosome. His reasoning can be followed in **Figure 15.4**. For a male, a single copy of the mutant allele would confer white eyes; since a male has only one X chromosome, there can be no wild-type allele (w^+) present to offset the recessive allele. On the other hand, a female could have white eyes only if both her X chromosomes carried the recessive mutant allele (*w*). This was impossible for the F_2 females in Morgan's experiment because all the F_1 fathers had red eyes.

Morgan's finding of the correlation between a particular trait and an individual's sex provided support for the chromosome theory of inheritance: namely, that a specific gene is carried on a specific chromosome (in this case, an eye-color gene

▼ Figure 15.4 Inquiry

In a cross between a wild-type female fruit fly and a mutant white-eyed male, what color eyes will the F_1 and F_2 offspring have?

EXPERIMENT Thomas Hunt Morgan wanted to analyze the behavior of two alleles of a fruit fly eye-color gene. In crosses similar to those done by Mendel with pea plants, Morgan and his colleagues mated a wild-type (red-eyed) female with a mutant white-eyed male.

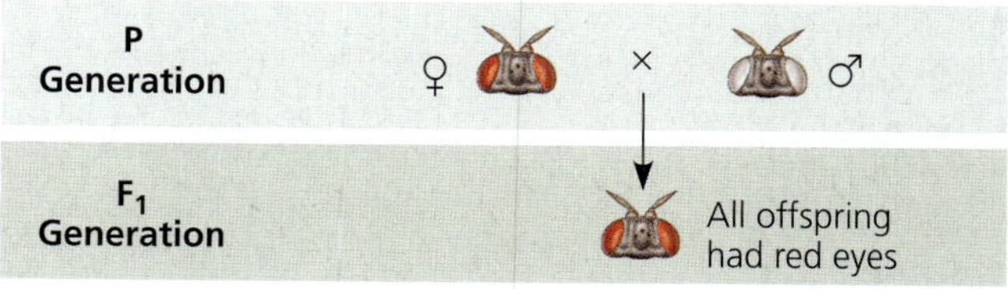

Morgan then bred an F_1 red-eyed female to an F_1 red-eyed male to produce the F_2 generation.

RESULTS The F_2 generation showed a typical Mendelian ratio of 3 red-eyed flies : 1 white-eyed fly. However, no females displayed the white-eye trait; all white-eyed flies were males.

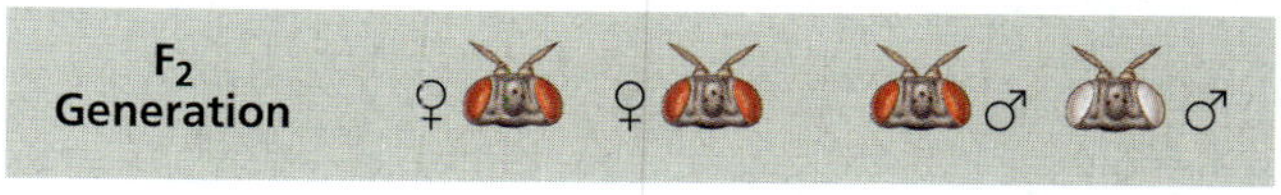

CONCLUSION All F_1 offspring had red eyes, so the mutant white-eye trait (w) must be recessive to the wild-type red-eye trait (w^+). Since the recessive trait—white eyes—was expressed only in males in the F_2 generation, Morgan deduced that this eye-color gene is located on the X chromosome and that there is no corresponding locus on the Y chromosome.

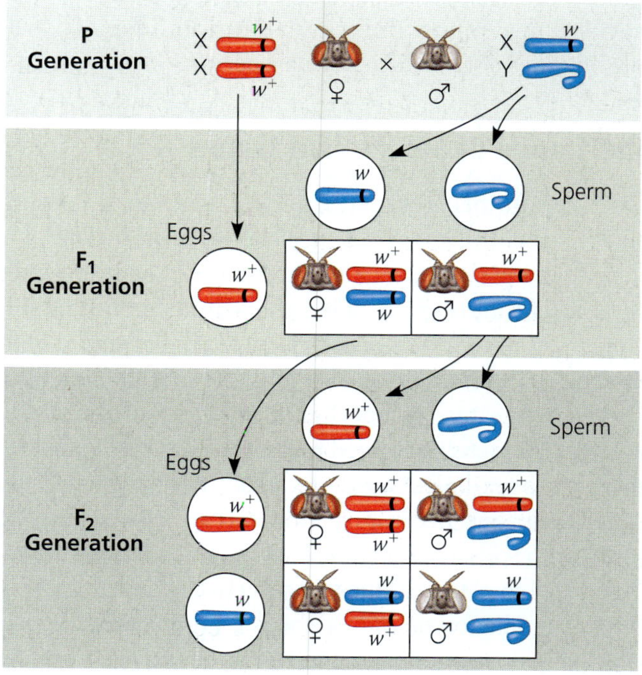

SOURCE T. H. Morgan, Sex-limited inheritance in *Drosophila*, *Science* 32:120–122 (1910).

WHAT IF? Suppose this eye-color gene were located on an autosome. Predict the phenotypes (including gender) of the F_2 flies in this hypothetical cross. (*Hint:* Draw a Punnett square.)

on the X chromosome). In addition, Morgan's work indicated that genes located on a sex chromosome exhibit unique inheritance patterns, which we will discuss in the next section. Recognizing the importance of Morgan's early work, many bright students were attracted to his fly room.

CONCEPT CHECK 15.1

1. Which one of Mendel's laws relates to the inheritance of alleles for a single character? Which law relates to the inheritance of alleles for two characters in a dihybrid cross?
2. What is the physical basis of Mendel's laws?
3. **WHAT IF?** Propose a possible reason that the first naturally occurring mutant fruit fly Morgan saw involved a gene on a sex chromosome.

For suggested answers, see Appendix A.

CONCEPT 15.2

Sex-linked genes exhibit unique patterns of inheritance

As you just learned, Morgan's discovery of a trait (white eyes) that correlated with the sex of flies was a key episode in the development of the chromosome theory of inheritance. Because the identity of the sex chromosomes in an individual could be inferred by observing the sex of the fly, the behavior of the two members of the pair of sex chromosomes could be correlated with the behavior of the two alleles of the eye-color gene. In this section, we consider the role of sex chromosomes in inheritance in more detail. We begin by reviewing the chromosomal basis of sex determination in humans and some other animals.

The Chromosomal Basis of Sex

Whether we are male or female is one of our more obvious phenotypic characters. Although the anatomical and physiological differences between women and men are numerous, the chromosomal basis for determining sex is rather simple. In humans and other mammals, there are two varieties of sex chromosomes, designated X and Y. The Y chromosome is much smaller than the X chromosome (**Figure 15.5**). A person who inherits two X chromosomes, one from each parent, usually develops as a female. A male develops from a zygote containing one X chromosome and one Y chromosome

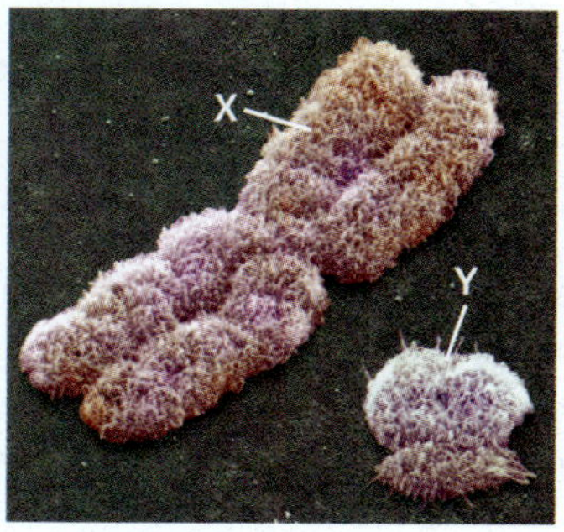

▲ Figure 15.5 Human sex chromosomes.

(Figure 15.6a). Short segments at either end of the Y chromosome are the only regions that are homologous with corresponding regions of the X. These homologous regions allow the X and Y chromosomes in males to pair and behave like homologous chromosomes during meiosis in the testes.

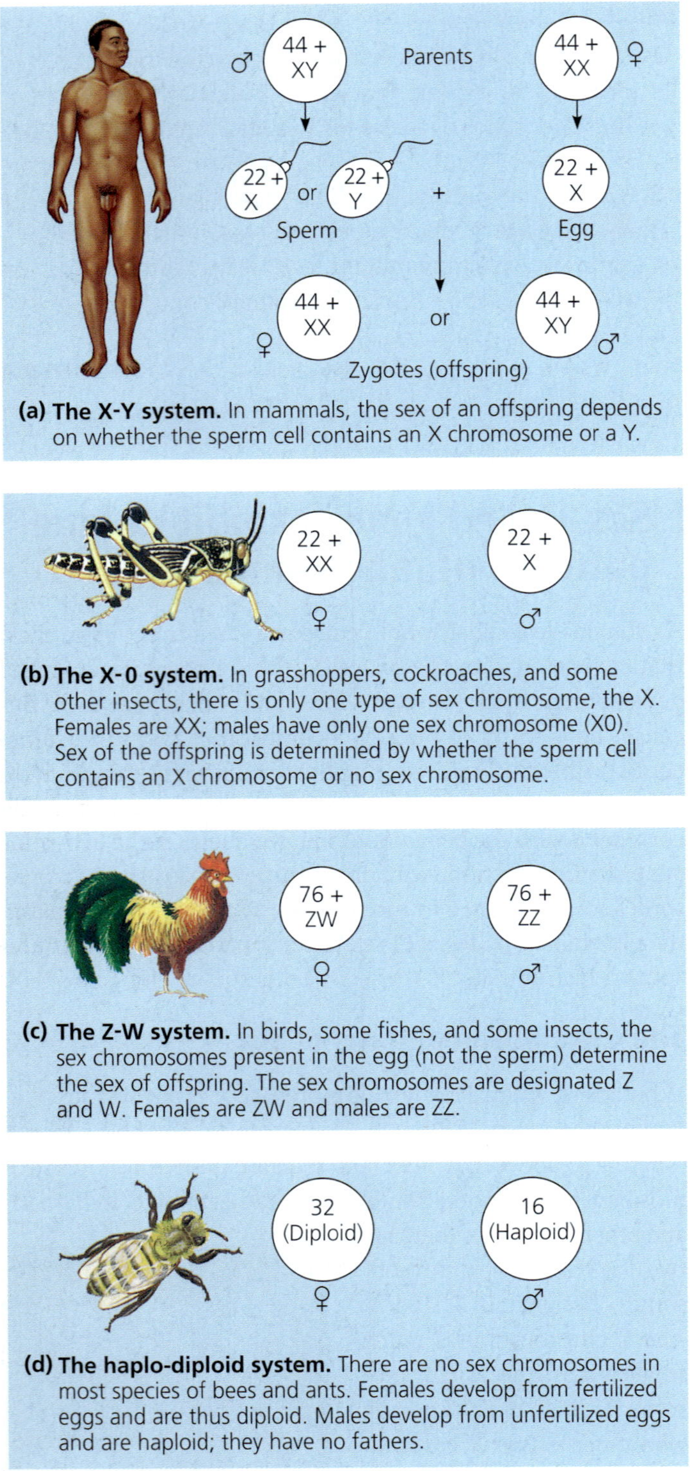

(a) The X-Y system. In mammals, the sex of an offspring depends on whether the sperm cell contains an X chromosome or a Y.

(b) The X-0 system. In grasshoppers, cockroaches, and some other insects, there is only one type of sex chromosome, the X. Females are XX; males have only one sex chromosome (X0). Sex of the offspring is determined by whether the sperm cell contains an X chromosome or no sex chromosome.

(c) The Z-W system. In birds, some fishes, and some insects, the sex chromosomes present in the egg (not the sperm) determine the sex of offspring. The sex chromosomes are designated Z and W. Females are ZW and males are ZZ.

(d) The haplo-diploid system. There are no sex chromosomes in most species of bees and ants. Females develop from fertilized eggs and are thus diploid. Males develop from unfertilized eggs and are haploid; they have no fathers.

▲ **Figure 15.6 Some chromosomal systems of sex determination.** Numerals indicate the number of autosomes in the species pictured. In *Drosophila*, males are XY, but sex depends on the ratio between the number of X chromosomes and the number of autosome sets, not simply on the presence of a Y chromosome.

In both testes and ovaries, the two sex chromosomes segregate during meiosis, and each gamete receives one. Each egg contains one X chromosome. In contrast, sperm fall into two categories: Half the sperm cells a male produces contain an X chromosome, and half contain a Y chromosome. We can trace the sex of each offspring to the moment of conception: If a sperm cell bearing an X chromosome happens to fertilize an egg, the zygote is XX, a female; if a sperm cell containing a Y chromosome fertilizes an egg, the zygote is XY, a male (see Figure 15.6a). Thus, sex determination is a matter of chance—a fifty-fifty chance. Besides the mammalian X-Y system, three other chromosomal systems for determining sex are shown in **Figure 15.6b–d**.

In humans, the anatomical signs of sex begin to emerge when the embryo is about 2 months old. Before then, the rudiments of the gonads are generic—they can develop into either testes or ovaries, depending on whether or not a Y chromosome is present. In 1990, a British research team identified a gene on the Y chromosome required for the development of testes. They named the gene *SRY*, for sex-determining region of Y. In the absence of *SRY*, the gonads develop into ovaries. The biochemical, physiological, and anatomical features that distinguish males and females are complex, and many genes are involved in their development. In fact, *SRY* codes for a protein that regulates other genes.

Researchers have sequenced the human Y chromosome and have identified 78 genes, which code for about 25 proteins (some genes are duplicates). About half of these genes are expressed only in the testis, and some are required for normal testicular functioning. In their absence, an XY individual is male but does not produce normal sperm.

Inheritance of Sex-Linked Genes

In addition to their role as carriers of genes that determine sex, the sex chromosomes, especially X chromosomes, have genes for many characters unrelated to sex. A gene located on either sex chromosome is called a **sex-linked gene**, although in humans the term has historically referred specifically to a gene on the X chromosome. Sex-linked genes in humans follow the same pattern of inheritance that Morgan observed for the eye-color locus he studied in *Drosophila* (see Figure 15.4). Fathers pass sex-linked alleles to all of their daughters but to none of their sons. In contrast, mothers can pass sex-linked alleles to both sons and daughters, as shown in **Figure 15.7**.

If a sex-linked trait is due to a recessive allele, a female will express the phenotype only if she is a homozygote. Because males have only one locus, the terms *homozygous* and *heterozygous* lack meaning for describing their sex-linked genes; the term *hemizygous* is used in such cases. Any male receiving the recessive allele from his mother will express the trait. For this reason, far more males than females have sex-linked recessive disorders. However, even though the chance of a female inheriting a double dose of the mutant allele is much less

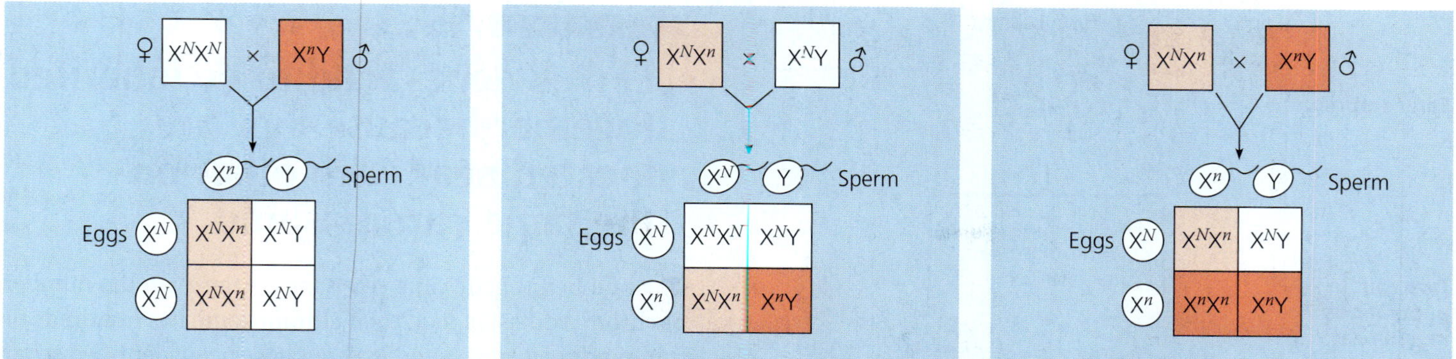

(a) A color-blind father will transmit the mutant allele to all daughters but to no sons. When the mother is a dominant homozygote, the daughters will have the normal phenotype but will be carriers of the mutation.

(b) If a carrier mates with a male who has normal color vision, there is a 50% chance that each daughter will be a carrier like her mother and a 50% chance that each son will have the disorder.

(c) If a carrier mates with a color-blind male, there is a 50% chance that each child born to them will have the disorder, regardless of sex. Daughters who have normal color vision will be carriers, whereas males who have normal color vision will be free of the recessive allele.

▲ **Figure 15.7 The transmission of sex-linked recessive traits.** In this diagram, color blindness is used as an example. The superscript *N* represents the dominant allele for normal color vision carried on the X chromosome, and the superscript *n* represents the recessive allele, which has a mutation causing color blindness. White boxes indicate unaffected individuals, light orange boxes indicate carriers, and dark orange boxes indicate color-blind individuals.

? *If a color-blind woman married a man who had normal color vision, what would be the probable phenotypes of their children?*

than the probability of a male inheriting a single dose, there *are* females with sex-linked disorders. For instance, color blindness is a mild disorder inherited as a sex-linked trait. A color-blind daughter may be born to a color-blind father whose mate is a carrier (see Figure 15.7c). Because the sex-linked allele for color blindness is relatively rare, though, the probability that such a man and woman will mate is low.

A number of human sex-linked disorders are much more serious than color blindness. An example is **Duchenne muscular dystrophy**, which affects about one out of every 3,500 males born in the United States. The disease is characterized by a progressive weakening of the muscles and loss of coordination. Affected individuals rarely live past their early 20s. Researchers have traced the disorder to the absence of a key muscle protein called dystrophin and have mapped the gene for this protein to a specific locus on the X chromosome.

Hemophilia is a sex-linked recessive disorder defined by the absence of one or more of the proteins required for blood clotting. When a person with hemophilia is injured, bleeding is prolonged because a firm clot is slow to form. Small cuts in the skin are usually not a problem, but bleeding in the muscles or joints can be painful and can lead to serious damage. Today, people with hemophilia are treated as needed with intravenous injections of the missing protein.

X Inactivation in Female Mammals

Since female mammals, including humans, inherit two X chromosomes, you may wonder whether females make twice as much of the proteins encoded by genes on the X chromosome, compared to the amounts in males. In fact, one X chromosome in each cell in females becomes almost completely inactivated during embryonic development. As a result, the cells of females and males have the same effective dose (one copy) of these genes. The inactive X in each cell of a female condenses into a compact object called a **Barr body**, which lies along the inside of the nuclear envelope. Most of the genes of the X chromosome that forms the Barr body are not expressed. In the ovaries, Barr-body chromosomes are reactivated in the cells that give rise to eggs, so every female gamete has an active X.

British geneticist Mary Lyon demonstrated that selection of which X chromosome will form the Barr body occurs randomly and independently in each embryonic cell present at the time of X inactivation. As a consequence, females consist of a *mosaic* of two types of cells: those with the active X derived from the father and those with the active X derived from the mother. After an X chromosome is inactivated in a particular cell, all mitotic descendants of that cell have the same inactive X. Thus, if a female is heterozygous for a sex-linked trait, about half her cells will express one allele, while the others will express the alternate allele. **Figure 15.8**, on the next page, shows how this mosaicism results in the mottled coloration of a tortoiseshell cat. In humans, mosaicism can be observed in a recessive X-linked mutation that prevents the development of sweat glands. A woman who is heterozygous for this trait has patches of normal skin and patches of skin lacking sweat glands.

Inactivation of an X chromosome involves modification of the DNA, including attachment of methyl groups ($—CH_3$) to one of the nitrogenous bases of DNA nucleotides. (The regulatory role of DNA methylation is discussed further in Chapter 18.) Researchers also have discovered an X chromosome gene called *XIST* (for X-inactive specific transcript) that is active

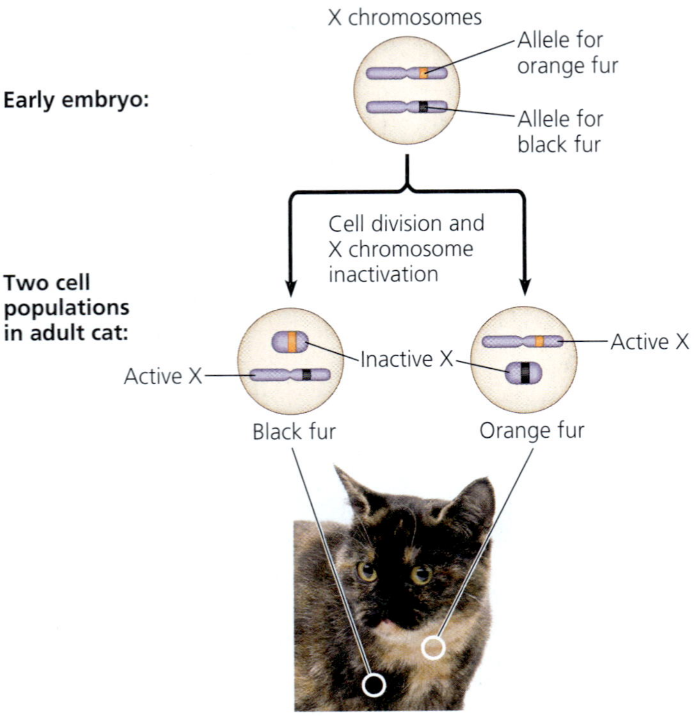

▲ **Figure 15.8 X inactivation and the tortoiseshell cat.** The tortoiseshell gene is on the X chromosome, and the tortoiseshell phenotype requires the presence of two different alleles, one for orange fur and one for black fur. Normally, only females can have both alleles, because only they have two X chromosomes. If a female is heterozygous for the tortoiseshell gene, she is tortoiseshell. Orange patches are formed by populations of cells in which the X chromosome with the orange allele is active; black patches have cells in which the X chromosome with the black allele is active. ("Calico" cats also have white areas, which are determined by yet another gene.)

only on the Barr-body chromosome. Multiple copies of the RNA product of this gene apparently attach to the X chromosome on which they are made, eventually almost covering it. Interaction of this RNA with the chromosome seems to initiate X inactivation, and regulation of this process by other genes is currently an active area of research.

CONCEPT CHECK 15.2

1. A white-eyed female *Drosophila* is mated with a red-eyed (wild-type) male, the reciprocal cross of the one shown in Figure 15.4. What phenotypes and genotypes do you predict for the offspring?
2. Neither Tim nor Rhoda has Duchenne muscular dystrophy, but their firstborn son does have it. What is the probability that a second child of this couple will have the disease? What is the probability if the second child is a boy? A girl?
3. **WHAT IF?** During early embryonic development of female carriers for color blindness, the normal allele is inactivated by chance in about half the cells. Why, then, aren't 50% of female carriers color-blind?

For suggested answers, see Appendix A.

CONCEPT **15.3**

Linked genes tend to be inherited together because they are located near each other on the same chromosome

The number of genes in a cell is far greater than the number of chromosomes; in fact, each chromosome has hundreds or thousands of genes. (Only 78 genes have been identified so far on the Y chromosome, but more may be found as methods for analyzing genetic sequences become more refined.) Genes located on the same chromosome that tend to be inherited together in genetic crosses are said to be **linked genes.** (Note the distinction between the terms *sex-linked gene*, referring to a single gene on a sex chromosome, and *linked genes*, referring to two or more genes on the same chromosome that tend to be inherited together.) When geneticists follow linked genes in breeding experiments, the results deviate from those expected from Mendel's law of independent assortment.

How Linkage Affects Inheritance

To see how linkage between genes affects the inheritance of two different characters, let's examine another of Morgan's *Drosophila* experiments. In this case, the characters are body color and wing size, each with two different phenotypes. Wild-type flies have gray bodies and normal-sized wings. In addition to these flies, Morgan had doubly mutant flies with black bodies and wings much smaller than normal, called vestigial wings. The mutant alleles are recessive to the wild-type alleles, and neither gene is on a sex chromosome. In his investigation of these two genes, Morgan carried out the crosses shown in **Figure 15.9**.

In these crosses, Morgan observed a much higher proportion of parental phenotypes than would be expected if the two genes assorted independently. Based on these results, he concluded that body color and wing size are usually inherited together in specific combinations (the parental combinations) because the genes for these characters are on the same chromosome:

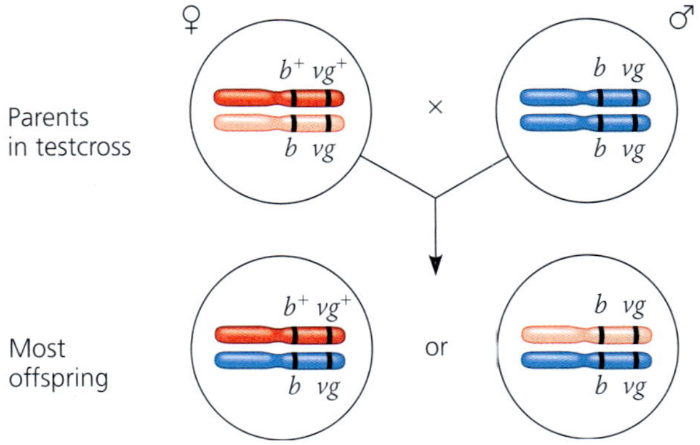

▼ Figure 15.9 Inquiry

How does linkage between two genes affect inheritance of characters?

EXPERIMENT Morgan wanted to know whether the genes for body color and wing size were on the same chromosome, and if so, how this affected their inheritance. The alleles for body color are b^+ (gray) and b (black), and those for wing size are vg^+ (normal) and vg (vestigial).

Morgan first mated true-breeding wild-type flies with black, vestigial-winged flies to produce heterozygous F_1 dihybrids ($b^+ \, b \, vg^+ \, vg$), all of which are wild-type in appearance.

He then mated wild-type F_1 dihybrid females with black, vestigial-winged males.

The male's sperm contributes only recessive alleles, so the phenotype of the offspring reflects the genotype of the female's eggs.

Note: Although only females (with pointed abdomen) are shown, half the offspring in each class would be male (with rounded abdomen).

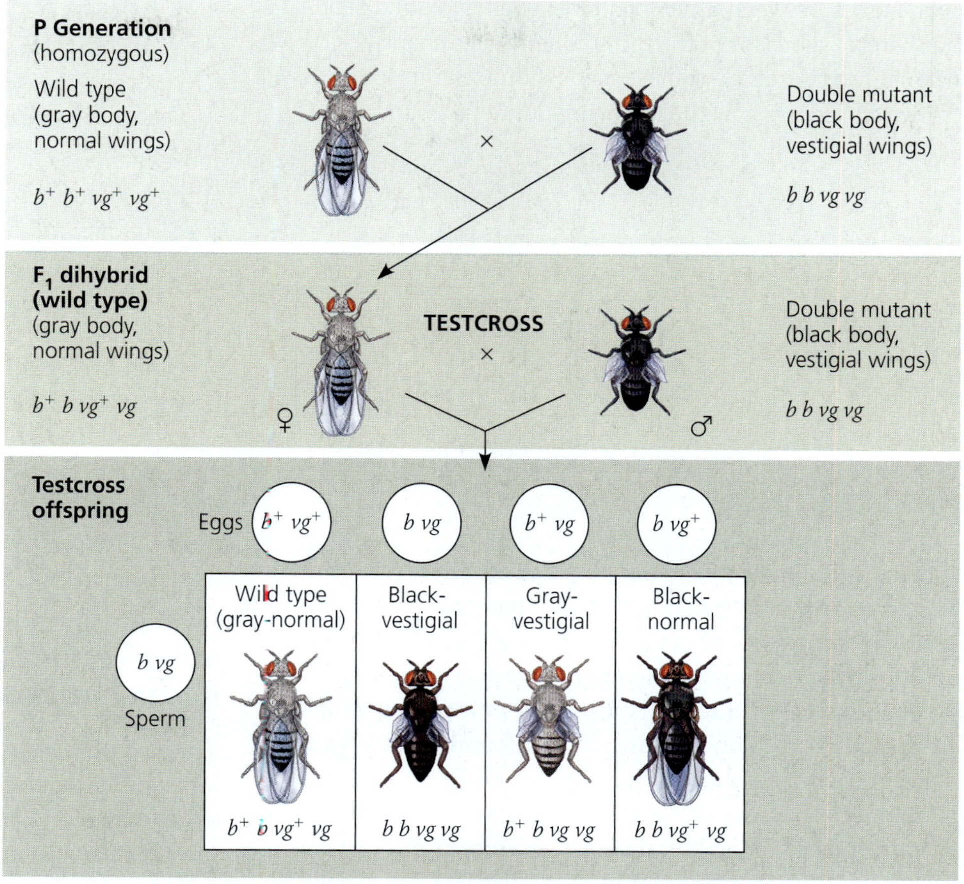

PREDICTED RATIOS

If genes are located on different chromosomes:	1	:	1	:	1	:	1
If genes are located on the same chromosome *and* parental alleles are always inherited together:	1	:	1	:	0	:	0
RESULTS	965	:	944	:	206	:	185

CONCLUSION Since most offspring had a parental phenotype, Morgan concluded that the genes for body color and wing size are located on the same chromosome. However, the production of a relatively small number of offspring with nonparental phenotypes indicated that some mechanism occasionally breaks the linkage between specific alleles of genes on the same chromosome.

SOURCE T. H. Morgan, The explanation of a new sex ratio in *Drosophila*, *Science* 36:718–720 (1912).

WHAT IF? If the parental flies had been true-breeding for gray body with vestigial wings and black body with normal wings, which phenotypic class(es) would be largest among the testcross offspring?

However, as Figure 15.9 shows, both of the nonparental phenotypes were also produced in Morgan's experiments, suggesting that the body-color and wing-size genes are only partially linked genetically. To understand this conclusion, we need to further explore **genetic recombination**, the production of offspring with combinations of traits that differ from those found in either parent.

Genetic Recombination and Linkage

In Chapter 13, you learned that meiosis and random fertilization generate genetic variation among offspring of sexually reproducing organisms. Here we will examine the chromosomal basis of recombination in relation to the genetic findings of Mendel and Morgan.

Recombination of Unlinked Genes: Independent Assortment of Chromosomes

Mendel learned from crosses in which he followed two characters that some offspring have combinations of traits that do not match those of either parent. For example, we can represent the cross between a pea plant with yellow-round seeds that is heterozygous for both seed color and seed shape (*YyRr*) and a plant with green-wrinkled seeds (homozygous for both recessive alleles, *yyrr*) by the following Punnett square

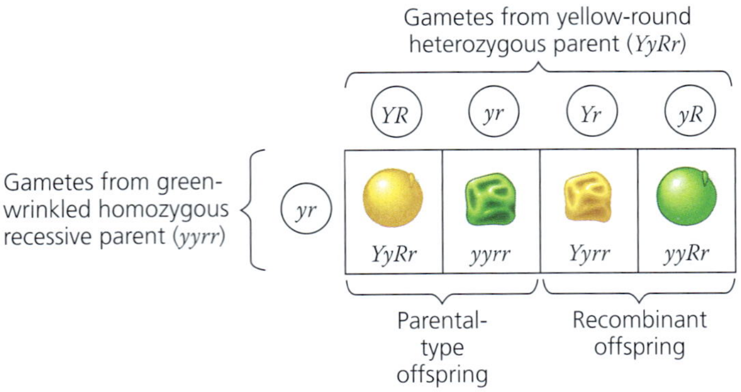

Notice in this Punnett square that one-half of the offspring are expected to inherit a phenotype that matches one of the parental phenotypes. These offspring are called **parental types**. But two nonparental phenotypes are also found among the offspring. Because these offspring have new combinations of seed shape and color, they are called **recombinant types**, or **recombinants** for short. When 50% of all offspring are recombinants, as in this example, geneticists say that there is a 50% frequency of recombination. The predicted phenotypic ratios among the offspring are similar to what Mendel actually found in *YyRr* × *yyrr* crosses (a type of testcross).

A 50% frequency of recombination in such testcrosses is observed for any two genes that are located on different chromosomes and are thus unlinked. The physical basis of recombination between unlinked genes is the random orientation of homologous chromosomes at metaphase I of meiosis, which leads to the independent assortment of the two unlinked genes (see Figure 13.11 and the question in the Figure 15.2 legend).

Recombination of Linked Genes: Crossing Over

Now let's return to Morgan's fly room to see how we can explain the results of the *Drosophila* testcross illustrated in Figure 15.9. Recall that most of the offspring from the testcross for body color and wing size had parental phenotypes. That suggested that the two genes were on the same chromosome, since the occurrence of parental types with a frequency greater than 50% indicates that the genes are linked. About 17% of offspring, however, were recombinants.

Faced with these results, Morgan proposed that some process must occasionally break the physical connection between specific alleles of genes on the same chromosome. Subsequent experiments demonstrated that this process, now called **crossing over**, accounts for the recombination of linked genes. In crossing over, which occurs while replicated homologous chromosomes are paired during prophase of meiosis I, a set of proteins orchestrates an exchange of corresponding segments of one maternal and one paternal chromatid (see Figure 13.12). In effect, end portions of two nonsister chromatids trade places each time a crossover occurs.

The recombinant chromosomes resulting from crossing over may bring alleles together in new combinations, and the subsequent events of meiosis distribute the recombinant chromosomes to gametes. **Figure 15.10** shows how crossing over in a dihybrid female fly resulted in recombinant eggs and ultimately recombinant offspring in Morgan's testcross. Most of the eggs had a chromosome with either the $b^+\ vg^+$ or *b vg* parental genotype for body color and wing size, but some eggs had a recombinant chromosome ($b^+\ vg$ or $b\ vg^+$). Fertilization of these various classes of eggs by homozygous recessive sperm (*b vg*) produced an offspring population in which 17% exhibited a nonparental, recombinant phenotype. As we discuss next, the percentage of recombinant offspring, the *recombination frequency*, is related to the distance between linked genes.

Mapping the Distance Between Genes Using Recombination Data: *Scientific Inquiry*

The discovery of linked genes and recombination due to crossing over led one of Morgan's students, Alfred H. Sturtevant, to a method for constructing a **genetic map**, an ordered list of the genetic loci along a particular chromosome.

Sturtevant hypothesized that recombination frequencies calculated from experiments like the one in Figures 15.9 and 15.10 depend on the distances between genes on a chromosome. He assumed that crossing over is a random event, with the chance of crossing over approximately equal at all points along a chromosome. Based on these assumptions, Sturtevant predicted that *the farther apart two genes are, the higher the probability that a crossover will occur between them and therefore the higher the recombination frequency.* His reasoning was simple: The greater the distance between two genes, the more points there are between them where crossing over can occur. Using recombination data from various fruit fly crosses, Sturtevant proceeded to assign relative positions to genes on the same chromosomes—that is, to *map* genes.

A genetic map based on recombination frequencies is called a **linkage map**. **Figure 15.11**, on page 296, shows Sturtevant's linkage map of three genes: the body-color (*b*) and wing-size (*vg*) genes depicted in Figure 15.10 and a third gene, called cinnabar (*cn*). Cinnabar is one of many *Drosophila* genes affecting eye color. Cinnabar eyes, a mutant phenotype, are a brighter red than the wild-type color. The recombination frequency between *cn* and *b* is 9%; that between *cn* and *vg*,

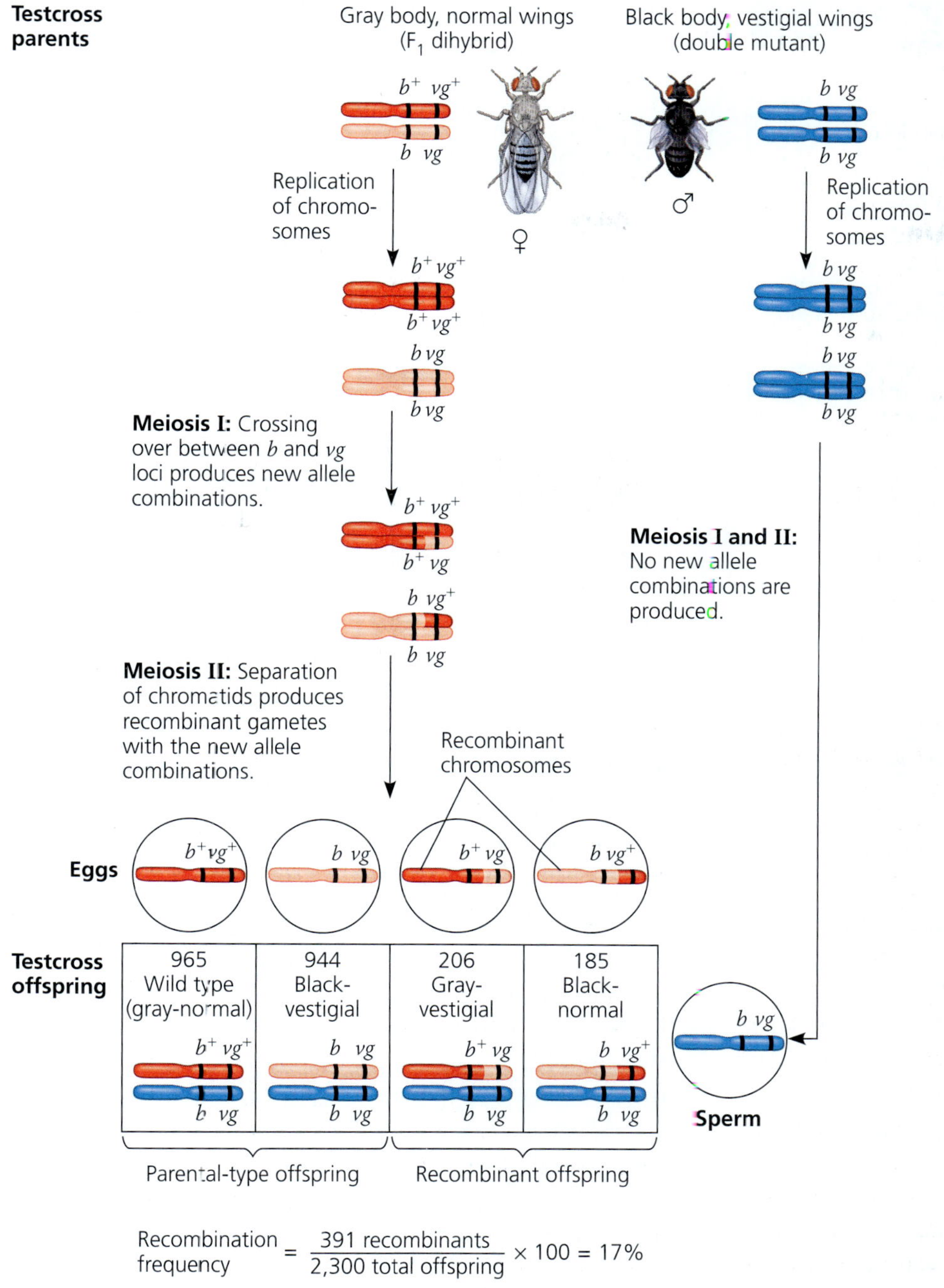

Figure 15.10 Chromosomal basis for recombination of linked genes. In these diagrams re-creating the testcross in Figure 15.9, we track chromosomes as well as genes. The maternal chromosomes are color-coded red and pink to distinguish one homolog from the other before any meiotic crossing over has taken place. Because crossing over between the *b* and *vg* loci occurs in some, but not all, egg-producing cells, more eggs with parental-type chromosomes than with recombinant ones are produced in the mating females. Fertilization of the eggs by sperm of genotype *b vg* gives rise to some recombinant offspring. The recombination frequency is the percentage of recombinant flies in the total pool of offspring.

DRAW IT *Suppose, as in the question at the bottom of Figure 15.9, that the parental flies were true-breeding for gray body with vestigial wings and black body with normal wings. Draw the chromosomes in each of the four possible kinds of eggs from an F_1 female, and label each chromosome as "parental" or "recombinant."*

9.5%; and that between *b* and *vg*, 17%. In other words, crossovers between *cn* and *b* and between *cn* and *vg* are about half as frequent as crossovers between *b* and *vg*. Only a map that locates *cn* about midway between *b* and *vg* is consistent with these data, as you can prove to yourself by drawing alternative maps. Sturtevant expressed the distances between genes in **map units**, defining one map unit as equivalent to a 1% recombination frequency. Today, map units are often called *centimorgans* in honor of Morgan.

In practice, the interpretation of recombination data is more complicated than this example suggests. For example, some genes on a chromosome are so far from each other that a crossover between them is virtually certain. The observed frequency of recombination in crosses involving two such genes can have a maximum value of 50%, a result indistinguishable from that for genes on different chromosomes. In this case, the physical connection between genes on the same chromosome is not reflected in the results of genetic crosses. Despite being on the same chromosome and thus being *physically linked*, the genes are *genetically unlinked*; alleles of such genes assort independently, as if they were on different chromosomes. In fact, several of the genes for pea characters that Mendel studied are now known to be on the same chromosome, but the distance between them is so great that linkage is not observed in genetic crosses. Consequently, they behaved as if they were on different chromosomes in Mendel's experiments. Genes located far

▼ **Figure 15.11 Research Method**

Constructing a Linkage Map

APPLICATION A linkage map shows the relative locations of genes along a chromosome.

TECHNIQUE A linkage map is based on the assumption that the probability of a crossover between two genetic loci is proportional to the distance separating the loci. The recombination frequencies used to construct a linkage map for a particular chromosome are obtained from experimental crosses, such as the cross depicted in Figures 15.9 and 15.10. The distances between genes are expressed as map units (centimorgans), with one map unit equivalent to a 1% recombination frequency. Genes are arranged on the chromosome in the order that best fits the data.

RESULTS In this example, the observed recombination frequencies between three *Drosophila* gene pairs (*b–cn* 9%, *cn–vg* 9.5%, and *b–vg* 17%) best fit a linear order in which *cn* is positioned about halfway between the other two genes:

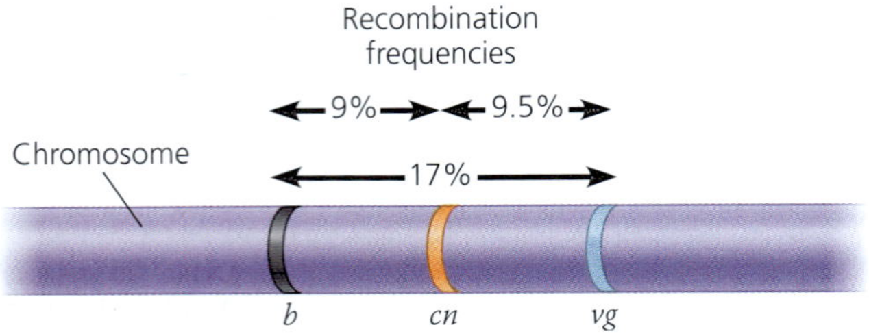

The *b–vg* recombination frequency (17%) is slightly less than the sum of the *b–cn* and *cn–vg* frequencies (9 + 9.5 = 18.5%) because of the few times that a crossover occurs between *b* and *cn* and an additional crossover occurs between *cn* and *vg*. The second crossover would "cancel out" the first, reducing the observed *b–vg* recombination frequency while contributing to the frequency between each of the closer pairs of genes. The value of 18.5% (18.5 map units) is closer to the actual distance between the genes, so a geneticist would add the smaller distances in constructing a map.

apart on a chromosome are mapped by adding the recombination frequencies from crosses involving a set of closer pairs of genes lying between the two distant genes.

Using recombination data, Sturtevant and his colleagues were able to map numerous *Drosophila* genes in linear arrays. They found that the genes clustered into four groups of linked genes. Because light microscopy had revealed four pairs of chromosomes in *Drosophila* cells, the linkage map thus provided additional evidence that genes are located on chromosomes. Each chromosome has a linear array of specific genes, each with its own locus **(Figure 15.12)**.

Because a linkage map is based only on recombination frequencies, it gives only an approximate picture of a chromosome. The frequency of crossing over is not actually uniform over the length of a chromosome, as Sturtevant assumed, and therefore map units do not correspond to actual physical distances (in nanometers, for instance). A linkage map does portray the order of genes along a chromosome, but it does not accurately portray the precise locations of those genes. Other methods enable geneticists to construct **cytogenetic maps** of chromosomes, which

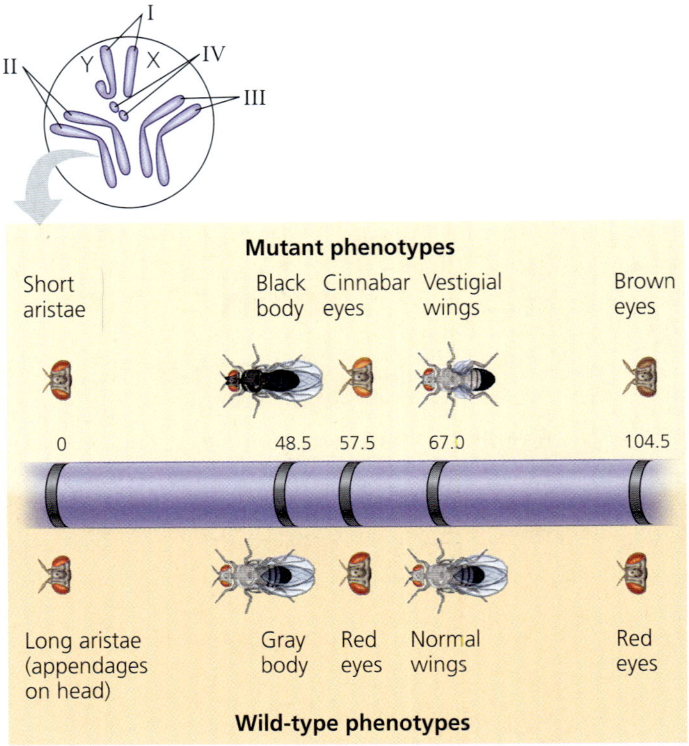

▲ **Figure 15.12 A partial genetic (linkage) map of a *Drosophila* chromosome.** This simplified map shows just a few of the genes that have been mapped on *Drosophila* chromosome II. The number at each gene locus indicates the number of map units between that locus and the locus for arista length (left). Notice that more than one gene can affect a given phenotypic characteristic, such as eye color. Also, note that in contrast to the homologous autosomes (II–IV), the X and Y sex chromosomes (I) have distinct shapes.

locate genes with respect to chromosomal features, such as stained bands, that can be seen in the microscope. The ultimate maps, which we will discuss in Chapter 21, show the physical distances between gene loci in DNA nucleotides. Comparing a linkage map with such a physical map or with a cytogenetic map of the same chromosome, we find that the linear order of genes is identical in all the maps, but the spacing between genes is not.

CONCEPT CHECK 15.3

1. When two genes are located on the same chromosome, what is the physical basis for the production of recombinant offspring in a testcross between a dihybrid parent and a double-mutant (recessive) parent?
2. For each type of offspring of the testcross in Figure 15.9, explain the relationship between its phenotype and the alleles contributed by the female parent.
3. **WHAT IF?** Genes *A*, *B*, and *C* are located on the same chromosome. Testcrosses show that the recombination frequency between *A* and *B* is 28% and between *A* and *C* is 12%. Can you determine the linear order of these genes? Explain.

For suggested answers, see Appendix A.

CONCEPT 15.4

Alterations of chromosome number or structure cause some genetic disorders

As you have learned so far in this chapter, the phenotype of an organism can be affected by small-scale changes involving individual genes. Random mutations are the source of all new alleles, which can lead to new phenotypic traits.

Large-scale chromosomal changes can also affect an organism's phenotype. Physical and chemical disturbances, as well as errors during meiosis, can damage chromosomes in major ways or alter their number in a cell. Large-scale chromosomal alterations often lead to spontaneous abortion (miscarriage) of a fetus, and individuals born with these types of genetic defects commonly exhibit various developmental disorders. In plants, such genetic defects may be tolerated to a greater extent than in animals.

Abnormal Chromosome Number

Ideally, the meiotic spindle distributes chromosomes to daughter cells without error. But there is an occasional mishap, called a **nondisjunction**, in which the members of a pair of homologous chromosomes do not move apart properly during meiosis I or sister chromatids fail to separate during meiosis II (**Figure 15.13**). In these cases, one gamete receives two of the same type of chromosome and another gamete receives no copy. The other chromosomes are usually distributed normally.

If either of the aberrant gametes unites with a normal one at fertilization, the zygote will also have an abnormal number of a chromosome, a condition known as **aneuploidy**. (Aneuploidy may involve more than one chromosome.) Fertilization involving a gamete that has no copy of a particular chromosome will lead to a missing chromosome in the zygote (so that the cell has $2n - 1$ chromosomes); the aneuploid zygote is said to be **monosomic** for that chromosome. If a chromosome is present in triplicate in the zygote (so that the cell has $2n + 1$ chromosomes), the aneuploid cell is **trisomic** for that chromosome. Mitosis will subsequently transmit the anomaly to all embryonic cells. If the organism survives, it usually has a set of traits caused by the abnormal dose of the genes associated with the extra or missing chromosome. Down syndrome is an example of trisomy in humans that will be discussed later. Nondisjunction can also occur during mitosis. If such an error takes place early in embryonic development, then the aneuploid condition is passed along by mitosis to a large number of cells and is likely to have a substantial effect on the organism.

Some organisms have more than two complete chromosome sets in all somatic cells. The general term for this chromosomal alteration is **polyploidy**; the specific terms ***triploidy*** ($3n$) and ***tetraploidy*** ($4n$) indicate three or four chromosomal sets, respectively. One way a triploid cell may arise is by the fertilization of an abnormal diploid egg produced by nondisjunction of all its chromosomes. Tetraploidy could result from the failure of a $2n$ zygote to divide after replicating its chromosomes. Subsequent normal mitotic divisions would then produce a $4n$ embryo.

Polyploidy is fairly common in the plant kingdom. As we will see in Chapter 24, the spontaneous origin of polyploid individuals plays an important role in the evolution of plants. Many of the plant species we eat are polyploid; for example, bananas are triploid and wheat is hexaploid ($6n$). In the animal kingdom, polyploid species are much less common, although they are known to occur among fishes and amphibians. Researchers in Chile were the first to identify a polyploid mammal, a rodent whose cells are tetraploid (**Figure 15.14**, on the next page); a closely related species also appears to be tetraploid. In general, polyploids are more nearly normal in appearance than aneuploids. One extra (or missing) chromosome apparently disrupts genetic balance more than does an entire extra set of chromosomes.

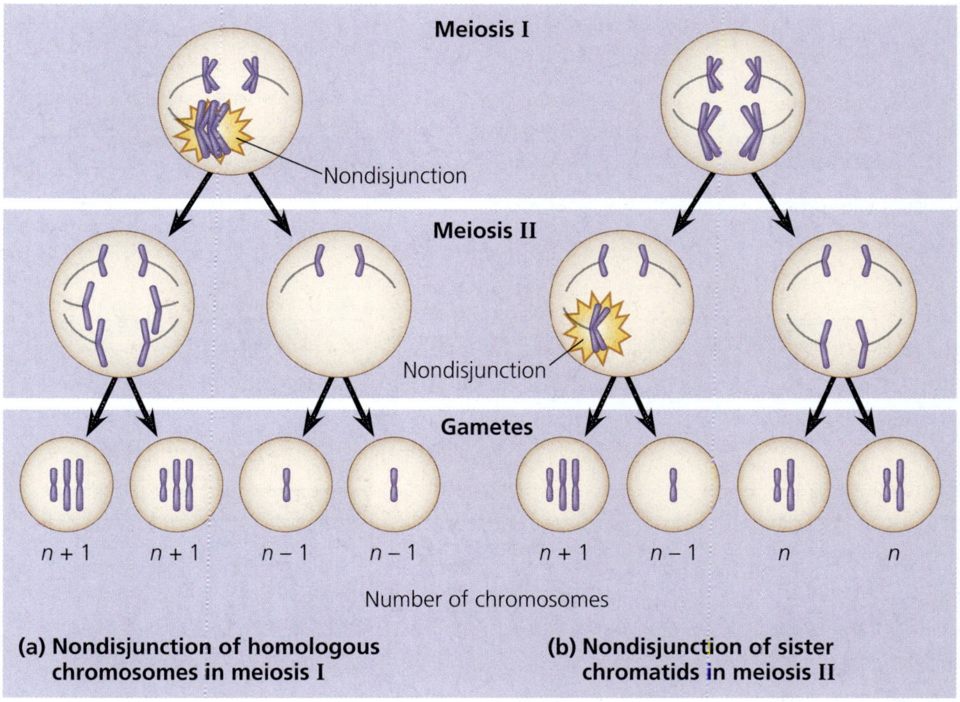

▲ **Figure 15.13 Meiotic nondisjunction.** Gametes with an abnormal chromosome number can arise by nondisjunction in either meiosis I or meiosis II.

▲ **Figure 15.14 A tetraploid mammal.** The somatic cells of this burrowing rodent, *Tympanoctomys barrerae*, have about twice as many chromosomes as those of closely related species. Interestingly, its sperm's head is unusually large, presumably a necessity for holding all that genetic material. Scientists think that this tetraploid species may have arisen when an ancestor doubled its chromosome number, presumably by errors in mitosis or meiosis within the animal's reproductive organs.

Alterations of Chromosome Structure

Errors in meiosis or damaging agents such as radiation can cause breakage of a chromosome, which can lead to four types of changes in chromosome structure **(Figure 15.15)**. A **deletion** occurs when a chromosomal fragment is lost. The affected chromosome is then missing certain genes. (If the centromere is deleted, the entire chromosome will be lost.) The "deleted" fragment may become attached as an extra segment to a sister chromatid, producing a **duplication**. Alternatively, a detached fragment could attach to a nonsister chromatid of a homologous chromosome. In that case, though, the "duplicated" segments might not be identical because the homologs could carry different alleles of certain genes. A chromosomal fragment may also reattach to the original chromosome but in the reverse orientation, producing an **inversion**. A fourth possible result of chromosomal breakage is for the fragment to join a nonhomologous chromosome, a rearrangement called a **translocation**.

Deletions and duplications are especially likely to occur during meiosis. In crossing over, nonsister chromatids sometimes exchange unequal-sized segments of DNA, so that one partner gives up more genes than it receives. The products of such a *nonreciprocal* crossover are one chromosome with a deletion and one chromosome with a duplication.

A diploid embryo that is homozygous for a large deletion (or has a single X chromosome with a large deletion, in a male) is usually missing a number of essential genes, a condition that is ordinarily lethal. Duplications and translocations also tend to be harmful. In reciprocal translocations, in which segments are exchanged between nonhomologous chromosomes, and in inversions, the balance of genes is not abnormal—all genes are present in their normal doses. Nevertheless, translocations and inversions can alter phenotype because a gene's expression can be influenced by its location among neighboring genes; such events sometimes have devastating effects.

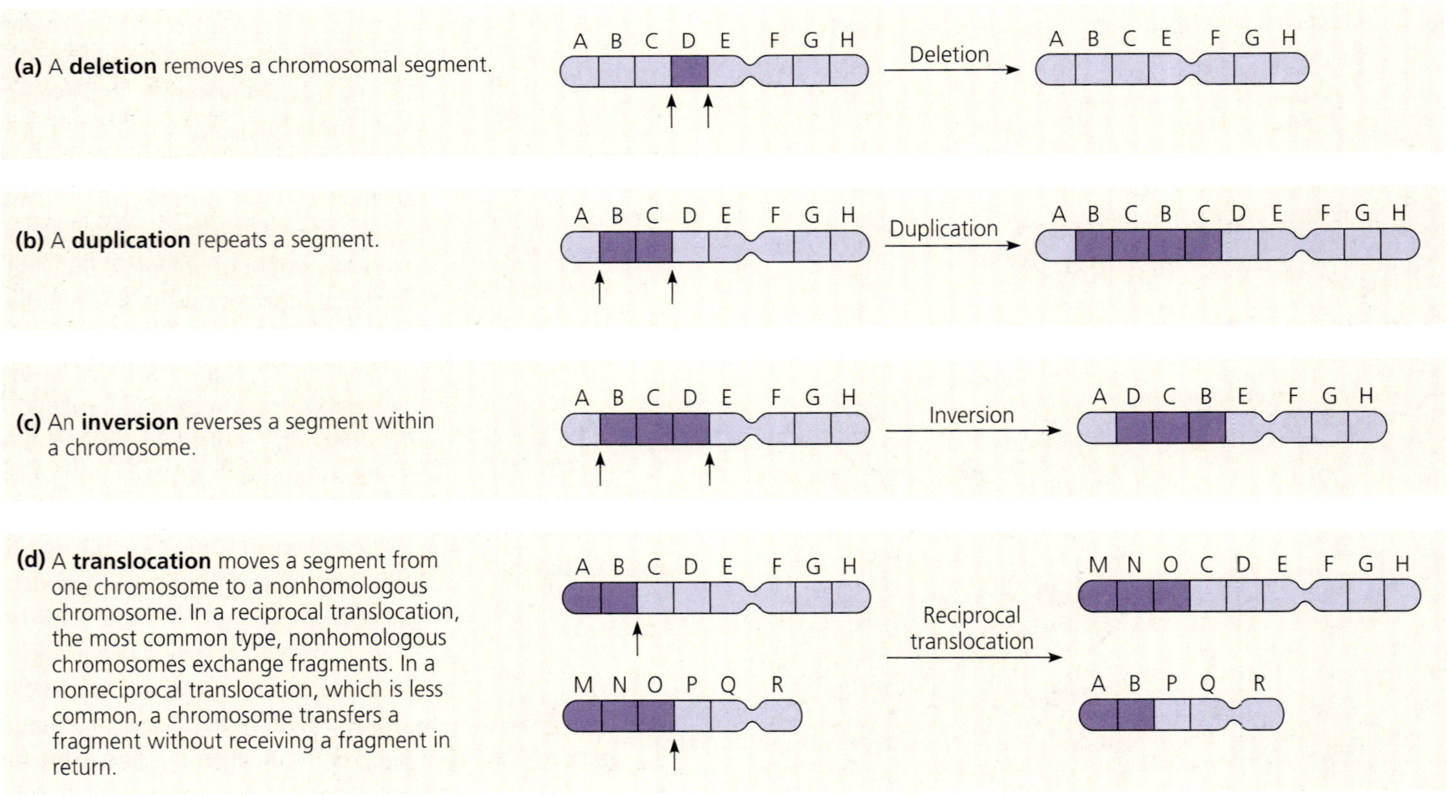

▲ **Figure 15.15 Alterations of chromosome structure.** Vertical arrows indicate breakage points. Dark purple highlights the chromosomal parts affected by the rearrangements.

Human Disorders Due to Chromosomal Alterations

Alterations of chromosome number and structure are associated with a number of serious human disorders. As described earlier, nondisjunction in meiosis results in aneuploidy in gametes and any resulting zygotes. Although the frequency of aneuploid zygotes may be quite high in humans, most of these chromosomal alterations are so disastrous to development that the embryos are spontaneously aborted long before birth. However, some types of aneuploidy appear to upset the genetic balance less than others, with the result that individuals with certain aneuploid conditions can survive to birth and beyond. These individuals have a set of traits—a *syndrome*—characteristic of the type of aneuploidy. Genetic disorders caused by aneuploidy can be diagnosed before birth by fetal testing (see Figure 14.18).

Down Syndrome (Trisomy 21)

One aneuploid condition, **Down syndrome**, affects approximately one out of every 700 children born in the United States **(Figure 15.16)**. Down syndrome is usually the result of an extra chromosome 21, so that each body cell has a total of 47 chromosomes. Because the cells are trisomic for chromosome 21, Down syndrome is often called *trisomy 21*. Down syndrome includes characteristic facial features, short stature, heart defects, susceptibility to respiratory infection, and mental retardation. Furthermore, individuals with Down syndrome are prone to developing leukemia and Alzheimer's disease. Although people with Down syndrome, on average, have a life span shorter than normal, some live to middle age or beyond. Most are sexually underdeveloped and sterile.

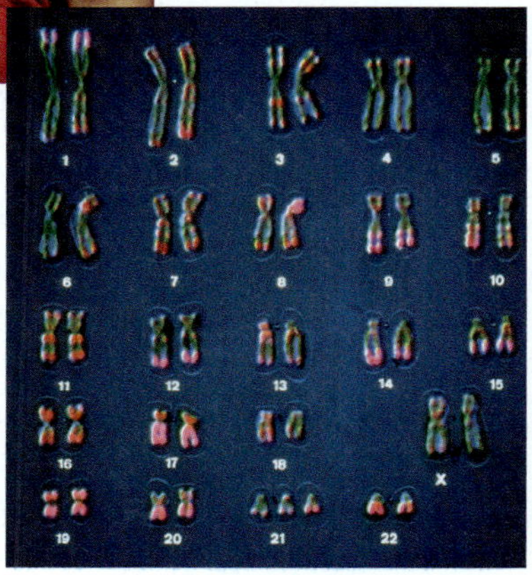

▲ **Figure 15.16 Down syndrome.** The child exhibits the facial features characteristic of Down syndrome. The karyotype shows trisomy 21, the most common cause of this disorder.

The frequency of Down syndrome increases with the age of the mother. While the disorder occurs in just 0.04% of children born to women under age 30, the risk climbs to 0.92% for mothers at age 40 and is even higher for older mothers. The correlation of Down syndrome with maternal age has not yet been explained. Most cases result from nondisjunction during meiosis I, and some research points to an age-dependent abnormality in a meiosis checkpoint that normally delays anaphase until all the kinetochores are attached to the spindle (like the M phase checkpoint of the mitotic cell cycle; see Chapter 12). Trisomies of some other chromosomes also increase in incidence with maternal age, although infants with other autosomal trisomies rarely survive for long. Prenatal screening for trisomies in the embryo is now offered to all pregnant women since the possible benefits outweigh any risks.

Aneuploidy of Sex Chromosomes

Nondisjunction of sex chromosomes produces a variety of aneuploid conditions. Most of these conditions appear to upset genetic balance less than aneuploid conditions involving autosomes. This may be because the Y chromosome carries relatively few genes and because extra copies of the X chromosome become inactivated as Barr bodies in somatic cells.

An extra X chromosome in a male, producing XXY, occurs approximately once in every 2,000 live births. People with this disorder, called *Klinefelter syndrome*, have male sex organs, but the testes are abnormally small and the man is sterile. Even though the extra X is inactivated, some breast enlargement and other female body characteristics are common. Affected individuals may have subnormal intelligence. Males with an extra Y chromosome (XYY) do not exhibit any well-defined syndrome, but they tend to be somewhat taller than average.

Females with trisomy X (XXX), which occurs once in approximately 1,000 live births, are healthy and cannot be distinguished from XX females except by karyotype. Monosomy X, called *Turner syndrome*, occurs about once in every 5,000 births and is the only known viable monosomy in humans. Although these X0 individuals are phenotypically female, they are sterile because their sex organs do not mature. When provided with estrogen replacement therapy, girls with Turner syndrome do develop secondary sex characteristics. Most have normal intelligence.

Disorders Caused by Structurally Altered Chromosomes

Many deletions in human chromosomes, even in a heterozygous state, cause severe problems. One such syndrome, known as *cri du chat* ("cry of the cat"), results from a specific deletion in chromosome 5. A child born with this deletion is mentally retarded, has a small head with unusual facial features, and has a cry that sounds like the mewing of a distressed cat. Such individuals usually die in infancy or early childhood.

Chromosomal translocations have been implicated in certain cancers, including *chronic myelogenous leukemia* (*CML*). This disease occurs when a reciprocal translocation happens during mitosis of cells that will become white blood cells. In these cells, the exchange of a large portion of chromosome 22 with a small fragment from a tip of chromosome 9 produces a much shortened, easily recognized chromosome 22, called the *Philadelphia chromosome* **(Figure 15.17)**. We will discuss how such an exchange can cause cancer in Chapter 18.

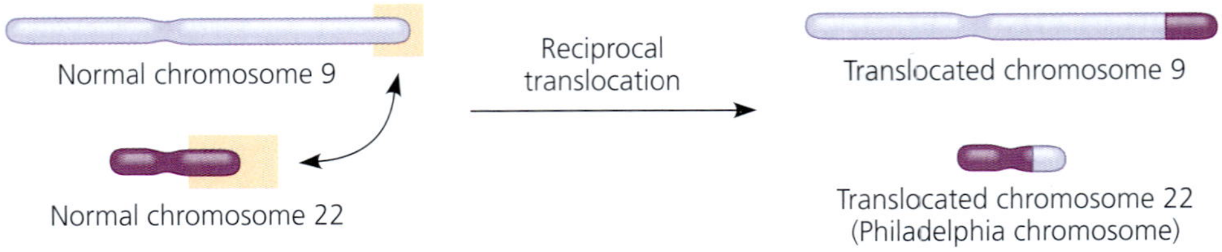

▲ **Figure 15.17 Translocation associated with chronic myelogenous leukemia (CML).** The cancerous cells in nearly all CML patients contain an abnormally short chromosome 22, the so-called Philadelphia chromosome, and an abnormally long chromosome 9. These altered chromosomes result from the translocation shown here, which presumably occurred in a single white blood cell precursor undergoing mitosis and was then passed along to all descendant cells.

CONCEPT CHECK 15.4

1. More common than completely polyploid animals are mosaic polyploids, animals that are diploid except for patches of polyploid cells. How might a mosaic tetraploid—an animal with some cells containing four sets of chromosomes—arise?
2. About 5% of individuals with Down syndrome have a chromosomal translocation in which a third copy of chromosome 21 is attached to chromosome 14. If this translocation occurred in a parent's gonad, how could it lead to Down syndrome in a child?
3. **WHAT IF?** The ABO blood type locus has been mapped on chromosome 9. A father who has type AB blood and a mother who has type O blood have a child with trisomy 9 and type A blood. Using this information, can you tell in which parent the nondisjunction occurred? Explain your answer.

For suggested answers, see Appendix A.

CONCEPT **15.5**

Some inheritance patterns are exceptions to the standard chromosome theory

In the previous section, you learned about deviations from the usual patterns of chromosomal inheritance due to abnormal events in meiosis and mitosis. We conclude this chapter by describing two normally occurring exceptions to Mendelian genetics, one involving genes located in the nucleus and the other involving genes located outside the nucleus. In both cases, the sex of the parent contributing an allele is a factor in the pattern of inheritance.

Genomic Imprinting

Throughout our discussions of Mendelian genetics and the chromosomal basis of inheritance, we have assumed that a given allele will have the same effect whether it was inherited from the mother or the father. This is probably a safe assumption most of the time. For example, when Mendel crossed purple-flowered pea plants with white-flowered pea plants, he observed the same results regardless of whether the purple-flowered parent supplied the eggs or the sperm. In recent years, however, geneticists have identified two to three dozen traits in mammals that depend on which parent passed along the alleles for those traits. Such variation in phenotype depending on whether an allele is inherited from the male or female parent is called **genomic imprinting**. (Note that this phenomenon is different from sex linkage; most imprinted genes are on autosomes.)

Genomic imprinting occurs during the formation of gametes and results in the silencing of one allele of certain genes. Because these genes are imprinted differently in sperm and eggs, a zygote expresses only one allele of an imprinted gene, either the allele inherited from the female parent or the allele inherited from the male parent. The imprints are transmitted to all the body cells during development, so either the maternal or paternal allele of a given imprinted gene is expressed in every cell of that organism. In each generation, the old imprints are "erased" in gamete-producing cells, and the chromosomes of the developing gametes are newly imprinted according to the sex of the individual forming the gametes. In a given species, the imprinted genes are always imprinted in the same way. For instance, a gene imprinted for maternal allele expression is always imprinted for maternal allele expression, generation after generation.

Consider, for example, the mouse gene for insulin-like growth factor 2 (*Igf2*), one of the first imprinted genes to be identified. Although this growth factor is required for normal prenatal growth, only the paternal allele is expressed **(Figure 15.18a)**. Evidence that the *Igf2* gene is imprinted initially came from crosses between wild-type mice and dwarf mice homozygous for a recessive mutation in the *Igf2* gene. The phenotypes of heterozygous offspring (which had one normal allele and one

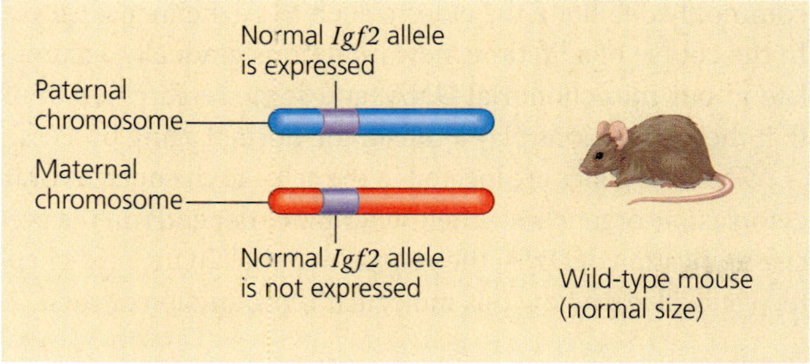

(a) Homozygote. A mouse homozygous for the wild-type *Igf2* allele is normal sized. Only the paternal allele of this gene is expressed.

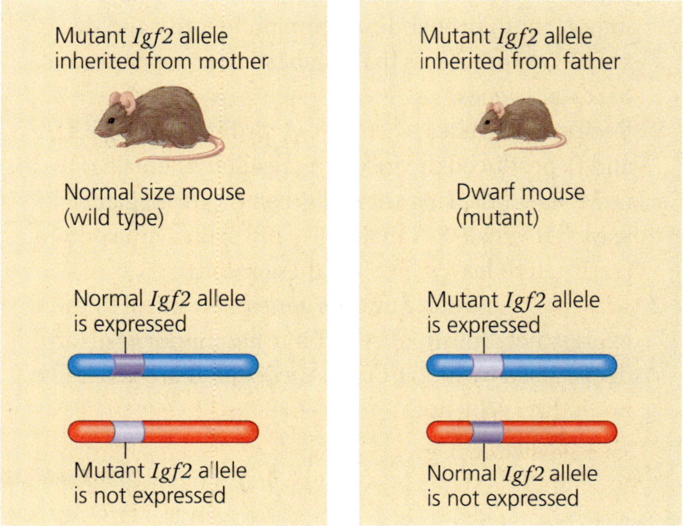

(b) Heterozygotes. Matings between wild-type mice and those homozygous for the recessive mutant *Igf2* allele produce heterozygous offspring. The mutant phenotype is seen only when the father contributed the mutant allele because the maternal allele is not expressed.

▲ **Figure 15.18 Genomic imprinting of the mouse *Igf2* gene.**

mutant) differed, depending on whether the mutant allele came from the mother or the father **(Figure 15.18b)**.

What exactly is a genomic imprint? In many cases, it seems to consist of methyl (—CH_3) groups that are added to cytosine nucleotides of one of the alleles. Such methylation may directly silence the allele, an effect consistent with evidence that heavily methylated genes are usually inactive (see Chapter 18). However, for a few genes, methylation has been shown to *activate* expression of the allele. This is the case for the *Igf2* gene: Methylation of certain cytosines on the paternal chromosome leads to expression of the paternal *Igf2* allele.

Genomic imprinting is thought to affect only a small fraction of the genes in mammalian genomes, but most of the known imprinted genes are critical for embryonic development. In experiments with mice, for example, embryos engineered to inherit both copies of certain chromosomes from the same parent usually die before birth, whether that parent is male or female. In 2004, however, scientists in Japan combined the genetic material from two eggs in a zygote while allowing expression of the *Igf2* gene from only one of the egg nuclei. The zygote developed into an apparently healthy mouse, shown in the photograph at the right. Apparently, normal development requires that embryonic cells have exactly one active copy—not zero, not two—of certain genes. The association of aberrant imprinting with abnormal development and certain cancers has stimulated numerous studies of how different genes are imprinted.

Inheritance of Organelle Genes

Although our focus in this chapter has been on the chromosomal basis of inheritance, we end with an important amendment: Not all of a eukaryotic cell's genes are located on nuclear chromosomes, or even in the nucleus. Some genes are located in organelles in the cytoplasm; because they are outside the nucleus, these genes are sometimes called *extranuclear genes* or *cytoplasmic genes.* Mitochondria, as well as chloroplasts and other plant plastids, contain small circular DNA molecules that carry a number of genes. These organelles reproduce themselves and transmit their genes to daughter organelles. Organelle genes are not distributed to offspring according to the same rules that direct the distribution of nuclear chromosomes during meiosis, so they do not display Mendelian inheritance.

The first hint that extranuclear genes exist came from studies by the German scientist Karl Correns on the inheritance of yellow or white patches on the leaves of an otherwise green plant. In 1909, he observed that the coloration of the offspring was determined only by the maternal parent (the source of eggs) and not by the paternal parent (the source of sperm). Subsequent research showed that such coloration patterns, or variegation, are due to mutations in plastid genes that control pigmentation **(Figure 15.19)**. In most plants, a zygote receives all its plastids from the cytoplasm of the egg and none from the sperm, which contributes little more than a haploid set of chromosomes. As the zygote develops, plastids containing

▶ **Figure 15.19 Variegated leaves from *Croton dioicus*.** Variegated (striped or spotted) leaves result from mutations in pigment genes located in plastids, which generally are inherited from the maternal parent.

wild-type or mutant pigment genes are distributed randomly to daughter cells. The pattern of leaf coloration exhibited by a plant depends on the ratio of wild-type to mutant plastids in its various tissues.

Similar maternal inheritance is also the rule for mitochondrial genes in most animals and plants, because almost all the mitochondria passed on to a zygote come from the cytoplasm of the egg. The products of most mitochondrial genes help make up the protein complexes of the electron transport chain and ATP synthase (see Chapter 9). Defects in one or more of these proteins, therefore, reduce the amount of ATP the cell can make and have been shown to cause a number of rare human disorders. Because the parts of the body most susceptible to energy deprivation are the nervous system and the muscles, most mitochondrial diseases primarily affect these systems. For example, *mitochondrial myopathy* causes weakness, intolerance of exercise, and muscle deterioration. Another mitochondrial disorder is *Leber's hereditary optic neuropathy*, which can produce sudden blindness in people as young as their 20s or 30s. The four mutations found thus far to cause this disorder affect oxidative phosphorylation during cellular respiration, a crucial function for the cell.

In addition to the rare diseases clearly caused by defects in mitochondrial DNA, mitochondrial mutations inherited from a person's mother may contribute to at least some cases of diabetes and heart disease, as well as to other disorders that commonly debilitate the elderly, such as Alzheimer's disease. In the course of a lifetime, new mutations gradually accumulate in our mitochondrial DNA, and some researchers think that these mutations play a role in the normal aging process.

Wherever genes are located in the cell—in the nucleus or in cytoplasmic organelles—their inheritance depends on the precise replication of DNA, the genetic material. In the next chapter, you will learn how this molecular reproduction occurs.

CONCEPT CHECK 15.5

1. Gene dosage, the number of active copies of a gene, is important to proper development. Identify and describe two processes that establish the proper dosage of certain genes.
2. Reciprocal crosses between two primrose varieties, A and B, produced the following results: A female × B male ⟶ offspring with all green (nonvariegated) leaves. B female × A male ⟶ offspring with spotted (variegated) leaves. Explain these results.
3. **WHAT IF?** Mitochondrial genes are critical to the energy metabolism of cells, but mitochondrial disorders caused by mutations in these genes are generally not lethal. Why not?

For suggested answers, see Appendix A.

Chapter 15 Review

MEDIA Go to the Study Area at **www.masteringbio.com** for BioFlix 3-D Animations, MP3 Tutors, Videos, Practice Tests, an eBook, and more.

SUMMARY OF KEY CONCEPTS

CONCEPT 15.1

Mendelian inheritance has its physical basis in the behavior of chromosomes (pp. 286–289)

- The behavior of chromosomes during meiosis accounts for Mendel's laws of segregation and independent assortment.
- **Morgan's Experimental Evidence: *Scientific Inquiry*** Morgan's discovery that transmission of the X chromosome in *Drosophila* correlates with inheritance of an eye-color trait was the first solid evidence indicating that a specific gene is associated with a specific chromosome.

MEDIA

MP3 Tutor Chromosomal Basis of Inheritance

CONCEPT 15.2

Sex-linked genes exhibit unique patterns of inheritance (pp. 289–292)

- **The Chromosomal Basis of Sex** Sex is an inherited phenotypic character usually determined by which sex chromosomes are present. Humans and other mammals have an X-Y system in which sex is determined by whether a Y chromosome is present. Other systems are found in birds, fishes, and insects.
- **Inheritance of Sex-Linked Genes** The sex chromosomes carry genes for some traits that are unrelated to sex characteristics. For instance, recessive alleles causing color blindness are carried on the X chromosome. Fathers transmit this and other sex-linked alleles to all daughters but to no sons. Any male who inherits such an allele from his mother will express the trait.
- **X Inactivation in Female Mammals** In mammalian females, one of the two X chromosomes in each cell is randomly inactivated during early embryonic development. If a female is heterozygous for a particular gene located on the X chromosome, she will be mosaic for that character, with about half her cells expressing the maternal allele and about half expressing the paternal allele.

MEDIA

Activity Sex-Linked Genes
Investigation What Can Fruit Flies Reveal About Inheritance?

CONCEPT 15.3

Linked genes tend to be inherited together because they are located near each other on the same chromosome (pp. 292–296)

- **How Linkage Affects Inheritance**

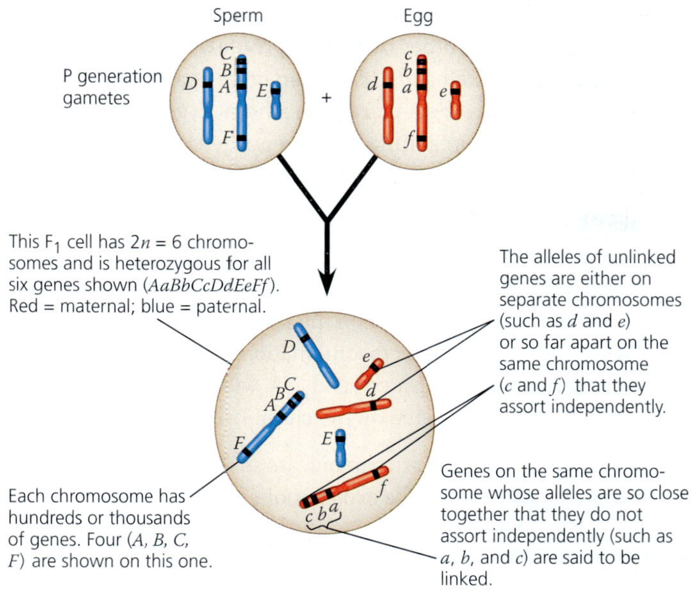

▶ **Genetic Recombination and Linkage** Recombinant offspring exhibit new combinations of traits inherited from two parents. Because of the independent assortment of chromosomes, unlinked genes exhibit a 50% frequency of recombination in the gametes. For linked genes, crossing over between nonsister chromatids during meiosis I accounts for the observed recombinants, always less than 50% of the total.

▶ **Mapping the Distance Between Genes Using Recombination Data: *Scientific Inquiry*** The order of genes on a chromosome and the relative distances between them can be deduced from recombination frequencies observed in genetic crosses. The farther apart genes are, the more likely their allele combinations will be recombined during crossing over.

MEDIA

Activity Linked Genes and Crossing Over
Biology Labs On-Line FlyLab
Biology Labs On-Line PedigreeLab

CONCEPT **15.4**

Alterations of chromosome number or structure cause some genetic disorders (pp. 297–300)

▶ **Abnormal Chromosome Number** Aneuploidy can result from nondisjunction during meiosis. When a normal gamete unites with one containing two copies or no copies of a particular chromosome, the resulting zygote and its descendant cells either have one extra copy of that chromosome (trisomy, $2n + 1$) or are missing a copy (monosomy, $2n - 1$). Polyploidy (more than two complete sets of chromosomes) can result from complete nondisjunction during gamete formation.

▶ **Alterations of Chromosome Structure** Chromosome breakage can result in deletions, inversions, duplications, and translocations.

▶ **Human Disorders Due to Chromosomal Alterations** Changes in the number of chromosomes per cell or in the structure of individual chromosomes can affect phenotype. Such alterations cause Down syndrome (usually due to trisomy of chromosome 21), certain cancers associated with chromosomal translocations, and various other human disorders.

MEDIA

Activity Polyploid Plants

CONCEPT **15.5**

Some inheritance patterns are exceptions to the standard chromosome theory (pp. 300–302)

▶ **Genomic Imprinting** In mammals, the phenotypic effects of certain genes depend on which allele is inherited from each parent. Imprints are formed during gamete production, with the result that one allele (either maternal or paternal) is not expressed in offspring.

▶ **Inheritance of Organelle Genes** The inheritance of traits controlled by the genes present in mitochondria and plastids depends solely on the maternal parent because the zygote's cytoplasm comes from the egg. Some diseases affecting the nervous and muscular systems are caused by defects in mitochondrial genes that prevent cells from making enough ATP.

TESTING YOUR KNOWLEDGE

GENETICS PROBLEMS

1. A man with hemophilia (a recessive, sex-linked condition) has a daughter of normal phenotype. She marries a man who is normal for the trait. What is the probability that a daughter of this mating will be a hemophiliac? That a son will be a hemophiliac? If the couple has four sons, what is the probability that all four will be born with hemophilia?

2. Pseudohypertrophic muscular dystrophy is an inherited disorder that causes gradual deterioration of the muscles. It is seen almost exclusively in boys born to apparently normal parents and usually results in death in the early teens. Is this disorder caused by a dominant or a recessive allele? Is its inheritance sex-linked or autosomal? How do you know? Explain why this disorder is almost never seen in girls.

3. Red-green color blindness is caused by a sex-linked recessive allele. A color-blind man marries a woman with normal vision whose father was color-blind. What is the probability that they will have a color-blind daughter? What is the probability that their first son will be color-blind? (Note the different wording in the two questions.)

4. A wild-type fruit fly (heterozygous for gray body color and normal wings) is mated with a black fly with vestigial wings. The offspring have the following phenotypic distribution: wild type, 778; black-vestigial, 785; black-normal, 158; gray-vestigial, 162. What is the recombination frequency between these genes for body color and wing size?

5. In another cross, a wild-type fruit fly (heterozygous for gray body color and red eyes) is mated with a black fruit fly with purple eyes. The offspring are as follows: wild type, 721; black-purple, 751; gray-purple, 49; black-red, 45. What is the recombination frequency between these genes for body color and eye color? Using information from problem 4, what fruit flies (genotypes and phenotypes) would you mate to determine the sequence of the body-color, wing-size, and eye-color genes on the chromosome?

6. **DRAW IT** A fruit fly that is true-breeding for gray body with vestigial wings ($b^+ b^+ vg\, vg$) is mated with one that is true-breeding for black body with normal wings ($b\, b\, vg^+ vg^+$).
   a. Draw the chromosomes for the P generation flies, using red for the gray fly and pink for the black one. Show the position of each allele.
   b. Draw the chromosomes and label the alleles of an F_1 fly.
   c. Suppose an F_1 female is testcrossed. Draw the chromosomes of the resulting offspring in a Punnett square.
   d. Knowing that the distance between these two genes is 17 map units, predict the phenotypic ratios of these offspring.

7. What pattern of inheritance would lead a geneticist to suspect that an inherited disorder of cell metabolism is due to a defective mitochondrial gene?

8. Women born with an extra X chromosome (XXX) are healthy and phenotypically indistinguishable from normal XX women. What is a likely explanation for this finding? How could you test this explanation?

9. Determine the sequence of genes along a chromosome based on the following recombination frequencies: *A–B*, 8 map units; *A–C*, 28 map units; *A–D*, 25 map units; *B–C*, 20 map units; *B–D*, 33 map units.

10. Assume that genes *A* and *B* are linked and are 50 map units apart. An animal heterozygous at both loci is crossed with one that is homozygous recessive at both loci. What percentage of the offspring will show phenotypes resulting from crossovers? If you did not know that genes *A* and *B* were linked, how would you interpret the results of this cross?

11. A space probe discovers a planet inhabited by creatures that reproduce with the same hereditary patterns seen in humans. Three phenotypic characters are height (T = tall, t = dwarf), head appendages (A = antennae, a = no antennae), and nose morphology (S = upturned snout, s = downturned snout). Since the creatures are not "intelligent," Earth scientists are able to do some controlled breeding experiments using various heterozygotes in testcrosses. For tall heterozygotes with antennae, the offspring are: tall-antennae, 46; dwarf-antennae, 7; dwarf-no antennae, 42; tall-no antennae, 5. For heterozygotes with antennae and an upturned snout, the offspring are: antennae-upturned snout, 47; antennae-downturned snout, 2; no antennae-downturned snout, 48; no antennae-upturned snout, 3. Calculate the recombination frequencies for both experiments.

12. Using the information from problem 11, scientists do a further testcross using a heterozygote for height and nose morphology. The offspring are: tall-upturned snout, 40; dwarf-upturned snout, 9; dwarf-downturned snout, 42; tall-downturned snout, 9. Calculate the recombination frequency from these data; then use your answer from problem 11 to determine the correct sequence of the three linked genes.

13. Two genes of a flower, one controlling blue (B) versus white (b) petals and the other controlling round (R) versus oval (r) stamens, are linked and are 10 map units apart. You cross a homozygous blue-oval plant with a homozygous white-round plant. The resulting F_1 progeny are crossed with homozygous white-oval plants, and 1,000 F_2 progeny are obtained. How many F_2 plants of each of the four phenotypes do you expect?

14. You design *Drosophila* crosses to provide recombination data for gene *a*, which is located on the chromosome shown in Figure 15.12. Gene *a* has recombination frequencies of 14% with the vestigial-wing locus and 26% with the brown-eye locus. Where is *a* located on the chromosome?

15. Banana plants, which are triploid, are seedless and therefore sterile. Propose a possible explanation.

For Genetics Problems answers, see Appendix A.

MEDIA Visit the Study Area at **www.masteringbio.com** for a Practice Test.

EVOLUTION CONNECTION

16. You have seen that crossing over, or recombination, is thought to be evolutionarily advantageous because it continually shuffles genetic alleles into novel combinations, allowing evolutionary processes to occur. Until recently, it was thought that the genes on the Y chromosome might degenerate because they lack homologous genes on the X chromosome with which to recombine. However, when the Y chromosome was sequenced, eight large regions were found to be internally homologous to each other, and quite a few of the 78 genes represent duplicates. (Y chromosome researcher David Page has called it a "hall of mirrors.") What might be a benefit of these regions?

SCIENTIFIC INQUIRY

17. Butterflies have an X-Y sex determination system that is different from that of flies or humans. Female butterflies may be either XY or XO, while butterflies with two or more X chromosomes are males. This photograph shows a tiger swallowtail *gynandromorph*, an individual that is half male (left side) and half female (right side). Given that the first division of the zygote divides the embryo into the future right and left halves of the butterfly, propose a hypothesis that explains how nondisjunction during the first mitosis might have produced this unusual-looking butterfly.

16 The Molecular Basis of Inheritance

▲ **Figure 16.1 How was the structure of DNA determined?**

KEY CONCEPTS

16.1 DNA is the genetic material
16.2 Many proteins work together in DNA replication and repair
16.3 A chromosome consists of a DNA molecule packed together with proteins

OVERVIEW

Life's Operating Instructions

In April 1953, James Watson and Francis Crick shook the scientific world with an elegant double-helical model for the structure of deoxyribonucleic acid, or DNA. **Figure 16.1** shows Watson (left) and Crick admiring their DNA model, which they built from tin and wire. Over the past 50 years, their model has evolved from a novel proposition to an icon of modern biology. DNA, the substance of inheritance, is the most celebrated molecule of our time. Mendel's heritable factors and Morgan's genes on chromosomes are, in fact, composed of DNA. Chemically speaking, your genetic endowment is the DNA contained in the 46 chromosomes you inherited from your parents and in the mitochondria passed along by your mother.

Of all nature's molecules, nucleic acids are unique in their ability to direct their own replication from monomers. Indeed, the resemblance of offspring to their parents has its basis in the precise replication of DNA and its transmission from one generation to the next. Hereditary information is encoded in the chemical language of DNA and reproduced in all the cells of your body. It is this DNA program that directs the development of your biochemical, anatomical, physiological, and, to some extent, behavioral traits. In this chapter, you will learn how biologists deduced that DNA is the genetic material and how Watson and Crick discovered its structure. You will also see how DNA is replicated—the molecular basis of inheritance—and how cells repair their DNA. Finally, you will explore how a molecule of DNA is packed together with proteins in a chromosome.

CONCEPT **16.1**

DNA is the genetic material

Today, even schoolchildren have heard of DNA, and scientists routinely manipulate DNA in the laboratory, often to change the heritable traits of cells in their experiments. Early in the 20th century, however, the identification of the molecules of inheritance loomed as a major challenge to biologists.

The Search for the Genetic Material: *Scientific Inquiry*

Once T. H. Morgan's group showed that genes are located along chromosomes (described in Chapter 15), the two chemical components of chromosomes—DNA and protein—became the candidates for the genetic material. Until the 1940s, the case for proteins seemed stronger, especially since biochemists had identified them as a class of macromolecules with great heterogeneity and specificity of function, essential requirements for the hereditary material. Moreover, little was known about nucleic acids, whose physical and chemical properties seemed far too uniform to account for the multitude of specific inherited traits exhibited by every organism. This view gradually changed as experiments with microorganisms yielded unexpected results. As with the work of Mendel and Morgan, a key factor in determining the identity of the genetic material was the choice of appropriate experimental organisms. The role of DNA in heredity was first worked out by studying bacteria and the viruses that infect them, which are far simpler than pea plants, fruit flies, or humans. In this section, we will trace the search for the genetic material in some detail as a case study in scientific inquiry.

Evidence That DNA Can Transform Bacteria

We can trace the discovery of the genetic role of DNA back to 1928. While attempting to develop a vaccine against pneumonia, a British medical officer named Frederick Griffith studied *Streptococcus pneumoniae*, a bacterium that causes pneumonia in mammals. Griffith had two strains (varieties) of the bacterium, one pathogenic (disease-causing) and one nonpathogenic (harmless). He was surprised to find that when he killed the pathogenic bacteria with heat and then mixed the cell remains with living bacteria of the nonpathogenic strain, some of the living cells became pathogenic **(Figure 16.2)**. Furthermore, this newly acquired trait of pathogenicity was inherited by all the descendants of the transformed bacteria. Clearly, some chemical component of the dead pathogenic cells caused this heritable change, although the identity of the substance was not known. Griffith called the phenomenon **transformation**, now defined as a change in genotype and phenotype due to the assimilation of external DNA by a cell. (This use of the word *transformation* should not be confused with the conversion of a normal animal cell to a cancerous one, discussed in Chapter 12.)

▼ **Figure 16.2 Inquiry**

Can a genetic trait be transferred between different bacterial strains?

EXPERIMENT Frederick Griffith studied two strains of the bacterium *Streptococcus pneumoniae*. Bacteria of the S (smooth) strain can cause pneumonia in mice; they are pathogenic because they have a capsule that protects them from an animal's defense system. Bacteria of the R (rough) strain lack a capsule and are nonpathogenic. To test for the trait of pathogenicity, Griffith injected mice with the two strains as shown below:

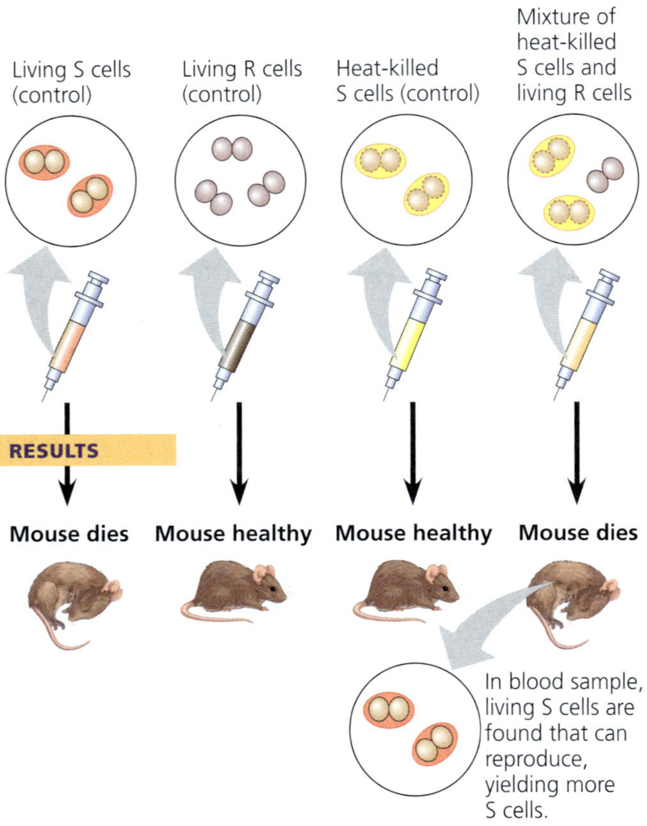

CONCLUSION Griffith concluded that the living R bacteria had been transformed into pathogenic S bacteria by an unknown, heritable substance from the dead S cells that allowed the R cells to make capsules.

SOURCE F. Griffith, The significance of pneumococcal types, *Journal of Hygiene* 27:113–159 (1928).

WHAT IF? How did this experiment rule out the possibility that the R cells could have simply used the capsules of the dead S cells to become pathogenic?

Griffith's work set the stage for a 14-year search by American bacteriologist Oswald Avery for the identity of the transforming substance. Avery focused on the three main candidates: DNA, RNA (the other nucleic acid), and protein. Avery broke open the heat-killed pathogenic bacteria and extracted the cellular contents. In separate samples, he used specific treatments that inactivated each of the three types of molecules. He then tested each treated sample for its ability to transform live nonpathogenic bacteria. Only when DNA was allowed to remain active did transformation occur. In 1944, Avery and his colleagues Maclyn McCarty and Colin MacLeod announced that the transforming agent was DNA. Their discovery was greeted with interest but considerable skepticism, in part because of the lingering belief that proteins were better candidates to be the genetic material. Moreover, many biologists were not convinced that the genes of bacteria would be similar in composition and function to those of more complex organisms. But the major reason for the continued doubt was that so little was known about DNA.

Evidence That Viral DNA Can Program Cells

Additional evidence for DNA as the genetic material came from studies of viruses that infect bacteria **(Figure 16.3)**. These

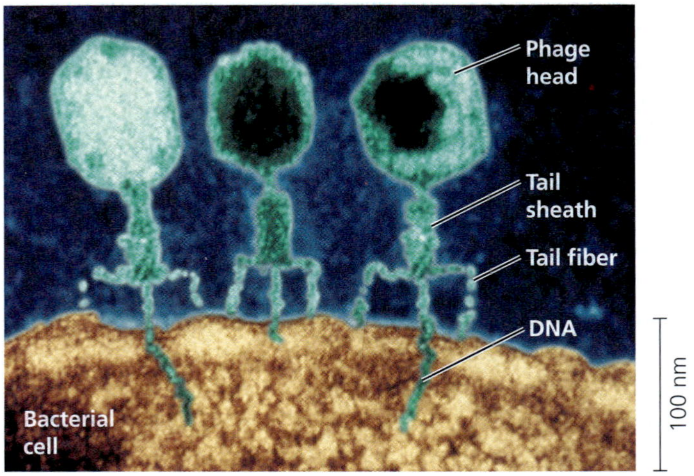

▲ **Figure 16.3 Viruses infecting a bacterial cell.** T2 and related phages attach to the host cell and inject their genetic material through the plasma membrane, while the head and tail parts remain on the outer bacterial surface (colorized TEM).

viruses are called **bacteriophages** (meaning "bacteria-eaters"), or just **phages**. Viruses are much simpler than cells. A virus is little more than DNA (or sometimes RNA) enclosed by a protective coat, which is often simply protein. To reproduce, a virus must infect a cell and take over the cell's metabolic machinery.

Phages have been widely used as tools by researchers in molecular genetics. In 1952, Alfred Hershey and Martha Chase performed experiments showing that DNA is the genetic material of a phage known as T2. This is one of many phages that infect *Escherichia coli* (*E. coli*), a bacterium that normally lives in the intestines of mammals. At that time, biologists already knew that T2, like many other phages, was composed almost entirely of DNA and protein. They also knew that the T2 phage could quickly turn an *E. coli* cell into a T2-producing factory that released many copies when the cell ruptured. Somehow, T2 could reprogram its host cell to produce viruses. But which viral component—protein or DNA—was responsible?

Hershey and Chase answered this question by devising an experiment showing that only one of the two components of T2 actually enters the *E. coli* cell during infection **(Figure 16.4)**. In

▼ Figure 16.4 Inquiry

Is protein or DNA the genetic material of phage T2?

EXPERIMENT Alfred Hershey and Martha Chase used radioactive sulfur and phosphorus to trace the fates of protein and DNA, respectively, of T2 phages that infected bacterial cells. They wanted to see which of these molecules entered and could reprogram the cells to make more phages.

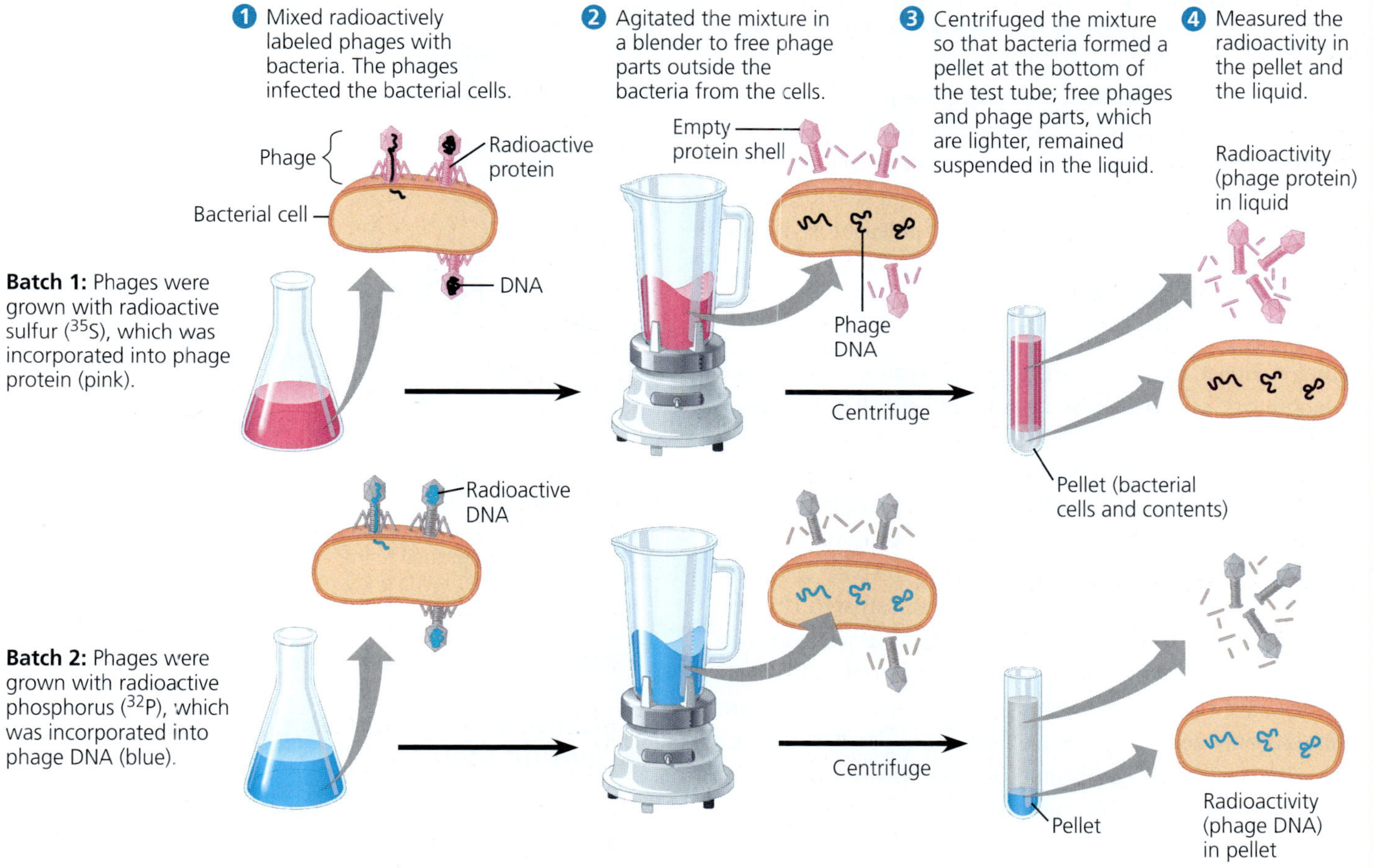

RESULTS When proteins were labeled (batch 1), radioactivity remained outside the cells; but when DNA was labeled (batch 2), radioactivity was found inside the cells. Bacterial cells with radioactive phage DNA released new phages with some radioactive phosphorus.

CONCLUSION Phage DNA entered bacterial cells, but phage proteins did not. Hershey and Chase concluded that DNA, not protein, functions as the genetic material of phage T2.

SOURCE A. D. Hershey and M. Chase, Independent functions of viral protein and nucleic acid in growth of bacteriophage, *Journal of General Physiology* 36:39–56 (1952).

WHAT IF? How would the results have differed if proteins carried the genetic information?

their experiment, they used a radioactive isotope of sulfur to tag protein in one batch of T2 and a radioactive isotope of phosphorus to tag DNA in a second batch. Because protein, but not DNA, contains sulfur, radioactive sulfur atoms were incorporated only into the protein of the phage. In a similar way, the atoms of radioactive phosphorus labeled only the DNA, not the protein, because nearly all the phage's phosphorus is in its DNA. In the experiment, separate samples of nonradioactive *E. coli* cells were allowed to be infected by the protein-labeled and DNA-labeled batches of T2. The researchers then tested the two samples shortly after the onset of infection to see which type of molecule—protein or DNA—had entered the bacterial cells and would therefore be capable of reprogramming them.

Hershey and Chase found that the phage DNA entered the host cells but the phage protein did not. Moreover, when these bacteria were returned to a culture medium, the infection ran its course, and the *E. coli* released phages that contained some radioactive phosphorus, further showing that the DNA inside the cell played an ongoing role during the infection process.

Hershey and Chase concluded that the DNA injected by the phage must be the molecule carrying the genetic information that makes the cells produce new viral DNA and proteins. The Hershey-Chase experiment was a landmark study because it provided powerful evidence that nucleic acids, rather than proteins, are the hereditary material, at least for viruses.

Additional Evidence That DNA Is the Genetic Material

Further evidence that DNA is the genetic material came from the laboratory of biochemist Erwin Chargaff. It was already known that DNA is a polymer of nucleotides, each consisting of three components: a nitrogenous (nitrogen-containing) base, a pentose sugar called deoxyribose, and a phosphate group **(Figure 16.5)**. The base can be adenine (A), thymine (T), guanine (G), or cytosine (C). Chargaff analyzed the base composition of DNA from a number of different organisms. In 1950, he reported that the base composition of DNA varies from one species to another. For example, 30.3% of human DNA nucleotides have the base A, whereas DNA from the bacterium *E. coli* has only 26.0% A. This evidence of molecular diversity among species, which had been presumed absent from DNA, made DNA a more credible candidate for the genetic material.

Chargaff also noticed a peculiar regularity in the ratios of nucleotide bases within a single species. In the DNA of each species he studied, the number of adenines approximately equaled the number of thymines, and the number of guanines approximately equaled the number of cytosines. In human DNA, for example, the four bases are present in these percentages: A = 30.3% and T = 30.3%; G = 19.5% and C = 19.9%. The equivalences for any given species between the number of A and T bases and the number of G and C bases became known as *Chargaff's rules*. The basis for these rules remained unexplained until the discovery of the double helix.

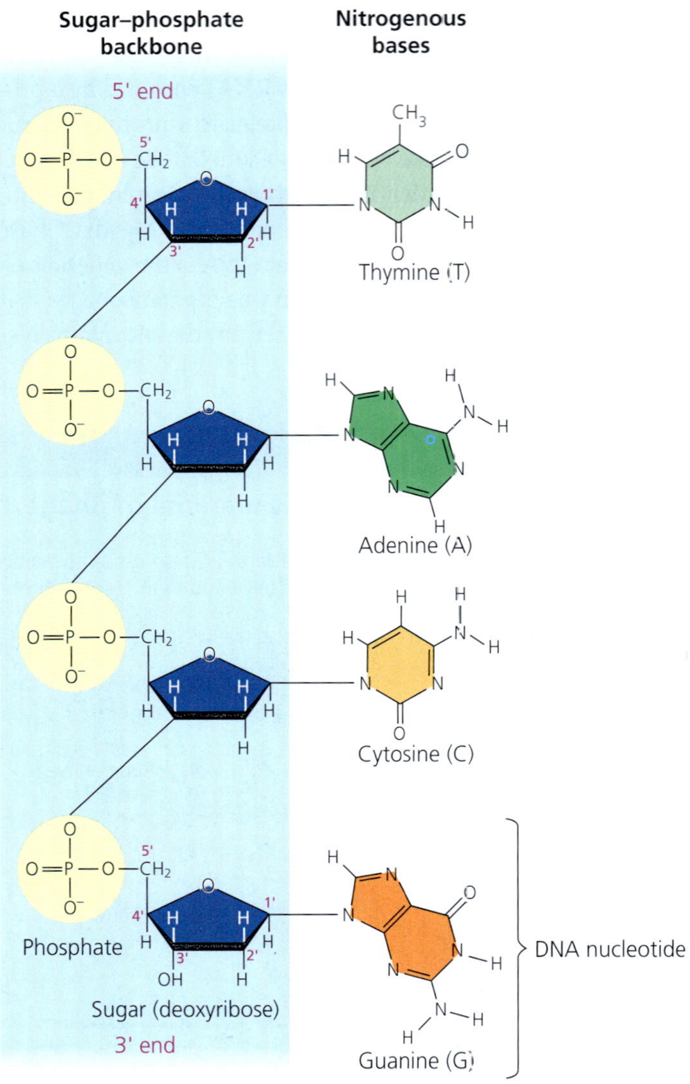

▲ **Figure 16.5 The structure of a DNA strand.** Each nucleotide monomer consists of a nitrogenous base (T, A, C, or G), the sugar deoxyribose (blue), and a phosphate group (yellow). The phosphate of one nucleotide is attached to the sugar of the next, resulting in a "backbone" of alternating phosphates and sugars from which the bases project. The polynucleotide strand has directionality, from the 5′ end (with the phosphate group) to the 3′ end (with the —OH group). 5′ and 3′ refer to the numbers assigned to the carbons in the sugar ring.

Building a Structural Model of DNA: *Scientific Inquiry*

Once most biologists were convinced that DNA was the genetic material, the challenge was to determine how the structure of DNA could account for its role in inheritance. By the early 1950s, the arrangement of covalent bonds in a nucleic acid polymer was well established (see Figure 16.5), and researchers focused on discovering the three-dimensional structure of DNA. Among the scientists working on the problem were Linus Pauling, at the California Institute of Technology, and Maurice Wilkins and Rosalind Franklin, at King's College in London. First to come up with the correct answer, however, were two scientists who were relatively unknown at the time—the American James Watson and the Englishman Francis Crick.

The brief but celebrated partnership that solved the puzzle of DNA structure began soon after Watson journeyed to Cambridge University, where Crick was studying protein structure with a technique called X-ray crystallography (see Figure 5.25). While visiting the laboratory of Maurice Wilkins, Watson saw an X-ray diffraction image of DNA produced by Wilkins's accomplished colleague Rosalind Franklin **(Figure 16.6a)**. Images produced by X-ray crystallography are not actually pictures of molecules. The spots and smudges in **Figure 16.6b** were produced by X-rays that were diffracted (deflected) as they passed through aligned fibers of purified DNA. Crystallographers use mathematical equations to translate such patterns into information about the three-dimensional shapes of molecules, and Watson was familiar with the types of patterns that helical molecules produce. A careful study of Franklin's X-ray diffraction photo of DNA not only told him that DNA was helical in shape, but also enabled him to approximate the width of the helix and the spacing of the nitrogenous bases along it. The width of the helix suggested that it was made up of two strands, contrary to a three-stranded model that Linus Pauling had proposed a short time earlier. The presence of two strands accounts for the now-familiar term **double helix** **(Figure 16.7)**.

Watson and Crick began building models of a double helix that would conform to the X-ray measurements and what was then known about the chemistry of DNA. Having also read an unpublished annual report summarizing Franklin's work, they knew she had concluded that the sugar-phosphate backbones were on the outside of the double helix. This arrangement was appealing because it put the relatively hydrophobic nitrogenous

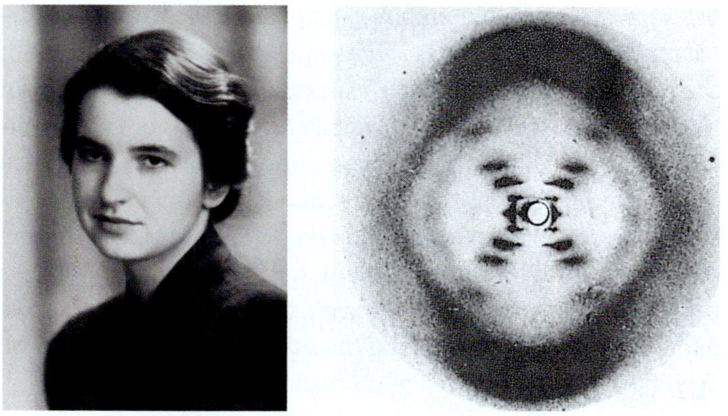

(a) Rosalind Franklin **(b) Franklin's X-ray diffraction photograph of DNA**

▲ **Figure 16.6 Rosalind Franklin and her X-ray diffraction photo of DNA.** Franklin, a very accomplished X-ray crystallographer, conducted critical experiments resulting in the photograph that allowed Watson and Crick to deduce the double-helical structure of DNA. Franklin died of cancer in 1958, when she was only 38. Her colleague Maurice Wilkins received the Nobel Prize in 1962 along with Watson and Crick.

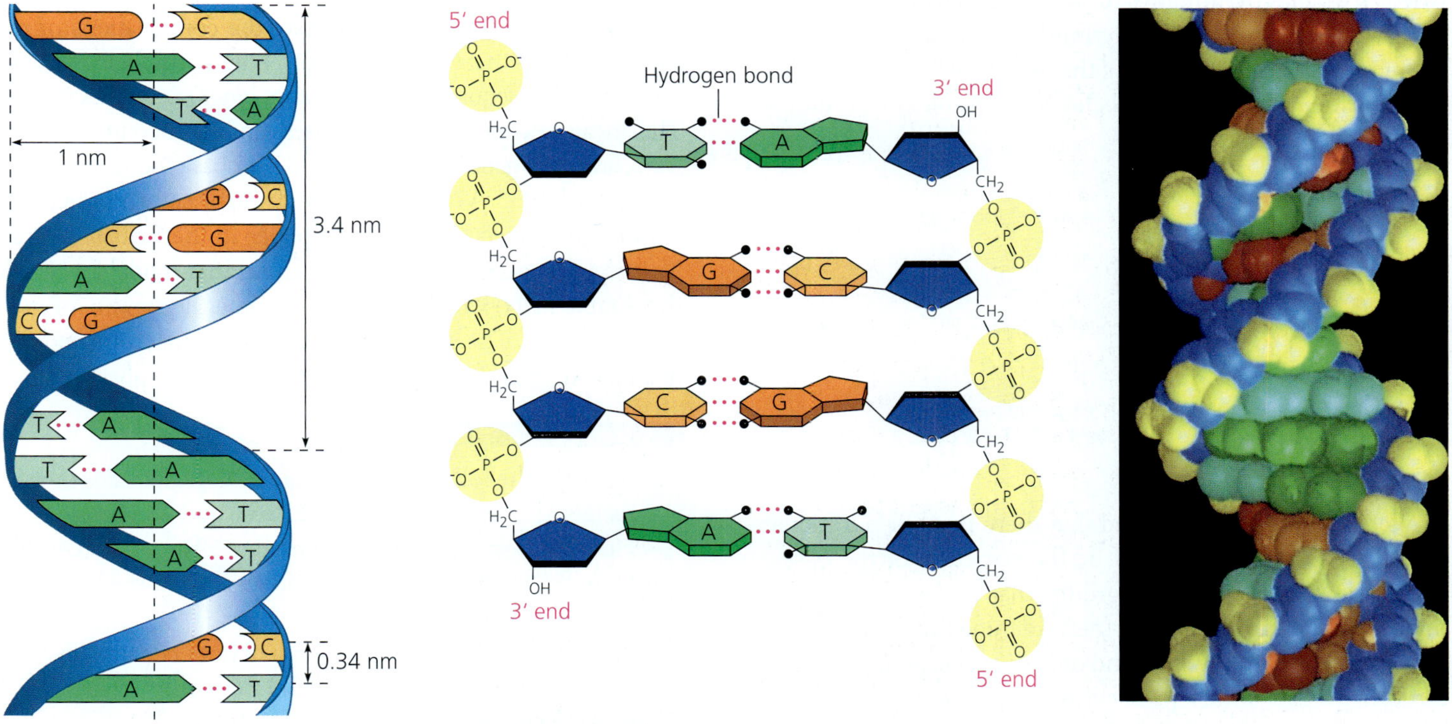

(a) Key features of DNA structure **(b) Partial chemical structure** **(c) Space-filling model**

▲ **Figure 16.7 The double helix. (a)** The "ribbons" in this diagram represent the sugar-phosphate backbones of the two DNA strands. The helix is "right-handed," curving up to the right. The two strands are held together by hydrogen bonds (dotted lines) between the nitrogenous bases, which are paired in the interior of the double helix. **(b)** For clarity, the two DNA strands are shown untwisted in this partial chemical structure. Strong covalent bonds link the units of each strand, while weaker hydrogen bonds hold one strand to the other. Notice that the strands are antiparallel, meaning that they are oriented in opposite directions. **(c)** The tight stacking of the base pairs is clear in this computer model. Van der Waals attractions between the stacked pairs play a major role in holding the molecule together (see Chapter 2).

bases in the molecule's interior and thus away from the surrounding aqueous solution. Watson constructed a model with the nitrogenous bases facing the interior of the double helix. In this model, the two sugar-phosphate backbones are antiparallel—that is, their subunits run in opposite directions to each other (see Figure 16.7). You can imagine the overall arrangement as a rope ladder with rigid rungs. The side ropes are the equivalent of the sugar-phosphate backbones, and the rungs represent pairs of nitrogenous bases. Now imagine holding one end of the ladder and twisting the other end, forming a spiral. Franklin's X-ray data indicated that the helix makes one full turn every 3.4 nm along its length. With the bases stacked just 0.34 nm apart, there are ten layers of base pairs, or rungs of the ladder, in each full turn of the helix.

The nitrogenous bases of the double helix are paired in specific combinations: adenine (A) with thymine (T), and guanine (G) with cytosine (C). It was mainly by trial and error that Watson and Crick arrived at this key feature of DNA. At first, Watson imagined that the bases paired like with like—for example, A with A and C with C. But this model did not fit the X-ray data, which suggested that the double helix had a uniform diameter. Why is this requirement inconsistent with like-with-like pairing of bases? Adenine and guanine are purines, nitrogenous bases with two organic rings. In contrast, cytosine and thymine belong to the family of nitrogenous bases known as pyrimidines, which have a single ring. Thus, purines (A and G) are about twice as wide as pyrimidines (C and T). A purine-purine pair is too wide and a pyrimidine-pyrimidine pair too narrow to account for the 2-nm diameter of the double helix. Always pairing a purine with a pyrimidine, however, results in a uniform diameter:

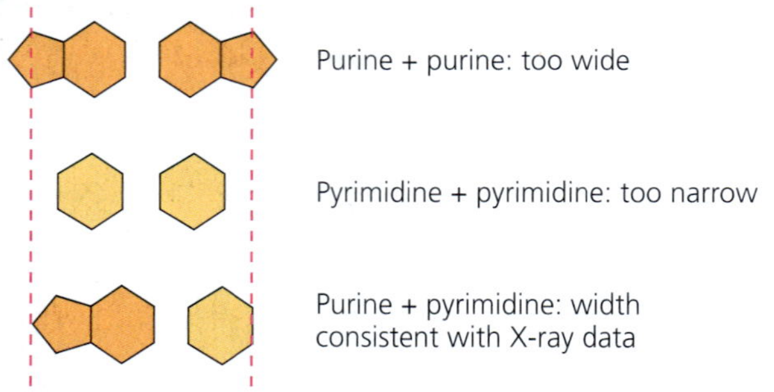

Watson and Crick reasoned that there must be additional specificity of pairing dictated by the structure of the bases. Each base has chemical side groups that can form hydrogen bonds with its appropriate partner: Adenine can form two hydrogen bonds with thymine and only thymine; guanine forms three hydrogen bonds with cytosine and only cytosine. In shorthand, A pairs with T, and G pairs with C (**Figure 16.8**).

The Watson-Crick model explained the basis for Chargaff's rules. Wherever one strand of a DNA molecule has an A, the partner strand has a T. And a G in one strand is always paired with a C in the complementary strand. Therefore, in the DNA of any organism, the amount of adenine equals the amount of thymine, and the amount of guanine equals the amount of cytosine. Although the base-pairing rules dictate the combinations of nitrogenous bases that form the "rungs" of the double helix, they do not restrict the sequence of nucleotides *along* each DNA strand. The linear sequence of the four bases can be varied in countless ways, and each gene has a unique order, or base sequence.

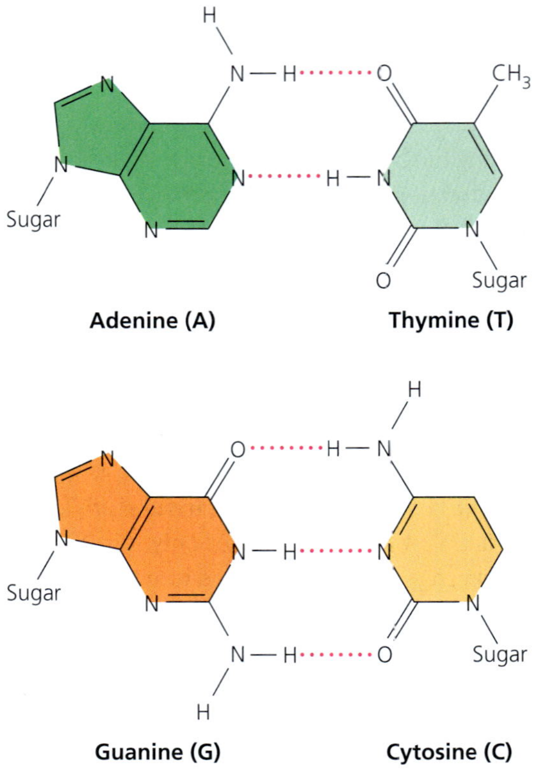

▲ **Figure 16.8 Base pairing in DNA.** The pairs of nitrogenous bases in a DNA double helix are held together by hydrogen bonds, shown here as pink dotted lines.

In April 1953, Watson and Crick surprised the scientific world with a succinct, one-page paper in the British journal *Nature*.* The paper reported their molecular model for DNA: the double helix, which has since become the symbol of molecular biology. The beauty of the model was that the structure of DNA suggested the basic mechanism of its replication.

CONCEPT CHECK 16.1

1. A fly has the following percentages of nucleotides in its DNA: 27.3% A, 27.6% T, 22.5% G, and 22.5% C. How do these numbers demonstrate Chargaff's rules?
2. How did Watson and Crick's model explain the basis for Chargaff's rules?
3. **WHAT IF?** If transformation had not occurred in Griffith's experiment, how would the results have differed? Explain.

For suggested answers, see Appendix A.

* J. D. Watson and F. H. C. Crick, Molecular structure of nucleic acids: a structure for deoxyribose nucleic acids, *Nature* 171:737–738 (1953).

CONCEPT **16.2**

Many proteins work together in DNA replication and repair

The relationship between structure and function is manifest in the double helix. The idea that there is specific pairing of nitrogenous bases in DNA was the flash of inspiration that led Watson and Crick to the correct double helix. At the same time, they saw the functional significance of the base-pairing rules. They ended their classic paper with this wry statement: "It has not escaped our notice that the specific pairing we have postulated immediately suggests a possible copying mechanism for the genetic material." In this section, you will learn about the basic principle of DNA replication, as well as some important details of the process.

The Basic Principle: Base Pairing to a Template Strand

In a second paper, Watson and Crick stated their hypothesis for how DNA replicates:

> Now our model for deoxyribonucleic acid is, in effect, a pair of templates, each of which is complementary to the other. We imagine that prior to duplication the hydrogen bonds are broken, and the two chains unwind and separate. Each chain then acts as a template for the formation onto itself of a new companion chain, so that eventually we shall have two pairs of chains, where we only had one before. Moreover, the sequence of the pairs of bases will have been duplicated exactly.*

* F. H. C. Crick and J. D. Watson, The complementary structure of deoxyribonucleic acid, *Proceedings of the Royal Society of London A* 223:80 (1954).

Figure 16.9 illustrates Watson and Crick's basic idea. To make it easier to follow, we show only a short section of double helix in untwisted form. Notice that if you cover one of the two DNA strands of Figure 16.9a, you can still determine its linear sequence of bases by referring to the uncovered strand and applying the base-pairing rules. The two strands are complementary; each stores the information necessary to reconstruct the other. When a cell copies a DNA molecule, each strand serves as a template for ordering nucleotides into a new, complementary strand. Nucleotides line up along the template strand according to the base-pairing rules and are linked to form the new strands. Where there was one double-stranded DNA molecule at the beginning of the process, there are soon two, each an exact replica of the "parent" molecule. The copying mechanism is analogous to using a photographic negative to make a positive image, which can in turn be used to make another negative, and so on.

This model of DNA replication remained untested for several years following publication of the DNA structure. The requisite experiments were simple in concept but difficult to perform. Watson and Crick's model predicts that when a double helix replicates, each of the two daughter molecules will have one old strand, derived from the parent molecule, and one newly made strand. This **semiconservative model** can be distinguished from a conservative model of replication, in which the two parent strands somehow come back together after the process (that is, the parent molecule is conserved). In yet a third model, called the dispersive model, all four strands of DNA following replication have a mixture of old and new DNA (**Figure 16.10**, on the next page). Although mechanisms for conservative or dispersive DNA replication are not easy to devise, these models remained possibilities until they could be ruled out. After two years of preliminary work in the late

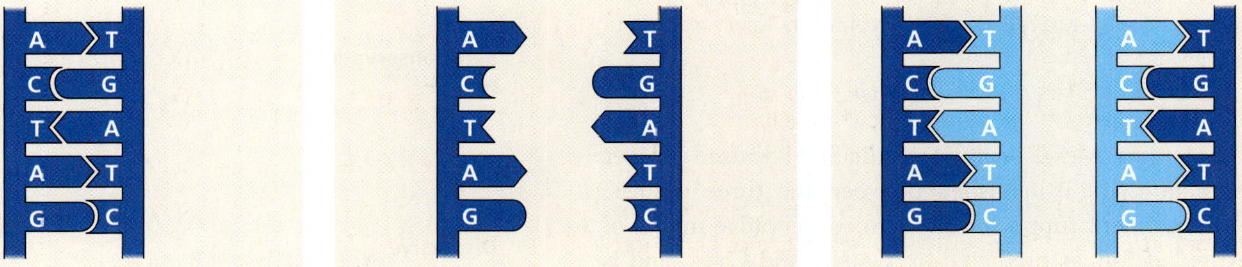

(a) The parent molecule has two complementary strands of DNA. Each base is paired by hydrogen bonding with its specific partner, A with T and G with C.

(b) The first step in replication is separation of the two DNA strands. Each parental strand can now serve as a template that determines the order of nucleotides along a new, complementary strand.

(c) The complementary nucleotides line up and are connected to form the sugar-phosphate backbones of the new strands. Each "daughter" DNA molecule consists of one parental strand (dark blue) and one new strand (light blue).

▲ **Figure 16.9 A model for DNA replication: the basic concept.** In this simplified illustration, a short segment of DNA has been untwisted into a structure that resembles a ladder. The rails of the ladder are the sugar-phosphate backbones of the two DNA strands; the rungs are the pairs of nitrogenous bases. Simple shapes symbolize the four kinds of bases. Dark blue represents DNA strands present in the parent molecule; light blue represents newly synthesized DNA.

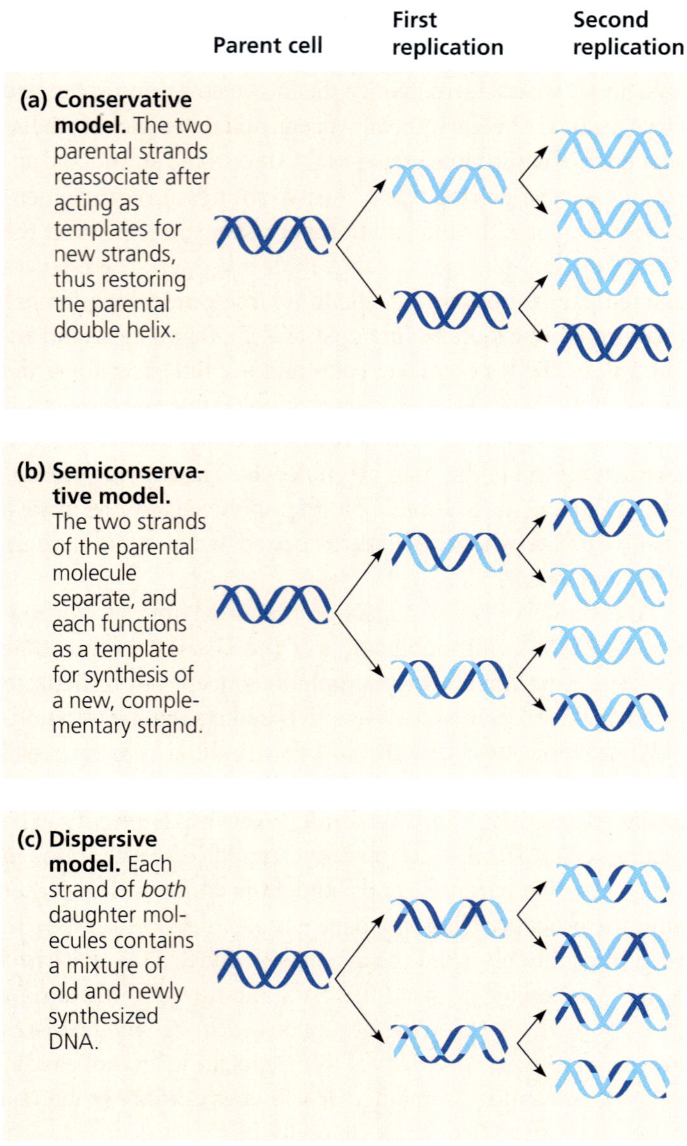

▲ **Figure 16.10 Three alternative models of DNA replication.** Each short segment of double helix symbolizes the DNA within a cell. Beginning with a parent cell, we follow the DNA for two generations of cells—two rounds of DNA replication. Newly made DNA is light blue.

1950s, Matthew Meselson and Franklin Stahl devised a clever experiment that distinguished between the three models. Their experiment supported the semiconservative model of DNA replication, as predicted by Watson and Crick, and is widely acknowledged among biologists to be a classic example of elegant experimental design **(Figure 16.11)**.

The basic principle of DNA replication is conceptually simple. However, the actual process involves some complicated biochemical gymnastics, as we will now see.

DNA Replication: *A Closer Look*

The bacterium *E. coli* has a single chromosome of about 4.6 million nucleotide pairs. In a favorable environment, an *E. coli* cell

▼ **Figure 16.11 Inquiry**

Does DNA replication follow the conservative, semiconservative, or dispersive model?

EXPERIMENT At the California Institute of Technology, Matthew Meselson and Franklin Stahl cultured *E. coli* for several generations in a medium containing nucleotide precursors labeled with a heavy isotope of nitrogen, ^{15}N. The scientists then transferred the bacteria to a medium with only ^{14}N, a lighter isotope. Two DNA samples were taken from this flask, one at 20 minutes and one at 40 minutes, after the first and second replications, respectively. Meselson and Stahl could distinguish DNA of different densities by centrifuging DNA extracted from the bacteria.

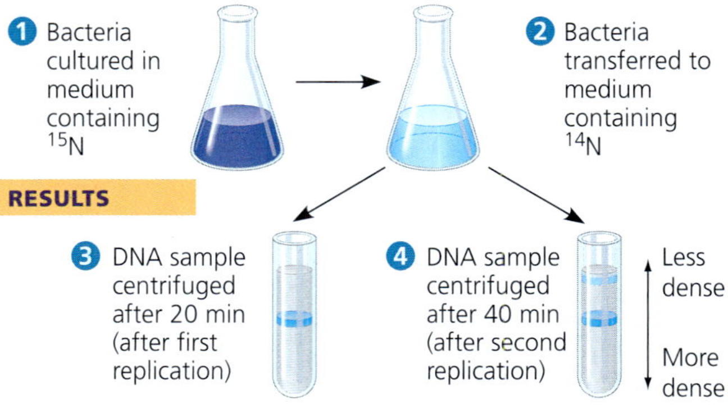

RESULTS

CONCLUSION Meselson and Stahl compared their results to those predicted by each of the three models in Figure 16.10, as shown below. The first replication in the ^{14}N medium produced a band of hybrid (^{15}N-^{14}N) DNA. This result eliminated the conservative model. The second replication produced both light and hybrid DNA, a result that refuted the dispersive model and supported the semiconservative model. They therefore concluded that DNA replication is semiconservative.

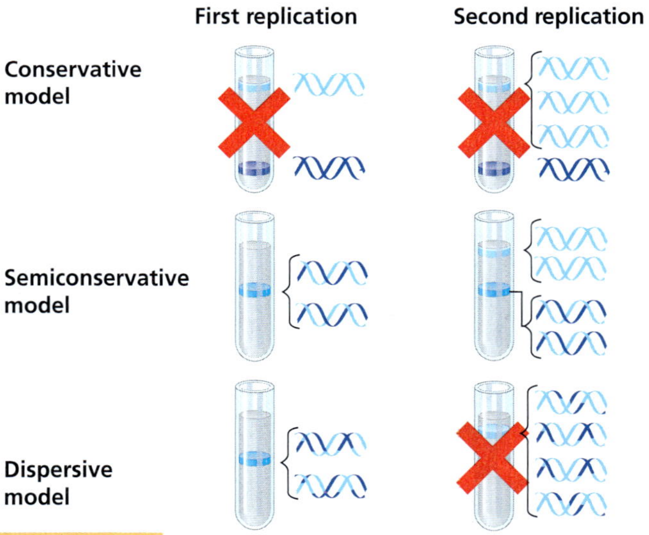

SOURCE M. Meselson and F. W. Stahl, The replication of DNA in *Escherichia coli, Proceedings of the National Academy of Sciences USA* 44:671–682 (1958).

Inquiry in Action Read and analyze the original paper in *Inquiry in Action: Interpreting Scientific Papers.*

WHAT IF? If Meselson and Stahl had first grown the cells in ^{14}N-containing medium and then moved them into ^{15}N-containing medium before taking samples, what would have been the result?

can copy all this DNA and divide to form two genetically identical daughter cells in less than an hour. Each of *your* cells has 46 DNA molecules in its nucleus, one long double-helical molecule per chromosome. In all, that represents about 6 billion base pairs, or over a thousand times more DNA than is found in a bacterial cell. If we were to print the one-letter symbols for these bases (A, G, C, and T) the size of the letters you are now reading, the 6 billion base pairs of information in a diploid human cell would fill about 1,200 books as thick as this text. Yet it takes a cell just a few hours to copy all of this DNA. This replication of an enormous amount of genetic information is achieved with very few errors—only about one per 10 billion nucleotides. The copying of DNA is remarkable in its speed and accuracy.

More than a dozen enzymes and other proteins participate in DNA replication. Much more is known about how this "replication machine" works in bacteria (such as *E. coli*) than in eukaryotes, and we will describe the basic steps of the process for *E. coli*, except where otherwise noted. What scientists have learned about eukaryotic DNA replication suggests, however, that most of the process is fundamentally similar for prokaryotes and eukaryotes.

Getting Started

The replication of a DNA molecule begins at special sites called **origins of replication**, short stretches of DNA having a specific sequence of nucleotides. The *E. coli* chromosome, like many other bacterial chromosomes, is circular and has a single origin. Proteins that initiate DNA replication recognize this sequence and attach to the DNA, separating the two strands and opening up a replication "bubble." Replication of DNA then proceeds in both directions until the entire molecule is copied **(Figure 16.12a)**. In

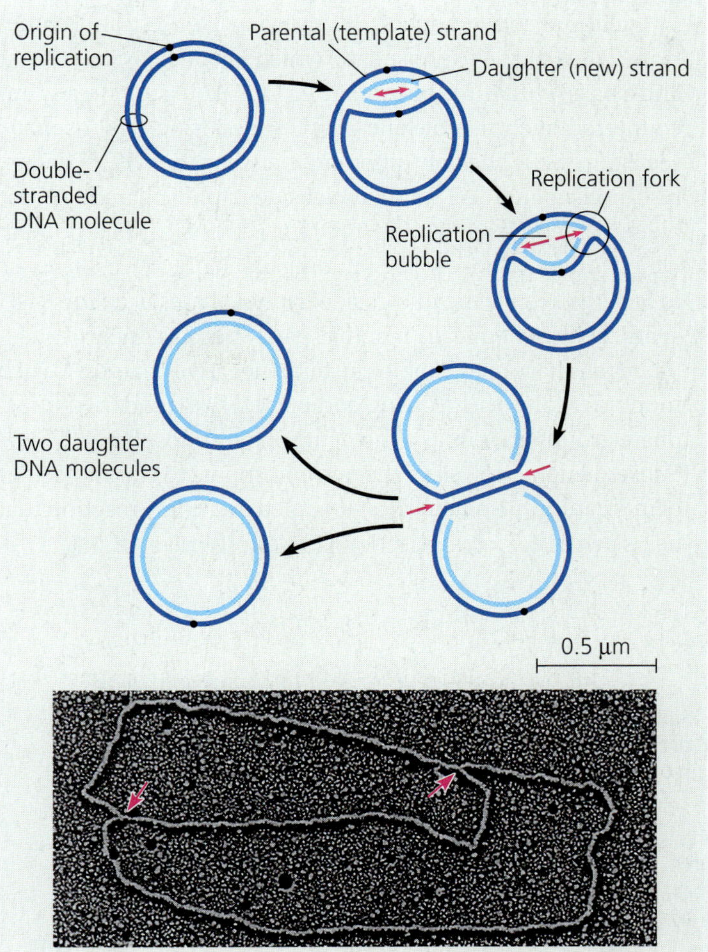

(a) In the circular chromosome of *E. coli* and many other bacteria, only one origin of replication is present. The parental strands separate at the origin, forming a replication bubble with two forks. Replication proceeds in both directions until the forks meet on the other side, resulting in two daughter DNA molecules. The TEM shows a bacterial chromosome with a replication bubble.

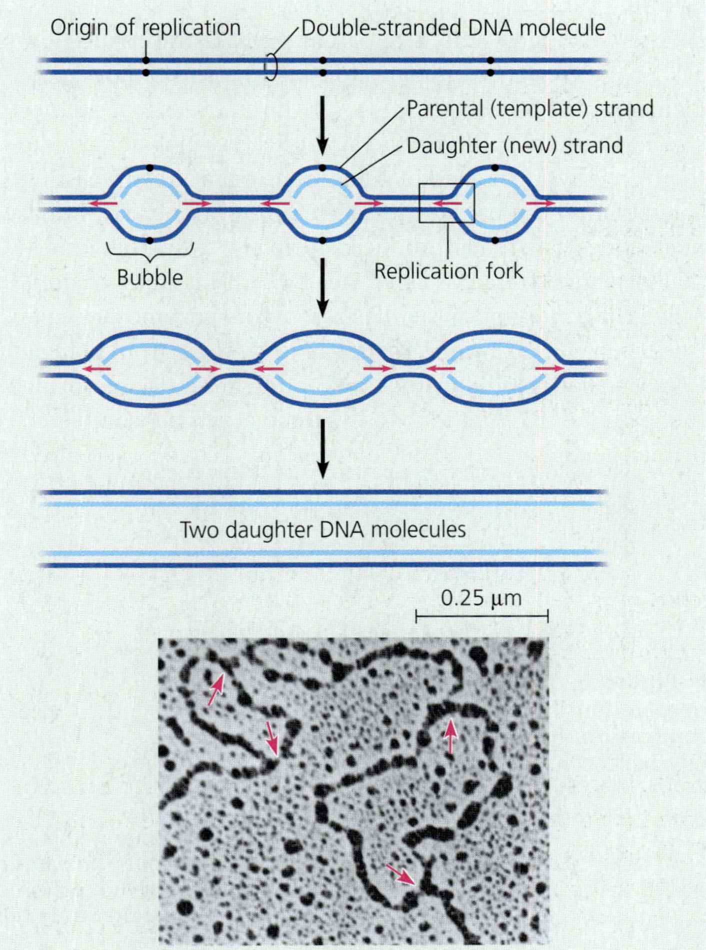

(b) In each linear chromosome of eukaryotes, DNA replication begins when replication bubbles form at many sites along the giant DNA molecule. The bubbles expand as replication proceeds in both directions. Eventually, the bubbles fuse and synthesis of the daughter strands is complete. The TEM shows three replication bubbles along the DNA of a cultured Chinese hamster cell.

▲ **Figure 16.12 Origins of replication in *E. coli* and eukaryotes.** The red arrows indicate the movement of the replication forks and thus the overall directions of DNA replication within each bubble.

DRAW IT *In the TEM in (b), add arrows for the third bubble.*

contrast to a bacterial chromosome, a eukaryotic chromosome may have hundreds or even a few thousand replication origins. Multiple replication bubbles form and eventually fuse, thus speeding up the copying of the very long DNA molecules **(Figure 16.12b)**. As in bacteria, eukaryotic DNA replication proceeds in both directions from each origin.

At each end of a replication bubble is a **replication fork**, a Y-shaped region where the parental strands of DNA are being unwound. Several kinds of proteins participate in the unwinding **(Figure 16.13)**. **Helicases** are enzymes that untwist the double helix at the replication forks, separating the two parental strands and making them available as template strands. After parental strand separation, **single-strand binding proteins** bind to the unpaired DNA strands, stabilizing them. The untwisting of the double helix causes tighter twisting and strain ahead of the replication fork. **Topoisomerase** helps relieve this strain by breaking, swiveling, and rejoining DNA strands.

The unwound sections of parental DNA strands are now available to serve as templates for the synthesis of new complementary DNA strands. However, the enzymes that synthesize DNA cannot *initiate* the synthesis of a polynucleotide; they can only add nucleotides to the end of an already existing chain that is base-paired with the template strand. The initial nucleotide chain that is produced during DNA synthesis is actually a short stretch of RNA, not DNA. This RNA chain is called a **primer** and is synthesized by the enzyme **primase** (see Figure 16.13). Primase starts an RNA chain from a single RNA nucleotide, adding RNA nucleotides one at a time, using the parental DNA strand as a template. The completed primer, generally 5 to 10 nucleotides long, is thus base-paired to the template strand. The new DNA strand will start from the 3′ end of the RNA primer.

Synthesizing a New DNA Strand

Enzymes called **DNA polymerases** catalyze the synthesis of new DNA by adding nucleotides to a preexisting chain. In *E. coli*, there are several different DNA polymerases, but two appear to play the major roles in DNA replication: DNA polymerase III and DNA polymerase I. The situation in eukaryotes is more complicated, with at least 11 different DNA polymerases discovered so far; however, the general principles are the same.

Most DNA polymerases require a primer and a DNA template strand, along which complementary DNA nucleotides line up. In *E. coli*, DNA polymerase III (abbreviated DNA pol III) adds a DNA nucleotide to the RNA primer and then continues adding DNA nucleotides, complementary to the parental DNA template strand, to the growing end of the new DNA strand. The rate of elongation is about 500 nucleotides per second in bacteria and 50 per second in human cells.

Each nucleotide added to a growing DNA strand comes from a nucleoside triphosphate, which is a nucleoside (a sugar and a base) with three phosphate groups. You have already encountered such a molecule—ATP (adenosine triphosphate; see Figure 8.8). The only difference between the ATP of energy metabolism and dATP, the nucleoside triphosphate that supplies an adenine nucleotide to DNA, is the sugar component, which is deoxyribose in the building block of DNA, but ribose in ATP. Like ATP, the nucleoside triphosphates used for DNA synthesis are chemically reactive, partly because their triphosphate tails have an unstable cluster of negative charge. As each monomer joins the growing end of a DNA strand, two phosphate groups are lost as a molecule of pyrophosphate Ⓟ—Ⓟ$_i$. Subsequent hydrolysis of the pyrophosphate to two molecules of inorganic phosphate Ⓟ$_i$ is a coupled exergonic reaction that helps drive the polymerization reaction **(Figure 16.14)**.

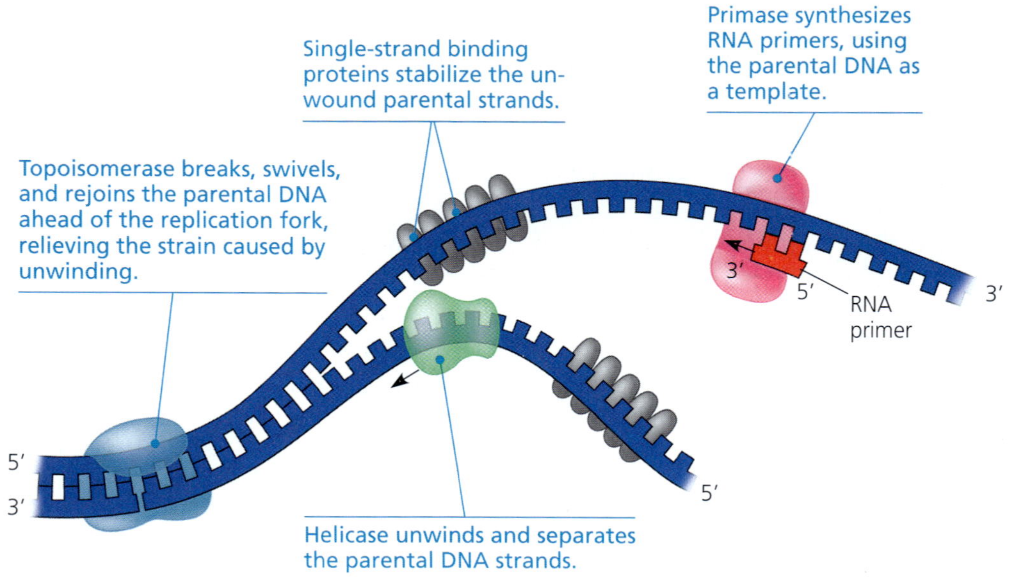

▶ **Figure 16.13 Some of the proteins involved in the initiation of DNA replication.** The same proteins function at both replication forks in a replication bubble. For simplicity, only one fork is shown.

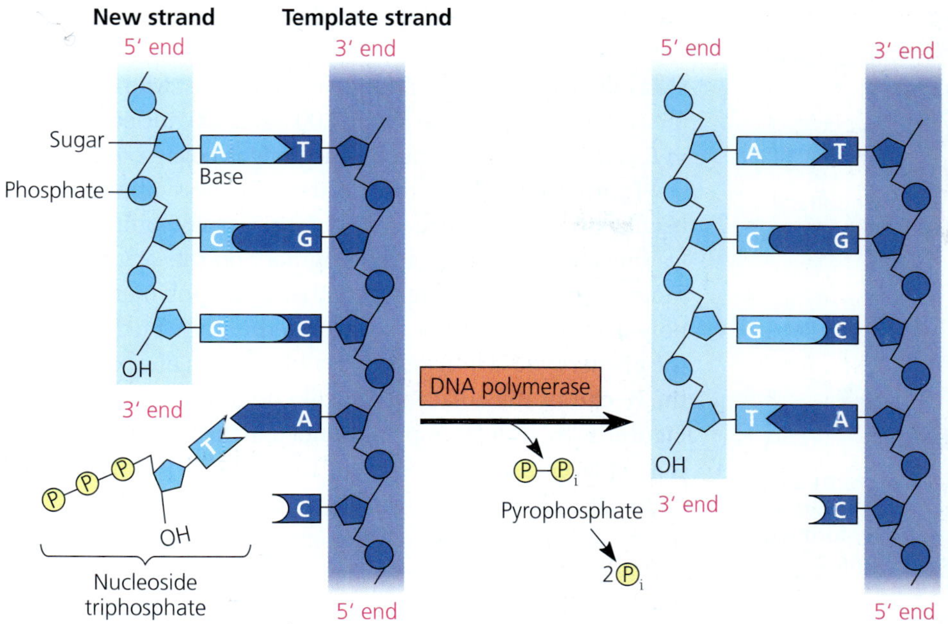

◀ **Figure 16.14 Incorporation of a nucleotide into a DNA strand.** DNA polymerase catalyzes the addition of a nucleoside triphosphate to the 3′ end of a growing DNA strand, with the release of two phosphates.

? *Use this diagram to explain what we mean when we say that each DNA strand has directionality.*

Antiparallel Elongation

As we have noted previously, the two ends of a DNA strand are different, giving each strand directionality, like a one-way street (see Figure 16.5). In addition, the two strands of DNA in a double helix are antiparallel, meaning that they are oriented in opposite directions to each other, like a divided highway (see Figure 16.14). Clearly, the two new strands formed during DNA replication must also be antiparallel to their template strands.

How does the antiparallel arrangement of the double helix affect replication? Because of their structure, DNA polymerases can add nucleotides only to the free 3′ end of a primer or growing DNA strand, never to the 5′ end (see Figure 16.14). Thus, a new DNA strand can elongate only in the 5′→3′ direction. With this in mind, let's examine a replication fork **(Figure 16.15)**. Along one template strand, DNA polymerase III can synthesize a complementary strand continuously by elongating the new DNA in the mandatory 5′→3′ direction. DNA pol III simply nestles in the replication fork on that template strand and continuously adds nucleotides to the new complementary strand as the fork progresses. The DNA strand made by this mechanism is called the **leading strand**. Only one primer is required for DNA pol III to synthesize the leading strand (see Figure 16.15).

To elongate the other new strand of DNA in the mandatory 5′→3′ direction, DNA pol III must work along the other template strand in the direction *away from* the replication fork. The DNA strand elongating in this direction is called the **lagging strand**.* In contrast to the leading strand, which elongates continuously,

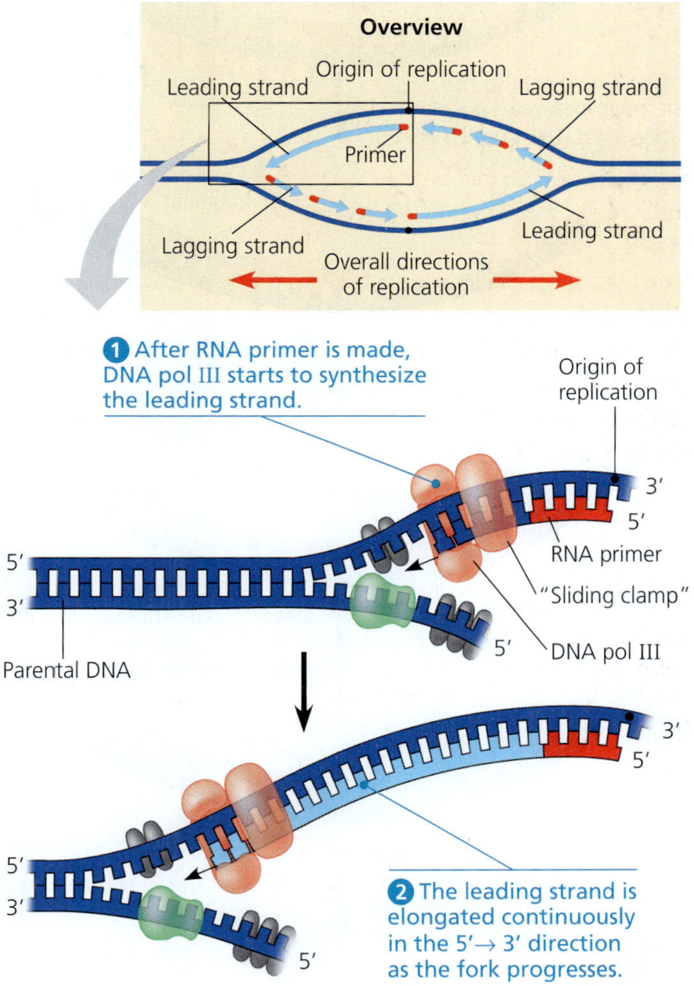

▲ **Figure 16.15 Synthesis of the leading strand during DNA replication.** This diagram focuses on the left replication fork shown in the overview box. DNA polymerase III (DNA pol III), shaped like a cupped hand, is closely associated with a protein called the "sliding clamp" that encircles the newly synthesized double helix like a doughnut. The sliding clamp moves DNA pol III along the DNA template strand.

* Synthesis of the leading strand and synthesis of the lagging strand occur concurrently and at the same rate. The lagging strand is so named because its synthesis is delayed slightly relative to synthesis of the leading strand; each new fragment of the lagging strand cannot be started until enough template has been exposed at the replication fork.

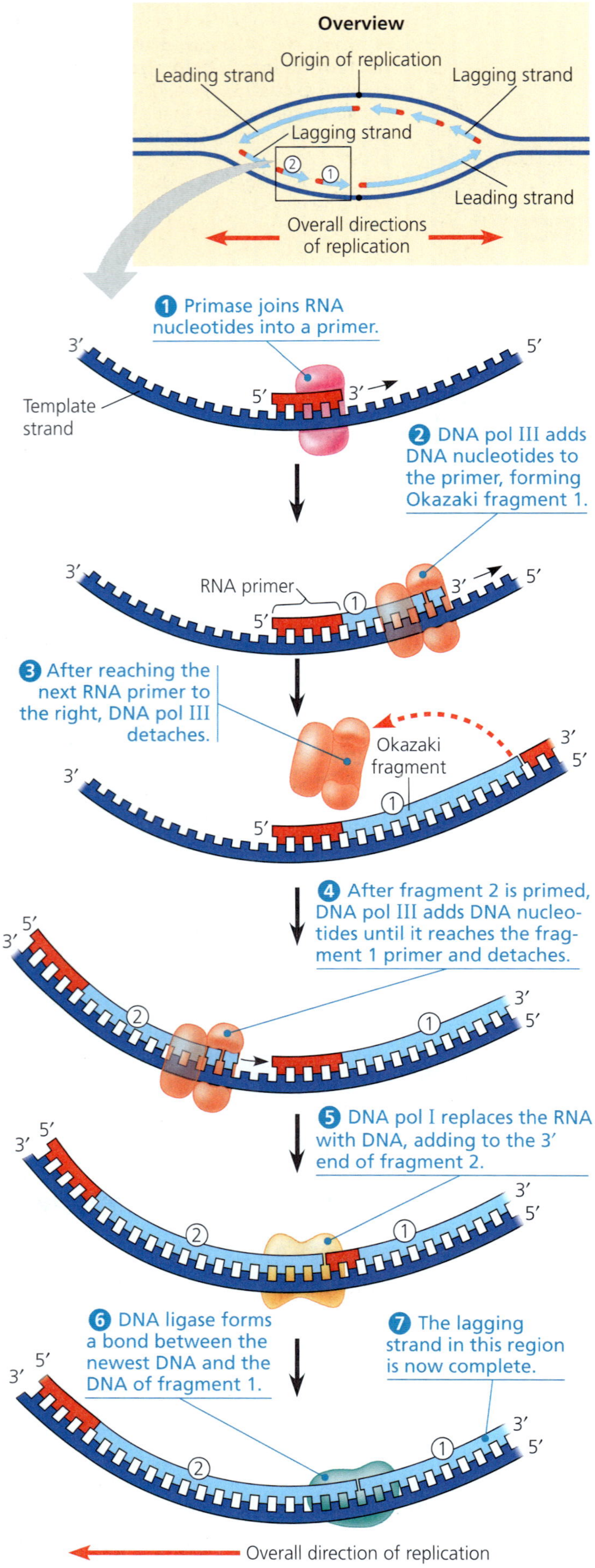

▲ **Figure 16.16 Synthesis of the lagging strand.**

the lagging strand is synthesized discontinuously, as a series of segments. These segments of the lagging strand are called **Okazaki fragments**, after the Japanese scientist who discovered them. The fragments are about 1,000 to 2,000 nucleotides long in *E. coli* and 100 to 200 nucleotides long in eukaryotes.

Figure 16.16 illustrates the steps in the synthesis of the lagging strand. Whereas only one primer is required on the leading strand, each Okazaki fragment on the lagging strand must be primed separately. Another DNA polymerase, DNA polymerase I (DNA pol I), replaces the RNA nucleotides of the primers with DNA versions, adding them one by one onto the 3′ end of the adjacent Okazaki fragment (fragment 2 in Figure 16.16). But DNA pol I cannot join the final nucleotide of this replacement DNA segment to the first DNA nucleotide of the Okazaki fragment whose primer was just replaced (fragment 1 in Figure 16.16). Another enzyme, **DNA ligase**, accomplishes this task, joining the sugar-phosphate backbones of all the Okazaki fragments into a continuous DNA strand.

Figure 16.17 and **Table 16.1**, on the next page, summarize DNA replication. Study them carefully before proceeding.

The DNA Replication Complex

It is traditional—and convenient—to represent DNA polymerase molecules as locomotives moving along a DNA "railroad track," but such a model is inaccurate in two important ways. First, the various proteins that participate in DNA replication actually form a single large complex, a "DNA replication machine." Many protein-protein interactions facilitate the efficiency of this complex. For example, by interacting with other proteins at the fork, primase apparently acts as a molecular brake, slowing progress of the replication fork and coordinating the rate of replication on the leading and lagging strands. Second, the DNA replication complex does not move along the DNA; rather, the DNA moves through the complex during the replication process. In eukaryotic cells, multiple copies of the complex, perhaps grouped into "factories," may be anchored to the nuclear matrix, a framework of fibers extending through the interior of the nucleus. Recent studies support a model in which two DNA polymerase molecules, one on each template strand, "reel in" the parental DNA and extrude newly made daughter DNA molecules. Additional evidence suggests that the lagging strand is looped back through the complex, so that when a DNA polymerase completes synthesis of an Okazaki fragment and dissociates, it doesn't have far to travel to reach the primer for the next fragment, near the replication fork. This looping of the lagging strand enables more Okazaki fragments to be synthesized in less time.

Proofreading and Repairing DNA

We cannot attribute the accuracy of DNA replication solely to the specificity of base pairing. Although errors in the completed DNA molecule amount to only one in 10 billion nucleotides, initial pairing errors between incoming nucleotides and those in

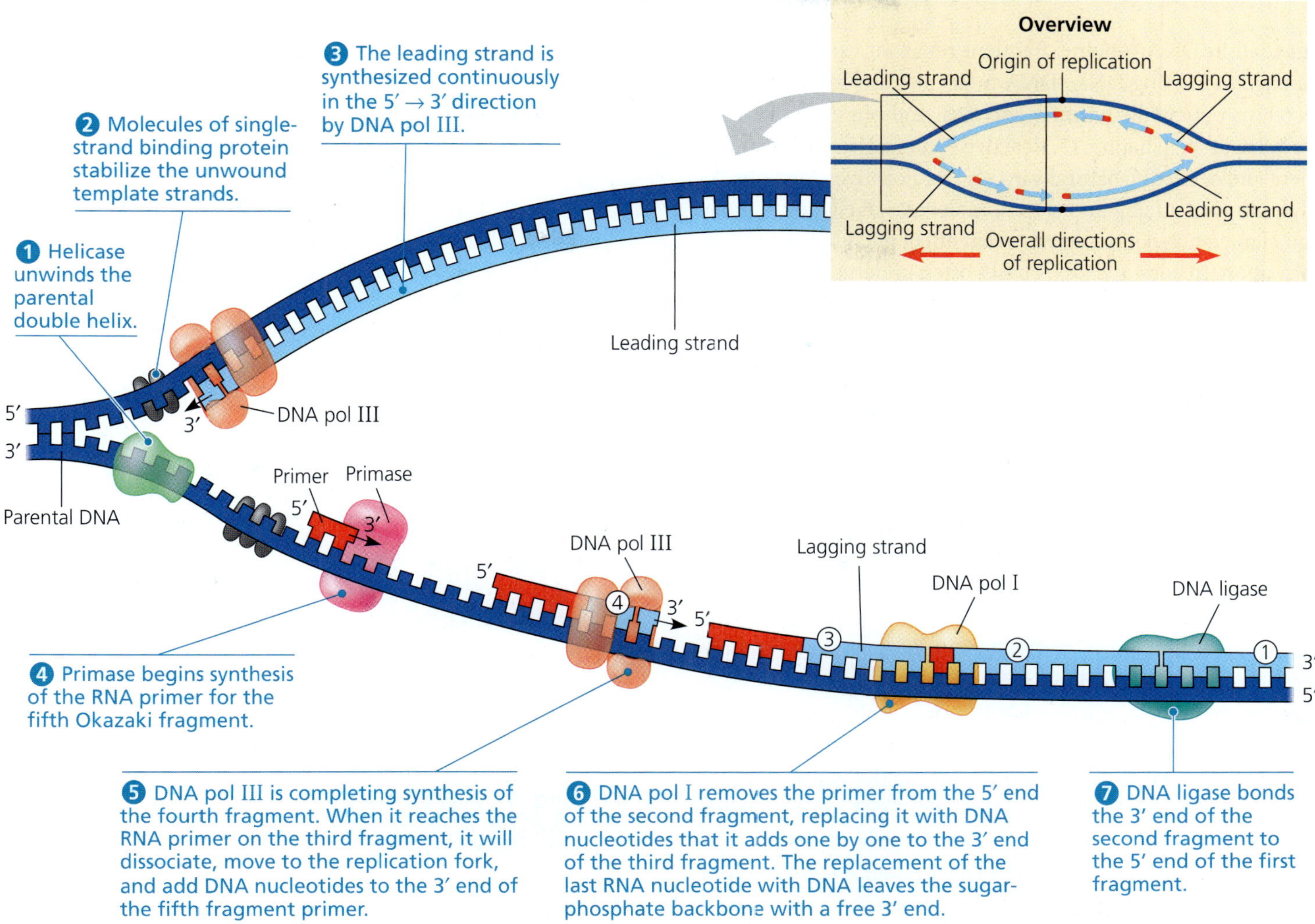

▲ **Figure 16.17 A summary of bacterial DNA replication.** The detailed diagram shows one replication fork, but as indicated in the overview (upper right), replication usually occurs simultaneously at two forks, one at either end of a replication bubble. Viewing each daughter strand in its entirety in the overview, you can see that half of it is made continuously as the leading strand, while the other half (on the other side of the origin) is synthesized in fragments as the lagging strand.

the template strand are 100,000 times more common—an error rate of one in 100,000 nucleotides. During DNA replication, DNA polymerases proofread each nucleotide against its template as soon as it is added to the growing strand. Upon finding an incorrectly paired nucleotide, the polymerase removes the nucleotide and then resumes synthesis. (This action is similar to fixing a word processing error by using the "delete" key and then entering the correct letter.)

Mismatched nucleotides sometimes evade proofreading by a DNA polymerase. In **mismatch repair**, enzymes remove and replace incorrectly paired nucleotides that have resulted from replication errors. Researchers spotlighted the importance of such enzymes when they found that a hereditary defect in one of them is associated with a form of colon cancer. Apparently, this defect allows cancer-causing errors to accumulate in the DNA at a faster rate than normal.

Incorrectly paired or altered nucleotides can also arise after replication. In fact, maintenance of the genetic information

Table 16.1 Bacterial DNA Replication Proteins and Their Functions

Protein	Function
Helicase	Unwinds parental double helix at replication forks
Single-strand binding protein	Binds to and stabilizes single-stranded DNA until it can be used as a template
Topoisomerase	Relieves "overwinding" strain ahead of replication forks by breaking, swiveling, and rejoining DNA strands
Primase	Synthesizes an RNA primer at 5′ end of leading strand and of each Okazaki fragment of lagging strand
DNA pol III	Using parental DNA as a template, synthesizes new DNA strand by covalently adding nucleotides to the 3′ end of a pre-existing DNA strand or RNA primer
DNA pol I	Removes RNA nucleotides of primer from 5′ end and replaces them with DNA nucleotides
DNA ligase	Joins 3′ end of DNA that replaces primer to rest of leading strand and joins Okazaki fragments of lagging strand

encoded in DNA requires frequent repair of various kinds of damage to existing DNA. DNA molecules are constantly subjected to potentially harmful chemical and physical agents, as we'll discuss in Chapter 17. Reactive chemicals (in the environment and occurring naturally in cells), radioactive emissions, X-rays, ultraviolet light, and certain molecules in cigarette smoke can change nucleotides in ways that affect encoded genetic information. In addition, DNA bases often undergo spontaneous chemical changes under normal cellular conditions. However, these changes in DNA are usually corrected before they become mutations perpetuated through successive replications. Each cell continuously monitors and repairs its genetic material. Because repair of damaged DNA is so important to the survival of an organism, it is no surprise that many different DNA repair enzymes have evolved. Almost 100 are known in *E. coli*, and about 130 have been identified so far in humans.

Most cellular systems for repairing incorrectly paired nucleotides, whether they are due to DNA damage or to replication errors, use a mechanism that takes advantage of the base-paired structure of DNA. Often, a segment of the strand containing the damage is cut out (excised) by a DNA-cutting enzyme—a **nuclease**—and the resulting gap is then filled in with nucleotides, using the undamaged strand as a template. The enzymes involved in filling the gap are a DNA polymerase and DNA ligase. One such DNA repair system is called **nucleotide excision repair** **(Figure 16.18)**.

An important function of the DNA repair enzymes in our skin cells is to repair genetic damage caused by the ultraviolet rays of sunlight. One type of damage, shown in Figure 16.18, is the covalent linking of thymine bases that are adjacent on a DNA strand. Such *thymine dimers* cause the DNA to buckle and interfere with DNA replication. The importance of repairing this kind of damage is underscored by the disorder xeroderma pigmentosum, which in most cases is caused by an inherited defect in a nucleotide excision repair enzyme. Individuals with this disorder are hypersensitive to sunlight; mutations in their skin cells caused by ultraviolet light are left uncorrected and cause skin cancer.

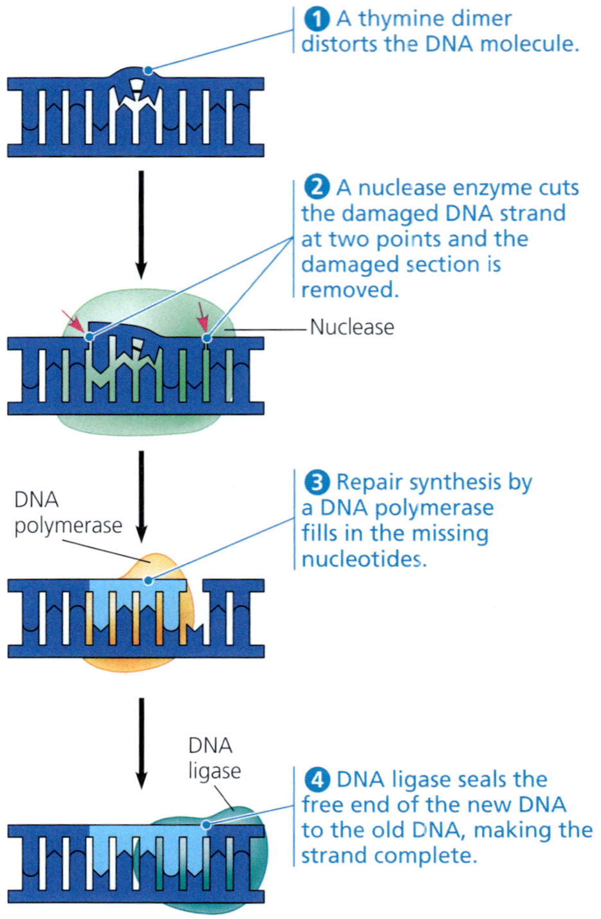

▲ **Figure 16.18 Nucleotide excision repair of DNA damage.** A team of enzymes detects and repairs damaged DNA. This figure shows DNA containing a thymine dimer, a type of damage often caused by ultraviolet radiation. A nuclease enzyme cuts out the damaged region of DNA, and a DNA polymerase (in bacteria, DNA pol I) replaces it with nucleotides complementary to the undamaged strand. DNA ligase completes the process by closing the remaining break in the sugar-phosphate backbone.

Replicating the Ends of DNA Molecules

In spite of the impressive capabilities of DNA polymerases, there is a small portion of the cell's DNA that DNA polymerases can neither replicate nor repair. For linear DNA, such as the DNA of eukaryotic chromosomes, the fact that a DNA polymerase can add nucleotides only to the 3′ end of a preexisting polynucleotide leads to an apparent problem. The usual replication machinery provides no way to complete the 5′ ends of daughter DNA strands. Even if an Okazaki fragment can be started with an RNA primer bound to the very end of the template strand, once that primer is removed, it cannot be replaced with DNA because there is no 3′ end available for nucleotide addition **(Figure 16.19)**. As a result, repeated rounds of replication produce shorter and shorter DNA molecules with uneven ("staggered") ends.

The shortening of DNA does not occur in most prokaryotes because their DNA is circular and therefore has no ends. But what protects the genes of eukaryotes from being eroded away during successive rounds of DNA replication? It turns out that eukaryotic chromosomal DNA molecules have special nucleotide sequences called **telomeres** at their ends **(Figure 16.20)**. Telomeres do not contain genes; instead, the DNA typically consists of multiple repetitions of one short nucleotide sequence. In each human telomere, for example, the six-nucleotide sequence TTAGGG is repeated between 100 and 1,000 times. Telomeric DNA protects the organism's genes. In addition, specific proteins associated with telomeric DNA prevent the staggered ends of the daughter molecule from activating the cell's systems for monitoring DNA damage. (Staggered ends of a DNA molecule, which often result from double-strand breaks, can trig-

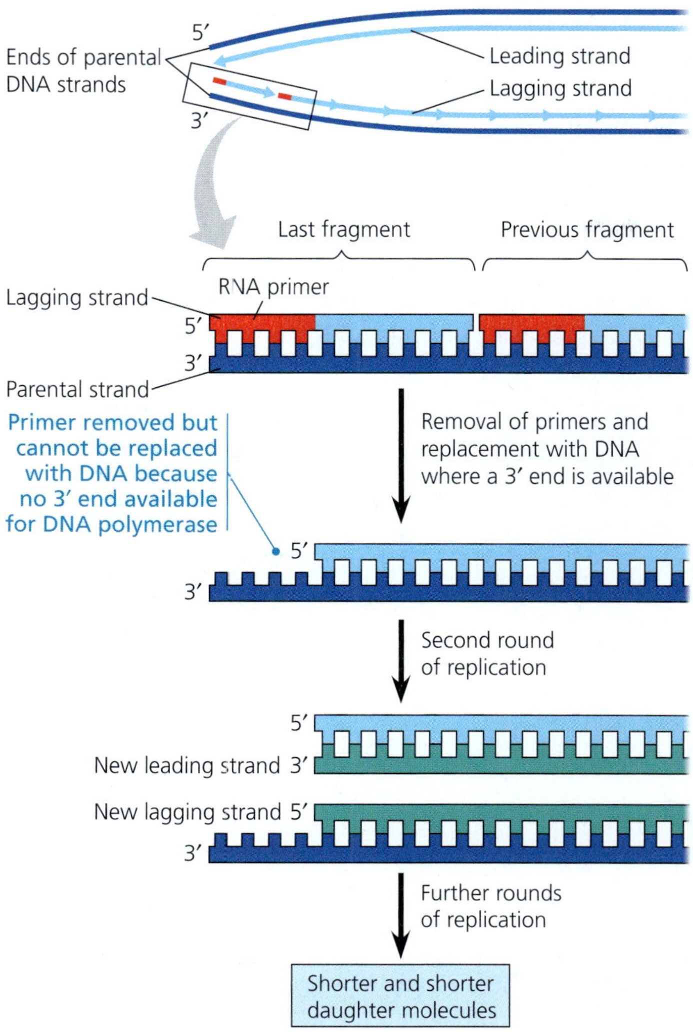

▲ **Figure 16.19 Shortening of the ends of linear DNA molecules.** Here we follow the end of one strand of a DNA molecule through two rounds of replication. After the first round, the new lagging strand is shorter than its template. After a second round, both the leading and lagging strands have become shorter than the original parental DNA. Although not shown here, the other ends of these DNA molecules also become shorter.

ger signal transduction pathways leading to cell cycle arrest or cell death.)

Telomeres do not prevent the shortening of DNA molecules due to successive rounds of replication; they just postpone the erosion of genes near the ends of DNA molecules. As shown in Figure 16.19, telomeres become shorter during every round of replication. As we would expect, telomeric DNA does tend to be shorter in dividing somatic cells of older individuals and in cultured cells that have divided many times. It has been proposed that shortening of telomeres is somehow connected to the aging process of certain tissues and even to aging of the organism as a whole.

But what about the cells whose genomes must persist unchanged from an organism to its offspring over many generations? If the chromosomes of germ cells (which give rise to gametes) became shorter in every cell cycle, essential genes

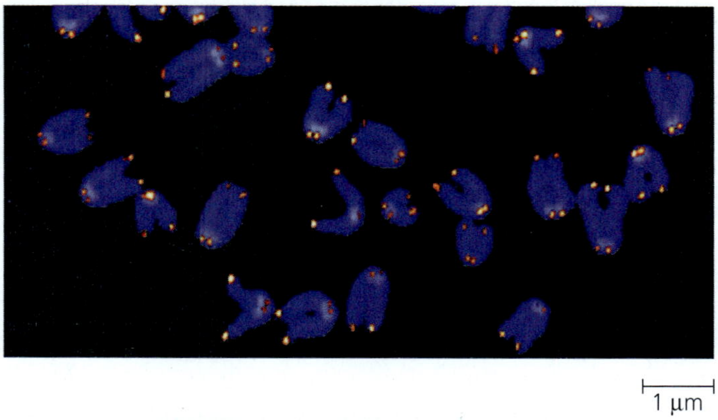

▲ **Figure 16.20 Telomeres.** Eukaryotes have repetitive, noncoding sequences called telomeres at the ends of their DNA. Telomeres are stained orange in these mouse chromosomes (LM).

would eventually be missing from the gametes they produce. However, this does not occur: An enzyme called **telomerase** catalyzes the lengthening of telomeres in eukaryotic germ cells, thus restoring their original length and compensating for the shortening that occurs during DNA replication. Telomerase is not active in most human somatic cells, but its activity in germ cells results in telomeres of maximum length in the zygote.

Normal shortening of telomeres may protect organisms from cancer by limiting the number of divisions that somatic cells can undergo. Cells from large tumors often have unusually short telomeres, as one would expect for cells that have undergone many cell divisions. Further shortening would presumably lead to self-destruction of the tumor cells. Intriguingly, researchers have found telomerase activity in cancerous somatic cells, suggesting that its ability to stabilize telomere length may allow these cancer cells to persist. Many cancer cells do seem capable of unlimited cell division, as do immortal strains of cultured cells (see Chapter 12). If telomerase is indeed an important factor in many cancers, it may provide a useful target for both cancer diagnosis and chemotherapy.

Thus far in this chapter, you have learned about the structure and replication of a DNA molecule. In the next section, we'll examine how DNA is packaged into chromosomes, the structures that carry the genetic information.

CONCEPT CHECK 16.2

1. What role does complementary base pairing play in the replication of DNA?
2. Identify two major functions of DNA pol III in DNA replication.
3. **WHAT IF?** If the DNA pol I in a given cell were nonfunctional, how would that affect the synthesis of a *leading* strand? In the overview box in Figure 16.17, point out where DNA pol I would normally function on the top leading strand.

For suggested answers, see Appendix A.

CONCEPT **16.3**

A chromosome consists of a DNA molecule packed together with proteins

The main component of the genome in most bacteria is one double-stranded, circular DNA molecule that is associated with a small amount of protein. Although we refer to this structure as the *bacterial chromosome*, it is very different from a eukaryotic chromosome, which consists of one linear DNA molecule associated with a large amount of protein. In *E. coli*, the chromosomal DNA consists of about 4.6 million nucleotide pairs, representing about 4,400 genes. This is 100 times more DNA than is found in a typical virus, but only about one-thousandth as much DNA as in a human somatic cell. Still, it is a lot of DNA to be packaged in such a small container.

Stretched out, the DNA of an *E. coli* cell would measure about a millimeter in length, 500 times longer than the cell.

▼ **Figure 16.21**

Exploring Chromatin Packing in a Eukaryotic Chromosome

This series of diagrams and transmission electron micrographs depicts a current model for the progressive levels of DNA coiling and folding. The illustration zooms out from a single molecule of DNA to a metaphase chromosome, which is large enough to be seen with a light microscope.

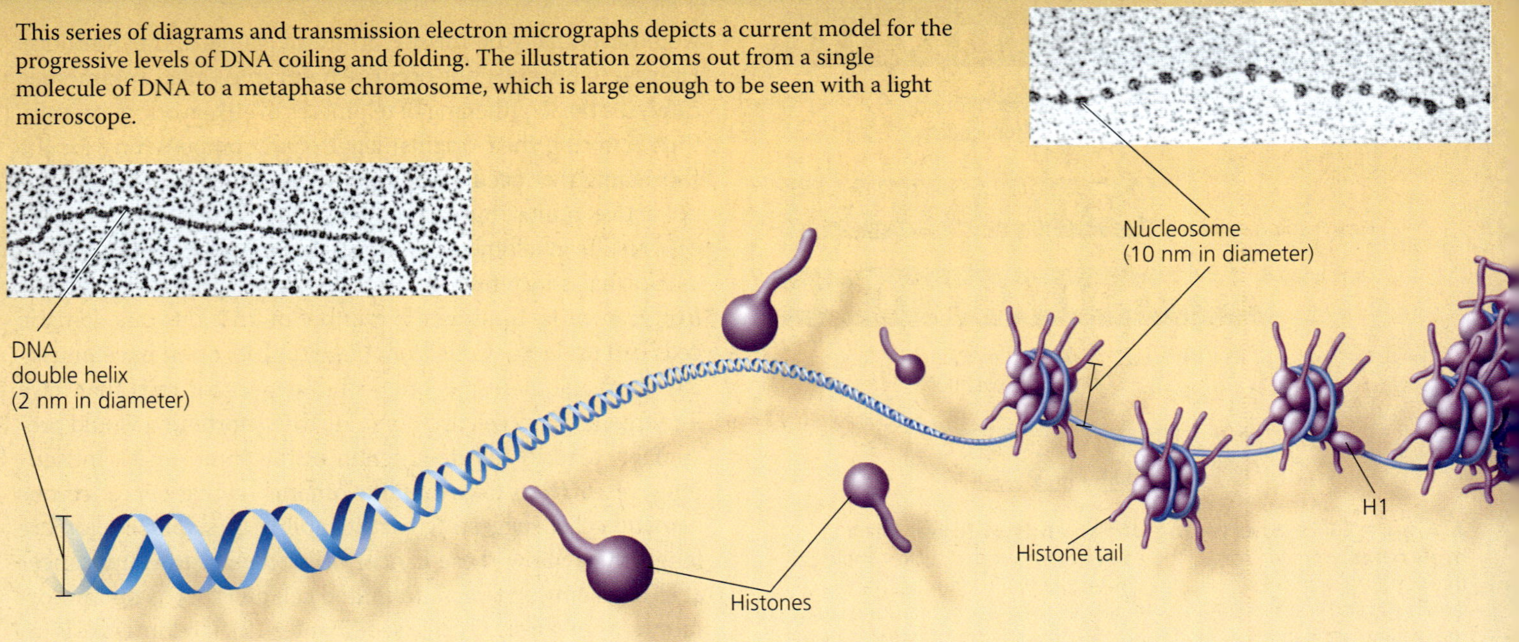

1 DNA, the double helix

Shown here is a ribbon model of DNA, with each ribbon representing one of the sugar-phosphate backbones. As you will recall from Figure 16.7, the phosphate groups along the backbone contribute a negative charge along the outside of each strand. The TEM shows a molecule of naked DNA; the double helix alone is 2 nm across.

2 Histones

Proteins called histones are responsible for the first level of DNA packing in chromatin. Although each histone is small—containing about 100 amino acids—the total mass of histone in chromatin approximately equals the mass of DNA. More than a fifth of a histone's amino acids are positively charged (lysine or arginine) and bind tightly to the negatively charged DNA.

Four types of histones are most common in chromatin: H2A, H2B, H3, and H4. The histones are very similar among eukaryotes; for example, all but two of the amino acids in cow H4 are identical to those in pea H4. The apparent conservation of histone genes during evolution probably reflects the pivotal role of histones in organizing DNA within cells.

The four main types of histones are critical to the next level of DNA packing. (A fifth type of histone, called H1, is involved in a further stage of packing.)

3 Nucleosomes, or "beads on a string" (10-nm fiber)

In electron micrographs, unfolded chromatin is 10 nm in diameter (the *10-nm fiber*). Such chromatin resembles beads on a string (see the TEM). Each "bead" is a *nucleosome*, the basic unit of DNA packing; the "string" between beads is called *linker DNA*.

A nucleosome consists of DNA wound twice around a protein core composed of two molecules each of the four main histone types. The amino end (N-terminus) of each histone (the *histone tail*) extends outward from the nucleosome.

In the cell cycle, the histones leave the DNA only briefly during DNA replication. Generally, they do the same during gene expression, another process that requires access to the DNA by the cell's molecular machinery. Chapter 18 will discuss some recent findings about the role of histone tails and nucleosomes in the regulation of gene expression.

Within a bacterium, however, certain proteins cause the chromosome to coil and "supercoil," densely packing it so that it fills only part of the cell. Unlike the nucleus of a eukaryotic cell, this dense region of DNA in a bacterium, called the **nucleoid**, is not bounded by membrane (see Figure 6.6).

Eukaryotic chromosomes each contain a single linear DNA double helix that, in humans, averages about 1.5×10^8 nucleotide pairs. This is an enormous amount of DNA relative to a chromosome's condensed length. If completely stretched out, such a DNA molecule would be about 4 cm long, thousands of times the diameter of a cell nucleus—and that's not even considering the DNA of the other 45 human chromosomes!

In the cell, eukaryotic DNA is precisely combined with a large amount of protein. Together, this complex of DNA and protein, called **chromatin**, fits into the nucleus through an elaborate, multilevel system of DNA packing. Our current view of the successive levels of DNA packing in a chromosome is outlined in **Figure 16.21**. Study this figure carefully before reading further.

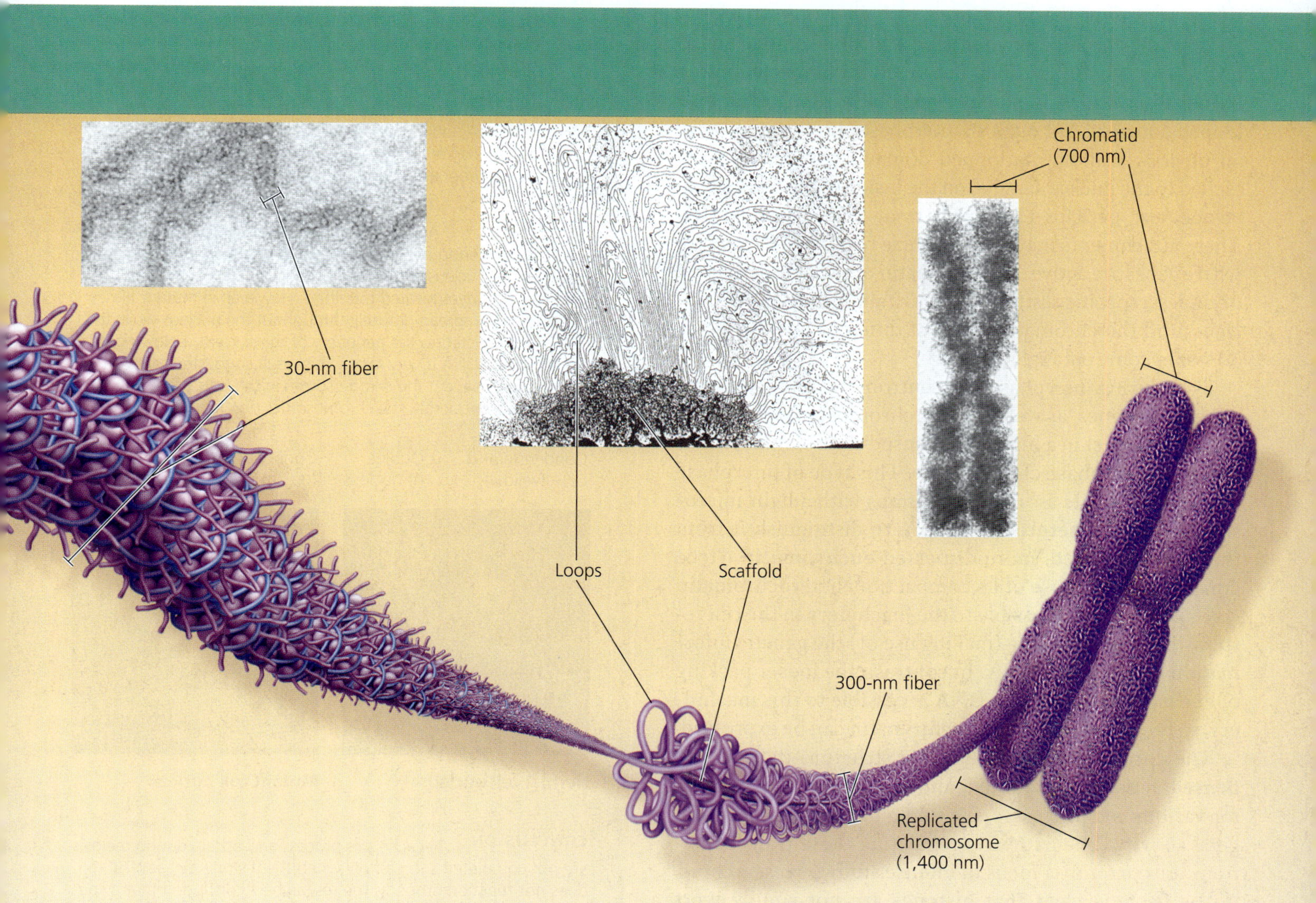

4 30-nm fiber

The next level of packing is due to interactions between the histone tails of one nucleosome and the linker DNA and nucleosomes on either side. A fifth histone, H1, is involved at this level. These interactions cause the extended 10-nm fiber to coil or fold, forming a chromatin fiber roughly 30 nm in thickness, the *30-nm fiber*. Although the 30-nm fiber is quite prevalent in the interphase nucleus, the packing arrangement of nucleosomes in this form of chromatin is still a matter of some debate.

5 Looped domains (300-nm fiber)

The 30-nm fiber, in turn, forms loops called *looped domains* attached to a chromosome scaffold made of proteins, thus making up a *300-nm fiber*. The scaffold is rich in one type of topoisomerase, and H1 molecules also appear to be present.

6 Metaphase chromosome

In a mitotic chromosome, the looped domains themselves coil and fold in a manner not yet fully understood, further compacting all the chromatin to produce the characteristic metaphase chromosome shown in the micrograph above. The width of one chromatid is 700 nm. Particular genes always end up located at the same places in metaphase chromosomes, indicating that the packing steps are highly specific and precise.

Chromatin undergoes striking changes in its degree of packing during the course of the cell cycle (see Figure 12.6). In interphase cells stained for light microscopy, the chromatin usually appears as a diffuse mass within the nucleus, suggesting that the chromatin is highly extended. As a cell prepares for mitosis, its chromatin coils and folds up (condenses), eventually forming a characteristic number of short, thick metaphase chromosomes that are distinguishable from each other with the light microscope.

Though interphase chromatin is generally much less condensed than the chromatin of mitotic chromosomes, it shows several of the same levels of higher-order packing. Some of the chromatin comprising a chromosome seems to be present as a 10-nm fiber, but much is compacted into a 30-nm fiber, which in some regions is further folded into looped domains. Although an interphase chromosome lacks an obvious scaffold, its looped domains appear to be attached to the nuclear lamina, on the inside of the nuclear envelope, and perhaps also to fibers of the nuclear matrix. These attachments may help organize regions of chromatin where genes are active. The chromatin of each chromosome occupies a specific restricted area within the interphase nucleus, and the chromatin fibers of different chromosomes do not become entangled.

Even during interphase, the centromeres and telomeres of chromosomes, as well as other chromosomal regions in some cells, exist in a highly condensed state similar to that seen in a metaphase chromosome. This type of interphase chromatin, visible as irregular clumps with a light microscope, is called **heterochromatin**, to distinguish it from the less compacted, more dispersed **euchromatin** ("true chromatin"). Because of its compaction, heterochromatin DNA is largely inaccessible to the machinery in the cell responsible for expressing (making use of) the genetic information coded in the DNA. In contrast, the looser packing of euchromatin makes its DNA accessible to this machinery, so the genes present in euchromatin can be expressed.

The chromosome is a dynamic structure that is condensed, loosened, modified, and remodeled as necessary for various cell processes, including mitosis, meiosis, and gene activity. The pathways regulating these transformations are currently a focus of intense study by researchers. It has become clear that histones are not simply inert spools around which the DNA is wrapped. Instead, they can undergo chemical modifications that result in changes in chromatin organization. Terry Orr-Weaver, the scientist interviewed at the beginning of this unit (pp. 246–247), has long been interested in the molecular mechanisms of chromosome dynamics during mitosis and meiosis. Using a genetic approach in *Drosophila*, she and her colleagues showed that phosphorylation of a specific amino acid on a histone tail plays a crucial role in chromosome behavior during prophase I of meiosis **(Figure 16.22)**.

▼ Figure 16.22 Inquiry

What role does histone phosphorylation play in chromosome behavior during meiosis?

EXPERIMENT Terry Orr-Weaver and colleagues at the Massachusetts Institute of Technology mutagenized fruit flies and looked for mutations that caused sterility, reasoning that such mutations might be found in genes coding for proteins that play important roles during meiosis. They found a mutation in the *nhk-1* gene that caused sterility in *Drosophila* females. They knew that the gene product, nucleosomal histone kinase-1, or NHK-1, is an enzyme that phosphorylates a specific amino acid on the tail of histone H2A. They hypothesized that sterility is caused by unsuccessful meiosis due to abnormal chromosome behavior when this enzyme does not function properly.

To test this hypothesis, they observed chromosome behavior closely during meiosis in ovarian cells of normal and mutant flies. In one experiment, they used a red fluorescent dye to mark the location of DNA and a green fluorescent dye to mark the location of the protein *condensin*, which normally coats the chromosomes at the end of prophase I and helps them condense.

RESULTS At the end of prophase I in ovarian cells of normal flies, condensin and DNA were both localized to a very small region in the nucleus (below left; the yellow color is the result of green and red dyes located together). However, in mutant flies, condensin was spread diffusely throughout the nucleus while the DNA was restricted to the periphery of the nucleus (below right; the red of the DNA is very faint). This result suggested that condensin does not coat the chromosomes in the cells of mutant flies and that, consequently, the chromosomes don't condense.

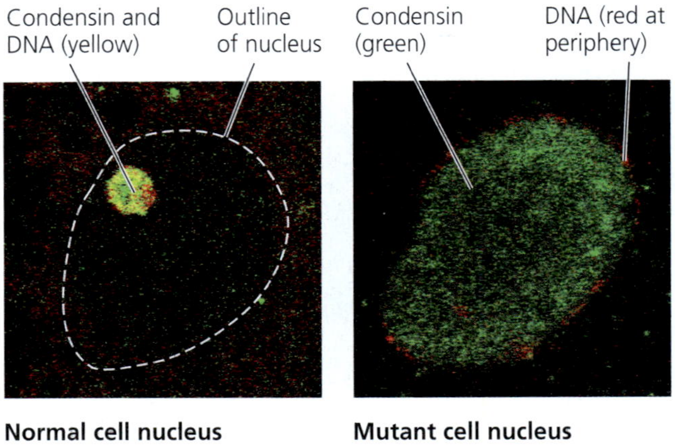

CONCLUSION Because this process and others during meiosis are not completed successfully when the histone kinase NHK-1 does not function properly, the researchers concluded that a specific phosphorylation of histone H2A is necessary for normal chromosome behavior during meiosis.

SOURCE I. Ivanovska, T. Khandan, T. Ito, and T. L. Orr-Weaver, A histone code in meiosis: the histone kinase, NHK-1, is required for proper chromosomal architecture in *Drosophila* oocytes, *Genes and Development* 19:2571–2582 (2005).

WHAT IF? Suppose a researcher discovered a mutant fly in which the histone H2A tail was missing the specific amino acid usually phosphorylated by the histone kinase NHK-1. How would this mutation likely affect chromosome behavior during meiosis in ovarian cells?

Phosphorylation and other chemical modifications of histones also have multiple effects on gene activity, as you will see in Chapter 18.

In this chapter, you have learned how DNA molecules are arranged in chromosomes and how DNA replication provides the copies of genes that parents pass to offspring. However, it is not enough that genes be copied and transmitted; the information they carry must be used by the cell. In other words, genes must also be "expressed." In the next chapter, we will examine how the cell translates genetic information encoded in DNA.

CONCEPT CHECK 16.3

1. Describe the structure of a nucleosome, the basic unit of DNA packing in eukaryotic cells.
2. What two properties distinguish heterochromatin from euchromatin?
3. **WHAT IF?** Although the proteins that cause the *E. coli* chromosome to coil are not histones, what property would you expect them to share with histones, given their ability to bind to DNA?

For suggested answers, see Appendix A.

Chapter 16 Review

MEDIA Go to the Study Area at **www.masteringbio.com** for BioFlix 3-D Animations, MP3 Tutors, Videos, Practice Tests, an eBook, and more.

SUMMARY OF KEY CONCEPTS

CONCEPT 16.1

DNA is the genetic material (pp. 305–310)

▶ **The Search for the Genetic Material: *Scientific Inquiry*** Experiments with bacteria and with phages provided the first strong evidence that the genetic material is DNA.

▶ **Building a Structural Model of DNA: *Scientific Inquiry*** Watson and Crick deduced that DNA is a double helix. Two antiparallel sugar-phosphate chains wind around the outside of the molecule; the nitrogenous bases project into the interior, where they hydrogen-bond in specific pairs, A with T, G with C.

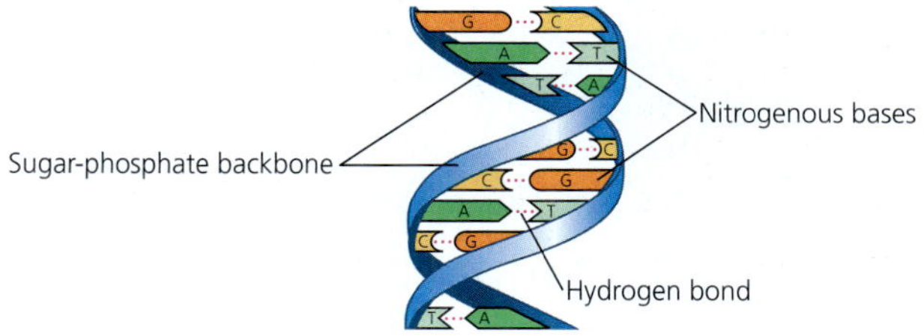

MEDIA
Activity The Hershey-Chase Experiment
Activity DNA and RNA Structure
Activity DNA Double Helix

CONCEPT 16.2

Many proteins work together in DNA replication and repair (pp. 311–319)

▶ **The Basic Principle: Base Pairing to a Template Strand** The Meselson-Stahl experiment showed that DNA replication is semiconservative: The parent molecule unwinds, and each strand then serves as a template for the synthesis of a new strand according to base-pairing rules.

▶ **DNA Replication: *A Closer Look***

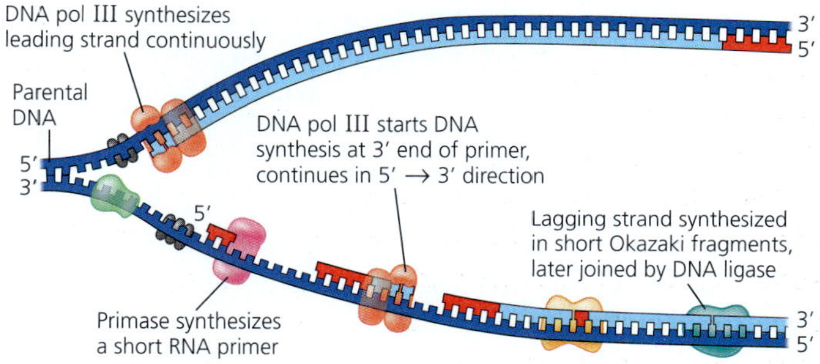

▶ **Proofreading and Repairing DNA** DNA polymerases proofread new DNA, replacing incorrect nucleotides. In mismatch repair, enzymes correct errors that persist. Nucleotide excision repair is a general process by which enzymes cut out and replace damaged stretches of DNA.

▶ **Replicating the Ends of DNA Molecules** The ends of eukaryotic chromosomal DNA get shorter with each round of replication. The presence of telomeres, repetitive sequences at the ends of linear DNA molecules, postpones the erosion of genes. Telomerase catalyzes the lengthening of telomeres in germ cells.

MEDIA
Activity DNA Replication: An Overview
Investigation What Is the Correct Model for DNA Replication?
Activity DNA Replication: A Closer Look
Activity DNA Replication: A Review

CONCEPT 16.3

A chromosome consists of a DNA molecule packed together with proteins (pp. 320–323)

▶ The bacterial chromosome is usually a circular DNA molecule with some associated proteins. Eukaryotic chromatin making up a chromosome is composed mostly of DNA, histones, and other proteins. The histones bind to each other and to the DNA to form nucleosomes, the most basic units of DNA packing. Histone tails extend outward from each bead-like nucleosome core. Additional folding leads ultimately to the highly condensed chromatin of the metaphase chromosome. In interphase cells, most chromatin is less compacted (euchromatin), but some remains highly condensed (heterochromatin). Histone modifications may influence the state of chromatin condensation.

MEDIA

Activity DNA Packing

TESTING YOUR KNOWLEDGE

SELF-QUIZ

1. In his work with pneumonia-causing bacteria and mice, Griffith found that
   a. the protein coat from pathogenic cells was able to transform nonpathogenic cells.
   b. heat-killed pathogenic cells caused pneumonia.
   c. some substance from pathogenic cells was transferred to nonpathogenic cells, making them pathogenic.
   d. the polysaccharide coat of bacteria caused pneumonia.
   e. bacteriophages injected DNA into bacteria.

2. *E. coli* cells grown on ^{15}N medium are transferred to ^{14}N medium and allowed to grow for two more generations (two rounds of DNA replication). DNA extracted from these cells is centrifuged. What density distribution of DNA would you expect in this experiment?
   a. one high-density and one low-density band
   b. one intermediate-density band
   c. one high-density and one intermediate-density band
   d. one low-density and one intermediate-density band
   e. one low-density band

3. A biochemist isolates and purifies molecules needed for DNA replication. When she adds some DNA, replication occurs, but each DNA molecule consists of a normal strand paired with numerous segments of DNA a few hundred nucleotides long. What has she probably left out of the mixture?
   a. DNA polymerase
   b. DNA ligase
   c. nucleotides
   d. Okazaki fragments
   e. primase

4. What is the basis for the difference in how the leading and lagging strands of DNA molecules are synthesized?
   a. The origins of replication occur only at the 5′ end.
   b. Helicases and single-strand binding proteins work at the 5′ end.
   c. DNA polymerase can join new nucleotides only to the 3′ end of a growing strand.
   d. DNA ligase works only in the 3′→5′ direction.
   e. Polymerase can work on only one strand at a time.

5. In analyzing the number of different bases in a DNA sample, which result would be consistent with the base-pairing rules?
   a. A = G
   b. A + G = C + T
   c. A + T = G + T
   d. A = C
   e. G = T

6. The elongation of the leading strand during DNA synthesis
   a. progresses away from the replication fork.
   b. occurs in the 3′→5′ direction.
   c. produces Okazaki fragments.
   d. depends on the action of DNA polymerase.
   e. does not require a template strand.

7. The spontaneous loss of amino groups from adenine results in hypoxanthine, an uncommon base, opposite thymine in DNA. What combination of molecules could repair such damage?
   a. nuclease, DNA polymerase, DNA ligase
   b. telomerase, primase, DNA polymerase
   c. telomerase, helicase, single-strand binding protein
   d. DNA ligase, replication fork proteins, adenylyl cyclase
   e. nuclease, telomerase, primase

8. In a nucleosome, the DNA is wrapped around
   a. polymerase molecules.
   b. ribosomes.
   c. histones.
   d. a thymine dimer.
   e. satellite DNA.

For Self-Quiz answers, see Appendix A.

MEDIA Visit the Study Area at **www.masteringbio.com** for a Practice Test.

EVOLUTION CONNECTION

9. Some bacteria may be able to respond to environmental stress by increasing the rate at which mutations occur during cell division. How might this be accomplished? Might there be an evolutionary advantage of this ability? Explain.

SCIENTIFIC INQUIRY

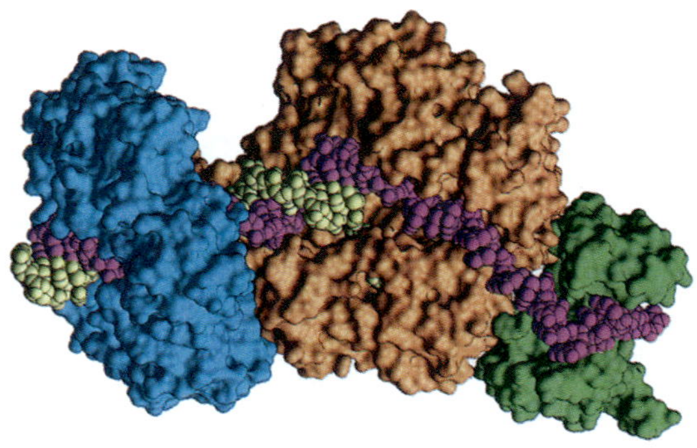

10. **DRAW IT** Model building can be an important part of the scientific process. The illustration above is a computer-generated model of a DNA replication complex. The parental and newly synthesized DNA strands are color-coded differently, as are each of the following three proteins: DNA pol III, the sliding clamp, and single-strand binding protein. Use what you've learned in this chapter to clarify this model by labeling each DNA strand and each protein and showing the overall direction of DNA replication.

17 From Gene to Protein

▲ **Figure 17.1 How does a single faulty gene result in the dramatic appearance of an albino deer?**

KEY CONCEPTS

- **17.1** Genes specify proteins via transcription and translation
- **17.2** Transcription is the DNA-directed synthesis of RNA: *a closer look*
- **17.3** Eukaryotic cells modify RNA after transcription
- **17.4** Translation is the RNA-directed synthesis of a polypeptide: *a closer look*
- **17.5** Point mutations can affect protein structure and function
- **17.6** While gene expression differs among the domains of life, the concept of a gene is universal

OVERVIEW

The Flow of Genetic Information

In 2006, a young albino deer seen frolicking with several brown deer in the mountains of eastern Germany elicited a public outcry **(Figure 17.1)**. A local hunting organization said the albino deer suffered from a "genetic disorder" and should be shot. Some people felt the deer should merely be prevented from mating with other deer in order to safeguard the gene pool of the population. Others favored relocating the albino deer to a nature reserve because they worried that it might be more noticeable to predators if left in the wild. A German rock star even held a benefit concert to raise funds for the relocation. What led to the striking phenotype of this deer, the cause of this lively debate?

You learned in Chapter 14 that inherited traits are determined by genes and that the trait of albinism is caused by a recessive allele of a pigmentation gene. The information content of genes is in the form of specific sequences of nucleotides along strands of DNA, the genetic material. But how does this information determine an organism's traits? Put another way, what does a gene actually say? And how is its message translated by cells into a specific trait, such as brown hair, type A blood, or, in the case of an albino deer, a total lack of pigment? The albino deer has a faulty version of a key protein, an enzyme required for pigment synthesis, and this protein is faulty because the gene that codes for it contains incorrect information.

This example illustrates the main point of this chapter: The DNA inherited by an organism leads to specific traits by dictating the synthesis of proteins and of RNA molecules involved in protein synthesis. In other words, proteins are the link between genotype and phenotype. **Gene expression** is the process by which DNA directs the synthesis of proteins (or, in some cases, just RNAs). The expression of genes that code for proteins includes two stages: transcription and translation. This chapter describes the flow of information from gene to protein in detail and explains how genetic mutations affect organisms through their proteins. Gene expression involves similar processes in all three domains of life. Understanding these processes will allow us to revisit the concept of the gene in more detail at the end of the chapter.

CONCEPT **17.1**

Genes specify proteins via transcription and translation

Before going into the details of how genes direct protein synthesis, let's step back and examine how the fundamental relationship between genes and proteins was discovered.

Evidence from the Study of Metabolic Defects

In 1909, British physician Archibald Garrod was the first to suggest that genes dictate phenotypes through enzymes that catalyze specific chemical reactions in the cell. Garrod postulated that the symptoms of an inherited disease reflect a person's inability to make a particular enzyme. He referred to such diseases as "inborn errors of metabolism." Garrod gave as one example the hereditary condition called alkaptonuria, in which the urine is black because it contains the chemical alkapton, which darkens upon exposure to air. Garrod reasoned that most people have an enzyme that metabolizes alkapton, whereas people with alkaptonuria have inherited an inability to make that enzyme.

Garrod may have been the first person to recognize that Mendel's principles of heredity apply to humans as well as peas. Garrod's realization was ahead of its time, but research conducted several decades later supported his hypothesis that a gene dictates the production of a specific enzyme. Biochemists accumulated much evidence that cells synthesize and degrade most organic molecules via metabolic pathways, in which each chemical reaction in a sequence is catalyzed by a specific enzyme (see p. 142). Such metabolic pathways lead, for instance, to the synthesis of the pigments that give fruit flies (*Drosophila*) their eye color (see Figure 15.3). In the 1930s, George Beadle and Boris Ephrussi speculated that in *Drosophila*, each of the various mutations affecting eye color blocks pigment synthesis at a specific step by preventing production of the enzyme that catalyzes that step. However, neither the chemical reactions nor the enzymes that catalyze them were known at the time.

Nutritional Mutants in Neurospora*: Scientific Inquiry*

A breakthrough in demonstrating the relationship between genes and enzymes came a few years later, when Beadle and Edward Tatum began working with a bread mold, *Neurospora crassa*. Using a treatment shown in the 1920s to cause genetic changes, they bombarded *Neurospora* with X-rays and then looked among the survivors for mutants that differed in their nutritional needs from the wild-type mold. Wild-type *Neurospora* has modest food requirements. It can survive in the laboratory on a moist support medium called agar, mixed only with inorganic salts, glucose, and the vitamin biotin. From this *minimal medium*, the mold cells use their metabolic pathways to produce all the other molecules they need. Beadle and Tatum identified mutants that could not survive on minimal medium, apparently because they were unable to synthesize certain essential molecules from the minimal ingredients. To ensure survival of these nutritional mutants, Beadle and Tatum allowed them to grow on a *complete growth medium*, which consisted of minimal medium supplemented with all 20 amino acids and a few other nutrients. The complete growth medium could support any mutant that couldn't synthesize one of the supplements.

To characterize the metabolic defect in each nutritional mutant, Beadle and Tatum took samples from the mutant growing on complete medium and distributed them to a number of different vials. Each vial contained minimal medium plus a single additional nutrient. The particular supplement that allowed growth indicated the metabolic defect. For example, if the only supplemented vial that supported growth of the mutant was the one fortified with the amino acid arginine, the researchers could conclude that the mutant was defective in the biochemical pathway that wild-type cells use to synthesize arginine.

Beadle and Tatum went on to pin down each mutant's defect more specifically. **Figure 17.2** shows how they used additional tests to distinguish among three classes of arginine-requiring mutants. Mutants in each class required a different set of compounds along the arginine-synthesizing pathway, which has three steps. Based on their results, the researchers reasoned that each class must be blocked at a different step in this pathway because mutants in that class lacked the enzyme that catalyzes the blocked step.

Because each mutant was defective in a single gene, Beadle and Tatum's results provided strong support for the *one gene–one enzyme hypothesis*, as they dubbed it, which states that the function of a gene is to dictate the production of a specific enzyme. Further support for this hypothesis came from experiments that identified the specific enzymes lacking in the mutants. Beadle and Tatum shared a Nobel Prize in 1958 for "their discovery that genes act by regulating definite chemical events" (in the words of the Nobel committee).

The Products of Gene Expression: A Developing Story

As researchers learned more about proteins, they made revisions to the one gene–one enzyme hypothesis. First of all, not all proteins are enzymes. Keratin, the structural protein of animal hair, and the hormone insulin are two examples of nonenzyme proteins. Because proteins that are not enzymes are nevertheless gene products, molecular biologists began to think in terms of one gene–one protein. However, many proteins are constructed from two or more different polypeptide chains, and each polypeptide is specified by its own gene. For example, hemoglobin, the oxygen-transporting protein of vertebrate red blood cells, is built from two kinds of polypeptides, and thus two genes code for this protein (see Figure 5.21). Beadle and Tatum's idea was therefore restated as the *one gene–one polypeptide hypothesis*. Even this description is not entirely accurate, though. First, many eukaryotic genes can code for a set of closely related polypeptides in a process called alternative splicing, which you will learn about later in this chapter. Second, quite a few genes code for RNA molecules that have important functions in cells even though they are never translated into protein. For now, we will focus on genes that do code for polypeptides. (Note that it is common to refer to these gene products as proteins, rather than more precisely as polypeptides—a practice you will encounter in this book.)

▼ Figure 17.2 Inquiry

Do individual genes specify the enzymes that function in a biochemical pathway?

EXPERIMENT Working with the mold *Neurospora crassa*, George Beadle and Edward Tatum, then at Stanford University, isolated mutants that required arginine in their growth medium. The researchers showed that these mutants fell into three classes, each defective in a different gene. From other considerations, they suspected that the metabolic pathway of arginine biosynthesis involved a precursor nutrient and the intermediate molecules ornithine and citrulline. Their most famous experiment, shown here, tested both their one gene–one enzyme hypothesis and their postulated arginine-synthesizing pathway. In this experiment, they grew their three classes of mutants under the four different conditions shown in the Results section below. They included minimal medium (MM) as a control because they knew that wild-type cells could grow on MM but mutant cells could not. (See test tubes on the right.)

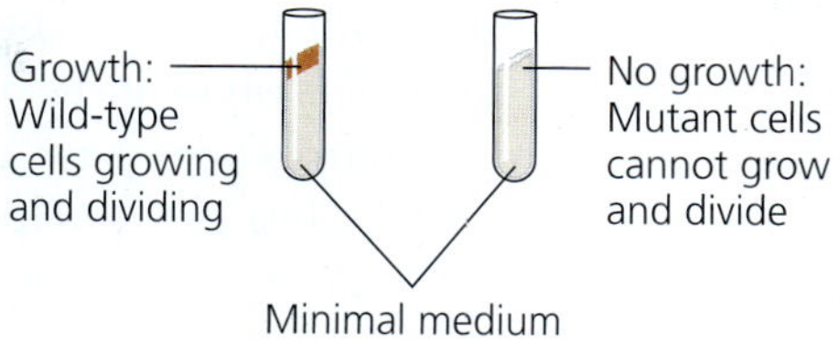

RESULTS The wild-type strain was capable of growth under all experimental conditions, requiring only the minimal medium. The three classes of mutants each had a specific set of growth requirements. For example, class II mutants could not grow when ornithine alone was added but could grow when either citrulline or arginine was added.

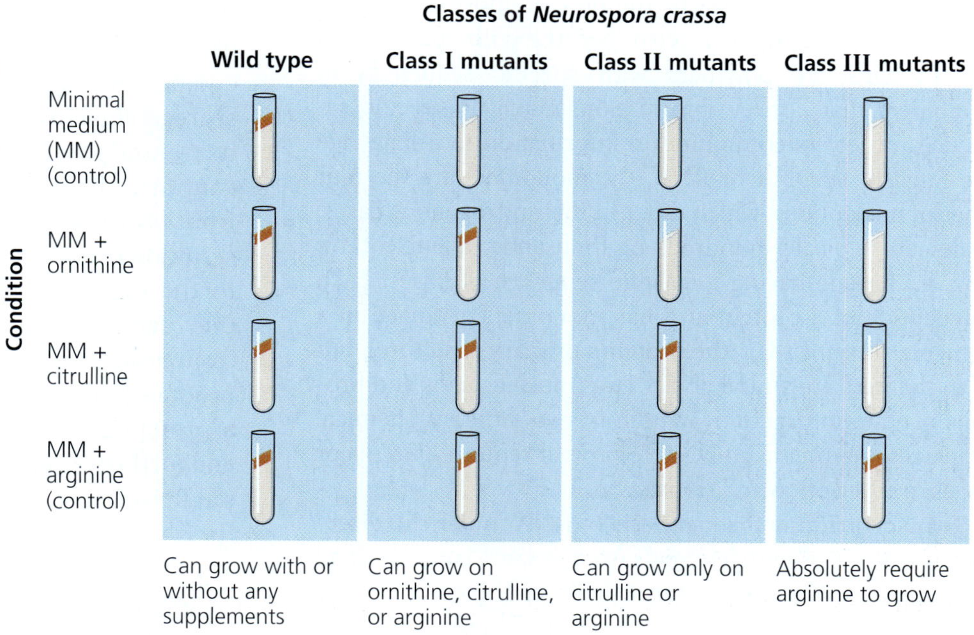

CONCLUSION From the growth requirements of the mutants, Beadle and Tatum deduced that each class of mutant was unable to carry out one step in the pathway for synthesizing arginine, presumably because it lacked the necessary enzyme. Because each of their mutants was mutated in a single gene, they concluded that each mutated gene must normally dictate the production of one enzyme. Their results supported the one gene–one enzyme hypothesis and also confirmed the arginine pathway. (Notice in the Results that a mutant can grow only if supplied with a compound made *after* the defective step, because this bypasses the defect.)

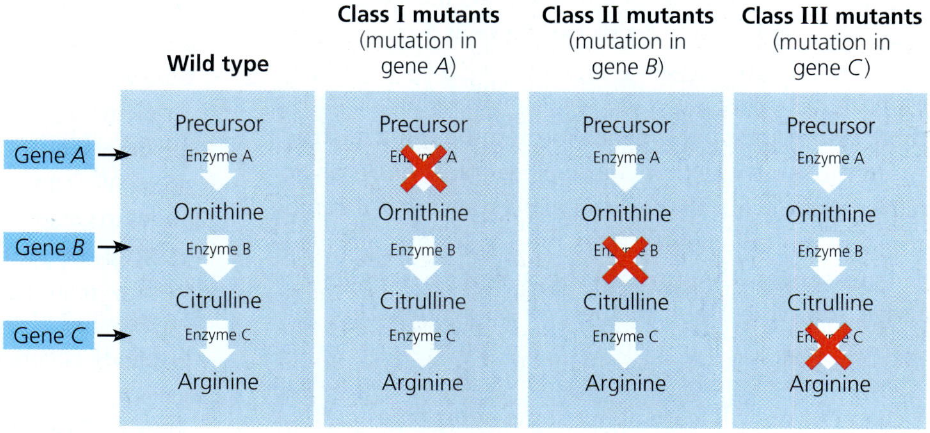

SOURCE G. W. Beadle and E. L. Tatum, Genetic control of biochemical reactions in *Neurospora*, *Proceedings of the National Academy of Sciences* 27:499–506 (1941).

WHAT IF? Suppose the experiment had shown that class I mutants could grow only in MM supplemented by ornithine or arginine and that class II mutants could grow in MM supplemented by citrulline, ornithine, or arginine. What conclusions would Beadle and Tatum have drawn from those results regarding the biochemical pathway and the defect in class I and class II mutants?

Basic Principles of Transcription and Translation

Genes provide the instructions for making specific proteins. But a gene does not build a protein directly. The bridge between DNA and protein synthesis is the nucleic acid RNA. You learned in Chapter 5 that RNA is chemically similar to DNA, except that it contains ribose instead of deoxyribose as its sugar and has the nitrogenous base uracil rather than thymine (see Figure 5.27). Thus, each nucleotide along a DNA strand has A, G, C, or T as its base, and each nucleotide along an RNA strand has A, G, C, or U as its base. An RNA molecule usually consists of a single strand.

It is customary to describe the flow of information from gene to protein in linguistic terms because both nucleic acids and proteins are polymers with specific sequences of monomers that convey information, much as specific sequences of letters communicate information in a language like English. In DNA or RNA, the monomers are the four types of nucleotides, which differ in their nitrogenous bases. Genes are typically hundreds or thousands of nucleotides long, each gene having a specific sequence of bases. Each polypeptide of a protein also has monomers arranged in a particular linear order (the protein's primary structure), but its monomers are amino acids. Thus, nucleic acids and proteins contain information written in two different chemical languages. Getting from DNA to protein requires two major stages: transcription and translation.

Transcription is the synthesis of RNA under the direction of DNA. Both nucleic acids use the same language, and the information is simply transcribed, or copied, from one molecule to the other. Just as a DNA strand provides a template for the synthesis of a new complementary strand during DNA replication, it also can serve as a template for assembling a complementary sequence of RNA nucleotides. For a protein-coding gene, the resulting RNA molecule is a faithful transcript of the gene's protein-building instructions, in the same way that your college transcript is an accurate record of your grades, and like a transcript, it can be sent out in multiple copies. This type of RNA molecule is called **messenger RNA (mRNA)** because it carries a genetic message from the DNA to the protein-synthesizing machinery of the cell. (Transcription is the general term for the synthesis of *any* kind of RNA on a DNA template. Later in this chapter, you will learn about some other types of RNA produced by transcription.)

Translation is the synthesis of a polypeptide, which occurs under the direction of mRNA. During this stage, there is a change in language: The cell must translate the base sequence of an mRNA molecule into the amino acid sequence of a polypeptide. The sites of translation are **ribosomes**, complex particles that facilitate the orderly linking of amino acids into polypeptide chains.

Transcription and translation occur in all organisms. Recall from Chapter 1 that there are three domains of life: Bacteria, Archaea, and Eukarya. Organisms in the first two domains are grouped as prokaryotes because their cells lack a membrane-bounded nucleus—a defining feature of eukaryotic cells. Most studies of transcription and translation have been done on bacteria and eukaryotes, which are therefore our main focus in this chapter. Although our understanding of these processes in archaea lags behind, in the last section we will discuss a few aspects of archaeal gene expression.

The basic mechanics of transcription and translation are similar for bacteria and eukaryotes, but there is an important difference in the flow of genetic information within the cells. Because bacteria do not have nuclei, their DNA is not segregated from ribosomes and the other protein-synthesizing equipment **(Figure 17.3a)**. As you will see later, this lack of segregation allows translation of an mRNA to begin while its transcription is still in progress. In a eukaryotic cell, by contrast, the nuclear envelope separates transcription from translation in space and time **(Figure 17.3b)**. Transcription occurs in the nucleus, and mRNA is transported to the cytoplasm, where translation occurs. But before they can leave the nucleus, eukaryotic RNA transcripts from protein-coding genes are modified in various ways to produce the final, functional mRNA. The transcription of a protein-coding eukaryotic gene results in *pre-mRNA*, and further processing yields the finished mRNA. The initial RNA transcript from any gene, including those coding for RNA that is not translated into protein, is more generally called a **primary transcript**.

Let's summarize: Genes program protein synthesis via genetic messages in the form of messenger RNA. Put another way, cells are governed by a molecular chain of command with a directional flow of genetic information: DNA → RNA → protein. This concept was dubbed the *central dogma* by Francis Crick in 1956. How has the concept held up over time? In the 1970s, scientists were surprised to discover that some RNA molecules can act as templates for DNA, a process you'll read about in Chapter 19. However, this rare exception does not invalidate the idea that, in general, genetic information flows from DNA to RNA to protein. In the next section, we discuss how the instructions for assembling amino acids into a specific order are encoded in nucleic acids.

The Genetic Code

When biologists began to suspect that the instructions for protein synthesis were encoded in DNA, they recognized a problem: There are only four nucleotide bases to specify 20 amino acids. Thus, the genetic code cannot be a language like Chinese, where each written symbol corresponds to a word. How many bases, then, correspond to an amino acid?

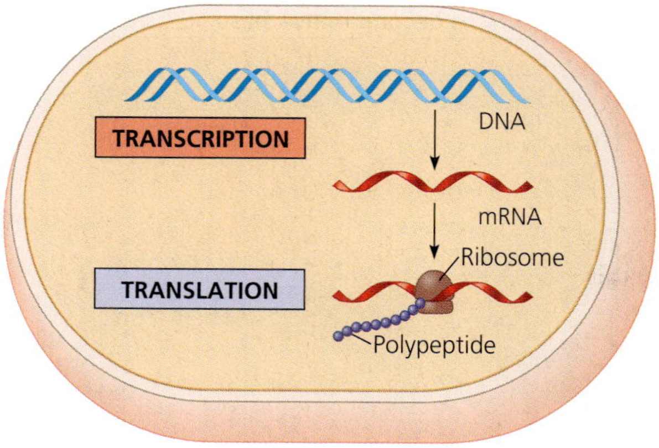

(a) Bacterial cell. In a bacterial cell, which lacks a nucleus, mRNA produced by transcription is immediately translated without additional processing.

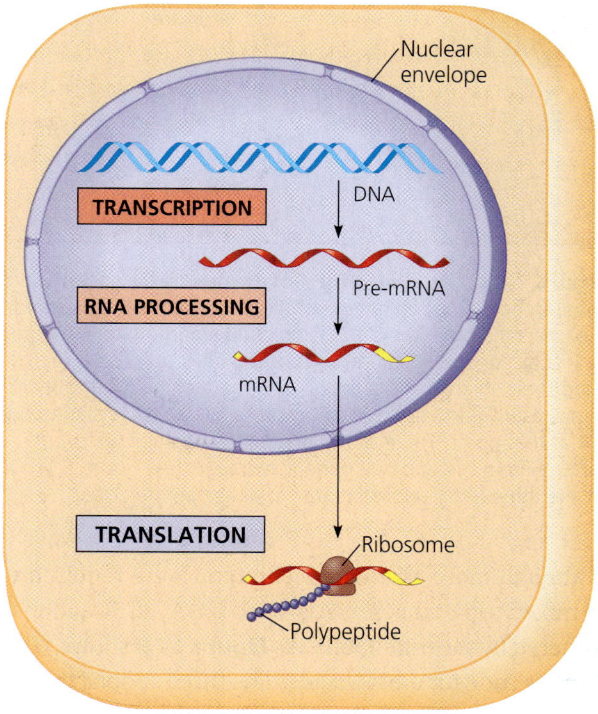

(b) Eukaryotic cell. The nucleus provides a separate compartment for transcription. The original RNA transcript, called pre-mRNA, is processed in various ways before leaving the nucleus as mRNA.

▲ **Figure 17.3 Overview: the roles of transcription and translation in the flow of genetic information.** In a cell, inherited information flows from DNA to RNA to protein. The two main stages of information flow are transcription and translation. A miniature version of part (a) or (b) accompanies several figures later in the chapter as an orientation diagram to help you see where a particular figure fits into the overall scheme.

Codons: Triplets of Bases

If each nucleotide base were translated into an amino acid, only 4 of the 20 amino acids could be specified. Would a language of two-letter code words suffice? The two-base sequence AG, for example, could specify one amino acid, and GT could specify another. Since there are four possible bases in each position, this would give us 16 (that is, 4^2) possible arrangements—still not enough to code for all 20 amino acids.

Triplets of nucleotide bases are the smallest units of uniform length that can code for all the amino acids. If each arrangement of three consecutive bases specifies an amino acid, there can be 64 (that is, 4^3) possible code words—more than enough to specify all the amino acids. Experiments have verified that the flow of information from gene to protein is based on a **triplet code**: The genetic instructions for a polypeptide chain are written in the DNA as a series of nonoverlapping, three-nucleotide words. For example, the base triplet AGT at a particular position along a DNA strand results in the placement of the amino acid serine at the corresponding position of the polypeptide being produced.

During transcription, the gene determines the sequence of bases along the length of an mRNA molecule **(Figure 17.4)**. For each gene, only one of the two DNA strands is transcribed. This strand is called the **template strand** because it provides the pattern, or template, for the sequence of nucleotides in an RNA transcript. A given DNA strand is the template strand for some genes along a DNA molecule, while for other genes the complementary strand functions as the template. Note that

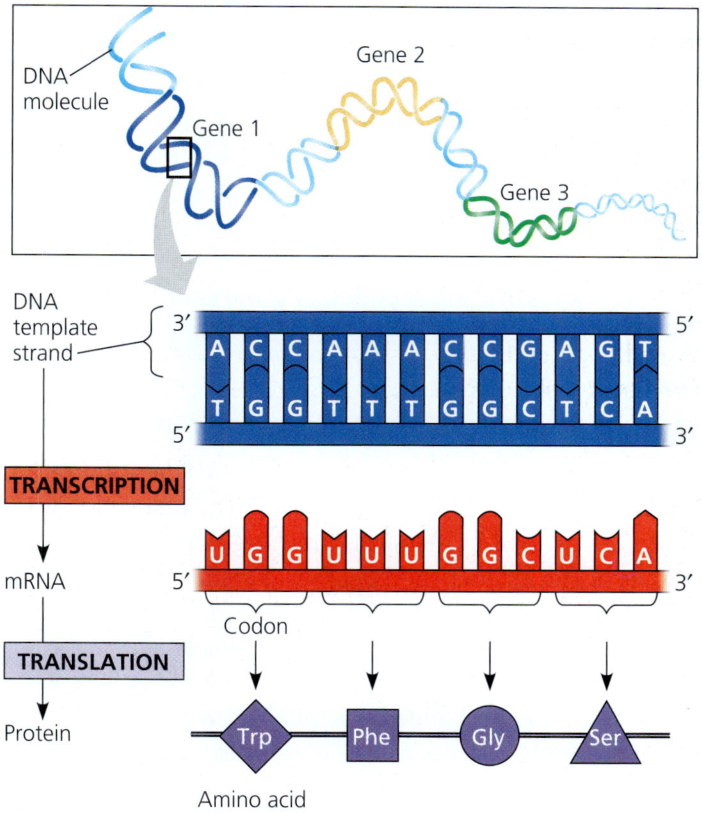

▲ **Figure 17.4 The triplet code.** For each gene, one DNA strand functions as a template for transcription. The base-pairing rules for DNA synthesis also guide transcription, but uracil (U) takes the place of thymine (T) in RNA. During translation, the mRNA is read as a sequence of base triplets, called codons. Each codon specifies an amino acid to be added to the growing polypeptide chain. The mRNA is read in the 5′ → 3′ direction.

for a particular gene, the same strand is used as the template every time it is transcribed.

An mRNA molecule is complementary rather than identical to its DNA template because RNA bases are assembled on the template according to base-pairing rules. The pairs are similar to those that form during DNA replication, except that U, the RNA substitute for T, pairs with A and the mRNA nucleotides contain ribose instead of deoxyribose. Like a new strand of DNA, the RNA molecule is synthesized in an antiparallel direction to the template strand of DNA. (To review what is meant by "antiparallel" and the 5′ and 3′ ends of a nucleic acid chain, see Figure 16.7.) For example, the base triplet ACC along the DNA (written as 3′-ACC-5′) provides a template for 5′-UGG-3′ in the mRNA molecule. The mRNA base triplets are called **codons**, and they are customarily written in the 5′ → 3′ direction. In our example, UGG is the codon for the amino acid tryptophan (abbreviated Trp). The term *codon* is also used for the DNA base triplets along the *nontemplate* strand. These codons are complementary to the template strand and thus identical in sequence to the mRNA except that they have T instead of U. (For this reason, the nontemplate DNA strand is sometimes called the "coding strand.")

During translation, the sequence of codons along an mRNA molecule is decoded, or translated, into a sequence of amino acids making up a polypeptide chain. The codons are read by the translation machinery in the 5′ → 3′ direction along the mRNA. Each codon specifies which one of the 20 amino acids will be incorporated at the corresponding position along a polypeptide. Because codons are base triplets, the number of nucleotides making up a genetic message must be three times the number of amino acids in the protein product. For example, it takes 300 nucleotides along an mRNA strand to code for the amino acids in a polypeptide that is 100 amino acids long.

Cracking the Code

Molecular biologists cracked the code of life in the early 1960s when a series of elegant experiments disclosed the amino acid translations of each of the RNA codons. The first codon was deciphered in 1961 by Marshall Nirenberg, of the National Institutes of Health, and his colleagues. Nirenberg synthesized an artificial mRNA by linking identical RNA nucleotides containing uracil as their base. No matter where this message started or stopped, it could contain only one codon in repetition: UUU. Nirenberg added this "poly-U" to a test-tube mixture containing amino acids, ribosomes, and the other components required for protein synthesis. His artificial system translated the poly-U into a polypeptide containing many units of the amino acid phenylalanine (Phe), strung together as a long polyphenylalanine chain. Thus, Nirenberg determined that the mRNA codon UUU specifies the amino acid phenylalanine. Soon, the amino acids specified by the codons AAA, GGG, and CCC were also determined.

Second mRNA base

First mRNA base (5′ end of codon)	U	C	A	G	Third mRNA base (3′ end of codon)
U	UUU Phe	UCU Ser	UAU Tyr	UGU Cys	U
U	UUC Phe	UCC Ser	UAC Tyr	UGC Cys	C
U	UUA Leu	UCA Ser	UAA Stop	UGA Stop	A
U	UUG Leu	UCG Ser	UAG Stop	UGG Trp	G
C	CUU Leu	CCU Pro	CAU His	CGU Arg	U
C	CUC Leu	CCC Pro	CAC His	CGC Arg	C
C	CUA Leu	CCA Pro	CAA Gln	CGA Arg	A
C	CUG Leu	CCG Pro	CAG Gln	CGG Arg	G
A	AUU Ile	ACU Thr	AAU Asn	AGU Ser	U
A	AUC Ile	ACC Thr	AAC Asn	AGC Ser	C
A	AUA Ile	ACA Thr	AAA Lys	AGA Arg	A
A	AUG Met or start	ACG Thr	AAG Lys	AGG Arg	G
G	GUU Val	GCU Ala	GAU Asp	GGU Gly	U
G	GUC Val	GCC Ala	GAC Asp	GGC Gly	C
G	GUA Val	GCA Ala	GAA Glu	GGA Gly	A
G	GUG Val	GCG Ala	GAG Glu	GGG Gly	G

▲ **Figure 17.5 The dictionary of the genetic code.** The three bases of an mRNA codon are designated here as the first, second, and third bases, reading in the 5′ → 3′ direction along the mRNA. (Practice using this dictionary by finding the codons in Figure 17.4.) The codon AUG not only stands for the amino acid methionine (Met) but also functions as a "start" signal for ribosomes to begin translating the mRNA at that point. Three of the 64 codons function as "stop" signals, marking the end of a genetic message. See Figure 5.17 for a list of the three-letter abbreviations for all the amino acids.

Although more elaborate techniques were required to decode mixed triplets such as AUA and CGA, all 64 codons were deciphered by the mid-1960s. As **Figure 17.5** shows, 61 of the 64 triplets code for amino acids. The three codons that do not designate amino acids are "stop" signals, or termination codons, marking the end of translation. Notice that the codon AUG has a dual function: It codes for the amino acid methionine (Met) and also functions as a "start" signal, or initiation codon. Genetic messages begin with the mRNA codon AUG, which signals the protein-synthesizing machinery to begin translating the mRNA at that location. (Because AUG also stands for methionine, polypeptide chains begin with methionine when they are synthesized. However, an enzyme may subsequently remove this starter amino acid from the chain.)

Notice in Figure 17.5 that there is redundancy in the genetic code, but no ambiguity. For example, although codons GAA and GAG both specify glutamic acid (redundancy), neither of them ever specifies any other amino acid (no ambiguity). The redundancy in the code is not altogether random. In many cases, codons that are synonyms for a particular amino acid differ only in the third base of the triplet. We will consider a possible benefit of this redundancy later in the chapter.

Our ability to extract the intended message from a written language depends on reading the symbols in the correct groupings—that is, in the correct **reading frame**. Consider this statement: "The red dog ate the bug." Group the letters incorrectly by starting at the wrong point, and the result will probably be gibberish: for example, "her edd oga tet heb ug." The reading frame is also important in the molecular language of cells. The short stretch of polypeptide shown in Figure 17.4, for instance, will be made correctly only if the mRNA nucleotides are read from left to right (5′ → 3′) in the groups of three shown in the figure: UGG UUU GGC UCA. Although a genetic message is written with no spaces between the codons, the cell's protein-synthesizing machinery reads the message as a series of nonoverlapping three-letter words. The message is *not* read as a series of overlapping words—UGGUUU, and so on—which would convey a very different message.

Evolution of the Genetic Code

The genetic code is nearly universal, shared by organisms from the simplest bacteria to the most complex plants and animals. The RNA codon CCG, for instance, is translated as the amino acid proline in all organisms whose genetic code has been examined. In laboratory experiments, genes can be transcribed and translated after being transplanted from one species to another, sometimes with quite striking results, as shown in **Figure 17.6**! Bacteria can be programmed by the insertion of human genes to synthesize certain human proteins for medical use, such as insulin. Such applications have produced many exciting developments in the area of biotechnology (see Chapter 20).

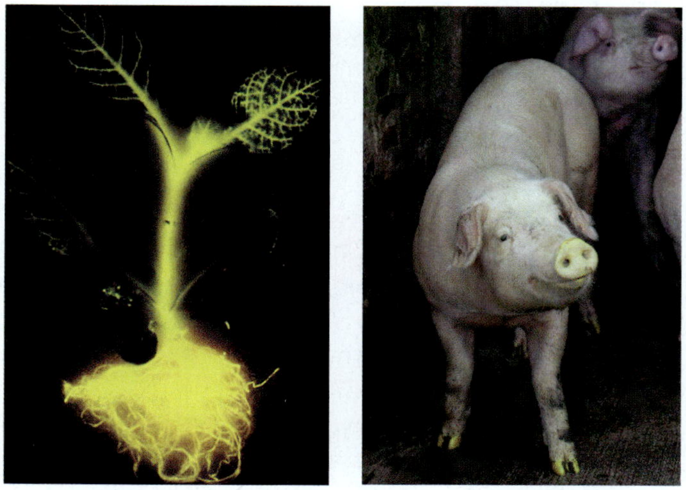

(a) Tobacco plant expressing a firefly gene. The yellow glow is produced by a chemical reaction catalyzed by the protein product of the firefly gene.

(b) Pig expressing a jellyfish gene. Researchers injected the gene for a fluorescent protein into fertilized pig eggs. One of the eggs developed into this fluorescent pig.

▲ **Figure 17.6 Expression of genes from different species.** Because diverse forms of life share a common genetic code, one species can be programmed to produce proteins characteristic of a second species by introducing DNA from the second species into the first.

Exceptions to the universality of the genetic code include translation systems in which a few codons differ from the standard ones. Slight variations in the genetic code exist in certain unicellular eukaryotes and in the organelle genes of some species. There are also exceptions in which stop codons can be translated into one of two amino acids not found in most organisms. Although one of these amino acids (pyrrolysine) has been detected thus far only in archaea, the other (selenocysteine) is a component of some bacterial proteins and even some human enzymes. Despite these exceptions, the evolutionary significance of the code's *near* universality is clear. A language shared by all living things must have been operating very early in the history of life—early enough to be present in the common ancestor of all present-day organisms. A shared genetic vocabulary is a reminder of the kinship that bonds all life on Earth.

CONCEPT CHECK 17.1

1. What polypeptide product would you expect from a poly-G mRNA that is 30 nucleotides long?
2. **DRAW IT** The template strand of a gene contains the sequence 3′-TTCAGTCGT-5′. Draw the nontemplate sequence and the mRNA sequence, indicating 5′ and 3′ ends of each. Compare the two sequences.
3. **WHAT IF?** Imagine that the nontemplate sequence in question 2 was transcribed instead of the template sequence. Draw the mRNA sequence and translate it using Figure 17.5. (Be sure to pay attention to the 5′ and 3′ ends.) Predict how well the protein synthesized from the nontemplate strand would function, if at all.

For suggested answers, see Appendix A.

CONCEPT 17.2

Transcription is the DNA-directed synthesis of RNA: *a closer look*

Now that we have considered the linguistic logic and evolutionary significance of the genetic code, we are ready to reexamine transcription, the first stage of gene expression, in more detail.

Molecular Components of Transcription

Messenger RNA, the carrier of information from DNA to the cell's protein-synthesizing machinery, is transcribed from the template strand of a gene. An enzyme called an **RNA polymerase** pries the two strands of DNA apart and joins the RNA nucleotides as they base-pair along the DNA template **(Figure 17.7)**. Like the DNA polymerases that function in DNA replication, RNA polymerases can assemble a polynucleotide only in its 5′ → 3′ direction. Unlike DNA polymerases, however, RNA polymerases are able to start a chain from scratch; they don't need a primer.

Specific sequences of nucleotides along the DNA mark where transcription of a gene begins and ends. The DNA sequence where RNA polymerase attaches and initiates transcription is known as the **promoter**; in bacteria, the sequence that signals the end of transcription is called the **terminator**. (The termination mechanism is different in eukaryotes; we'll describe it later.) Molecular biologists refer to the direction of transcription as "downstream" and the other direction as "upstream." These terms are also used to describe the positions of nucleotide sequences within the DNA or RNA. Thus, the promoter sequence in DNA is said to be upstream from the terminator. The stretch of DNA that is transcribed into an RNA molecule is called a **transcription unit**.

Bacteria have a single type of RNA polymerase that synthesizes not only mRNA but also other types of RNA that function in protein synthesis, such as ribosomal RNA. In contrast, eukaryotes have at least three types of RNA polymerase in their nuclei. The one used for mRNA synthesis is called RNA polymerase II. The other RNA polymerases transcribe RNA molecules that are not translated into protein. In the discussion of transcription that follows, we start with the features of mRNA synthesis common to both bacteria and eukaryotes and then describe some key differences.

Synthesis of an RNA Transcript

The three stages of transcription, as shown in Figure 17.7 and described next, are initiation, elongation, and termination of the RNA chain. Study Figure 17.7 to familiarize yourself with the stages and the terms used to describe them.

RNA Polymerase Binding and Initiation of Transcription

The promoter of a gene includes within it the transcription start point (the nucleotide where RNA synthesis actually begins) and typically extends several dozen nucleotide pairs

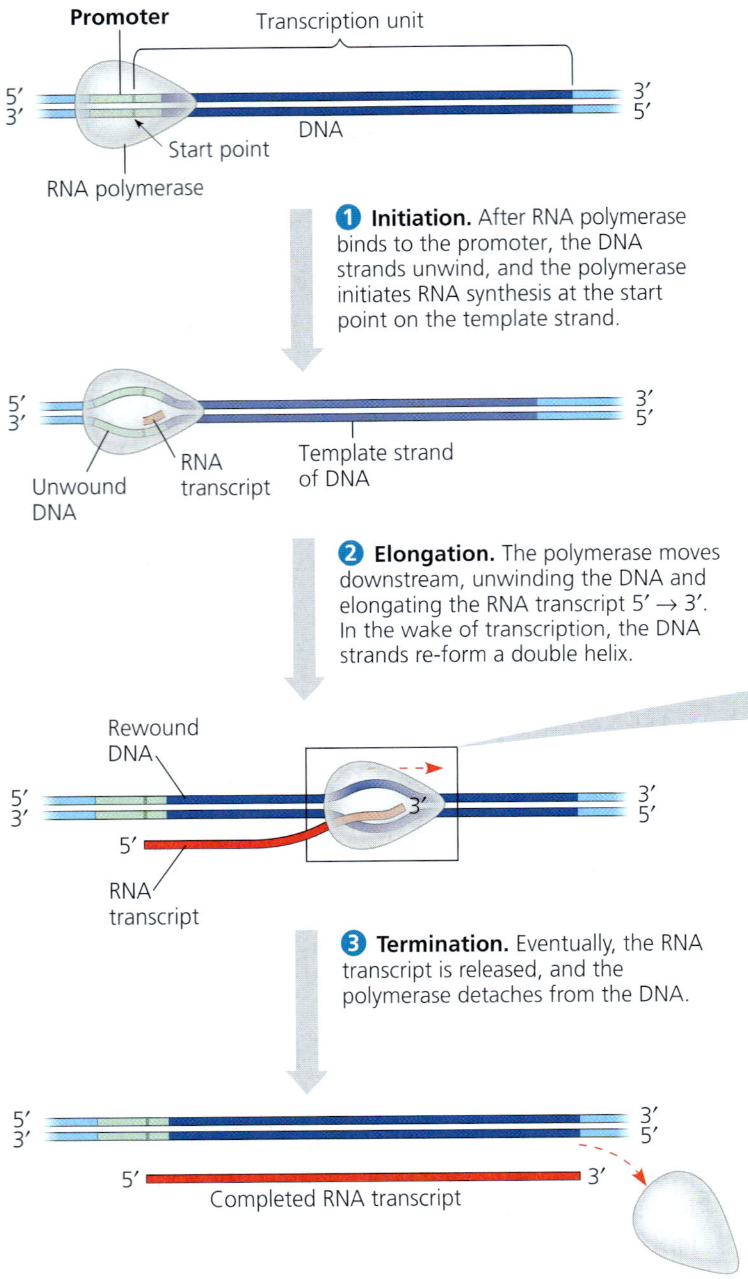

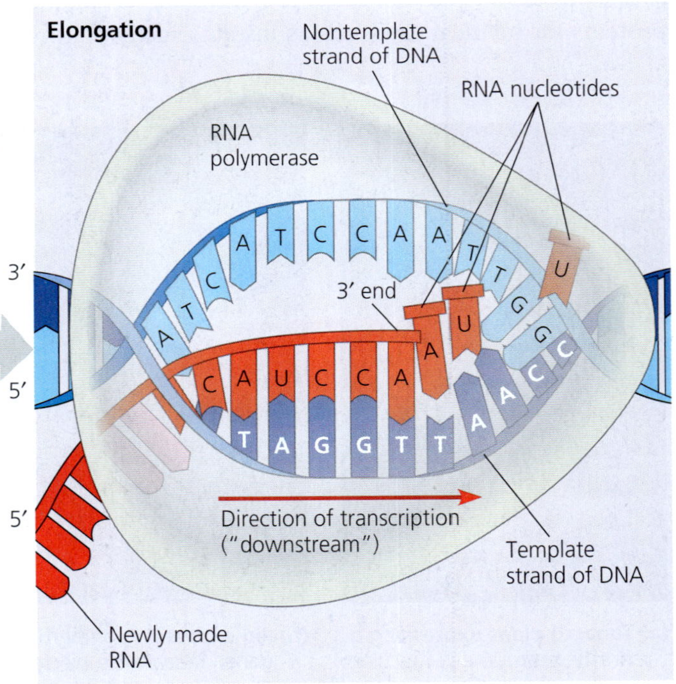

▲ **Figure 17.7 The stages of transcription: initiation, elongation, and termination.** This general depiction of transcription applies to both bacteria and eukaryotes, but the details of termination differ, as described in the text. Also, in a bacterium, the RNA transcript is immediately usable as mRNA; in a eukaryote, the RNA transcript must first undergo processing.

upstream from the start point. In addition to serving as a binding site for RNA polymerase and determining where transcription starts, the promoter determines which of the two strands of the DNA helix is used as the template.

Certain sections of a promoter are especially important for binding RNA polymerase. In bacteria, the RNA polymerase itself specifically recognizes and binds to the promoter. In eukaryotes, a collection of proteins called **transcription factors** mediate the binding of RNA polymerase and the initiation of transcription. Recall from Chapter 16 that the DNA of a eukaryotic chromosome is complexed with histones and other proteins in the form of chromatin. The roles of these proteins in making the DNA accessible to transcription factors will be discussed in Chapter 18. Only after certain transcription factors are attached to the promoter does RNA polymerase II bind to it. The whole complex of transcription factors and RNA polymerase II bound to the promoter is called a **transcription initiation complex**. **Figure 17.8** shows the role of transcription factors and a crucial promoter DNA sequence called a **TATA box** in forming the initiation complex at a eukaryotic promoter.

The interaction between eukaryotic RNA polymerase II and transcription factors is an example of the importance of protein-protein interactions in controlling eukaryotic transcription. Once the polymerase is firmly attached to the promoter DNA, the two DNA strands unwind there, and the enzyme starts transcribing the template strand.

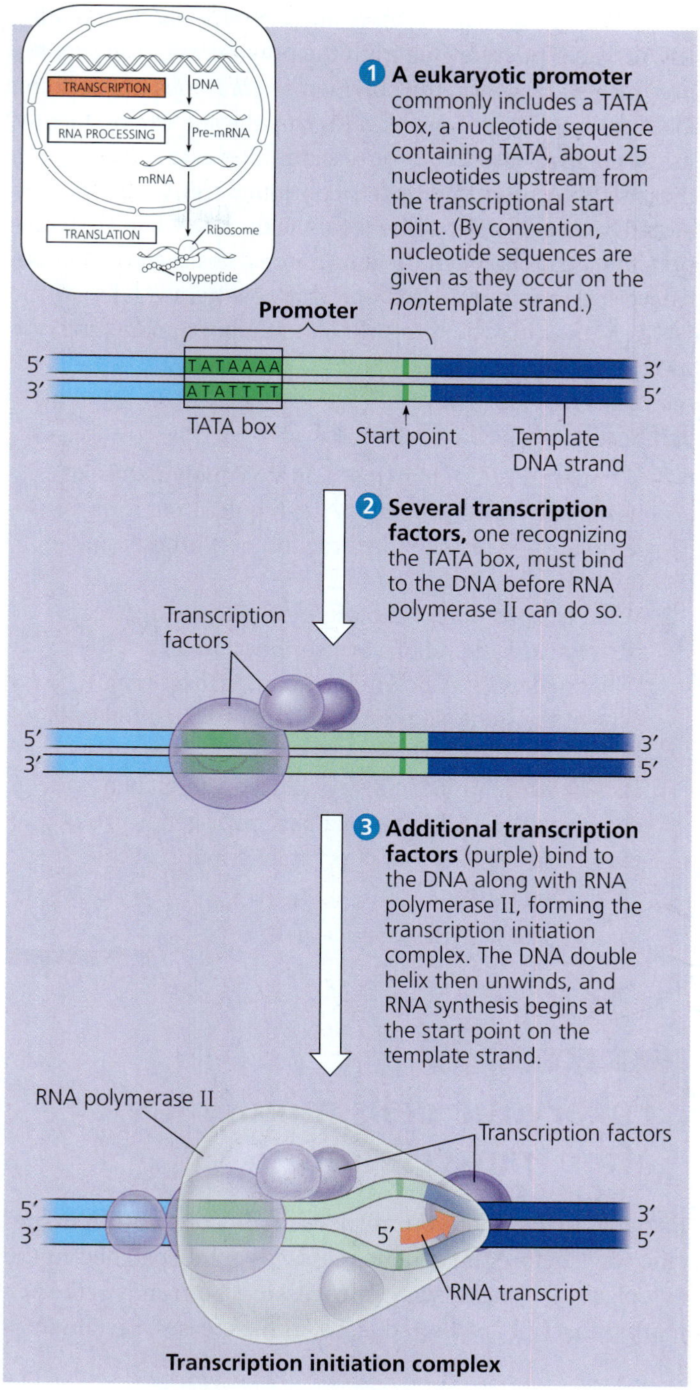

▲ **Figure 17.8 The initiation of transcription at a eukaryotic promoter.** In eukaryotic cells, proteins called transcription factors mediate the initiation of transcription by RNA polymerase II.

? *Explain how the interaction of RNA polymerase with the promoter would differ if the figure showed transcription initiation for bacteria.*

Elongation of the RNA Strand

As RNA polymerase moves along the DNA, it continues to untwist the double helix, exposing about 10 to 20 DNA bases at a time for pairing with RNA nucleotides (see Figure 17.7). The enzyme adds nucleotides to the 3′ end of the growing RNA molecule as it continues along the double helix. In the wake of this advancing wave of RNA synthesis, the new RNA molecule peels away from its DNA template and the DNA double helix re-forms. Transcription progresses at a rate of about 40 nucleotides per second in eukaryotes.

A single gene can be transcribed simultaneously by several molecules of RNA polymerase following each other like trucks in a convoy. A growing strand of RNA trails off from each polymerase, with the length of each new strand reflecting how far along the template the enzyme has traveled from the start point (see the mRNA molecules in Figure 17.24). The congregation of many polymerase molecules simultaneously transcribing a single gene increases the amount of mRNA transcribed from it, which helps the cell make the encoded protein in large amounts.

Termination of Transcription

The mechanism of termination differs between bacteria and eukaryotes. In bacteria, transcription proceeds through a terminator sequence in the DNA. The transcribed terminator (an RNA sequence) functions as the termination signal, causing the polymerase to detach from the DNA and release the transcript, which is available for immediate use as mRNA. In eukaryotes, RNA polymerase II transcribes a sequence on the DNA called the polyadenylation signal sequence, which codes for a polyadenylation signal (AAUAAA) in the pre-mRNA. Then, at a point about 10 to 35 nucleotides downstream from the AAUAAA signal, proteins associated with the growing RNA transcript cut it free from the polymerase, releasing the pre-mRNA. However, the polymerase continues transcribing DNA for hundreds of nucleotides past the site where the pre-mRNA was released. Recent research on yeast cells suggests that the RNA produced by this continued transcription is digested by an enzyme that moves along the RNA. The data support the idea that when the enzyme reaches the polymerase, transcription is terminated and the polymerase falls off the DNA. Meanwhile, the pre-mRNA undergoes processing, the topic of the next section.

CONCEPT CHECK 17.2

1. Compare DNA polymerase and RNA polymerase in terms of how they function, the requirement for a template and primer, the direction of synthesis, and the type of nucleotides used.
2. What is a promoter, and is it located at the upstream or downstream end of a transcription unit?
3. What makes RNA polymerase start transcribing a gene at the right place on the DNA in a bacterial cell? In a eukaryotic cell?
4. **WHAT IF?** Suppose X-rays caused a sequence change in the TATA box of a particular gene's promoter. How would that affect transcription of the gene? (See Figure 17.8.)

For suggested answers, see Appendix A.

CONCEPT **17.3**

Eukaryotic cells modify RNA after transcription

Enzymes in the eukaryotic nucleus modify pre-mRNA in specific ways before the genetic messages are dispatched to the cytoplasm. During this **RNA processing**, both ends of the primary transcript are altered. Also, in most cases, certain interior sections of the RNA molecule are cut out and the remaining parts spliced together. These modifications produce an mRNA molecule ready for translation.

Alteration of mRNA Ends

Each end of a pre-mRNA molecule is modified in a particular way **(Figure 17.9)**. The 5′ end is synthesized first; it receives a **5′ cap**, a modified form of a guanine (G) nucleotide added onto the 5′ end after transcription of the first 20 to 40 nucleotides. The 3′ end of the pre-mRNA molecule is also modified before the mRNA exits the nucleus. Recall that the pre-mRNA is released soon after the polyadenylation signal, AAUAAA, is transcribed. At the 3′ end, an enzyme adds 50 to 250 more adenine (A) nucleotides, forming a **poly-A tail**. The 5′ cap and poly-A tail share several important functions. First, they seem to facilitate the export of the mature mRNA from the nucleus. Second, they help protect the mRNA from degradation by hydrolytic enzymes. And third, they help ribosomes attach to the 5′ end of the mRNA once the mRNA reaches the cytoplasm. Figure 17.9 shows a diagram of a eukaryotic mRNA molecule with cap and tail. The figure also shows the untranslated regions (UTRs) at the 5′ and 3′ ends of the mRNA (referred to as the 5′ UTR and 3′ UTR). The UTRs are parts of the mRNA that will not be translated into protein, but they have other functions, such as ribosome binding.

Split Genes and RNA Splicing

A remarkable stage of RNA processing in the eukaryotic nucleus is the removal of large portions of the RNA molecule that is initially synthesized—a cut-and-paste job called **RNA splicing**, similar to editing a video **(Figure 17.10)**. The average length of a transcription unit along a human DNA molecule is about 27,000 base pairs, so the primary RNA transcript is also that long. However, it takes only 1,200 nucleotides in RNA to code for the average-sized protein of 400 amino acids. (Remember, each amino acid is encoded by a *triplet* of nucleotides.) This means that most eukaryotic genes and their RNA transcripts have long noncoding stretches of nucleotides, regions that are not

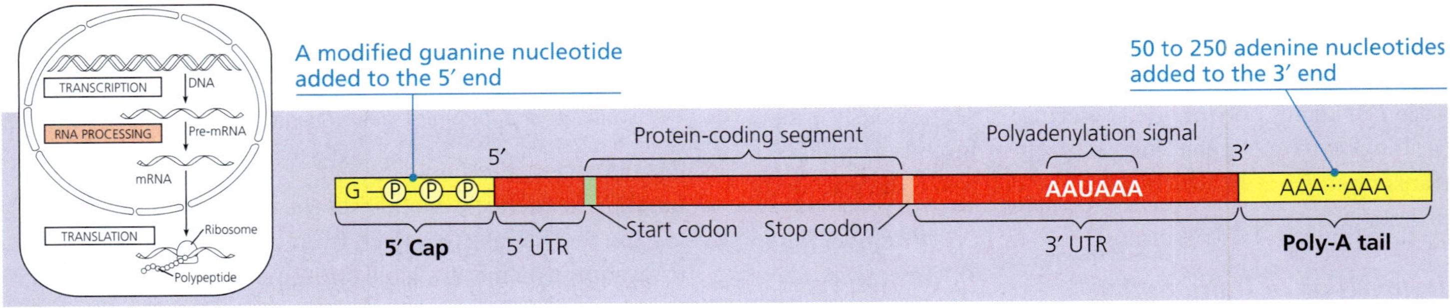

▲ **Figure 17.9 RNA processing: addition of the 5′ cap and poly-A tail.** Enzymes modify the two ends of a eukaryotic pre-mRNA molecule. The modified ends may promote the export of mRNA from the nucleus, and they help protect the mRNA from degradation. When the mRNA reaches the cytoplasm, the modified ends, in conjunction with certain cytoplasmic proteins, facilitate ribosome attachment. The 5′ cap and poly-A tail are not translated into protein, nor are the regions called the 5′ untranslated region (5′ UTR) and 3′ untranslated region (3′ UTR).

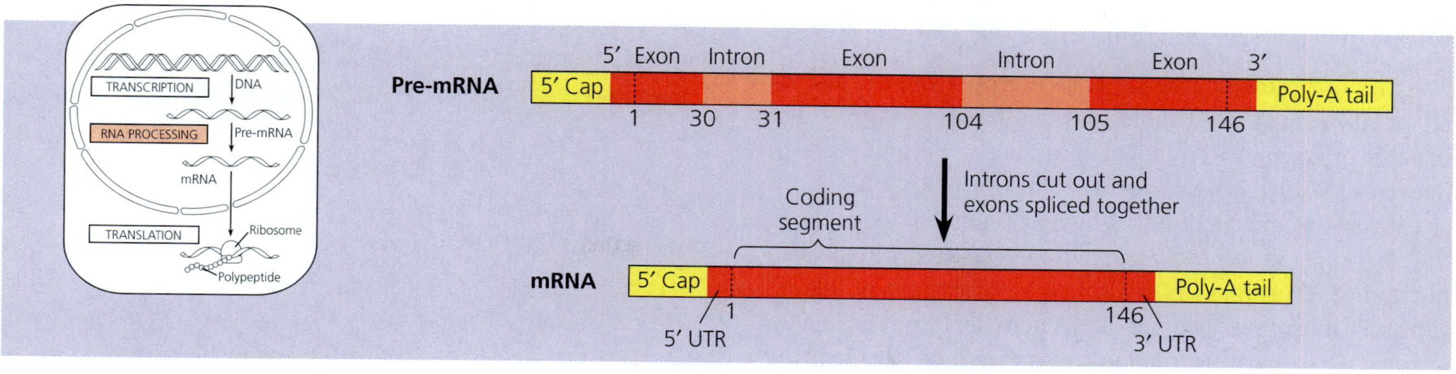

▲ **Figure 17.10 RNA processing: RNA splicing.** The RNA molecule shown here codes for β-globin, one of the polypeptides of hemoglobin. The numbers under the RNA refer to codons; β-globin is 146 amino acids long. The β-globin gene and its pre-mRNA transcript have three exons, corresponding to sequences that will leave the nucleus as mRNA. (The 5′ UTR and 3′ UTR are parts of exons because they are included in the mRNA; however, they do not code for protein.) During RNA processing, the introns are cut out and the exons spliced together. In many genes, the introns are much larger relative to the exons than they are in the β-globin gene. (The pre-mRNA is not drawn to scale.)

translated. Even more surprising is that most of these noncoding sequences are interspersed between coding segments of the gene and thus between coding segments of the pre-mRNA. In other words, the sequence of DNA nucleotides that codes for a eukaryotic polypeptide is usually not continuous; it is split into segments. The noncoding segments of nucleic acid that lie between coding regions are called intervening sequences, or **introns**. The other regions are called **exons**, because they are eventually expressed, usually by being translated into amino acid sequences. (Exceptions include the UTRs of the exons at the ends of the RNA, which make up part of the mRNA but are not translated into protein. Because of these exceptions, you may find it helpful to think of exons as sequences of RNA that *exit* the nucleus.) The terms *intron* and *exon* are used for both RNA sequences and the DNA sequences that encode them.

In making a primary transcript from a gene, RNA polymerase II transcribes both introns and exons from the DNA, but the mRNA molecule that enters the cytoplasm is an abridged version. The introns are cut out from the molecule and the exons joined together, forming an mRNA molecule with a continuous coding sequence. This is the process of RNA splicing.

How is pre-mRNA splicing carried out? Researchers have learned that the signal for RNA splicing is a short nucleotide sequence at each end of an intron. Particles called *small nuclear ribonucleoproteins*, abbreviated *snRNPs* (pronounced "snurps"), recognize these splice sites. As the name implies, snRNPs are located in the cell nucleus and are composed of RNA and protein molecules. The RNA in a snRNP particle is called a *small nuclear RNA (snRNA)*; each molecule is about 150 nucleotides long. Several different snRNPs join with additional proteins to form an even larger assembly called a **spliceosome**, which is almost as big as a ribosome. The spliceosome interacts with certain sites along an intron, releasing the intron and joining together the two exons that flanked the intron **(Figure 17.11)**. There is strong evidence that snRNAs catalyze these processes, as well as participating in spliceosome assembly and splice site recognition.

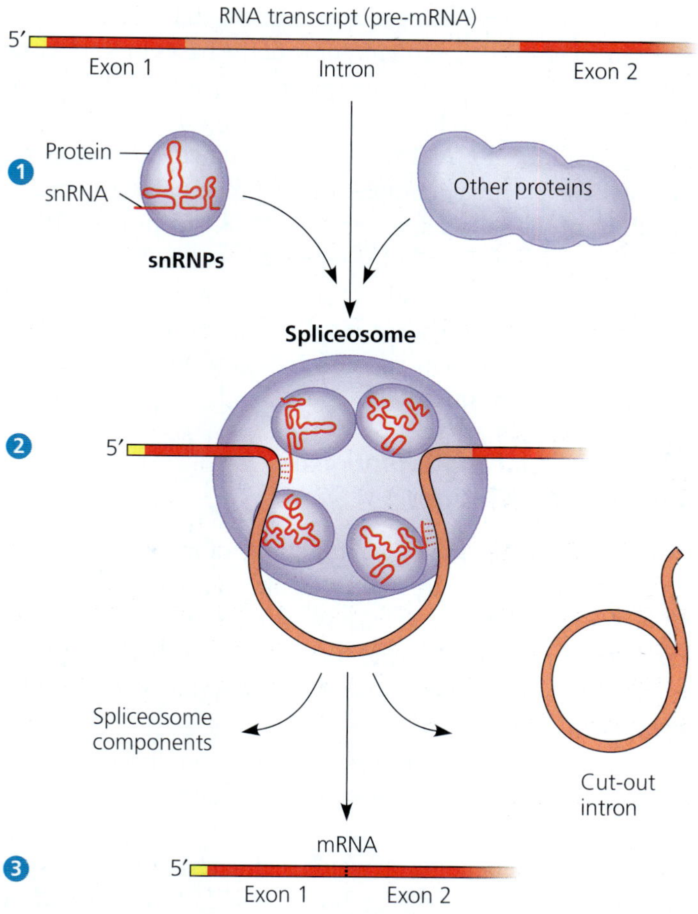

▲ **Figure 17.11 The roles of snRNPs and spliceosomes in pre-mRNA splicing.** The diagram shows only a portion of the pre-mRNA transcript; additional introns and exons lie downstream from the ones pictured here. **1** Small nuclear ribonucleoproteins (snRNPs) and other proteins form a molecular complex called a spliceosome on a pre-mRNA molecule containing exons and introns. **2** Within the spliceosome, snRNA base-pairs with nucleotides at specific sites along the intron. **3** The spliceosome cuts the pre-mRNA, releasing the intron, and at the same time splices the exons together. The spliceosome then comes apart, releasing mRNA, which now contains only exons.

Ribozymes

The idea of a catalytic role for snRNA arose from the discovery of **ribozymes**, RNA molecules that function as enzymes. In some organisms, RNA splicing can occur without proteins or even additional RNA molecules: The intron RNA functions as a ribozyme and catalyzes its own excision! For example, in the ciliate protist *Tetrahymena*, self-splicing occurs in the production of ribosomal RNA (rRNA), a component of the organism's ribosomes. The pre-rRNA actually removes its own introns. The discovery of ribozymes rendered obsolete the idea that all biological catalysts are proteins.

Three properties of RNA enable some RNA molecules to function as enzymes. First, because RNA is single-stranded, a region of an RNA molecule may base-pair with a complementary region elsewhere in the same molecule, which gives the molecule a particular three-dimensional structure. A specific structure is essential to the catalytic function of ribozymes, just as it is for enzymatic proteins. Second, like certain amino acids in an enzymatic protein, some of the bases in RNA contain functional groups that may participate in catalysis. Third, the ability of RNA to hydrogen-bond with other nucleic acid molecules (either RNA or DNA) adds specificity to its catalytic activity. For example, complementary base pairing between the RNA of the spliceosome and the RNA of a primary RNA transcript precisely locates the region where the ribozyme catalyzes splicing. Later in this chapter, you will see how these properties of RNA also allow it to perform important noncatalytic roles in the cell, such as recognition of the three-nucleotide codons on mRNA.

The Functional and Evolutionary Importance of Introns

What could be the biological functions of introns and RNA splicing? While specific functions may not have been identified for most introns, at least some contain sequences that regulate gene activity. And the splicing process itself is necessary for the passage of mRNA from the nucleus to the cytoplasm.

One consequence of the presence of introns in genes is that a single gene can encode more than one kind of polypeptide. Many genes are known to give rise to two or more different polypeptides, depending on which segments are treated as exons during RNA processing; this is called **alternative RNA splicing** (see Figure 18.11). For example, sex differences in fruit flies are largely due to differences in how males and females splice the RNA transcribed from certain genes. Results from the Human Genome Project (discussed in Chapter 21) suggest that alternative RNA splicing is one reason humans can get along with a relatively small number of genes—about one and a half times as many as a fruit fly. Because of alternative splicing, the number of different protein products an organism produces can be much greater than its number of genes.

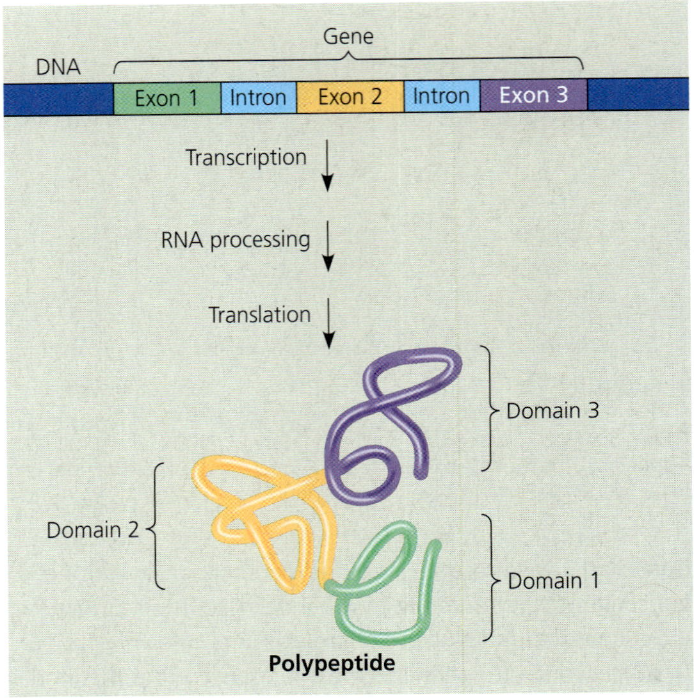

▲ **Figure 17.12 Correspondence between exons and protein domains.**

Proteins often have a modular architecture consisting of discrete structural and functional regions called **domains**. One domain of an enzymatic protein, for instance, might include the active site, while another might attach the protein to a cellular membrane. In quite a few cases, different exons code for the different domains of a protein (**Figure 17.12**).

The presence of introns in a gene may facilitate the evolution of new and potentially useful proteins as a result of a process known as *exon shuffling*. Introns increase the probability of potentially beneficial crossing over between the exons of alleles—simply by providing more terrain for crossovers without interrupting coding sequences. We can also imagine the occasional mixing and matching of exons between completely different (nonallelic) genes. Exon shuffling of either sort could lead to new proteins with novel combinations of functions. While most of the shuffling would result in nonbeneficial changes, occasionally a beneficial variant might arise.

CONCEPT CHECK 17.3

1. How does alteration of the 5′ and 3′ ends of pre-mRNA affect the mRNA that exits the nucleus?
2. How is RNA splicing similar to editing a video?
3. **WHAT IF?** In nematode worms, a gene that codes for an ATPase has two alternatives for exon 4 and three alternatives for exon 7. How many different forms of the protein could be made from this gene?

For suggested answers, see Appendix A.

CONCEPT 17.4

Translation is the RNA-directed synthesis of a polypeptide: *a closer look*

We will now examine in greater detail how genetic information flows from mRNA to protein—the process of translation. As we did for transcription, we'll concentrate on the basic steps of translation that occur in both bacteria and eukaryotes, while pointing out key differences.

Molecular Components of Translation

In the process of translation, a cell interprets a genetic message and builds a polypeptide accordingly. The message is a series of codons along an mRNA molecule, and the interpreter is called **transfer RNA (tRNA)**. The function of tRNA is to transfer amino acids from the cytoplasmic pool of amino acids to a ribosome. A cell keeps its cytoplasm stocked with all 20 amino acids, either by synthesizing them from other compounds or by taking them up from the surrounding solution. The ribosome adds each amino acid brought to it by tRNA to the growing end of a polypeptide chain **(Figure 17.13)**.

Molecules of tRNA are not all identical. The key to translating a genetic message into a specific amino acid sequence is that each type of tRNA molecule translates a particular mRNA codon into a particular amino acid. As a tRNA molecule arrives at a ribosome, it bears a specific amino acid at one end. At the other end of the tRNA is a nucleotide triplet called an **anticodon**, which base-pairs with a complementary codon on mRNA. For example, consider the mRNA codon UUU, which is translated as the amino acid phenylalanine. The tRNA that base-pairs with this codon by hydrogen bonding has AAA as its anticodon and carries phenylalanine at its other end (see the middle tRNA on the ribosome in Figure 17.13). As an mRNA molecule is moved through a ribosome, phenylalanine will be added to the polypeptide chain whenever the codon UUU is presented for translation. Codon by codon, the genetic message is translated as tRNAs deposit amino acids in the order prescribed, and the ribosome joins the amino acids into a chain. The tRNA molecule is a translator because it can read a nucleic acid word (the mRNA codon) and interpret it as a protein word (the amino acid).

Translation is simple in principle but complex in its biochemistry and mechanics, especially in the eukaryotic cell. In dissecting translation, we'll concentrate on the slightly less complicated version of the process that occurs in bacteria. Let's first look at the major components in this cellular process. Then we will see how they act together to make a polypeptide.

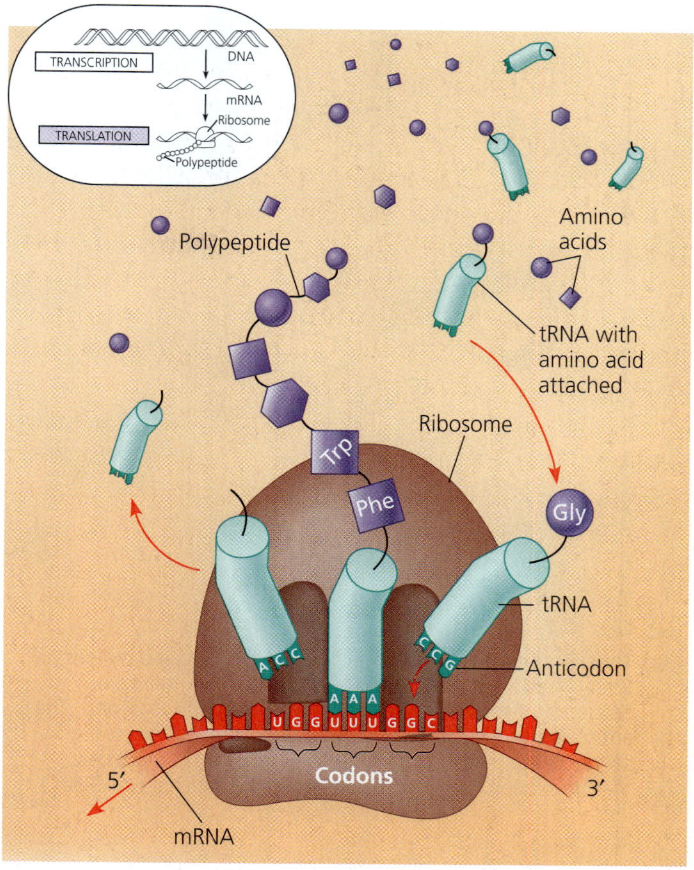

▲ **Figure 17.13 Translation: the basic concept.** As a molecule of mRNA is moved through a ribosome, codons are translated into amino acids, one by one. The interpreters are tRNA molecules, each type with a specific anticodon at one end and a corresponding amino acid at the other end. A tRNA adds its amino acid cargo to a growing polypeptide chain when the anticodon hydrogen-bonds to a complementary codon on the mRNA. The figures that follow show some of the details of translation in a bacterial cell.

MEDIA **BioFlix** Visit **www.campbellbiology.com** for the BioFlix 3-D Animation on Protein Synthesis.

The Structure and Function of Transfer RNA

Like mRNA and other types of cellular RNA, transfer RNA molecules are transcribed from DNA templates. In a eukaryotic cell, tRNA, like mRNA, is made in the nucleus and must travel from the nucleus to the cytoplasm, where translation occurs. In both bacterial and eukaryotic cells, each tRNA molecule is used repeatedly, picking up its designated amino acid in the cytosol, depositing this cargo onto a polypeptide chain at the ribosome, and then leaving the ribosome, ready to pick up another amino acid.

A tRNA molecule consists of a single RNA strand that is only about 80 nucleotides long (compared to hundreds of nucleotides for most mRNA molecules). Because of the presence of complementary stretches of bases that can hydrogen-bond to each other, this single strand can fold back upon itself and form a molecule with a three-dimensional structure. Flattened into one plane to reveal this base pairing, a tRNA molecule

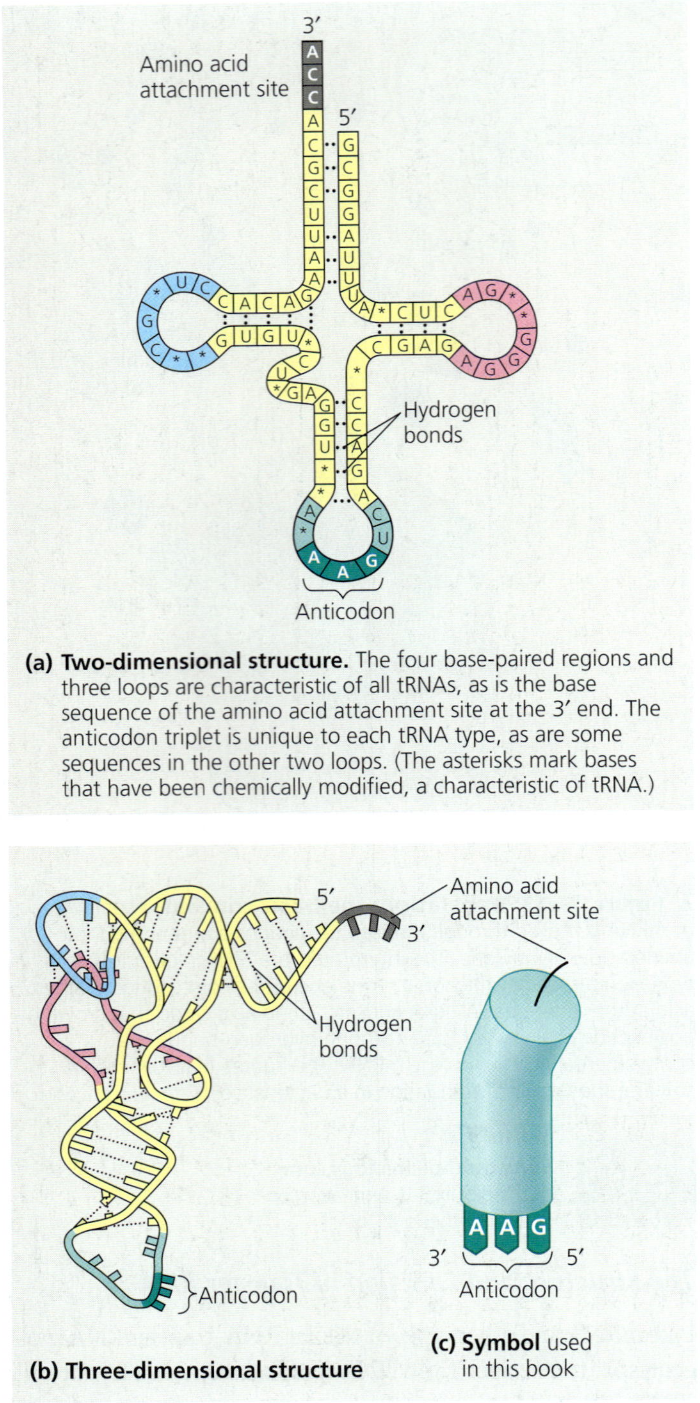

(a) Two-dimensional structure. The four base-paired regions and three loops are characteristic of all tRNAs, as is the base sequence of the amino acid attachment site at the 3′ end. The anticodon triplet is unique to each tRNA type, as are some sequences in the other two loops. (The asterisks mark bases that have been chemically modified, a characteristic of tRNA.)

(b) Three-dimensional structure

(c) Symbol used in this book

▲ **Figure 17.14 The structure of transfer RNA (tRNA).** Anticodons are conventionally written 3′ → 5′ to align properly with codons written 5′ → 3′ (see Figure 17.13). For base pairing, RNA strands must be antiparallel, like DNA. For example, anticodon 3′-AAG-5′ pairs with mRNA codon 5′-UUC-3′.

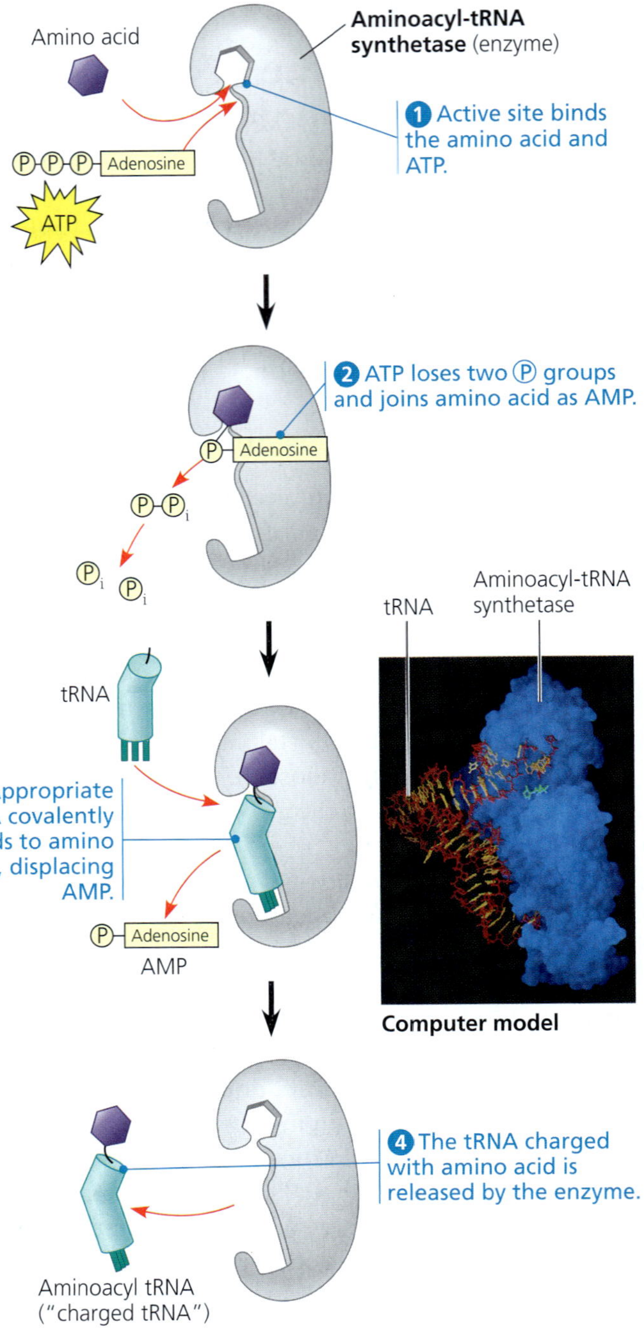

▲ **Figure 17.15 An aminoacyl-tRNA synthetase joining a specific amino acid to a tRNA.** Linkage of the tRNA and amino acid is an endergonic process that occurs at the expense of ATP. The ATP loses two phosphate groups, becoming AMP (adenosine monophosphate).

looks like a cloverleaf **(Figure 17.14a)**. The tRNA actually twists and folds into a compact three-dimensional structure that is roughly L-shaped **(Figure 17.14b)**. The loop extending from one end of the L includes the anticodon, the particular base triplet that base-pairs to a specific mRNA codon. From the other end of the L-shaped tRNA molecule protrudes its 3′ end, which is the attachment site for an amino acid. Thus, the structure of a tRNA molecule fits its function.

The accurate translation of a genetic message requires two processes that involve molecular recognition. First, a tRNA that binds to an mRNA codon specifying a particular amino acid must carry that amino acid, and no other, to the ribosome. The correct matching up of tRNA and amino acid is carried out by a family of related enzymes called **aminoacyl-tRNA synthetases** **(Figure 17.15)**. The active site of each type of aminoacyl-tRNA synthetase fits only a specific combination of amino acid and tRNA. There are 20 different synthetases, one for each amino acid; each synthetase is able to bind all the different tRNAs that code for its particular amino acid. The synthetase catalyzes the covalent attachment of the amino acid to its tRNA in a process driven by the hydrolysis of ATP. The resulting aminoacyl tRNA, also called a charged tRNA, is released from the enzyme and is then available to deliver its amino acid to a growing polypeptide chain on a ribosome.

The second recognition process involves matching up the tRNA anticodon with the appropriate mRNA codon. If one tRNA variety existed for each mRNA codon that specifies an amino acid, there would be 61 tRNAs (see Figure 17.5). In fact, there are only about 45, signifying that some tRNAs must be able to bind to more than one codon. Such versatility is possible because the rules for base pairing between the third base of a codon and the corresponding base of a tRNA anticodon are relaxed compared to those at other codon positions. For example, the base U at the 5′ end of a tRNA anticodon can pair with either A or G in the third position (at the 3′ end) of an mRNA codon. The flexible base pairing at this codon position is called **wobble**. Wobble explains why the synonymous codons for a given amino acid can differ in their third base, but usually not in their other bases. For example, a tRNA with the anticodon 3′-UCU-5′ can base-pair with either the mRNA codon 5′-AGA-3′ or 5′-AGG-3′, both of which code for arginine (see Figure 17.5).

Ribosomes

Ribosomes facilitate the specific coupling of tRNA anticodons with mRNA codons during protein synthesis. A ribosome is made up of two subunits, called the large and small subunits **(Figure 17.16)**. The ribosomal subunits are constructed of proteins and RNA molecules named **ribosomal RNAs**, or **rRNAs**. In eukaryotes, the subunits are made in the nucleolus. Ribosomal RNA genes on the chromosomal DNA are transcribed, and the RNA is processed and assembled with proteins imported from the cytoplasm. The resulting ribosomal subunits are then exported via nuclear pores to the cytoplasm. In both bacteria and eukaryotes, large and small subunits join to form a functional ribosome only when they attach to an mRNA molecule. About two-thirds of the mass of a ribosome consists of rRNAs, either three molecules (in bacteria) or four (in eukaryotes). Because most cells contain thousands of ribosomes, rRNA is the most abundant type of cellular RNA.

Although the ribosomes of bacteria and eukaryotes are very similar in structure and function, those of eukaryotes are slightly larger and differ somewhat from bacterial ribosomes in their molecular composition. The differences are medically significant. Certain antibiotic drugs can inactivate bacterial ribosomes without inhibiting the ability of eukaryotic ribosomes to make proteins. These drugs, including tetracycline and streptomycin, are used to combat bacterial infections.

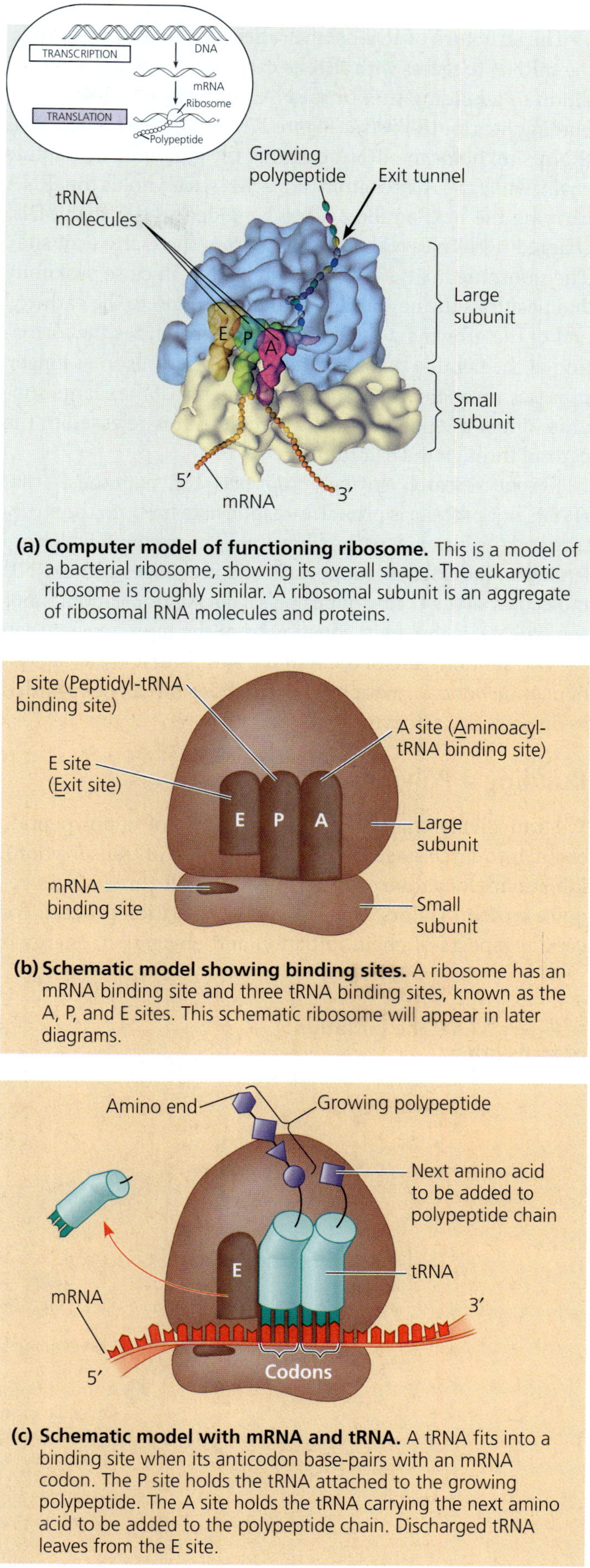

(a) Computer model of functioning ribosome. This is a model of a bacterial ribosome, showing its overall shape. The eukaryotic ribosome is roughly similar. A ribosomal subunit is an aggregate of ribosomal RNA molecules and proteins.

(b) Schematic model showing binding sites. A ribosome has an mRNA binding site and three tRNA binding sites, known as the A, P, and E sites. This schematic ribosome will appear in later diagrams.

(c) Schematic model with mRNA and tRNA. A tRNA fits into a binding site when its anticodon base-pairs with an mRNA codon. The P site holds the tRNA attached to the growing polypeptide. The A site holds the tRNA carrying the next amino acid to be added to the polypeptide chain. Discharged tRNA leaves from the E site.

▲ **Figure 17.16 The anatomy of a functioning ribosome.**

The structure of a ribosome reflects its function of bringing mRNA together with tRNAs carrying amino acids. In addition to a binding site for mRNA, each ribosome has three binding sites for tRNA (see Figure 17.16). The **P site** (peptidyl-tRNA site) holds the tRNA carrying the growing polypeptide chain, while the **A site** (aminoacyl-tRNA site) holds the tRNA carrying the next amino acid to be added to the chain. Discharged tRNAs leave the ribosome from the **E site** (exit site). The ribosome holds the tRNA and mRNA in close proximity and positions the new amino acid for addition to the carboxyl end of the growing polypeptide. It then catalyzes the formation of the peptide bond. As the polypeptide becomes longer, it passes through an *exit tunnel* in the ribosome's large subunit. When the polypeptide is complete, it is released to the cytosol through the exit tunnel.

Recent research strongly supports the hypothesis that rRNA, not protein, is primarily responsible for both the structure and the function of the ribosome. The proteins, which are largely on the exterior, support the shape changes of the rRNA molecules as they carry out catalysis during translation. Ribosomal RNA is the main constituent of the interface between the two subunits and of the A and P sites, and it is the catalyst of peptide bond formation. Thus, a ribosome can be regarded as one colossal ribozyme!

Building a Polypeptide

We can divide translation, the synthesis of a polypeptide chain, into three stages (analogous to those of transcription): initiation, elongation, and termination. All three stages require protein "factors" that aid in the translation process. For certain aspects of chain initiation and elongation, energy is also required. It is provided by the hydrolysis of GTP (guanosine triphosphate), a molecule closely related to ATP.

Ribosome Association and Initiation of Translation

The initiation stage of translation brings together mRNA, a tRNA bearing the first amino acid of the polypeptide, and the two subunits of a ribosome **(Figure 17.17)**. First, a small ribosomal subunit binds to both mRNA and a specific initiator tRNA, which carries the amino acid methionine. In bacteria, the small subunit can bind these two in either order; it binds the mRNA at a specific RNA sequence, just upstream of the start codon, AUG. In eukaryotes, the small subunit, with the initiator tRNA already bound, binds to the 5′ cap of the mRNA and then moves, or *scans*, downstream along the mRNA until it reaches the start codon, and the initiator tRNA hydrogen-bonds to it. In either case, the start codon signals the start of translation; this is important because it establishes the codon reading frame for the mRNA.

The union of mRNA, initiator tRNA, and a small ribosomal subunit is followed by the attachment of a large ribosomal subunit, completing the *translation initiation complex*. Proteins called *initiation factors* are required to bring all these components together. The cell also expends energy in the form of a GTP molecule to form the initiation complex. At the completion of the initiation process, the initiator tRNA sits in the P site of the ribosome, and the vacant A site is ready for the next aminoacyl tRNA. Note that a polypeptide is always synthesized in one direction, from the initial methionine at the amino end, also called the N-terminus, toward the final amino acid at the carboxyl end, also called the C-terminus (see Figure 5.18).

▶ **Figure 17.17 The initiation of translation.**

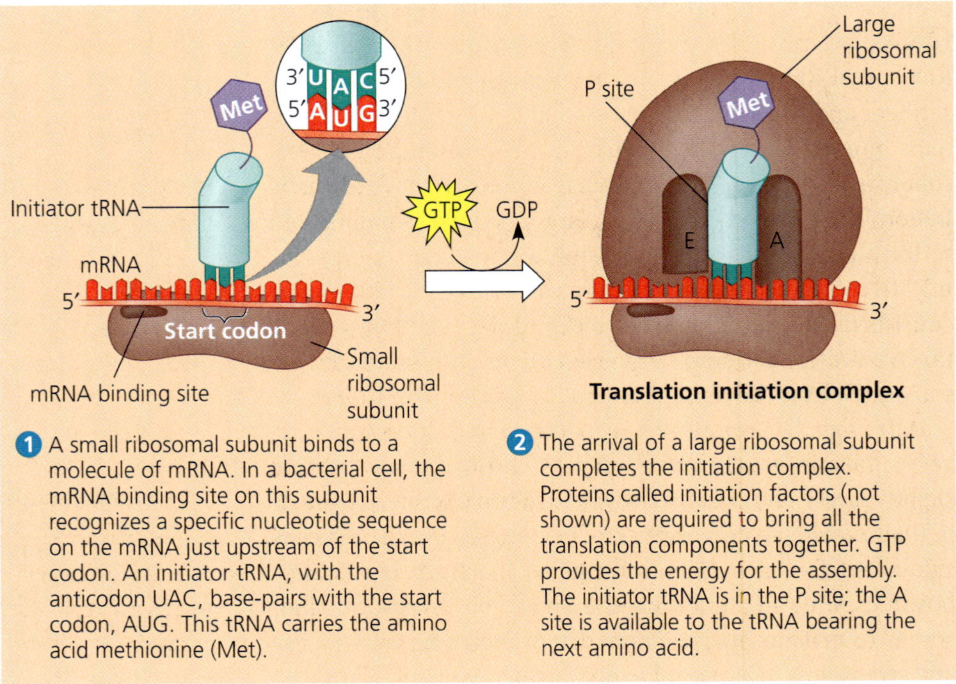

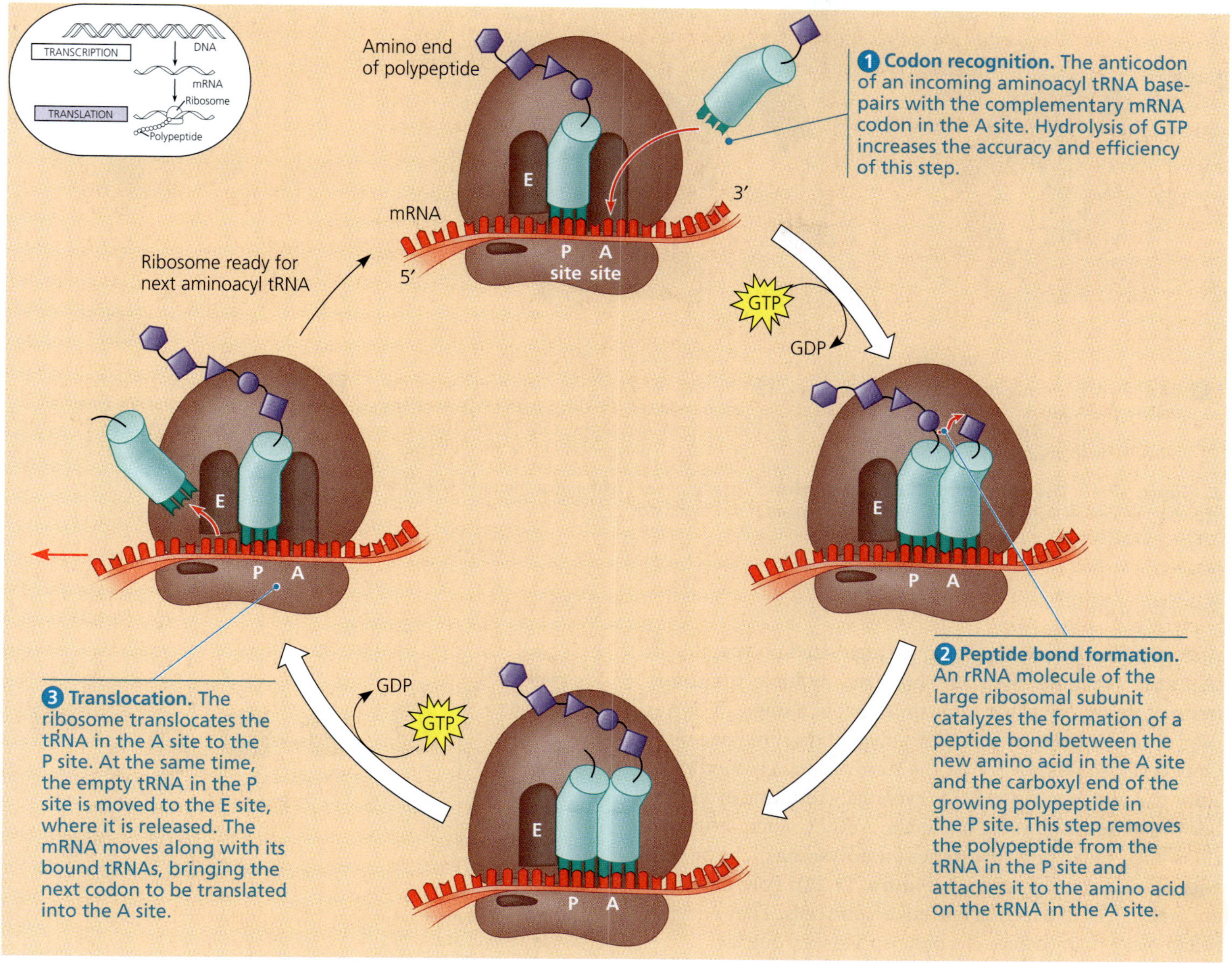

▲ **Figure 17.18 The elongation cycle of translation.** The hydrolysis of GTP plays an important role in the elongation process. Not shown are the proteins called elongation factors.

Elongation of the Polypeptide Chain

In the elongation stage of translation, amino acids are added one by one to the preceding amino acid. Each addition involves the participation of several proteins called *elongation factors* and occurs in a three-step cycle described in **Figure 17.18**. Energy expenditure occurs in the first and third steps. Codon recognition requires hydrolysis of one molecule of GTP, which increases the accuracy and efficiency of this step. One more GTP is hydrolyzed to provide energy for the translocation step.

The mRNA is moved through the ribosome in one direction only, 5′ end first; this is equivalent to the ribosome moving 5′ → 3′ on the mRNA. The important point is that the ribosome and the mRNA move relative to each other, unidirectionally, codon by codon. The elongation cycle takes less than a tenth of a second in bacteria and is repeated as each amino acid is added to the chain until the polypeptide is completed.

Termination of Translation

The final stage of translation is termination (**Figure 17.19**, on the next page). Elongation continues until a stop codon in the mRNA reaches the A site of the ribosome. The base triplets UAG, UAA, and UGA do not code for amino acids but instead act as signals to stop translation. A protein called a *release factor* binds directly to the stop codon in the A site. The release factor causes the addition of a water molecule instead of an amino acid to the polypeptide chain. This reaction breaks (hydrolyzes) the bond between the completed polypeptide and the tRNA in the P site, releasing the polypeptide through the exit tunnel of the ribosome's large subunit (see Figure 17.16a). The remainder of the translation assembly then comes apart in a multistep process, aided by other protein factors. Breakdown of the translation assembly requires the hydrolysis of two more GTP molecules.

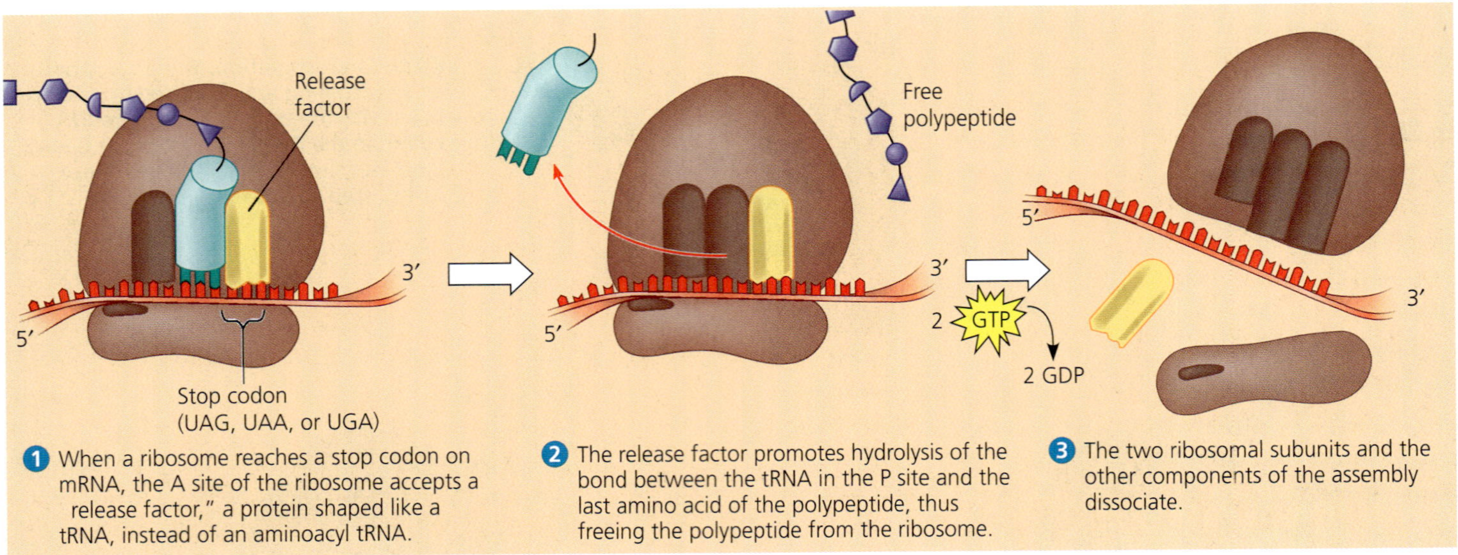

▲ **Figure 17.19 The termination of translation.** Like elongation, termination requires GTP hydrolysis as well as additional protein factors, which are not shown here.

Polyribosomes

A single ribosome can make an average-sized polypeptide in less than a minute. Typically, however, multiple ribosomes translate an mRNA at the same time; that is, a single mRNA is used to make many copies of a polypeptide simultaneously. Once a ribosome moves past the start codon, a second ribosome can attach to the mRNA, eventually resulting in a number of ribosomes trailing along the mRNA. Such strings of ribosomes, called **polyribosomes** (or **polysomes**), can be seen with an electron microscope **(Figure 17.20)**. Polyribosomes are found in both bacterial and eukaryotic cells. They enable a cell to make many copies of a polypeptide very quickly.

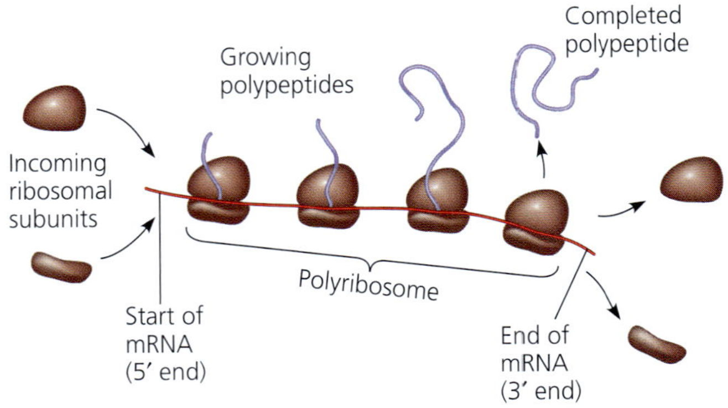

(a) An mRNA molecule is generally translated simultaneously by several ribosomes in clusters called polyribosomes.

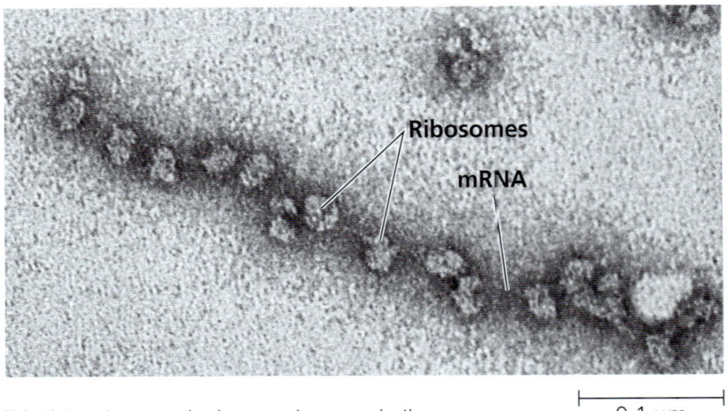

(b) This micrograph shows a large polyribosome in a bacterial cell (TEM).

▲ **Figure 17.20 Polyribosomes.**

Completing and Targeting the Functional Protein

The process of translation is often not sufficient to make a functional protein. In this section, you will learn about modifications that polypeptide chains undergo after the translation process as well as some of the mechanisms used to target completed proteins to specific sites in the cell.

Protein Folding and Post-Translational Modifications

During its synthesis, a polypeptide chain begins to coil and fold spontaneously as a consequence of its amino acid sequence (primary structure), forming a protein with a specific shape: a three-dimensional molecule with secondary and tertiary structure (see Figure 5.21). Thus, a gene determines primary structure, and primary structure in turn determines shape. In many cases, a chaperone protein (chaperonin) helps the polypeptide fold correctly (see Figure 5.24).

Additional steps—*post-translational modifications*—may be required before the protein can begin doing its particular job in the cell. Certain amino acids may be chemically modified by the attachment of sugars, lipids, phosphate groups, or other additions. Enzymes may remove one or more amino acids from the leading (amino) end of the polypeptide chain. In some cases, a polypeptide chain may be enzymatically cleaved into two or more pieces. For example, the protein insulin is first synthesized as a single polypeptide chain but becomes active only after an enzyme cuts out a central part of the chain, leaving a protein made up of two polypeptide chains connected by disulfide bridges. In other cases, two or more polypeptides that are synthesized separately may come together, becoming the subunits of a protein that has quaternary structure. A familiar example is hemoglobin (see Figure 5.21).

Targeting Polypeptides to Specific Locations

In electron micrographs of eukaryotic cells active in protein synthesis, two populations of ribosomes (and polyribosomes) are evident: free and bound (see Figure 6.11). Free ribosomes are suspended in the cytosol and mostly synthesize proteins that stay in the cytosol and function there. In contrast, bound ribosomes are attached to the cytosolic side of the endoplasmic reticulum (ER) or to the nuclear envelope. Bound ribosomes make proteins of the endomembrane system (the nuclear envelope, ER, Golgi apparatus, lysosomes, vacuoles, and plasma membrane) as well as proteins secreted from the cell, such as insulin. The ribosomes themselves are identical and can switch their status from free to bound.

What determines whether a ribosome will be free in the cytosol or bound to rough ER at any given time? Polypeptide synthesis always begins in the cytosol, when a free ribosome starts to translate an mRNA molecule. There the process continues to completion—*unless* the growing polypeptide itself cues the ribosome to attach to the ER. The polypeptides of proteins destined for the endomembrane system or for secretion are marked by a **signal peptide**, which targets the protein to the ER **(Figure 17.21)**. The signal peptide, a sequence of about 20 amino acids at or near the leading (amino) end of the polypeptide, is recognized as it emerges from the ribosome by a protein-RNA complex called a **signal-recognition particle (SRP)**. This particle functions as an adapter that brings the ribosome to a receptor protein built into the ER membrane. This receptor is part of a multiprotein translocation complex. Polypeptide synthesis continues there, and the growing polypeptide snakes across the membrane into the ER lumen via a protein pore. The signal peptide is usually removed by an enzyme. The rest of the completed polypeptide, if it is to be secreted from the cell, is released into solution within the ER lumen (as in Figure 17.21). Alternatively, if the polypeptide is to be a membrane protein, it remains partially embedded in the ER membrane.

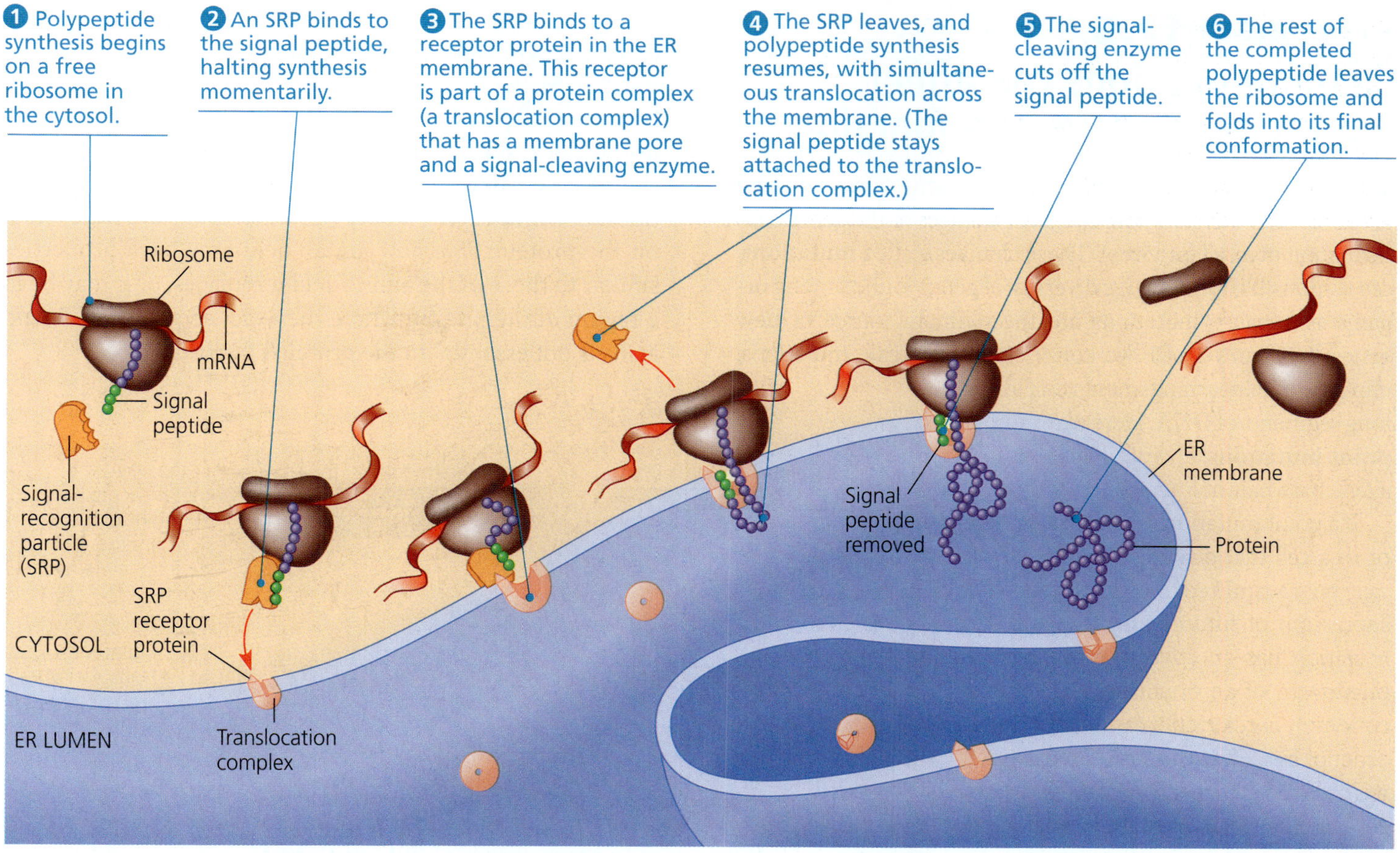

▲ **Figure 17.21 The signal mechanism for targeting proteins to the ER.** A polypeptide destined for the endomembrane system or for secretion from the cell begins with a signal peptide, a series of amino acids that targets it for the ER. This figure shows the synthesis of a secretory protein and its simultaneous import into the ER. In the ER and then in the Golgi, the protein will be processed further. Finally, a transport vesicle will convey it to the plasma membrane for release from the cell (see Figure 7.10).

Other kinds of signal peptides are used to target polypeptides to mitochondria, chloroplasts, the interior of the nucleus, and other organelles that are not part of the endomembrane system. The critical difference in these cases is that translation is completed in the cytosol before the polypeptide is imported into the organelle. The mechanisms of translocation also vary, but in all cases studied to date, the "zip codes" that address proteins for secretion or to cellular locations are signal peptides of some sort. Bacteria also employ signal peptides to target proteins for secretion.

CONCEPT CHECK 17.4

1. What two processes ensure that the correct amino acid is added to a growing polypeptide chain?
2. Describe how the formation of polyribosomes can benefit the cell.
3. Describe how a polypeptide to be secreted is transported to the endomembrane system.
4. **WHAT IF?** Discuss the ways in which rRNA structure likely contributes to ribosomal function.

For suggested answers, see Appendix A.

CONCEPT **17.5**

Point mutations can affect protein structure and function

Now that you have explored the process of gene expression, you are ready to understand the effects of changes to the genetic information of a cell (or virus). These changes, called **mutations**, are responsible for the huge diversity of genes found among organisms because mutations are the ultimate source of new genes. In Figure 15.15, we considered large-scale mutations, chromosomal rearrangements that affect long segments of DNA. Here we examine **point mutations**, chemical changes in a single base pair of a gene.

If a point mutation occurs in a gamete or in a cell that gives rise to gametes, it may be transmitted to offspring and to a succession of future generations. If the mutation has an adverse effect on the phenotype of an organism, the mutant condition is referred to as a genetic disorder or hereditary disease. For example, we can trace the genetic basis of sickle-cell disease to the mutation of a single base pair in the gene that encodes the β-globin polypeptide of hemoglobin. The change of a single nucleotide in the DNA's template strand leads to the production of an abnormal protein (**Figure 17.22**; also see Figure 5.22). In individuals who are homozygous for the mutant allele, the sickling of red blood cells caused by the altered hemoglobin produces the multiple symptoms associated with sickle-cell disease (see Chapter 14). Another example is a heart condition responsible for some incidents of sudden death in young athletes, called familial cardiomyopathy. Point mutations in several genes have been identified, each of which can lead to this disorder.

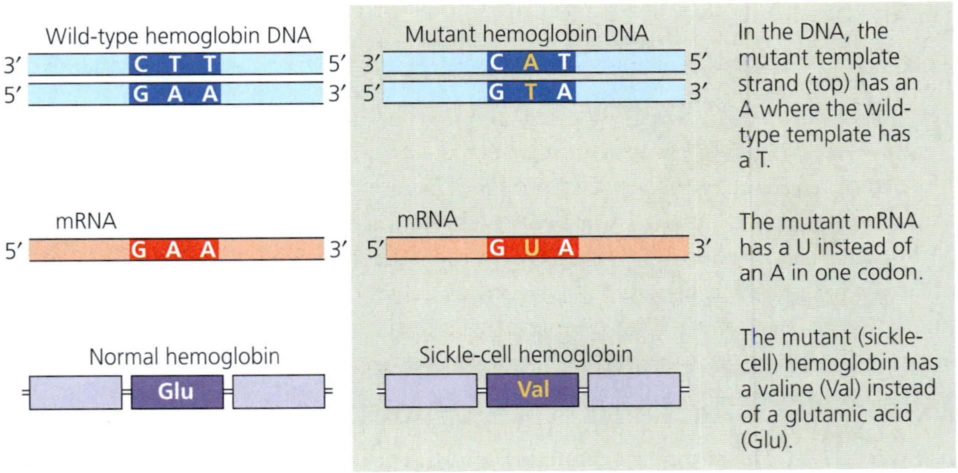

▲ **Figure 17.22 The molecular basis of sickle-cell disease: a point mutation.** The allele that causes sickle-cell disease differs from the wild-type (normal) allele by a single DNA base pair.

Types of Point Mutations

Point mutations within a gene can be divided into two general categories: base-pair substitutions and base-pair insertions or deletions. Let's now consider how these mutations affect proteins.

Substitutions

A **base-pair substitution** is the replacement of one nucleotide and its partner with another pair of nucleotides (**Figure 17.23a**). Some substitutions are called *silent mutations* because, owing to the redundancy of the genetic code, they have no effect on the encoded protein. In other words, a change in a base pair may transform one codon into another that is translated into the same amino acid. For example, if 3′-CCG-5′ on the template strand mutated to 3′-CCA-5′, the mRNA codon that used to be GGC would become GGU, but a glycine would still be inserted at the proper location in the protein (see Figure 17.5). Substitutions that change one amino acid to another one are called **missense mutations**. Such a mutation may have little effect on the protein: The new amino acid may have properties similar to those of the amino acid it replaces, or it may be in a region of the protein where the exact sequence of amino acids is not essential to the protein's function.

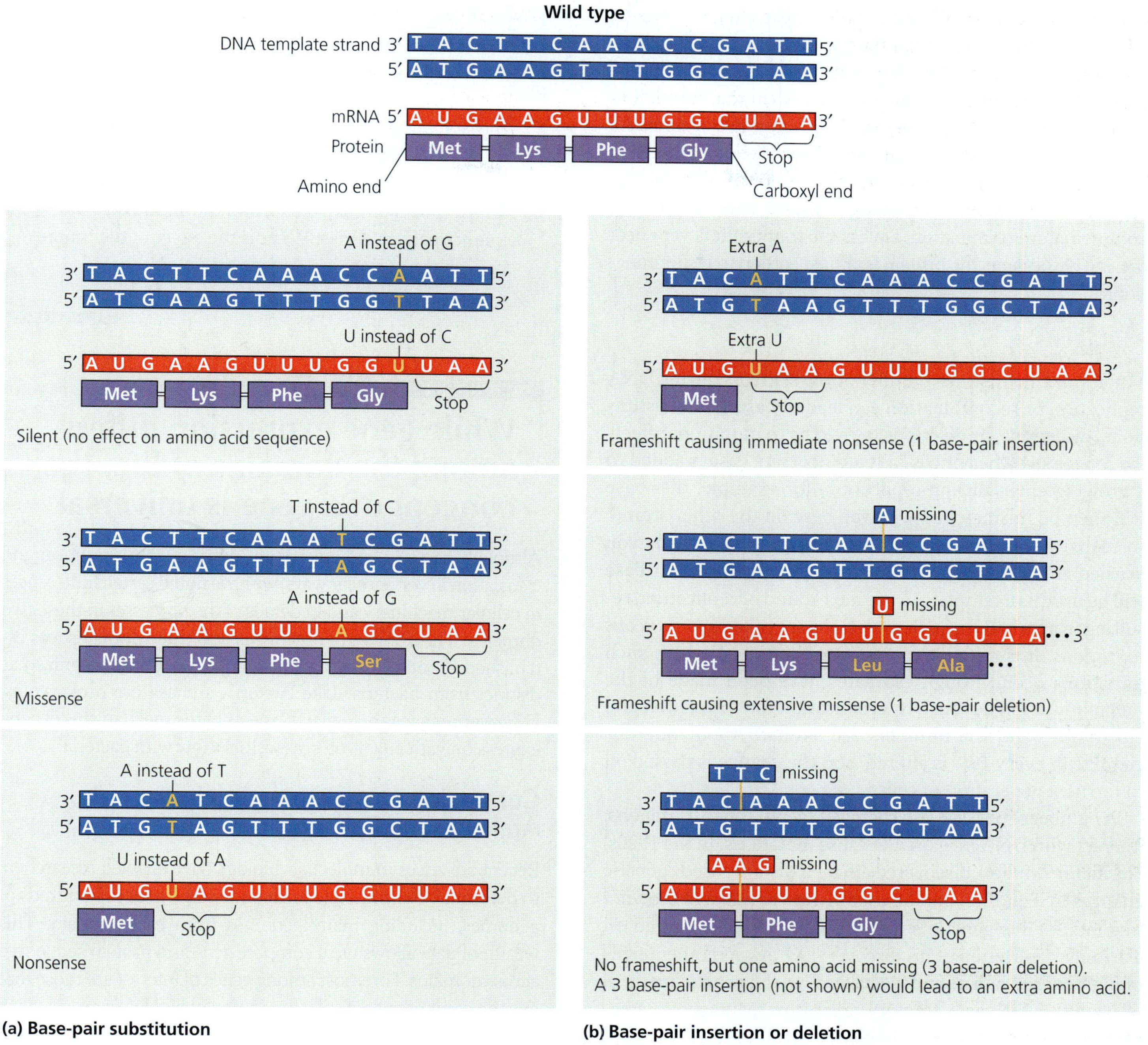

▲ **Figure 17.23 Types of point mutations.** Mutations are changes in DNA that result in changes in the mRNA.

However, the base-pair substitutions of greatest interest are those that cause a major change in a protein. The alteration of a single amino acid in a crucial area of a protein—such as in the part of hemoglobin shown in Figure 17.22 or in the active site of an enzyme—will significantly alter protein activity. Occasionally, such a mutation leads to an improved protein or one with novel capabilities, but much more often such mutations are detrimental, leading to a useless or less active protein that impairs cellular function.

Substitution mutations are usually missense mutations; that is, the altered codon still codes for an amino acid and thus makes sense, although not necessarily the *right* sense. But a point mutation can also change a codon for an amino acid into a stop codon. This is called a **nonsense mutation**, and it causes translation to be terminated prematurely; the resulting polypeptide will be shorter than the polypeptide encoded by the normal gene. Nearly all nonsense mutations lead to nonfunctional proteins.

Insertions and Deletions

Insertions and **deletions** are additions or losses of nucleotide pairs in a gene **(Figure 17.23b)**. These mutations have a disastrous effect on the resulting protein more often than substitutions do. Insertion or deletion of nucleotides may alter the reading frame of the genetic message, the triplet grouping of bases on the mRNA that is read during translation. Such a mutation, called a **frameshift mutation**, will occur whenever the number of nucleotides inserted or deleted is not a multiple of three. All the nucleotides that are downstream of the deletion or insertion will be improperly grouped into codons, and the result will be extensive missense, usually ending sooner or later in nonsense and premature termination. Unless the frameshift is very near the end of the gene, the protein is almost certain to be nonfunctional.

Mutagens

Mutations can arise in a number of ways. Errors during DNA replication or recombination can lead to base-pair substitutions, insertions, or deletions, as well as to mutations affecting longer stretches of DNA. If an incorrect base is added to a growing chain during replication, for example, that base will then be mismatched with the base on the other strand. In many cases, the error will be corrected by systems you learned about in Chapter 16. Otherwise, the incorrect base will be used as a template in the next round of replication, resulting in a mutation. Such mutations are called *spontaneous mutations*. It is difficult to calculate the rate at which such mutations occur. Rough estimates have been made of the rate of mutation during DNA replication for both *E. coli* and eukaryotes, and the numbers are similar: About one nucleotide in every 10^{10} is altered, and the change is passed on to the next generation of cells.

A number of physical and chemical agents, called **mutagens**, interact with DNA in ways that cause mutations. In the 1920s, Hermann Muller discovered that X-rays caused genetic changes in fruit flies, and he used X-rays to make *Drosophila* mutants for his genetic studies. But he also recognized an alarming implication of his discovery: X-rays and other forms of high-energy radiation pose hazards to the genetic material of people as well as laboratory organisms. Mutagenic radiation, a physical mutagen, includes ultraviolet (UV) light, which can cause disruptive thymine dimers in DNA (see Figure 16.18).

Chemical mutagens fall into several categories. Base analogs are chemicals that are similar to normal DNA bases but that pair incorrectly during DNA replication. Some other chemical mutagens interfere with correct DNA replication by inserting themselves into the DNA and distorting the double helix. Still other mutagens cause chemical changes in bases that change their pairing properties.

Researchers have developed various methods to test the mutagenic activity of chemicals. A major application of these tests is the preliminary screening of chemicals to identify those that may cause cancer. This approach makes sense because most carcinogens (cancer-causing chemicals) are mutagenic, and conversely, most mutagens are carcinogenic.

CONCEPT CHECK 17.5

1. What happens when one nucleotide pair is lost from the middle of the coding sequence of a gene?
2. **WHAT IF?** A gene whose template strand contains the sequence 3′-TACTTGTCCGATATC-5′ is mutated to 3′-TACTTGTCCAATATC-5′. For both normal and mutant genes, draw the double-stranded DNA, the resulting mRNA, and the amino acid sequence each encodes. What is the effect of the mutation on the amino acid sequence?

For suggested answers, see Appendix A.

CONCEPT **17.6**

While gene expression differs among the domains of life, the concept of a gene is universal

Although bacteria and eukaryotes carry out transcription and translation in very similar ways, we have noted certain differences in cellular machinery and in details of the processes in these two domains. The division of organisms into three domains was established about 40 years ago, when archaea were recognized as distinct from bacteria. Like bacteria, archaea are prokaryotes. However, archaea share many aspects of the mechanisms of gene expression with eukaryotes, as well as a few with bacteria.

Comparing Gene Expression in Bacteria, Archaea, and Eukarya

Recent advances in molecular biology have enabled researchers to determine the complete nucleotide sequences of hundreds of genomes, including many genomes from each domain. This wealth of data allows us to compare gene and protein sequences across domains. Foremost among genes of interest are those that encode components of such fundamental biological processes as transcription and translation.

Bacterial and eukaryotic RNA polymerases differ significantly from each other. In contrast, the single RNA polymerase of archaea resembles the three eukaryotic RNA polymerases, and archaea and eukaryotes use a complex set of transcription factors, unlike bacteria. Transcription is terminated differently in bacteria and eukaryotes. Here again, the little that is known about archaeal transcription termination suggests it may be more like the eukaryotic process.

As far as translation is concerned, bacterial and eukaryotic ribosomes are slightly different. Archaeal ribosomes are the same size as bacterial ribosomes, but their sensitivity to chemical inhibitors most closely matches that of eukaryotic ribosomes. We mentioned earlier that initiation of translation is slightly different in bacteria and eukaryotes. In this respect, the archaeal process is more like that of bacteria.

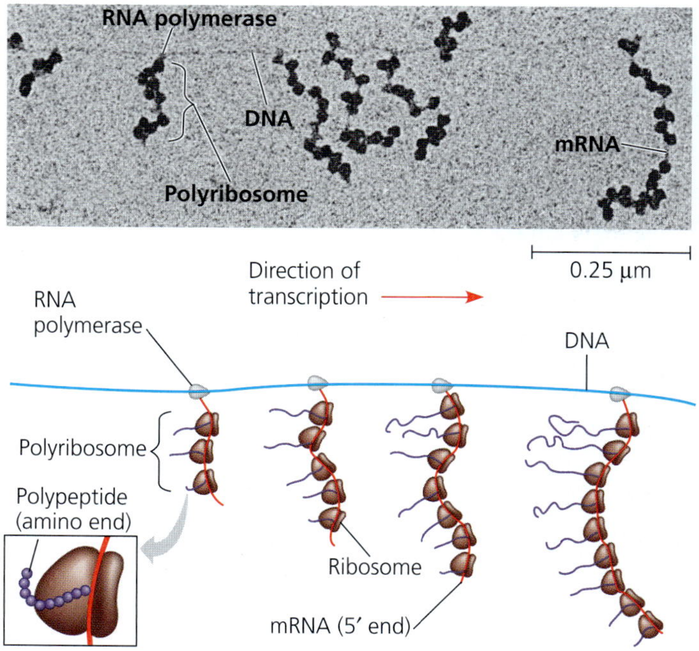

▲ **Figure 17.24 Coupled transcription and translation in bacteria.** In bacterial cells, the translation of mRNA can begin as soon as the leading (5′) end of the mRNA molecule peels away from the DNA template. The micrograph (TEM) shows a strand of *E. coli* DNA being transcribed by RNA polymerase molecules. Attached to each RNA polymerase molecule is a growing strand of mRNA, which is already being translated by ribosomes. The newly synthesized polypeptides are not visible in the micrograph but are shown in the diagram.

? *Which one of the mRNA molecules started transcription first? On that mRNA, which ribosome started translating first?*

The most important differences between bacteria and eukaryotes with regard to gene expression arise from the bacterial cell's relative absence of compartmental organization. Like a one-room workshop, a bacterial cell ensures a streamlined operation. In the absence of a nucleus, it can simultaneously transcribe and translate the same gene **(Figure 17.24)**, and the newly made protein can quickly diffuse to its site of function. Little is currently known about whether the processes of transcription and translation are coupled like this in archaeal cells, but most researchers suspect that they are, since archaea lack a nuclear envelope. In contrast, the eukaryotic cell's nuclear envelope segregates transcription from translation and provides a compartment for extensive RNA processing. This processing stage includes additional steps whose regulation can help coordinate the eukaryotic cell's elaborate activities (see Chapter 18). Finally, eukaryotic cells have complicated mechanisms for targeting proteins to the appropriate cellular compartment (organelle).

Learning more about the proteins and RNAs involved in archaeal transcription and translation will tell us much about the evolution of these processes in all three domains. In spite of the differences in gene expression cataloged here, however, the idea of the gene itself is a unifying concept among all forms of life.

What Is a Gene? *Revisiting the Question*

Our definition of a gene has evolved over the past few chapters, as it has through the history of genetics. We began with the Mendelian concept of a gene as a discrete unit of inheritance that affects a phenotypic character (Chapter 14). We saw that Morgan and his colleagues assigned such genes to specific loci on chromosomes (Chapter 15). We went on to view a gene as a region of specific nucleotide sequence along the length of a DNA molecule in a chromosome (Chapter 16). Finally, in this chapter, we have considered a functional definition of a gene as a DNA sequence that codes for a specific polypeptide chain. (**Figure 17.25**, on the next page, summarizes the path from gene to polypeptide in a eukaryotic cell.) All these definitions are useful, depending on the context in which genes are being studied.

Clearly, the statement that a gene codes for a polypeptide is too simple. Most eukaryotic genes contain noncoding segments (introns), so large portions of these genes have no corresponding segments in polypeptides. Molecular biologists also often include promoters and certain other regulatory regions of DNA within the boundaries of a gene. These DNA sequences are not transcribed, but they can be considered part of the functional gene because they must be present for transcription to occur. Our molecular definition of a gene must also be broad enough to include the DNA that is transcribed into rRNA, tRNA, and other RNAs that are not translated. These genes have no polypeptide products but play crucial roles in the cell. Thus, we arrive at the following definition: *A gene is a region of DNA that can be expressed to produce a final functional product that is either a polypeptide or an RNA molecule.*

When considering phenotypes, however, it is often useful to start by focusing on genes that code for polypeptides. In this chapter, you have learned in molecular terms how a typical gene is expressed—by transcription into RNA and then translation into a polypeptide that forms a protein of specific structure and function. Proteins, in turn, bring about an organism's observable phenotype.

A given type of cell expresses only a subset of its genes. This is an essential feature in multicellular organisms: You'd be in trouble if the lens cells in your eyes started expressing the genes for hair proteins, which are normally expressed only in hair follicle cells! Gene expression is precisely regulated. We'll explore gene regulation in the next chapter, beginning with the simpler case of bacteria and continuing with eukaryotes.

CONCEPT CHECK 17.6

1. Would the coupling of processes shown in Figure 17.24 be found in a eukaryotic cell? Explain.
2. **WHAT IF?** In eukaryotic cells, mRNAs have been found to have a circular arrangement in which the poly-A tail is held by proteins near the 5′ end cap. How might this increase translation efficiency?

For suggested answers, see Appendix A.

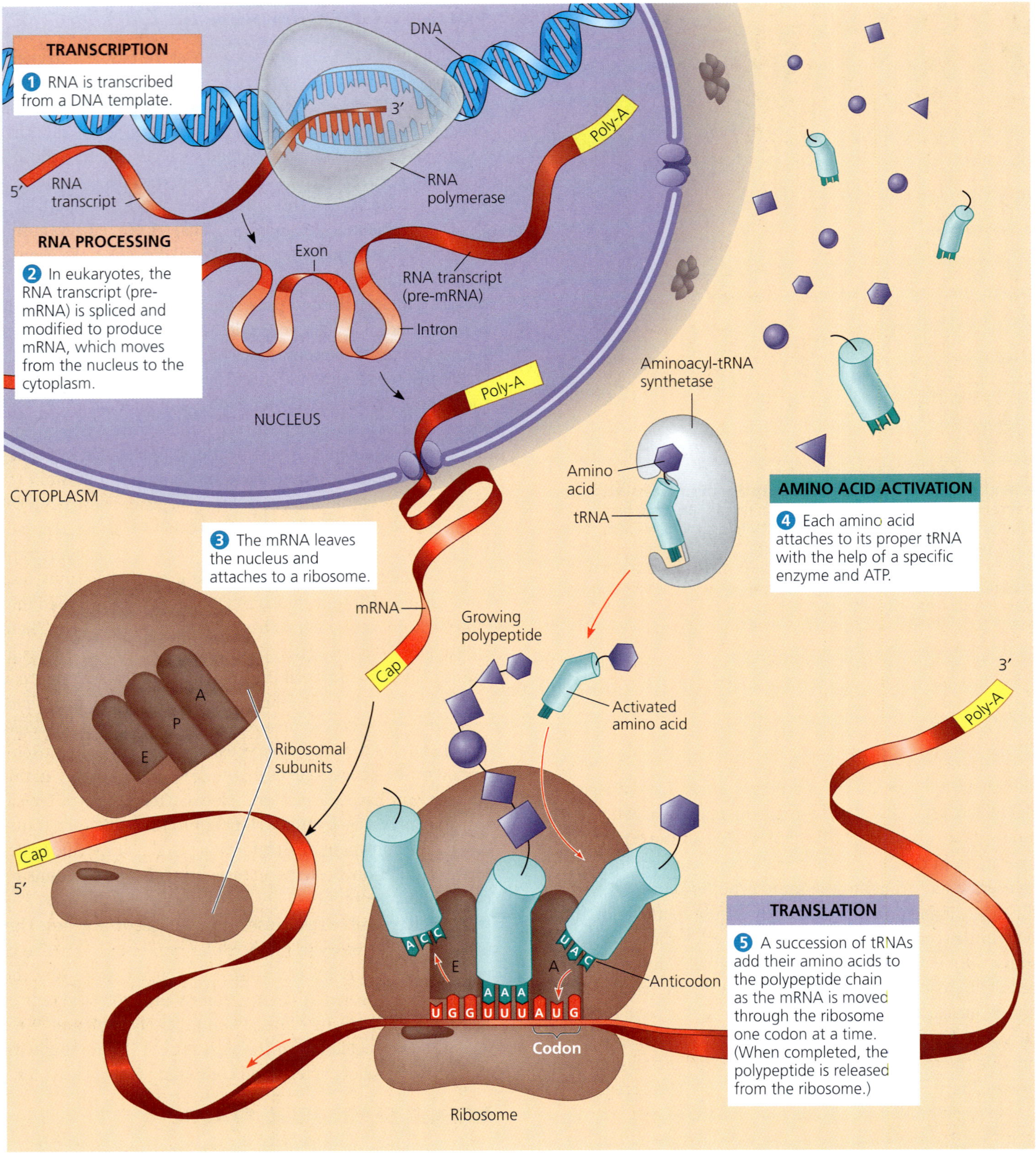

▲ **Figure 17.25 A summary of transcription and translation in a eukaryotic cell.** This diagram shows the path from one gene to one polypeptide. Keep in mind that each gene in the DNA can be transcribed repeatedly into many identical RNA molecules and that each mRNA can be translated repeatedly to yield many identical polypeptide molecules. (Also, remember that the final products of some genes are not polypeptides but RNA molecules, including tRNA and rRNA.) In general, the steps of transcription and translation are similar in bacterial, archaeal, and eukaryotic cells. The major difference is the occurrence of RNA processing in the eukaryotic nucleus. Other significant differences are found in the initiation stages of both transcription and translation and in the termination of transcription.

Chapter 17 Review

MEDIA Go to the Study Area at **www.masteringbio.com** for BioFlix 3-D Animations, MP3 Tutors, Videos, Practice Tests, an eBook, and more.

SUMMARY OF KEY CONCEPTS

CONCEPT 17.1

Genes specify proteins via transcription and translation (pp. 325–331)

- **Evidence from the Study of Metabolic Defects** DNA controls metabolism by directing cells to make specific enzymes and other proteins. Beadle and Tatum's experiments with mutant strains of *Neurospora* supported the one gene–one enzyme hypothesis. Genes code for polypeptide chains or for RNA molecules.
- **Basic Principles of Transcription and Translation** Transcription is the nucleotide-to-nucleotide transfer of information from DNA to RNA, while translation is the informational transfer from nucleotide sequence in RNA to amino acid sequence in a polypeptide.
- **The Genetic Code** Genetic information is encoded as a sequence of nonoverlapping base triplets, or codons. A codon in messenger RNA (mRNA) either is translated into an amino acid (61 of the 64 codons) or serves as a stop signal (3 codons). Codons must be read in the correct reading frame.

MEDIA

Investigation How Is a Metabolic Pathway Analyzed?
MP3 Tutor DNA to RNA to Protein
Activity Overview of Protein Synthesis

CONCEPT 17.2

Transcription is the DNA-directed synthesis of RNA: *a closer look* (pp. 331–334)

- **Molecular Components of Transcription** RNA synthesis is catalyzed by RNA polymerase. It follows the same base-pairing rules as DNA replication, except that in RNA, uracil substitutes for thymine.

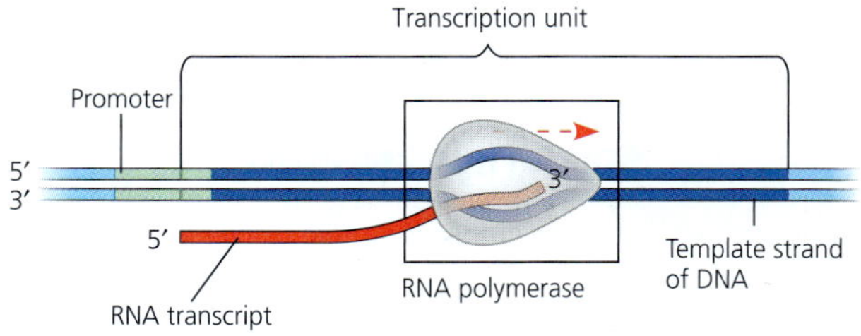

- **Synthesis of an RNA Transcript** The three stages of transcription are initiation, elongation, and termination. Promoters signal the initiation of RNA synthesis. Transcription factors help eukaryotic RNA polymerase recognize promoter sequences. The mechanisms of termination are different in bacteria and eukaryotes.

MEDIA

Activity Transcription

CONCEPT 17.3

Eukaryotic cells modify RNA after transcription (pp. 334–336)

- **Alteration of mRNA Ends** Eukaryotic mRNA molecules are processed before leaving the nucleus by modification of their ends and by RNA splicing. The 5′ end receives a modified nucleotide cap, and the 3′ end a poly-A tail.

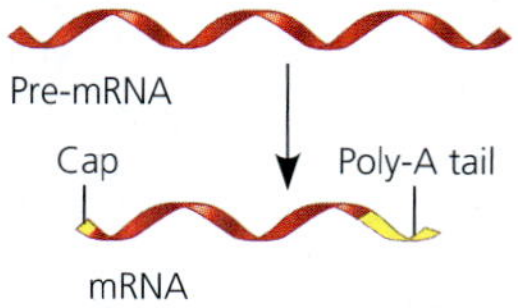

- **Split Genes and RNA Splicing** Most eukaryotic genes have introns interspersed among the coding regions, the exons. In RNA splicing, introns are removed and exons joined. RNA splicing is typically carried out by spliceosomes, but in some cases, RNA alone catalyzes its own splicing. The catalytic ability of some RNA molecules, called ribozymes, derives from the inherent properties of RNA. The presence of introns allows for alternative RNA splicing.

MEDIA

Activity RNA Processing

CONCEPT 17.4

Translation is the RNA-directed synthesis of a polypeptide: *a closer look* (pp. 337–344)

- **Molecular Components of Translation** A cell translates an mRNA message into protein using transfer RNAs (tRNAs). After binding specific amino acids, tRNAs line up via their anticodons at complementary codons on mRNA. Ribosomes help facilitate this coupling with binding sites for mRNA and tRNA.
- **Building a Polypeptide** Ribosomes coordinate the three stages of translation: initiation, elongation, and termination. The formation of peptide bonds between amino acids is catalyzed by rRNA. A number of ribosomes can translate a single mRNA molecule simultaneously, forming a polyribosome.

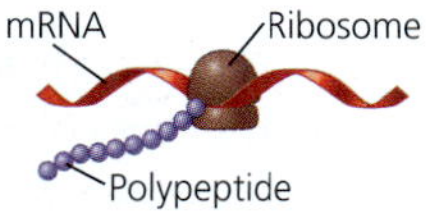

- **Completing and Targeting the Functional Protein** After translation, modifications to proteins can affect their three-dimensional shape. Free ribosomes in the cytosol initiate synthesis of all proteins, but proteins destined for the endomembrane system or for secretion are transported into the ER. Such proteins have a signal peptide to which a signal-recognition particle (SRP) binds, enabling the translating ribosome to bind to the ER.

MEDIA

BioFlix 3-D Animation Protein Synthesis
Activity Translation
Biology Labs On-Line TranslationLab

CONCEPT 17.5

Point mutations can affect protein structure and function (pp. 344–346)

- **Types of Point Mutations** A point mutation is a change in one DNA base pair, which may lead to production of a nonfunctional protein. Base-pair substitutions can cause missense or nonsense mutations. Base-pair insertions or deletions may produce frameshift mutations.

▶ **Mutagens** Spontaneous mutations can occur during DNA replication, recombination, or repair. Chemical and physical mutagens cause DNA damage that can alter genes.

CONCEPT 17.6

While gene expression differs among the domains of life, the concept of a gene is universal (pp. 346–348)

▶ **Comparing Gene Expression in Bacteria, Archaea, and Eukarya** Because bacterial cells lack a nuclear envelope, translation can begin while transcription is still in progress. Archaeal cells show similarities to both eukaryotic and bacterial cells in their processes of gene expression. In a eukaryotic cell, the nuclear envelope separates transcription from translation, and extensive RNA processing occurs in the nucleus.

▶ **What Is a Gene? *Revisiting the Question*** A gene is a region of DNA whose final functional product is either a polypeptide or an RNA molecule.

TESTING YOUR KNOWLEDGE

SELF-QUIZ

1. In eukaryotic cells, transcription cannot begin until
  a. the two DNA strands have completely separated and exposed the promoter.
  b. several transcription factors have bound to the promoter.
  c. the 5′ caps are removed from the mRNA.
  d. the DNA introns are removed from the template.
  e. DNA nucleases have isolated the transcription unit.

2. Which of the following is *not* true of a codon?
  a. It consists of three nucleotides.
  b. It may code for the same amino acid as another codon.
  c. It never codes for more than one amino acid.
  d. It extends from one end of a tRNA molecule.
  e. It is the basic unit of the genetic code.

3. The anticodon of a particular tRNA molecule is
  a. complementary to the corresponding mRNA codon.
  b. complementary to the corresponding triplet in rRNA.
  c. the part of tRNA that bonds to a specific amino acid.
  d. changeable, depending on the amino acid that attaches to the tRNA.
  e. catalytic, making the tRNA a ribozyme.

4. Which of the following is *not* true of RNA processing?
  a. Exons are cut out before mRNA leaves the nucleus.
  b. Nucleotides may be added at both ends of the RNA.
  c. Ribozymes may function in RNA splicing.
  d. RNA splicing can be catalyzed by spliceosomes.
  e. A primary transcript is often much longer than the final RNA molecule that leaves the nucleus.

5. Using Figure 17.5, identify a 5′ → 3′ sequence of nucleotides in the DNA template strand for an mRNA coding for the polypeptide sequence Phe-Pro-Lys.
  a. 5′-UUUGGGAAA-3′
  b. 5′-GAACCCCTT-3′
  c. 5′-AAAACCTTT-3′
  d. 5′-CTTCGGGAA-3′
  e. 5′-AAACCCUUU-3′

6. Which of the following mutations would be *most* likely to have a harmful effect on an organism?
  a. a base-pair substitution
  b. a deletion of three nucleotides near the middle of a gene
  c. a single nucleotide deletion in the middle of an intron
  d. a single nucleotide deletion near the end of the coding sequence
  e. a single nucleotide insertion downstream of, and close to, the start of the coding sequence

7. Which component is *not* directly involved in translation?
  a. mRNA
  b. DNA
  c. tRNA
  d. ribosomes
  e. GTP

8. DRAW IT Review the roles of RNA by filling in the following table:

Type of RNA	Functions
Messenger RNA (mRNA)	
Transfer RNA (tRNA)	
	Plays catalytic (ribozyme) roles and structural roles in ribosomes
Primary transcript	
Small nuclear RNA (snRNA)	

For Self-Quiz answers, see Appendix A.

MEDIA Visit the Study Area at **www.masteringbio.com** for a Practice Test.

EVOLUTION CONNECTION

9. The genetic code (see Figure 17.5) is rich with evolutionary implications. For instance, notice that the 20 amino acids are not randomly scattered; most amino acids are coded for by a similar set of codons. What evolutionary explanations can be given for this pattern? (*Hint:* There is one explanation relating to historical ancestry, and some less obvious ones of a "form-fits-function" type.)

SCIENTIFIC INQUIRY

10. Knowing that the genetic code is almost universal, a scientist uses molecular biological methods to insert the human β-globin gene (shown in Figure 17.10) into bacterial cells, hoping the cells will express it and synthesize functional β-globin protein. Instead, the protein produced is nonfunctional and is found to contain many fewer amino acids than does β-globin made by a eukaryotic cell. Explain why.

Biological Inquiry: A Workbook of Investigative Cases Explore translation and use of sequence data in testing hypotheses with the case "The Doctor's Dilemma."

18 Regulation of Gene Expression

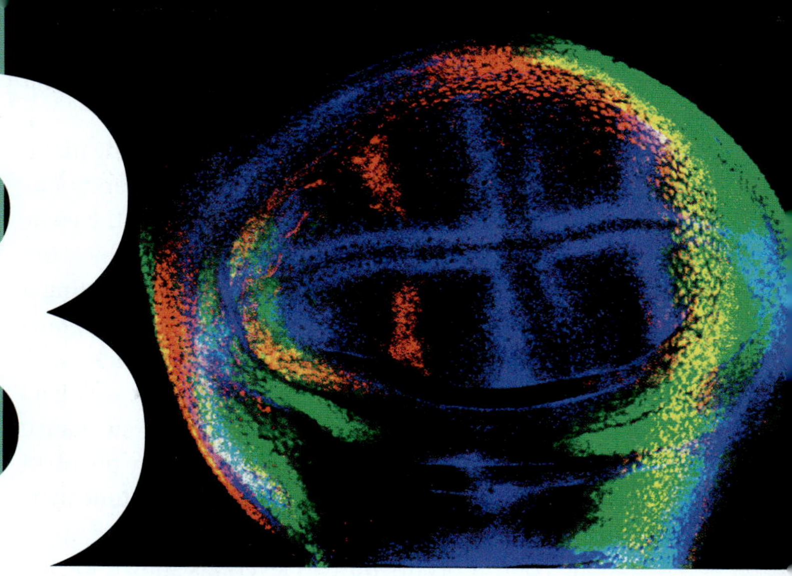

▲ **Figure 18.1 What regulates the precise pattern of expression of different genes?**

KEY CONCEPTS

18.1 Bacteria often respond to environmental change by regulating transcription

18.2 Eukaryotic gene expression can be regulated at any stage

18.3 Noncoding RNAs play multiple roles in controlling gene expression

18.4 A program of differential gene expression leads to the different cell types in a multicellular organism

18.5 Cancer results from genetic changes that affect cell cycle control

OVERVIEW

Conducting the Genetic Orchestra

An oboe squawks loudly, several violins squeak shrilly, and a tuba adds its rumble to the noisy chaos. Then the conductor's baton rises, pauses, and begins a series of elaborate movements, directing specific instruments to join in and others to raise or lower their volume at exact moments. Properly balanced and timed, discordant sounds are thus transformed into a beautiful symphony that enraptures the audience.

In a similar way, cells intricately and precisely regulate their gene expression. Both prokaryotes and eukaryotes must alter their patterns of gene expression in response to changes in environmental conditions. Multicellular eukaryotes must also develop and maintain multiple cell types. Each cell type contains the same genome but expresses a different subset of genes, a significant challenge in gene regulation.

An adult fruit fly, for example, develops from a single fertilized egg, passing through a wormlike stage called a larva. At every stage, gene expression is carefully regulated, ensuring that the right genes are expressed only at the correct time and place. In the larva, the adult wing forms in a disk-shaped pocket of several thousand cells, shown in **Figure 18.1**. The tissue in this image has been treated to reveal the mRNA for three genes—labeled red, blue, and green—using techniques covered in Chapter 20. (Red and green together appear yellow.) The intricate pattern of expression for each gene is the same from larva to larva at this stage, and it provides a graphic display of the precision of gene regulation. But what is the molecular basis for this pattern? Why is one particular gene expressed only in the few hundred cells that appear blue in this image and not in the other cells?

In this chapter, we first explore how bacteria regulate expression of their genes in response to different environmental conditions. We then examine how eukaryotes regulate gene expression to maintain different cell types. Gene expression in eukaryotes, as in bacteria, is often regulated at the stage of transcription, but control at other levels of gene expression is also important. In recent years, researchers have been surprised to discover the many roles played by RNA molecules in regulating eukaryotic gene expression, a topic we cover next. Putting together these aspects of gene regulation, we then consider how a carefully orchestrated program of gene regulation can allow a single cell—the fertilized egg—to become a fully functioning organism made up of many different cell types. Finally, we investigate how disruptions in gene regulation can lead to cancer. Orchestrating proper gene expression by all cells is crucial to the functions of life.

CONCEPT **18.1**

Bacteria often respond to environmental change by regulating transcription

Bacterial cells that can conserve resources and energy have a selective advantage over cells that are unable to do so. Thus, natural selection has favored bacteria that express only the genes whose products are needed by the cell.

Consider, for instance, an individual *E. coli* cell living in the erratic environment of a human colon, dependent for its nutrients on the whimsical eating habits of its host. If the environment is lacking in the amino acid tryptophan, which the bacterium needs to survive, the cell responds by activating a metabolic pathway that makes tryptophan from another compound. Later, if the human host eats a tryptophan-rich meal, the bacterial cell stops producing tryptophan, thus saving itself from squandering its resources to produce a substance that is available from the surrounding solution in prefabricated form. This is just one example of how bacteria tune their metabolism to changing environments.

Metabolic control occurs on two levels, as shown for the synthesis of tryptophan in **Figure 18.2**. First, cells can adjust the activity of enzymes already present. This is a fairly fast response, which relies on the sensitivity of many enzymes to chemical cues that increase or decrease their catalytic activity (see Chapter 8). The activity of the first enzyme in the tryptophan synthesis pathway is inhibited by the pathway's end product (Figure 18.2a). Thus, if tryptophan accumulates in a cell, it shuts down the synthesis of more tryptophan by inhibiting enzyme activity. Such *feedback inhibition*, typical of anabolic (biosynthetic) pathways, allows a cell to adapt to short-term fluctuations in the supply of a substance it needs.

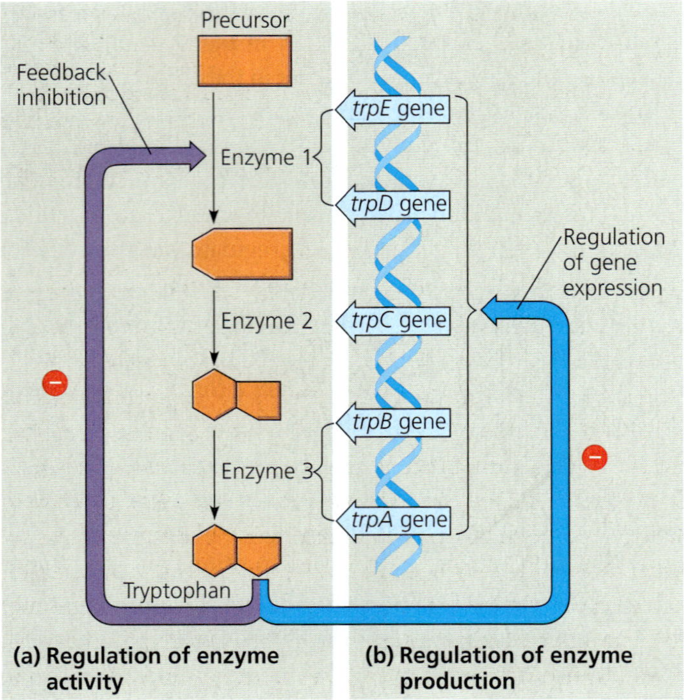

▲ **Figure 18.2 Regulation of a metabolic pathway.** In the pathway for tryptophan synthesis, an abundance of tryptophan can both **(a)** inhibit the activity of the first enzyme in the pathway (feedback inhibition), a rapid response, and **(b)** repress expression of the genes encoding all subunits of the enzymes in the pathway, a longer-term response. Genes *trpE* and *trpD* encode the two subunits of enzyme 1, and genes *trpB* and *trpA* encode the two subunits of enzyme 3. (The genes were named before the order in which they functioned in the pathway was determined.) The ⊖ symbol stands for inhibition.

Second, cells can adjust the production level of certain enzymes; that is, they can regulate the expression of the genes encoding the enzymes. If, in our example, the environment provides all the tryptophan the cell needs, the cell stops making the enzymes that catalyze the synthesis of tryptophan (Figure 18.2b). In this case, the control of enzyme production occurs at the level of transcription, the synthesis of messenger RNA coding for these enzymes. More generally, many genes of the bacterial genome are switched on or off by changes in the metabolic status of the cell. The basic mechanism for this control of gene expression in bacteria, described as the *operon model*, was discovered in 1961 by François Jacob and Jacques Monod at the Pasteur Institute in Paris. Let's see what an operon is and how it works, using the control of tryptophan synthesis as our first example.

Operons: The Basic Concept

E. coli synthesizes the amino acid tryptophan from a precursor molecule in the multistep pathway shown in Figure 18.2. Each reaction in the pathway is catalyzed by a specific enzyme, and the five genes that code for the subunits of these enzymes are clustered together on the bacterial chromosome. A single promoter serves all five genes, which together constitute a transcription unit. (Recall from Chapter 17 that a promoter is a site where RNA polymerase can bind to DNA and begin transcription.) Thus, transcription gives rise to one long mRNA molecule that codes for the five polypeptides making up the enzymes in the tryptophan pathway. The cell can translate this one mRNA into five separate polypeptides because the mRNA is punctuated with start and stop codons that signal where the coding sequence for each polypeptide begins and ends.

A key advantage of grouping genes of related function into one transcription unit is that a single on-off "switch" can control the whole cluster of functionally related genes; in other words, these genes are under *coordinate control*. When an *E. coli* cell must make tryptophan for itself because the nutrient medium lacks this amino acid, all the enzymes for the metabolic pathway are synthesized at one time. The switch is a segment of DNA called an **operator**. Both its location and name suit its function: Positioned within the promoter or, in some cases, between the promoter and the enzyme-coding genes, the operator controls the access of RNA polymerase to the genes. All together, the operator, the promoter, and the genes they control—the entire stretch of DNA required for enzyme production for the tryptophan pathway—constitute an **operon**. The *trp* operon (*trp* for tryptophan) is one of many operons in the *E. coli* genome **(Figure 18.3)**.

If the operator is the switch for controlling transcription, how does this switch work? By itself, the *trp* operon is turned on; that

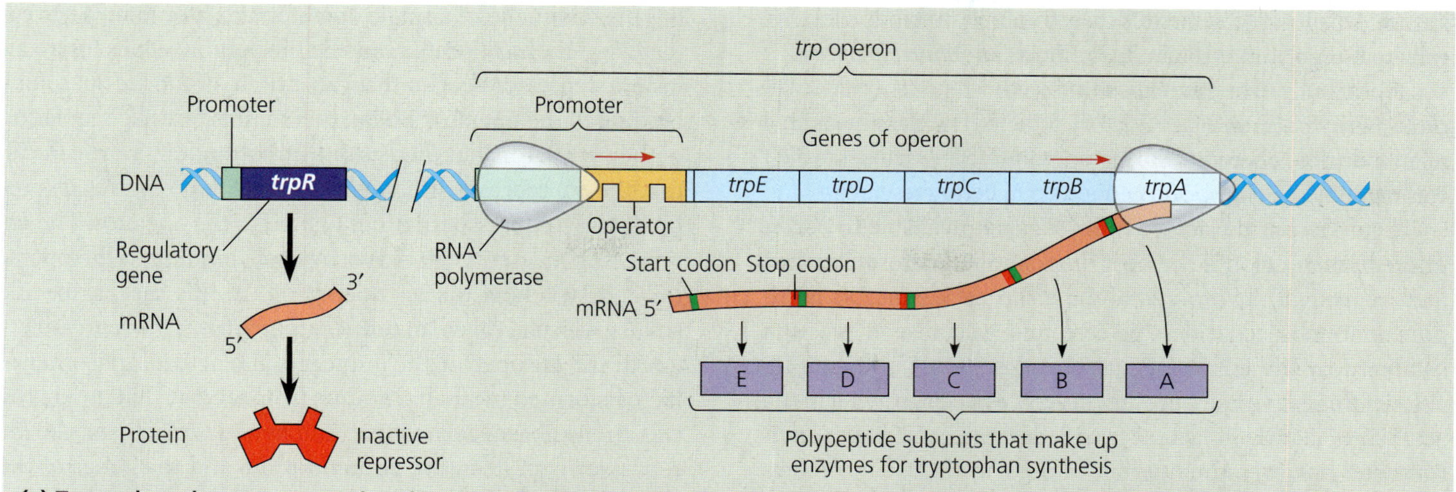

(a) Tryptophan absent, repressor inactive, operon on. RNA polymerase attaches to the DNA at the promoter and transcribes the operon's genes.

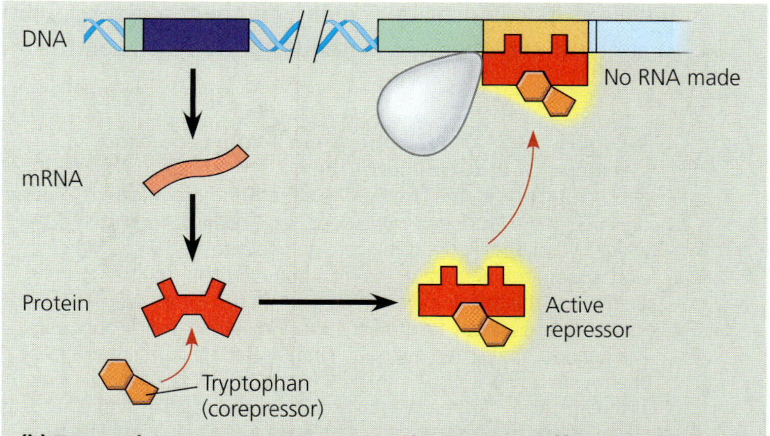

(b) Tryptophan present, repressor active, operon off. As tryptophan accumulates, it inhibits its own production by activating the repressor protein, which binds to the operator, blocking transcription.

▲ **Figure 18.3 The *trp* operon in *E. coli*: regulated synthesis of repressible enzymes.** Tryptophan is an amino acid produced by an anabolic pathway catalyzed by repressible enzymes. **(a)** The five genes encoding the polypeptide subunits of the enzymes in this pathway (see Figure 18.2) are grouped, along with a promoter, into the *trp* operon. The *trp* operator (the repressor-binding site) is located within the *trp* promoter (the RNA polymerase-binding site). **(b)** Accumulation of tryptophan, the end product of the pathway, represses transcription of the *trp* operon, thus blocking synthesis of all the enzymes in the pathway.

? *Describe what happens to the* trp *operon as the cell uses up its store of tryptophan.*

is, RNA polymerase can bind to the promoter and transcribe the genes of the operon. The operon can be switched off by a protein called the *trp* **repressor**. The repressor binds to the operator and blocks attachment of RNA polymerase to the promoter, preventing transcription of the genes. A repressor protein is specific for the operator of a particular operon. For example, the repressor that switches off the *trp* operon by binding to the *trp* operator has no effect on other operons in the *E. coli* genome.

The *trp* repressor is the product of a **regulatory gene** called *trpR*, which is located some distance from the operon it controls and has its own promoter. Regulatory genes are expressed continuously, although at a low rate, and a few *trp* repressor molecules are always present in *E. coli* cells. Why, then, is the *trp* operon not switched off permanently? First, the binding of repressors to operators is reversible. An operator vacillates between two states: one without the repressor bound and one with the repressor bound. The relative duration of each state depends on the number of active repressor molecules around. Second, the *trp* repressor, like most regulatory proteins, is an allosteric protein, with two alternative shapes, active and inactive (see Figure 8.20). The *trp* repressor is synthesized in an inactive form with little affinity for the *trp* operator. Only if tryptophan binds to the *trp* repressor at an allosteric site does the repressor protein change to the active form that can attach to the operator, turning the operon off.

Tryptophan functions in this system as a **corepressor**, a small molecule that cooperates with a repressor protein to switch an operon off. As tryptophan accumulates, more tryptophan molecules associate with *trp* repressor molecules, which can then bind to the *trp* operator and shut down production of the tryptophan pathway enzymes. If the cell's tryptophan level drops, transcription of the operon's genes resumes. This is one example of how gene expression can respond to changes in the cell's internal and external environment.

Repressible and Inducible Operons: Two Types of Negative Gene Regulation

The *trp* operon is said to be a *repressible operon* because its transcription is usually on but can be inhibited (repressed) when a specific small molecule (in this case, tryptophan) binds allosterically to a regulatory protein. In contrast, an *inducible operon* is usually off but can be stimulated (induced) when a specific small molecule interacts with a regulatory protein. The classic example of an inducible operon is the *lac* operon (*lac* for lactose), which was the subject of Jacob and Monod's pioneering research.

The disaccharide lactose (milk sugar) is available to *E. coli* in the human colon if the host drinks milk. Lactose metabolism begins with hydrolysis of the disaccharide into its component monosaccharides, glucose and galactose, a reaction catalyzed by the enzyme β-galactosidase. Only a few molecules of this enzyme are present in an *E. coli* cell growing in the absence of lactose. If lactose is added to the bacterium's environment, however, the number of β-galactosidase molecules in the cell increases a thousandfold within about 15 minutes.

The gene for β-galactosidase is part of the *lac* operon, which includes two other genes coding for enzymes that function in lactose utilization. The entire transcription unit is under the command of a single operator and promoter. The regulatory gene, *lacI*, located outside the operon, codes for an allosteric repressor protein that can switch off the *lac* operon by binding to the operator. So far, this sounds just like regulation of the *trp* operon, but there is one important difference. Recall that the *trp* repressor is inactive by itself and requires tryptophan as a corepressor in order to bind to the operator. The *lac* repressor, in contrast, is active by itself, binding to the operator and switching the *lac* operon off. In this case, a specific small molecule, called an **inducer**, *inactivates* the repressor.

For the *lac* operon, the inducer is allolactose, an isomer of lactose formed in small amounts from lactose that enters the cell. In the absence of lactose (and hence allolactose), the *lac* repressor is in its active configuration, and the genes of the *lac* operon are silenced **(Figure 18.4a)**. If lactose is added to the cell's surroundings, allolactose binds to the *lac* repressor and alters its conformation, nullifying the repressor's ability to attach to the operator. Without bound repressor, the *lac*

▼ **Figure 18.4 The *lac* operon in *E.coli*: regulated synthesis of inducible enzymes.** *E. coli* uses three enzymes to take up and metabolize lactose. The genes for these three enzymes are clustered in the *lac* operon. One gene, *lacZ*, codes for β-galactosidase, which hydrolyzes lactose to glucose and galactose. The second gene, *lacY*, codes for a permease, the membrane protein that transports lactose into the cell. The third gene, *lacA*, codes for an enzyme called transacetylase, whose function in lactose metabolism is still unclear. The gene for the *lac* repressor, *lacI*, happens to be adjacent to the *lac* operon, an unusual situation. The function of the darker green region at the upstream (left) end of the promoter is revealed in Figure 18.5.

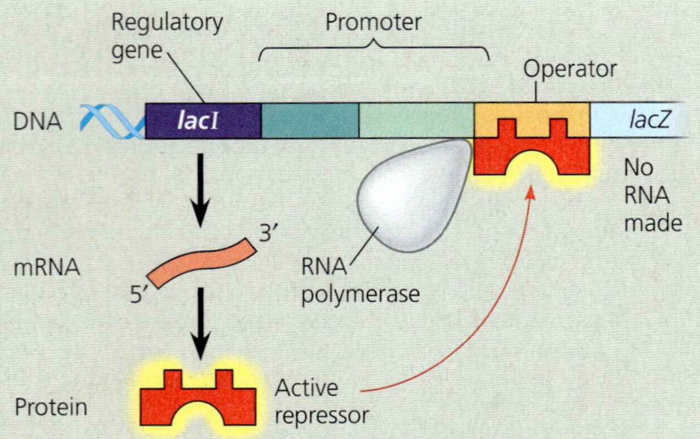

(a) Lactose absent, repressor active, operon off. The *lac* repressor is innately active, and in the absence of lactose it switches off the operon by binding to the operator.

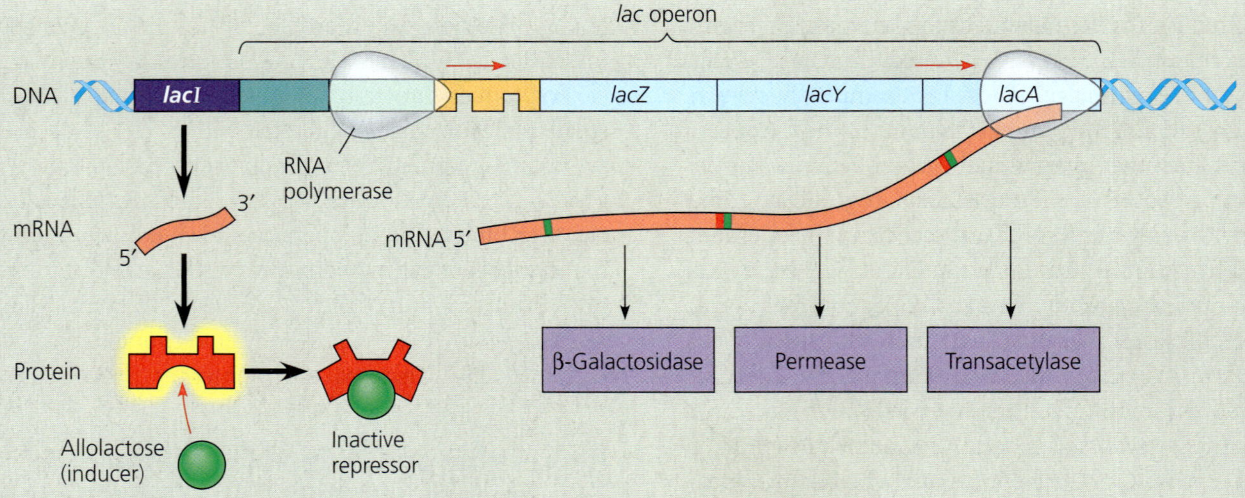

(b) Lactose present, repressor inactive, operon on. Allolactose, an isomer of lactose, derepresses the operon by inactivating the repressor. In this way, the enzymes for lactose utilization are induced.

operon is transcribed into mRNA for the lactose-utilizing enzymes **(Figure 18.4b)**.

In the context of gene regulation, the enzymes of the lactose pathway are referred to as *inducible enzymes* because their synthesis is induced by a chemical signal (allolactose, in this case). Analogously, the enzymes for tryptophan synthesis are said to be repressible. *Repressible enzymes* generally function in anabolic pathways, which synthesize essential end products from raw materials (precursors). By suspending production of an end product when it is already present in sufficient quantity, the cell can allocate its organic precursors and energy for other uses. In contrast, inducible enzymes usually function in catabolic pathways, which break down a nutrient to simpler molecules. By producing the appropriate enzymes only when the nutrient is available, the cell avoids wasting energy and precursors making proteins that are not needed.

Regulation of both the *trp* and *lac* operons involves the *negative* control of genes, because the operons are switched off by the active form of the repressor protein. It may be easier to see this for the *trp* operon, but it is also true for the *lac* operon. Allolactose induces enzyme synthesis not by acting directly on the genome, but by freeing the *lac* operon from the negative effect of the repressor. Gene regulation is said to be *positive* only when a regulatory protein interacts directly with the genome to switch transcription on. Let's look at an example of positive control of genes, again involving the *lac* operon.

Positive Gene Regulation

When glucose and lactose are both present in its environment, *E. coli* preferentially uses glucose. The enzymes for glucose breakdown in glycolysis (see Figure 9.9) are continually present. Only when lactose is present *and* glucose is in short supply does *E. coli* use lactose as an energy source, and only then does it synthesize appreciable quantities of the enzymes for lactose breakdown.

How does the *E. coli* cell sense the glucose concentration and relay this information to the genome? Again, the mechanism depends on the interaction of an allosteric regulatory protein with a small organic molecule, in this case **cyclic AMP (cAMP)**, which accumulates when glucose is scarce (see Figure 11.10 for the structure of cAMP). The regulatory protein, called *catabolite activator protein (CAP)*, is an **activator**, a protein that binds to DNA and stimulates transcription of a gene. When cAMP binds to this regulatory protein, CAP assumes its active shape and can attach to a specific site at the upstream end of the *lac* promoter **(Figure 18.5a)**. This attachment increases the affinity of RNA polymerase for the promoter, which is actually rather low even when no repressor is bound to the operator. By facilitating the binding of RNA polymerase to the promoter and thereby increasing the rate of transcription, the attachment of CAP to the promoter directly stimulates gene expression. Therefore, this mechanism qualifies as positive regulation.

If the amount of glucose in the cell increases, the cAMP concentration falls, and without cAMP, CAP detaches from the operon. Because CAP is inactive, RNA polymerase binds less efficiently to the promoter, and transcription of the *lac* operon proceeds at only a low level, even in the presence of lactose **(Figure 18.5b)**. Thus, the *lac* operon is under dual control: negative control by the *lac* repressor and positive control by CAP. The state of the *lac* repressor (with or without bound allolactose) determines whether or not transcription of the *lac* operon's genes occurs at all; the state of CAP (with or without bound cAMP) controls the *rate* of transcription if the operon is repressor-free. It is as though the operon has both an on-off switch and a volume control.

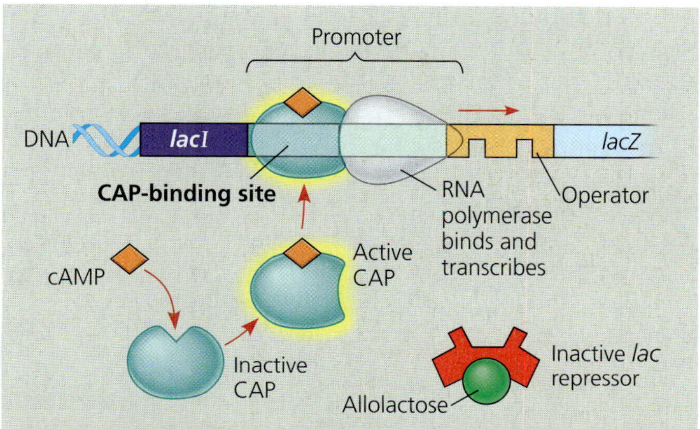

(a) Lactose present, glucose scarce (cAMP level high): abundant *lac* mRNA synthesized. If glucose is scarce, the high level of cAMP activates CAP, and the *lac* operon produces large amounts of mRNA coding for the enzymes in the lactose pathway.

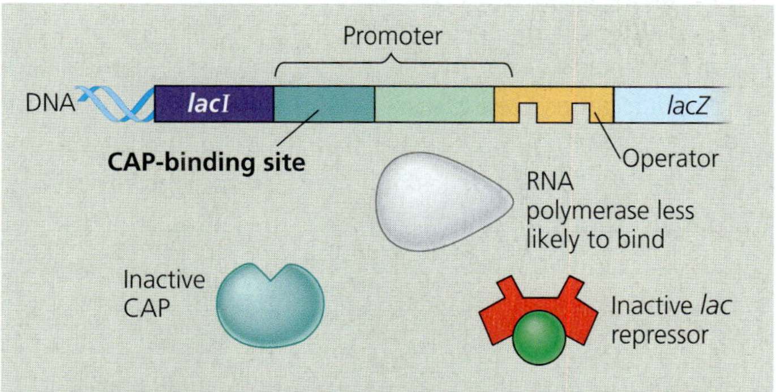

(b) Lactose present, glucose present (cAMP level low): little *lac* mRNA synthesized. When glucose is present, cAMP is scarce, and CAP is unable to stimulate transcription.

▲ **Figure 18.5 Positive control of the *lac* operon by catabolite activator protein (CAP).** RNA polymerase has high affinity for the *lac* promoter only when catabolite activator protein (CAP) is bound to a DNA site at the upstream end of the promoter. CAP attaches to its DNA site only when associated with cyclic AMP (cAMP), whose concentration in the cell rises when the glucose concentration falls. Thus, when glucose is present, even if lactose also is available, the cell preferentially catabolizes glucose and makes very little of the lactose-utilizing enzymes.

In addition to the *lac* operon, CAP helps regulate other operons that encode enzymes used in catabolic pathways. All told, it may affect the expression of more than 100 genes in *E. coli*. When glucose is plentiful and CAP is inactive, the synthesis of enzymes that catabolize compounds other than glucose generally slows down. The ability to catabolize other compounds, such as lactose, enables a cell deprived of glucose to survive. The compounds present in the cell at the moment determine which operons are switched on—the result of simple interactions of activator and repressor proteins with the promoters of the genes in question.

CONCEPT CHECK 18.1

1. How does binding of the *trp* corepressor and the *lac* inducer to their respective repressor proteins alter repressor function and transcription in each case?
2. A certain mutation in *E. coli* changes the *lac* operator so that the active repressor cannot bind. How would this affect the cell's production of β-galactosidase?
3. **WHAT IF?** Describe binding of RNA polymerase, repressors, and activators to the *lac* operon when both lactose and glucose are scarce. What would be the effect on transcription? How might the transcription of other genes outside the *lac* operon be regulated if another sugar were present?

For suggested answers, see Appendix A.

CONCEPT 18.2

Eukaryotic gene expression can be regulated at any stage

All organisms, whether prokaryotes or eukaryotes, must regulate which genes are expressed at any given time. Both unicellular organisms and the cells of multicellular organisms must continually turn genes on and off in response to signals from their external and internal environments. Regulation of gene expression is also essential for cell specialization in multicellular organisms, which are made up of different types of cells, each with a distinct role. To perform its role, each cell type must maintain a specific program of gene expression in which certain genes are expressed and others are not.

Differential Gene Expression

A typical human cell probably expresses about 20% of its genes at any given time. Highly differentiated cells, such as muscle or nerve cells, express an even smaller fraction of their genes. Almost all the cells in an organism contain an identical genome. (Cells of the immune system are one exception. During their differentiation, rearrangement of the immunoglobulin genes results in a change in the genome, as you will see in Chapter 43.) However, the subset of genes expressed in the cells of each type is unique, allowing these cells to carry out their specific function. The differences between cell types, therefore, are due not to different genes being present, but to **differential gene expression**, the expression of different genes by cells with the same genome.

The genomes of eukaryotes may contain tens of thousands of genes, but for quite a few species, only a small amount of the DNA—about 1.5% in humans—codes for protein. The rest of the DNA either codes for RNA products, such as tRNAs, or isn't transcribed at all. The transcription factors of a cell must locate the right genes at the right time, a task on a par with finding a needle in a haystack. When gene expression goes awry, serious imbalances and diseases, including cancer, can arise.

Figure 18.6 summarizes the entire process of gene expression in a eukaryotic cell, highlighting key stages in the expression of a protein-coding gene. Each stage depicted in Figure 18.6 is a potential control point at which gene expression can be turned on or off, accelerated, or slowed down.

Only 40 years ago, an understanding of the mechanisms that control gene expression in eukaryotes seemed almost hopelessly out of reach. Since then, new research methods, notably advances in DNA technology (see Chapter 20), have enabled molecular biologists to uncover many of the details of eukaryotic gene regulation. In all organisms, a common control point for gene expression is at transcription; regulation at this stage is often in response to signals coming from outside the cell, such as hormones or other signaling molecules. For that reason, the term *gene expression* is often equated with transcription for both bacteria and eukaryotes. While this is most often the case for bacteria, the greater complexity of eukaryotic cell structure and function provides opportunities for regulating gene expression at many additional stages (see Figure 18.6). In the remainder of this section, we'll examine some of the important control points of eukaryotic gene expression more closely.

Regulation of Chromatin Structure

Recall that the DNA of eukaryotic cells is packaged with proteins in an elaborate complex known as chromatin, the basic unit of which is the nucleosome (see Figure 16.21). The structural organization of chromatin not only packs a cell's DNA into a compact form that fits inside the nucleus but also helps regulate gene expression in several ways. The location of a gene's promoter relative to nucleosomes and to the sites where the DNA attaches to the chromosome scaffold or nuclear lamina can affect whether the gene is transcribed. In addition, genes within heterochromatin, which is highly condensed, are usually not expressed. The repressive effect of heterochromatin has been seen in experiments in which a transcriptionally active gene was inserted into a region of heterochromatin in yeast cells; the inserted gene was no longer expressed. Lastly, as revealed in a flurry of recent research, certain chemical modifications to the histones and DNA

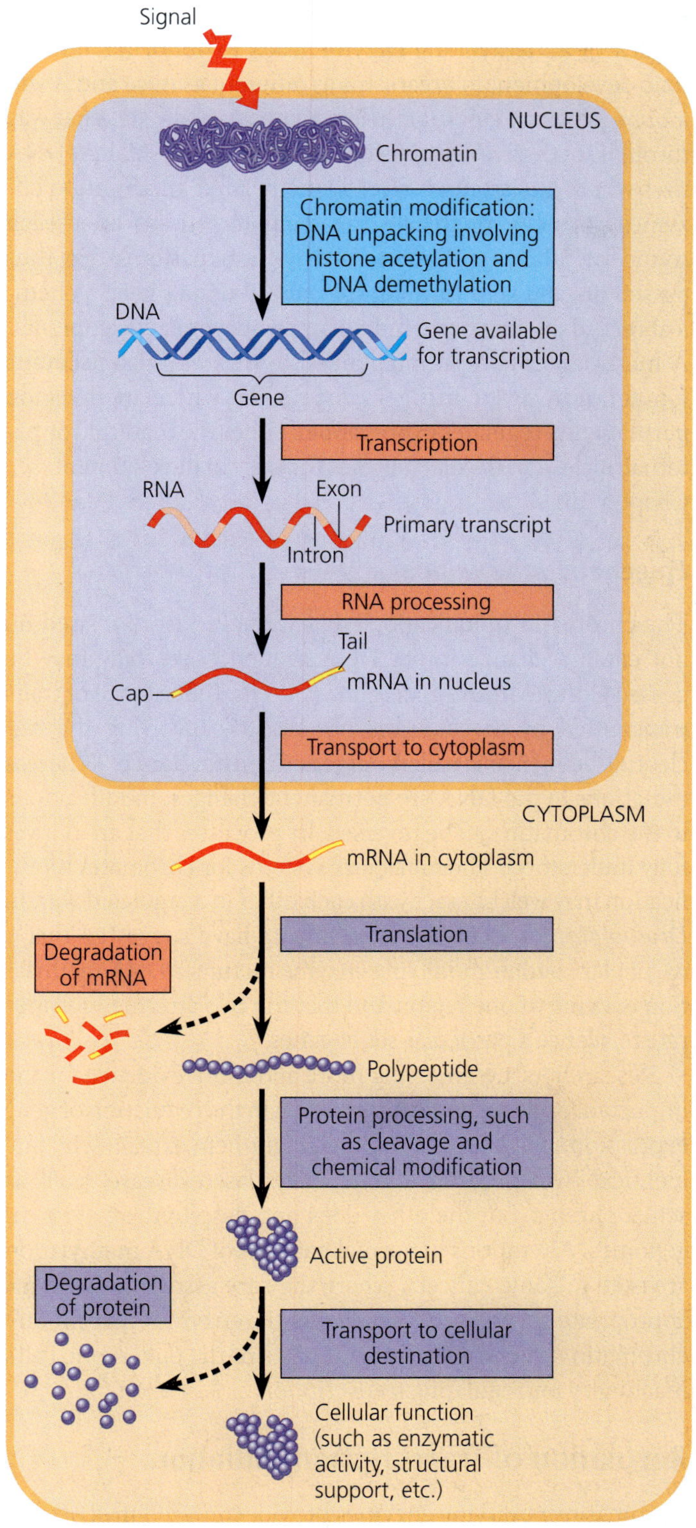

▲ **Figure 18.6 Stages in gene expression that can be regulated in eukaryotic cells.** In this diagram, the colored boxes indicate the processes most often regulated; each color indicates the type of molecule that is affected (blue = DNA, orange = RNA, purple = protein). The nuclear envelope separating transcription from translation in eukaryotic cells offers an opportunity for post-transcriptional control in the form of RNA processing that is absent in prokaryotes. In addition, eukaryotes have a greater variety of control mechanisms operating before transcription and after translation. The expression of any given gene, however, does not necessarily involve every stage shown; for example, not every polypeptide is cleaved.

of chromatin can influence both chromatin structure and gene expression. Here we examine the effects of these modifications, which are catalyzed by specific enzymes.

Histone Modifications

There is mounting evidence that chemical modifications to histones, the proteins around which the DNA is wrapped in nucleosomes, play a direct role in the regulation of gene transcription. The N-terminus of each histone molecule in a nucleosome protrudes outward from the nucleosome **(Figure 18.7a)**. These

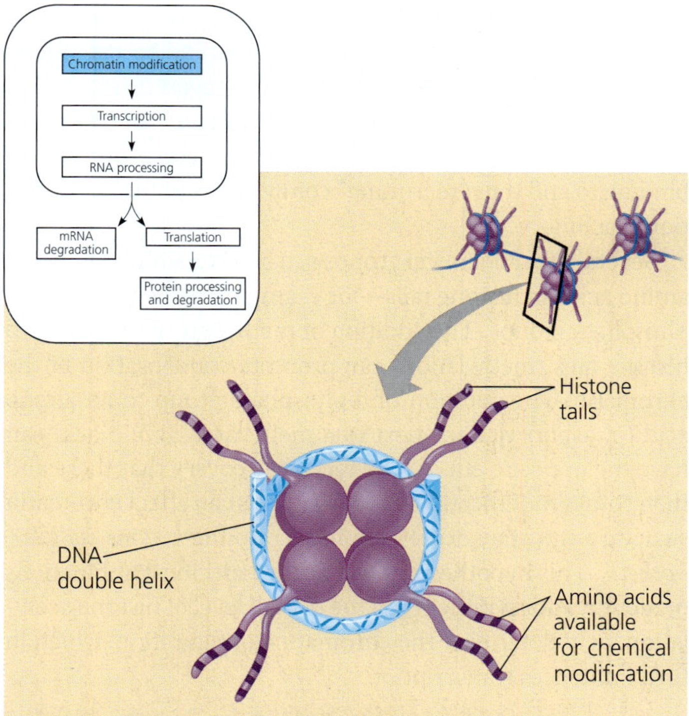

(a) Histone tails protrude outward from a nucleosome. This is an end view of a nucleosome. The amino acids in the N-terminal tails are accessible for chemical modification.

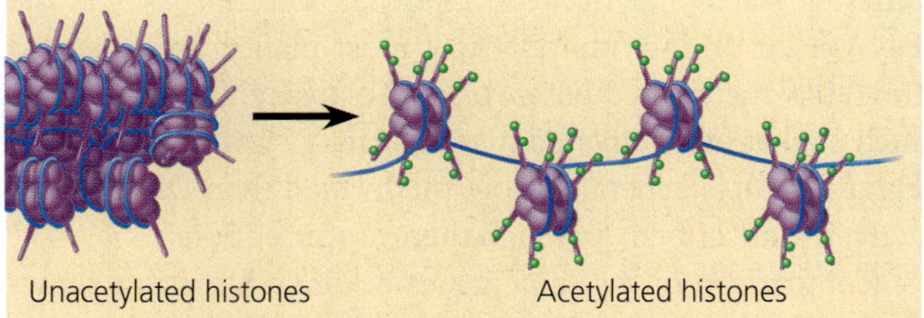

(b) Acetylation of histone tails promotes loose chromatin structure that permits transcription. A region of chromatin in which nucleosomes are unacetylated forms a compact structure (left) in which the DNA is not transcribed. When nucleosomes are highly acetylated (right), the chromatin becomes less compact, and the DNA is accessible for transcription.

▲ **Figure 18.7 A simple model of histone tails and the effect of histone acetylation.** In addition to acetylation, histones can undergo several other types of modifications that also help determine the chromatin configuration in a region.

histone tails are accessible to various modifying enzymes, which catalyze the addition or removal of specific chemical groups.

In **histone acetylation,** acetyl groups (—$COCH_3$) are attached to lysines in histone tails; deacetylation is the removal of acetyl groups. When the lysines are acetylated, their positive charges are neutralized and the histone tails no longer bind to neighboring nucleosomes **(Figure 18.7b)**. Recall that such binding promotes the folding of chromatin into a more compact structure; when this binding does not occur, chromatin has a looser structure. As a result, transcription proteins have easier access to genes in an acetylated region. Researchers have shown that some enzymes that acetylate or deacetylate histones are closely associated with or even components of the transcription factors that bind to promoters (see Figure 17.8). These observations suggest that histone acetylation enzymes may promote the initiation of transcription not only by remodeling chromatin structure, but also by binding to and thus "recruiting" components of the transcription machinery.

Several other chemical groups can be reversibly attached to amino acids in histone tails—for example, methyl groups and phosphate groups. The addition of methyl groups (—CH_3) to histone tails (methylation) can promote condensation of the chromatin. The addition of a phosphate group to an amino acid (phosphorylation) next to a methylated amino acid can have the opposite effect. The recent discovery that these and many other modifications to histone tails can affect chromatin structure and gene expression has led to the *histone code hypothesis*. This hypothesis proposes that specific combinations of modifications, rather than the overall level of histone acetylation, help determine the chromatin configuration, which in turn influences transcription.

DNA Methylation

While some enzymes methylate the tails of histone proteins, a different set of enzymes can methylate certain bases in the DNA itself. In fact, the DNA of most plants, animals, and fungi has methylated bases, usually cytosine. Inactive DNA, such as that of inactivated mammalian X chromosomes (see Figure 15.8), is generally more methylated than DNA that is actively transcribed, although there are exceptions.

Comparison of the same genes in different tissues shows that the genes are usually more heavily methylated in cells in which they are not expressed. Removal of the extra methyl groups can turn on some of these genes. Moreover, researchers have discovered proteins that bind to methylated DNA and recruit histone deacetylation enzymes. Thus, a dual mechanism, involving both DNA methylation and histone deacetylation, can repress transcription.

At least in some species, DNA methylation seems to be essential for the long-term inactivation of genes that occurs during normal cell differentiation in the embryo. For instance, experiments have shown that deficient DNA methylation due to lack of a methylating enzyme leads to abnormal embryonic development in organisms as different as mice and *Arabidopsis* (a plant). Once methylated, genes usually stay that way through successive cell divisions in a given individual. At DNA sites where one strand is already methylated, methylation enzymes correctly methylate the daughter strand after each round of DNA replication. Methylation patterns are thus passed on, and cells forming specialized tissues keep a chemical record of what occurred during embryonic development. A methylation pattern maintained in this way also accounts for **genomic imprinting** in mammals, where methylation permanently regulates expression of either the maternal or paternal allele of particular genes at the start of development (see Chapter 15).

Epigenetic Inheritance

The chromatin modifications that we have just discussed do not entail a change in the DNA sequence, yet they may be passed along to future generations of cells. Inheritance of traits transmitted by mechanisms not directly involving the nucleotide sequence is called **epigenetic inheritance**. Whereas mutations in the DNA are permanent changes, modifications to the chromatin can be reversed, by processes that are not yet fully understood. The molecular systems for chromatin modification may well interact with each other in a regulated way. In *Drosophila*, for example, experiments have suggested that a particular histone-modifying enzyme recruits a DNA methylation enzyme to one region and that the two enzymes collaborate to silence a particular set of genes.

Researchers are amassing more and more evidence for the importance of epigenetic information in the regulation of gene expression. Epigenetic variations might help explain why one identical twin acquires a genetically based disease, such as schizophrenia, but the other does not, despite their identical genomes. Alterations in normal patterns of DNA methylation are seen in some cancers, where they are associated with inappropriate gene expression. Evidently, enzymes that modify chromatin structure are integral parts of the eukaryotic cell's machinery for regulating transcription.

Regulation of Transcription Initiation

Chromatin-modifying enzymes provide initial control of gene expression by making a region of DNA either more or less able to bind the transcription machinery. Once the chromatin of a gene is optimally modified for expression, the initiation of transcription is the next major step at which gene expression is regulated. As in bacteria, the regulation of transcription initiation in eukaryotes involves proteins that bind to DNA and either facilitate or inhibit binding of RNA polymerase. The process is more complicated in eukaryotes, however. Before looking at how eukaryotic cells control their transcription, let's review the structure of a typical eukaryotic gene and its transcript.

Organization of a Typical Eukaryotic Gene

A eukaryotic gene and the DNA elements (segments) that control it are typically organized as shown in **Figure 18.8**, which extends what you learned about eukaryotic genes in Chapter 17. Recall that a cluster of proteins called a *transcription initiation complex* assembles on the promoter sequence at the "upstream" end of the gene. One of these proteins, RNA polymerase II, then proceeds to transcribe the gene, synthesizing a primary RNA transcript (pre-mRNA). RNA processing includes enzymatic addition of a 5′ cap and a poly-A tail, as well as splicing out of introns, to yield a mature mRNA. Associated with most eukaryotic genes are multiple **control elements**, segments of noncoding DNA that help regulate transcription by binding certain proteins. These control elements and the proteins they bind are critical to the precise regulation of gene expression seen in different cell types.

The Roles of Transcription Factors

To initiate transcription, eukaryotic RNA polymerase requires the assistance of proteins called transcription factors. Some transcription factors, such as those illustrated in Figure 17.8, are essential for the transcription of *all* protein-coding genes; therefore, they are often called *general transcription factors.* Only a few general transcription factors independently bind a DNA sequence, such as the TATA box within the promoter; the others primarily bind proteins, including each other and RNA polymerase II. Protein-protein interactions are crucial to the initiation of eukaryotic transcription. Only when the complete initiation complex has assembled can the polymerase begin to move along the DNA template strand, producing a complementary strand of RNA.

The interaction of general transcription factors and RNA polymerase II with a promoter usually leads to only a low rate of initiation and production of few RNA transcripts. In eukaryotes, high levels of transcription of particular genes at the appropriate time and place depend on the interaction of control elements with another set of proteins, which can be thought of as *specific transcription factors.*

Enhancers and Specific Transcription Factors As you can see in Figure 18.8, some control elements, named *proximal control elements,* are located close to the promoter. (Although some biologists consider proximal control elements part of the promoter, we do not.) The more distant *distal control elements,* groupings of which are called **enhancers**, may be thousands of nucleotides upstream or downstream of a gene

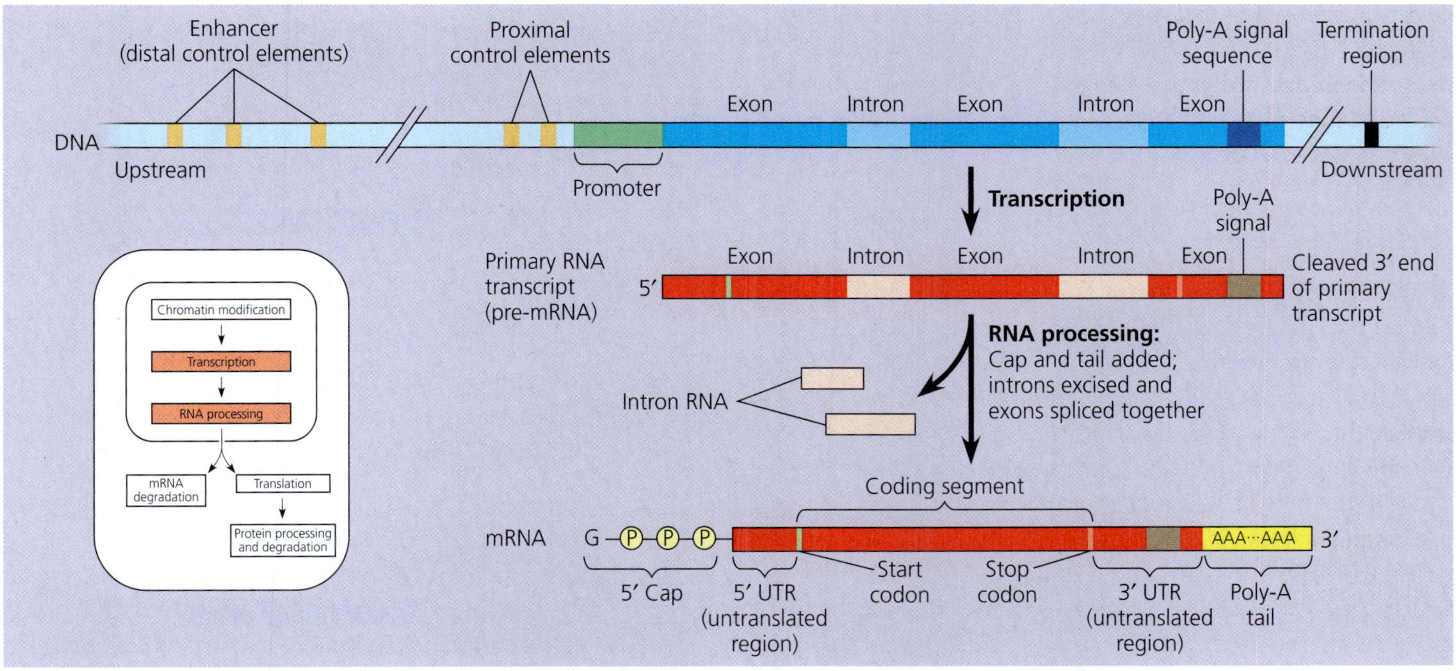

▲ **Figure 18.8 A eukaryotic gene and its transcript.** Each eukaryotic gene has a promoter, a DNA sequence where RNA polymerase binds and starts transcription, proceeding "downstream." A number of control elements (gold) are involved in regulating the initiation of transcription; these are DNA sequences located near (proximal to) or far from (distal to) the promoter. Distal control elements can be grouped together as enhancers, one of which is shown for this gene. A polyadenylation (poly-A) signal sequence in the last exon of the gene is transcribed into an RNA sequence that signals where the transcript is cleaved and the poly-A tail added. Transcription may continue for hundreds of nucleotides beyond the poly-A signal before terminating. RNA processing of the primary transcript into a functional mRNA involves three steps: addition of the 5′ cap, addition of the poly-A tail, and splicing. In the cell, the 5′ cap is added soon after transcription is initiated; splicing and poly-A tail addition may also occur while transcription is still under way (see Figure 17.9).

or even within an intron. A given gene may have multiple enhancers, each active at a different time or in a different cell type or location in the organism. Each enhancer, however, is associated with only that gene and no other.

In eukaryotes, the rate of gene expression can be strongly increased or decreased by the binding of proteins, either activators or repressors, to the control elements of enhancers. **Figure 18.9** shows a current model for how binding of activators to an enhancer located far from the promoter can influence transcription. Protein-mediated bending of the DNA is thought to bring the bound activators in contact with a group of so-called *mediator proteins*, which in turn interact with proteins at the promoter. These multiple protein-protein interactions help assemble and position the initiation complex on the promoter. Support for this model includes a study showing that the proteins regulating a mouse globin gene contact both the gene's promoter and an enhancer located about 50,000 nucleotides upstream. Clearly, these two regions in the DNA must be brought together in a very specific fashion for this interaction to occur.

Hundreds of transcription activators have been discovered in eukaryotes. Researchers have identified two common structural elements in a large number of activator proteins: a DNA-binding domain—a part of the protein's three-dimensional structure that binds to DNA—and one or more activation domains. Activation domains bind other regulatory proteins or components of the transcription machinery, facilitating a sequence of protein-protein interactions that result in transcription of a given gene.

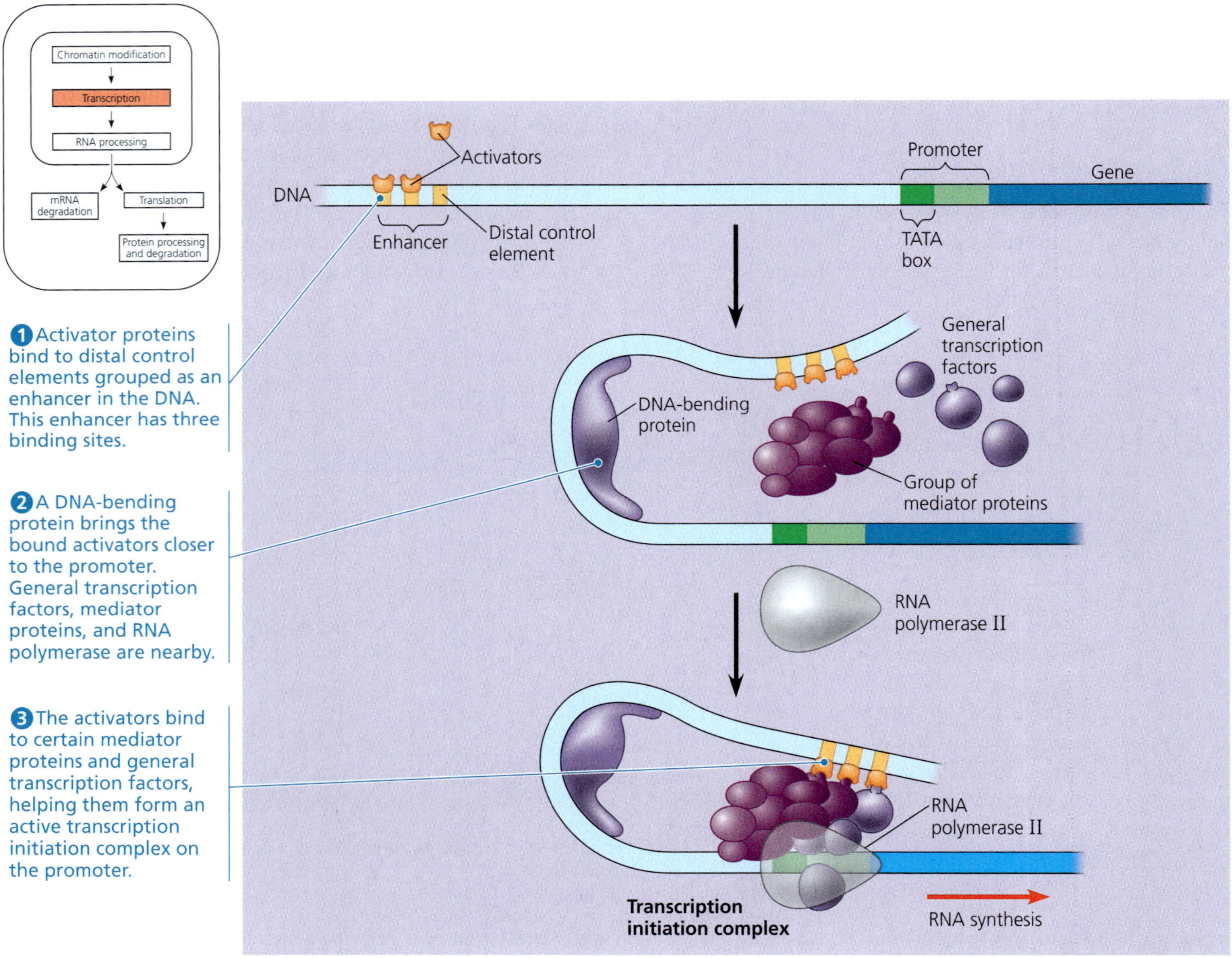

▲ **Figure 18.9 A model for the action of enhancers and transcription activators.** Bending of the DNA by a protein enables enhancers to influence a promoter hundreds or even thousands of nucleotides away. Specific transcription factors called activators bind to the enhancer DNA sequences and then to a group of mediator proteins, which in turn bind to general transcription factors, assembling the transcription initiation complex. These protein-protein interactions facilitate the correct positioning of the complex on the promoter and the initiation of RNA synthesis. Only one enhancer (with three orange control elements) is shown here, but a gene may have several enhancers that act at different times or in different cell types.

Specific transcription factors that function as repressors can inhibit gene expression in several different ways. Some repressors bind directly to control element DNA (in enhancers or elsewhere), blocking activator binding or, in some cases, turning off transcription even when activators are bound. Other repressors block the binding of activators to proteins that allow the activators to bind to DNA.

In addition to influencing transcription directly, some activators and repressors act indirectly by affecting chromatin structure. Studies using yeast and mammalian cells show that some activators recruit proteins that acetylate histones near the promoters of specific genes, thus promoting transcription (see Figure 18.7). Similarly, some repressors recruit proteins that deacetylate histones, leading to reduced transcription, a phenomenon referred to as *silencing*. Indeed, recruitment of chromatin-modifying proteins seems to be the most common mechanism of repression in eukaryotes.

Combinatorial Control of Gene Activation In eukaryotes, the precise control of transcription depends largely on the binding of activators to DNA control elements. Considering the great number of genes that must be regulated in a typical animal or plant cell, the number of completely different nucleotide sequences found in control elements is surprisingly small. A dozen or so short nucleotide sequences appear again and again in the control elements for different genes. On average, each enhancer is composed of about ten control elements, each of which can bind only one or two specific transcription factors. The particular *combination* of control elements in an enhancer associated with a gene turns out to be more important than the presence of a single unique control element in regulating transcription of the gene.

Even with only a dozen control element sequences available, a very large number of combinations are possible. A particular combination of control elements will be able to activate transcription only when the appropriate activator proteins are present, which may occur at a precise time during development or in a particular cell type. **Figure 18.10** illustrates how the use of different combinations of just a few control elements can allow differential regulation of transcription in two cell types.

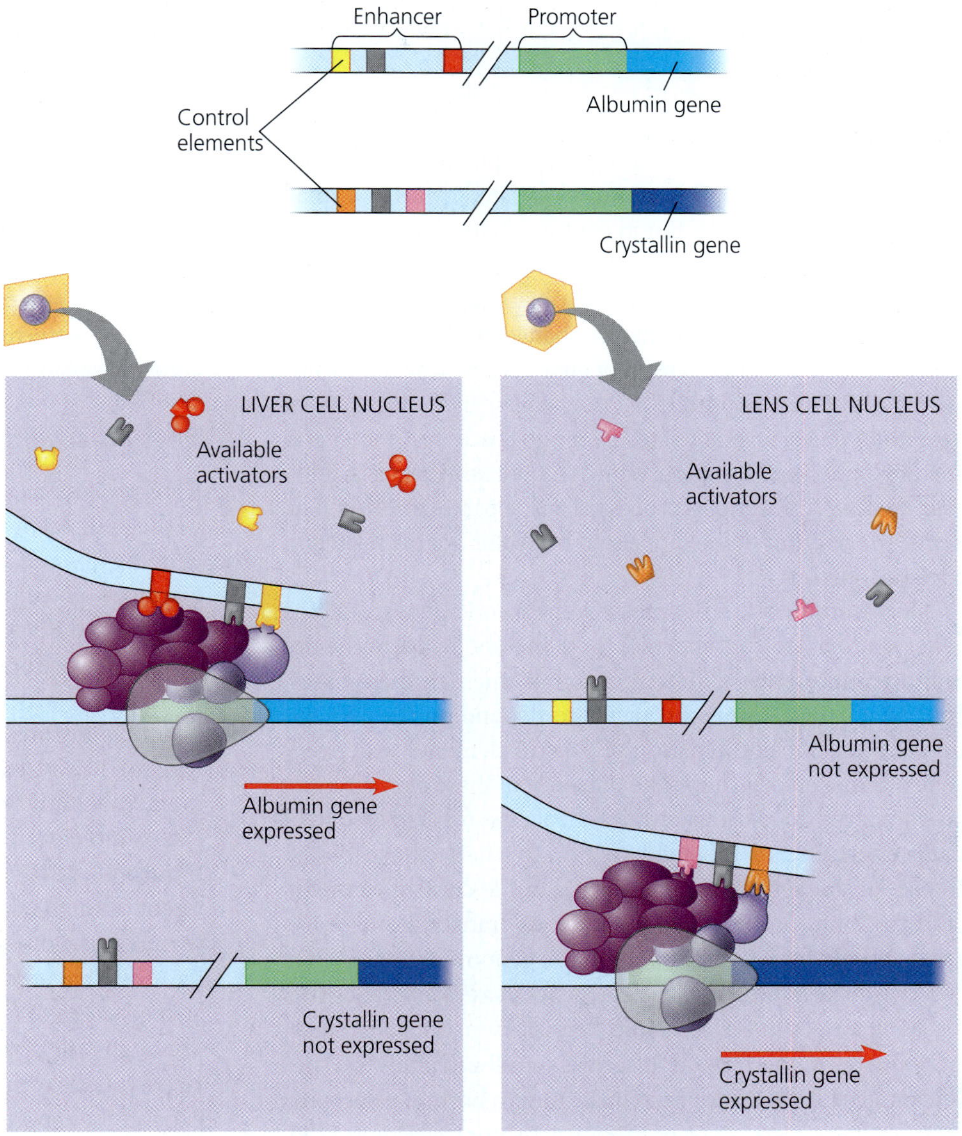

(a) Liver cell. The albumin gene is expressed, and the crystallin gene is not.

(b) Lens cell. The crystallin gene is expressed, and the albumin gene is not.

▲ **Figure 18.10 Cell type–specific transcription.** Both liver cells and lens cells have the genes for making the proteins albumin and crystallin, but only liver cells make albumin (a blood protein) and only lens cells make crystallin (the main protein of lenses). The specific transcription factors made in a cell determine which genes are expressed. In this example, the genes for albumin and crystallin are shown at the top, each with an enhancer made up of three different control elements. Although the enhancers for the two genes share one control element (gray), each enhancer has a unique combination of elements. All the activators required for high-level expression of the albumin gene are present only in liver cells (a), whereas the activators needed for expression of the crystallin gene are present only in lens cells (b). For simplicity, we consider only the role of activators here, although the presence or absence of repressors may also influence transcription in certain cell types.

? *Describe the enhancer for the albumin gene in each cell. How would the nucleotide sequence of this enhancer in the liver cell compare with that in the lens cell?*

Coordinately Controlled Genes in Eukaryotes

How does the eukaryotic cell deal with genes of related function that need to be turned on or off at the same time? Earlier in this chapter, you learned that in bacteria, such coordinately controlled genes are often clustered into an operon, which is regulated by a single promoter and transcribed into a single mRNA molecule. Thus, the genes are expressed together, and the encoded proteins are produced concurrently. Operons that work in this way have not been found in eukaryotic cells, with some exceptions.

Analysis of the genomes of several eukaryotic species has revealed some co-expressed genes that are clustered near one another on the same chromosome. Examples include certain genes in the testis of the fruit fly and muscle-related genes in a small worm called a nematode. But in contrast to the genes of bacterial operons, each gene in such a cluster has its own promoter and is individually transcribed. The coordinate regulation of these clustered genes is thought to involve changes in chromatin structure that make the entire group of genes either available or unavailable for transcription. In other cases, including 15% of nematode genes, several related genes do share a promoter and are transcribed into a single pre-mRNA. Unlike in bacteria, however, the RNA transcript is processed into separate mRNAs. The nematode operons do not appear to be evolutionarily related to bacterial operons.

More commonly, co-expressed eukaryotic genes, such as genes coding for the enzymes of a metabolic pathway, are found scattered over different chromosomes. In these cases, coordinate gene expression seems to depend on the association of a specific combination of control elements with every gene of a dispersed group. The presence of these elements can be compared to the raised flags on a few mailboxes out of many, signaling to the mail carrier to check those boxes. Copies of the activators that recognize the control elements bind to them, promoting simultaneous transcription of the genes, no matter where they are in the genome.

Coordinate control of dispersed genes in a eukaryotic cell often occurs in response to chemical signals from outside the cell. A steroid hormone, for example, enters a cell and binds to a specific intracellular receptor protein, forming a hormone-receptor complex that serves as a transcription activator (see Figure 11.8). Every gene whose transcription is stimulated by a particular steroid hormone, regardless of its chromosomal location, has a control element recognized by that hormone-receptor complex. This is how estrogen activates a group of genes that stimulate cell division in uterine cells, preparing the uterus for pregnancy.

Many signal molecules, such as nonsteroid hormones and growth factors, bind to receptors on a cell's surface and never actually enter the cell. Such molecules can control gene expression indirectly by triggering signal transduction pathways that lead to activation of particular transcription activators or repressors (see Figure 11.14). The principle of coordinate regulation is the same as in the case of steroid hormones: Genes with the same control elements are activated by the same chemical signals. Systems for coordinating gene regulation probably arose early in evolutionary history and evolved by the duplication and distribution of control elements within the genome.

Mechanisms of Post-Transcriptional Regulation

Transcription alone does not constitute gene expression. The expression of a protein-coding gene is ultimately measured by the amount of functional protein a cell makes, and much happens between the synthesis of the RNA transcript and the activity of the protein in the cell. Researchers are discovering more and more regulatory mechanisms that operate at various stages after transcription (see Figure 18.6). These mechanisms allow a cell to fine-tune gene expression rapidly in response to environmental changes without altering its transcription patterns. Here we discuss how cells can regulate gene expression once a gene has been transcribed.

RNA Processing

RNA processing in the nucleus and the export of mature RNA to the cytoplasm provide several opportunities for regulating gene expression that are not available in prokaryotes. One example of regulation at the RNA-processing level is **alternative RNA splicing**, in which different mRNA molecules are produced from the same primary transcript, depending on which RNA segments are treated as exons and which as introns. Regulatory proteins specific to a cell type control intron-exon choices by binding to regulatory sequences within the primary transcript.

A simple example of alternative RNA splicing is shown in **Figure 18.11** for the troponin T gene, which encodes two different (though related) proteins. Other genes offer possibilities for greater numbers of products. For instance, researchers have discovered a fruit fly gene that has enough alternatively spliced exons to generate more than 38,000 different proteins, although only a small number have been found to be synthesized. It is clear that alternative RNA splicing can significantly expand the repertoire of a eukaryotic genome.

mRNA Degradation

The life span of mRNA molecules in the cytoplasm is important in determining the pattern of protein synthesis in a cell. Bacterial mRNA molecules typically are degraded by enzymes within a few minutes of their synthesis. This short life span of mRNAs is one reason bacteria can change their patterns of protein synthesis so quickly in response to environmental changes. In contrast, mRNAs in multicellular eukaryotes typically survive for hours, days, or even weeks. For instance, the mRNAs for the hemoglobin polypeptides

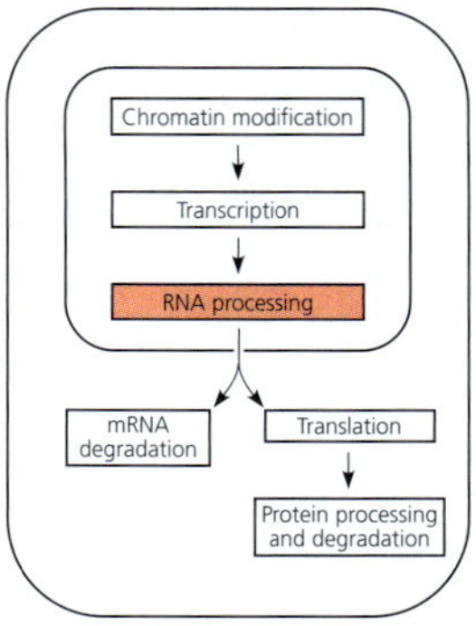

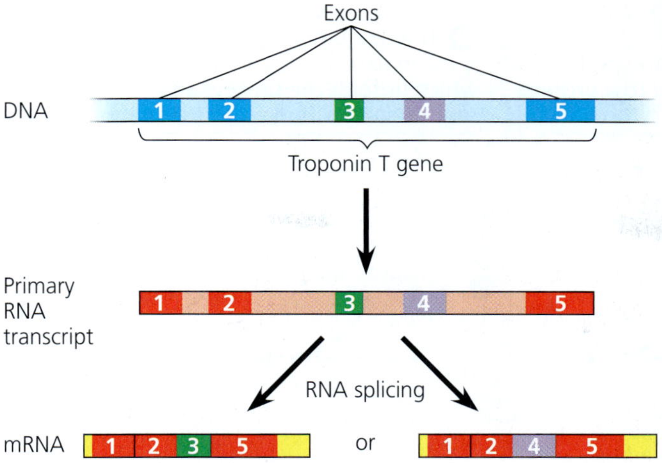

◀ **Figure 18.11 Alternative RNA splicing of the troponin T gene.** The primary transcript of this gene can be spliced in more than one way, generating different mRNA molecules. Notice that one mRNA molecule has ended up with exon 3 (green) and the other with exon 4 (purple). These two mRNAs are translated into different but related muscle proteins.

(α-globin and β-globin) in developing red blood cells are unusually stable, and these long-lived mRNAs are translated repeatedly in these cells.

Research on yeast species suggests that a common pathway of mRNA breakdown begins with the enzymatic shortening of the poly-A tail (see Figure 18.8). This helps trigger the action of enzymes that remove the 5′ cap (the two ends of the mRNA may be briefly held together by the proteins involved). Removal of the cap, a critical step, is also regulated by particular nucleotide sequences within the mRNA. Once the cap is removed, nuclease enzymes rapidly chew up the mRNA.

Nucleotide sequences that affect how long an mRNA remains intact are often found in the untranslated region (UTR) at the 3′ end of the molecule (see Figure 18.8). In one experiment, researchers transferred such a sequence from the short-lived mRNA for a growth factor to the 3′ end of a normally stable globin mRNA. The globin mRNA was quickly degraded.

During the past few years, other mechanisms that degrade or block expression of mRNA molecules have come to light. These mechanisms involve an important group of newly discovered RNA molecules that regulate gene expression at several levels, and we will discuss them later in this chapter.

Initiation of Translation

Translation presents another opportunity for regulating gene expression; such regulation occurs most commonly at the initiation stage (see Figure 17.17). The initiation of translation of some mRNAs can be blocked by regulatory proteins that bind to specific sequences or structures within the untranslated region at the 5′ end (5′ UTR) of the mRNA, preventing the attachment of ribosomes. (Recall from Chapter 17 that both the 5′ cap and the poly-A tail of an mRNA molecule are important for ribosome binding.) A different mechanism for blocking translation is seen in a variety of mRNAs present in the eggs of many organisms: Initially, these stored mRNAs lack poly-A tails of sufficient length to allow translation initiation. At the appropriate time during embryonic development, however, a cytoplasmic enzyme adds more adenine (A) nucleotides, prompting translation to begin.

Alternatively, translation of *all* the mRNAs in a cell may be regulated simultaneously. In a eukaryotic cell, such "global" control usually involves the activation or inactivation of one or more of the protein factors required to initiate translation. This mechanism plays a role in starting translation of mRNAs that are stored in eggs. Just after fertilization, translation is triggered by the sudden activation of translation initiation factors. The response is a burst of synthesis of the proteins encoded by the stored mRNAs. Some plants and algae store mRNAs during periods of darkness; light then triggers the reactivation of the translational apparatus.

Protein Processing and Degradation

The final opportunities for controlling gene expression occur after translation. Often, eukaryotic polypeptides must be processed to yield functional protein molecules. For instance, cleavage of the initial insulin polypeptide (pro-insulin) forms the active hormone. In addition, many proteins undergo chemical modifications that make them functional. Regulatory proteins are commonly activated or inactivated by the reversible addition of phosphate groups, and proteins destined for the surface of animal cells acquire sugars. Cell-surface proteins and many others must also be transported to target destinations in the cell in order to function. Regulation might occur at any of the steps involved in modifying or transporting a protein.

Finally, the length of time each protein functions in the cell is strictly regulated by means of selective degradation. Many proteins, such as the cyclins involved in regulating the cell cycle, must be relatively short-lived if the cell is to function appropriately (see Figure 12.17). To mark a particular protein for destruction, the cell commonly attaches molecules of a small protein called ubiquitin to the protein. Giant protein complexes called **proteasomes** then recognize the ubiquitin-tagged

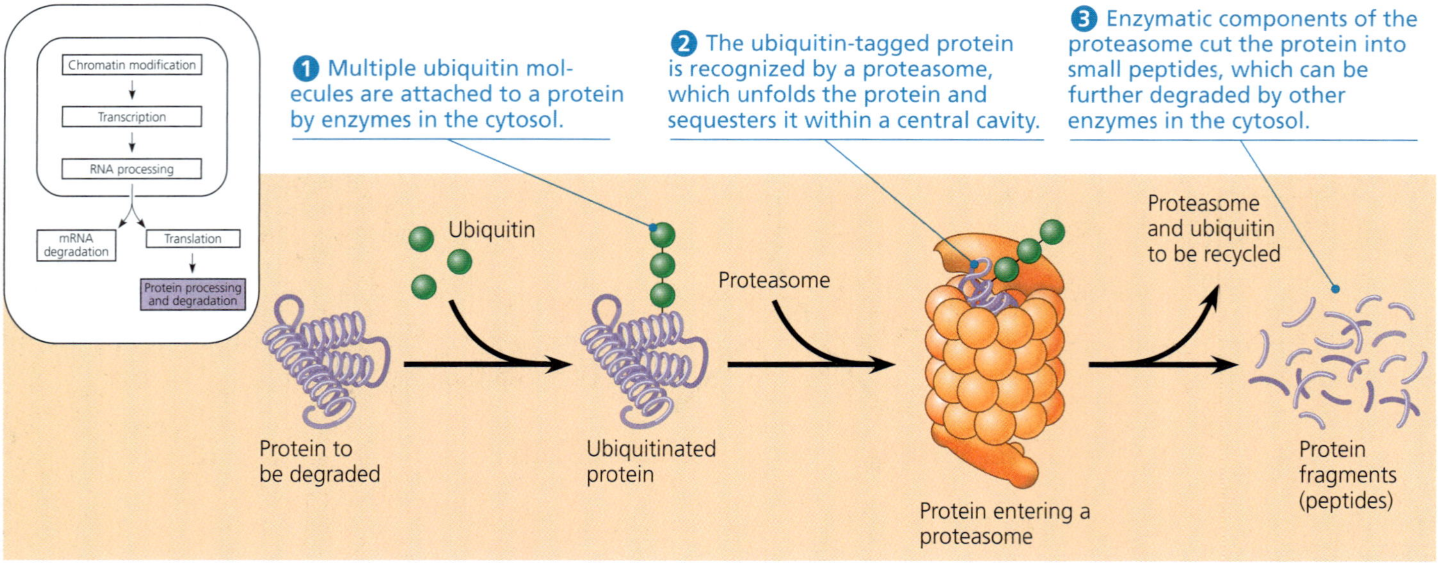

▲ **Figure 18.12 Degradation of a protein by a proteasome.** A proteasome, an enormous protein complex shaped like a trash can, chops up unneeded proteins in the cell. In most cases, the proteins attacked by a proteasome have been tagged with short chains of ubiquitin, a small protein. Steps 1 and 3 require ATP. Eukaryotic proteasomes are as massive as ribosomal subunits and are distributed throughout the cell. Their shape somewhat resembles that of chaperone proteins, which protect protein structure rather than destroy it (see Figure 5.24).

proteins and degrade them **(Figure 18.12)**. The importance of proteasomes is underscored by the finding that mutations making specific cell cycle proteins impervious to proteasome degradation can lead to cancer.

CONCEPT CHECK 18.2

1. In general, what is the effect of histone acetylation and DNA methylation on gene expression?
2. Compare the roles of general and specific transcription factors in regulating gene expression.
3. Suppose you compared the nucleotide sequences of the distal control elements in the enhancers of three genes that are expressed only in muscle tissue. What would you expect to find? Why?
4. Once mRNA encoding a particular protein reaches the cytoplasm, what are four mechanisms that can regulate the amount of the protein that is active in the cell?
5. **WHAT IF?** Examine Figure 18.10 and suggest a mechanism by which the yellow activator protein comes to be present in the liver cell but not in the lens cell.

For suggested answers, see Appendix A.

CONCEPT 18.3

Noncoding RNAs play multiple roles in controlling gene expression

Recall that only 1.5% of the human genome—and a similarly small percentage of the genome of many other multicellular eukaryotes—codes for proteins. Of the remainder, a very small fraction consists of genes for small RNAs, such as ribosomal RNA and transfer RNA. Until recently, most of the rest of the DNA was assumed to be untranscribed. The general idea was that since it didn't code for proteins or the few known types of RNA, such DNA didn't contain meaningful genetic information. However, a flood of recent data has contradicted this idea. For example, a study of two human chromosomes showed that ten times as much of the genome was transcribed as was predicted by the number of protein-coding exons present. Introns accounted for some of this transcribed, nontranslated RNA, but only a small fraction of the total. These and other results suggest that a significant amount of the genome may be transcribed into non-protein-coding RNAs (also called *noncoding RNAs*), including a variety of small RNAs. While many questions about the functions of these RNAs remain unanswered, researchers are uncovering more evidence of their biological roles every day.

Biologists are excited about these recent discoveries, which hint at a large, diverse population of RNA molecules in the cell that play crucial roles in regulating gene expression—and have gone largely unnoticed until now. Clearly, we must revise our long-standing view that because they code for proteins, mRNAs are the most important RNAs functioning in the cell. It's as if we've been so focused on the famous ruler of a country that we've completely overlooked the many advisors and cabinet members working behind the scenes.

Regulation by noncoding RNAs is known to occur at two points in the pathway of gene expression: mRNA translation and chromatin configuration. We will focus on several types of small RNAs that have been extensively studied in the past few years; the importance of these RNAs was recognized when they were the focus of the 2006 Nobel Prize in Physiology or Medicine.

Effects on mRNAs by MicroRNAs and Small Interfering RNAs

Since 1993, a number of research studies have uncovered small single-stranded RNA molecules, called **microRNAs (miRNAs)**, that are capable of binding to complementary sequences in mRNA molecules. The miRNAs are formed from longer RNA precursors that fold back on themselves, forming one or more short double-stranded hairpin structures, each held together by hydrogen bonds **(Figure 18.13)**. After each hairpin is cut away from the precursor, it is trimmed by an enzyme (fittingly called Dicer) into a short double-stranded fragment of about 20 nucleotide pairs. One of the two strands is degraded, while the other strand, which is the miRNA, forms a complex with one or more proteins; the miRNA allows the complex to bind to any mRNA molecule with the complementary sequence. The miRNA-protein complex then either degrades the target mRNA or blocks its translation. It has been estimated that expression of up to one-third of all human genes may be regulated by miRNAs, a remarkable figure given that the existence of miRNAs was unknown a mere two decades ago.

A growing understanding of the miRNA pathway provided an explanation for a perplexing observation: Researchers had found that injecting double-stranded RNA molecules into a cell somehow turned off expression of a gene with the same sequence as the RNA. They called this experimental phenomenon **RNA interference (RNAi)**. It was later shown to be due to **small interfering RNAs (siRNAs)**, which are similar in size and function to miRNAs. In fact, subsequent research showed that the same cellular machinery generates miRNAs and siRNAs and that both can associate with the same proteins, producing similar results. The distinction between miRNAs and siRNAs is based on the nature of the precursor molecule for each. While an miRNA is usually formed from a single hairpin in a precursor RNA (see Figure 18.13), siRNAs are formed from much longer double-stranded RNA molecules, each of which gives rise to many siRNAs.

We mentioned that laboratory investigators had injected double-stranded RNAs into cells, and you may wonder whether such molecules are ever found naturally. As you will learn in Chapter 19, some viruses have double-stranded RNA genomes. Because the cellular RNAi pathway can lead to the destruction of

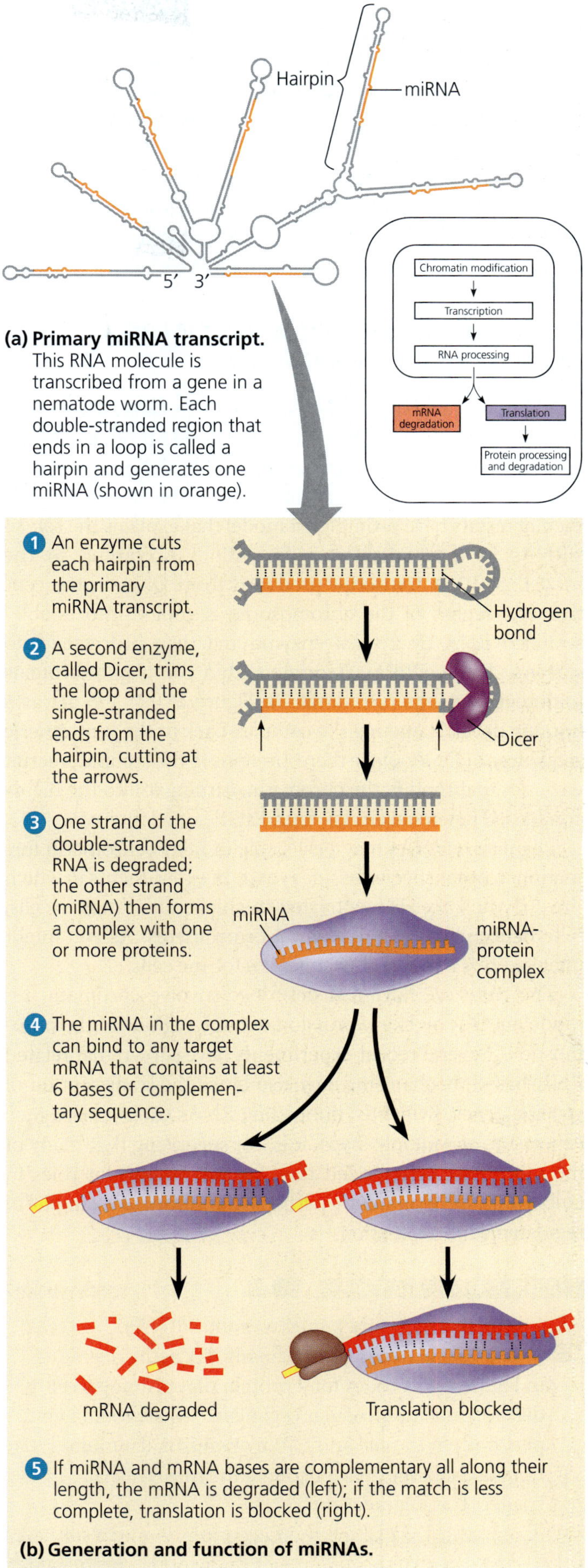

▶ **Figure 18.13 Regulation of gene expression by miRNAs.** RNA transcripts are processed into miRNAs, which prevent expression of mRNAs containing complementary sequences.

RNAs with sequences complementary to those found in double-stranded RNAs, this pathway may have evolved as a natural defense against infection by such viruses. However, the fact that the RNAi pathway can also affect the expression of nonviral cellular genes may reflect a different evolutionary origin for the RNAi pathway. Moreover, some species apparently produce their own long double-stranded RNA precursors to small RNAs such as siRNAs. Once produced, these RNAs can interfere with gene expression at stages other than translation, as we'll discuss next.

Chromatin Remodeling and Silencing of Transcription by Small RNAs

In addition to affecting mRNAs, small RNAs can cause remodeling of chromatin structure. In yeast, siRNAs produced by the yeast cells themselves appear to be crucial for the formation of heterochromatin at the centromeres of chromosomes. Experimental results have prompted a model that explains the role of siRNAs in heterochromatin formation. According to the model, an RNA transcript produced from DNA in the centromeric region of the chromosome is copied into double-stranded RNA by a yeast enzyme and then processed into siRNAs. These siRNAs associate with a complex of proteins (different from the one shown in Figure 18.13) and act as a homing device, targeting the complex back to the centromeric sequences of DNA. Once there, proteins in the complex recruit enzymes that modify the chromatin, turning it into the highly condensed heterochromatin found at the centromere.

Regulatory RNAs may play a role in heterochromatin formation in other species besides yeast. In experiments in which the enzyme Dicer is inactivated in chicken and mouse cells, heterochromatin fails to form at centromeres. As you might imagine, this has dire consequences for the cells.

The cases we have just described involve chromatin remodeling that blocks expression of large regions of the chromosome. Several recent experiments have shown that related RNA-based mechanisms may also block the transcription of specific genes. Evidently, noncoding RNAs can regulate gene expression at multiple steps. It is not surprising that many of the miRNAs characterized thus far play important roles in embryonic development—perhaps the ultimate example of regulated gene expression.

CONCEPT CHECK 18.3

1. Compare and contrast miRNAs and siRNAs.
2. **WHAT IF?** Imagine that the mRNA being degraded in Figure 18.13 codes for a protein that promotes cell division in a multicellular organism. What would happen, both in the cell and to the organism, if a mutation disabled the gene encoding the miRNA that triggers this degradation?

For suggested answers, see Appendix A.

CONCEPT 18.4

A program of differential gene expression leads to the different cell types in a multicellular organism

In the embryonic development of multicellular organisms, a fertilized egg (a zygote) gives rise to cells of many different types, each with a different structure and corresponding function. Typically, cells are organized into tissues, tissues into organs, organs into organ systems, and organ systems into the whole organism. Thus, any developmental program must produce cells of different types that form higher-level structures arranged in a particular way in three dimensions. The processes that occur during development in plants and animals are detailed in Chapters 35 and 47, respectively. In this chapter, we focus instead on the program of regulation of gene expression that orchestrates development, using a few animal species as examples.

A Genetic Program for Embryonic Development

The photos in **Figure 18.14** illustrate the dramatic difference between a zygote and the organism it becomes. This remarkable transformation results from three interrelated processes: cell division, cell differentiation, and morphogenesis. Through a succession of mitotic cell divisions, the zygote gives rise to a large number of cells. Cell division alone, however, would produce only a great ball of identical cells, nothing like a tadpole. During embryonic development, cells not only increase in number, but also undergo **cell differentiation**, the process by which cells become specialized in structure and function. Moreover, the different kinds of cells are not randomly distributed but are organized into tissues and organs in a particular three-dimensional

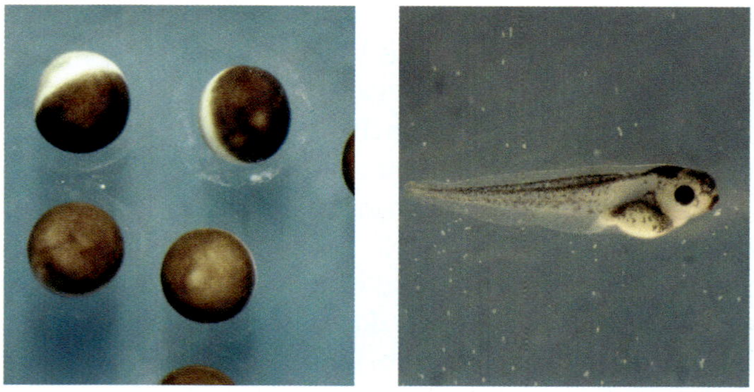

(a) Fertilized eggs of a frog **(b) Newly hatched tadpole**

▲ **Figure 18.14 From fertilized egg to animal: What a difference four days makes.** It takes just four days for cell division, differentiation, and morphogenesis to transform each of the fertilized frog eggs shown in (a) into a tadpole like the one in (b).

arrangement. The physical processes that give an organism its shape constitute **morphogenesis**, meaning "creation of form."

All three processes have their basis in cellular behavior. Even morphogenesis, the shaping of the organism, can be traced back to changes in the shape, motility, and other characteristics of the cells that make up various regions of the embryo. As you have seen, the activities of a cell depend on the genes it expresses and the proteins it produces. Almost all cells in an organism have the same genome; therefore, differential gene expression results from the genes being regulated differently in each cell type.

In Figure 18.10, you saw a simplified view of how differential gene expression occurs in two cell types, a liver cell and a lens cell. Each of these fully differentiated cells has a particular mix of specific activators that turn on the collection of genes whose products are required in the cell. The fact that both cells arose through a series of mitoses from a common fertilized egg inevitably leads to a question: How do different sets of activators come to be present in the two cells?

It turns out that materials placed into the egg by the mother set up a sequential program of gene regulation that is carried out as cells divide, and this program makes the cells become different from each other in a coordinated fashion. To understand how this works, we will consider two basic developmental processes: First, we'll explore how cells that arise from early embryonic mitoses develop the differences that start each cell along its own differentiation pathway. Second, we'll see how cellular differentiation leads to one particular cell type, using muscle development as an example.

Cytoplasmic Determinants and Inductive Signals

What generates the first differences among cells in an early embryo? And what controls the differentiation of all the various cell types as development proceeds? By this point in the chapter, you can probably deduce the answer: The specific genes expressed in any particular cell of a developing organism determine its path. Two sources of information, used to varying extents in different species, "tell" a cell which genes to express at any given time during embryonic development.

One important source of information early in development is the egg's cytoplasm, which contains both RNA and proteins encoded by the mother's DNA. The cytoplasm of an unfertilized egg is not homogeneous. Messenger RNA, proteins, other substances, and organelles are distributed unevenly in the unfertilized egg, and this unevenness has a profound impact on the development of the future embryo in many species. Maternal substances in the egg that influence the course of early development are called **cytoplasmic determinants** (Figure 18.15a). After fertilization, early

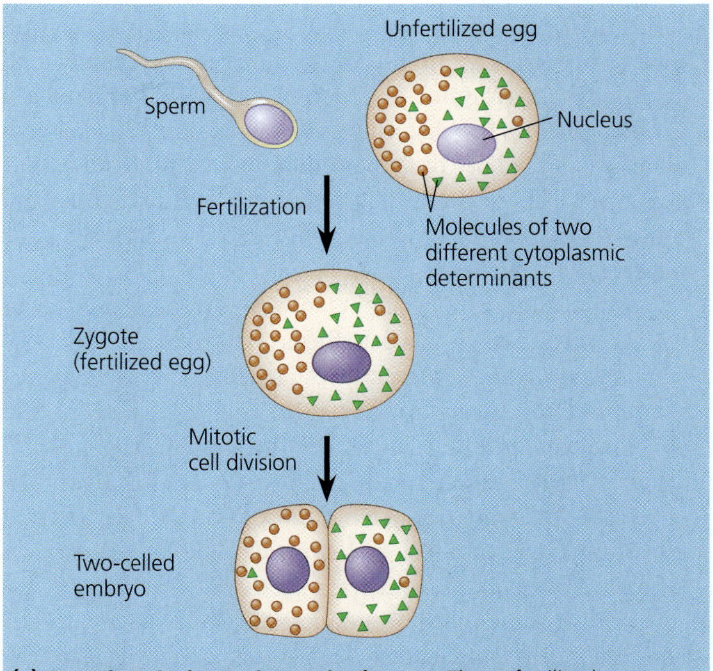

(a) Cytoplasmic determinants in the egg. The unfertilized egg has molecules in its cytoplasm, encoded by the mother's genes, that influence development. Many of these cytoplasmic determinants, like the two shown here, are unevenly distributed in the egg. After fertilization and mitotic division, the cell nuclei of the embryo are exposed to different sets of cytoplasmic determinants and, as a result, express different genes.

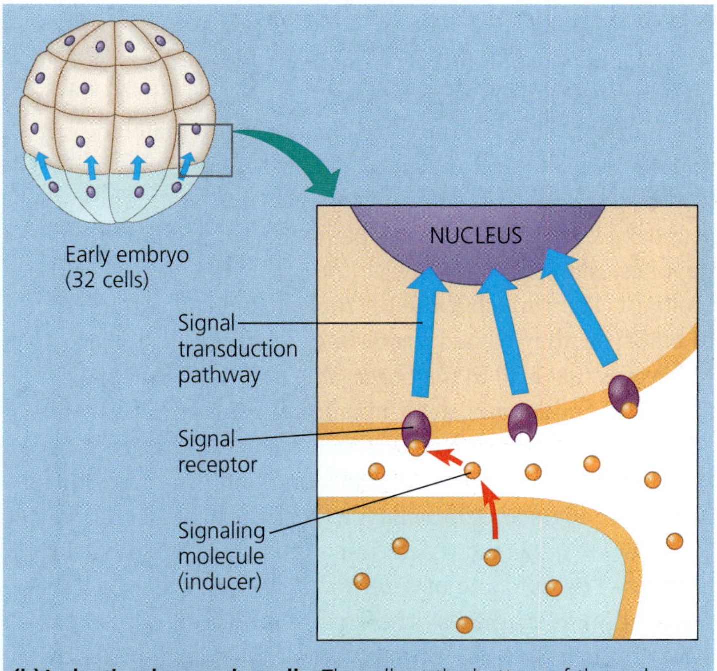

(b) Induction by nearby cells. The cells at the bottom of the early embryo depicted here are releasing chemicals that signal nearby cells to change their gene expression.

▲ **Figure 18.15 Sources of developmental information for the early embryo.**

mitotic divisions distribute the zygote's cytoplasm into separate cells. The nuclei of these cells may thus be exposed to different cytoplasmic determinants, depending on which portions of the zygotic cytoplasm a cell received. The combination of cytoplasmic determinants in a cell helps determine its developmental fate by regulating expression of the cell's genes during the course of cell differentiation.

The other major source of developmental information, which becomes increasingly important as the number of embryonic cells increases, is the environment around a particular cell. Most influential are the signals impinging on an embryonic cell from other embryonic cells in the vicinity, including contact with cell-surface molecules on neighboring cells and the binding of growth factors secreted by neighboring cells. Such signals cause changes in the target cells, a process called **induction** **(Figure 18.15b)**. The molecules conveying these signals within the target cell are cell-surface receptors and other proteins expressed by the embryo's own genes. In general, the signaling molecules send a cell down a specific developmental path by causing changes in its gene expression that eventually result in observable cellular changes. Thus, interactions between embryonic cells help induce differentiation of the many specialized cell types making up a new organism.

Sequential Regulation of Gene Expression During Cellular Differentiation

As the tissues and organs of an embryo develop and their cells differentiate, the cells become noticeably different in structure and function. These observable changes are actually the outcome of a cell's developmental history beginning at the first mitotic division of the zygote, as we have just seen. The earliest changes that set a cell on a path to specialization are subtle ones, showing up only at the molecular level. Before biologists knew much about the molecular changes occurring in embryos, they coined the term **determination** to refer to the events that lead to the observable differentiation of a cell. Once it has undergone determination, an embryonic cell is irreversibly committed to its final fate. If a committed cell is experimentally placed in another location in the embryo, it will still differentiate into the cell type that is its normal fate.

Today we understand determination in terms of molecular changes. The outcome of determination, observable cell differentiation, is marked by the expression of genes for *tissue-specific proteins*. These proteins are found only in a specific cell type and give the cell its characteristic structure and function. The first evidence of differentiation is the appearance of mRNAs for these proteins. Eventually, differentiation is observable with a microscope as changes in cellular structure. On the molecular level, different sets of genes are sequentially expressed in a regulated manner as new cells arise from division of their precursors. A number of the steps in gene expression may be regulated during differentiation, with transcription among the most important. In the fully differentiated cell, transcription remains the principal regulatory point for maintaining appropriate gene expression.

Differentiated cells are specialists at making tissue-specific proteins. For example, as a result of transcriptional regulation, liver cells specialize in making albumin, and lens cells specialize in making crystallin (see Figure 18.10). Skeletal muscle cells in vertebrates are another instructive example. Each of these cells is a long fiber containing many nuclei within a single plasma membrane. Skeletal muscle cells have high concentrations of muscle-specific versions of the contractile proteins myosin and actin, as well as membrane receptor proteins that detect signals from nerve cells.

Muscle cells develop from embryonic precursor cells that have the potential to develop into a number of cell types, including cartilage cells and fat cells, but particular conditions commit them to becoming muscle cells. Although the committed cells appear unchanged under the microscope, determination has occurred, and they are now *myoblasts*. Eventually, myoblasts start to churn out large amounts of muscle-specific proteins and fuse to form mature, elongated, multinucleate skeletal muscle cells (**Figure 18.16**, left).

Researchers have worked out what happens at the molecular level during muscle cell determination by growing myoblasts in culture and analyzing them using molecular biological techniques you will learn about in Chapter 20. In a series of experiments, they isolated different genes, caused each to be expressed in a separate embryonic precursor cell, and then looked for differentiation into myoblasts and muscle cells. In this way, they identified several so-called "master regulatory genes" whose protein products commit the cells to becoming skeletal muscle. Thus, in the case of muscle cells, the molecular basis of determination is the expression of one or more of these master regulatory genes.

To understand more about how commitment occurs in muscle cell differentiation, let's focus on the master regulatory gene called *myoD* (**Figure 18.16**, right). This gene encodes MyoD protein, a transcription factor that binds to specific control elements in the enhancers of various target genes and stimulates their expression. Some target genes for MyoD encode still other muscle-specific transcription factors. MyoD also stimulates expression of the *myoD* gene itself, thus perpetuating its effect in maintaining the cell's differentiated state. Presumably, all the genes activated by MyoD have enhancer control elements recognized by MyoD and are thus coordinately controlled. Finally, the secondary transcription factors activate the genes for proteins such as myosin and actin that confer the unique properties of skeletal muscle cells.

The MyoD protein deserves its designation as a master regulatory gene. Researchers have shown that it is even capable of changing some kinds of fully differentiated nonmuscle cells, such as fat cells and liver cells, into muscle cells. Why doesn't

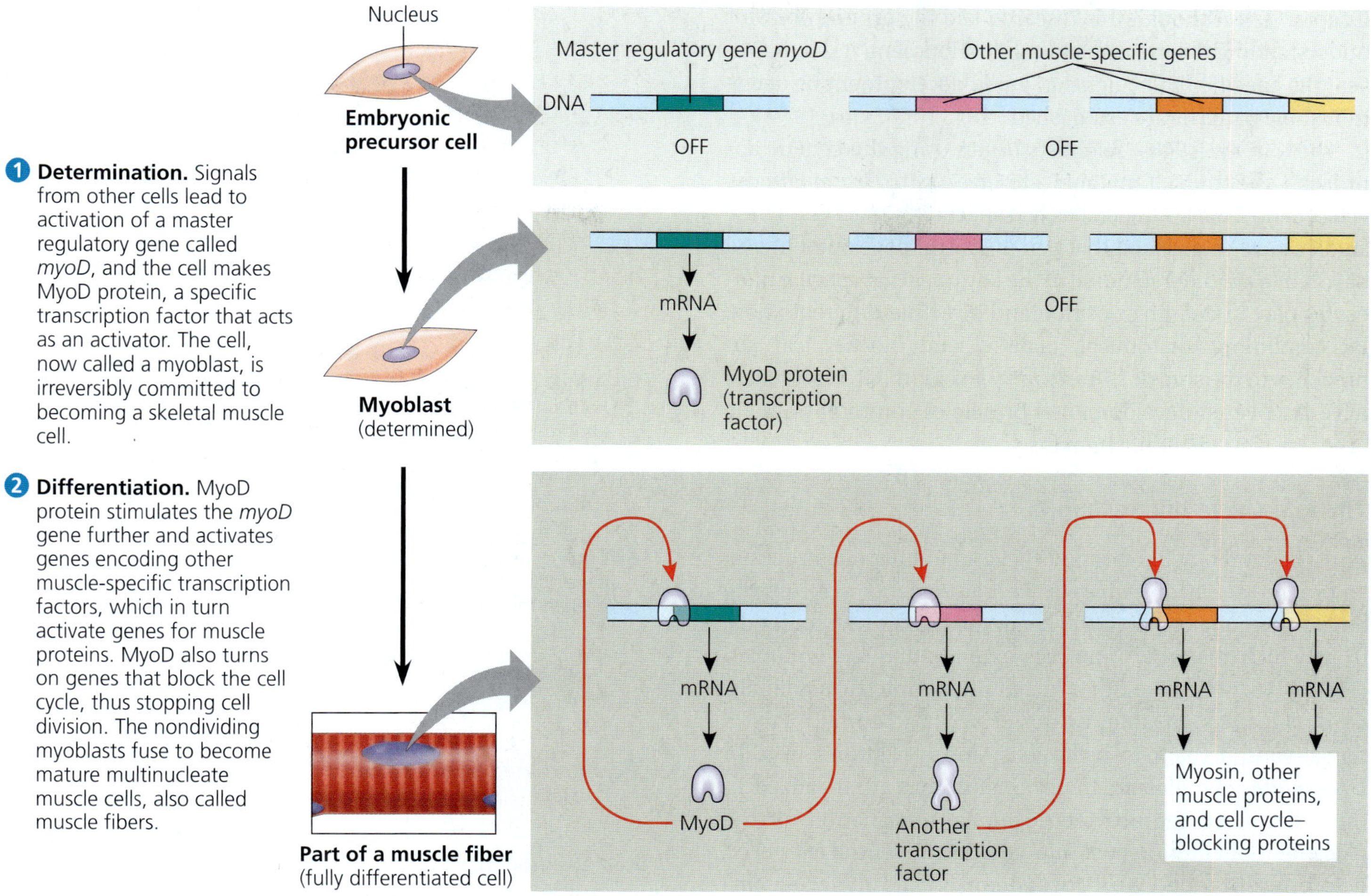

▲ **Figure 18.16 Determination and differentiation of muscle cells.** Skeletal muscle cells arise from embryonic cells as a result of changes in gene expression. (In this depiction, the process of gene activation is greatly simplified.)

WHAT IF? *What would happen if a mutation in the* myoD *gene resulted in a MyoD protein that could not activate the* myoD *gene?*

it work on *all* kinds of cells? One likely explanation is that activation of the muscle-specific genes is not solely dependent on MyoD but requires a particular *combination* of regulatory proteins, some of which are lacking in cells that do not respond to MyoD. The determination and differentiation of other kinds of tissues may play out in a similar fashion.

We have now seen how different programs of gene expression that are activated in the fertilized egg can result in differentiated cells and tissues. But for the tissues to function effectively in the organism as a whole, the organism's *body plan*—its overall three-dimensional arrangement—must be established and superimposed on the differentiation process. Next we'll investigate the molecular basis for the establishment of the body plan, using the well-studied *Drosophila* as an example.

Pattern Formation: Setting Up the Body Plan

Cytoplasmic determinants and inductive signals both contribute to the development of a spatial organization in which the tissues and organs of an organism are all in their characteristic places. This process is called **pattern formation**.

Pattern formation in animals begins in the early embryo, when the major axes of an animal are established. Before construction begins on a new building, the locations of the front, back, and sides are determined. In the same way, before the tissues and organs of a bilaterally symmetrical animal appear, the relative positions of the animal's head and tail, right and left sides, and back and front are set up, thus establishing the three major body axes. The molecular cues that control pattern formation, collectively called **positional information**, are provided by cytoplasmic determinants and inductive signals (see Figure 18.15). These cues tell a cell its location relative to the body axes and to neighboring cells and determine how the cell and its progeny will respond to future molecular signals.

During the first half of the 20th century, classical embryologists made detailed anatomical observations of embryonic development in a number of species and performed experiments in which they manipulated embryonic tissues (see

Chapter 47). Although this research laid the groundwork for understanding the mechanisms of development, it did not reveal the specific molecules that guide development or determine how patterns are established.

Then, in the 1940s, scientists began using the genetic approach—the study of mutants—to investigate *Drosophila* development. That approach has had spectacular success. These studies have established that genes control development and have led to an understanding of the key roles that specific molecules play in defining position and directing differentiation. By combining anatomical, genetic, and biochemical approaches to the study of *Drosophila* development, researchers have discovered developmental principles common to many other species, including humans.

The Life Cycle of Drosophila

Fruit flies and other arthropods have a modular construction, an ordered series of segments. These segments make up the body's three major parts: the head, the thorax (the midbody, from which the wings and legs extend), and the abdomen (**Figure 18.17a**). Like other bilaterally symmetrical animals, *Drosophila* has an anterior-posterior (head-to-tail) axis, a dorsal-ventral (back-to-belly) axis, and a right-left axis. In *Drosophila*, cytoplasmic determinants that are localized in the unfertilized egg provide positional information for the placement of anterior-posterior and dorsal-ventral axes even before fertilization. We'll focus here on the molecules involved in establishing the anterior-posterior axis.

The *Drosophila* egg develops in the female's ovary, surrounded by ovarian cells called nurse cells and follicle cells (**Figure 18.17b**, top). These support cells supply the egg with nutrients, mRNAs, and other substances needed for development and make the egg shell. After fertilization and laying of the egg, embryonic development results in the formation of a segmented larva, which goes through three larval stages. Then, in a process much like that by which a caterpillar becomes a butterfly, the fly larva forms a cocoon in which it metamorphoses into the adult fly pictured in Figure 18.17a.

Genetic Analysis of Early Development: Scientific Inquiry

Edward B. Lewis was a visionary American biologist who, in the 1940s, first showed the value of the genetic approach to studying embryonic development in *Drosophila*. Lewis studied bizarre mutant flies with developmental defects that led to extra wings or legs in the wrong places (**Figure 18.18**). He located the mutations on the fly's genetic map, thus connecting the developmental abnormalities to specific genes. This research supplied the first concrete evidence that genes somehow direct the developmental processes studied by embryologists. The genes Lewis discovered, called **homeotic genes**, control pattern formation in the late embryo, larva, and adult.

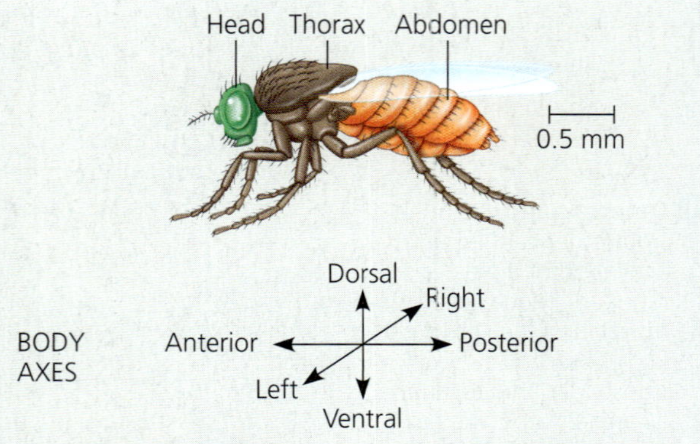

(a) Adult. The adult fly is segmented, and multiple segments make up each of the three main body parts—head, thorax, and abdomen. The body axes are shown by arrows.

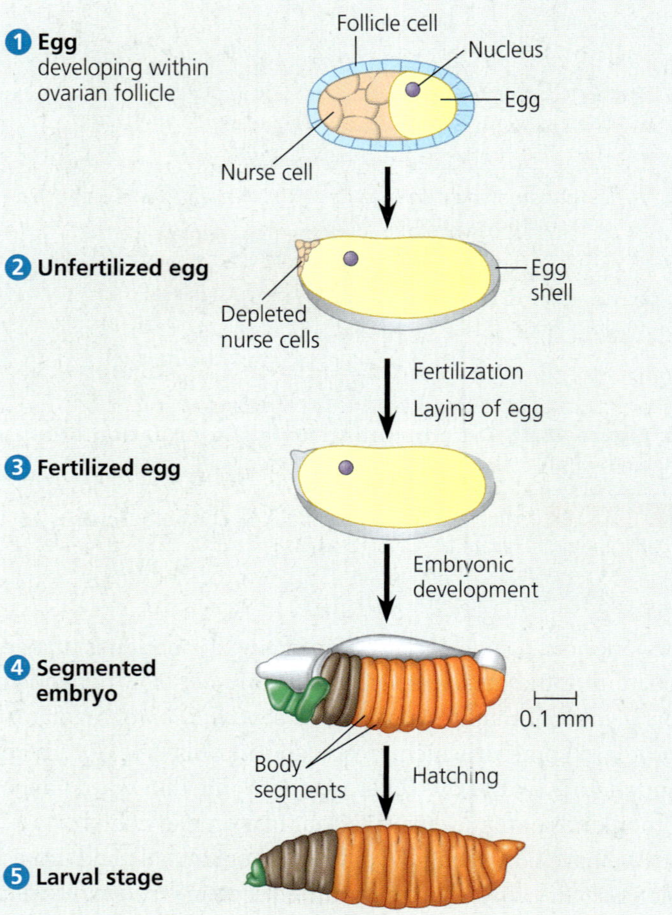

(b) Development from egg to larva. ① The yellow egg is surrounded by other cells that form a structure called the follicle within one of the mother's ovaries. ② The nurse cells shrink as they supply nutrients and mRNAs to the developing egg, which grows larger. Eventually, the mature egg fills the egg shell that is secreted by the follicle cells. ③ The egg is fertilized within the mother and then laid. ④ Embryonic development forms ⑤ a larva, which goes through three stages. The third stage forms a cocoon (not shown), within which the larva metamorphoses into the adult shown in (a).

▲ **Figure 18.17 Key developmental events in the life cycle of *Drosophila*.**

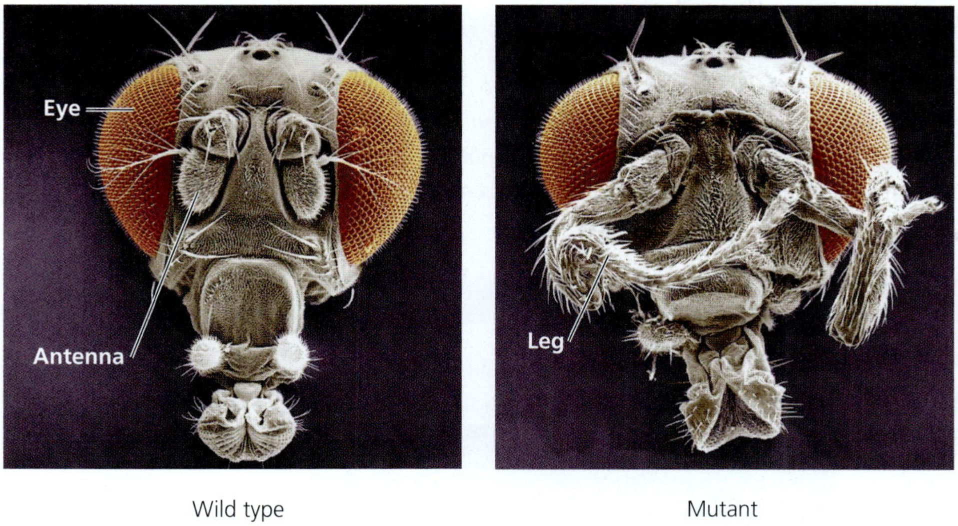

◀ Figure 18.18 Abnormal pattern formation in *Drosophila*. Mutations in certain regulatory genes, called homeotic genes, cause a misplacement of structures in an animal. These micrographs contrast the head of a wild-type fly, bearing a pair of small antennae, with that of a homeotic mutant, bearing a pair of legs in place of antennae (SEMs).

Insight into pattern formation during early embryonic development did not come for another 30 years, when two researchers in Germany, Christiane Nüsslein-Volhard and Eric Wieschaus, set out to identify *all* the genes that affect segment formation in *Drosophila*. The project was daunting for three reasons. The first was the sheer number of *Drosophila* genes, now known to total about 13,700. The genes affecting segmentation might be just a few needles in a haystack or might be so numerous and varied that the scientists would be unable to make sense of them. Second, mutations affecting a process as fundamental as segmentation would surely be **embryonic lethals**, mutations with phenotypes causing death at the embryonic or larval stage. Because organisms with embryonic lethal mutations never reproduce, they cannot be bred for study. The researchers dealt with this problem by looking for recessive mutations, which can be propagated in heterozygous flies. Third, cytoplasmic determinants in the egg were known to play a role in axis formation, and therefore the researchers knew they would have to study the mother's genes as well as those of the embryo. It is the mother's genes that we will discuss further as we focus on how the anterior-posterior body axis is set up in the developing egg.

Nüsslein-Volhard and Wieschaus began their search for segmentation genes by exposing flies to a mutagenic chemical that affected the flies' gametes. They mated the mutagenized flies and then scanned their descendants for dead embryos or larvae with abnormal segmentation or other defects. For example, to find genes that might set up the anterior-posterior axis, they looked for embryos or larvae with abnormal ends, such as two heads or two tails, predicting that such abnormalities would arise from mutations in maternal genes required for correctly setting up the offspring's head or tail end.

Using this approach, Nüsslein-Volhard and Wieschaus eventually identified about 1,200 genes essential for pattern formation during embryonic development. Of these, about 120 were essential for normal segmentation. Over several years, the researchers were able to group these segmentation genes by general function, to map them, and to clone many of them for further study in the lab. The result was a detailed molecular understanding of the early steps in pattern formation in *Drosophila*.

When the results of Nüsslein-Volhard and Wieschaus were combined with Lewis's earlier work, a coherent picture of *Drosophila* development emerged. In recognition of their discoveries, the three researchers were awarded a Nobel Prize in 1995.

Let's consider further the genes that Nüsslein-Volhard, Wieschaus, and co-workers found for cytoplasmic determinants deposited in the egg by the mother. These genes set up the initial pattern of the embryo by regulating gene expression in broad regions of the early embryo.

Axis Establishment

As we mentioned earlier, cytoplasmic determinants in the egg are the substances that initially establish the axes of the *Drosophila* body. These substances are encoded by genes of the mother, fittingly called maternal effect genes. A **maternal effect gene** is a gene that, when mutant in the mother, results in a mutant phenotype in the offspring, regardless of the offspring's own genotype. In fruit fly development, the mRNA or protein products of maternal effect genes are placed in the egg while it is still in the mother's ovary. When the mother has a mutation in such a gene, she makes a defective gene product (or none at all), and her eggs are defective; when these eggs are fertilized, they fail to develop properly.

Because they control the orientation (polarity) of the egg and consequently of the fly, maternal effect genes are also called **egg-polarity genes**. One group of these genes sets up the anterior-posterior axis of the embryo, while a second group establishes the dorsal-ventral axis. Like mutations in segmentation genes, mutations in maternal effect genes are generally embryonic lethals.

Bicoid: A Morphogen Determining Head Structures To see how maternal effect genes determine the body axes of the offspring, we will focus on one such gene, called ***bicoid***, a term meaning "two-tailed." An embryo whose mother has a mutant *bicoid* gene lacks the front half of its body and has posterior structures at both ends **(Figure 18.19)**. This phenotype suggested to Nüsslein-Volhard and her colleagues that the product of the mother's *bicoid* gene is essential for setting up the anterior end of the fly and might be concentrated at the future anterior end of the embryo. This hypothesis is a specific example of the *morphogen gradient hypothesis* first proposed by embryologists a century ago; in this hypothesis, gradients of

▼ Figure 18.19 Inquiry

Is Bicoid a morphogen that determines the anterior end of a fruit fly?

EXPERIMENT Using a genetic approach to study *Drosophila* development, Christiane Nüsslein-Volhard and colleagues at the European Molecular Biology Laboratory in Heidelberg, Germany, obtained many embryos and larvae with defects in their body patterns, some due to mutations in the mother's genes. One such gene was called *bicoid*, meaning "two-tailed," because its mutation resulted in larvae with two tails and no head. Subsequent studies analyzed expression of the *bicoid* gene.

The researchers hypothesized that *bicoid* normally codes for a morphogen that specifies the head (anterior) end of the embryo. To test this hypothesis, they used molecular techniques to determine where the mRNA and protein encoded by this gene were found in the fertilized egg and early embryo.

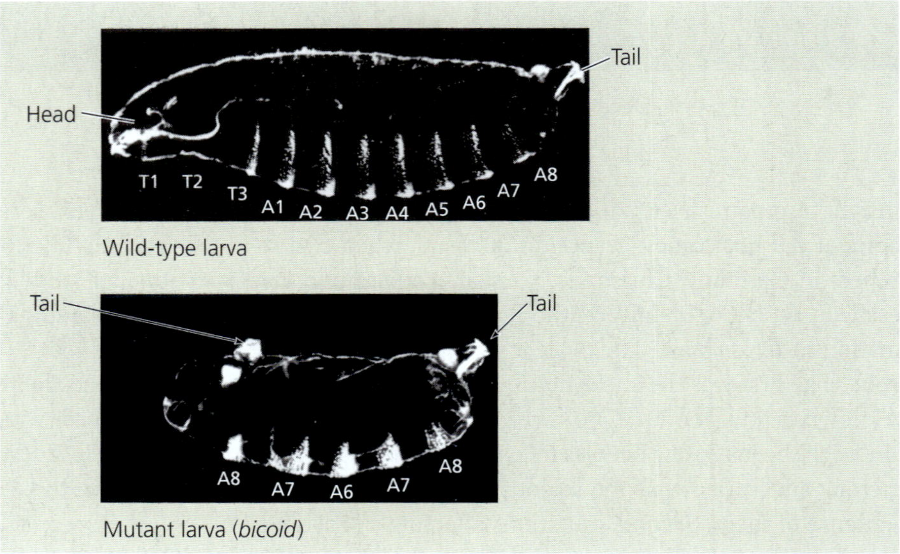

RESULTS *Bicoid* mRNA (dark blue) was confined to the anterior end of the unfertilized egg. Later in development, Bicoid protein was seen to be concentrated in cells at the anterior end of the embryo.

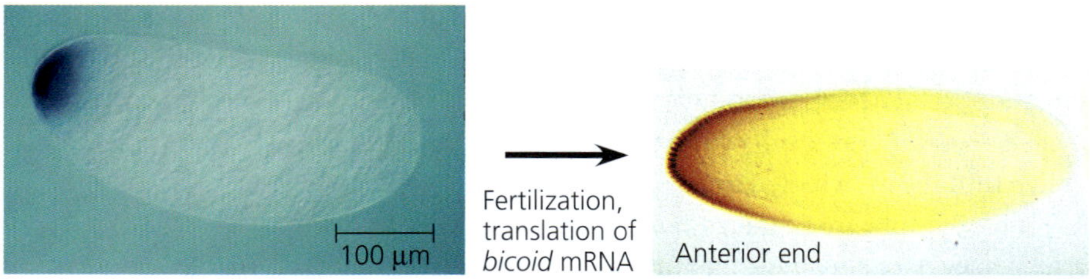

Bicoid mRNA in mature unfertilized egg

Bicoid protein in early embryo

CONCLUSION The results support the hypothesis that Bicoid protein is a morphogen specifying formation of head-specific structures.

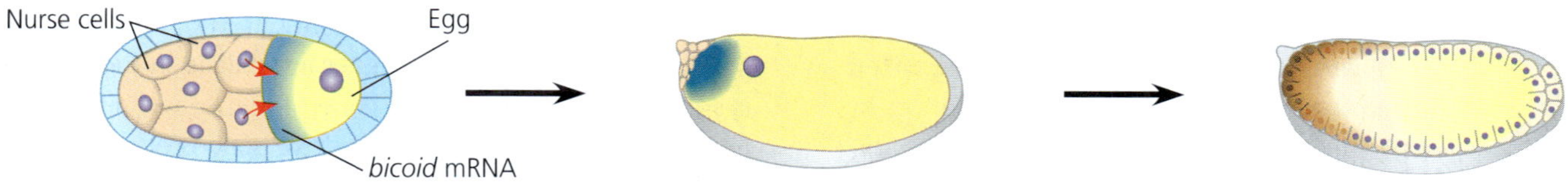

Developing egg

Bicoid mRNA in mature unfertilized egg

Bicoid protein in early embryo

SOURCE C. Nüsslein-Volhard et al., Determination of anteroposterior polarity in *Drosophila*, *Science* 238:1675–1681 (1987). W. Driever and C. Nüsslein-Volhard, A gradient of *bicoid* protein in *Drosophila* embryos, *Cell* 54:83–93 (1988). T. Berleth et al., The role of localization of *bicoid* RNA in organizing the anterior pattern of the *Drosophila* embryo, *EMBO Journal* 7:1749–1756 (1988).

WHAT IF? If the hypothesis is correct, predict what would happen if you injected *bicoid* mRNA into the anterior end of an egg from a female mutant for *bicoid*.

substances called **morphogens** establish an embryo's axes and other features of its form.

DNA technology and other modern biochemical methods enabled the researchers to test whether the *bicoid* product is in fact a morphogen that determines the anterior end of the fly. The first question they asked was whether the mRNA and protein products of these genes are located in the egg in a position consistent with the hypothesis. They found that *bicoid* mRNA is highly concentrated at the extreme anterior end of the mature egg, as predicted by the hypothesis (see Figure 18.19). The mRNA is produced in nurse cells, transferred to the egg via cytoplasmic bridges, and anchored to the cytoskeleton at the anterior end of the egg. After the egg is fertilized, the mRNA is translated into protein. The Bicoid protein then diffuses from the anterior end toward the posterior, resulting in a gradient of protein within the early embryo, with the highest concentration at the anterior end. These results are consistent with the hypothesis that Bicoid protein is responsible for specifying the fly's anterior end. To test the hypothesis more specifically, scientists injected pure *bicoid* mRNA into various regions of early embryos. The protein that resulted from its translation caused anterior structures to form at the injection sites.

The *bicoid* research was groundbreaking for several reasons. First, it led to the identification of a specific protein required for some of the earliest steps in pattern formation. It thus helped us understand how different regions of the egg can give rise to cells that go down different developmental pathways. Second, it increased our understanding of the mother's critical role in the initial phases of embryonic development. (As one developmental biologist has put it, "Mom tells Junior which way is up.") Finally, the principle that a gradient of morphogens can determine polarity and position has proved to be a key developmental concept for a number of species, just as early embryologists had thought.

In *Drosophila*, gradients of specific proteins determine the posterior end as well as the anterior and also are responsible for establishing the dorsal-ventral axis. Later, positional information operating on an ever finer scale establishes a specific number of correctly oriented segments and finally triggers the formation of each segment's characteristic structures. When the genes operating in this final step are abnormal, the pattern of the adult is abnormal, as you saw in Figure 18.18.

In this section, we have seen how a carefully orchestrated program of sequential gene regulation controls the transformation of a fertilized egg into a multicellular organism. The program is carefully balanced between turning on the genes for differentiation in the right place and turning off other genes. Even when an organism is fully developed, gene expression is regulated in a similarly fine-tuned manner. In the final section of the chapter, we'll consider how fine this tuning is, by looking at how specific changes in expression of one or a few genes can lead to the development of cancer.

CONCEPT CHECK 18.4

1. As you learned in Chapter 12, mitosis gives rise to two daughter cells that are genetically identical to the parent cell. Yet you, the product of many mitotic divisions, are not composed of identical cells. Why?
2. The signaling molecules released by an embryonic cell can induce changes in a neighboring cell without entering the cell. How?
3. Why are fruit fly maternal effect genes also called egg-polarity genes?
4. **WHAT IF?** In the blowup box in Figure 18.15b, the lower cell is synthesizing signaling molecules, whereas the upper cell is expressing signal receptors. In terms of gene regulation, explain how these cells came to have different functions.

For suggested answers, see Appendix A.

CONCEPT 18.5

Cancer results from genetic changes that affect cell cycle control

In Chapter 12, we considered cancer as a set of diseases in which cells escape from the control mechanisms that normally limit their growth. Now that we have discussed the molecular basis of gene expression and its regulation, we are ready to look at cancer more closely. The gene regulation systems that go wrong during cancer turn out to be the very same systems that play important roles in embryonic development, the immune response, and many other biological processes. Thus, research into the molecular basis of cancer has both benefited from and informed many other fields of biology.

Types of Genes Associated with Cancer

The genes that normally regulate cell growth and division during the cell cycle include genes for growth factors, their receptors, and the intracellular molecules of signaling pathways. (To review the cell cycle, see Chapter 12.) Mutations that alter any of these genes in somatic cells can lead to cancer. The agent of such change can be random spontaneous mutation. However, it is likely that many cancer-causing mutations result from environmental influences, such as chemical carcinogens, X-rays and other high-energy radiation, and certain viruses.

An early breakthrough in understanding cancer came in 1911, when Peyton Rous, an American pathologist, discovered a virus that causes cancer in chickens. Since then, scientists have recognized a number of *tumor viruses* that cause cancer in various animals, including humans (see Table 19.1). The Epstein-Barr virus, which causes infectious mononucleosis, has been linked to several types of cancer, notably Burkitt's lymphoma. Papillomaviruses are associated with cancer of the cervix, and a virus called HTLV-1 causes a type of adult leukemia. Worldwide, viruses seem to play a role in about 15% of the cases of human cancer.

Oncogenes and Proto-Oncogenes

Research on tumor viruses led to the discovery of cancer-causing genes called **oncogenes** (from the Greek *onco*, tumor) in certain retroviruses (see Chapter 19). Subsequently, close counterparts of these oncogenes were found in the genomes of humans and other animals. The normal versions of the cellular genes, called **proto-oncogenes**, code for proteins that stimulate normal cell growth and division.

How might a proto-oncogene—a gene that has an essential function in normal cells—become an oncogene, a cancer-causing gene? In general, an oncogene arises from a genetic change that leads to an increase either in the amount of the proto-oncogene's protein product or in the intrinsic activity of each protein molecule. The genetic changes that convert proto-oncogenes to oncogenes fall into three main categories: movement of DNA within the genome, amplification of a proto-oncogene, and point mutations in a control element or in the proto-oncogene itself **(Figure 18.20)**.

Cancer cells are frequently found to contain chromosomes that have broken and rejoined incorrectly, translocating fragments from one chromosome to another (see Figure 15.15). Now that you have learned how gene expression is regulated, you can understand the possible consequences of such translocations. If a translocated proto-oncogene ends up near an especially active promoter (or other control element), its transcription may increase, making it an oncogene. The second main type of genetic change, amplification, increases the number of copies of the proto-oncogene in the cell. The third possibility is a point mutation either (1) in the promoter or an enhancer that controls a proto-oncogene, causing an increase in its expression, or (2) in the coding sequence, changing the gene's product to a protein that is more active or more resistant to degradation than the normal protein. All these mechanisms can lead to abnormal stimulation of the cell cycle and put the cell on the path to malignancy.

Tumor-Suppressor Genes

In addition to genes whose products normally promote cell division, cells contain genes whose normal products *inhibit* cell division. Such genes are called **tumor-suppressor genes** because the proteins they encode help prevent uncontrolled cell growth. Any mutation that decreases the normal activity of a tumor-suppressor protein may contribute to the onset of cancer, in effect stimulating growth through the absence of suppression.

The protein products of tumor-suppressor genes have various functions. Some tumor-suppressor proteins normally repair damaged DNA, a function that prevents the cell from accumulating cancer-causing mutations. Other tumor-suppressor proteins control the adhesion of cells to each other or to the extracellular matrix; proper cell anchorage is crucial in normal tissues—and often absent in cancers. Still other tumor-suppressor proteins are components of cell-signaling pathways that inhibit the cell cycle.

Interference with Normal Cell-Signaling Pathways

The proteins encoded by many proto-oncogenes and tumor-suppressor genes are components of cell-signaling pathways **(Figure 18.21)**. Let's take a closer look at how such proteins function in normal cells and what goes wrong with their function in cancer cells. We will focus on the products of two key

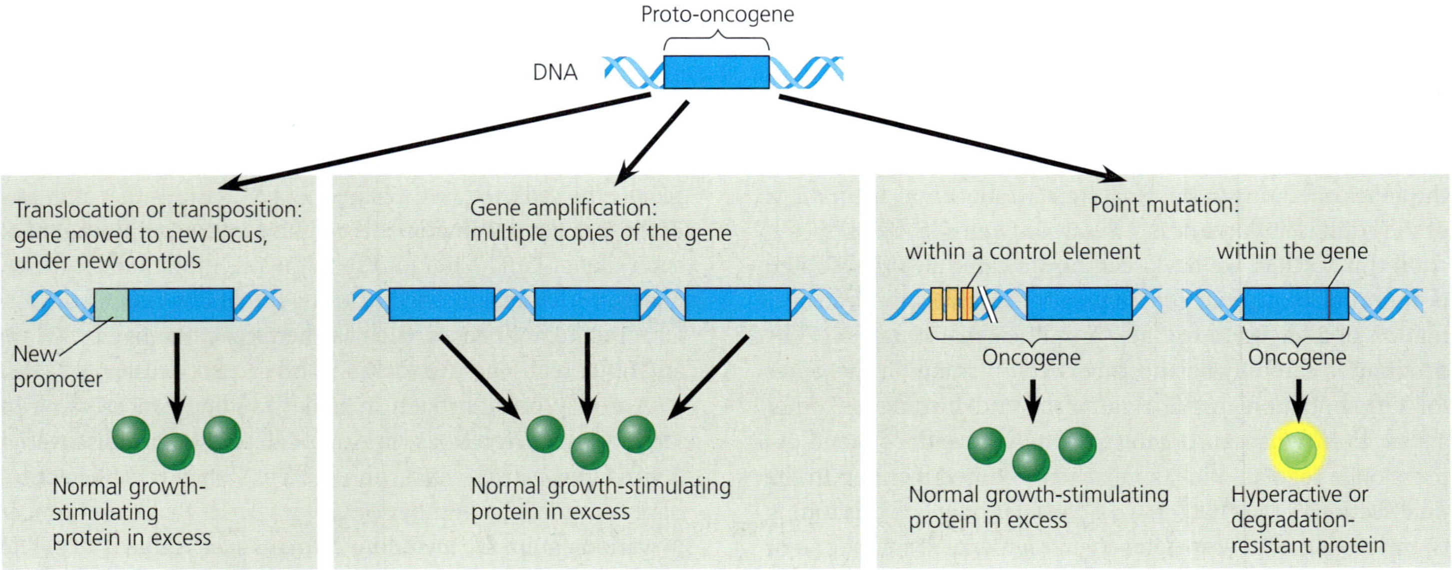

▲ **Figure 18.20 Genetic changes that can turn proto-oncogenes into oncogenes.**

(a) Cell cycle–stimulating pathway. This pathway is triggered by **1** a growth factor that binds to **2** its receptor in the plasma membrane. The signal is relayed to **3** a G protein called Ras. Like all G proteins, Ras is active when GTP is bound to it. Ras passes the signal to **4** a series of protein kinases. The last kinase activates **5** a transcription activator that turns on one or more genes for proteins that stimulate the cell cycle. If a mutation makes Ras or any other pathway component abnormally active, excessive cell division and cancer may result.

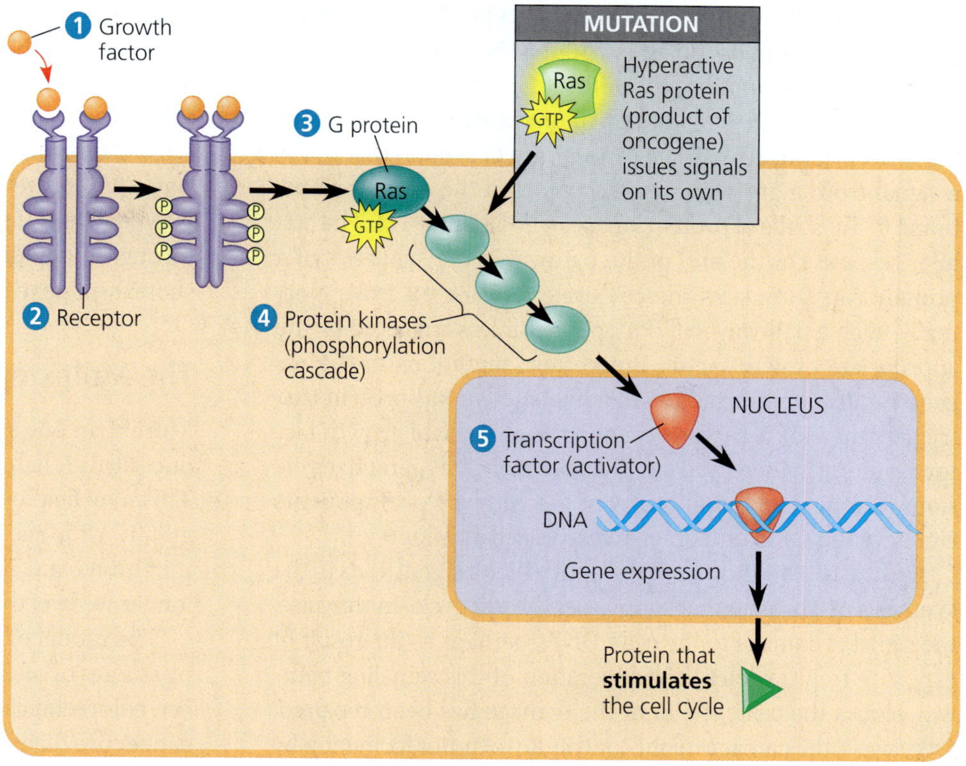

(b) Cell cycle–inhibiting pathway. In this pathway, **1** DNA damage is an intracellular signal that is passed via **2** protein kinases and leads to activation of **3** p53. Activated p53 promotes transcription of the gene for a protein that inhibits the cell cycle. The resulting suppression of cell division ensures that the damaged DNA is not replicated. Mutations causing deficiencies in any pathway component can contribute to the development of cancer.

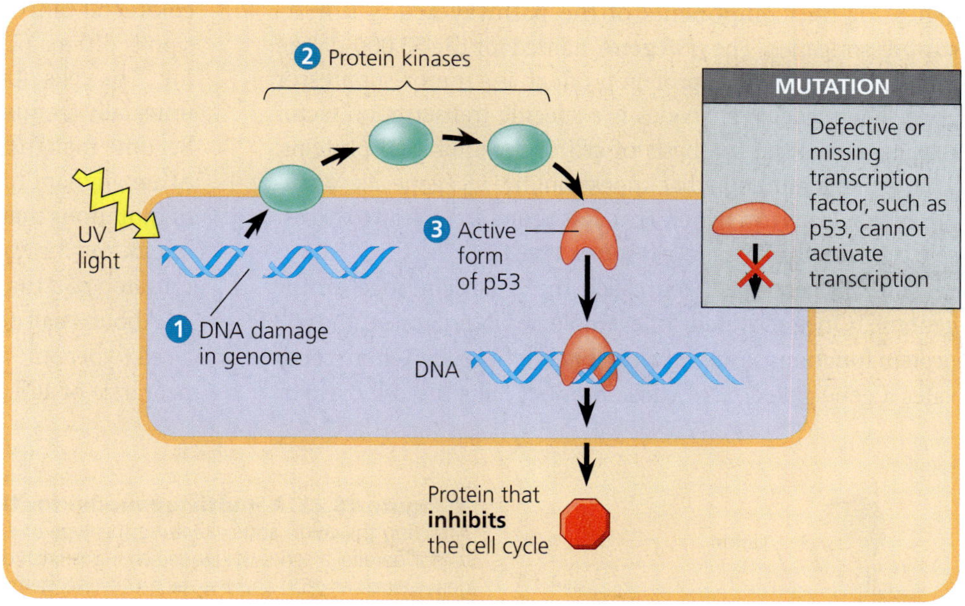

(c) Effects of mutations. Increased cell division, possibly leading to cancer, can result if the cell cycle is overstimulated, as in (a), or not inhibited when it normally would be, as in (b).

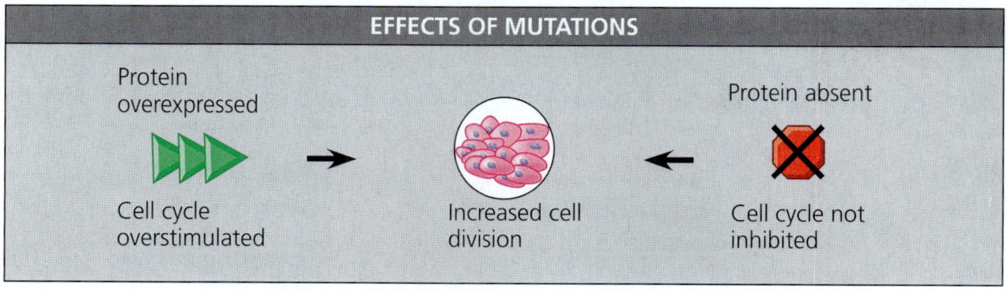

▲ **Figure 18.21 Signaling pathways that regulate cell division.** Both stimulatory and inhibitory pathways regulate the cell cycle, commonly by influencing transcription. Cancer can result from aberrations in such pathways, which may be caused by mutations, either spontaneous or environmentally triggered.

? *Looking at the pathway in (b), explain whether a cancer-causing mutation in a tumor-suppressor gene, such as p53, is more likely to be a recessive or a dominant mutation.*

genes, the *ras* proto-oncogene and the *p53* tumor-suppressor gene. Mutations in *ras* occur in about 30% of human cancers, mutations in *p53* in more than 50%.

The Ras protein, encoded by the ***ras* gene** (named for rat sarcoma, a connective tissue cancer), is a G protein that relays a signal from a growth factor receptor on the plasma membrane to a cascade of protein kinases (see Chapter 11). The cellular response at the end of the pathway is the synthesis of a protein that stimulates the cell cycle **(Figure 18.21a)**. Normally, such a pathway will not operate unless triggered by the appropriate growth factor. But certain mutations in the *ras* gene can lead to production of a hyperactive Ras protein that triggers the kinase cascade even in the absence of growth factor, resulting in increased cell division. In fact, hyperactive versions or excess amounts of any of the pathway's components can have the same outcome: excessive cell division.

Figure 18.21b shows a pathway in which a signal leads to the synthesis of a protein that suppresses the cell cycle. In this case, the signal is damage to the cell's DNA, perhaps as the result of exposure to ultraviolet light. Operation of this signaling pathway blocks the cell cycle until the damage has been repaired. Otherwise, the damage might contribute to tumor formation by causing mutations or chromosomal abnormalities. Thus, the genes for the components of the pathway act as tumor-suppressor genes. The ***p53* gene**, named for the 53,000-dalton molecular weight of its protein product, is a tumor-suppressor gene. The protein it encodes is a specific transcription factor that promotes the synthesis of cell cycle–inhibiting proteins. That is why a mutation that knocks out the *p53* gene, like a mutation that leads to a hyperactive Ras protein, can lead to excessive cell growth and cancer **(Figure 18.21c)**.

The *p53* gene has been called the "guardian angel of the genome." Once activated, for example by DNA damage, the p53 protein functions as an activator for several genes. Often it activates a gene called *p21*, whose product halts the cell cycle by binding to cyclin-dependent kinases, allowing time for the cell to repair the DNA; the p53 protein can also turn on genes directly involved in DNA repair. When DNA damage is irreparable, p53 activates "suicide" genes, whose protein products cause cell death by apoptosis (see Figure 11.20). Thus, in at least three ways, p53 prevents a cell from passing on mutations due to DNA damage. If mutations do accumulate and the cell survives through many divisions—as is more likely if the *p53* tumor-suppressor gene is defective or missing—cancer may ensue.

The Multistep Model of Cancer Development

More than one somatic mutation is generally needed to produce all the changes characteristic of a full-fledged cancer cell. This may help explain why the incidence of cancer increases greatly with age. If cancer results from an accumulation of mutations and if mutations occur throughout life, then the longer we live, the more likely we are to develop cancer.

The model of a multistep path to cancer is well supported by studies of one of the best-understood types of human cancer, colorectal cancer. About 135,000 new cases of colorectal cancer are diagnosed each year in the United States, and the disease causes 60,000 deaths each year. Like most cancers, colorectal cancer develops gradually **(Figure 18.22)**. The first sign is often a polyp, a small, benign growth in the colon lining. The cells of the polyp look normal, although they divide unusually frequently. The tumor grows and may eventually become malignant, invading other tissues. The development of a malignant tumor is paralleled by a gradual accumulation of mutations that convert proto-oncogenes to oncogenes and knock out tumor-suppressor genes. A *ras* oncogene and a mutated *p53* tumor-suppressor gene are often involved.

About a half dozen changes must occur at the DNA level for a cell to become fully cancerous. These usually include the appearance of at least one active oncogene and the mutation or

▼ **Figure 18.22 A multistep model for the development of colorectal cancer.** Affecting the colon and/or rectum, this type of cancer is one of the best understood. Changes in a tumor parallel a series of genetic changes, including mutations affecting several tumor-suppressor genes (such as *p53*) and the *ras* proto-oncogene. Mutations of tumor-suppressor genes often entail loss (deletion) of the gene. *APC* stands for "adenomatous polyposis coli," and *DCC* stands for "deleted in colorectal cancer." Other mutation sequences can also lead to colorectal cancer.

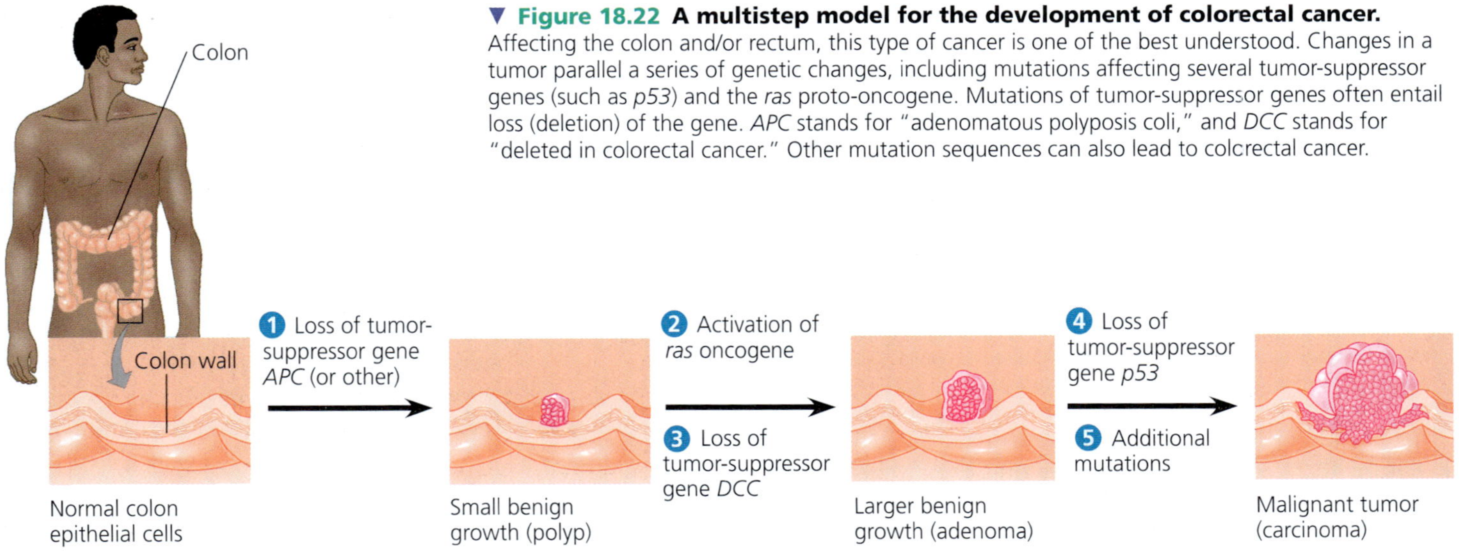

loss of several tumor-suppressor genes. Furthermore, since mutant tumor-suppressor alleles are usually recessive, in most cases mutations must knock out *both* alleles in a cell's genome to block tumor suppression. (Most oncogenes, on the other hand, behave as dominant alleles.) In many malignant tumors, the gene for telomerase is activated. This enzyme reverses the shortening of chromosome ends during DNA replication (see Figure 16.19). Production of telomerase in cancer cells removes a natural limit on the number of times the cells can divide.

Inherited Predisposition and Other Factors Contributing to Cancer

The fact that multiple genetic changes are required to produce a cancer cell helps explain the observation that cancers can run in families. An individual inheriting an oncogene or a mutant allele of a tumor-suppressor gene is one step closer to accumulating the necessary mutations for cancer to develop than is an individual without any such mutations.

Geneticists are devoting much effort to identifying inherited cancer alleles so that predisposition to certain cancers can be detected early in life. About 15% of colorectal cancers, for example, involve inherited mutations. Many of these affect the tumor-suppressor gene called *adenomatous polyposis coli*, or *APC* (see Figure 18.22). This gene has multiple functions in the cell, including regulation of cell migration and adhesion. Even in patients with no family history of the disease, the *APC* gene is mutated in 60% of colorectal cancers. In these individuals, new mutations must occur in both *APC* alleles before the gene's function is lost. Since only 15% of colorectal cancers are associated with known inherited mutations, researchers continue in their efforts to identify "markers" that could predict the risk of developing this type of cancer.

There is evidence of a strong inherited predisposition in 5–10% of patients with breast cancer. This is the second most common type of cancer in the United States, striking over 180,000 women (and some men) annually and killing 40,000 each year. Mutations in the *BRCA1* or *BRCA2* gene are found in at least half of inherited breast cancers **(Figure 18.23)**. (*BRCA* stands for *br*east *ca*ncer.) A woman who inherits one mutant *BRCA1* allele has a 60% probability of developing breast cancer before the age of 50, compared with only a 2% probability for an individual homozygous for the normal allele. Both *BRCA1* and *BRCA2* are considered tumor-suppressor genes because their wild-type alleles protect against breast cancer and their mutant alleles are recessive. Apparently, the BRCA1 and BRCA2 proteins both function in the cell's DNA damage repair pathway. More is known about BRCA2, which, in association with another protein, helps to repair breaks that occur in both strands of DNA, crucial for maintaining undamaged DNA in a cell's nucleus.

Because DNA breakage can contribute to cancer, it makes sense that the risk of cancer can be lowered by minimizing exposure to DNA-damaging agents, such as the ultraviolet radiation in sunlight and chemicals found in cigarette smoke. Novel methods for early diagnosis and treatment of specific cancers are being developed that rely on new techniques for analyzing, and perhaps interfering with, gene expression in tumors. Ultimately, such approaches may lower the death rate from cancer.

▲ **Figure 18.23 Tracking the molecular basis of breast cancer.** In 1990, after 16 years of research, geneticist Mary-Claire King convincingly demonstrated that mutations in one gene—*BRCA1*—are associated with increased susceptibility to breast cancer, a finding that flew in the face of medical opinion at the time. Her lab is currently working to identify the environmental conditions that may affect the timing of cancer development in people who carry mutations in *BRCA1* and another breast cancer gene, *BRCA2*.

The study of genes associated with cancer, inherited or not, increases our basic understanding of how disruption of normal gene regulation results in this disease. We've come a long way in our understanding since Peyton Rous's discovery. We now know that viruses can contribute to cancer development in several ways if they integrate their genetic material into the DNA of infected cells. Viral integration may donate an oncogene to the cell, disrupt a tumor-suppressor gene, or convert a proto-oncogene to an oncogene. In addition, some viruses produce proteins that inactivate p53 and other tumor-suppressor proteins, thus making the cell more prone to becoming cancerous. Although viruses are little more than a nucleic acid surrounded by a protective coat, they are powerful biological agents. You'll learn more about how viruses function in the next chapter.

CONCEPT CHECK 18.5

1. Compare the usual functions of proteins encoded by proto-oncogenes with those encoded by tumor-suppressor genes.
2. Under what circumstances is cancer considered to have a hereditary component?
3. **WHAT IF?** Explain how the types of mutations that lead to cancer are different for a proto-oncogene and a tumor-suppressor gene, in terms of the effect of the mutation on the activity of the gene product.

For suggested answers, see Appendix A.

Chapter 18 Review

MEDIA Go to the Study Area at **www.masteringbio.com** for BioFlix 3-D Animations, MP3 Tutors, Videos, Practice Tests, an eBook, and more.

SUMMARY OF KEY CONCEPTS

CONCEPT 18.1

Bacteria often respond to environmental change by regulating transcription (pp. 351–356)

▶ **Operons: The Basic Concept** Cells control metabolism by regulating enzyme activity or the expression of genes coding for enzymes. In bacteria, genes are often clustered into operons, with one promoter serving several adjacent genes. An operator site on the DNA switches the operon on or off, resulting in coordinate regulation of the genes.

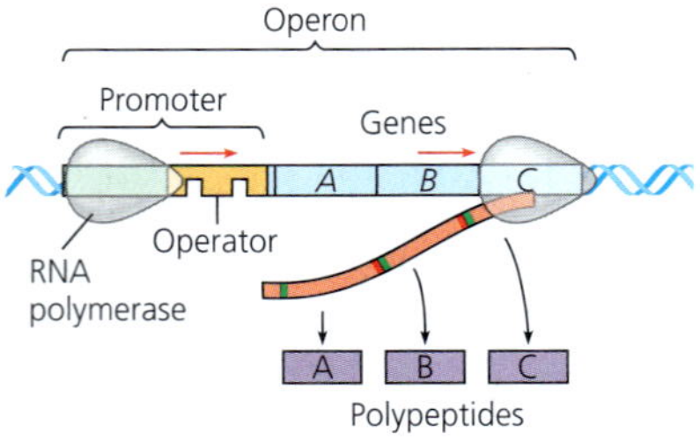

▶ **Repressible and Inducible Operons: Two Types of Negative Gene Regulation** In either type of operon, binding of a specific repressor protein to the operator shuts off transcription. (The repressor is encoded by a separate regulatory gene.) In a repressible operon, the repressor is active when bound to a corepressor, usually the end product of an anabolic pathway.

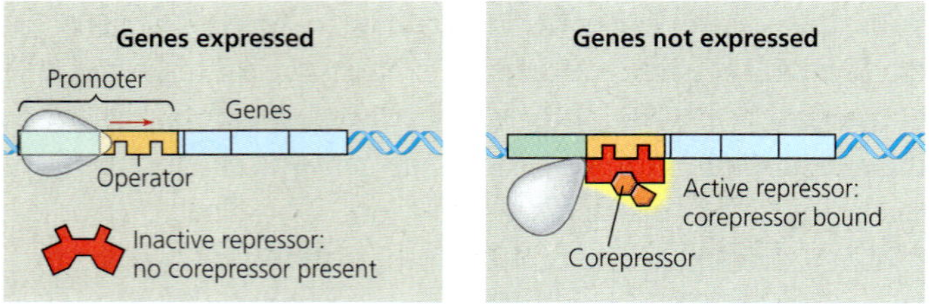

In an inducible operon, binding of an inducer to an innately active repressor inactivates the repressor and turns on transcription. Inducible enzymes usually function in catabolic pathways.

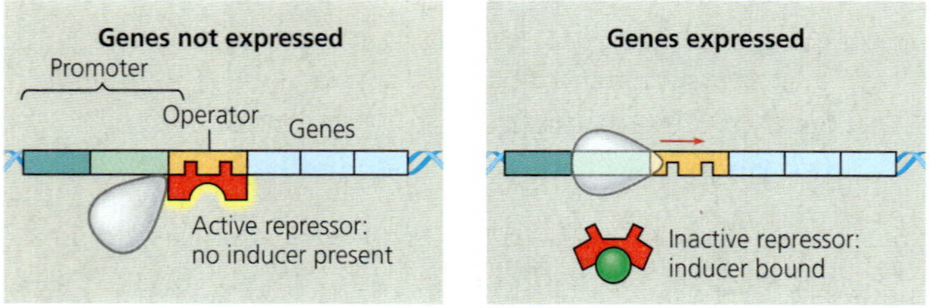

▶ **Positive Gene Regulation** Some operons are also subject to positive control via a stimulatory activator protein, such as catabolite activator protein (CAP), which promotes transcription when bound to a site within the promoter.

MEDIA

MP3 Tutor Control of Gene Expression

Activity The *lac* Operon in *E. coli*

CONCEPT 18.2

Eukaryotic gene expression can be regulated at any stage (pp. 356–364)

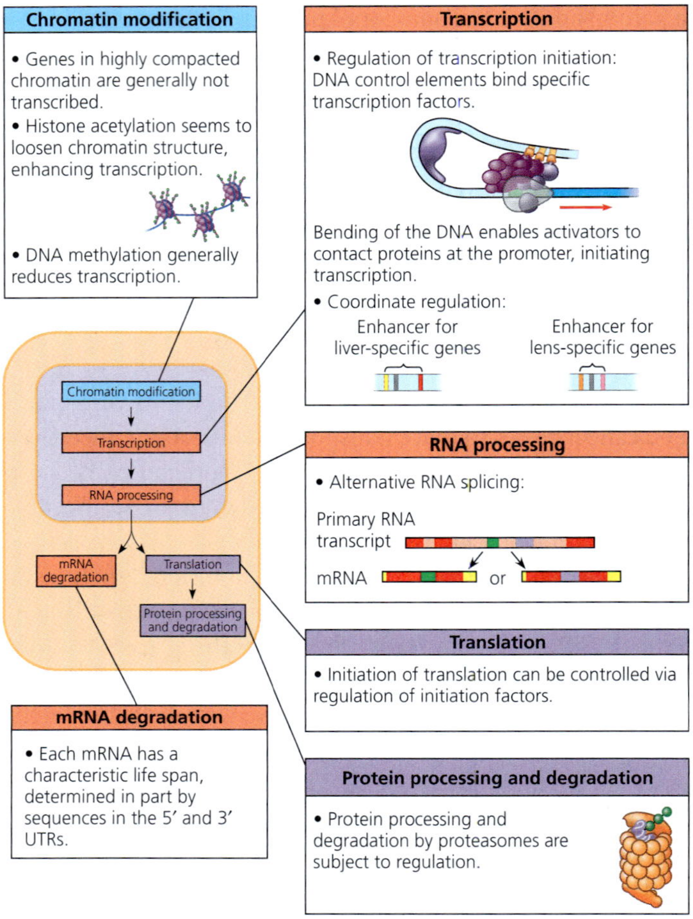

CONCEPT 18.3

Noncoding RNAs play multiple roles in controlling gene expression (pp. 364–366)

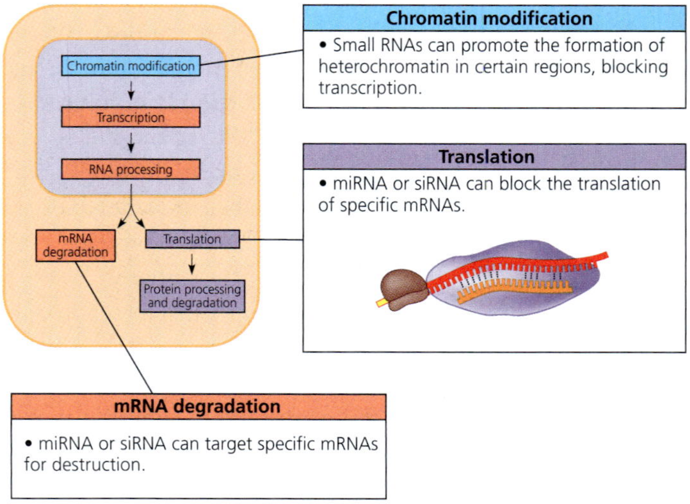

MEDIA

Activity Overview: Control of Gene Expression
Activity Control of Transcription
Activity Post-Transcriptional Control Mechanisms
Activity Review: Control of Gene Expression
Investigation How Do You Design a Gene Expression System?

CONCEPT 18.4

A program of differential gene expression leads to the different cell types in a multicellular organism (pp. 366–373)

- **A Genetic Program for Embryonic Development** Embryonic cells undergo differentiation, becoming specialized in structure and function. Morphogenesis encompasses the processes that give shape to the organism and its various parts. Cells differ in structure and function not because they contain different genes but because they express different portions of a common genome.
- **Cytoplasmic Determinants and Inductive Signals** Cytoplasmic determinants in the unfertilized egg regulate the expression of genes in the zygote that affect the developmental fate of embryonic cells. In the process called induction, signaling molecules from embryonic cells cause transcriptional changes in nearby target cells.
- **Sequential Regulation of Gene Expression During Cellular Differentiation** Differentiation is heralded by the appearance of tissue-specific proteins, which enable differentiated cells to carry out their specialized roles.
- **Pattern Formation: Setting Up the Body Plan** In animals, pattern formation, the development of a spatial organization of tissues and organs, begins in the early embryo. Positional information, the molecular cues that control pattern formation, tell a cell its location relative to the body's axes and to other cells. In *Drosophila*, gradients of morphogens encoded by maternal effect genes determine the body axes. For example, the gradient of Bicoid protein determines the anterior-posterior axis.

MEDIA

Activity Signal Transduction Pathways
Activity Role of *bicoid* Gene in *Drosophila* Development
Investigation How Do *bicoid* Mutations Alter Development?

CONCEPT 18.5

Cancer results from genetic changes that affect cell cycle control (pp. 373–377)

- **Types of Genes Associated with Cancer** The products of proto-oncogenes and tumor-suppressor genes control cell division. A DNA change that makes a proto-oncogene excessively active converts it to an oncogene, which may promote excessive cell division and cancer. A tumor-suppressor gene encodes a protein that inhibits abnormal cell division. A mutation in such a gene that reduces the activity of its protein product may also lead to excessive cell division and possibly to cancer.
- **Interference with Normal Cell-Signaling Pathways** Many proto-oncogenes and tumor-suppressor genes encode components of growth-stimulating and growth-inhibiting signaling pathways, respectively. A hyperactive version of a protein in a stimulatory pathway, such as Ras (a G protein), functions as an oncogene protein. A defective version of a protein in an inhibitory pathway, such as p53 (a transcription activator), fails to function as a tumor suppressor.
- **The Multistep Model of Cancer Development** Normal cells are converted to cancer cells by the accumulation of mutations affecting proto-oncogenes and tumor-suppressor genes.
- **Inherited Predisposition and Other Factors Contributing to Cancer** Individuals who inherit a mutant oncogene or tumor-suppressor allele have an increased risk of developing cancer. Certain viruses promote cancer by integration of viral DNA into a cell's genome.

MEDIA

Activity Causes of Cancer

TESTING YOUR KNOWLEDGE

SELF-QUIZ

1. If a particular operon encodes enzymes for making an essential amino acid and is regulated like the *trp* operon, then
   a. the amino acid inactivates the repressor.
   b. the enzymes produced are called inducible enzymes.
   c. the repressor is active in the absence of the amino acid.
   d. the amino acid acts as a corepressor.
   e. the amino acid turns on transcription of the operon.

2. Muscle cells differ from nerve cells mainly because they
   a. express different genes.
   b. contain different genes.
   c. use different genetic codes.
   d. have unique ribosomes.
   e. have different chromosomes.

3. What would occur if the repressor of an inducible operon were mutated so it could not bind the operator?
   a. irreversible binding of the repressor to the promoter
   b. reduced transcription of the operon's genes
   c. buildup of a substrate for the pathway controlled by the operon
   d. continuous transcription of the operon's genes
   e. overproduction of catabolite activator protein (CAP)

4. The functioning of enhancers is an example of
   a. transcriptional control of gene expression.
   b. a post-transcriptional mechanism for editing mRNA.
   c. the stimulation of translation by initiation factors.
   d. post-translational control that activates certain proteins.
   e. a eukaryotic equivalent of prokaryotic promoter functioning.

5. Absence of *bicoid* mRNA from a *Drosophila* egg leads to the absence of anterior larval body parts and mirror-image duplication of posterior parts. This is evidence that the product of the *bicoid* gene
   a. is transcribed in the early embryo.
   b. normally leads to formation of tail structures.
   c. normally leads to formation of head structures.
   d. is a protein present in all head structures.
   e. leads to programmed cell death.

6. Which of the following statements about the DNA in one of your brain cells is true?
   a. Most of the DNA codes for protein.
   b. The majority of genes are likely to be transcribed.
   c. Each gene lies immediately adjacent to an enhancer.
   d. Many genes are grouped into operon-like clusters.
   e. It is the same as the DNA in one of your heart cells.

7. Cell differentiation always involves
   a. the production of tissue-specific proteins, such as muscle actin.
   b. the movement of cells.
   c. the transcription of the *myoD* gene.
   d. the selective loss of certain genes from the genome.
   e. the cell's sensitivity to environmental cues, such as light or heat.

8. Which of the following is an example of post-transcriptional control of gene expression?
   a. the addition of methyl groups to cytosine bases of DNA
   b. the binding of transcription factors to a promoter
   c. the removal of introns and splicing together of exons
   d. gene amplification contributing to cancer
   e. the folding of DNA to form heterochromatin

9. Within a cell, the amount of protein made using a given mRNA molecule depends partly on
   a. the degree of DNA methylation.
   b. the rate at which the mRNA is degraded.
   c. the presence of certain transcription factors.
   d. the number of introns present in the mRNA.
   e. the types of ribosomes present in the cytoplasm.

10. Proto-oncogenes can change into oncogenes that cause cancer. Which of the following best explains the presence of these potential time bombs in eukaryotic cells?
   a. Proto-oncogenes first arose from viral infections.
   b. Proto-oncogenes normally help regulate cell division.
   c. Proto-oncogenes are genetic "junk."
   d. Proto-oncogenes are mutant versions of normal genes.
   e. Cells produce proto-oncogenes as they age.

11. **DRAW IT** The diagram below shows five genes (with their enhancers) from the genome of a certain species. Imagine that orange, blue, green, black, red, and purple activator proteins exist that can bind to the appropriately color-coded control elements in the enhancers of these genes.

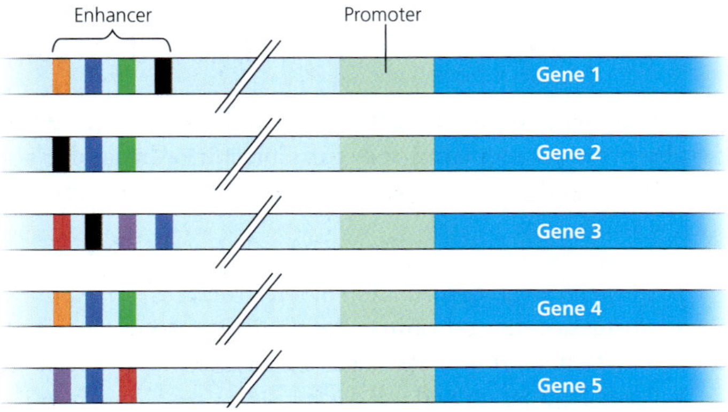

   a. Draw an X above enhancer elements (of all the genes) that would have activators bound in a cell in which only gene 5 is transcribed. Which colored activators would be present?
   b. Draw a dot above all enhancer elements that would have activators bound in a cell in which the green, blue, and orange activators are present. Which gene(s) would be transcribed?
   c. Imagine that genes 1, 2, and 4 code for nerve-specific proteins, and genes 3 and 5 are skin specific. Which activators would have to be present in each cell type to ensure transcription of the appropriate genes?

For Self-Quiz answers, see Appendix A.

MEDIA Visit the Study Area at **www.masteringbio.com** for a Practice Test.

EVOLUTION CONNECTION

12. DNA sequences can act as "tape measures of evolution" (see Chapter 5). Scientists analyzing the human genome sequence were surprised to find that some of the regions of the human genome that are most highly conserved (similar to comparable regions in other species) don't code for proteins. Propose a possible explanation for this observation.

SCIENTIFIC INQUIRY

13. Prostate cells usually require testosterone and other androgens to survive. But some prostate cancer cells thrive despite treatments that eliminate androgens. One hypothesis is that estrogen, often considered a female hormone, may be activating genes normally controlled by an androgen in these cancer cells. Describe one or more experiments to test this hypothesis. (See Figure 11.8 to review the action of these steroid hormones.)

Biological Inquiry: A Workbook of Investigative Cases Explore gene regulation by the hedgehog pathway with the case "Shh: Silencing the Hedgehog Pathway."

SCIENCE, TECHNOLOGY, AND SOCIETY

14. Trace amounts of dioxin were present in Agent Orange, a defoliant sprayed on vegetation during the Vietnam War. Animal tests suggest that dioxin can cause birth defects, cancer, liver and thymus damage, and immune system suppression, sometimes leading to death. But the animal tests are equivocal; a hamster is not affected by a dose that can kill a guinea pig. Dioxin acts somewhat like a steroid hormone, entering a cell and binding to a receptor protein that then attaches to the cell's DNA. How might this mechanism help explain the variety of dioxin's effects on different body systems and in different animals? How might you determine whether a type of illness is related to dioxin exposure? How might you determine whether a particular individual became ill as a result of exposure to dioxin? Which would be more difficult to demonstrate? Why?

Viruses 19

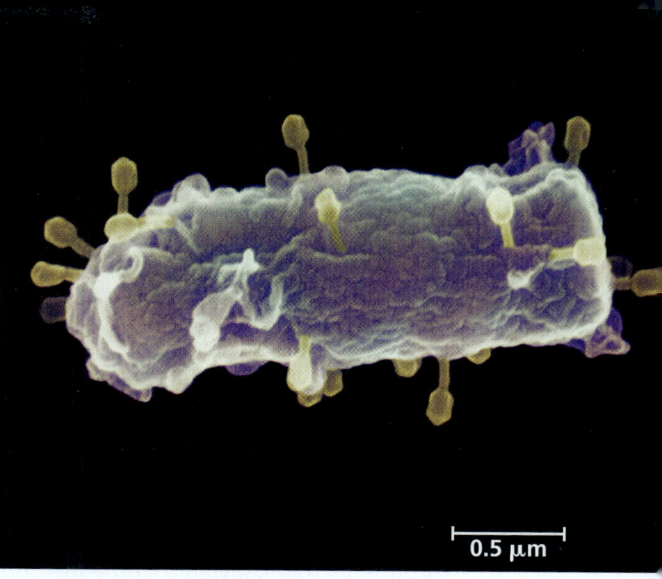

▲ **Figure 19.1 Are the tiny viruses infecting this *E. coli* cell alive?**

KEY CONCEPTS

19.1 A virus consists of a nucleic acid surrounded by a protein coat

19.2 Viruses reproduce only in host cells

19.3 Viruses, viroids, and prions are formidable pathogens in animals and plants

OVERVIEW

A Borrowed Life

The photo in **Figure 19.1** shows a remarkable event: the attack of a bacterial cell by numerous structures that resemble miniature lollipops. These structures, a type of virus called T4 bacteriophage, are seen infecting the bacterium *Escherichia coli* in this colorized SEM. By injecting its DNA into the cell, the virus sets in motion a genetic takeover of the bacterium, recruiting cellular machinery to mass-produce many new viruses.

Recall that bacteria and other prokaryotes are cells much smaller and more simply organized than those of eukaryotes, such as plants and animals. Viruses are smaller and simpler still. Lacking the structures and metabolic machinery found in cells, most viruses are little more than genes packaged in protein coats.

Are viruses living or nonliving? Early on, they were considered biological chemicals; in fact, the Latin root for the word *virus* means "poison." Because viruses are capable of causing a wide variety of diseases and can be spread between organisms, researchers in the late 1800s saw a parallel with bacteria and proposed that viruses were the simplest of living forms. However, viruses cannot reproduce or carry out metabolic activities outside of a host cell. Most biologists studying viruses today would probably agree that they are not alive but exist in a shady area between life-forms and chemicals. The simple phrase used recently by two researchers describes them aptly enough: Viruses lead "a kind of borrowed life."

To a large extent, molecular biology was born in the laboratories of biologists studying viruses that infect bacteria. Experiments with viruses provided important evidence that genes are made of nucleic acids, and they were critical in working out the molecular mechanisms of the fundamental processes of DNA replication, transcription, and translation.

Beyond their value as experimental systems, viruses have unique genetic mechanisms that are interesting in their own right and that also help us understand how viruses cause disease. In addition, the study of viruses has led to the development of techniques that enable scientists to manipulate genes and transfer them from one organism to another. These techniques play an important role in basic research, biotechnology, and medical applications. For instance, viruses are used as agents of gene transfer in gene therapy (see Chapter 20).

In this chapter, we will explore the biology of viruses. We will begin with the structure of these simplest of all genetic systems and then describe their reproductive cycles. Next, we will discuss the role of viruses as disease-causing agents, or pathogens, and conclude by considering some even simpler infectious agents, viroids and prions.

CONCEPT **19.1**

A virus consists of a nucleic acid surrounded by a protein coat

Scientists were able to detect viruses indirectly long before they were actually able to see them. The story of how viruses were discovered begins near the end of the 19th century.

The Discovery of Viruses: *Scientific Inquiry*

Tobacco mosaic disease stunts the growth of tobacco plants and gives their leaves a mottled, or mosaic, coloration. In 1883, Adolf Mayer, a German scientist, discovered that he could transmit the disease from plant to plant by rubbing sap extracted from diseased leaves onto healthy plants. After an unsuccessful search for an infectious microbe in the sap, Mayer suggested that the disease was caused by unusually small bacteria that were invisible under a microscope. This hypothesis was tested a decade later by Dimitri Ivanowsky, a Russian biologist who passed sap from infected tobacco leaves through a filter designed to remove bacteria. After filtration, the sap still produced mosaic disease.

But Ivanowsky clung to the hypothesis that bacteria caused tobacco mosaic disease. Perhaps, he reasoned, the bacteria were small enough to pass through the filter or made a toxin that could do so. The second possibility was ruled out when the Dutch botanist Martinus Beijerinck carried out a classic series of experiments that showed that the infectious agent in the filtered sap could reproduce **(Figure 19.2)**.

In fact, the pathogen reproduced only within the host it infected. In further experiments, Beijerinck showed that unlike bacteria used in the lab at that time, the mysterious agent of mosaic disease could not be cultivated on nutrient media in test tubes or petri dishes. Beijerinck imagined a reproducing particle much smaller and simpler than a bacterium, and he is generally credited with being the first scientist to voice the concept of a virus. His suspicions were confirmed in 1935 when the American scientist Wendell Stanley crystallized the infectious particle, now known as tobacco mosaic virus (TMV). Subsequently, TMV and many other viruses were actually seen with the help of the electron microscope.

▼ **Figure 19.2 Inquiry**

What causes tobacco mosaic disease?

EXPERIMENT In the late 1800s, Martinus Beijerinck, of the Technical School in Delft, the Netherlands, investigated the properties of the agent that causes tobacco mosaic disease (then called spot disease).

RESULTS When the filtered sap was rubbed on healthy plants, they became infected. Their sap, when extracted and filtered, could then act as the source of infection for another group of plants. Each successive group of plants developed the disease to the same extent as earlier groups.

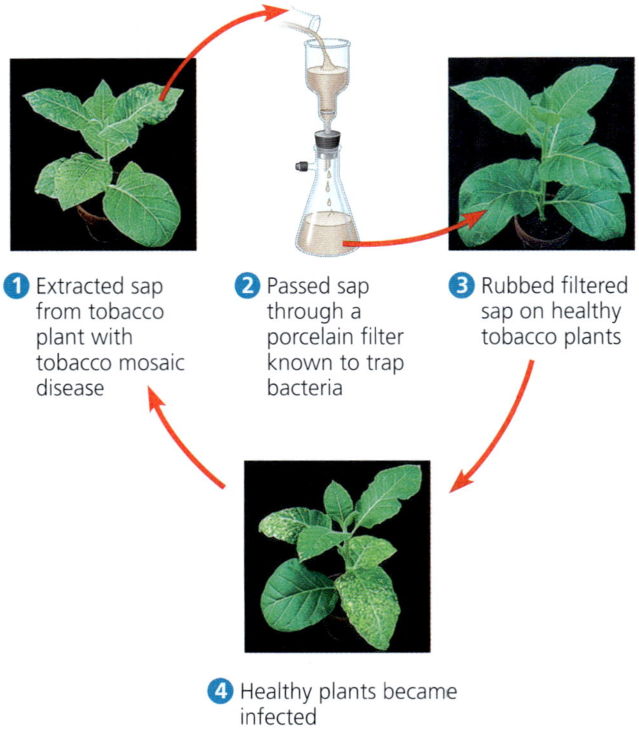

CONCLUSION The infectious agent was apparently not a bacterium because it could pass through a bacterium-trapping filter. The pathogen must have been reproducing in the plants because its ability to cause disease was undiluted after several transfers from plant to plant.

SOURCE M. J. Beijerinck, Concerning a *contagium vivum fluidum* as cause of the spot disease of tobacco leaves, *Verhandelingen der Koninkyke akademie Wettenschappen te Amsterdam* 65:3–21 (1898). Translation published in English as Phytopathological Classics Number 7 (1942), American Phytopathological Society Press, St. Paul, MN.

WHAT IF? If Beijerinck had observed that the infection of each group was weaker than that of the previous group and that ultimately the sap could no longer cause disease, what might he have concluded?

Structure of Viruses

The tiniest viruses are only 20 nm in diameter—smaller than a ribosome. Millions could easily fit on a pinhead. Even the largest known virus, which has a diameter of several hundred nanometers, is barely visible in the light microscope. Stanley's discovery that some viruses could be crystallized was exciting and puzzling news. Not even the simplest of cells can aggregate into regular crystals. But if viruses are not cells, then what are they? Examining the structure of viruses more closely reveals that they are infectious particles consisting of nucleic acid enclosed in a protein coat and, in some cases, a membranous envelope.

Viral Genomes

We usually think of genes as being made of double-stranded DNA—the conventional double helix—but many viruses defy this convention. Their genomes may consist of double-stranded DNA, single-stranded DNA, double-stranded RNA, or single-stranded RNA, depending on the kind of virus. A virus is called a DNA virus or an RNA virus, according to the kind of nucleic acid that makes up its genome. In either case, the genome is usually organized as a single linear or circular molecule of nucleic acid, although the genomes of some viruses consist of multiple molecules of nucleic acid. The smallest viruses known have only four genes in their genome, while the largest have several hundred to a thousand. For comparison, bacterial genomes contain about 200 to a few thousand genes.

Capsids and Envelopes

The protein shell enclosing the viral genome is called a **capsid**. Depending on the type of virus, the capsid may be rod-shaped, polyhedral, or more complex in shape (like T4). Capsids are built from a large number of protein subunits called *capsomeres*, but the number of different *kinds* of proteins in a capsid is usually small. Tobacco mosaic virus has a rigid, rod-shaped capsid made from over a thousand molecules of a single type of protein arranged in a helix; rod-shaped viruses are commonly called *helical viruses* for this reason **(Figure 19.3a)**. Adenoviruses, which infect the respiratory tracts of animals, have 252 identical protein molecules arranged in a polyhedral capsid with 20 triangular facets—an icosahedron; thus, these and other similarly shaped viruses are referred to as *icosahedral viruses* **(Figure 19.3b)**.

Some viruses have accessory structures that help them infect their hosts. For instance, a membranous envelope surrounds the capsids of influenza viruses and many other viruses found in animals **(Figure 19.3c)**. These **viral envelopes**, which are derived from the membranes of the host cell, contain host cell phospholipids and membrane proteins. They also contain proteins and glycoproteins of viral origin. (Glycoproteins are proteins with carbohydrates covalently attached.) Some viruses carry a few viral enzyme molecules within their capsids.

Many of the most complex capsids are found among the viruses that infect bacteria, called **bacteriophages**, or simply **phages**. The first phages studied included seven that infect *E. coli*. These seven phages were named type 1 (T1), type 2 (T2), and so forth, in the order of their discovery. The three T-even phages (T2, T4, and T6) turned out to be very similar in structure. Their capsids have elongated icosahedral heads enclosing their DNA. Attached to the head is a protein tail piece with fibers by which the phages attach to a bacterium **(Figure 19.3d)**. In the next section, we'll examine how these few viral parts function together with cellular components to produce large numbers of viral progeny.

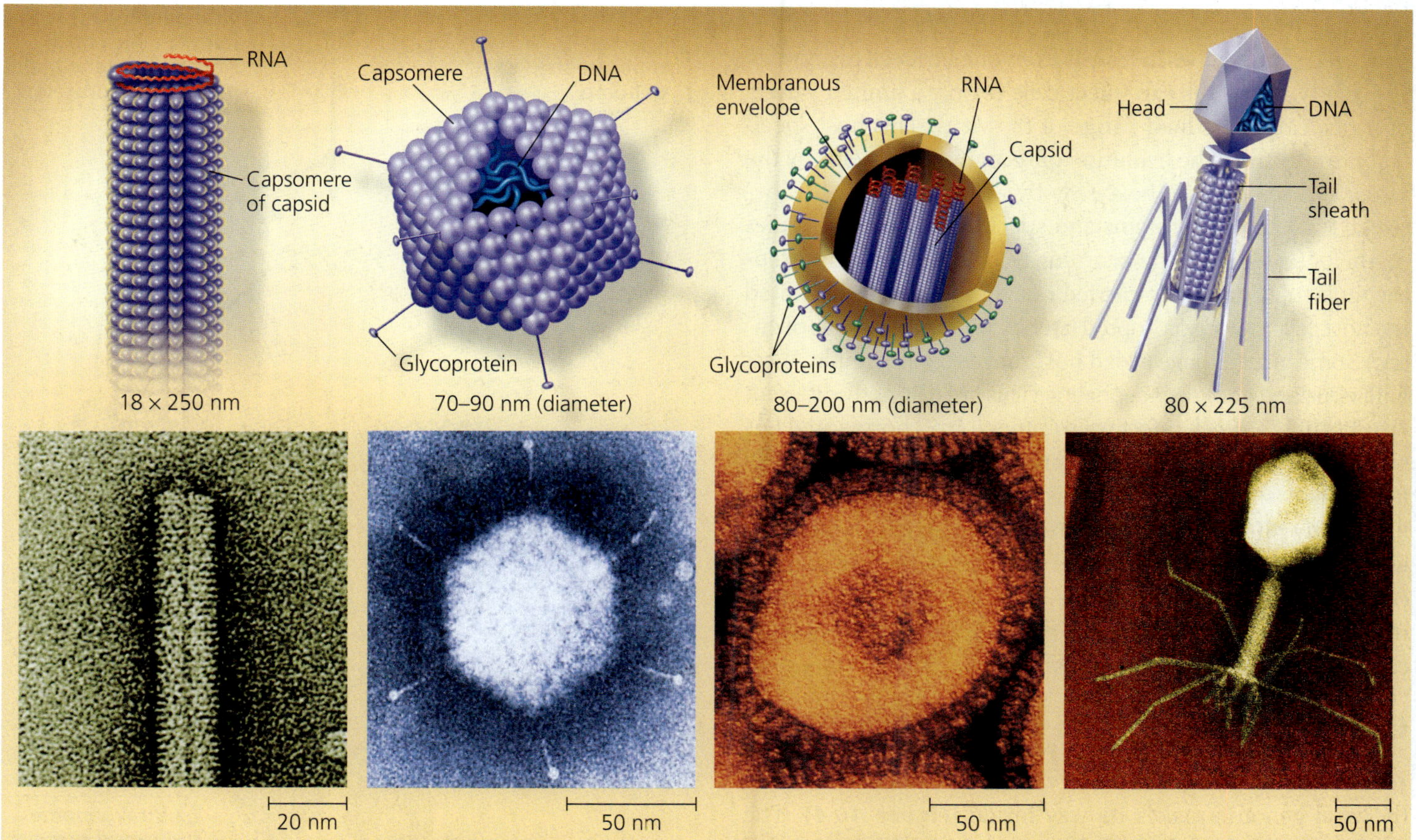

(a) Tobacco mosaic virus has a helical capsid with the overall shape of a rigid rod.

(b) Adenoviruses have an icosahedral capsid with a glycoprotein spike at each vertex.

(c) Influenza viruses have an outer envelope studded with glycoprotein spikes. The genome consists of eight different RNA molecules, each wrapped in a helical capsid.

(d) Bacteriophage T4, like other "T-even" phages, has a complex capsid consisting of an icosahedral head and a tail apparatus.

▲ **Figure 19.3 Viral structure.** Viruses are made up of nucleic acid (DNA or RNA) enclosed in a protein coat (the capsid) and sometimes further wrapped in a membranous envelope. The individual protein subunits making up the capsid are called capsomeres. Although diverse in size and shape, viruses have common structural features, most of which appear in the four examples shown here. (All the micrographs are colorized TEMs.)

CONCEPT CHECK 19.1

1. Compare the structures of tobacco mosaic virus (TMV) and influenza virus (see Figure 19.3).
2. WHAT IF? In 2005, scientists discovered a virus that could, under certain conditions, develop pointed projections at each end while outside a host cell. How does this observation fit with the characterization of viruses as nonliving?

For suggested answers, see Appendix A.

CONCEPT 19.2 Viruses reproduce only in host cells

Viruses lack metabolic enzymes and equipment for making proteins, such as ribosomes. They are obligate intracellular parasites; in other words, they can reproduce only within a host cell. It is fair to say that viruses in isolation are merely packaged sets of genes in transit from one host cell to another.

Each type of virus can infect cells of only a limited variety of hosts, called the **host range** of the virus. This host specificity results from the evolution of recognition systems by the virus. Viruses identify host cells by a "lock-and-key" fit between viral surface proteins and specific receptor molecules on the outside of cells. (According to one model, such receptor molecules originally carried out functions that benefited the host cell but were co-opted later by viruses as portals of entry.) Some viruses have broad host ranges. For example, West Nile virus and equine encephalitis virus are distinctly different viruses that can each infect mosquitoes, birds, horses, and humans. Other viruses have host ranges so narrow that they infect only a single species. Measles virus, for instance, can infect only humans. Furthermore, viral infection of multicellular eukaryotes is usually limited to particular tissues. Human cold viruses infect only the cells lining the upper respiratory tract, and the AIDS virus binds to receptors present only on certain types of white blood cells.

General Features of Viral Reproductive Cycles

A viral infection begins when a virus binds to a host cell and the viral genome makes its way inside **(Figure 19.4)**. The mechanism of genome entry depends on the type of virus and the type of host cell. For example, T-even phages use their elaborate tail apparatus to inject DNA into a bacterium (see Figure 19.3d). Other viruses are taken up by endocytosis or, in the case of enveloped viruses, by fusion of the viral envelope with the plasma membrane. Once the viral genome is inside, the proteins it encodes can commandeer the host, reprogramming the cell to copy the viral nucleic acid and manufacture viral proteins. The host provides the nucleotides for making viral nucleic acids, as well as enzymes, ribosomes, tRNAs, amino acids, ATP, and other components needed for making the viral proteins. Most DNA viruses use the DNA polymerases of the host cell to synthesize new genomes along the templates provided by the viral DNA. In contrast, to replicate their genomes, RNA viruses use virally encoded polymerases that can use RNA as a template. (Uninfected cells generally make no enzymes for carrying out this process.)

After the viral nucleic acid molecules and capsomeres are produced, they spontaneously self-assemble into new viruses. In fact, researchers can separate the RNA and capsomeres of TMV and then reassemble complete viruses simply by mixing the components together under the right conditions. The simplest type of viral reproductive cycle ends with the exit of

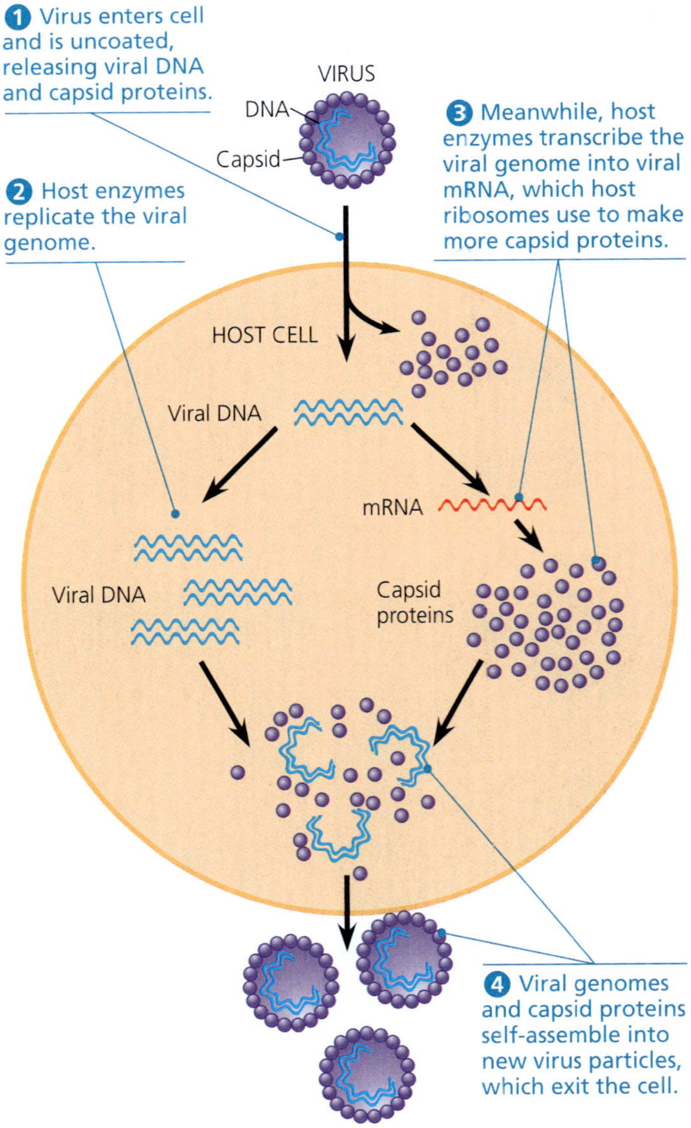

▲ **Figure 19.4 A simplified viral reproductive cycle.** A virus is an obligate intracellular parasite that uses the equipment and small molecules of its host cell to reproduce. In this simplest of viral cycles, the parasite is a DNA virus with a capsid consisting of a single type of protein.

? *Label each of the straight black arrows with one word representing the name of the process that is occurring.*

hundreds or thousands of viruses from the infected host cell, a process that often damages or destroys the cell. Such cellular damage and death, as well as the body's responses to this destruction, cause many of the symptoms associated with viral infections. The viral progeny that exit a cell have the potential to infect additional cells, spreading the viral infection.

There are many variations on the simplified viral reproductive cycle we have traced in this general description. We will now take a closer look at some of these variations in bacterial viruses (phages) and animal viruses; later in the chapter, we will consider plant viruses.

Reproductive Cycles of Phages

Phages are the best understood of all viruses, although some of them are also among the most complex. Research on phages led to the discovery that some double-stranded DNA viruses can reproduce by two alternative mechanisms: the lytic cycle and the lysogenic cycle.

The Lytic Cycle

A phage reproductive cycle that culminates in death of the host cell is known as a **lytic cycle**. The term refers to the last stage of infection, during which the bacterium lyses (breaks open) and releases the phages that were produced within the cell. Each of these phages can then infect a healthy cell, and a few successive lytic cycles can destroy an entire bacterial population in just a few hours. A phage that reproduces only by a lytic cycle is a **virulent phage**. **Figure 19.5** illustrates the major steps in the lytic cycle of T4, a typical virulent phage. The figure and legend describe the process, which you should study before proceeding.

After reading about the lytic cycle, you may wonder why phages haven't exterminated all bacteria. In fact, phage treatments have been used medically in some countries to help control bacterial infections in patients. Solutions containing bacteriophages have also been sprayed on chicken carcasses, significantly reducing bacterial contamination of poultry on the way to the marketplace. Bacteria are not defenseless, however. First, natural selection favors bacterial mutants with receptors that are no longer recognized by a particular type of phage. Second, when phage DNA successfully enters a bacterium, the DNA often is identified as foreign and cut up by cellular enzymes called **restriction enzymes**, which are so named because their activity *restricts* the ability of the phage to infect the bacterium. The bacterial cell's own DNA is methylated in a

▶ **Figure 19.5 The lytic cycle of phage T4, a virulent phage.** Phage T4 has almost 300 genes, which are transcribed and translated using the host cell's machinery. One of the first phage genes translated after the viral DNA enters the host cell codes for an enzyme that degrades the host cell's DNA (step 2); the phage DNA is protected from breakdown because it contains a modified form of cytosine that is not recognized by the enzyme. The entire lytic cycle, from the phage's first contact with the cell surface to cell lysis, takes only 20–30 minutes at 37°C.

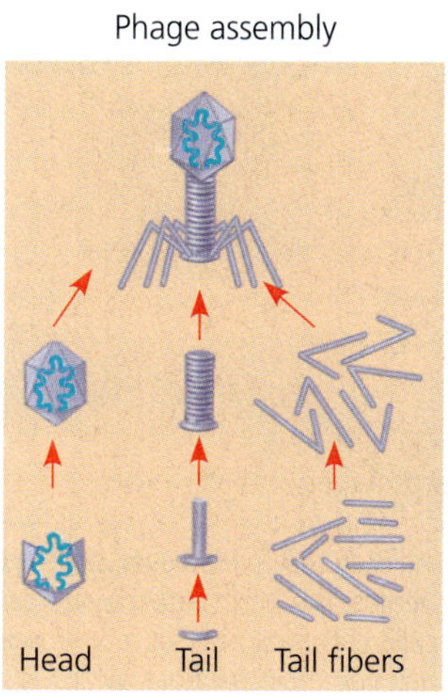

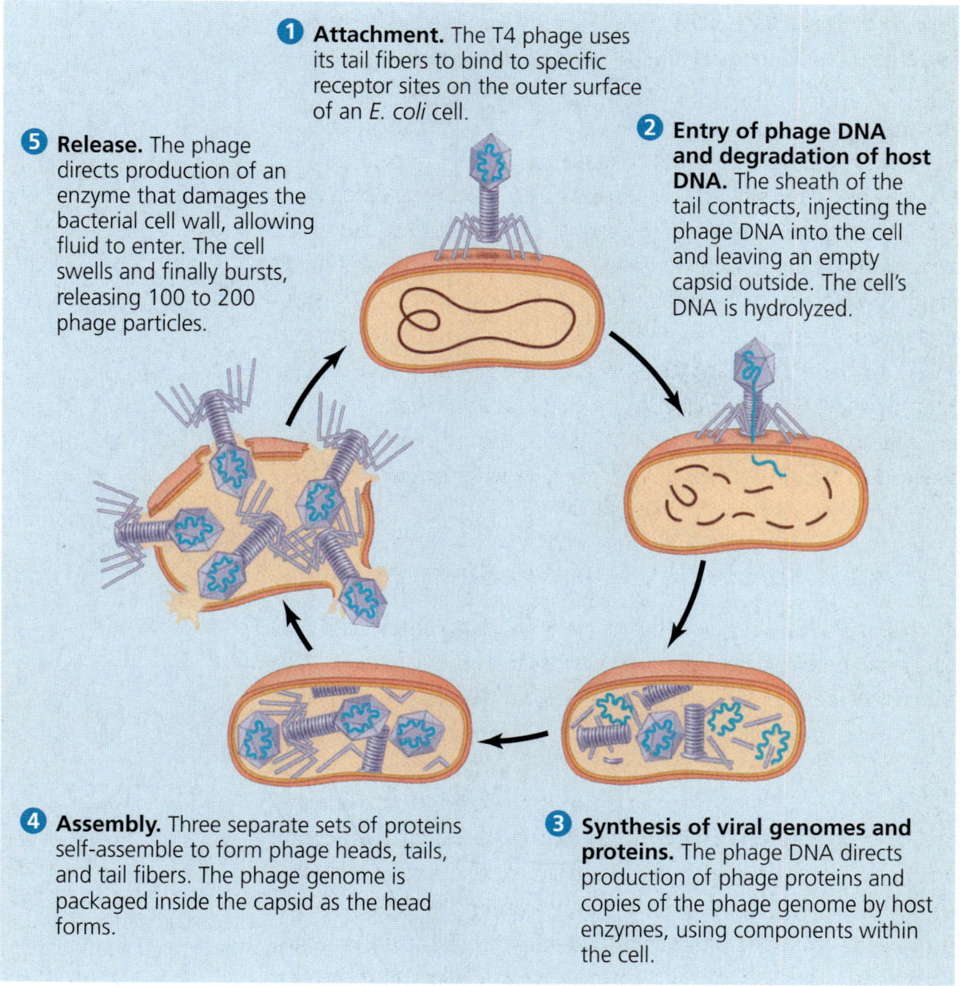

way that prevents attack by its own restriction enzymes. But just as natural selection favors bacteria with mutant receptors or effective restriction enzymes, it also favors phage mutants that can bind the altered receptors or are resistant to particular restriction enzymes. Thus, the parasite-host relationship is in constant evolutionary flux.

There is yet a third important reason bacteria have been spared from extinction as a result of phage activity. Instead of lysing their host cells, many phages coexist with them in a state called lysogeny, which we'll now discuss.

The Lysogenic Cycle

In contrast to the lytic cycle, which kills the host cell, the **lysogenic cycle** allows replication of the phage genome without destroying the host. Phages capable of using both modes of reproducing within a bacterium are called **temperate phages**. A temperate phage called lambda, written with the Greek letter λ, is widely used in biological research. Phage λ resembles T4, but its tail has only one, short tail fiber.

Infection of an *E. coli* cell by phage λ begins when the phage binds to the surface of the cell and injects its linear DNA genome **(Figure 19.6)**. Within the host, the λ DNA molecule forms a circle. What happens next depends on the reproductive mode: lytic cycle or lysogenic cycle. During a lytic cycle, the viral genes immediately turn the host cell into a λ-producing factory, and the cell soon lyses and releases its viral products. During a lysogenic cycle, however, the λ DNA molecule is incorporated into a specific site on the *E. coli* chromosome by viral proteins that break both circular DNA molecules and join them to each other. When integrated into the bacterial chromosome in this way, the viral DNA is known as a **prophage**. One prophage gene codes for a protein that prevents transcription of most of the other prophage genes. Thus, the phage genome is mostly silent within the bacterium. Every time the *E. coli* cell prepares to divide, it replicates the phage DNA along with its own and passes the copies on to daughter cells. A single infected cell can quickly give rise to a large population of bacteria carrying the virus in prophage form. This mechanism enables viruses to propagate without killing the host cells on which they depend.

The term *lysogenic* implies that prophages are capable of generating active phages that lyse their host cells. This occurs when the λ genome is induced to exit the bacterial chromosome and initiate a lytic cycle. An environmental signal, such as a certain chemical or high-energy radiation, usually triggers the switchover from the lysogenic to the lytic mode.

In addition to the gene for the transcription-preventing protein, a few other prophage genes may be expressed during lysogeny. Expression of these genes may alter the host's phenotype, a phenomenon that can have important medical

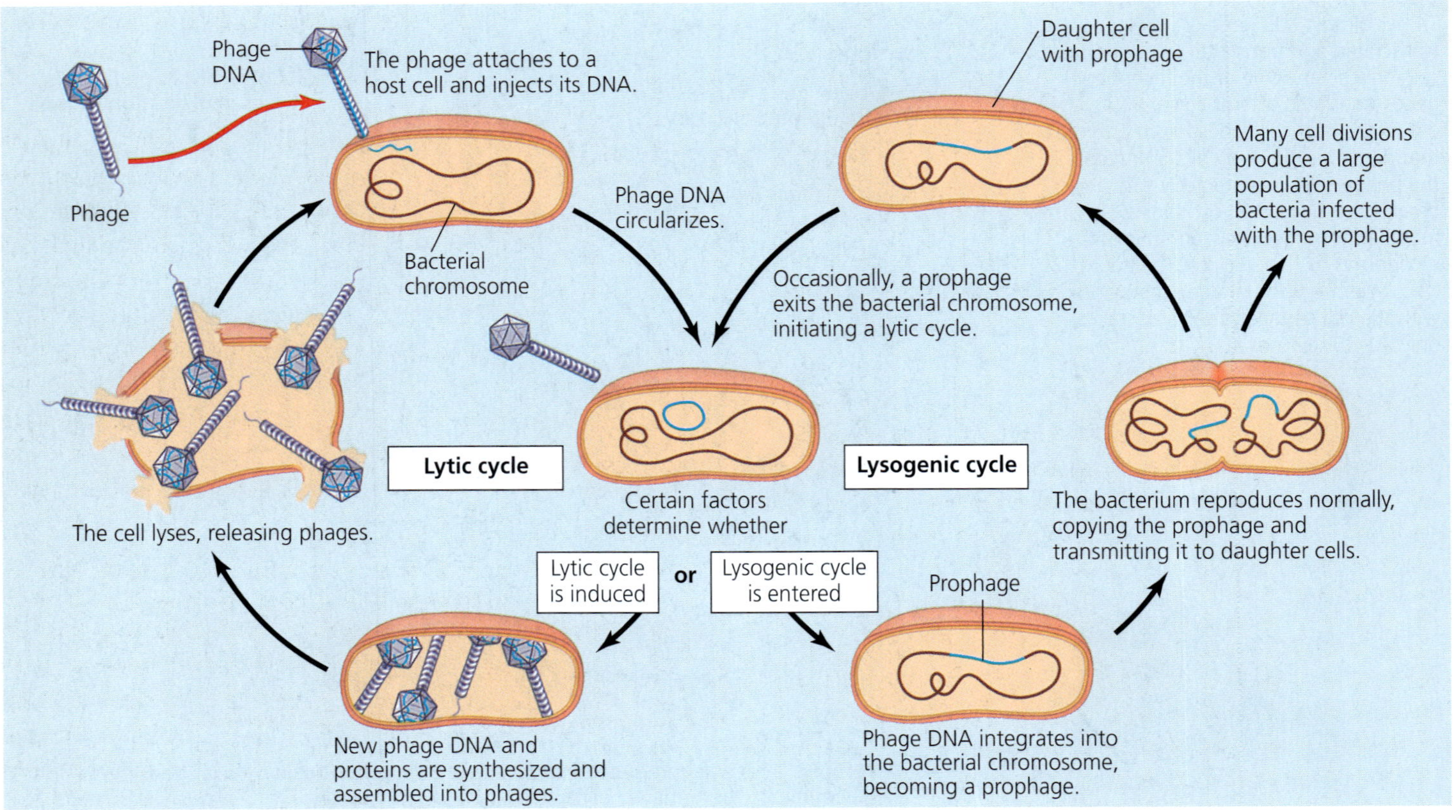

▲ **Figure 19.6 The lytic and lysogenic cycles of phage λ, a temperate phage.** After entering the bacterial cell and circularizing, the λ DNA can immediately initiate the production of a large number of progeny phages (lytic cycle) or integrate into the bacterial chromosome (lysogenic cycle). In most cases, phage λ follows the lytic pathway, which is similar to that detailed in Figure 19.5. However, once a lysogenic cycle begins, the prophage may be carried in the host cell's chromosome for many generations. Phage λ has one main tail fiber, which is short.

Table 19.1 Classes of Animal Viruses

Class/Family	Envelope	Examples/Disease
I. Double-stranded DNA (dsDNA)		
Adenovirus (see Figure 19.3b)	No	Respiratory diseases; tumors
Papovavirus	No	Papillomavirus (warts, cervical cancer); polyomavirus (tumors)
Herpesvirus	Yes	Herpes simplex I and II (cold sores, genital sores); varicella zoster (shingles, chicken pox); Epstein-Barr virus (mononucleosis, Burkitt's lymphoma)
Poxvirus	Yes	Smallpox virus; cowpox virus
II. Single-stranded DNA (ssDNA)		
Parvovirus	No	B19 parvovirus (mild rash)
III. Double-stranded RNA (dsRNA)		
Reovirus	No	Rotavirus (diarrhea); Colorado tick fever virus
IV. Single-stranded RNA (ssRNA); serves as mRNA		
Picornavirus	No	Rhinovirus (common cold); poliovirus, hepatitis A virus, and other enteric (intestinal) viruses
Coronavirus	Yes	Severe acute respiratory syndrome (SARS)
Flavivirus	Yes	Yellow fever virus; West Nile virus; hepatitis C virus
Togavirus	Yes	Rubella virus; equine encephalitis viruses
V. ssRNA; template for mRNA synthesis		
Filovirus	Yes	Ebola virus (hemorrhagic fever)
Orthomyxovirus (see Figures 19.3c and 19.9b)	Yes	Influenza virus
Paramyxovirus	Yes	Measles virus; mumps virus
Rhabdovirus	Yes	Rabies virus
VI. ssRNA; template for DNA synthesis		
Retrovirus (see Figure 19.8)	Yes	HIV, human immunodeficiency virus (AIDS); RNA tumor viruses (leukemia)

significance. For example, the three species of bacteria that cause the human diseases diphtheria, botulism, and scarlet fever would not be so harmful to humans without certain prophage genes that cause the host bacteria to make toxins. And the difference between the *E. coli* strain that resides in our intestines and the O157:H7 strain that has caused several deaths by food poisoning appears to be the presence of prophages in the O157:H7 strain.

Reproductive Cycles of Animal Viruses

Everyone has suffered from viral infections, whether cold sores, influenza, or the common cold. Like all viruses, those that cause illness in humans and other animals can reproduce only inside host cells. Many variations on the basic scheme of viral infection and reproduction are represented among the animal viruses. One key variable is the nature of the viral genome: Is it composed of DNA or RNA? Is it double-stranded or single-stranded? The nature of the genome is the basis for the common classification of viruses shown in **Table 19.1**. Single-stranded RNA viruses are further classified into three classes (IV–VI) according to how the RNA genome functions in a host cell.

Whereas few bacteriophages have an envelope or RNA genome, many animal viruses have both. In fact, nearly all animal viruses with RNA genomes have an envelope, as do some with DNA genomes (see Table 19.1). Rather than consider all the mechanisms of viral infection and reproduction, we will focus on the roles of viral envelopes and on the functioning of RNA as the genetic material of many animal viruses.

Viral Envelopes

An animal virus equipped with an envelope—that is, an outer membrane—uses it to enter the host cell. Protruding from the outer surface of this envelope are viral glycoproteins that bind to specific receptor molecules on the surface of a host cell. **Figure 19.7**, on the next page, outlines the events in the reproductive cycle of an enveloped virus with an RNA genome. The protein parts of envelope glycoproteins are made by ribosomes bound to the endoplasmic reticulum (ER) of the host cell; cellular enzymes in the ER and Golgi apparatus then add the sugars. The resulting viral glycoproteins, embedded in host cell-derived membrane, are transported to the cell surface. In a process much like exocytosis, new viral capsids are wrapped in membrane as they bud from the cell. In other words, the viral envelope is derived from the host cell's plasma membrane, although some of the molecules of this membrane are specified by viral genes. The enveloped viruses are now free to infect other cells. This reproductive cycle does not necessarily kill the host cell, in contrast to the lytic cycles of phages.

Some viruses have envelopes that are not derived from plasma membrane. Herpesviruses, for example, are temporarily cloaked in membrane derived from the nuclear envelope of the host; they then shed this membrane in the cytoplasm and acquire a new envelope made from membrane of the Golgi apparatus. These

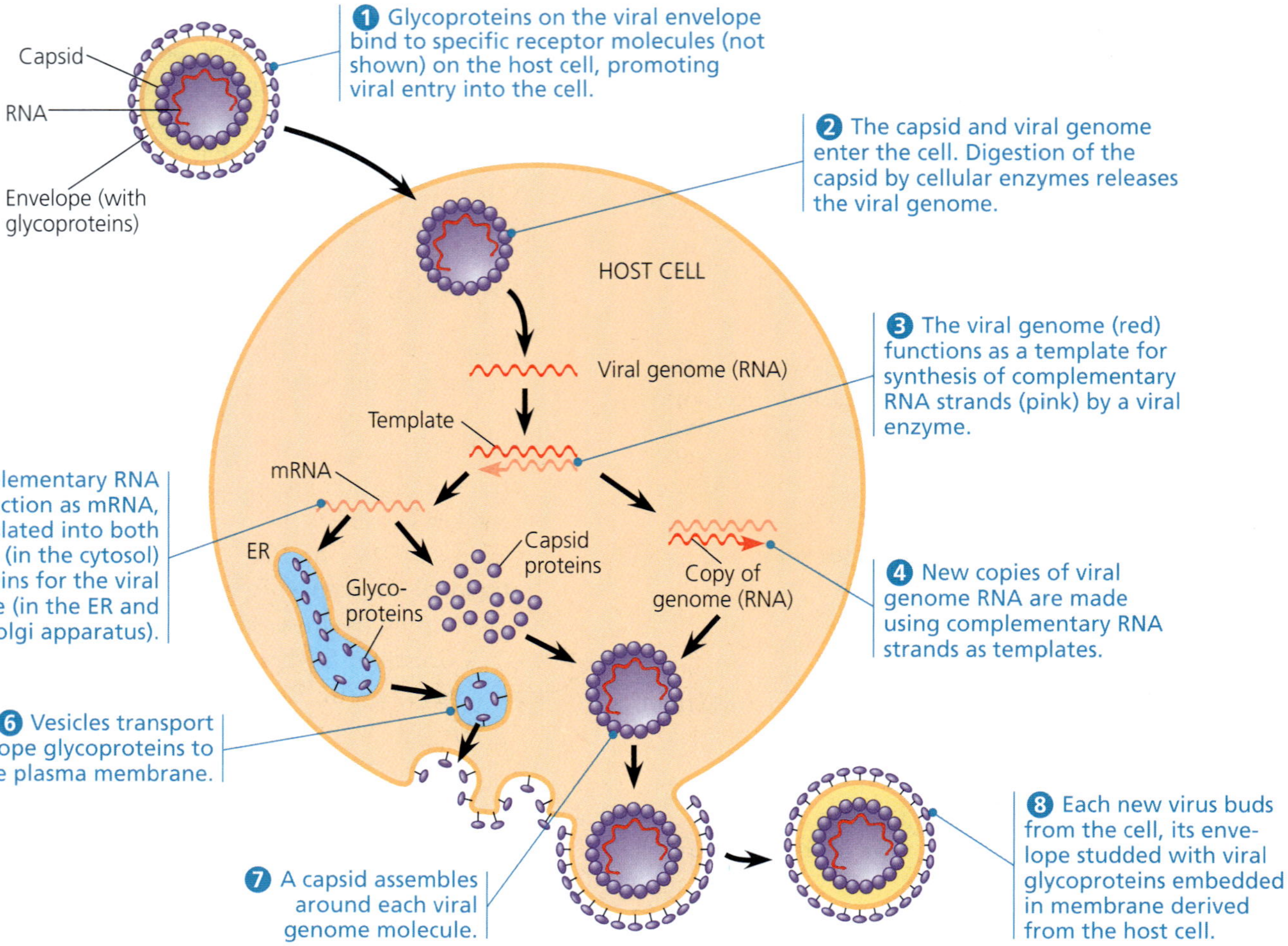

▲ **Figure 19.7 The reproductive cycle of an enveloped RNA virus.** Shown here is a virus with a single-stranded RNA genome that functions as a template for synthesis of mRNA. Some enveloped viruses enter the host cell by fusion of the envelope with the cell's plasma membrane; others enter by endocytosis. For all enveloped RNA viruses, the formation of new envelopes for progeny viruses occurs by the mechanism depicted in this figure.

? *Name a virus that has infected you and has a reproductive cycle matching this one. (Hint: See Table 19.1.)*

viruses have a double-stranded DNA genome and reproduce within the host cell nucleus, using a combination of viral and cellular enzymes to replicate and transcribe their DNA. In the case of herpesviruses, copies of the viral DNA can remain behind as mini-chromosomes in the nuclei of certain nerve cells. There they remain latent until some sort of physical or emotional stress triggers a new round of active virus production. The infection of other cells by these new viruses causes the blisters characteristic of herpes, such as cold sores or genital sores. Once someone acquires a herpesvirus infection, flare-ups may recur throughout the person's life.

RNA as Viral Genetic Material

Although some phages and most plant viruses are RNA viruses, the broadest variety of RNA genomes is found among the viruses that infect animals. Among the three types of single-stranded RNA genomes found in animal viruses, the genome of class IV viruses can directly serve as mRNA and thus can be translated into viral protein immediately after infection. Figure 19.7 shows a virus of class V, in which the RNA genome serves as a *template* for mRNA synthesis. The RNA genome is transcribed into complementary RNA strands, which function both as mRNA and as templates for the synthesis of additional copies of genomic RNA. All viruses that require RNA → RNA synthesis to make mRNA use a viral enzyme capable of carrying out this process; there are no such enzymes in most cells. The viral enzyme is packaged with the genome inside the viral capsid.

The RNA animal viruses with the most complicated reproductive cycles are the **retroviruses** (class VI). These viruses are equipped with an enzyme called **reverse transcriptase**, which transcribes an RNA template into DNA, providing an RNA → DNA information flow, the opposite of the usual direction. This unusual phenomenon is the source of the name retroviruses (*retro* means "backward"). Of particular medical importance is **HIV (human immunodeficiency virus)**, the retrovirus that causes **AIDS (acquired immunodeficiency syndrome)**. HIV and other retroviruses are enveloped viruses that contain two identical molecules of single-stranded RNA and two molecules of reverse transcriptase.

▼ Figure 19.8 The reproductive cycle of HIV, the retrovirus that causes AIDS. Note in step 4 that DNA synthesized from the viral RNA genome is integrated into the host cell chromosomal DNA, a characteristic unique to retroviruses. The photos on the left (artificially colored TEMs) show HIV entering and leaving a human white blood cell.

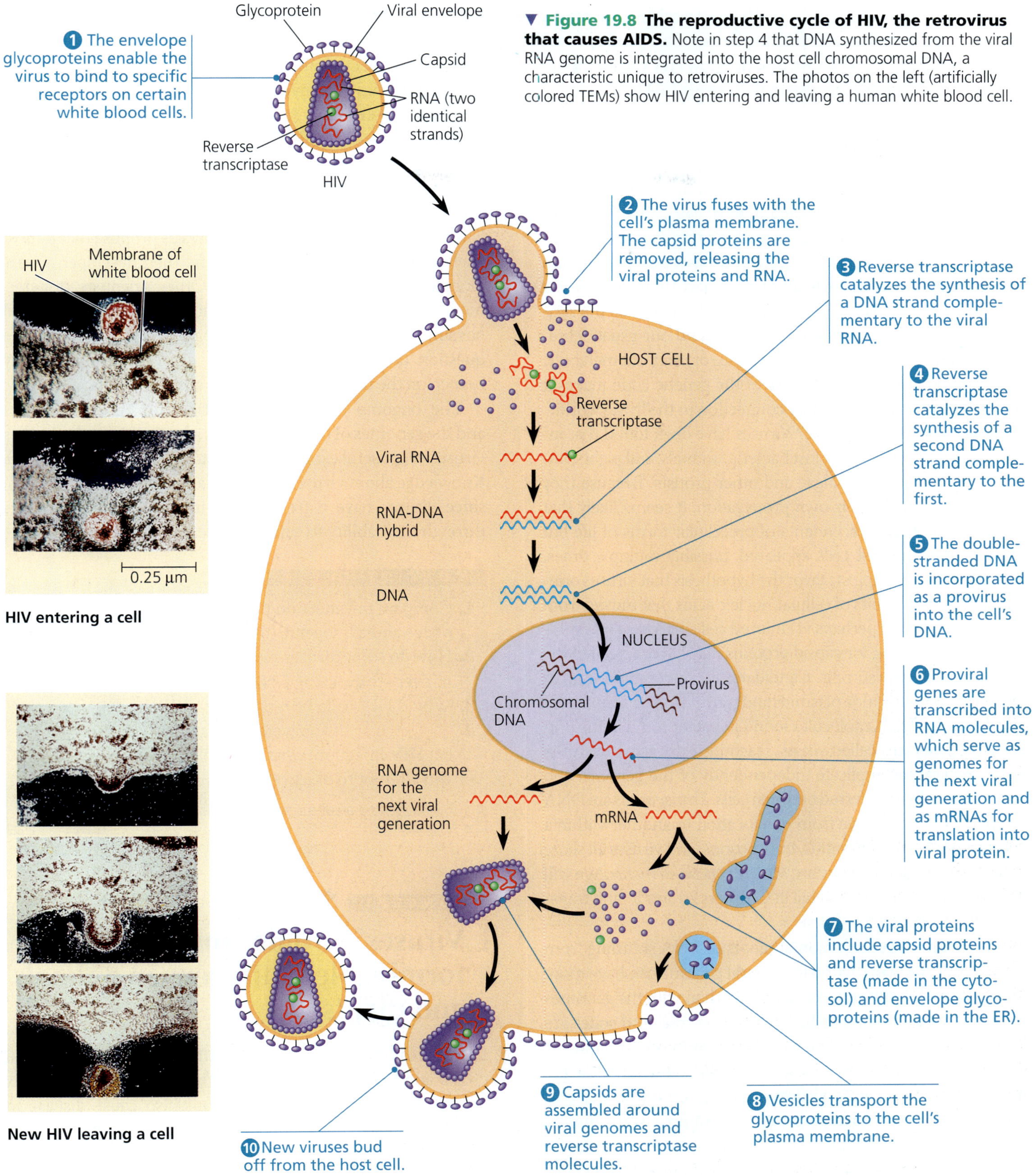

Figure 19.8 traces the HIV reproductive cycle, which is typical of a retrovirus. After HIV enters a host cell, its reverse transcriptase molecules are released into the cytoplasm, where they catalyze synthesis of viral DNA. The newly made viral DNA then enters the cell's nucleus and integrates into the DNA of a chromosome. The integrated viral DNA, called a **provirus**, never leaves the host's genome, remaining a permanent resident of the cell. (Recall that a prophage, in contrast, leaves the host's genome at the start of a lytic cycle.) The host's RNA polymerase transcribes the proviral DNA into RNA molecules, which can

function both as mRNA for the synthesis of viral proteins and as genomes for the new viruses that will be assembled and released from the cell. In Chapter 43, we describe how HIV causes the deterioration of the immune system that occurs in AIDS.

Evolution of Viruses

We began this chapter by asking whether or not viruses are alive. Viruses do not really fit our definition of living organisms. An isolated virus is biologically inert, unable to replicate its genes or regenerate its own supply of ATP. Yet it has a genetic program written in the universal language of life. Do we think of viruses as nature's most complex associations of molecules or as the simplest forms of life? Either way, we must bend our usual definitions. Although viruses cannot reproduce or carry out metabolic activities independently, their use of the genetic code makes it hard to deny their evolutionary connection to the living world.

How did viruses originate? Viruses have been found that infect every form of life—not just bacteria, animals, and plants, but also archaea, fungi, and algae and other protists. Because they depend on cells for their own propagation, it seems likely that viruses are not the descendants of precellular forms of life but evolved *after* the first cells appeared, possibly multiple times. Most molecular biologists favor the hypothesis that viruses originated from naked bits of cellular nucleic acids that moved from one cell to another, perhaps via injured cell surfaces. The evolution of genes coding for capsid proteins may have facilitated the infection of uninjured cells. Candidates for the original sources of viral genomes include plasmids and transposons. *Plasmids* are small, circular DNA molecules found in bacteria and in the unicellular eukaryotes called yeasts. Plasmids exist apart from the cell's genome, can replicate independently of the genome, and are occasionally transferred between cells. *Transposons* are DNA segments that can move from one location to another within a cell's genome. Thus, plasmids, transposons, and viruses all share an important feature: They are *mobile genetic elements.* We will discuss plasmids in more detail in Chapters 20 and 27 and transposons in Chapter 21.

Consistent with this vision of pieces of DNA shuttling from cell to cell is the observation that a viral genome can have more in common with the genome of its host than with the genomes of viruses that infect other hosts. Indeed, some viral genes are essentially identical to genes of the host. On the other hand, recent sequencing of many viral genomes has shown that the genetic sequences of some viruses are quite similar to those of seemingly distantly related viruses; for example, some animal viruses share similar sequences with plant viruses. This genetic similarity may reflect the persistence of groups of viral genes that were favored by natural selection during the early evolution of viruses and the eukaryotic cells that served as their hosts.

The debate about the origin of viruses has been reinvigorated recently by reports of mimivirus, the largest virus yet discovered. Mimivirus is a double-stranded DNA virus with an icosahedral capsid that is 400 nm in diameter. (The beginning of its name is short for *mi*micking *mi*crobe because the virus is the size of a small bacterium.) Its genome contains 1.2 million bases (about 100 times as many as the influenza virus genome) and an estimated 1,000 genes. Perhaps the most surprising aspect of mimivirus, however, is that some of the genes appear to code for products previously thought to be hallmarks of cellular genomes. These products include proteins involved in translation, DNA repair, protein folding, and polysaccharide synthesis. The researchers who described mimivirus propose that it most likely evolved *before* the first cells and then developed an exploitative relationship with them. Other scientists disagree, maintaining that the virus evolved more recently than cells and has simply been efficient at scavenging genes from its hosts. The question of whether some viruses deserve their own early branch on the tree of life may not be answered for some time.

The ongoing evolutionary relationship between viruses and the genomes of their host cells is an association that makes viruses very useful experimental systems in molecular biology. Knowledge about viruses also has many practical applications, since viruses have a tremendous impact on all organisms through their ability to cause disease.

CONCEPT CHECK 19.2

1. Compare the effect on the host cell of a lytic (virulent) phage and a lysogenic (temperate) phage.
2. How do some viruses reproduce without possessing or ever synthesizing DNA?
3. Why is HIV called a retrovirus?
4. **WHAT IF?** If you were a researcher trying to combat HIV infection, what molecular processes could you attempt to block? (See Figure 19.8.)

For suggested answers, see Appendix A.

CONCEPT **19.3**

Viruses, viroids, and prions are formidable pathogens in animals and plants

Diseases caused by viral infections afflict humans, agricultural crops, and livestock worldwide. Other smaller, less complex entities known as viroids and prions also cause disease in plants and animals, respectively.

Viral Diseases in Animals

A viral infection can produce symptoms by a number of different routes. Viruses may damage or kill cells by causing the release of hydrolytic enzymes from lysosomes. Some viruses cause infected cells to produce toxins that lead to disease symptoms, and some have molecular components that are toxic, such as envelope proteins. How much damage a virus causes depends partly on the ability of the infected tissue to regenerate by cell division. People usually recover completely from colds because the epithelium of the respiratory tract, which the viruses infect, can efficiently repair itself. In contrast, damage inflicted by poliovirus to mature nerve cells is permanent, because these cells do not divide and usually cannot be replaced. Many of the temporary symptoms associated with viral infections, such as fever and aches, actually result from the body's own efforts at defending itself against infection rather than from cell death caused by the virus.

The immune system is a complex and critical part of the body's natural defenses (see Chapter 43). It is also the basis for the major medical tool for preventing viral infections—vaccines. A **vaccine** is a harmless variant or derivative of a pathogen that stimulates the immune system to mount defenses against the harmful pathogen. Smallpox, a viral disease that was at one time a devastating scourge in many parts of the world, was eradicated by a vaccination program carried out by the World Health Organization. The very narrow host range of the smallpox virus—it infects only humans—was a critical factor in the success of this program. Similar worldwide vaccination campaigns are currently under way to eradicate polio and measles. Effective vaccines are also available against rubella, mumps, hepatitis B, and a number of other viral diseases.

Although vaccines can prevent certain viral illnesses, medical technology can do little, at present, to cure most viral infections once they occur. The antibiotics that help us recover from bacterial infections are powerless against viruses. Antibiotics kill bacteria by inhibiting enzymes specific to bacteria but have no effect on eukaryotic or virally encoded enzymes. However, the few enzymes that are encoded by viruses have provided targets for other drugs. Most antiviral drugs resemble nucleosides and as a result interfere with viral nucleic acid synthesis. One such drug is acyclovir, which impedes herpesvirus reproduction by inhibiting the viral polymerase that synthesizes viral DNA. Similarly, azidothymidine (AZT) curbs HIV reproduction by interfering with the synthesis of DNA by reverse transcriptase. In the past two decades, much effort has gone into developing drugs against HIV. Currently, multidrug treatments, sometimes called "cocktails," have been found to be most effective. Such treatments commonly include a combination of two nucleoside mimics and a protease inhibitor, which interferes with an enzyme required for assembly of the viruses.

Emerging Viruses

Viruses that appear suddenly or are new to medical scientists are often referred to as *emerging viruses*. HIV, the AIDS virus, is a classic example: This virus appeared in San Francisco in the early 1980s, seemingly out of nowhere, although later studies uncovered a case in the Belgian Congo that occurred as early as 1959. The deadly Ebola virus, recognized initially in 1976 in central Africa, is one of several emerging viruses that cause *hemorrhagic fever*, an often fatal syndrome (set of symptoms) characterized by fever, vomiting, massive bleeding, and circulatory system collapse. A number of other dangerous emerging viruses cause encephalitis, inflammation of the brain. One example is the West Nile virus, which appeared in North America for the first time in 1999 and has spread to all 48 contiguous states in the United States.

Severe acute respiratory syndrome (*SARS*) first appeared in southern China in November 2002. A global outbreak that occurred during the following eight months infected about 8,000 people and killed more than 700. Researchers quickly identified the infectious agent as a *coronavirus*, a virus with a single-stranded RNA genome (class IV) that had not previously been known to cause disease in humans. Public health workers responded rapidly, isolating patients and quarantining those who had come in contact with them. Because of low infectivity and other characteristics of the SARS virus, this rapid response succeeded in quelling the outbreak before it could infect a much larger population.

How do such viruses burst on the human scene, giving rise to harmful diseases that were previously rare or even unknown? Three processes contribute to the emergence of viral diseases. The first, and perhaps most important, is the mutation of existing viruses. RNA viruses tend to have an unusually high rate of mutation because errors in replicating their RNA genomes are not corrected by proofreading. Some mutations change existing viruses into new genetic varieties (strains) that can cause disease, even in individuals who are immune to the ancestral virus. For instance, general outbreaks of flu, or flu **epidemics**, are caused by new strains of influenza virus genetically different enough from earlier strains that people have little immunity to them.

A second process that can lead to the emergence of viral diseases is the dissemination of a viral disease from a small, isolated human population. For instance, AIDS went unnamed and virtually unnoticed for decades before it began to spread around the world. In this case, technological and social factors, including affordable international travel, blood transfusions, sexual promiscuity, and the abuse of intravenous drugs, allowed a previously rare human disease to become a global scourge.

A third source of new viral diseases in humans is the spread of existing viruses from other animals. Scientists estimate that about three-quarters of new human diseases originate in this way. Animals that harbor and can transmit a particular virus but are generally unaffected by it are said to act as a natural reservoir for that virus. For example, a species of bat has been identified as the likely natural reservoir of the SARS virus. Bats are sold as food in China, and their dried feces are even sold for medicinal uses; either of these practices could provide a route for transmission of the virus to humans.

Flu epidemics provide an instructive example of the effects of viruses moving between species. There are three types of influenza virus: types B and C, which infect only humans and have never caused an epidemic, and type A, which infects a wide range of animals, including birds, pigs, horses, and humans. Influenza A strains have caused three major flu epidemics among humans in the last 100 years. The worst was the "Spanish flu" **pandemic** (a global epidemic) of 1918–1919, which killed about 40 million people, including many World War I soldiers **(Figure 19.9a)**. Evidence points to birds as the source of the 1918 flu pandemic.

A likely scenario for that pandemic and others is that they began when the virus mutated as it passed from one host species to another. When an animal is infected with more than one strain of flu virus, the different strains can undergo genetic recombination if the RNA molecules making up their genomes mix and match during viral assembly. Coupled with mutation, these changes can lead to the emergence of a viral strain that is capable of infecting human cells. Having never been exposed to that particular strain before, humans will lack immunity, and the recombinant virus has the potential to be highly pathogenic. If such a flu virus recombines with viruses that circulate widely among humans, it may acquire the ability to spread easily from person to person, dramatically increasing the potential for a major human outbreak.

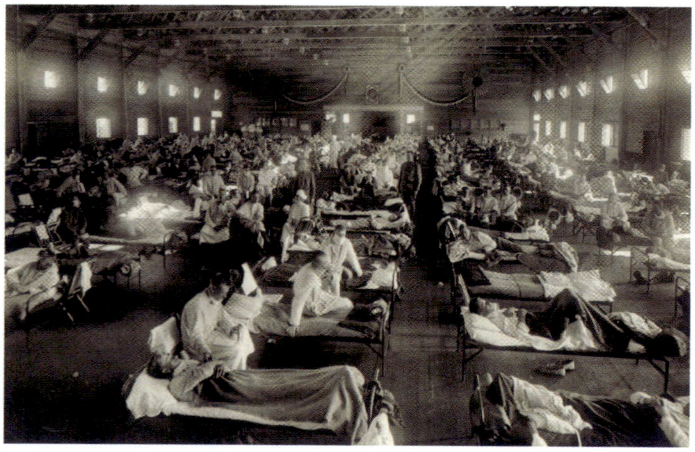

(a) The 1918 flu pandemic. Many of those infected were treated in large makeshift hospitals, such as this one.

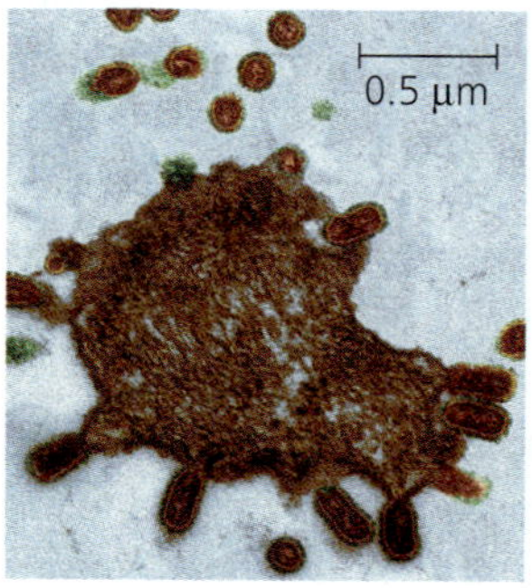

(b) Influenza A H5N1 virus. Virus particles are seen budding from an infected cell in this colorized TEM.

(c) Vaccinating ducks. Veterinarians administer vaccinations in a region of China reporting cases of avian flu, caused by strain H5N1.

▲ **Figure 19.9 Influenza in humans and other animals.**

Different strains of influenza A are given standardized names; for example, the strain that caused the 1918 flu is called H1N1. The name identifies which forms of two viral surface proteins are present: hemagglutinin (H) and neuraminidase (N). There are 16 different types of hemagglutinin, a protein that helps the flu virus attach to host cells, and 9 types of neuraminidase, an enzyme that helps release new virus particles from infected cells. Water birds have been found that carry viruses with all possible combinations of H and N.

In 1997, at least 18 people in Hong Kong were infected with an H5N1 virus **(Figure 19.9b)**; six of these people subsequently died. The same strain, previously seen only in wild birds, had killed several thousand chickens earlier that year, presumably passed along from wild birds or other species. A mass culling of all of Hong Kong's 1.5 million domestic birds appeared to stop that outbreak. Beginning in 2002, however, new cases of H5N1 human infection began to crop up around southeast Asia. By 2007, the disease caused by this virus, now called "avian flu," had killed about 160 people. Perhaps even more alarming is the overall mortality rate, which is greater than 50%. More than 100 million birds have either died from the disease or been killed to prevent the spread of infection; efforts are under way to vaccinate birds of several species **(Figure 19.9c)**.

The geographical and host ranges of avian flu virus continue to expand. It has shown up in wild or domestic birds in Africa and Europe, as well as in pigs, tigers, and domestic cats and dogs. The expanding host range provides increasing opportunities for different strains of virus to reassort their genetic material and for new strains to emerge. If the H5N1 avian flu virus evolves so that it can spread easily from person to person, it could bring about a major human outbreak. Human-to-human transmission is strongly suspected in several cases where the disease has clustered in families, but so far the disease has not spread beyond small groups to cause an epidemic. For those studying emerging viruses and their ability to give rise to a human pandemic, avian flu provides a sobering lesson in progress.

As we have seen, emerging viruses are generally not new; rather, they are existing viruses that mutate, disseminate more widely in the current host species, or spread to new host species. Changes in host behavior or environmental changes can increase the viral traffic responsible for emerging diseases. For example, new roads through remote areas can allow viruses to spread between previously isolated human populations. Also, the destruction of forests to expand cropland can bring humans into contact with other animals that may host viruses capable of infecting humans.

Viral Diseases in Plants

More than 2,000 types of viral diseases of plants are known, and together they account for an estimated annual loss of $15 billion worldwide due to their destruction of agricultural and horticultural crops. Common signs of viral infection include bleached or brown spots on leaves and fruits, stunted growth, and damaged flowers or roots, all tending to diminish the yield and quality of crops **(Figure 19.10)**.

Plant viruses have the same basic structure and mode of reproduction as animal viruses. Most plant viruses discovered thus far, including tobacco mosaic virus (TMV), have an RNA genome. Many have a helical capsid, like TMV (see Figure 19.3a); others have an icosahedral capsid.

Viral diseases of plants spread by two major routes. In the first route, called *horizontal transmission*, a plant is infected from an external source of the virus. Because the invading virus must get past the plant's outer protective layer of cells (the epidermis), a plant becomes more susceptible to viral infections if it has been damaged by wind, injury, or herbivores. Herbivores, especially insects, pose a double threat because they can also act as carriers of viruses, transmitting disease from plant to plant. Farmers and gardeners may transmit plant viruses inadvertently on pruning shears and other tools. The other route of viral infection is *vertical transmission*, in which a plant inherits a viral infection from a parent. Vertical transmission can occur in asexual propagation (for example, cuttings) or in sexual reproduction via infected seeds.

Once a virus enters a plant cell and begins reproducing, viral genomes and associated proteins can spread throughout the plant by means of plasmodesmata, the cytoplasmic connections that penetrate the walls between adjacent plant cells (see Figure 6.28). The passage of viral macromolecules from cell to cell is facilitated by virally encoded proteins that cause enlargement of plasmodesmata. Scientists have not yet devised cures for most viral plant diseases. Consequently, their efforts are focused largely on reducing the transmission of such diseases and on breeding resistant varieties of crop plants.

▲ **Figure 19.10 Viral infection of plants.** Infection with particular viruses causes irregular brown patches on tomatoes (left), black blotching on squash (center), and streaking in tulips due to redistribution of pigment granules (right).

Viroids and Prions: The Simplest Infectious Agents

As small and simple as viruses are, they dwarf another class of pathogens: **viroids**. These are circular RNA molecules, only a few hundred nucleotides long, that infect plants. Viroids do not encode proteins but can replicate in host plant cells, apparently using host cell enzymes. These small RNA molecules seem to cause errors in the regulatory systems that control plant growth, and the typical signs of viroid diseases are abnormal development and stunted growth. One viroid disease, called cadang-cadang, has killed more than 10 million coconut palms in the Philippines.

An important lesson from viroids is that a single molecule can be an infectious agent that spreads a disease. But viroids are nucleic acid, whose ability to be replicated is well known. Even more surprising is the evidence for infectious *proteins*, called **prions**, which appear to cause a number of degenerative brain diseases in various animal species. These diseases include scrapie in sheep; mad cow disease, which has plagued the European beef industry in recent years; and Creutzfeldt-Jakob disease in humans, which has caused the death of some 150 people in Great Britain over the past decade. Prions are most likely transmitted in food, as may occur when people eat prion-laden beef from cattle with mad cow disease. Kuru, another human disease caused by prions, was identified in the early 1900s among the South Fore natives of New Guinea. A kuru epidemic peaked there in the 1960s, puzzling scientists, who at first thought the disease had a genetic basis. Eventually, however, anthropological investigations ferreted out how the disease was spread: ritual cannibalism, a widespread practice among South Fore natives at that time.

Two characteristics of prions are especially alarming. First, prions act very slowly, with an incubation period of at least ten years before symptoms develop. The lengthy incubation period prevents sources of infection from being identified until long after the first cases appear, allowing many more infections to occur. Second, prions are virtually indestructible; they are not destroyed or deactivated by heating to normal cooking temperatures. To date, there is no known cure for

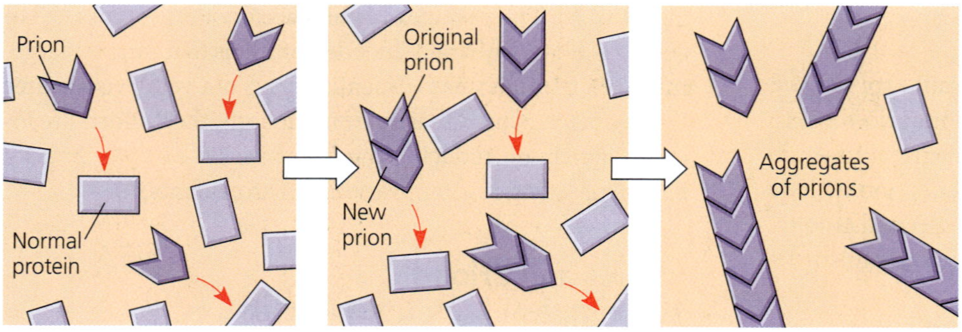

Figure 19.11 Model for how prions propagate. Prions are misfolded versions of normal brain proteins. When a prion contacts a normally folded version of the same protein, it may induce the normal protein to assume the abnormal shape. The resulting chain reaction may continue until high levels of prion aggregation cause cellular malfunction and eventual degeneration of the brain.

prion diseases, and the only hope for developing effective treatments lies in understanding the process of infection.

How can a protein, which cannot replicate itself, be a transmissible pathogen? According to the leading model, a prion is a misfolded form of a protein normally present in brain cells. When the prion gets into a cell containing the normal form of the protein, the prion somehow converts normal protein molecules to the misfolded prion versions. Several prions then aggregate into a complex that can convert other normal proteins to prions, which join the chain (**Figure 19.11**). Prion aggregation interferes with normal cellular functions and causes disease symptoms. This model was greeted with much skepticism when it was first proposed by Stanley Prusiner in the early 1980s, but it is now widely accepted. Prusiner was awarded the Nobel Prize in 1997 for his work on prions.

CONCEPT CHECK 19.3

1. Describe two ways a preexisting virus can become an emerging virus.
2. Contrast horizontal and vertical transmission of viruses in plants.
3. TMV has been isolated from virtually all commercial tobacco products. Why, then, is TMV infection not an additional hazard for smokers?
4. **WHAT IF?** How might the H5N1 avian flu virus have spread from Asia to Africa and Europe? Is it likely that human air travel could have spread this virus? How could you test your hypothesis?

For suggested answers, see Appendix A.

Chapter 19 Review

MEDIA Go to the Study Area at **www.masteringbio.com** for BioFlix 3-D Animations, MP3 Tutors, Videos, Practice Tests, an eBook, and more.

SUMMARY OF KEY CONCEPTS

CONCEPT 19.1

A virus consists of a nucleic acid surrounded by a protein coat (pp. 381–384)

- **The Discovery of Viruses: *Scientific Inquiry*** Researchers discovered viruses in the late 1800s by studying a plant disease, tobacco mosaic disease.
- **Structure of Viruses** A virus is a small nucleic acid genome enclosed in a protein capsid and sometimes a membranous envelope containing viral proteins that help viruses enter cells. The genome may be single- or double-stranded DNA or RNA.

CONCEPT 19.2

Viruses reproduce only in host cells (pp. 384–390)

- **General Features of Viral Reproductive Cycles** Viruses use enzymes, ribosomes, and small molecules of host cells to synthesize progeny viruses. Each type of virus has a characteristic host range.
- **Reproductive Cycles of Phages** Phages (viruses that infect bacteria) can reproduce by two alternative mechanisms: the lytic cycle and the lysogenic cycle.

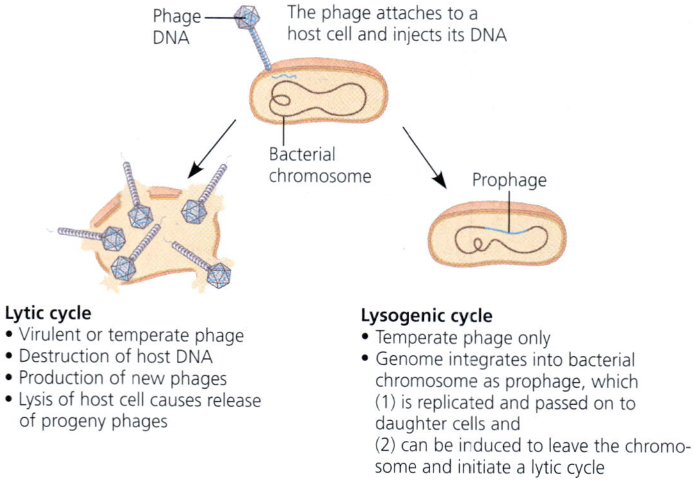

▶ **Reproductive Cycles of Animal Viruses** Many animal viruses have an envelope. Retroviruses (such as HIV) use the enzyme reverse transcriptase to copy their RNA genome into DNA, which can be integrated into the host genome as a provirus.

▶ **Evolution of Viruses** Since viruses can reproduce only within cells, they probably evolved after the first cells appeared, perhaps as packaged fragments of cellular nucleic acid. The origin of viruses is still being debated.

MEDIA

Activity Simplified Viral Reproductive Cycle
Activity Phage Lytic Cycle
Activity Phage Lysogenic and Lytic Cycles
Activity Retrovirus (HIV) Reproductive Cycle

CONCEPT 19.3

Viruses, viroids, and prions are formidable pathogens in animals and plants (pp. 390–394)

▶ **Viral Diseases in Animals** Symptoms may be caused by direct viral harm to cells or by the body's immune response. Vaccines stimulate the immune system to defend the host against specific viruses.

▶ **Emerging Viruses** Outbreaks of "new" viral diseases in humans are usually caused by existing viruses that expand their host territory. The H5N1 avian flu virus is being closely monitored for its potential to cause a serious flu pandemic.

▶ **Viral Diseases in Plants** Viruses enter plant cells through damaged cell walls (horizontal transmission) or are inherited from a parent (vertical transmission).

▶ **Viroids and Prions: The Simplest Infectious Agents** Viroids are naked RNA molecules that infect plants and disrupt their growth. Prions are slow-acting, virtually indestructible infectious proteins that cause brain diseases in mammals.

MEDIA

Investigation What Causes Infections in AIDS Patients?
Investigation Why Do AIDS Rates Differ Across the U.S.?

TESTING YOUR KNOWLEDGE

SELF-QUIZ

1. A bacterium is infected with an experimentally constructed bacteriophage composed of the T2 phage protein coat and T4 phage DNA. The new phages produced would have
   a. T2 protein and T4 DNA.
   b. T2 protein and T2 DNA.
   c. a mixture of the DNA and proteins of both phages.
   d. T4 protein and T4 DNA.
   e. T4 protein and T2 DNA.

2. RNA viruses require their own supply of certain enzymes because
   a. host cells rapidly destroy the viruses.
   b. host cells lack enzymes that can replicate the viral genome.
   c. these enzymes translate viral mRNA into proteins.
   d. these enzymes penetrate host cell membranes.
   e. these enzymes cannot be made in host cells.

3. Which of the following characteristics, structures, or processes is common to both bacteria and viruses?
   a. metabolism
   b. ribosomes
   c. genetic material composed of nucleic acid
   d. cell division
   e. independent existence

4. Emerging viruses arise by
   a. mutation of existing viruses.
   b. the spread of existing viruses to new host species.
   c. the spread of existing viruses more widely within their host species.
   d. all of the above
   e. none of the above

5. To cause a human pandemic, the H5N1 avian flu virus would have to
   a. spread to primates such as chimpanzees.
   b. develop into a virus with a different host range.
   c. become capable of human-to-human transmission.
   d. arise independently in chickens in North and South America.
   e. become much more pathogenic.

6. **DRAW IT** Redraw Figure 19.7 to show the reproductive cycle of a virus with a single-stranded genome that can function as mRNA (a class IV virus).

For Self-Quiz answers, see Appendix A.

MEDIA Visit the Study Area at **www.masteringbio.com** for a Practice Test.

EVOLUTION CONNECTION

7. The success of some viruses lies in their ability to evolve rapidly within the host. Such a virus evades the host's defenses by mutating and producing many altered progeny viruses before the body can mount an attack. Thus, the viruses present late in infection differ from those that initially infected the body. Discuss this as an example of evolution in microcosm. Which viral lineages tend to predominate?

SCIENTIFIC INQUIRY

8. When bacteria infect an animal, the number of bacteria in the body increases in an exponential fashion (graph A). After infection by a virulent animal virus with a lytic reproductive cycle, there is no evidence of infection for a while. Then, the number of viruses rises suddenly and subsequently increases in a series of steps (graph B). Explain the difference in the curves.

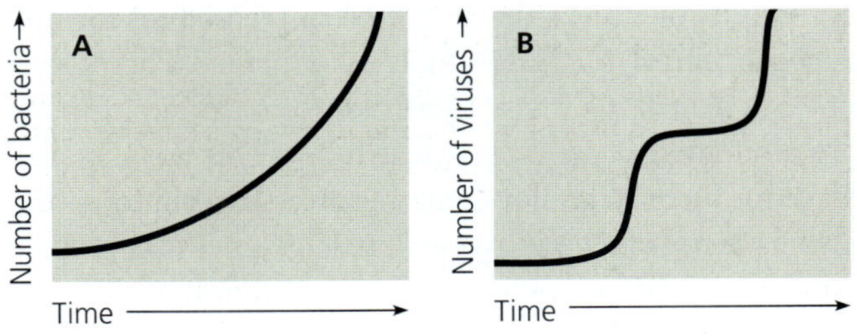

Biological Inquiry: A Workbook of Investigative Cases Explore West Nile virus in the case "The Donor's Dilemma." Explore the immune response to flu pathogens with the case "Pandemic Flu (Past and Possible)."

20 Biotechnology

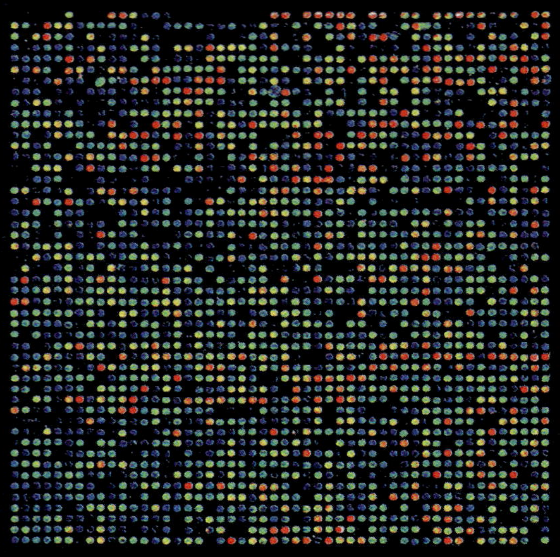

▲ **Figure 20.1 How can this array of spots be used to compare normal and cancerous tissues?**

KEY CONCEPTS

20.1 DNA cloning yields multiple copies of a gene or other DNA segment

20.2 DNA technology allows us to study the sequence, expression, and function of a gene

20.3 Cloning organisms may lead to production of stem cells for research and other applications

20.4 The practical applications of DNA technology affect our lives in many ways

OVERVIEW

The DNA Toolbox

In 1995, a major scientific milestone was announced: For the first time, researchers had sequenced the entire genome of a free-living organism, the bacterium *Haemophilus influenzae*. This news electrified the scientific community. Few among them would have dared to dream that a mere 12 years later, genome sequencing would be under way for more than 2,000 species. By 2007, researchers had completely sequenced hundreds of prokaryotic genomes and dozens of eukaryotic ones, including all 3 billion base pairs of the human genome.

Ultimately, these achievements can be attributed to advances in DNA technology—methods of working with and manipulating DNA—that had their roots in the 1970s. A key accomplishment was the invention of techniques for making **recombinant DNA**, DNA molecules formed when segments of DNA from two different sources—often different species—are combined *in vitro* (in a test tube). This advance set the stage for further development of powerful techniques for analyzing genes and gene expression. How scientists prepare recombinant DNA and use DNA technology to answer fundamental biological questions are one focus of this chapter.

Another focus of the chapter is how our lives are affected by **biotechnology**, the manipulation of organisms or their components to make useful products. Biotechnology has a long history that includes such early practices as selective breeding of farm animals and using microorganisms to make wine and cheese. Today, biotechnology also encompasses **genetic engineering**, the direct manipulation of genes for practical purposes. Genetic engineering has launched a revolution in biotechnology, greatly expanding the scope of its potential applications. Tools from the DNA toolbox are now applied in ways that were unthinkable only a decade ago, affecting everything from agriculture to criminal law to medical research. For instance, on the DNA microarray in **Figure 20.1**, the colored spots represent the relative level of expression of 2,400 human genes. Using microarray analysis, researchers can quickly compare gene expression in different samples, such as those obtained from normal and cancerous tissues. The knowledge gained from such gene expression studies is making a significant contribution to the study of cancer and other diseases.

In this chapter, we first describe the main techniques for manipulating DNA and analyzing gene expression and function. Next, we explore parallel advances in cloning organisms and producing stem cells, techniques that have both expanded our basic understanding of biology and enhanced our ability to apply this understanding to global problems. Finally, we survey the practical applications of biotechnology and consider some of the social and ethical issues that arise as modern biotechnology becomes more pervasive in our lives.

CONCEPT **20.1**

DNA cloning yields multiple copies of a gene or other DNA segment

The molecular biologist studying a particular gene faces a challenge. Naturally occurring DNA molecules are very long, and a single molecule usually carries many genes. Moreover, in many eukaryotic genomes, genes occupy only a small proportion of the chromosomal DNA, the rest being noncoding nucleotide sequences. A single human gene, for example, might constitute only 1/100,000 of a chromosomal DNA molecule. As a further complication, the distinctions between a gene and the surrounding DNA are subtle, consisting only of differences in nucleotide sequence. To work directly with specific genes, scientists have developed methods for preparing well-defined segments of DNA in multiple identical copies, a process called *DNA cloning.*

DNA Cloning and Its Applications: *A Preview*

Most methods for cloning pieces of DNA in the laboratory share certain general features. One common approach uses bacteria, most often *Escherichia coli.* Recall from Chapter 16 that the *E. coli* chromosome is a large circular molecule of DNA. In addition, *E. coli* and many other bacteria have **plasmids**, small circular DNA molecules that replicate separately from the bacterial chromosome. A plasmid has only a small number of genes; these genes may be useful when the bacterium is in a particular environment but may not be required for survival or reproduction under most conditions.

To clone pieces of DNA in the laboratory, researchers first isolate a plasmid from a bacterial cell and insert DNA from another source ("foreign" DNA) into it **(Figure 20.2)**. The resulting plasmid is now a recombinant DNA molecule, combining DNA from two sources. The plasmid is then returned to a bacterial cell, producing a *recombinant bacterium*. This single cell reproduces through repeated cell divisions to form a clone of cells, a population of genetically identical cells. Because the dividing bacteria replicate the recombinant plasmid and pass it on to their descendants, the foreign DNA and any genes it carries are cloned at the same time. The production of multiple copies of a single gene is called **gene cloning**.

Gene cloning is useful for two basic purposes: to make many copies of a particular gene and to produce a protein product. Researchers can isolate copies of a cloned gene from bacteria for use in basic research or to endow an organism with a new metabolic capability, such as pest resistance. For example, a resistance gene present in one crop species might be cloned and transferred into plants of another species. Alternatively, a protein with medical uses, such as human growth hormone, can be harvested in large quantities from cultures of bacteria carrying the cloned gene for the protein.

Most protein-coding genes exist in only one copy per haploid genome—roughly one part per million in the case of human DNA—so the ability to prepare large amounts of

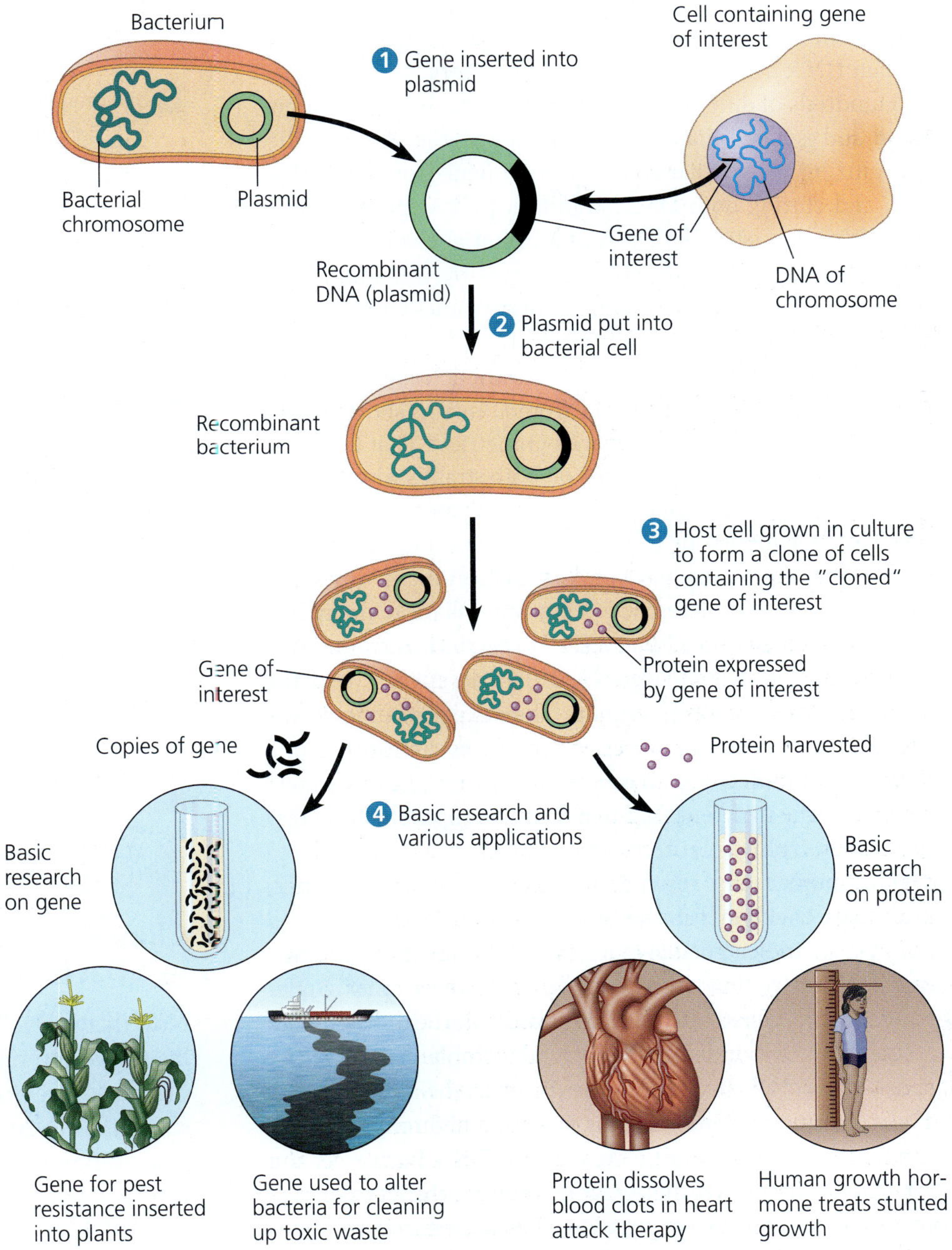

▲ **Figure 20.2 A preview of gene cloning and some uses of cloned genes.** In this simplified diagram of gene cloning, we start with a plasmid isolated from a bacterial cell and a gene of interest from another organism. Only one copy of the plasmid and one copy of the gene of interest are shown at the top of the figure, but the starting materials would include many copies of each.

such rare DNA fragments is crucial for any application involving a single gene. In the rest of this chapter, you will learn more about the techniques outlined in Figure 20.2 and related methods.

Using Restriction Enzymes to Make Recombinant DNA

Gene cloning and genetic engineering rely on the use of enzymes that cut DNA molecules at a limited number of specific locations. These enzymes, called restriction endonucleases, or **restriction enzymes**, were discovered in the late 1960s by researchers studying bacteria. As you learned in Chapter 19, restriction enzymes protect the bacterial cell by cutting up foreign DNA from other organisms or phages.

Hundreds of different restriction enzymes have been identified and isolated. Each restriction enzyme is very specific, recognizing a particular short DNA sequence, or **restriction site**, and cutting both DNA strands at precise points within this restriction site. The DNA of a bacterial cell is protected from the cell's own restriction enzymes by the addition of methyl groups (—CH_3) to adenines or cytosines within the sequences recognized by the enzymes.

The top of **Figure 20.3** illustrates a restriction site recognized by a particular restriction enzyme from *E. coli*. As shown in this example, most restriction sites are symmetrical. That is, the sequence of nucleotides is the same on both strands when read in the 5′ → 3′ direction. Most restriction enzymes recognize sequences containing four to eight nucleotides. Because any sequence this short usually occurs (by chance) many times in a long DNA molecule, a restriction enzyme will make many cuts in a DNA molecule, yielding a set of **restriction fragments**. All copies of a particular DNA molecule always yield the same set of restriction fragments when exposed to the same restriction enzyme. In other words, a restriction enzyme cuts a DNA molecule in a reproducible way. (Later you will learn how the different fragments can be separated and distinguished from each other.)

The most useful restriction enzymes cleave the sugar-phosphate backbones in the two DNA strands in a staggered manner, as indicated in Figure 20.3. The resulting double-stranded restriction fragments have at least one single-stranded end, called a **sticky end**. These short extensions can form hydrogen-bonded base pairs with complementary sticky ends on any other DNA molecules cut with the same enzyme. The associations formed in this way are only temporary but can be made permanent by the enzyme **DNA ligase**. As you learned in Chapter 16, this enzyme catalyzes the formation of covalent bonds that close up the sugar-phosphate backbones of DNA strands; for example, it joins Okazaki fragments during replication. You can see at the bottom of Figure 20.3 that the ligase-catalyzed joining of DNA from two different sources produces a stable recombinant DNA molecule.

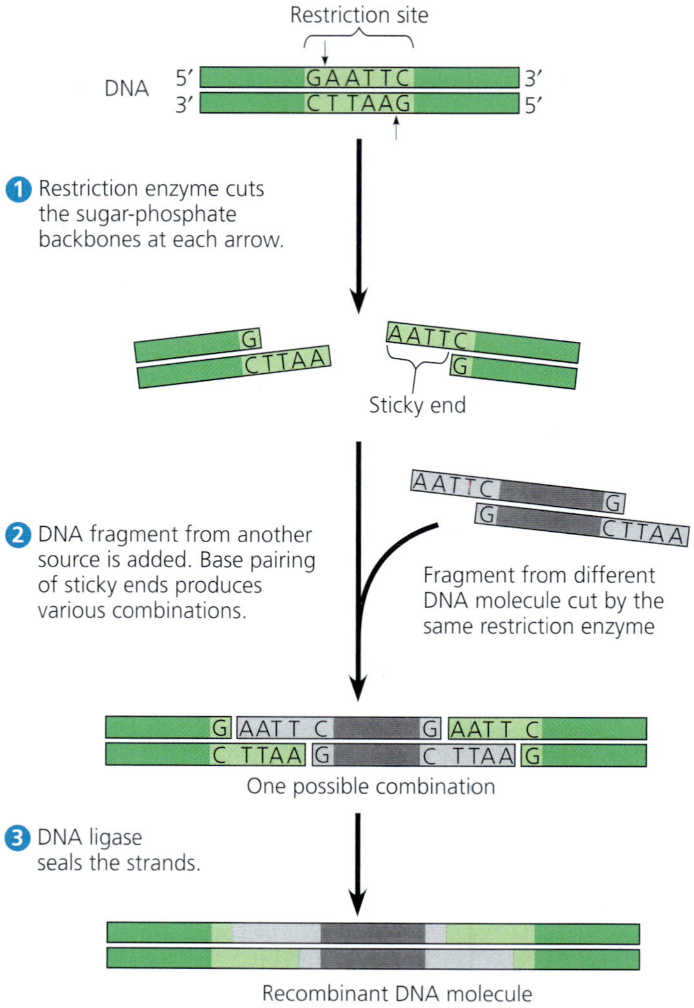

▲ **Figure 20.3 Using a restriction enzyme and DNA ligase to make recombinant DNA.** The restriction enzyme in this example (called *Eco*RI) recognizes a specific six-base-pair sequence, the restriction site, and makes staggered cuts in the sugar-phosphate backbones within this sequence, producing fragments with sticky ends. Any fragments with complementary sticky ends can base-pair, including the two original fragments; if the fragments come from different DNA molecules, the product is recombinant DNA.

DRAW IT *The restriction enzyme* HindIII *recognizes the sequence 5′-AAGCTT-3′, cutting between the two 'A's. Draw the double-stranded sequence before and after the enzyme cuts.*

Cloning a Eukaryotic Gene in a Bacterial Plasmid

Now that you've learned about restriction enzymes and DNA ligase, we can take a closer look at how genes are cloned in plasmids. The original plasmid is called a **cloning vector**, defined as a DNA molecule that can carry foreign DNA into a host cell and replicate there. Bacterial plasmids are widely used as cloning vectors for several reasons. They can be easily isolated from bacteria, manipulated to form recombinant plasmids by insertion of foreign DNA *in vitro*, and then reintroduced into bacterial cells. Moreover, recombinant bacterial plasmids (and the foreign DNA they carry) multiply rapidly owing to the high reproductive rate of their host cells.

Producing Clones of Cells Carrying Recombinant Plasmids

Let's say we are researchers interested in studying the β-globin gene in a particular species of hummingbird to see if this oxygen-carrying protein is different from its counterpart in other, less metabolically active species. **Figure 20.4** details one method for cloning hummingbird genes using a bacterial plasmid as the cloning vector.

1 We begin by isolating hummingbird genomic DNA from hummingbird cells. We also isolate our chosen vector, a particular bacterial plasmid from *E. coli* cells. The plasmid has been engineered to carry two genes that will later prove useful: *amp*^{R}, which makes *E. coli* cells resistant to the antibiotic ampicillin, and *lacZ*, which encodes an enzyme called β-galactosidase that hydrolyzes the sugar lactose. This enzyme can also hydrolyze a similar synthetic molecule called X-gal to form a blue product. The plasmid contains only one copy of the restriction site recognized by the restriction enzyme used in the next step; that site is within the *lacZ* gene.

2 Both the plasmid and the hummingbird DNA are cut with the same restriction enzyme, and then **3** the fragments are mixed together, allowing base pairing between their complementary sticky ends. We then add DNA ligase, which covalently bonds the sugar-phosphate backbones of the fragments whose sticky ends have base-paired. Some of the resulting recombinant plasmids contain hummingbird DNA fragments like the three shown in Figure 20.4; one of the three carries the β-globin gene. This step will also generate other products, such as plasmids containing several hummingbird DNA fragments, a combination of two plasmids, or a rejoined, nonrecombinant version of the original plasmid.

4 The DNA mixture is then added to bacteria that have a mutation in the *lacZ* gene on their own chromosome, making them unable to hydrolyze lactose or X-gal. Under suitable experimental conditions, the cells take up foreign DNA by

▼ **Figure 20.4 Research Method**

Cloning Genes in Bacterial Plasmids

APPLICATION Gene cloning is a process that produces many copies of a gene of interest. These copies can be used in sequencing the gene, in producing its encoded protein, or in basic research or other applications.

TECHNIQUE In this example, hummingbird genes are inserted into plasmids from *E. coli*. Only three plasmids and three hummingbird DNA fragments are shown, but millions of copies of the plasmid and a mixture of millions of different hummingbird DNA fragments would be present in the samples.

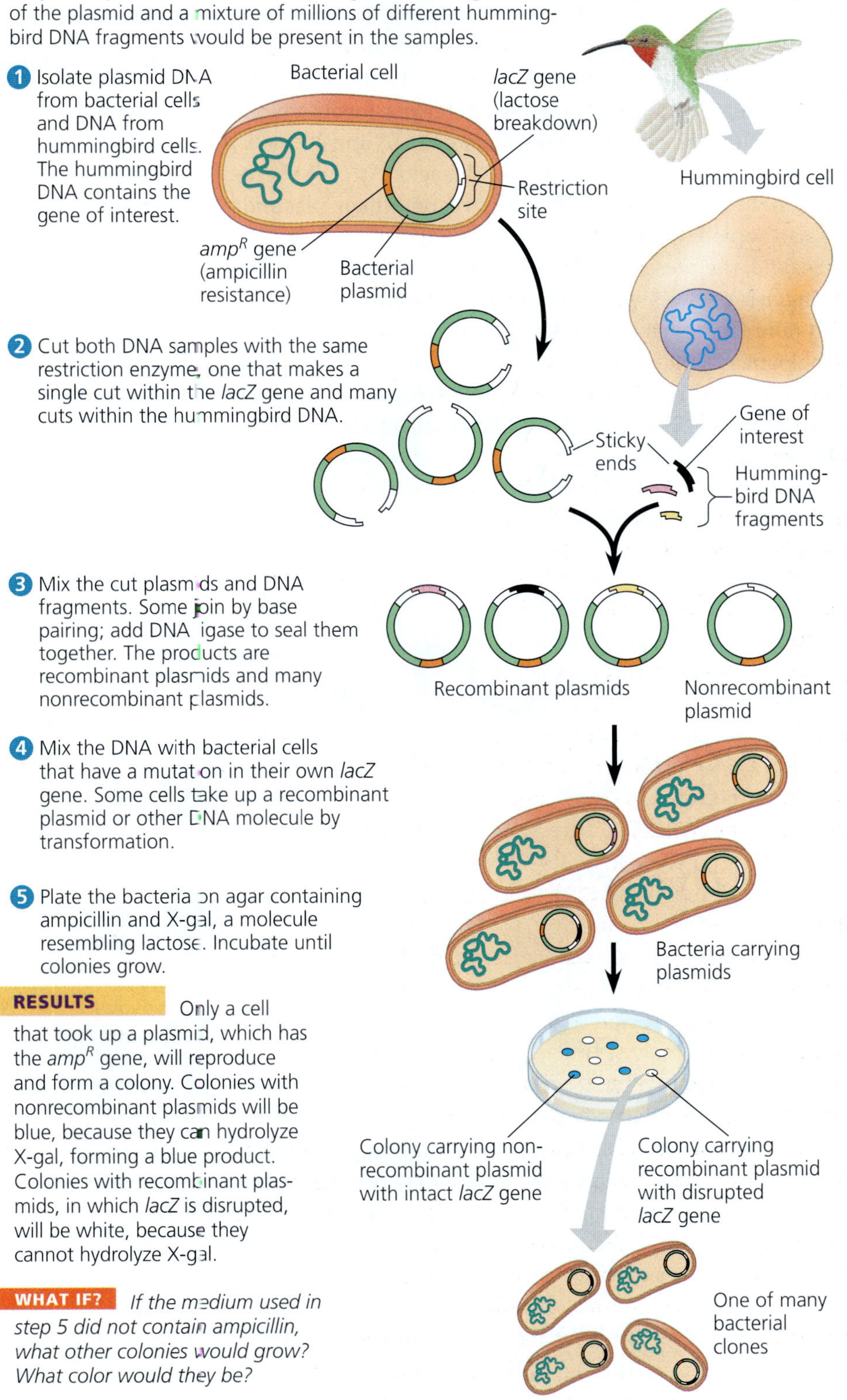

transformation (see pp. 306 and 561). Some cells acquire a recombinant plasmid carrying a gene, while others may take up a nonrecombinant plasmid, a fragment of noncoding hummingbird DNA, or nothing at all. The *amp*^{R} and *lacZ* genes on the plasmid can help us sort out these possibilities.

5 First, plating out all the bacteria on solid nutrient medium (agar) containing ampicillin allows us to distinguish the cells that have taken up plasmids, whether recombinant or not, from the other cells. Under these conditions, only cells with a plasmid will reproduce because only they have the *amp*^{R} gene conferring resistance to the ampicillin in the medium. Each reproducing bacterium forms a clone of cells. Once the clone contains between 10^5 and 10^8 cells, it is visible as a mass, or *colony*, on the agar. As cells reproduce, any foreign genes carried by recombinant plasmids are also copied (cloned).

Second, the presence of X-gal in the medium allows us to distinguish colonies with recombinant plasmids from those with nonrecombinant plasmids. Colonies containing nonrecombinant plasmids have the *lacZ* gene intact and will produce functional β-galactosidase. These colonies will be blue because the enzyme hydrolyzes the X-gal in the medium, forming a blue product. In contrast, no functional β-galactosidase is produced in colonies containing recombinant plasmids with foreign DNA inserted into the *lacZ* gene; these colonies will therefore be white.

The procedure to this point will have cloned many different hummingbird DNA fragments, not just the β-globin gene that interests us. In fact, taken together, the white colonies should represent all the DNA sequences from the hummingbird genome, including noncoding regions as well as genes. How do we find the colony (cell clone) carrying the β-globin gene among the many clones carrying other pieces of hummingbird DNA? Before addressing this question, we must consider how the clones are stored.

Storing Cloned Genes in DNA Libraries

The cloning procedure in Figure 20.4, which starts with a mixture of fragments from the entire genome of an organism, is called a "shotgun" approach; no single gene is targeted for cloning. Thousands of different recombinant plasmids are produced in step 3, and a clone of cells carrying each type of plasmid ends up as a white colony in step 5. The complete set of plasmid-containing cell clones, each carrying copies of a particular segment from the initial genome, is referred to as a **genomic library** **(Figure 20.5a)**. Each "plasmid clone" in the library is like a book containing specific information. Today, scientists often obtain such libraries (or even particular cloned genes) from another researcher, a commercial source, or a sequencing center.

Certain bacteriophages have also been used as cloning vectors for making genomic libraries. Fragments of foreign DNA

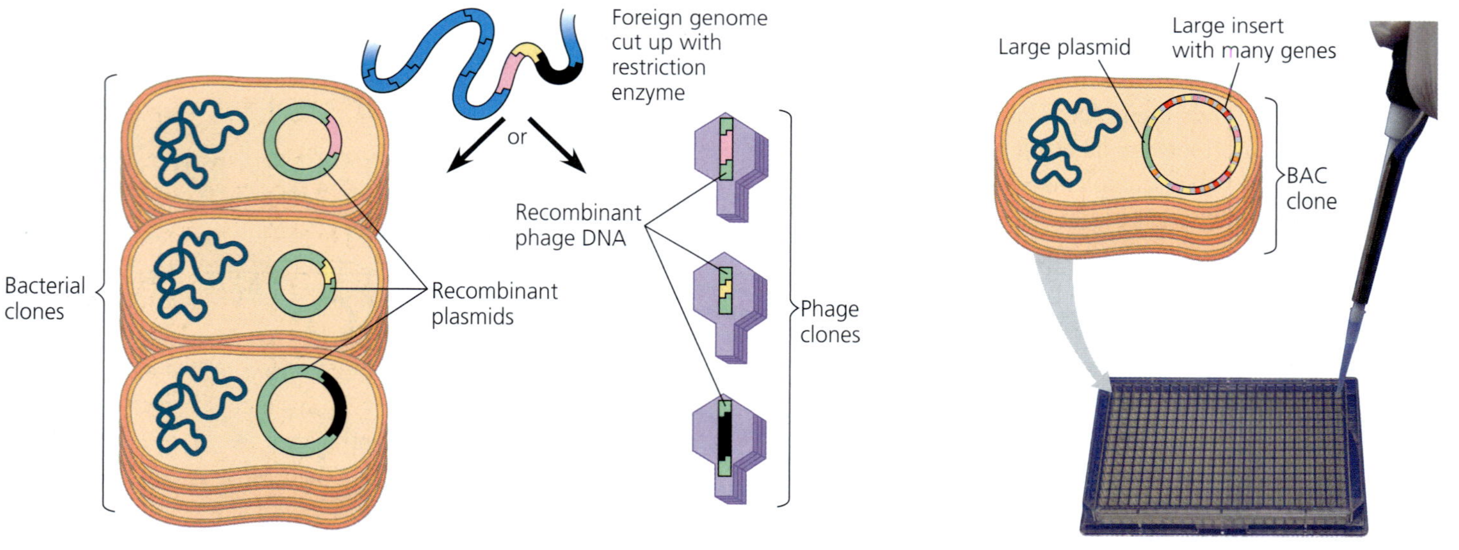

(a) Plasmid library. Shown are three of the thousands of "books" in a plasmid library. Each "book" is a clone of bacterial cells, which contain copies of a particular foreign genome fragment (pink, yellow, black segments) in their recombinant plasmids.

(b) Phage library. The same three foreign genome segments are shown in three "books" of a phage library.

(c) A library of bacterial artificial chromosome (BAC) clones. Such libraries are usually stored in a "multiwell" plastic plate; a 384-well plate is shown here. Each clone occupies one well. (The library of an entire genome would require many such plates.) Plasmid libraries can also be stored this way.

▲ **Figure 20.5 Genomic libraries.** A genomic library is a collection of many bacterial or phage clones. Each clone carries copies of a particular DNA segment from a foreign genome, integrated into an appropriate DNA vector, such as a plasmid or a trimmed-down phage genome. In a complete genomic library, the foreign DNA segments cover the entire genome of an organism. Note that the bacterial chromosomes in (a) and (c) are not drawn to scale; they are actually about 1,000 times as large as the vectors.

can be spliced into a trimmed-down version of a phage genome, as into a plasmid, by using a restriction enzyme and DNA ligase. The normal infection process allows production of many new phage particles, each carrying the foreign DNA. An advantage of using phages as vectors is that while a standard plasmid can carry a DNA insert no larger than 12,000 base pairs (12 kb), a phage can carry an insert of about 25 kb. A genomic library made using phages is stored as a collection of phage clones **(Figure 20.5b)**. Because restriction enzymes do not recognize gene boundaries, some genes in either type of genomic library will be cut and divided up among two or more clones.

Another type of vector used in library construction is a **bacterial artificial chromosome (BAC)**. In spite of the name, these are simply large plasmids, trimmed down so they contain just the genes necessary to ensure replication and capable of carrying inserts of 100–300 kb. The very large insert size minimizes the number of clones needed to make up the genomic library, but it also makes them more challenging to work with in the lab. BAC clones are usually stored in multiwelled plastic plates, with one clone per well **(Figure 20.5c)**. This orderly storage of clones, identified by their location in the plate, makes screening for the gene of interest very efficient, as we will see.

Researchers can make another kind of DNA library by starting with mRNA extracted from cells **(Figure 20.6)**. The enzyme reverse transcriptase (obtained from retroviruses) is used *in vitro* to make single-stranded DNA transcripts of the mRNA molecules. Recall that the 3′ end of the mRNA has a stretch of adenine (A) ribonucleotides called a poly-A tail. This feature allows use of a short strand of thymine (dT) deoxyribonucleotides as a primer for the reverse transcriptase. Following enzymatic degradation of the mRNA, a second DNA strand, complementary to the first, is synthesized by DNA polymerase. The resulting double-stranded DNA is called **complementary DNA (cDNA)**. For creating a library, the cDNA is modified by the addition of restriction enzyme recognition sequences at each end. Then the cDNA is inserted into vector DNA in a manner similar to the insertion of genomic DNA fragments. The extracted mRNA is a mixture of all the mRNA molecules in the original cells, transcribed from many different genes. Therefore, the cDNAs that are cloned make up a **cDNA library** containing a collection of genes. However, a cDNA library represents only part of the genome—only the subset of genes that were transcribed in the cells from which the mRNA was isolated.

Genomic and cDNA libraries each have advantages, depending on what is being studied. If you want to clone a gene but don't know what cell type expresses it or cannot obtain enough cells of the appropriate type, a genomic library is almost certain to contain the gene. Also, if you are interested in the regulatory sequences or introns associated with a gene, a genomic library is required because these sequences are absent from the fully processed mRNAs used in making a cDNA library. For this very reason, if you are interested only in the coding sequence of a gene, you can obtain a stripped-down version of the gene from a cDNA library. In our example of the hummingbird β-globin protein, a cDNA library would serve us well. We know which cells express the gene, so we could start making the library by isolating mRNA from hummingbird red blood cells. A cDNA library is also useful for studying the set of genes responsible for the specialized functions of a particular cell type, such as brain or liver cells. Finally, by making cDNA from cells of the same type at different times in the life of an organism, researchers can trace changes in patterns of gene expression during development.

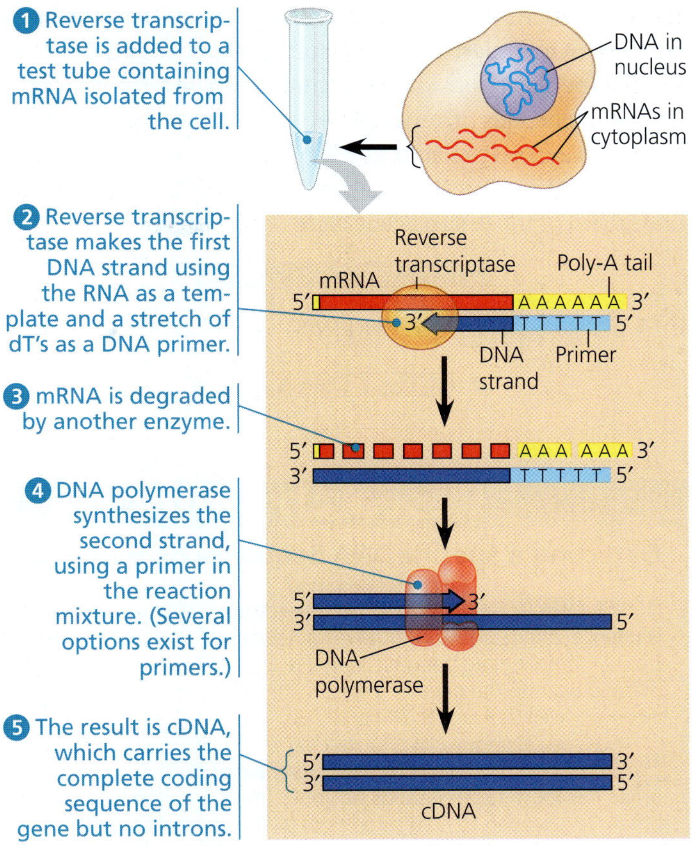

▲ **Figure 20.6 Making complementary DNA (cDNA) for a eukaryotic gene.** Complementary DNA is DNA made *in vitro* using mRNA as a template for the first strand. Because the mRNA contains only exons, the resulting double-stranded cDNA carries the complete coding sequence of the gene but no introns. Although only one mRNA is shown here, the final collection of cDNAs would reflect all the mRNAs that were present in the cell.

Screening a Library for Clones Carrying a Gene of Interest

Now, returning to the results in Figure 20.4, we're ready to screen all the colonies with recombinant plasmids (the white colonies) for a clone of cells containing the hummingbird β-globin gene. We can detect this gene's DNA by its ability to base-pair with a complementary sequence on another nucleic acid molecule, using **nucleic acid hybridization**. The complementary molecule, a short, single-stranded nucleic acid that can be either RNA or DNA, is called a **nucleic acid probe**. If we know at least part of the nucleotide sequence of the gene of interest (perhaps from knowing the amino acid sequence of the protein it encodes or, as in our case, the gene's nucleotide sequence in a closely related species), we can synthesize a probe complementary to it. For example, if part of the sequence on one strand of the desired gene were

5′ ···GGCTAACTTAGC··· 3′

then we would synthesize this probe:

3′ CCGATTGAATCG 5′

Each probe molecule, which will hydrogen-bond specifically to a complementary sequence in the desired gene, is labeled with a radioactive isotope or a fluorescent tag so we can track it.

Recall that the clones in our hummingbird genomic library have been stored in a multiwell plate (see Figure 20.5c). If we transfer a few cells from each well to a defined location on a membrane made of nylon or nitrocellulose, we can screen a large number of clones simultaneously for the presence of DNA complementary to our DNA probe **(Figure 20.7)**.

Once we've identified the location of a clone carrying the β-globin gene, we can grow some cells from that colony in liquid

▼ Figure 20.7 Research Method

Detecting a Specific DNA Sequence by Hybridization with a Nucleic Acid Probe

APPLICATION Hybridization with a complementary nucleic acid probe detects a specific DNA sequence within a mixture of DNA molecules. In this example, a collection of bacterial clones from a hummingbird genomic library is screened to identify clones that carry a recombinant plasmid bearing the gene of interest. The library is stored in many multiwell plates, with one clone per well (see Figure 20.5c).

TECHNIQUE Cells from each clone are applied to a special nylon membrane. Each membrane has room for thousands of clones (many more than are shown here) so only a few membranes are needed to hold samples of all the clones in the library. This set of membranes is an *arrayed library* that can be screened for a specific gene using a labeled probe. Here the label is a radioactive nucleotide, but other commonly used labels include enzymes, covalently linked to the probe nucleotides, that can produce either a colored or luminescent product.

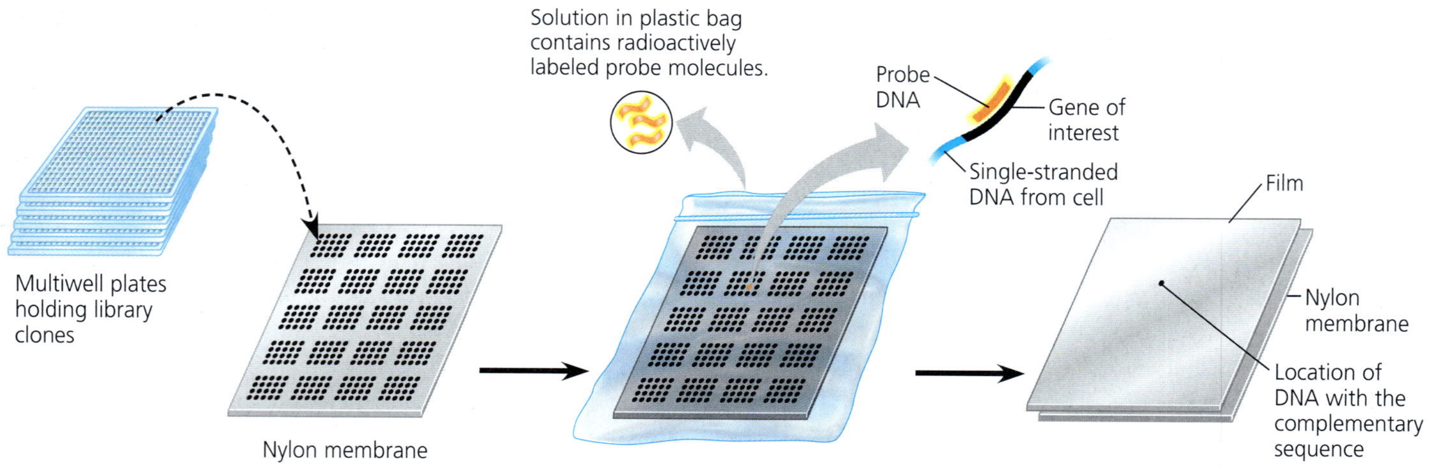

1 Plate by plate, cells from each well, representing one clone, are transferred to a defined spot on a special nylon membrane. The nylon membrane is treated to break open the cells and denature their DNA; the resulting single-stranded DNA molecules stick to the membrane.

2 The membrane is then incubated in a solution of radioactive probe molecules complementary to the gene of interest. Because the DNA immobilized on the membrane is single-stranded, the single-stranded probe can base-pair with any complementary DNA on the membrane. Excess DNA is then rinsed off. (One spot with radioactive probe–DNA hybrids is shown here in orange but would not be visible yet.)

3 The membrane is laid under photographic film, allowing any radioactive areas to expose the film (autoradiography). Black spots on the film correspond to the locations on the membrane of DNA that has hybridized to the probe. This location can be traced back to the original well containing the bacterial clone that holds the gene of interest.

RESULTS The location of the black spot on the photographic film identifies the clone containing the gene of interest. By using probes with different nucleotide sequences, researchers can screen the collection of bacterial clones for different genes.

culture in a large tank and then easily isolate many copies of the gene for our studies. We can also use the cloned gene as a probe to identify similar or identical genes in DNA from other sources, such as other species of birds.

Expressing Cloned Eukaryotic Genes

Once a particular gene has been cloned in host cells, its protein product can be produced in large amounts for research purposes or valuable practical applications, which we'll explore in Concept 20.4. Cloned genes can be expressed as protein in either bacterial or eukaryotic cells; each option has advantages and disadvantages.

Bacterial Expression Systems

Getting a cloned eukaryotic gene to function in bacterial host cells can be difficult because certain aspects of gene expression are different in eukaryotes and bacteria. To overcome differences in promoters and other DNA control sequences, scientists usually employ an **expression vector**, a cloning vector that contains a highly active bacterial promoter just upstream of a restriction site where the eukaryotic gene can be inserted in the correct reading frame. The bacterial host cell will recognize the promoter and proceed to express the foreign gene now linked to that promoter. Such expression vectors allow the synthesis of many eukaryotic proteins in bacterial cells.

Another problem with expressing cloned eukaryotic genes in bacteria is the presence of noncoding regions (introns) in most eukaryotic genes. Introns can make a eukaryotic gene very long and unwieldy, and they prevent correct expression of the gene by bacterial cells, which do not have RNA-splicing machinery. This problem can be surmounted by using a cDNA form of the gene, which includes only the exons.

Eukaryotic Cloning and Expression Systems

Molecular biologists can avoid eukaryotic-bacterial incompatibility by using eukaryotic cells such as yeasts, rather than bacteria, as hosts for cloning and/or expressing eukaryotic genes of interest. Yeasts, single-celled fungi, offer two advantages: They are as easy to grow as bacteria, and they have plasmids, a rarity among eukaryotes. Scientists have even constructed recombinant plasmids that combine yeast and bacterial DNA and can replicate in either type of cell. Another useful tool for cloning eukaryotic genes are **yeast artificial chromosomes (YACs)**, which combine the essentials of a eukaryotic chromosome—an origin for DNA replication, a centromere, and two telomeres—with foreign DNA. These chromosome-like vectors behave like ordinary chromosomes during mitosis, cloning the foreign DNA as the yeast cell divides. Because a YAC can carry a much longer DNA segment than can a plasmid vector, a cloned fragment is more likely to contain an entire gene rather than just a portion of it.

Another reason to use eukaryotic host cells for expressing a cloned eukaryotic gene is that many eukaryotic proteins will not function unless they are modified after translation, for example by the addition of carbohydrate or lipid groups. Bacterial cells cannot carry out these modifications, and if the gene product requiring such processing is from a mammal, even yeast cells may not be able to modify the protein correctly. The use of host cells from an animal cell culture may therefore be necessary.

Scientists have developed a variety of methods for introducing recombinant DNA into eukaryotic cells. In **electroporation**, a brief electrical pulse applied to a solution containing cells creates temporary holes in their plasma membranes, through which DNA can enter. (This technique is now commonly used for bacteria as well.) Alternatively, scientists can inject DNA directly into single eukaryotic cells using microscopically thin needles. To get DNA into plant cells, the soil bacterium *Agrobacterium* can be used, as well as other methods you will learn about later. If the introduced DNA is incorporated into a cell's genome by genetic recombination, then it may be expressed by the cell.

Amplifying DNA *in Vitro*: The Polymerase Chain Reaction (PCR)

DNA cloning in cells remains the best method for preparing large quantities of a particular gene or other DNA sequence. However, when the source of DNA is scanty or impure, the **polymerase chain reaction**, or **PCR**, is quicker and more selective. In this technique, any specific target segment within one or many DNA molecules can be quickly amplified (copied many times) in a test tube. With automation, PCR can make billions of copies of a target segment of DNA in a few hours, significantly faster than the days it would take to obtain the same number of copies by screening a DNA library for a clone with the desired gene and letting it replicate within host cells. In fact, PCR is being used increasingly to make enough of a specific DNA fragment to insert it directly into a vector, entirely skipping the steps of making and screening a library. To continue our literary analogy, PCR is like photocopying just one page rather than checking out all the books in a library.

In the PCR procedure, a three-step cycle brings about a chain reaction that produces an exponentially growing population of identical DNA molecules. During each cycle, the reaction mixture is heated to denature (separate) the DNA strands and then cooled to allow annealing (hydrogen bonding) of short, single-stranded DNA primers complementary to sequences on opposite strands at each end of the target sequence; finally, a heat-stable DNA polymerase extends the primers in the $5' \rightarrow 3'$ direction. If a standard DNA polymerase were used, the protein would be denatured along with the DNA during the first heating step and would have to be replaced after each cycle. The key to automating PCR was the discovery of an unusual heat-stable DNA polymerase, first isolated from cells of a bacterial species living in hot springs, that could withstand the heat at the start of each cycle. **Figure 20.8** illustrates the steps in PCR.

Just as impressive as the speed of PCR is its specificity. Only minute amounts of DNA need be present in the starting material, and this DNA can be in a partially degraded state, as long as a few molecules contain the complete target sequence. The key to this high specificity is the primers, which hydrogen-bond *only* to sequences at opposite ends of the target segment. (For high specificity, the primers must be at least 15 or so nucleotides long.) By the end of the third cycle, one-fourth of the molecules are identical to the target segment, with both strands the appropriate length. With each successive cycle, the number of target segment molecules of the correct length doubles, soon greatly outnumbering all other DNA molecules in the reaction.

Despite its speed and specificity, PCR amplification cannot substitute for gene cloning in cells when large amounts of a gene are desired. Occasional errors during PCR replication impose limits on the number of good copies that can be made by this method. When PCR is used to provide the specific DNA fragment for cloning, the resulting clones are sequenced to select clones with error-free inserts.

Devised in 1985, PCR has had a major impact on biological research and biotechnology. PCR has been used to amplify DNA from a wide variety of sources: fragments of ancient DNA from a 40,000-year-old frozen woolly mammoth; DNA from fingerprints or from tiny amounts of blood, tissue, or semen found at crime scenes; DNA from single embryonic cells for rapid prenatal diagnosis of genetic disorders; and DNA of viral genes from cells infected with viruses that are difficult to detect, such as HIV. We'll return to applications of PCR later in the chapter.

▼ Figure 20.8 Research Method

The Polymerase Chain Reaction (PCR)

APPLICATION With PCR, any specific segment—the target sequence—within a DNA sample can be copied many times (amplified), completely *in vitro*.

TECHNIQUE PCR requires double-stranded DNA containing the target sequence, a heat-resistant DNA polymerase, all four nucleotides, and two 15- to 20-nucleotide DNA strands that serve as primers. One primer is complementary to one end of the target sequence on one strand; the second primer is complementary to the other end of the sequence on the other strand.

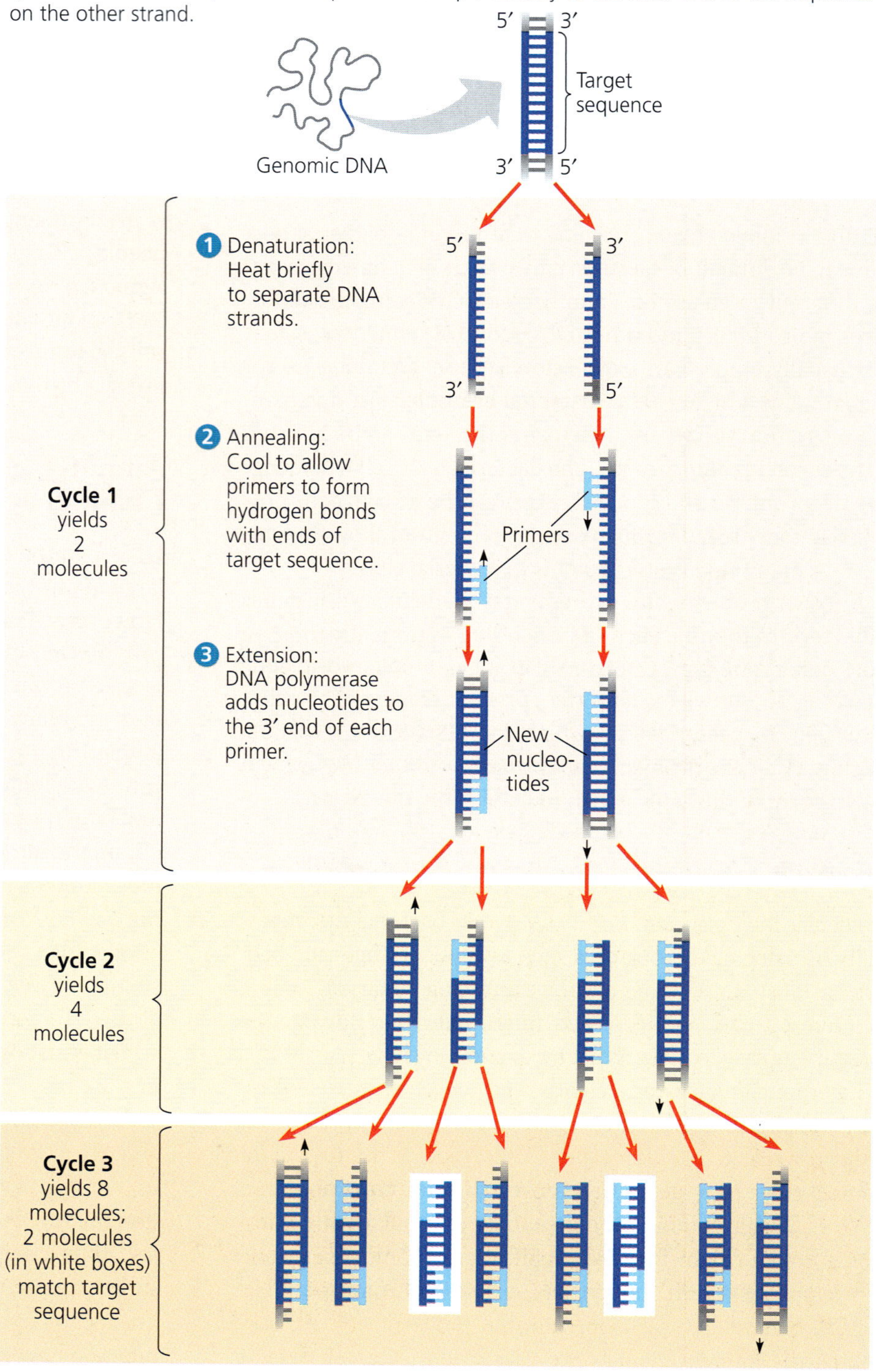

RESULTS After 3 cycles, one-fourth of the molecules match the target sequence exactly. After 30 cycles, more than 99.99% of the molecules match the target sequence.

CONCEPT CHECK 20.1

1. The restriction site for an enzyme called *PvuI* is the following sequence:

   5′-C G A T C G-3′
   3′-G C T A G C-5′

   Staggered cuts are made between the T and C on each strand. What type of bonds are being cleaved?
2. **DRAW IT** One strand of a DNA molecule has the following sequence: 5′-CCTTGACGATCGTTACCG-3′. Draw the other strand. Will *PvuI* cut this molecule? If so, draw the products.
3. What are some potential difficulties in using plasmid vectors and bacterial host cells to produce large quantities of proteins from cloned eukaryotic genes?
4. **WHAT IF?** If you wanted to amplify the double-stranded DNA fragment you drew for question 2, show the sequence of the primers you would use, assuming each is four nucleotides long. (In reality, primers are much longer.) Be sure to label the 5′ and 3′ ends.

For suggested answers, see Appendix A.

CONCEPT 20.2

DNA technology allows us to study the sequence, expression, and function of a gene

Once DNA cloning has provided us with large quantities of specific DNA segments, we can tackle some interesting questions about a particular gene and its function. For example, does the sequence of the hummingbird β-globin gene suggest a protein structure that can carry oxygen more efficiently than its counterpart in less metabolically active species? Does a particular human gene differ from person to person, and are certain alleles of that gene associated with a hereditary disorder? Where in the body and when is a given gene expressed? And, ultimately, what role does a certain gene play in an organism?

Before we can begin to address such compelling questions, we must consider a few standard laboratory techniques that are used to analyze the DNA of genes.

Gel Electrophoresis and Southern Blotting

Many approaches for studying DNA molecules involve **gel electrophoresis**. This technique uses a gel made of a polymer, such as a polysaccharide. The gel acts as a molecular sieve to separate nucleic acids or proteins on the basis of size, electrical charge, and other physical properties **(Figure 20.9)**. Because nucleic acid molecules carry negative charges on their phosphate groups, they all travel toward the positive pole in

▼ Figure 20.9 Research Method

Gel Electrophoresis

APPLICATION Gel electrophoresis is used for separating nucleic acids or proteins that differ in size, electrical charge, or other physical properties. DNA molecules are separated by gel electrophoresis in restriction fragment analysis of both cloned genes (see Figure 20.10) and genomic DNA (see Figure 20.11).

TECHNIQUE Gel electrophoresis separates macromolecules on the basis of their rate of movement through a polymeric gel in an electric field: The distance a DNA molecule travels is inversely proportional to its length. A mixture of DNA molecules, usually fragments produced by restriction enzyme digestion (cutting) or PCR amplification, is separated into bands. Each band contains thousands of molecules of the same length.

1 Each sample, a mixture of DNA molecules, is placed in a separate well near one end of a thin slab of gel. The gel is set into a small plastic support and immersed in an aqueous solution in a tray with electrodes at each end.

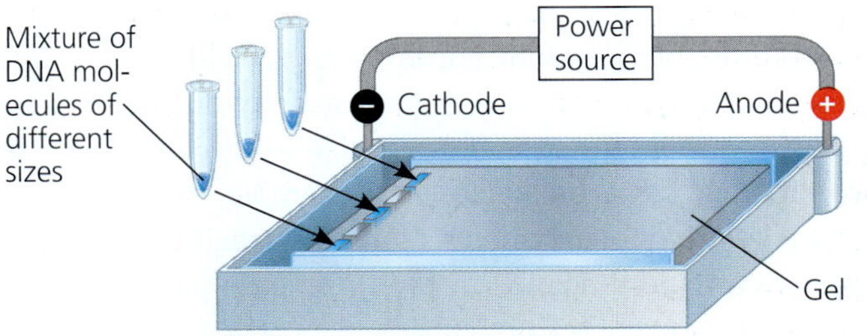

2 When the current is turned on, the negatively charged DNA molecules move toward the positive electrode, with shorter molecules moving faster than longer ones. Bands are shown here in blue, but on an actual gel, the bands would not be visible at this time.

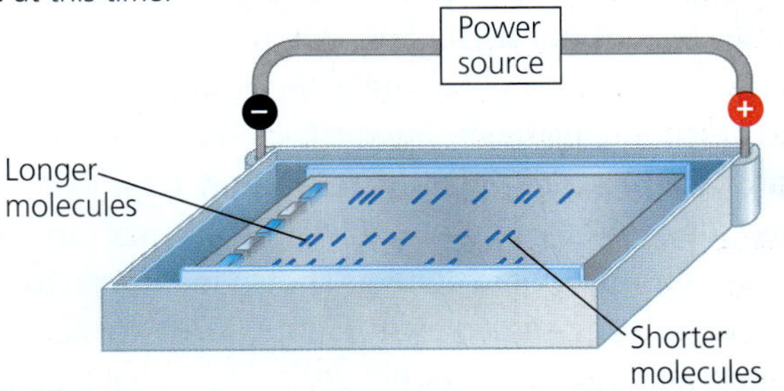

RESULTS After the current is turned off, a DNA-binding dye is added. This dye fluoresces pink in ultraviolet light, revealing the separated bands to which it binds. In the gel below, the pink bands correspond to DNA fragments of different lengths separated by electrophoresis. If all the samples were initially cut with the same restriction enzyme, then the different band patterns indicate that they came from different sources.

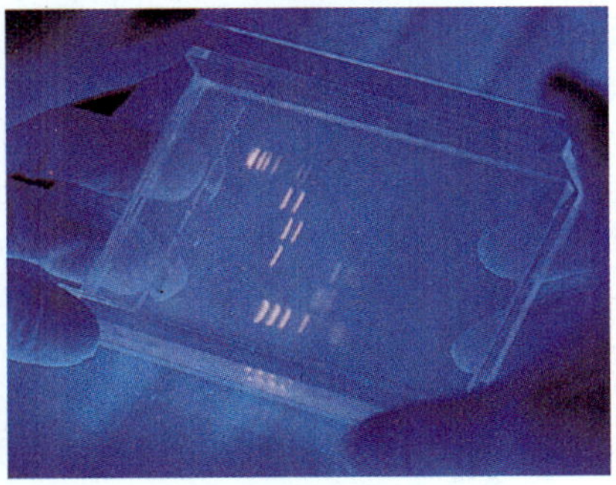

an electric field. As they move, the thicket of polymer fibers impedes longer molecules more than it does shorter ones, separating them by length. Thus, gel electrophoresis separates a mixture of linear DNA molecules into bands, each consisting of DNA molecules of the same length.

One useful application of this technique is *restriction fragment analysis,* which can rapidly provide useful information about DNA sequences. In this type of analysis, the DNA fragments produced by restriction enzyme digestion (cutting) of a DNA molecule are sorted by gel electrophoresis. When the mixture of restriction fragments undergoes electrophoresis, it yields a band pattern characteristic of the starting molecule and the restriction enzyme used. In fact, the relatively small DNA molecules of viruses and plasmids can be identified simply by their restriction fragment patterns. Because DNA can be recovered undamaged from gels, the procedure also provides a way to prepare pure samples of individual fragments—if the bands can be clearly resolved. (Very large DNA molecules, such as those of eukaryotic chromosomes, yield so many fragments that they appear as a smear instead of distinct bands.)

Restriction fragment analysis is also useful for comparing two different DNA molecules—for example, two alleles of a gene. A restriction enzyme recognizes a specific sequence of nucleotides, and a change in even one base pair of that sequence will prevent it from cutting at a particular site. Therefore, if the nucleotide differences between two alleles occur within a restriction site, digestion with the enzyme that recognizes the site will produce a different mixture of fragments from each allele. Each mixture will give its own band pattern in gel electrophoresis. For example, sickle-cell disease is caused by mutation of a single nucleotide located within a restriction sequence in the human β-globin gene (see Figure 17.22 and p. 278). Consequently, restriction fragment analysis by electrophoresis can distinguish the normal and sickle-cell alleles of the β-globin gene, as shown in **Figure 20.10**.

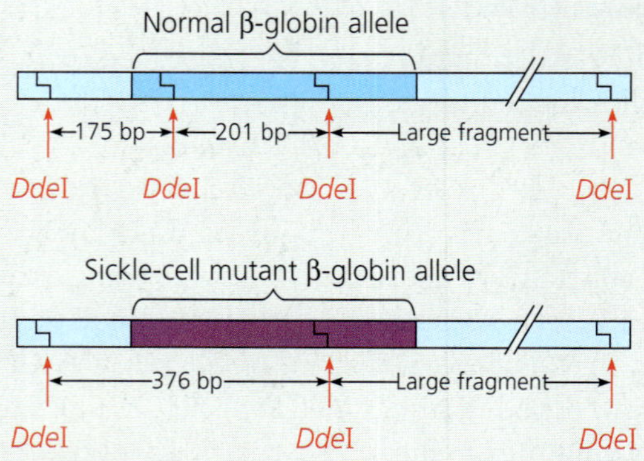

(a) *Dde*I restriction sites in normal and sickle-cell alleles of β-globin gene. Shown here are the cloned alleles, separated from the vector DNA but including some DNA next to the coding sequence. The normal allele contains two sites within the coding sequence recognized by the *Dde*I restriction enzyme. The sickle-cell allele lacks one of these sites.

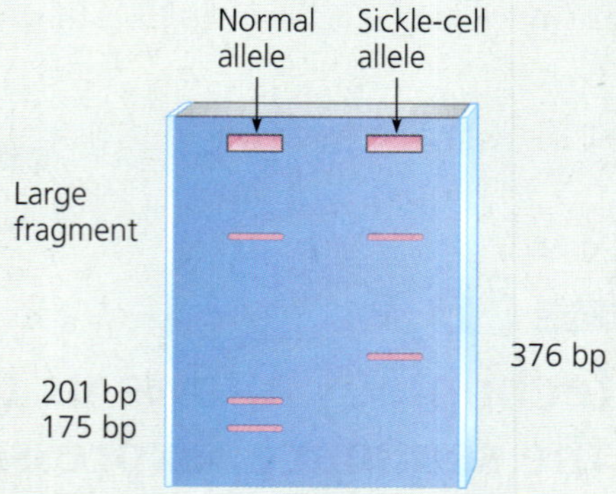

(b) Electrophoresis of restriction fragments from normal and sickle-cell alleles. Samples of each purified allele were cut with the *Dde*I enzyme and then subjected to gel electrophoresis, resulting in three bands for the normal allele and two bands for the sickle-cell allele. (The tiny fragments on the ends of both initial DNA molecules are identical and are not seen here.)

▲ **Figure 20.10 Using restriction fragment analysis to distinguish the normal and sickle-cell alleles of the human β-globin gene. (a)** The sickle-cell mutation destroys one of the *Dde*I restriction sites within the gene. **(b)** As a result, digestion with the *Dde*I enzyme generates different fragments from the normal and sickle-cell alleles.

WHAT IF? *Given recombinant bacterial clones carrying each of these alleles, how would you isolate the DNA of the alleles to provide the pure samples run on the gel in (b)? (Hint: Study Figures 20.4 and 20.9.)*

The starting materials in Figure 20.10 are samples of the cloned and purified β-globin alleles. But what can we do if we don't have purified alleles to start with? Suppose we wanted to determine whether a person is a heterozygous carrier of the mutant allele for sickle-cell disease. In this case, we would directly compare the genomic DNA from that person with DNA from both a person who has sickle-cell disease (and is homozygous for the mutant allele) and a person who is homozygous for the normal allele. As we mentioned already, electrophoresis of genomic DNA digested with a restriction enzyme and stained with a DNA-binding dye yields too many bands to distinguish them individually. However, a method called **Southern blotting** (developed by British biochemist Edwin Southern), which combines gel electrophoresis and nucleic acid hybridization, allows us to detect just those bands that include parts of the β-globin gene. The principle is the same as in nucleic acid hybridization for screening bacterial clones (see Figure 20.7). In Southern blotting, the probe is usually a radioactive single-stranded DNA molecule that is complementary to the gene of interest. **Figure 20.11** outlines the entire procedure and demonstrates how it can differentiate a heterozygote (in this case, for the sickle-cell allele) from an individual homozygous for the normal allele.

The identification of carriers of mutant alleles associated with genetic diseases is only one of the ways Southern blotting has been used. In fact, this technique has been a laboratory workhorse for many years. Recently, however, it has been supplanted by more rapid methods, often involving PCR amplification of the specific parts of genomes that may differ.

▼ **Figure 20.11** **Research Method**

Southern Blotting of DNA Fragments

APPLICATION Researchers can detect specific nucleotide sequences within a complex DNA sample with this method. In particular, Southern blotting is useful for comparing the restriction fragments produced from different samples of genomic DNA.

TECHNIQUE In this example, we compare genomic DNA samples from three individuals: a homozygote for the normal β-globin allele (I), a homozygote for the mutant sickle-cell allele (II), and a heterozygote (III). As in Figure 20.7, we show a radioactively labeled probe, but other methods of probe labeling and detection are also used.

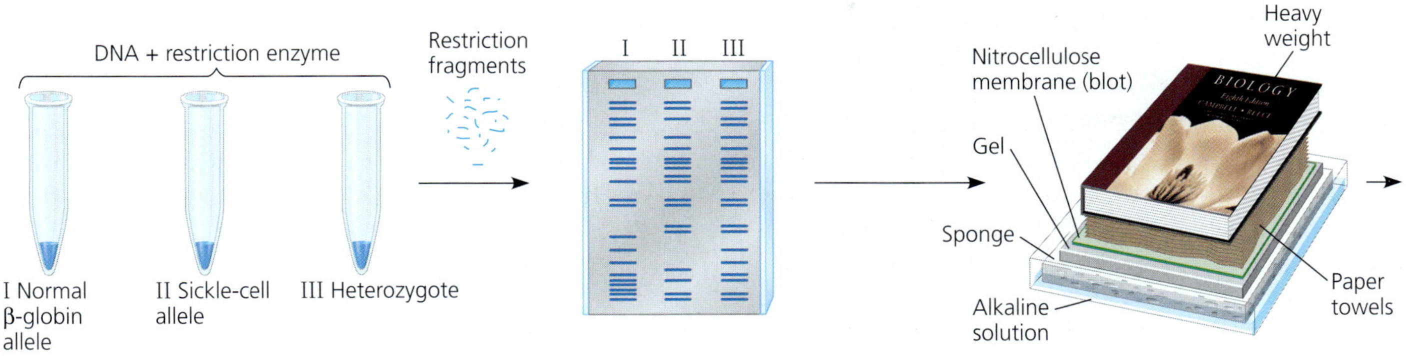

1 Preparation of restriction fragments. Each DNA sample is mixed with the same restriction enzyme, in this case *Dde*I. Digestion of each sample yields a mixture of thousands of restriction fragments.

2 Gel electrophoresis. The restriction fragments in each sample are separated by electrophoresis, forming a characteristic pattern of bands. (In reality, there would be many more bands than shown here, and they would be invisible unless stained.)

3 DNA transfer (blotting). With the gel arranged as shown above, capillary action pulls the alkaline solution upward through the gel, transferring the DNA to a nitrocellulose membrane, producing the blot; the DNA is denatured in the process. (Another version of this method uses an electrical current to hasten the DNA transfer.) The single strands of DNA stuck to the nitrocellulose are positioned in bands corresponding to those on the gel.

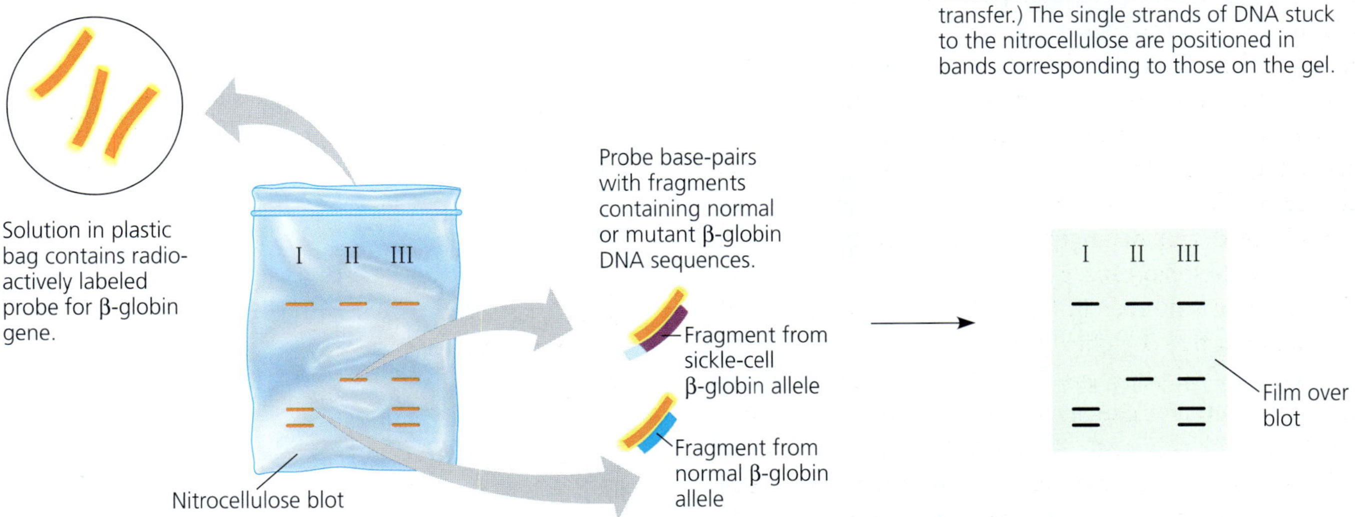

4 Hybridization with radioactive probe. The nitrocellulose blot is exposed to a solution containing a radioactively labeled probe. In this example, the probe is single-stranded DNA complementary to the β-globin gene. Probe molecules attach by base-pairing to any restriction fragments containing a part of the β-globin gene. (The bands would not be visible yet.)

5 Probe detection. A sheet of photographic film is laid over the blot. The radioactivity in the bound probe exposes the film to form an image corresponding to those bands containing DNA that base-paired with the probe.

RESULTS Because the band patterns for the three samples are clearly different, this method can be used to identify heterozygous carriers of the sickle-cell allele (III), as well as those with the disease, who have two mutant alleles (II), and unaffected individuals, who have two normal alleles (I). The band patterns for samples I and II resemble those observed for the purified normal and mutant alleles, respectively, seen in Figure 20.10b. The band pattern for the sample from the heterozygote (III) is a combination of the patterns for the two homozygotes (I and II).

▼ Figure 20.12 Research Method

Dideoxy Chain Termination Method for Sequencing DNA

APPLICATION The sequence of nucleotides in any cloned DNA fragment up to about 800 base pairs in length can be determined rapidly with machines that carry out sequencing reactions and separate the labeled reaction products by length.

TECHNIQUE This method synthesizes a set of DNA strands complementary to the original DNA fragment. Each strand starts with the same primer and ends with a dideoxyribonucleotide (ddNTP), a modified nucleotide. Incorporation of a ddNTP terminates a growing DNA strand because it lacks a 3′—OH group, the site for attachment of the next nucleotide (see Figure 16.14). In the set of strands synthesized, each nucleotide position along the original sequence is represented by strands ending at that point with the complementary ddNTP. Because each type of ddNTP is tagged with a distinct fluorescent label, the identity of the ending nucleotides of the new strands, and ultimately the entire original sequence, can be determined.

1. The fragment of DNA to be sequenced is denatured into single strands and incubated in a test tube with the necessary ingredients for DNA synthesis: a primer designed to base pair with the known 3′ end of the template strand, DNA polymerase, the four deoxyribonucleotides, and the four dideoxyribonucleotides, each tagged with a specific fluorescent molecule.

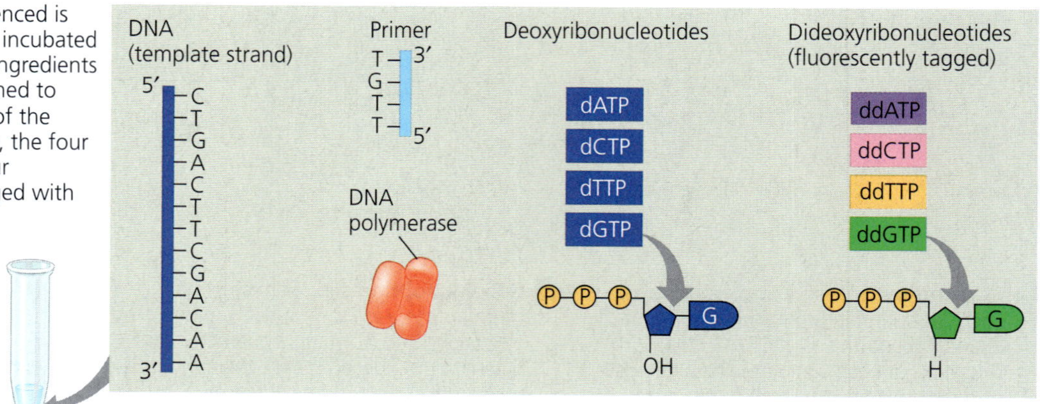

2. Synthesis of each new strand starts at the 3′ end of the primer and continues until a dideoxyribonucleotide is inserted, at random, instead of the normal equivalent deoxyribonucleotide. This prevents further elongation of the strand. Eventually, a set of labeled strands of various lengths is generated, with the color of the tag representing the last nucleotide in the sequence.

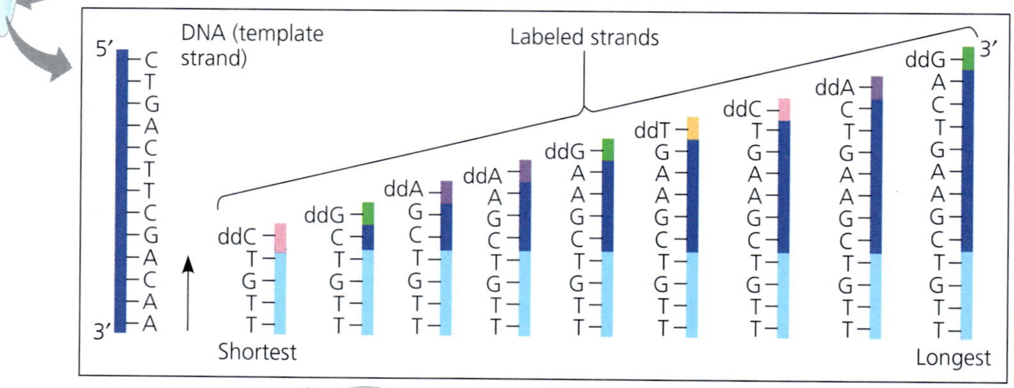

3. The labeled strands in the mixture are separated by passage through a polyacrylamide gel, with shorter strands moving through faster. For DNA sequencing, the gel is formed in a capillary tube rather than a slab like that shown in Figure 20.9. The small size of the tube allows a fluorescence detector to sense the color of each fluorescent tag as the strands come through. Strands differing in length by as little as one nucleotide can be distinguished from each other.

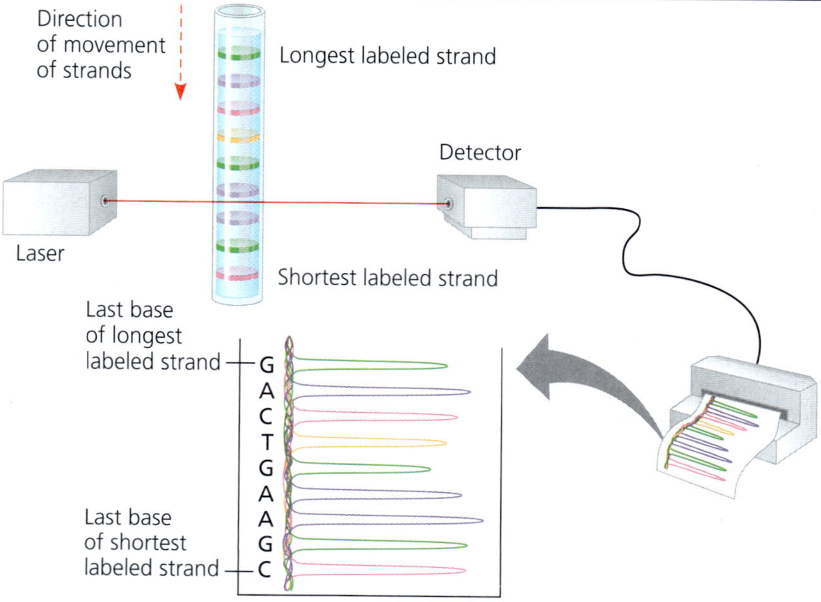

RESULTS The color of the fluorescent tag on each strand indicates the identity of the nucleotide at its end. The results can be printed out as a spectrogram, and the sequence, which is complementary to the template strand, can then be read from bottom (shortest strand) to top (longest strand). (Notice that the sequence here begins after the primer.)

DNA Sequencing

Once a gene is cloned, its complete nucleotide sequence can be determined. Today, sequencing is automated, carried out by sequencing machines. The automated procedure is based on a technique called the *dideoxyribonucleotide* (or *dideoxy*, for short) *chain termination method*, for reasons you can see in **Figure 20.12**. This method was developed by British biochemist Frederick Sanger, who received the Nobel Prize in 1980 for this accomplishment. (One of only four people to win two Nobel Prizes, Sanger also won one in 1975 for determining the amino acid sequence of insulin.)

Knowing the sequence of a gene allows researchers to compare it directly with genes in other species, where the function of the gene product may be known. If two genes from different species are quite similar in sequence, it is reasonable to suppose that their gene products perform similar functions. In this way, sequence comparisons provide clues to a gene's function, a topic we'll return to shortly. Another set of clues is provided by experimental approaches that analyze when and where a gene is expressed.

▼ Figure 20.13 **Research Method**

RT-PCR Analysis of the Expression of Single Genes

APPLICATION RT-PCR uses the enzyme reverse transcriptase (RT) in combination with PCR and gel electrophoresis. RT-PCR can be used to compare gene expression between samples—for instance, in different embryonic stages, in different tissues, or in the same type of cell under different conditions.

TECHNIQUE In this example, samples containing mRNAs from six embryonic stages of hummingbird were processed as shown below. (The mRNA from only one stage is shown.)

1 **cDNA synthesis** is carried out by incubating the mRNAs with reverse transcriptase and other necessary components.

2 **PCR amplification** of the sample is performed using primers specific to the hummingbird β-globin gene.

3 **Gel electrophoresis** will reveal amplified DNA products only in samples that contained mRNA transcribed from the β-globin gene.

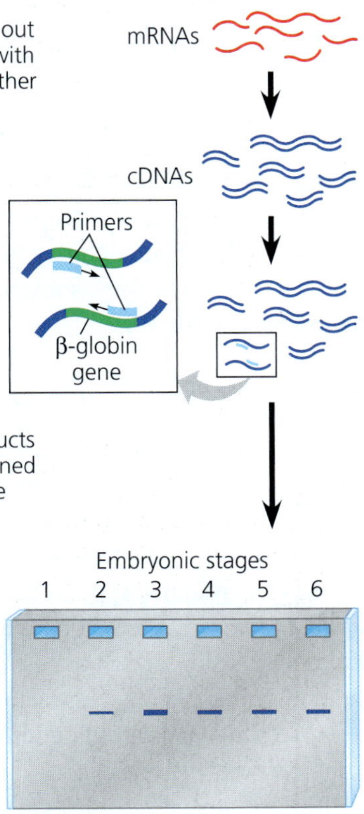

RESULTS The mRNA for this gene first is expressed at stage 2 and continues to be expressed through stage 6. The size of the amplified fragment depends on the distance between the primers that were used.

Analyzing Gene Expression

Having cloned a given gene, researchers can make labeled nucleic acid probes that will hybridize with mRNAs transcribed from the gene. The probes can provide information about when or where in the organism the gene is transcribed. Transcription levels are commonly used as a measure of gene expression.

Studying the Expression of Single Genes

Suppose we want to find out how the expression of the β-globin gene changes during the embryonic development of the hummingbird. There are at least two ways to do this.

The first is called **Northern blotting** (a play on words based on this method's similarity to Southern blotting). In this method, we carry out gel electrophoresis on samples of mRNA from hummingbird embryos at different stages of development, transfer the samples to a nitrocellulose membrane, and then allow the mRNAs on the membrane to hybridize with a labeled probe recognizing β-globin mRNA. If we expose a film to the membrane, the resulting image will look similar to the Southern blot in Figure 20.11, with one band of a given size showing up in each sample. If the mRNA band is seen at a particular stage, we can hypothesize that the protein functions during events taking place at that stage. Like Southern blotting, Northern blotting has been a mainstay over the years, but it is being supplanted in many labs by other techniques.

A method that is quicker and more sensitive (and therefore becoming more widely used) is the **reverse transcriptase–polymerase chain reaction**, or **RT-PCR (Figure 20.13)**. Analysis of hummingbird β-globin gene expression with RT-PCR begins similarly to Northern blotting, with the isolation of mRNAs from different developmental stages of hummingbird embryos. Reverse transcriptase is added next to make cDNA, which then serves as a template for PCR amplification using primers from the β-globin gene. When the products are run on a gel, copies of the amplified region will be observed as bands only in samples that originally contained the β-globin mRNA. In the case of hummingbird β-globin, for instance, we might expect to see a band appear at the stage when red blood cells begin forming, with all subsequent stages showing the same band. RT-PCR can also be carried out with mRNAs collected from different tissues at one time to discover which tissue is producing a specific mRNA.

An alternative way to determine which tissues or cells are expressing certain genes is to track down the location of specific mRNAs using labeled probes in place, or *in situ*, in the in-

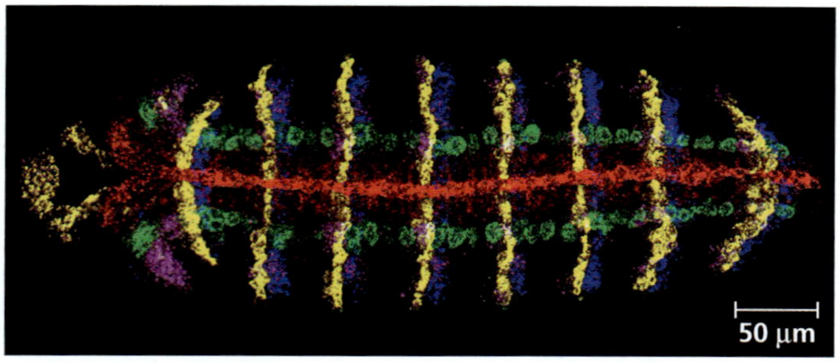

▲ **Figure 20.14 Determining where genes are expressed by *in situ* hybridization analysis.** This *Drosophila* embryo was incubated in a solution containing probes for five different mRNAs, each probe labeled with a different fluorescently colored tag. The embryo was then viewed using fluorescence microscopy. Each color marks where a specific gene is expressed as mRNA.

tact organism. This technique, called ***in situ* hybridization**, is most often carried out with probes labeled by attachment of fluorescent dyes (see Chapter 6). Different probes can be labeled with different dyes, sometimes with strikingly beautiful results **(Figure 20.14)**.

Studying the Expression of Interacting Groups of Genes

A major goal of biologists is to learn how genes act together to produce and maintain a functioning organism. Now that the entire genomes of a number of organisms have been sequenced, it is possible to study the expression of large groups of genes—a systems approach. Researchers use genome sequences as probes to investigate which genes are transcribed in different situations, such as in different tissues or at different stages of development. They also look for groups of genes that are expressed in a coordinated manner, with the aim of identifying networks of gene expression across an entire genome.

The basic strategy in such global (genome-wide) expression studies is to isolate the mRNAs made in particular cells, use these molecules as templates for making the corresponding cDNAs by reverse transcription, and then employ nucleic acid hybridization to compare this set of cDNAs with a collection of DNA fragments representing all or part of the genome. The results identify the subset of genes in the genome that are being expressed at a given time or under certain conditions. DNA technology makes such studies possible; with automation, they are easily performed on a large scale. Scientists can now measure the expression of thousands of genes at one time.

Genome-wide expression studies are made possible by **DNA microarray assays**. A DNA microarray consists of tiny amounts of a large number of single-stranded DNA fragments representing different genes fixed to a glass slide in a tightly spaced array, or grid (see Figure 20.1). (The microarray is also called a *DNA chip* by analogy to a computer chip.) Ideally, these fragments represent all the genes of an organism. **Figure 20.15** outlines how the DNA fragments on a microarray are tested for

▼ **Figure 20.15 Research Method**

DNA Microarray Assay of Gene Expression Levels

APPLICATION With this method, researchers can test thousands of genes simultaneously to determine which ones are expressed in a particular tissue, under different environmental conditions, in various disease states, or at different developmental stages. They can also look for coordinated gene expression.

TECHNIQUE

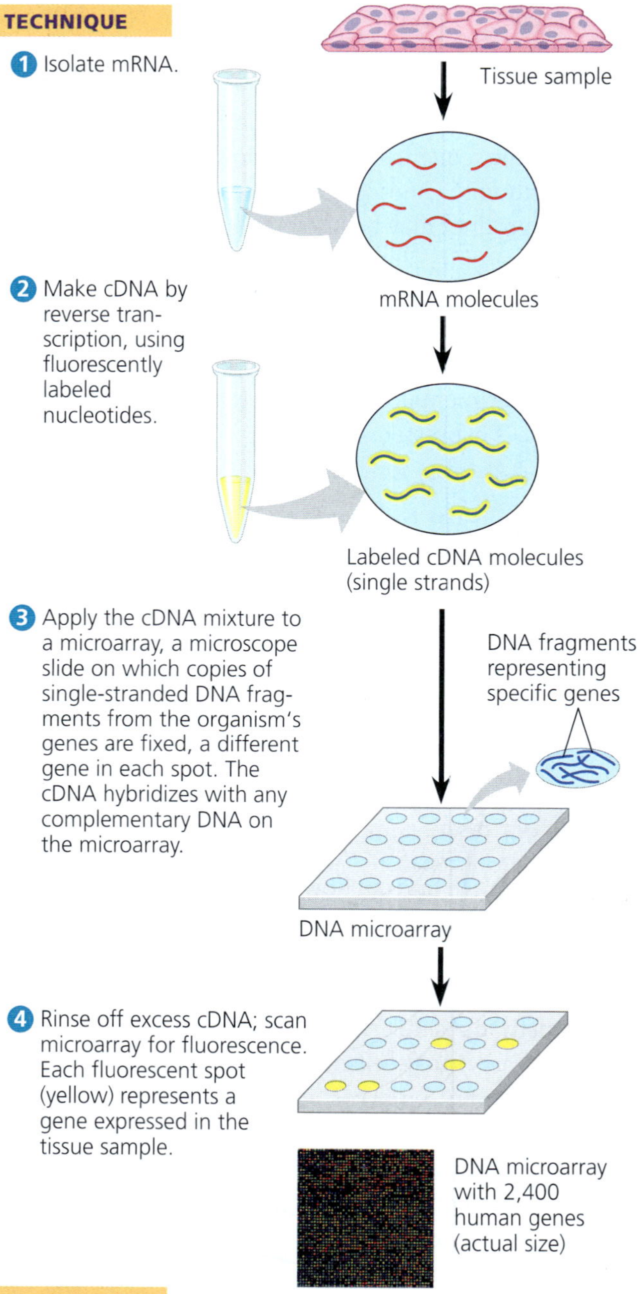

RESULTS The intensity of fluorescence at each spot is a measure of the expression in the tissue sample of the gene represented by that spot. Most often, as in the actual microarray above, two different samples are tested together by labeling the cDNAs prepared from each sample with labels of different colors, often green and red. The resulting color at a spot reveals the relative levels of expression of a particular gene in the two samples: Green indicates expression in one sample, red in the other, yellow in both, and black in neither. (See Figure 20.1 for a larger view.)

hybridization with cDNA molecules that have been prepared from the mRNAs in particular cells of interest and labeled with fluorescent dyes.

Using this technique, researchers have performed DNA microarray assays on more than 90% of the genes of the nematode *Caenorhabditis elegans,* during every stage of its life cycle. The results show that expression of nearly 60% of the genes changes dramatically during development and that many genes are expressed in a sex-specific pattern. This study supports the model held by most developmental biologists that embryonic development involves a complex and elaborate program of gene expression, rather than simply the expression of a small number of important genes. This example illustrates the ability of DNA microarrays to reveal general profiles of gene expression over the lifetime of an organism.

In addition to uncovering gene interactions and providing clues to gene function, DNA microarray assays may contribute to a better understanding of diseases and suggest new diagnostic techniques or therapies. For instance, comparing patterns of gene expression in breast cancer tumors and noncancerous breast tissue has already resulted in more informed and effective treatment protocols. Ultimately, information from DNA microarray assays should provide a grander view of how ensembles of genes interact to form an organism and maintain its vital systems.

Determining Gene Function

How do scientists determine the function of a gene identified by the techniques described thus far in the chapter? Perhaps the most common approach is to disable the gene and then observe the consequences in the cell or organism. In one application of this approach, called ***in vitro* mutagenesis**, specific mutations are introduced into a cloned gene, and then the mutated gene is returned to a cell in such a way that it disables ("knocks out") the normal cellular copies of the same gene. If the introduced mutations alter or destroy the function of the gene product, the phenotype of the mutant cell may help reveal the function of the missing normal protein. Using molecular and genetic techniques worked out in the 1980s, researchers can even generate mice with any given gene disabled, in order to study the role of that gene in development and in the adult. Mario Capecchi, Martin Evans, and Oliver Smithies received the 2007 Nobel Prize in Medicine for first accomplishing this feat. Several projects currently underway aim to disable every one of the over 20,000 genes in the mouse genome. These ambitious efforts will enhance our understanding of mammalian gene function.

A newer method for silencing expression of selected genes exploits the phenomenon of **RNA interference (RNAi)**, described in Chapter 18. This experimental approach uses synthetic double-stranded RNA molecules matching the sequence of a particular gene to trigger breakdown of the gene's messenger RNA or to block its translation. To date, the RNAi technique has been used successfully to reduce ("knock down") the expression of specific genes in mammalian cells, including human cells in culture, and plans are under way to try RNAi for the treatment of certain human disorders, including macular degeneration of the eye. In other organisms, such as the nematode and the fruit fly, RNAi has already proved valuable for analyzing the functions of genes on a large scale. In one study, RNAi was used to prevent expression of 86% of the genes in early nematode embryos, one gene at a time. Analysis of the phenotypes of the worms that developed from these embryos allowed the researchers to classify most of the genes into a small number of groups by function. This type of genome-wide analysis of gene function is sure to become more common as research focuses on the importance of interactions between genes in the system as a whole—the basis of systems biology (see Chapter 1).

The techniques and experimental approaches you have learned about thus far have already taught us a great deal about genes and the functions of their products. This DNA research is now being augmented by the development of powerful techniques for cloning whole multicellular organisms. An aim of this work is to obtain special types of cells, called stem cells, that give rise to all the different kinds of tissues. On a basic level, stem cells would allow scientists to use the DNA methods previously discussed to study the process of cell differentiation. On a more applied level, recombinant DNA techniques could be used to alter stem cells for the treatment of disease. Methods involving the cloning of organisms and production of stem cells are the subject of the next section.

CONCEPT CHECK 20.2

1. Suppose you carried out electrophoresis on a sample of genomic DNA isolated from an individual and treated with a restriction enzyme. After staining the gel with a DNA-binding dye, what would you see? Explain.
2. Describe the roles played by complementary base pairing between nucleic acid strands in each of the following techniques: Southern blotting, DNA sequencing, Northern blotting, RT-PCR, and microarray analysis.
3. **WHAT IF?** Look at the microarray in Figure 20.1, which is a larger version of the one in Figure 20.15. If one sample, labeled with a green fluorescent dye, was from normal tissue, and the other, labeled red, was from cancerous tissue, what does this mean for a spot that is green? Red? Yellow? Black? Based on their expression patterns, which genes would you be interested in examining further if you were studying cancer? Explain.

For suggested answers, see Appendix A.

CONCEPT 20.3

Cloning organisms may lead to production of stem cells for research and other applications

In parallel with advances in DNA technology, scientists have been developing and refining methods for cloning whole multicellular organisms from single cells. In this context, cloning produces one or more organisms genetically identical to the "parent" that donated the single cell. This is often called *organismal cloning* to differentiate it from gene cloning and, more significantly, from cell cloning—the division of an asexually reproducing cell into a collection of genetically identical cells. (The common theme for all types of cloning is that the product is genetically identical to the parent. In fact, the word *clone* comes from the Greek *klon*, meaning "twig.") The current interest in organismal cloning arises primarily from its potential to generate stem cells, which can in turn generate many different tissues.

The cloning of plants and animals was first attempted over 50 years ago in experiments designed to answer basic biological questions. For example, researchers wondered if all the cells of an organism have the same genes (a concept called *genomic equivalence*) or if cells lose genes during the process of differentiation (see Chapter 18). One way to answer this question is to see whether a differentiated cell can generate a whole organism—in other words, whether cloning an organism is possible. Let's discuss these early experiments before we consider more recent progress in organismal cloning and procedures for producing stem cells.

Cloning Plants: Single-Cell Cultures

The successful cloning of whole plants from single differentiated cells was accomplished during the 1950s by F. C. Steward and his students at Cornell University, who worked with carrot plants **(Figure 20.16)**. They found that differentiated cells taken from the root (the carrot) and incubated in culture medium could grow into normal adult plants, each genetically identical to the parent plant. These results showed that differentiation does not necessarily involve irreversible changes in the DNA. In plants, at least, mature cells can "dedifferentiate" and then give rise to all the specialized cell types of the organism. Any cell with this potential is said to be **totipotent**.

Plant cloning is now used extensively in agriculture. For some plants, such as orchids, cloning is the only commercially practical means of reproducing plants. In other cases, cloning has been used to reproduce a plant with valuable characteristics, such as the ability to resist a plant pathogen. In fact, you yourself may be a plant cloner: If you have ever grown a new plant from a cutting, you have practiced cloning!

▼ Figure 20.16 Inquiry

Can a differentiated plant cell develop into a whole plant?

EXPERIMENT In 1958, F. C. Steward and colleagues at Cornell University tested individual cells from a fully formed carrot for their ability to form a plant.

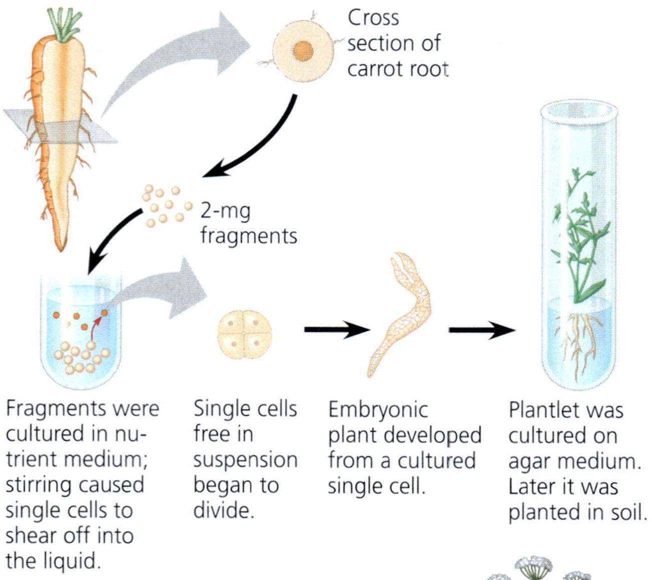

RESULTS A single somatic (nonreproductive) carrot cell developed into a mature carrot plant. The new plant was a genetic duplicate (clone) of the parent plant.

CONCLUSION At least some differentiated somatic cells in plants are totipotent, able to reverse their differentiation and then give rise to all the cell types in a mature plant.

SOURCE F. C. Steward et al., Growth and organized development of cultured cells. II. Organization in cultures grown from suspended cells, *American Journal of Botany* 45:705–708 (1958).

WHAT IF? If each individual cell had developed into only one type of tissue instead of a whole plant, what might the researchers have concluded?

Cloning Animals: Nuclear Transplantation

Differentiated cells from animals generally do not divide in culture, much less develop into the multiple cell types of a new organism. Therefore, early researchers had to use a different approach to the question of whether differentiated animal cells can be totipotent. Their approach was to remove the nucleus of an unfertilized or fertilized egg and replace it with the nucleus of a differentiated cell, a procedure called *nuclear transplantation.* If the nucleus from the differentiated donor cell retains its full genetic capability, then it should be able to direct development of the recipient cell into all the tissues and organs of an organism.

Such experiments were conducted on frogs by Robert Briggs and Thomas King in the 1950s and by John Gurdon in the 1970s. These researchers transplanted a nucleus from an embryonic or tadpole cell into an enucleated (nucleus-lacking) egg of the same species. In Gurdon's experiments, the transplanted nucleus was often able to support normal development of the egg into a tadpole (**Figure 20.17**). However, he found that the potential of a transplanted nucleus to direct normal development was inversely related to the age of the donor: the older the donor nucleus, the lower the percentage of normally developing tadpoles.

From these results, Gurdon concluded that something in the nucleus *does* change as animal cells differentiate. In frogs and most other animals, nuclear potential tends to be restricted more and more as embryonic development and cell differentiation progress.

Reproductive Cloning of Mammals

In addition to cloning frogs, researchers have long been able to clone mammals using nuclei or cells from a variety of early embryos. But it was not known whether a nucleus from a fully differentiated cell could be reprogrammed to succeed in acting as a donor nucleus. In 1997, however, Scottish researchers captured newspaper headlines when they announced the birth of Dolly, a lamb cloned from an adult sheep by nuclear transplantation from a differentiated cell (**Figure 20.18**, on the next page). These researchers achieved the necessary dedifferentiation of donor nuclei by culturing mammary cells in nutrient-poor medium. They then fused these cells with enucleated sheep eggs. The resulting diploid cells divided to form early embryos, which were implanted into surrogate mothers. Out of several hundred implanted embryos, one successfully completed normal development, and Dolly was born.

Later analyses showed that Dolly's chromosomal DNA was indeed identical to that of the nucleus donor. (Her mitochondrial DNA came from the egg donor, as expected.) In 2003, at age 6, Dolly suffered complications from a lung disease usually seen only in much older sheep and was euthanized. Dolly's premature death, as well as her arthritic condition, led to speculation that her cells were in some way not quite as healthy as those of a normal sheep, possibly reflecting incomplete reprogramming of the original transplanted nucleus.

Since 1997, researchers have cloned numerous other mammals, including mice, cats, cows, horses, mules, pigs, and dogs. In most cases, their goal has been the production of new individuals; this is known as *reproductive cloning.* We have already learned a lot from such experiments. For example, cloned animals of the same species do *not* always look or behave identically. In a herd of cows cloned from the same line of cultured cells, certain cows are dominant in behavior and others are

▼ Figure 20.17 Inquiry

Can the nucleus from a differentiated animal cell direct development of an organism?

EXPERIMENT John Gurdon and colleagues at Oxford University, in England, destroyed the nuclei of frog eggs by exposing the eggs to ultraviolet light. They then transplanted nuclei from cells of frog embryos and tadpoles into the enucleated eggs.

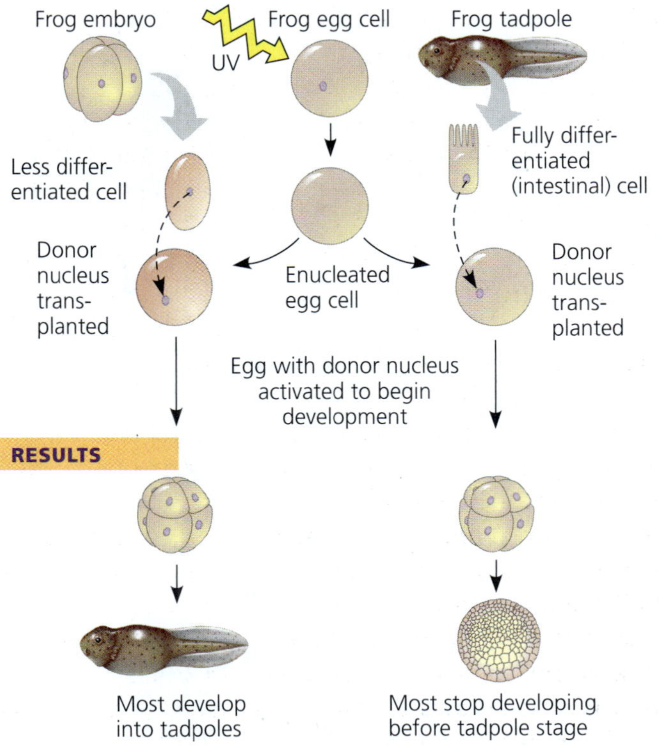

When the transplanted nuclei came from an early embryo, whose cells are relatively undifferentiated, most of the recipient eggs developed into tadpoles. But when the nuclei came from the fully differentiated intestinal cells of a tadpole, fewer than 2% of the eggs developed into normal tadpoles, and most of the embryos stopped developing at a much earlier stage.

CONCLUSION The nucleus from a differentiated frog cell can direct development of a tadpole. However, its ability to do so decreases as the donor cell becomes more differentiated, presumably because of changes in the nucleus.

SOURCE J. B. Gurdon et al., The developmental capacity of nuclei transplanted from keratinized cells of adult frogs, *Journal of Embryology and Experimental Morphology* 34:93–112 (1975).

WHAT IF? If each cell in a four-cell embryo was already so specialized that it was not totipotent, what results would you predict for the experiment on the left side of the figure?

▼ **Figure 20.18 Research Method**

Reproductive Cloning of a Mammal by Nuclear Transplantation

APPLICATION This method is used to produce cloned animals whose nuclear genes are identical to those of the animal supplying the nucleus.

TECHNIQUE Shown here is the procedure used to produce Dolly, the first reported case of a mammal cloned using the nucleus of a differentiated cell.

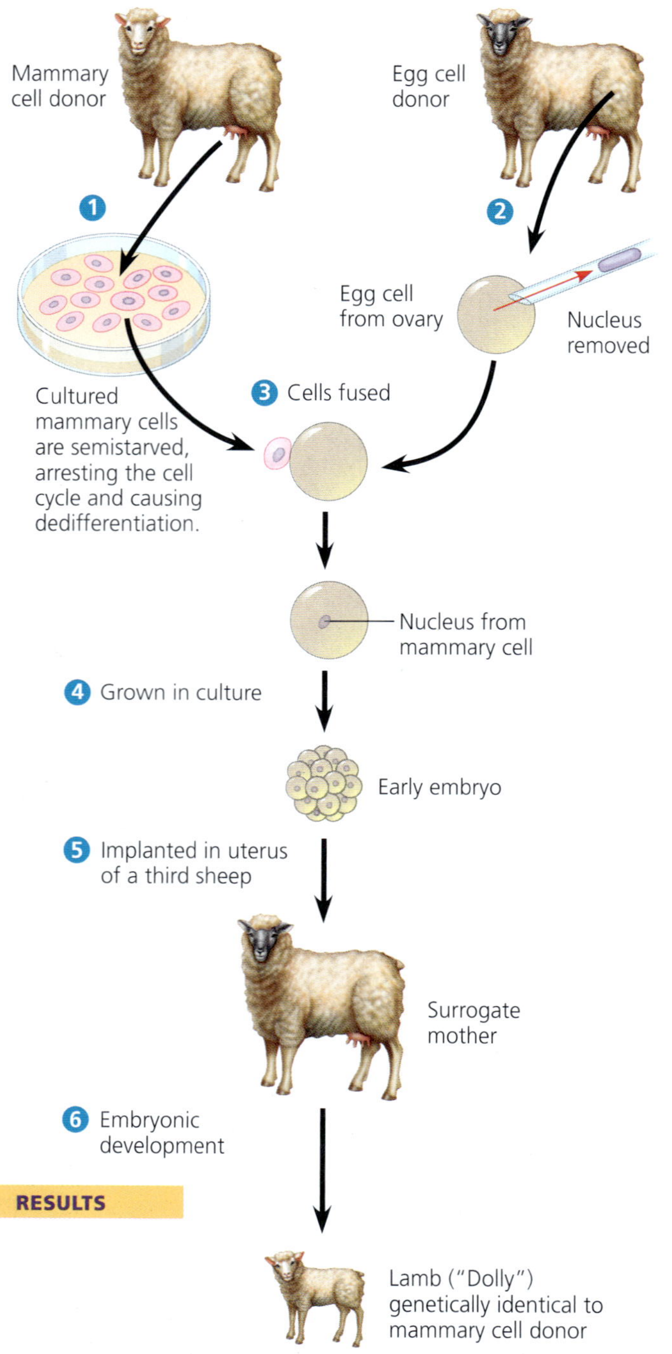

The genetic makeup of the cloned animal is identical to that of the animal supplying the nucleus but differs from that of the egg donor and surrogate mother. (The latter two are "Scottish blackface" sheep, with dark faces.)

▲ **Figure 20.19 CC, the first cloned cat, and her single parent.** Rainbow (left) donated the nucleus in a cloning procedure that resulted in CC (right). However, the two cats are not identical: Rainbow is a classic calico cat with orange patches on her fur and has a "reserved personality," while CC has a gray and white coat and is more playful.

more submissive. Another example of nonidentity in clones is the first cloned cat, named CC for Carbon Copy **(Figure 20.19)**. She has a calico coat, like her single female parent, but the color and pattern are different because of random X chromosome inactivation, which is a normal occurrence during embryonic development (see Figure 15.8). And identical human twins, which are naturally occurring "clones," are always slightly different. Clearly, environmental influences and random phenomena can play a significant role during development.

The successful cloning of so many mammals has heightened speculation about the cloning of humans. Scientists in several labs around the world have tackled the first steps of this feat. In the most common approach, nuclei from differentiated human cells are transplanted into unfertilized enucleated eggs, and the eggs are stimulated to divide. In 2001, a research group in Massachusetts observed a few early cell divisions in such an experiment. A few years later, South Korean researchers reported cloning embryos to an early stage called the blastocyst stage. This apparent success of human cloning techniques raised unprecedented ethical issues, igniting a heated debate. The later retraction of the Korean report has bought us a little more time to consider such issues.

Problems Associated with Animal Cloning

In most nuclear transplantation studies thus far, only a small percentage of cloned embryos develop normally to birth. And like Dolly, many cloned animals exhibit defects. Cloned mice, for instance, are prone to obesity, pneumonia, liver failure, and premature death. Scientists assert that even cloned animals that appear normal are likely to have subtle defects.

In recent years, we have begun to uncover some reasons for the low efficiency of cloning and the high incidence of abnormalities. In the nuclei of fully differentiated cells, a small subset of genes is turned on and expression of the rest is repressed. This regulation often is the result of epigenetic changes in chromatin, such as acetylation of histones or methylation of DNA (see Figure 18.7). During the nuclear transfer procedure, many of these changes must be reversed in the later-stage nucleus from a donor animal for genes to be expressed or repressed appropriately in early stages of development. Researchers have found that the DNA in embryonic cells from cloned embryos, like that of differentiated cells, often has more methyl groups than does the DNA in equivalent cells from uncloned embryos of the same species. This finding suggests that the reprogramming of donor nuclei requires chromatin restructuring, which occurs incompletely during cloning procedures. Because DNA methylation helps regulate gene expression, misplaced methyl groups in the DNA of donor nuclei may interfere with the pattern of gene expression necessary for normal embryonic development. In fact, the success of a cloning attempt may depend in large part on whether or not the chromatin structure in the donor nucleus can be restored to that of a newly fertilized egg.

Stem Cells of Animals

The major goal of cloning human embryos is not reproduction, but the production of stem cells for treating human diseases. A **stem cell** is a relatively unspecialized cell that can both reproduce itself indefinitely and, under appropriate conditions, differentiate into specialized cells of one or more types. Thus, stem cells are able both to replenish their own population and to generate cells that travel down specific differentiation pathways.

Many early animal embryos contain stem cells capable of giving rise to differentiated embryonic cells of any type. Stem cells can be isolated from early embryos at a stage called the blastula stage or its human equivalent, the blastocyst stage **(Figure 20.20)**. In culture, these *embryonic stem (ES) cells* reproduce indefinitely; and depending on culture conditions, they can be made to differentiate into a variety of specialized cells, including even eggs and sperm.

The adult body also has stem cells, which serve to replace nonreproducing specialized cells as needed. In contrast to ES cells, *adult stem cells* are not able to give rise to all cell types in the organism, though they can generate multiple types. For example, one of the several types of stem cells in bone marrow can generate all the different kinds of blood cells (see Figure 20.20), and another can differentiate into bone, cartilage, fat, muscle, and the linings of blood vessels. To the surprise of many, the adult brain has been found to contain stem cells that continue to produce certain kinds of nerve cells there. And recently, researchers have reported finding stem cells in skin, hair, eyes, and dental pulp. Although adult animals have only tiny numbers of stem cells, scientists are learning to identify and isolate these cells from various tissues and, in some cases, to grow them in culture. With the right culture conditions (for instance, the addition of specific growth factors), cultured stem cells from adult animals have been made to differentiate into multiple types of specialized cells.

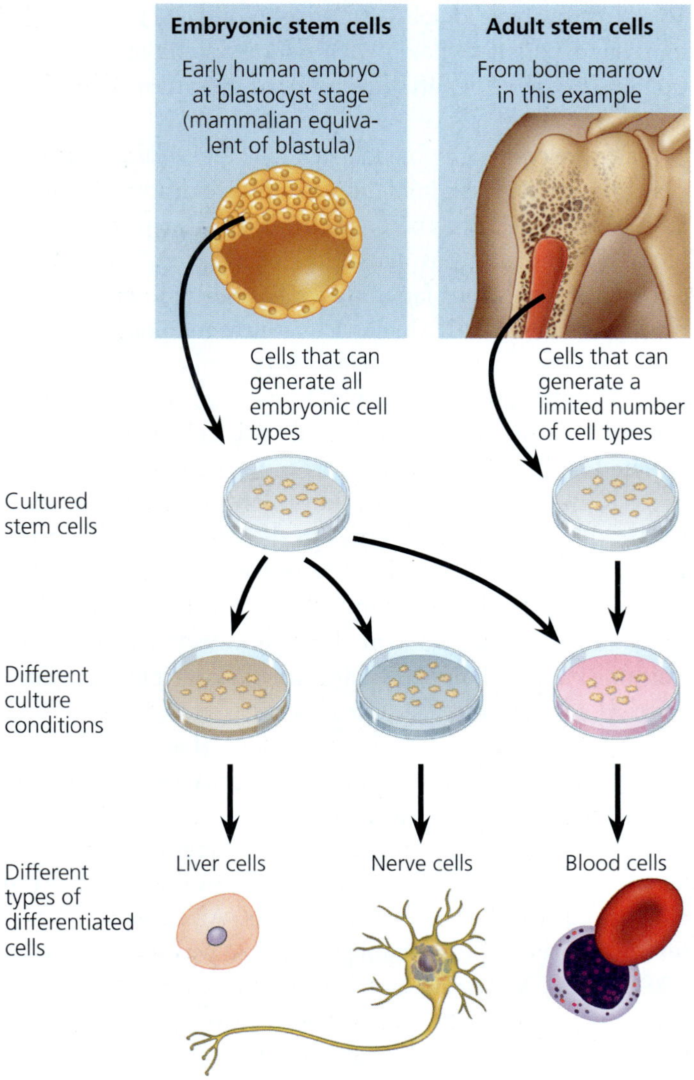

▲ **Figure 20.20 Working with stem cells.** Animal stem cells, which can be isolated from early embryos or adult tissues and grown in culture, are self-perpetuating, relatively undifferentiated cells. Embryonic stem cells are easier to grow than adult stem cells and can theoretically give rise to *all* types of cells in an organism. The range of cell types that can arise from adult stem cells is not yet fully understood.

Research with embryonic or adult stem cells is a source of valuable data about differentiation and has enormous potential for medical applications. The ultimate aim is to supply cells for the repair of damaged or diseased organs: for example, insulin-producing pancreatic cells for people with diabetes or certain kinds of brain cells for people with Parkinson's disease or Huntington's disease. Adult stem cells from bone marrow have long been used as a source of immune system cells in patients whose own immune systems are nonfunctional because of genetic disorders or radiation treatments for cancer.

The developmental potential of adult stem cells is limited to certain tissues. ES cells hold more promise than adult stem cells for medical applications because ES cells are **pluripotent**, capable of differentiating into many different cell types. The only way to obtain ES cells thus far, however, has been to harvest them from human embryos, which raises ethical and political issues.

ES cells are currently obtained from embryos donated by patients undergoing infertility treatment or from long-term cell cultures originally established with cells isolated from donated embryos. If scientists are able to clone human embryos to the blastocyst stage, they might be able to use such clones as the source of ES cells in the future. Furthermore, with a donor nucleus from a person with a particular disease, they might be able to produce ES cells for treatment that match the patient and are thus not rejected by his or her immune system. When the main aim of cloning is to produce ES cells to treat disease, the process is called *therapeutic cloning.* Although most people believe that reproductive cloning of humans is unethical, opinions vary about the morality of therapeutic cloning.

Resolving the debate now seems less imperative due to recent reports that ES cells can be acquired by turning back the clock in fully differentiated adult cells. Accomplishing this feat, which posed formidable obstacles, was announced in 2007, first by labs using mouse skin cells and then by others using human skin cells. In all cases, researchers transformed the skin cells into ES cells by using retroviruses to introduce extra cloned copies of four "stem cell" master regulatory genes. All tests carried out so far indicate that the transformed cells, which are known as *induced pluripotent stem (iPS) cells*, can do everything ES cells can do. Fine-tuning of techniques to direct iPS cells to become specific cell types is an area of intense research that has already seen some success. These cell types could eventually provide tailor-made "replacement" cells for patients without using any human eggs or embryos, thus circumventing most ethical objections.

CONCEPT CHECK 20.3

1. If you clone a carrot, will all the progeny plants ("clones") look identical? Why or why not?
2. Based on current knowledge, how would you explain the difference in the percentage of tadpoles that developed from the two kinds of donor nuclei in Figure 20.17?
3. **WHAT IF?** If you were a doctor who wanted to use iPS cells to treat a patient with severe diabetes, what new technique would have to be developed?

For suggested answers, see Appendix A.

CONCEPT 20.4 The practical applications of DNA technology affect our lives in many ways

DNA technology is in the news almost every day. Most often, the topic is a new and promising application in medicine, but this is just one of numerous fields benefiting from DNA technology and genetic engineering.

Medical Applications

One important use of DNA technology is the identification of human genes whose mutation plays a role in genetic diseases. These discoveries may lead to ways of diagnosing, treating, and even preventing such conditions. DNA technology is also contributing to our understanding of "nongenetic" diseases, from arthritis to AIDS, since a person's genes influence susceptibility to these diseases. Furthermore, diseases of all sorts involve changes in gene expression within the affected cells and often within the patient's immune system. By using DNA microarray assays or other techniques to compare gene expression in healthy and diseased tissues, as seen in Figure 20.1, researchers hope to find many of the genes that are turned on or off in particular diseases. These genes and their products are potential targets for prevention or therapy.

Diagnosis of Diseases

A new chapter in the diagnosis of infectious diseases has been opened by DNA technology, in particular the use of PCR and labeled nucleic acid probes to track down pathogens. For example, because the sequence of the RNA genome of HIV is known, RT-PCR can be used to amplify, and thus detect, HIV RNA in blood or tissue samples (see Figure 20.13). RT-PCR is often the best way to detect an otherwise elusive infective agent.

Medical scientists can now diagnose hundreds of human genetic disorders by using PCR with primers that target the genes associated with these disorders. The amplified DNA product is then sequenced to reveal the presence or absence of the disease-causing mutation. Among the genes for human diseases that have been identified are those for sickle-cell disease, hemophilia, cystic fibrosis, Huntington's disease, and Duchenne muscular dystrophy. Individuals afflicted with such diseases can often be identified before the onset of symptoms, even before birth. PCR can also be used to identify symptomless carriers of potentially harmful recessive alleles, essentially replacing Southern blotting for this purpose.

For some genetic disorders, medical scientists can detect an abnormal disease-causing allele by testing for *genetic markers* that are known to be very close (linked) to the allele. A genetic marker is a DNA sequence that varies in a population; in a gene, such sequence variation is the basis of different alleles. Just like coding sequences, noncoding DNA at a specific locus on a chromosome may exhibit small nucleotide differences among individuals. Variations in DNA sequence are called *polymorphisms* (from the Greek for "many forms").

Among the most useful of these genetic markers are single base-pair variations in the genomes of the human population. A single base-pair site where variation is found in at least 1% of the population is called a **single nucleotide polymorphism** (**SNP**, pronounced "snip"). SNPs occur on average about once in 100 to 300 base pairs in the human genome and are found in both coding and noncoding DNA sequences. (About 99% of our genome doesn't code for protein, as you will learn in Chapter 21.)

Some SNPs alter the sequence recognized by a restriction enzyme, as is the case for the single-nucleotide difference between the alleles for normal and sickle-cell β-globin. Such alterations change the lengths of the restriction fragments formed by digestion with that enzyme (see Figure 20.10). This type of sequence change, which can occur in either coding or noncoding regions, is called a **restriction fragment length polymorphism** (**RFLP**, pronounced "Rif-lip"). Scientists figured out how to use Southern blotting to detect RFLPs almost 30 years ago, realizing that RFLPs could serve as useful genetic markers. It is not necessary to actually sequence the DNA of multiple individuals to find SNPs; today they can be detected by very sensitive microarray analysis or by PCR.

But how can RFLPs or other SNPs help us diagnose a genetic disorder? The presence of an abnormal allele can be diagnosed with reasonable accuracy if a closely linked SNP marker has been found **(Figure 20.21)**. Alleles for Huntington's disease and a number of other genetic diseases were first detected by means of RFLPs in this indirect way. If the marker and the gene itself are close enough, crossing over between the marker and the gene is very unlikely during gamete formation. Therefore, the marker and gene will almost always be inherited together, even though the marker is not part of the gene. The same principle applies to all kinds of markers.

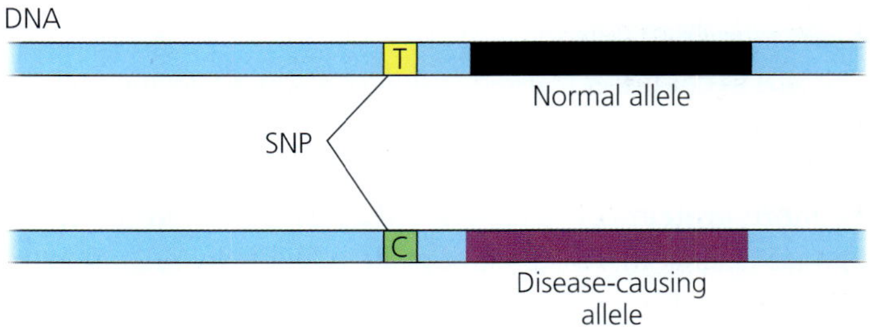

▲ **Figure 20.21 Single nucleotide polymorphisms (SNPs) as genetic markers for disease-causing alleles.** This diagram depicts homologous segments of DNA from a family in which some members have a genetic disease. In this family, unaffected family members have a T at a particular SNP locus. If a family member has a C at that locus, there is a high probability that the individual has also inherited the disease-causing allele. (Here, only a single strand is shown for each DNA molecule.)

Human Gene Therapy

Gene therapy—introducing genes into an afflicted individual for therapeutic purposes—holds great potential for treating disorders traceable to a single defective gene. In theory, a normal allele of the defective gene could be inserted into the somatic cells of the tissue affected by the disorder.

For gene therapy of somatic cells to be permanent, the cells that receive the normal allele must be ones that multiply throughout the patient's life. Bone marrow cells, which include the stem cells that give rise to all the cells of the blood and immune system, are prime candidates. **Figure 20.22** outlines one possible procedure for gene therapy of an individual whose bone marrow cells do not produce a vital enzyme because of a single defective gene. One type of severe combined immunodeficiency (SCID) is caused by just this kind of defect. If the treatment is successful, the patient's bone marrow cells will begin producing the missing protein, and the patient will be cured.

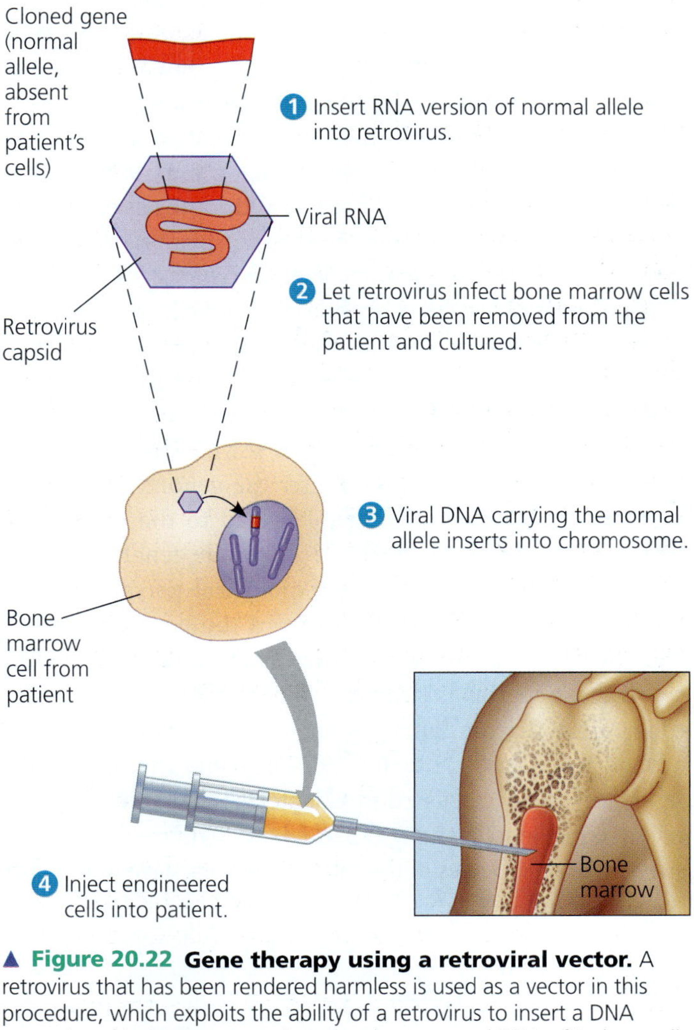

▲ **Figure 20.22 Gene therapy using a retroviral vector.** A retrovirus that has been rendered harmless is used as a vector in this procedure, which exploits the ability of a retrovirus to insert a DNA transcript of its RNA genome into the chromosomal DNA of its host cell (see Figure 19.8). If the foreign gene carried by the retroviral vector is expressed, the cell and its descendants will possess the gene product, and the patient may be cured. Cells that reproduce throughout life, such as bone marrow cells, are ideal candidates for gene therapy.

The procedure shown in Figure 20.22 has been used in gene therapy trials for SCID. In a trial begun in France in 2000, ten young children with SCID were treated by the same procedure. Nine of these patients showed significant, definitive improvement after two years, the first indisputable success of gene therapy. However, three of the patients subsequently developed leukemia, a type of blood cell cancer, and one of them died. Researchers discovered that in two cases, the retroviral vector used to carry the normal allele into bone marrow cells had inserted near a gene involved in the proliferation and development of blood cells, which could have led to the leukemia. In a recent experiment on mice, a replacement of the mouse version of the same gene led to a high incidence of lymphoma, also a blood cancer. These results suggest that an unknown function of the gene itself may be responsible.

Gene therapy also raises several other technical questions. For example, how can the activity of the transferred gene be controlled so that cells make appropriate amounts of the gene product at the right time and in the right place? How can we be sure that the insertion of the therapeutic gene does not harm some other necessary cell function? As more is learned about DNA control elements and gene interactions, researchers may be able to answer such questions.

In addition to technical challenges, gene therapy raises ethical questions. Some critics believe that tampering with human genes in any way is immoral. Other observers see no fundamental difference between the transplantation of genes into somatic cells and the transplantation of organs.

Treatment of human germ-line cells in the hope of correcting a defect in future generations raises further ethical issues. Such genetic engineering is routinely done in laboratory mice, and the technical problems relating to similar genetic engineering in humans will eventually be solved. Under what circumstances, if any, should we alter the genomes of human germ lines? Would this inevitably lead to the practice of eugenics, a deliberate effort to control the genetic makeup of human populations? From a biological perspective, the elimination of unwanted alleles from the gene pool could backfire. Genetic variation is a necessary ingredient for the survival of a species as environmental conditions change with time. Alleles that are damaging under some conditions may be advantageous under other conditions (one example is the sickle-cell allele, discussed in Chapter 14). Are we willing to risk making genetic changes that could be detrimental to the survival of our species in the future? We may have to face this question soon.

Pharmaceutical Products

The pharmaceutical industry derives significant benefit from advances in DNA technology and genetic research, applying them to the development of useful drugs to treat diseases. Pharmaceutical products are synthesized using methods of either organic chemistry or biotechnology, depending on the nature of the product.

Synthesis of Small Molecules for Use as Drugs An exciting recent development has been the synthesis of small molecules that are tailored to combat certain cancers by blocking the function of a protein crucial for the tumor cells' survival. One drug, imatinib (trade name Gleevec), is a small molecule that inhibits a specific receptor tyrosine kinase (see Figure 11.7). The overexpression of this receptor, resulting from a chromosomal translocation, is instrumental in causing chronic myelogenous leukemia (CML; see Figure 15.17). Patients in the early stages of CML who are treated with imatinib have exhibited nearly complete, sustained remission from the cancer. Drugs that work like this have also been used with success to treat a few types of lung and breast cancers. This approach is feasible only for cancers for which the molecular basis is fairly well understood.

Pharmaceutical products that are proteins can be synthesized on a large scale, using cells or whole organisms. Cell cultures are more widely used at present.

Protein Production in Cell Cultures You learned earlier in the chapter about DNA cloning and gene expression systems for producing large quantities of proteins that are present naturally in only minute amounts. The host cells used in such expression systems can even be engineered to secrete a protein as it is made, thereby simplifying the task of purifying it by traditional biochemical methods.

Among the first pharmaceutical products "manufactured" in this way were human insulin and human growth hormone (HGH). Some 2 million people with diabetes in the United States depend on insulin treatment to control their disease. Human growth hormone has been a boon to children born with a form of dwarfism caused by inadequate amounts of HGH. Another important pharmaceutical product produced by genetic engineering is tissue plasminogen activator (TPA). If administered shortly after a heart attack, TPA helps dissolve blood clots and reduces the risk of subsequent heart attacks.

Cells in culture can also be used to produce vaccines, which stimulate the immune system to defend the body against specific pathogens (see Chapter 43). A pathogen typically has one or more proteins on its surface that trigger an immune response against it. This type of protein, made by recombinant DNA techniques, can function as a vaccine against the pathogen. Alternatively, genetic engineering can be used to modify a pathogen's genome, resulting in a weakened pathogen that can then serve as a "live" vaccine.

Protein Production by "Pharm" Animals and Plants In some cases, instead of using cell systems *in vitro* to produce large quantities of protein products, pharmaceutical scientists can use whole animals. Their techniques are those we alluded to when we discussed the possibility of genetically engineering human germ-line cells, but they are performed on animals other than humans. Using these methods, scientists can introduce a gene from an animal of one genotype into the genome of another individual, often of a different species. This individual is then called a **transgenic** animal. To do this, they first remove eggs from a female of the recipient species and fertilize them *in vitro*. Meanwhile, they have cloned the desired gene from the donor organism. They then inject the cloned DNA directly into the nuclei of the fertilized eggs. Some of the cells integrate the foreign DNA, the *transgene*, into their genomes and are able to express the foreign gene. The engineered embryos are then surgically implanted in a surrogate mother. If an embryo develops successfully, the result is a transgenic animal, containing a gene from a third "parent," which may be of a different species.

Assuming that the introduced gene encodes a protein desired in large quantities, these transgenic animals can act as pharmaceutical "factories." For example, a transgene for a human blood protein such as antithrombin can be inserted into the genome of a goat in such a way that the transgene's product is secreted in the animal's milk **(Figure 20.23)**. The protein can then be purified from the milk, usually more easily than from a cell culture. Researchers have also engineered transgenic chickens that express large amounts of the transgene's product in eggs. Biotechnology companies consider the characteristics of candidate animals in deciding which to use for engineering. For example, goats reproduce faster than cows, and it is possible to harvest more protein from goat milk than from the milk of other rapidly reproducing mammals, such as rabbits.

Human proteins produced in transgenic farm animals for use in humans may differ in some ways from the naturally produced human proteins, possibly because of subtle differences in protein modification. Therefore, such proteins must be tested very carefully to ensure that they (or contaminants from the farm animals) will not cause allergic reactions or other adverse effects in patients who receive them.

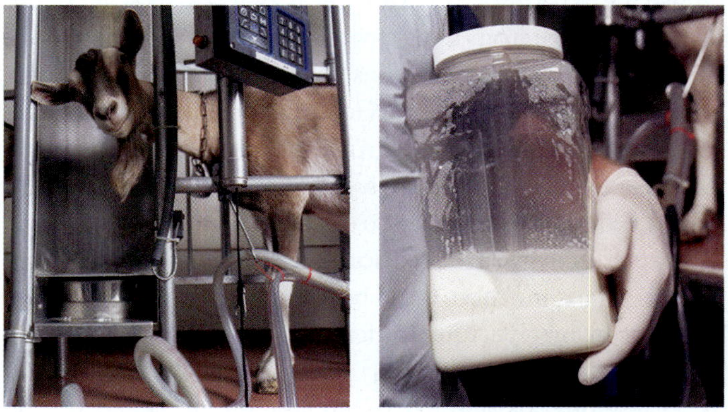

▲ **Figure 20.23 Goats as "pharm" animals.** This transgenic goat carries a gene for a human blood protein, antithrombin, which she secretes in her milk. Patients with a rare hereditary disorder in which this protein is lacking suffer from formation of blood clots in their blood vessels. Easily purified from the goat's milk, the protein is currently under evaluation as an anticlotting agent.

In a novel twist, the pharmaceutical industry is beginning to develop "pharm" plants, analogous to "pharm" animals. Although natural plants have long been sources of drugs, researchers are now creating plants that make human proteins for medical use and viral proteins for use as vaccines. In 2007, the U.S. Food and Drug Administration (FDA) approved a plan to plant over 3,000 acres in Kansas with rice harboring genes for milk proteins. The harvested proteins would be used in rehydration formulas to treat infant diarrhea, a serious problem in developing countries.

Forensic Evidence and Genetic Profiles

In violent crimes, body fluids or small pieces of tissue may be left at the scene or on the clothes or other possessions of the victim or assailant. If enough blood, semen, or tissue is available, forensic laboratories can determine the blood type or tissue type by using antibodies to detect specific cell-surface proteins. However, such tests require fairly fresh samples in relatively large amounts. Also, because many people have the same blood or tissue type, this approach can only exclude a suspect; it cannot provide strong evidence of guilt.

DNA testing, on the other hand, can identify the guilty individual with a high degree of certainty, because the DNA sequence of every person is unique (except for identical twins). Genetic markers that vary in the population can be analyzed for a given person to determine that individual's unique set of genetic markers, or **genetic profile**. (This term is preferred over "DNA fingerprint" by forensic scientists, who want to emphasize the heritable aspect of these markers rather than simply the fact that they produce a pattern on a gel that, like a fingerprint, is visually recognizable.) The FBI started applying DNA technology in forensics in 1988, using RFLP analysis by Southern blotting to detect similarities and differences in DNA samples. This method required much smaller samples of blood or tissue than earlier methods—only about 1,000 cells.

Today, forensic scientists use an even more sensitive method that takes advantage of variations in length of genetic markers called **short tandem repeats (STRs)**. These are tandemly repeated units of 2- to 5-base sequences in specific regions of the genome. The number of repeats present in these regions is highly variable from person to person (polymorphic). For example, one individual may have the sequence ACAT repeated 30 times at one genome locus, whereas another individual may have 18 repeats at this locus. PCR is used to amplify particular STRs, using sets of primers that are labeled with different-colored fluorescent tags; the length of the region, and thus the number of repeats, can then be determined by electrophoresis. Because Southern blotting is not required, this method is quicker than RFLP analysis. And the PCR step allows use of the method even when the DNA is in poor condition or available only in minute quantities. A tissue sample containing as few as 20 cells can be sufficient for PCR amplification.

In a murder case, for example, this method can be used to compare DNA samples from the suspect, the victim, and a small amount of blood found at the crime scene. The forensic scientist tests only a few selected portions of the DNA—usually 13 STR markers. However, even this small set of markers can provide a forensically useful genetic profile because the probability that two people (who are not identical twins) would have exactly the same set of STR markers is vanishingly small. The Innocence Project, a nonprofit organization dedicated to overturning wrongful convictions, uses STR analysis of archived samples from crime scenes to revisit old cases. As of 2006, 18 innocent people had been released from prison as a result of forensic and legal work by this group **(Figure 20.24)**.

Genetic profiles can also be useful for other purposes. A comparison of the DNA of a mother, her child, and the purported father can conclusively settle a question of paternity. Sometimes paternity is of historical interest: Genetic profiles provided strong evidence that Thomas Jefferson or one of his close male relatives fathered at least one of the children of his slave Sally Hemings. Genetic profiles can also identify victims of mass casualties. The largest such effort occurred after the attack on the World Trade Center in 2001; more than 10,000 samples of victims' remains were compared with DNA samples from personal items, such as toothbrushes, provided by families. Ultimately, forensic scientists succeeded in identifying almost 3,000 victims using these methods.

Just how reliable is a genetic profile? The greater the number of markers examined in a DNA sample, the more likely it is that the profile is unique to one individual. In forensic cases using STR analysis with 13 markers, the probability of two people having identical DNA profiles is somewhere between one chance in 10 billion and one in several trillion. (For comparison, the world's population in 2007 was about 6.6 billion.) The exact probability depends on the frequency of those markers in the general population. Information on how common various markers are in different ethnic groups is critical because these marker frequencies may vary considerably among ethnic groups and between a particular ethnic group and the population as a whole. With the increasing availability of frequency data, forensic scientists can make extremely accurate statistical calculations. Thus, despite problems that can still arise from insufficient data, human error, or flawed evidence, genetic profiles are now accepted as compelling evidence by legal experts and scientists alike.

(a) In 1984, Earl Washington was convicted and sentenced to death for the 1982 rape and murder of Rebecca Williams. His sentence was commuted to life in prison in 1993 due to new doubts about the evidence. In 2000, STR analysis by forensic scientists associated with The Innocence Project showed conclusively that he was innocent. This photo shows Washington just before his release in 2001, after 17 years in prison.

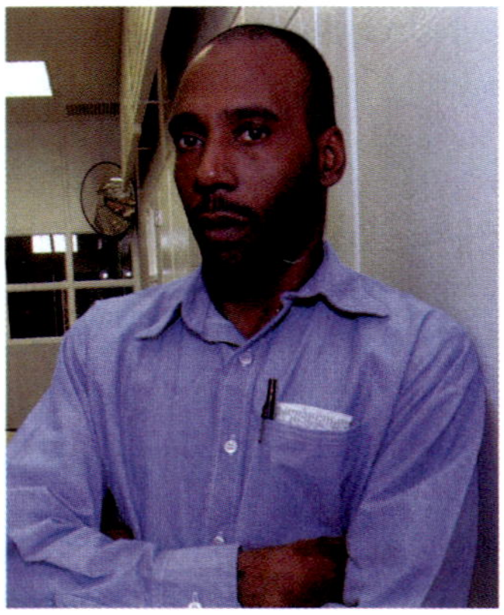

Source of sample	STR marker 1	STR marker 2	STR marker 3
Semen on victim	17, 19	13, 16	12, 12
Earl Washington	16, 18	14, 15	11, 12
Kenneth Tinsley	17, 19	13, 16	12, 12

(b) In STR analysis, selected STR markers in a DNA sample are amplified by PCR, and the PCR products are separated by electrophoresis. The procedure reveals how many repeats are present for each STR locus in the sample. An individual has two alleles per STR locus, each with a certain number of repeats. This table shows the number of repeats for three STR markers in three samples: from semen found on the victim, from Washington, and from another man named Kenneth Tinsley, who was in prison because of an unrelated conviction. These and other STR data (not shown) exonerated Washington and led Tinsley to plead guilty to the murder.

▲ **Figure 20.24 STR analysis used to release an innocent man from prison.**

Environmental Cleanup

Increasingly, the remarkable ability of certain microorganisms to transform chemicals is being exploited for environmental cleanup. If the growth needs of such microbes make them unsuitable for direct use, scientists can now transfer the genes for their valuable metabolic capabilities into other microorganisms, which can then be used to treat environmental problems. For example, many bacteria can extract heavy metals, such as copper, lead, and nickel, from their environments and incorporate the metals into compounds such as copper sulfate or lead sulfate, which are readily recoverable. Genetically engineered microbes may become important in both mining minerals (especially as ore reserves are depleted) and cleaning up highly toxic mining wastes. Biotechnologists are also trying to engineer microbes that can degrade chlorinated hydrocarbons and other harmful compounds. These microbes could be used in wastewater treatment plants or by manufacturers before the compounds are ever released into the environment.

In a more "old-fashioned" application of biotechnology, *biofuels* from crops such as corn, soybeans, and cassava have been proposed as a supplement or even replacement for fossil fuels. In this case, no genetic engineering is involved. To produce "bioethanol," the starch made naturally by the plants is simply converted to sugar and then fermented by microorganisms; a similar process can produce "biodiesel" from plant oils. Either product can be mixed with gasoline or used alone to power vehicles. Proponents say these fuels have less impact on the environment than fossil fuels, but others disagree, asserting that the environmental effects of growing these crops make their total impact greater than that of fossil fuels.

Agricultural Applications

Scientists are working to learn more about the genomes of agriculturally important plants and animals. For a number of years, they have been using DNA technology in an effort to improve agricultural productivity.

Animal Husbandry

The selective breeding of livestock, or animal husbandry, has exploited naturally occurring mutations and genetic recombination for centuries. As we described earlier, DNA technology enables scientists to produce transgenic animals, which speeds up the selective breeding process. The goals of creating a transgenic animal are often the same as the goals of traditional breeding—for instance, to make a sheep with better quality wool, a pig with leaner meat, or a cow that will mature in a shorter time. Scientists might, for example, identify and clone a gene that causes the development of larger muscles (muscles make up most of the meat we eat) in one breed of cattle and transfer it to other cattle or even to sheep.

However, problems such as low fertility or increased susceptibility to disease are not uncommon among farm animals carrying genes from humans and other foreign species. Animal health and welfare are important issues to consider when developing transgenic animals.

Genetic Engineering in Plants

Agricultural scientists have already endowed a number of crop plants with genes for desirable traits, such as delayed ripening and resistance to spoilage and disease. In one striking way, plants are easier to genetically engineer than most animals. For many plant species, a single tissue cell grown in culture can give rise to an adult plant (see Figure 20.16). Thus, genetic manipulations can be performed on an ordinary somatic cell and the cell then used to generate an organism with new traits.

The most commonly used vector for introducing new genes into plant cells is a plasmid, called the **Ti plasmid**, from the soil bacterium *Agrobacterium tumefaciens*. This plasmid integrates a segment of its DNA, known as T DNA, into the chromosomal DNA of its host plant cells. For vector purposes, researchers work with versions of the plasmid that do not cause disease (unlike the wild-type version) and that have been engineered to carry genes of interest within the borders of the T DNA. **Figure 20.25** outlines one method for using the Ti plasmid to produce transgenic plants. Scientists can introduce recombinant Ti plasmids into plant cells by electroporation. Alternatively, the recombinant plasmid can be put back into *Agrobacterium*; susceptible plants or plant cells growing in culture are then infected with bacteria that contain the recombinant plasmid.

▼ **Figure 20.25 Research Method**

Using the Ti Plasmid to Produce Transgenic Plants

APPLICATION Genes conferring useful traits, such as pest resistance, herbicide resistance, delayed ripening, and increased nutritional value, can be transferred from one plant variety or species to another using the Ti plasmid as a vector.

TECHNIQUE

1. The Ti plasmid is isolated from the bacterium *Agrobacterium tumefaciens*. The segment of the plasmid that integrates into the genome of host cells is called T DNA.
2. The foreign gene of interest is inserted into the middle of the T DNA using methods shown in Figure 20.4.
3. Recombinant plasmids can be introduced into cultured plant cells by electroporation. Or plasmids can be returned to *Agrobacterium*, which is then applied as a liquid suspension to the leaves of susceptible plants, infecting them. Once a plasmid is taken into a plant cell, its T DNA integrates into the cell's chromosomal DNA.

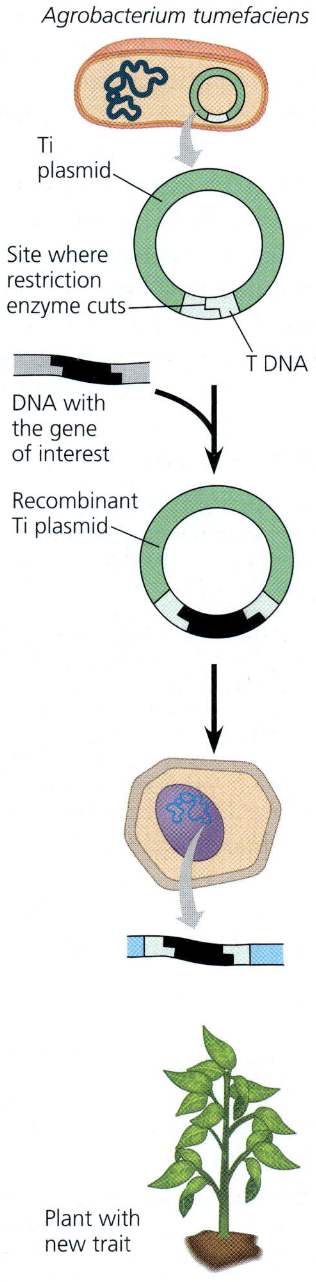

RESULTS Transformed cells carrying the transgene of interest can regenerate complete plants that exhibit the new trait conferred by the transgene.

Genetic engineering is rapidly replacing traditional plant-breeding programs, especially for useful traits, such as herbicide or pest resistance, determined by one or a few genes. Crops engineered with a bacterial gene making the plants resistant to herbicides can grow while weeds are destroyed, and genetically engineered crops that can resist destructive insects reduce the need for chemical insecticides. In India, the insertion of a salinity resistance gene from a coastal mangrove plant into the genomes of several rice varieties has resulted in rice plants that can grow in water three times as salty as seawater. The research foundation that carried out this feat of genetic engineering estimates that one-third of all irrigated land has high salinity owing to overirrigation and intensive use of chemical fertilizers, representing a serious threat to the food supply. Thus, salinity-resistant crop plants would be enormously valuable worldwide.

Genetic engineering also has great potential for improving the nutritional value of crop plants. For instance, scientists have developed transgenic rice plants that produce yellow rice grains containing beta-carotene, which our body uses to make vitamin A (see Figure 38.18). This "golden" rice could help prevent vitamin A deficiency in the half of the world's population that depends on rice as a staple food. Currently, large numbers of young children in Southeast Asia suffer from vitamin A deficiency, which leads to vision impairment and increases susceptibility to disease.

Safety and Ethical Questions Raised by DNA Technology

Early concerns about potential dangers associated with recombinant DNA technology focused on the possibility that hazardous new pathogens might be created. What might happen, for instance, if cancer cell genes were transferred into bacteria or viruses? To guard against such rogue microbes, scientists developed a set of guidelines that were adopted as formal government regulations in the United States and some other countries. One safety measure is a set of strict laboratory procedures designed to protect researchers from infection by engineered microbes and to prevent the microbes from accidentally leaving the laboratory. In addition, strains of microorganisms to be used in recombinant DNA experiments are genetically crippled to ensure that they cannot survive outside the laboratory. Finally, certain obviously dangerous experiments have been banned.

Today, most public concern about possible hazards centers not on recombinant microbes but on **genetically modified (GM) organisms** used as food. In common language, a GM organism is one that has acquired by artificial means one or more genes from another species or even from another variety of the same species. Some salmon, for example, have been genetically modified by addition of a more active salmon growth hormone gene. However, the majority of the GM organisms that contribute to our food supply are not animals, but crop plants.

GM crops are widespread in the United States, Argentina, and Brazil; together these countries account for over 80% of the world's acreage devoted to such crops. In the United States, the majority of corn, soybeans, and canola are GM crops, and GM products are not required to be labeled. However, the same foods are an ongoing subject of controversy in Europe, where the GM revolution has met with strong opposition. Many Europeans are concerned about the safety of GM foods and the possible environmental consequences of growing GM plants. Early in 2000, negotiators from 130 countries (including the United States) agreed on a Biosafety Protocol that requires exporters to identify GM organisms present in bulk food shipments and allows importing countries to decide whether the products pose environmental or health risks. Since then, European countries have, on occasion, refused crops from the United States and other countries, leading to trade disputes. Although a small number of GM crops have been grown on European soil, these products have generally failed in local markets, and the future of GM crops in Europe is uncertain.

Advocates of a cautious approach toward GM crops fear that transgenic plants might pass their new genes to close relatives in nearby wild areas. We know that lawn and crop grasses, for example, commonly exchange genes with wild relatives via pollen transfer. If crop plants carrying genes for resistance to herbicides, diseases, or insect pests pollinated wild ones, the offspring might become "super weeds" that are very difficult to control. Another possible hazard, suggested by one laboratory-based study, is that a transgene encoding a pesticide-like protein might cause plants to produce pollen toxic to butterflies. However, scientists with the Agricultural Research Service concluded from a two-year field study that butterflies were unlikely to be exposed to toxic levels of pollen.

As for the risks to human health from GM foods, some people fear that the protein products of transgenes might lead to allergic reactions. Although there is some evidence that this could happen, advocates claim that these proteins could be tested in advance to avoid producing ones that cause allergic reactions.

Today, governments and regulatory agencies throughout the world are grappling with how to facilitate the use of biotechnology in agriculture, industry, and medicine while ensuring that new products and procedures are safe. In the United States, such applications of biotechnology are evaluated for potential risks by various regulatory agencies, including the FDA, the Environmental Protection Agency, the National Institutes of Health, and the Department of Agriculture. Meanwhile, these same agencies and the public must consider the ethical implications of biotechnology.

Advances in biotechnology have allowed us to obtain complete genome sequences for humans and many other species, providing a vast treasure trove of information about genes. We can ask how certain genes differ from species to species, as well as how genes and, ultimately, entire genomes have evolved.

(These are the subjects of Chapter 21.) At the same time, the increasing speed and falling cost of determining the genome sequences of individuals are raising significant ethical questions. Who should have the right to examine someone else's genetic information? How should that information be used? Should a person's genome be a factor in determining eligibility for a job or insurance? Ethical considerations, as well as concerns about potential environmental and health hazards, will likely slow some applications of biotechnology. There is always a danger that too much regulation will stifle basic research and its potential benefits. However, the power of DNA technology and genetic engineering—our ability to profoundly and rapidly alter species that have been evolving for millennia—demands that we proceed with humility and caution.

CONCEPT CHECK 20.4

1. What is the advantage of using stem cells for gene therapy?
2. List at least three different properties that have been acquired by crop plants via genetic engineering.
3. **WHAT IF?** As a physician, you have a patient with symptoms that suggest a hepatitis A infection. The symptoms come and go, but you have not been able to detect viral proteins in the blood. Knowing that hepatitis A is an RNA virus, what lab tests could you perform to support your diagnosis? Explain what the results would mean.

For suggested answers, see Appendix A.

Chapter 20 Review

MEDIA Go to the Study Area at **www.masteringbio.com** for BioFlix 3-D Animations, MP3 Tutors, Videos, Practice Tests, an eBook, and more.

SUMMARY OF KEY CONCEPTS

CONCEPT 20.1

DNA cloning yields multiple copies of a gene or other DNA segment (pp. 396–405)

- **DNA Cloning and Its Applications: *A Preview*** DNA cloning and other techniques, collectively termed DNA technology, can be used to manipulate and analyze DNA and to produce useful new products and organisms.
- **Using Restriction Enzymes to Make Recombinant DNA** Bacterial restriction enzymes cut DNA molecules within short, specific nucleotide sequences to yield a set of double-stranded DNA fragments with single-stranded sticky ends. The sticky ends on fragments from one DNA source can base-pair with complementary sticky ends on fragments from other DNA molecules; sealing of the base-paired fragments with DNA ligase produces recombinant DNA molecules.
- **Cloning a Eukaryotic Gene in a Bacterial Plasmid**

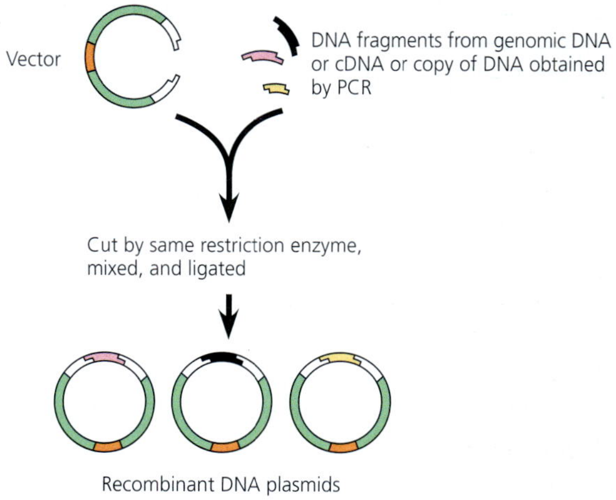

Recombinant plasmids are returned to host cells, each of which divides to form a clone of cells. Collections of clones are stored as libraries.

- **Expressing Cloned Eukaryotic Genes** Several technical difficulties hinder the expression of cloned eukaryotic genes in bacterial host cells. The use of cultured eukaryotic cells as host cells and yeast artificial chromosomes (YACs) as vectors helps avoid these problems.
- **Amplifying DNA *in Vitro*: The Polymerase Chain Reaction (PCR)** PCR can produce many copies of a specific target segment of DNA, using primers that bracket the desired sequence and a heat-resistant DNA polymerase.

MEDIA
Activity Applications of DNA Technology
Activity Restriction Enzymes
Activity Cloning a Gene in Bacteria
Investigation How Can Antibiotic-Resistant Plasmids Transform *E. coli*?

CONCEPT 20.2

DNA technology allows us to study the sequence, expression, and function of a gene (pp. 405–411)

- **Gel Electrophoresis and Southern Blotting** DNA restriction fragments of different lengths can be separated by gel electrophoresis. Specific fragments can be identified by Southern blotting, using labeled probes that hybridize to the DNA immobilized on a "blot" of the gel.
- **DNA Sequencing** Relatively short DNA fragments can be sequenced by the dideoxy chain termination method, which can be performed in automated sequencing machines.
- **Analyzing Gene Expression** Expression of a gene can be investigated using hybridization with labeled probes to look for specific mRNAs, either on a gel (Northern blotting) or in a whole organism (*in situ* hybridization). Also, RNA can be transcribed into cDNA by reverse transcriptase and the cDNA amplified with specific primers (RT-PCR). Microarrays allow researchers to compare the expression of many genes at once in different tissues, at different times, or under different conditions.

▶ **Determining Gene Function** For a gene of unknown function, experimental inactivation of the gene and observation of the resulting phenotypic effects can provide clues to its function.

MEDIA

Activity Gel Electrophoresis of DNA
Activity Analyzing DNA Fragments Using Gel Electrophoresis
Investigation How Can Gel Electrophoresis Be Used to Analyze DNA?

CONCEPT 20.3

Cloning organisms may lead to production of stem cells for research and other applications (pp. 412–416)

▶ Studies showing genomic equivalence (that an organism's cells have the same genome) were the first examples of organismal cloning.

▶ **Cloning Plants: Single-Cell Cultures** Differentiated cells from mature plants are often capable of generating a complete new plant.

▶ **Cloning Animals: Nuclear Transplantation** The nucleus from a differentiated animal cell can sometimes give rise to a new animal if transplanted to an enucleated egg.

▶ **Stem Cells of Animals** Certain stem cells from animal embryos or adult tissues can reproduce and differentiate *in vitro* as well as *in vivo*, offering the potential for medical use.

CONCEPT 20.4

The practical applications of DNA technology affect our lives in many ways (pp. 416–423)

▶ **Medical Applications** DNA technology is increasingly being used in the diagnosis of genetic and other diseases and offers potential for better treatment of genetic disorders or even permanent cures. Single nucleotide polymorphisms (SNPs) and restriction fragment length polymorphisms (RFLPs) are useful as genetic markers linked to disease-causing alleles. Large-scale production of protein hormones and other proteins with therapeutic uses, including safer vaccines, is possible with DNA technology. Some therapeutic proteins are being produced in transgenic "pharm" animals and plants.

▶ **Forensic Evidence and Genetic Profiles** Analysis of genetic markers such as short tandem repeats (STRs) in DNA isolated from tissue or body fluids found at crime scenes can provide definitive evidence that a suspect is guilty or innocent. Such analysis is also useful in parenthood disputes and in identifying the remains of crime victims.

▶ **Environmental Cleanup** Genetically engineered microorganisms can be used to extract minerals from the environment or degrade various types of potentially toxic waste materials.

▶ **Agricultural Applications** The aim of developing transgenic plants and animals is to improve agricultural productivity and food quality.

▶ **Safety and Ethical Questions Raised by DNA Technology** The potential benefits of genetic engineering must be carefully weighed against the potential hazards of creating products or developing procedures that are harmful to humans or the environment.

MEDIA

Activity DNA Fingerprinting
Activity Making Decisions About DNA Technology: Golden Rice

TESTING YOUR KNOWLEDGE

SELF-QUIZ

1. Which of the following tools of recombinant DNA technology is *incorrectly* paired with its use?
a. restriction enzyme—production of RFLPs
b. DNA ligase—enzyme that cuts DNA, creating the sticky ends of restriction fragments
c. DNA polymerase—used in a polymerase chain reaction to amplify sections of DNA
d. reverse transcriptase—production of cDNA from mRNA
e. electrophoresis—separation of DNA fragments

2. Which of the following would *not* be true of cDNA produced using human brain tissue as the starting material?
a. It could be amplified by the polymerase chain reaction.
b. It could be used to create a complete genomic library.
c. It is produced from mRNA using reverse transcriptase.
d. It could be used as a probe to detect genes expressed in the brain.
e. It lacks the introns of the human genes.

3. Plants are more readily manipulated by genetic engineering than are animals because
a. plant genes do not contain introns.
b. more vectors are available for transferring recombinant DNA into plant cells.
c. a somatic plant cell can often give rise to a complete plant.
d. genes can be inserted into plant cells by microinjection.
e. plant cells have larger nuclei.

4. A paleontologist has recovered a bit of tissue from the 400-year-old preserved skin of an extinct dodo (a bird). The researcher would like to compare a specific region of the DNA from the sample with DNA from living birds. Which of the following would be most useful for increasing the amount of dodo DNA available for testing?
a. RFLP analysis
b. polymerase chain reaction (PCR)
c. electroporation
d. gel electrophoresis
e. Southern blotting

5. Expression of a cloned eukaryotic gene in a bacterial cell involves many challenges. The use of mRNA and reverse transcriptase is part of a strategy to solve the problem of
a. post-transcriptional processing.
b. electroporation.
c. post-translational processing.
d. nucleic acid hybridization.
e. restriction fragment ligation.

6. DNA technology has many medical applications. Which of the following is *not* done routinely at present?
a. production of hormones for treating diabetes and dwarfism
b. production of viral proteins for vaccines
c. introduction of genetically engineered genes into human gametes
d. prenatal identification of genetic disease genes
e. genetic testing for carriers of harmful alleles

7. Which of the following sequences in double-stranded DNA is most likely to be recognized as a cutting site for a restriction enzyme?

a. AAGG	b. AGTC	c. GGCC	d. ACCA	e. AAAA
TTCC	TCAG	CCGG	TGGT	TTTT

8. In recombinant DNA methods, the term *vector* can refer to
   a. the enzyme that cuts DNA into restriction fragments.
   b. the sticky end of a DNA fragment.
   c. a RFLP marker.
   d. a plasmid used to transfer DNA into a living cell.
   e. a DNA probe used to identify a particular gene.

9. **DRAW IT** You are making a genomic library for the aardvark, using a bacterial plasmid as a vector. The green diagram below shows the plasmid, which contains the restriction site for the enzyme used in Figure 20.3. Above the plasmid is a region of linear aardvark DNA. Diagram your cloning procedure, showing what would happen to these two molecules during each step. Use one color for the aardvark DNA and its bases and another color for those of the plasmid. Label each step.

Aardvark DNA

Plasmid

10. **WHAT IF?** Imagine you want to study human crystallins, proteins present in the lens of the eye. To obtain a sufficient amount of the protein, you decide to clone the crystallin gene. Would you construct a genomic library or a cDNA library? What material would you use as a source of DNA or RNA?

For Self-Quiz answers, see Appendix A.

MEDIA Visit the Study Area at **www.masteringbio.com** for a Practice Test.

EVOLUTION CONNECTION

11. If DNA-based technologies become widely used, how might they change the way evolution proceeds, as compared with the natural evolutionary mechanisms of the past 4 billion years?

SCIENTIFIC INQUIRY

12. You hope to study a gene that codes for a neurotransmitter protein in human brain cells. You know the amino acid sequence of the protein. Explain how you might (a) identify the genes expressed in a specific type of brain cell, (b) identify the gene for the neurotransmitter, (c) produce multiple copies of the gene for study, and (d) produce a quantity of the neurotransmitter for evaluation as a potential medication.

SCIENCE, TECHNOLOGY, AND SOCIETY

13. Is there danger of discrimination based on testing for "harmful" genes? What policies can you suggest that would prevent such abuses?

14. Government funding of embryonic stem cell research has been a contentious political issue. Why has this debate been so heated? Summarize the arguments for and against embryonic stem cell research, and explain your own position on the issue.

21 Genomes and Their Evolution

▲ **Figure 21.1 What genomic information makes a human or a chimpanzee?**

KEY CONCEPTS

21.1 New approaches have accelerated the pace of genome sequencing

21.2 Scientists use bioinformatics to analyze genomes and their functions

21.3 Genomes vary in size, number of genes, and gene density

21.4 Multicellular eukaryotes have much noncoding DNA and many multigene families

21.5 Duplication, rearrangement, and mutation of DNA contribute to genome evolution

21.6 Comparing genome sequences provides clues to evolution and development

OVERVIEW

Reading the Leaves from the Tree of Life

The woman in **Figure 21.1** and her chimpanzee companions are sharing a laugh—or are they? Does each of them perceive a "joke" and react with this joyful facial expression and its accompanying vocalization? With the advent of recent techniques for rapidly sequencing complete genomes, we can now start to address the genetic basis of intriguing questions like this.

The chimpanzee (*Pan troglodytes*) is our closest living relative on the evolutionary tree of life. Its genome was sequenced in 2005, two years after sequencing of the human genome was largely completed. Now that we can compare our genome with that of the chimpanzee base by base, we can tackle the more general issue of what differences in the genetic information account for the distinct characteristics of these two species of primates.

In addition to determining the sequences of the human and chimpanzee genomes, researchers have obtained complete genome sequences for *E. coli* and numerous other prokaryotes, as well as many eukaryotes, including *Saccharomyces cerevisiae* (brewer's yeast), *Caenorhabditis elegans* (a nematode), *Drosophila melanogaster* (fruit fly), *Mus musculus* (house mouse), and *Macaca mulatta* (rhesus macaque). Fragments of DNA have been sequenced even from extinct species, such as the cave bear and the woolly mammoth! These whole and partial genomes are of great interest in their own right and are also providing important insights into evolution and other biological processes. Broadening the human-chimpanzee comparison to the genomes of other primates and more distantly related animals should reveal the sets of genes that control group-defining characteristics. Beyond that, comparisons with the genomes of bacteria, archaea, fungi, protists, and plants should enlighten us about the long evolutionary history of shared ancient genes and their products.

With the genomes of many species fully sequenced, scientists can study whole sets of genes and their interactions, an approach called **genomics**. The sequencing efforts that feed this approach have generated, and continue to generate, enormous volumes of data. The need to deal with this ever-increasing flood of information has spawned the new field of **bioinformatics**, the application of computational methods to the storage and analysis of biological data.

We will begin this chapter by discussing two approaches to genome sequencing and some of the advances in and applications of bioinformatics. We will then summarize what has been learned from the genomes that have been sequenced thus far. Next, we will describe the composition of the human genome as a representative genome of a complex multicellular eukaryote. Finally, we will explore current ideas about how genomes evolve and about how the evolution of developmental mechanisms could have generated the great diversity of life found on Earth today.

CONCEPT 21.1

New approaches have accelerated the pace of genome sequencing

Sequencing of the human genome, an ambitious undertaking, officially began as the **Human Genome Project** in 1990. Organized by an international, publicly funded consortium of scientists at universities and research institutes, the project involved 20 large sequencing centers in six countries plus a host of other labs working on small projects.

After sequencing of the human genome was largely completed in 2003, the sequence of each chromosome was carefully analyzed and described in a series of papers, the last of which covered chromosome 1 and was published in 2006. With this refinement, researchers termed the sequencing "virtually complete." To reach these milestones, the project proceeded through three stages that provided progressively more detailed views of the human genome: linkage mapping, physical mapping, and DNA sequencing.

Three-Stage Approach to Genome Sequencing

Even before the Human Genome Project began, earlier research had sketched a rough picture of the organization of the genomes of many organisms. For instance, the karyotyping of many species had revealed their chromosome numbers and banding patterns (see Figure 13.3). And some genes had already been located on a particular region of a chromosome by fluorescence *in situ* hybridization (FISH), a method in which fluorescently labeled probes are allowed to hybridize to an immobilized array of whole chromosomes (see Figure 15.1). Cytogenetic maps based on this type of information provided the starting point for more detailed mapping.

With these cytogenetic maps of the chromosomes in hand, the initial stage in sequencing the human genome was to construct a **linkage map** (a type of genetic map; see Chapter 15) of several thousand genetic markers spaced throughout the chromosomes (**Figure 21.2** stage 1). The order of the markers and the relative distances between them on such a map are based on recombination frequencies (see Figure 15.11). The markers can be genes or any other identifiable sequences in the DNA, such as RFLPs or the short tandem repeats (STRs) discussed in Chapter 20. By 1992, researchers had compiled a human linkage map with some 5,000 markers. Such a map enabled them to locate other markers, including genes, by testing for genetic linkage to the known markers. It was also valuable as a framework for organizing more detailed maps of particular regions.

The next stage was the physical mapping of the human genome. In a **physical map**, the distances between markers are expressed by some physical measure, usually the number of base pairs along the DNA. For whole-genome mapping, a physical map is made by cutting the DNA of each chromosome into a number of restriction fragments and then determining the original order of the fragments in the chromosomal DNA. The key is to make fragments that overlap and then use probes or automated nucleotide sequencing of the ends to find the overlaps (Figure 21.2, stage 2). In this way, fragments can be assigned to a sequential order that corresponds to their order in a chromosome.

Supplies of the DNA fragments used for physical mapping are prepared by DNA cloning. In working with large genomes, researchers carry out several rounds of DNA cutting, cloning, and physical mapping. The first cloning vector is often a yeast artificial chromosome (YAC), which can carry inserted fragments a million base pairs long, or a bacterial artificial chromosome (BAC), which typically carries inserts of 100,000 to 300,000 base pairs. After such long fragments are put in order,

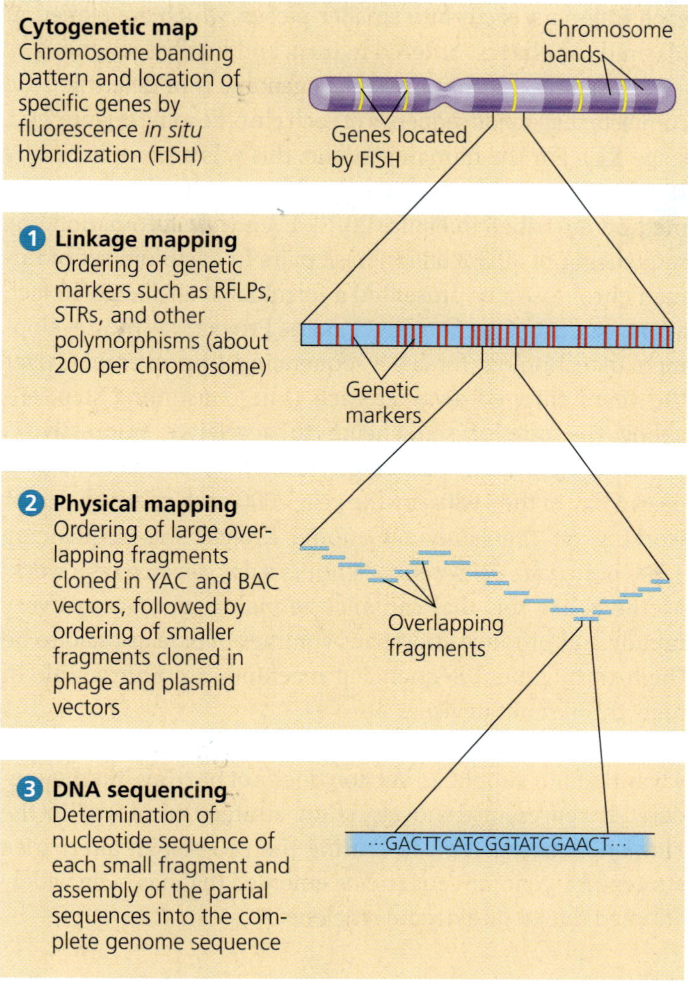

▲ **Figure 21.2 Three-stage approach to sequencing an entire genome.** Starting with a cytogenetic map of each chromosome, researchers with the Human Genome Project proceeded through three stages to reach the ultimate goal, the virtually complete nucleotide sequence of every chromosome.

each fragment is cut into smaller pieces, which are cloned in plasmids or phages, ordered in turn, and finally sequenced.

The ultimate goal in mapping a genome is to determine the complete nucleotide sequence of each chromosome (Figure 21.2, stage 3). For the human genome, this was accomplished by sequencing machines, using the dideoxy chain-termination method described in Figure 20.12. Even with automation, the sequencing of all 3.2 billion base pairs in a haploid set of human chromosomes presented a formidable challenge. In fact, a major thrust of the Human Genome Project was the development of technology for faster sequencing. Improvements over the years chipped away at each time-consuming step, enabling the rate of sequencing to accelerate impressively: Whereas a productive lab could typically sequence 1,000 base pairs a day in the 1980s, by the year 2000 each research center working on the Human Genome Project was sequencing 1,000 base pairs *per second*, 24 hours a day, seven days a week. Methods like this that can analyze biological materials very rapidly and produce enormous volumes of data are said to be "high-throughput." Sequencing machines are an example of high-throughput devices.

In practice, the three stages shown in Figure 21.2 overlap in a way that our simplified version does not portray, but they accurately represent the overarching strategy employed in the Human Genome Project. During the project, an alternative strategy for genome sequencing emerged that has been widely adopted due to its extreme efficiency.

Whole-Genome Shotgun Approach to Genome Sequencing

In 1992, emboldened by advances in sequencing and computer technology, molecular biologist J. Craig Venter devised an alternative approach to the sequencing of whole genomes. Called the *whole-genome shotgun approach*, it essentially skips the linkage mapping and physical mapping stages and starts directly with the sequencing of random DNA fragments. Powerful computer programs then assemble the resulting very large number of overlapping short sequences into a single continuous sequence **(Figure 21.3)**.

Despite the skepticism of many scientists, the value of Venter's approach became clear in 1995 when he and colleagues reported the first complete genome sequence of an organism, the bacterium *Haemophilus influenzae*. In 1998, he set up a company, Celera Genomics, and declared his intention to sequence the entire human genome. Five years later, Celera Genomics and the public consortium jointly announced that sequencing of the human genome was largely complete, two years before the original target date of the Human Genome Project.

Representatives of the public consortium point out that Celera's accomplishment relied heavily on the consortium's maps and sequence data and that the infrastructure established by their approach was a tremendous aid to Celera's efforts. Venter, on the other hand, argues for the efficiency and economy of Celera's methods, and indeed, the public consortium has made some use of them as well. Clearly, both approaches are valuable and have contributed to the rapid completion of genome sequencing for quite a few species.

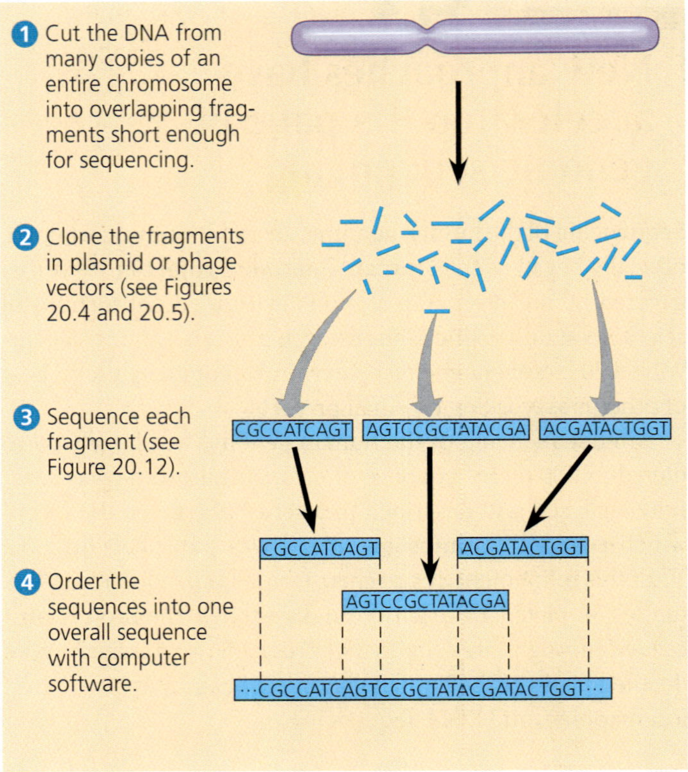

▲ **Figure 21.3 Whole-genome shotgun approach to sequencing.** In this approach, developed by Craig Venter and colleagues at the company he founded, Celera Genomics, random DNA fragments are sequenced and then ordered relative to each other. Compare this approach with the hierarchical, three-stage approach shown in Figure 21.2.

? *The fragments in stage 2 of this figure are depicted as scattered, whereas those in stage 2 of Figure 21.2 are drawn in a much more orderly fashion. How do these depictions reflect the two approaches?*

Today, the whole-genome shotgun approach is widely used. Typically, the fragments are cloned into three different vectors, each of which takes a defined size of insert. The known distance between the ends of the inserted DNA is one more piece of information the computer can use to assemble the sequences properly. A recent study comparing the two approaches has shown that whole-genome shotgun sequencing can miss some duplicated sequences, thus underestimating the size of the genome and missing some genes in those regions. The hybrid approach that ended up being used for the human genome, with the more rapid shotgun sequencing being augmented by some mapping of clones, may in fact prove to be most useful in the long run.

As of 2007, there were still some small sections of the human genome that had not been sequenced. Because of the presence of repetitive DNA and for other poorly understood reasons, certain parts of the chromosomes of multicellular organisms resist detailed sequencing by the usual methods.

At first glance, genome sequences of humans and other organisms are simply dry lists of nucleotide bases—millions of A's, T's, C's, and G's in mind-numbing succession. Crucial to making sense of this massive amount of data have been new analytical approaches, which we discuss next.

CONCEPT CHECK 21.1

1. What is the major difference between a linkage map and a physical map of a chromosome?
2. In general, how does the approach to genome mapping used in the Human Genome Project differ from the whole-genome shotgun approach?
3. **WHAT IF?** Imagine you decide to sequence the genome of a field mouse, a species closely related to the lab mouse whose genome has already been sequenced. Why might having the lab mouse sequence available lead you to use the whole-genome shotgun approach instead of the three-stage approach?

For suggested answers, see Appendix A.

CONCEPT **21.2**

Scientists use bioinformatics to analyze genomes and their functions

Each of the 20 or so sequencing centers around the world working on the Human Genome Project churned out voluminous amounts of DNA sequence day after day. As the data began to accumulate, the need to coordinate efforts to keep track of all the sequences became clear. Thanks to the foresight of scientists and government officials involved in the Human Genome Project, its goals included the establishment of banks of data, or databases, and the refining of analytical software. These databases and software programs would then be centralized and made readily accessible on the Internet. Accomplishing this aim has accelerated progress in DNA sequence analysis by making bioinformatics resources available to researchers worldwide and by speeding up the dissemination of information.

Centralized Resources for Analyzing Genome Sequences

Government-funded agencies carried out their mandate to establish databases and provide software with which scientists could analyze the sequence data. For example, in the United States, a joint endeavor between the National Library of Medicine and the National Institutes of Health (NIH) created the National Center for Biotechnology Information (NCBI), which maintains a website (www.ncbi.nlm.nih.gov) with extensive bioinformatics resources. On this site are links to databases, software, and a wealth of information about genomics and related topics. Similar websites have also been established by the European Molecular Biology Laboratory and the DNA Data Bank of Japan, two genome centers with which the NCBI collaborates. These large, comprehensive websites are complemented by others maintained by individual or small groups of laboratories. Smaller websites often provide databases and software designed for a narrower purpose, such as studying genetic and genomic changes in one particular type of cancer.

The NCBI database of sequences is called Genbank. As of August 2007, it included the sequences of 76 million fragments of genomic DNA, totaling 80 billion base pairs! Genbank is constantly updated, and the amount of data it contains is estimated to double approximately every 18 months. Any sequence in the database can be retrieved and analyzed using software from the NCBI website or elsewhere.

One software program available on the NCBI website, called BLAST, allows the visitor to compare a DNA sequence with every sequence in Genbank, base by base, to look for similar regions. Another program allows comparison of predicted protein sequences. Yet a third can search any protein sequence for common stretches of amino acids (domains) for which a function is known or suspected, and it can show a three-dimensional model of the domain alongside other relevant information (**Figure 21.4**, on the next page). There is even a software program that can compare a collection of sequences, either nucleic acids or polypeptides, and diagram them in the form of an evolutionary tree based on the sequence relationships. (You'll learn more about such diagrams in Chapter 26.)

The NCBI website also maintains a database that includes all three-dimensional protein structures that have been determined. (To review how protein structures are analyzed, see Figure 5.25.) These structures can be rotated by the viewer to see all sides of the protein. Let's say a researcher has an amino acid sequence for all or part of an unknown protein, and it is similar to an amino acid sequence for which the structure has been worked out. In this event, the possible structure of the unknown protein can be predicted by one software program and compared with all other known protein structures using another program. This information can aid the researcher in determining the function of the unknown protein.

There is a vast array of resources available for researchers anywhere in the world to use. Let us now consider the types of questions scientists can address using these resources.

Identifying Protein-Coding Genes Within DNA Sequences

Using available DNA sequences, geneticists can study genes directly, without having to infer genotype from phenotype, as in classical genetics. But this approach poses a new challenge:

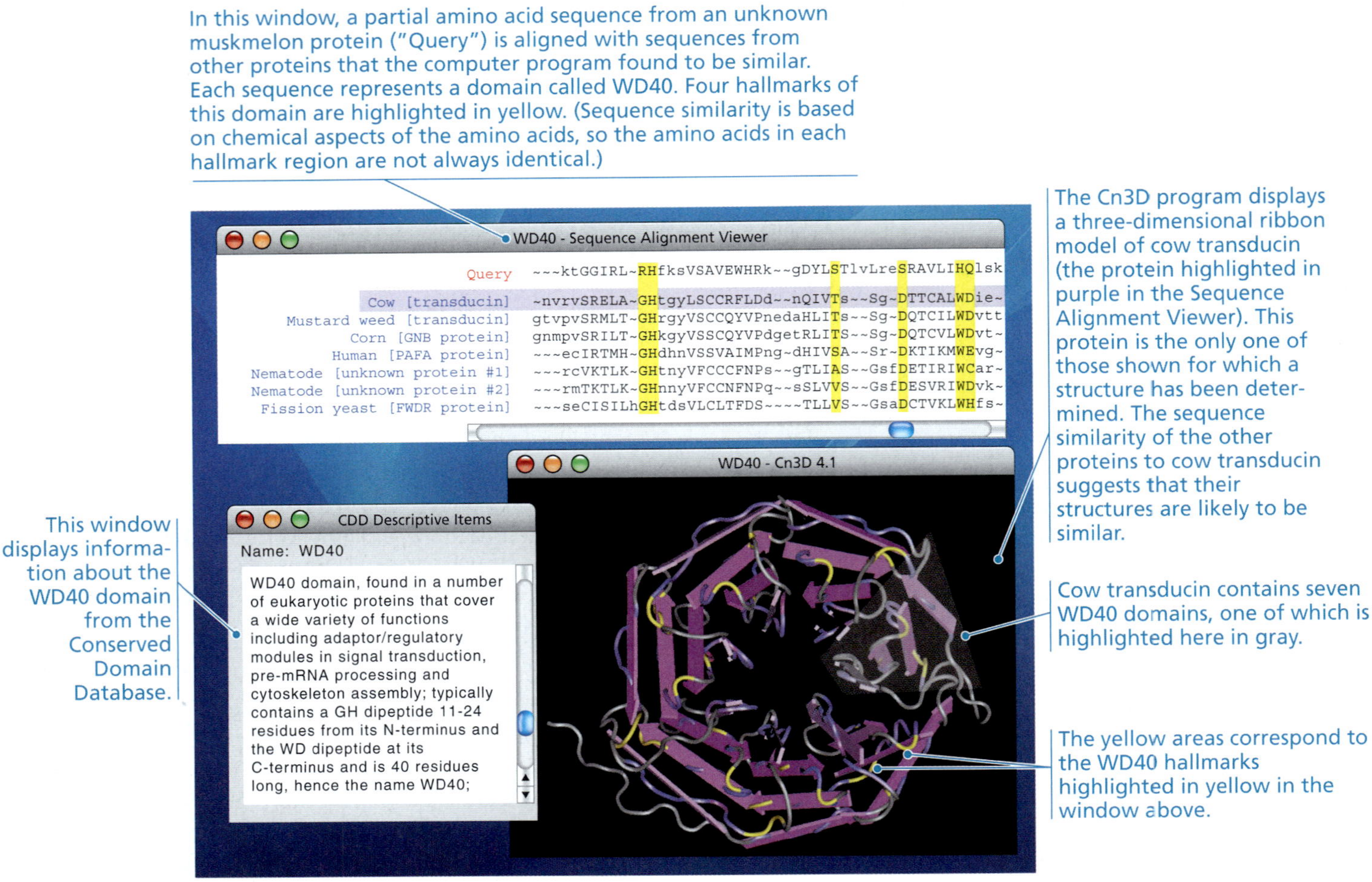

▲ **Figure 21.4 Bioinformatics tools available on the Internet.** A website maintained by the National Center for Biotechnology Information allows scientists and the public to access DNA and protein sequences and other stored data. The site includes a link to a protein structure database (Conserved Domain Database, CDD) that can find and describe similar domains in related proteins, as well as software (Cn3D, "see in 3D") that displays three-dimensional models of domains for which the structure has been determined. Some results are shown from a search for regions of proteins similar to an amino acid sequence in a muskmelon protein.

determining the phenotype from the genotype. Given a long DNA sequence from a database such as Genbank, how can we recognize as-yet unknown protein-coding genes and determine their function?

The usual approach is to use software to scan these stored sequences for transcriptional and translational start and stop signals, for RNA-splicing sites, and for other telltale signs of protein-coding genes. The software also looks for certain short sequences that correspond to sequences present in known mRNAs. Thousands of such sequences, called *expressed sequence tags*, or *ESTs*, have been collected from cDNA sequences and are cataloged in computer databases. This type of analysis identifies sequences that may be previously unknown protein-coding genes.

The identities of about half of the human genes were known before the Human Genome Project began. But what about the others, the previously unknown genes revealed by analysis of DNA sequences? Clues about their identities come from comparing the sequences of gene candidates with those of known genes from other organisms, using the software described previously. Due to redundancy in the genetic code, the DNA sequence itself may vary more than the protein sequence does. Thus, scientists interested in proteins often compare the predicted amino acid sequence of the protein to that of other proteins.

Sometimes a newly identified sequence will match, at least partially, the sequence of a gene or protein whose function is well known. For example, part of a new gene may match a known gene that encodes an important signaling pathway protein such as a protein kinase (see Chapter 11), suggesting that the new gene does, too. Alternatively, the new gene sequence may be similar to a previously encountered sequence whose function is still unknown. Another possibility is that the sequence is entirely unlike anything ever seen before. This was true for about a third of the genes of *E. coli* when its genome was sequenced. In the last case, protein function is usually deduced through a combination of biochemical and functional studies. The biochemical approach aims to determine the three-dimensional structure of the protein as well as other attributes, such as potential binding sites for other molecules. Functional studies usually involve blocking or disabling the gene to see what effect that has on the phenotype. RNAi, described in Chapter 20, is an example of an experimental technique used to block gene function.

Understanding Genes and Their Products at the Systems Level

The impressive computational power provided by the tools of bioinformatics allows the study of whole sets of genes and their interactions, as well as the comparison of genomes from different species. Genomics is a rich source of new insights into fundamental questions about genome organization, regulation of gene expression, growth and development, and evolution.

The success in sequencing genomes and studying entire sets of genes has encouraged scientists to attempt similar systematic study of the full protein sets (*proteomes*) encoded by genomes, an approach called **proteomics**. Proteins, not the genes that encode them, actually carry out most of the activities of the cell. Therefore, we must study when and where proteins are produced in an organism, as well as how they interact in networks, if we are to understand the functioning of cells and organisms.

How Systems Are Studied: An Example

Genomics and proteomics are enabling biologists to approach the study of life from an increasingly global perspective. Using the tools we have described, biologists have begun to compile catalogs of genes and proteins—listings of all the "parts" that contribute to the operation of cells, tissues, and organisms. With such catalogs in hand, researchers have shifted their attention from the individual parts to their functional integration in biological systems. As you may recall, in Chapter 1 we discussed this systems biology approach, which aims to model the dynamic behavior of whole biological systems.

One basic application of the systems biology approach is to define gene circuits and protein interaction networks. To map the protein interaction network in *Drosophila* that was mentioned in Chapter 1, for instance, researchers started with over 10,000 predicted RNA transcripts. They then used molecular techniques to test interactions between the whole or partial protein products of these transcripts. Using statistical tests to select the interactions for which the data were strongest, they ended up with approximately 4,700 proteins that appeared to participate in more than 4,000 interactions. A subset of these interactions is displayed graphically in **Figure 21.5**; the details are more easily seen in the two successive blowups in the figure. To process the vast number of protein-protein interactions generated by this experiment and integrate them into the graphic model required powerful computers, mathematical tools, and newly developed software. Thus, the systems biology approach has really been made possible by advances in bioinformatics.

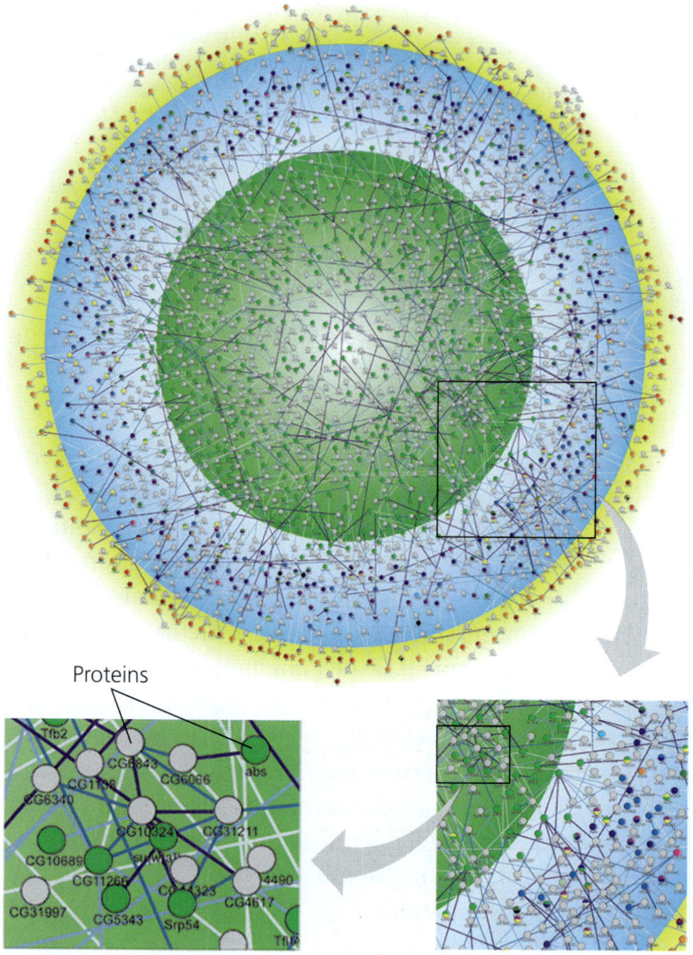

▲ **Figure 21.5 The systems biology approach to protein interactions.** This global protein interaction map shows a subset of the most statistically likely interactions (lines) among about 2,300 proteins (small circles) in *Drosophila*. The three background colors in the map correspond to the general location of each protein: Green is the nucleus, blue the cytoplasm, and yellow the plasma membrane. The proteins are also color-coded according to their specific subcellular localization; for example, green circles represent nuclear proteins.

Application of Systems Biology to Medicine

The Cancer Genome Atlas is another example of systems biology in which a large group of interacting genes and gene products are analyzed together. This project, under the joint leadership of the National Cancer Institute and the NIH, aims to understand how changes in biological systems lead to cancer. In a three-year pilot project running from 2007 to 2010, three types of cancer—lung cancer, ovarian cancer, and glioblastoma of the brain—are being analyzed by comparing gene sequences and patterns of gene expression in cancer cells with those of normal cells. A set of approximately 2,000 genes from the cancer cells will be sequenced at several different times during the progression of the disease to monitor changes due to mutations and rearrangements. If this approach proves fruitful, it will be extended to other types of cancer.

◀ **Figure 21.6 A human gene microarray chip.** Tiny dots of DNA arranged in a grid on this silicon wafer represent almost all of the genes in the human genome. Using this chip, researchers can analyze expression patterns for all human genes at the same time, minimizing the amount of reagents used and ensuring uniform conditions for all the genes.

Systems biology has tremendous potential in human medicine that is just starting to be explored. Silicon and glass "chips" have already been developed that hold a microarray of most of the known human genes **(Figure 21.6)**. Such chips are being used to analyze gene expression patterns in patients suffering from various cancers and other diseases, with the eventual aim of tailoring their treatment to their unique genetic makeup and the specifics of their cancers. This approach has already had modest success in characterizing subsets of several cancers.

Ultimately, every person may carry with their medical records a catalog of their DNA sequence, a sort of genetic bar code, with regions highlighted that predispose them to specific diseases. The potential application for disease prevention and treatment is substantial.

Systems biology is a highly efficient way to study emergent properties at the molecular level. Recall from Chapter 1 that according to the concept of emergent properties, novel properties arise at each successive level of biological complexity as a result of the arrangement of building blocks at the underlying level. The more we are able to learn about the arrangement and interactions of the components of genetic systems, the deeper will be our understanding of organisms. The remaining sections of this chapter will survey what we've learned from genomic studies thus far.

CONCEPT CHECK 21.2

1. What role does the Internet play in current genomics and proteomics research?
2. Explain the advantage of the systems biology approach to studying cancer versus the approach of studying a single gene at a time.
3. **WHAT IF?** Let's say you used the classic genetic approach to study a certain inherited trait in *Drosophila*—that is, you mutated flies and selected those with a phenotype of interest to you. Assuming you could use molecular biological tools to obtain the region of DNA that is mutated, how would you analyze it further to determine how it is involved in the trait of interest?

For suggested answers, see Appendix A.

CONCEPT **21.3**

Genomes vary in size, number of genes, and gene density

By early 2008, the sequencing of over 700 genomes had been completed and that of more than 2,700 genomes was in progress. In the completely sequenced group, about 600 are genomes of bacteria, and 50 are archaeal genomes. Among the 65 eukaryotic species in the group are vertebrates, invertebrates, protists, fungi, and plants. The accumulated genome sequences contain a wealth of information that we are just beginning to mine. What have we learned so far by comparing the genomes that have now been sequenced? In this section, we will examine the characteristics of genome size, number of genes, and gene density. Because these characteristics are so broad, we will focus on general trends, for which there are often exceptions.

Genome Size

Comparing the three domains (Bacteria, Archaea, and Eukarya), we find a general difference in genome size between prokaryotes and eukaryotes **(Table 21.1)**. While there are some exceptions, most bacterial genomes have between 1 and 6 million base pairs (Mb); the genome of *E. coli*, for instance, has 4.6 Mb. Genomes of archaea are, for the most part, within the size range of bacterial genomes. (Keep in mind, however, that many fewer archaeal genomes have been completely sequenced, so this picture may change.) Eukaryotic genomes tend to be larger: The genome of the single-celled yeast *Saccharomyces cerevisiae* (a fungus) has about 13 Mb, while most animals and plants, which are multicellular, have genomes of at least 100 Mb. There are 180 Mb in the fruit fly genome, while humans have 3,200 Mb, about 500 to 3,000 times as many as a typical bacterium.

Aside from this general difference between prokaryotes and eukaryotes, a comparison of genome sizes among eukaryotes fails to reveal any systematic relationship between genome size and the organism's phenotype. For instance, the genome of *Fritillaria assyriaca*, a flowering plant in the lily family, contains 120 billion base pairs (120,000 Mb), about 40 times the size of the human genome. Even more striking, there is a single-celled amoeba, *Amoeba dubia*, whose genome has 670,000 Mb. (This genome has not yet been sequenced.) On a finer scale, comparing two insect species, the cricket (*Anabrus simplex*) genome turns out to have 11 times as many base pairs as the *Drosophila melanogaster* genome. There is a wide range of genome sizes within the groups of protists, insects, amphibians, and plants and less of a range within mammals and reptiles.

Number of Genes

A similar difference holds true for the number of genes: Bacteria and archaea, in general, have fewer genes than eukaryotes.

Free-living bacteria and archaea have from 1,500 to 7,500 genes, while the number of genes in eukaryotes ranges from about 5,000 for unicellular fungi to at least 40,000 for some multicellular eukaryotes (see Table 21.1).

Within the eukaryotes, the number of genes in a species is often lower than expected from simply considering the size of its genome. Looking at Table 21.1, you can see that the genome of the nematode *C. elegans* is 100 Mb in size and contains roughly 20,000 genes. The *Drosophila* genome, in comparison, is almost twice as big (180 Mb) but has about two-thirds the number of genes—only 13,700 genes.

Considering an example closer to home, we noted that the human genome contains 3,200 Mb, well over ten times the size of either the *Drosophila* or *C. elegans* genome. At the outset of the Human Genome Project, biologists expected somewhere between 50,000 and 100,000 genes to be identified in the completed sequence, based on the number of known human proteins. As the project progressed, the estimate was revised downward several times, and, as of 2007, the most reliable count places the number at 20,488. This relatively low number, similar to the number of genes in the nematode *C. elegans*, has surprised biologists, who had clearly expected many more human genes.

What genetic attributes allow humans (and other vertebrates) to get by with no more genes than nematodes? An important factor is that vertebrate genomes "get more bang for the buck" from their coding sequences because of extensive alternative splicing of RNA transcripts. Recall that this process generates more than one functional protein from a single gene (see Figure 18.11). For instance, nearly all human genes contain multiple exons, and an estimated 75% of these multi-exon genes are spliced in at least two different ways. If we assume that each alternatively spliced human gene on average specifies three different polypeptides, then the total number of different human polypeptides would be about 75,000. Additional polypeptide diversity could result from post-translational modifications such as cleavage or addition of carbohydrate groups in different cell types or at different developmental stages.

Table 21.1 Genome Sizes and Estimated Numbers of Genes*

Organism	Haploid Genome Size (Mb)	Number of Genes	Genes per Mb
Bacteria			
Haemophilus influenzae	1.8	1,700	940
Escherichia coli	4.6	4,400	950
Archaea			
Archaeoglobus fulgidus	2.2	2,500	1,130
Methanosarcina barkeri	4.8	3,600	750
Eukaryotes			
Saccharomyces cerevisiae (yeast, a fungus)	13	6,200	480
Caenorhabditis elegans (nematode)	100	20,000	200
Arabidopsis thaliana (mustard family plant)	118	25,500	215
Drosophila melanogaster (fruit fly)	180	13,700	76
Oryza sativa (rice)	390	40,000	140
Danio rerio (zebrafish)	1,700	23,000	13
Mus musculus (house mouse)	2,600	22,000	11
Homo sapiens (human)	3,200	20,500	7
Fritillaria assyriaca (lily family plant)	120,000	ND	ND

*Some values given here are likely to be revised as genome analysis continues. Mb = million base pairs. ND = not determined.

Gene Density and Noncoding DNA

In addition to genome size and number of genes, we can compare gene density in different species—in other words, how many genes there are in a given length of DNA. When we compare the genomes of bacteria, archaea, and eukaryotes, we see that eukaryotes generally have larger genomes but fewer genes in a given number of base pairs. Humans have hundreds or thousands of times as many base pairs in their genome as most bacteria, as we already noted, but they have only 5 to 15 times as many genes; thus, gene density is lower in humans (see Table 21.1). Even unicellular eukaryotes, such as yeasts, have fewer genes per million base pairs than bacteria and archaea. Among the genomes that have been sequenced completely thus far, humans and other mammals have the lowest gene density.

In all bacterial genomes studied so far, most of the DNA consists of genes for protein, tRNA, or rRNA; the small amount of other DNA consists mainly of nontranscribed regulatory sequences, such as promoters. The sequence of nucleotides along a bacterial protein-coding gene proceeds from start to finish without interruption by noncoding sequences (introns). In eukaryotic genomes, by contrast, most of the DNA neither encodes protein nor is transcribed into known functional RNA molecules (such as tRNAs), and the DNA includes more complex regulatory sequences. In fact, humans have 10,000 times as much noncoding DNA as bacteria. Some of this DNA in multicellular eukaryotes is present as introns within genes. Indeed, introns account for most of the difference in average length between human genes (27,000 base pairs) and bacterial genes (1,000 base pairs).

In addition to introns, multicellular eukaryotes have a vast amount of noncoding DNA between genes. In the next section, we will describe the composition and arrangement of these great stretches of DNA in the human genome.

CONCEPT CHECK 21.3

1. According to the best current estimate, the human genome contains about 20,500 genes. However, there is evidence that human cells produce more than 20,500 different polypeptides. What processes might explain this discrepancy?
2. The number of sequenced genomes is constantly being updated. Use the website www.genomesonline.org to find the current number of completed genomes within each domain as well as the number of genomes whose sequencing is in progress. (*Hint:* Click on "GOLD Tables," and then click on "Published Complete Genomes" for extra information.)
3. **WHAT IF?** What evolutionary processes might account for prokaryotes having smaller genomes than eukaryotes?

For suggested answers, see Appendix A.

CONCEPT **21.4**

Multicellular eukaryotes have much noncoding DNA and many multigene families

We have spent most of this chapter, and indeed this unit, focusing on genes that code for proteins. Yet the coding regions of these genes and the genes for RNA products such as rRNA, tRNA, and microRNA (miRNA) make up only a small portion of the genomes of most multicellular eukaryotes. The bulk of most eukaryotic genomes consists of DNA sequences that neither code for proteins nor are transcribed to produce known RNAs; this noncoding DNA was often described in the past as "junk DNA." However, much evidence is accumulating that this DNA plays important roles in the cell, an idea supported by its persistence in diverse genomes over many hundreds of generations. For example, comparison of the genomes of humans, rats, and mice has revealed the presence of almost 500 regions of noncoding DNA that are identical in sequence in all three species. This is a higher level of sequence conservation than is seen for protein-coding regions in these species, strongly suggesting that the noncoding regions have important functions. In this section, we examine how genes and noncoding DNA sequences are organized within genomes of multicellular eukaryotes, using the human genome as our main example. Genome organization tells us much about how genomes have evolved and continue to evolve, the next subject we'll consider.

Once the sequencing of the human genome was completed, it became clear that only a tiny part—1.5%—codes for proteins or is transcribed into rRNAs or tRNAs. **Figure 21.7** shows what is known about the makeup of the remaining 98.5%. Gene-related regulatory sequences and introns account for 24% of the human genome; the rest, located between functional genes, includes some unique noncoding DNA, such as gene fragments and **pseudogenes**, former genes that have accumulated mutations over a long time and have become nonfunctional. Most intergenic DNA, however, is **repetitive DNA**, which consists of sequences that are present in multiple copies in the genome. Somewhat surprisingly, about three-fourths of this repetitive DNA (44% of the entire human genome) is made up of units called transposable elements and sequences related to them.

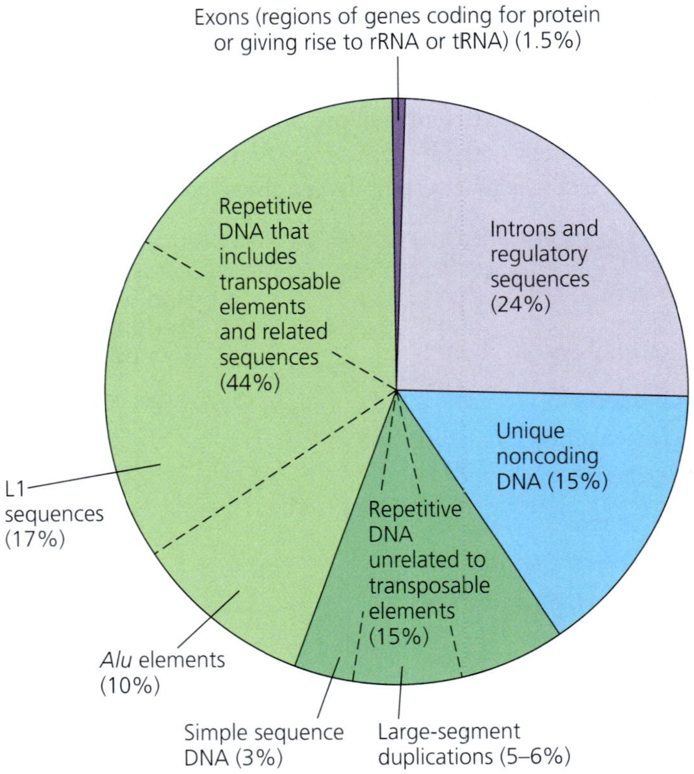

▲ **Figure 21.7 Types of DNA sequences in the human genome.** The gene sequences that code for proteins or are transcribed into rRNA or tRNA molecules make up only about 1.5% of the human genome (dark purple in the pie chart), while introns and regulatory sequences associated with genes (light purple) make up about a quarter. The vast majority of the human genome does not code for proteins or give rise to known RNAs, and much of it is repetitive DNA (dark and light green). Because repetitive DNA is the most difficult to sequence and analyze, classification of some portions is tentative, and the percentages given here may shift slightly as genome analysis proceeds. The genes that are transcribed into miRNAs, which were recently discovered, are found among unique noncoding DNA sequences and within introns; thus, they are included in two segments of this chart.

Transposable Elements and Related Sequences

Both prokaryotes and eukaryotes have stretches of DNA that can move from one location to another within the genome. These stretches are known as *transposable genetic elements*, or simply **transposable elements**. During the process called *transposition*, a transposable element moves from one site in a cell's DNA to a different target site by a type of recombination process. Transposable elements are sometimes called "jumping genes," but the phrase is misleading because they never completely detach from the cell's DNA. (The original and new DNA sites are brought together by DNA bending.)

The first evidence for wandering DNA segments came from American geneticist Barbara McClintock's breeding experiments with Indian corn (maize) in the 1940s and 1950s **(Figure 21.8)**. As she tracked corn plants through multiple generations, McClintock identified changes in the color of corn kernels that made sense only if she postulated the existence of genetic elements capable of moving from other locations in the genome into the genes for kernel color, disrupting the genes so that the kernel color was changed. McClintock's discovery was met with great skepticism and virtually discounted at the time. Her careful work and insightful ideas were finally validated many years later when transposable elements were found in bacteria and microbial geneticists learned more about the molecular basis of transposition.

Movement of Transposons and Retrotransposons

Eukaryotic transposable elements are of two types. The first type are **transposons**, which move within a genome by means of a DNA intermediate. Transposons can move by a "cut-and-paste" mechanism, which removes the element from the original site, or by a "copy-and-paste" mechanism, which leaves a copy behind **(Figure 21.9a)**.

Most transposable elements in eukaryotic genomes are of the second type, **retrotransposons**, which move by means of an RNA intermediate that is a transcript of the retrotransposon DNA. Retrotransposons always leave a copy at the original site during transposition, since they are initially transcribed into an RNA intermediate **(Figure 21.9b)**. To insert at another site, the RNA intermediate is first converted back to DNA by reverse transcriptase, an enzyme encoded in the retrotransposon itself. Thus, reverse transcriptase can be present in cells not infected with retroviruses. (In fact, retroviruses, which were discussed in Chapter 19, may have evolved from retrotransposons.) A cellular enzyme catalyzes insertion of the reverse-transcribed DNA at a new site.

▲ **Figure 21.8 The effect of transposable elements on corn kernel color.** Barbara McClintock first proposed the idea of mobile genetic elements after observing variegations in corn kernel color. Although her idea was met with skepticism when she proposed it in the 1940s, it was later fully validated. She received the Nobel Prize in 1983, at the age of 81, for her pioneering research.

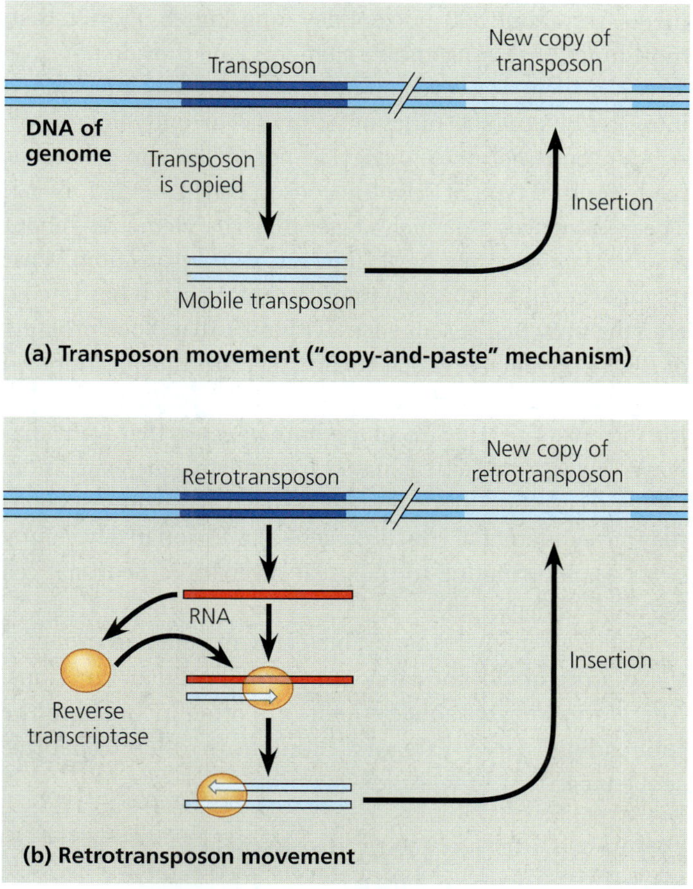

▲ **Figure 21.9 Movement of eukaryotic transposable elements. (a)** Movement of transposons by either the cut-and-paste mechanism or the copy-and-paste mechanism (shown here) involves a double-stranded DNA intermediate that is inserted into the genome. **(b)** Movement of retrotransposons begins with formation of a single-stranded RNA intermediate. The remaining steps are essentially identical to part of the retrovirus reproductive cycle (see Figure 19.8). In the movement of transposons by the copy-and-paste mechanism and in the movement of retrotransposons, the DNA sequence remains in the original site as well as appearing in a new site.

? *How would part (a) differ if it showed the cut-and-paste mechanism?*

Sequences Related to Transposable Elements

Multiple copies of transposable elements and sequences related to them are scattered throughout eukaryotic genomes. A single unit is usually hundreds to thousands of base pairs long, and the dispersed "copies" are similar but usually not identical to each other. Some of these are transposable elements that can move; the enzymes required for this movement may be encoded by any transposable element, including the one that is moving. Others are related sequences that have lost the ability to move altogether. Transposable elements and related sequences make up 25–50% of most mammalian genomes (see Figure 21.7) and even higher percentages in amphibians and many plants.

In humans and other primates, a large portion of transposable element–related DNA consists of a family of similar sequences called *Alu elements*. These sequences alone account for approximately 10% of the human genome. *Alu* elements are about 300 nucleotides long, much shorter than most functional transposable elements, and they do not code for any protein. However, many *Alu* elements are transcribed into RNA; its cellular function, if any, is currently unknown.

An even larger percentage (17%) of the human genome is made up of a type of retrotransposon called *LINE-1*, or *L1*. These sequences are much longer than *Alu* elements—about 6,500 base pairs—and have a low rate of transposition. What might account for this low rate? Recent research has uncovered the presence of sequences within L1 that block progress of RNA polymerase, which is necessary for transposition. An accompanying genomic analysis found L1 sequences within the introns of nearly 80% of the human genes that were analyzed, suggesting that L1 may help regulate gene expression. Other researchers have proposed that L1 retrotransposons may have differential effects on gene expression in developing neurons, contributing to the great diversity of neuronal cell types (see Chapter 48).

Although many transposable elements encode proteins, these proteins do not carry out normal cellular functions. Therefore, transposable elements are often included in the "noncoding" DNA category, along with other repetitive sequences.

Other Repetitive DNA, Including Simple Sequence DNA

Repetitive DNA that is not related to transposable elements probably arises due to mistakes during DNA replication or recombination. Such DNA accounts for about 15% of the human genome (see Figure 21.7). About a third of this (5–6% of the human genome) consists of duplications of long stretches of DNA, with each unit ranging from 10,000 to 300,000 base pairs. The large segments seem to have been copied from one chromosomal location to another site on the same or a different chromosome.

In contrast to scattered copies of long sequences, **simple sequence DNA** contains many copies of tandemly repeated short sequences, as in the following example (showing one DNA strand only):

. . . GTTACGTTACGTTACGTTACGTTACGTTAC . . .

In this case, the repeated unit (GTTAC) consists of five nucleotides. Repeated units may contain as many as 500 nucleotides, but often contain fewer than 15 nucleotides, as seen for this example. When the unit contains 2 to 5 nucleotides, the series of repeats is called a **short tandem repeat**, or **STR**; we discussed the use of STR analysis in preparing genetic profiles in Chapter 20. The number of copies of the repeated unit can vary from site to site within a given genome. There could be as many as several hundred thousand repetitions of the GTTAC unit at one site, but only half that number at another. The repeat number can also vary from person to person, producing the variation represented in the genetic profiles that result from STR analysis. Altogether, simple sequence DNA makes up 3% of the human genome.

The nucleotide composition of simple sequence DNA is often different enough from the rest of the cell's DNA to have an intrinsically different density. If genomic DNA is cut into pieces and centrifuged at high speed, segments of different density migrate to different positions in the centrifuge tube. Repetitive DNA isolated in this way was originally called *satellite DNA* because it appeared as a "satellite" band in the centrifuge tube, separate from the rest of the DNA. Now the term is often used interchangeably with simple sequence DNA.

Much of a genome's simple sequence DNA is located at chromosomal telomeres and centromeres, suggesting that this DNA plays a structural role for chromosomes. The DNA at centromeres is essential for the separation of chromatids in cell division (see Chapter 12). Centromeric DNA, along with simple sequence DNA located elsewhere, may also help organize the chromatin within the interphase nucleus. The simple sequence DNA located at telomeres, at the tips of chromosomes, prevents genes from being lost as the DNA shortens with each round of replication (see Chapter 16). Telomeric DNA also binds proteins that protect the ends of a chromosome from degradation and from joining to other chromosomes.

Genes and Multigene Families

We finish our discussion of the various types of DNA sequences in eukaryotic genomes with a closer look at genes. Recall that DNA sequences that code for proteins or give rise to tRNA or rRNA compose a mere 1.5% of the human genome (see Figure 21.7). If we include introns and regulatory sequences associated with genes, the total amount of gene-related DNA—coding and noncoding—constitutes about 25%

of the human genome. Put another way, only about 6% (1.5% out of 25%) of the length of the average gene is represented in the final gene product.

Like the genes of bacteria, many eukaryotic genes are present as unique sequences, with only one copy per haploid set of chromosomes. But in the human genome and the genomes of many other animals and plants, such solitary genes make up less than half of the total transcribed DNA. The rest occurs in **multigene families**, collections of two or more identical or very similar genes.

In multigene families that consist of *identical* DNA sequences, those sequences are usually clustered tandemly and, with the notable exception of the genes for histone proteins, have RNAs as their final products. An example is the family of identical DNA sequences that are the genes for the three largest rRNA molecules (**Figure 21.10a**). These rRNA molecules are transcribed from a single transcription unit that is repeated tandemly hundreds to thousands of times in one or several clusters in the genome of a multicellular eukaryote. The many copies of this rRNA transcription unit help cells to quickly make the millions of ribosomes needed for active protein synthesis. The primary transcript is cleaved to yield the three rRNA molecules. These are then combined with proteins and one other kind of rRNA (5S rRNA) to form ribosomal subunits.

The classic examples of multigene families of *nonidentical* genes are two related families of genes that encode globins, a group of proteins that include the α and β polypeptide subunits of hemoglobin. One family, located on chromosome 16 in humans, encodes various forms of α-globin; the other, on chromosome 11, encodes forms of β-globin (**Figure 21.10b**). The different forms of each globin subunit are expressed at different times in development, allowing hemoglobin to function effectively in the changing environment of the developing animal. In humans, for example, the embryonic and fetal forms of hemoglobin have a higher affinity for oxygen than the adult forms, ensuring the efficient transfer of oxygen from mother to fetus. Also found in the globin gene family clusters are several pseudogenes.

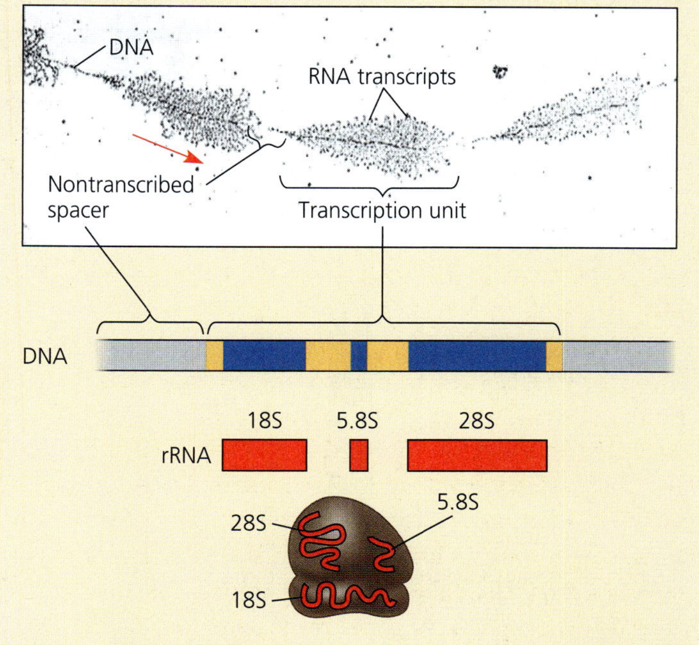

(a) Part of the ribosomal RNA gene family. Three of the hundreds of copies of rRNA transcription units in a salamander genome are shown at the top (TEM). Each "feather" corresponds to a single transcription unit being transcribed by about 100 molecules of RNA polymerase (the dark dots along the DNA), moving left to right. The growing RNA transcripts extend out from the DNA. In the diagram below the TEM, one transcription unit is shown. It includes the genes for three types of rRNA (blue), adjacent to regions that are transcribed but later removed (yellow). A single transcript is made and then processed to yield one molecule of each of the three rRNAs, which are crucial components of the ribosome. A fourth rRNA (5S rRNA) is also found in the ribosome, but the gene encoding it is not part of this transcription unit.

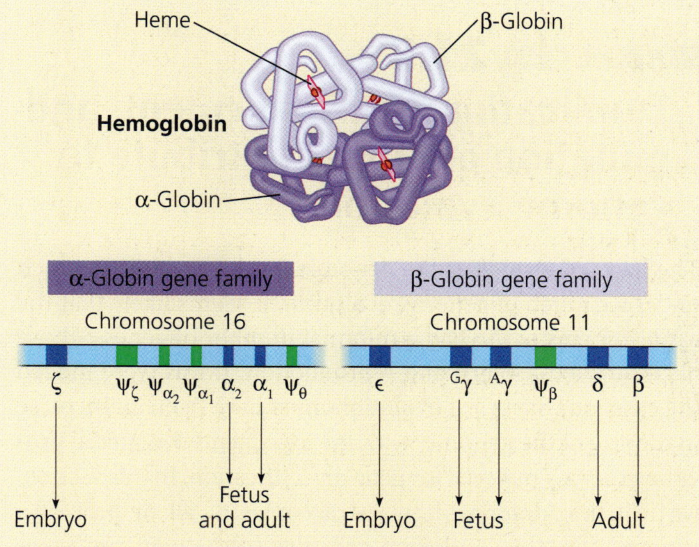

(b) The human α-globin and β-globin gene families. Hemoglobin is composed of two α-globin and two β-globin polypeptide subunits. The genes (dark blue) encoding α- and β-globins are found in two families, organized as shown here. The noncoding DNA separating the functional genes within each family cluster includes pseudogenes (green), nonfunctional versions of the functional genes. Genes and pseudogenes are named with Greek letters.

▲ **Figure 21.10 Gene families.**

? *In (a), how could you determine the direction of transcription if it wasn't indicated by the red arrow?*

The arrangement of the genes in gene families has given biologists insight into the evolution of genomes. We will consider some of the processes that have shaped the genomes of different species over evolutionary time in the next section.

CONCEPT CHECK 21.4

1. Discuss the characteristics of mammalian genomes that make them larger than prokaryotic genomes.
2. How are introns, transposable elements, and simple sequence DNA distributed differently in the genome?
3. Discuss the differences in the organization of the rRNA gene family and the globin gene families. For each, explain how the existence of a family of genes benefits the organism.
4. **WHAT IF?** Imagine you have found a DNA sequence similar to that of a known gene, but with quite a few nucleotide differences. How would you determine whether the "gene" you found is functional?

For suggested answers, see Appendix A.

CONCEPT **21.5**

Duplication, rearrangement, and mutation of DNA contribute to genome evolution

The basis of change at the genomic level is mutation, which underlies much of genome evolution. It seems likely that the earliest forms of life had a minimal number of genes—those necessary for survival and reproduction. If this were indeed the case, one aspect of evolution must have been an increase in the size of the genome, with the extra genetic material providing the raw material for gene diversification. In this section, we will first describe how extra copies of all or part of a genome can arise and then consider subsequent processes that can lead to the evolution of proteins (or RNA products) with slightly different or entirely new functions.

Duplication of Entire Chromosome Sets

An accident in meiosis can result in one or more extra sets of chromosomes, a condition known as polyploidy. Although such accidents would most often be lethal, in rare cases they could facilitate the evolution of genes. In a polyploid organism, one set of genes can provide essential functions for the organism. The genes in the one or more extra sets can diverge by accumulating mutations; these variations may persist if the organism carrying them survives and reproduces. In this way, genes with novel functions can evolve. As long as one copy of an essential gene is expressed, the divergence of another copy can lead to its encoded protein acting in a novel way, thereby changing the organism's phenotype. The outcome of this accumulation of mutations may be the branching off of a new species, as happens often in plants (see Chapter 24). Polyploid animals also exist, but they are rare.

Alterations of Chromosome Structure

Scientists have long known that sometime in the last 6 million years, when the ancestors of humans and chimpanzees diverged as species, the fusion of two ancestral chromosomes in the human line led to different haploid numbers for humans ($n = 23$) and chimpanzees ($n = 24$). With the recent explosion in genomic sequence information, we can now compare the chromosomal organization of many different species on a much finer scale. This information allows us to make inferences about the evolutionary processes that shape chromosomes and may drive speciation.

In one study, for example, researchers compared the DNA sequence of each human chromosome with the whole genome sequence of the mouse. **Figure 21.11** shows the results of this comparison for human chromosome 16: Large blocks of genes on this chromosome are found on four mouse chromosomes, indicating that the genes in each block stayed together during the evolution of the mouse and human lineages. Performing the same comparative analysis between chromosomes of humans

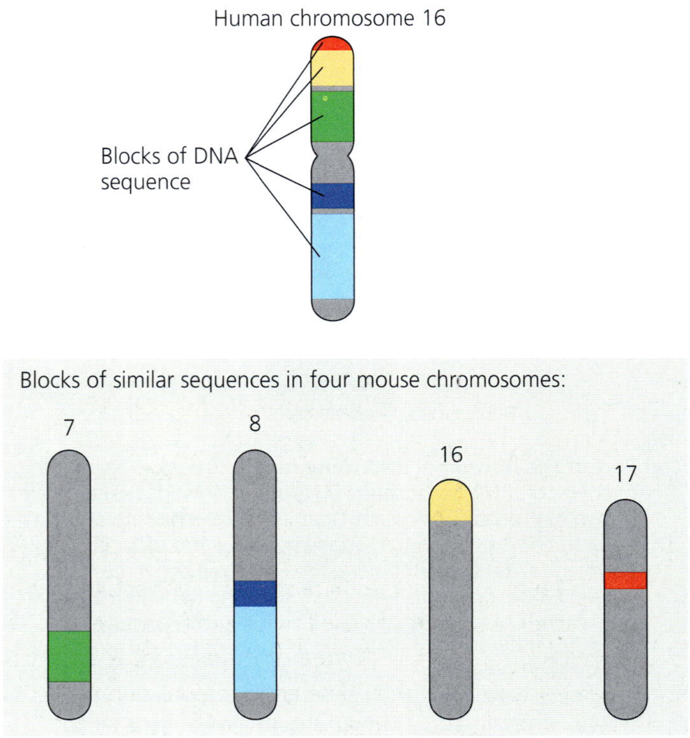

▲ **Figure 21.11 Similar blocks of sequences on human and mouse chromosomes.** DNA sequences very similar to large blocks of human chromosome 16 are found on mouse chromosomes 7, 8, 16, and 17. This suggests that the DNA sequence in each block has stayed together in the mouse and human lineages since the time they diverged from a common ancestor.

and six other mammalian species allowed the researchers to reconstruct the evolutionary history of chromosomal rearrangements in these eight species. They found many duplications and inversions of large portions of chromosomes, the result of mistakes during meiotic recombination in which the DNA broke and was rejoined incorrectly. The rate of these events seems to have accelerated about 100 million years ago, around the time large dinosaurs became extinct and the number of mammalian species increased rapidly. The apparent coincidence is interesting because chromosomal rearrangements are thought to contribute to the generation of new species. Although two individuals with different arrangements could still mate and produce offspring, the offspring would have two nonequivalent sets of chromosomes, making meiosis inefficient or even impossible. Thus, chromosomal rearrangements would lead to two populations that could not successfully mate with each other, a step on the way to their becoming two separate species. (You'll learn more about this in Chapter 24.)

Somewhat unexpectedly, the same study also unearthed a pattern with medical relevance. Analysis of the chromosomal breakage points associated with the rearrangements showed that they were not randomly distributed, but that specific sites were used over and over again. A number of these recombination "hot spots" correspond to locations of chromosomal rearrangements within the human genome that are associated with congenital diseases. Researchers are, of course, looking at the other sites as well for their possible association with as yet unidentified diseases.

Duplication and Divergence of Gene-Sized Regions of DNA

Errors during meiosis can also lead to the duplication of chromosomal regions that are smaller than the ones we've just discussed, including segments the length of individual genes. Unequal crossing over during prophase I of meiosis, for instance, can result in one chromosome with a deletion and another with a duplication of a particular gene. As illustrated in **Figure 21.12**, transposable elements in the genome can provide sites where nonsister chromatids can cross over, even when their homologous sequences are not correctly aligned.

Also, slippage can occur during DNA replication, such that the template shifts with respect to the new complementary strand, and a part of the template strand is either skipped by the replication machinery or used twice as a template. As a result, a segment of DNA is deleted or duplicated. It is easy to imagine how such errors could occur in regions of repeats like the simple sequence DNA described previously. The variable number of repeated units of simple sequence DNA at a given site, used for STR analysis, is probably due to errors like these. Evidence that unequal crossing over and template slippage during DNA replication lead to duplication of genes is found in the existence of multigene families.

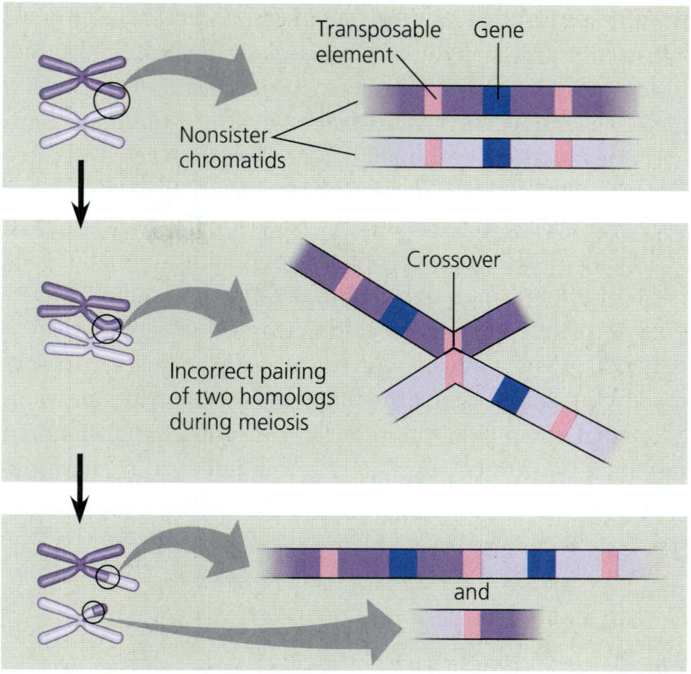

▲ **Figure 21.12 Gene duplication due to unequal crossing over.** One mechanism by which a gene (or other DNA segment) can be duplicated is recombination during meiosis between copies of a transposable element flanking the gene. Such recombination between misaligned nonsister chromatids of homologous chromosomes produces one chromatid with two copies of the gene and one chromatid with no copy.

Evolution of Genes with Related Functions: The Human Globin Genes

Duplication events can lead to the evolution of genes with related functions, such as those of the α-globin and β-globin gene families (see Figure 21.10b). A comparison of gene sequences within a multigene family can suggest the order in which the genes arose. This approach to re-creating the evolutionary history of the globin genes indicates that they all evolved from one common ancestral globin gene that underwent duplication and divergence into the α-globin and β-globin ancestral genes about 450–500 million years ago (**Figure 21.13**, on the next page). Each of these genes was later duplicated several times, and the copies then diverged from each other in sequence, yielding the current family members. In fact, the common ancestral globin gene also gave rise to the oxygen-binding muscle protein myoglobin and to the plant protein leghemoglobin. The latter two proteins function as monomers, and their genes are included in a "globin superfamily."

After the duplication events, the differences between the genes in the globin families undoubtedly arose from mutations that accumulated in the gene copies over many generations. The current model is that the necessary function provided by an α-globin protein, for example, was fulfilled by one gene, while other copies of the α-globin gene accumulated random mutations. Many mutations may have had an adverse effect on the organism and others may have had no effect, but a few mutations must have altered the function of the protein product in a way that was advantageous to the organism at a particular life stage without substantially changing the protein's oxygen-carrying function. Presumably, natural selection acted on these altered genes, maintaining them in the population.

The similarity in the amino acid sequences of the various α-globin and β-globin polypeptides supports this model of gene duplication and mutation **(Table 21.2)**. The amino acid sequences of the β-globins, for instance, are much more similar to each other than to the α-globin sequences. The existence of several pseudogenes among the functional globin genes provides additional evidence for this model (see Figure 21.10b): Random mutations in these "genes" over evolutionary time have destroyed their function.

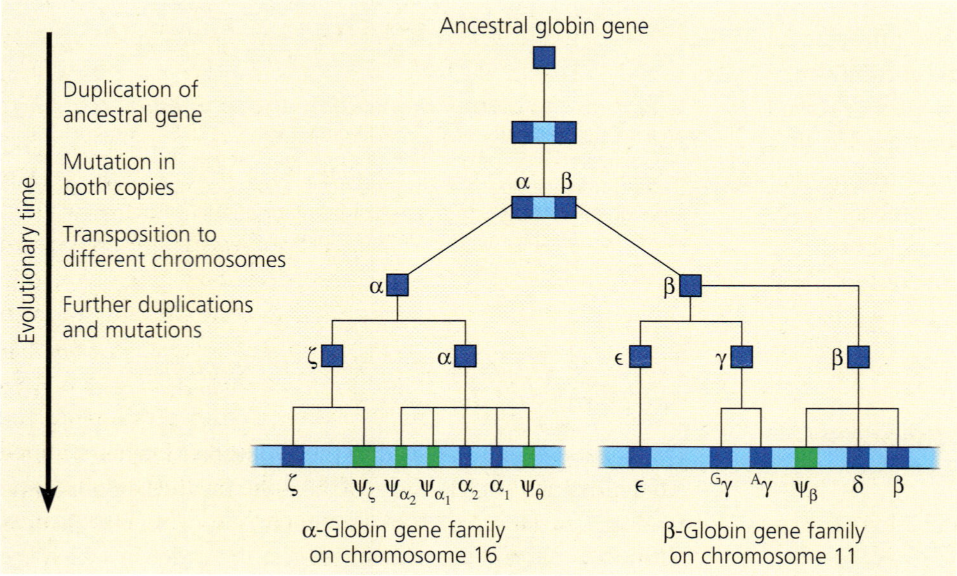

▲ **Figure 21.13 A model for the evolution of the human α-globin and β-globin gene families from a single ancestral globin gene.**

? *The green elements are pseudogenes. Explain how they could have arisen after gene duplication.*

Table 21.2 Percentage of Similarity in Amino Acid Sequence Between Human Globin Proteins

		α-Globins		β-Globins		
		α	ζ	β	γ	ϵ
α-Globins	α	—	58	42	39	37
	ζ	58	—	34	38	37
β-Globins	β	42	34	—	73	75
	γ	39	38	73	—	80
	ϵ	37	37	75	80	—

Evolution of Genes with Novel Functions

In the evolution of the globin gene families, gene duplication and subsequent divergence produced family members whose protein products performed similar functions (oxygen transport). Alternatively, one copy of a duplicated gene can undergo alterations that lead to a completely new function for the protein product. The genes for lysozyme and α-lactalbumin are good examples.

Lysozyme is an enzyme that helps protect animals against bacterial infection by hydrolyzing bacterial cell walls; α-lactalbumin is a nonenzymatic protein that plays a role in milk production in mammals. The two proteins are quite similar in their amino acid sequences and three-dimensional structures. Both genes are found in mammals, whereas only the lysozyme gene is present in birds. These findings suggest that at some time after the lineages leading to mammals and birds had separated, the lysozyme gene underwent a duplication event in the mammalian lineage but not in the avian lineage. Subsequently, one copy of the duplicated lysozyme gene evolved into a gene encoding α-lactalbumin, a protein with a completely different function.

Rearrangements of Parts of Genes: Exon Duplication and Exon Shuffling

Rearrangement of existing DNA sequences within genes has also contributed to genome evolution. The presence of introns in most genes of multicellular eukaryotes may have promoted the evolution of new and potentially useful proteins by facilitating the duplication or repositioning of exons in the genome. Recall from Chapter 17 that an exon often codes for a domain, a distinct structural or functional region of a protein.

We've already seen that unequal crossing over during meiosis can lead to duplication of a gene on one chromosome and its loss from the homologous chromosome (see Figure 21.12). By a similar process, a particular exon within a gene could be duplicated on one chromosome and deleted from the other. The gene with the duplicated exon would code for a protein containing a second copy of the encoded domain. This change in the protein's structure could augment its function by increasing its stability, enhancing

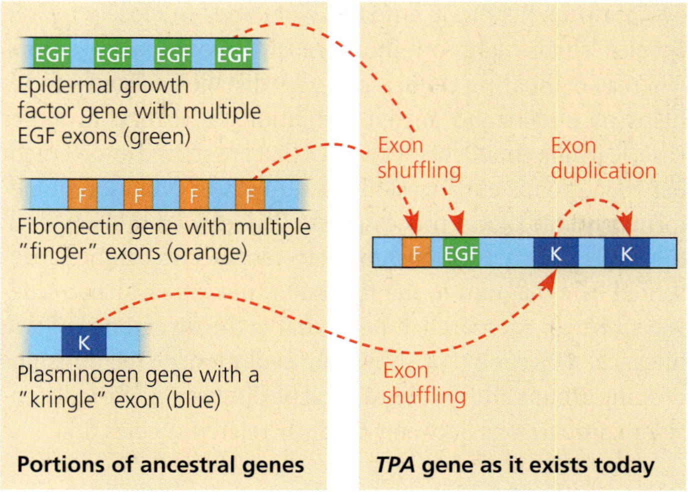

▲ **Figure 21.14 Evolution of a new gene by exon shuffling.** Exon shuffling could have moved exons from ancestral forms of the genes for epidermal growth factor, fibronectin, and plasminogen (left) into the evolving gene for tissue plasminogen activator, TPA (right). The order in which these events might have occurred is unknown. Duplication of the "kringle" exon from the plasminogen gene after its movement could account for the two copies of this exon in the *TPA* gene. Each type of exon encodes a particular domain in the TPA protein.

? *How could the presence of transposable elements in introns have facilitated the exon shuffling shown here?*

its ability to bind a particular ligand, or altering some other property. Quite a few protein-coding genes have multiple copies of related exons, which presumably arose by duplication and then diverged. The gene encoding the extracellular matrix protein collagen is a good example. Collagen is a structural protein with a highly repetitive amino acid sequence, which is reflected in the repetitive pattern of exons in the collagen gene.

Alternatively, we can imagine the occasional mixing and matching of different exons either within a gene or between two nonallelic genes owing to errors in meiotic recombination. This process, termed *exon shuffling*, could lead to new proteins with novel combinations of functions. As an example, let's consider the gene for tissue plasminogen activator (TPA). The TPA protein is an extracellular protein that helps control blood clotting. It has four domains of three types, each encoded by an exon; one exon is present in two copies. Because each type of exon is also found in other proteins, the gene for TPA is thought to have arisen by several instances of exon shuffling and duplication **(Figure 21.14)**.

How Transposable Elements Contribute to Genome Evolution

The persistence of transposable elements as a large fraction of some eukaryotic genomes is consistent with the idea that they play an important role in shaping a genome over evolutionary time. These elements can contribute to the evolution of the genome in several ways. They can promote recombination, disrupt cellular genes or control elements, and carry entire genes or individual exons to new locations.

Transposable elements of similar sequence scattered throughout the genome facilitate recombination between different chromosomes by providing homologous regions for crossing over. Most such alterations are probably detrimental, causing chromosomal translocations and other changes in the genome that may be lethal to the organism. But over the course of evolutionary time, an occasional recombination event of this sort may be advantageous to the organism.

The movement of a transposable element can also have direct consequences. For instance, if a transposable element "jumps" into the middle of a protein-coding sequence, it will prevent the production of a normal transcript of the gene. If a transposable element inserts within a regulatory sequence, the transposition may lead to increased or decreased production of one or more proteins. Transposition caused both types of effects on the genes coding for pigment-synthesizing enzymes in McClintock's corn kernels. Again, while such changes may usually be harmful, in the long run some may prove beneficial by providing a survival advantage.

During transposition, a transposable element may carry along a gene or group of genes to a new position in the genome. This mechanism probably accounts for the location of the α-globin and β-globin gene families on different human chromosomes, as well as the dispersion of the genes of certain other gene families. By a similar tag-along process, an exon from one gene may be inserted into another gene in a mechanism similar to that of exon shuffling during recombination. For example, an exon may be inserted by transposition into the intron of a protein-coding gene. If the inserted exon is retained in the RNA transcript during RNA splicing, the protein that is synthesized will have an additional domain, which may confer a new function on the protein.

A recent study reveals yet another way that transposable elements can bring about new coding sequences. This work shows that an *Alu* element may hop into an intron in a way that creates a weak alternative splice site in the RNA transcript. During processing of the transcript, the regular splice sites are used more often, so that the original protein is made. On occasion, however, splicing occurs at the new weak site, with the result that some of the *Alu* element ends up in the mRNA, coding for a new portion of the protein. In this way, alternative genetic combinations may be "tried out" while the function of the original gene product is retained.

Clearly, all the processes discussed in this section most often produce either harmful effects, which may be lethal, or no effect at all. In a few cases, however, small beneficial changes may occur. Over many generations, the resulting genetic diversity provides valuable raw material for natural selection. Diversification of genes and their products is an important factor in the evolution of new species. Thus, the accumulation of changes in the genome of each species provides a record of its evolutionary history. To read this record, we must be able to identify genomic changes. Comparing the genomes of different species allows us to do that and has increased our understanding of how genomes evolve. You will learn more about these topics in the final section.

CONCEPT CHECK 21.5

1. Describe three examples of errors in cellular processes that lead to DNA duplications.
2. Explain how multiple exons might have arisen in the ancestral EGF and fibronectin genes shown in Figure 21.14 (left).
3. What are three ways that transposable elements are thought to contribute to genome evolution?
4. **WHAT IF?** In 2005, Icelandic scientists reported finding a large chromosomal inversion present in 20% of northern Europeans, and they noted that Icelandic women with this inversion had significantly more children than women without it. What would you expect to happen to the frequency of this inversion in the Icelandic population in future generations?

For suggested answers, see Appendix A.

CONCEPT **21.6**

Comparing genome sequences provides clues to evolution and development

One researcher has likened the current state of biology to the Age of Exploration in the 15th century after major improvements in navigation and the building of faster ships. In the last 20 years, we have seen rapid advances in genome sequencing and data collection, new techniques for assessing gene activity across the whole genome, and refined approaches for understanding how genes and their products work together in complex systems. We are truly poised on the brink of a new world.

Comparisons of genome sequences from different species reveal much about the evolutionary history of life, from very ancient to more recent. Similarly, comparative studies of the genetic programs that direct embryonic development in different species are beginning to clarify the mechanisms that generated the great diversity of life-forms present today. In this final section of the chapter, we will discuss what has been learned from these two approaches.

Comparing Genomes

The more similar in sequence the genes and genomes of two species are, the more closely related those species are in their evolutionary history. Comparing genomes of closely related species sheds light on more recent evolutionary events, whereas comparing genomes of very distantly related species helps us understand ancient evolutionary history. In either case, learning about characteristics that are shared or divergent between groups enhances our picture of the evolution of lifeforms and biological processes. As you learned in Chapter 1, the evolutionary relationships between species can be represented by a diagram in the form of a tree (often turned sideways), where each branch point marks the divergence of two lineages. **Figure 21.15** shows the evolutionary relationships of some groups and species we will be discussing. We will consider comparisons between distantly related species first.

Comparing Distantly Related Species

Analyzing which genes have remained similar—that is, are *highly conserved*—in distantly related species can help clarify evolutionary relationships among species that diverged from each other long ago. Indeed, comparisons of the complete genome sequences of bacteria, archaea, and eukaryotes indicate that these three groups diverged between 2 and 4 billion years ago and strongly support the theory that they are the fundamental domains of life (see Figure 21.15).

In addition to their value in evolutionary biology, comparative genomic studies confirm the relevance of research on model organisms to our understanding of biology in general and human biology in particular. Genes that evolved a very long time ago can still be surprisingly similar in disparate species. As a case in point, several genes in yeast are so similar to certain human disease genes that researchers have deduced the functions of the disease genes by studying their yeast counterparts. This striking similarity underscores the common origin of these two distantly related species.

Comparing Closely Related Species

The genomes of two closely related species are likely to be organized similarly because of their relatively recent divergence. As we mentioned earlier, this allows the fully sequenced genome of one species to be used as a scaffold for assembling the genomic sequences of a closely related species, accelerating mapping of the second genome. For instance, using the human genome sequence as a guide, researchers were able to quickly sequence the mouse genome.

The recent divergence of two closely related species also underlies the small number of gene differences that are found when their genomes are compared. The particular genetic differences can therefore be more easily correlated with phenotypic differences between the two species. An exciting application of this type of analysis is seen as researchers compare the human genome with the genomes of the chimpanzee, mouse, rat, and other mammals. Identifying the genes shared by all of these species but not by nonmammals should give clues about what it

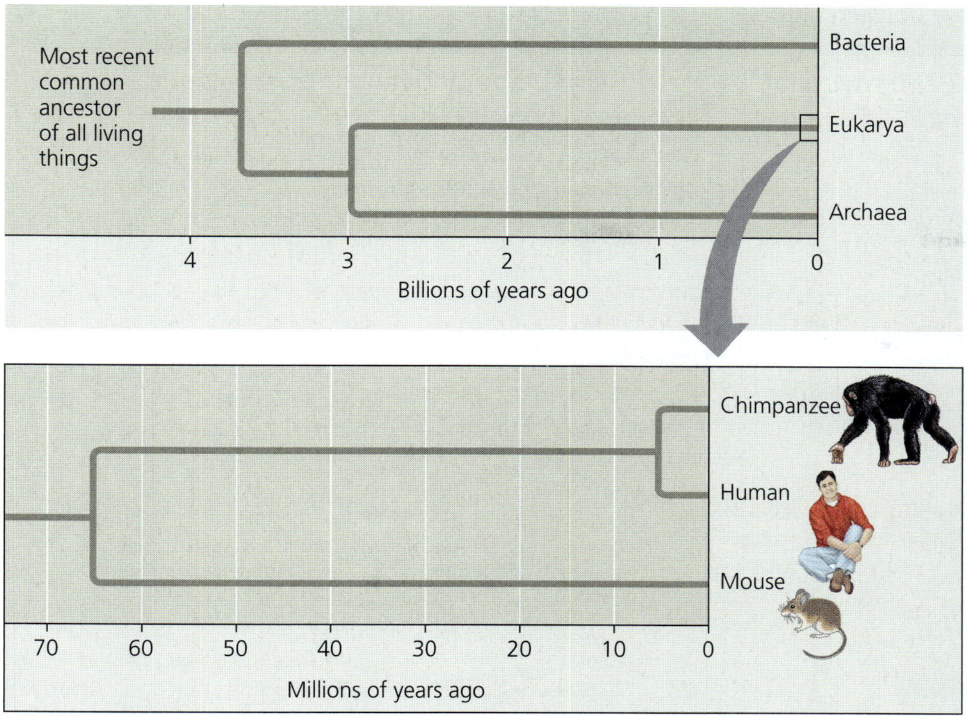

▲ **Figure 21.15 Evolutionary relationships of the three domains of life.** This tree diagram shows the ancient divergence of bacteria, archaea, and eukaryotes. A portion of the eukaryote lineage is expanded in the inset to show the more recent divergence of three mammalian species discussed in this chapter.

takes to make a mammal, while finding the genes shared by chimpanzees and humans but not by rodents should tell us something about primates. And, of course, comparing the human genome with that of the chimpanzee should help us answer the tantalizing question we asked at the beginning of the chapter: What genomic information makes a human or a chimpanzee?

An analysis of the overall composition of the human and chimpanzee genomes, which are thought to have diverged only about 6 million years ago (see Figure 21.15), reveals some general differences. Considering single base substitutions, the two genomes differ by only 1.2%. When researchers looked at longer stretches of DNA, however, they were surprised to find a further 2.7% difference due to insertions or deletions of larger regions in the genome of one or the other species; many of the insertions were duplications or other repetitive DNA. In fact, a third of the human duplications are not present in the chimpanzee genome, and some of these duplications contain regions associated with human diseases. There are more *Alu* elements in the human genome than in that of the chimpanzee, and the latter contains many copies of a retroviral provirus not present in humans. All of these observations provide clues to the forces that might have swept the two genomes along different paths, but we don't have a complete picture yet. We also don't know how these differences might account for the distinct characteristics of each species.

To discover the basis for the phenotypic differences between the two species, biologists are studying specific genes and types of genes that differ between humans and chimpanzees and comparing them with their counterparts in other mammals. This approach has revealed a number of genes that are apparently changing (evolving) faster in the human than in either the chimpanzee or the mouse. Among them are genes involved in defense against malaria and tuberculosis and at least one gene that regulates brain size. When genes are classified by function, the genes that seem to be evolving the fastest are those that code for transcription factors. This is exciting news because transcription factors regulate gene expression and thus play a key role in orchestrating the overall genetic program.

One transcription factor whose gene shows evidence of rapid change in the human lineage is called FOXP2. Several lines of evidence suggest that the *FOXP2* gene functions in vocalization in vertebrates. For one thing, mutations in this gene can produce severe speech and language impairment in humans. Moreover, the *FOXP2* gene is expressed in the brains of zebra finches and canaries at the time when these songbirds are learning their songs. But perhaps the strongest evidence comes from a "knock-out" experiment in which Joseph Buxbaum and colleagues disrupted the *FOXP2* gene in mice and analyzed the resulting phenotype (**Figure 21.16**, on the next page). The homozygous mutant mice had malformed brains and failed to emit normal ultrasonic vocalizations, and mice with one faulty copy of the gene also showed significant problems with vocalization. These results support the idea that the *FOXP2* gene product turns on genes involved in vocalization.

Expanding on this notion, researchers are exploring the question of whether differences between the human and chimpanzee FOXP2 proteins could account for the ability of humans, but not chimpanzees, to communicate by speech. There are only two amino acid differences between the human and chimpanzee FOXP2 proteins, and the effect of these differences on the function of the human protein is an intriguing question, whose answer is not yet known.

The *FOXP2* story is an excellent example of how different approaches complement each other in uncovering biological phenomena of widespread importance. The experiment described in Figure 21.16 uses mice as a model for humans because it would be unethical (as well as impractical) to carry out such an experiment in humans. Mice and humans diverged about 65.5 million years ago (see Figure 21.15) and share about 85% of their genes. This genetic similarity can be exploited in

▼ Figure 21.16 Inquiry

What is the function of a gene (*FOXP2*) that is rapidly evolving in the human lineage?

EXPERIMENT Several lines of evidence support a role for the *FOXP2* gene in the development of speech and language in humans and of vocalization in other vertebrates. In 2005, Joseph Buxbaum and collaborators at the Mount Sinai School of Medicine and several other institutions set out to test the function of *FOXP2*. They used the mouse, a model organism in which genes can be easily knocked out, as a representative vertebrate that vocalizes. Mice produce ultrasonic squeaks, referred to as whistles, to communicate stress. The researchers applied genetic engineering techniques to produce mice in which one or both copies of *FOXP2* were disrupted.

Wild type: two normal copies of *FOXP2*

Heterozygote: one copy of *FOXP2* disrupted

Homozygote: both copies of *FOXP2* disrupted

They then compared the phenotypes of these mice. Two of the characters they examined are included here: brain anatomy and vocalization.

Experiment 1: Researchers cut thin sections of brain and stained them with reagents, allowing visualization of brain anatomy in a UV fluorescence microscope.

Experiment 2: Researchers separated each newborn pup from its mother and recorded the number of ultrasonic whistles produced by the pup.

RESULTS

Experiment 1: Disruption of both copies of *FOXP2* led to brain abnormalities in which the cells were disorganized. Phenotypic effects on the brain of heterozygotes, with one disrupted copy, were less severe.

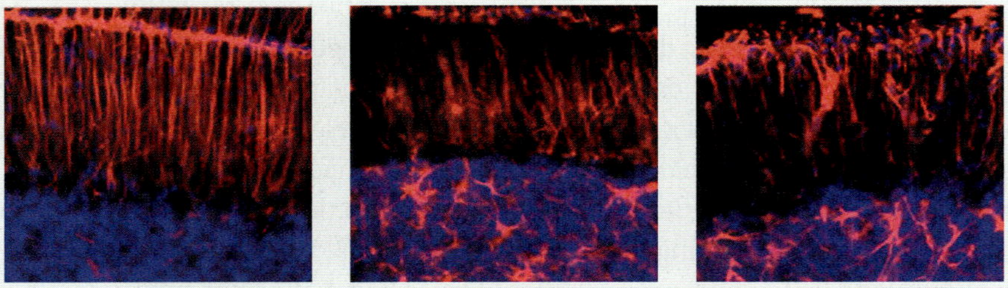

Wild type | Heterozygote | Homozygote

Experiment 2: Disruption of both copies of *FOXP2* led to an absence of ultrasonic vocalization in response to stress. The effect on vocalization in the heterozygote was also extreme.

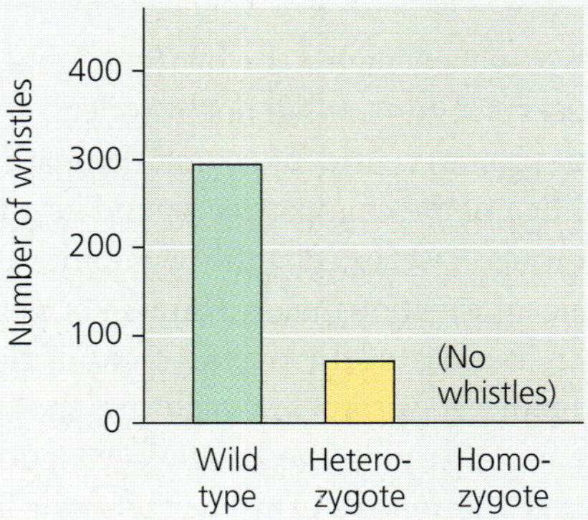

CONCLUSION *FOXP2* plays a significant role in the development of functional communication systems in mice. The results augment evidence from studies of birds and humans, supporting the hypothesis that *FOXP2* may act similarly in diverse organisms.

SOURCE W. Shu et al., Altered ultrasonic vocalization in mice with a disruption in the *Foxp2* gene, *Proceedings of the National Academy of Sciences* 102:9643–9648 (2005).

WHAT IF? Since the results support a role for mouse *FOXP2* in vocalization, you might wonder whether the human FOXP2 protein is a key regulator of speech. If you were given the amino acid sequences of wild-type and mutant human FOXP2 proteins and the wild-type chimpanzee FOXP2 protein, how would you investigate this question? What further clues could you obtain by comparing these sequences to that of the mouse FOXP2 protein?

studying other human genetic disorders. If researchers know the organ or tissue that is affected by a particular genetic disorder, they can look for genes that are expressed in these locations in experiments using mice. This approach has revealed several human genes of interest, including one that may contribute to Down syndrome.

Further research efforts are under way to extend genomic studies to many more microbial species, additional primates, and neglected species from diverse branches of the tree of life. These studies will advance our understanding of all aspects of biology, including health and ecology as well as evolution.

Comparing Genomes Within a Species

Another exciting prospect that stems from our ability to analyze genomes is increasing our understanding of the spectrum of genetic variation in humans. Because the history of the human species is so short—probably about 200,000 years—the amount of DNA variation among humans is small compared to that of many other species. Much of our diversity seems to be in the form of single nucleotide polymorphisms (SNPs, described in Chapter 20), usually detected by DNA sequencing. In the human genome, SNPs occur on average about once in 100 to 300 base pairs.

Scientists have already identified the location of several million SNP sites in the human genome and continue to find more. In the course of this search, they have also found other variations—including inversions, deletions, and duplications—that occur without apparent ill effect on the individual carrying them. These variations, as well as SNPs, will be useful genetic markers for studying human evolution, the differences between human populations, and the migratory routes of human populations throughout history. Such polymorphisms in human DNA will also be valuable markers for identifying genes that cause diseases or affect our health in more subtle ways. As well as informing us about evolution, analysis of the differences in individual genomes is likely to change the practice of medicine later in the 21st century.

Comparing Developmental Processes

Biologists in the field of evolutionary developmental biology, or **evo-devo** as it is often called, compare developmental processes of different multicellular organisms. Their aim is to understand how these processes have evolved and how changes in them can modify existing organismal features or lead to new ones. With the advent of molecular techniques and the recent flood of genomic information, we are beginning to realize that the genomes of related species with strikingly different forms may have only minor differences in gene sequence or regulation. Discovering the molecular basis underlying these differences in turn helps us understand the origin of the myriad diverse forms that cohabit this planet, thus informing our study of evolution.

Widespread Conservation of Developmental Genes Among Animals

In Chapter 18, you learned about the homeotic genes in *Drosophila*, which specify the identity of body segments in the fruit fly (see Figure 18.18). Molecular analysis of the homeotic genes in *Drosophila* has shown that they all include a 180-nucleotide sequence called a **homeobox**, which specifies a 60-amino-acid *homeodomain* in the encoded proteins. An identical or very similar nucleotide sequence has been discovered in the homeotic genes of many invertebrates and vertebrates. The sequences are so similar between humans and fruit flies, in fact, that one researcher has whimsically referred to flies as "little people with wings." The resemblance even extends to the organization of these genes: The vertebrate genes homologous to the homeotic genes of fruit flies have kept the same chromosomal arrangement **(Figure 21.17)**. Homeobox-containing sequences have also been found in regulatory genes of much more distantly related eukaryotes, including plants and yeasts. From these similarities, we can deduce that the homeobox DNA sequence evolved very early in the history of

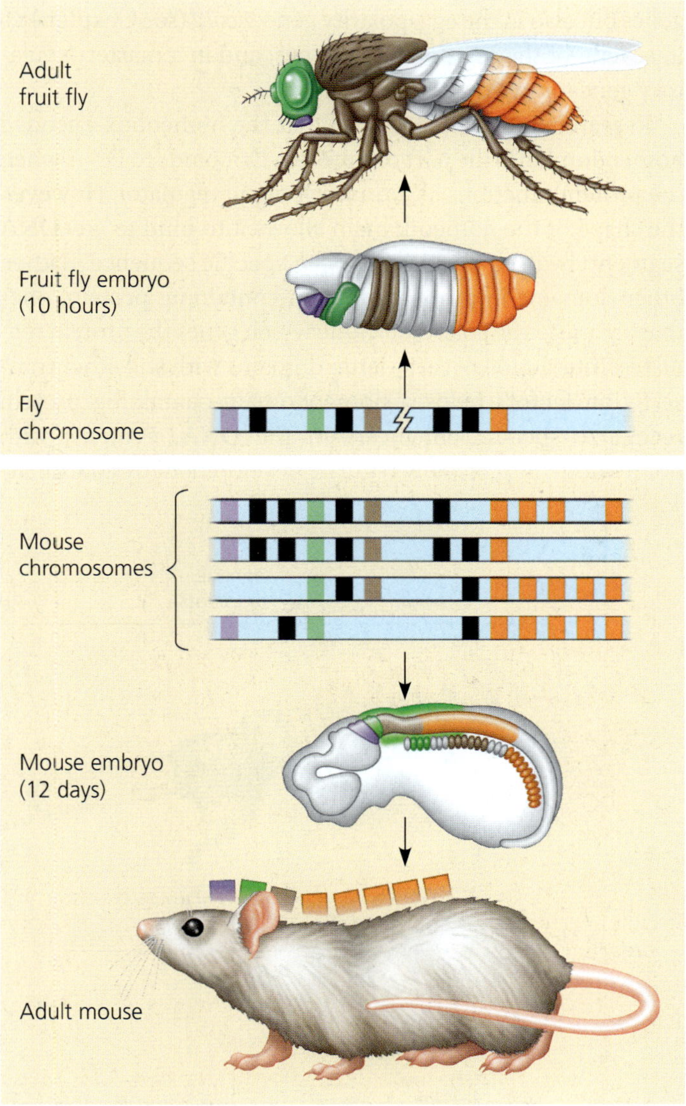

▲ **Figure 21.17 Conservation of homeotic genes in a fruit fly and a mouse.** Homeotic genes that control the form of anterior and posterior structures of the body occur in the same linear sequence on chromosomes in *Drosophila* and mice. Each colored band on the chromosomes shown here represents a homeotic gene. In fruit flies, all homeotic genes are found on one chromosome. The mouse and other mammals have the same or similar sets of genes on four chromosomes. The color code indicates the parts of the embryos in which these genes are expressed and the adult body regions that result. All of these genes are essentially identical in flies and mice, except for those represented by black bands, which are less similar in the two animals.

life and was sufficiently valuable to organisms to have been conserved in animals and plants virtually unchanged for hundreds of millions of years.

Homeotic genes in animals were named *Hox* genes, short for *h*omeob*ox*-containing genes, because homeotic genes were the first genes found to have this sequence. Other homeobox-containing genes were later found that do not act as homeotic genes; that is, they do not directly control the identity of body parts. However, most of these genes, in animals at least, are associated with development, suggesting their ancient and fundamental importance in that process. In *Drosophila*, for example, homeoboxes are present not only in the homeotic genes but also in the egg-polarity gene *bicoid* (see Chapter 18), in several of the segmentation genes, and in a master regulatory gene for eye development.

Researchers have discovered that the homeobox-encoded homeodomain is the part of a protein that binds to DNA when the protein functions as a transcriptional regulator. However, the shape of the homeodomain allows it to bind to any DNA segment; by itself it cannot select a specific sequence. Rather, other domains in a homeodomain-containing protein, ones that are more variable, determine which genes the protein regulates. Interaction of these latter domains with still other transcription factors helps a homeodomain-containing protein recognize specific enhancers in the DNA. Proteins with homeodomains probably regulate development by coordinating the transcription of batteries of developmental genes, switching them on or off. In embryos of *Drosophila* and other animal species, different combinations of homeobox genes are active in different parts of the embryo. This selective expression of regulatory genes, varying over time and space, is central to pattern formation.

Developmental biologists have found that in addition to homeotic genes, many other genes involved in development are highly conserved from species to species. These include numerous genes encoding components of signaling pathways. The extraordinary similarity among particular developmental genes in different animal species raises a question: How can the same genes be involved in the development of animals whose forms are so very different from each other?

Current studies are suggesting answers to this question. In some cases, small changes in regulatory sequences of particular genes cause changes in gene expression patterns that can lead to major changes in body form. For example, the differing patterns of expression of the *Hox* genes along the body axis in insects and crustaceans can explain the variation in number of leg-bearing segments among these segmented animals **(Figure 21.18)**. Also, recent research suggests that the same *Hox* gene product may have subtly dissimilar effects in different species, turning on new genes or turning on the same genes at higher or lower levels. In other cases, similar genes direct distinct developmental processes in different organisms, resulting in diverse body shapes. Several *Hox* genes, for instance, are expressed in the embryonic and larval stages of the sea urchin, a nonsegmented animal that has a body plan quite different from those of insects and mice. Sea urchin adults make the pincushion-shaped shells you may have seen on the beach. They are among the organisms long used in classical embryological studies (see Chapter 47).

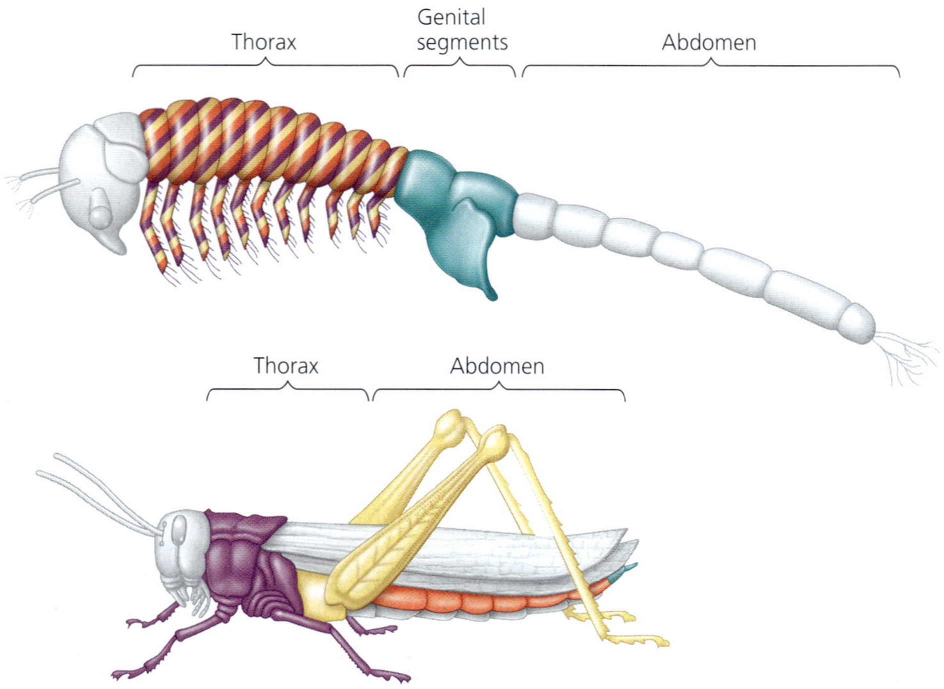

▲ **Figure 21.18 Effect of differences in *Hox* gene expression during development in crustaceans and insects.** Changes in the expression patterns of four *Hox* genes have occurred over evolutionary time. These changes account in part for the different body plans of the brine shrimp *Artemia*, a crustacean (top), and the grasshopper, an insect. Shown here are regions of the adult body color-coded for expression of the *Hox* genes that determine formation of particular body parts during embryonic development. Each color represents a specific *Hox* gene.

Comparison of Animal and Plant Development

The last common ancestor of animals and plants was probably a single-celled eukaryote that lived hundreds of millions of years ago, so the processes of development must have evolved independently in the two multicellular lineages of organisms. Plants evolved with rigid cell walls, which rule out the morphogenetic movements of cells and tissues that are so important in animals. Instead, morphogenesis in plants relies primarily on differing planes of cell division and on selective cell enlargement. (You will learn about these processes in

Chapter 35.) But despite the differences between animals and plants, there are similarities in the molecular mechanisms of development, which are legacies of their shared unicellular origin.

In both animals and plants, development relies on a cascade of transcriptional regulators turning on or turning off genes in a finely tuned series. For example, work on a small flowering plant in the mustard family, *Arabidopsis thaliana*, has shown that establishing the radial pattern of flower parts, like setting up the head-to-tail axis in *Drosophila*, involves a cascade of transcription factors. The genes that direct these processes, however, differ considerably in animals and plants. While quite a few of the master regulatory switches in *Drosophila* are homeobox-containing *Hox* genes, those in *Arabidopsis* belong to a completely different family of genes, called the *Mads-box* genes. And although homeobox-containing genes can be found in plants and *Mads-box* genes in animals, in neither case do they perform the same major roles in development that they do in the other group. Thus, molecular evidence supports the supposition that developmental programs evolved separately in animals and plants.

In this final chapter of the genetics unit, you have learned how studying genomic composition and comparing the genomes of different species can disclose much about how genomes evolve. Further, comparing developmental programs, we can see that the unity of life is reflected in the similarity of molecular and cellular mechanisms used to establish body pattern, although the genes directing development may differ among organisms. The similarities between genomes reflect the common ancestry of life on Earth. But the differences are also crucial, for they have created the huge diversity of organisms that have evolved. In the remainder of the book, we expand our perspective beyond the level of molecules, cells, and genes to explore this diversity on the organismal level.

CONCEPT CHECK 21.6

1. Would you expect the genome of the macaque (a monkey) to be more similar to the mouse genome or the human genome? Why?
2. The DNA sequences called homeoboxes, which help homeotic genes in animals direct development, are common to flies and mice. Given this similarity, explain why these animals are so different.
3. **WHAT IF?** There are three times as many *Alu* elements in the human genome as in the chimpanzee genome. How do you think these extra *Alu* elements arose in the human genome? Propose a role they might have played in the divergence of these two species.

For suggested answers, see Appendix A.

Chapter 21 Review

MEDIA Go to the Study Area at **www.masteringbio.com** for BioFlix 3-D Animations, MP3 Tutors, Videos, Practice Tests, an eBook, and more.

SUMMARY OF KEY CONCEPTS

CONCEPT 21.1
New approaches have accelerated the pace of genome sequencing (pp. 427–429)

- **Three-Stage Approach to Genome Sequencing** In linkage mapping, the order of genes and other inherited markers in the genome and the relative distances between them can be determined from recombination frequencies. Next, physical mapping uses overlaps between DNA fragments to order the fragments and determine the distance in base pairs between markers. Finally, the ordered fragments are sequenced, providing the finished genome sequence.
- **Whole-Genome Shotgun Approach to Genome Sequencing** The whole genome is cut into many small, overlapping fragments that are sequenced; computer software then assembles the complete sequence. Correct assembly is made easier when mapping information is also available.

MEDIA
Activity The Human Genome Project: Genes on Human Chromosome 17

CONCEPT 21.2
Scientists use bioinformatics to analyze genomes and their functions (pp. 429–432)

- **Centralized Resources for Analyzing Genome Sequences** Websites on the Internet provide centralized access to genome sequence databases, analytical tools, and genome-related information.
- **Identifying Protein-Coding Genes Within DNA Sequences** Computer analysis of genome sequences helps researchers identify sequences likely to encode proteins. Comparison of the sequences of "new" genes with those of known genes in other species may help identify functions of new genes. For a gene of unknown function, experimental inactivation of the gene and observation of the resulting phenotypic effects can provide clues to its function.
- **Understanding Genes and Their Products at the Systems Level** Using the computer-based tools of bioinformatics, scientists can compare genomes and study sets of genes and proteins as whole systems (genomics and proteomics). Studies include large-scale analyses of protein interactions.

CONCEPT **21.3**

Genomes vary in size, number of genes, and gene density (pp. 432–434)

	Bacteria	Archaea	Eukarya
Genome size	Most are 1–6 Mb		Most are 10–4,000 Mb, but a few are much larger
Number of genes	1,500–7,500		5,000–40,000
Gene density	Higher than in eukaryotes		Lower than in prokaryotes (Within eukaryotes, lower density is correlated with larger genomes.)
Introns	None in protein-coding genes	Present in some genes	Unicellular eukaryotes: present, but prevalent only in some species Multicellular eukaryotes: present in most genes
Other noncoding DNA	Very little		Can be large amounts; generally more repetitive noncoding DNA in multicellular eukaryotes

CONCEPT **21.4**

Multicellular eukaryotes have much noncoding DNA and many multigene families (pp. 434–438)

- Only 1.5% of the human genome codes for proteins or gives rise to rRNAs or tRNAs; the rest is noncoding DNA, including repetitive DNA.
- **Transposable Elements and Related Sequences** The most abundant type of repetitive DNA in multicellular eukaryotes consists of transposable elements and related sequences. Two types of transposable elements occur in eukaryotes: transposons, which move via a DNA intermediate, and retrotransposons, which are more prevalent and move via an RNA intermediate. Each element may be hundreds or thousands of base pairs long, and similar but usually not identical copies are dispersed throughout the genome.
- **Other Repetitive DNA, Including Simple Sequence DNA** Short noncoding sequences that are tandemly repeated thousands of times (simple sequence DNA, which includes STRs) are especially prominent in centromeres and telomeres, where they probably play structural roles in the chromosome.
- **Genes and Multigene Families** Though many eukaryotic genes are present in one copy per haploid chromosome set, others (most, in some species) are members of a family of related genes. The transcription unit corresponding to the three largest rRNAs is tandemly repeated hundreds to thousands of times at one or several chromosomal sites, enabling the cell to quickly make the rRNA for millions of ribosomes. The multiple, slightly different genes in the two globin gene families encode polypeptides used at different developmental stages of an animal.

CONCEPT **21.5**

Duplication, rearrangement, and mutation of DNA contribute to genome evolution (pp. 438–442)

- **Duplication of Entire Chromosome Sets** Accidents in cell division can lead to extra copies of all or part of a genome, which may then diverge if one set accumulates sequence changes.
- **Alterations of Chromosome Structure** The chromosomal organization of genomes can be compared among species, providing information about evolutionary relationships. Within a given species, rearrangements of chromosomes are thought to contribute to the emergence of new species.
- **Duplication and Divergence of Gene-Sized Regions of DNA** The genes encoding the various globin proteins evolved from one common ancestral globin gene, which duplicated and diverged into α-globin and β-globin ancestral genes. Subsequent duplication and random mutation gave rise to the present globin genes, all of which code for oxygen-binding proteins. The copies of some duplicated genes have diverged so much that the functions of their encoded proteins are now substantially different.
- **Rearrangements of Parts of Genes: Exon Duplication and Exon Shuffling** Rearrangement of exons within and between genes during evolution has led to genes containing multiple copies of similar exons and/or several different exons derived from other genes.
- **How Transposable Elements Contribute to Genome Evolution** Movement of transposable elements or recombination between copies of the same element occasionally generates new sequence combinations that are beneficial to the organism. Such mechanisms can alter the functions of genes or their patterns of expression and regulation.

MEDIA

Biology Labs On-Line HemoglobinLab

CONCEPT **21.6**

Comparing genome sequences provides clues to evolution and development (pp. 442–447)

- **Comparing Genomes** Comparative studies of genomes from widely divergent and closely related species provides valuable information about ancient and more recent evolutionary history, respectively. Human and chimpanzee sequences show about 4% difference, mostly due to insertions, deletions, and duplications in one lineage. Along with nucleotide variations in specific genes (such as *FOXP2*, a gene affecting speech), these differences may account for the distinct characteristics of the two species. Single nucleotide polymorphisms among individuals in a species can also yield information about the history of that species.
- **Comparing Developmental Processes** Homeotic genes and some other genes associated with animal development contain a homeobox region, whose sequence is identical or similar in diverse species. Related sequences are present in the genes of plants and yeasts. Other developmental genes also are highly conserved among animal species, but they may play different roles in the development of different species. During embryonic development in both plants and animals, a cascade of transcription regulators turns genes on or off in a carefully regulated sequence. However, the genes that direct analogous developmental processes differ considerably in sequence in plants and animals as a result of their remote ancestry.

TESTING YOUR KNOWLEDGE

SELF-QUIZ

1. Bioinformatics includes all of the following *except*
   a. using computer programs to align DNA sequences.
   b. analyzing protein interactions in a species.
   c. using molecular biology to combine DNA from two different sources in a test tube.
   d. development of computer-based tools for genome analysis.
   e. use of mathematical tools to make sense of biological systems.

2. Which of the following has the largest genome and the fewest genes per million base pairs?
   a. *Haemophilus influenzae* (bacterium)
   b. *Saccharomyces cerevisiae* (yeast)
   c. *Arabidopsis thaliana* (plant)
   d. *Drosophila melanogaster* (fruit fly)
   e. *Homo sapiens* (human)

3. One of the characteristics of retrotransposons is that
   a. they code for an enzyme that synthesizes DNA using an RNA template.
   b. they are found only in animal cells.
   c. they generally move by a cut-and-paste mechanism.
   d. they contribute a significant portion of the genetic variability seen within a population of gametes.
   e. their amplification is dependent on a retrovirus.

4. Multigene families are
   a. groups of enhancers that control transcription.
   b. usually clustered at the telomeres.
   c. equivalent to the operons of prokaryotes.
   d. sets of genes that are coordinately controlled.
   e. sets of identical or similar genes that have evolved by gene duplication.

5. Two eukaryotic proteins have one domain in common but are otherwise very different. Which of the following processes is most likely to have contributed to this similarity?
   a. gene duplication
   b. RNA splicing
   c. exon shuffling
   d. histone modification
   e. random point mutations

6. Homeotic genes
   a. encode transcription factors that control the expression of genes responsible for specific anatomical structures.
   b. are found only in *Drosophila* and other arthropods.
   c. are the only genes that contain the homeobox domain.
   d. encode proteins that form anatomical structures in the fly.
   e. are responsible for patterning during plant development.

7. **DRAW IT** At the top of the next column are the amino acid sequences (using the single-letter code; see Figure 5.17) of four short segments of the FOXP2 protein from six species: chimpanzee, orangutan, gorilla, rhesus macaque, mouse, and human. These segments contain all of the amino acid differences between the FOXP2 proteins of these species.

1. ATETI...PKSSD...TSSTT...NARRD
2. ATETI...PKSSE...TSSTT...NARRD
3. ATETI...PKSSD...TSSTT...NARRD
4. ATETI...PKSSD...TSSNT...SARRD
5. ATETI...PKSSD...TSSTT...NARRD
6. VTETI...PKSSD...TSSTT...NARRD

Use a highlighter to color any amino acid that varies among the species. (Color that amino acid in all sequences.) Then answer the following questions.
   a. The chimpanzee, gorilla, and rhesus macaque (C, G, R) sequences are identical. Which lines correspond to those sequences?
   b. The human sequence differs from that of the C, G, R species at two amino acids. Which line corresponds to the human sequence? Underline the two differences.
   c. The orangutan sequence differs from the C, G, R sequence at one amino acid (having valine instead of alanine) and from the human sequence at three amino acids. Which line corresponds to the orangutan sequence?
   d. How many amino acid differences are there between the mouse and the C, G, R species? Circle the amino acid(s) that differ(s) in the mouse. How many amino acid differences are there between the mouse and the human? Draw a square around the amino acid(s) that differ(s) in the mouse.
   e. Primates and rodents diverged between 60 and 100 million years ago, and chimpanzees and humans diverged about 6 million years ago. Knowing that, what can you conclude by comparing the amino acid differences between the mouse and the C, G, R species with the differences between the human and the C, G, R species?

For Self-Quiz answers, see Appendix A.

MEDIA Visit the Study Area at **www.masteringbio.com** for a Practice Test.

EVOLUTION CONNECTION

8. Genes important in the embryonic development of animals, such as homeobox-containing genes, have been relatively well conserved during evolution; that is, they are more similar among different species than are many other genes. Why is this?

SCIENTIFIC INQUIRY

9. The scientists mapping the SNPs in the human genome noticed that groups of SNPs tended to be inherited together, in blocks known as haplotypes, ranging in length from 5,000 to 200,000 base pairs. There are as few as four or five commonly occurring combinations of SNPs per haplotype. Propose an explanation for this observation, integrating what you've learned throughout this chapter and this unit.

UNIT 4 Mechanisms of Evolution

AN INTERVIEW WITH

Scott V. Edwards

Birds—and the birds of Australia in particular—might seem a surprising focus for a scientist who grew up in New York City, but they are the main subjects of Scott Edwards's research on evolution. A graduate of Harvard College, with a Ph.D. from the University of California, Berkeley, and postdoctoral work at the University of Florida, Dr. Edwards was at the University of Washington until he returned to Harvard in 2003 as Professor of Organismic and Evolutionary Biology. Jane Reece and Michael Cain interviewed Dr. Edwards at Harvard's Museum of Comparative Zoology, where he is Curator of Ornithology and the head of an active research group.

How big is the bird collection here?

Since the Museum of Comparative Zoology was founded in 1859, its collection of bird specimens has grown to 350,000 specimens, the largest university collection in the world. Each specimen is tagged with data about the location where it was obtained, the date, often the bird's sex and weight, and other pertinent information. The collection provides a remarkable record of how bird species have changed over the years, as environments have changed. Using the oldest specimens here, we can now compare the sizes, shapes, and genes of birds from populations separated by more than a century. What's exciting now is that all the information is being digitized, so that we can easily look at large amounts of data from multiple museums.

Were you interested in birds as a child?

When I was about six years old, we moved from an urban neighborhood to Riverdale, in the northwest corner of the Bronx. There were actually trees there, and we were close to the Hudson River. A few years later, a neighbor took me bird watching—and the rest is history. I'll never forget how impressed I was with my first Northern flicker; it was just remarkable to me that this bizarrely colored woodpecker lived right in my backyard.

Later, as an undergraduate biology major, I really needed a break after taking organic chemistry, and I was able to take a year off. I volunteered at the Smithsonian for several months and then at national parks in Hawaii and California. It was that first exposure to fieldwork that showed me what biologists do. And when I came back to college, I was much more focused and motivated. I think it should be mandatory for biology majors to work in the field or in a lab for a few months, because that's how you find out what science is all about.

How did you get interested in evolution?

I was impressed with the precision that molecular tools seemed to bring to the study of evolution. I realize now that things are not as precise as I thought, but the DNA code is still a remarkable yardstick for comparing different species on the same scale. I actually didn't like biochemistry or molecular biology until I could connect them with evolution. But coming back from my year off, I worked in a lab that helped me make the connection. The world is very diverse biologically—there are millions of species—and the fact that these species can be compared at the DNA level was a revelation to me.

What led you to study Australian birds?

After entering grad school, I volunteered for a research project on birds of paradise in New Guinea; these are flamboyant songbirds that live in the rain forest. I was hunting around for a project of my own. Several ornithologists directed me to a group of songbirds called babblers, which mostly live in Australia. These birds are fascinating: They live in family groups, using large, domed nests with a hole on the side. Eight or nine of them will clamber into one of these nests. It was very interesting to see the organization of their family groups in the wild and also to study the birds on the DNA level. Were they all related to each other, or not? How different were different families in a single locality? So I ended up doing my dissertation on babblers, comparing individuals within a family, families within a region, and populations in different parts of the continent.

As a postdoc in Florida, you studied the evolution of genes involved in disease resistance in birds. What were these genes?

They're called MHC genes, for Major Histocompatibility Complex. These genes are important components of the immune systems of all vertebrates. In humans, they are the genes you try to match when looking for a compatible donor for an organ transplant. The MHC genes encode proteins that bind to fragments of pathogens and other foreign cells that have been phagocytized. The MHC proteins then move to the cell surface and present the fragments to the rest of the immune system, saying, "Hey, I found something foreign." The MHC is fascinating from an evolutionary standpoint, because pathogens and parasites seem to be major drivers of evolutionary change. They're constantly playing cat and mouse with the host. In the pathogens, more efficient ways of infecting the host keep evolving, while evolving defenses keep pace in their hosts. So MHC genes are under strong selection by pathogens, and we can see signs of that selection when we look at the DNA sequences. In mammals, these genes are wildly diverse compared to typical "housekeeping" genes. I was curious to see if they were as diverse in birds as they are in mammals.

How did so much variation evolve in MHC genes?

There are a couple of different hypotheses. One comes from the idea that the primary drivers of MHC diversity are pathogens. It says that if you are heterozygous—if you have two different alleles for each of these MHC genes—you have a better surveillance system for dealing with the pathogenic world. You're going to be able to recognize at least twice as many pathogens compared to a homozygous individual. When you have a situation where heterozygotes are more fit because they can combat disease, it results in a lot of diversity. The other hypothesis focuses on mating preferences, though it's not inconsistent with the

MEDIA
Activity Conservation Biology Review
GraphIt! Global Freshwater Resources
GraphIt! Prospects for Renewable Energy

TESTING YOUR KNOWLEDGE

SELF-QUIZ

1. Ecologists conclude there is a biodiversity crisis because
a. biophilia causes humans to feel ethically responsible for protecting other species.
b. scientists have at last discovered and counted most of Earth's species and can now accurately calculate the current extinction rate.
c. current extinction rates are very high and many species are threatened or endangered.
d. many potential life-saving medicines are being lost as species evolve.
e. there are too few biodiversity hot spots.

2. Which of the following would be considered an example of bioremediation?
a. adding nitrogen-fixing microorganisms to a degraded ecosystem to increase nitrogen availability
b. using a bulldozer to regrade a strip mine
c. identifying a new biodiversity hot spot
d. reconfiguring the channel of a river
e. adding seeds of a chromium-accumulating plant to soil contaminated by chromium

3. What is the effective population size (N_e) of a population of 50 strictly monogamous swans (40 males and 10 females) if every female breeds successfully?
a. 50 b. 40 c. 30 d. 20 e. 10

4. One characteristic that distinguishes a population in an extinction vortex from most other populations is that
a. its habitat is fragmented.
b. it is a rare, top-level predator.
c. its effective population size is much lower than its total population size.
d. its genetic diversity is very low.
e. it is not well adapted to edge conditions.

5. The discipline that applies ecological principles to returning degraded ecosystems to more natural states is known as
a. population viability analysis.
b. landscape ecology.
c. conservation ecology.
d. restoration ecology.
e. resource conservation.

6. What is the single greatest threat to biodiversity?
a. overexploitation of commercially important species
b. introduced species that compete with or prey on native species
c. pollution of Earth's air, water, and soil
d. disruption of trophic relationships as more and more prey species become extinct
e. habitat alteration, fragmentation, and destruction

7. Which of the following strategies would most rapidly increase the genetic diversity of a population in an extinction vortex?
a. Capture all remaining individuals in the population for captive breeding followed by reintroduction to the wild.
b. Establish a reserve that protects the population's habitat.
c. Introduce new individuals transported from other populations of the same species.
d. Sterilize the least fit individuals in the population.
e. Control populations of the endangered population's predators and competitors.

8. Of the following statements about protected areas that have been established to preserve biodiversity, which one is *not* correct?
a. About 25% of Earth's land area is now protected.
b. National parks are one of many types of protected areas.
c. Most protected areas are too small to protect species.
d. Management of a protected area should be coordinated with management of the land surrounding the area.
e. It is especially important to protect biodiversity hot spots.

For Self-Quiz answers, see Appendix A.

MEDIA Visit the Study Area at **www.masteringbio.com** for a Practice Test.

EVOLUTION CONNECTION

9. One factor favoring rapid population growth by an introduced species is the absence of the predators, parasites, and pathogens that controlled its population in the region where it evolved. Over the long term, how should evolution by natural selection influence the rate at which the native predators, parasites, and pathogens in a region of introduction attack an introduced species?

SCIENTIFIC INQUIRY

10. **DRAW IT** Suppose that you are in charge of planning a forest reserve, and one of your goals is to help sustain local populations of woodland birds suffering from parasitism by the brown-headed cowbird. Reading research reports, you note that female cowbirds are usually reluctant to penetrate more than about 100 m into a forest and that nest parasitism is reduced for woodland birds nesting in denser, more central forest regions. The forested area you have to work with extends about 6,000 m from east to west and 1,000 m from north to south. Intact forest surrounds the reserve everywhere but on the west side, where the reserve borders deforested pastureland, and in the southwest corner, where it borders an agricultural field for 500 m. Your plan must include space for a small maintenance building, which you estimate to take up about 100 m^2. It will also be necessary to build a road, 10 m by 1,000 m, from the north to the south side of the reserve. Draw a map of the reserve, showing where you would construct the road and the building to minimize cowbird intrusion along edges. Explain your reasoning.

APPENDIX A Answers

CHAPTER 1

Figure Questions

Figure 1.3 Of the properties shown in this figure, the lawn mower shows only order, regulation, and energy processing. **Figure 1.6** The arrangement of fingers and opposable thumb in the human hand, combined with fingernails and a complex system of nerves and muscles, allows the hand to grasp and manipulate objects with great dexterity. **Figure 1.13** Substance B would be made continuously and would accumulate in large amounts. Neither C nor D would be made. **Figure 1.27** The percentage of brown artificial snakes attacked would probably be higher than the percentage of artificial kingsnakes attacked in all areas (whether or not inhabited by coral snakes). **Figure 1.28** The hole would allow some mixing of blood between the two ventricles. As a result, some of the blood pumped from the left ventricle to the body would not have received oxygen in the lungs, and some of the blood pumped to the lungs would already carry oxygen.

Concept Check 1.1

1. Examples: A molecule consists of *atoms* bonded together. Each organelle has an orderly arrangement of *molecules*. Photosynthetic plant cells contain *organelles* called chloroplasts. A tissue consists of a group of similar *cells*. Organs such as the heart are constructed from several *tissues*. A complex multicellular organism, such as a plant, has several types of *organs*, such as leaves and roots. A population is a set of *organisms* of the same species. A community consists of *populations* of the various species inhabiting a specific area. An ecosystem consists of a biological *community* along with the nonliving factors important to life, such as air, soil, and water. The biosphere is made up of all of Earth's *ecosystems*. **2.** (a) Structure and function are correlated. (b) Cells are an organism's basic units, *and* the continuity of life is based on heritable information in the form of DNA. (c) Organisms interact with their environments, exchanging matter and energy. **3.** Some possible answers: *Evolution:* All plants have chloroplasts, indicating their descent from a common ancestor. *Emergent properties:* The ability of a human heart to pump blood requires an intact heart; it is not a capability of any of the heart's tissues or cells working alone. *Exchange of matter and energy with the environment:* A mouse eats food, then uses the nutrients for growth and the generation of energy for its activities; some of the food material is expelled in urine and feces, and some of the energy returns to the environment as heat. *Structure and function:* The strong, sharp teeth of a wolf are well suited to grasping and dismembering its prey. *Cells:* The digestion of food is made possible by chemicals (chiefly enzymes) made by cells of the digestive tract. *DNA:* Human eye color is determined by the combination of genes inherited from the two parents. *Feedback regulation:* When your stomach is full, it signals your brain to decrease your appetite.

Concept Check 1.2

1. An address pinpoints a location by tracking from broader to narrower categories—a state, city, zip, street, and building number. This is analogous to the groups-subordinate-to-groups structure of biological taxonomy. **2.** Natural selection starts with the naturally occurring heritable variation in a population and then "edits" the population as individuals with heritable traits better suited to the environment survive and reproduce more successfully than others.
3.

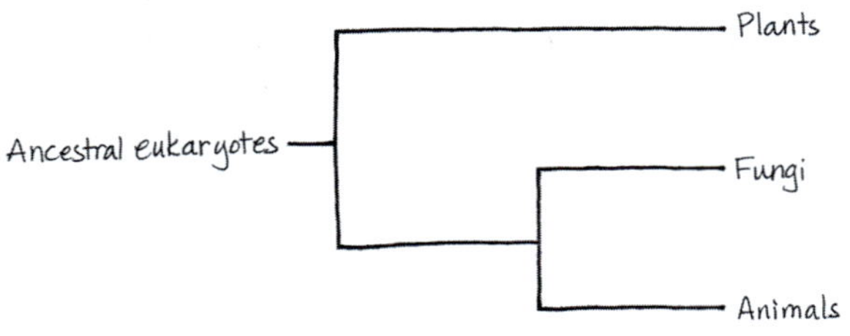

Concept Check 1.3

1. Inductive reasoning derives generalizations from specific cases; deductive reasoning predicts specific outcomes from general premises. **2.** Compared to a hypothesis, a scientific theory is usually more general and substantiated by a much greater amount of evidence. Natural selection is an explanatory idea that applies to all kinds of organisms and is supported by vast amounts of evidence of various kinds. **3.** Based on the results shown in Figure 1.27, you might predict that the colorful artificial snakes would be attacked more often than the brown ones, simply because they are easier to see. This prediction assumes that the area in Virginia where you are working has predators that attack snakes but no poisonous snakes that resemble the colorful artificial snakes.

Self-Quiz

1. b **2.** d **3.** a **4.** c **5.** c **6.** c **7.** c **8.** d **9.** b **10.** c
11. Your figure should show: (1) For the biosphere, the Earth with an arrow coming out of a tropical ocean; (2) for the ecosystem, a distant view of a coral reef; (3) for the community, a collection of reef animals and algae, with corals, fish, some seaweed, and any other organisms you can think of; (4) for the population, a group of fish of the same species; (5) for the organism, one fish from your population; (6) for the organ, the fish's stomach, and for the organ system, the whole digestive tract (see Chapter 41 for help); (7) for a tissue, a group of similar cells from the stomach; (8) for a cell, one cell from the tissue, showing its nucleus and a few other organelles; (9) for an organelle, the nucleus, where most of the cell's DNA is located; and (10) for a molecule, a DNA double helix. Your sketches can be very rough!

CHAPTER 2

Figure Questions

Figure 2.2 The most significant difference in the results would be that the two *Cedrela* saplings inside each garden would show similar amounts of dying leaf tissue because a poisonous chemical released from the *Duroia* trees would presumably reach the saplings via the air or soil and would not be blocked by the insect barrier. The *Cedrela* saplings planted outside the gardens would not show damage unless *Duroia* trees were nearby. Also, any ants present on the unprotected *Cedrela* saplings inside the gardens would probably not be observed making injections into the leaves. However, formic acid would likely still be found in the ants' glands, as for most species of ants. **Figure 2.9** Atomic number = 12; 12 protons, 12 electrons; three electron shells; 2 electrons in the valence shell
Figure 2.16

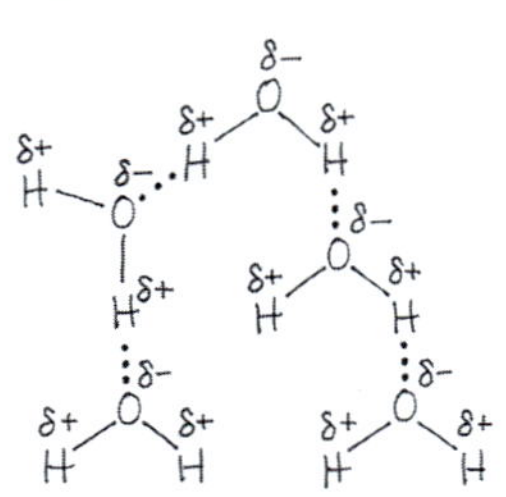

Figure 2.19 The plant is submerged in water (H_2O), in which the CO_2 is dissolved. The sun's energy is used to make sugar, which is found in the plant and can act as food for the plant itself, as well as for animals that eat the plant. The oxygen (O_2) is present in the bubbles.

Concept Check 2.1

1. Table salt is made up of sodium and chlorine. We are able to eat the compound, showing that it has different properties from those of a metal and a poisonous gas. **2.** Yes, because an organism requires trace elements, even though only in small amounts. **3.** A person with an iron deficiency will probably show effects of low oxygen in the blood, such as fatigue. (The condition is called anemia and can also result from too few red blood cells or abnormal hemoglobin.)

Concept Check 2.2

1. 7 **2.** $^{15}_{7}N$ **3.** 9 electrons; two electron shells; 1*s*, 2*s*, 2*p* (three orbitals); 1 electron is needed to fill the valence shell. **4.** The elements in a row all have the same number of electron shells. In a column, all the elements have the same number of electrons in their valence shells.

Concept Check 2.3

1. Each carbon atom has only three covalent bonds instead of the required four. **2.** The attractions between oppositely charged ions form ionic bonds. **3.** If researchers can synthesize molecules that mimic these shapes, they may be able to treat diseases or conditions caused by the inability of affected individuals to synthesize such molecules.

Concept Check 2.4

1.

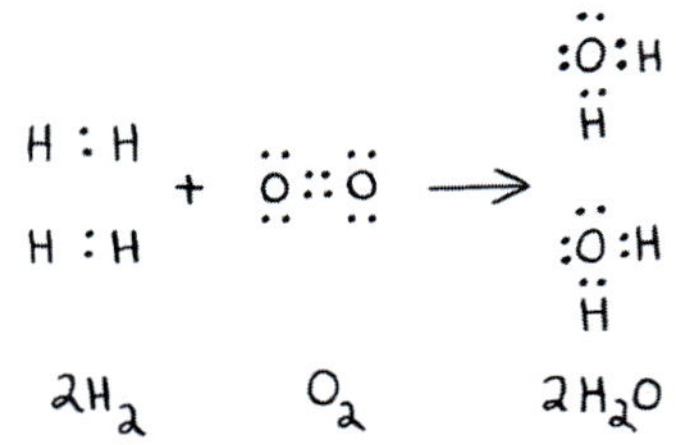

2. At equilibrium, the forward and reverse reactions occur at the same rate. **3.** $C_6H_{12}O_6 + 6\,O_2 \rightarrow 6\,CO_2 + 6\,H_2O$ + Energy. Glucose and oxygen react to form carbon dioxide, water, and energy. We breathe in oxygen because we need it for this reaction to occur, and we breathe out carbon dioxide because it is a by-product of this reaction. By the way, this reaction is called cellular respiration, and you will learn more about it in Chapter 9.

Self-Quiz

1. a **2.** b **3.** b **4.** c **5.** b **6.** a **7.** b **8.** b

9.

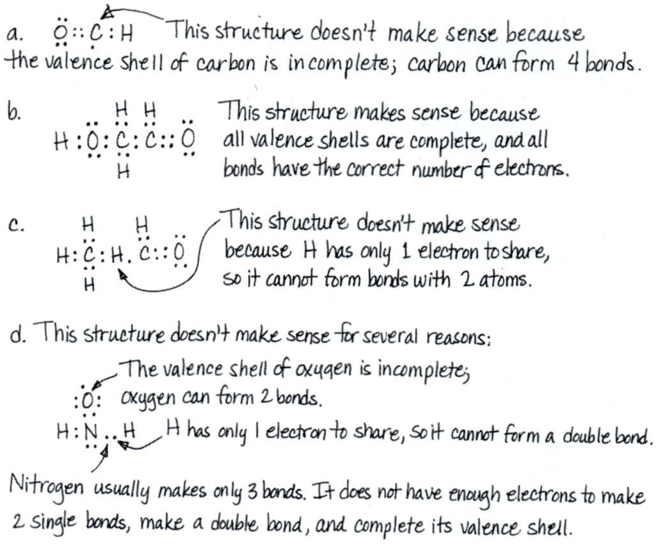

CHAPTER 3

Figure Questions

Figure 3.6 Without hydrogen bonds, water would behave like other small molecules, and the solid phase (ice) would be denser than liquid water. The ice would sink to the bottom, and because it would no longer insulate the whole body of water, it could freeze. Freezing would take a longer time because the Antarctic is an ocean (the Southern Ocean), not a pond or lake, but the average annual temperature at the South Pole is −50°C, so eventually it would freeze. The krill could not survive. **Figure 3.7** Heating the solution would cause the water to evaporate faster than it is evaporating at room temperature. At a certain point, there wouldn't be enough water molecules to solubilize the salt ions. The salt would start coming out of solution and re-forming crystals. Eventually, all the water would evaporate, leaving behind a pile of salt like the original pile. **Figure 3.11** Given that Ca^{2+} and $CO_3{}^{2-}$ must interact to form $CaCO_3$, you would predict that $[Ca^{2+}]$ would also have an effect on the calcification rate, and this result is observed in the current study. Under natural conditions in the oceans, the $[Ca^{2+}]$ remains relatively constant, so the $[CO_3{}^{2-}]$ has a much more important effect on calcification rate.

Concept Check 3.1

1. Electronegativity is the attraction of an atom for the electrons of a covalent bond. Because oxygen is more electronegative than hydrogen, the oxygen atom in H_2O pulls electrons toward itself, resulting in a partial negative charge on the oxygen atom and partial positive charges on the hydrogen atoms. Oppositely charged ends of water molecules are attracted to each other, forming a hydrogen bond. **2.** The hydrogen atoms of one molecule, with their partial positive charges, would repel the hydrogen atoms of the adjacent molecule. **3.** Water molecules would not be polar, and they would not form hydrogen bonds with each other.

Concept Check 3.2

1. Hydrogen bonds hold neighboring water molecules together. This cohesion helps the molecules resist gravity. Adhesion between water molecules and the walls of water-conducting cells also counters gravity. As water evaporates from leaves, the chain of water molecules in water-conducting cells moves upward. **2.** High humidity hampers cooling by suppressing the evaporation of sweat. **3.** As water freezes, it expands because water molecules move farther apart in forming ice crystals. When there is water in a crevice of a boulder, expansion due to freezing may crack the rock. **4.** The molecular mass of NaCl is 58.5 daltons. A mole would have a mass of 58.5 g, so you would measure out 0.5 mol, or 29.3 g, of NaCl and gradually add water, stirring until it is dissolved. You would add water to bring the final volume to 1 L. **5.** The hydrophobic substance repels water, perhaps helping to keep the ends of the legs from becoming coated with water and breaking through the surface. If the legs were coated with a hydrophilic substance, water would be drawn up them, possibly making it more difficult for the water strider to walk on water.

Concept Check 3.3

1. 10^5, or 100,000 **2.** $[H^+] = 0.01\ M = 10^{-2}\ M$, so pH = 2 **3.** $CH_3COOH \rightleftharpoons CH_3COO^- + H^+$. CH_3COOH is the acid (the H^+ donor) and CH_3COO^- is the base (the H^+ acceptor). **4.** The pH of the water should go from 7 to about 2; the pH of the acetic acid solution will only decrease a small amount, because the reaction shown for question 3 will shift to the left, with CH_3COO^- accepting the influx of H^+ and becoming CH_3COOH molecules.

Self-Quiz

1. d **2.** c **3.** b **4.** c **5.** c **6.** d **7.** c **8.** c

9.

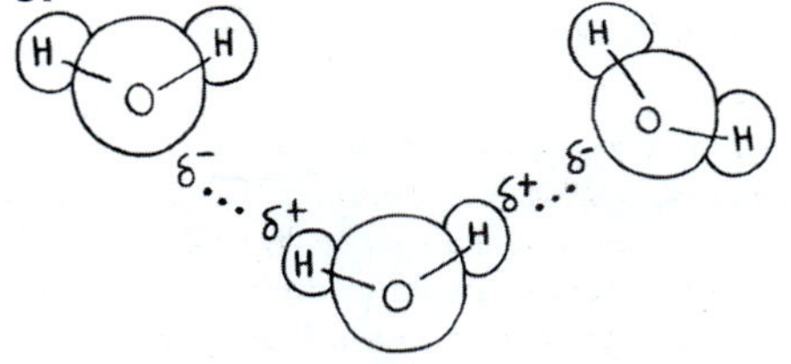

CHAPTER 4

Figure Questions

Figure 4.2 Because the concentration of the reactants influences the equilibrium (as discussed in Chapter 2), there might be more HCN relative to CH_2O, since there would be a higher concentration of the reactant gas that contains nitrogen.

Figure 4.4

Figure 4.7

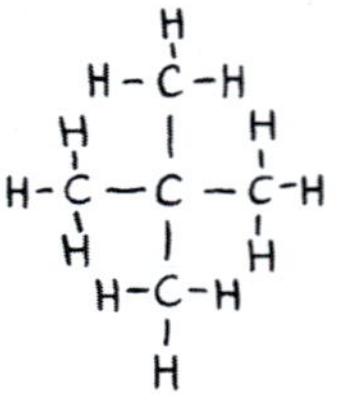

Figure 4.10 Molecule b, because there are not only the two electronegative oxygens of the carboxyl group, but also an oxygen on the next (carbonyl) carbon. All of these oxygens help make the bond between the O and H of the —OH group more polar, thus making the dissociation of H^+ more likely.

Concept Check 4.1

1. Amino acids are essential molecules for living organisms. Their synthesis from gases of the primitive atmosphere on Earth demonstrated that life's molecules could initially have been synthesized from nonliving molecules. **2.** The spark provides energy needed for the inorganic molecules in the atmosphere to react with each other. (You'll learn more about energy and chemical reactions in Chapter 8.)

Concept Check 4.2

1.

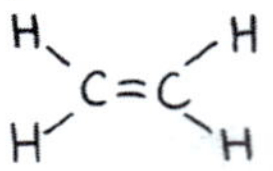

2. The forms of C_4H_{10} in (b) are structural isomers, as are the butenes in (c). **3.** Both consist largely of hydrocarbon chains. **4.** No. There is not enough diversity in the atoms. It can't form structural isomers because there is only one way for three carbons to attach to each other (in a line). There are no double bonds, so geometric isomers are not possible. Each carbon has at least two hydrogens attached to it, so the molecule is symmetrical and cannot have enantiomeric isomers.

Concept Check 4.3

1. It has both an amino group (—NH_2), which makes it an amine, and a carboxyl group (—COOH), which makes it a carboxylic acid. **2.** The ATP molecule loses a phosphate, becoming ADP.

3.

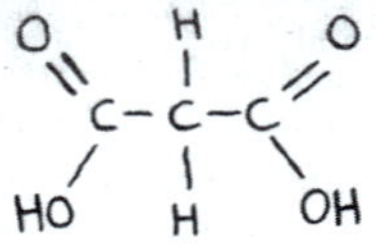

A chemical group that can act as a base has been replaced with a group that can act as an acid, increasing the acidic properties of the molecule. The shape of the molecule would also change, likely changing the molecules with which it can interact.

Self-Quiz

1. b **2.** d **3.** a **4.** b **5.** b **6.** a **7.** d

8.

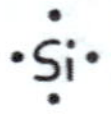

Si has four valence electrons, the same number as carbon. Therefore, silicon would be able to form long chains, including branches, that could act as skeletons for organic molecules. It would clearly do this much better than neon (with no valence electrons) or aluminum (with three valence electrons).

CHAPTER 5

Figure Questions

Figure 5.4

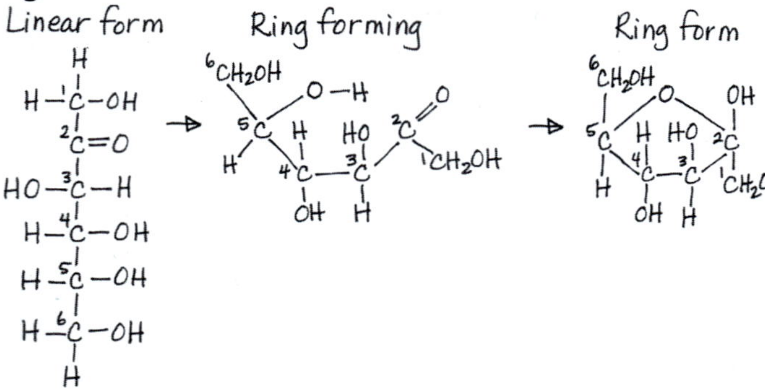

Four carbons are in the fructose ring, and two are not. (The latter two carbons are hanging off carbons 2 and 5, which are in the ring.) This form differs from glucose, which has five carbons in the ring and one that is not. (Note that the orientation of this fructose molecule is flipped relative to the one in Figure 5.5b.)

Figure 5.18

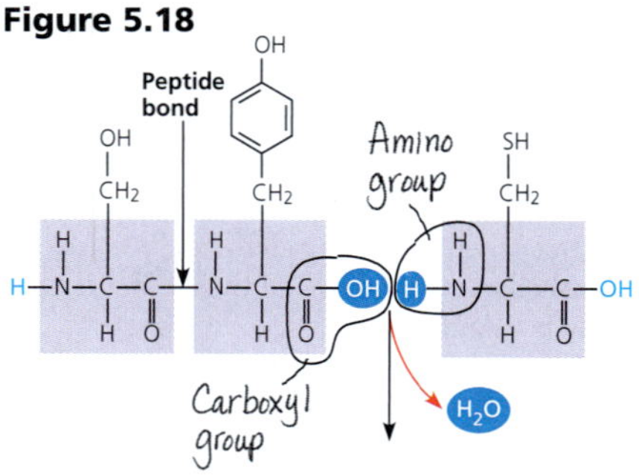

Figure 5.25 The green spiral is an α helix.

Concept Check 5.1

1. Proteins, carbohydrates, lipids, and nucleic acids **2.** Nine, with one water required to hydrolyze each connected pair of monomers **3.** The amino acids in the green bean protein are released in hydrolysis reactions and incorporated into other proteins in dehydration reactions.

Concept Check 5.2

1. $C_3H_6O_3$ **2.** $C_{12}H_{22}O_{11}$ **3.** The absence of these prokaryotes would hamper the cow's ability to obtain energy from food and could lead to weight loss and possibly death.

Concept Check 5.3

1. Both have a glycerol molecule attached to fatty acids. The glycerol of a fat has three fatty acids attached, whereas the glycerol of a phospholipid is attached to two fatty acids and one phosphate group. **2.** Human sex hormones are steroids, a type of hydrophobic compound. **3.** The oil droplet membrane could consist of a single layer of phospholipids rather than a bilayer, because an arrangement in which the hydrophobic tails of the membrane phospholipids were in contact with the hydrocarbon regions of the oil molecules would be more stable.

Concept Check 5.4

1. The function of a protein is a consequence of its specific shape, which is lost when a protein becomes denatured. **2.** Secondary structure involves hydrogen bonds between atoms of the polypeptide backbone. Tertiary structure involves bonding between atoms of the R groups of the amino acid subunits. **3.** Primary structure, the amino acid sequence, affects the secondary structure, which affects the tertiary structure, which affects the quaternary structure (if any). In short, the amino acid sequence affects the shape of the protein. Because the function of a protein depends on its shape, a change in primary structure can destroy a protein's function.

Concept Check 5.5

1. **5′ end**

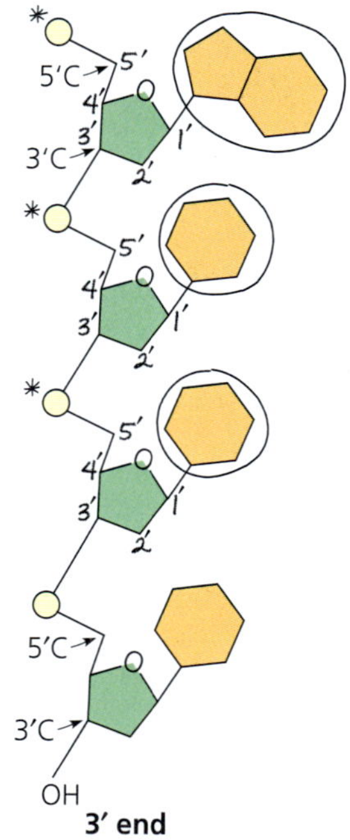

3′ end

2. 5′-TAGGCCT–3′
3′–ATCCGGA–5′

3. (a)

Mismatch

5′–T A (A) G C C T–3′
3′–A T (C) C G G A–5′

(b)

3′–A T T C G G A–5′

Self-Quiz

1. d **2.** c **3.** a **4.** b **5.** a **6.** d **7.** b

8.

	Monomers or Components	Polymer or larger molecule	Type of linkage
Sugars	Monosaccharides	Polysaccharides	Glycosidic linkages
Lipids	Fatty acids	Triacylglycerols	Ester linkages
Proteins	Amino acids	Polypeptides	Peptide bonds
Nucleic acids	Nucleotides	Polynucleotides	Phosphodiester linkages

9.

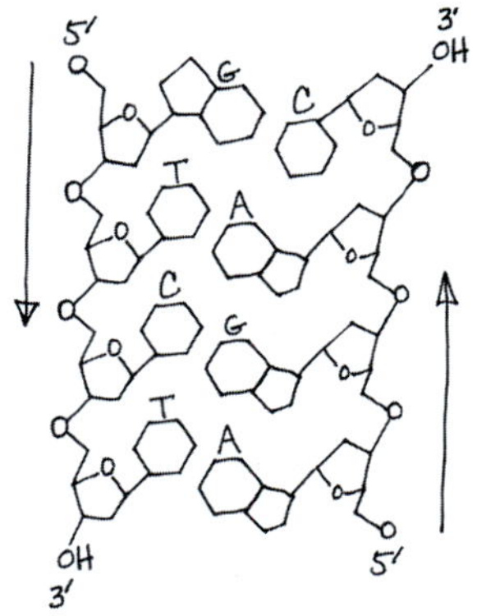

Original strand Complementary strand

CHAPTER 6

Figure Questions

Figure 6.7 A phospholipid is a lipid, consisting of a glycerol molecule joined to two fatty acids and one phosphate group. Together, the glycerol and phosphate end of the phospholipid form is the "head," which is hydrophilic, while the hydrocarbon chains on the fatty acids form hydrophobic "tails." The presence in a single molecule of both a hydrophilic and a hydrophobic region makes the molecule ideal as the main building block of a membrane. **Figure 6.22** Each centriole has 9 sets of 3 microtubules, so the entire centrosome has 54. Each microtubule consists of a helical array of tubulin dimers (as shown in Table 6.1).

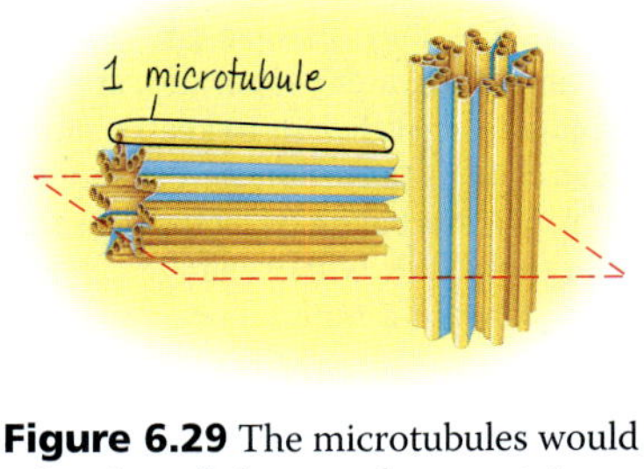

Figure 6.29 The microtubules would reorient, and based on the earlier results, the cellulose synthase proteins would also change their path, orienting along the repositioned microtubules. (This is, in fact, what was observed.)

Concept Check 6.1

1. Stains used for light microscopy are colored molecules that bind to cell components, affecting the light passing through, while stains used for electron microscopy involve heavy metals that affect the beams of electrons passing through. **2.** (a) Light microscope, (b) scanning electron microscope, (c) transmission electron microscope

Concept Check 6.2

1. See Figure 6.9.
2.

This cell would have the same volume as the cells in columns 2 and 3 but proportionally more surface area than that in column 2 and less than that in column 3. Thus, the surface-to-volume ratio should be greater than 1.2 but less than 6. To obtain the surface area, you'd have to add the area of the six sides (the top, bottom, sides, and ends): $125 + 125 + 125 + 125 + 1 + 1 = 502$. The surface-to-volume ratio equals 502 divided by a volume of 125, or 4.0.

Concept Check 6.3

1. Ribosomes in the cytoplasm translate the genetic message, carried from the DNA in the nucleus by mRNA, into a polypeptide chain. **2.** Nucleoli consist of DNA and the ribosomal RNA (rRNA) made according to its instructions, as well as proteins imported from the cytoplasm. Together, the rRNA and proteins are assembled into large and small ribosomal subunits. (These are exported through nuclear pores to the cytoplasm, where they will participate in polypeptide synthesis.) **3.** The information in a gene (on a chromosome in the nucleus) is used to synthesize an mRNA that is then transported through a nuclear pore to the cytoplasm. There it is translated into protein, which is transported back through a nuclear pore into the nucleus, where it joins other proteins and DNA, forming chromatin.

Concept Check 6.4

1. The primary distinction between rough and smooth ER is the presence of bound ribosomes on the rough ER. While both types of ER make phospholipids, membrane proteins and secretory proteins are all produced on the ribosomes of the rough ER. The smooth ER also functions in detoxification, carbohydrate metabolism, and storage of calcium ions. **2.** Transport vesicles move membranes and substances they enclose between other components of the endomembrane system. **3.** The mRNA is synthesized in the nucleus and then passes out through a nuclear pore to be translated on a bound ribosome, attached to the rough ER. The protein is synthesized into the lumen of the ER and perhaps modified there. A transport vesicle carries the protein to the Golgi apparatus. After further modification in the Golgi, another transport vesicle carries it back to the ER, where it will perform its cellular function.

Concept Check 6.5

1. Both organelles are involved in energy transformation, mitochondria in cellular respiration and chloroplasts in photosynthesis. They both have multiple membranes that separate their interiors into compartments. In both organelles, the innermost membranes—cristae, or infoldings of the inner membrane, in mitochondria, and the thylakoid membranes in chloroplasts—have large surface areas with embedded enzymes that carry out their main functions. **2.** Mitochondria, chloroplasts, and peroxisomes are not derived from the ER, nor are they connected physically or via transport vesicles to organelles of the endomembrane system. Mitochondria and chloroplasts are structurally quite different from vesicles derived from the ER, which are bounded by a single membrane.

Concept Check 6.6

1. Both systems of movement involve long filaments that are moved in relation to each other by motor proteins that grip, release, and grip again adjacent polymers. **2.** Dynein arms, powered by ATP, move neighboring doublets of microtubules relative to one another. Because they are anchored within the organelle and with respect to each other, the doublets bend instead of sliding past one another. **3.** Such individuals have defects in the microtubule-based movement of cilia and flagella. Thus, the sperm can't move because of malfunctioning flagella; the airways are compromised; and signaling events during embryogenesis do not occur correctly due to malfunctioning cilia.

Concept Check 6.7

1. The most obvious difference is the presence of direct cytoplasmic connections between cells of plants (plasmodesmata) and animals (gap junctions). These connections result in the cytoplasm being continuous between adjacent cells. **2.** The cell would not be able to function properly and would probably soon die, as the cell wall or ECM must be permeable to allow the exchange of matter between the cell and its external environment. Molecules involved with energy production and use must be allowed entry, as well as those that provide information about the cell's environment. Other molecules, such as products synthesized by the cell for export and the by-products of cellular respiration, must be allowed to exit.

Self-Quiz

1. c **2.** b **3.** d **4.** d **5.** b **6.** c **7.** e **8.** a
9. See Figure 6.9.

CHAPTER 7

Figure Questions

Figure 7.6 You couldn't rule out movement of proteins within the cell membrane of the same species. You might speculate that the membrane lipids and proteins from one species weren't able to mingle with those from the other species because of some incompatibility. **Figure 7.9** A transmembrane protein like the integrin dimer in (f) might change its shape upon binding to a particular ECM molecule. The new shape might enable the interior portion of the protein to bind to a second, cytoplasmic protein that would relay the message to the inside of the cell, as shown in (c). **Figure 7.12** The orange solute would be evenly distributed throughout the solution on both sides of the membrane. The solution levels would not be affected because the orange solute can diffuse through the membrane and equalize its concentration. Thus, no additional osmosis of water would take place in either direction.

Concept Check 7.1

1. They are on the inner side of the transport vesicle membrane. **2.** Plants adapted to cold environments would be expected to have more unsaturated fatty acids in their membranes because those remain fluid at lower temperatures. Plants adapted to hot environments would be expected to have more saturated fatty acids, which would allow the fatty acids to "stack" more closely, making the membranes less fluid and therefore helping them to stay intact at higher temperatures.

Concept Check 7.2

1. O_2 and CO_2 are both small nonpolar molecules that can easily pass through the hydrophobic core of a membrane. **2.** Water is a polar molecule, so it cannot pass very rapidly through the hydrophobic region in the middle of a phospholipid bilayer. **3.** The hydronium ion is charged, while glycerol is not. Charge is probably more significant than size as a basis for exclusion by the aquaporin channel.

Concept Check 7.3

1. CO_2 is a small nonpolar molecule that can diffuse through the plasma membrane. As long as it diffuses away so the concentration remains low outside the cell, it will continue to exit the cell in this way. (This is the opposite of the case for O_2, described in this section.) **2.** The water is hypotonic to the plant cells, so they take up water and the cells of the vegetable remain turgid, rather than plasmolyzing. The vegetable (for example, lettuce or spinach) remains crisp and not wilted. **3.** The activity of the *Paramecium's* contractile vacuole will decrease. The vacuole pumps out excess water that flows into the cell; this flow occurs only in a hypotonic environment.

Concept Check 7.4

1. The pump uses ATP. To establish a voltage, ions have to be pumped against their gradients, which requires energy. **2.** Each ion is being transported against its electrochemical gradient. If either ion were flowing down its electrochemical gradient, this *would* be considered cotransport. **3.** Even if proton pumps were still using ATP and moving protons, no proton gradient would become established. This would have serious consequences for the cells, because processes like the cotransport of sucrose (as well as synthesis of ATP) depend on establishment of a proton gradient.

Concept Check 7.5

1. Exocytosis. When a transport vesicle fuses with the plasma membrane, the vesicle membrane becomes part of the plasma membrane. **2.** Receptor-mediated endocytosis. In this case, one specific kind of molecule needs to be taken up at a particular time; pinocytosis takes up substances in a nonspecific manner.

Self-Quiz

1. b **2.** c **3.** a **4.** d **5.** b
6. a.

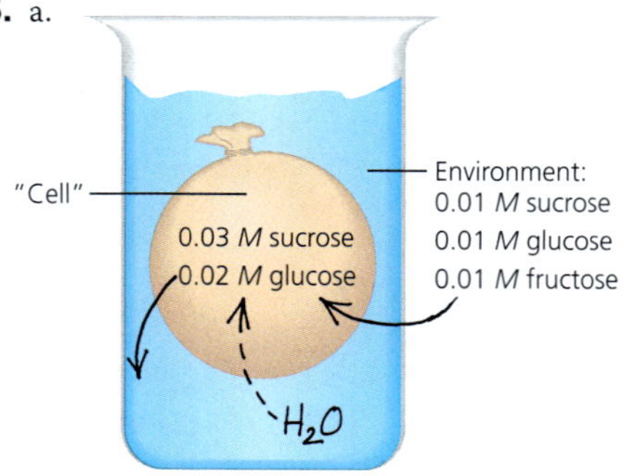

b. The solution outside is hypotonic. It has less sucrose, which is a nonpenetrating solute.
c. See answer for (a).
d. The artificial cell will become more turgid.
e. Eventually, the two solutions will have the same solute concentrations. Even though sucrose can't move through the membrane, water flow (osmosis) will lead to isotonic conditions.

CHAPTER 8

Figure Questions

Figure 8.14

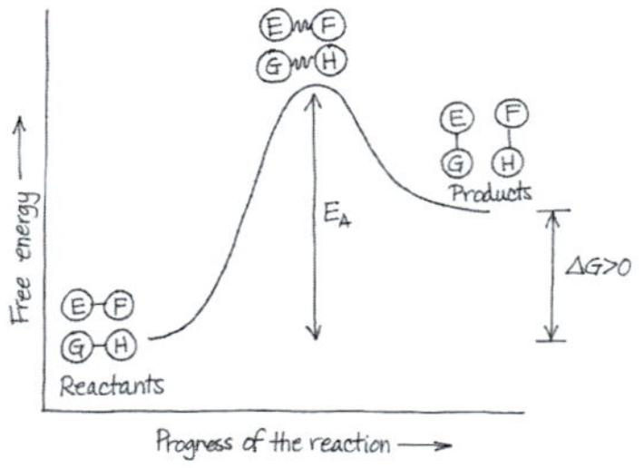

Figure 8.18

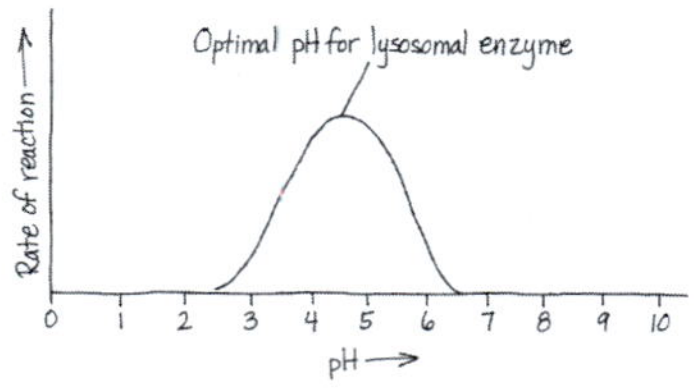

Figure 8.21 Because the affinity of the caspase for the inhibitor is very low (as is expected of an allosterically inhibited enzyme), the inhibitor is likely to diffuse away. Because no additional source of the inhibitory compound is present (the concentration of inhibitor is very low), the inhibitor is unlikely to bind again to the enzyme once the covalent linkage is broken. Thus, normal activity of the enzyme would most likely not be affected. (This test was performed by the researchers, and enzyme activity was observed to be normal upon release of the inhibitor.)

Concept Check 8.1

1. The second law is the trend toward randomness. Equal concentrations of a substance on both sides of a membrane is a more random distribution than unequal concentrations. Diffusion of a substance to a region where it is initially less concentrated increases entropy, as described by the second law. **2.** The apple has potential energy in its position hanging on the tree, and the sugars and other nutrients it contains have chemical energy. The apple has kinetic energy as it falls from the tree to the ground. Finally, when the apple is digested and its molecules broken down, some of the chemical energy is used to do work, and the rest is lost as thermal energy. **3.** The sugar crystals become less ordered (entropy increases) as they dissolve and become randomly spread out in the water. Over time, the water evaporates, and the crystals form again because the water volume is insufficient to keep them in solution. While the reappearance of sugar crystals may represent a "spontaneous" increase in order (decrease in entropy), it is balanced by the decrease in order (increase in entropy) of the water molecules, which changed from a relatively compact arrangement in liquid water to a much more dispersed and disordered form in water vapor.

Concept Check 8.2

1. Cellular respiration is a spontaneous and exergonic process. The energy released from glucose is used to do work in the cell or is lost as heat. **2.** Hydrogen ions can perform work only if their concentrations on each side of a membrane differ. When the H^+ concentrations are the same, the system is at equilibrium and can do no work. **3.** The reaction is exergonic because it releases energy—in this case, in the form of light. (This is a chemical version of the bioluminescence seen in Figure 8.1.)

Concept Check 8.3

1. ATP transfers energy to endergonic processes by phosphorylating (adding phosphate groups to) other molecules. (Exergonic processes phosphorylate ADP to regenerate ATP.) **2.** A set of coupled reactions can transform the first combination into the second. Since, overall, this is an exergonic process, ΔG is negative and the first group must have more free energy. (See Figure 8.10.)

Concept Check 8.4

1. A spontaneous reaction is a reaction that is exergonic. However, if it has a high activation energy that is rarely attained, the rate of the reaction may be low. **2.** Only the specific substrate(s) will fit properly into the active site of an enzyme, the part of the enzyme that carries out catalysis. **3.** Increase the concentration of the normal substrate (succinate) and see whether the rate of reaction increases. If it does, malonate is a competitive inhibitor.

Concept Check 8.5

1. The activator binds in such a way that it stabilizes the active form of an enzyme, whereas the inhibitor stabilizes the inactive form. **2.** You might choose to screen chemical compounds that bind allosterically to the enzyme, because allosteric regulatory sites are less likely to share similarity between different enzymes.

Self-Quiz

1. b **2.** c **3.** b **4.** a **5.** e **6.** c **7.** c

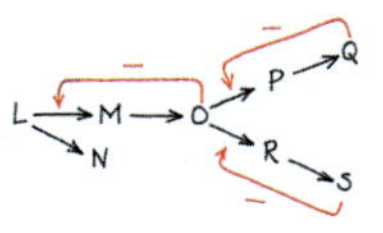

Scientific Inquiry

9.

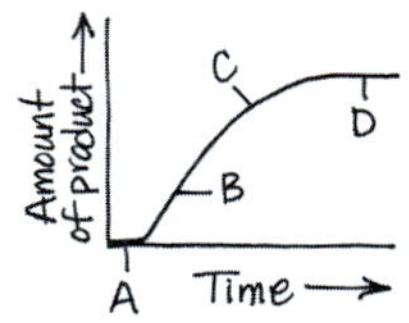

A: The substrate molecules are entering the cells, so no product is made yet.
B: There is sufficient substrate, so the reaction is proceeding at a maximum rate.
C: As the substrate is used up, the rate falls.
D: The line is flat because no new substrate remains and thus no new product appears.

CHAPTER 9

Figure Questions

Figure 9.7 Because an enzyme is catalyzing this reaction and there is no external source of energy, it must be exergonic, and the reactants must be at a higher energy level than the products. **Figure 9.9** It would probably stop glycolysis, or at least slow it down, since this would tend to push the equilibrium for this step toward the left. If less (or no) glyceraldehyde-3-phosphate were made, step 6 would slow down (or be unable to occur). **Figure 9.15** Rotation in the opposite direction (blue bars) would be expected to hydrolyze some of the ATP present, lowering ATP concentration below the background level. Thus, the blue bars would be expected to be lower than the gray bars, which is not what the researchers observed. (A possible explanation: In the article, the researchers explained that when they enclosed ATP synthases in a chamber for this assay, a number of the complexes adhered to the chamber ceiling instead of the nickel plate. The enzymes adhering to the chamber ceiling would be expected to spin in an opposite direction to those on the nickel plate on the floor. When the floor-based enzymes produce ATP during a particular spin (yellow bars), those on the ceiling would be expected to consume ATP, which would make the yellow bars lower than they would have been if all enzymes were floor-based. The opposite is also true: When the floor-based enzymes hydrolyze ATP (blue bars), the ceiling-based enzymes would be synthesizing ATP, which would make the blue bars higher than if all enzymes were floor-based. Evidence of this phenomenon is shown in the graph: The spins expected to hydrolyze ATP (blue bars) result in higher ATP levels than those with no rotation (gray bars), suggesting that there are probably some "upside-down" ceiling-based complexes generating ATP while the rest are floor-based and are hydrolyzing ATP.) **Figure 9.16** At first, some ATP could be made, since electron transport could proceed as far as complex III, and a small H^+ gradient could be built up. Soon, however, no more electrons could be passed to complex III because it could not be reoxidized by passing its electrons to complex IV.

Concept Check 9.1

1. Both processes include glycolysis, the citric acid cycle, and oxidative phosphorylation. In aerobic respiration, the final electron acceptor is molecular oxygen (O_2), whereas in anaerobic respiration, the final electron acceptor is a different substance. **2.** $C_4H_6O_5$ would be oxidized and NAD^+ would be reduced.

Concept Check 9.2

1. NAD^+ acts as the oxidizing agent in step 6, accepting electrons from glyceraldehyde-3-phosphate, which thus acts as the reducing agent. **2.** Since the overall process of glycolysis results in net production of ATP, it would make sense for the process to slow down when ATP levels have increased substantially. Thus we would expect ATP to allosterically inhibit phosphofructokinase.

Concept Check 9.3

1. NADH and $FADH_2$; they will donate electrons to the electron transport chain. **2.** CO_2 is released from the pyruvate that is formed during glycolysis, and CO_2 is also released during the citric acid cycle. **3.** In both cases, the precursor molecule loses a CO_2 molecule and then donates electrons to an electron carrier in an oxidation step. Also, the product has been activated due to the attachment of a CoA group.

Concept Check 9.4

1. Oxidative phosphorylation would stop entirely, resulting in no ATP production by this process. Without oxygen to "pull" electrons down the electron transport chain, H^+ would not be pumped into the mitochondrion's intermembrane space and chemiosmosis would not occur. **2.** Decreasing the pH is the addition of H^+. It would establish a proton gradient even without the function of the electron transport chain, and we would expect ATP synthase to function and synthesize ATP. (In fact, it was experiments like this that provided support for chemiosmosis as an energy-coupling mechanism.)

Concept Check 9.5

1. A derivative of pyruvate—such as acetaldehyde during alcohol fermentation—or pyruvate itself during lactic acid fermentation; oxygen. **2.** The cell would need to consume glucose at a rate about 19 times the consumption rate in the aerobic environment (2 ATP are generated by fermentation versus up to 38 ATP by cellular respiration).

Concept Check 9.6

1. The fat is much more reduced; it has many —CH_2— units, and in all these bonds the electrons are equally shared. The electrons present in a carbohydrate molecule are already somewhat oxidized (shared unequally in bonds), as quite a few of them are bound to oxygen. **2.** When we consume more food than necessary for metabolic processes, our body synthesizes fat as a way of storing energy for later use. **3.** AMP will accumulate, stimulating phosphofructokinase, which increases the rate of glycolysis. Since oxygen is not present, the cell will convert pyruvate to lactate in lactic acid fermentation, providing a supply of ATP.

Self-Quiz

1. b **2.** d **3.** c **4.** c **5.** a **6.** a **7.** d **8.** b **9.** b

10.

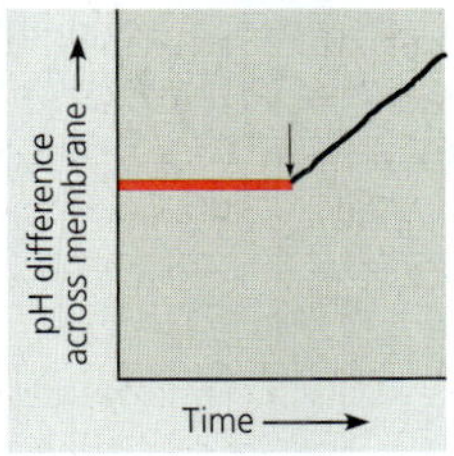

CHAPTER 10

Figure Questions

Figure 10.9 Red, but not violet-blue, wavelengths would pass through the filter, so the bacteria would not congregate where the violet-blue light normally comes through. Therefore, the left "peak" of bacteria would not be present, but the right peak would be observed because the red wavelengths passing through the filter would be used for photosynthesis. **Figure 10.11** In the leaf, most of the chlorophyll electrons excited by photon absorption are used to power the reactions of photosynthesis.

Figure 10.18

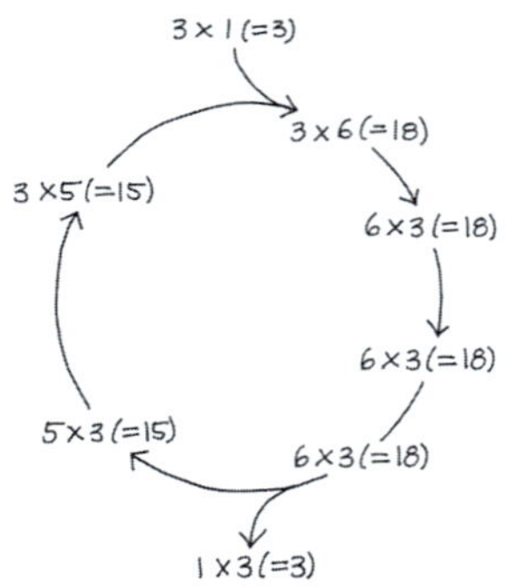

Three carbon atoms enter the cycle, one by one, as individual CO_2 molecules, and leave the cycle in one three-carbon molecule (G3P) per three turns of the cycle.

Concept Check 10.1

1. CO_2 enters leaves via stomata, and water enters via roots and is carried to leaves through veins. **2.** Using ^{18}O, a heavy isotope of oxygen, as a label, van Niel was able to show that the oxygen produced during photosynthesis originates in water, not in carbon dioxide. **3.** The light reactions could *not* keep producing NADPH and ATP without the $NADP^+$, ADP, and Ⓟ$_i$ that the Calvin cycle generates. The two cycles are interdependent.

Concept Check 10.2

1. Green, because green light is mostly transmitted and reflected—not absorbed—by photosynthetic pigments **2.** In chloroplasts, light-excited electrons are trapped by a primary electron acceptor, which prevents them from dropping back to the ground state. In isolated chlorophyll, there is no electron acceptor, so the photoexcited electrons immediately drop back down to the ground state, with the emission of light and heat. **3.** Water (H_2O) is the initial electron donor; $NADP^+$ accepts electrons at the end of the electron transport chain, becoming reduced to NADPH. **4.** In this experiment, the rate of ATP synthesis would slow and eventually stop. Because the added compound would not allow a proton gradient to build up across the membrane, ATP synthase could not catalyze ATP production.

Concept Check 10.3

1. 6, 18, 12 **2.** The more potential energy a molecule stores, the more energy and reducing power is required for the formation of that molecule. Glucose is a valuable energy source because it is highly reduced, storing lots of potential energy in its electrons. To reduce CO_2 to glucose, much energy and reducing power are required in the form of large numbers of ATP and NADPH molecules, respectively. **3.** The light reactions require ADP and $NADP^+$, which would not be formed in sufficient quantities from ATP and NADPH if the Calvin cycle stopped.

Concept Check 10.4

1. Photorespiration decreases photosynthetic output by adding oxygen, instead of carbon dioxide, to the Calvin cycle. As a result, no sugar is generated (no carbon is fixed), and O_2 is used rather than generated. **2.** Without PS II, no O_2 is generated in bundle-sheath cells. This avoids the problem of O_2 competing with CO_2 for binding to rubisco in these cells. **3.** C_4 and CAM species would replace many of the C_3 species.

Self-Quiz

1. d **2.** b **3.** b **4.** c **5.** d **6.** d **7.** c

Scientific Inquiry

9.

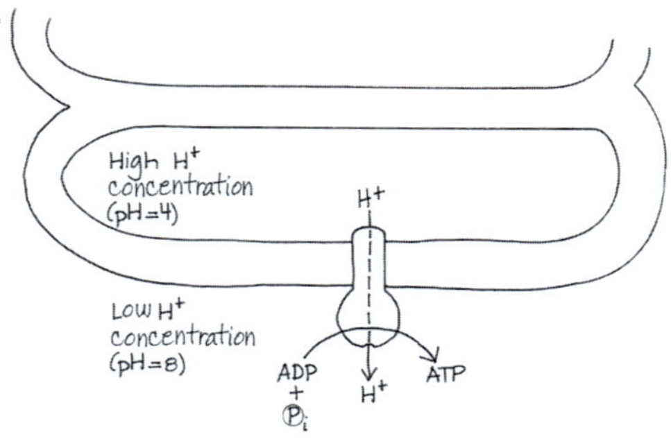

The ATP would end up outside the thylakoid. The chloroplasts were able to make ATP in the dark because the researchers set up an artificial proton concentration gradient across the thylakoid membrane; thus, the light reactions were not necessary to establish the H^+ gradient required for ATP synthesis by ATP synthase.

CHAPTER 11

Figure Questions

Figure 11.6 Epinephrine is a signaling molecule; presumably it binds to a cell-surface receptor protein. **Figure 11.8** The testosterone molecule is hydrophobic and can therefore pass directly through the lipid bilayer of the plasma membrane into the cell. (Hydrophilic molecules cannot do this.) **Figure 11.9** The active form of protein kinase 2 **Figure 11.10** The signaling molecule (cAMP) would remain in its active form and would continue to signal. **Figure 11.16** In the model, the directionality of growth is determined by the association of Fus3 with the membrane near the site of receptor activation. Thus, the development of shmoos would be severely compromised, and the affected cell would likely resemble the ΔFus3 and Δformin cells.

Concept Check 11.1

1. The two cells of opposite mating type (**a** and **α**) each secrete a certain signaling molecule, which can only be bound by receptors carried on cells of the opposite mating type. Thus, the **a** mating factor cannot bind to another **a** cell and cause it to grow toward the first **a** cell. Only an **α** cell can "receive" the signaling molecule and respond by directed growth (see Figure 11.16 for more information). **2.** The secretion of neurotransmitter molecules at a synapse is an example of local signaling. The electrical signal that travels along a very long nerve cell and is passed to the next nerve cell can be considered an example of long-distance signaling. (Note, however, that local signaling at the synapse between two cells is necessary for the signal to pass from one cell to the next.) **3.** Glucose-1-phosphate is not generated, because the activation of the enzyme requires an intact cell, with an intact receptor in the membrane and an intact signal transduction pathway. The enzyme cannot be activated directly by interaction with the signaling molecule in the test tube. **4.** Glycogen phosphorylase acts in the third stage, the response to epinephrine signaling.

Concept Check 11.2

1. The water-soluble NGF molecule cannot pass through the lipid membrane to reach intracellular receptors, as steroid hormones can. Therefore, you'd expect the NGF receptor to be in the plasma membrane—which is, in fact, the case. **2.** The cell with the faulty receptor would not be able to respond appropriately to the signaling molecule when it was present. This would most likely have dire consequences for the cell, since regulation of the cell's activities by this receptor would not occur appropriately.

Concept Check 11.3

1. A protein kinase is an enzyme that transfers a phosphate group from ATP to a protein, usually activating that protein (often a second type of protein kinase). Many signal transduction pathways include a series of such interactions, in which each phosphorylated protein kinase in turn phosphorylates the next protein kinase in the series. Such phosphorylation cascades carry a signal from outside the cell to the cellular protein(s) that will carry out the response. **2.** Protein phosphatases reverse the effects of the kinases. **3.** Information is transduced by way of sequential protein-protein interactions that change protein shapes, causing them to function in a way that passes the signal along. **4.** The IP_3-gated channel opens, allowing calcium ions to flow out of the ER, which raises the cytosolic Ca^{2+} concentration.

Concept Check 11.4

1. At each step in a cascade of sequential activations, one molecule or ion may activate numerous molecules functioning in the next step. **2.** Scaffolding proteins hold molecular components of signaling pathways in a complex with each other. Different scaffolding proteins would assemble different collections of proteins, leading to different cellular responses in the two cells.

Concept Check 11.5

1. In formation of the hand or paw in mammals, cells in the regions between the digits are programmed to undergo apoptosis. This serves to shape the digits of the hand or paw so that they are not webbed. **2.** If a receptor protein for a death-signaling molecule was defective so that it was activated for signaling even in the absence of the death signal, this would lead to apoptosis when it wouldn't normally occur. Similar defects in any of the proteins in the signaling pathway, which would activate these relay or response proteins in the absence of interaction with the previous protein or second messenger in the pathway, would have the same effect. Conversely, if any protein in the pathway were defective in its ability to respond to an interaction with an early protein or other molecule or ion, apoptosis would not occur when it normally should. For example, a receptor protein for a death-signaling ligand might not be able to be activated, even when ligand was bound. This would stop the signal from being transduced into the cell.

Self-Quiz

1. c **2.** d **3.** a **4.** c **5.** c **6.** b **7.** a **8.** d
9. This is one possible drawing of the pathway. (Similar drawings would also be correct.)

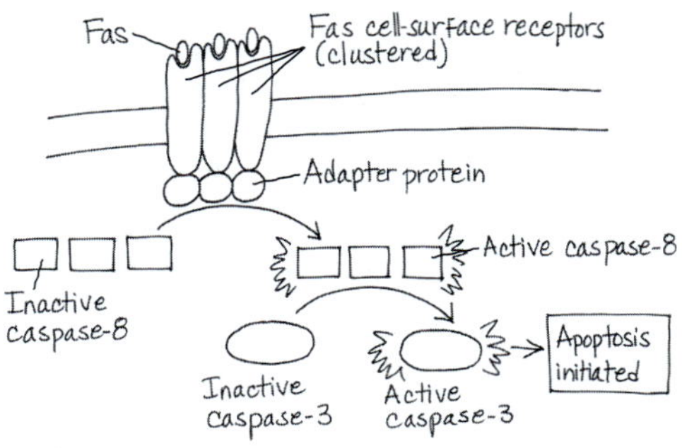

CHAPTER 12

Figure Questions

Figure 12.4

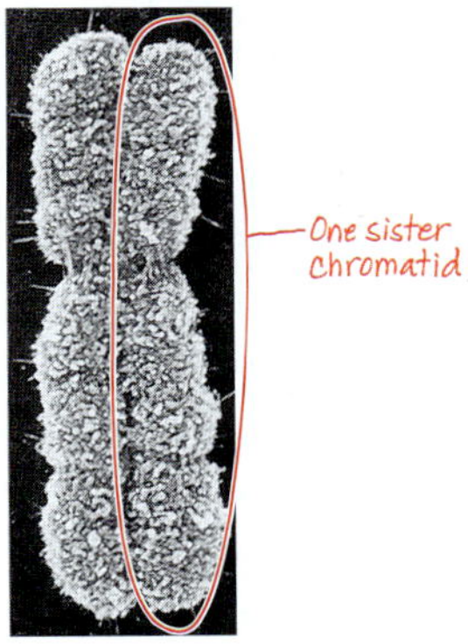

Circling the other chromatid instead would also be correct. The chromosome has four arms. **Figure 12.6** 12; 2; 2; 1

Figure 12.7

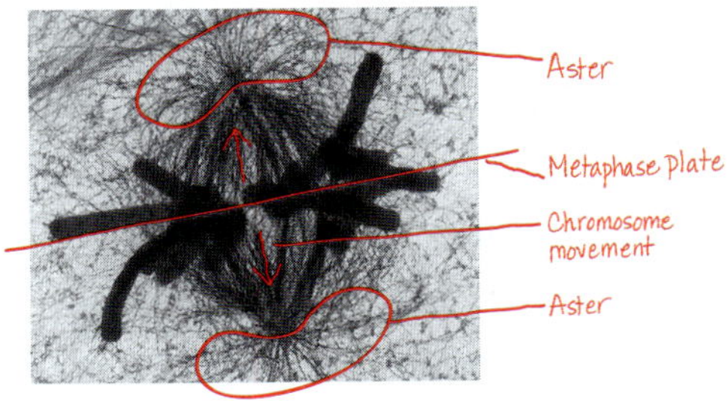

Figure 12.8 The mark would have moved toward the nearer pole. The lengths of fluorescent microtubules between that pole and the mark would have decreased, while the lengths between the chromosomes and the mark would have remained the same. **Figure 12.13** In both cases, the G_1 nucleus would have remained in G_1 until the time it normally would have entered the S phase. Chromosome condensation and spindle formation would not have occurred until the S and G_2 phases had been completed. **Figure 12.15** The cell would divide under conditions where it was inappropriate to do so. If the daughter cells and their descendants also ignored the checkpoint and divided, there would soon be an abnormal mass of cells. (This type of inappropriate cell division can contribute to the development of cancer.) **Figure 12.16** Given that control experiments showed that the *cdc2* protein kinase was the primary source of kinase activity detected in this experiment, there would be virtually no kinase activity. The percentage of cells dividing would be zero because the cells would be unable to undergo mitosis without the *cdc2* kinase. **Figure 12.18** The cells in the vessel with PDGF would not be able to respond to the growth factor signal and thus would not divide. The culture would resemble that without the added PDGF.

Concept Check 12.1

1. 32 cells **2.** 2 **3.** 39; 39; 78

Concept Check 12.2

1. 6; 12 **2.** Cytokinesis results in two genetically identical daughter cells in both plant cells and animal cells, but the mechanism of dividing the cytoplasm is different in animals and plants. In an animal cell, cytokinesis occurs by cleavage, which divides the parent cell in two with a contractile ring of actin filaments. In a plant cell, a cell plate forms in the middle of the cell and grows until its membrane fuses with the plasma membrane of the parent cell. A new cell wall grows inside the cell plate. **3.** They elongate the cell during anaphase. **4.** Sample answer: Each type of chromosome consists of a single molecule of DNA with attached proteins. If stretched out, the molecules of DNA would be many times longer than the cells in which they reside. During cell division, the two copies of each type of chromosome actively move apart, and one copy ends up in each of the two daughter cells. Chromosome movement in both types of cells may involve similar cytoskeletal proteins. **5.** During eukaryotic cell division, tubulin is involved in spindle formation and chromosome movement, while actin functions during cytokinesis. In bacterial binary fission, it's the opposite: Tubulin-like molecules are thought to act in daughter cell separation, and actin-like molecules are thought to move the daughter bacterial chromosomes to opposite ends of the cell. **6.** From the end of S phase in interphase through the end of metaphase in mitosis

Concept Check 12.3

1. The nucleus on the right was originally in the G_1 phase; therefore, it had not yet duplicated its chromosome. The nucleus on the left was in the M phase, so it had already duplicated its chromosome. **2.** A sufficient amount of MPF has to build up for a cell to pass the G_2 checkpoint. **3.** Most body cells are in a nondividing state called G_0. **4.** Both types of tumors consist of abnormal cells. A benign tumor stays at the original site and can usually be surgically removed. Cancer cells from a malignant tumor spread from the original site by metastasis and may impair the functions of one or more organs. **5.** The cells might divide even in the absence of PDGF, in which case they would not stop when the surface of the culture vessel was covered; they would continue to divide, piling on top of one another.

Self-Quiz

1. b **2.** a **3.** a **4.** c **5.** c **6.** e **7.** a **8.** b
9. See Figure 12.6 for a description of major events.

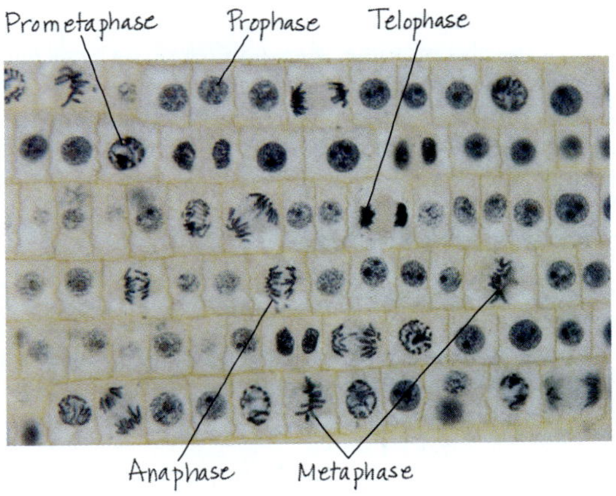

Answers

10.

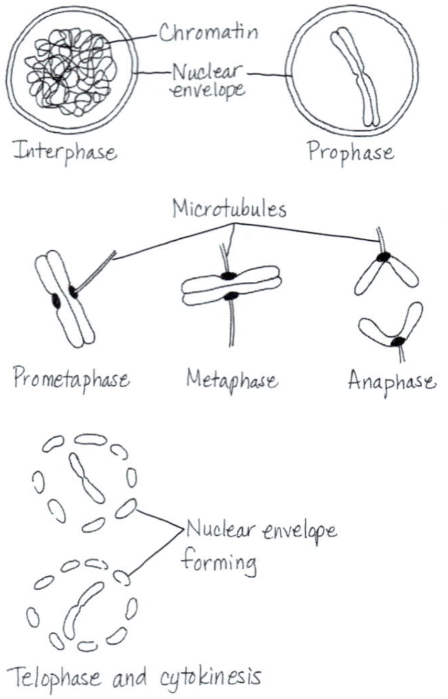

CHAPTER 13

Figure Questions

Figure 13.4 The haploid number, *n*, is 3. A set is always haploid. **Figure 13.7** A short strand of DNA is shown here for simplicity, but each chromosome or chromatid contains a very long coiled and folded DNA molecule.

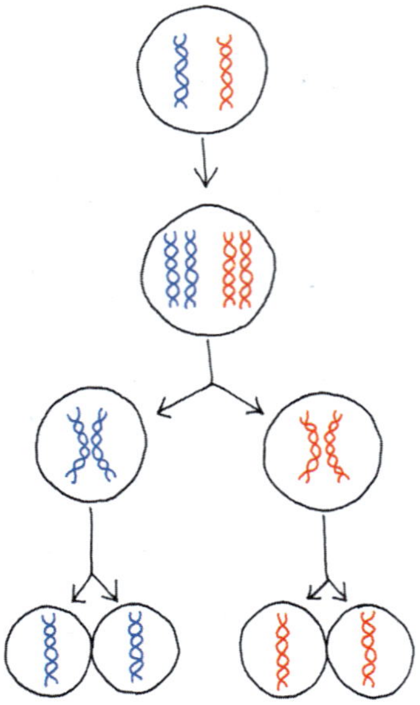

Figure 13.9 Yes. Each of the chromosomes shown in telophase I has one nonrecombinant chromatid and one recombinant chromatid. Therefore, eight possible sets of chromosomes can be generated for the cell on the left and eight for the cell on the right. **Figure 13.10** The chromosomes arising from chromatids of the unlabeled chromosome would be expected to behave exactly like those of the labeled chromosome. Therefore, the graph would look identical to the one shown in the figure.

Concept Check 13.1

1. Parents pass genes to their offspring; the genes program cells to make specific enzymes and other proteins, whose cumulative action produces an individual's inherited traits. **2.** Such organisms reproduce by mitosis, which generates offspring whose genomes are exact copies of the parent's genome (in the absence of mutation). **3.** She should clone it. Breeding it would generate offspring that have additional variation, which she no longer desires now that she has obtained her ideal orchid.

Concept Check 13.2

1. A female has two X chromosomes; a male has an X and a Y. **2.** In meiosis, the chromosome count is reduced from diploid to haploid; the union of two haploid gametes in fertilization restores the diploid chromosome count. **3.** The haploid number (n) is 7; the diploid number ($2n$) is 14. **4.** This organism has the life cycle shown in Figure 13.6c. Therefore, it must be a fungus or a protist, perhaps an alga.

Concept Check 13.3

1. The chromosomes are similar in that each is composed of two sister chromatids, and the individual chromosomes are positioned similarly on the metaphase plate. The chromosomes differ in that in a mitotically dividing cell, sister chromatids of each chromosome are genetically identical, but in a meiotically dividing cell, sister chromatids are genetically distinct because of crossing over in meiosis I. Moreover, the chromosomes in metaphase of mitosis can be a diploid set or a haploid set, but the chromosomes in metaphase of meiosis II always consist of a haploid set. **2.** If crossing over did not occur, the two homologs would not be associated in any way. This might result in incorrect arrangement of homologs during metaphase I and ultimately in formation of gametes with an abnormal number of chromosomes.

Concept Check 13.4

1. Mutations in a gene lead to the different versions (alleles) of that gene. **2.** Without crossing over, independent assortment of chromosomes during meiosis I theoretically can generate 2^n possible haploid gametes, and random fertilization can produce $2^n \times 2^n$ possible diploid zygotes. Because the haploid number (n) of grasshoppers is 23 and that of fruit flies is 4, two grasshoppers would be expected to produce a greater variety of zygotes than would two fruit flies. **3.** If the segments of the maternal and paternal chromatids that undergo crossing over are genetically identical and thus have the same two alleles for every gene, then the recombinant chromosomes will be genetically equivalent to the parental chromosomes. Crossing over contributes to genetic variation only when it involves the rearrangement of different alleles.

Self-Quiz

1. a **2.** d **3.** b **4.** a **5.** d **6.** c **7.** d **8.** This cell must be undergoing meiosis because homologous chromosomes are associated with each other; this does not occur in mitosis. **9.** Metaphase I
10.

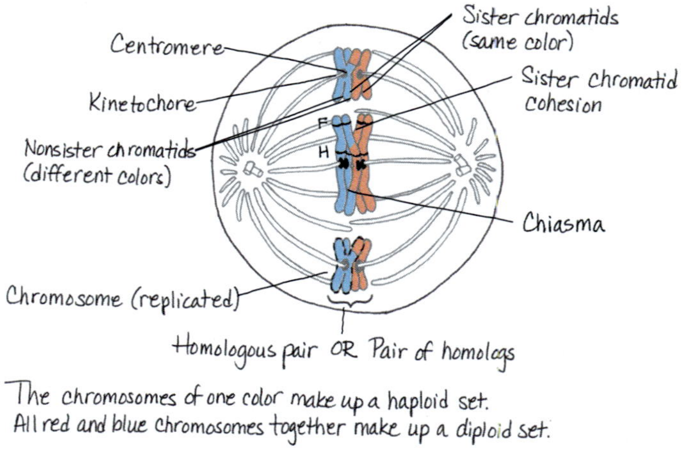

CHAPTER 14

Figure Questions

Figure 14.3 All offspring would have purple flowers. (The ratio would be one purple to zero white.) The P generation plants are true-breeding, so mating two purple-flowered plants produces the same result as self-pollination: All the offspring have the same trait.

Figure 14.8

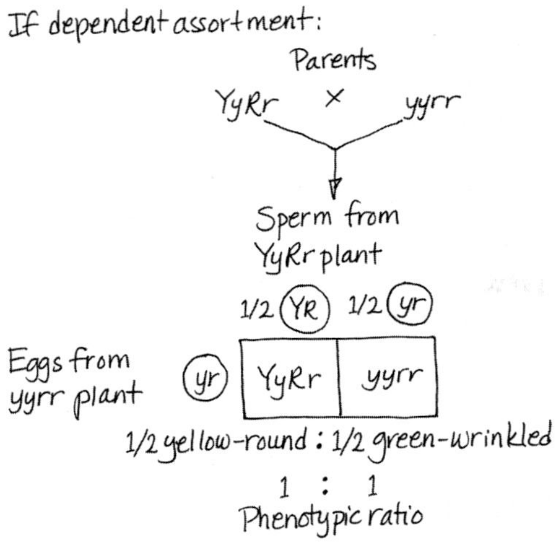

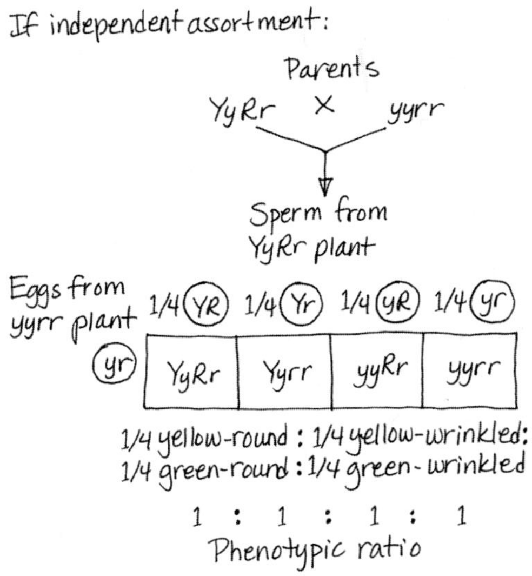

Yes, this cross would also have allowed Mendel to make different predictions for the two hypotheses, thereby allowing him to distinguish the correct one. **Figure 14.10** Your classmate would probably point out that the F_1 generation hybrids show an intermediate phenotype between those of the homozygous parents, which supports the blending hypothesis. You could respond that crossing the F_1 hybrids results in the reappearance of the white phenotype, rather than identical pink offspring, which fails to support the idea of blending traits during inheritance. **Figure 14.11** Both the I^A and I^B alleles are dominant to the *i* allele, which results in no attached carbohydrate. The I^A and I^B alleles are codominant; both are expressed in the phenotype of $I^A I^B$ heterozygotes, who have type AB blood.

Figure 14.13

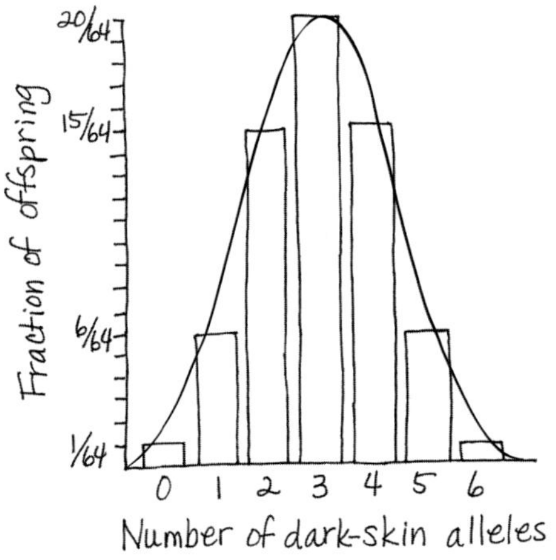

The majority of individuals have intermediate phenotypes (skin color in the middle range), while fewer individuals have phenotypes at either end (very dark or very light skin). (As you may know, this is called a "bell curve" and represents a "normal distribution.") **Figure 14.16** In the Punnett square, two of the three individuals with normal coloration are carriers, so the probability is ⅔.

Concept Check 14.1

1. A cross of *Ii* × *ii* would yield offspring with a genotypic ratio of 1 *Ii* : 1 *ii* (2:2 is an equivalent answer) and a phenotypic ratio of 1 inflated : 1 constricted (2:2 is equivalent).

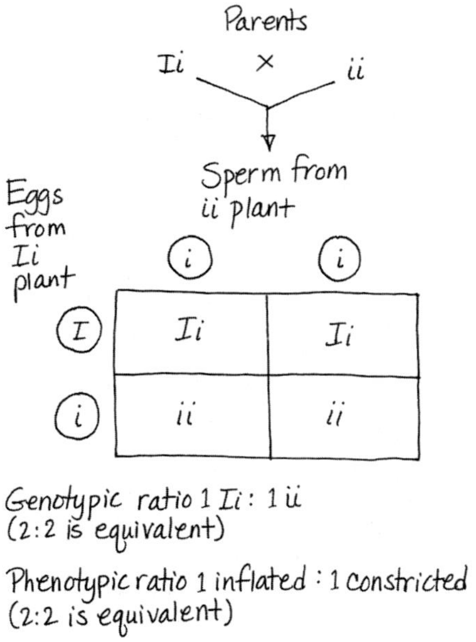

2. According to the law of independent assortment, 25 plants (1⁄16 of the offspring) are predicted to be *aatt*, or recessive for both characters. The actual result is likely to differ slightly from this value.

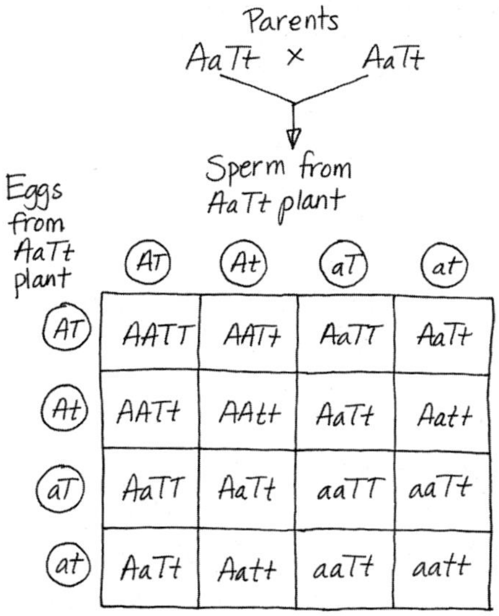

3. The plant could make eight different gametes (*YRI, YRi, YrI, Yri, yRI, yRi, yrI,* and *yri*). To fit all the possible gametes in a self-pollination, a Punnett square would need 8 rows and 8 columns. It would have spaces for the 64 possible unions of gametes in the offspring.

Concept Check 14.2

1. ½ homozygous dominant (*CC*), 0 homozygous recessive (*cc*), and ½ heterozygous (*Cc*) **2.** ¼ *BBDD*; ¼ *BbDD*; ¼ *BBDd*; ¼ *BbDd* **3.** The genotypes that fulfill this condition are *ppyyIi, ppYyii, Ppyyii, ppYYii,* and *ppyyii.* Use the multiplication rule to find the probability of getting each genotype, and then use the addition rule to find the overall probability of meeting the conditions of this problem:

$$ppyyIi \quad \tfrac{1}{2}\,(\text{probability of } pp) \times \tfrac{1}{4}\,(yy) \times \tfrac{1}{2}\,(ii) = \tfrac{1}{16}$$
$$ppYyii \quad \tfrac{1}{2}\,(pp) \times \tfrac{1}{2}\,(Yy) \times \tfrac{1}{2}\,(ii) = \tfrac{2}{16}$$
$$Ppyyii \quad \tfrac{1}{2}\,(Pp) \times \tfrac{1}{4}\,(yy) \times \tfrac{1}{2}\,(ii) = \tfrac{1}{16}$$
$$ppYYii \quad \tfrac{1}{2}\,(pp) \times \tfrac{1}{4}\,(YY) \times \tfrac{1}{2}\,(ii) = \tfrac{1}{16}$$
$$ppyyii \quad \tfrac{1}{2}\,(pp) \times \tfrac{1}{4}\,(yy) \times \tfrac{1}{2}\,(ii) = \tfrac{1}{16}$$

$$\text{Fraction predicted to have at least two recessive traits} = \tfrac{6}{16} \text{ or } \tfrac{3}{8}$$

Concept Check 14.3

1. Incomplete dominance describes the relationship between two alleles of a single gene, whereas epistasis relates to the genetic relationship between two genes (and the respective alleles of each). **2.** Half of the children would be expected to have type A blood and half type B blood. **3.** The black and white alleles are incompletely dominant, with heterozygotes being gray in color. A cross between a gray rooster and a black hen should yield approximately equal numbers of gray and black offspring.

Concept Check 14.4

1. 1/9 (Since cystic fibrosis is caused by a recessive allele, Beth and Tom's siblings who have CF must be homozygous recessive. Therefore, each parent must be a carrier of the recessive allele. Since neither Beth nor Tom has CF, this means they each have a 2/3 chance of being a carrier. If they are both carriers, there is a 1/4 chance that they will have a child with CF. 2/3 × 2/3 × 1/4 = 1/9); 0 (Both Beth and Tom would have to be carriers to produce a child with the disease.) **2.** Joan's genotype is *Dd*. Because the allele for polydactyly (*D*) is dominant to the allele for five digits per appendage (*d*), the trait is expressed in people with either the *DD* or *Dd* genotype. But because Joan's father does not have polydactyly, his genotype must be *dd*, which means Joan inherited a *d* allele from him. Therefore Joan, who does have the trait, must be heterozygous. **3.** Since polydactyly is a dominant trait, one of the parents of an affected individual should show the trait. Therefore, this must be an extremely rare case of a mutation that occurred during formation of one of the gametes involved in the fertilization that created Peter.

Genetics Problems

1. Gene, l. Allele, e. Character, g. Trait, b. Dominant allele, j. Recessive allele, a. Genotype, k. Phenotype, h. Homozygous, c. Heterozygous, f. Testcross, i. Monohybrid cross, d.

2.

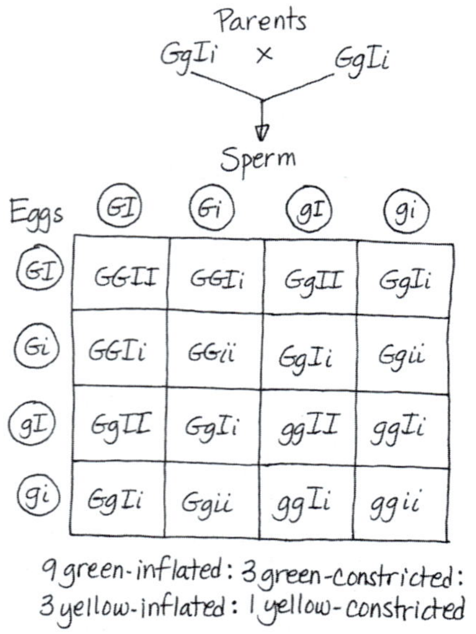

3. Parental cross is $AAC^RC^R \times aaC^WC^W$. F_1 genotype is AaC^RC^W, phenotype is all axial-pink. F_2 genotypes are 1 AAC^RC^R : 2 AAC^RC^W : 1 AAC^WC^W : 2 AaC^RC^R : 4 AaC^RC^W : 2 AaC^WC^W : 1 aaC^RC^R : 2 aaC^RC^W : 1 aaC^WC^W. F_2 phenotypes are 3 axial-red : 6 axial-pink : 3 axial-white : 1 terminal-red : 2 terminal-pink : 1 terminal-white.

4. a. 1/64
b. 1/64
c. 1/8
d. 1/32

5. Albino (*b*) is a recessive trait; black (*B*) is dominant. First cross: parents *BB* × *bb*; gametes *B* and *b*; offspring all *Bb* (black coat). Second cross: parents *Bb* × *bb*; gametes ½ *B* and ½ *b* (heterozygous parent) and *b*; offspring ½ *Bb* and ½ *bb*.

6. a. *PPLl* × *PPLl*, *PPLl* × *PpLl*, or *PPLl* × *ppLl*.
b. *ppLl* × *ppLl*.
c. *PPLL* × any of the 9 possible genotypes or *PPll* × *ppLL*.
d. *PpLl* × *Ppll*.
e. *PpLl* × *PpLl*.

7. Man I^Ai; woman I^Bi; child *ii*. Other genotypes for children are ¼ I^AI^B, ¼ I^Ai, ¼ I^Bi.

8. a. ¾ × ¾ × ¾ = 27/64
b. 1 − 27/64 = 37/64
c. ¼ × ¼ × ¼ = 1/64
d. 1 − 1/64 = 63/64

9. a. 1/256
b. 1/16
c. 1/256
d. 1/64
e. 1/128

10. a. 1
b. 1/32
c. 1/8
d. 1/2

11. 1/9

12. Matings of the original mutant cat with true-breeding noncurl cats will produce both curl and noncurl F_1 offspring if the curl allele is dominant, but only noncurl offspring if the curl allele is recessive. You would obtain some true-breeding offspring homozygous for the curl allele from matings between the F_1 cats resulting from the original curl × noncurl crosses whether the curl trait is dominant or recessive. You know that cats are true-breeding when curl × curl matings produce only curl offspring. As it turns out, the allele that causes curled ears is dominant. **13.** 1/16 **14.** 25% will be cross-eyed; all of the cross-eyed offspring will also be white. **15.** The dominant allele *I* is epistatic to the *P/p* locus, and thus the genotypic ratio for the F_1 generation will be 9 *I_P_* (colorless) : 3 *I_pp* (colorless) : 3 *iiP_* (purple) : 1 *iipp* (red). Overall, the phenotypic ratio is 12 colorless : 3 purple : 1 red. **16.** Recessive. All affected individuals (Arlene, Tom, Wilma, and Carla) are homozygous recessive *aa*. George is *Aa*, since some of his children with Arlene are affected. Sam, Ann, Daniel, and Alan are each *Aa*, since they are all unaffected children with one affected parent. Michael also is *Aa*, since he has an affected child (Carla) with his heterozygous wife Ann. Sandra, Tina, and Christopher can each have the *AA* or *Aa* genotype. **17.** 1/2 **18.** 1/6 **19.** 9 *B_A_* (agouti) : 3 *B_aa* (black) : 3 *bbA_* (white) : 1 *bbaa* (white). Overall, 9 agouti : 3 black : 4 white.

CHAPTER 15

Figure Questions

Figure 15.2 The ratio would be 1 yellow-round : 1 green-round : 1 yellow-wrinkled : 1 green-wrinkled. **Figure 15.4** About ¾ of the F_2 offspring would have red eyes and about ¼ would have white eyes. About half of the white-eyed flies would be female and half would be male; about half of the red-eyed flies would be female. **Figure 15.7** All the males would be color-blind, and all the females would be carriers. **Figure 15.9** The two largest classes would still be the parental-type offspring, but now they would be gray-vestigial and black-normal because those were the specific allele combinations in the P generation. **Figure 15.10** The two chromosomes below, left are like the two chromosomes inherited by the F_1 female, one from each P generation fly. They are passed by the F_1 female intact to the offspring and thus could be called "parental" chromosomes. The other two chromosomes result from crossing over during meiosis in the F_1 female. Because they have combinations of alleles not seen in either of the F_1 female's chromosomes, they can be called "recombinant" chromosomes.

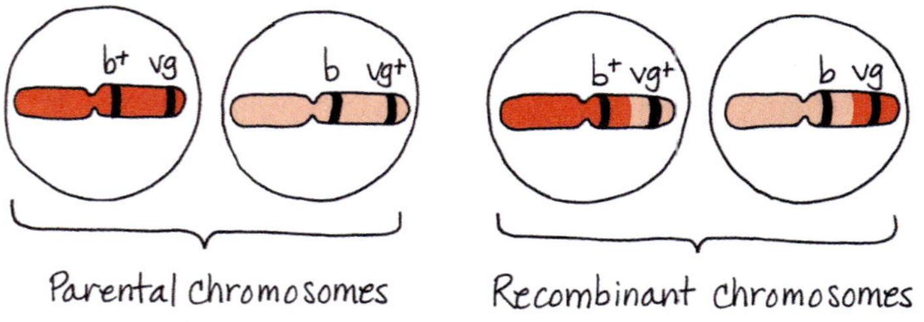

Concept Check 15.1

1. The law of segregation relates to the inheritance of alleles for a single character. The law of independent assortment of alleles relates to the inheritance of alleles for two characters. **2.** The physical basis for the law of segregation is the separation of homologs in anaphase I. The physical basis for the law of independent assortment is the alternative arrangements of homologous chromosome pairs in metaphase I. **3.** To show the mutant phenotype, a male needs to possess only one mutant allele. If this gene had been on a pair of autosomes, *two* mutant alleles would have had to be present for an individual to show the mutant phenotype, a much less probable situation.

Concept Check 15.2

1. Because the gene for this eye-color character is located on the X chromosome, all female offspring will be red-eyed and heterozygous ($X^{w^+}X^w$); all male offspring will inherit a Y chromosome from the father and be white-eyed (X^wY). **2.** $\frac{1}{4}$; $\frac{1}{2}$ chance that the child will inherit a Y chromosome from the father and be male × $\frac{1}{2}$ chance that he will inherit the X carrying the disease allele from his mother. If the child is a boy, there is a $\frac{1}{2}$ chance he will have the disease; a female would have zero chance (but $\frac{1}{2}$ chance of being a carrier). **3.** The cells in the eye responsible for color vision must come from multiple cells in the early embryo. The descendants of half of those cells express the allele for normal color vision and half the allele for color blindness. Having half the number of mature eye cells expressing the normal allele must be sufficient for normal color vision.

Concept Check 15.3

1. Crossing over during meiosis I in the heterozygous parent produces some gametes with recombinant genotypes for the two genes. Offspring with a recombinant phenotype arise from fertilization of the recombinant gametes by homozygous recessive gametes from the double-mutant parent. **2.** In each case, the alleles contributed by the female parent determine the phenotype of the offspring because the male contributes only recessive alleles in this cross. **3.** No. The order could be *A-C-B* or *C-A-B*. To determine which possibility is correct, you need to know the recombination frequency between *B* and *C*.

Concept Check 15.4

1. At some point during development, one of the embryo's cells may have failed to carry out mitosis after duplicating its chromosomes. Subsequent normal cell cycles would produce genetic copies of this tetraploid cell. **2.** In meiosis, a combined 14-21 chromosome will behave as one chromosome. If a gamete receives the combined 14-21 chromosome and a normal copy of chromosome 21, trisomy 21 will result when this gamete combines with a normal gamete during fertilization. **3.** No. The child can be either I^AI^Ai or I^Aii. A sperm of genotype I^AI^A could result from nondisjunction in the father during meiosis II, while an egg with the genotype *ii* could result from nondisjunction in the mother during either meiosis I or meiosis II.

Concept Check 15.5

1. Inactivation of an X chromosome in females and genomic imprinting. Because of X inactivation, the effective dose of genes on the X chromosome is the same in males and females. As a result of genomic imprinting, only one allele of certain genes is phenotypically expressed. **2.** The genes for leaf coloration are located in plastids within the cytoplasm. Normally, only the maternal parent transmits plastid genes to offspring. Since variegated offspring are produced only when the female parent is of the B variety, we can conclude that variety B contains both the wild-type and mutant alleles of pigment genes, producing variegated leaves. **3.** The situation is similar to that for chloroplasts. Each cell contains numerous mitochondria, and in affected individuals, most cells contain a variable mixture of normal and mutant mitochondria. The normal mitochondria carry out enough cellular respiration for survival.

Genetics Problems

1. 0; $\frac{1}{2}$, $\frac{1}{16}$ **2.** Recessive; if the disorder were dominant, it would affect at least one parent of a child born with the disorder. The disorder's inheritance is sex-linked because it is seen only in boys. For a girl to have the disorder, she would have to inherit recessive alleles from *both* parents. This would be very rare, since males with the recessive allele on their X chromosome die in their early teens. **3.** $\frac{1}{4}$ for each daughter ($\frac{1}{2}$ chance that child will be female × $\frac{1}{2}$ chance of a homozygous recessive genotype); $\frac{1}{2}$ for first son. **4.** 17% **5.** 6%. Wild type (heterozygous for normal wings and red eyes) × recessive homozygote with vestigial wings and purple eyes

6. a.

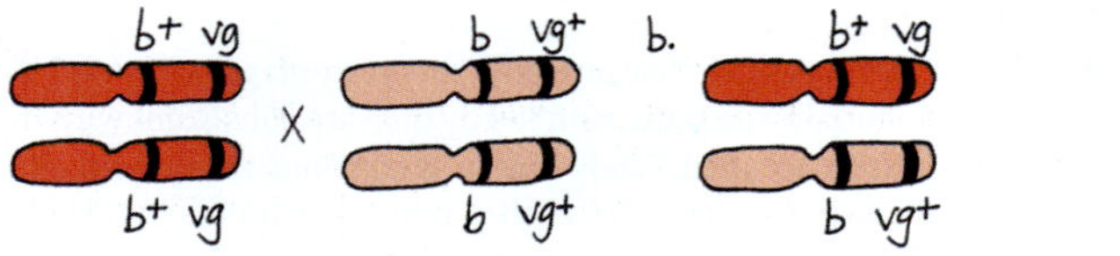

c.

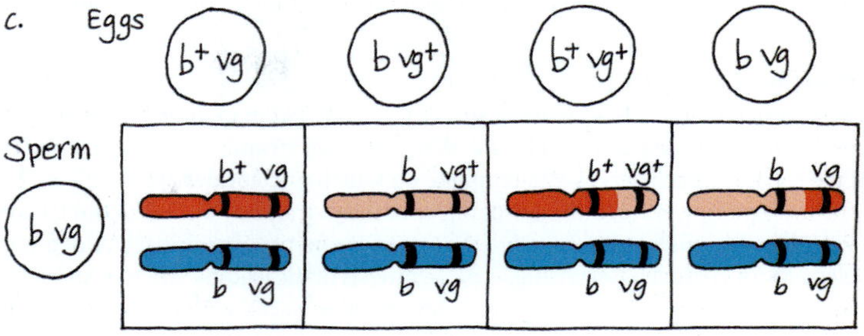

d. 41.5% gray body, vestigial wings
41.5% black body, normal wings
8.5% gray body, normal wings
8.5% black body, vestigial wings

7. The disorder would always be inherited from the mother. **8.** The inactivation of two X chromosomes in XXX women would leave them with one genetically active X, as in women with the normal number of chromosomes. Microscopy should reveal two Barr bodies in XXX women. **9.** *D–A–B–C* **10.** Fifty percent of the offspring would show phenotypes that resulted from crossovers. These results would be the same as those from a cross where *A* and *B* were not linked. Further crosses involving other genes on the same chromosome would reveal the linkage and map distances. **11.** Between *T* and *A*, 12%; between *A* and *S*, 5% **12.** Between *T* and *S*, 18%; sequence of genes is *T-A-S* **13.** 450 each of blue-oval and white-round (parentals) and 50 each of blue-round and white-oval (recombinants) **14.** About one-third of the distance from the vestigial-wing locus to the brown-eye locus **15.** Because bananas are triploid, homologous pairs cannot line up during meiosis. Therefore, it is not possible to generate gametes that can fuse to produce a zygote with the triploid number of chromosomes.

CHAPTER 16

Figure Questions

Figure 16.2 The living S cells found in the blood sample were able to reproduce to yield more S cells, indicating that the S trait is a permanent, heritable change, rather than just a one-time use of the dead S cells' capsules. **Figure 16.4** The radioactivity would have been found in the pellet when proteins were labeled (batch 1) because proteins would have had to enter the bacterial cells to program them with genetic instructions. It's hard for us to imagine now, but the DNA might have played a structural role that allowed some of the proteins to be injected while it remained outside the bacterial cell (thus no radioactivity in the pellet in batch 2). **Figure 16.11** The tube from the first replication would look the same, with a middle band of hybrid ^{15}N-^{14}N DNA, but the second tube would not have the upper band of two light blue strands. Instead it would have a bottom band of two dark blue strands, like the bottom band in the result predicted after one replication in the conservative model. **Figure 16.12** In the bubble at the top in (b), arrows should be drawn pointing left and right to indicate the two replication forks. **Figure 16.14** Looking at any of the DNA strands, we see that one end is called the 5′ end and the other the 3′ end. If we proceed from the 5′ end to the 3′ end on the left-most strand, for example, we list the components in this order: phosphate group → 5′ C of the sugar → 3′ C → phosphate → 5′ C → 3′ C. Going in the opposite direction on the same strand, the components proceed in the reverse order: 3′ C → 5′ C → phosphate. Thus, the two directions are distinguishable, which is what we mean when we say that the strands have directionality. (Review Figure 16.5 if necessary.) **Figure 16.22** The cells in the mutant would probably have the same defects in meiosis that were seen in this experiment, such as the failure of condensin to be concentrated in a small region in the nucleus. The defect in the two mutants is essentially the same: In the mutant described in the experiment, the kinase doesn't function properly; in the newly discovered mutant, the kinase could not phosphorylate the correct amino acid because that amino acid is missing.

Answers

Concept Check 16.1

1. Chargaff's rules state that in DNA, the percentages of A and T and of G and C are essentially the same, and the fly data are consistent with those rules. (Slight variations are most likely due to limitations of analytical technique.) **2.** In the Watson-Crick model, each A hydrogen-bonds to a T, so in a DNA double helix, their numbers are equal; the same is true for G and C. **3.** The mouse injected with the mixture of heat-killed S cells and living R cells would have survived, since neither type of cell alone could have killed the mouse.

Concept Check 16.2

1. Complementary base pairing ensures that the two daughter molecules are exact copies of the parent molecule. When the two strands of the parent molecule separate, each serves as a template on which nucleotides are arranged, by the base-pairing rules, into new complementary strands. **2.** DNA pol III covalently adds nucleotides to new DNA strands and proofreads each added nucleotide for correct base pairing. **3.** Synthesis of the leading strand is initiated by an RNA primer, which must be removed and replaced with DNA, a task that could not be performed if the cell's DNA pol I were nonfunctional. In the overview box in Figure 16.17, just to the left of the top origin of replication, a functional DNA pol I would replace the RNA primer of the leading strand (shown in red) with DNA nucleotides (blue).

Concept Check 16.3

1. A nucleosome is made up of eight histone proteins, two each of four different types, around which DNA is wound. Linker DNA runs from one nucleosome to the next. **2.** Euchromatin is chromatin that becomes less compacted during interphase and is accessible to the cellular machinery responsible for gene activity. Heterochromatin, on the other hand, remains quite condensed during interphase and contains genes that are largely inaccessible to this machinery. **3.** Like histones, the *E. coli* proteins would be expected to contain many basic (positively charged) amino acids, such as lysine and arginine, which can form weak bonds with the negatively charged phosphate groups on the sugar-phosphate backbone of the DNA molecule.

Self-Quiz

1. c **2.** d **3.** b **4.** c **5.** b **6.** d **7.** a **8.** c

10.

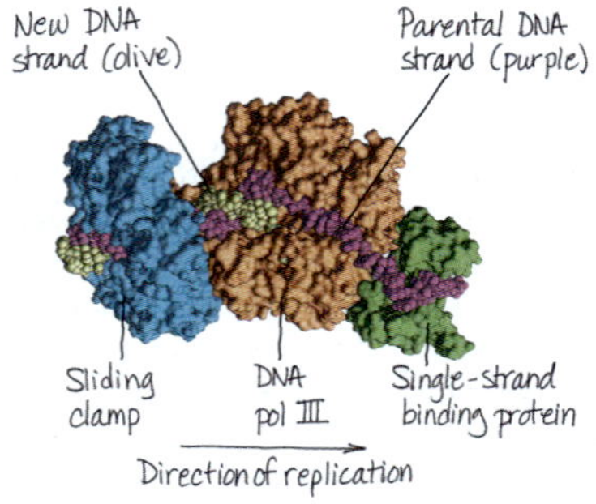

CHAPTER 17

Figure Questions

Figure 17.2 The previously presumed pathway would have been wrong. The new results would support this pathway: precursor → citrulline → ornithine → arginine. They would also indicate that class I mutants have a defect in the second step and class II mutants have a defect in the first step. **Figure 17.8** The RNA polymerase would bind directly to the promoter, rather than depending on the previous binding of other factors. **Figure 17.24** The mRNA on the right (the longest one) started transcription first. The ribosome at the top, closest to the DNA, started translating first and thus has the longest polypeptide.

Concept Check 17.1

1. A polypeptide made up of 10 Gly (glycine) amino acids

2.

Template sequence (from problem):	3′-TTCAGTCGT-5′
Nontemplate sequence:	5′-AAGTCAGCA-3′
mRNA sequence:	5′-AAGUCAGCA-3′

The nontemplate and mRNA base sequences are the same, except there is T in the nontemplate strand of DNA wherever there is U in the mRNA.

3.

"Template sequence" (from nontemplate sequence in problem, written 3′ → 5′):	3′-ACGACTGAA-5′
mRNA sequence:	5′-UGCUGACUU-3′
Translated:	Cys-STOP-Leu

(Remember that the mRNA is antiparallel to the DNA strand.) A protein translated from the nontemplate sequence would have a completely different amino acid sequence and would surely be nonfunctional. (It would also be shorter because of the stop signal shown in the mRNA sequence above—and possibly others earlier in the mRNA sequence.)

Concept Check 17.2

1. Both assemble nucleic acid chains from monomer nucleotides whose order is determined by complementary base pairing to a template strand. Both synthesize in the 5′ → 3′ direction, antiparallel to the template. DNA polymerase requires a primer, but RNA polymerase can start a nucleotide chain from scratch. DNA polymerase uses nucleotides with the sugar deoxyribose and the base T, whereas RNA polymerase uses nucleotides with the sugar ribose and the base U. **2.** The promoter is the region of DNA to which RNA polymerase binds to begin transcription, and it is at the upstream end of the gene (transcription unit). **3.** In a bacterial cell, RNA polymerase recognizes the gene's promoter and binds to it. In a eukaryotic cell, transcription factors mediate the binding of RNA polymerase to the promoter. **4.** The transcription factor that recognizes the TATA sequence would be unable to bind, so RNA polymerase could not bind and transcription of that gene probably would not occur.

Concept Check 17.3

1. The 5′ cap and poly-A tail facilitate mRNA export from the nucleus, prevent the mRNA from being degraded by hydrolytic enzymes, and facilitate ribosome attachment. **2.** In editing a video, segments are cut out and discarded (like introns), and the remaining segments are joined together (like exons) so that the regions of joining ("splicing") are not noticeable. **3.** Six different forms could be made because alternative splicing could generate six different mRNAs (two possibilities for exon 4 × three possibilities for exon 7).

Concept Check 17.4

1. First, each aminoacyl-tRNA synthetase specifically recognizes a single amino acid and attaches it only to an appropriate tRNA. Second, a tRNA charged with its specific amino acid binds only to an mRNA codon for that amino acid. **2.** Polyribosomes enable the cell to produce multiple copies of a polypeptide very quickly. **3.** A signal peptide on the leading end of the polypeptide being synthesized is recognized by a signal-recognition particle that brings the ribosome to the ER membrane. There the ribosome attaches and continues to synthesize the polypeptide, depositing it in the ER lumen. **4.** The structure and function of the ribosome seem to depend more on the rRNAs than on the ribosomal proteins. Because it is single-stranded, an RNA molecule can hydrogen-bond with itself and with other RNA molecules. RNA molecules make up the interface between the two ribosomal subunits, so presumably RNA-RNA binding helps hold the ribosome together. The binding site for mRNA in the ribosome could include rRNA that can bind the mRNA. (In fact, this turns out to be the case.) Also, complementary bonding within an RNA molecule allows it to assume a particular three-dimensional shape, and, along with the RNA's functional groups, presumably enables rRNA to catalyze peptide bond formation during translation.

Concept Check 17.5

1. In the mRNA, the reading frame downstream from the deletion is shifted, leading to a long string of incorrect amino acids in the polypeptide and, in most cases, a stop codon will arise, leading to premature termination. The polypeptide will most likely be nonfunctional.

2. Normal DNA sequence (template strand is on top): 3′-TACTTGTCCGATATC-5′
5′-ATGAACAGGCTATAG-3′

mRNA sequence: 5′-AUGAACAGGCUAUAG-3′

Amino acid sequence: Met-Asn-Arg-Leu-STOP

Mutated DNA sequence (template strand is on top): 3′-TACTTGTCCAATATC-5′
5′-ATGAACAGGTTATAG-3′

mRNA sequence: 5′-AUGAACAGGUUAUAG-3′

Amino acid sequence: Met-Asn-Arg-Leu-STOP

The amino acid sequence is Met-Asn-Arg-Leu both before and after the mutation because the mRNA codons 5′-CUA-3′ and 5′-UUA-3′ both code for Leu. (The fifth codon is a stop codon.)

Concept Check 17.6

1. No, transcription and translation are separated in space and time in a eukaryotic cell, a result of the eukaryotic cell's nuclear compartment. **2.** When one ribosome terminates translation and dissociates, the two subunits would be very close to the cap. This could facilitate their rebinding and initiating synthesis of a new polypeptide, thus increasing the efficiency of translation.

Self-Quiz

1. b **2.** d **3.** a **4.** a **5.** d **6.** e **7.** b

8.

Type of RNA	Functions
Messenger RNA (mRNA)	Carries information specifying amino acid sequences of proteins from DNA to ribosomes.
Transfer RNA (tRNA)	Serves as adapter molecule in protein synthesis; translates mRNA codons into amino acids.
Ribosomal RNA (rRNA)	Plays catalytic (ribozyme) roles and structural roles in ribosomes.
Primary transcript	Is a precursor to mRNA, rRNA, or tRNA, before being processed. Some intron RNA acts as a ribozyme, catalyzing its own splicing.
Small nuclear RNA (snRNA)	Plays structural and catalytic roles in spliceosomes, the complexes of protein and RNA that splice pre-mRNA.

CHAPTER 18

Figure Questions

Figure 18.3 As the concentration of tryptophan in the cell falls, eventually there will be none bound to repressor molecules. These will then take on their inactive shapes and dissociate from the operator, allowing transcription of the operon to resume. The enzymes for tryptophan synthesis will be made, and they will begin to synthesize tryptophan again in the cell. **Figure 18.10** The albumin gene enhancer has the three control elements colored yellow, gray, and red. The sequences in the two cells would be identical, since the cells are in the same organism. **Figure 18.16** Even if the mutant MyoD protein couldn't activate the *myoD* gene, it could still turn on genes for the other proteins in the pathway (other transcription factors, which would turn on the genes for muscle-specific proteins, for example). Therefore, some differentiation would occur. But unless there were other activators that could compensate for the loss of the MyoD protein's activation of the *myoD* gene, the cell would not be able to maintain its differentiated state. **Figure 18.19** Normal Bicoid protein would be made in the anterior end and compensate for the presence of mutant *bicoid* mRNA put into the egg by the mother. Development should be normal, with a head present. **Figure 18.21** The mutation is likely to be recessive because it is more likely to have an effect if both copies of the gene are mutated and code for nonfunctional proteins. If one normal copy of the gene is present, its product could inhibit the cell cycle. (However, there are also known cases of dominant *p53* mutations.)

Concept Check 18.1

1. Binding by the *trp* corepressor (tryptophan) activates the *trp* repressor, shutting off transcription of the *trp* operon; binding by the *lac* inducer (allolactose) inactivates the *lac* repressor, leading to transcription of the *lac* operon. **2.** The cell would continuously produce β-galactosidase and the two other enzymes for lactose utilization, even in the absence of lactose, thus wasting cell resources. **3.** With glucose scarce, cAMP would be bound to CAP and CAP would be bound to the promoter, favoring the binding of RNA polymerase. However, in the absence of lactose, the repressor would be bound to the operator, blocking RNA polymerase binding to the promoter. The operon genes would therefore not be transcribed. If another sugar were present and the genes encoding enzymes for its breakdown were in an operon regulated like the *lac* operon, we might expect to find active transcription of those genes.

Concept Check 18.2

1. Histone acetylation is generally associated with gene expression, while DNA methylation is generally associated with lack of expression. **2.** General transcription factors function in assembling the transcription initiation complex at the promoters for all genes. Specific transcription factors bind to control elements associated with a particular gene and, once bound, either increase (activators) or decrease (repressors) transcription of that gene. **3.** The three genes should have some similar or identical sequences in the control elements of their enhancers. Because of this similarity, the same specific transcription factors could bind to the enhancers of all three genes and stimulate their expression coordinately. **4.** Degradation of the mRNA, regulation of translation, activation of the protein (by chemical modification, for example), and protein degradation **5.** Expression of the gene encoding the yellow activator (YA) must be regulated at one of the steps shown in Figure 18.6. The YA gene might be transcribed only in liver cells because the necessary activators for the enhancer of the YA gene are found only in liver cells.

Concept Check 18.3

1. Both miRNAs and siRNAs are small, single-stranded RNAs that associate with a complex of proteins and then can base-pair with mRNAs that have a complementary sequence. This base pairing leads to either degradation of the mRNA or blockage of its translation. Some siRNAs, in association with other proteins, can bind back to the chromatin in a certain region, causing chromatin changes that affect transcription. Both miRNAs and siRNAs are processed from double-stranded RNA precursors by the enzyme Dicer. However, miRNAs are encoded by genes in the cell's genome, and the single transcript folds back on itself to form one or more double-stranded hairpins, each of which is processed into an miRNA. In contrast, siRNAs arise from a longer stretch of double-stranded RNA, which may be introduced into the cell by a virus or an experimenter. In some cases, a cellular gene codes for one RNA strand of the precursor molecule, and an enzyme then synthesizes the complementary strand. **2.** The mRNA would persist and be translated into the cell division–promoting protein, and the cell would probably divide. If the intact miRNA is necessary for inhibition of cell division, then division of this cell might be inappropriate. Uncontrolled cell division could lead to formation of a mass of cells (tumor) that prevents proper functioning of the organism.

Concept Check 18.4

1. Cells undergo differentiation during embryonic development, becoming different from each other; in the adult organism, there are many highly specialized cell types. **2.** By binding to a receptor on the receiving cell's surface and triggering a signal transduction pathway that affects gene expression **3.** Because their products, made and deposited into the egg by the mother, determine the head and tail ends, as well as the back and belly, of the embryo (and eventually the adult fly) **4.** The lower cell is synthesizing signaling molecules because the gene encoding them is activated, meaning that the appropriate specific transcription factors are binding to the gene's enhancer. The genes encoding these specific transcription factors are also being expressed in this cell because the transcriptional activators that can turn them on were expressed in the precursor to this cell. A similar explanation also applies to the cells expressing the receptor proteins. This scenario began with specific cytoplasmic determinants localized in specific regions of the egg. These cytoplasmic determinants were distributed unevenly to daughter cells, resulting in cells going down different developmental pathways.

Concept Check 18.5

1. The protein product of a proto-oncogene is usually involved in a pathway that stimulates cell division. The protein product of a tumor-suppressor gene is usually involved in a pathway that inhibits cell division. **2.** When an individual has inherited an oncogene or a mutant allele of a tumor-suppressor gene **3.** A cancer-causing mutation in a proto-oncogene usually makes the gene product overactive, whereas a cancer-causing mutation in a tumor-suppressor gene usually makes the gene product nonfunctional.

Self-Quiz

1. d **2.** a **3.** d **4.** a **5.** c **6.** e **7.** a **8.** c **9.** b **10.** b

11. a.

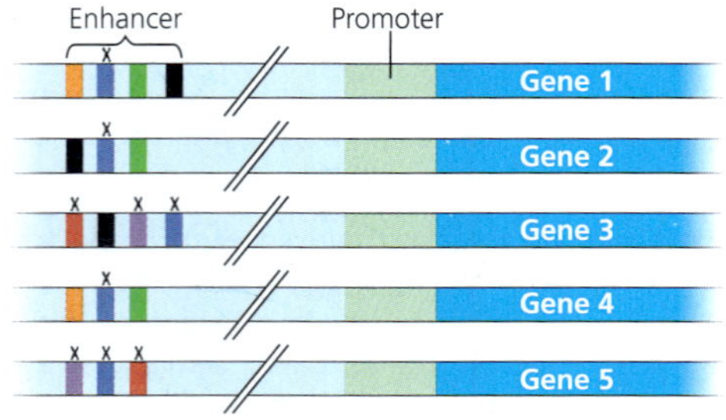

The purple, blue, and red activator proteins would be present.

b.

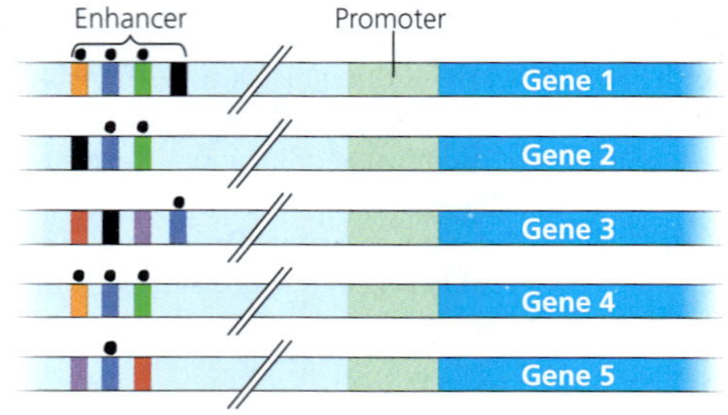

Only gene 4 would be transcribed.

c. In nerve cells, the orange, blue, green, and black activators would have to be present, thus activating transcription of genes 1, 2, and 4. In skin cells, the red, black, purple, and blue activators would have to be present, thus activating genes 3 and 5.

CHAPTER 19

Figure Questions

Figure 19.2 Beijerinck might have concluded that the agent was a toxin produced by the plant that was able to pass through a filter but that became more and more dilute. In this case, he would have concluded that the infectious agent could not reproduce. **Figure 19.4** Top vertical arrow: Infection. Left upper arrow: Replication. Right upper arrow: Transcription. Right middle arrow: Translation. Lower left and right arrows: Self-assembly. Bottom middle arrow: Exit. **Figure 19.7** Any class V virus, including the viruses that cause influenza (flu), measles, and mumps.

Concept Check 19.1

1. TMV consists of one molecule of RNA surrounded by a helical array of proteins. The influenza virus has eight molecules of RNA, each surrounded by a helical array of proteins, similar to the arrangement of the single RNA molecule in TMV. Another difference is that the influenza virus has an outer envelope. **2.** One of the arguments for regarding viruses as nonliving is that they cannot perform any activity characteristic of living organisms unless they are inside a host cell. This virus challenges that generalization because the virus can change its shape without having access to host cell proteins. (Further analysis suggested that the projections contain proteins related to intermediate filaments that may polymerize spontaneously under certain conditions.)

Concept Check 19.2

1. Lytic phages can only carry out lysis of the host cell, whereas lysogenic phages may either lyse the host cell or integrate into the host chromosome. In the latter case, the viral DNA (prophage) is simply replicated along with the host chromosome. Under certain conditions, a prophage may exit the host chromosome and initiate a lytic cycle. **2.** The genetic material of these viruses is RNA, which is replicated inside the infected cell by enzymes encoded by the virus. The viral genome (or a complementary copy of it) serves as mRNA for the synthesis of viral proteins. **3.** Because it synthesizes DNA from its RNA genome. This is the reverse ("retro") of the usual DNA → RNA information flow. **4.** There are many steps that could be interfered with: binding of the virus to the cell, reverse transcriptase function, integration into the host cell chromosome, genome synthesis (in this case, transcription of RNA from the integrated provirus), assembly of the virus inside the cell, and budding of the virus. (Many, if not all, of these are targets of actual medical strategies to block progress of the infection in HIV-infected people.)

Concept Check 19.3

1. Mutations can lead to a new strain of a virus that can no longer be effectively fought by the immune system, even if an animal had been exposed to the original strain; a virus can jump from one species to a new host; and a rare virus can spread if a host population becomes less isolated. **2.** In horizontal transmission, a plant is infected from an external source of virus, which could enter through a break in the plant's epidermis due to damage by herbivores. In vertical transmission, a plant inherits viruses from its parent either via infected seeds (sexual reproduction) or via an infected cutting (asexual reproduction). **3.** Humans are not within the host range of TMV, so they can't be infected by the virus. **4.** It is unlikely that human air travel could have spread the virus, since existing strains of the virus do not seem to be transmissible from human to human. It is conceivable but unlikely that an infected human traveling from Asia passed the virus to birds in Africa and Europe. It is possible that domestic birds carried the virus, perhaps in shipments of poultry. The likeliest scenario of all may be that migratory wild birds carried the virus during their migrations and passed it to domestic and wild birds in the new locations. To test these latter hypotheses, the timing of the outbreaks should be analyzed to see if they correlate with recent poultry shipments or known wild bird migrations. Any such migratory birds should be tested for the presence of the African or European strain of the virus, based on the nucleotide sequences of their genomes.

Self-Quiz

1. d **2.** b **3.** c **4.** d **5.** c

6. As shown below, the viral genome would be translated into capsid proteins and envelope glycoproteins directly, rather than after a complementary RNA copy was made. A complementary RNA strand would still be made, however, that could be used as a template for many new copies of the viral genome.

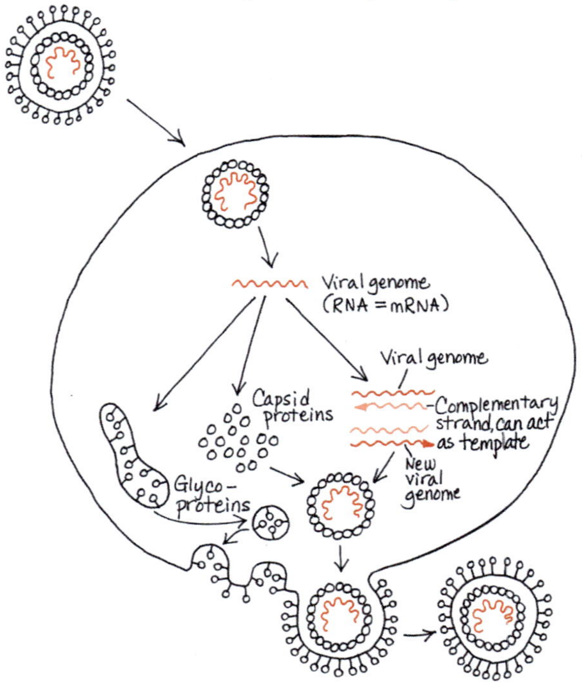

CHAPTER 20

Figure Questions

Figure 20.3

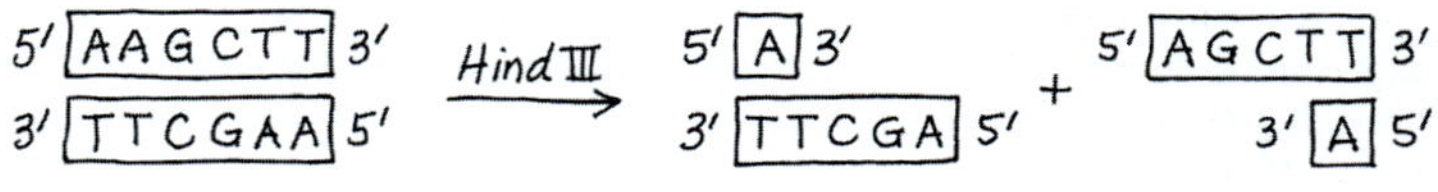

Figure 20.4 Cells containing no plasmid at all would be able to grow; these colonies would be white because they would lack functional *lacZ* genes. **Figure 20.10** Grow each clone of cells in culture. Isolate the plasmids from each and cut them with the restriction enzyme originally used to make the clone (see Figure 20.4). Run each sample on an electrophoretic gel, and recover the DNA of the insert from the gel band. **Figure 20.16** The researchers might have concluded that differentiated cells are irreversibly changed so that they can make only one type of tissue in the plant. (This result would support the idea that cloning isn't possible.) **Figure 20.17** None of the eggs with the transplanted nuclei would have developed into a tadpole. Also, the result might include only some of the tissues of a tadpole which might differ depending on which nucleus was transplanted. (This assumes that there was some way to tell the four cells apart, as one can in some frog species.)

Concept Check 20.1

1. The covalent sugar-phosphate bonds of the DNA strands **2.** Yes, *Pvu*I will cut the molecule

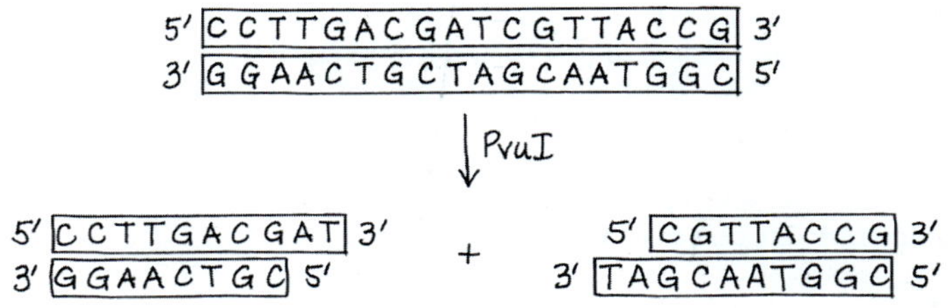

3. Some human genes are too large to be incorporated into bacterial plasmids. Bacterial cells lack the means to process RNA transcripts into mRNA, and even if the need for RNA processing is avoided by using cDNA, bacteria lack enzymes to catalyze the post-translational processing that many human proteins require to function properly. **4.** 5′-CGGT-3′ and 5′-CCTT-3′

Concept Check 20.2

1. Any restriction enzyme will cut genomic DNA in many places, generating such a large number of fragments that they would appear as a smear rather than distinct bands when the gel is stained after electrophoresis. **2.** In Southern blotting, Northern blotting, and microarray analysis, the labeled probe binds only to the specific target sequence owing to complementary nucleic acid hybridization (DNA-DNA hybridization in Southern blotting and microarray analysis, DNA-RNA hybridization in Northern blotting). In DNA sequencing, primers base-pair to the template, allowing DNA synthesis to start. In RT-PCR, the primers must base-pair with their target sequences in the DNA mixture. **3.** If a spot is green, the gene represented on that spot is expressed only in normal tissue. If red, the gene is expressed only in cancerous tissue. If yellow, the gene is expressed in both. And if black, the gene is expressed in neither type of tissue. As a researcher interested in cancer development, you would want to study genes represented by spots that are green or red because these are genes for which the expression level differs between the two types of tissues. Some of these genes may be expressed differently as a result of cancer, but others might play a role in causing cancer.

Concept Check 20.3

1. No, primarily because of subtle (and perhaps not so subtle) differences in their environments **2.** The state of chromatin modification in the nucleus from the intestinal cell was undoubtedly less similar to that of a nucleus from a fertilized egg, explaining why many fewer of these nuclei were able to be reprogrammed. In contrast, the chromatin in a nucleus from a cell at the four-cell stage would have been much more like that of a nucleus in a fertilized egg and therefore much more easily programmed to direct development. **3.** A technique would have to be worked out for turning a human iPS cell into a pancreatic cell (probably by inducing expression of pancreas-specific regulatory genes in the cell).

Concept Check 20.4

1. Stem cells continue to reproduce themselves. **2.** Herbicide resistance, pest resistance, disease resistance, salinity resistance, delayed ripening, and improved nutritional value **3.** Because hepatitis A is an RNA virus, you could isolate RNA from the blood and try to detect copies of hepatitis A RNA by one of three methods. First, you could run the RNA on a gel and then do a Northern blot using probes complementary to hepatitis A genome sequences. A second approach would be to use reverse transcriptase to make cDNA from the RNA in the blood, run the cDNA on a gel, and do a Southern blot using the same probe. However, neither of these methods would be as sensitive as RT-PCR, in which you would reverse transcribe the blood RNA into cDNA and then use PCR to amplify the cDNA, using primers specific to hepatitis A sequences. If you then ran the products on an electrophoretic gel, the presence of a band would support your hypothesis.

Self-Quiz

1. b **2.** b **3.** c **4.** b **5.** a **6.** c **7.** c **8.** d
9.

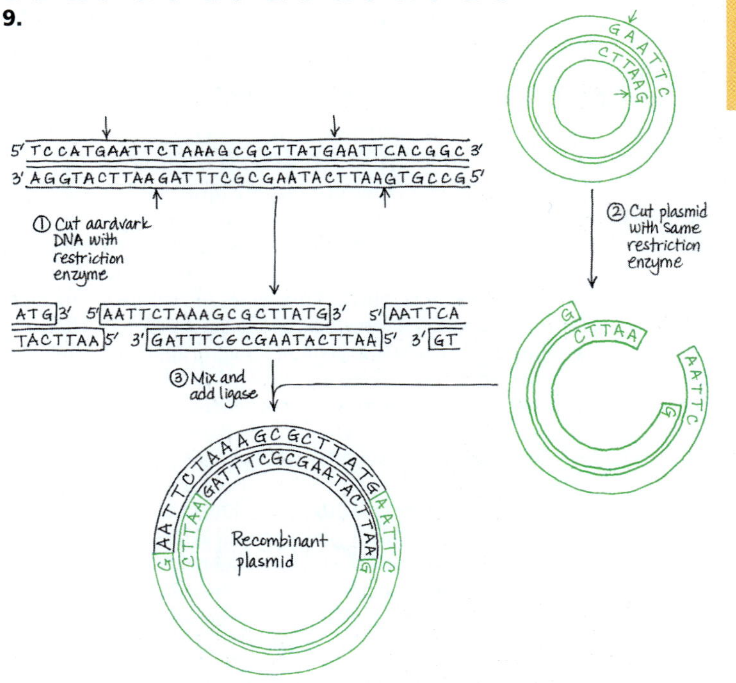

10. A cDNA library, made using mRNA from human lens cells, which would be expected to contain many copies of crystallin mRNAs

CHAPTER 21

Figure Questions

Figure 21.3 The fragments in stage 2 of this figure are like those in stage 2 of Figure 21.2, but in this figure their order relative to each other is not known and will be determined later by computer. The order of the fragments in Figure 21.2 is completely known before sequencing begins. (Determining the order takes longer but makes the eventual sequence assembly much easier.) **Figure 21.9** The transposon would be cut out of the DNA at the original site rather than copied, so part (a) would show the original stretch of DNA without the transposon after the mobile transposon had been cut out. **Figure 21.10** The RNA transcripts extending from the DNA in each transcription unit are shorter on the left and longer on the right. This means that RNA polymerase must be starting on the left end of the unit and moving toward the right. **Figure 21.13** Pseudogenes are nonfunctional. They could have arisen by any mutations in the second copy that made the gene product unable to function. Examples would be base changes that introduce stop codons in the sequence, alter amino acids, or change a region of the gene promoter so that the gene can no longer be expressed. **Figure 21.14** Let's say a transposable element (TE) existed in the intron to the left of the indicated EGF exon in the *EGF* gene, and the same TE was present in the intron to the left of the indicated F exon

Answers

in the fibronectin gene. During meiotic recombination, these TEs could cause nonsister chromatids on the same chromosome to pair up incorrectly, as seen in Figure 21.12. One gene might end up with an F exon next to an EGF exon. Further mistakes in pairing over many generations might result in these two exons being separated from the rest of the gene and placed next to a single or duplicated K exon. In general, the presence of repeated sequences in introns and between genes facilitates these processes because it allows incorrect pairing of nonsister chromatids, leading to novel exon combinations. **Figure 21.16** Since you know that chimpanzees do not speak but humans do, you'd probably want to know how many amino acid differences there are between the human wild-type FOXP2 protein and that of the chimpanzee and whether these changes affect the function of the protein. (As we explain later in the text, there are two amino acid differences.) You know that humans with mutations in this gene have severe language impairment. You would want to learn more about the human mutations by checking whether they affect the same amino acids in the gene product that the chimpanzee sequence differences affect. If so, those amino acids might play an important role in the function of the protein in language. Going further, you could analyze the differences between the chimpanzee and mouse FOXP2 proteins. You might ask: Are they more similar than the chimpanzee and human proteins? (It turns out that the chimpanzee and mouse proteins have only one amino acid difference and thus are more similar than the chimpanzee and human proteins, which have two differences, and than the human and mouse proteins, which have three differences.)

Concept Check 21.1

1. In a linkage map, genes and other markers are ordered with respect to each other, but only the relative distances between them are known. In a physical map, the actual distances between markers, expressed in base pairs, are known. **2.** The three-stage approach employed in the Human Genome Project involves linkage mapping, physical mapping, and then sequencing of short, overlapping fragments that previously have been ordered relative to each other (see Figure 21.2). The whole-genome shotgun approach eliminates the linkage mapping and physical mapping stages; instead, short fragments generated by multiple restriction enzymes are sequenced and then ordered by computer programs that identify overlapping regions (see Figure 21.3). **3.** Because the two mouse species are very closely related, their genome sequences are expected to be very similar. This means that the field mouse genome fragments could be compared with the assembled lab mouse genome, providing valuable information to use in placing the field mouse genome fragments in the correct order. In a sense, the lab mouse genome could be used as a rough map for the field mouse genome, removing the necessity to carry out complete genetic and physical mapping for the field mouse.

Concept Check 21.2

1. The Internet allows centralization of databases such as GenBank and software resources such as BLAST, making them freely accessible. Having all the data in a central database, easily accessible on the Internet, minimizes the possibility of errors and of researchers working with different data. It streamlines the process of science, since all researchers are able to use the same software programs, rather than each having to obtain their own software. It speeds up dissemination of data and ensures as much as possible that errors are corrected in a timely fashion. These are just a few answers; you can probably think of more. **2.** Cancer is a disease caused by multiple factors. To focus on a single gene or a single defect would ignore other factors that may influence the cancer and even the behavior of the single gene being studied. The systems approach, because it takes into account many factors at the same time, is more likely to lead to an understanding of the causes and most useful treatments for cancer. **3.** The DNA would first be sequenced and analyzed for whether the mutation is in the coding region for a gene or in a promoter or enhancer, affecting the expression of a gene. In either case, the nature of the gene product could be explored by searching the protein database for similar proteins. If similar proteins have known functions, that would provide a clue about the function of your protein. Otherwise, biochemical and other methods could provide some ideas about possible function. Software could be used to compare what is known about your protein and similar proteins.

Concept Check 21.3

1. Alternative splicing of RNA transcripts from a gene and post-translational processing of polypeptides **2.** The total number of completed genomes is found by clicking on "Published Complete Genomes." Add the figures for bacterial, archaeal, and eukaryotic "ongoing genomes" to get the number "in progress." Finally, look at the top of the Published Complete Genomes page to get numbers of completed genomes for each domain. (*Note:* You can click on the "Size" column and the table will be re-sorted by genome size. Scroll down to get an idea of relative sizes of genomes in the three domains. Remember, though, that most of the sequenced genomes are bacterial.) **3.** Prokaryotes are generally smaller cells than eukaryotic cells, and they reproduce by binary fission. The evolutionary process involved is natural selection for more quickly reproducing cells: The faster they can replicate their DNA and divide, the more likely they will be able to dominate a population of prokaryotes. The less DNA they have to replicate, then, the faster they will reproduce.

Concept Check 21.4

1. The number of genes is higher in mammals, and the amount of noncoding DNA is greater. Also, the presence of introns in mammalian genes makes them longer, on average, than prokaryotic genes. **2.** Introns are interspersed within the coding sequences of genes. Many copies of each transposable element are scattered throughout the genome. Simple sequence DNA is concentrated at the centromeres and telomeres and is clustered in other locations. **3.** In the rRNA gene family, identical transcription units for the three different RNA products are present in long, tandemly repeated arrays. The large number of copies of the rRNA genes enable organisms to produce the rRNA for enough ribosomes to carry out active protein synthesis, and the single transcription unit ensures that the relative amounts of the different rRNA molecules produced are correct. Each globin gene family consists of a relatively small number of nonidentical genes. The differences in the globin proteins encoded by these genes result in production of hemoglobin molecules adapted to particular developmental stages of the organism. **4.** First, you could check the sequence by translating it into a predicted amino acid sequence and see if there are multiple stop codons. If there aren't, the next step would be to see whether the gene is expressed, probably by carrying out a Northern blot or *in situ* hybridization to look for the mRNA in the cells that express the gene.

Concept Check 21.5

1. If meiosis is faulty, two copies of the entire genome can end up in a single cell. Errors in crossing over during meiosis can lead to one segment being duplicated while another is deleted. During DNA replication, slippage backward along the template strand can result in a duplication. **2.** For either gene, a mistake in crossing over during meiosis could have occurred between the two copies of that gene, such that one ended up with a duplicated exon. This could have happened several times, resulting in the multiple copies of a particular exon in each gene. **3.** Homologous transposable elements scattered throughout the genome provide sites where recombination can occur between different chromosomes. Movement of these elements into coding or regulatory sequences may change expression of genes. Transposable elements also can carry genes with them, leading to dispersion of genes and in some cases different patterns of expression. Transport of an exon during transposition and its insertion into a gene may add a new functional domain to the originally encoded protein, a type of exon shuffling. **4.** Because more offspring are born to women who have this inversion, it must provide some advantage. It would be expected to persist and spread in the population. (In fact, evidence in the study allowed the researchers to conclude that it has been increasing in proportion in the population. You'll learn more about population genetics in the next unit.)

Concept Check 21.6

1. Because both humans and macaques are primates, their genomes are expected to be more similar than the macaque and mouse genomes are. The mouse lineage diverged from the primate lineage before the human and macaque lineages diverged. **2.** Homeotic genes differ in their *non*homeobox sequences, which determine the interactions of homeotic gene products with other transcription factors and hence which genes are regulated by the homeotic genes. These nonhomeobox sequences differ in the two organisms, as do the expression patterns of the homeobox genes. **3.** *Alu* elements must have undergone transposition more actively in the human genome for some reason. Their increased numbers may have then allowed more recombination errors in the human genome, resulting in more or different duplications. The divergence of the organization and content of the two genomes presumably accelerated divergence of the two species by making matings less and less likely to result in fertile offspring.

Self-Quiz

1. c **2.** e **3.** a **4.** e **5.** c **6.** a

7.
1. ATETI...PKSSD...TSSTT...NARRD
2. ATETI...PKSSE...TSSIT...NARRD
3. ATETI...PKSSD...TSSTT...NARRD
4. ATETI...PKSSD...TSSNT...SARRD
5. ATETI...PKSSD...TSSTT...NARRD
6. VTETI...PKSSD...TSSTT...NARRD

a. Lines 1, 3, and 5 are the C, G, R species.
b. Line 4 is the human sequence.
c. Line 6 is the orangutan sequence.
d. There is one amino acid difference between the mouse (line 2) and the C, G, R species; there are three amino acid differences between the mouse and the human.
e. Because only one amino acid difference arose during the 60–100 million years since the mouse and C, G, R species diverged, it is somewhat surprising that two additional amino acid differences resulted during the 6 million years since chimpanzees and humans diverged. This indicates that the *FOXP2* gene has been evolving faster in the human lineage than in the lineages of other primates.

CHAPTER 22

Figure Questions

Figure 22.8 More than 5.5 million years ago. **Figure 22.13** The original pool of the transplanted guppy population contains pike-cichlids, a potent predator of adult guppies. Brightly colored adult males would be at a disadvantage in this pool. Thus, it is likely that color patterns in the guppy population would become more drab if they were returned to their original pool. **Figure 22.19** Based on this evolutionary tree, crocodiles are more closely related to birds than to lizards because they share a more recent common ancestor with birds (ancestor 5) than with lizards (ancestor 4).

Concept Check 22.1

1. Hutton and Lyell proposed that events in the past were caused by the same processes operating today. This principle suggested that Earth must be much older than a few thousand years, the age that was widely accepted at that time. Hutton and Lyell also thought that geologic change occurs gradually, stimulating Darwin to reason that the slow accumulation of small changes could ultimately produce the profound changes documented in the fossil record. In this context, the age of Earth was important to Darwin, because unless Earth was very old, he could not envision how there would have been enough time for evolution to occur. **2.** By these criteria, Cuvier's explanation of the fossil record and Lamarck's hypothesis of evolution are both scientific. Cuvier suggested that catastrophes and the resulting extinctions were usually confined to local regions, and that such regions were later repopulated by a different set of species that immigrated from other areas. These assertions can be tested against the fossil record (they have been found to be false). With respect to Lamarck, his principle of use and disuse can be used to make testable predictions for fossils of groups such as whale ancestors as they adapt to a new habitat. Lamarck's principle of the inheritance of acquired characteristics can be tested directly in living organisms (it has been found to be false).

Concept Check 22.2

1. Organisms share characteristics (the unity of life) because they share common ancestors; the great diversity of life occurs because new species have repeatedly formed when descendant organisms gradually adapted to different environments, becoming different from their ancestors. **2.** All species have the potential to produce more offspring (overreproduce) than can be supported by the environment. This ensures there will be what Darwin called a "struggle for existence" in which many of the offspring are eaten, starved, diseased, or unable to reproduce for a variety of other reasons. Members of a population exhibit a range of heritable variations, some of which make it likely that their bearers will leave more offspring than other individuals (for example, the bearer may escape predators more effectively or be more tolerant of the physical conditions of the environment). Over time, natural selection imposed by factors such as predators, lack of food, or the physical conditions of the environment can increase the proportion of individuals with favorable traits in a population (evolutionary adaptation). **3.** The fossil mammal species (or its ancestors) would most likely have colonized the Andes from within South America, whereas ancestors of mammals currently found in African mountains would most likely have colonized those mountains from other parts of Africa. As a result, the Andes fossil species would share a more recent common ancestor with South American mammals than with mammals in Africa. Thus, for many of its traits, the fossil mammal species would probably more closely resemble mammals that live in South American jungles than mammals that live on African mountains.

Concept Check 22.3

1. An environmental factor such as a drug does not create new traits such as drug resistance, but rather selects for traits among those that are already present in the population. **2.** (a) Despite their different functions, the forelimbs of different mammals are structurally similar because they all represent modifications of a structure found in the common ancestor. (b) Convergent evolution: The similarities between the sugar glider and flying squirrel indicate that similar environments selected for similar adaptations despite different ancestry.

3. At the time that dinosaurs originated, Earth's landmasses formed a single large continent, Pangaea. Because many dinosaurs were large and mobile, it is likely that early members of these groups lived on many different parts of Pangaea. When Pangaea broke apart, fossils of these organisms would have moved with the rocks in which they were deposited. As a result, we would predict that fossils of early dinosaurs would have a broad geographic distribution (this prediction has been upheld).

Self-Quiz

1. b **2.** c **3.** d **4.** d **5.** a **6.** d

8. (a)

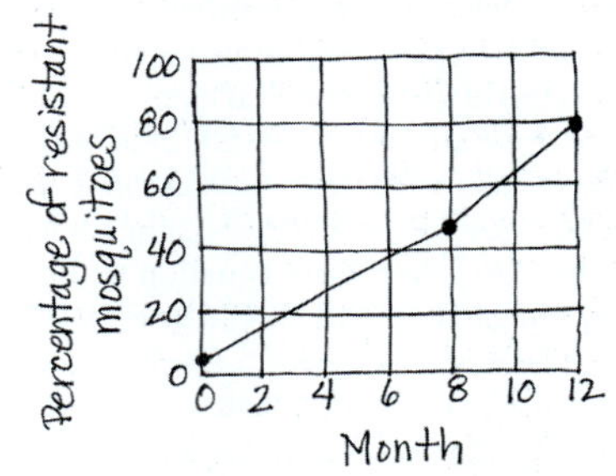

(b) The rapid rise in the percentage of mosquitoes resistant to DDT was most likely caused by natural selection in which mosquitoes resistant to DDT could survive and reproduce while other mosquitoes could not. (c) In India—where DDT resistance first appeared—natural selection would have caused the frequency of resistant mosquitoes to increase over time. If resistant mosquitoes then migrated from India (for example, transported by wind or in planes, trains, or ships) to other parts of the world, the frequency of DDT resistance would increase there as well.

CHAPTER 23

Figure Questions

Figure 23.7 The predicted frequencies are 36% C^RC^R, 48% C^RC^W, and 16% C^WC^W. **Figure 23.12** Because such a shift in prevailing winds would increase the transfer of alleles (gene flow) from plants living on mine soils to plants living at the location marked by the arrow, the change would probably lead to an increase in the index of copper tolerance of plants at that location. **Figure 23.16** Crossing a single female's eggs with both an SC and an LC male's sperm allowed the researchers to directly compare the effects of the males' contribution to the next generation, since both batches of offspring had the same maternal contribution. This isolation of the male's impact enabled researchers to draw conclusions about differences in genetic "quality" between the SC and LC males. **Figure 23.18** The researchers measured the percentages of successfully reproducing adults out of the breeding adult population that had each phenotype. This approach of determining which phenotype was favored by selection assumes that reproduction was a sufficient indicator of relative fitness (as opposed to counting the number of eggs laid or offspring hatched, for example) and that mouth phenotype was the driving factor determining the fishes' ability to reproduce.

Concept Check 23.1

1. (a) Within a population, genetic differences among individuals provide the raw material on which natural selection and other mechanisms can act. Without such differences, allele frequencies could not change over time—and hence the population could not evolve. (b) Genetic variation among populations can arise by natural selection if selection favors different alleles in different populations; this might occur, for example, if the different populations experienced different environmental conditions. Genetic variation among populations can also arise by genetic drift when the genetic differences between populations are selectively neutral. **2.** Many mutations occur in somatic cells that do not produce gametes and so are lost when the organism dies. Of mutations that do occur in cell lines that produce gametes, many do not have a phenotypic effect on which natural selection can act. Others have a harmful effect and are thus unlikely to increase in frequency because they decrease the reproductive success of their bearers. **3.** Its genetic variation (whether measured at the level of the gene or at the level of nucleotide sequences) would probably drop over time. During meiosis, crossing over and the independent assortment of chromosomes produce many new combinations of alleles. In addition, a population contains a vast number of possible mating combinations, and fertilization brings together the gametes of individuals with different genetic backgrounds. Thus, via crossing over, independent assortment of chromosomes, and fertilization, sexual reproduction reshuffles alleles into fresh combinations each generation. Without sexual reproduction, new sources of genetic variation would be reduced, causing the overall amount of genetic variation to drop.

Concept Check 23.2

1. 750. Half the loci (250) are fixed, meaning only one allele exists for each locus: $250 \times 1 = 250$. There are two alleles each for the other loci: $250 \times 2 = 500$. $250 + 500 = 750$. **2.** $p^2 + 2pq$; p^2 represents homozygotes with two *A* alleles, and $2pq$ represents heterozygotes with one *A* allele. **3.** There are 120 individuals in the population, so there are 240 alleles. Of these, there are 124 *A* alleles—32 from the 16 *AA* individuals and 92 from the 92 *Aa* individuals. Thus, the frequency of the *A* allele is $p = 124/240 = 0.52$; hence, the frequency of the *a* allele is $q = 0.48$. Based on the Hardy-Weinberg equation, if the population were not evolving, the frequency of genotype *AA* should be $p^2 = 0.52 \times 0.52 = 0.27$; the frequency of genotype *Aa* should be $2pq = 2 \times 0.52 \times 0.48 = 0.5$; and the frequency of genotype *aa* should be $q^2 = 0.48 \times 0.48 = 0.23$. In a population of 120 individuals, these expected genotype frequencies lead us to predict that there would be 32 *AA* individuals (0.27×120), 60 *Aa* individuals (0.5×120), and 28 *aa* individuals (0.23×120). The actual numbers for the population (16 *AA*, 92 *Aa*, 12 *aa*) deviate from these expectations (fewer homozygotes and more heterozygotes than expected). This suggests that the population is not in Hardy-Weinberg equilibrium and hence is evolving.

Concept Check 23.3

1. Natural selection is more "predictable" in that it alters allele frequencies in a nonrandom way: It tends to increase the frequency of alleles that increase the organism's reproductive success in its environment and decrease the frequency of alleles that decrease the organism's reproductive success. Alleles subject to genetic drift increase or decrease in frequency by chance alone, whether or not they are advantageous. **2.** Genetic drift results from chance events that cause allele frequencies to fluctuate at random from generation to generation; within a population, this process tends to decrease genetic variation over time. Gene flow is the exchange of alleles between populations; a process that can introduce new alleles to a population and hence may increase its genetic variation (albeit slightly, since rates of gene flow are often low). **3.** Selection is not important at this locus; furthermore, the populations are reasonably large, and hence the effects of genetic drift should not be pronounced. Gene flow is occurring via the movement of pollen and seeds. Thus, allele and genotype frequencies in these populations should become more similar over time as a result of gene flow.

Concept Check 23.4

1. Zero, because fitness includes reproductive contribution to the next generation, and a sterile mule cannot produce offspring. **2.** Although both gene flow and genetic drift can increase the frequency of advantageous alleles in a population, they can also decrease the frequency of advantageous alleles or increase the frequency of harmful alleles. Only natural selection *consistently* results in an increase in the frequency of alleles that enhance survival or reproduction. Thus, natural selection is the only mechanism that consistently causes adaptive evolution. **3.** The three modes of natural selection (directional, stabilizing, and disruptive) are defined in terms of the selective advantage of different *phenotypes,* not different genotypes. Thus, the type of selection represented by heterozygote advantage depends on the phenotype of the heterozygotes. In this question, because heterozygous individuals have a more extreme phenotype than either homozygote, heterozygote advantage represents directional selection.

Self-Quiz

1. d **2.** a **3.** e
4. b **5.** b
7. The frequency of the lap^{94} allele forms a cline, decreasing as one moves from southwest to northeast across Long Island Sound.

Site	1	2	3	4	5	6
lap^{94}%	13	16	16	25	36	37

Site	7	8	9	10	11
lap^{94}%	39	55	59	59	59

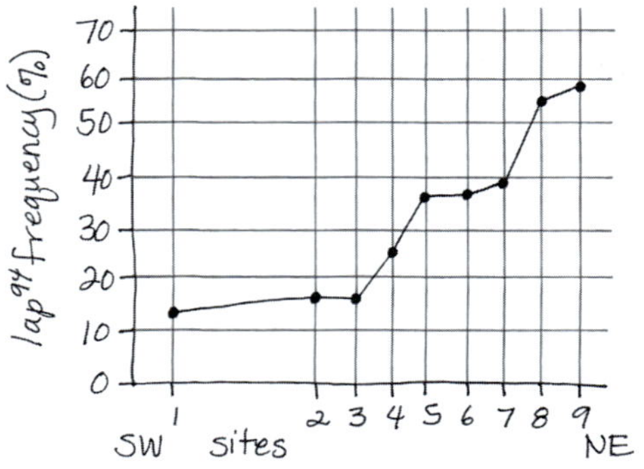

A hypothesis that explains the cline and accounts for the observations stated in the question is that the cline is maintained by an interaction between selection and gene flow. Under this hypothesis, in the southwest portion of the Sound, salinity is relatively low, and selection against the lap^{94} allele is strong. Moving toward the northeast and into the open ocean, where salinity is relatively high, selection favors a high frequency of the lap^{94} allele. However, because mussel larvae disperse long distances, gene flow prevents the lap^{94} allele from becoming fixed in the open ocean or from declining to zero in the southwestern portion of Long Island Sound.

CHAPTER 24

Figure Questions

Figure 24.3 Allele 1 (found in some birds in Population B) is more closely related to alleles found in Population A than to other alleles found in Population B. This implies that the ancestral allele from which allele 1 descended existed in Population A. Hence, the direction of gene flow was from Population A to Population B. **Figure 24.9** This change would have the effect of increasing gene flow between the populations, which would make the evolution of reproductive isolation more difficult. **Figure 24.12** Such results would suggest that mate choice based on coloration does not provide a reproductive barrier between these two cichlid species. **Figure 24.14** Because the populations had only just begun to diverge from one another at this point in the process, it is likely that any existing barriers to reproduction would weaken over time. **Figure 24.20** The presence of *M. cardinalis* plants that carry the *M. lewisii yup* allele would make it more likely that bumblebees would transfer pollen between the two monkey flower species. As a result, we would expect the number of hybrid offspring to increase.

Concept Check 24.1

1. (a) All except the biological species concept can be applied to both asexual and sexual species because they define species on the basis of characteristics other than ability to reproduce. In contrast, the biological species concept can be applied only to sexual species. (b) The easiest species concept to apply in the field would be the morphological species concept because it is based only on the appearance of the organism. Additional information about its ecological habits, evolutionary history, and reproduction are not required. **2.** Because these birds live in fairly similar environments and can breed successfully in captivity, the reproductive barrier in nature is probably prezygotic; given the species differences in habitat preference, this barrier could result from habitat isolation.

Concept Check 24.2

1. In allopatric speciation, a new species forms while in geographic isolation from its parent species; in sympatric speciation, a new species forms in the absence of geographic isolation. Geographic isolation greatly reduces gene flow between populations, whereas ongoing gene flow is more likely in sympatry. As a result, sympatric speciation is less common than allopatric speciation. **2.** Gene flow between subsets of a population that live in the same area can be reduced in a variety of ways. In some species—especially plants—changes in chromosome number can block gene flow and establish reproductive isolation in a single generation. Gene flow can also be reduced in sympatric populations by habitat differentiation (as seen in the apple maggot fly, *Rhagoletis*) and sexual selection (as seen in Lake Victoria cichlids). **3.** Allopatric speciation would be less likely to occur on a nearby island than on an isolated island of the same size. The reason we expect this result is that continued gene flow between mainland populations and those on a nearby island reduces the chance that enough genetic divergence will take place for allopatric speciation to occur.

Concept Check 24.3

1. Hybrid zones are regions in which members of different species meet and mate, producing some offspring of mixed ancestry. Such regions are "natural laboratories" in which to study speciation because scientists can directly observe factors that cause (or fail to cause) reproductive isolation. **2.** (a) If hybrids consistently survive and reproduce poorly compared to the offspring of intraspecific matings, it is possible that reinforcement would occur. If it did, natural selection would cause prezygotic barriers to reproduction between the parent species to strengthen over time, decreasing the production of unfit hybrids and leading to a completion of the speciation process. (b) If hybrid offspring survive and reproduce as well as the offspring of intraspecific matings, indiscriminate mating between the parent species would lead to the production of large numbers of hybrid offspring. As these hybrids mated with each other and with members of both parent species, the gene pools of the parent species could fuse over time, reversing the speciation process.

Concept Check 24.4

1. The time between speciation events includes (1) the length of time that it takes for populations of a newly formed species to begin diverging reproductively from one another and (2) the time it takes for speciation to be complete once this divergence begins. Although speciation can occur rapidly once populations have begun to diverge from one another, it may take millions of years for that divergence to begin. **2.** Investigators transferred alleles at the *yup* locus (which influences flower color) from each parent species to the other. *M. lewisii* plants with an *M. cardinalis yup* allele received many more visits from hummingbirds than usual; hummingbirds usually pollinate *M. cardinalis* but avoid *M. lewisii*. Similarly, *M. cardinalis* plants with an *M. lewisii yup* allele received many more visits from bumblebees than usual; bumblebees usually pollinate *M. lewisii* and avoid *M. cardinalis*. Thus, alleles at the *yup* locus can influence pollinator choice, which in these species provides the primary barrier to interspecific mating. Nevertheless, the experiment does not prove that the *yup* locus alone controls barriers to reproduction between *M. lewisii* and *M. cardinalis*; other genes might enhance the effect of the *yup* locus (by modifying flower color) or cause entirely different barriers to reproduction (for example, gametic isolation or a postzygotic barrier).

Self-Quiz

1. b **2.** a **3.** c **4.** e **5.** d **6.** c
8. One possible process is

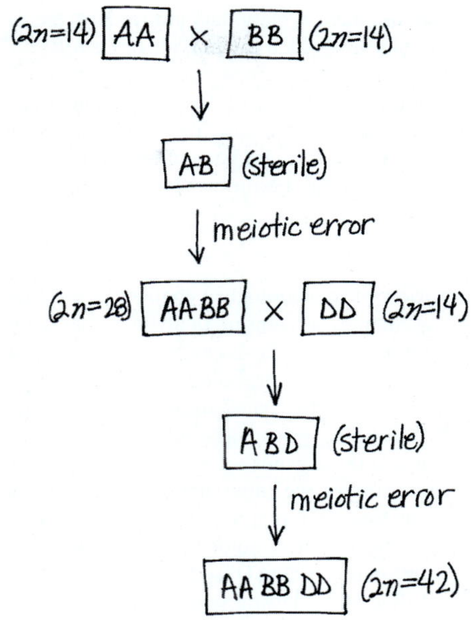

CHAPTER 25

Figure Questions

Figure 25.5 Because uranium-238 has a half-life of 4.5 billion years, the *x*-axis would be relabeled (in billions of years) as: 4.5, 9, 13.5, and 18. **Figure 25.23** The coding sequence of the *Pitx1* gene would differ between the marine and lake populations, but patterns of gene expression would not.

Concept Check 25.1

1. The hypothesis that conditions on early Earth could have permitted the synthesis of organic molecules from inorganic ingredients **2.** In contrast to random mingling of molecules in an open solution, segregation of molecular systems by membranes could concentrate organic molecules, assisting biochemical reactions. **3.** No. Such a result would only show that life *could* have begun as in the experiment.

Concept Check 25.2

1. 22,920 years (four half-lives: $5{,}730 \times 4$) **2.** The fossil record shows that different groups of organisms dominated life on Earth at different points in time and that many organisms once alive are now extinct; specific examples of these points can be found in Figure 25.4. The fossil record also indicates that new groups of organisms can arise via the gradual modification of previously existing organisms, as illustrated by fossils that document the origin of mammals from cynodont ancestors. **3.** The discovery of such a (hypothetical) fossil organism would indicate that aspects of our current understanding of the origin of mammals are not correct because mammals are thought to have originated much more recently (see Figure 25.6). For example, such a discovery could suggest that the dates of previous fossil discoveries are not correct or that the lineages shown in Figure 25.6 shared features with mammals but were not their direct ancestors. Such a discovery would also suggest that radical changes in multiple aspects of the skeletal structure of organisms could arise suddenly—an idea that is not supported by the known fossil record.

Concept Check 25.3

1. Free oxygen attacks chemical bonds and can inhibit enzymes and damage cells. **2.** All eukaryotes have mitochondria or remnants of these organelles, but not all eukaryotes have plastids. **3.** A fossil record of life today would include many organisms with hard body parts (such as vertebrates and many marine invertebrates), but might not include some species we are very familiar with, such as those that have small geographic ranges and/or small population sizes (for example, all five rhinoceros species).

Concept Check 25.4

1. Continental drift alters the physical geography and climate of Earth, as well as the extent to which organisms are geographically isolated. Because these factors affect extinction and speciation rates, continental drift has a major impact on life on Earth. **2.** Mass extinctions; major evolutionary innovations; the diversification of another group of organisms (which can provide new sources of food); migration to new locations where few competitor species exist **3.** Their fossils should be present right up to the time of the catastrophic event, then disappear. Reality is a bit more complicated because the fossil record is not perfect. So the most recent fossil for a species might be a million years before the mass extinction, even if the species did not become extinct until the mass extinction.

Concept Check 25.5

1. Heterochrony can cause a variety of morphological changes. For example, if the onset of sexual maturity changes, a retention of juvenile characteristics (paedomorphosis) may result. Paedomorphosis can be caused by small genetic changes that result in large changes in morphology, as seen in the axolotl salamander. **2.** In animal embryos, *Hox* genes influence the development of structures such as limbs or feeding appendages. As a result, changes in these genes—or in the regulation of these genes—are likely to have major effects on morphology. **3.** From genetics, we know that gene regulation is altered by how well transcription factors bind to noncoding DNA sequences called control elements. Thus, if changes in morphology are often caused by changes in gene regulation, portions of noncoding DNA that contain control elements are likely to be strongly affected by natural selection.

Concept Check 25.6

1. Complex structures do not evolve all at once, but in increments, with natural selection selecting for adaptive variants of the earlier versions. **2.** Although the myxoma virus is highly lethal, initially some of the rabbits are resistant (0.2% of infected rabbits are not killed). Thus, assuming resistance is an inherited trait, we would expect the rabbit population to show a trend for increased resistance to the virus. We would also expect the virus to show an evolutionary trend toward reduced lethality. We would expect this trend because a rabbit infected with a less lethal virus would be more likely to live long enough for a mosquito to bite it and hence potentially transmit the virus to another rabbit. (A virus that kills its rabbit host before a mosquito transmits it to another rabbit dies with its host.)

Self-Quiz

1. c **2.** a **3.** e **4.** b **5.** c **6.** d **7.** b

8.

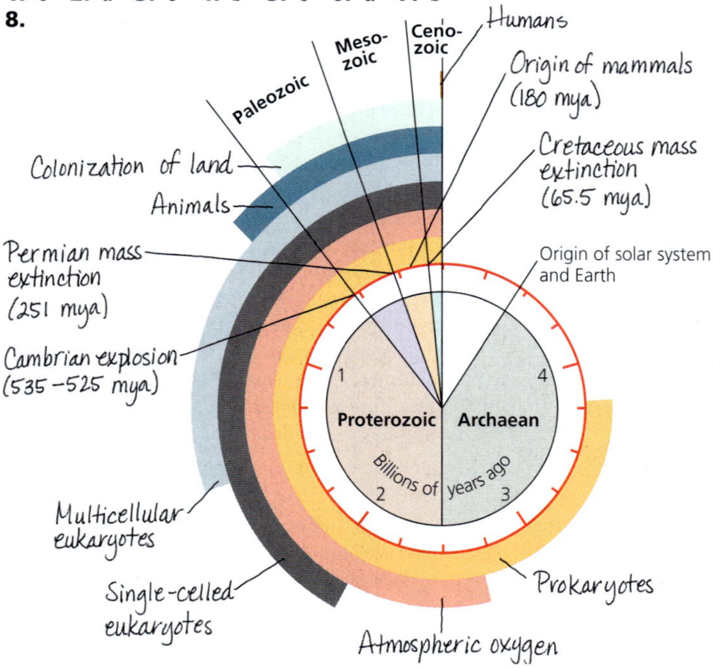

CHAPTER 26

Figure Questions

Figure 26.5 This new version does not alter any of the evolutionary relationships shown in Figure 26.5. For example, B and C remain sister taxa, taxon A is still as closely related to taxon B as it is to taxon C, and so on.

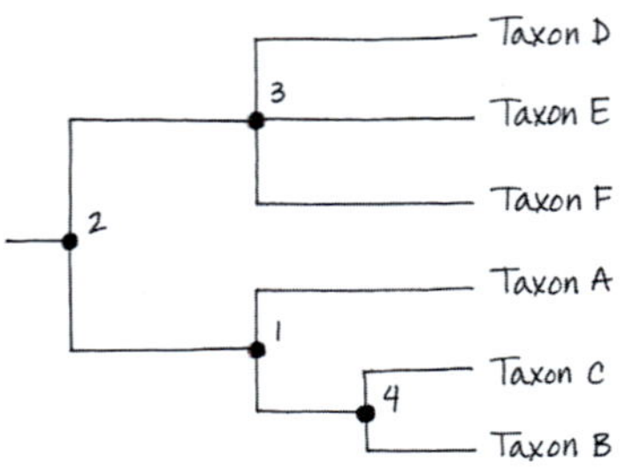

Figure 26.6 Mistakes can occur while performing the experiment (such as errors in DNA sequencing) and analyzing the results (such as misaligning the DNA sequences of different species). But an erroneous conclusion—such as concluding that a sample was from a humpback whale when in fact it came from a gray whale—could be reached even if no such errors were made. A particular humpback whale might, for example, happen to have a DNA sequence that was rare for its species, yet common for another species. To reduce the chance that such events could lead to an erroneous conclusion, gene trees could be constructed for multiple genes; if similar results emerged from all of these gene trees, there would be little reason to doubt the conclusions. **Figure 26.9** There are four possible bases (A, C, G, T) at each nucleotide position. If the base at each position depends on chance, not common descent, we would expect roughly one out of four (25%) of them to be the same. **Figure 26.12** The zebrafish lineage; of the five vertebrate lineages shown, its branch length is the longest. **Figure 26.19** The molecular clock indicates that the divergence time is roughly 45–50 million years. **Figure 26.21** Bacteria was the first to emerge. Archaea is the sister domain to Eukarya.

Concept Check 26.1

1. We are classified the same down to the class level; both the leopard and human are mammals. Leopards belong to order Carnivora, whereas humans do not. **2.** The branching pattern of the tree indicates that the badger and the wolf share a common ancestor that is more recent than the ancestor that these two animals share with the leopard. **3.** The tree in (c) shows a different pattern of evolutionary relationships. In (c), C and B are sister taxa, whereas C and D are sister taxa in (a) and (b).

4.

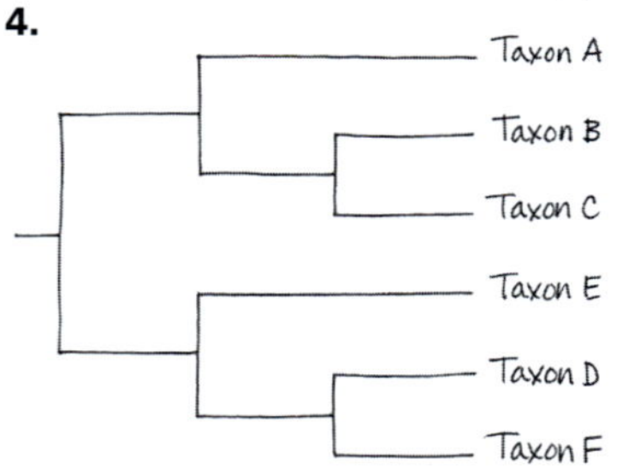

Concept Check 26.2

1. (a) Analogy, since porcupines and cacti are not closely related and since most other animals and plants do not have similar structures; (b) homology, since cats and humans are both mammals and have homologous forelimbs, of which the hand and paw are the lower part; (c) analogy, since owls and hornets are not closely related and since the structure of their wings is very different. **2.** Species 2 and 3 are more likely to be closely related. Small genetic changes (as between species 2 and 3) can produce divergent physical appearances, whereas if genes have diverged greatly (as in species 1 and 2), that suggests that the lineages have been separate for a long time.

Concept Check 26.3

1. No; hair is a shared ancestral character common to all mammals and thus is not helpful in distinguishing different mammalian subgroups. **2.** The principle of maximum parsimony states that the hypothesis about nature we investigate first should be the simplest explanation found to be consistent with the facts. Actual evolutionary relationships may differ from those inferred by parsimony owing to complicating factors such as convergent evolution. **3.** The traditional classification provides a poor match to evolutionary history, thus violating the basic principle of cladistics—that classification should be based on common descent. Both birds and mammals originated from groups traditionally designated as reptiles, making reptiles as traditionally delineated a paraphyletic group. These problems can be addressed by removing *Dimetrodon* and cynodonts from the reptiles, and by considering birds as a group of reptiles (specifically, as a group of dinosaurs).

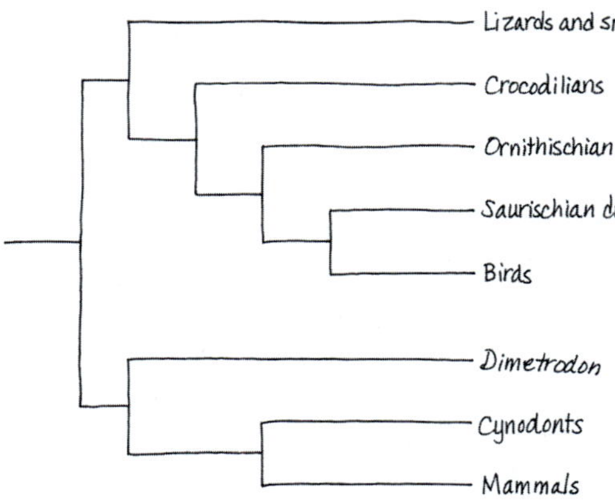

Concept Check 26.4

1. Proteins are gene products. Their amino acid sequences are determined by the nucleotide sequences of the DNA that codes for them. Thus, differences between comparable proteins in two species reflect underlying genetic differences. **2.** These observations suggest that the evolutionary lineages leading to species 1 and species 2 diverged from one another before a gene duplication event in species 1 produced gene B from gene A.

Concept Check 26.5

1. A molecular clock is a method of estimating the actual time of evolutionary events based on numbers of base changes in orthologous genes. It is based on the assumption that the regions of genomes being compared evolve at constant rates. **2.** There are many portions of the genome that do not code for genes; many base changes in these regions could accumulate through drift without affecting an organism's fitness. Even in coding regions of the genome, some mutations may not have a critical effect on genes or proteins. **3.** The gene (or genes) used for the molecular clock may have evolved more slowly in these two taxa than in the species used to calibrate the clock; as a result, the clock would underestimate the time at which the taxa diverged from one another.

Concept Check 26.6

1. The kingdom Monera included bacteria and archaea, but we now know that these organisms are in separate domains. Kingdoms are subsets of domains, so a single kingdom (like Monera) that includes taxa from different domains is not valid (it is polyphyletic). **2.** Because of horizontal gene transfer, some genes in eukaryotes are more closely related to bacteria, while others are more closely related to archaea; thus, depending on which genes are used, phylogenetic trees constructed from DNA data can yield conflicting results.
3.

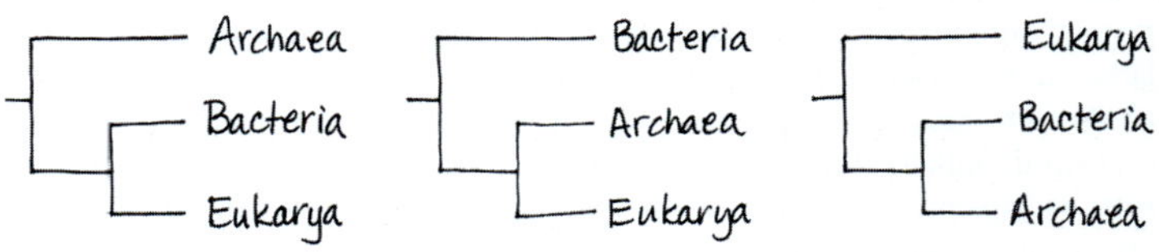

The third tree, in which the eukaryotic lineage diverged first, is not likely to receive support from genetic data because the fossil record shows that prokaryotes originated long before eukaryotes.

Self-Quiz

1. b **2.** d **3.** a **4.** d **5.** c **6.** d **7.** d

9.

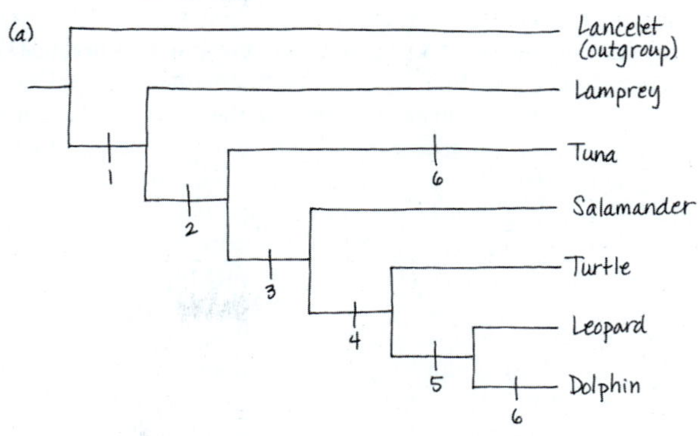

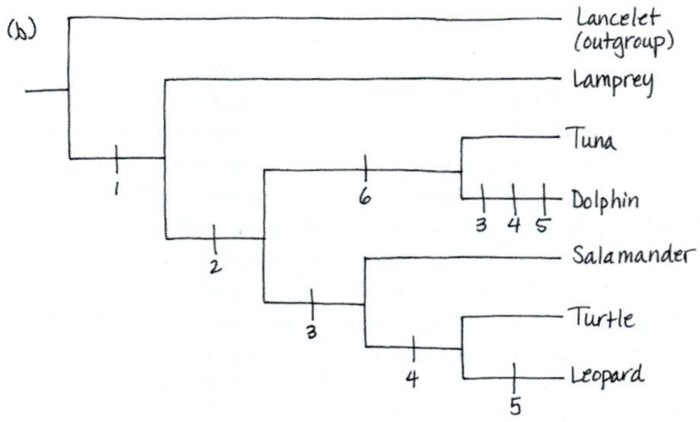

(c) The tree in (a) requires seven evolutionary changes, while the tree in (b) requires nine evolutionary changes. Thus, the tree in (a) is the most parsimonious, since it requires fewer evolutionary changes.

CHAPTER 27

Figure Questions

Figure 27.10 It is likely that the expression or sequence of genes that affect glucose metabolism may have changed; genes for metabolic processes no longer needed by the cell also may have changed. **Figure 27.12** The population that included individuals capable of conjugation would probably be more successful, since some of its members could form recombinant cells whose new gene combinations might be advantageous in a novel environment. **Figure 27.17** Thermophiles live in very hot environments, so it is likely that their enzymes can continue to function normally at much higher temperatures than do the enzymes of other organisms. At low temperatures, however, the enzymes of thermophiles may not function as well as the enzymes of other organisms. **Figure 27.19** From the graph, plant uptake can be estimated as 0.7, 0.6, and 0.95 (mg K) for strains 1, 2, and 3, respectively. These values average to 0.75 mg K. If bacteria had no effect, the average plant uptake of potassium for strains 1, 2, and 3 should be close to 0.5 mg K, the value observed for plants grown in bacteria-free soil.

Concept Check 27.1

1. Adaptations include the capsule (shields prokaryotes from host's immune system) and endospores (enable cells to survive harsh conditions and to revive when the environment becomes favorable). **2.** Prokaryotic cells generally lack the internal compartmentalization of eukaryotic cells. Prokaryotic genomes have much less DNA than eukaryotic genomes, and most of this DNA is contained in a single ring-shaped chromosome located in the nucleoid rather than within a true membrane-bounded nucleus. In addition, many prokaryotes also have plasmids, small ring-shaped DNA molecules containing a few genes. **3.** Because prokaryotic populations evolve rapidly in response to their environment, it is likely that bacteria from endospores that formed 40 years ago would already be adapted to the polluted conditions. Hence, at least initially, these bacteria would probably grow better than bacteria from endospores that formed 150 years ago, when the lake was not polluted.

Concept Check 27.2

1. Prokaryotes have extremely large population sizes, in part because they have short generation times. The large number of individuals in prokaryotic populations makes it likely that in each generation there will be thousands of individuals that have new mutations at any particular gene, thereby adding considerable genetic diversity to the population. **2.** In transformation, naked, foreign DNA from the environment is taken up by a bacterial cell. In transduction, phages carry bacterial genes from one bacterial cell to another. In conjugation, a bacterial cell directly transfers plasmid or chromosomal DNA to another cell via a mating bridge that temporarily connects the two cells. **3.** Yes. Genes for antibiotic resistance could be transferred (by transformation, transduction, or conjugation) from the nonpathogenic bacterium to a pathogenic bacterium; this could make the pathogen an even greater threat to human health. In general, transformation, transduction, and conjugation tend to increase the spread of resistance genes.

Concept Check 27.3

1. A phototroph derives its energy from light, while a chemotroph gets its energy from chemical sources. An autotroph derives its carbon from inorganic sources (often CO_2), while a heterotroph gets its carbon from organic sources. Thus, there are four nutritional modes: photoautotrophic, photoheterotrophic (unique to prokaryotes), chemoautotrophic (unique to prokaryotes), and chemoheterotrophic. **2.** Chemoheterotrophy; the bacterium must rely on chemical sources of energy, since it is not exposed to light, and it must be a heterotroph if it requires an organic source of carbon rather than CO_2 (or another inorganic source, like bicarbonate). **3.** If humans could fix nitrogen, we could build proteins using atmospheric N_2 and hence would not need to eat high-protein foods such as meat or fish. Our diet would, however, need to include a source of carbon, along with minerals and water. Thus, a typical meal might consist of carbohydrates as a carbon source, along with fruits and vegetables to provide essential minerals (and additional carbon).

Concept Check 27.4

1. Before molecular systematics, taxonomists classified prokaryotes according to phenotypic characters that did not clarify evolutionary relationships. Molecular comparisons—of DNA in particular—indicate key divergences in prokaryotic lineages. **2.** By not requiring that organisms be cultured in the laboratory, genetic prospecting has revealed an immense diversity of previously unknown prokaryotic species. Over time, the ongoing discovery of new species by genetic prospecting is likely to alter our understanding of prokaryotic phylogeny greatly. **3.** At present, all known methanogens are archaea in the clade Euryarchaeota; this suggests that this unique metabolic pathway arose in ancestral species within Euryarchaeota. Since Bacteria and Archaea have been separate evolutionary lineages for billions of years, the discovery of a methanogen from the domain Bacteria would suggest that adaptations that enabled the use of CO_2 to oxidize H_2 evolved at least twice—once in Archaea (within Euryarchaeota) and once in Bacteria.

Concept Check 27.5

1. Although prokaryotes are small, their large numbers and metabolic abilities enable them to play key roles in ecosystems by decomposing wastes, recycling chemicals, and affecting the concentrations of nutrients available to other organisms. **2.** *Bacteroides thetaiotaomicron*, which lives inside the human intestine, benefits by obtaining nutrients from the digestive system and by receiving protection from competing bacteria from host-produced antimicrobial compounds to which it is not sensitive. The human host benefits because the bacterium manufactures carbohydrates, vitamins, and other nutrients. **3.** Some of the many different species of prokaryotes that live in the human gut compete with one another for resources (in the food you eat). Because different prokaryotic species have different adaptations, a change in diet may alter which species can grow most rapidly, thus altering species abundance.

Concept Check 27.6

1. Sample answers: eating fermented foods such as yogurt, sourdough bread, or cheese; receiving clean water from sewage treatment; taking medicines produced by bacteria. **2.** No. If the poison is secreted as an exotoxin, live bacteria could be transmitted to another person. But the same is true if the poison is an endotoxin—only in this case, the live bacteria that are transmitted may be descendants of the (now-dead) bacteria that produced the poison. **3.** Strain K-12 may have lost genes by deletion mutations. A phylogenetic analysis would help distinguish between these hypotheses—if some of the genes found in O157:H7 but not in K-12 are present in the common ancestor of the two strains, that would suggest that strain K-12 lost these genes over the course of its evolution.

Self-Quiz

1. e **2.** a **3.** d **4.** d **5.** b **6.** a
8. (a)

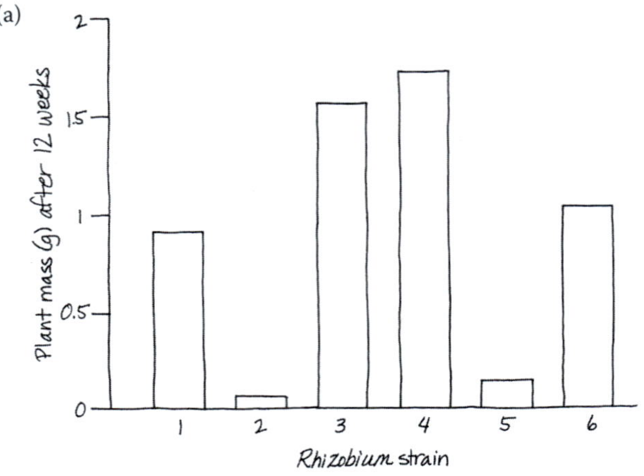

(b) Some *Rhizobium* strains are much more effective at promoting plant growth than are other *Rhizobium* strains; the most ineffective strains have little positive effect (plant growth with these strains differs little from plant growth in the absence of *Rhizobium*). The ineffective strains may transfer relatively little nitrogen to their plant host, hence limiting plant growth.

CHAPTER 28

Figure Questions

Figure 28.10 Merozoites are produced by the asexual (mitotic) cell division of haploid sporozoites; similarly, gametocytes are produced by the asexual cell division of merozoites. Hence, it is likely that individuals in these three stages have the same complement of genes and that morphological differences between them result from changes in gene expression. **Figure 28.22** The following stage should be circled: step 6, where a mature cell undergoes mitosis and forms four or more daughter cells. In step 7, the zoospores eventually grow into mature haploid cells, but they do not produce new daughter cells. Likewise, in step 2, a mature cell develops into a gamete, but it does not produce new daughter cells. **Figure 28.23** If the assumption is correct, then their results indicate that the DHFR-TS gene fusion may be a derived trait shared by members of four supergroups of eukaryotes (Excavata, Chromalveolata, Rhizaria, and Archaeplastida). However, if the assumption is not correct, the presence or absence of the gene fusion may tell little about phylogenetic history. For example, if the genes fused multiple times, groups could share the trait because of convergent evolution rather than common descent. If the genes were secondarily split, a group with such a split could be placed (incorrectly) in Unikonta rather than its correct placement in one of the other four supergroups.

Concept Check 28.1

1. Sample response: Protists include unicellular, colonial, and multicellular organisms; photoautotrophs, heterotrophs, and mixotrophs; species that reproduce asexually, sexually, or both ways; and organisms with diverse physical forms and adaptations. **2.** Strong evidence shows that eukaryotes acquired mitochondria after an early eukaryote first engulfed and then formed an endosymbiotic association with an alpha proteobacterium. Similarly, chloroplasts in red and green algae appear to have descended from a photosynthetic cyanobacterium that was engulfed by an ancient heterotrophic eukaryote. Secondary endosymbiosis also played an important role: Various protist lineages acquired plastids by engulfing unicellular red or green algae. **3.** The modified tree would look as follows:

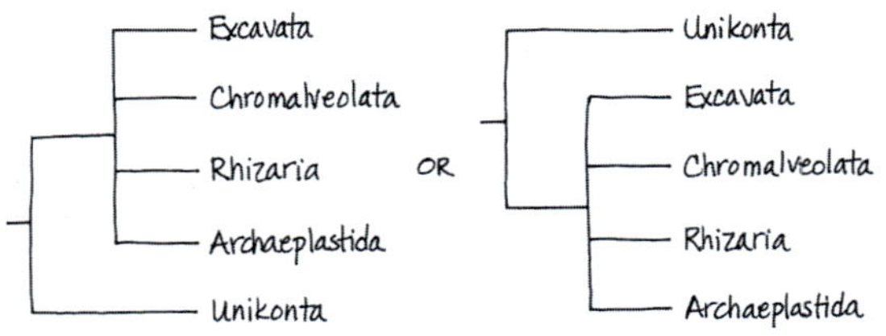

Concept Check 28.2

1. Their mitochondria do not have an electron transport chain and so cannot function in aerobic respiration. **2.** Since the unknown protist is more closely related to diplomonads than to euglenids, it must have evolved after the diplomonads and parabasalids diverged from the euglenozoans. In addition, since the unknown species has fully functional mitochondria—yet both diplomonads and parabasalids do not—it is likely that the unknown species evolved *before* the last common ancestor of the diplomonads and parabasalids.

Concept Check 28.3

1. Some DNA data indicate that Chromalveolata is a monophyletic group, but other DNA data fail to support this result. In support of monophyly, for many species in the group, the structure of their plastids and the sequence of their plastid DNA suggest that the group originated by a secondary endosymbiosis event (in which a red alga was engulfed). However, other species in the group lack plastids entirely, making the secondary endosymbiosis hypothesis difficult to test. **2.** Figure 13.6b. Algae and plants with alternation of generations have a multicellular haploid stage *and* a multicellular diploid stage. In the other two life cycles, either the haploid stage or the diploid stage is unicellular. **3.** The plastid DNA would likely be more similar to the chromosomal DNA of cyanobacteria based on the well-supported hypothesis that eukaryotic plastids (such as those found in the eukaryotic groups listed) originated by an endosymbiosis event in which a eukaryote engulfed a cyanobacterium. If the plastid is derived from the cyanobacterium, its DNA would be derived from the bacterial DNA.

Concept Check 28.4

1. Because foram tests are hardened with calcium carbonate, they form long-lasting fossils in marine sediments and sedimentary rocks. **2.** Convergent evolution. The different organisms have come to display similar morphological adaptations over time owing to their similar lifestyles.

Concept Check 28.5

1. Many red algae contain an accessory pigment called phycoerythrin, which gives them a reddish color and allows them to carry out photosynthesis in relatively deep coastal water. Also unlike brown algae, red algae have no flagellated stages in their life cycle and must depend on water currents to bring gametes together for fertilization. **2.** *Ulva*'s thallus contains many cells and is differentiated into leaflike blades and a rootlike holdfast. *Caulerpa*'s thallus is composed of multinucleate filaments without cross-walls, so it is essentially one large cell. **3.** Red algae have no flagellated stages in their life cycle and hence must depend on water currents to bring their gametes together. This feature of their biology might increase the difficulty of reproducing on land. In contrast, the gametes of green algae are flagellated, making it possible for them to swim in thin films of water. In addition, a variety of green algae contain compounds in their cytoplasm, cell wall, or zygote coat that protect against intense sunlight and other terrestrial conditions. Such compounds may have increased the chance that descendants of green algae could survive on land.

Concept Check 28.6

1. Amoebozoans have lobe-shaped pseudopodia, whereas forams have thread-like pseudopodia. **2.** Slime molds are fungus-like in that they produce fruiting bodies that aid in the dispersal of spores, and they are animal-like in that they are motile and ingest food. However, slime molds are more closely related to gymnamoebas and entamoebas than to fungi or animals. **3.** Support. Unikonts lack the unique cytoskeletal features shared by many excavates (see Concept 28.2). Thus, if the unikonts were the first group of eukaryotes to diverge from other eukaryotes (as shown in Figure 28.23), it would be unlikely that the eukaryote common ancestor had the cytoskeletal features found today in many excavates. Such a result would strengthen the case that many excavates share cytoskeletal features because they are members of a monophyletic group, the Excavata.

Concept Check 28.7

1. Because photosynthetic protists lie at the base of aquatic food webs, many aquatic organisms depend on them for food, either directly or indirectly. (In addition, a substantial percentage of the oxygen produced in photosynthesis on Earth is made by photosynthetic protists.) **2.** Protists form mutualistic and parasitic associations with other organisms. Examples include parabasalids that form a mutualistic symbiosis with termites, as well as the oomycete *Phytophthora ramorum*, a parasite of oak trees. **3.** Corals depend on their dinoflagellate symbionts for nourishment, so coral bleaching would be expected to cause the corals to die. As the corals die, less food will be available for fishes and other species that eat coral. As a result, populations of these species may decline, and that, in turn, might cause populations of their predators to decline.

Self-Quiz

1. d **2.** b **3.** c **4.** d **5.** e **6.** d
7.

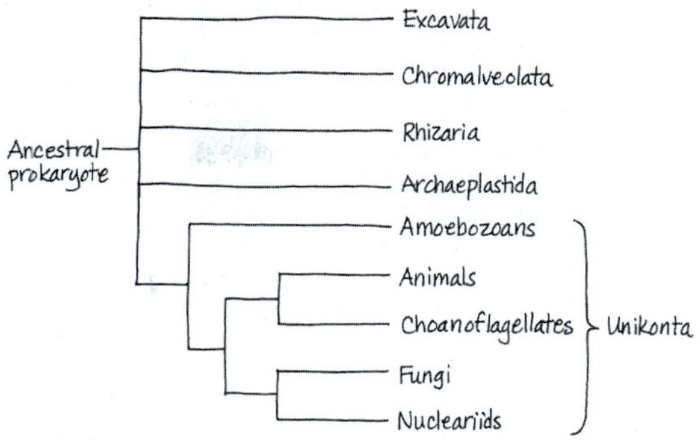

Pathogens that share a relatively recent common ancestor with humans should also share metabolic and structural characteristics with humans. Because drugs target the pathogen's metabolism or structure, developing drugs that harm the pathogen but not the patient should be most difficult for pathogens with whom we share the most recent evolutionary history. Working backward in time, we can use the phylogenetic tree to determine the order in which humans shared a common ancestor with pathogens in different taxa. This process leads to the prediction that it should be hardest to develop drugs to combat animal pathogens, followed by choanoflagellate pathogens, fungal and nucleariid pathogens, amoebozoans, other protists, and finally prokaryotes.

CHAPTER 29

Figure Questions

Figure 29.7

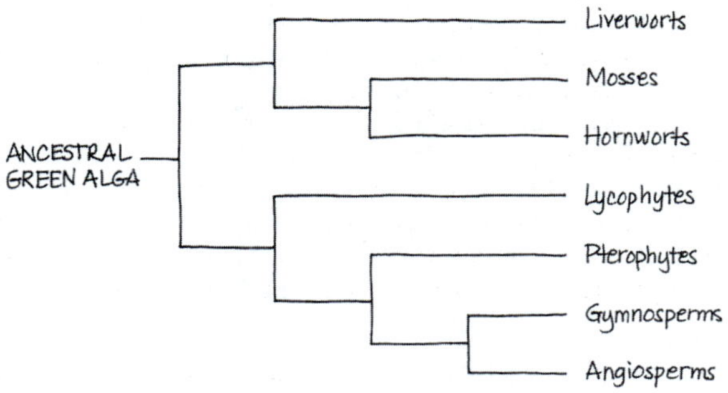

Figure 29.10 Because the moss reduces nitrogen loss from the ecosystem, species that typically colonize the soils after the moss probably experience higher soil nitrogen levels than they otherwise would—an effect that may benefit these species, since nitrogen is an essential nutrient that often is in short supply. **Figure 29.13** A fern that had wind-dispersed sperm would not require water for fertilization, thus removing a difficulty that ferns face when they live in arid environments. The fern would also be under strong selection to produce sperm above ground (as opposed to the current situation, where some fern gametophytes are located below ground).

Concept Check 29.1

1. Land plants share some key traits only with charophytes: rosette cellulose-synthesizing complexes, presence of peroxisome enzymes, similarity in sperm structure, and the formation of a phragmoplast in cell division. Comparisons of nuclear and chloroplast genes also point to a common ancestry. **2.** Spore walls toughened by sporopollenin (protects against harsh environmental conditions); multicellular, dependent embryos (provides nutrients and protection to the developing embryo); cuticle (reduces water loss). **3.** The multicellular diploid stage of the life cycle would not reproduce sexually. Instead, both males and females would produce haploid spores by meiosis. These spores would give rise to multicellular male and female haploid stages—a major change from the single-celled haploid stages (sperm and eggs) that we actually have. The multicellular haploid stages would produce gametes and reproduce sexually. An individual at the multicellular haploid stage of the human life cycle might look like us, or it might look completely different.

Concept Check 29.2

1. Bryophytes do not have an extensive vascular transport system, and their life cycle is dominated by gametophytes rather than sporophytes. **2.** Answers may include the following: Large surface area of protonema enhances absorption of water and minerals; the vase-shaped archegonia protect eggs during fertilization and transport nutrients to the embryos via placental transfer cells; the stalk-like seta conducts nutrients from the gametophyte to the capsule, where spores are produced; the peristome enables gradual spore discharge; stomata enable CO_2/O_2 exchange while minimizing water loss; lightweight spores are readily dispersed by wind.
3.

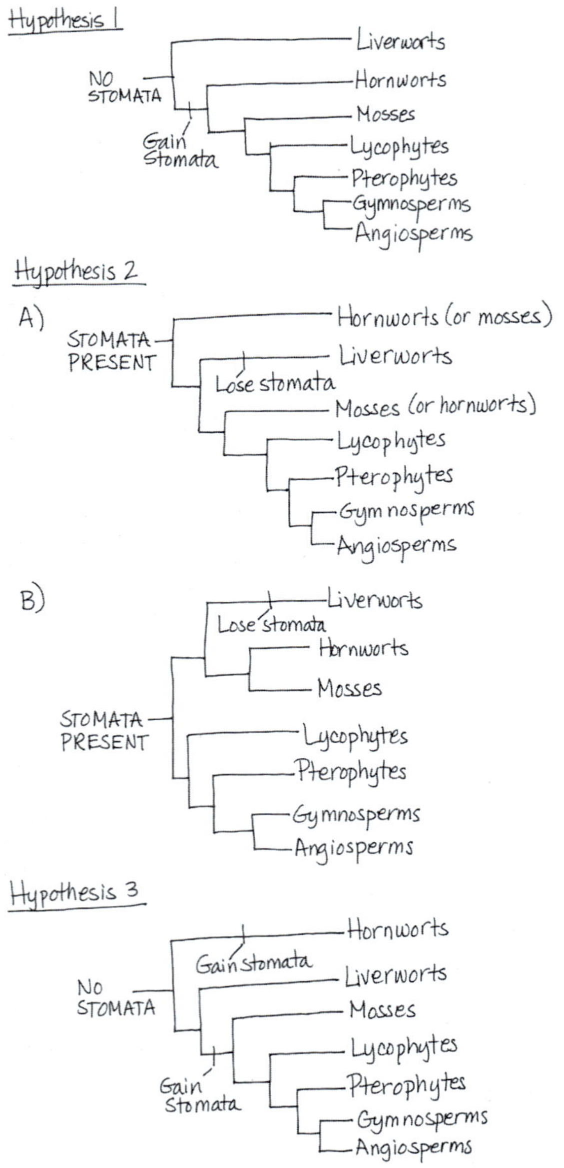

Concept Check 29.3

1. Lycophytes have microphylls, whereas seed plants and pterophytes (ferns and their relatives) have megaphylls. Pterophytes and seed plants also share other traits not found in lycophytes, such as overtopping growth and the initiation of new root branches at various points along the length of an existing root. **2.** Both seedless vascular plants and bryophytes have flagellated sperm that require moisture for fertilization; this shared similarity poses challenges for these species in arid regions. With respect to key differences, seedless vascular plants have lignified, well-developed vascular tissue, a trait that enables the sporophyte to grow tall and that has transformed life on Earth (via the formation of forests). Seedless vascular plants also have true leaves and roots, which, when compared to bryophytes, provides increased surface area for photosynthesis and improves their ability to extract nutrients from soil. **3.** If lycophytes and pterophytes formed a clade, the traits shared by pterophytes and seed plants might have been present in the common ancestor of all vascular plants, but lost in the lycophytes. Alternatively, the common ancestor of all vascular plants may have lacked the traits shared by pterophytes and seed plants; in this case, pterophytes and seed plants would share these traits as a result of convergent evolution.

Self-Quiz

1. b **2.** e **3.** a **4.** d **5.** c **6.** b **7.** c
8. a. diploid; b. haploid; c. haploid; d. diploid; e. haploid
9. Based on our current understanding of the evolution of major plant groups, the phylogeny has the four branch points shown here:

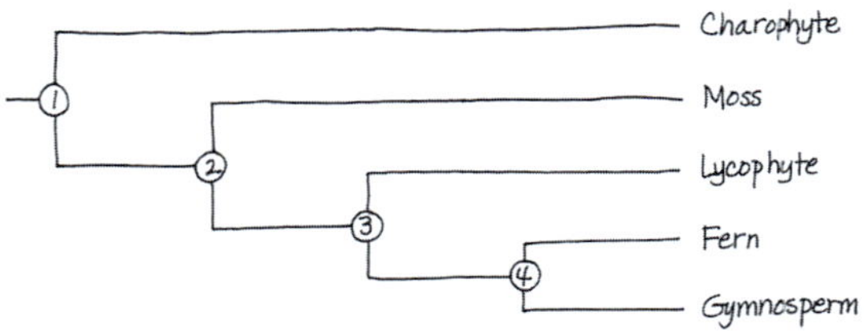

Derived characters unique to the charophyte and land plant clade (indicated by branch point 1) include rosette cellulose-synthesizing complexes, peroxisome enzymes, flagellated sperm structure, and a phragmoplast. Derived characters unique to the land plant clade (branch point 2) include apical meristems, alternation of generations, walled spores produced in sporangia, and multicellular gametangia. Derived characters unique to the vascular plant clade (branch point 3) include life cycles with dominant sporophytes, complex vascular systems (xylem and phloem), and well-developed roots and leaves. Derived characters unique to the pterophyte and seed plant clade (branch point 4) include megaphylls and overtopping growth.

CHAPTER 30

Figure Questions

Figure 30.3 Three: (1) the current sporophyte (cells of ploidy $2n$, found in the integument, or seed coat); (2) the female gametophyte (cells of ploidy n, found in the food supply); and (3) the sporophyte of the next generation (cells of ploidy $2n$, found in the embryo). **Figure 30.12** No. The branching order shown could still be correct if *Amborella* and other early angiosperms had originated prior to 150 million years ago but angiosperm fossils of that age had not yet been discovered. In such a situation, the 140-million-year-old date for the origin of the angiosperms shown on the phylogeny would be incorrect. **Figure 30.14** This study establishes a correlation between the type of floral symmetry and the rate of plant speciation—but it is possible that floral symmetry is correlated with another factor that was the actual cause of the observed results. Note, however, that floral symmetry was associated with increased speciation rates in a variety of different plant lineages. This suggests—but does not establish—that differences in floral symmetry cause differences in speciation rates. In general, strong evidence for causation can come from controlled, manipulative experiments, but such experiments are usually not possible for studies of past evolutionary events.

Concept Check 30.1

1. To have any chance of reaching the eggs, the flagellated sperm of seedless vascular plants must swim through a film of water, usually over a distance of no more than a few centimeters. In contrast, the sperm of seed plants do not require water because they are produced within pollen grains that can be transported long distances by wind or by animal pollinators. Although flagellated in some species, the sperm of seed plants do not require mobility because pollen tubes convey them from the point at which the pollen grain is deposited (near the ovules) directly to the eggs. **2.** The reduced gametophytes of seed plants are nurtured by sporophytes and protected from stress, such as drought conditions and UV radiation. Pollen grains have tough protective walls. Seeds have one or two layers of protective tissue, the seed coat, that improve survival by providing more protection from environmental stresses than do the walls of spores. Seeds also contain a stored supply of food, which enables seeds to live longer than spores and provides developing embryos with nourishment for growth. **3.** If seed plants were homosporous, only one type of spore would be produced—as opposed to the actual situation in which microspores give rise to sperm cells within pollen grains, and megaspores give rise to eggs within ovules. Thus, if structures like pollen grains and seeds were produced, they would arise in a very different way from how they now are formed.

Concept Check 30.2

1. Although gymnosperms are similar in not having their seeds enclosed in ovaries and fruits, their seed-bearing structures vary greatly. For instance, cycads have large cones, whereas some gymnosperms, such as *Ginkgo* and *Gnetum*, have small cones that look somewhat like berries, even though they are not fruits. Leaf shape also varies greatly, from the needles of many conifers to the palmlike leaves of cycads to *Gnetum* leaves that look like those of flowering plants. **2.** The life cycle illustrates heterospory, as ovulate cones produce megaspores and pollen cones produce microspores. The reduced gametophytes are evident in the form of the microscopic pollen grains and the microscopic female gametophyte within the megaspore. The egg is shown developing within an ovule, and a pollen tube is shown conveying the sperm. The figure also shows the protective and nutritive features of a seed. **3.** No. Fossil evidence indicates that gymnosperms originated at least 305 million years ago, but this does not mean that angiosperms are that old—only that the most recent common ancestor of gymnosperms and angiosperms must be that old.

Concept Check 30.3

1. In the oak's life cycle, the tree (the sporophyte) produces flowers, which contain gametophytes in pollen grains and ovules; the eggs in ovules are fertilized; the mature ovaries develop into dry fruits called acorns. We can view the oak's life cycle as starting when the acorn seeds germinate, resulting in embryos giving rise to seedlings and finally to mature trees, which produce flowers—and then more acorns. **2.** Pine cones and flowers both have sporophylls, modified leaves that produce spores. Pine trees have separate pollen cones (with pollen grains) and ovulate cones (with ovules inside cone scales). In flowers, pollen grains are produced by the anthers of stamens, and ovules are within the ovaries of carpels. Unlike pine cones, many flowers produce both pollen and ovules. **3.** Such a discovery would remove support for the idea, based on 125-million-year-old *Archaefructus sinensis* fossils, that the earliest angiosperms may have been herbaceous, aquatic plants.

Concept Check 30.4

1. Because extinction is irreversible, it decreases the total diversity of plants, many of which may have brought important benefits to humans. **2.** A detailed phylogeny of the seed plants would identify many different monophyletic groups of seed plants. Using this phylogeny, researchers could look for clades that contained species in which medicinally useful compounds had already been discovered. Identification of such clades would allow researchers to concentrate their search for new medicinal compounds among clade members—as opposed to searching for new compounds in species that were selected at random from the more than 250,000 existing species of seed plants.

Self-Quiz

1. d **2.** a **3.** b **4.** a **5.** d

6.

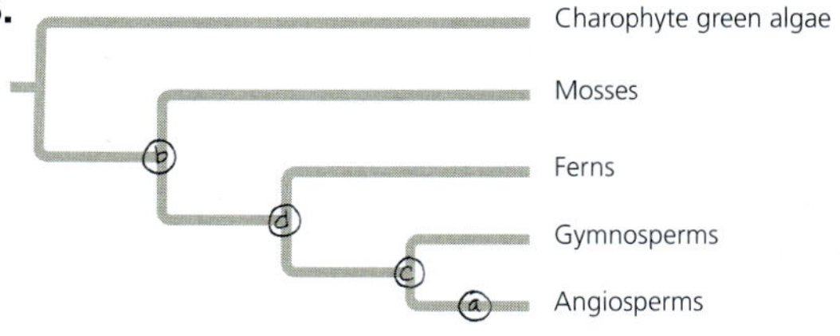

8. (a)

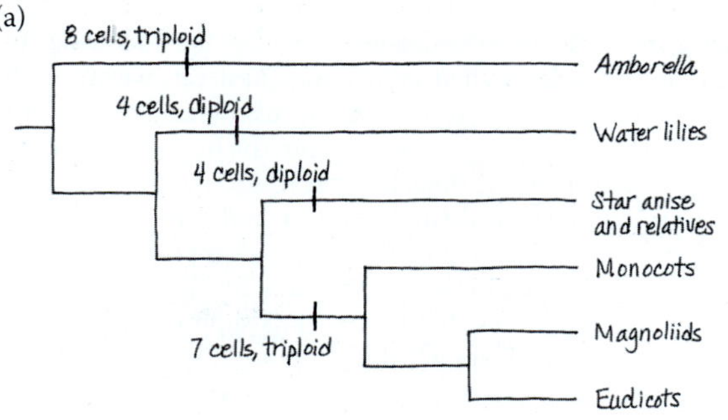

(b) The phylogeny indicates that basal angiosperms differed from other angiosperms in terms of the number of cells in female gametophytes and the ploidy of the endosperm. The ancestral state of the angiosperms cannot be determined from these data alone. It is possible that the common ancestor of angiosperms had seven-celled female gametophytes and triploid endosperm and hence that the eight-celled and four-celled conditions found in basal angiosperms represent derived traits for those lineages. Alternatively, either the eight-celled or four-celled condition may represent the ancestral state.

CHAPTER 31

Figure Questions

Figure 31.2 DNA from each of these mushrooms would be identical if each mushroom is part of a single hyphal network, as is likely. **Figure 31.16** One or both of the following would apply to each species: DNA analyses would reveal that it is a member of the ascomycetes clade, or aspects of its sexual life cycle would indicate that it is an ascomycete (for example, it would produce asci and ascospores). **Figure 31.21** Two possible controls would be E−P− and E+P−. Results from an E−P− control could be compared with results from the E−P+ experiment, and results from an E+P− control could be compared with results from the E+P+ experiment; together, these two comparisons would indicate whether the addition of the pathogen causes an increase in leaf mortality. Results from an E−P− experiment could also be compared with results from the second control (E+P−) to determine whether adding the endophytes has a negative effect on the plant.

Concept Check 31.1

1. Both a fungus and a human are heterotrophs. Many fungi digest their food externally by secreting enzymes into the food and then absorbing the small molecules that result from digestion. Other fungi absorb such small molecules directly from their environment. In contrast, humans (and most other animals) ingest relatively large pieces of food and digest the food within their bodies. **2.** The ancestors of such a mutualist most likely secreted powerful enzymes to digest the body of their insect host. Since such enzymes would harm a living host, it is likely that the mutualist would not produce such enzymes or would restrict their secretion and use.

Concept Check 31.2

1. The majority of the fungal life cycle is spent in the haploid stage, whereas the majority of the human life cycle is spent in the diploid stage. **2.** The two mushrooms might be reproductive structures of the same mycelium (the same organism). Or they might be parts of two separate organisms that have arisen from a single parent organism through asexual reproduction and thus carry the same genetic information.

Concept Check 31.3

1. DNA evidence indicates that fungi, animals, and their protistan relatives form a clade, the opisthokonts. Furthermore, an early-diverging fungal lineage, the chytrids, have posterior flagella, as do most other opisthokonts. This suggests that other fungal lineages lost their flagella after diverging from chytrids. **2.** This indicates that fungi had already established mutualistic relationships with plants by the date the fossils of the earliest vascular plants had formed. **3.** Fungi are heterotrophs. Prior to the colonization of land by plants, terrestrial fungi could have lived only where other organisms (or their remains) were present and provided a source of food. Thus, if fungi had colonized land before plants, they could have fed on any prokaryotes or protists that lived on land or by the water's edge—but not on the plants or animals on which many fungi feed today.

Concept Check 31.4

1. Flagellated spores **2.** Possible answers include the following: In zygomycetes, the sturdy, thick-walled zygosporangium can withstand harsh conditions and then undergo karyogamy and meiosis when the environment is favorable for reproduction. In glomeromycetes, the hyphae have a specialized morphology that enables the fungi to form arbuscular mycorrhizae with plant roots. In ascomycetes, the asexual spores (conidia) are often produced in chains or clusters at the tips of conidiophores, where they are easily dispersed by wind. The often cup-shaped ascocarps house the sexual spore-forming asci. In basidiomycetes, the basidiocarp supports and protects a large surface area of basidia, from which spores are dispersed. **3.** Such a change to the life cycle of an ascomycete would reduce the number and genetic diversity of ascospores that result from a mating event. Ascospore number would drop because a mating event would lead to the formation of only one ascus. Ascospore genetic diversity would also drop because in ascomycetes, one mating event leads to the formation of asci by many different dikaryotic cells. As a result, genetic recombination and meiosis occurs independently many different times—which could not happen if only a single ascus was formed. It is also likely that if such an ascomycete formed an ascocarp, the shape of the ascocarp would differ considerably from that found in its close relatives.

Concept Check 31.5

1. A suitable environment for growth, retention of water and minerals, protection from intense sunlight, and protection from being eaten **2.** A hardy spore stage enables dispersal to host organisms through a variety of mechanisms; their ability to grow rapidly in a favorable new environment enables them to capitalize on the host's resources. **3.** Many different outcomes might have occurred. Organisms that currently form mutualisms with fungi might have gained the ability to perform the tasks currently done by their fungal partners, or they might have formed similar mutualisms with other organisms (such as bacteria). Alternatively, organisms that currently form mutualisms with fungi might be less effective at living in their present environments. For example, the colonization of land by plants might have been more difficult. And if plants did eventually colonize land without fungal mutualists, natural selection might have favored plants that formed more highly divided and extensive root systems (in part replacing mycorrhizae).

Self-Quiz

1. b **2.** c **3.** d **4.** e **5.** b **6.** a
8.

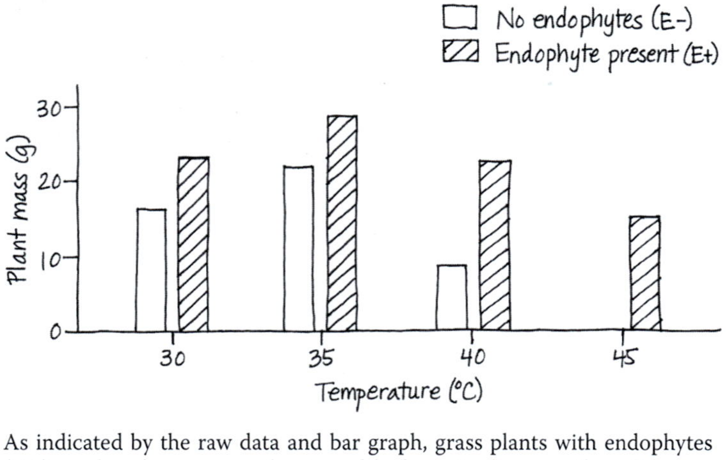

As indicated by the raw data and bar graph, grass plants with endophytes (E+) produced more new shoots and had greater biomass than did grass plants that lacked endophytes (E−). These differences were especially pronounced at the highest soil temperature, where E− grass plants produced no new shoots and had a biomass of zero (indicating they were dead).

CHAPTER 32

Figure Questions

Figure 32.3 As described in 1 and 2, choanoflagellates and a broad range of animals have collar cells. Since collar cells have never been observed in plants, fungi, or non-choanoflagellate protists, this suggests that choanoflagellates may be more closely related to animals than to other eukaryotes. If choanoflagellates are more closely related to animals than is any other group of eukaryotes, choanoflagellates and animals should share other traits that are not found in other eukaryotes. The data described in 3 are consistent with this prediction. **Figure 32.6** The sea anemone embryos could be infused with a protein that can bind to β-catenin's DNA-binding site, thereby limiting the extent to which β-catenin activates the transcription of genes necessary for gastrulation. Such an experiment would provide an independent check of the results shown in step 4. **Figure 32.10** Ctenophora is the sister phylum in this figure, while Cnidaria is the sister phylum in Figure 32.11.

Concept Check 32.1

1. In most animals, the zygote undergoes cleavage, which leads to the formation of a blastula. Next, in gastrulation, one end of the embryo folds inward, producing layers of embryonic tissue. As the cells of these layers differentiate, a wide variety of animal forms result. Despite the diversity of animal forms, animal development is controlled by a similar set of *Hox* genes across a broad range of taxa. **2.** The imaginary plant would require tissues composed of cells that were analogous to the muscle and nerve cells found in animals: "muscle" tissue would be necessary for the plant to chase prey, and "nerve" tissue would be required for the plant to coordinate its movements when chasing prey. To digest captured prey, the plant would need to either secrete enzymes into one or more digestive cavities (which could be modified leaves, as in a Venus' flytrap), or secrete enzymes outside of its body and feed by absorption. To extract nutrients from the soil—yet be able to chase prey—the plant would need something other than fixed roots, perhaps retractable "roots" or a way to ingest soil. To conduct photosynthesis, the plant would require chloroplasts. Overall, such an imaginary plant would be very similar to an animal that had chloroplasts and retractable roots.

Concept Check 32.2

1. c, b, a, d **2.** We cannot infer whether animals originated before or after fungi. If correct, the date provided for the most recent common ancestor of fungi and animals would indicate that animals originated some time within the last billion years. The fossil record indicates that animals originated at least 565 million years ago. Thus, we could conclude only that animals originated some time between 565 million years ago and 1 billion years ago.

Concept Check 32.3

1. Grade-level characteristics are those that multiple lineages share regardless of evolutionary history. Some grade-level characteristics may have evolved multiple times independently. Features that unite clades are derived characteristics that originated in a common ancestor and were passed on to the various descendants. **2.** A snail has a spiral and determinate cleavage pattern; a human has radial, indeterminate cleavage. In a snail, the coelomic cavity is formed by splitting of mesoderm masses; in a human, the coelom forms from folds of archenteron. In a snail, the mouth forms from the blastopore; in a human, the anus develops from the blastopore. **3.** Most coelomate triploblasts have two openings to their digestive tract, a mouth and an anus. As such, their bodies have a structure that is analogous to that of a doughnut: The digestive tract (the hole of the doughnut) runs from the mouth to the anus and is surrounded by various tissues (the solid part of the doughnut). The doughnut analogy is most obvious at early stages of development (see Figure 32.9c).

Concept Check 32.4

1. Cnidarians possess true tissues, while sponges do not. Also unlike sponges, cnidarians exhibit body symmetry, though it is radial and not bilateral as in other animal phyla. **2.** The morphology-based tree divides Bilateria into two major clades: Deuterostomia and Protostomia. The molecular-based tree recognizes three major clades: Deuterostomia, Ecdysozoa, and Lophotrochozoa. **3.** Both statements could be correct. Figure 32.11 shows that the lineage leading to Deuterostomia was the first to diverge from the other two main bilaterian lineages (those leading to Lophotrochozoa and Ecdysozoa). By itself, however, this information does not indicate whether the most recent common ancestor of the Deuterostomia lived before or after the first arthropods. For example, the ancestors of Deuterostomia could have diverged from the ancestors of Lophotrochozoa and Ecdysozoa 570 million years ago; it could have then taken 35 million years for the clade Deuterostomia to originate, but only 10 million years for first Ecdysozoa and then the arthropod clades to originate.

Self-Quiz

1. a **2.** d **3.** b **4.** e **5.** c **6.** e

8.

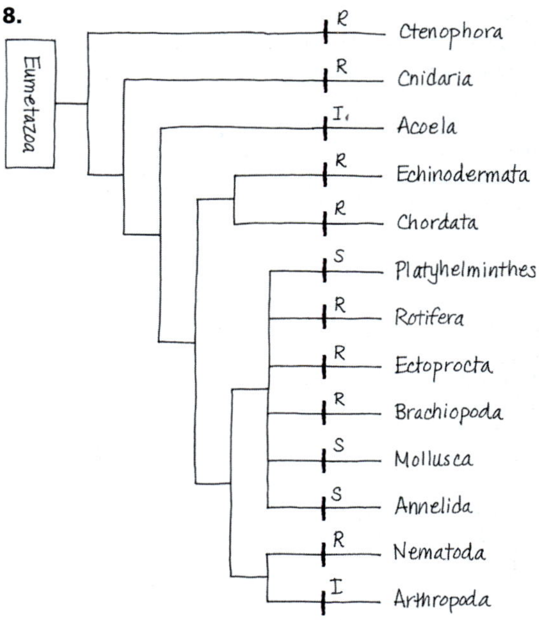

From the phylogeny, it appears that radial cleavage is the ancestral condition for eumetazoans. However, because relationships within Lophotrochozoa are not resolved, we cannot estimate the precise number of times that cleavage patterns have changed over the course of evolution. If, for example, Platyhelminthes, Mollusca, and Annelida form a clade, it would be reasonable to infer three cleavage pattern changes (one in Acoela, one in the ancestor of this hypothetical clade, and one in Arthropoda). Various other possible relationships among lophotrochozoans lead to other estimates.

CHAPTER 33

Figure Questions

Figure 33.8 Within a reproductive polyp, a cell that gives rise to a medusa would have to divide by meiosis. A resulting haploid cell would then divide repeatedly (by mitosis), forming a haploid medusa. Later, cells in the medusa's gonads would divide by mitosis, forming the haploid eggs and sperm. **Figure 33.11** Adding fertilizer to the water supply would probably increase the abundance of algae. This, in turn, might increase the abundance of both snails (which eat algae) and blood flukes (which require snails as an intermediate host). As a result, the occurrence of schistosomiasis might increase. **Figure 33.28** Such a result would be consistent with the *Ubx* and *abd-A Hox* genes having played a major role in the evolution of increased body segment diversity in arthropods. However, by itself, such a result would simply show that the presence of the *Ubx* and *abd-A Hox* genes was *correlated with* an increase in body segment diversity in arthropods; it would not provide direct experimental evidence that the acquisition of the *Ubx* and *adb-A* genes *caused* an increase in arthropod body segment diversity.

Concept Check 33.1

1. The flagella of choanocytes draw water through their collars, which trap food particles. The particles are engulfed by phagocytosis and digested, either by choanocytes or by amoebocytes. **2.** The collar cells of sponges (and other animals—see Chapter 32) bear a striking resemblance to a choanoflagellate cell. This suggests that the last common ancestor of animals and their protist sister group may have resembled a choanoflagellate. Nevertheless, mesomycetozoans could still be the sister group of animals. If this is the case, the lack of collar cells in mesomycetozoans would indicate that over time their structure evolved in ways that caused it to no longer resemble a choanoflagellate cell.

Concept Check 33.2

1. Both the polyp and the medusa are composed of an outer epidermis and an inner gastrodermis separated by a gelatinous layer, the mesoglea. The polyp is a cylindrical form that adheres to the substrate by its aboral end; the medusa is a flattened, mouth-down form that moves freely in the water. **2.** Cnidarian stinging cells (cnidocytes) function in defense and prey capture. They contain capsule-like organelles (cnidae), which in turn contain coiled threads. The threads either inject poison or stick to and entangle small prey. **3.** This would suggest that the life cycle of basal cnidarians was probably dominated by the medusa stage. Over time, the polyp stage came to be increasingly important in some groups, such as Hydrozoa, which alternate between medusa and polyp stages, and Anthozoa, which lack the medusa stage entirely.

Concept Check 33.3

1. Tapeworms can absorb food from their environment and release ammonia into their environment through their body surface because their body is very flat, due in part to the lack of a coelom. **2.** The function of the foot reflects the locomotion required in each class. Gastropods use their foot as a holdfast or to move slowly on the substrate. In cephalopods, the foot functions as a siphon and tentacles. **3.** The inner tube is the alimentary canal, which runs the length of the body. The outer tube is the body wall. The two tubes are separated by the coelom. **4.** Many lophotrochozoans lack skeletons or other structures that could support their soft bodies against the force of gravity, making it difficult for them to live above the surface of the soil. Some species, such as ectoprocts (bryozoans), have a sturdy exoskeleton, but they are stationary and so would find it difficult to capture food on land. (Note that those lophotrochozoans that do live above the soil surface, such as slugs, have some form of hydrostatic skeleton.)

Concept Check 33.4

1. Nematodes lack body segments and a true coelom; annelids have both. **2.** Arthropod mouthparts are modified appendages, which are bilaterally paired. **3.** The arthropod exoskeleton, which had already evolved in the ocean, allowed terrestrial species to retain water and support their bodies on land. Wings allowed them to disperse quickly to new habitats and to find food and mates. The tracheal system allows for efficient gas exchange despite the presence of an exoskeleton. **4.** Yes. Under the traditional hypothesis, we would expect body segmentation to be controlled by similar *Hox* genes in annelids and arthropods. However, if annelids are in Lophotrochozoa and arthropods are in Ecdysozoa, body segmentation may have evolved independently in these two groups. In such a case, we might expect that different *Hox* genes would control the development of body segmentation in the two clades.

Concept Check 33.5

1. Each tube foot consists of an ampulla and a podium. When the ampulla squeezes, it forces water into the podium, which causes the podium to expand and contact the substrate. Adhesive chemicals are then secreted from the base of the podium, thereby attaching the podium to the substrate. **2.** These two organisms look very different from one another, but they share features found in all echinoderms, such as a water vascular system and tube feet. Hence, their shared characteristics probably result from homology, not analogy. **3.** Both insects and nematodes are members of Ecdysozoa, one of the three major clades of bilaterians. Therefore, a characteristic shared by *Drosophila* and *Caenorhabditis* may be informative for other members of their clade—but not necessarily for members of Deuterostomia. Instead, Figure 33.2 suggests that a species within Echinodermata or Chordata might be a more appropriate invertebrate model organism from which to draw inferences about humans and other vertebrates.

Self-Quiz

1. c **2.** a **3.** d **4.** e **5.** b **6.** e
7.

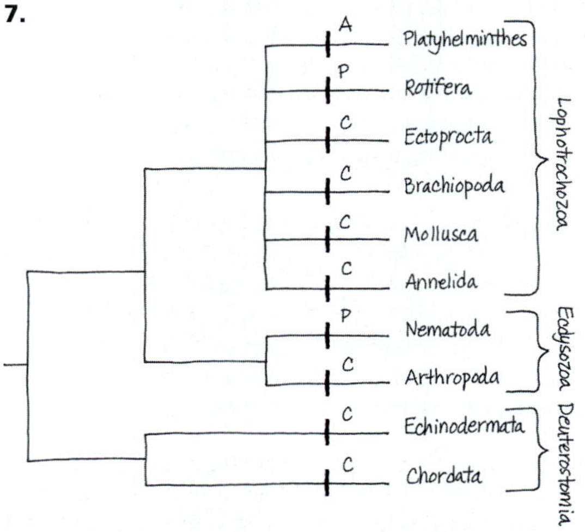

(a) Both phyla in Deuterostomia are coelomates, suggesting that their most recent common ancestor had a true coelom. Lophotrochozoa contains one phylum of acoelomates (Platyhelminthes), one phylum of pseudocoelomates (Rotifera), and four phyla of coelomates (Ectoprocta, Brachiopoda, Mollusca, Annelida); thus, we cannot from this information alone infer the condition of the most recent common ancestor shared by these phyla. Similarly, since Ecdysozoa contains one phylum of pseudocoelomates (Nematoda) and one phylum of coelomates (Arthropoda), we cannot infer whether their most recent common ancestor had a true coelom or not. (b) Depending on whether or not the last common ancestor of Bilateria had a true coelom, the presence of a true coelom has either been lost or gained multiple times during the evolutionary history of bilaterians. Thus, the presence of a true coelom appears to change over the course of evolution.

CHAPTER 34

Figure Questions

Figure 34.20 Amphibians must have originated some time between the date that the most recent common ancestor of *Hynerpeton* and later tetropods originated (380 mya) and the date of the earliest known fossils of amphibians (shown in the figure as 340 mya). **Figure 34.37** The phylogeny shows humans as the sister group to the genus *Pan*. This relationship is consistent with humans being placed in *Pan* along with its two living members, chimpanzees and bonobos. **Figure 34.43** It is not likely that these two sources of error significantly influenced the results. We can conclude this in part because the results were reproducible: similar sequences were found for mtDNA obtained from two different Neanderthal fossils and sequenced by two different research teams. In addition, the close relationship of the two Neanderthal mtDNA sequences to each other would not be expected if the fossil DNA had broken down considerably. Similarly, the fact that Europeans and other living humans formed a sister group to the Neanderthals, and that chimpanzees formed a sister group to the human/Neanderthal clade also would not be expected had the DNA broken down greatly—nor would these results be expected if the fossil DNA sequences were contaminated (for example, by DNA from microorganisms or from living humans).

Concept Check 34.1

1. As water passes through the slits, food particles are filtered from the water and transported to the digestive system. **2.** In humans, these characters are present only in the embryo. The notochord becomes disks between the vertebrae, the tail is almost completely lost, and the pharyngeal clefts develop into various adult structures. **3.** Not necessarily. It would be possible that the chordate common ancestor had this gene, which was then lost in the lancelet lineage and retained in other chordates. However, it would also be possible that the chordate common ancestor lacked this gene—this could occur if the gene originated after lancelets diverged from other chordates yet before tunicates diverged from other chordates.

Concept Check 34.2

1. Hagfishes have a head and skull made of cartilage, plus a small brain, sensory organs, and tooth-like structures. They have a neural crest, gill slits, and more extensive organ systems. In addition, hagfishes have slime glands that ward off predators and may repel competing scavengers. **2.** *Myllokunmingia*. Fossils of this organism provide evidence of ear capsules and eye capsules; these structures are part of the skull. Thus, *Myllokunmingia* is considered a craniate, as are humans. *Haikouella* did not have a skull. **3.** Such a finding suggests that early organisms with a head were favored by natural selection in several different evolutionary lineages. However, while a logical argument can be made that having a head was advantageous, fossils alone do not constitute proof.

Concept Check 34.3

1. Lampreys have a round, rasping mouth, which they use to attach to fish. Conodonts had two sets of mineralized dental elements, which may have been used to impale prey and cut it into smaller pieces. **2.** In armored jawless vertebrates, bone served as external armor that may have provided protection from predators. Some species also had mineralized mouthparts, which could be used for either predation or scavenging. Still others had mineralized fin rays, which may have enabled them to swim more rapidly and with greater steering control.

Concept Check 34.4

1. Both are gnathostomes and have jaws, four clusters of *Hox* genes, enlarged forebrains, and lateral line systems. Shark skeletons consist mainly of cartilage, whereas tuna have bony skeletons. Sharks also have a spiral valve. Tuna have an operculum and a swim bladder, as well as flexible rays supporting their fins. **2.** Aquatic gnathostomes have jaws (an adaptation for feeding) and paired fins and a tail (adaptations for swimming). Aquatic gnathostomes also typically have streamlined bodies for efficient swimming and swim bladders or other mechanisms (such as oil storage in sharks) for buoyancy. **3.** Yes, that could have happened. The paired appendages of aquatic gnathostomes other than the lobe-fins could have served as a starting point for the evolution of limbs. The colonization of land by aquatic gnathostomes other than the lobe-fins might have been facilitated in lineages that possessed lungs, as that would have enabled those organisms to breathe air.

Concept Check 34.5

1. Tetrapods are thought to have originated about 360 million years ago when the fins of some lobe-fins evolved into the limbs of tetrapods. In addition to their four limbs—a key derived trait for which the group is named—other derived traits of tetrapods include a neck (consisting of vertebrae that separate the head from the rest of the body), a pelvic girdle that is fused to the backbone, and a lack of gill slits. **2.** Some fully aquatic species are paedomorphic, retaining larval features for life in water as adults. Species that live in dry environments may avoid dehydration by burrowing or living under moist leaves, and they protect their eggs with foam nests, viviparity, and other adaptations. **3.** Many amphibians spend part of their life cycle in aquatic environments and part on land. Thus, they may be exposed to a wide range of environmental problems, including water and air pollution and the loss or degradation of aquatic and/or terrestrial habitats. In addition, amphibians have highly permeable skin, providing relatively little protection from external conditions, and their eggs do not have a protective shell.

Concept Check 34.6

1. The amniotic egg provides protection to the embryo and allows the embryo to develop on land, eliminating the necessity of a watery environment for reproduction. Another key adaptation is rib cage ventilation, which improves the efficiency of air intake and may have allowed early amniotes to dispense with breathing through their skin. And not breathing through their skin allowed amniotes to develop relatively impermeable skin, thereby conserving water. **2.** Birds have weight-saving modifications, including the absence of teeth, a urinary bladder, and a second ovary in females. The wings and feathers are adaptations that facilitate flight, and so are efficient respiratory and circulatory systems that support a high metabolic rate.

3.

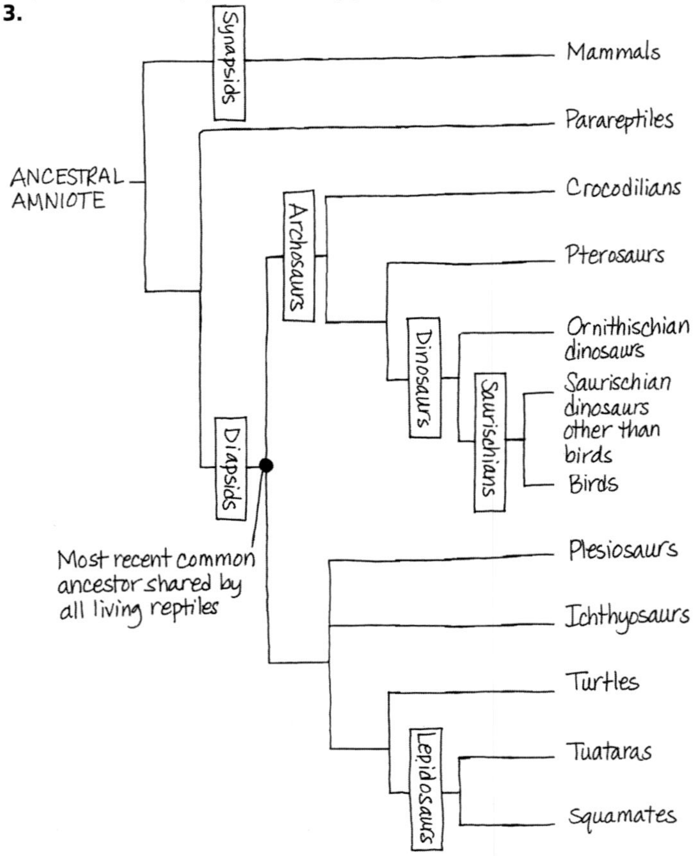

Under this convention, the reptiles would consist of all groups in Figure 34.24 except parareptiles and mammals.

Concept Check 34.7

1. Monotremes lay eggs. Marsupials give birth to very small live young that attach to a nipple in the mother's pouch, where they complete development. Eutherians give birth to more developed live young. **2.** Hands and feet adapted for grasping, flat nails, large brain, forward-looking eyes on a flat face, parental care, moveable big toe and thumb. **3.** Mammals are endothermic, enabling them to live in a wide range of habitats. Milk provides young with a balanced set of nutrients, and hair and a layer of fat under the skin help mammals retain heat. Mammals have differentiated teeth, enabling them to eat many different kinds of food. Mammals also have relatively large brains, and many species are capable learners. Following the mass extinction at the end of the Cretaceous period, the absence of large terrestrial dinosaurs may have opened many new ecological niches to mammals, promoting their adaptive radiation. Continental drift also isolated many groups of mammals from one another, promoting the formation of many new species.

Concept Check 34.8

1. Hominins are a clade within the ape clade that includes humans and all species more closely related to humans than other apes. The derived characters of hominins include bipedal locomotion and relatively larger brains. **2.** In hominins, bipedal locomotion evolved long before large brain size. *Homo ergaster*, for example, was fully upright, bipedal, and as tall as modern humans, but its brain was significantly smaller than that of modern humans. **3.** Yes, both can be correct. *Homo sapiens* may have established populations outside of Africa as early as 115,000 years ago, as indicated by the fossil record. However, those populations may have left few or no descendants today. Instead, all living humans may have descended from Africans that spread from Africa roughly 50,000 years ago, as indicated by genetic data.

Self-Quiz

1. e **2.** c **3.** a **4.** d **5.** b **6.** c **7.** c

9. (a) Because brain size tends to increase consistently in such lineages, we can conclude that natural selection favored the evolution of larger brains and hence that the benefits outweighed the costs. (b) As long as the benefits of brains that are large relative to body size are greater than the costs, large brains can evolve. Natural selection might favor the evolution of brains that are large relative to body size because such brains confer an advantage in obtaining mates and/or an advantage in survival.

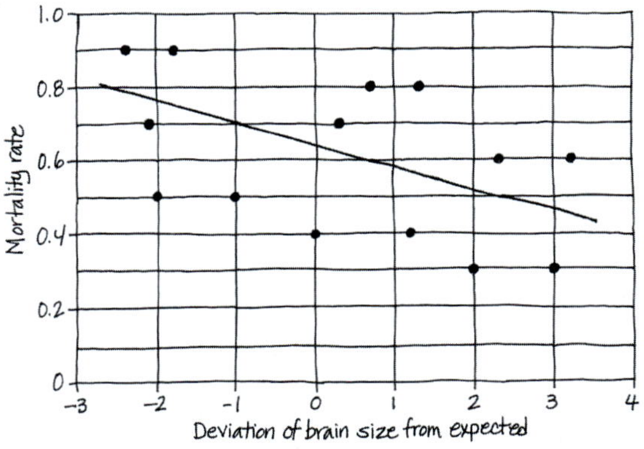

(c) Adult mortality tends to be lower in birds with larger brains.

CHAPTER 35

Figure Questions

Figure 35.9 The finding might suggest that the tawny-colored trichomes deter the beetles by some means other than physically obstructing the beetles. Perhaps they contain a chemical that is harmful or distasteful to the beetles, or their color is a deterrent. **Figure 35.17** Pith and cortex are defined, respectively, as ground tissue that is internal and external to vascular tissue. Since vascular bundles of monocot stems are scattered throughout the ground tissue, there is no clear distinction between internal and external relative to the vascular tissue. **Figure 35.19** The vascular cambium produces growth that increases the diameter of a stem or root. The tissues that are exterior to the vascular cambium cannot keep pace with the growth because their cells no longer divide. As a result, these tissues rupture. **Figure 35.31** Every root epidermal cell would develop a root hair.

Concept Check 35.1

1. The vascular tissue system connects leaves and roots, allowing sugars to move from leaves to roots in the phloem and allowing water and minerals to move to the leaves in the xylem. **2.** (a) large axillary buds; (b) petioles; (c) storage leaves; (d) storage roots **3.** The dermal tissue system is the leaf's protective covering. The vascular tissue system consists of the transport tissues xylem and phloem. The ground tissue system performs metabolic functions such as photosynthesis. **4.** Here are a few examples: The tubular, hollow structures of the tracheids and vessel elements of the xylem and the sieve plates in the sieve-tube elements of the phloem facilitate transport. Root hairs aid in absorption of water and nutrients. The cuticle in leaves and stems protects these structures from desiccation and pathogens. Leaf trichomes protect from herbivores and pathogens. Collenchyma and sclerenchyma cells have thick walls that provide support for plants. **5.** To get sufficient energy from photosynthesis, we would need lots of surface area exposed to the sun. This large surface-to-volume ratio, however, would create a new problem—evaporative water loss. We would have to be permanently connected to a water source—the soil, also our source of minerals. In short, we would probably look and behave very much like plants.

Concept Check 35.2

1. Primary growth arises from apical meristems and involves production and elongation of organs. Secondary growth arises from lateral meristems and adds to the girth of roots and stems. **2.** Your dividing cells are normally limited in the types of cells they can form. In contrast, the products of cell division in a plant meristem can differentiate into all the types of plant cells. **3.** The largest, oldest leaves would be lowest on the shoot. Since they would probably be heavily shaded, they would not photosynthesize much regardless of their size. **4.** No, the radish roots will probably be smaller at the end of the second year because the food stored in the root will be used to produce flowers, fruits, and seeds.

Concept Check 35.3

1. Lateral roots emerge from the root's interior (from the pericycle), pushing through cortical and epidermal cells. In contrast, shoot branches arise on the exterior of a shoot (from axillary buds). **2.** In roots, primary growth occurs in three successive stages, moving away from the tip of the root: the zones of cell division, elongation, and differentiation. In shoots, it occurs at the tip of apical buds, with leaf primordia arising along the sides of an apical meristem. Most growth in length occurs in older internodes below the shoot tip. **3.** Grazing animals that crop plants close to the ground have more of a detrimental effect on eudicots than on monocots because the removal of the lowest axillary buds prevents the eudicot from recovering. In contrast, the underground stems of grasses and the intercalary meristems of their leaves are less affected by grazing. Thus, the presence of grazing animals selects for the survival of grasses. **4.** No. Because vertically oriented leaves can capture light equally on both sides of the leaf, you would expect them to have mesophyll cells that are not differentiated into palisade and spongy layers. This is typically the case. Also, vertical leaves usually have stomata on both leaf surfaces.

Concept Check 35.4

1. The sign will still be 2 m above the ground because this part of the tree is no longer growing in length (primary growth); it is now growing only in thickness (secondary growth). **2.** Stomata must be able to close because evaporation is much more intensive from leaves than from the trunks of woody trees as a result of the higher surface-to-volume ratio in leaves. **3.** The growth rings of a tree from the tropics would be difficult to discern unless the tree came from an area that had pronounced wet and dry seasons. **4.** Girdling removes an entire ring of secondary phloem (part of the bark), completely preventing transport of sugars and starches from the shoots to the roots.

Concept Check 35.5

1. *Arabidopsis* is a small, easy-to-grow plant with a small genome and a short generation time. **2.** Differential gene expression **3.** In *fass* mutants, the arrangement of microtubules is disrupted so that the preprophase band does not form. This results in random planes of cell division, rather than the ordered planes of division that normally occur. Disruption of microtubule organization also prevents the alignment of cellulose microfibrils that sets the plane of cell elongation. Because of this randomness, directional growth is disrupted, and the plant becomes stubby. **4.** In theory, tepals could arise if *B* gene activity was present in all three of the outer whorls of the flower.

Self-Quiz

1. d **2.** c **3.** c **4.** d **5.** a **6.** e **7.** d **8.** d **9.** b **10.** b

11.

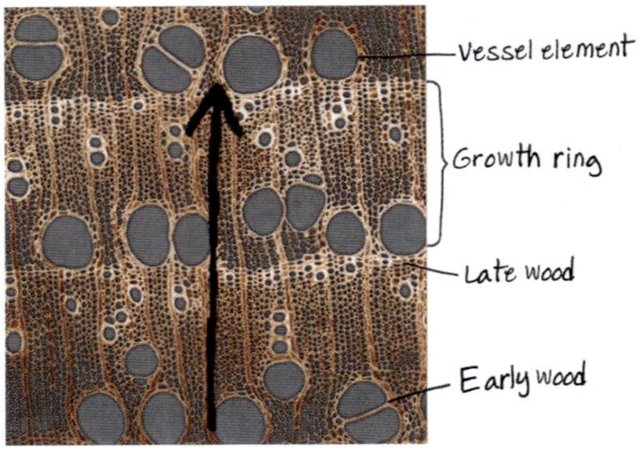

CHAPTER 36

Figure Questions

Figure 36.3 The leaves are being produced in a counterclockwise spiral. **Figure 36.4** A higher leaf area index will not necessarily increase photosynthesis because of upper leaves shading lower leaves. **Figure 36.12** The Casparian strip blocks water and minerals from moving between endodermal cells or moving around an endodermal cell via the cell's wall. Therefore, water and minerals must pass through an endodermal cell's plasma membrane. **Figure 36.21** Because the xylem is under negative pressure (tension), an isolated stylet inserted into a tracheid or vessel element would probably introduce air into the cell. No xylem sap would exude unless pressure was predominant. **Figure 36.22** Such a finding (although not considered likely) would cast doubts on the interpretation of the experiment. If the small fluorescent molecule was cleaved off the larger molecular probe, this small molecule could move through the plasmodesmata without them being dilated.

Concept Check 36.1

1. Vascular plants must transport minerals and water absorbed by the root to all the other parts of the plant. They must also transport sugars from sites of production to sites of use. **2.** Many features of plant architecture affect self-shading, including leaf arrangement, leaf orientation, and leaf area index. **3.** The fungicide may kill the mycorrhizal fungi that help the plants absorb phosphate and other minerals. **4.** Increased stem elongation would raise the plant's upper leaves. Erect leaves and reduced lateral branching would make the plant less subject to shading by the encroaching neighbors. **5.** As described in Chapter 35, pruning the shoot tips will end their apical dominance, allowing axillary buds to grow into lateral shoots (branches). This branching will produce a bushier plant with a higher leaf area index.

Concept Check 36.2

1. The cell's ψ_P is 0.7 MPa. In a solution with a ψ of -0.4 MPa, the cell's ψ_P at equilibrium would be 0.3 MPa. **2.** The cells would still adjust to changes in their osmotic environment, but their responses would be slower. Although aquaporins do not affect the water potential gradient across membranes, they allow for more rapid osmotic adjustments. **3.** If tracheids and vessel elements were living cells, their cytoplasm would impede water movement, preventing rapid long-distance transport. **4.** The protoplasts would burst. Because the cytoplasm has many dissolved solutes, water would enter the protoplast continuously without reaching equilibrium. (When present, the cell wall prevents rupturing by excessive expansion of the protoplast.)

Concept Check 36.3

1. Because water-conducting xylem cells are dead at maturity and form essentially hollow tubes, they offer little resistance to water flow, and their thick walls prevent the cells from collapsing from the negative pressure inside. **2.** At dawn, a drop is exuded because the xylem is under positive pressure due to root pressure. At noon, the xylem is under negative pressure potential due to transpiration and the root pressure cannot keep pace with the increased rate of transpiration. **3.** The endodermis regulates the passage of water-soluble solutes by requiring all such molecules to cross a selectively permeable membrane. Presumably the inhibitor never reaches the plant's photosynthetic cells. **4.** Perhaps greater root mass helps compensate for the lower water permeability of the plasma membranes.

Concept Check 36.4

1. Stomatal aperture is controlled by drought, light, CO_2 concentrations, a circadian rhythm, and the plant hormone abscisic acid. **2.** The activation of the proton pump of stomatal cells would cause the guard cells to take up K^+. The increased turgor of the guard cells would lock the stomata open and lead to extreme evaporation from the leaf. **3.** After the flowers are cut, transpiration from any leaves and from the petals (which are modified leaves) will continue to draw water up the xylem. If cut flowers are transferred directly to a vase, air pockets in xylem vessels prevent delivery of water from the vase to the flowers. Cutting stems again underwater, a few centimeters from the original cut, will sever the xylem above the air pocket. The water droplets prevent another air pocket from forming while placing the flowers in a vase.

Concept Check 36.5

1. In both cases, the long-distance transport is a bulk flow driven by a pressure difference at opposite ends of tubes. Pressure is generated at the source end of a sieve tube by the loading of sugar and resulting osmotic flow of water into the phloem, and this pressure *pushes* sap from the source end to the sink end of the tube. In contrast, transpiration generates a negative pressure potential (tension) as a force that *pulls* the ascent of xylem sap. **2.** The main sources are fully grown leaves (by photosynthesis) and fully developed storage organs (by breakdown of starch). Roots, buds, stems, expanding leaves, and fruits are powerful sinks because they are actively growing. A storage organ may be a sink in the summer when accumulating carbohydrates, but a source in the spring when breaking down starch into sugar for growing shoot tips. **3.** Positive pressure, whether it be in the xylem when root pressure predominates, or in the sieve-tube elements of the phloem, requires active transport. Most long-distance transport in the xylem depends on bulk flow driven by negative pressure potential generated ultimately by the evaporation of water from the leaf and does not require living cells. **4.** The spiral slash prevents optimal bulk flow of the phloem sap to the root sinks. Therefore, more phloem sap can move from the source leaves to the fruit sinks, making them sweeter.

Concept Check 36.6

1. Voltage between cells, cytoplasmic pH, cytoplasmic calcium, and movement proteins all affect symplastic communication, as do developmental changes in plasmodesmatal number. **2.** Plasmodesmata, unlike gap junctions, have the ability to pass RNA, proteins, and viruses from cell to cell. **3.** Although this strategy would eliminate the systemic spread of viral infections, it would also severely impact the development of the plants.

Self-Quiz

1. d **2.** e **3.** c **4.** b **5.** a **6.** d **7.** c **8.** c **9.** b **10.** c

11.

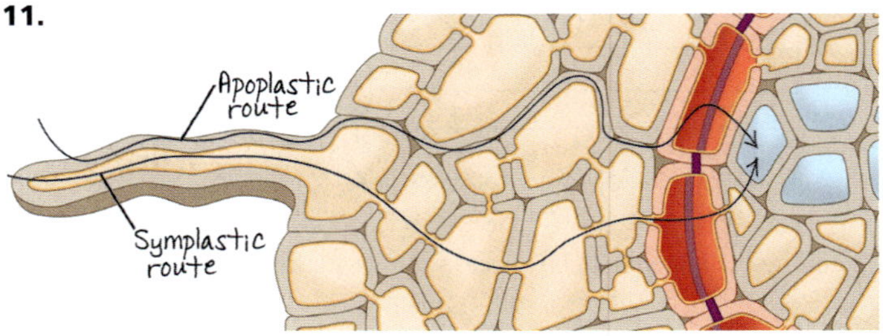

CHAPTER 37

Figure Questions

Figure 37.3 Anions. Because cations are bound to soil particles, they are less likely to be lost from the soil following heavy rains. **Figure 37.10** The legume plants benefit because the bacteria fix nitrogen that is absorbed by their roots. The bacteria benefit because they acquire photosynthetic products from the plants. **Figure 37.11** All three plant tissue systems are affected. Root hairs (dermal tissue) are modified to allow rhizobial penetration. The cortex (ground tissue) and pericycle (vascular tissue) proliferate during nodule formation. The vascular tissue of the nodule links up with the vascular cylinder of the root to allow for efficient nutrient exchange. **Figure 37.13** If phosphate were the only limiting mineral, then native tree growth would be less severely impacted by the reduction in mycorrhizal associations in soils invaded by garlic mustard.

Concept Check 37.1

1. Overwatering deprives roots of oxygen. Overfertilizing is wasteful and can lead to soil salinization and water pollution. **2.** As lawn clippings decompose, they restore mineral nutrients to the soil. If they are removed, the minerals lost from the soil must be replaced by fertilization. **3.** Because of their small size and negative charge, clay particles would increase the number of binding sites for cations and water molecules and would therefore increase cation exchange and water retention in the soil.

Concept Check 37.2

1. Table 37.1 shows that CO_2 is the source of 90% of a plant's dry weight, supporting Hales's view that plants are nourished mostly by air. **2.** No, because even though macronutrients are required in greater amounts, all essential elements are necessary for the plant to complete its life cycle. **3.** No. Most plants can complete their life cycles in the absence of silicon. Therefore, by definition, it is not an essential nutrient.

Concept Check 37.3

1. The rhizosphere is a narrow zone in the soil immediately adjacent to living roots. This zone is especially rich in both organic and inorganic nutrients and has a microbial population that is many times greater than the bulk of the soil. **2.** Soil bacteria and mycorrhizae enhance plant nutrition by making certain minerals more available for plants. For example, many types of soil bacteria are involved in the nitrogen cycle, whereas the hyphae of mycorrhizae provide a large surface area for the absorption of nutrients, particularly phosphate ions. **3.** Saturating rainfall may deplete the soil of oxygen. A lack of soil oxygen would inhibit nitrogen fixation by the peanut root nodules and decrease the nitrogen available to the plant. Alternatively, heavy rain may leach nitrate from the soil. A symptom of nitrogen deficiency is yellowing of older leaves.

Self-Quiz

1. b **2.** b **3.** c **4.** b **5.** b **6.** c **7.** d **8.** a **9.** d **10.** b
11.

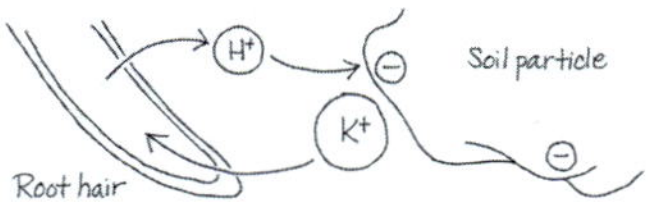

CHAPTER 38

Figure Questions

Figure 38.4 Having a specific pollinator is more efficient because less pollen gets delivered to flowers of the wrong species. However, it is also a risky strategy: If the pollinator population suffers to an unusual degree from predation, disease, or climate change, then the plant may not be able to produce seeds. **Figure 38.6** An inability to produce GABA would also prevent the establishment of a GABA gradient to help direct pollen tube growth. Thus, these mutants would be sterile also. **Figure 38.9** Beans use a hypocotyl hook to push through the soil. The delicate leaves and shoot apical meristem are also protected by being sandwiched between two large cotyledons. The coleoptile of maize seedlings helps protect the emerging leaves.

Concept Check 38.1

1. In angiosperms, pollination is the transfer of pollen from an anther to a stigma. Fertilization is the fusion of the egg and sperm to form the zygote; it cannot occur until after the growth of the pollen tube from the pollen grain. **2.** Seed dormancy prevents the premature germination of seeds. A seed will germinate only when the environmental conditions are optimal for the survival of its embryo as a young seedling. **3.** The fruit types are not completely separate categories because the term *accessory fruit* applies to any fruit that develops not only from one or more carpels but also from additional floral parts. Therefore, a simple, aggregate, or multiple fruit can also be an accessory fruit. The terms *simple, aggregate,* and *multiple* refer only to the number of carpels and flowers from which the fruit develops. **4.** Long styles help to weed out pollen grains that are genetically inferior and not capable of successfully growing long pollen tubes.

Concept Check 38.2

1. Sexual reproduction produces genetic variety, which may be advantageous in an unstable environment. The likelihood is better that at least one offspring of sexual reproduction will survive in a changed environment. Asexual reproduction can be advantageous in a stable environment because individual plants that are well suited to that environment pass on all their genes to offspring. Asexual reproduction also generally results in offspring that are less fragile than the seedlings produced by sexual reproduction. However, sexual reproduction offers the advantage of dispersal of tough seeds. **2.** Asexually propagated crops lack genetic diversity. Genetically diverse populations are less likely to become extinct in the face of an epidemic because there is a greater likelihood that a few individuals in the population are resistant. **3.** In the short term, selfing may be advantageous in a population that is so dispersed and sparse that pollen delivery is unreliable. In the long term, however, selfing is an evolutionary dead end because it leads to a loss of genetic diversity that may preclude adaptive evolution. **4.** This might be possible, but satisfactory results would be very unlikely. Both tubers and fruits are tremendous energy sinks. Each plant has only a finite amount of energy to divide between sexual and asexual reproduction. Although a tomato-potato hybrid could, in theory, produce an offspring that makes fruits and tubers equally, these fruits and tubers would be of inferior quality or low yielding.

Concept Check 38.3

1. Traditional breeding and genetic engineering both involve artificial selection for desired traits. However, genetic engineering techniques facilitate faster gene transfer and are not limited to transferring genes between closely related varieties or species. **2.** GM crops may be more nutritious and less susceptible to insect damage or pathogens that invade insect-damaged plants. They also may not require as much chemical spraying. However, unknown risks may include adverse effects on human health and nontarget organisms and the possibility of transgene escape. **3.** *Bt* maize suffers less insect damage; therefore, *Bt* maize plants are less likely to be infected by fumonisin-producing fungi that infect plants through wounds. **4.** In such species, engineering the transgene into the chloroplast DNA would not prevent its escape in pollen; such a method requires that the chloroplast DNA be found only in the egg. An entirely different method of preventing transgene escape would therefore be needed, such as male sterility, apomixis, or self-pollinating closed flowers.

Self-Quiz

1. d **2.** c **3.** a **4.** c **5.** d **6.** c **7.** e **8.** a **9.** c **10.** e
11.

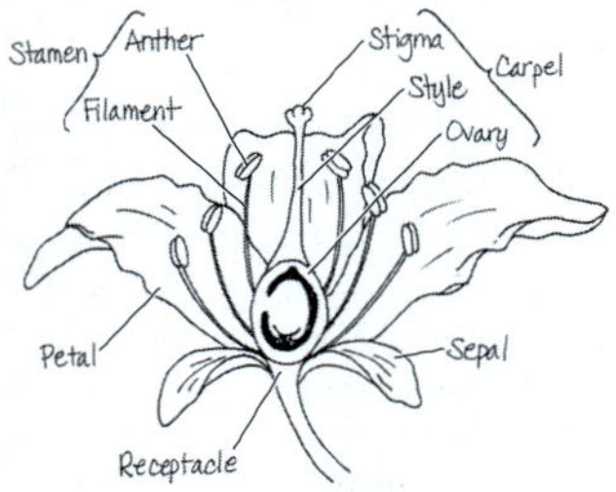

CHAPTER 39

Figure Questions

Figure 39.5 To determine which wavelengths of light are most effective in phototropism, one could use a glass prism to split white light into its component colors and see which colors cause the quickest bending (the answer is blue; see Figure 39.16). **Figure 39.6** The coleoptile would bend toward the side with the TIBA-containing agar bead. **Figure 39.7** No. Polar auxin transport depends on the polar distribution of auxin transport proteins. **Figure 39.17** Yes. The white light, which contains red light, would stimulate seed germination in all treatments. **Figure 39.22** The short-day plant would not flower. The long-day plant would flower. **Figure 39.23** If this were true, florigen would be an inhibitor of flowering, not an inducer.

Concept Check 39.1

1. Dark-grown seedlings have long stems, underdeveloped root systems, and unexpanded leaves, and their shoots lack chlorophyll. **2.** Etiolated growth is beneficial to seeds sprouting under the dark conditions they would encounter underground. By devoting more energy to stem elongation and less to leaf expansion and root growth, a plant increases the likelihood that the shoot will reach the sunlight before its stored foods run out. **3.** Cycloheximide should inhibit de-etiolation by preventing the synthesis of new proteins necessary for de-etiolation. **4.** No. Applying Viagra, like injecting cyclic GMP as described in the text, should cause only a partial de-etiolation response. Full de-etiolation would require activation of the calcium branch of the signal transduction pathway.

Concept Check 39.2

1. The release of ethylene by the damaged apple stimulates ripening in the other apples. **2.** Because cytokinins delay leaf senescence and floral parts are modified leaves, cytokinins also delay the senescence of cut flowers. **3.** Fusicoccin's ability to cause an increase in plasma H^+ pump activity is similar to an effect of auxin and leads to an auxin-like effect, a promotion of stem cell elongation. **4.** The plant will exhibit a constitutive triple response. Because the kinase that normally prevents the triple response is dysfunctional, the plant will undergo the triple response regardless of whether ethylene is present or the ethylene receptor is functional.

Concept Check 39.3

1. Not necessarily. Many environmental factors, such as temperature and light, change over a 24-hour period in the field. To determine whether the enzyme is under circadian control, the scientist would have to demonstrate that its activity oscillates even when environmental conditions are held constant. **2.** Flowering of the species may have been day-neutral or required multiple exposures to short nights. **3.** You might determine which wavelengths of light are most effective and plot an action spectrum. If the action spectrum indicates phytochrome, you could do further experiments to test for red/far-red photosensitivity. **4.** It is impossible to say. To establish that this species is a short-day plant, it would be necessary to establish the critical night length for flowering and that this species only flowers when the night is longer than the critical night length.

Concept Check 39.4

1. A plant that overproduces ABA would undergo less evaporative cooling because its stomata would not open as widely. **2.** Plants close to the aisles may be more subject to mechanical stresses caused by passing workers and air currents. The plants nearer to the center of the bench may also be taller as a result of shading and less evaporative stress. **3.** Like drought stress, freezing leads to cellular dehydration. Any process that helps mitigate drought stress will tend also to reduce freezing stress. **4.** No. Because root caps are involved in sensing gravity, roots that have their root caps removed are almost completely insensitive to gravity.

Concept Check 39.5

1. Some insects increase plants' productivity by eating harmful insects or aiding in pollination. **2.** Mechanical damage breaches a plant's first line of defense against infection, its protective dermal tissue. **3.** No. Pathogens that kill their hosts would soon run out of victims and might themselves go extinct. **4.** Perhaps the breeze dilutes the local concentration of a volatile defense compound that the plants produce.

Self-Quiz

1. a **2.** c **3.** d **4.** b **5.** e **6.** b **7.** b **8.** c **9.** e **10.** b
11.

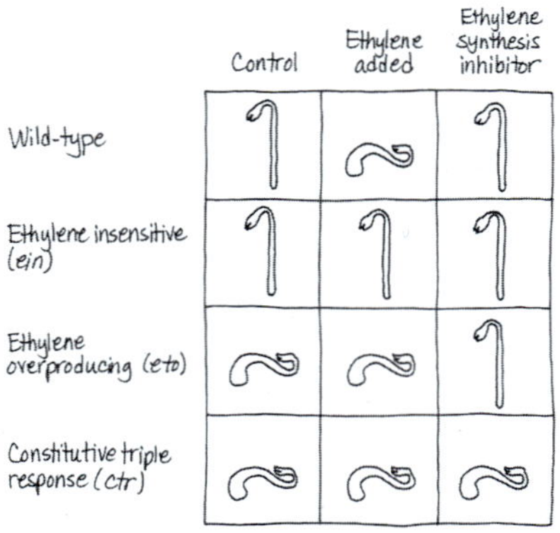

CHAPTER 40

Figure Questions

Figure 40.4 Such exchange surfaces are internal in the sense that they are inside the body. However, they are also continuous with openings on the external body surface that contact the environment. **Figure 40.8** The air conditioner would form a second negative-feedback loop, cooling the house when air temperature exceeded the set point. Such opposing, or antagonistic, pairs of negative-feedback loops increase the effectiveness of a homeostatic mechanism. **Figure 40.14** When a female Burmese python is not incubating eggs, her oxygen consumption will decrease with decreasing temperature, as for any other ectotherm. **Figure 40.21** If falling temperatures triggered hibernation, you would predict hibernation would begin earlier than normal. If another seasonal change, such as day length, controlled hibernation, its timing should be unaffected. By controlling these environmental variables in the laboratory, scientists have shown that lowering temperature without a change in day length is sufficient to induce ground squirrel hibernation.

Concept Check 40.1

1. Epithelial cells line a surface, are tightly packed, are situated on top of a basal lamina, and form an active and protective interface with the external environment. **2.** By flattening its ears along its body, the jackrabbit can reduce the total surface area of its body and hence the amount of heat absorbed when environmental temperatures are high, or the amount of heat lost when environmental temperatures are low. **3.** You need the nervous system to perceive the danger and provoke a split-second muscular response to keep from falling. The nervous system, however, does not make a direct connection with blood vessels or liver cells. Instead, the nervous system triggers the release of a hormone (called epinephrine or adrenaline) by the endocrine system, bringing about a change in these tissues in just a few seconds.

Concept Check 40.2

1. No; even though an animal regulates some aspects of its internal environment, the internal environment fluctuates slightly around set points. Homeostasis is a dynamic state. Furthermore, there are sometimes programmed changes in set points, such as those resulting in radical increases in hormone levels at particular times in development. **2.** In negative feedback, a change triggers control mechanisms that counteract further change in that direction. In positive feedback, a change triggers mechanisms that amplify the change. **3.** You would want to locate a thermostat close to where you would be spending time, protected from environmental perturbations, such as direct sunshine, and not right in the path of the output of the heating system. Similarly, the sensors for homeostasis located in the human brain are separated from environmental influences and can monitor conditions in a vital and sensitive tissue.

Concept Check 40.3

1. "Wind chill" involves heat loss through convection. **2.** The hummingbird, being a very small endotherm, has a very high metabolic rate. If by absorbing sunlight certain flowers warm their nectar, a hummingbird feeding on these flowers is saved the metabolic expense of warming the nectar to its body temperature. **3.** The ice water would cool tissues in your head, including blood that would then circulate throughout the body. This effect would accelerate the return to a normal body temperature. If, however, the ice water reached the eardrum and cooled the hypothalamic thermostat, the perceived drop in temperature would inhibit sweating and blood vessel dilation, slowing cooling elsewhere in the body.

Concept Check 40.4

1. The mouse would consume oxygen at a higher rate because it is an endotherm and therefore its basal metabolic rate is higher than the ectothermic lizard's standard metabolic rate. **2.** The house cat; the smaller an animal is, the higher its metabolic rate and its demand for food per unit of body mass are. **3.** Although penguins do not grow as adults, they increase and decrease in size as they repeatedly form and use energy stores. A significant amount of energy might be stored in fat during part of the year but be missing from the pie chart because it is used later in the year.

Self-Quiz

1. b **2.** e **3.** c **4.** e **5.** a **6.** d

7.

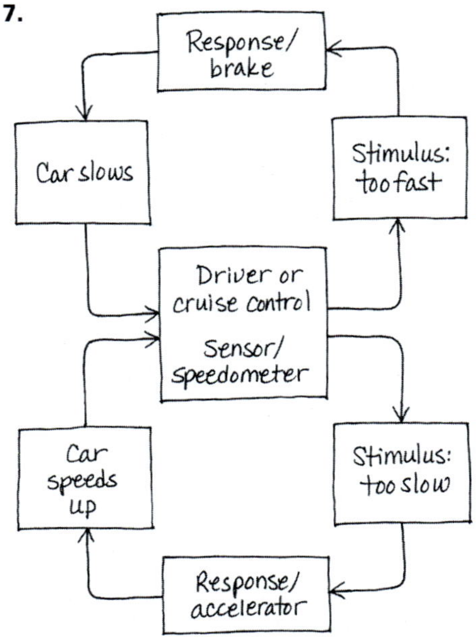

CHAPTER 41

Figure Questions

Figure 41.5 As in the described study, they needed a sample size large enough that they could expect a significant number of neural tube defects in the control group. The information needed to determine the appropriate sample size was the frequency of neural tube defects in first-time pregnancies in the general population. **Figure 41.13** Since enzymes are proteins, and proteins are hydrolyzed in the small intestine, the digestive enzymes in that compartment need to be resistant to cleavage by proteases other than the cleavage required for enzyme activation. **Figure 41.15** None. Since digestion is completed in the small intestine, tapeworms simply absorb predigested nutrients through their large body surface area. **Figure 41.24** The *db* mouse would have higher leptin levels. The wild-type mouse produces leptin after a meal. As the mouse depletes its fat stores, leptin production drops. The mouse eventually regains its appetite, eats another meal, and makes another burst of leptin. Because the *db* mouse cannot respond to leptin, its fat stores are constantly replenished through excessive consumption. As a result, leptin is produced continuously and builds up to a high concentration in the blood.

Concept Check 41.1

1. The only essential amino acids are those that an animal can't synthesize from other molecules containing carbon and nitrogen. **2.** Carbohydrates are needed throughout the body as a source of energy and carbon in the biosynthesis of cellular components, whereas vitamins typically serve as reusable enzyme cofactors or as raw materials for certain specialized cell structures. **3.** To identify the essential nutrient missing from an animal's diet, a researcher could supplement the diet with particular nutrients and determine which nutrient eliminates the signs of malnutrition.

Concept Check 41.2

1. A gastrovascular cavity is a digestive sac with a single opening that functions in both ingestion and elimination; an alimentary canal is a digestive tube with a separate mouth and anus at opposite ends. **2.** As long as nutrients are within the cavity of the alimentary canal, they are in a compartment that is continuous with the outside environment via the mouth and anus and have not yet crossed a membrane to enter the body. **3.** Just as food remains outside the body in a digestive tract, gasoline moves from the fuel tank to the engine and the exhaust without ever entering the passenger compartment of the car. In addition, gasoline, like food, is broken down in a specialized compartment, so that the rest of the body is protected from disassembly. In both cases, high-energy fuels are consumed and waste products are eliminated.

Concept Check 41.3

1. By peristalsis, which can squeeze food through the esophagus even without the help of gravity. **2.** Fats can cross the membranes of epithelial cells by diffusion, whereas proteins and sugars, which are not lipid-soluble, require transport or exchange proteins. **3.** Proteins would be denatured and digested into peptides. Further digestion, to individual amino acids, would require enzymatic secretions found in the small intestine.

Concept Check 41.4

1. The increased time for transit allows for more extensive processing, and the increased surface area provides greater opportunity for absorption. **2.** Mutualistic microbes in the intestines of vertebrates have an environment that is protected against other microbes by saliva and gastric juice, that is held at a constant temperature conducive to enzyme action, and that provides a steady source of nutrients. **3.** For the yogurt treatment to be effective, the bacteria from yogurt would have to establish a mutualistic relationship with the small intestine, where disaccharides are broken down and sugars are absorbed. Conditions in the small intestine are likely to be very different than in a yogurt culture. The bacteria might be killed before they reach the small intestine, or they might not be able to grow there in sufficient numbers to aid in digestion.

Concept Check 41.5

1. Over the long term, the body converts excess calories to fat, whether those calories are consumed as fat, carbohydrate, or protein. **2.** Both hormones have appetite-suppressing effects on the brain's satiety center. During the course of a day, PYY, secreted by the intestine, suppresses appetite after meals. Over the longer term, leptin, produced by adipose tissue, normally reduces appetite as fat storage increases. **3.** In normal individuals, leptin levels decline during fasting. The group with low levels of leptin are likely to be defective in leptin production, so leptin levels would remain low regardless of food intake. The group with high leptin levels are likely to be defective in responding to leptin, but they still should shut off leptin production as fat stores are used up.

Self-Quiz

1. e **2.** a **3.** c **4.** c **5.** c **6.** d **7.** e **8.** b

9.

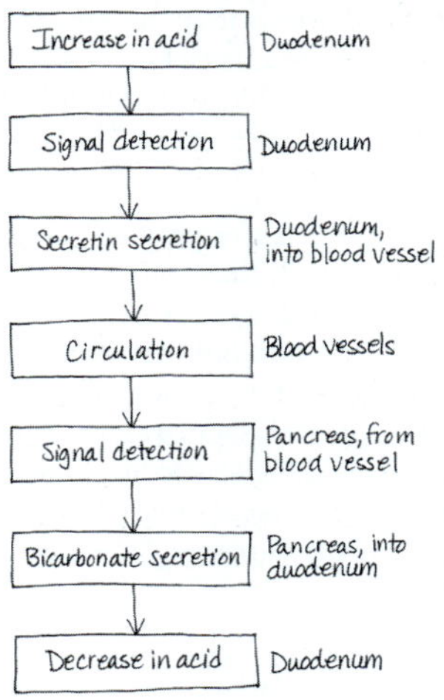

CHAPTER 42

Figure Questions

Figure 42.2 As the name indicates, a gastrovascular cavity functions in both digestion and circulation. Although gas exchange might be improved by a steady, one-way flow of fluid, there would likely be inadequate time for food to be digested and nutrients absorbed if fluids flowed through the cavity in this manner. **Figure 42.12** Because endothelin regulates the smooth muscle cells in blood vessels, you would expect it to be secreted from the basal surface of endothelial cells, a prediction that has been confirmed experimentally. **Figure 42.28** The resulting increase in tidal volume would enhance ventilation within the lungs, increasing P_{O_2} in the alveoli. **Figure 42.30** Some CO_2 is dissolved in plasma, some is bound to hemoglobin, and some is converted to bicarbonate ion (HCO_3^-), which is dissolved in plasma. **Figure 42.31** You might find some fast runners in those with the highest V_{O_2} max, but you might also find some sloths. There are two principal factors that contribute to V_{O_2} max, genetics and exercise. Elite athletes have a very high V_{O_2} max, reflecting not only the strengthening of their cardiovascular system through training but also the genetic circumstance of their having a particular combination of alleles.

Concept Check 42.1

1. In both an open circulatory system and a fountain, fluid is pumped through a tube and then returns to the pump after collecting in a pool. **2.** The ability to shut off blood supply to the lungs when the animal is submerged **3.** The O_2 content would be abnormally low because some oxygen-depleted blood returned to the right ventricle from the systemic circuit would mix with the oxygen-rich blood in the left ventricle.

Concept Check 42.2

1. The pulmonary veins carry blood that has just passed through capillary beds in the lungs, where it accumulated O_2. The venae cavae carry blood that has just passed through capillary beds in the rest of the body, where it lost O_2 to the tissues. **2.** The delay allows the atria to empty completely, filling ventricles fully before they contract. **3.** The heart, like any other muscle, becomes stronger through regular exercise. You would expect a stronger heart to have a greater stroke volume, which would allow for the decrease in heart rate.

Concept Check 42.3

1. The large total cross-sectional area of the capillaries **2.** An increase in blood pressure and cardiac output combined with the diversion of more blood to the skeletal muscles would increase the capacity for action by increasing the rate of blood circulation and delivering more O_2 and nutrients to the skeletal muscles. **3.** Additional hearts could be used to improve blood return from the legs. However, it might be difficult to coordinate the activity of multiple hearts and to maintain adequate blood flow to hearts far from the gas exchange organs.

Concept Check 42.4

1. An increase in the number of white blood cells (leukocytes) may indicate that the person is combating an infection. **2.** Clotting factors do not initiate clotting but are essential steps in the clotting process. Also, the clots that form a thrombus typically result from an inflammatory response to an atherosclerotic plaque, not from clotting at a wound site. **3.** The chest pain results from inadequate blood flow in coronary arteries. Vasodilation promoted by nitric oxide from nitroglycerin increases blood flow, providing the heart muscle with additional oxygen and thus relieving the pain.

Concept Check 42.5

1. Their interior position helps them stay moist. If the respiratory surfaces of lungs extended out into the terrestrial environment, they would quickly dry out, and diffusion of O_2 and CO_2 across these surfaces would stop. **2.** Earthworms need to keep their skin moist for gas exchange, but they need air outside this moist layer. If they stay in their waterlogged tunnels after a heavy rain, they will suffocate because they cannot get as much O_2 from water as from air. **3.** Since exhalation is largely passive, the recoil of the elastic fibers in alveoli helps force air out of the lungs. When alveoli lose their elasticity, as occurs in the disease emphysema, the volume of each breath decreases, lowering the efficiency of gas exchange.

Concept Check 42.6

1. An increase in blood CO_2 concentration causes an increase in the rate of CO_2 diffusion into the cerebrospinal fluid, where the CO_2 combines with water to form carbonic acid. Dissociation of carbonic acid releases hydrogen ions, decreasing the pH of the cerebrospinal fluid. **2.** Increased heart rate increases the rate at which CO_2-rich blood is delivered to the lungs, where CO_2 is removed. **3.** A hole would allow air to enter the space between the inner and outer layers of the double membrane, resulting in a condition called a pneumothorax. The two layers would no longer stick together, and the lung on the side with the hole would collapse and cease functioning.

Concept Check 42.7

1. Differences in partial pressure; gases diffuse from a region of higher partial pressure to a region of lower partial pressure. **2.** The Bohr shift causes hemoglobin to release more O_2 at a lower pH, such as found in the vicinity of tissues with high rates of cellular respiration and CO_2 release. **3.** The doctor is assuming that the rapid breathing is the body's response to low blood pH. Metabolic acidosis, the lowering of blood pH, can have many causes, including complications of certain types of diabetes, shock (extremely low blood pressure), and poisoning.

Self-Quiz

1. c **2.** b **3.** d **4.** e **5.** b **6.** c **7.** a **8.** a

9.

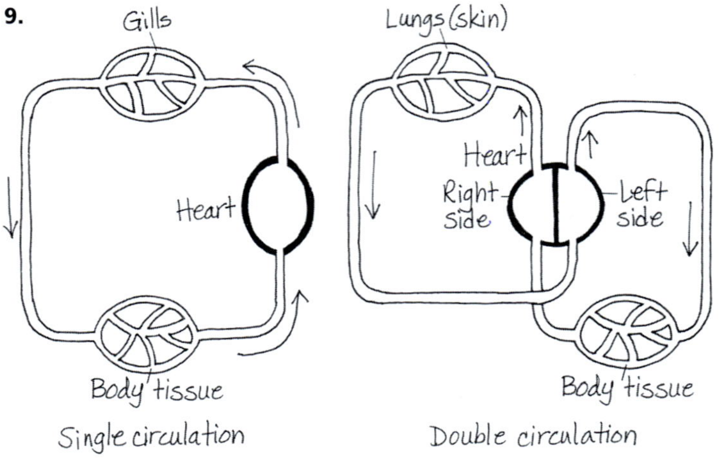

CHAPTER 43

Figure Questions

Figure 43.5 The seemingly inactive peptides might offer protection against pathogens other than those studied. Also, some antimicrobial peptides might work best in combination. **Figure 43.6** Cell surface TLRs recognize pathogens identifiable by surface molecules, whereas TLRs in vesicles recognize pathogens identifiable by internal molecules after the pathogens are broken down. **Figure 43.16** Primary response: arrows extending from Antigen (1st exposure), Antigen-presenting cell, Helper T cell, B cell, Plasma cells, Cytotoxic T cell, and Active Cytotoxic T cells; secondary response: arrows extending from Antigen (2nd exposure), Memory Helper T cells, Memory B cells, and Memory Cytotoxic T cells. **Figure 43.19** Before mounting a secondary immune response, memory cells must become activated by displaying antigen on the cell surface to a helper T cell.

Concept Check 43.1

1. A physical barrier often provides a very effective defense against infection. However, it is necessarily incomplete because animals need openings in their bodies for exchange with the environment. **2.** Because pus contains white blood cells, fluid, and cell debris, it indicates an active and at least partially successful inflammatory response against invading microbes. **3.** A microbe that grew optimally at low pH would be able to colonize the skin or stomach more readily. At the same time, it would not be well adapted to growth in other parts of the body.

Concept Check 43.2

1. See Figure 43.9a. All of the functions shared among receptors map to C regions, whereas the antigen-binding site maps to the V regions. **2.** Generating memory cells ensures both that a receptor specific for a particular epitope will be present and that there will be more lymphocytes with this specificity than in a host that had never encountered the antigen. **3.** If each B cell produced two different light and heavy chains for its antigen receptor, different combinations would make four different receptors. If any one was self-reactive, the lymphocyte would be eliminated in the generation of self-tolerance. For this reason, many more B cells would be eliminated, and those that could respond to a foreign antigen would be less effective at doing so due to the variety of receptors (and antibodies) they express.

Concept Check 43.3

1. A child lacking a thymus would have no functional T cells. Without helper T cells to help activate B cells, the child would be unable to produce antibodies against extracellular bacteria. Furthermore, without cytotoxic T cells or helper T cells, the child's immune system would be unable to kill virus-infected cells. **2.** Since the antigen-binding site is intact, the antibody fragments could neutralize viruses and opsonize bacteria. **3.** If the handler developed immunity to proteins in the antivenin, another injection could provoke a severe immune response. The handler's immune system might also now produce antibodies that could neutralize the venom.

Concept Check 43.4

1. Myasthenia gravis is considered an autoimmune disease because the immune system produces antibodies against self molecules (acetylcholine receptors).

2. A person with a cold is likely to produce oral and nasal secretions that facilitate viral transfer. In addition, since sickness can cause incapacitation or death, a virus that is programmed to exit the host when there is a physiological stress has the opportunity to find a new host at a time when the current host may cease to function. **3.** A person with a macrophage deficiency would have frequent infections. The causes would be poor innate responses, due to diminished phagocytosis and inflammation, and poor acquired responses, due to the lack of macrophages to present antigens to helper T cells.

Self-Quiz

1. b **2.** d **3.** c **4.** b **5.** c **6.** d **7.** b
8. One possible answer:

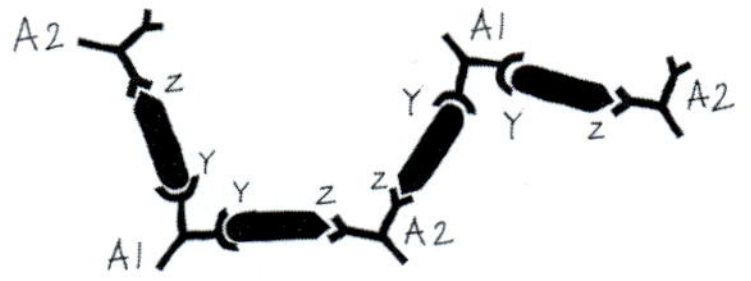

CHAPTER 44

Figure Questions

Figure 44.7 The body fluids of osmoconformers, such as most marine invertebrates, have the same salt concentration as sea water. Any such animals in the diet of the bird would add to the salt load to be eliminated. In contrast, marine fishes that osmoregulate maintain body fluids that have a much lower salt concentration than the surrounding ocean. By eating such fish, marine birds can obtain nutrients and water without adding as much to their salt load. **Figure 44.15** Tubule cells in the medulla are in contact with extracellular fluid of very high osmolarity. By producing solutes that keep intracellular osmolarity high, these cells achieve homeostasis with regard to volume. **Figure 44.16** Furosemide increases urine volume. The absence of ion transport in the ascending limb leaves the filtrate too concentrated for substantial volume reduction in the distal tubule and collecting duct. **Figure 44.20** The ADH levels would likely be elevated in both sets of patients with mutations because either defect prevents the recapture of water that restores blood osmolarity to normal levels.

Concept Check 44.1

1. Because the salt is moved against its concentration gradient, from low concentration (freshwater) to high concentration (blood) **2.** A freshwater osmoconformer would have body fluids too dilute to carry out life's processes. **3.** Without a layer of insulating fur, the camel must use the cooling effect of evaporative water loss to maintain body temperature, thus linking thermoregulation and osmoregulation.

Concept Check 44.2

1. Because uric acid is largely insoluble in water, it can be excreted as a semisolid paste, thereby reducing an animal's water loss. **2.** Humans produce uric acid from purine breakdown, and reducing purines in the diet often lessens the severity of gout. Birds, however, produce uric acid as a waste product of general nitrogen metabolism. They would therefore need a diet low in all nitrogen-containing compounds, not just purines.

Concept Check 44.3

1. In flatworms, ciliated cells draw interstitial fluids containing waste products into protonephridia. In earthworms, waste products pass from interstitial fluids into the coelom. From there they enter metanephridia by beating of cilia in a funnel surrounding an internal opening. In insects, the Malphigian tubules pump fluids from the hemolymph, which receives waste products during exchange with interstitial fluids in the course of circulation. **2.** Filtration produces a fluid for exchange processes that is free of cells and large molecules which are of benefit to the animal and could not readily be reabsorbed. **3.** The presence of Na^+ and other ions (electrolytes) in the dialysate would limit the extent to which they would be removed from the filtrate during dialysis. Adjusting the electrolytes in the starting dialysate can thus lead to the restoration of proper electrolyte concentrations in the plasma. Similarly, the absence of urea and other waste products in the starting dialysate results in their efficient removal from the filtrate.

Concept Check 44.4

1. The numerous nephrons and well-developed glomeruli of freshwater fishes produce urine at a high rate, while the small numbers of nephrons and smaller glomeruli of marine fishes produce urine at a low rate. **2.** The kidney medulla would absorb less water and thus the drug would increase the amount of water lost in the urine. **3.** A decline in blood pressure in the afferent arteriole would reduce the rate of filtration by moving less material through the vessels.

Concept Check 44.5

1. Alcohol inhibits the release of ADH, causing an increase in urinary water loss and increasing the chance of dehydration. **2.** The consumption of a large amount of water in a very short period of time, coupled with an absence of solute intake, can reduce sodium levels in the blood below tolerable levels. This condition, called hyponatremia, leads to disorientation and, sometimes, respiratory distress. It has been seen in marathon runners who drink water rather than sports drinks. (It has also caused the death of a fraternity pledge as a consequence of a water hazing ritual and the death of a contestant in a water-drinking competition.) **3.** High blood pressure

Self-Quiz

1. d **2.** b **3.** e **4.** d **5.** a **6.** c **7.** b
8.

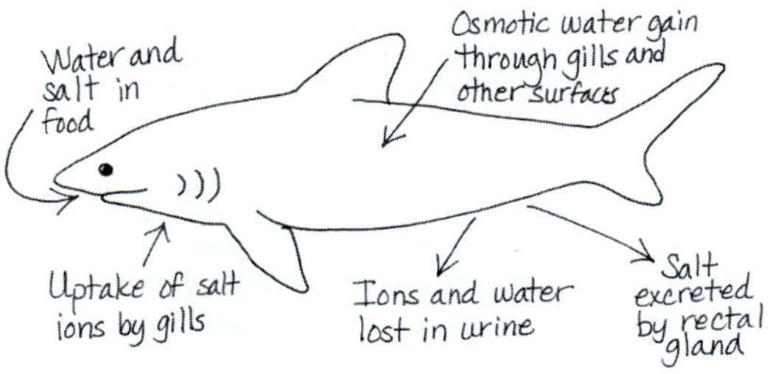

CHAPTER 45

Figure Questions

Figure 45.4 Because lipid-soluble hormones can freely diffuse across the lipid bilayers of cell membranes, you would expect to see biological activity after microinjection either into cells or into the interstitial space. **Figure 45.5** The hormone is water-soluble and has a cell-surface receptor. Such receptors, unlike those for lipid-soluble hormones, can cause observable changes in cells without hormone-dependent gene transcription. **Figure 45.18** Both diagnoses could be correct. In one case, the thyroid gland may produce excess thyroid hormone despite normal hormonal input from the hypothalamus and anterior pituitary. In the other, abnormally elevated hormonal input may be the cause of the overactive thyroid gland. **Figure 45.22** The result of the surgery would have been the same for both sexes—an absence of sexual differentiation in the genitals.

Concept Check 45.1

1. Water-soluble hormones, which cannot penetrate the plasma membrane, bind to cell-surface receptors. This interaction triggers an intracellular signal transduction pathway that ultimately alters the activity of a preexisting cytoplasmic protein and/or changes transcription of specific genes in the nucleus. Steroid hormones are lipid-soluble and can cross the plasma membrane into the cell interior, where they bind to receptors located in the cytosol or nucleus. The hormone-receptor complex then functions directly as a transcription factor that binds to the cell's DNA and activates or inhibits transcription of specific genes. **2.** Prostaglandins in semen that induce contractions in the uterus are aiding reproduction as signaling molecules that are transferred from one individual to another of the same species, like pheromones. **3.** Whether in different tissues or species, a particular hormone may cause diverse responses in target cells having different receptors for the hormone, different signal transduction pathways, and/or different proteins for carrying out the response.

Concept Check 45.2

1. In a healthy person, insulin released in response to the initial rise in blood glucose stimulates uptake of glucose by body cells. In a person with diabetes, however, inadequate production of insulin or nonresponsiveness of target cells decreases the body's ability to clear excess glucose from the blood. The initial increase in blood glucose is therefore greater in a person with diabetes, and it remains high for a prolonged period. **2.** A pathway governed by a short-lived stimulus would be less dependent on negative feedback. **3.** Since patients with type 2 diabetes produce insulin but fail to maintain normal glucose levels, you might predict that there could be mutations in the genes for the insulin receptor or the signal transduction pathway it activates. Such mutations have in fact been found in type 2 patients.

Concept Check 45.3

1. The posterior pituitary, an extension of the hypothalamus that contains the axons of neurosecretory cells, is the storage and release site for two neurohormones, oxytocin and antidiuretic hormone (ADH). The anterior pituitary, derived from tissues of the embryonic mouth, contains endocrine cells that make at least six different hormones. Secretion of anterior pituitary hormones is controlled by hypothalamic hormones that travel via portal vessels to the anterior pituitary. **2.** Because oxytocin responses involve positive feedback from suckling, the pathway does not require a sustained hormonal input stimulus. **3.** The hypothalamus and pituitary glands function in many different endocrine pathways. Many defects in these glands, such as those affecting growth or organization, would therefore disrupt many hormone pathways. Only a very specific defect, such as a mutation affecting a particular hormone receptor, would alter just one endocrine pathway. The situation is quite different for the final gland in a pathway, such as the thyroid gland. In this case, a wide range of defects that disrupt gland function would disrupt only the one pathway or small set of pathways in which that gland functions.

Concept Check 45.4

1. The adrenal medulla is derived from neural tissue during development. Reflecting this origin, it is an endocrine organ that produces two molecules—epinephrine and norepinephrine—that act both as hormones and as neurotransmitters. **2.** The levels of these hormones in the blood would become very high. This would be due to the diminished negative feedback on the hypothalamic neurons that secrete the releasing hormone that stimulates the secretion of ACTH by the anterior pituitary. **3.** By applying glucocorticoids to tissue by local injection, you in principle exploit their anti-inflammatory activity. Local injection avoids the effects on glucose metabolism that would occur if glucocorticoids were taken orally and transported throughout the body in the bloodstream.

Self-Quiz

1. c **2.** d **3.** d **4.** c **5.** b **6.** b **7.** c **8.** a

9.

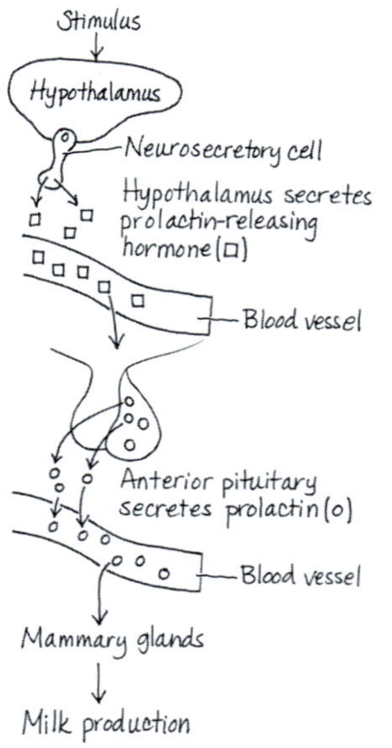

CHAPTER 46

Figure Questions

Figure 46.9 According to the graph, about one-third of the females rid themselves of all sperm from the first mating. Thus, two-thirds retain some sperm from the first mating. We would therefore predict that two-thirds of the females would have some offspring exhibiting the small eye phenotype of the dominant mutation carried by the males with which the females mated first. **Figure 46.16** Testosterone can pass from fetal blood to maternal blood via the placental circulation, temporarily upsetting the hormonal balance in the mother. **Figure 46.18** Oxytocin would most likely induce labor and start a positive-feedback loop that would direct labor to completion. Synthetic oxytocin is in fact frequently used to induce labor when prolonged pregnancy might endanger the mother or fetus.

Concept Check 46.1

1. The offspring of sexual reproduction are more genetically diverse. However, asexual reproduction can produce more offspring over multiple generations. **2.** Unlike other forms of asexual reproduction, parthenogenesis involves gamete production. By controlling whether or not haploid eggs are fertilized, species such as honeybees can readily switch between asexual and sexual reproduction. **3.** No. Owing to random assortment of chromosomes during meiosis, the offspring may receive the same copy or different copies of a particular parental chromosome from the sperm and the egg. Furthermore, genetic recombination during meiosis will result in reassortment of genes between pairs of parental chromosomes.

Concept Check 46.2

1. Internal fertilization allows the sperm to reach the egg without either gamete drying out. **2.** (a) Animals with external fertilization tend to release many gametes at once, resulting in the production of enormous numbers of zygotes. This increases the chances that some will survive to adulthood. (b) Animals with internal fertilization produce fewer offspring but generally exhibit greater care of the embryos and the young. **3.** The antimicrobial peptide might serve to protect the sperm before mating, the females with which the male mates, or the eggs those females produce. In all three cases, the reproductive success of the male would be enhanced, providing a mechanism for selection for peptide production over the course of evolution. You might want to think about how you might determine which function is most critical.

Concept Check 46.3

1. Primarily the penis and clitoris, but also the testes, labia, breasts, and outer third of the vagina **2.** Spermatogenesis occurs normally only when the testicles are cooler than normal body temperature. Extensive use of a hot tub (or of very tight-fitting underwear) can cause a decrease in sperm quality and number. **3.** The only effect of sealing off each vas deferens is an absence of sperm in the ejaculate. Sexual response and ejaculate volume are unchanged. The cutting and sealing off of these ducts, a *vasectomy*, is a common surgical procedure for men who do not wish to produce any (more) offspring.

Concept Check 46.4

1. The small size and lack of cytoplasm characteristic of a sperm are adaptations well suited to its function as a delivery vehicle for DNA. The large size and rich cytoplasmic contents of eggs support the growth and development of the embryo. **2.** In humans, the secondary oocyte combines with a sperm before it finishes the second meiotic division. Thus, oogenesis is completed after, not before, fertilization. **3.** The analysis would be informative because the polar bodies contain all of the maternal chromosomes that don't end up in the mature egg. For example, finding two copies of the disease gene in the polar bodies would indicate its absence in the egg. This method of genetic testing is sometimes carried out when oocytes collected from a female are fertilized with sperm in a laboratory dish.

Concept Check 46.5

1. In the testis, FSH stimulates the Sertoli cells, which nourish developing sperm. LH stimulates the production of androgens (mainly testosterone), which in turn stimulate sperm production. In both females and males, FSH encourages the growth of cells that support and nourish developing gametes (follicle cells in females and Sertoli cells in males), and LH stimulates the production of sex hormones that promote gametogenesis (estrogens, primarily estradiol, in females and androgens, especially testosterone, in males). **2.** In estrous cycles, which occur in most female mammals, the endometrium is reabsorbed (rather than shed) if fertilization does not occur. Estrous cycles often occur just one or a few times a year, and the female is usually receptive to copulation only during the period around ovulation. Menstrual cycles are found only in humans and some other primates. **3.** The combination of estradiol and progesterone would have a negative-feedback effect on the hypothalamus, blocking release of GnRH. This would interfere with LH secretion by the pituitary, thus preventing ovulation. This is in fact one basis of action of the most common hormonal contraceptives.

Concept Check 46.6

1. hCG secreted by the early embryo stimulates the corpus luteum to make progesterone, which helps maintain the pregnancy. During the second trimester, however, hCG production drops, the corpus luteum disintegrates, and the placenta completely takes over progesterone production. **2.** Both tubal ligation and vasectomy block the movement of gametes from the gonads to a site where fertilization could take place. **3.** By introducing a spermatid nucleus directly into an oocyte, ICSI bypasses the sperm's acquisition of motility in the epididymis, its swimming to meet the egg in the oviduct, and its fusion with the egg.

Self-Quiz

1. d **2.** b **3.** a **4.** b **5.** c **6.** c **7.** a **8.** d

9.

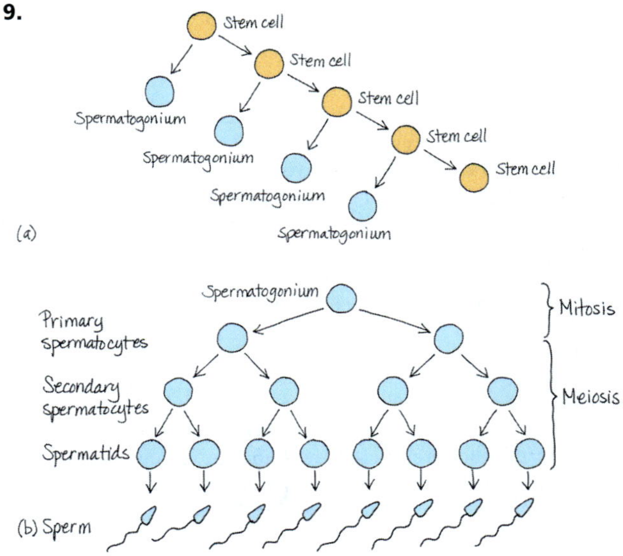

(c) The supply of stem cells would be used up and spermatogenesis would not be able to continue.

CHAPTER 47

Figure Questions

Figure 47.4 You could inject the compound into an unfertilized egg, expose the egg to sperm, and see whether the fertilization envelope forms. **Figure 47.7** The researchers allowed normal cortical rotation to occur, resulting in activation of the "back-forming" determinants. Then they forced the opposite rotation to occur, which established the back on the opposite side as well. Because the molecules on the normal side were already activated, forcing the opposite rotation apparently did not "cancel out" the establishment of the back side by the first rotation. **Figure 47.14** Given that these regions form from ectoderm but are just inside the body, you might propose that they form by an inpocketing of the ectoderm that then meets and fuses with the endoderm. And you would be right! **Figure 47.19** Cadherin is required to hold the cells of the blastula together, and extracellular calcium is required for cadherin function, so in the absence of calcium in the water, you'd expect to see a disorganized embryo like the one shown in the experimental SEM. **Figure 47.20** You could cut out the same tissues from control and injected embryos as was done in experiment 2 and place them between cover slips coated with the artificial fibronectin (FN) matrix. If convergent extension occurred in the tissues from both injected and control embryos, that would support the hypothesis that convergent extension can occur on a preexisting FN matrix in the embryo. **Figure 47.23** In Spemann's control, the two blastomeres were physically separated, and each grew into a whole embryo. In Roux's experiment, remnants of the dead blastomere were still contacting the live blastomere, which developed into a half-embryo. Therefore, molecules present in the dead cell's remnants may have been signaling to the live cell, inhibiting it from making all the embryonic structures. **Figure 47.24** You could inject the isolated protein or an mRNA encoding it into ventral cells of an earlier gastrula. If dorsal structures formed on the ventral side, that would support the idea that the protein is the signaling molecule secreted or presented by the dorsal lip. You should also do a control experiment to make sure the injection process alone did not cause dorsal structures to form. **Figure 47.26** You could remove the AER and look for Sonic hedgehog mRNA or protein as a marker of the ZPA. If either was absent, that would support your hypothesis. You could also block FGF function and see whether the ZPA formed (by looking for Sonic hedgehog).

Concept Check 47.1

1. The fertilization envelope forms after cortical granules release their contents outside the egg, causing the vitelline membrane to rise and harden. The fertilization envelope serves as a barrier to fertilization by more than one sperm. **2.** During cleavage in frogs and many other animals, the cell cycle is modified so that it virtually skips G_1 and G_2, the growth phases. As a result, the early cleavage divisions divide the zygote's cytoplasm into smaller and smaller cells as the embryo's size remains nearly the same. **3.** Cleavage transforms the single-celled zygote into an embryo consisting of many cells; cleavage does not involve cell or tissue movement. During gastrulation, the cells and tissues of a blastula are extensively rearranged, so that by the late gastrula stage there are three tissue layers positioned in new relationships to each other. **4.** The neural tube forms when a band of ectodermal tissue on the dorsal side along the anterior-posterior axis, called the neural plate, rolls into a tube and pinches off from the rest of the ectoderm. Neural crest cells arise as groups of cells in the regions between the edges of the neural tube and the surrounding ectoderm migrate away from the neural tube. **5.** The increased Ca^{2+} concentration in the egg would cause the cortical granules to fuse with the plasma membrane, releasing their contents and causing a fertilization envelope to form, even though no sperm had entered. This would prevent fertilization. **6.** Conjoined twins develop from monozygotic twins that separate quite late, after part of the embryo has already formed. (This part is shared by the twins.) By this time, both the chorion and amnion have formed, so there is only one of each.

Concept Check 47.2

1. Microtubules elongate, lengthening the cell along one axis, while microfilaments oriented crosswise at one end of the cell contract, making that end smaller and the whole cell wedge-shaped. **2.** The cells of the notochord migrate toward the midline of the embryo (converge), rearranging themselves so there are fewer cells across the notochord, which thus becomes longer overall (extends; see Figure 47.18). **3.** Because microfilaments would not be able to contract and decrease the size of one end of the cell, both the inward bending in the middle of the neural tube and the outward bending of the hinge regions at the edges would be blocked. Therefore, the neural tube probably would not form.

Concept Check 47.3

1. Once the first two axes are specified, the third one is automatically determined. (Think of your own body: If you know where your anterior and posterior ends are and where your left and right sides are, you automatically know which sides are your front and back.) Of course, there still must be a mechanism for determining where asymmetrically placed organs must go, such as the vertebrate stomach or appendix. **2.** Yes, a second embryo could develop because inhibiting BMP-4 activity would have the same effect as transplanting an organizer. **3.** The limb that developed probably would have a mirror-image duplication, with the most posterior digits in the middle and the most anterior digits at either end.

Self-Quiz

1. a **2.** b **3.** e **4.** c **5.** a **6.** c **7.** e

8.

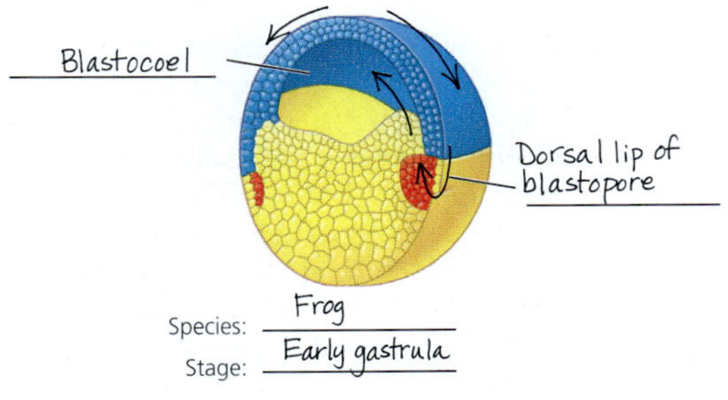

CHAPTER 48

Figure Questions

Figure 48.7 Adding chloride channels makes the membrane potential less negative. Adding sodium or potassium channels would have no effect, because sodium movement is already at equilibrium and there are no potassium ions present. **Figure 48.15** The production and transmission of action potentials would be unaffected. However, action potentials arriving at chemical synapses would be unable to trigger release of neurotransmitter. Signaling at such synapses would thus be blocked. **Figure 48.17** In theory, the results would be similar because the binding studies indicate that both opiates and naloxone, an opiate antagonist, bind directly to the receptor.

Concept Check 48.1

1. Sensors in your ear transmit information to your brain. There the activity of interneurons in processing centers enables you to recognize your name. In response, signals transmitted via motor neurons cause contraction of muscles that turn your neck. **2.** The nervous system is required for control of vital functions, such as circulation and gas exchange, and the transmission of information occurs on a very short time scale. **3.** It would prevent information from being transmitted away from the cell body along the axon.

Concept Check 48.2

1. Ions can flow against a chemical concentration gradient if there is an opposing electrical gradient of greater magnitude. **2.** A decrease in permeability to K^+, an increase in permeability to Na^+, or both. **3.** The activity of the sodium-potassium pump is essential to maintain the resting potential. With the pump inactivated, the sodium and potassium concentration gradients would gradually disappear, and so would the resting potential.

Concept Check 48.3

1. A graded potential has a magnitude that varies with stimulus strength, whereas an action potential has an all-or-none magnitude that is independent of stimulus strength. **2.** Loss of the insulation provided by myelin sheaths leads to a disruption of action potential propagation along axons. Voltage-gated sodium channels are restricted to the nodes of Ranvier, and without the insulating effect of myelin, the inward current produced at one node during an action potential cannot depolarize the membrane to the threshold at the next node. **3.** The maximum frequency would decrease because the refractory period would be extended.

Concept Check 48.4

1. It can bind to different types of receptors, each triggering a specific response in postsynaptic cells. **2.** These toxins would prolong the EPSPs that acetylcholine produces because the neurotransmitter would remain longer in the synaptic cleft. **3.** Such a drug might act as a sedative, decreasing the general level of activity in the brain and hence in the person.

Self-Quiz

1. c **2.** b **3.** c **4.** a **5.** c **6.** e

7. As shown in this pair of drawings, a pair of action potentials would move outward in both directions from each electrode. (Action potentials are unidirectional only if they begin at one end of an axon.) However, because of the refractory period, the two action potentials between the electrodes both stop where they meet. Thus, only one action potential reaches the synaptic terminals.

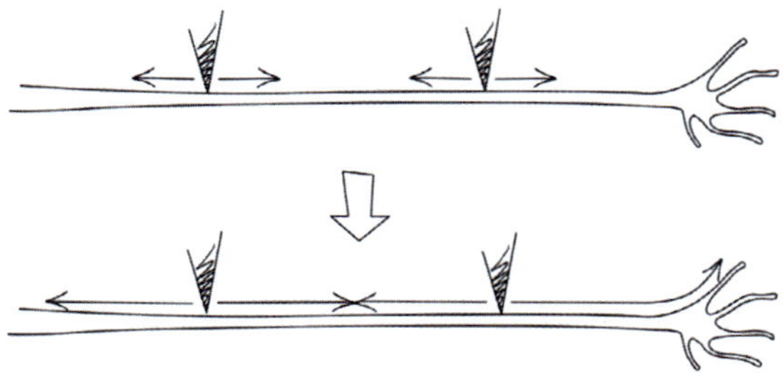

CHAPTER 49

Figure Questions

Figure 49.8 Neurosecretory cells of the adrenal medulla secrete epinephrine in response to preganglionic input from sympathetic neurons. Epinephrine travels in the circulation to reach target tissues throughout the body. Thus, the rapid changes in body tissues required for the "fight-or-flight" response rely on direct input from the nervous system as well as indirect input via neurohormone products of the adrenal medulla. **Figure 49.12** If the new mutation disrupted only pacemaker function, you should be able to restore rhythmic activity by removing the SCN and replacing it with an SCN transplant from either a wild-type or τ mutant hamster. Using the new mutant as the donor would not be as informative, since both failed transplants and successful ones would result in a lack of rhythmic activity. **Figure 49.22** The depolarization should mimic natural stimulation of the brain reward system, resulting in positive and perhaps pleasurable sensations.

Concept Check 49.1

1. The sympathetic division, which mediates the "fight-or-flight" response in stressful situations **2.** The preganglionic neurons use the same neurotransmitter and function similarly in each division (to activate postganglionic neurons). The postganglionic neurons generally have opposing functions and use different neurotransmitters. **3.** Nerves contain bundles of axons, some of which belong to motor neurons that send signals outward from the CNS, and some that belong to sensory neurons that bring signals into the CNS. Therefore, you would expect effects on both motor control and sensation.

Concept Check 49.2

1. The cerebral cortex on the left side of the brain control initiates voluntary movement of the right side of the body. **2.** Alcohol diminishes function of the cerebellum. **3.** Paralysis reflects an inability to carry out motor functions transmitted from the cerebrum to the spinal cord. You would expect these patients to have injuries below the reticular formation. A coma reflects a disruption in the cycles of sleep and arousal regulated by communication between the reticular formation and the cerebrum. You would expect these patients to have injuries at or above the reticular formation.

Concept Check 49.3

1. Brain lesions that disrupt behavior, cognition, memory, or other functions provide evidence that the portion of the brain affected by the damage is important for the normal activity that is blocked or altered. **2.** Broca's area, which is active during the generation of speech, is located near the part of the primary motor cortex that controls muscles in the face. Wernicke's area, which is active when speech is heard, is located near the part of the temporal lobe that is involved in hearing. **3.** Each cerebral hemisphere is specialized for different parts of this task—the right for face recognition and the left for language. Without an intact corpus callosum, neither hemisphere can take advantage of the other's processing abilities.

Concept Check 49.4

1. There can be an increase in the number of synapses between the neurons or an increase in the strength of existing synaptic connections. **2.** If consciousness is an emergent property resulting from the interaction of many different regions of the brain, then it is unlikely that localized brain damage will have a discrete effect on consciousness. **3.** The hippocampus is responsible for organizing newly acquired information. Without hippocampal function, the links necessary to retrieve information from the neocortex will be lacking and no functional memory, short- or long-term, will be formed.

Concept Check 49.5

1. Both are progressive brain diseases whose risk increases with advancing age. Both result from the death of brain neurons and are associated with the accumulation of peptide or protein aggregates. **2.** The symptoms of schizophrenia can be mimicked by a drug that stimulates dopamine-releasing neurons. The brain reward system, which is involved in addiction, is comprised of dopamine-releasing neurons that connect the ventral tegmental area to regions in the cerebrum. Parkinson's disease results from the death of dopamine-releasing neurons. **3.** Not necessarily. It might be that the plaques, tangles, and missing regions of the brain seen at death reflect secondary effects, the consequence of other unseen changes that are actually responsible for the alterations in brain function.

Self-Quiz

1. c **2.** a **3.** d **4.** d **5.** e **6.** c

7. (a)

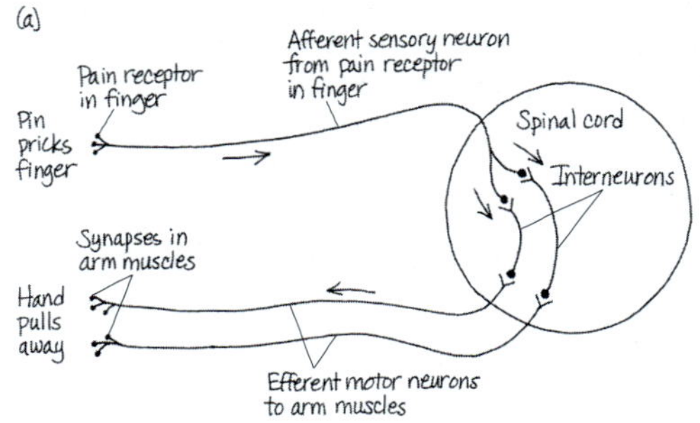

(b)

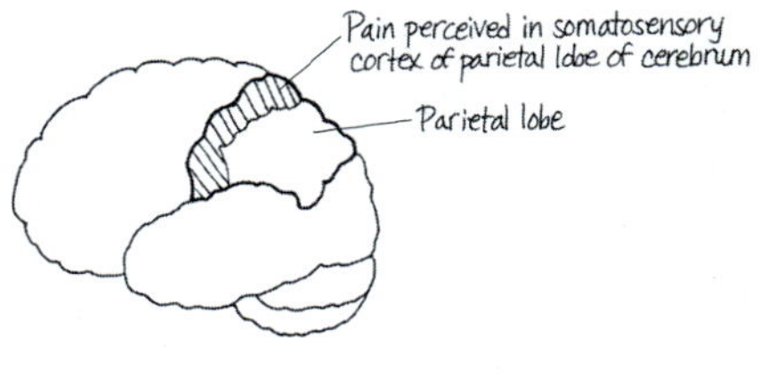

CHAPTER 50

Figure Questions

Figure 50.10 In the brain. Each note is detected separately in the ear, with each causing vibration of the basilar membrane and deflection of hair cells in a distinct location. Sensory neurons in each location provide output in the form of action potentials that travel along distinct axons in the auditory nerve. It is not until the information reaches the brain that the individual notes are detected and the perception of the chord is generated. **Figure 50.14** The results of the experiment would have been identical. What matters is the activation of particular sets of neurons, not the manner in which they are activated. Any signal from a bitter cell will be interpreted by the brain as a bitter taste, regardless of the nature of the compound and the receptor involved. **Figure 50.15** Only perception. Binding of an odorant to its receptor will cause action potentials to be sent to the brain. Although an excess of that odorant might cause a diminished response through adaptation, another odorant can mask the first only at the level of perception in the brain. **Figure 50.22** Each of the three types of cones is most sensitive to a distinct wavelength of light. A cone might be fully depolarized when there is light present if the light is of a wavelength far from its optimum. **Figure 50.27** Hundreds of myosin heads participate in sliding each pair of thick and thin filaments past each other. Because cross-bridge formation and breakdown are not synchronized, many myosin heads are exerting force on the thin filaments at all times during muscle contraction. **Figure 50.37** Since a duck is more specialized for flying than for swimming, you might expect that it would consume more energy per unit body mass and distance in swimming than would, for example, a fish. (In fact, if the value for a 10^3 g swimming duck were plotted on this graph, it would appear well above the line for swimmers and just above the line for runners.)

Concept Check 50.1

1. Electromagnetic receptors in general detect only external stimuli. Nonelectromagnetic receptors, such as chemoreceptors or mechanoreceptors, can act as either internal or external sensors. **2.** The capsaicin present in the spice mix activates the thermoreceptor for high temperatures. In response to the perceived high temperature, the nervous system triggers sweating to achieve evaporative cooling. **3.** You would perceive the electrical stimulus as if the sensory receptors that regulate that neuron had been activated. For example, electrical stimulation of the sensory neuron controlled by the thermoreceptor activated by menthol would likely be perceived as a local cooling.

Concept Check 50.2

1. Statocysts detect the animal's orientation with respect to gravity, providing information that is essential in environments such as these, where light cues are absent. **2.** As a sound that changes gradually from a very low to a very high pitch **3.** The stapes and the other middle ear bones transmit vibrations from the tympanic membrane to the oval window. Fusion of these bones, as occurs in otosclerosis, would block this transmission and result in hearing loss.

Concept Check 50.3

1. Both taste cells and olfactory cells have receptor proteins in their plasma membrane that bind certain substances, leading to membrane depolarization through a signal transduction pathway involving a G protein. However, olfactory cells are sensory neurons, whereas taste cells are not. **2.** Since animals rely on chemical signals for behaviors that include finding mates, marking territories, and avoiding dangerous substances, it is adaptive for the olfactory system to have a robust response to a very small number of molecules of a particular odorant. **3.** Because the sweet, bitter, and umami tastes involve GPCR proteins but the sour taste does not, you might predict that the mutation is in a molecule that acts in the signal transduction pathway common to the different GPCR receptors.

Concept Check 50.4

1. Planarians have ocelli that cannot form images but can sense the intensity and direction of light, providing enough information to enable the animals to find protection in shaded places. Flies have compound eyes that form images and excel at detecting movement. **2.** The person can focus on distant objects but not close objects (without glasses) because close focusing requires the lens to become almost spherical. This problem is common after age 50. **3.** Close each eye in turn. An object floating on the surface of an eyeball will appear only when that eye is open.

Concept Check 50.5

1. By causing all of the motor neurons that control the muscle to generate action potentials at a rate high enough to produce tetanus in all of the muscle fibers **2.** In a skeletal muscle fiber, Ca^{2+} binds to the troponin complex, which moves tropomyosin away from the myosin-binding sites on actin and allows cross-bridges to form. In a smooth muscle cell, Ca^{2+} binds to calmodulin, which activates an enzyme that phosphorylates the myosin head and thus enables cross-bridge formation. **3.** *Rigor mortis*, a Latin phrase meaning "stiffness of death," results from the complete depletion of ATP in skeletal muscle. Since ATP is required for release of myosin from actin and to pump Ca^{2+} out of the cytosol, muscles become chronically contracted beginning about 3 or 4 hours after death.

Concept Check 50.6

1. Septa provide the divisions of the coelom that allow for peristalsis, a form of locomotion requiring independent control of different body segments. **2.** The main problem in swimming is drag; a fusiform body minimizes drag. The main problem in flying is overcoming gravity; wings shaped like airfoils provide lift, and adaptations such as air-filled bones reduce body mass. **3.** You could start by standing with your upper arm against your side and your lower arm extended out at a ninety degree angle from your hip. You could then slowly lower your hand toward the ground. Because you are holding the weight of your hand and lower arm against gravity, you need to maintain tetanus in the biceps muscle. As you lower your hand, you are gradually decreasing the number of motor units in the biceps that are contracted. The triceps is not involved because gravity is providing the force for arm extension.

Self Quiz

1. e **2.** d **3.** b **4.** a **5.** c **6.** b
7.

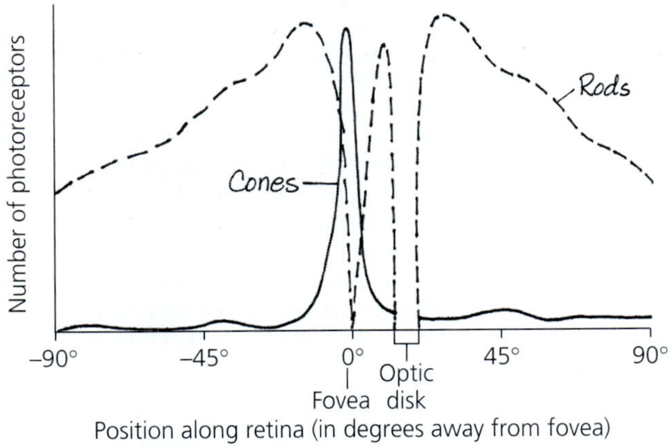

The answer shows the actual distribution of rods and cones in the human eye. Your graph may differ but should have the following properties: Only cones at the fovea; fewer cones and more rods at both ends of the x-axis; no photoreceptors in the optic disk.

CHAPTER 51

Figure Questions

Figure 51.3 The fixed action pattern based on the sign stimulus of a red belly ensures that the male will chase away any invading males of his species. By chasing away such males, the defender decreases the chance that eggs laid in his nesting territory will be fertilized by another male. **Figure 51.10** There should be no effect. Imprinting is an innate behavior that is carried out anew in each generation. Assuming the nest was not disturbed, the offspring of the Lorenz followers would imprint on the mother goose. **Figure 51.11** Perhaps the wasp doesn't use visual cues. It might also be that wasps recognize objects native to their environment, but not foreign objects, such as the pinecones. Tinbergen addressed these ideas before carrying out the pinecone study. When he swept away the pebbles and sticks around the nest, the wasps could no longer find their nests. If he shifted the natural objects in their natural arrangement, the shift in the landmarks caused a shift in the site to which the wasps returned. Finally, if the natural objects around the nest site were replaced with pinecones while the wasp was in the burrow, the wasp nevertheless found her way back to the nest site. **Figure 51.14** Courtship song generation must be coupled to courtship song recognition. Unless the genes that control generation of particular song elements also control recognition, the hybrids might be unlikely to find mating partners, depending on what aspects of the songs are important for mate recognition and acceptance. **Figure 51.15** It might be that the birds require stimuli during flight to exhibit their migratory preference. If this were true, the birds would show the same orientation in the funnel experiment despite their distinct genetic programming. **Figure 51.28** It holds true for some, but not all individuals. If a parent has more than one reproductive partner, the offspring of different partners will have a coefficient of relatedness less than 0.5.

Concept Check 51.1

1. It is an example of a fixed action pattern. The proximate explanation might be that nudging and rolling are released by the sign stimulus of an object outside the nest, and the behavior is carried to completion once initiated. The ultimate explanation might be that ensuring that eggs remain in the nest increases the chance of producing healthy offspring. **2.** Circannual rhythms are typically based on the cycles of light and dark in the environment. As the global climate changes, animals that migrate in response to these rhythms may shift to a location before or after local environmental conditions are optimal for reproduction and survival. **3.** There might be selective pressure for other prey fish to detect an injured fish because the source of the injury might threaten them as well. There might be selection for predators to be attracted to the alarm substance because they would be more likely to encounter crippled prey than would be predators that can't respond. Fish with adequate defenses might show no change because they have a selective advantage if they do not waste energy responding to the alarm substance.

Concept Check 51.2

1. Natural selection would tend to favor convergence in color pattern because a predator learning to associate a pattern with a sting or bad taste would avoid all other individuals with that same color pattern, regardless of species. **2.** Forgetting the location of some caches, which consist of pine seeds buried in the ground, might benefit the nutcracker by increasing the number of pines growing in its habitat. This example points out one of the difficulties in making simplistic assumptions about the purpose of a behavior. **3.** You might move objects around to establish an abstract rule, such as "past landmark A, the same distance as A is from the starting point," while maintaining a minimum of fixed metric relationships, that is, avoiding having the food directly adjacent to or a set distance from a landmark. As you might surmise, designing an informative experiment of this kind is not easy.

Concept Check 51.3

1. Because this geographic variation corresponds to differences in prey availability between two garter snake habitats, it seems likely that snakes with characteristics enabling them to feed on the abundant prey in their locale would have had increased survival and reproductive success, and thus natural selection would have resulted in the divergent foraging behaviors. **2.** Courtship is easier to study because it is essential for reproduction, but not for growth, development, and survival. Mutations disrupting many other behaviors would be lethal. **3.** You would need to know the percentage of time that unrelated individuals behave identically when performing this behavior.

Concept Check 51.4

1. Certainty of paternity is higher with external fertilization. **2.** Natural selection acts on genetic variation in the population. **3.** Because females would now be present in much larger numbers than males, all three types of males should have some reproductive success. Nevertheless, since the advantage that the blue-throats rely on—a limited number of females in their territory—will be absent, the yellow-throats are likely to increase in frequency in the short term.

Concept Check 51.5

1. Reciprocal altruism, the exchange of helpful behaviors for future similar behaviors, can explain cooperative behaviors between unrelated animals, though often the behavior has some potential benefit to the benefactor as well. **2.** Yes. Kin selection does not require any recognition or awareness of relatedness. **3.** The older individual cannot be the beneficiary because he or she cannot have extra offspring. However, the cost is low for an older individual performing the altruistic act because that person has already reproduced (but perhaps is still caring for a child or grandchild). There can therefore be selection for an altruistic act by a postreproductive individual that benefits a young relative.

Self-Quiz

1. d **2.** a **3.** c **4.** a **5.** c **6.** b **7.** c

8.

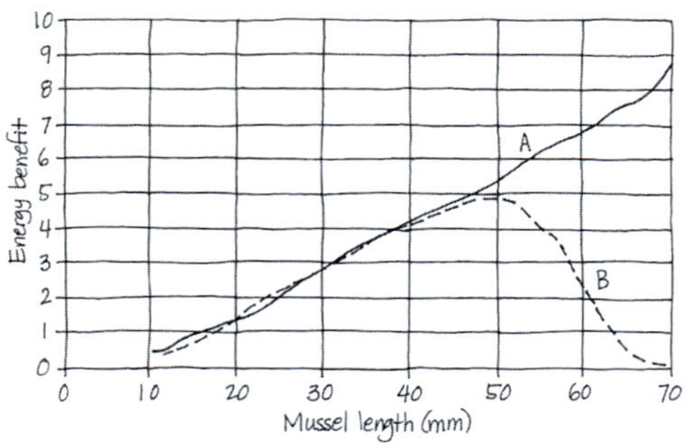

You could measure the size of mussels that oystercatchers successfully open and compare that with the size distribution in the habitat.

CHAPTER 52

Figure Questions

Figure 52.6 Some factors, such as fire, are relevant only for terrestrial systems. At first glance, water availability is primarily a terrestrial factor, too. However, species living along the intertidal zone of oceans or along the edge of lakes suffer desiccation as well. Salinity stress is important for species in some aquatic and terrestrial systems. Oxygen availability is an important factor primarily for species in some aquatic systems and in soils and sediments. **Figure 52.8** When only urchins were removed, limpets may have increased in abundance and reduced seaweed cover somewhat (the difference between the purple and blue lines on the graph). **Figure 52.14** Dispersal limitations, the activities of people (such as a broad-scale conversion of forests to agriculture or selective harvesting), or other factors listed in Figure 52.6

Concept Check 52.1

1. *Ecology* is the scientific study of the interactions between organisms and their environment; *environmentalism* is advocacy for the environment. Ecology provides scientific understanding that can inform decision making about environmental issues. **2.** Interactions in ecological time that affect the survival or reproduction of organisms can result in changes to the population's gene pool and ultimately result in a change in the population on an evolutionary time scale. **3.** If the fungicides are used together, fungi will likely evolve resistance to all four much more quickly than if the fungicides are used individually at different times.

Concept Check 52.2

1. a. Humans could transplant a species to a new area that it could not previously reach because of a geographic barrier (dispersal change). b. Humans could change a species' biotic interactions by eliminating a predator or herbivore species, such as sea urchins, from an area. **2.** The sun's unequal heating of Earth's surface produces temperature variations between the warmer tropics and colder polar regions, and it influences the movement of air masses and thus the distribution of moisture at different latitudes. **3.** One test would be to build a fence around a plot of land in an area that has trees of that species, excluding all deer from the plot. You could then compare the abundance of tree seedlings inside and outside the fenced plot over time.

Concept Check 52.3

1. Rapid changes in salinity can cause salt stress in many organisms. **2.** In the oceanic pelagic zone, the ocean bottom lies below the photic zone, so there is too little light to support benthic algae or rooted plants. **3.** In a river below a dam, the fish are more likely to be species that prefer colder water. In summer, the deep layers of a reservoir are colder than the surface layers, so a river below a dam will be colder than an undammed river.

Concept Check 52.4

1. Higher average temperature in deserts **2.** Answers will vary by location but should be based on the information and maps in Figure 52.21. How much your local area has been altered from its natural state will influence how much it reflects the expected characteristics of your biome, particularly the expected plants and animals. **3.** Northern coniferous forest is likely to replace tundra along the boundary between these biomes. To see why, note that northern coniferous forest is adjacent to tundra throughout North America, northern Europe, and Asia (see Figure 52.19) and that the temperature range for northern coniferous forest is just above that for tundra (see Figure 52.20).

Self-Quiz

1. c **2.** d **3.** a **4.** d **5.** b **6.** d **7.** d **8.** e **9.** a
10.

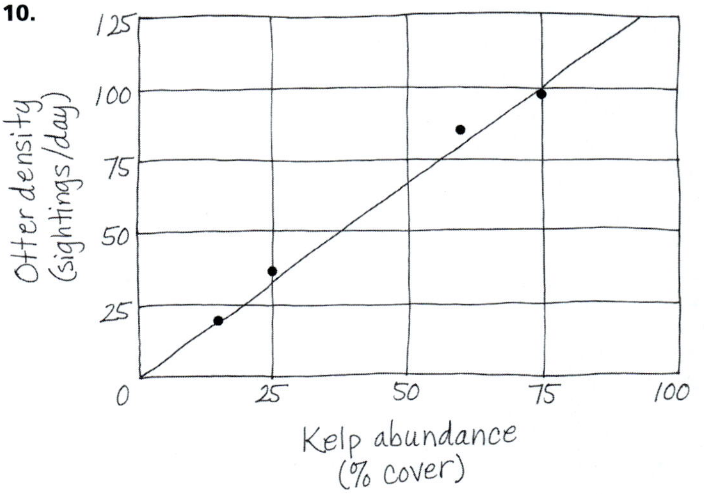

Based on what you learned from Figure 52.8 and on the positive relationship you observed in the field between kelp abundance and otter density, you could hypothesize that otters lower sea urchin density, reducing feeding of the urchins on kelp.

CHAPTER 53

Figure Questions

Figure 53.4 The dispersion of the penguins would likely appear clumped as you flew over densely populated islands and sparsely populated ocean. **Figure 53.8** If male European kestrels provided no parental care, brood size should not affect their survival. Therefore, the three bars representing male survival in Figure 53.8 should have similar heights. In contrast, female survival should still decline with increasing brood size, as shown in the current figure. **Figure 53.19** The moose population grew quickly because food was abundant and predators were absent. During this period, the population experienced exponential growth. **Figure 53.20** Hare numbers typically peaked slightly before lynx numbers did. The lynx depend on the hares for food, but there is a delay between increased food availability and increased reproduction by the lynx.

Concept Check 53.1

1. The territorial species likely has a uniform pattern of dispersion, since the interactions between individuals will maintain constant space between them. The flocking species is probably clumped, since most individuals probably live in one of the clumps (flocks).
2.

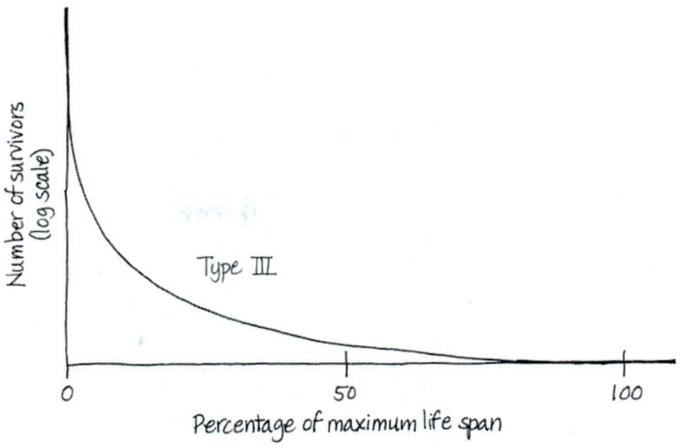

A type III survivorship curve is most likely because very few of the young probably survive. **3.** If an animal is captured by attracting it with food, it may be more likely to be recaptured if it seeks the same food. The number of marked animals recaptured (x) would be an overestimate, and because the population size $(N) = mn/x$, N would be an underestimate. Alternatively, if an animal has a negative experience during capture and learns from that experience, it may be less likely to be recaptured. In this case, x would be an underestimate and N would be an overestimate.

Concept Check 53.2

1. The constant, spring-fed stream. In more constant physical conditions, where populations are more stable and competition for resources is more likely, larger well-provisioned young, which are more typical of iteroparous species, have a better chance of surviving. **2.** By preferentially investing in the eggs it lays in the nest, the peacock wrasse increases their probability of survival. The eggs it disperses widely and does not provide care for are less likely to survive, at least some of the time, but require a lower investment by the adults. (In this sense, the adults avoid the risk of placing all their eggs in one basket.) **3.** If a parent's survival is compromised greatly by bearing young during times of stress, the animal's fitness may increase if it abandons its current young and survives to produce healthier young at a later time.

Concept Check 53.3

1. Though r_{max} is constant, N, the population size, is increasing. As r_{max} is applied to an increasingly large N, population growth ($r_{max}N$) accelerates, producing the J-shaped curve. **2.** On the new island. The first plants that found suitable habitat on the island would encounter an abundance of space, nutrients, and light. In the rain forest, competition among plants for these resources is intense. **3.** The net population growth is $\Delta N/\Delta t = bN - dN$. The annual per capita birth rate, b, equals 14/1,000, or 0.014, and the per capita death rate, d, equals 8/1,000, or 0.008. Therefore, the net population growth in 2006 is

$$\frac{\Delta N}{\Delta t} = (0.014 \times 300{,}000{,}000) - (0.008 \times 300{,}000{,}000)$$

or 1.8 million people. A population is growing exponentially only if its per capita rate of increase equals its maximum rate. That is not the case for the United States currently.

Concept Check 53.4

1. When N (population size) is small, there are relatively few individuals producing offspring. When N is large, near the carrying capacity, per capita growth is relatively small because it is limited by available resources. The steepest part of the logistic growth curve corresponds to a population with a number of reproducing individuals that is substantial but not yet near carrying capacity. **2.** r-selected. Weeds that colonize an abandoned field face little competition, and their initial populations are well below carrying capacity. These are characteristics of environments that favor r-selected species. **3.** Using a population size of 1,600 as an example,

$$\frac{dN}{dt} = r_{max}N\frac{(K-N)}{K} = \frac{1(1{,}600)(1{,}500 - 1{,}600)}{1{,}500}$$

and the population "growth" rate is −107 individuals per year. The population shrinks even faster when N is farther from the carrying capacity; when N equals 1,750 and 2,000 individuals, the population shrinks by 292 and 667 individuals per year, respectively. These negative growth rates correspond most closely to the time when the *Daphnia* population has overshot its carrying capacity and is shrinking, about days 65–100 in Figure 53.13b.

Concept Check 53.5

1. Competition for resources and space can negatively impact population growth by limiting reproductive output. Diseases that are transmitted more easily in crowded populations can exert negative feedback on increasing population size. Some predators feed preferentially on species at higher population densities, since those prey are easier to find than are prey in less dense populations. In crowded populations, toxic metabolic wastes can build up and poison the organisms. **2.** Three attributes are the size, quality, and isolation of patches. A patch that is larger or of higher quality is more likely to attract individuals and to be a source of individuals for other patches. A patch that is relatively isolated will undergo less exchange of individuals with other patches. **3.** You would need to study the population for more than one cycle (longer than 10 years and probably at least 20) before having sufficient data to examine changes through time. Otherwise, it would be impossible to know whether an observed decrease in the population size reflected a long-term trend or was part of the normal cycle.

Concept Check 53.6

1. A bottom-heavy age structure, with a disproportionate number of young people, portends continuing growth of the population as these young people begin reproducing. In contrast, a more evenly distributed age structure predicts a more stable population size. **2.** The growth rate of Earth's human population has dropped by half since the 1960s, from 2.2% in 1962 to 1.15% today. Nonetheless, growth has not slowed much because the smaller growth rate is counterbalanced by increased population size; the number of extra people on Earth each year remains enormous—approximately 75 million. **3.** Each of us influences our ecological footprint by how we live—what we eat, how much energy we use, and the amount of waste we generate—as well as by how many children we have. Making choices that reduce our demand for resources makes our ecological footprint smaller.

Self-Quiz

1. c **2.** c **3.** c **4.** d **5.** d **6.** b **7.** c **8.** d **9.** c **10.** d
11.

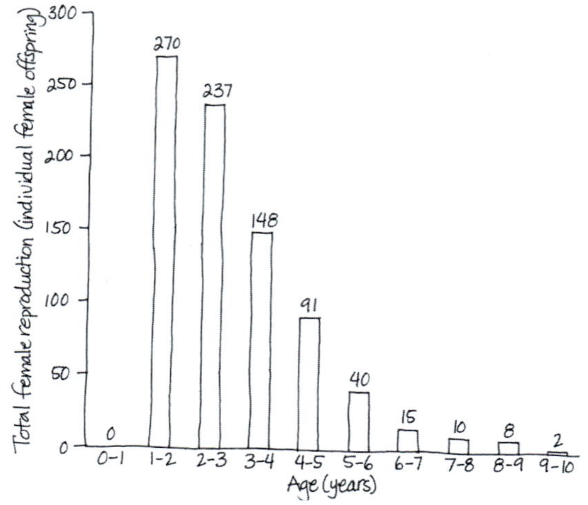

The total number of female offspring produced is greatest in females 1–2 years of age. Sample calculation for females of this age group: 252 indiv. × 1.07 female offspring/indiv. = 270 female offspring.

CHAPTER 54

Figure Questions

Figure 54.3 Its realized and fundamental niches would be similar. **Figure 54.4** If both species were feeding on seeds of the same size, you would expect differences in beak size to disappear over evolutionary time. The species could not specialize on seeds of different sizes. **Figure 54.14** The low-productivity treatment had the shortest food chain, so that food chain should be the most stable. **Figure 54.15** The death of individuals of *Mytilus*, a dominant species, should open up space for other species and increase species richness even in the absence of *Pisaster*. **Figure 54.19** Because the abundance of the second predatory species is unaffected by soil warming, there would be a less dramatic decrease in total predator numbers. Therefore, if the top-down model applies in this community, you would expect a smaller increase in *S. lindsayae* density than was actually observed. **Figure 54.28** Other factors not included in the model must contribute to the unexplained variation in the results.

Concept Check 54.1

1. Interspecific competition has negative effects on both species (−/−). In predation, the predator population benefits at the expense of the prey population (+/−). Mutualism is a symbiosis in which both species benefit (+/+). **2.** One of the competing species will become locally extinct because of the greater reproductive success of the more efficient competitor. **3.** Examples of relevant interactions include competition between weeds and food crops; predation by humans on herbivores, such as cattle; herbivory by humans on leafy vegetables, such as lettuce or spinach; and the planting of symbiotic nitrogen-fixing plants, such as beans or peas.

Concept Check 54.2

1. Species richness, the number of species in the community, and relative abundance, the proportions of the community represented by the various species, both contribute to species diversity. Compared to a community with a very high proportion of one species, one with a more even proportion of species is considered to be more diverse. **2.** The energetic hypothesis suggests that the length of a food chain is limited by the inefficiency of energy transfer along the chain, while the dynamic stability hypothesis proposes that long food chains are less stable than short chains. The energetic hypothesis predicts that food chains will be longer in habitats with higher primary productivity. The dynamic stability hypothesis predicts that food chains will be longer in more predictable environments. **3.** According to the bottom-up model, adding extra predators would have little effect on lower trophic levels, particularly vegetation. If the top-down model applied, increased bobcat numbers would decrease raccoon numbers, increase snake numbers, decrease grasshopper numbers, and increase plant biomass.

Concept Check 54.3

1. High levels of disturbance are generally so disruptive that they eliminate many species from communities, leaving the community dominated by a few tolerant species. Low levels of disturbance permit competitively dominant species to exclude other species from the community. In contrast, moderate levels of disturbance can facilitate coexistence of a greater number of species in a community by preventing competitively dominant species from becoming abundant enough to eliminate other species from the community. **2.** Early successional species can facilitate the arrival of other species in many ways, including increasing the fertility or water-holding capacity of soils or providing shelter to seedlings from wind and intense sunlight. **3.** The absence of fire for 100 years would represent a change to a low level of disturbance. According to the intermediate disturbance hypothesis, this change should cause diversity to decline as competitively dominant species gain sufficient time to exclude less competitive species.

Concept Check 54.4

1. Ecologists propose that the greater species richness of tropical regions is the result of their longer evolutionary history and the greater solar energy input and water availability in tropical regions. **2.** Immigration of species to islands declines with distance from the mainland and increases with island area. Extinction of species is lower on larger islands and on less isolated islands. Since the number of species on islands is largely determined by the difference between rates of immigration and extinction, the number of species will be highest on large islands near the mainland and lowest on small islands far from the mainland. **3.** Because of their greater mobility, birds disperse to islands more often than snakes or mammals, so birds should have greater richness.

Concept Check 54.5

1. Pathogens are microorganisms, viruses, viroids, or prions that cause disease. **2.** If the parasite requires contact with a human and another animal, the parasite might be an especially likely vector for the pathogens that cause zoonotic diseases. **3.** If you can identify the host or hosts of the disease as well as any intermediate vectors, such as mosquitoes or fleas, you can reduce the rate of infection by decreasing the abundance of the host and vector or by reducing their contact with people.

Self-Quiz

1. c **2.** d **3.** b **4.** c **5.** c **6.** d **7.** b **8.** c
9. Community 1: $H = -[(0.05)(\ln 0.05) + (0.05)(\ln 0.05) + (0.85)(\ln 0.85) + (0.05)(\ln 0.05)] = 0.59$. Community 2: $H = -[(0.30)(\ln 0.30) + (0.40)(\ln 0.40) + (0.30)(\ln 0.30)] = 1.1$. Community 2 is more diverse.
10.

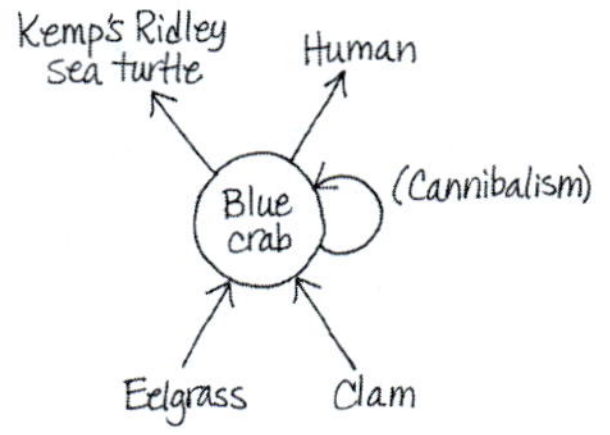

Crab numbers should increase, reducing the abundance of eelgrass.

CHAPTER 55

Figure Questions

Figure 55.6 Wetlands, coral reefs, and coastal zones cover areas too small to show up clearly on global maps. **Figure 55.7** If the new duck farms made nitrogen available in rich supply, as phosphorus already is, then adding extra nitrogen in the experiment would not increase phytoplankton density. **Figure 55.13** By dissolving minerals, the fungi move nutrients from reservoir D (inorganic materials unavailable as nutrients) to reservoir C (inorganic materials available as nutrients). **Figure 55.15** Water availability is probably another factor that varied across the sites. Such factors not included in the experimental design could make the results more difficult to interpret. Multiple factors can also covary in nature, so ecologists must be careful that the factor they are studying is actually causing the observed response and is not just correlated with it.

Concept Check 55.1

1. Energy passes through an ecosystem, entering as sunlight and leaving as heat. It is not recycled within the ecosystem. **2.** The second law states that in any energy transfer or transformation, some of the energy is dissipated to the surroundings as heat. This "escape" of energy from an ecosystem is offset by the continuous influx of solar radiation. **3.** You would need to know how much biomass the wildebeests ate from your plot and how much nitrogen was contained in that biomass. You would also need to know how much nitrogen they deposited in urine or feces.

Concept Check 55.2

1. Only a fraction of solar radiation strikes plants or algae, only a portion of that fraction is of wavelengths suitable for photosynthesis, and much energy is lost as a result of reflection or heating of plant tissue. **2.** By manipulating the level of the factors of interest, such as phosphorus availability or soil moisture, and measuring responses by primary producers **3.** The student is missing the plant biomass eaten by herbivores and the production allocated to plant roots and other belowground tissues.

Concept Check 55.3

1. 20 J; 40% **2.** Nicotine protects the plant from herbivores. **3.** There are many things they could do to reduce their production efficiency. For example, exercising vigorously will use energy that might otherwise go to biomass, and keeping the house cool will force their bodies to use energy to stay warm.

Concept Check 55.4

1. For example, for the carbon cycle:

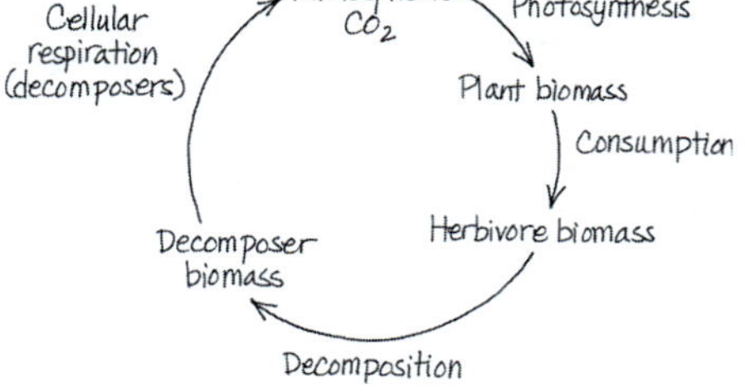

Cycling of a carbon atom

2. Removal of the trees stops nitrogen uptake from the soil, allowing nitrate to accumulate there. The nitrate is washed away by precipitation and enters the streams. **3.** Most of the nutrients in a tropical rain forest are contained in the trees, so removing the trees by logging rapidly depletes nutrients from the ecosystem. The nutrients that remain in the soil are quickly carried away into streams and groundwater by the abundant precipitation.

Concept Check 55.5

1. Adding nutrients causes population explosions of algae and the organisms that feed on them. Increased respiration by algae and consumers, including detritivores, depletes the lake's oxygen, which the fish require. **2.** At a lower trophic level, because biological magnification increases the concentration of toxins up the food chain **3.** Because higher temperatures lead to faster decomposition, organic matter in these soils could be quickly decomposed to CO_2, speeding up global warming.

Self-Quiz

1. c **2.** b **3.** d **4.** c **5.** e **6.** a **7.** d
8.

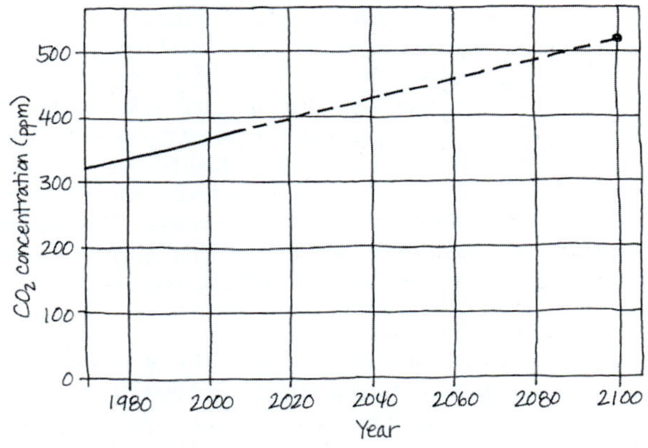

Between 1974 and 2007, Earth's atmospheric CO_2 concentration increased from approximately 330 ppm to 385 ppm. If this rate of increase of 1.7 ppm/yr continues, the concentration in 2100 will be about 540 ppm. The actual rise in CO_2 concentration could be larger or smaller, depending on Earth's human population, per capita energy use, and the extent to which societies take steps to reduce CO_2 emissions, including replacing fossil fuels with renewable or nuclear fuels. Additional scientific data will be important for many reasons, including determining how quickly greenhouse gases such as CO_2 are removed from the atmosphere by the biosphere.

CHAPTER 56

Figure Questions

Figure 56.4 You would need to know the complete range of the species and that it is missing across all of that range. You would also need to be certain that the species isn't hidden, as might be the case for an animal that is hibernating underground or a plant that is present in the form of seeds or spores. **Figure 56.11** Because the population of Illinois birds has a different genetic makeup than birds in other regions, you would want to maintain to the greatest extent possible the frequency of beneficial genes or alleles found only in that population. In restoration, preserving genetic diversity in a species is as important as increasing organism numbers. **Figure 56.13** The natural disturbance regime in this habitat included frequent fires that cleared undergrowth but did not kill mature pine trees. Without these fires, the undergrowth quickly fills in and the habitat becomes unsuitable for red-cockaded woodpeckers.

Concept Check 56.1

1. In addition to species loss, the biodiversity crisis includes the loss of genetic diversity within populations and species and the degradation of entire ecosystems. **2.** Habitat destruction, such as deforestation, channelizing of rivers, or conversion of natural ecosystems to agriculture or cities, deprives species of places to live. Introduced species, which are transported by humans to regions outside their native range, where they are not controlled by their natural pathogens or predators, often reduce the population sizes of native species through competition or predation. Overexploitation has reduced populations of plants and animals or driven them to extinction. **3.** If both populations breed separately, then gene flow between the populations would not occur and genetic differences between them would be greater. As a result, the loss of genetic diversity would be greater than if the populations interbreed.

Concept Check 56.2

1. Reduced genetic variation decreases the capacity of a population to evolve in the face of change. **2.** The effective population size, N_e, is $4(35 \times 10)/(35 + 10) = 31$ birds. **3.** Because millions of people use the greater Yellowstone ecosystem each year, it would be impossible to eliminate all contact between people and bears. Instead, you might try to reduce the kinds of encounters where bears are killed. You might recommend lower speed limits on roads in the park, adjust the timing or location of hunting seasons (where hunting is allowed outside the park) to minimize contact with mother bears and cubs, and provide financial incentives for livestock owners to try alternative means (such as guard dogs) of protecting livestock.

Concept Check 56.3

1. A small area supporting an exceptionally large number of endemic species as well as a disproportionate number of endangered and threatened species **2.** Zoned reserves may provide sustained supplies of forest products, water, hydroelectric power, educational opportunities, and income from ecotourism. **3.** Habitat corridors can increase the rate of movement or dispersal of organisms between habitat patches and thus the rate of gene flow between subpopulations. They thus help prevent a decrease in fitness attributable to inbreeding. They can also minimize interactions between organisms and humans as the organisms disperse; in cases involving potential predators, such as bears or large cats, minimizing such interactions is desirable.

Concept Check 56.4

1. The main goal is to restore degraded ecosystems to a more natural state. **2.** Bioremediation uses organisms, generally prokaryotes, fungi, or plants, to detoxify or remove pollutants from ecosystems. Biological augmentation uses organisms, such as nitrogen-fixing plants, to add essential materials to degraded ecosystems. **3.** The Kissimmee River project returns the flow of water to the original channel and restores natural flow, a self-sustaining outcome. Ecologists at the Maungatautari reserve will need to maintain the integrity of the fence indefinitely, an outcome that is not self-sustaining in the long term.

Concept Check 56.5

1. Sustainable development is an approach to development that works toward the long-term prosperity of human societies and the ecosystems that support them, which requires linking the biological sciences with the social sciences, economics, and humanities. **2.** Biophilia, our sense of connection to nature and other forms of life, may act as a significant motivation for the development of an environmental ethic that resolves not to allow species to become extinct or ecosystems to be destroyed. Such an ethic is necessary if we are to become more attentive and effective custodians of the environment. **3.** At a minimum, you would want to know the size of the population and the average reproductive rate of individuals in it. To develop the fishery sustainably, you would seek a harvest rate that maintains the population near its original size and maximizes its harvest in the long term rather than the short term.

Self-Quiz

1. c **2.** e **3.** d **4.** d **5.** d **6.** e **7.** c **8.** a
10.

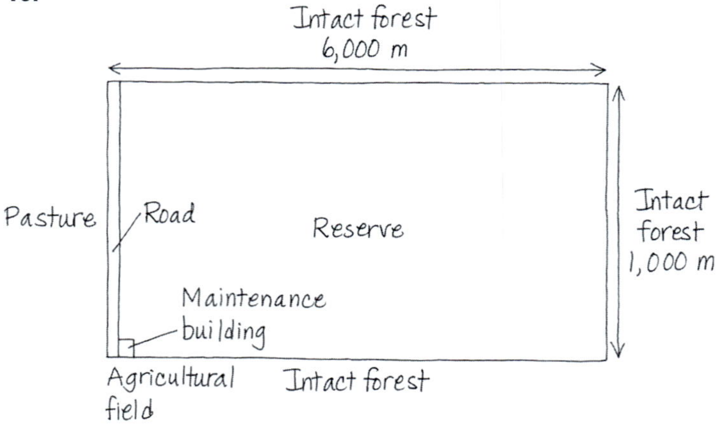

To minimize the area of forest into which the cowbirds penetrate, you should locate the road along one edge of the reserve. Any other location would increase the area of affected habitat. Similarly, the maintenance building should be in a corner of the reserve to minimize the area susceptible to cowbirds.

APPENDIX

B Periodic Table of the Elements

Representative elements: Group 1A (Alkali metals), Group 2A (Alkaline earth metals), Groups 3A–7A (Group 7A: Halogens), Group 8A (Noble gases)

Transition elements: Groups 3B–2B

Period number	1 Group 1A	2 Group 2A	3 3B	4 4B	5 5B	6 6B	7 7B	8 8B	9 8B	10 8B	11 1B	12 2B	13 Group 3A	14 Group 4A	15 Group 5A	16 Group 6A	17 Group 7A	18 Group 8A
1	1 H 1.008																	2 He 4.003
2	3 Li 6.941	4 Be 9.012											5 B 10.81	6 C 12.01	7 N 14.01	8 O 16.00	9 F 19.00	10 Ne 20.18
3	11 Na 22.99	12 Mg 24.31											13 Al 26.98	14 Si 28.09	15 P 30.97	16 S 32.07	17 Cl 35.45	18 Ar 39.95
4	19 K 39.10	20 Ca 40.08	21 Sc 44.96	22 Ti 47.87	23 V 50.94	24 Cr 52.00	25 Mn 54.94	26 Fe 55.85	27 Co 58.93	28 Ni 58.69	29 Cu 63.55	30 Zn 65.41	31 Ga 69.72	32 Ge 72.64	33 As 74.92	34 Se 78.96	35 Br 79.90	36 Kr 83.80
5	37 Rb 85.47	38 Sr 87.62	39 Y 88.91	40 Zr 91.22	41 Nb 92.91	42 Mo 95.94	43 Tc (98)	44 Ru 101.1	45 Rh 102.9	46 Pd 106.4	47 Ag 107.9	48 Cd 112.4	49 In 114.8	50 Sn 118.7	51 Sb 121.8	52 Te 127.6	53 I 126.9	54 Xe 131.3
6	55 Cs 132.9	56 Ba 137.3	57* La 138.9	72 Hf 178.5	73 Ta 180.9	74 W 183.8	75 Re 186.2	76 Os 190.2	77 Ir 192.2	78 Pt 195.1	79 Au 197.0	80 Hg 200.6	81 Tl 204.4	82 Pb 207.2	83 Bi 209.0	84 Po (209)	85 At (210)	86 Rn (222)
7	87 Fr (223)	88 Ra (226)	89† Ac (227)	104 Rf (261)	105 Db (262)	106 Sg (266)	107 Bh (264)	108 Hs (269)	109 Mt (268)	110 Ds (271)	111 — (272)	112 — (285)	113 — (284)	114 — (289)	115 — (288)			

*Lanthanides	58 Ce 140.1	59 Pr 140.9	60 Nd 144.2	61 Pm (145)	62 Sm 150.4	63 Eu 152.0	64 Gd 157.3	65 Tb 158.9	66 Dy 162.5	67 Ho 164.9	68 Er 167.3	69 Tm 168.9	70 Yb 173.0	71 Lu 175.0
†Actinides	90 Th 232.0	91 Pa 231.0	92 U 238.0	93 Np (237)	94 Pu (244)	95 Am (243)	96 Cm (247)	97 Bk (247)	98 Cf (251)	99 Es 252	100 Fm 257	101 Md 258	102 No 259	103 Lr 260

Metals　Metalloids　Nonmetals

Name (Symbol)	Atomic Number								
Actinium (Ac)	89	Cobalt (Co)	27	Iodine (I)	53	Osmium (Os)	76	Silicon (Si)	14
Aluminum (Al)	13	Copper (Cu)	29	Iridium (Ir)	77	Oxygen (O)	8	Silver (Ag)	47
Americium (Am)	95	Curium (Cm)	96	Iron (Fe)	26	Palladium (Pd)	46	Sodium (Na)	11
Antimony (Sb)	51	Darmstadtium (Ds)	110	Krypton (Kr)	36	Phosphorus (P)	15	Strontium (Sr)	38
Argon (Ar)	18	Dubnium (Db)	105	Lanthanum (La)	57	Platinum (Pt)	78	Sulfur (S)	16
Arsenic (As)	33	Dysprosium (Dy)	66	Lawrencium (Lr)	103	Plutonium (Pu)	94	Tantalum (Ta)	73
Astatine (At)	85	Einsteinium (Es)	99	Lead (Pb)	82	Polonium (Po)	84	Technetium (Tc)	43
Barium (Ba)	56	Erbium (Er)	68	Lithium (Li)	3	Potassium (K)	19	Tellurium (Te)	52
Berkelium (Bk)	97	Europium (Eu)	63	Lutetium (Lu)	71	Praseodymium (Pr)	59	Terbium (Tb)	65
Beryllium (Be)	4	Fermium (Fm)	100	Magnesium (Mg)	12	Promethium (Pm)	61	Thallium (Tl)	81
Bismuth (Bi)	83	Fluorine (F)	9	Manganese (Mn)	25	Protactinium (Pa)	91	Thorium (Th)	90
Bohrium (Bh)	107	Francium (Fr)	87	Meitnerium (Mt)	109	Radium (Ra)	88	Thulium (Tm)	69
Boron (B)	5	Gadolinium (Gd)	64	Mendelevium (Md)	101	Radon (Rn)	86	Tin (Sn)	50
Bromine (Br)	35	Gallium (Ga)	31	Mercury (Hg)	80	Rhenium (Re)	75	Titanium (Ti)	22
Cadmium (Cd)	48	Germanium (Ge)	32	Molybdenum (Mo)	42	Rhodium (Rh)	45	Tungsten (W)	74
Calcium (Ca)	20	Gold (Au)	79	Neodymium (Nd)	60	Rubidium (Rb)	37	Uranium (U)	92
Californium (Cf)	98	Hafnium (Hf)	72	Neon (Ne)	10	Ruthenium (Ru)	44	Vanadium (V)	23
Carbon (C)	6	Hassium (Hs)	108	Neptunium (Np)	93	Rutherfordium (Rf)	104	Xenon (Xe)	54
Cerium (Ce)	58	Helium (He)	2	Nickel (Ni)	28	Samarium (Sm)	62	Ytterbium (Yb)	70
Cesium (Cs)	55	Holmium (Ho)	67	Niobium (Nb)	41	Scandium (Sc)	21	Yttrium (Y)	39
Chlorine (Cl)	17	Hydrogen (H)	1	Nitrogen (N)	7	Seaborgium (Sg)	106	Zinc (Zn)	30
Chromium (Cr)	24	Indium (In)	49	Nobelium (No)	102	Selenium (Se)	34	Zirconium (Zr)	40

APPENDIX C

The Metric System

Measurement	Unit and Abbreviation	Metric Equivalent	Metric-to-English Conversion Factor	English-to-Metric Conversion Factor
Length	1 kilometer (km)	= 1000 (10^3) meters	1 km = 0.62 mile	1 mile = 1.61 km
	1 meter (m)	= 100 (10^2) centimeters = 1000 millimeters	1 m = 1.09 yards 1 m = 3.28 feet 1 m = 39.37 inches	1 yard = 0.914 m 1 foot = 0.305 m
	1 centimeter (cm)	= 0.01 (10^{-2}) meter	1 cm = 0.394 inch	1 foot = 30.5 cm 1 inch = 2.54 cm
	1 millimeter (mm)	= 0.001 (10^{-3}) meter	1 mm = 0.039 inch	
	1 micrometer (μm) (formerly micron, μ)	= 10^{-6} meter (10^{-3} mm)		
	1 nanometer (nm) (formerly millimicron, mμ)	= 10^{-9} meter (10^{-3} μm)		
	1 angstrom (Å)	= 10^{-10} meter (10^{-4} μm)		
Area	1 hectare (ha)	= 10,000 square meters	1 ha = 2.47 acres	1 acre = 0.405 ha
	1 square meter (m^2)	= 10,000 square centimeters	1 m^2 = 1.196 square yards 1 m^2 = 10.764 square feet	1 square yard = 0.8361 m^2 1 square foot = 0.0929 m^2
	1 square centimeter (cm^2)	= 100 square millimeters	1 cm^2 = 0.155 square inch	1 square inch = 6.4516 cm^2
Mass	1 metric ton (t)	= 1000 kilograms	1 t = 1.103 tons	1 ton = 0.907 t
	1 kilogram (kg)	= 1000 grams	1 kg = 2.205 pounds	1 pound = 0.4536 kg
	1 gram (g)	= 1000 milligrams	1 g = 0.0353 ounce 1 g = 15.432 grains	1 ounce = 28.35 g
	1 milligram (mg)	= 10^{-3} gram	1 mg = approx. 0.015 grain	
	1 microgram (μg)	= 10^{-6} gram		
Volume (solids)	1 cubic meter (m^3)	= 1,000,000 cubic centimeters	1 m^3 = 1.308 cubic yards 1 m^3 = 35.315 cubic feet	1 cubic yard = 0.7646 m^3 1 cubic foot = 0.0283 m^3
	1 cubic centimeter (cm^3 or cc)	= 10^{-6} cubic meter	1 cm^3 = 0.061 cubic inch	1 cubic inch = 16.387 cm^3
	1 cubic millimeter (mm^3)	= 10^{-9} cubic meter (10^{-3}cubic centimeter)		
Volume (liquids and gases)	1 kiloliter (kl or kL)	= 1000 liters	1 kL = 264.17 gallons	1 gallon = 3.785 L
	1 liter (l or L)	= 1000 milliliters	1 L = 0.264 gallons 1 L = 1.057 quarts	1 quart = 0.946 L
	1 milliliter (ml or mL)	= 10^{-3} liter = 1 cubic centimeter	1 mL = 0.034 fluid ounce 1 mL = approx. 1/4 teaspoon 1 mL = approx. 15-16 drops (gtt.)	1 quart = 946 mL 1 pint = 473 mL 1 fluid ounce = 29.57 mL 1 teaspoon = approx. 5 mL
	1 microliter (μl or μL)	= 10^{-6} liter (10^{-3} milliliters)		
Time	1 second (s)	= 1/60 minute		
	1 millisecond (ms)	= 10^{-3} second		
Temperature	Degrees Celsius (°C) (Absolute zero, when all molecular motion ceases, is − 273°C. The Kelvin [K] scale, which has the same size degrees as Celsius, has its zero point at absolute zero. Thus, 0°K = −273°C.)		°F = 9/5°C + 32	°C = 5/9 (°F − 32)

APPENDIX D

A Comparison of the Light Microscope and the Electron Microscope

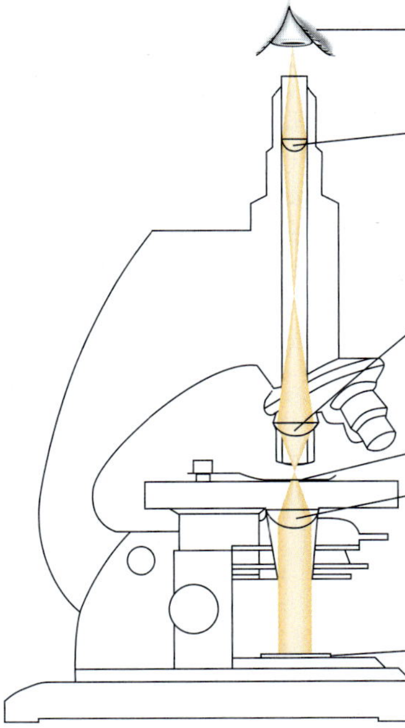

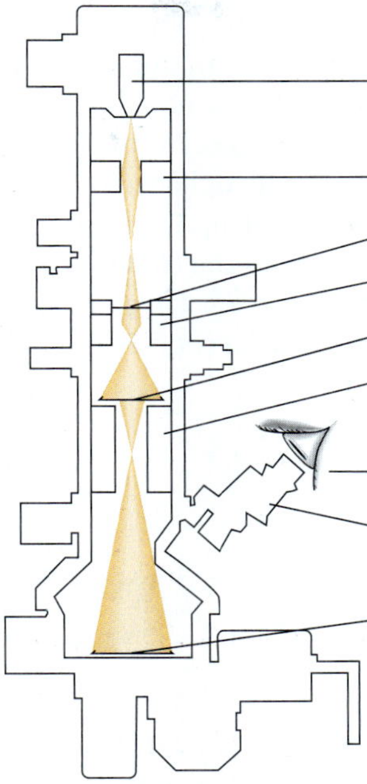

Light Microscope

In light microscopy, light is focused on a specimen by a glass condenser lens; the image is then magnified by an objective lens and an ocular lens, for projection on the eye, digital camera, digital video camera, or photographic film.

Electron Microscope

In electron microscopy, a beam of electrons (top of the microscope) is used instead of light, and electromagnets are used instead of glass lenses. The electron beam is focused on the specimen by a condenser lens; the image is magnified by an objective lens and a projector lens for projection on a digital detector, fluorescent screen, or photographic film.

APPENDIX E Classification of Life

This appendix presents a taxonomic classification for the major extant groups of organisms discussed in this text; not all phyla are included. The classification presented here is based on the three-domain system, which assigns the two major groups of prokaryotes, bacteria and archaea, to separate domains (with eukaryotes making up the third domain). This classification contrasts with the traditional five-kingdom system, which groups all prokaryotes in a single kingdom, Monera. Systematists no longer recognize the kingdom Monera because it would have members in two different domains (see Chapter 26).

Various alternative classification schemes are discussed in Unit Five of the text. The taxonomic turmoil includes debates about the number and boundaries of kingdoms and about the alignment of the Linnaean classification hierarchy with the findings of modern cladistic analysis. In this review, asterisks (*) indicate currently recognized phyla thought by some systematists to be paraphyletic.

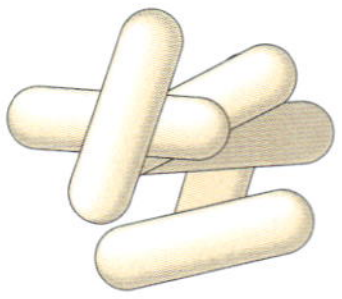

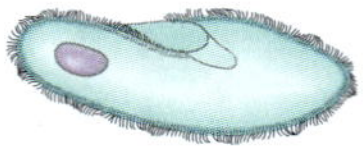

DOMAIN BACTERIA

- **Proteobacteria**
- **Chlamydia**
- **Spirochetes**
- **Gram-positive Bacteria**
- **Cyanobacteria**

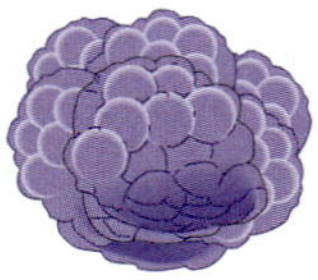

DOMAIN ARCHAEA

- **Korarchaeota**
- **Euryarchaeota**
- **Crenarchaeota**
- **Nanoarchaeota**

DOMAIN EUKARYA

In the phylogenetic hypothesis we present in Chapter 28, major clades of eukaryotes are grouped together in the five "supergroups" listed below in bold type. The traditional five-kingdom classification scheme united all the eukaryotes generally called protists in a single kingdom, Protista. However, advances in systematics have made it clear that Protista is in fact polyphyletic: Some protists are more closely related to plants, fungi, or animals than they are to other protists. As a result, the kingdom Protista has been abandoned. In contrast, the kingdoms Plantae (land plants), Fungi, and Animalia (animals) have survived from the five-kingdom system.

Excavata

- Diplomonadida (diplomonads)
- Parabasala (parabasalids)
- Euglenozoa (euglenozoans)
  - Euglenophyta (euglenids)
  - Kinetoplastida (kinetoplastids)

Chromalveolata

- Alveolata (alveolates)
  - Dinoflagellata (dinoflagellates)
  - Apicomplexa (apicomplexans)
  - Ciliophora (ciliates)
- Stramenopila (stramenopiles)
  - Bacillariophyta (diatoms)
  - Chrysophyta (golden algae)
  - Phaeophyta (brown algae)
  - Oomycota (water molds)

Archaeplastida

- Rhodophyta (red algae)
- Chlorophyta (green algae: chlorophytes)
- Charophyceae (green algae: charophyceans)
- Plantae

Phylum Hepatophyta (liverworts)	Bryophytes (nonvascular plants)	
Phylum Anthocerophyta (hornworts)	Bryophytes (nonvascular plants)	
Phylum Bryophyta (mosses)	Bryophytes (nonvascular plants)	
Phylum Lycophyta (lycophytes)	Seedless vascular plants	
Phylum Pterophyta (ferns, horsetails, whisk ferns)	Seedless vascular plants	
Phylum Ginkgophyta (ginkgo)	Gymnosperms	Seed plants
Phylum Cycadophyta (cycads)	Gymnosperms	Seed plants
Phylum Gnetophyta (gnetophytes)	Gymnosperms	Seed plants
Phylum Coniferophyta (conifers)	Gymnosperms	Seed plants
Phylum Anthophyta (flowering plants)	Angiosperms	Seed plants

Rhizaria
- ▶ Chlorarachniophyta (chlorarachniophytes)
- ▶ Foraminifera (forams)
- ▶ Radiolaria (radiolarians)

Unikonta
- ▶ Amoebozoa (amoebozoans)
  - Myxogastrida (plasmodial slime molds)
  - Dictyostelida (cellular slime molds)
  - Gymnamoeba (gymnamoebas)
  - Entamoeba (entamoebas)
  - Nucleariida (nucleariids)

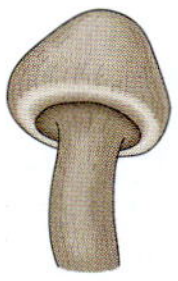

- ▶ Fungi
  - *Phylum Chytridiomycota (chytrids)
  - Phylum Zygomycota (zygomycetes)
  - Phylum Glomeromycota (glomeromycetes)
  - Phylum Ascomycota (sac fungi)
  - Phylum Basidiomycota (club fungi)
- ▶ Choanoflagellata (choanoflagellates)

- ▶ Animalia
  - Phylum Calcarea } (sponges)
  - Phylum Silicea } (sponges)
  - Phylum Cnidaria (cnidarians)
    - Class Hydrozoa (hydrozoans)
    - Class Scyphozoa (jellies)
    - Class Cubozoa (box jellies and sea wasps)
    - Class Anthozoa (sea anemones and most corals)
  - Phylum Ctenophora (comb jellies)
  - Phylum Acoela (acoel flatworms)
  - Lophotrochozoa (lophotrochozoans)
    - Phylum Placozoa (placozoans)
    - Phylum Kinorhyncha (kinorhynchs)
    - Phylum Platyhelminthes (flatworms)
      - Class Turbellaria (free-living flatworms)
      - Class Trematoda (flukes)
      - Class Monogenea (monogeneans)
      - Class Cestoda (tapeworms)
    - Phylum Nemertea (proboscis worms)
    - Phylum Ectoprocta (ectoprocts)
    - Phylum Phoronida (phoronids)
    - Phylum Brachiopoda (brachiopods)
    - Phylum Rotifera (rotifers)
    - Phylum Cycliophora (cycliophorans)
    - Phylum Mollusca (molluscs)
      - Class Polyplacophora (chitons)
      - Class Gastropoda (gastropods)
      - Class Bivalvia (bivalves)
      - Class Cephalopoda (cephalopods)
    - Phylum Annelida (segmented worms)
      - Class Oligochaeta (oligochaetes)
      - Class Polychaeta (polychaetes)
      - Class Hirudinea (leeches)
    - Phylum Acanthocephala (spiny-headed worms)
  - Ecdysozoa (ecdysozoans)
    - Phylum Loricifera (loriciferans)
    - Phylum Priapula (priapulans)
    - Phylum Nematoda (roundworms)
    - Phylum Arthropoda (This survey groups arthropods into a single phylum, but some zoologists now split the arthropods into multiple phyla.)
      - Subphylum Cheliceriformes (horseshoe crabs, arachnids)
      - Subphylum Myriapoda (millipedes, centipedes)
      - Subphylum Hexapoda (insects, springtails)
      - Subphylum Crustacea (crustaceans)
    - Phylum Tardigrada (tardigrades)
    - Phylum Onychophora (velvet worms)
  - Deuterostomia (deuterostomes)
    - Phylum Hemichordata (hemichordates)
    - Phylum Echinodermata (echinoderms)
      - Class Asteroidea (sea stars)
      - Class Ophiuroidea (brittle stars)
      - Class Echinoidea (sea urchins and sand dollars)
      - Class Crinoidea (see lilies)
      - Class Concentricycloidea (sea daisies)
      - Class Holothuroidea (sea cucumbers)
    - Phylum Chordata (chordates)
      - Subphylum Cephalochordata (cephalochordates: lancelets)
      - Subphylum Urochrodata (urochordates: tunicates)
      - Subphylum Craniata (craniates)
        - Class Myxini (hagfishes)
        - Vertebrates:
          - Class Cephalaspidomorphi (lampreys)
          - Class Chondrichthyes (sharks, rays, chimaeras)
          - Class Acinopterygii (ray-finned fishes)
          - Class Actinistia (coelacanths)
          - Class Dipnoi (lungfishes)
          - Class Amphibia (amphibians)
          - Class Reptilia (tuataras, lizards, snakes, turtles, crocodilians, birds)
          - Class Mammalia (mammals)

Credits

PHOTO CREDITS

Cover Image Chris Fox/Corbis

Unit Opening Interviews **Unit One** Stuart Brinin; **Unit Two** Zach Veilleux, Rockefeller University; **Unit Three** Maria Nemchuk; **Unit Four** Justin Ide; **Unit Five** Brent Nicastro; **Unit Six** Noah Berger Photography; **Unit Seven** Skeeter Hagler; **Unit Eight** William A. Cotton, Colorado State University.

Detailed Contents **Unit One** David A. Bushnell. **Unit Two** Albert Tousson, High Resolution Imaging Facility, University of Alabama at Birmingham. **Unit Three** Kosman D, Mizutani CM, Lemons D, Cox WG, McGinnis W, Bier E. Multiplex detection of RNA expression in Drosophila embryos. Science. 2004 Aug 6;305(5685):846; Fig 1. **Unit Four** Olivier Grunewald. **Unit Five** Ray Watson, canadianphotography.ca. **Unit Six** Pierre-Michel Blais, orchidees.provence.free.fr. **Unit Seven** Michael Nichols/National Geographic/Getty Images. **Unit Eight** Galen Rowell/Corbis.

Chapter 1 **1.1** Chris Fox/Corbis; **1.2** Henri and Käthi Oertli; **1.3.1** Fred Bavendam/Minden Pictures; **1.3.2** Kim Taylor and Jane Burton/Dorling Kindersley; **1.3.3** Frans Lanting/Minden Pictures; **1.3.4** Frans Lanting/Minden Pictures; **1.3.5** Michael & Patricia Fogden/Corbis; **1.3.6** Joe McDonald/Corbis; **1.3.7** ImageState/International Stock Photography Ltd.; **1.4.1** WorldSat International/Photo Researchers, Inc.; **1.4.2** Bill Brooks/Alamy; **1.4.3** Mark Raycroft/Minden Pictures, and Dave McShaffrey, Marietta College, Ohio; **1.4.4** Michael Orton/Photographer's Choice/Getty Images; **1.4.5** Ross M. Horowitz/Iconica/Getty Images; **1.4.6** Photodisc/Getty Images; **1.4.7** Jeremy Burgess/SPL/Photo Researchers; **1.4.8** Manfred Kage/Peter Arnold, Inc.; **1.4.9** E.H. Newcomb & W.P. Wergin/Biological Photo Service; **1.5** Shin Yoshino/Minden Pictures; **1.6a** PhotoLink/Photodisc/Getty Images; **1.6b** Janice Sheldon; **1.6c** T. D. Parsons, D. Kleinfeld, F. Raccuia-Behling and B. Salzberg. Biophysical Journal, July 1989. Photo courtesy of Brian Salzberg; **1.7** Conly L. Rieder, Ph.D.; **1.8.1** S. C. Holt/Biological Photo Service; **1.8.2** Visuals Unlimited; **1.9** Camille Tokerud/Stone/Getty Images; **1.10a** Photodisc/Getty Images; **1.11** Roy Kaltschmidt, Lawrence Berkeley National Laboratory; **1.12** Giot L, Bader JS, Brouwer C, Chaudhuri A, Kuang B, Li Y, Hao YL, Ooi CE, Godwin B, Vitols E, Vijayadamodar G, Pochart P, Machineni H, Welsh M, Kong Y, Zerhusen B, Malcolm R, Varrone Z, Collis A, Minto M, Burgess S, McDaniel L, Stimpson E, Spriggs F, Williams J, Neurath K, Ioime N, Agee M, Voss E, Furtak K, Renzulli R, Aanensen N, Carrolla S, Bickelhaupt E, Lazovatsky Y, DaSilva A, Zhong J, Stanyon CA, Finley RL Jr, White KP, Braverman M, Jarvie T, Gold S, Leach M, Knight J, Shimkets RA, McKenna MP, Chant J, Rothberg JM. A protein interaction map of Drosophila melanogaster. Science. 2003 Dec 5; 302 (5651): 1727-36; Fig. 4b. Image supplied by Joel S. Bader; **1.15a** Oliver Meckes/Nicole Ottawa/Photo Researchers; **1.15b** Ralph Robinson/Visuals Unlimited; **1.15c left to right** D. P. Wilson/Photo Researchers; Peter Lilja/Taxi/Getty Images; Kunst & Scheidulin/AGE Fotostock; Anup Shah/Nature Picture Library; **1.16.1** VVG/SPL/Photo Researchers; **1.16.2** Omikron/Photo Researchers; **1.16.3** W. L. Dentler, University of Kansas/Biological Photo Services; **1.17** Mike Hettwer/Project Exploration; **1.18** Archiv/Photo Researchers; **1.19.1** Michael P. Fogden/Bruce Coleman Inc./Alamy; **1.19.2** Michael & Patricia Fogden/Minden Pictures **1.19.3** Hal Horwitz/Corbis; **1.21** Frank Greenaway/Dorling Kindersley; **1.23.1** Karl Ammann/Corbis; **1.23.2** Tim Ridley/Dorling Kindersley; **1.25.1** Breck P. Kent; **1.25.2** E R Degginger; **1.26** David Pfennig; **1.29** Gloria M. Coruzzi, Professor of Biology, New York University; **1.30** Stone/Getty Images. **Chapter 2** **2.1** Martin Dohrn/BBC Natural History Unit; **2.2** Martin Dohrn/BBC Natural History Unit; **2.3** Chip Clark; **2.4a** Grant Heilman/Grant Heilman Photography; **2.4b** Ivan Polunin/Bruce Coleman Inc.; **2.6** Clayton T Hamilton, Stanford University; **2.7** CTI Molecular Imaging, Inc.; **p. 41** Jerry Young/Dorling Kindersley; **2.19** Runk/Schoenberg/Grant Heilman, Inc.; **p. 45** E. R. Degginger/Color-Pic, Inc. **Chapter 3** **3.1** NASA/Johnson Space Center; **3.3** R. Kessel-Shih/Visuals Unlimited; **3.4** Charles S. Lewallen; **3.6** Flip Nicklin/Minden Pictures; **3.10** Oliver Strewe/Stone Allstock/Getty Images; **3.11** Raena M. Cota; **inset** DeLois Galleciez; **p. 57** Kargel JS. Planetary science. Proof for water, hints of life? Science. 2004 Dec 3; 306(5702):1689-91; Fig. E and O. **Chapter 4** **4.1** Mark Moffett/Minden Pictures; **4.6a** Manfred Kage/Peter Arnold, Inc. **Chapter 5** **5.1** Lester Lefkowitz/Corbis; **5.6a** John N. A. Lott/Biological Photo Service; **5.6b** H. Shio and P. B. Lazarow; **5.8.1** Photodisc/Getty Images; **5.8.2** Manfred Kage/Peter Arnold, Inc.; **5.8.3** J. Litray/Visuals Unlimited; **5.9.1** Photodisc/Getty Images **5.9.2** T. J. Beveridge/Visuals Unlimited; **5.10b** F. Collet/Photo Researchers; **5.10c** Corbis; **5.12a** Dorling Kindersley; **5.12b** Corbis; **5.20** Tulip WR, Varghese JN, Laver WG, Webster RG, Colman PM. Refined crystal structure of the influenza virus N9 neuraminidase-NC41 Fab complex. J Mol Biol. 1992 Sep 5;227(1):122-48; **5.21** Wolfgang Kaehler; **5.22** Eye of Science/Photo Researchers; **5.24** Reprinted by permission from Nature. P. B. Sigler from Z. Xu, A. L. Horwich, and P. B. Sigler. 388:741-750 Copyright (c) 1997 Macmillan Magazines Limited; **5.25.1** Cramer P, Bushnell DA, Fu J, Gnatt AL, Maier-Davis B, Thompson NE, Burgess RR, Edwards AM, David PR, Kornberg RD. Architecture of RNA polymerase II and implications for the transcription mechanism.Science. 2000 Apr 28;288(5466):640-9; Fig. 3. Image supplied by David Bushnell; **5.25.2** David A. Bushnell. **Chapter 6** **6.1** Albert Tousson, High Resolution Imaging Facility, University of Alabama at Birmingham; **6.3.1** Biophoto Associates/Photo Researchers; **6.3.2** Ed Reschke; **6.3.3** David M. Phillips/Visuals Unlimited; **6.3.4** David M. Phillips/Visuals Unlimited; **6.3.5** Molecular Probes, Inc.; **6.3.6** Karl Garsha; **6.3.7** Karl Garsha; **6.4** William Dentler/Biological Photo Service; **6.6b** S. C. Holt, University of Texas Health Center/Biological Photo Service; **6.7a** Daniel S. Friend; **6.10.1** Reproduced by permission from L.Orci and A.Perelet, Freeze-Etch Histology. (Heidelberg: Springer-Veerlag, 1975). Copyright (c)1975 by Springer-Verlag GmbH & Co KG; **6.10.2** Reproduced by permission from A. C. Faberge, Cell Tiss. Res. 151 Copyright (c) 1974 by Springer-Verlag GmbH & Co KG; **6.10.3** Reprinted by permission from Nature 323. U. Aebi et al. Copyright (c) 1996: 560-564, figure 1a. Used with permission. Macmillan Magazines Limited; **6.11** D. W. Fawcett/Photo Researchers; **6.12** R. Bolender, D. Fawcett/Photo Researchers; **6.13** Don Fawcett/Visuals Unlimited; **6.14** Daniel S. Friend; **6.15** E. H. Newcomb; **6.17** Daniel S. Friend; **6.18** E.H. Newcomb & W.P. Wergin/Biological Photo Service; **6.19** From S. E. Fredrick and E. H. Newcomb, The Journal of Cell Biology 43 (1969): 343; **6.20** John E. Heuser, M.D., Washington University School of Medicine, St. Louis, Missouri; **6.21b** B. J. Schapp et al., 1985, Cell 40:455; p. 113 **left to right** Dr. Mary Osborn, Dr. Frank Solomon, Mark S. Ladinsky, and J. Richard McIntosh, University of Colorado; **6.22** Kent L. McDonald; **6.23a** Biophoto Associates/Photo Researchers; **6.23b** Meckes/Ottawa/Eye of Science/Photo Researchers, Inc.; **6.24.1** Omikron/Science Source/Photo Researchers; **6.24.2** W. L. Dentler/Biological Photo Service; **6.24.3** Linck RW, Stephens RE. Functional protofilament numbering of ciliary, flagellar, and centriolar microtubules. Cell Motil Cytoskeleton. 2007 Jul;64(7):489-95; cover. Micrograph by D. Woodrum Hensley; **6.26** From Hirokawa Nobutaka, The Journal of Cell Biology 94 (1982): 425. Fig.1. Reproduced by copyright permission of The Rockefeller University Press; **6.28** G. F. Leedale/Photo Researchers; **6.29** Paredez AR, Somerville CR, Ehrhardt DW. Visualization of Cellulose Synthase Demonstrates Functional Association with Microtubules. Science. 2006 Jun 9; 312 (5779): 1491-5; Fig. 2; **6.31** Micrograph by W. P. Wergin, provided by E. H. Newcomb; **6.32.1** From Douglas J. Kelly, The Journal of Cell Biology 28 (1966): 51. Fig.17. Reproduced by copyright permission of The Rockefeller University Press; **6.32.2** Reproduced by permission from L. Orci and A. Perrelet, Freeze-Etch Histology. (Heidelberg: Springer-Verlag). Copyright (c) 1975 by Springer-Verlag GmbH & Co KG; **6.32.3** From C. Peracchia and A. F. Dulhunty, The Journal of Cell Biology 70 (1976): 419. Fig. 6 Reproduced by copyright permission of The Rockefeller University Press; **6.33** Lennart Nilsson/Albert Bonniers Forlag; **p 123** E. H. Newcomb; p. 123 From S. E. Fredrick and E. H. Newcomb, The Journal of Cell Biology 43 (1969): 343. **Chapter 7** **7.1** Long SB, Campbell EB, Mackinnon R. Crystal structure of a mammalian voltage-dependent Shaker family K^+ channel. Science. 2005 Aug 5;309 (5736):897-903; cover; **7.4** D. W. Fawcett/Photo Researchers, Inc.; **7.14** Cabisco/Visuals Unlimited; **7.20.1** H. S. Pankratz, T.C. Beaman & P. Gerhardt/Biological Photo Service; **7.20.2** D. W. Fawcett/Photo Researchers, Inc.; **7.20.3** M. M. Perry & A. B. Gilbert, J. Cell Science 39(1979):257 (c) 1979 by the Company of Biologists Ltd; **7.20.4** M. M. Perry & A. B. Gilbert, J. Cell Science 39 (1979):257 (c) 1979 by the Company of Biologists Ltd. **Chapter 8** **8.1** Jean-Marie Bassot/Photo Researchers, Inc.; **8.2** David W. Hamilton/Getty Images, Inc.; **8.3a** Joe McDonald/Corbis; **8.3b** Manoj Shah/Getty Images, Inc.; **8.4** Brian Capon; **8.16** Thomas A. Steitz, Yale University; **8.21** Scheer JM, Romanowski MJ, Wells JA. A common allosteric site and mechanism in caspases. Proc Natl Acad Sci U S A. 2006 May 16;103(20): 7595-600; Fig. 4a; **8.23** Nicolae Simionescu. **Chapter 9** **9.1** Frans Lanting/Minden Pictures. **Chapter 10** **10.1** Bob Rowan, Progressive Image/Corbis;

10.2a Jim Brandenburg/Minden Pictures; **10.2b** Bob Evans/Peter Arnold, Inc.; **10.2c** Michael Abbey/Visuals Unlimited; **10.2d** Susan M. Barns, Ph.D.; **10.2e** National Library of Medicine; **10.3.1** M. Eichelberger/Visuals Unlimited; **10.3.2** Courtesy of W.P. Wergin and E.H. Newcomb, University of Wisconsin/BPS; **10.11** Christine Case, Skyline College; **10.20a** David Muench/Corbis; **10.20b** Dave Bartruff/Corbis. **Chapter 11** **11.1** CrystalGenomics, Inc.; **11.3.1** Dale Kaiser; **11.3.2** Kuner JM, Kaiser D. Fruiting body morphogenesis in submerged cultures of Myxococcus xanthus. J Bacteriol. 1982 Jul;151(1):458-61. By permission of ASM; **11.16** Matheos D, Metodiev M, Muller E, Stone D, Rose MD.Pheromone-induced polarization is dependent on the Fus3p MAPK acting through the formin Bni1p. J Cell Biol. 2004 Apr;165(1):99-109; Fig. 9; **11.19** Dr. Gopal Murti/Visuals Unlimited; **11.21** Development 127, 5245-5252 (2000). Mesenchymal cells engulf and clear apoptotic footplate cells in macrophageless PU. 1 null mouse embryos. William Wood, Mark Turmaine, Roberta Weber, Victoria Camp, Richard A. Maki, Scott R. McKercher and Paul Martin. **Chapter 12** **12.1** Jan-Michael Peters/Silke Hauf; **12.2a** Biophoto Associates/Photo Researchers, Inc.; **12.2b** C.R. Wyttenback/Biological Photo Service; **12.2c** Biophoto/Science Source/Photo Researchers, Inc.; **12.3** John Murray; **12.4** Biophoto/Photo Researchers, Inc.; **12.6** Conly L. Rieder; **12.7.1** J. Richard McIntosh, University of Colorado at Boulder; **12.7.2** Reproduced by permission from Matthew Schibler, from Protoplasma 137. Copyright (c) 1987: 29-44 by Springer-Verlag GmbH & Co KG; **12.9a** David M. Phillips/Visuals Unlimited; **12.9b** B. A. Palevitz, Courtesy of E. H. Newcomb, University of Wisconsin; **12.10** Carolina Biological Supply/Phototake NYC; **12.18** Guenter Albrecht-Buehler, Ph.D.; **12.19** Lan Bo Chen; p. 245 Carolina Biological Supply/Phototake NYC. **Chapter 13** **13.1** Steve Granitz/WireImage; **13.2a** Roland Birke/OKAPIA/ Photo Researchers; **13.2b** Robert Nessler; **13.3.1** Veronique Burger/Phanie Agency/Photo Researchers; **13.3.2** CNRI/Photo Researchers; **13.12** Petronczki M, Siomos MF, Nasmyth K. Un menage a quatre: the molecular biology of chromosome segregation in meiosis. Cell. 2003 Feb 21;112(4):423-40; Fig. 5a. **Chapter 14** **14.1** Mendel Museum, Augustinian Abbey, Brno; **14.14.1** altrendo nature/Getty Images; **14.14.2** Corbis; **14.15 top** Photodisc/Getty Images; **14.15 bottom** Anthony Loveday; **14.16** Rick Guidotti for Positive Exposure; **14.17** Dick Zimmerman/Shooting Star; **p 285** Norma Jubinville. **Chapter 15** **15.1** Peter Lichter and David Ward, Science 247 (1990). Copyright 1990 American Association for the Advancemcent of Science; **15.3** Cabisco/Visuals Unlimited; **15.5** Andrew Syred/Photo Researchers; **15.8** Dave King/Dorling Kindersley; **15.14** Milton H. Gallardo; **15.16.1** Lauren Shear/SPL/Photo Researchers; **15.16.2** CNRI/SPL/Photo Researchers; **p 301** Koichi Kamoshida/Getty Images; **15.19** Ken Wagner/Phototake NYC. **p 304** James K. Adams, Biology, Dalton State College, Dalton, Georgia. **Chapter 16** **16.1** National Institute of Health; **16.3** Oliver Meckes/Photo Researchers; **16.6a** Courtesy of the Library of Congress; **16.6b** From the Double Helix: by James D. Watson, Atheneum Press, N. Y., 1968, p. 215. (c) 1968. Courtesy CSHL Archive; **16.7c** Richard Wagner; **16.12a** Jerome Vinograd; **16.12b** From D. J. Burks and P. J. Stambrook, The Journal of Cell Biology 77 (1978). 762, fig.6 by copyright permission of The Rockefeller University Press. Photo provided by P. J. Stambrook; **16.20** Peter Lansdorp/Terry Fox Laboratory; **16.21.1** S. C. Holt. University of Texas, Health Science Center, San Antonio/Biological Photo Service; **16.21.2** Dr. Victoria E. Foe; **16.21.3** Barbara Hamkalo; **16.21.4** From J. R. Paulsen and U. K. Laemmli, Cell 12 (1977):817-828; **16.21.5** National Museum of Health and Medicine/Armed Forces Institute of Pathology; **16.22** Ivanovska I, Khandan T, Ito K and Orr-Weaver TL. A histone code in meiosis: the histone kinase, NHK-1, is required for proper chromosomal architecture in Drosophila oocytes. Genes & Development. 2005; 192571-2582; **p 324** Courtesy of J. Wang and T. A. Steitz, modified from J. Wang et al., Cell (1997) 89:1087-1099. (c)Cell Press. **Chapter 17** **17.1** DDP; **17.6a** Keith V. Wood, University of California, San Diego; **17.6b** Associated Press; **17.15** M. A. Rould, J.J. Perona, P. Vogt, and T.A. Steitz, Science 246 (1 December 1989): cover. Copyright 1989 by the American Association for the Advancement of Science. Thomas A. Steitz, Yale University; **17.16a** Joachim Frank; **17.20** B. Hamkalo and O. Miller, Jr; **17.24** Reproduced with permission from O. L. Miller, Jr., B. A. Hamkalo, and C. A. Thomas, Jr., Science 169 (1970): 392. Copyright (c)1970 American Association for the Advancement of Science. Fig# 3. **Chapter 18** **18.1** Cook O, Biehs B, Bier E. Brinker and optomotor-blind act coordinately to initiate development of the L5 wing vein primordium in Drosophila. Development. 2004 May;131(9):2113-24; **18.14a** Carolina Biological/ Visuals Unlimited; **18.14b** Hans Pfletschinger/Peter Arnold, Inc.; **18.18** F. Rudolf Turner, Indiana University; **18.19.1** Wolfgang Driever, University of Freiburg, Freiburg, Germany; **18.19.2** Wolfgang Driever, University of Freiburg, Freiburg, Germany; **18.19.3** Dr. Ruth Lahmann, The Whitehead Institution; **18.19.4** Dr. Ruth Lahmann, The Whitehead Institution; **18.23** University of Washington. **Chapter 19** **19.1** Science Photo Library/Photo Researchers; **19.2** Peter von Sengbusch/Botanik; **19.3.1** Robley C. Williams/Biological Photo Service; **19.3.2** R.C. Valentine and H.G. Pereira, J. Mol. Biol/Biological Photo Service; **19.3.3** K.G. Murti/Visuals Unlimited; **19.3.4** Robley C. Williams/Biological Photo Service; **19.8** C. Dauguet/Institute Pasteur/Photo Researchers; **19.9.1** National Museum of Health and Medicine, Armed Forces Institute of Pathology, Washington, DC (NCP 1603); **19.9.2** Eye of Science/Photo Researchers, Inc.; **19.9.3** Ryan Pyle/Corbis; **19.10.1** Thomas A. Zitter; **19.10.2** Dennis E. Mayhew; **19.10.3** Science VU/Wayside/Visuals Unlimited. **Chapter 20** **20.1** Reproduced with permission from R.F. Service, Science (1998) 282:396-3999. Copyright 1998 American Association for the Advancement of Science. Incyte Pharmaceuticals, Inc., Palo Alto, CA; **20.5c** L. Brent Selinger, Department of Biological Sciences, University of Lethbridge, Alberta, Canada; **20.9** Repligen Corporation; **20.14** Kosman D, Mizutani CM, Lemons D, Cox WG, McGinnis W, Bier E. Multiplex detection of RNA expression in Drosophila embryos. Science. 2004 Aug 6;305(5685):846; Fig 1; **20.15** Reproduced with permission from R.F. Service, Science (1998) 282:396-3999. Copyright 1998 American Association for the Advancement of Science. Incyte Pharmaceuticals, Inc., Palo Alto, CA; **20.19** Pat Sullivan/Associated Press; **20.23** Brad DeCecco; **20.24a** Steve Helber/ Associated Press. **Chapter 21** **21.1** Yann Arthus-Bertrand/Corbis; **21.5** Giot L, Bader JS, Brouwer C, Chaudhuri A, Kuang B, Li Y, Hao YL, Ooi CE, Godwin B, Vitols E, Vijayadamodar G, Pochart P, Machineni H, Welsh M, Kong Y, Zerhusen B, Malcolm R, Varrone Z, Collis A, Minto M, Burgess S, McDaniel L, Stimpson E, Spriggs F, Williams J, Neurath K, Ioime N, Agee M, Voss E, Furtak K, Renzulli R, Aanensen N, Carrolla S, Bickelhaupt E, Lazovatsky Y, DaSilva A, Zhong J, Stanyon CA, Finley RL Jr, White KP, Braverman M, Jarvie T, Gold S, Leach M, Knight J, Shimkets RA, McKenna MP, Chant J, Rothberg JM. A protein interaction map of Drosophila melanogaster. Science. 2003 Dec 5; 302 (5651): 1727-36, Fig. 4b. Image supplied by Joel S. Bader; **21.6** Affymetrix; **21.8.1** AP Wide World Photos; **21.8.2** Courtesy of Virginia Walbot, Stanford University; **21.10a** Courtesy of O. L. Miller Jr., Dept. of Biology, University of Virginia; **21.16** Shu W, Cho JY, Jiang Y, Zhang M, Weisz D, Elder GA, Schmeidler J, De Gasperi R, Sosa MA, Rabidou D, Santucci AC, Perl D, Morrisey E, Buxbaum JD. Altered ultrasonic vocalization in mice with a disruption in the Foxp2 gene. : Proc Natl Acad Sci U S A. 2005 Jul 5;102(27):9643-8; Fig. 3. Image supplied by Joseph Buxbaum. **Chapter 22** **22.1** Olivier Grunewald; **22.4** Michael S. Yamashita/Corbis; **22.5.1** Archiv/Photo Researchers; **22.5.2** National Maritime Museum Picture Library, London, England; **22.6 all** Tui De Roy/Minden Pictures; **22.7** Darwin's Tree of life sketch, MS.DAR.121:p.36. Reproduced with the permission of the Cambridge University Library; **22.9** Jack Wilburn/Animals Animals/Earth Scenes; **22.10** Wil Meinderts/FOTO NATURA/Minden Pictures; **22.11** Richard Packwood/Oxfor Scientific/Jupiter Images; **22.12a** Edward S. Ross, California Academy of Sciences; **22.12b** Michael & Patricia Fogden/Minden Pictures; **22.15** Mark Webster, Department of the Geophysical Sciences. University of Chicago; **22.18.1** Dwight R. Kuhn Photography; **22.18.2** Lennart Nilsson/Albert Bonniers Forlag AB. **Chapter 23** **23.1** Grant PR, Grant BR. Evolution of character displacement in Darwin's finches. Science. 2006 Jul 14;313(5784):224-6; Fig. 1a; **23.2a** Greene, E. A diet-induced developmental polymorphism in a caterpillar. Science. 1989. 243:643-646. Photo by Erick Greene; **23.2b** Erick Greene; **23.3a** Steve Gorton/Dorling Kindersley; **23.3b** Janice Britton-Davidian, ISEM, UMR 5554 CNRS, Universite Montpellier II. Reprinted by permission from Nature, Vol.403, 13 January 2000, p. 158 Copyright (c) 2000 Macmillan Magazines Limited; **23.4** New York State Department of Environmental Conservation; **23.5.1** Michio Hosino/Minden Pictures; **23.5.2** James L. Davis/ProWildlife; **23.10a** William Ervin/Photo Researchers, Inc.; **23.11** 1993 Time Magazine cover/TimePix; **23.12** J. Antonovics/Visuals Unlimited; **23.14a** Fred Bavendam/Minden Pictures; **23.14b** John Visser/Bruce Coleman USA; **23.15** Frans Lanting/Minden Pictures; **23.16a** Allison M. Welch, College of Charleston, South Carolina; **23.18** Dr. Michio Hori, Kyoto University; **23.19** Merlin D. Tuttle, Bat Conservation International. **Chapter 24** **24.1** George Harrison/Grant Heilman Photography, Inc.; **24.2a1** John Shaw/Tom Stack & Associates, Inc.; **24.2a2** Don & Pat Valenti/Tom Stack & Associates, Inc.; **24.2b1** Zefa/Corbis; **24.2b2** Photodisc/Getty Images; **24.2b3** Photodisc/Getty Images; **24.2b4** Photodisc/Getty Images; **24.2b5** Photodisc/Getty Images; **24.2b6** Masterfile; **24.3** Graeme Chapman; **24.4a** Joe McDonald/Bruce Coleman Inc.; **24.4b** Joe McDonald/Corbis; **24.4c** USDA/APHIS/Animal and Plant Health Inspection Service; **24.4d** Stephen Krasemann/Photo Researchers; **24.4e** Barbara Gerlach/Tom Stack & Associates, Inc.; **24.4f** Ueshima R, Asami T. Evolution: single-gene speciation by left-right reversal. Nature. 2003 Oct 16;425(6959):679; Fig. 1; **24.4g** William E. Ferguson; **24.4h** Charles W. Brown; **24.4i** Photodisc/Getty Images; **24.4j** Ralph A. Reinhold/Animals Animals/Earth Scenes; **24.4k** Grant Heilman/Grant Heilman Photography, Inc.; **24.4l** Kazutoshi Okuno; **24.6.1** Corbis; **24.6.2** John Shaw/Bruce Coleman, Inc.; **24.6.3** Michael Fogden/Bruce Coleman, Inc.; **24.12** Ole Seehausen; **24.13.1** Stephen Dalton/ Animals Animals - Earth Scenes; **24.13.2** ChristopheCorteau/naturepl.com; **24.15.1** Juan Martín Simón; **24.15.2** Melvin Grey/NHPA; **24.16** Ole Seehausen; **24.18a** Jason Rick; **24.19** Ueshima R, Asami T. Evolution: single-gene speciation by left-right reversal. Nature. 2003 Oct 16;425(6959):679; Fig. 1 **24.20** Bradshaw HD, Schemske DW. Allele substitution at a flower colour locus produces a pollinator shift in monkeyflowers. Nature. 2003 Nov 13;426(6963):176-8. **Chapter 25** **25.1** Gerhard Boeggemann; **p 507** William R. Hammer and

ReBecca Hunt, Department of Geology, Augustana College, Rock Island, Illinois; 25.2 George Luther, University of Delaware Graduate College of Marine Studies; 25.3a Courtesy of F. M. Menger and Kurt Gabrielson, Emory University; 25.4.1a Mitsuaki Iwago/Minden Pictures; 25.4.1b S. M. Awramik/Biological Photo Service; 25.4.2 Andrew H. Knoll; 25.4.3 Lisa-Ann Gershwin/University of California-Berkeley, Museum of Paleontology; 25.4.4 Chip Clark, National Museum of Natural History, Smithsonian Institution; 25.4.5 http://www.fossils.eu.com; 25.4.6 Chip Clark; 25.4.7 Seelevel.com; 25.4.8 Specimen No. 12478, Markus Moser, Staatliches Museum für Naturkunde Stuttgart; 25.8 Theodore J. Bornhorst, Michigan Technological University; 25.11 Shuhai Xiao, Tulane University; 25.18.1 Gerald D. Carr;25.18.2 Gerald D. Carr; 25.18.3 Gerald D. Carr; 25.18.4 Gerald D. Carr; 25.18.5 Gerald D. Carr; 25.18.6 Bruce G. Baldwin; 25.20 Stephen Dalton/Minden Pictures; 25.23 Shapiro MD, Marks ME, Peichel CL, Blackman BK, Nereng KS, Jonsson B, Schluter D, Kingsley DM. Genetic and developmental basis of evolutionary pelvic reduction in threespine sticklebacks.Nature. Erratum. 2006 Feb 23;439(7079):1014; Fig. 1. Chapter 26 26.1 Michael & Patricia Fogden/Minden Pictures; 26.2.1 Ryan McVay/Photodisc/Getty Images; 26.2.2 Neil Fletcher/Dorling Kindersley; 26.2.3 Dorling Kindersley; 26.17a Courtesy Dept. of Library Services, American Museum of Natural History; 26.17b EdHeck.com; 26.20 John W. Karapelou, CMI/Phototake. Chapter 27 27.1 Wayne P. Armstrong; 27.2a Dr. Dennis Kunkel/Visuals Unlimited; 27.2b Dr. Dennis Kunkel/Visuals Unlimited; 27.2c Stem Jems/Photo Researchers; 27.3 Jack Bostrack/Visuals Unlimited; 27.4 Dr. Immo Rantala/SPL/Photo Researchers; 27.5 Fran Heyl Associates; 27.6 Julius Adler; 27.7a S. W. Watson. ©Journal of Bacteriology, American Society of Microbiology; 27.7b N.J. Lang/Biological Photo Service; 27.8 Huntington Potter, Byrd Alzheimer's Institute and University of South Florida and David Dressler, Oxford University and Balliol College; 27.9 H.S. Pankratz, T.C. Beaman/Biological Photo Service; 27.12 Dennis Kunkel/Phototake NYC; 27.14 Susan M. Barns, Ph.D.; 27.15 Dr. Tony Brain/Science Photo Library/Photo Researchers; 27.17 Jack Dykinga/Stone/Getty Images; 27.18.1 L. Evans/Biological Photo Service; 27.18.2 Yuichi Suwa; 27.18.3 National Library of Medicine; 27.18.4 Phototake NYC; 27.18.5 Alfred Pasieka/Peter Arnold, Inc.; 27.18.6 Photo Researchers; 27.18.7 Moredon Animal Health/SPL/Photo Researchers; 27.18.8 CNRI/SPL/Photo Researchers; 27.18.9 T.E. Adams/Visuals Unlimited; 27.18.10 Frederick P. Mertz/Visuals Unlimited; 27.18.11 David M. Phillips/Visuals Unlimited; 27.19 Pascale Frey-Klett, Tree-Microbes Interaction Joint Unit, Centre INRA de Nancy; 27.20 Ken Lucas/Biological Photo Service; 27.21.1 Scott Camazine/Photo Researchers; 27.21.2 David M. Phillips/Photo Researchers; 27.21.3 James Marshall/The Image Works; 27.22.1 Seelevel.com; 27.22.2 Mirel™ natural plastics/Metabolix; 27.22.3 Courtesy of Exxon Mobil Corporation. Chapter 28 28.1 Moreira D, López-García P. The molecular ecology of microbial eukaryotes unveils a hidden world. Trends Microbiol. 2002 Jan;10(1):31-8; Fig. 4. Photo by Brian S. Leander; 28.3a Jerome Paulin/Visuals Unlimited; 28.3b Eric Condliffe/Visuals Unlimited; 28.3c1 Manfred Kage/Peter Arnold, Inc.; 28.3c2 Visuals Unlimited; 28.3d1 Manfred Kage/Peter Arnold, Inc.; 28.3d2 David J. Patterson/microscope; 28.3e Wim van Egmond/Getty Images; 28.4 David M. Phillips/Visuals Unlimited; 28.5 David J. Patterson; 28.6 Meckes/Ottawa/Photo Researchers, Inc.; 28.7 Michael Abbey/Visuals Unlimited; 28.8 Guy Brugerolle, Universitad Clearmont Ferrand; 28.9 Virginia Institute of Marine Science; 28.10 Masamichi Aikawa, Tokai University School of Medicine, Japan; 28.11 Mike Abbey/Visuals Unlimited; 28.12 Centers for Disease Control & Prevention; 28.13 Eric Condliffe/Visuals Unlimited; 28.14 Stephen Durr; 28.15 Colin Bates, http://www.coastalimageworks.com; 28.16 J.R. Waaland/Biological Photo Service; 28.17 Fred Rhoades; 28.18 Robert Brons/Biological Photo Service; 28.19.1 D. P. Wilson, Eric & David Hosking/Photo Researchers, Inc.; 28.19.2 Michael D. Guiry; 28.19.3 Biophoto Associates/Photo Researchers, Inc.; 28.19.4 Michael Yamashita/IPN/Aurora & Quanta Productions Inc; 28.19.5 David Murray/Dorling Kindersley; 28.20 Gerald and Buff Corsi/Visuals Unlimited; 28.21.1 Laurie Campbell/NHPA; 28.21.2 David L. Ballantine, Department of Marine Sciences, University of Puerto Rico; 28.22 William L. Dentler; 28.24.1 George Barron; 28.24.2 R. Calentine/Visuals Unlimited; 28.25 Robert Kay, MRC Cambridge; 28.26 Kevin Carpenter and Patrick Keeling. Chapter 29 29.1 Martin Rugner/AGE Fotostock America, Inc.; 29.2 S. C. Mueller and R. M. Brown, Jr.; 29.3a Natural Visions; 29.3b Linda Graham, University of Wisconsin-Madison; 29.5.1 Linda Graham, University of Wisconsin-Madison; 29.5.2 Photo courtesy Karen S. Renzaglia; 29.5.3 Alan S. Heilman; 29.5.4 Michael Clayton; 29.5.5 Barry Runk/Stan/Grant Heilman Photography, Inc.; 29.5.6 Ed Reschke; 29.5.7 Centers for Disease Control & Prevention; 29.6 Charles H. Wellman; 29.8 R. Kessel-Shih/Visuals Unlimited; 29.9.1 Runk/Schoenberger/Grant Heilman Photography, Inc.; 29.9.2 Linda Graham, University of Wisconsin-Madison; 29.9.3 Hidden Forest; 29.9.4 Hidden Forest; 29.9.5 Tony Wharton, Frank Lane Picture Agency/Corbis; 29.11a Brian Lightfoot/AGE Fotostock America, Inc.; 29.11b Chris Lisle/Corbis; 29.15.1 Jane Grushow/Grant Heilman Photography, Inc.; 29.15.2 Murray Fagg, Australian National Botanic Gardens; 29.15.3 Helga & Kurt Rasbach; 29.15.4 Barry Runk/Stan/Grant Heilman Photography, Inc.; 29.15.5 Milton Rand/Tom Stack & Associates, Inc.; 29.15.6 Michael Viard/Peter Arnold, Inc.; 29.16 The Open University. Chapter 30 30.1 National Museum of Natural History, Smithsonian Institution; 30.5.1 George Louin/Visuals Unlimited; 30.5.3 Grant Heilman Photography, Inc.; 30.5.4 Michael and Patricia Fogden/Minden Pictures; 30.5.5 Thomas Schoepke; 30.5.6 Michael Clayton; 30.5.7 Doug Sokell/Visuals Unlimited; 30.5.8 Raymond Gehman/Corbis; 30.5.9 Adam Jones/Getty Images, Inc.; 30.5.10 David Muench/Corbis; 30.5.11 Gunter Marx, Photography/Corbis; 30.5.12 Jaime Plaza/Wildlight Photo Agency; 30.5.13 Royal Botanic Gardens Sydney; 30.5.14 Kent, Breck P./Animals Animals/Earth Scenes; 30.8.1 Dave King/Dorling Kindersley; 30.8.2 Andy Crawford/Dorling Kindersley; 30.8.3 Dave King/Dorling Kindersley; 30.8.4 Bill Steele/Stone/Getty Images Inc.; 30.8.5 Roger Phillips/Dorling Kindersley; 30.9.1 C. P. George/Visuals Unlimited; 30.9.2 Hans Doeter Brandl, Frank Lane Picture Agency/Corbis; 30.9.3 Scott Camazine/Photo Researchers, Inc.; 30.9.4 Derek Hall/Dorling Kindersley; 30.11a David L. Dilcher; 30.13.1 Howard Rice/Dorling Kindersley; 30.13.2 Bob & Ann Simpson/Visuals Unlimited; 30.13.3 Stephen McCabe; 30.13.4 Andrew Butler/Dorling Kindersley; 30.13.5 Eric Crichton/Dorling Kindersley; 30.13.6 John Dransfield; 30.13.7 Dorling Kindersley; 30.13.8 Terry W. Eggers/Corbis; 30.13.9 Ed Reschke/Peter Arnold, Inc.; 30.13.10 Matthew Ward/Dorling Kindersley; 30.13.11 Tony Wharton, Frank Lane Picture Agency/Corbis; 30.13.12 Howard Rice/Dorling Kindersley; 30.13.13 Dr. Gerald D. Carr, PhD; p 633 Dorling Kindersley. Chapter 31 31.1 Georg Müller/www.pilzepilze.de; 31.2.1 Hans Reinhard/Taxi/Getty Images; 31.2.2 Fred Rhoades/Mycena Consulting; 31.2.3 Elmer Koneman/Visuals Unlimited; 31.4a N. Allin & G.L. Barron, University of Guelph/Biological Photo Service; 31.6.1 Jack M. Bostack/Visuals Unlimited; 31.6.2 David Scharf/Peter Arnold, Inc.; 31.7 Stephen J. Kron; 31.9 Dirk Redecker, Robin Kodner, and Linda E. Graham. Glomalean Fungi from the Ordovician. Science 15. September 2000; 289:1920-1921; 31.10 Centers for Disease Control & Prevention; 31.11.1 John Taylor; 31.11.2 Ray Watson, anadianphotography.ca; 31.11.3 Kiers ET, van der Heijden MG. Mutualistic stability in the arbuscular mycorrhizal symbiosis: exploring hypotheses of evolutionary cooperation. Ecology. 2006 Jul;87(7):1627-36; Fig. 1a. Image by Marcel van der Heijden, Swiss Federal Research Station for Agroecology and Agriculture; 31.11.4 Frank Young/Papilio/Corbis; 31.11.5 Phil Dotson/Photo Researchers, Inc.; 31.12 William E. Barstow; 31.13.1 Barry Runk/Stan/Grant Heilman Photography, Inc.; 31.13.2 Barry Runk/Stan/Grant Heilman Photography, Inc.; 31.13.3 Ed Reschke/Peter Arnold, Inc.; 31.13.4 George Barron; 31.14 G.L. Barron, University of Guelph/Biological Photo Service; 31.15 M. F. Brown/Biological Photo Service; 31.16.1 David M. Dennis/Animals Animals/Earth Scenes; 31.16.2 Viard/Jacana/Photo Researchers, Inc.; 31.16c Matt Springer; 31.17 Fred Spiegel; 31.18.1 Fletcher and Baylis/Photo Researchers, Inc.; 31.18.2 Michael Fogden/DRK Photo; 31.18.3 Konrad Wothe/Minden Pictures; 31.19 Biophoto Associates/Photo Researchers, Inc.; 31.20 Rob Simpson/Visuals Unlimited; 31.22 Mark Moffett/Minden Pictures; 31.23.1 Gerald & Buff Corsi/Visuals Unlimited; 31.23.2 Fritz Polking/Visuals Unlimited; 31.23.3 David Sieren/Visuals Unlimited; 31.24 V. Ahmadijian/Visuals Unlimited; 31.25a Brad Mogen/Visuals Unlimited; 31.25b Peter Chadwick/Dorling Kindersley; 31.25c Robert Calentine/Visuals Unlimited; 31.26 Christine Case/Skyline College. Chapter 32 32.1 Jeff Hunter/Image Bank/Getty Images; 32.4a Dennis Rice, South Australian Museum; 32.4b James G. Gehling, South Australian Museum; 32.5 J Sibbick, The Natural History Museum, London; 32.6 Wikramanayake AH, Hong M, Lee PN, Pang K, Byrum CA, Bince JM, Xu R, Martindale MQ. An ancient role for nuclear beta-catenin in the evolution of axial polarity and germ layer segregation. Nature. 2003 Nov 27;426(6965): 446-50; Fig. 2, 3 and 4; 32.12 Kent Wood/Photo Researchers, Inc.; 32.13a Carolina Biological/Visuals Unlimited. Chapter 33 33.1 C. Wolcott Henry III/National Geographic/Getty Images; 33.3.1 Andrew J. Martinez/Photo Researchers, Inc.; 33.3.2 Robert Brons/Biological Photo Service; 33.3.3 Stephen Dellaporta; 33.3.4 Gregory G. Dimijian/Photo Researchers, Inc.; 33.3.5 Ed Robinson/Pacific Stock/Photolibrary; 33.3.6 W. I. Walker/Photo Researchers, Inc.; 33.3.7 Colin Milkins/Oxford Scientific Films/Animals Animals/Earth Scenes; 33.3.8 Fred Bavendam/Peter Arnold, Inc.; 33.3.9 Lauritz Jensen/Visuals Unlimited; 33.3.10 Peter Funch; 33.3.11 Erling Svensen/UWPhoto ANS; 33.3.12 Robert Pickett/Papilio/Alamy Images; 33.3.13 Peter Batson/Image Quest Marine; 33.3.14 Reinhart Mobjerg Kristensen; 33.3.15 Erling Svensen/UWPhoto ANS; 33.3.16 Andrew Syred/Photo Researchers, Inc.; 33.3.17 Thomas Stromberg; 33.3.18 Reproduced with permission from A. Eizinger and R. Sommer, Max Planck Institut fur entwicklungsbiologie, Tubingen. Copyright 2000 American Association for the Advancement of Science. Cover 278(5337) 17 Oct 97; 33.3.19 Tim Flach/Stone/Getty Images; 33.3.20 Heather Angel/Natural Visions; 33.3.21 Robert Harding World Imagery/Alamy Images; 33.3.22 Robert Brons/Biological Photo Service; 33.3.23 Wim van Egmond/Visuals Unlimited; 33.4 Andrew J. Martinez/Photo Researchers, Inc.; 33.7a Andrew J. Martinez/Photo Researchers, Inc.; 33.7b Robert Brons/Biological Photo Service; 33.7c Great Barrier Reef Marine Park Authority; 33.7d Neil G. McDaniel/Photo Researchers, Inc.; 33.8 Robert Brons/Biological Photo Service; 33.9 Ed Robinson/Pacific Stock/Photolibrary;

33.11 Centers for Disease Control and Prevention (CDC); **33.12** Stanley Fleger/Visuals Unlimited; **33.13** W. I. Walker/Photo Researchers, Inc.; **33.14a** Colin Milkins, Oxford Scientific Films/Animals Animals/Earth Scenes; **33.14b** Fred Bavendam/Peter Arnold, Inc.; **33.16** Jeff Foott/Tom Stack & Associates, Inc.; **33.17a** Gerry Ellis/Minden Pictures; **33.17b** Corbis; **33.19** Harold W. Pratt/Biological Photo Service; **33.21a** Robert Pickett/Papilio/Alamy Images; **33.21b** Mark Conlin/Image Quest Marine; **33.21c** Jonathan Blair/Corbis; **33.22** A.N.T./Photoshot/NHPA Limited; **33.23** Peter Batson/Image Quest Marine; **33.24** Astrid & Hanns-Frieder Michler/Photo Researchers, Inc.; **33.25** Reproduced with permission from A. Eizinger and R. Sommer, Max Planck Institut fur entwicklungsbiologie, Tubingen. Copyright 2000 American Association for the Advancement of Science. Cover 278(5337) 17 Oct 97; **33.26** Science Photo Library/Photo Researchers, Inc.; **33.27** Dan Cooper, http://www.isotelus.com; **33.28** Grenier JK, Garber TL, Warren R, Whitington PM, Carroll S. Evolution of the entire arthropod Hox gene set predated the origin and radiation of the onychophoran/arthropod clade. Curr Biol. 1997 Aug 1;7(8):547-53; Fig. 3c; **33.30** Milton Tierney, Jr./Visuals Unlimited; **33.31.1** Tim Flach/Stone/Getty Images; **33.31.2** Andrew Syred/Photo Researchers, Inc.; **33.31.3** Peter Arnold/Peter Arnold, Inc.; **33.33** Premaphotos/Nature Picture Library; **33.34** Tom McHugh/Photo Researchers, Inc.; **33.36 all** John Shaw/Tom Stack & Associates, Inc.; **33.38a** Maximilian Weinzierl/Alamy Images; **33.38b** Peter Herring/Image Quest Marine; **33.38c** Peter Parks/Image Quest Marine; **33.40a** Fred Bavendam/Minden Pictures; **33.40b** Jeff Rotman/Photo Researchers, Inc.; **33.40c** Robert Harding World Imagery/Alamy Images; **33.40d** Jurgen Freund/Nature Picture Library; **33.40e** Hal Beral/Corbis; **33.40f** Daniel Janies. **Chapter 34** **34.1** Xian-guang H, Aldridge RJ, Siveter DJ, Siveter DJ, Xiang-hong F. New evidence on the anatomy and phylogeny of the earliest vertebrates. Proceedings of the Royal Society of London-B-Biological Sciences, Sept. 22, 2002; 269 (1503) 1865-1869; Fig. 1c; **34.4** Runk/Schoenberger/Grant Heilman Photography, Inc.; **34.5a** Robert Bron/Biological Photo Service; **34.8** Nanjing Institute of Geology and Palaeontology; **34.9** Tom McHugh/Photo Researchers, Inc.; **34.10** Breck P. Kent/Animals Animals/Earth Scenes; **34.11** Mark A. Purnell, Department of Geology, University of Leicester, UK; **34.14** The Field Museum; **34.15a** Carlos Villoch/Image Quest Marine; **34.15b** Masa Ushioda/Image Quest Marine; **34.15c** Jeff Mondragon/Mondragon Photo; **34.17a** James D. Watt/Image Quest Marine; **34.17b** Fred Bavendam/Minden Pictures; **34.17c** Marevision/AGE Fotostock America, Inc.; **34.17d** Fred McConnaughey/Photo Researchers, Inc.; **34.18** Arnaz Mehta; **34.21a** Alberto Fernandez/AGE Fotostock America, Inc.; **34.21b** Michael Fogden/Bruce Coleman Inc.; **34.21c** Michael Fogden/Bruce Coleman Inc.; **34.22a** John Cancalosi/Peter Arnold, Inc.; **34.22b** Stephen Dalton/Photo Researchers, Inc.; **34.22c** Hans Pfletschinger/Peter Arnold, Inc.; **34.23** Michael & Patricia Fogden/Minden Pictures; **34.26** Michael & Patricia Fogden/Minden Pictures; **34.27a** Doug Wechsler; **34.27b** Matt T. Lee; **34.27c** Michael & Patricia Fogden/Minden Pictures; **34.27d** Medford Taylor/National Geographic Image Collection; **34.27e** Carl & Ann Purcell/Corbis; **34.28a** Stephen J. Kraseman/DRK Photo; **34.28b** Janice Sheldon; **34.30a** Russell Mountford/Alamy Images; **34.30b** Corbis; **34.30c** Frans Lanting/Minden Pictures; **34.30d** franzfoto.com/Alamy Images; **34.32.1** D. Parer and E. Parer Cook/Auscape International Proprietary Ltd.; **34.32.2** Mervyn Griffiths/Commonwealth Scientific and Industrial Research Organization; **34.33a** ARDEA/Retna Ltd.; **34.33b** Fritz Prenzel/Animals Animals/Earth Scenes; **34.36** Frans Lanting/Minden Pictures; **34.38a** Kevin Schafer/AGE Fotostock America, Inc.; **34.38b** Frans Lanting/Minden Pictures; **34.39a** Morales/AGE Fotostock America, Inc.; **34.39b** Anup Shah/Image State/Alamy Images; **34.39c** T. J. Rick/Nature Picture Library; **34.39d** E. A. Janes/AGE Fotostock America, Inc.; **34.39e** Frans Lanting/Minden Pictures; **34.41a** The Cleveland Museum of Natural History; **34.41b** John Reader/SPL/Photo Researchers, Inc.; **34.41c** John Gurche Studios; **34.42.1** Alan Walker @ National Museums of Kenya. Printed with permission; **34.42.2** J.H. Matternes; **34.44** David L. Brill/Brill Atlanta; **34.45** C. Henshilwood & F. d'Errico. **Chapter 35** **35.3** Robert & Linda Mitchell/Robert & Linda Mitchell Photography; **35.4.1** James Strawser/Grant Heilman Photography, Inc.; **35.4.2** Rob Walls/Alamy; **35.4.3** Drew Weiner; **35.4.4** Robert Holmes/Corbis; **35.4.5** Geoff Tompkinson/Science Photo Library/Photo Researchers, Inc.; **35.5.1** Lee W. Wilcox; **35.5.2** Gusto Production/Science Photo Library/Photo Researchers, Inc.; **35.5.3** Dorling Kindersley; **35.5.4** Barry Runk/Stan/Grant Heilman Photography, Inc.; **35.7.1** Scott Camazine/Photo Researchers, Inc.; **35.7.2** Fritz Polking/Visuals Unlimited; **35.7.3** Mike Zens/Corbis; **35.7.4** Jerome Wexler/Photo Researchers, Inc.; **35.7.5** James Strawser/Grant Heilman Photography, Inc.; **35.9** Purdue Extension Entomology; **35.10.1** Brian Capon; **35.10.2** Ed Reschke/Peter Arnold, Inc.; **35.10.3** Graham Kent, Benjamin Cummings; **35.10.4** Graham Kent, Benjamin Cummings; **35.10.5** Richard Kessel and Gene Shih/Visuals Unlimited; **35.10.6** Reproduced with permission from Plant Cell Biology on CD, by B E S Gunning, http://www.plantcellbiologyonCD.com; **35.10.7** Ray F. Evert; **35.10.8** Dr. Richard Kessel & Dr. Gene Shih/Visuals Unlimited; **35.13** Carolina Biological Supply/Phototake NYC; **35.14a1** Ed Reschke; **35.14a2** Carolina Biological Supply/Phototake NYC; **35.14b** Ed Reschke; **35.15** Michael Clayton; **35.16** Michael Clayton; **35.17** Ed Reschke; **35.18** Ed Reschke; **35.19b1** Michael Clayton; **35.19b2** Alison W. Roberts; **35.21** Edward Cook, Lamont-Doherty Earth Observatory, Columbia University, Palisades, NY; **35.23** California Historical Society Collection (CHS-1177), University of Southern California on behalf of the USC Specialized Libraries and Archival Collections; **35.24** Reproduced by permission from Janet Braam, Cell 60 (9 February 1990): Cover. Copyright (c)1990 Cell Press. Image courtesy of Elsevier Sciences Ltd; **35.26** Susan Wick, University of Minnesota; **35.27** B. Wells and K. Roberts; **35.28a-b** From figure 1 in B. Scheres et al, Development 121:53. (c) 1995 The Company of Biologists Ltd; **35.28c** From figure 6c in R. Torres Ruiz and G. Jurgens, Development 120:2967-2978. (c) 1994 The Company of Biologists Ltd; **35.29** From figure 1a in U. Mayer et al, Development 117 (1): 149-162. (c) 1993 The Company of Biologists Ltd; **35.30** Reproduced by permission from figure 1 in D. Hareven et al, Cell 84 (5): 735-744. Copyright (c) 1996. by Elsevier Science Ltd; **35.31** Reproduced by permission from Figure 2g in Hung et al, Plant Physiology 117:73-84. Copyright (c) 1998 by the American Society of Plant Biologists. Image courtesy of John Schiefelbein/Univesity of Michigan; **35.32** Dr. Gerald D. Carr, PhD; **35.33** Dr. E. M. Meyerowitz and John Bowman, Development 112 1991:1-231.2, Division of Biology, California Institute of Technology; **p 763** Kitin PB, Fujii T, Abe H and Funada R. Anatomy of the vessel network within and between tree rings of Fraxinus lanuginosa (Oleaceae). American Journal of Botany. 2004;91:779-788; Fig. 1. **Chapter 36** **36.1** Peggy Heard/FLPA/Alamy Images; **36.3** Rolf Rutishauser; **36.5** Dana Richter/Visuals Unlimited; **36.10** Nigel Cattlin/Holt Studios International/Photo Researchers, Inc.; **36.13** Scott Camazine/Photo Researchers, Inc.; **36.16** Jeremy Burgess /Science Photo Library/Photo Researchers, Inc.; **36.18.1** C. C. Lockwood/Animals Animals - Earth Scenes; **36.18.2** Kate Shane; **36.18.3** Frans Lanting/Minden Pictures; **36.18.4** John D. Cunningham/Visuals Unlimited; **36.18.5** Andrew de Lory/Dorling Kindersley; **36.18.6** Maddie Thornhill/Garden World Images Ltd/Alamy; **36.21** M. H. Zimmermann, courtesy of Professor P. B. Tomlinson, Harvard University; **36.22** Kim I, Hempel FD, Sha K, Pfluger J, Zambryski PC. Identification of a developmental transition in plasmodesmatal function during embryogenesis in Arabidopsis thaliana. Development. 2002 Mar;129(5):1261-72; Fig. 3g, 6f, 8b and 8c. Images supplied by Patricia Zambryski. **Chapter 37** 37.1 NOAA Photo Library, Historic NWS collection; **37.2** Agricultural Research Service/USDA; **37.4.1** U.S. Geological Survey, Denver; **37.4.2** USGS/Menlo Park; **37.5** Kevin Horan/Stone/Getty Images Inc.; **37.7** Maurice Reece. From the Country Gentleman, courtesy of the Curtis Publishing Co.; **37.8** White et al, Plant Physiology, June 2003; **37.10a** Breck P. Kent/Animals Animals/Earth Scenes; **37.10b** E. H. Newcomb and S. R. Tandon/Biological Photo Service; **37.12a** Gerald Van Dyke/Visuals Unlimited; **37.12b** Carolina Biological Supply/Phototake NYC; **37.13** Elizabeth J. Czarapata; **37.14.1** Wolfgang Kaehler/Corbis; **37.14.2** James Strawser/Grant Heilman Photography, Inc.; **37.14.3** Kevin Schafer/Corbis; **37.14.4** Andrew Syred/Science Photo Library/Photo Researchers, Inc.; **37.14.5** Gary W. Carter/Corbis; **37.14.6** Kim Taylor and Jane Burton/Dorling Kindersley; **37.14.7** Biophoto Associates/Photo Researchers, Inc.; **37.14.8** Philip Blenkinsop/Dorling Kindersley; **37.14.9** Dr. Paul A. Zahl/Photo Researchers, Inc.; **37.14.10** Fritz Polking, Frank Lane Picture Agency/Corbis; **37.15** R. Ronacordi/Visuals Unlimited. **Chapter 38** **38.1** Pierre-Michel Blais, orchidees.provence.free.fr; **38.2.1** Craig Lovell/Corbis; **38.2.2** John Cancalosi/Nature Picture Library; **38.3.1** Ed Reschke; **38.3.2** David Scharf/Peter Arnold, Inc.; **38.3.3** Ed Reschke; **38.4.1** Marianne Wiora, gartenspaziergang.de; **38.4.2** Stephen Dalton/NHPA; **38.4.3** Bjørn Rørslett, naturfotograf.com; **38.4.4** Doug Backlund; **38.4.5** Martin Heigan, anti-matter-3d.com; **38.4.6** Michael & Patricia Fogden/Minden Pictures; **38.4.7** Merlin D. Tuttle, Bat Conservation International; **38.6** Palanivelu R, Brass L, Edlund AF, Preuss D. Pollen tube growth and guidance is regulated by POP2, an Arabidopsis gene that controls GABA levels. Cell. 2003 Jul 11; 114(1):47-59; Fig. 5ab; **38.11.1** BIOS Klein & Hubert/Peter Arnold, Inc.; **38.11.2** Satoshi Kuribayashi/Nature Production; **38.11.3** Brian Gordon Green/National Geographic Image Collection; **38.11.4** Steve Bloom Images/Alamy; **38.11.5** Aaron McCoy/Botanica/Photolibrary; **38.11.6** California Department of Food and Agriculture's Plant Health and Pest Prevention Services; **38.11.7** Steve Shattuck/CSIRO Division Entomology; **38.11.8** Kim A. Cabrera, http://www.beartracker.com; **38.11.9** Alan Williams/Alamy; **38.12** David Cavagnaro/DRK Photo; **38.13.1** Marcel E. Dorken; **38.13.2** Marcel E. Dorken; **38.13.3** Nobumitsu Kawakubo, Gifu University, Japan; **38.14** Bruce Iverson, Photomicrography; **38.15** Sinclair Stammers/Photo Researchers, Inc.; **38.16** Andrew McRobb/Dorling Kindersley; **38.17** Richard M. Manshardt; **38.18.1** Peter Berger; **38.18.2** Adrian Arbib/Peter Arnold, Inc. **Chapter 39** **39.1** Plant Physiol. 1999 Jul;120(3): Cover. By permission of the American Society of Plant Physiologists. Illustration by Niemeyer MI and Fernandez MC; **39.2** Natalie B. Bronstein, Mercy College, Dobbs Ferry, New York; **39.7** Regulation of Polar Auxin transport ATPIN1 in Arabidopsis Vascular Tissue," by Leo Galweiler, et al) Science 18 DEC 1998, vol. 282; **39.9** Malcolm B. Wilkins, University of Glasgow, Glasgow, Scotland, U.K.; **39.10a** Mark Doyle/Department of Biochemistry, University of Wisconsin, Madison; **39.10b** Fred Jensen; **39.12.1** Mia Molvray, http://www.molvray.com; **39.12.2** Karen E. Koch; **39.14a** Kurt Stepnitz, DOE Plant Research Laboratory, Michigan State University; **39.14b** Joe Kieber, University of North Carolina; **39.15** Ed Reschke; **39.16** Malcolm B. Wilkins, University of Glasgow, Glasgow, Scotland, U.K.; **39.17** Malcolm B. Wilkins, University of Glasgow, Glasgow, Scotland, U.K.; **39.20** Malcolm B. Wilkins, University of Glasgow, Glasgow, Scotland, U.K.; **39.24** Michael Evans, Ohio State University; **39.25** Reproduced by permission from Janet Braam, Cell 60 (9 February 1990: Cover. Copyright (c)1990 Cell Press. Image courtesy of Elesevier Sciences Ltd; **39.26a-b** David Sieren/Visuals Unlimited; **39.26c** From K. Esau. Anatomy of Seed Plants, 2nd ed. (New York: John Wiley and sons, 1977), fig. 19.4, p.358; **39.27** J. L. Basq and M. C. Drew. **Chapter 40 40.1** Joe McDonald/ McDonald Wildlife Photography; **40.2a** Flip Nicklin/Minden Pictures; **40.2b** Tui De Roy/Minden Pictures; **40.2c** Norbert Wu/Minden Pictures; **40.4.1** G. Shih-R. Kessel/Visuals Unlimited; **40.4.2** D. M. Phillips/Visuals Unlimited; **40.4.3** Visuals Unlimited; **40.5.1** CNRI/SPL/Photo Researchers; **40.5.3** Chuck Brown/ Photo Researchers; **40.5.4** Science VU/Visuals Unlimited; **40.5.5** Nina Zanetti; **40.5.6** Nina Zanetti; **40.5.7** Dr. Gopal Murti/SPL/Photo Researchers; **40.5.8** Nina Zanetti; **40.5.9** Manfred Kage/Peter Arnold, Inc.; **40.5.10** Gladden Willis, M. D./Visuals Unlimited; **40.5.11** Thomas Deerinck/National Center for Microscopy and Imaging Research, University of California, San Diego; **40.5.12** Ulrich Gärtner, Paul Flechsig Institute for Brain Research, Dept. Neuroanatomy, Universität Leipzig, Germany; **40.9a** Patricio Robles Gil/naturepl.com; **40.9b** Jim Brandenburg/Minden Pictures; **40.13** Robert Ganz; **40.18** Stan Lindstedt; **40.21** John Gerlach/Visuals Unlimited. **Chapter 41 41.1** Michael deYoung/ Corbis; **41.2** Corbis; **41.3** Roland Seitre/Peter Arnold, Inc.; **41.4** Thomas Mangelsen/Minden Pictures; **41.6.1** Brandon Cole; **41.6.2** Thomas Eisner; **41.6.3** Photo Lennart Nilsson/Albert Bonniers Forlag AB; **41.6.4** Gunter Ziesler/Peter Arnold, Inc.; **41.12** Fred E. Hossler/Visuals Unlimited; **41.17** Centers for Disease Control & Prevention; **41.19.1** Photodisc/Getty Images; **41.19.2** EyeWire Collection/Photodisc/Getty Images; **41.22** Susumu Nishinaga/ SPL/Photo Researchers; **41.24** Photo courtesy of The Jackson Laboratory, Bar Harbor, Maine; **41.25** Wolfgang Kaehler/Corbis. **Chapter 42 42.1** Jane Burton/Bruce Coleman USA; **42.2a** © 2002 Norbert Wu /www.norbertwu.com; **42.2b** Dr. John D. Cunningham/Visuals Unlimited; **42.10.1** Dr. Richard Kessel & Dr. Randy Kardon/Visuals Unlimited; **42.10.2** Photo Lennart Nilsson/Albert Bonniers Forlag; **42.17** Nina Zanetti; **42.18** Science Source/Photo Researchers; **42.19** Nina Zanetti; **42.20a** Ed Reschke; **42.20b** W. Ober/Visuals Unlimited; **42.21a** Peter Batson/Image Quest 3-D; **42.21b** Dave Haas/The Image Finders; **42.21c** Frans Lanting/Minden Pictures; **42.23** Prepared by Dr. Hong Y. Yan, University of Kentcky and Dr. Peng Chai, University of Texas; **42.24.1** Dr. Richard Kessel & Dr. Randy Kardon/Visuals Unlimited; **42.24.2** CNRI/SPL/ Photo Researchers; **42.26** Hans-Rainer Duncker, University of Giessen, Germany. **Chapter 43 43.1** Biology Media/Science Source/Photo Researchers; **43.4** Ferrandon D, Jung AC, Criqui M, Lemaitre B, Uttenweiler-Joseph S, Michaut L, Reichhart J, Hoffmann JA. A drosomycin-GFP reporter transgene reveals a local immune response in Drosophila that is not dependent on the Toll pathway. EMBO J. 1998 Aug 10;17(5):1217-27; Fig. 1a; **43.17** David Scharf/Peter Arnold, Inc.; **43.19** Gopal Murti/Phototake NYC; **43.22** Alamy Images; **43.24** Alain Pol/ISM/PhototakeUSA. **Chapter 44 44.1** Felix Heintzenberg, Biofokus Nature Photography; **44.3** Brandon Cole; **44.5** Dr. John Crowe, University of California, Davis; **44.14b** Lise Bankir; **44.14d** Visuals Unlimited; **44.17** Gerry Ellis/Minden Pictures; **44.18** Michael & Patricia Fodgen/Minden Pictures. **Chapter 45 45.1.1** Ralph A. Clevenger/Corbis; **45.1.2** Stuart Wilson/Photo Researchers, Inc.; **45.9.1** Breck P. Kent/Animals Animals - Earth Scenes; **45.9.2** Leonard Rue Enterprises/Animals Animals - Earth Scenes; **45.19** ISM/Phototake. **Chapter 46 46.1** Robin Chittenden/Corbis; **46.2** David Wrobel; **46.4a** P. de Vries, courtesy of David Crews; **46.5** Dwight Kuhn/Dwight R. Kuhn Photography **46.6** William E. Ferguson; **46.17** Photo Lennart Nilsson/Albert Bonniers Forlag AB. A Child is Born, Dell Publishing. **Chapter 47 47.1** Photo Lennart Nilsson/Albert Bonniers Forlag AB. Gunilla Hedesund; **47.2** Historical Collections, College of Physicians; **47.4** Exp Vacquier VD, Payne JE. Methods for quantitating sea urchin sperm-egg binding. Exp Cell Res. 1973 Nov; 82(1):227-35; **47.4** Res Hafner, M., Petzelt, C., Nobiling, R., Pawley, J., Kramp, D. and G. Schatten. Wave of Free Calcium at Fertilization in the Sea Urchin Egg Visualized with Fura-2 Cell Motil. Cytoskel., 9:271-277 (1988); **47.6** George von Dassow; **47.8** R. Kessel & G. Shih/Visuals Unlimited; **47.9** Charles A. Ettensohn; **47.12a** P. Huw Williams and Jim Smith, The Wellcome Trust/Cancer Research UK Gurdon Institute; **47.12c** Thomas Poole, SUNY Health Science Center; **47.13b** Carolina Biological Supply/Phototake; **47.19** Dr. Janet Heasman; **47.20** Marsden M, DeSimone DW. Integrin-ECM interactions regulate cadherin-dependent cell adhesion and are required for convergent extension in Xenopus. Curr Biol. 2003 Jul 15;13(14):1182-91; **47.21** Hiroki Nishida, Developmental Biology 121 (1987): 526. Reprinted by permission of Academic Press; **47.22** J. E. Sulston and H. R. Horvitz, Dev. Biol. 56 (1977):110-156; **47.25** Kathryn W. Tosney, University of Michigan; **47.26** Dennis Summerbell; **47.27** CNRI/Photo Researchers, Inc. **Chapter 48 48.1** Marine Themes Pty Ltd, marinethemes.com; **48.2** David Fleetham/Alamy; **48.5b** Thomas Deerinck/National Center for Microscopy and Imaging Research, University of California, San Diego; **48.12** Bear, Connors, and Paradiso, Neuroscience: Exploring the Brain (c) 1996, p. 43; **48.14** Edwin R. Lewis, University of California at Berkeley. **Chapter 49 49.1** G. Fesl MD, Dep. of Neuroradiology, LMU Munich, Germany; **49.6b** N. Kedersha/Photo Researchers, Inc.; **49.17** Marcus E. Raichle, M.D., Washington University Medical Center; **49.23** Martin M. Rotker/Photo Researchers, Inc.; **49.24** Fred H. Gage, The Salk Institute, Laboratory of Genetics. **Chapter 50 50.1** Stephen Dalton/Photoshot/NHPA Limited; **50.4.1** OSF/Animals Animals/Earth Scenes; **50.4.2** R. A. Steinbrecht; **50.5a** Joe McDonald/Animals Animals/Earth Scenes; **50.5b** Flip Nicklin/Minden Pictures; **50.7** From Richard Elzinga, Fundamentals of Entomology 3 ed. (c)1987, p. 185. Reprinted by permission of Prentice-Hall, Upper Saddle River, NJ; **50.8** Peter Gillespie; **50.17a** USDA/APHIS Animal And Plant Health Inspection Service; **50.25** Clara Franzini-Armstrong, University of Pennsylvania; **50.26** Courtesy of Dr. H. E. Huxley; **50.35** Dave Watts/NHPA/Photo Researchers, Inc.; **50.36** Vance A. Tucker. **Chapter 51 51.1** Steve Kaufman/ Corbis; **51.2** Michael & Patricia Fogden/Minden Pictures; **51.5** Michael Quinton/Minden Pictures; **51.6** Michael Nichols/National Geographic/Getty Images; **51.8** Kenneth Lorenzen, UC Davis; **51.10a** Thomas McAvoy/Life Magazine/Getty Images; **51.10b** Operation Migration Inc.; **51.12** Harry Engels/Animals Animals/Earth Scenes; **51.13** Clive Bromhall/OSF/Animals Animals/Earth Scenes; **51.14** Robert Pickett/Corbis; **51.16** Rory Doolin; **51.17** Lowell L. Getz and Lisa Davis; **51.20a** Thomas Mangelsen/Minden Pictures; **51.20b** Robinson, James H./Animals Animals/Earth Scenes; **51.20c** Bill Schomoker; **51.21** Fred Bavendam/Minden Pictures; **51.22** Courtesy of Gerald S. Wilkinson; from G.S. Wilkinson and G. N. Dodson, in J. Choe and B. Crespi (eds)., The Evolution of Mating Systems in Insects and Arachnids, Cambridge University Press, Cambridge (1997), pp. 310-328; **51.23** Cyril Laubscher/Dorling Kindersley; **51.25** Mitsuaki Iwago/Minden Pictures; **51.26** Erik Svensson, Lund University, Sweden; **51.27** Jennifer Jarvis; **51.29** Stephen J. Krasemann/Peter Arnold, Inc.; **51.31.1** Richard Wrangham; **51.31.2** Alissa Crandall/Corbis; **51.32** Ulf Wallin/Riser/Getty Images. **Chapter 52 52.1** Flip Nicklin/Minden Pictures; **52.2.1** Science Faction/Getty Images; **52.2.2** Joe McDonald/Corbis; **52.2.3** Tom Bean/Corbis; **52.2.4** B. Tharp/Photo Researchers, Inc.; **52.2.5** Yann Arthus-Bertrand/Corbis; **52.2.6** Reto Stöckli, Alan Nelson, Fritz Hasler, Laboratory for Atmospheres, Goddard Space Flight Center, NASA; **52.4** Erich Hartmann/ Magnum Photos; **52.5** Geoff Dann/Dorling Kindersley; **52.7** Peter Llewellyn/ Alamy; **52.9** Tim Fitzharris/Minden Pictures; **52.18.1** Allen Russell/Index Stock Imagery, Inc.; **52.18.2** Gerry Ellis/Minden Pictures; **52.18.3** David Muench/ Corbis; **52.18.4** Ron Watts/Corbis; **52.18.5** Charles McDowell/Grant Heilman Photography, Inc.; **52.18.6** James Randklev/Image Bank/Getty Images; **52.18.7** Stuart Westmorland/Corbis; **52.18.8** Stuart Westmorland/Corbis; **52.18.9** Digital Vision/Getty Images; **52.18.10** William Lange/Woods Hole Oceanographic Institution; **52.21.1** Frans Lanting/Minden Pictures; **52.21.2** Gordon Whitten/ Corbis; **52.21.3** Wolfgang Kaehler/Corbis; **52.21.4** John D. Cunningham/Visuals Unlimited; **52.21.5** Tom Bean/Image Bank/Getty Images; **52.21.6** Bill Ross/Corbis; **52.21.7** Kennan Ward/Corbis; **52.21.8** Darrell Gulin/Corbis; **Chapter 53 53.1** David Tipling/Nature Picture Library; **53.2** Brandon Cole, http://www.brandoncole.com; **53.4a** Bernard Castelein/Nature Picture Library/ Alamy; **53.4b** Frans Lanting/Minden Pictures; **53.4c** Niall Benvie/Corbis; **53.7** Tom Bean/Corbis; **53.8** H. Willcox/Wildlife Picture/Peter Arnold, Inc.; **53.9a** Jean Louis Batt/Taxi/Getty Images, Inc.; **53.9b** Christine Osborne/Corbis; **53.14** Gerry Ellis/Minden Pictures; **53.17a** Adrian Bailey/ Aurora & Quanta Productions Inc.; **53.17b** Wolfgang Kaehler/Corbis; **53.20** Joe McDonald/ Corbis; **53.21** Niclas Fritzén; **53.27** Marc L. Imhoff, Lahouari Bounoua, Taylor Ricketts, Colby Loucks, Robert Harriss and William T. Lawrence/NASA's Goddard Space Flight Center. **Chapter 54 54.1** Phil Degginger/Alamy; **54.2.1** Kevin deQueiroz, National Museum of Natural History, Smithsonian Institution, Smithsonian Institution; **54.2.2** Joseph T. Collins/Photo Researchers, Inc.; **54.5a** C. Allan Morgan/Peter Arnold, Inc.; **54.5b** Fogden/Corbis; **54.5c1** Stephen J. Krasemann/Photo Researchers, Inc.; **54.5c2** Michael & Patricia Fodgen/Minden Pictures; **54.5d1** Edward S. Ross, California Academy of Sciences; **54.5d2** Runk/ Schoenberger/Grant Heilman Photography, Inc.; **54.6** Fred Bavendam/Minden Pictures; **54.7a** Fogden/Corbis; **54.7b** Daniel H. Janzen; **54.8** Peter Johnson/ Corbis; **54.15** Nancy Rotenberg/ Animals Animals - Earth Scenes; **54.17** Tom and Pat Leeson; **54.18** Sally D. Hacker; **54.20** Frank Gilliam; **54.21a** Michael Quinton/Minden Pictures; **54.21b** Scott T. Smith/Corbis; **54.22.1** Charles Mauzy/Corbis; **54.22.2** Tom Bean/DRK Photo; **54.22.3** Terry Donnelly, Donnelly-Austin Photography; **54.22.4** Glacier Bay National Park Photo/Glacier Bay National Park and Preserve; **54.24.1** R. Grant Gilmore, Dynamac Corporation; **54.24.2** Lance Horn, National Undersea Research Center, University of North Carolina-Wilmington; **54.29** Andy Bruckner, NOAA Fisheries; **54.30** Josh Spice, http://www.wildimages.blogspot.com. **Chapter 55 55.1** Galen Rowell/Corbis; **55.2** Annette Summers Engel, Department of Geology and Geophysics, Louisiana State University; **55.3** Fritz Poelking/AGE Fotostock; **55.6** Earth Observatory NASA; **55.7** Reprinted with permission from

D. W. Schindler, Science 184 (24 May 1974): 897, Figure 1.49.20. Copyright 1974 American Association for the Advancement of Science; **55.9** TK; **55.12** Thomas Del Brase/Photographer's Choice/Getty Images; **55.16a** Hubbard Brook Research Foundation; **55.16b** USDA Forest Service; **55.17** Arthur C. Smith III/Grant Heilman Photography; **55.18** NASA; **55.22** Prof. William H. Schlesinger; **55.25** NASA. **Chapter 56 56.1** Stephen J Richards; **56.2** Wayne Lawler/Ecoscene/ Corbis; **56.4a** Neil Lucas/Nature Picture Library; **56.4b** Mark Carwardine/Still Pictures/Peter Arnold, Inc.; **56.4c** Nazir Foead; **56.5** Merlin D. Tuttle, Bat Conservation International; **56.6** Scott Camazine/Photo Researchers, Inc.; **56.7** Michael Edwards/Getty Images; **56.8a** Michael Fodgen/Animals Animals/Earth Scenes; **56.8b** Robert Ginn/PhotoEdit Inc.; **56.9** Richard Vogel/Liaison/Getty Images, Inc.; **56.11** William Ervin/Photo Researchers, Inc.; **56.12** Lance Craighead/The Craighead Environmental Research Institute; **56.13a1** Tim Thompson/Corbis; **56.13a2** David Sieren/Visuals Unlimited; **56.13b** Blanche Haning/The Lamplighter; **56.14a** Yann Arthus-Bertrand/ Corbis; **56.14b** James P. Blair/National Geographic Image Collection; **56.15** R. O. Bierregaard, Jr., Biology Dept., University of North Carolina, Charlotte; **56.16** SPL/Photo Researchers, Inc.; **56.19b** Frans Lanting/Minden Pictures; **56.20** Mark Chiappone and Steven Miller, Center for Marine Science, University of North Carolina-Wilmington, Key Largo, Florida; **56.21** Princeton Hydro, LLC, Ringoes, NJ; **56.22** U.S. Department of Energy; **56.23.1** Stewart Rood, University of Lethbridge; **56.23.2** Daniel H. Janzen, University of Pennsylvania; **56.23.3** Photo provided by Kissimmee Division staff, South Florida Water Management District (WPB; **56.23.4** Tim Day, Xcluder Pest Proof Fencing Company; **56.23.5** Bert Boekhoven; **56.23.6** Jean Hall/Holt Studio/Photo Researchers, Inc.; **56.23.7** Kenji Morita/Environment Division, Tokyo Kyuei Co., Ltd; **56.25a** Serge de Sazo/Photo Researchers, Inc.; **56.25b** AP Photo/Hilde Jensen, University of Tübingen/Nature Magazine; **56.25c** Frans Lanting/Minden Pictures.

ILLUSTRATION CREDITS

The following figures are adapted from C. K. Matthews and K. E. van Holde, *Biochemistry*, 2nd ed. Copyright © 1996 Pearson Education, Inc., publishing as Pearson Benjamin Cummings: **4.6b, 9.9, 17.16b** and **c**. The following figures are adapted from W. M. Becker, J. B. Reece, and M. F. Poenie, *The World of the Cell*, 3rd ed. Copyright © 1996 Pearson Education, Inc., publishing as Pearson Benjamin Cummings: **4.7, 6.7b, 7.8, 11.7, 11.11, 17.10, 18.22, 20.8.,** and **21.9. Figures 6.9** and **6.23a** and cell organelle drawings in **6.12, 6.13, 6.14,** and **6.20** are adapted from illustrations by Tomo Narashima in E. N. Marieb, *Human Anatomy and Physiology*, 5th ed. **6.12a, 50.10,** and **50.11** are also from *Human Anatomy and Physiology*, 5th ed. Copyright © 2001 Pearson Education, Inc., publishing as Pearson Benjamin Cummings. The following figures are adapted from Gerard J. Tortora, Berdell R. Funke, and Christine L. Case. 1998. *Microbiology: An Introduction*, 6th ed. Copyright © 1998 Pearson Education, Inc., publishing as Pearson Benjamin Cummings: **27.6a** and **43.8**. The following figures are adapted from M. W. Nabors, *Introduction to Botany*, Copyright © 2004 Pearson Education, Inc., publishing as Pearson Benjamin Cummings: **30.4, 30.13j, 39.13,** and **41.2 (center)**. The following figures are adapted from L. G. Mitchell, J. A. Mutchmor, and W. D. Dolphin. *Zoology*. Copyright ©1988 Pearson Education, Inc., publishing as Pearson Benjamin Cummings: **41.8, 44.9,** and **51.11**. The following figures are adapted from E. N. Marieb, *Human Anatomy and Physiology*, 4th ed. Copyright © 1998 Pearson Education, Inc., publishing as Pearson Benjamin Cummings: **46.16, 49.8, 49.10 , 50.25,** and **50.29**. **Chapter 1 1.12** and **21.5** From Figure 4B from L. Giot et al., "A Protein Interaction Map of *Drosophila melanogaster*," *Science*, Dec. 5, 2003, p. 1733. Copyright © 2003 AAAS. Reprinted with permission from the American Association for the Advancement of Science; **1.25** Map provided courtesy of David W. Pfennig, University of North Carolina at Chapel Hill; **1.27** Data in bar graph based on D. W. Pfennig et al. 2001. Frequency-dependent Batesian mimicry. *Nature* 410: 323. **Chapter 2 2.2** (bottom) Graph adapted from M.E. Frederickson et al. 'Devil's gardens' bedevilled by ants, *Nature*, 437: 495, 9/22/05. Reprinted by permission of Macmillan Publishers, Ltd. **Chapter 3 3.8a** Adapted from *Scientific American*, Nov. 1998, p.102. **Chapter 5 5.13** Adapted from *Biology: The Science of Life*, 4/e by Robert Wallace et al. Copyright © 1991. Reprinted by permission of Pearson Education, Inc.; **5.19** Adapted from D. W. Heinz et al. 1993. How amino-acid insertions are allowed in an alpha-helix of T4 lysozyme. *Nature* 361: 561; **5.21** Collagen and hemoglobin art: © Illustration, Irving Geis. Images from Irving Geis collection/Howard Hughes Medical Institute. Rights owned by Howard Hughes Medical Institute. Not to be reproduced without permission. **Chapter 6 Table 6.1a** Adapted from W. M. Becker, L. J. Kleinsmith, and J. Hardin, *The World of the Cell*, 4th ed. p. 753. Copyright © 2000 Pearson Education, Inc., publishing as Pearson Benjamin Cummings. **Chapter 8 8.21** Adapted from J. M. Scheer et al. 2006. A common allosteric site and mechanism in caspases. *Proceedings of the National Academy of Sciences of the United States of America* 103: 7595-7600. **Chapter 9 9.5a, b** Copyright © 2002 from *Molecular Biology of the Cell*, 4th ed. by Bruce Alberts et al., fig. 2.69, p. 92. Reproduced by permission of Garland Science/Taylor & Francis Books, Inc.; **9.15** H. Itoh et al. 2004. Mechanically driven ATP synthesis by F_1-ATPase. *Nature* 427: 465-468. **Chapter 10 10.14** Adapted from Richard and David Walker. *Energy, Plants and Man*, Fig. 4.1, p. 69. Sheffield: University of Sheffield. Oxygraphics http://www.oxygraphics.co.uk © Richard Walker. Used with permission courtesy of Richard Walker. **Chapter 11 11.16** Adapted from D. Matheos et al. 2004. Pheromone-induced polarization is dependent on the Fus3p MAPK acting through the formin Bni1p, *Journal of Cell Biology* 165: 99-109. **Chapter 12 12.12** Copyright © 2002 from *Molecular Biology of the Cell*, 4th ed., by Bruce Alberts et al., fig. 18.41, p. 1059. Garland Science/Taylor & Francis Books, Inc.; **12.16** Adapted from S. Moreno et al. 1989. Regulation of $p34^{cdc2}$ protein kinase during mitosis. *Cell* 58: 361-372. Copyright © 1989. Used by permission. **Chapter 13 13.10** Adapted from T. S. Kitajima et al. 2004. The conserved kinetochore protein shugoshin protects centromeric cohesion during meiosis. *Nature* 427: 510-517, 2004. Copyright © 2004. Reprinted by permission of Macmillan Publishers, Ltd. **Chapter 17 17.12** Adapted from L. J. Kleinsmith and V. M. Kish. 1995. *Principles of Cell and Molecular Biology*, 2nd ed. New York, NY: HarperCollins. Reprinted by permission of Addison Wesley Educational Publishers. **Chapter 18 18.13a** Adapted from Fig. 1d in N. C. Lau et al. An abundant class of tiny RNAs with probable regulatory roles in *Caenorhabditis elegans. Science* 294: 858-862, 10/26/2001. Copyright © 2001. Reprinted with permission from AAAS. **Chapter 20 20.10** Adapted from Peter Russell, *Genetics*, 5th ed., fig. 15.24, p. 481. Copyright © 1998 Pearson Education, Inc., publishing as Pearson Benjamin Cummings. **Chapter 21 21.2** Adapted from a figure by Chris A. Kaiser and Erica Beade; **21.10:** Hemoglobin art: © Illustration, Irving Geis. Images from Irving Geis collection/Howard Hughes Medical Institute. Rights owned by Howard Hughes Medical Institute. Not to be reproduced without permission. **21.17** Adapted from an illustration by William McGinnis; **21.18** Adapted from M. Akam, "*Hox* genes and the evolution of diverse body plans," *Philosophical Transactions Biological Sciences*, Vol. 349, No. 1329, pp. 313-319, September 1995. Copyright © 1995 Royal Society of London. Used with permission. **Chapter 22 22.8** © Utako Kikutani, 2007. Used with permission; **22.14** Adapted from R. Shurman et al. 1995. *Journal of Infectious Diseases* 171: 1411; **22.16a** Adapted from J. G. M. Thewissen et al. 2001. Skeletons of terrestrial cetaceans and the relationship of whales to artiodactyls. *Nature* 413: 277-281, fig. 2a; **22.16b** Adapted from P. D. Gingerich et al. 2001. Origin of whales from early artiodactyls: Hands and feet of eocene protocetidae from Pakistan. *Science* 293: 2239-2242, fig. 3; **22.16c** and **d** Adapted from C. de Muizon. 2001. Walking with whales, *Nature* 413: 259-260, fig. 1. **Chapter 23 23.4** Graph adapted from D. A. Powers et al. 1991. Genetic mechanisms for adapting to a changing environment. *Annual Review of Genetics* 25: 629-659; **23.10a** and **b** J. L. Bouzat et al. 1998. The ghost of genetic diversity past: Historical DNA analysis of the greater prairie chicken. *The American Naturalist*, 152: 1-6; **23.12a** Adapted from D. Futuyma. 1998. *Evolutionary Biology* 3rd ed. Sinauer Associates, fig. 13.19. Copyright © 1998. Reprinted by permission of Sinauer Associates, Inc.; **23.14b** Art adapted from D Futuyma, 2005. *Evolution*, 1st ed. Sinauer. fig. 11.3; **23.16** Adapted from A. M. Welch et al.1998. Call duration as an indicator of genetic quality in male gray tree frogs. *Science* 280: 1928-1930; **23.17** Adapted from A. C. Allison. 1961. Abnormal hemoglobin and erythrocyte enzyme-deficiency traits. In *Genetic Variation in Human Populations*, ed. G.A. Harrison. Oxford: Elsevier Science; **Un 23.2** R. K. Koehn and T. J. Hilbish. 1987. The adaptive importance of genetic variation. *American Scientist* 75: 134-141. **Chapter 24 24.3** Based on data from S. V. Edwards, 1993. Long-distance gene flow in a cooperative breeder detected in genealogies of mitochondrial DNA sequences. Proceedings of the Royal Society of London. Series B, *Biological Sciences* 252: 177-185; **24.7** Adapted from fig. 1 of F. Bossuyt and M. C. Milinkovitch. Amphibians as indicator of early tertiary "out-of-India" dispersal vertebrates. *Science* 292: 93-95, 4/6/01. Copyright © 2001. Reprinted with permission from AAAS; **24.8** Graph adapted from figure 2 in "Correspondence between sexual isolation and allozyme differentiation" *Proceedings of the National Academy of Science*, 87: 2715-2719, 1990, p. 2718. Copyright © 1990 Stephen G. Tilley, Paul A. Verrell, Steven J. Arnold. Used with permission; **24.9** Adapted from D. M. B. Dodd, 1989. Reproductive isolation as a consequence of adaptive divergence in *Drosophila pseudoobscura*. *Evolution* 43: 1308-1311; **24.13** Map and graphs adapted from J. M. Szymura, 1993. Analysis of hybrid zones with bombina. In *Hybrid Zone and the Evolutionary Process* by R. G. Harrison, ed. Oxford University Press, NY; **24.15** Adapted from figure 2, from G.P. Saetre et al., "A sexually selected character displacement in flycatchers reinforces premating isolation" *Nature* 387: 589-591, June 5, 1997. Copyright © 1997. Reprinted by permission of Macmillan Publishers, Ltd.; **24.18b** Adapted from figure 2 in L. H. Rieseberg et al., Role of gene interactions in hybrid speciation: Evidence from ancient and experimental hybrids. *Science* 272: 741-745, 1996. Copyright © 1996. Reprinted with permission from AAAS. **Chapter 25 25.5** Adapted from D. J. Futuyma. 1998. *Evolutionary Biology*, 3rd ed., p. 128. Sunderland, MA: Sinauer Associates; **25.6a—d** Adapted from D. J. Futuyma. 2005. *Evolution*, 1st ed., fig. 4.10. Sunderland, MA: Sinauer Associates; **25.6e** Adapted from Luo et al. 2001. A new mammalia form from the Early

Jurassic and evolution of mammalian characteristics. *Science* 292: 1535; **25.7** Adapted from D. J. Des Marais. September 8, 2000. When did photosynthesis emerge on Earth? *Science* 289: 1703—1705; **25.10** Data from A. H. Knoll and S. B. Carroll, June 25, 1995. *Science* 284: 2129—2137; **25.12** Map adapted from http://gelology.er.usgs.gov/eastern/plates.html; **25.14** Graph created from D. M. Raup and J. J. Sepkoski, Jr. 1982. Mass extinctions in the marine fossil record. *Science.* 215: 1501-1503 and J. J. Sepkoski, Jr. 1984. A kinetic model of Phanerozoic taxonomic diversity. III. Post-Paleozoic families and mass extinctions. *Paleobiology.* Vol 10, No. 2, pp. 246-267 in D. J. Futuyma, fig. 7.3a, p. 143 and fig. 7.6, p. 145, Sunderland, MA: Sinauer Associates; **25.16** Graph from R. K. Bambach et al. 2002. Anatomical and ecological constraints on Phanerozoic animal diversity in the marien realm. *PNAS* 99: 6854-6859; **25.17** Adapted from Hickman, Roberts, and Larson.1997, *Zoology,* 10th ed, Wm. C. Brown, fig. 31.1; **25.22** Adapted from M. Ronshaugen et al. Feb. 21, 2002. Hox protein mutation and macroevolution of the insect body plan. *Nature* 415: 914-917, fig. 1a; **25.24** Adapted from M. Strickberger, 1990. *Evolution,* Boston: Jones & Bartlett. **Chapter 26** **26.6** Adapted from C. S. Baker and S. R. Palumbi. 1994. Which whales are hunted? A molecular genetic approach to monitoring whaling. *Science* 265: 1538-1539, fig. 1. Copyright © 1994. Reprinted with permission from AAAS; **26.12** Adapted from S. M. Shimeld. 1999. The evolution of the hedgehog gene family in chordates: Insights from amphioxus hedgehog. *Development Genes and Evolution* 209: 40-47, fig. 3; **26.19** Adapted from J. C. Avise, 2004. *Molecular Markers, Natural History, and Evolution,* 2/e, fig. 4.3c. Copyright © Sinauer Associates, Inc. Used with permission; **26.20a** Graph adapted from B. Korber et al. Timing the ancestor of the HIV-1 pandemic strains. *Science* 288: 1789-1796. June 9, 2000. Copyright © 2000. Reprinted with permission from AAAS. **26.21** Adapted from S. L. Baldauf et al. 2004. The Tree of Life: An Overview. In *Assembling the Tree of Life* by J. Cracraft and MJ Donoghue (eds), Oxford University Press, NY; **26.22** Adapted from S. Blair Hedges. The origin and evolution of model organisms. *Nature Reviews Genetics* 3: 838-848, fig. 1, p. 840; **26.23** Adapted from M. C. Rivera and J. A. Lake. 2004. The ring of life provides evidence for a genome fusion origin of eukaryotes. *Nature* 431: 152-155, fig. 3. Copyright © 2004. Reprinted by permission of Macmillan Publishers. Ltd. **Chapter 27** **27.10b** Graph adapted from V. S. Cooper and R. E. Lenski. 2000. The population genetics of ecological specialization in evolving *E. coli* populations. *Nature* 407: 736-739; **27.19** Graph created from data in C. Calvaruso et al. 2006. Root-associated bacteria contribute to mineral weathering and to mineral nutrition in trees: A budgeting analysis. *Applied and Environmental Microbiology* 72: 1258-1266. **Chapter 28** **28.2** From Archibald and Keeling, "Recycled Plastids," *Trends in Genetics,* Vol. 18, No. 11, 2002. Copyright © 2002, with permission from Elsevier; **28.11** Adapted from R. W. Bauman. 2004. *Microbiology,* fig. 12.7, p. 350. Copyright © 2004 Pearson Education, Inc., publishing as Pearson Benjamin Cummings; **28.23** Data from A. Stechman and T. Cavalier-Smith. 2002. Rooting the eukaryote tree by using a derived gene fusion, *Science* 297: 89-91; **28.27** Copyright © 2004 The Nature Conservancy. Map produced by N. Law – Global Priorities Group – 08/14/2004. **Chapter 29** **29.10** Source: R. D. Bowden. 1991. Inputs, outputs and accumulation of nitrogen in an early successional moss *(Polytrichum)* ecosystem. *Ecological Monographs* 61: 207-223; **29.14** Adapted from Raven et al. *Biology of Plants,* 6th ed., fig. 19.7. **Chapter 30** **30.12a** Adapted from Soltis et al., 2005. *Phylogeny and Evolution of Angiosperms,* Sinauer Associates, pp. 17, fig. 1.11 from Crane 1985, which is based on Harris, 1964; **30.12b** Adapted from Soltis et al., *Phylogeny and Evolution of Angiosperms,* p. 28, fig. 2.3. Copyright © 2005. Reprinted by permission of Sinauer Associates, Inc.; **Table 30.1** Adapted from Randy Moore et al., *Botany,* 2nd ed. Dubuque, IA: Brown, 1998, Table 2.2, p. 37. **Chapter 31** **31.21** Graphs adapted from A. E. Arnold et al. 2003. Fungal endophytes limit pathogen damage in a tropical tree. *PNAS* 100: 15649-15654, figs. 4 and 5. **Chapter 33** **33.28a** From J. K. Grenier et al. 1997. Evolution of the entire arthropod *Hox* gene set predated the origin and radiation of the onychophoran/arthropod clade, *Current Biology* 7: 547-553, fig. 3c, p. 551. Copyright © 1997, with permission from Elsevier, Inc. **Chapter 34** **34.8b** Adapted from J. Mallatt and J. Chen. 2003. Fossil sister group of craniates: Predicted and found. *Journal of Morphology* 258: 1-31, fig. 1, 5/15/03. Copyright © 2003. Reprinted with permission of Wiley-Liss, Inc., a subsidiary of John Wiley & Sons, Inc.; **34.12** Adapted from K. Kardong. 2001. *Vertebrates: Comparative Anatomy, Function, and Evolution,* 3/e. © 2001 McGraw-Hill Science/Engineering/Mathematics.; **34.19** and **34.20** From C. Zimmer. *At the Water's Edge.* Copyright © 1999 by Carl Zimmer. Reprinted by permission of Jankow & Nesbit; **34.31a** Adapted from D. J. Futuyma. 2005. *Evolution,* 1/e, fig. 4.10. Sunderland, MA: Sinauer Associates; **34.40** Drawn from photos of fossils: *O. tugenensis* photo in Michael Balter, Early hominid sows division, *ScienceNow,* Feb. 22, 2001, © 2001 American Association for the Advancement of Science. *A. ramidus kadabba* photo by Timothy White, 1999/Brill Atlanta. *A. anamensis, A. garhi,* and *H. neanderthalensis* adapted from *The Human Evolution Coloring Book. K platyops* drawn from photo in Meave Leakey et al., New hominid genus from eastern Africa shows diverse middle Pliocene lineages, *Nature,* March 22, 2001, 410: 433. *P. boisei* drawn from a photo by David Bill. *H. ergaster* drawn from a photo at www.inhandmuseum.com. *S. tchadensis* drawn from a photo in Michel Brunet et al., A new hominid from the Upper Miocene of Chad, Central Africa, *Nature,* July 11, 2002, 418: 147, fig. 1b.; **34.43a** and **b** Adapted from I. V. Ovchinnikov et al. 2000. Molecular analysis of Neanderthal DNA from the northern Caucasus. *Nature* 404: 492, fig.3a and b. **Chapter 35** **35.24 (right)** Pie chart adapted from *Nature,* Dec. 14, 2000, 408: 799. New data from The Arabidopsis Information Resource (TAIR). **Chapter 39** **39.16 (top)** Adapted from M. Wilkins. 1988. *Plant Watching,* Facts of File Publ.; **39.28** Reprinted with permission from Edward Farmer, 1997, *Science* 279: 912. Copyright © 1997 American Association for the Advancement of Science. **Chapter 40** **40.14** Adapted from fig. 2 in V. H. Hutchinson et al. Thermoregulation in a brooding female Indian python, *Python molurus bivittatus. Science* 151: 694-695, fig. 2. Copyright © 1966. Reprinted with permission from AAAS; **40.15** Adapted with permission from B. Heinrich. 1974. Thermoregulation in endothermic insects. *Science* 185: 747-756, fig. 7. © 1974 American Association for the Advancement of Science. **Chapter 41** **41.5** Source: R. W. Smithells et al. 1980. Possible prevention of neural-tube defects by periconceptual vitamins supplementation. *Lancet* 315: 339-340; **41.10** Adapted from R. A. Rhoades and R.G. Pflanzer, *Human Physiology,* 3/e, fig. 22-1, p. 666. Copyright © 1996. Reprinted by permission of Brooks/Cole, a division of Thomson Learning: www.thomsonrights.comFax 800 732-2215; **41.23** Adapted from J. Marx, "Cellular Warriors at the Battle of the Bulge," *Science,* Vol. 299, p. 846. Copyright © 2003 American Association for the Advancement of Science. Illustration: Katharine Sutliff. **Chapter 42** **42.31** Adapted from S. L. Lindstedt et al. 1991. Running energetics in the pronghorn antelope. *Nature* 353: 748–750. Copyright © 1991. Reprinted by permission of Macmillan Publishers, Ltd. **Chapter 43** **43.5** Adapted from Phoebe Tzou et al, "Constitutive expression of a single antimicrobial peptide can restore wild-type resistance to infection in immuno-deficient *Drosophila* mutants," *PNAS,* 99: 2152-2157, figs. 2a and 4a. Copyright © 2002 National Academy of Sciences, U.S.A. Used with permission. **Chapter 44** **44.6** Kangaroo rat data adapted from K. B. Schmidt-Nielson. 1990. *Animal Physiology: Adaptation and Environment,* 4th ed., p. 339. Cambridge: Cambridge University Press; **44.7** Adapted from K. B. Schmidt-Nielsen et al. 1958. Extrarenal salt excretion in birds. *American Journal of Physiology* 193: 101-107; **44.20** Table adapted from P. M. T. Deen et al. 1994. Requirement of human renal water channel aquaporin-2 for vasopressin-dependent concentration in urine. *Science* 264: 92-95, table 1. Copyright © 1994. Reprinted with permission from AAAS; **44 EOC** (visual summary) Adapted from W. S. Beck et al. 1991. *Life: An Introduction to Biology,* p. 649. Copyright © 1991 HarperCollins. Reprinted by permission of Pearson Education. **Chapter 45** **45.4** J. M. Horowitz, et al. 1980. The Response of Single Melanophores to Extracellular and Intracellular Iontophoretic Injection of Melanocyte-Stimulating Hormone. *Endocrinology* 106: 771, fig. B. © 1980 by The Endocrine Society; **45.22** A. Jost, Recherches sur la differenciation sexuelle de l'embryon de lapin (Studies on the sexual differentiation of the rabbit embryo). *Arch. Anat. Microsc. Morphol. Exp* (*Archives d'anatomie microscopique et de morphologie expérimentale*). 36: 271-316, 1947. **Chapter 46** **46.9** Figure adapted from R. R. Snook and D. J. Hosken. 2004. Sperm death and dumping in *Drosophila.. Nature* 428: 939—941, fig. 2. Copyright © 2004. Reprinted by permission of Macmillan Publishers, Ltd. **Chapter 47** **47.18** From Wolpert, et al. 1998. *Principles of Development,* fig. 8.25, p. 251 (right). Oxford: Oxford University Press. By permission of Oxford University Press; **47.21a** Copyright © 1989 from *Molecular Biology of the Cell,* 2nd ed. by Bruce Alberts et al. Reproduced by permission of Garland Science/Taylor & Francis Books, Inc.; **47.21b** From Hiroki Nishida, "Cell lineage analysis in ascidian embryos by intracellular injection of a tracer enzyme: III. Up to the tissue restricted stage," *Developmental Biology,* Vol. 121, p. 526, June, 1987. Copyright © 1987 with permission from Elsevier, Inc.; **47.22** Copyright © 2002 from *Molecular Biology of the Cell,* 4th ed. by Bruce Alberts et al., fig. 21.17, p. 1172. Reproduced by permission of Garland Science/Taylor & Francis Books, Inc.; **47.24 experiment** and **left side of results** From Wolpert et al. 1998. *Principles of Development,* fig. 1.10, Oxford: Oxford University Press. By permission of Oxford University Press; **47.24 right side of results** Figure 15.12, p. 604 from *Developmental Biology,* 5th ed. by Gilbert et al. Copyright © 1997 Sinauer Associates. Used with permission. **Chapter 48** **48.10** Adapted from G. Matthews, 1986. *Cellular Physiology of Nerve and Muscle,* Cambridge, MA: Blackwell Scientific Publications. © 1986 Blackwell Science. Used with permission of Blackwell Science, Inc.; **48.17** Adapted from C. B. Pert and S. H. Snyder. 1973. Opiate receptor: Demonstration in nervous tissue. *Science* 179: 1011-1014, Table 1. **Chapter 49** **49.11** Adapted from L. M. Mukhametov. 1984. Sleep in marine mammals. In *Sleep Mechanisms,* by A. A. Borbéley and J. L. Valatx (eds.). Munich: Springer-Verlag, pp 227-238; **49.12** Adapted from M. R. Ralph et al. 1990. Transplanted suprachiasmatic nucleus determines circadian period. *Science* 247: 975-978, fig. 2; **49.14** Adapted from E. D. Jarvis et al. 2005. Avian brains and a new understanding of vertebrate brain evolution. *Nature Reviews Neuroscience* 6: 151-159, fig. 1c. **Chapter 50** **50.14** Adapted from K. L. Mueller et al. 1975. The receptors and coding logic for bitter taste. *Nature* 434: 225-229;

50.19 Adapted from Bear et al. 2001. *Neuroscience: Exploring the Brain,* 2nd ed., figs. 11.8 and 11.9, pp. 281 and 283. Hagerstown,MD: Lippincott Williams & Wilkins. © 2001 Lippincott Williams and Wilkins; **50.22** Adapted from Shepherd. 1988. *Neurobiology,* 2nd ed., fig. 11.4, p. 227. Oxford University Press. (From V. G. Dethier. 1976. *The Hungry Fly.* Cambridge, MA: Harvard University Press.); **50.23** Adapted from Bear et al. 2001, Neuroscience: Exploring the Brain, 2nd ed., fig. 8.7, p. 196. Hagerstown,MD: Lippincott Williams & Wilkins. © 2001 Lippincott Williams and Wilkins; **50.32** Grasshopper adapted from Hickman et al. 1993. *Integrated Principles of Zoology,* 9th ed., Fig. 22.6, p. 518. New York: McGraw-Hill Higher Education. © 1993 The McGraw-Hill Companies; **50.37** K. Schmidt-Nielsen. 1972. Locomotion: Energy cost of swimming, flying, and running. *Science* 177: 222-228. **Chapter 51 51.3b** Adapted from N. Tinbergen. 1951. *The Study of Instinct.* Oxford: Oxford University Press. By permission of Oxford University Press; **51.7** Adapted from M. B. Sokolowski, 2001. *Drosophila:* Genetics meets behavior. *Nature Reviews: Genetics* 2: 881, fig. 1. Copyright © 2001 McMillan Publishing. Used with permission; **51.14 (left)** Adapted from C. S. Henry et al. 2002. The inheritance of mating songs in two cryptic, sibling lacewings species (Neuroptera: Chrysopidae: *Chrysoperla). Genetica* 116: 269-289, fig. 2. Copyright © 2002. Reprinted by permission of Springer Verlag; **51.15 (top)** Adapted from a photograph by Jonathan Blair in Alcock, 2002. *Animal Behavior,* 7th ed. Sinauer Associates, Inc., Publiishers. **51.15 (bottom)** Adapted from P. Berthold et al. 1992. Rapid microevolution of migratory behaviour in a wild bird species. *Nature* 360, 12/17,/92, p. 668, fig. 1. Copyright © 1992 Nature Publishing, Inc., used with permission; **51.18** Adapted from M. B. Sokolowski et al. 1997. Evolution of foraging behavior in *Drosophila* by density-dependent selection. Proceedings of the National Academy of Sciences of the United States of America. 94: 7373-7377. Copyright © 1997 National Academy of Sciences, U.S.A.; **51.24** K. Witte and N. Sawka. 2003. Sexual imprinting on a novel trait in the dimorphic zebra finch: sexes differ. *Animal Behaviour* 65: 195-203. Art adapted from http://www.uni-bielefeld.de/biologue/vhf/KW/Forschungsprojekte2.html. **Chapter 52 52.3** Adapted from figure at http://tde.ornl.gov. Project supported by the U.S. Dept. of Energy's Office of Science (BER); **52.5** Map adapted from G. Caughly et al. 1987. *Kangaroos: Their Ecology and Management in the Sheep Rangelands of Australia,* p. 12, fig. 1.2, Cambridge: Cambridge University Press. Copyright © 1987 Cambridge University Press. Used with permission; **52.7** Map adapted from R. L. Smith. 1974. *Ecology and Field Biology,* fig. 11.19, p. 353. Harper and Row Publishers. Map updated from D. A. Sibley. 2000. National Audubon Society *The Sibley Guide to Birds,* Alfred A. Knopf: New York; **52.8** Data from W. J. Fletcher. 1987. Interactions among subtidal Australian sea urchins, gastropods and algae: effects of experimental removals. *Ecological Monographs* 57: 89–109; **52.11** Map adapted from Physicists track great ocean conveyor belt. http://www.anl.gov/Media_Center/Frontiers/2003/d8ee.html; **52.14** Adapted from L. Roberts. 1989. How fast can trees migrate? *Science* 243: 736, fig. 2. © 1989 by the American Association for the Advancement of Science; **52.19** Adapted from Heinrich Walter and Siegmar-Walter Breckle. 2003. *Walter's Vegetation of the Earth,* fig. 16, p. 36. Springer-Verlag, © 2003; **Un. 52.2** Data from J. Clausen, D. D. Keck, and W. M. Hiesey. 1948. Experimental studies on the nature of species. III. Environmental responses of climatic races of *Achillea.* Carnegie Institution of Washington Publication 581. **Chapter 53 53.5** Adapted from P. W. Sherman and M. L. Morton, "Demography of Belding's ground squirrels," *Ecology,* Vol. 65, No. 5, p. 1622, fig. 1a, 1984. Copyright © 1984 Ecological Society of America. Used by pemission. **53.15** Adapted from J. T. Enright. 1976. Climate and population regulation: The biogeographer's dilemma. *Oecologia* 24: 295-310; **53.16** and **53.18** Adapted from T. Clutton-Brock and J. Pemberton. 2004. Soay Sheep: Dynamics and selection in an island population. Cambridge University Press. Used with permission; **53.19** Data courtesy of Rolf O. Peterson, Michigan Technological University; **53.23** Data from U. S. Census Bureau International Data Base; **53.24** Data from Population Reference Bureau 2000 and U. S. Census Bureau International Data Base, 2003; **53.25** Data from U. S. Census Bureau International Data Base; **53.26** Data from U. S. Census Bureau International Data Base 2003; **53.27** Adapted from M. L. Imhoff et al. 2004. Global patterns in human consumption of net primary production. *Nature* 429: 870-873, fig. 1a; **Tables 53.1** and **53.2** Data from P. W. Sherman and M. L. Morton, "Demography of Belding's ground squirrels," *Ecology,* Vol. 65, No. 5, p. 1622, fig. 1a, 1984. Copyright © 1984 Ecological Society of America. **Chapter 54 54.2** A. S. Rand and E. E. Williams. 1969. The anoles of La Palma: Aspects of their ecological relationships. *Breviora* 327. Museum of Comparative Zoology, Harvard University. Copyright © 1969 by the President and Fellows of Harvard College. Reprinted with permission from the Museum of Comparative Zoology; **54.10** Adapted from N. Fierer and R. B. Jackson. 2006. The diversity and biogeography of soil bacterial communities. *Proceedings of the National Academy of Sciences USA* 103: 626–631 fig. 1a. Copyright © 2006 National Academy of Sciences, U.S.A. Used with permission; **54.12** Adapted from E. A, Knox. 1970. Antarctic marine ecosystems. In *Antarctic Ecology,* ed. M. W. Holdgate, 69-96. London: Academic Press; **54.13** Adapted from D. L. Breitburg et al. 1997. Varying effects of low dissolved oxygen on trophic interactions in an estuarine food web. *Ecological Monographs* 67: 490. Copyright © 1997 Ecological Society of America; **54.14** Adapted from B. Jenkins. 1992. Productivity, disturbance and food web structure at a local spatial scale in experimental container habitats. *Oikos* 65: 252. Copyright © 1992 Oikos, Sweden; **54.15** Adapted from R. T. Paine. 1966. Food web complexity and species diversity. *American Naturalist* 100: 65-75; **54.16** Adapted from J. A. Estes et al. 1998. Killer whale predation on sea otters linking oceanic and nearshore ecosystems. *Science* 282: 474. Copyright © 1998 by the American Association for the Advancement of Science. Reprinted with permission from AAAS; **54.18** Data for graph from S. D. Hacker and M. D. Bertness. 1999. Experimental evidence for factors maintaining plant species diversity in a New England salt marsh. *Ecology* 80: 2064-2073; **54.19** Data from D. Wall Freckman and R. A. Virginia 1997. Low-diversity Antarctic soil nematode communities: distribution and response to disturbance. *Ecology* 78: 363-369; **54.20** Graph adapted from A. R. Townsend et al. 1997. The intermediate disturbance hypothesis, refugia, and diversity in streams. *Limnology and Oceanography* 42: 938-949. Copyright © 1997 by the American Society of Limnology and Oceanography, Inc. Used with permission; **54.22** Adapted from R. L. Crocker and J. Major. 1955. Soil Development in relation to vegetation and surface age at Glacier Bay, Alaska. *Journal of Ecology* 43: 427-448; **54.23** Data from F. S. Chapin, III, et al. 1994. Mechanisms of primary succession following deglaciation at Glacier Bay, Alaska. *Ecological Monographs* 64: 149-175; **54.25** Adapted from D. J. Currie. 1991. Energy and large-scale patterns of animal- and plant-species richness. *American Naturalist* 137: 27-49; **54.26** Adapted from F. W. Preston. 1960. Time and space and the variation of species. *Ecology* 41: 611-627; **54.28** Adapted from F. W. Preston. 1962. The canonical distribution of commonness and rarity. *Ecology* 43: 185-215, 410-432. **Chapter 55 55.4** and **Un 55.1** Adapted from D. L. DeAngelis. 1992. *Dynamics of Nutrient Cycling and Food Webs.* New York: Chapman & Hall; **55.7** Adapted from J. H. Ryther and W. M. Dunstan. 1971. Nitrogen, phosphorus, and eutrophication in the coastal marine environment. *Science* 171: 1008-1013. Copyright © 1971. Reprinted with permission from AAAS; **55.8** Data from M. L. Rosenzweig. 1968. New primary productivity of terrestrial environments: Predictions from climatologic data, *American Naturalist* 102: 67-74; **55.14a** Adapted from R. E. Ricklefs. 1997. *The Economy of Nature,* 4th ed. © 1997 by W. H. Freeman and Company. Used with permission; **55.15** Map adapted from Moore et al. 1999. Litter decomposition rates in Canadian forests, Global Change Biology 5: 75-82. Trofymow et al. 1998. Canadian Intersite Decomposition Experiment (CIDET): Project and site establishment. Information Report BC-X-378. NRCAN CFS Victoria. 126pp. Produced under license from Her Majesty the Queen in Right of Canada, with permission of Natural Resources Canada. The Canadian Intersite Decomposition Experiment (CIDET) - http://cfs.nrcan.gc.ca/subsite/cidet; **55.21** CO_2 data from C. D. Keeling and T. P. Whorf, Scripps Institution of Oceanography. Temperature data from www.earth-policy.org/Indicators/Temp/Temp_data.htm. **55.23** Data from ozonewatch.gsfc.nasa.gov/facts/history/htmml; **Table 55.1** Data from Menzel and Ryther. 1961. *Deep Sea Ranch* 7: 276-281. **Chapter 56 56.10** Adapted from C. J. Krebs. 2001. *Ecology,* 5th ed., fig. 19.1. Copyright © 2001 Pearson Education, Inc., publishing as Pearson Benjamin Cummings; **56.11** Adapted from R. L. Westemeiier et al. 1998. Tracking the long-term decline and recovery of an isolated population. *Science* 282: 1696. © 1998 by the American Association for the Advancement of Science; **56.17** Adapted from N. Myers et al., "Biodiversity hotspots for conservation priorities," *Nature,* Vol. 403, p. 853, 2/24/2000. Copyright © 2000 Nature Publishing, Inc. Used with permission. Updated with data from C. H. Roberts et al. 2002. Marine biodiversity hotspots and conservation priorities for tropical reefs. *Science* 295: 1280-1284; **56.18** Adapted from W.D. Newmark, "Legal and biotic boundaries of western North American national parks: A problem of congruence." *Biological Conservation* 33: 199, 1985. © 1985 Elsevier, with kind permission; **56.19a** Map adapted from W. Purves and G. Orians, *Life, The Science of Biology,* 5th ed., fig. 55.23, p. 1239. © 1998 by Sinauer Associates, Inc. Used with permission; **56.22b** Graph adapted from http://news-service.standford.edu/news/2006/may24/gcriddle_ponds.jpg; **56.24** Data from Instituto Nacional de Estadistica y Censos de Costa Rica and Centro Centroamericano de Poblacion, Universidad de Costa Rica.

Glossary

Pronunciation Key

Pronounce		
ā	as in	**a**ce
a		**a**sh
ch		**ch**ose
ē		m**ee**t
e/eh		b**e**t
g		**g**ame
ī		**i**ce
i		h**i**t
ks		bo**x**
kw		**qu**ick
ng		so**ng**
ō		r**o**be
o		**o**x
oy		b**oy**
s		**s**ay
sh		**sh**ell
th		**th**in
ū		b**oo**t
u/uh		**u**p
z		**z**oo

′ = primary accent
′ = secondary accent

5′ cap A modified form of guanine nucleotide added onto the nucleotide at the 5′ end of a pre-mRNA molecule.

A site One of a ribosome's three binding sites for tRNA during translation. The A site holds the tRNA carrying the next amino acid to be added to the polypeptide chain. (A stands for aminoacyl tRNA.)

ABC model A model of flower formation identifying three classes of organ identity genes that direct formation of the four types of floral organs.

abiotic (ā′-bī-ot′-ik) Nonliving; referring to physical and chemical properties of an environment.

abortion The termination of a pregnancy in progress.

abscisic acid (ABA) (ab-sis′-ik) A plant hormone that slows growth, often antagonizing actions of growth hormones. Two of its many effects are to promote seed dormancy and facilitate drought tolerance.

absorption The third stage of food processing in animals: the uptake of small nutrient molecules by an organism's body.

absorption spectrum The range of a pigment's ability to absorb various wavelengths of light; also a graph of such a range.

abyssal zone (uh-bis′-ul) The part of the ocean's benthic zone between 2,000 and 6,000 m deep.

acanthodian (ak′-an-thō′-d ē-un) Any of a group of ancient jawed aquatic vertebrates from the Devonian period.

accessory fruit A fruit, or assemblage of fruits, in which the fleshy parts are derived largely or entirely from tissues other than the ovary.

acclimatization (uh-klī′-muh-tī-zā′-shun) Physiological adjustment to a change in an environmental factor.

acetyl CoA Acetyl coenzyme A; the entry compound for the citric acid cycle in cellular respiration, formed from a fragment of pyruvate attached to a coenzyme.

acetylcholine (as′-uh-til-kō′-lēn) One of the most common neurotransmitters; functions by binding to receptors and altering the permeability of the postsynaptic membrane to specific ions, either depolarizing or hyperpolarizing the membrane.

acid A substance that increases the hydrogen ion concentration of a solution.

acid precipitation Rain, snow, or fog that is more acidic than pH 5.2.

acoelomate (uh-sē′-lō-māt) A solid-bodied animal lacking a cavity between the gut and outer body wall.

acquired immunity A vertebrate-specific defense that is mediated by B lymphocytes (B cells) and T lymphocytes (T cells). It exhibits specificity, memory, and self-nonself recognition. Also called adaptive immunity.

acrosomal reaction (ak′-ruh-sōm′-ul) The discharge of hydrolytic enzymes from the acrosome, a vesicle in the tip of a sperm, when the sperm approaches or contacts an egg.

acrosome (ak′-ruh-sōm) A vesicle in the tip of a sperm containing hydrolytic enzymes and other proteins that help the sperm reach the egg.

actin (ak′-tin) A globular protein that links into chains, two of which twist helically about each other, forming microfilaments (actin filaments) in muscle and other kinds of cells.

action potential A rapid change in the membrane potential of an excitable cell, caused by stimulus-triggered, selective opening and closing of voltage-sensitive gates in sodium and potassium ion channels.

action spectrum A graph that profiles the relative effectiveness of different wavelengths of radiation in driving a particular process.

activation energy The amount of energy that reactants must absorb before a chemical reaction will start; also called free energy of activation.

activator A protein that binds to DNA and stimulates gene transcription. In prokaryotes, activators bind in or near the promoter; in eukaryotes, activators bind to control elements in enhancers.

active immunity Long-lasting immunity conferred by the action of B cells and T cells and the resulting B and T memory cells specific for a pathogen. Active immunity can develop as a result of natural infection or immunization.

active site The specific portion of an enzyme that binds the substrate by means of multiple weak interactions and that forms the pocket in which catalysis occurs.

active transport The movement of a substance across a cell membrane, with an expenditure of energy, against its concentration or electrochemical gradient; mediated by specific transport proteins.

actual evapotranspiration The amount of water transpired by plants and evaporated from a landscape over a given period of time, usually measured in millimeters and estimated for a year.

adaptation Inherited characteristic of an organism that enhances its survival and reproduction in specific environments.

adaptive radiation Period of evolutionary change in which groups of organisms form many new species whose adaptations allow them to fill vacant ecological roles in their communities.

adenylyl cyclase (uh-den′-uh-lil) An enzyme that converts ATP to cyclic AMP in response to a signal.

adhesion The attraction between different kinds of molecules.

adipose tissue A connective tissue that insulates the body and serves as a fuel reserve; contains fat-storing cells called adipose cells.

adrenal gland (uh-drē′-nul) One of two endocrine glands located adjacent to the kidneys in mammals. Endocrine cells in the outer portion (cortex) respond to ACTH by secreting steroid hormones that help maintain homeostasis during long-term stress. Neurosecretory cells in the central portion (medulla) secrete epinephrine and norepinephrine in response to nervous inputs triggered by short-term stress.

adrenocorticotropic hormone (ACTH) A tropic hormone that is produced and secreted by the anterior pituitary and that stimulates the production and secretion of steroid hormones by the adrenal cortex.

aerobic respiration A catabolic pathway that consumes oxygen (O_2) and organic molecules, producing ATP. This is the most efficient catabolic pathway and is carried out in most eukaryotic cells and many prokaryotic organisms.

afferent arteriole (af′-er-ent) In the kidney, the blood vessel supplying a nephron.

age structure The relative number of individuals of each age in a population.

aggregate fruit A fruit derived from a single flower that has more than one carpel.

agonistic behavior (a′-gō-nis′-tik) In animals, an often ritualized contest that determines which competitor gains access to a resource, such as food or mates.

AIDS (acquired immunodeficiency syndrome) The symptoms and signs present during the late stages of HIV infection, defined by a specified reduction in the number of T cells and the appearance of characteristic secondary infections.

alcohol fermentation Glycolysis followed by the conversion of pyruvate to carbon dioxide and ethyl alcohol.

aldosterone (al-dos′-tuh-rōn) A steroid hormone that acts on tubules of the kidney to regulate the transport of sodium ions (Na^+) and potassium ions (K^+).

alimentary canal (al′-uh-men′-tuh-rē) A digestive tract consisting of a tube running between a mouth and an anus; also called a complete digestive tract.

allantois (al-an′-tō′-is) One of four extraembryonic membranes; serves as a repository for the embryo's nitrogenous waste and functions in gas exchange.

allele (uh-lē′-ul) Any of the alternative versions of a gene that produce distinguishable phenotypic effects.

allopatric speciation (al′-uh-pat′-rik) The formation of new species in populations that are geographically isolated from one another.

allopolyploid (al′-ō-pol′-ē-ployd) A fertile individual that has more than two chromosome sets as a result of two different species interbreeding and combining their chromosomes.

allosteric regulation The binding of a regulatory molecule to a protein at one site that affects the function of the protein at a different site.

alpha (α) helix (al′-fuh hē′-liks) A spiral shape constituting one form of the secondary structure of proteins, arising from a specific pattern of hydrogen bonding.

alternation of generations A life cycle in which there is both a multicellular diploid form, the sporophyte, and a multicellular haploid form, the gametophyte; characteristic of plants and some algae.

alternative RNA splicing A type of eukaryotic gene regulation at the RNA-processing level in which different mRNA molecules are produced from the same primary transcript, depending on which RNA segments are treated as exons and which as introns.

altruism (al′-trū-iz-um) Selflessness; behavior that reduces an individual's fitness while increasing the fitness of another individual.

alveolate (al-vē′-uh-let) A protist with membrane-bounded sacs (alveoli) located just under the plasma membrane.

alveolus (al-vē′-uh-lus) (plural, **alveoli**) One of the dead-end, multilobed air sacs where gas exchange occurs in a mammalian lung.

Alzheimer's disease (alts′-hī-merz) An age-related dementia (mental deterioration) characterized by confusion, memory loss, and other symptoms.

amacrine cell (am′-uh-krin) A neuron of the retina that helps integrate information before it is sent to the brain.

amino acid (uh-mēn′-ō) An organic molecule possessing both carboxyl and amino groups. Amino acids serve as the monomers of polypeptides.

amino group A chemical group consisting of a nitrogen atom bonded to two hydrogen atoms; can act as a base in solution, accepting a hydrogen ion and acquiring a charge of 1+.

aminoacyl-tRNA synthetase An enzyme that joins each amino acid to the appropriate tRNA.

ammonia A small, very toxic molecule (NH_3) produced by nitrogen fixation or as a metabolic waste product of protein and nucleic acid metabolism.

ammonite A member of a group of shelled cephalopods that were important marine predators for hundreds of millions of years until their extinction at the end of the Cretaceous period (65.5 mya).

amniocentesis (am′-nē-ō-sen-tē′-sis) A technique of prenatal diagnosis in which amniotic fluid, obtained by aspiration from a needle inserted into the uterus, is analyzed to detect certain genetic and congenital defects in the fetus.

amnion (am′-nē-on) One of four extraembryonic membranes. It surrounds a fluid-filled cavity that cushions the embryo.

amniote (am′-nē-ōt) Member of a clade of tetrapods named for a key derived character, the amniotic egg, which contains specialized membranes, including the fluid-filled amnion, that protect the embryo. Amniotes include mammals as well as birds and other reptiles.

amniotic egg A shelled egg in which an embryo develops within a fluid-filled amniotic sac and is nourished by yolk. Produced by reptiles (including birds) and egg-laying mammals, it enables them to complete their life cycles on dry land.

amoeba (uh-mē′-buh) A protist grade characterized by the presence of pseudopodia.

amoebocyte (uh-mē′-buh-sīt′) An amoeba-like cell that moves by pseudopodia and is found in most animals. Depending on the species, it may digest and distribute food, dispose of wastes, form skeletal fibers, fight infections, and change into other cell types.

amoebozoan (uh-mē′-buh-zō′-an) A protist in a clade that includes many species with lobe- or tube-shaped pseudopodia.

amphibian Member of the tetrapod class Amphibia, including salamanders, frogs, and caecilians.

amphipathic (am′-fē-path′-ik) Having both a hydrophilic region and a hydrophobic region.

amplification The strengthening of stimulus energy during transduction.

amygdala (uh-mig′-duh-luh) A structure in the temporal lobe of the vertebrate brain that has a major role in the processing of emotions.

amylase (am′-uh-lās′) An enzyme in saliva that hydrolyzes starch (a glucose polymer from plants) and glycogen (a glucose polymer from animals) into smaller polysaccharides and the disaccharide maltose.

anabolic pathway (an′-uh-bol′-ik) A metabolic pathway that consumes energy to synthesize a complex molecule from simpler compounds.

anaerobic respiration (an-er-ō′-bik) The use of inorganic molecules other than oxygen to accept electrons at the "downhill" end of electron transport chains.

analogous Having characteristics that are similar because of convergent evolution, not homology.

analogy (an-al′-uh-jē) Similarity between two species that is due to convergent evolution rather than to descent from a common ancestor with the same trait.

anaphase The fourth stage of mitosis, in which the chromatids of each chromosome have separated and the daughter chromosomes are moving to the poles of the cell.

anatomy The structure of an organism and its study.

anchorage dependence The requirement that a cell must be attached to a substratum in order to divide.

androgen (an′-drō-jen) Any steroid hormone, such as testosterone, that stimulates the development and maintenance of the male reproductive system and secondary sex characteristics.

aneuploidy (an′-yū-ploy′-dē) A chromosomal aberration in which one or more chromosomes are present in extra copies or are deficient in number.

angiosperm (an′-jē-ō-sperm) A flowering plant, which forms seeds inside a protective chamber called an ovary.

angiotensin II A peptide hormone that stimulates constriction of precapillary arterioles and increases reabsorption of NaCl and water by the proximal tubules of the kidney, increasing blood pressure and volume.

anhydrobiosis (an-hī′-drō-bī-ō′-sis) A dormant state involving loss of almost all body water.

animal pole The point at the end of an egg in the hemisphere where the least yolk is concentrated; opposite of vegetal pole.

Animalia The kingdom that consists of multicellular eukaryotes that ingest their food.

anion (an′-ī-on) A negatively charged ion.

annual A flowering plant that completes its entire life cycle in a single year or growing season.

anterior Pertaining to the front, or head, of a bilaterally symmetrical animal.

anterior pituitary Also called the adenohypophysis; portion of the pituitary that develops from nonneural tissue; consists of endocrine cells that synthesize and secrete several tropic and nontropic hormones.

anther In an angiosperm, the terminal pollen sac of a stamen, where pollen grains containing sperm-producing male gametophytes form.

antheridium (an-thuh-rid′-ē-um) (plural, **antheridia**) In plants, the male gametangium, a moist chamber in which gametes develop.

anthropoid (an′-thruh-poyd) Member of a primate group made up of the monkeys and the apes (gibbons, orangutans, gorillas, chimpanzees, bonobos, and humans).

antibody A protein secreted by plasma cells (differentiated B cells) that binds to a particular antigen; also called immunoglobulin. All antibody molecules have the same Y-shaped structure and in their monomer form consist of two identical heavy chains and two identical light chains.

anticodon (an′-tī-kō′-don) A nucleotide triplet at one end of a tRNA molecule that recognizes a particular complementary codon on an mRNA molecule.

antidiuretic hormone (ADH) (an′-tī-dī-yū-ret′-ik) A peptide hormone, also known as vasopressin, that promotes water retention by the kidneys. Produced in the hypothalamus and released from the posterior pituitary, ADH also has activities in the brain.

antigen (an′-ti-jen) A macromolecule that elicits an immune response by binding to receptors of B cells or T cells.

antigen presentation The process by which an MHC molecule binds to a fragment of an intracellular protein antigen and carries it to the cell surface, where it is displayed and can be recognized by a T cell.

antigen receptor The general term for a surface protein, located on B cells and T cells, that binds to antigens, initiating acquired immune responses. The antigen receptors on B cells are called B cell receptors, and the antigen receptors on T cells are called T cell receptors.

antigen-presenting cell A cell that upon ingesting pathogens or internalizing pathogen proteins generates peptide fragments that are bound by class II MHC molecules and subsequently displayed on the cell surface to T cells. Macrophages, dendritic cells, and B cells are the primary antigen-presenting cells.

antiparallel The opposite arrangement of the sugar-phosphate backbones in a DNA double helix.

aphotic zone (ā′-fō′-tik) The part of an ocean or lake beneath the photic zone, where light does not penetrate sufficiently for photosynthesis to occur.

apical bud (ā′-pik-ul) A bud at the tip of a plant stem; also called a terminal bud.

apical dominance Concentration of growth at the tip of a plant shoot, where a terminal bud partially inhibits axillary bud growth.

apical ectodermal ridge (AER) A thickened area of ectoderm at the tip of a limb bud that promotes outgrowth of the limb bud.

apical meristem (mār′-uh-stem) Embryonic plant tissue in the tips of roots and the buds of shoots. The dividing cells of an apical meristem enable the plant to grow in length.

apicomplexan (ap′-ē-kom-pleks′-un) A protist in a clade that includes many species that parasitize animals. Some apicomplexans cause human disease.

apomixis (ap′-uh-mik′-sis) The ability of some plant species to reproduce asexually through seeds without fertilization by a male gamete.

apoplast (ap′-ō-plast) In plants, the continuum of cell walls plus the extracellular spaces.

apoptosis (ā-puh-tō′-sus) A program of controlled cell suicide, which is brought about by signals that trigger the activation of a cascade of suicide proteins in the cell destined to die.

aposematic coloration (ap′-ō-si-mat′-ik) The bright coloration of animals with effective physical or chemical defenses that acts as a warning to predators.

appendix A small, finger-like extension of the vertebrate cecum; contains a mass of white blood cells that contribute to immunity.

aquaporin A channel protein in the plasma membrane of a plant, animal, or microorganism cell that specifically facilitates osmosis, the diffusion of water across the membrane.

aqueous humor Plasma-like liquid in the space between the lens and the cornea in the vertebrate eye; helps maintain the shape of the eye, supplies nutrients and oxygen to its tissues, and disposes of its wastes.

aqueous solution (ā′-kwē-us) A solution in which water is the solvent.

arachnid A member of a major arthropod group, the cheliceriforms. Arachnids include spiders, scorpions, ticks, and mites.

arbuscular mycorrhiza (ar-bus′-kyū-lur mī′-kō-rī′-zuh) Association of a fungus with a plant root system in which the fungus causes the invagination of the host (plant) cells' plasma membranes.

arbuscular mycorrhizal fungus A symbiotic fungus whose hyphae grow through the cell wall of plant roots and extend into the root cell (enclosed in tubes formed by invagination of the root cell plasma membrane).

Archaea (ar′-kē′-uh) One of two prokaryotic domains, the other being Bacteria.

archaean Member of the prokaryotic domain Archaea.

Archaeplastida (ar′-kē-plas′-tid-uh) One of five supergroups of eukaryotes proposed in a current hypothesis of the evolutionary history of eukaryotes. This monophyletic group, which includes red algae, green alage, and land plants, descended from an ancient protist ancestor that engulfed a cyanobacterium. *See also* Excavata, Chromalveolata, Rhizaria, and Unikonta.

archegonium (ar-ki-gō′-nē-um) (plural, **archegonia**) In plants, the female gametangium, a moist chamber in which gametes develop.

archenteron (ar-ken′-tuh-ron) The endoderm-lined cavity, formed during gastrulation, that develops into the digestive tract of an animal.

archosaur (ar′-kō-sōr) Member of the reptilian group that includes crocodiles, alligators, dinosaurs, and birds.

arteriole (ar-ter′-ē-ōl) A vessel that conveys blood between an artery and a capillary bed.

artery A vessel that carries blood away from the heart to organs throughout the body.

arthropod A segmented ecdysozoan with a hard exoskeleton and jointed appendages. Familiar examples include insects, spiders, millipedes, and crabs.

artificial selection The selective breeding of domesticated plants and animals to encourage the occurrence of desirable traits.

ascocarp The fruiting body of a sac fungus (ascomycete).

ascomycete (as′-kuh-mī′-sēt) Member of the fungal phylum Ascomycota, commonly called sac fungus. The name comes from the saclike structure in which the spores develop.

ascus (plural, **asci**) A saclike spore capsule located at the tip of a dikaryotic hypha of a sac fungus.

asexual reproduction The generation of offspring from a single parent that occurs without the fusion of gametes (by budding, division of a single cell, or division of the entire organism into two or more parts). In most cases, the offspring are genetically identical to the parent.

assisted reproductive technology A fertilization procedure that generally involves surgically removing eggs (secondary oocytes) from a woman's ovaries after hormonal stimulation, fertilizing the eggs, and returning them to the woman's body.

associative learning The acquired ability to associate one environmental feature (such as a color) with another (such as danger).

aster A radial array of short microtubules that extends from each centrosome toward the plasma membrane in an animal cell undergoing mitosis.

astrocyte A glial cell with diverse functions, including providing structural support for neurons, regulating the interstitial environment, facilitating synaptic transmission, and assisting in regulating the blood supply to the brain.

atherosclerosis A cardiovascular disease in which fatty deposits called plaques develop in the inner walls of the arteries, obstructing the arteries and causing them to harden.

atom The smallest unit of matter that retains the properties of an element.

atomic mass The total mass of an atom, which is the mass in grams of 1 mole of the atom.

atomic nucleus An atom's dense central core, containing protons and neutrons.

atomic number The number of protons in the nucleus of an atom, unique for each element and designated by a subscript to the left of the elemental symbol.

ATP (adenosine triphosphate) (a-den′-ō-sēn trī-fos′-fāt) An adenine-containing nucleoside triphosphate that releases free energy when its phosphate bonds are hydrolyzed. This energy is used to drive endergonic reactions in cells.

ATP synthase A complex of several membrane proteins that provide a port through which protons diffuse. This complex functions in chemiosmosis with adjacent electron transport chains, using the energy of a hydrogen ion (proton) concentration gradient to make ATP. ATP synthases are found in the inner mitochondrial membrane of eukaryotic cells and in the plasma membrane of prokaryotes.

atrial natriuretic peptide (ANP) (ā′-trē-ul na′-trē-yū-ret′-ik) A peptide hormone secreted by cells of the atria of the heart in response to high blood pressure. ANP's effects on the kidney alter ion and water movement and thereby reduce blood pressure.

atrioventricular (AV) node A region of specialized heart muscle tissue between the left and right atria where electrical impulses are delayed for about 0.1 second before spreading to both ventricles and causing them to contract.

atrioventricular (AV) valve A heart valve located between each atrium and ventricle that prevents a backflow of blood when the ventricle contracts.

atrium (ā′-trē-um) (plural, **atria**) A chamber of the vertebrate heart that receives blood from the veins and transfers blood to a ventricle.

autocrine Referring to a secreted molecule that acts on the cell that secreted it.

autoimmune disease An immunological disorder in which the immune system turns against self.

autonomic nervous system (ot′-ō-nom′-ik) An efferent branch of the vertebrate peripheral nervous system that regulates the internal environment; consists of the sympathetic, parasympathetic, and enteric divisions.

autopolyploid (ot′-ō-pol′-ē-ployd) An individual that has more than two chromosome sets that are all derived from a single species.

autosome (ot′-ō-sōm) A chromosome that is not directly involved in determining sex; not a sex chromosome.

autotroph (ot′-ō-trōf) An organism that obtains organic food molecules without eating other organisms or substances derived from other organisms. Autotrophs use energy from the sun or from the oxidation of inorganic substances to make organic molecules from inorganic ones.

auxin (ok′-sin) A term that primarily refers to indoleacetic acid (IAA), a natural plant hormone that has a variety of effects, including cell elongation, root formation, secondary growth, and fruit growth.

average heterozygosity (het′-er-ō-zī-gō′-si-tē) The percent, on average, of a population's loci that are heterozygous in members of the population.

avirulent Describing a pathogen that can only mildly harm, but not kill, the host.

axillary bud (ak′-sil-ār-ē) A structure that has the potential to form a lateral shoot, or branch. The bud appears in the angle formed between a leaf and a stem.

axon (ak′-son) A typically long extension, or process, of a neuron that carries nerve impulses away from the cell body toward target cells.

axon hillock The conical region of a neuron's axon where it joins the cell body; typically the region where nerve impulses are generated.

B cell receptor The antigen receptor on B cells: a Y-shaped, membrane-bound molecule consisting of two identical heavy chains and two identical light chains linked by disulfide bridges and containing two antigen-binding sites.

B cells The lymphocytes that complete their development in the bone marrow and become effector cells for the humoral immune response.

Bacteria One of two prokaryotic domains, the other being Archaea.

bacterial artificial chromosome (BAC) A large plasmid that acts as a bacterial chromosome and can carry inserts of 100,000 to 300,000 base pairs.

bacteriophage (bak-tēr′-ē-ō-fāj) A virus that infects bacteria; also called a phage.

bacterium Member of the prokaryotic domain Bacteria.

bacteroid A form of the bacterium *Rhizobium* contained within the vesicles formed by the root cells of a root nodule.

balancing selection Natural selection that maintains two or more phenotypic forms in a population.

bark All tissues external to the vascular cambium, consisting mainly of the secondary phloem and layers of periderm.

Barr body A dense object lying along the inside of the nuclear envelope in cells of female mammals, representing a highly condensed, inactivated X chromosome.

barrier method Contraception that relies on a physical barrier to block the passage of sperm. Examples include condoms and diaphragms.

basal angiosperm Member of a clade of three early-diverging lineages of flowering plants. Examples are *Amborella*, water lilies, and star anise and its relatives.

basal body (bā′-sul) A eukaryotic cell structure consisting of a 9 + 0 arrangement of microtubule triplets. The basal body may organize the microtubule assembly of a cilium or flagellum and is structurally very similar to a centriole.

basal metabolic rate (BMR) The metabolic rate of a resting, fasting, and nonstressed endotherm at a comfortable temperature.

base A substance that reduces the hydrogen ion concentration of a solution.

base-pair substitution A type of point mutation; the replacement of one nucleotide and its partner in the complementary DNA strand by another pair of nucleotides.

basidiocarp Elaborate fruiting body of a dikaryotic mycelium of a club fungus.

basidiomycete (buh-sid′-ē-ō-mī′-sēt) Member of the fungal phylum Basidiomycota, commonly called club fungus. The name comes from the club-like shape of the basidium.

basidium (plural, **basidia**) (buh-sid′-ē-um, buh-sid′-ē-ah) A reproductive appendage that produces sexual spores on the gills of mushrooms (club fungi).

Batesian mimicry (bāt′-zē-un mim′-uh-krē) A type of mimicry in which a harmless species looks like a species that is poisonous or otherwise harmful to predators.

behavior (in animals) Individually, an action carried out by muscles or glands under control of the nervous system in response to a stimulus; collectively, the sum of an animal's responses to external and internal stimuli.

behavioral ecology The study of the evolution of and ecological basis for animal behavior.

benign tumor A mass of abnormal cells that remains at the site of its origin.

benthic zone The bottom surface of an aquatic environment.

benthos (ben′-thōz) The communities of organisms living in the benthic zone of an aquatic biome.

beta (β) pleated sheet One form of the secondary structure of proteins in which the polypeptide chain folds back and forth. Two regions of the chain lie parallel to each other and are held together by hydrogen bonds.

beta oxidation A metabolic sequence that breaks fatty acids down to two-carbon fragments that enter the citric acid cycle as acetyl CoA.

bicoid A maternal effect gene that codes for a protein responsible for specifying the anterior end in *Drosophila*.

biennial (bī-en′-ē-ul) A flowering plant that requires two years to complete its life cycle.

big-bang reproduction Reproduction in which an organism produces all of its offspring in a single event; also known as semelparity.

bilateral symmetry Body symmetry in which a central longitudinal plane divides the body into two equal but opposite halves.

bilaterian (bī′-luh-ter′-ē-uhn) Member of a clade of animals with bilateral symmetry and three germ layers.

bile A mixture of substances that is produced in the liver but stored in the gallbladder and that enables formation of fat droplets in water as an aid in the digestion and absorption of fats.

binary fission A method of asexual reproduction by "division in half." In prokaryotes, binary fission does not involve mitosis; but in single-celled eukaryotes that undergo binary fission, mitosis is part of the process.

binomial The two-part latinized name of a species, consisting of the genus and specific epithet.

biodiversity hot spot A relatively small area with an exceptional concentration of endemic species and often a large number of endangered and threatened species.

bioenergetics (1) The overall flow and transformation of energy in an organism. (2) The study of how energy flows through organisms.

biofilm A surface-coating colony of one or more species of prokaryotes that engage in metabolic cooperation.

biofuel A fuel produced from dry organic matter or combustible oils produced by plants.

biogenic amine A neurotransmitter derived from an amino acid.

biogeochemical cycle Any of the various chemical cycles, which involve both biotic and abiotic components of ecosystems.

biogeography The study of the past and present distribution of species.

bioinformatics The use of computers, software, and mathematical models to process and integrate biological information from large data sets.

biological augmentation An approach to restoration ecology that uses organisms to add essential materials to a degraded ecosystem.

biological clock An internal timekeeper that controls an organism's biological rhythms. The biological clock marks time with or without environmental cues but often requires signals from the environment to remain tuned to an appropriate period. *See also* circadian rhythm.

biological magnification A process in which retained substances become more concentrated at each higher trophic level in a food chain.

biological species concept Definition of a species as a population or group of populations whose members have the potential to interbreed in nature and produce viable, fertile offspring, but do not produce viable, fertile offspring with members of other such groups.

biology The scientific study of life.

biomanipulation An approach that applies the top-down model of community organization to alter ecosystem characteristics. For example, ecologists can prevent algal blooms and eutrophication by altering the density of higher-level consumers in lakes instead of by using chemical treatments.

biomass The total mass of organic matter comprising a group of organisms in a particular habitat.

biome (bī′-ōm) Any of the world's major ecosystems, often classified according to the predominant vegetation and characterized by adaptations of organisms to that particular environment.

bioremediation The use of organisms to detoxify and restore polluted and degraded ecosystems.

biosphere The entire portion of Earth inhabited by life; the sum of all the planet's ecosystems.

biotechnology The manipulation of organisms or their components to produce useful products.

biotic (bī-ot′-ik) Pertaining to the living organisms in the environment.

bipolar cell A neuron that relays information between photoreceptors and ganglion cells in the retina.

bipolar disorder Depressive mental illness characterized by swings of mood from high to low; also called manic-depressive disorder.

birth control pill A chemical contraceptive that inhibits ovulation, retards follicular development, or alters a woman's cervical mucus to prevent sperm from entering the uterus.

blade (1) A leaflike structure of a seaweed that provides most of the surface area for photosynthesis. (2) The flattened portion of a typical leaf.

blastocoel (blas′-tuh-sēl) The fluid-filled cavity that forms in the center of a blastula.

blastocyst (blas′-tuh-sist) The blastula stage of mammalian embryonic development, consisting of an inner cell mass, a cavity, and an outer layer, the trophoblast. In humans, the blastocyst forms one week after fertilization.

blastomere An early embryonic cell arising during the cleavage stage of an early embryo.

blastopore (blas′-tō-pōr) In a gastrula, the opening of the archenteron that typically develops into the anus in deuterostomes and the mouth in protostomes.

blastula (blas′-tyū-luh) A hollow ball of cells that marks the end of the cleavage stage during early embryonic development in animals.

blood A connective tissue with a fluid matrix called plasma in which red blood cells, white blood cells, and cell fragments called platelets are suspended.

blood-brain barrier A specialized capillary arrangement in the brain that restricts the passage of most substances into the brain, thereby preventing dramatic fluctuations in the brain's environment.

blue-light photoreceptor A type of light receptor in plants that initiates a variety of responses, such as phototropism and slowing of hypocotyl elongation.

body cavity A fluid- or air-filled space between the digestive tract and the body wall.

body plan In animals, a set of morphological and developmental traits that are integrated into a functional whole—the living animal.

Bohr shift A lowering of the affinity of hemoglobin for oxygen, caused by a drop in pH. It facilitates the release of oxygen from hemoglobin in the vicinity of active tissues.

bolus A lubricated ball of chewed food.

bone A connective tissue consisting of living cells held in a rigid matrix of collagen fibers embedded in calcium salts.

book lung An organ of gas exchange in spiders, consisting of stacked plates contained in an internal chamber.

bottleneck effect Genetic drift that occurs when the size of a population is reduced, as by a natural disaster or human actions. Typically, the surviving population is no longer genetically representative of the original population.

bottom-up model A model of community organization in which mineral nutrients influence community organization by controlling plant or phytoplankton numbers, which in turn control herbivore numbers, which in turn control predator numbers.

Bowman's capsule (bō′-munz) A cup-shaped receptacle in the vertebrate kidney that is the initial, expanded segment of the nephron where filtrate enters from the blood.

brachiopod (bra′-kē-uh-pod′) A marine lophophorate with a shell divided into dorsal and ventral halves. Brachiopods are also called lamp shells.

brain Organ of the central nervous system where information is processed and integrated.

brainstem Collection of structures in the vertebrate brain, including the midbrain, the pons, and the medulla oblongata; functions in homeostasis, coordination of movement, and conduction of information to higher brain centers.

branch point The representation on a phylogenetic tree of the divergence of two or more taxa from a common ancestor. Most branch points are shown as dichotomies, in which a branch representing the ancestral lineage splits (at the branch point) into two branches, one for each of the two descendant taxa.

brassinosteroid A steroid hormone in plants that has a variety of effects, including cell elongation, retarding leaf abscission, and promoting xylem differentiation.

breathing Ventilation of the lungs through alternating inhalation and exhalation.

breathing control center A brain center that directs the activity of organs involved in breathing.

bronchiole (brong′-kē-ōl′) A fine branch of the bronchi that transports air to alveoli.

bronchus (brong′-kus) (plural, **bronchi**) One of a pair of breathing tubes that branch from the trachea into the lungs.

brown alga A multicellular, photosynthetic protist with a characteristic brown or olive color that results from carotenoids in its plastids. Most brown algae are marine, and some have a plantlike body (thallus).

bryophyte (brī′-uh-fīt) An informal name for a moss, liverwort, or hornwort; a nonvascular plant that lives on land but lacks some of the terrestrial adaptations of vascular plants.

budding Asexual reproduction in which outgrowths from the parent form and pinch off to live independently or else remain attached to eventually form extensive colonies.

buffer A substance that consists of acid and base forms in a solution and that minimizes changes in pH when extraneous acids or bases are added to the solution.

bulk feeder An animal that eats relatively large pieces of food.

bulk flow The movement of a fluid due to a difference in pressure between two locations.

bundle-sheath cell In C_4 plants, a type of photosynthetic cell arranged into tightly packed sheaths around the veins of a leaf.

C_3 plant A plant that uses the Calvin cycle for the initial steps that incorporate CO_2 into organic material, forming a three-carbon compound as the first stable intermediate.

C_4 plant A plant in which the Calvin cycle is preceded by reactions that incorporate CO_2 into a four-carbon compound, the end product of which supplies CO_2 for the Calvin cycle.

cadherin (kad-hēr′-in) A member of an important class of cell adhesion molecules that requires extracellular calcium ions for its function.

calcitonin (kal′-si-tō′-nin) A hormone secreted by the thyroid gland that lowers blood calcium levels by promoting calcium deposition in bone and calcium excretion from the kidneys; nonessential in adult humans.

callus A mass of dividing, undifferentiated cells at the cut end of a shoot.
calorie (cal) The amount of heat energy required to raise the temperature of 1 g of water by 1°C; also the amount of heat energy that 1 g of water releases when it cools by 1°C. The Calorie (with a capital C), usually used to indicate the energy content of food, is a kilocalorie.
Calvin cycle The second of two major stages in photosynthesis (following the light reactions), involving fixation of atmospheric CO_2 and reduction of the fixed carbon into carbohydrate.
CAM plant A plant that uses crassulacean acid metabolism, an adaptation for photosynthesis in arid conditions. In this process, carbon dioxide entering open stomata during the night is converted to organic acids, which release CO_2 for the Calvin cycle during the day, when stomata are closed.
Cambrian explosion A relatively brief time in geologic history when large, hard-bodied forms of animals with most of the major body plans known today appeared in the fossil record. This burst of evolutionary change occurred about 535–525 million years ago.
canopy The uppermost layer of vegetation in a terrestrial biome.
capillary (kap′-il-ār′-ē) A microscopic blood vessel that penetrates the tissues and consists of a single layer of endothelial cells that allows exchange between the blood and interstitial fluid.
capillary bed A network of capillaries in a tissue or organ.
capsid The protein shell that encloses a viral genome. It may be rod-shaped, polyhedral, or more complex in shape.
capsule (1) A sticky layer that surrounds the cell wall of some prokaryotes, protecting the cell surface and sometimes helping to glue the cell to surfaces. (2) The sporangium of a bryophyte (moss, liverwort, or hornwort).
carbohydrate (kar′-bō-hī′-drāt) A sugar (monosaccharide) or one of its dimers (disaccharides) or polymers (polysaccharides).
carbon fixation The initial incorporation of carbon from CO_2 into an organic compound by an autotrophic organism (a plant, another photosynthetic organism, or a chemoautotrophic prokaryote).
carbonyl group (kar′-buh-nēl′) A chemical group present in aldehydes and ketones and consisting of a carbon atom double-bonded to an oxygen atom.
carboxyl group (kar-bok′-sil) A chemical group present in organic acids and consisting of a single carbon atom double-bonded to an oxygen atom and also bonded to a hydroxyl group.
cardiac cycle (kar′-dē-ak) The alternating contractions and relaxations of the heart.
cardiac muscle A type of muscle that forms the contractile wall of the heart. Its cells are joined by intercalated disks that relay each heartbeat.
cardiac output The volume of blood pumped per minute by each ventricle of the heart.
cardiovascular system A closed circulatory system with a heart and branching network of arteries, capillaries, and veins. The system is characteristic of vertebrates.
carnivore An animal that mainly eats other animals.
carotenoid (kuh-rot′-uh-noyd′) An accessory pigment, either yellow or orange, in the chloroplasts of plants and in some prokaryotes. By absorbing wavelengths of light that chlorophyll cannot, carotenoids broaden the spectrum of colors that can drive photosynthesis.
carpel (kar′-pul) The ovule-producing reproductive organ of a flower, consisting of the stigma, style, and ovary.
carrier In genetics, an individual who is heterozygous at a given genetic locus, with one normal allele and one recessive allele. The heterozygote is phenotypically dominant for the character determined by the gene but can pass on the recessive allele to offspring.
carrying capacity The maximum population size that can be supported by the available resources, symbolized as *K*.
cartilage (kar′-til-ij) A flexible connective tissue with an abundance of collagenous fibers embedded in chondroitin sulfate.
Casparian strip (kas-pār′-ē-un) A water-impermeable ring of wax in the endodermal cells of plants that blocks the passive flow of water and solutes into the stele by way of cell walls.
catabolic pathway (kat′-uh-bol′-ik) A metabolic pathway that releases energy by breaking down complex molecules to simpler compounds.
catalyst (kat′-uh-list) A chemical agent that increases the rate of a reaction without being consumed by the reaction.
catastrophism (kuh-tas′-truh-fiz′-um) The principle that events in the past occurred suddenly and were caused by different mechanisms than those operating today. *See* uniformitarianism.
catecholamine (kat′-uh-kōl′-uh-mēn) Any of a class of neurotransmitters and hormones, including the hormones epinephrine and norepinephrine, that are synthesized from the amino acid tyrosine.
cation (cat′-ī-on) A positively charged ion.
cation exchange A process in which positively charged minerals are made available to a plant when hydrogen ions in the soil displace mineral ions from the clay particles.
CD4 A surface protein, present on most helper T cells, that binds to class II MHC molecules, enhancing the interaction between the T cell and an antigen-presenting cell.
CD8 A surface protein, present on most cytotoxic T cells, that binds to class I MHC molecules, enhancing the interaction between the T cell and a target cell.
cDNA library A gene library containing clones that carry complementary DNA (cDNA) inserts. The library includes only the genes that were transcribed in the cells whose mRNA was isolated to make the cDNA.
cecum (sē′-kum) (plural, **ceca**) The blind pouch at the beginning of the large intestine.
cell adhesion molecule (CAM) A transmembrane, cell-surface glycoprotein that binds to CAMs on other cells. The resulting cell-to-cell attachments contribute to stable tissue structure.
cell body The part of a neuron that houses the nucleus and most other organelles.
cell cycle An ordered sequence of events in the life of a cell, from its origin in the division of a parent cell until its own division into two; the eukaryotic cell cycle is composed of interphase (including G_1, S, and G_2 subphases) and M phase (including mitosis and cytokinesis).
cell cycle control system A cyclically operating set of molecules in the eukaryotic cell that both triggers and coordinates key events in the cell cycle.
cell differentiation The structural and functional divergence of cells as they become specialized during a multicellular organism's development. Cell differentiation depends on the control of gene expression.
cell division The reproduction of cells.
cell fractionation The disruption of a cell and separation of its parts by centrifugation.
cell plate A double membrane across the midline of a dividing plant cell, between which the new cell wall forms during cytokinesis.
cell wall A protective layer external to the plasma membrane in the cells of plants, prokaryotes, fungi, and some protists. Polysaccharides such as cellulose (in plants and some protists), chitin (in fungi), and peptidoglycan (in bacteria) are an important structural component of cell walls.
cell-mediated immune response The branch of acquired immunity that involves the activation of cytotoxic T cells, which defend against infected cells.
cellular respiration The catabolic pathways of aerobic and anaerobic respiration, which break down organic molecules for the production of ATP.
cellular slime mold A type of protist that has unicellular amoeboid cells and aggregated reproductive bodies in its life cycle.
cellulose (sel′-yū-lōs) A structural polysaccharide of plant cell walls, consisting of glucose monomers joined by β glycosidic linkages.
Celsius scale (sel′-sē-us) A temperature scale (°C) equal to $^5/_9$(°F – 32) that measures the freezing point of water at 0°C and the boiling point of water at 100°C.
central canal The narrow cavity in the center of the spinal cord that is continuous with the fluid-filled ventricles of the brain.
central vacuole A membranous sac in a mature plant cell with diverse roles in reproduction, growth, and development.
centriole (sen′-trē-ōl) A structure in the centrosome of an animal cell composed of a cylinder of microtubule triplets arranged in a 9 + 0 pattern. A centrosome has a pair of centrioles.

centromere (sen′-trō-mēr) The specialized region of the chromosome where two sister chromatids are most closely attached.

centrosome (sen′-trō-sōm) Structure present in the cytoplasm of animal cells, important during cell division; functions as a microtubule-organizing center. A centrosome has two centrioles.

cephalization (sef′-uh-luh-zā′-shun) An evolutionary trend toward the concentration of sensory equipment at the anterior end of the body.

cerebellum (sār′-ruh-bel′-um) Part of the vertebrate hindbrain located dorsally; functions in unconscious coordination of movement and balance.

cerebral cortex (suh-rē′-brul) The surface of the cerebrum; the largest and most complex part of the mammalian brain, containing nerve cell bodies of the cerebrum; the part of the vertebrate brain most changed through evolution.

cerebral hemisphere The right or left side of the cerebrum.

cerebrospinal fluid (suh-rē′-brō-spī′-nul) Blood-derived fluid that surrounds, protects against infection, nourishes, and cushions the brain and spinal cord.

cerebrum (suh-rē′-brum) The dorsal portion of the vertebrate forebrain, composed of right and left hemispheres; the integrating center for memory, learning, emotions, and other highly complex functions of the central nervous system.

cervix (ser′-viks) The neck of the uterus, which opens into the vagina.

chaparral A scrubland biome of dense, spiny evergreen shrubs found at midlatitudes along coasts where cold ocean currents circulate offshore; characterized by mild, rainy winters and long, hot, dry summers.

chaperonin (shap′-er-ō′-nin) A protein molecule that assists in the proper folding of other proteins.

character An observable heritable feature.

character displacement The tendency for characteristics to be more divergent in sympatric populations of two species than in allopatric populations of the same two species.

checkpoint A control point in the cell cycle where stop and go-ahead signals can regulate the cycle.

chelicera (kē-lih′-suh-ruh) (plural, **chelicerae**) One of a pair of clawlike feeding appendages characteristic of cheliceriforms.

cheliceriform (kē-lih-suh′-ri-form) An arthropod that has chelicerae and a body divided into a cephalothorax and an abdomen. Living cheliceriforms include sea spiders, horseshoe crabs, scorpions, ticks, and spiders.

chemical bond An attraction between two atoms, resulting from a sharing of outer-shell electrons or the presence of opposite charges on the atoms. The bonded atoms gain complete outer electron shells.

chemical energy Energy available in molecules for release in a chemical reaction; a form of potential energy.

chemical equilibrium In a chemical reaction, the state in which the rate of the forward reaction equals the rate of the reverse reaction, so that the relative concentrations of the reactants and products do not change with time.

chemical reaction The making and breaking of chemical bonds, leading to changes in the composition of matter.

chemiosmosis (kem′-ē-oz-mō′-sis) An energy-coupling mechanism that uses energy stored in the form of a hydrogen ion gradient across a membrane to drive cellular work, such as the synthesis of ATP. Most ATP synthesis in cells occurs by chemiosmosis.

chemoautotroph (kē′-mō-ot′-ō-trōf) An organism that needs only carbon dioxide as a carbon source but obtains energy by oxidizing inorganic substances.

chemoheterotroph (kē′-mō-het′-er-ō-trōf) An organism that must consume organic molecules for both energy and carbon.

chemoreceptor A sensory receptor that responds to a chemical stimulus, such as a solute or an odorant.

chiasma (plural, **chiasmata**) (kī-az′-muh, kī-az′-muh-tuh) The X-shaped, microscopically visible region where homologous nonsister chromatids have exchanged genetic material through crossing over during meiosis, the two homologs remaining associated due to sister chromatid cohesion.

chitin (kī′-tin) A structural polysaccharide, consisting of amino sugar monomers, found in many fungal cell walls and in the exoskeletons of all arthropods.

chlorophyll (klōr′-ō-fil) A green pigment located within the chloroplasts of plants and algae and in the membranes of certain prokaryotes. Chlorophyll *a* participates directly in the light reactions, which convert solar energy to chemical energy.

chlorophyll *a* A photosynthetic pigment that participates directly in the light reactions, which convert solar energy to chemical energy.

chlorophyll *b* An accessory photosynthetic pigment that transfers energy to chlorophyll *a*.

chloroplast (klōr′-ō-plast) An organelle found in plants and photosynthetic protists that absorbs sunlight and uses it to drive the synthesis of organic compounds from carbon dioxide and water.

choanocyte (kō-an′-uh-sīt) A flagellated feeding cell found in sponges. Also called a collar cell, it has a collar-like ring that traps food particles around the base of its flagellum.

cholesterol (kō-les′-tuh-rol) A steroid that forms an essential component of animal cell membranes and acts as a precursor molecule for the synthesis of other biologically important steroids, such as hormones.

chondrichthyan (kon-drik′-thē-an) Member of the class Chondrichthyes, vertebrates with skeletons made mostly of cartilage, such as sharks and rays.

chordate Member of the phylum Chordata, animals that at some point during their development have a notochord; a dorsal, hollow nerve cord; pharyngeal slits or clefts; and a muscular, post-anal tail.

chorion (kōr′-ē-on) The outermost of four extraembryonic membranes. It functions in gas exchange and contributes to the formation of the mammalian placenta.

chorionic villus sampling (CVS) (kōr′-ē-on′-ik vil′-us) A technique of prenatal diagnosis in which a small sample of the fetal portion of the placenta is removed and analyzed to detect certain genetic and congenital defects in the fetus.

choroid (kor′-oyd) A thin, pigmented inner layer of the vertebrate eye.

Chromalveolata One of five supergroups of eukaryotes proposed in a current hypothesis of the evolutionary history of eukaryotes. Chromalveolates may have originated by secondary endosymbiosis and include two large protist clades, the alveolates and the stramenopiles. *See also* Excavata, Rhizaria, Archaeplastida, and Unikonta.

chromatin (krō′-muh-tin) The complex of DNA and proteins that makes up a eukaryotic chromosome. When the cell is not dividing, chromatin exists in its dispersed form, as a mass of very long, thin fibers that are not visible with a light microscope.

chromosome (krō′-muh-sōm) A cellular structure carrying genetic material, found in the nucleus of eukaryotic cells. Each chromosome consists of one very long DNA molecule and associated proteins. (A bacterial chromosome usually consists of a single circular DNA molecule and associated proteins. It is found in the nucleoid region, which is not membrane bounded.) *See also* chromatin.

chromosome theory of inheritance A basic principle in biology stating that genes are located on chromosomes and that the behavior of chromosomes during meiosis accounts for inheritance patterns.

chylomicron (kī′-lō-mī′-kron) A small globule that transports lipids. Chylomicrons are composed of fats mixed with cholesterol and coated with proteins.

chyme (kīm) The mixture of partially digested food and digestive juices formed in the stomach.

chytrid (kī′-trid) Member of the fungal phylum Chytridiomycota, mostly aquatic fungi with flagellated zoospores that represent an early-diverging fungal lineage.

ciliary body A portion of the vertebrate eye associated with the lens. It produces the clear, watery aqueous humor that fills the anterior cavity of the eye.

ciliate (sil′-ē-it) A type of protist that moves by means of cilia.

cilium (sil′-ē-um) (plural, **cilia**) A short cellular appendage containing microtubules. A motile cilium is specialized for locomotion and is formed from a core of nine outer doublet microtubules and two inner single microtubules (the "9 + 2" arrangement) ensheathed in an extension of the plasma membrane. A primary cilium is usually nonmotile and plays a sensory and signaling role; it lacks the two inner microtubules (the "9 + 0" arrangement).

circadian rhythm (ser-kā′-dē-un) A physiological cycle of about 24 hours that is present in all eukaryotic organisms and that persists even in the absence of external cues.

citric acid cycle A chemical cycle involving eight steps that completes the metabolic breakdown of glucose molecules begun in glycolysis by oxidizing pyruvate to carbon dioxide; occurs within the mitochondrion in eukaryotic cells and in the cytosol of prokaryotes; the second major stage in cellular respiration.

clade (klayd) A group of species that includes an ancestral species and all its descendants.

cladistics (kluh-dis′-tiks) An approach to systematics in which organisms are placed into groups called clades based primarily on common descent.

class In classification, the taxonomic category above the level of order.

class I MHC molecule A type of MHC molecule found on the surface of nearly all nucleated cells and that functions in identification of infected cells by cytotoxic T cells.

class II MHC molecule A type of MHC molecule restricted to a few specialized immune cell types (dendritic cells, macrophages, and B cells) that serve as antigen-presenting cells.

classical conditioning A type of associative learning in which an arbitrary stimulus becomes associated with a particular outcome.

cleavage (1) The process of cytokinesis in animal cells, characterized by pinching of the plasma membrane. (2) The succession of rapid cell divisions without significant growth during early embryonic development that converts the zygote to a ball of cells.

cleavage furrow The first sign of cleavage in an animal cell; a shallow groove in the cell surface near the old metaphase plate.

climate The long-term prevailing weather conditions at a locality.

climograph A plot of the temperature and precipitation in a particular region.

cline A graded change in a character along a geographic axis.

clitoris (klit′-uh-ris) An organ at the upper intersection of the labia minora that engorges with blood and becomes erect during sexual arousal.

cloaca (klō-ā′-kuh) A common opening for the digestive, urinary, and reproductive tracts found in many nonmammalian vertebrates but in few mammals.

clonal selection The process by which an antigen selectively binds to and activates only those lymphocytes bearing receptors specific for the antigen. The selected lymphocytes proliferate and differentiate into a clone of effector cells and a clone of memory cells specific for the stimulating antigen.

clone (1) A lineage of genetically identical individuals or cells. (2) In popular usage, a single individual organism that is genetically identical to another individual. (3) As a verb, to make one or more genetic replicas of an individual or cell. *See also* gene cloning.

cloning vector In genetic engineering, a DNA molecule that can carry foreign DNA into a host cell and replicate there. Cloning vectors include plasmids that move recombinant DNA from a test tube back into a cell and viruses that transfer recombinant DNA by infection.

closed circulatory system A circulatory system in which blood is confined to vessels and is kept separate from the interstitial fluid.

club fungus *See* basidiomycete.

cnidocyte (nī′-duh-sīt) A specialized cell unique to the phylum Cnidaria; contains a capsule-like organelle housing a coiled thread that, when discharged, explodes outward and functions in prey capture or defense.

cochlea (kok′-lē-uh) The complex, coiled organ of hearing that contains the organ of Corti.

codominance The situation in which the phenotypes of both alleles are exhibited in the heterozygote because both alleles affect the phenotype in separate, distinguishable ways.

codon (kō′-don) A three-nucleotide sequence of DNA or mRNA that specifies a particular amino acid or termination signal; the basic unit of the genetic code.

coefficient of relatedness The fraction of genes that, on average, are shared by two individuals.

coelom (sē′-lōm) A body cavity lined by tissue derived only from mesoderm.

coelomate (sē′-lō-māt) An animal that possesses a true coelom (a body cavity lined by tissue completely derived from mesoderm).

coenocytic fungus (sē′-no-si′-tic) A fungus that lacks septa and hence whose body is made up of a continuous cytoplasmic mass that may contain hundreds or thousands of nuclei.

coenzyme (kō-en′-zīm) An organic molecule serving as a cofactor. Most vitamins function as coenzymes in metabolic reactions.

cofactor Any nonprotein molecule or ion that is required for the proper functioning of an enzyme. Cofactors can be permanently bound to the active site or may bind loosely with the substrate during catalysis.

cognition The process of knowing that may include awareness, reasoning, recollection, and judgment.

cognitive map A neural representation of the abstract spatial relationships between objects in an animal's surroundings.

cohesion The binding together of like molecules, often by hydrogen bonds.

cohort A group of individuals of the same age in a population.

coitus (kō′-uh-tus) The insertion of a penis into a vagina; also called sexual intercourse.

coleoptile (kō′-lē-op′-tul) The covering of the young shoot of the embryo of a grass seed.

coleorhiza (kō′-lē-uh-rī′-zuh) The covering of the young root of the embryo of a grass seed.

collagen A glycoprotein in the extracellular matrix of animal cells that forms strong fibers, found extensively in connective tissue and bone; the most abundant protein in the animal kingdom.

collecting duct The location in the kidney where processed filtrate, called urine, is collected from the renal tubules.

collenchyma cell (kō-len′-kim-uh) A flexible plant cell type that occurs in strands or cylinders that support young parts of the plant without restraining growth.

colloid A mixture made up of a liquid and particles that (because of their large size) remain suspended rather than dissolved in that liquid.

colon (kō′-len) The largest section of the vertebrate large intestine; functions in water absorption and formation of feces.

commensalism (kuh-men′-suh-lizm) A symbiotic relationship in which one organism benefits but the other is neither helped nor harmed.

communication In animal behavior, a process involving transmission of, reception of, and response to signals. The term is also used in connection with other organisms, as well as individual cells of multicellular organisms.

community All the organisms that inhabit a particular area; an assemblage of populations of different species living close enough together for potential interaction.

community ecology The study of how interactions between species affect community structure and organization.

companion cell A type of plant cell that is connected to a sieve-tube element by many plasmodesmata and whose nucleus and ribosomes may serve one or more adjacent sieve-tube elements.

competitive exclusion The concept that when populations of two similar species compete for the same limited resources, one population will use the resources more efficiently and have a reproductive advantage that will eventually lead to the elimination of the other population.

competitive inhibitor A substance that reduces the activity of an enzyme by entering the active site in place of the substrate whose structure it mimics.

complement system A group of about 30 blood proteins that may amplify the inflammatory response, enhance phagocytosis, or directly lyse extracellular pathogens.

complementary DNA (cDNA) A double-stranded DNA molecule made *in vitro* using mRNA as a template and the enzymes reverse transcriptase and DNA polymerase. A cDNA molecule corresponds to the exons of a gene.

complete digestive tract A digestive tube that runs between a mouth and an anus; also called an alimentary canal.

complete dominance The situation in which the phenotypes of the heterozygote and dominant homozygote are indistinguishable.

complete flower A flower that has all four basic floral organs: sepals, petals, stamens, and carpels.

complete metamorphosis The transformation of a larva into an adult that looks very different, and often functions very differently in its environment, than the larva.

compound A substance consisting of two or more different elements combined in a fixed ratio.

compound eye A type of multifaceted eye in insects and crustaceans consisting of up to several thousand light-detecting, focusing ommatidia; especially good at detecting movement.

concentration gradient A region along which the density of a chemical substance increases or decreases.

conception The fertilization of an egg by a sperm in humans.

condensation reaction A reaction in which two molecules become covalently bonded to each other through the loss of a small molecule, usually water, in which case it is also called a dehydration reaction.

condom A thin, latex rubber or natural membrane sheath that fits over the penis to collect semen.

conduction The direct transfer of thermal motion (heat) between molecules of objects in direct contact with each other.

cone A cone-shaped cell in the retina of the vertebrate eye, sensitive to color.

conformer An animal for which an internal condition conforms with a change in an environmental variable.

conidium (plural, **conidia**) A haploid spore produced at the tip of a specialized hypha in ascomycetes during asexual reproduction.

conifer Member of the largest gymnosperm phylum. Most conifers are cone-bearing trees, such as pines and firs.

conjugation (kon′-jū-gā′-shun) In prokaryotes, the direct transfer of DNA between two cells (of the same or different species) that are temporarily joined. In ciliates, a sexual process in which two cells exchange haploid micronuclei.

connective tissue Animal tissue that functions mainly to bind and support other tissues, having a sparse population of cells scattered through an extracellular matrix.

conodont An early, soft-bodied vertebrate with prominent eyes and dental elements.

conservation biology The integrated study of ecology, evolutionary biology, physiology, molecular biology, and genetics to sustain biological diversity at all levels.

continental drift The slow movement of the continental plates across Earth's surface.

contraception The deliberate prevention of pregnancy.

contractile vacuole A membranous sac that helps move excess water out of certain freshwater protists.

control element A segment of noncoding DNA that helps regulate transcription of a gene by binding a transcription factor. Multiple control elements are present in a eukaryotic gene's enhancer.

controlled experiment An experiment in which an experimental group is compared with a control group that varies only in the factor being tested.

convection The mass movement of warmed air or liquid to or from the surface of a body or object.

convergent evolution The evolution of similar features in independent evolutionary lineages.

convergent extension A process in which the cells of a tissue layer rearrange themselves, so that the sheet of cells becomes narrower (converges) and longer (extends).

cooperativity A kind of allosteric regulation whereby a shape change in one subunit of a protein caused by substrate binding is transmitted to all the others, facilitating binding of subsequent substrate molecules.

copepod (cō′-puh-pod) Any of a group of small crustaceans that are important members of marine and freshwater plankton communities.

coral reef Typically a warm-water, tropical ecosystem dominated by the hard skeletal structures secreted primarily by the resident cnidarians. Some reefs also exist in cold, deep waters.

corepressor A small molecule that binds to a bacterial repressor protein and changes its shape, allowing it to switch an operon off.

cork cambium (kam′-bē-um) A cylinder of meristematic tissue in woody plants that replaces the epidermis with thicker, tougher cork cells.

cornea (kor′-nē-uh) The transparent frontal portion of the sclera, which admits light into the vertebrate eye.

corpus callosum (kor′-pus kuh-lō′-sum) The thick band of nerve fibers that connects the right and left cerebral hemispheres in mammals, enabling the hemispheres to process information together.

corpus luteum (kor′-pus lū′-tē-um) A secreting tissue in the ovary that forms from the collapsed follicle after ovulation and produces progesterone.

cortex (1) The outer region of cytoplasm in a eukaryotic cell, lying just under the plasma membrane, that has a more gel-like consistency than the inner regions, due to the presence of multiple microfilaments. (2) In plants, ground tissue that is between the vascular tissue and dermal tissue in a root or eudicot stem.

cortical granule A vesicle containing enzymes and other macromolecules located in the cortex (the region just under the plasma membrane) of an egg. Cortical granules undergo exocytosis during the cortical reaction.

cortical nephron In mammals and birds, a nephron with a loop of Henle located almost entirely in the renal cortex.

cortical reaction Exocytosis of enzymes and other macromolecules from cortical granules in the egg cytoplasm during fertilization, leading to the formation of a fertilization envelope.

corticosteroid Any steroid hormone produced and secreted by the adrenal cortex.

cotransport The coupling of the "downhill" diffusion of one substance to the "uphill" transport of another against its own concentration gradient.

cotyledon (kot′-uh-lē′-dun) A seed leaf of an angiosperm embryo. Some species have one cotyledon, others two.

countercurrent exchange The exchange of a substance or heat between two fluids flowing in opposite directions. For example, blood in a fish gill flows in the opposite direction of water passing over the gill, maximizing diffusion of oxygen into and carbon dioxide out of the blood.

countercurrent multiplier system A countercurrent system in which energy is expended in active transport to facilitate exchange of materials and generate concentration gradients.

covalent bond (kō-vā′-lent) A type of strong chemical bond in which two atoms share one or more pairs of valence electrons.

cranial nerve A nerve that originates in the brain and terminates in an organ of the head or upper body.

craniate A chordate with a head.

crassulacean acid metabolism (CAM) An adaptation for photosynthesis in arid conditions, first discovered in the family Crassulaceae. In this process, a plant takes up CO_2 and incorporates it into a variety of organic acids at night; during the day, CO_2 is released from organic acids for use in the Calvin cycle.

crista (plural, **cristae**) (kris′-tuh, kris′-tē) An infolding of the inner membrane of a mitochondrion that houses electron transport chains and molecules of the enzyme catalyzing the synthesis of ATP (ATP synthase).

critical load The amount of added nutrient, usually nitrogen or phosphorus, that can be absorbed by plants without damaging ecosystem integrity.

crop rotation The practice of planting nonlegumes one year and legumes in alternating years to restore concentrations of fixed nitrogen in the soil.

cross-fostering study A behavioral study in which the young of one species are placed in the care of adults from another species.

crossing over The reciprocal exchange of genetic material between nonsister chromatids during prophase I of meiosis.

cross-pollination In angiosperms, the transfer of pollen from an anther of a flower on one plant to the stigma of a flower on another plant of the same species.

crustacean (kruh-stā′-shun) A member of a subphylum of mostly aquatic arthropods that includes lobsters, crayfishes, crabs, shrimps, and barnacles.

cryptic coloration Camouflage that makes a potential prey difficult to spot against its background.

culture A system of information transfer through social learning or teaching that influences the behavior of individuals in a population.

cuticle (kyū′-tuh-kul) (1) A waxy covering on the surface of stems and leaves that acts as an adaptation that prevents desiccation in terrestrial plants. (2) The exoskeleton of an arthropod, consisting of layers of protein and chitin that are variously modified for different functions. (3) A tough coat that covers the body of a nematode.

cyclic AMP (cAMP) Cyclic adenosine monophosphate, a ring-shaped molecule made from ATP that is a common intracellular signaling molecule (second messenger) in eukaryotic cells. It is also a regulator of some bacterial operons.

cyclic electron flow A route of electron flow during the light reactions of photosynthesis that involves only photosystem I and that produces ATP but not NADPH or O_2.

cyclin (sī′-klin) A cellular protein that occurs in a cyclically fluctuating concentration and that plays an important role in regulating the cell cycle.

cyclin-dependent kinase (Cdk) A protein kinase that is active only when attached to a particular cyclin.

cystic fibrosis (sis′-tik fī-brō′-sis) A human genetic disorder caused by a recessive allele for a chloride channel protein; characterized by an excessive secretion of mucus and consequent vulnerability to infection; fatal if untreated.

cytochrome (sī′-tō-krōm) An iron-containing protein that is a component of electron transport chains in the mitochondria and chloroplasts of eukaryotic cells and the plasma membranes of prokaryotic cells.

cytogenetic map A chart of a chromosome that locates genes with respect to chromosomal features distinguishable in a microscope.

cytokine (sī′-tō-kīn′) Any of a group of proteins secreted by a number of cell types, including macrophages and helper T cells, that regulate the function of lymphocytes and other cells of the immune system.

cytokinesis (sī′-tō-kuh-nē′-sis) The division of the cytoplasm to form two separate daughter cells immediately after mitosis, meiosis I, or meiosis II.

cytokinin (sī′-tō-kī′-nin) Any of a class of related plant hormones that retard aging and act in concert with auxin to stimulate cell division, influence the pathway of differentiation, and control apical dominance.

cytoplasm (sī′-tō-plaz′-um) The contents of the cell, exclusive of the nucleus and bounded by the plasma membrane.

cytoplasmic determinant A maternal substance, such as a protein or RNA, placed into an egg that influences the course of early development by regulating the expression of genes that affect the developmental fate of cells.

cytoplasmic streaming A circular flow of cytoplasm, involving myosin and actin filaments, that speeds the distribution of materials within cells.

cytoskeleton A network of microtubules, microfilaments, and intermediate filaments that branch throughout the cytoplasm and serve a variety of mechanical, transport, and signaling functions.

cytosol (sī′-tō-sol) The semifluid portion of the cytoplasm.

cytotoxic T cell A type of lymphocyte that, when activated, kills infected cells as well as certain cancer cells and transplanted cells.

dalton A measure of mass for atoms and subatomic particles; the same as the atomic mass unit, or amu.

data Recorded observations.

day-neutral plant A plant in which flower formation is not controlled by photoperiod or day length.

decapod A member of the group of crustaceans that includes lobsters, crayfishes, crabs, and shrimps.

decomposer An organism that absorbs nutrients from nonliving organic material such as corpses, fallen plant material, and the wastes of living organisms and converts them to inorganic forms; a detritivore.

deductive reasoning A type of logic in which specific results are predicted from a general premise.

deep-sea hydrothermal vent A dark, hot, oxygen-deficient environment associated with volcanic activity on or near the seafloor. The producers in a vent community are chemoautotrophic prokaryotes.

de-etiolation The changes a plant shoot undergoes in response to sunlight; also known informally as greening.

dehydration reaction A chemical reaction in which two molecules covalently bond to each other with the removal of a water molecule.

deletion (1) A deficiency in a chromosome resulting from the loss of a fragment through breakage. (2) A mutational loss of one or more nucleotide pairs from a gene.

demographic transition A shift from rapid population growth in which birth rate outpaces death rate to zero population growth characterized by low birth and death rates.

demography The study of statistics relating to births and deaths in populations.

denaturation (dē-nā′-chur-ā′-shun) In proteins, a process in which a protein unravels and loses its native shape, thereby becoming biologically inactive; in DNA, the separation of the two strands of the double helix. Denaturation occurs under extreme (noncellular) conditions of pH, salt concentration, and temperature.

dendrite (den′-drīt) One of usually numerous, short, highly branched extensions of a neuron that receive signals from other neurons.

dendritic cell An antigen-presenting cell, located mainly in lymphatic tissues and skin, that is particularly efficient in presenting antigens to helper T cells, thereby initiating a primary immune response.

density The number of individuals per unit area or volume.

density dependent Referring to any characteristic that varies according to an increase in population density.

density independent Referring to any characteristic that is not affected by population density.

density-dependent inhibition The phenomenon observed in normal animal cells that causes them to stop dividing when they come into contact with one another.

deoxyribonucleic acid (DNA) (dē-ok′-sē-rī′-bō-nū-klā′-ik) A double-stranded, helical nucleic acid molecule consisting of nucleotide monomers with a deoxyribose sugar and the nitrogenous bases adenine (A), cytosine (C), guanine (G), and thymine (T); capable of replicating and determining the inherited structure of a cell's proteins.

deoxyribose (dē-ok′-si-rī′-bōs) The sugar component of DNA nucleotides, having one fewer hydroxyl group than ribose, the sugar component of RNA nucleotides.

depolarization A change in a cell's membrane potential such that the inside of the membrane is made less negative relative to the outside. For example, a neuron membrane is depolarized if a stimulus decreases its voltage from the resting potential of –70 mV in the direction of zero voltage.

dermal tissue system The outer protective covering of plants.

desert A terrestrial biome characterized by very low precipitation.

desmosome A type of intercellular junction in animal cells that functions as a rivet.

determinate cleavage A type of embryonic development in protostomes that rigidly casts the developmental fate of each embryonic cell very early.

determinate growth A type of growth characteristic of most animals and some plant organs, in which growth stops after a certain size is reached.

determination The progressive restriction of developmental potential in which the possible fate of each cell becomes more limited as an embryo develops. At the end of determination, a cell is committed to its fate.

detritivore (deh-trī′-tuh-vōr) A consumer that derives its energy and nutrients from nonliving organic material such as corpses, fallen plant material, and the wastes of living organisms; a decomposer.

detritus (di-trī′-tus) Dead organic matter.

deuteromycete (dū′-tuh-rō-mī′-sēt) Traditional classification for a fungus with no known sexual stage. When a sexual stage for a so-called deuteromycete is discovered, the species is assigned to a phylum.

deuterostome development (dū′-tuh-rō-stōm′) In animals, a developmental mode distinguished by the development of the anus from the blastopore; often also characterized by radial cleavage and by the body cavity forming as outpockets of mesodermal tissue.

diabetes mellitus (dī′-uh-bē′-tis mel′-uh-tus) An endocrine disorder marked by inability to maintain glucose homeostasis. The type 1 form results from autoimmune destruction of insulin-secreting cells; treatment usually requires daily insulin injections. The type 2 form most commonly results from reduced responsiveness of target cells to insulin; obesity and lack of exercise are risk factors.

diacylglycerol (DAG) (dī-a′-sil-glis′-er-ol) A second messenger produced by the cleavage of a certain kind of phospholipid in the plasma membrane.

diaphragm (dī′-uh-fram′) (1) A sheet of muscle that forms the bottom wall of the thoracic cavity in mammals. Contraction of the diaphragm pulls air into the lungs. (2) A dome-shaped rubber cup fitted into the upper portion of the vagina before sexual intercourse. It serves as a physical barrier to the passage of sperm into the uterus.

diapsid (dī-ap′-sid) Member of an amniote clade distinguished by a pair of holes on each side of the skull. Diapsids include the lepidosaurs and archosaurs.

diastole (dī-as′-tō-lē) The stage of the cardiac cycle in which a heart chamber is relaxed and fills with blood.

diastolic pressure Blood pressure in the arteries when the ventricles are relaxed.

diatom (dī′-uh-tom) A unicellular photosynthetic alga with a unique glassy cell wall containing silica.

dicot A term traditionally used to refer to flowering plants that have two embryonic seed leaves, or cotyledons. Recent molecular evidence indicates that dicots do not form a clade; species once classified as dicots are now grouped into eudicots, magnoliids, and several lineages of basal angiosperms.

differential gene expression The expression of different sets of genes by cells with the same genome.

diffusion The spontaneous movement of a substance down its concentration gradient, from a region where it is more concentrated to a region where it is less concentrated.

digestion The second stage of food processing in animals: the breaking down of food into molecules small enough for the body to absorb.

dihybrid (dī′-hī′-brid) An organism that is heterozygous with respect to two genes of interest. All the offspring from a cross between parents doubly homozygous for different alleles are dihybrids. For example, parents of genotypes *AABB* and *aabb* produce a dihybrid of genotype *AaBb.*

dikaryotic (dī′-kār-ē-ot′-ik) Referring to a fungal mycelium with two haploid nuclei per cell, one from each parent.

dinoflagellate (dī′-nō-flaj′-uh-let) Member of a group of mostly unicellular photosynthetic algae with two flagella situated in perpendicular grooves in cellulose plates covering the cell.

dinosaur Member of an extremely diverse clade of reptiles varying in body shape, size, and habitat. Birds are the only extant dinosaurs.

dioecious (dī-ē′-shus) In plant biology, having the male and female reproductive parts on different individuals of the same species.

diploblastic Having two germ layers.

diploid cell (dip′-loyd) A cell containing two sets of chromosomes (*2n*), one set inherited from each parent.

diplomonad A protist that has modified mitochondria, two equal-sized nuclei, and multiple flagella.

directional selection Natural selection in which individuals at one end of the phenotypic range survive or reproduce more successfully than do other individuals.

disaccharide (dī-sak′-uh-rīd) A double sugar, consisting of two monosaccharides joined by a glycosidic linkage formed during dehydration synthesis.

discovery science The process of scientific inquiry that focuses on describing nature.

dispersal The movement of individuals (or gametes) away from their parent location. This movement sometimes expands the geographic range of a population or species.

dispersion The pattern of spacing among individuals within the boundaries of the geographic population.

disruptive selection Natural selection in which individuals on both extremes of a phenotypic range survive or reproduce more successfully than do individuals with intermediate phenotypes.

distal tubule In the vertebrate kidney, the portion of a nephron that helps refine filtrate and empties it into a collecting duct.

disturbance A natural or human-caused event that changes a biological community and usually removes organisms from it. Disturbances, such as fires and storms, play a pivotal role in structuring many communities.

disulfide bridge A strong covalent bond formed when the sulfur of one cysteine monomer bonds to the sulfur of another cysteine monomer.

DNA (deoxyribonucleic acid) (dē-ok′-sē-rī′-bō-nū-klā′-ik) A double-stranded, helical nucleic acid molecule, consisting of nucleotide monomers with a deoxyribose sugar and the nitrogenous bases adenine (A), cytosine (C), guanine (G), and thymine (T); capable of being replicated and determining the inherited structure of a cell's proteins.

DNA ligase (lī′-gās) A linking enzyme essential for DNA replication; catalyzes the covalent bonding of the 3′ end of one DNA fragment (such as an Okazaki fragment) to the 5′ end of another DNA fragment (such as a growing DNA chain).

DNA microarray assay A method to detect and measure the expression of thousands of genes at one time. Tiny amounts of a large number of single-stranded DNA fragments representing different genes are fixed to a glass slide and tested for hybridization with samples of labeled cDNA.

DNA polymerase (puh-lim′-er-ās) An enzyme that catalyzes the elongation of new DNA (for example, at a replication fork) by the addition of nucleotides to the 3′ end of an existing chain. There are several different DNA polymerases; DNA polymerase III and DNA polymerase I play major roles in DNA replication in prokaryotes.

domain (1) A taxonomic category above the kingdom level. The three domains are Archaea, Bacteria, and Eukarya. (2) An independently folding part of a protein.

dominant allele An allele that is fully expressed in the phenotype of a heterozygote.

dominant species A species with substantially higher abundance or biomass than other species in a community. Dominant species exert a powerful control over the occurrence and distribution of other species.

dopamine A neurotransmitter that is a catecholamine, like epinephrine and norepinephrine.

dormancy A condition typified by extremely low metabolic rate and a suspension of growth and development.

dorsal Pertaining to the top of an animal with radial or bilateral symmetry.

dorsal lip The region above the blastopore on the dorsal side of the amphibian embryo.

double bond A double covalent bond; the sharing of two pairs of valence electrons by two atoms.

double circulation A circulatory system consisting of separate pulmonary and systemic circuits, in which blood passes through the heart after completing each circuit.

double fertilization A mechanism of fertilization in angiosperms in which two sperm cells unite with two cells in the female gametophyte (embryo sac) to form the zygote and endosperm.

double helix The form of native DNA, referring to its two adjacent antiparallel polynucleotide strands wound around an imaginary axis into a spiral shape.

Down syndrome A human genetic disease caused by the presence of an extra chromosome 21; characterized by mental retardation and heart and respiratory defects.

Duchenne muscular dystrophy (duh-shen′) A human genetic disease caused by a sex-linked recessive allele; characterized by progressive weakening and a loss of muscle tissue.

duodenum (dū′-uh-dēn′-um) The first section of the small intestine, where chyme from the stomach mixes with digestive juices from the pancreas, liver, and gallbladder as well as from gland cells of the intestinal wall.

duplication An aberration in chromosome structure due to fusion with a fragment from a homologous chromosome, such that a portion of a chromosome is duplicated.

dynamic stability hypothesis The idea that long food chains are less stable than short chains.

dynein (dī′-nē-un) In cilia and flagella, a large contractile protein extending from one microtubule doublet to the adjacent doublet. ATP hydrolysis drives changes in dynein shape that lead to bending of cilia and flagella.

E site One of a ribosome's three binding sites for tRNA during translation. The E site is the place where discharged tRNAs leave the ribosome. (E stands for exit.)

ecdysone (ek′-duh-sōn) A steroid hormone, secreted by the prothoracic glands, that triggers molting in arthropods.

ecdysozoan Member of a group of animal phyla identified as a clade by molecular evidence. Many ecdysozoans are molting animals.

echinoderm (i-kī′-nō-derm) A slow-moving or sessile marine deuterostome with a water vascular system and, in larvae, bilateral symmetry. Echinoderms include sea stars, brittle stars, sea urchins, feather stars, and sea cucumbers.

ecological footprint The aggregate land and water area required by a person, city, or nation to produce all of the resources it consumes and to absorb all of the waste it generates.

ecological niche (nich) The sum of a species' use of the biotic and abiotic resources in its environment.

ecological species concept A definition of species in terms of ecological niche, the sum of how members of the species interact with the nonliving and living parts of their environment.

ecological succession Transition in the species composition of a community following a disturbance; the establishment of a community in an area virtually barren of life.

ecology The study of how organisms interact with each other and their environment.

ecosystem All the organisms in a given area as well as the abiotic factors with which they interact; one or more communities and the physical environment around them.

ecosystem ecology The study of energy flow and the cycling of chemicals among the various biotic and abiotic components in an ecosystem.

ecosystem service A function performed by an ecosystem that directly or indirectly benefits humans.

ecotone The transition from one type of habitat or ecosystem to another, such as the transition from a forest to a grassland.

ectoderm (ek′-tō-durm) The outermost of the three primary germ layers in animal embryos; gives rise to the outer covering and, in some phyla, the nervous system, inner ear, and lens of the eye.

ectomycorrhiza (ek′-tō-mī′-kō-rī′-zuh) Association of a fungus with a plant root system in which the fungus surrounds the roots but does not cause invagination of the host (plant) cells' plasma membranes.

ectomycorrhizal fungus A symbiotic fungus that forms sheaths of hyphae over the surface of plant roots and also grows into extracellular spaces of the root cortex.

ectoparasite A parasite that feeds on the external surface of a host.

ectopic Occurring in an abnormal location.

ectoproct A sessile, colonial lophophorate commonly called a bryozoan.

ectothermic Referring to organisms for which external sources provide most of the heat for temperature regulation.

Ediacaran biota (ē′-dē-uh-keh′-run bī-ō′-tuh) An early group of soft-bodied, multicellular eukaryotes known from fossils that range in age from 565 million to 545 million years old.

effective population size An estimate of the size of a population based on the numbers of females and males that successfully breed; generally smaller than the total population.

effector cell (1) A muscle cell or gland cell that performs the body's response to stimuli as directed by signals from the brain or other processing center of the nervous system. (2) A lymphocyte that has undergone clonal selection and is capable of mediating an acquired immune response.

efferent arteriole In the kidney, the blood vessel draining a nephron.

egg The female gamete.

egg-polarity gene A gene that helps control the orientation (polarity) of the egg; also called a maternal effect gene.

ejaculation The propulsion of sperm from the epididymis through the muscular vas deferens, ejaculatory duct, and urethra.

ejaculatory duct In mammals, the short section of the ejaculatory route formed by the convergence of the vas deferens and a duct from the seminal vesicle. The ejaculatory duct transports sperm from the vas deferens to the urethra.

electrocardiogram (ECG or EKG) A record of the electrical impulses that travel through heart muscle during the cardiac cycle.

electrochemical gradient The diffusion gradient of an ion, which is affected by both the concentration difference of the ion across a membrane (a chemical force) and the ion's tendency to move relative to the membrane potential (an electrical force).

electrogenic pump An ion transport protein that generates voltage across a membrane.

electromagnetic receptor A receptor of electromagnetic energy, such as visible light, electricity, or magnetism.

electromagnetic spectrum The entire spectrum of electromagnetic radiation ranging in wavelength from less than a nanometer to more than a kilometer.

electron A subatomic particle with a single negative electrical charge and a mass about $^{1}/_{2,000}$ that of a neutron or proton. One or more electrons move around the nucleus of an atom.

electron microscope (EM) A microscope that uses magnets to focus an electron beam on or through a specimen, resulting in resolving power a thousandfold greater than that of a light microscope. A transmission electron microscope (TEM) is used to study the internal structure of thin sections of cells. A scanning electron microscope (SEM) is used to study the fine details of cell surfaces.

electron shell An energy level of electrons at a characteristic average distance from the nucleus of an atom.

electron transport chain A sequence of electron carrier molecules (membrane proteins) that shuttle electrons during the redox reactions that release energy used to make ATP.

electronegativity The attraction of a given atom for the electrons of a covalent bond.

electroporation A technique to introduce recombinant DNA into cells by applying a brief electrical pulse to a solution containing the cells. The pulse creates temporary holes in the cells' plasma membranes, through which DNA can enter.

element Any substance that cannot be broken down to any other substance by chemical reactions.

elimination The fourth and final stage of food processing in animals: the passing of undigested material out of the digestive system.

embryo sac (em′-brē-ō) The female gametophyte of angiosperms, formed from the growth and division of the megaspore into a multicellular structure that typically has eight haploid nuclei.

embryonic lethal A mutation with a phenotype leading to death of an embryo or larva.

embryophyte Alternate name for land plants that refers to their shared derived trait of multicellular, dependent embryos.

emergent properties New properties that arise with each step upward in the hierarchy of life, owing to the arrangement and interactions of parts as complexity increases.

emigration The movement of individuals out of a population.

enantiomer (en-an′-tē-ō-mer) One of two compounds that are mirror images of each other.

endangered species A species that is in danger of extinction throughout all or a significant portion of its range.

endemic (en-dem′-ik) Referring to a species that is confined to a specific, relatively small geographic area.

endergonic reaction (en′-der-gon′-ik) A nonspontaneous chemical reaction, in which free energy is absorbed from the surroundings.

endocrine gland (en′-dō-krin) A ductless gland that secretes hormones directly into the interstitial fluid, from which they diffuse into the bloodstream.

endocrine system The internal system of communication involving hormones, the ductless glands that secrete hormones, and the molecular receptors on or in target cells that respond to hormones; functions in concert with the nervous system to effect internal regulation and maintain homeostasis.

endocytosis (en′-dō-sī-tō′-sis) Cellular uptake of biological molecules and particulate matter via formation of new vesicles from the plasma membrane.

endoderm (en′-dō-durm) The innermost of the three primary germ layers in animal embryos; lines the archenteron and gives rise to the liver, pancreas, lungs, and the lining of the digestive tract in species that have these structures.

endodermis The innermost layer of the cortex in plant roots; a cylinder one cell thick that forms the boundary between the cortex and the vascular cylinder.

endomembrane system The collection of membranes inside and around a eukaryotic cell, related either through direct physical contact or by the transfer of membranous vesicles; includes the smooth and rough endoplasmic reticulum, the Golgi apparatus, lysosomes, and vacuoles.

endometriosis (en′-dō-mē-trē-ō′-sis) The condition resulting from the presence of endometrial tissue outside of the uterus.

endometrium (en′-dō-mē′-trē-um) The inner lining of the uterus, which is richly supplied with blood vessels.

endoparasite A parasite that lives within a host.

endophyte A fungus that lives inside a leaf or other plant part without causing harm to the plant.

endoplasmic reticulum (ER) (en′-dō-plaz′-mik ruh-tik′-yū-lum) An extensive membranous network in eukaryotic cells, continuous with the outer nuclear membrane and composed of ribosome-studded (rough) and ribosome-free (smooth) regions.

endorphin (en-dōr′-fin) Any of several hormones produced in the brain and anterior pituitary that inhibits pain perception.

endoskeleton A hard skeleton buried within the soft tissues of an animal, such as the spicules of sponges, the plates of echinoderms, and the bony skeletons of vertebrates.

endosperm In angiosperms, a nutrient-rich tissue formed by the union of a sperm with two polar nuclei during double fertilization. The endosperm provides nourishment to the developing embryo in angiosperm seeds.

endospore A thick-coated, resistant cell produced by a bacterial cell exposed to harsh conditions.

endosymbiosis A process in which a unicellular organism (the "host") engulfs another cell, which lives within the host cell and ultimately becomes an organelle in the host cell; also refers to the hypothesis that mitochondria and plastids were formerly small prokaryotes that began living within larger cells.

endothelin A peptide produced by a blood vessel's endothelium that causes the vessel to constrict.

endothelium (en′-dō-thē′-lē-um) The simple squamous layer of cells lining the lumen of blood vessels.

endothermic Referring to organisms with bodies that are warmed by heat generated by metabolism. This heat is usually used to maintain a relatively stable body temperature higher than that of the external environment.

endotoxin A toxic component of the outer membrane of certain gram-negative bacteria that is released only when the bacteria die.

energetic hypothesis The concept that the length of a food chain is limited by the inefficiency of energy transfer along the chain.

energy The capacity to cause change, especially to do work (to move matter against an opposing force).

energy coupling In cellular metabolism, the use of energy released from an exergonic reaction to drive an endergonic reaction.

enhancer A segment of eukaryotic DNA containing multiple control elements, usually located far from the gene whose transcription it regulates.

enteric division Networks of neurons in the digestive tract, pancreas, and gallbladder; normally regulated by the sympathetic and parasympathetic divisions of the autonomic nervous system.

entropy A measure of disorder, or randomness.

enzymatic hydrolysis The process in digestion that splits macromolecules from food by the enzymatic addition of water.

enzyme (en′-zīm) A macromolecule serving as a catalyst, a chemical agent that changes the rate of a reaction without being consumed by the reaction.

enzyme-substrate complex A temporary complex formed when an enzyme binds to its substrate molecule(s).

eosinophil (ē′-ō-sin′-ō-fil) A type of white blood cell with low phagocytic activity that is thought to play a role in defense against parasitic worms by releasing enzymes toxic to these invaders.

epicotyl (ep′-uh-kot′-ul) In an angiosperm embryo, the embryonic axis above the point of attachment of the cotyledon(s) and below the first pair of miniature leaves.

epidemic A general outbreak of a disease.

epidermis (1) The dermal tissue system of nonwoody plants, usually consisting of a single layer of tightly packed cells. (2) The outermost layer of cells in an animal.

epididymis (ep′-uh-did′-uh-mus) A coiled tubule located adjacent to the mammalian testis where sperm are stored.

epigenetic inheritance Inheritance of traits transmitted by mechanisms not directly involving the nucleotide sequence of a genome.

epinephrine (ep′-i-nef′-rin) A catecholamine that, when secreted as a hormone by the adrenal medulla, mediates "fight-or-flight" responses to short-term stresses; also released by some neurons as a neurotransmitter; also known as adrenaline.

epiphyte (ep′-uh-fīt) A plant that nourishes itself but grows on the surface of another plant for support, usually on the branches or trunks of tropical trees.

epistasis (ep′-i-stā′-sis) A type of gene interaction in which one gene alters the phenotypic effects of another gene that is independently inherited.

epithelial tissue (ep′-uh-thē′-lē-ul) Sheets of tightly packed cells that line organs and body cavities as well as external surfaces.

epithelium An epithelial tissue.

epitope A small, accessible region of an antigen to which an antigen receptor or antibody binds; also called an antigenic determinant.

EPSP *See* excitatory postsynaptic potential.

equilibrium potential (E_{ion}) The magnitude of a cell's membrane voltage at equilibrium; calculated using the Nernst equation.

erythrocyte (eh-rith′-ruh-sīt) A blood cell that contains hemoglobin, which transports oxygen; also called a red blood cell.

erythropoietin (EPO) (eh-rith′-rō-poy′-uh-tin) A hormone that stimulates the production of erythrocytes. It is secreted by the kidney when body tissues do not receive enough oxygen.

esophagus (eh-sof′-uh-gus) A channel that conducts food, by peristalsis, from the pharynx to the stomach.

essential amino acid An amino acid that an animal cannot synthesize itself and must be obtained from food in prefabricated form. Eight amino acids are essential in the human adult.

essential element In plants, a chemical element required for the plant to grow from a seed and complete its life cycle, producing another generation in the form of seeds.

essential fatty acid An unsaturated fatty acid that an animal needs but cannot make.

essential nutrient A substance that an organism must absorb in preassembled form because it cannot be synthesized from any other material. In humans, there are essential vitamins, minerals, amino acids, and fatty acids.

estradiol (es′-truh-dī′-ol) A steroid hormone that stimulates the development and maintenance of the female reproductive system and secondary sex characteristics; the major estrogen in mammals.

estrogen (es′-trō-jen) Any steroid hormone, such as estradiol, that stimulates the development and maintenance of the female reproductive system and secondary sex characteristics.

estrous cycle (es′-trus) A reproductive cycle characteristic of female mammals except humans and certain other primates, in which the nonpregnant endometrium is reabsorbed rather than shed, and sexual response occurs only during mid-cycle at estrus.

estuary The area where a freshwater stream or river merges with the ocean.

ethology The scientific study of how animals behave, particularly in their natural environments.

ethylene (eth′-uh-lēn) The only gaseous plant hormone. Among its many effects are response to mechanical stress, programmed cell death, leaf abscission, and fruit ripening.

etiolation Plant morphological adaptations for growing in darkness.

euchromatin (yū-krō′-muh-tin) The less condensed form of eukaryotic chromatin that is available for transcription.

eudicot (yū-dī′-kot) Member of a clade consisting of the vast majority of flowering plants that have two embryonic seed leaves, or cotyledons.

euglenid (yū′-glen-id) A protist, such as *Euglena* or its relatives, characterized by an anterior pocket from which one or two flagella emerge.
euglenozoan Member of a diverse clade of flagellated protists that includes predatory heterotrophs, photosynthetic autotrophs, and pathogenic parasites.
Eukarya (yū-kar′-ē-uh) The domain that includes all eukaryotic organisms.
eukaryotic cell (yū′-ker-ē-ot′-ik) A type of cell with a membrane-enclosed nucleus and membrane-enclosed organelles. Organisms with eukaryotic cells (protists, plants, fungi, and animals) are called eukaryotes.
eumetazoan (yū′-met-uh-zō′-un) Member of a clade of animals with true tissues. All animals except sponges and a few other groups are eumetazoans.
euryhaline (yur′-i-hā′-līn) Referring to organisms that tolerate substantial changes in external osmolarity.
eurypterid (yur-ip′-tuh-rid) An extinct carnivorous cheliceriform also called a water scorpion.
Eustachian tube (yū-stā′-shun) The tube that connects the middle ear to the pharynx.
eutherian (yū-thēr′-ē-un) Placental mammal; mammal whose young complete their embryonic development within the uterus, joined to the mother by the placenta.
eutrophic lake (yū-trōf′-ik) A lake that has a high rate of biological productivity supported by a high rate of nutrient cycling.
eutrophication A process by which nutrients, particularly phosphorus and nitrogen, become highly concentrated in a body of water, leading to increased growth of organisms such as algae or cyanobacteria.
evaporation The process by which a liquid changes to a gas.
evaporative cooling The process in which the surface of an object becomes cooler during evaporation, owing to a change of the molecules with the greatest kinetic energy from the liquid to the gaseous state.
evapotranspiration The total evaporation of water from an ecosystem, including evaporation from soil and the outside of plants, as well as the transpiration of water from within plants through stomata.
evo-devo Evolutionary developmental biology; a field of biology that compares developmental processes of different multicellular organisms to understand how these processes have evolved and how changes can modify existing organismal features or lead to new ones.
evolution Descent with modification; the idea that living species are descendants of ancestral species that were different from the present-day ones; also defined more narrowly as the change in the genetic composition of a population from generation to generation.
evolutionary tree A branching diagram that reflects a hypothesis about evolutionary relationships among groups of organisms.
Excavata One of five supergroups of eukaryotes proposed in a current hypothesis of the evolutionary history of eukaryotes. Excavates have unique cytoskeletal features, and some species have an "excavated" feeding groove on one side of the cell body. *See also* Chromalveolata, Rhizaria, Archaeplastida, and Unikonta.
excitatory postsynaptic potential (EPSP) An electrical change (depolarization) in the membrane of a postsynaptic cell caused by the binding of an excitatory neurotransmitter from a presynaptic cell to a postsynaptic receptor; makes it more likely for a postsynaptic cell to generate an action potential.
excretion The disposal of nitrogen-containing metabolites and other waste products.
exergonic reaction (ek′-ser-gon′-ik) A spontaneous chemical reaction, in which there is a net release of free energy.
exocytosis (ek′-sō-sī-tō′-sis) The cellular secretion of biological molecules by the fusion of vesicles containing them with the plasma membrane.
exon A sequence within a primary transcript that remains in the RNA after RNA processing; also refers to the region of DNA from which this sequence was transcribed.
exoskeleton A hard encasement on the surface of an animal, such as the shell of a mollusc or the cuticle of an arthropod, that provides protection and points of attachment for muscles.
exotoxin (ek′-sō-tok′-sin) A toxic protein that is secreted by a prokaryote or other pathogen and that produces specific symptoms, even if the pathogen is no longer present.
expansin Plant enzyme that breaks the cross-links (hydrogen bonds) between cellulose microfibrils and other cell wall constituents, loosening the wall's fabric.
exponential population growth Growth of a population in an ideal, unlimited environment, represented by a J-shaped curve when population size is plotted over time.
expression vector A cloning vector that contains the requisite bacterial promoter just upstream of a restriction site where a eukaryotic gene can be inserted, allowing the gene to be expressed in a bacterial cell.
external fertilization The fusion of gametes that parents have discharged into the environment.
extinction vortex A downward population spiral in which inbreeding and genetic drift combine to cause a small population to shrink and, unless the spiral is reversed, to become extinct.
extracellular digestion The breakdown of food in compartments that are continuous with the outside of an animal's body.
extracellular matrix (ECM) The substance in which animal cells are embedded, consisting of protein and polysaccharides synthesized and secreted by cells.
extraembryonic membrane One of four membranes (yolk sac, amnion, chorion, and allantois) located outside the embryo that support the developing embryo in reptiles and mammals.
extreme halophile An organism that lives in a highly saline environment, such as the Great Salt Lake or the Dead Sea.
extreme thermophile An organism that thrives in hot environments (often 60–80°C or hotter).
extremophile An organism that lives in an environment whose conditions are so extreme that few other species can survive there. Extremophiles include extreme halophiles and extreme thermophiles.
F factor In bacteria, the DNA segment that confers the ability to form pili for conjugation and associated functions required for the transfer of DNA from donor to recipient. The F factor may exist as a plasmid or be integrated into the bacterial chromosome.
F plasmid The plasmid form of the F factor.
F_1 generation The first filial, or hybrid, offspring in a series of genetic crosses.
F_2 generation Offspring resulting from interbreeding of the hybrid F_1 generation.
facilitated diffusion The spontaneous passage of molecules or ions across a biological membrane with the assistance of specific transmembrane transport proteins.
facilitator A species that has a positive effect on the survival and reproduction of other species in a community and that influences community structure.
facultative anaerobe (fak′-ul-tā′-tiv an′-uh-rōb) An organism that makes ATP by aerobic respiration if oxygen is present but that switches to anaerobic respiration or fermentation if oxygen is not present.
family In classification, the taxonomic category above genus.
fast block to polyspermy The depolarization of the egg plasma membrane that begins within 1–3 seconds after a sperm binds to an egg membrane protein. The depolarization lasts about 1 minute and prevents additional sperm from fusing with the egg during that time.
fast-twitch fiber A muscle fiber used for rapid, powerful contractions.
fat A lipid consisting of three fatty acids linked to one glycerol molecule; also called a triacylglycerol or triglyceride.
fate map A territorial diagram of embryonic development that displays the future derivatives of individual cells and tissues.
fatty acid A long carbon chain carboxylic acid. Fatty acids vary in length and in the number and location of double bonds; three fatty acids linked to a glycerol molecule form a fat molecule, also known as a triacylglycerol or triglyceride.
feces (fē′-sēz) The wastes of the digestive tract.
feedback inhibition A method of metabolic control in which the end product of a metabolic pathway acts as an inhibitor of an enzyme within that pathway.
fermentation A catabolic process that makes a limited amount of ATP from glucose without

an electron transport chain and that produces a characteristic end product, such as ethyl alcohol or lactic acid.

fertilization (1) The union of haploid gametes to produce a diploid zygote. (2) The addition of mineral nutrients to the soil.

fertilization envelope The protective layer formed when the vitelline layer of an egg is pushed away from the plasma membrane and hardened after fertilization by molecules exocytosed during the cortical reaction.

fetus (fē′-tus) A developing mammal that has all the major structures of an adult. In humans, the fetal stage lasts from the 9th week of gestation until birth.

fiber A lignified cell type that reinforces the xylem of angiosperms and functions in mechanical support; a slender, tapered sclerenchyma cell that usually occurs in bundles.

fibrin (fī′-brin) The activated form of the blood-clotting protein fibrinogen. Fibrin aggregates into threads that form the fabric of the clot.

fibroblast (fī′-brō-blast) A type of cell in loose connective tissue that secretes the protein ingredients of the extracellular fibers.

fibronectin A glycoprotein that helps animal cells attach to the extracellular matrix.

filament In an angiosperm, the stalk portion of the stamen, the pollen-producing reproductive organ of a flower.

filtrate Cell-free fluid extracted from the body fluid by the excretory system.

filtration In excretory systems, the extraction of water and small solutes, including metabolic wastes, from the body fluid.

fimbria (plural, **fimbriae**) A short, hairlike appendage of a prokaryotic cell that helps it adhere to the substrate or to other cells; also known as an attachment pilus.

first law of thermodynamics The principle of conservation of energy: Energy can be transferred and transformed, but it cannot be created or destroyed.

fission The separation of an organism into two or more individuals of approximately equal size.

fixed action pattern In animal behavior, a sequence of unlearned acts that is essentially unchangeable and, once initiated, usually carried to completion.

flaccid (flas′-id) Limp. Lacking in stiffness or firmness, as in a plant cell in surroundings where there is no tendency for water to enter the cell.

flagellum (fluh-jel′-um) (plural, **flagella**) A long cellular appendage specialized for locomotion. Like motile cilia, eukaryotic flagella have a core with nine outer doublet microtubules and two inner single microtubules ensheathed in an extension of the plasma membrane. Prokaryotic flagella have a different structure.

florigen A flowering signal, not yet chemically identified, that may be a hormone or may be a change in relative concentrations of multiple hormones.

flower In an angiosperm, a short stem with up to four sets of modified leaves, bearing structures that function in sexual reproduction.

fluid feeder An animal that lives by sucking nutrient-rich fluids from another living organism.

fluid mosaic model The currently accepted model of cell membrane structure, which envisions the membrane as a mosaic of protein molecules drifting laterally in a fluid bilayer of phospholipids.

follicle (fol′-uh-kul) A microscopic structure in the ovary that contains the developing oocyte and secretes estrogens.

follicle-stimulating hormone (FSH) A tropic hormone that is produced and secreted by the anterior pituitary and that stimulates the production of eggs by the ovaries and sperm by the testes.

follicular phase That part of the ovarian cycle during which follicles are growing and oocytes maturing.

food chain The pathway along which food energy is transferred from trophic level to trophic level, beginning with producers.

food vacuole A membranous sac formed by phagocytosis of microorganisms or particles to be used as food by the cell.

food web The interconnected feeding relationships in an ecosystem.

foot (1) The portion of a bryophyte sporophyte that gathers sugars, amino acids, water, and minerals from the parent gametophyte via transfer cells. (2) One of the three main parts of a mollusc; a muscular structure usually used for movement. *See also* mantle, visceral mass.

foraging The seeking and obtaining of food.

foram (foraminiferan) An aquatic protist that secretes a hardened shell containing calcium carbonate and extends pseudopodia through pores in the shell.

foraminiferan *See* foram.

forebrain One of three ancestral and embryonic regions of the vertebrate brain; develops into the thalamus, hypothalamus, and cerebrum.

fossil A preserved remnant or impression of an organism that lived in the past.

founder effect Genetic drift that occurs when a few individuals become isolated from a larger population and form a new population whose gene pool composition is not reflective of that of the original population.

fovea (fō′-vē-uh) The place on the retina at the eye's center of focus, where cones are highly concentrated.

fragmentation A means of asexual reproduction whereby a single parent breaks into parts that regenerate into whole new individuals.

frameshift mutation A mutation occurring when the number of nucleotides inserted or deleted is not a multiple of three, resulting in the improper grouping of the subsequent nucleotides into codons.

free energy The portion of a biological system's energy that can perform work when temperature and pressure are uniform throughout the system. (The change in free energy of a system is calculated by the equation $\Delta G = \Delta H - T\Delta S$, where H is enthalpy [in biological systems, equivalent to total energy], T is absolute temperature, and S is entropy.)

frequency-dependent selection A decline in the reproductive success of individuals that have a phenotype that has become too common in a population.

fruit A mature ovary of a flower. The fruit protects dormant seeds and often aids in their dispersal.

functional group A specific configuration of atoms commonly attached to the carbon skeletons of organic molecules and usually involved in chemical reactions.

Fungi (fun′-jē) The eukaryotic kingdom that includes organisms that absorb nutrients after decomposing organic material.

G_0 phase A nondividing state occupied by cells that have left the cell cycle.

G_1 phase The first gap, or growth phase, of the cell cycle, consisting of the portion of interphase before DNA synthesis begins.

G_2 phase The second gap, or growth phase, of the cell cycle, consisting of the portion of interphase after DNA synthesis occurs.

gallbladder An organ that stores bile and releases it as needed into the small intestine.

game theory An approach to evaluating alternative strategies in situations where the outcome of a particular strategy depends on the strategies used by other individuals.

gametangium (gam′-uh-tan′-jē-um) (plural, **gametangia**) Multicellular plant structure in which gametes are formed. Female gametangia are called archegonia, and male gametangia are called antheridia.

gamete (gam′-ēt) A haploid reproductive cell, such as an egg or sperm. Gametes unite during sexual reproduction to produce a diploid zygote.

gametogenesis The process by which gametes are produced.

gametophore (guh-mē′-tō-fōr) The mature gamete-producing structure of a moss gametophyte.

gametophyte (guh-mē′-tō-fīt) In organisms (plants and some algae) that have alternation of generations, the multicellular haploid form that produces haploid gametes by mitosis. The haploid gametes unite and develop into sporophytes.

gamma-aminobutyric acid (GABA) An amino acid that functions as a CNS neurotransmitter in the central nervous system of vertebrates.

ganglion (gang′-glē-un) (plural, **ganglia**) A cluster (functional group) of nerve cell bodies in a centralized nervous system.

ganglion cell A type of neuron in the retina that synapses with bipolar cells and transmits action potentials to the brain via axons in the optic nerve.

gap junction A type of intercellular junction in animals that allows the passage of materials between cells.

gas exchange The uptake of molecular oxygen from the environment and the discharge of carbon dioxide to the environment.

gastric juice A digestive fluid secreted by the stomach.

gastrovascular cavity A central cavity with a single opening in the body of certain animals that functions in both the digestion and distribution of nutrients.

gastrula (gas′-trū-luh) An embryonic stage in animal development encompassing the formation of three layers: ectoderm, mesoderm, and endoderm.

gastrulation (gas′-trū-lā′-shun) In animal development, a series of cell and tissue movements in which the blastula-stage embryo folds inward, producing a three-layered embryo, the gastrula.

gated channel A transmembrane protein channel that opens or closes in response to a particular stimulus.

gated ion channel A gated channel for a specific ion. The opening or closing of such channels may alter the membrane potential.

gel electrophoresis (ē-lek′-trō-fōr-ē′-sis) A technique for separating nucleic acids or proteins on the basis of their size and electrical charge, both of which affect their rate of movement through an electric field in a gel.

gene A discrete unit of hereditary information consisting of a specific nucleotide sequence in DNA (or RNA, in some viruses).

gene cloning The production of multiple copies of a gene.

gene expression The process by which DNA directs the synthesis of proteins or, in some cases, just RNAs.

gene flow The transfer of alleles from one population to another, resulting from the movement of fertile individuals or their gametes.

gene pool The aggregate of all of the alleles for all of the loci in all individuals in a population. The term is also used in a more restricted sense as the aggregate of alleles for just one or a few loci in a population.

gene therapy The introduction of genes into an afflicted individual for therapeutic purposes.

gene-for-gene recognition A widespread form of plant disease resistance involving recognition of pathogen-derived molecules by the protein products of specific plant disease resistance genes.

genetic drift A process in which chance events cause unpredictable fluctuations in allele frequencies from one generation to the next. Effects of genetic drift are most pronounced in small populations.

genetic engineering The direct manipulation of genes for practical purposes.

genetic map An ordered list of genetic loci (genes or other genetic markers) along a chromosome.

genetic profile An individual's unique set of genetic markers, detected most often today by PCR or, previously, by electrophoresis and nucleic acid probes.

genetic recombination General term for the production of offspring with combinations of traits that differ from those found in either parent.

genetically modified (GM) organism An organism that has acquired one or more genes by artificial means; also known as a transgenic organism.

genetics The scientific study of heredity and hereditary variation.

genome (jē′-nōm) The genetic material of an organism or virus; the complete complement of an organism's or virus's genes along with its noncoding nucleic acid sequences.

genomic imprinting A phenomenon in which expression of an allele in offspring depends on whether the allele is inherited from the male or female parent.

genomic library A set of cell clones containing all the DNA segments from a genome, each within a plasmid, phage, or other cloning vector.

genomics (juh-nō′-miks) The study of whole sets of genes and their interactions.

genotype (jē′-nō-tīp) The genetic makeup, or set of alleles, of an organism.

genus (jē′-nus) (*plural*, **genera**) A taxonomic category above the species level, designated by the first word of a species' two-part scientific name.

geographic variation Differences between the gene pools of geographically separate populations or population subgroups.

geologic record The division of Earth's history into time periods, grouped into three eons—Archaean, Proterozoic, and Phanerozoic—and further subdivided into eras, periods, and epochs.

geometric isomer One of several compounds that have the same molecular formula and covalent arrangements but differ in the spatial arrangements of their atoms owing to the inflexibility of double bonds.

germ layer One of the three main layers in a gastrula that will form the various tissues and organs of an animal body.

gestation (jes-tā′-shun) Pregnancy; the state of carrying developing young within the female reproductive tract.

gibberellin (jib′-uh-rel′-in) Any of a class of related plant hormones that stimulate growth in the stem and leaves, trigger the germination of seeds and breaking of bud dormancy, and (with auxin) stimulate fruit development.

glans The rounded structure at the tip of the clitoris or penis that is involved in sexual arousal.

glial cells (glia) Supporting cells that are essential for the structural integrity of the nervous system and for the normal functioning of neurons.

global ecology The study of the functioning and distribution of organisms across the biosphere and how the regional exchange of energy and materials affects them.

glomeromycete (glō′-mer-ō-mī′-sēt) Member of the fungal phylum Glomeromycota, characterized by a distinct branching form of mycorrhizae (mutualistic relationships with plant roots) called arbuscular mycorrhizae.

glomerulus (glō-mār′-yū-lus) A ball of capillaries surrounded by Bowman's capsule in the nephron and serving as the site of filtration in the vertebrate kidney.

glucagon (glū′-kuh-gon) A hormone secreted by pancreatic alpha cells that raises blood glucose levels. It promotes glycogen breakdown and release of glucose by the liver.

glucocorticoid A steroid hormone that is secreted by the adrenal cortex and that influences glucose metabolism and immune function.

glutamate An amino acid that functions as a neurotransmitter in the central nervous system.

glyceraldehyde-3-phosphate (G3P) (glis′-er-al′-de-hīd) A three-carbon carbohydrate that is the direct product of the Calvin cycle; it is also an intermediate in glycolysis.

glycogen (glī′-kō-jen) An extensively branched glucose storage polysaccharide found in the liver and muscle of animals; the animal equivalent of starch.

glycolipid A lipid with covalently attached carbohydrate(s).

glycolysis (glī-kol′-uh-sis) The splitting of glucose into pyruvate. Glycolysis occurs in almost all living cells, serving as the starting point for fermentation or cellular respiration.

glycoprotein A protein with one or more carbohydrates covalently attached to it.

glycosidic linkage A covalent bond formed between two monosaccharides by a dehydration reaction.

gnathostome (na′-thu-stōm) Member of the vertebrate subgroup possessing jaws.

golden alga A biflagellated, photosynthetic protist named for its color, which results from its yellow and brown carotenoids.

Golgi apparatus (gol′-jē) An organelle in eukaryotic cells consisting of stacks of flat membranous sacs that modify, store, and route products of the endoplasmic reticulum and synthesize some products, notably noncellulose carbohydrates.

gonads (gō′-nadz) The male and female sex organs; the gamete-producing organs in most animals.

G protein A GTP-binding protein that relays signals from a plasma membrane signal receptor, known as a G protein-coupled receptor, to other signal transduction proteins inside the cell.

G protein-coupled receptor A signal receptor protein in the plasma membrane that responds to the binding of a signaling molecule by activating a G protein. Also called a G protein-linked receptor.

grade A group of organisms that share the same level of organizational complexity or share a key adaptation.

Gram stain A staining method that distinguishes between two different kinds of bacterial cell walls.

gram-negative Describing the group of bacteria that have a cell wall that is structurally more complex and contains less peptidoglycan than the cell wall of gram-positive bacteria. Gram-negative bacteria are often more toxic than gram-positive bacteria.

gram-positive Describing the group of bacteria that have a cell wall that is structurally less complex and contains more peptidoglycan than the cell wall of gram-negative bacteria. Gram-positive bacteria are usually less toxic than gram-negative bacteria.

granum (gran′-um) (plural, **grana**) A stack of membrane-bounded thylakoids in the chloroplast. Grana function in the light reactions of photosynthesis.

gravitropism (grav′-uh-trō′-pizm) A response of a plant or animal to gravity.

gray crescent A light gray, crescent-shaped region of cytoplasm that becomes exposed after cortical rotation, located near the equator of an egg on the side opposite sperm entry, marking the future dorsal side of the embryo.

gray matter Regions of dendrites and clustered neuron cell bodies within the CNS.

green alga A photosynthetic protist, named for green chloroplasts that are similar in structure and pigment composition to those of land plants. Green algae are a paraphyletic group, some of whose members are more closely related to land plants than they are to other green algae.

green world hypothesis The conjecture that terrestrial herbivores consume relatively little plant biomass because they are held in check by a variety of factors, including predators, parasites, and disease.

greenhouse effect The warming of Earth due to the atmospheric accumulation of carbon dioxide and certain other gases, which absorb reflected infrared radiation and reradiate some of it back toward Earth.

gross primary production (GPP) The total primary production of an ecosystem.

ground tissue system Plant tissues that are neither vascular nor dermal, fulfilling a variety of functions, such as storage, photosynthesis, and support.

growth factor (1) A protein that must be present in the extracellular environment (culture medium or animal body) for the growth and normal development of certain types of cells. (2) A local regulator that acts on nearby cells to stimulate cell proliferation and differentiation.

growth hormone (GH) A hormone that is produced and secreted by the anterior pituitary and that has both direct (nontropic) and tropic effects on a wide variety of tissues.

guard cells The two cells that flank the stomatal pore and regulate the opening and closing of the pore.

gustation The sense of taste.

guttation The exudation of water droplets, caused by root pressure in certain plants.

gymnosperm (jim′-nō-sperm) A vascular plant that bears naked seeds—seeds not enclosed in specialized chambers.

habituation A simple type of learning that involves a loss of responsiveness to stimuli that convey little or no new information.

half-life The amount of time it takes for 50% of a sample of a radioactive isotope to decay.

Hamilton's rule The principle that for natural selection to favor an altruistic act, the benefit to the recipient, devalued by the coefficient of relatedness, must exceed the cost to the altruist.

haploid cell (hap′-loyd) A cell containing only one set of chromosomes (n).

Hardy-Weinberg equilibrium The condition describing a nonevolving population (one that is in genetic equilibrium).

Hardy-Weinberg principle The principle that frequencies of alleles and genotypes in a population remain constant from generation to generation, provided that only Mendelian segregation and recombination of alleles are at work.

haustorium (plural, **haustoria**) (ho-stōr′-ē-um, ho-stōr′-ē-uh) In certain symbiotic fungi, a specialized hypha that can penetrate the tissues of host organisms.

heart A muscular pump that uses metabolic energy to elevate the hydrostatic pressure of the circulatory fluid (blood or hemolymph). The fluid then flows down a pressure gradient through the body and eventually returns to the heart.

heart attack The damage or death of cardiac muscle tissue resulting from prolonged blockage of one or more coronary arteries.

heart murmur A hissing sound that most often results from blood squirting backward through a leaky valve in the heart.

heart rate The frequency of heart contraction.

heat The total amount of kinetic energy due to the random motion of atoms or molecules in a body of matter; also called thermal energy. Heat is energy in its most random form.

heat of vaporization The quantity of heat a liquid must absorb for 1 g of it to be converted from the liquid to the gaseous state.

heat-shock protein A protein that helps protect other proteins during heat stress. Heat-shock proteins are found in plants, animals, and microorganisms.

heavy chain One of the two types of polypeptide chains that make up an antibody molecule and B cell receptor; consists of a variable region, which contributes to the antigen-binding site, and a constant region.

helicase An enzyme that untwists the double helix of DNA at the replication forks, separating the two strands and making them available as template strands.

helper T cell A type of T cell that, when activated, secretes cytokines that promote the response of B cells (humoral response) and cytotoxic T cells (cell-mediated response) to antigens.

hemoglobin (hē′-mō-glō′-bin) An iron-containing protein in red blood cells that reversibly binds oxygen.

hemolymph (hē′-mō-limf′) In invertebrates with an open circulatory system, the body fluid that bathes tissues.

hemophilia (hē′-muh-fil′-ē-uh) A human genetic disease caused by a sex-linked recessive allele resulting in the absence of one or more blood-clotting proteins; characterized by excessive bleeding following injury.

hepatic portal vein A large circulatory channel that conveys nutrient-laden blood from the small intestine to the liver, which regulates the blood's nutrient content.

herbivore (hur′-bi-vōr′) An animal that mainly eats plants or algae.

herbivory An interaction in which an organism eats parts of a plant or alga.

heredity The transmission of traits from one generation to the next.

hermaphrodite (hur-maf′-ruh-dīt′) An individual that functions as both male and female in sexual reproduction by producing both sperm and eggs.

hermaphroditism (hur-maf′-rō-dī-tizm) A condition in which an individual has both female and male gonads and functions as both a male and female in sexual reproduction by producing both sperm and eggs.

heterochromatin (het′-er-ō-krō′-muh-tin) Eukaryotic chromatin that remains highly compacted during interphase and is generally not transcribed.

heterochrony (het′-uh-rok′-ruh-nē) Evolutionary change in the timing or rate of an organism's development.

heterocyte (het′-er-ō-sīt) A specialized cell that engages in nitrogen fixation in some filamentous cyanobacteria; formerly called heterocyst.

heterokaryon (het′-er-ō-kār′-ē-un) A fungal mycelium that contains two or more haploid nuclei per cell.

heteromorphic (het′-er-ō-mōr′-fik) Referring to a condition in the life cycle of plants and certain algae in which the sporophyte and gametophyte generations differ in morphology.

heterosporous (het-er-os′-pōr-us) Referring to a plant species that has two kinds of spores: microspores, which develop into male gametophytes, and megaspores, which develop into female gametophytes.

heterotroph (het′-er-ō-trōf) An organism that obtains organic food molecules by eating other organisms or substances derived from them.

heterozygote advantage Greater reproductive success of heterozygous individuals compared with homozygotes; tends to preserve variation in a gene pool.

heterozygous (het′-er-ō-zī′-gus) Having two different alleles for a given gene.

hexapod An insect or closely related wingless, six-legged arthropod.

hibernation A physiological state in which metabolism decreases, the heart and respiratory system slow down, and body temperature is maintained at a lower level than normal.

high-density lipoprotein (HDL) A particle in the blood made up of cholesterol and other lipids surrounded by a single layer of phospholipids in which proteins are embedded. HDL carries less cholesterol than a related lipoprotein, LDL, and high HDL levels in the blood may be correlated with a decreased risk of blood vessel blockage.

hindbrain One of three ancestral and embryonic regions of the vertebrate brain; develops into the medulla oblongata, pons, and cerebellum.

histamine (his′-tuh-mēn) A substance released by mast cells that causes blood vessels to dilate and become more permeable in inflammatory and allergic responses.

histone (his′-tōn) A small protein with a high proportion of positively charged amino acids that binds to the negatively charged DNA and plays a key role in chromatin structure.

histone acetylation The attachment of acetyl groups to certain amino acids of histone proteins.

HIV (human immunodeficiency virus) The infectious agent that causes AIDS. HIV is a retrovirus.

holdfast A rootlike structure that anchors a seaweed.

holoblastic cleavage (hō′-lō-blas′-tik) A type of cleavage in which there is complete division of the egg; occurs in eggs that have little yolk (such as those of the sea urchin) or a moderate amount of yolk (such as those of the frog).

homeobox (hō′-mē-ō-boks′) A 180-nucleotide sequence within homeotic genes and some other developmental genes that is widely conserved in animals. Related sequences occur in plants and yeasts.

homeostasis (hō′-mē-ō-stā′-sis) The steady-state physiological condition of the body.

homeotic gene (ho′-mē-o′-tik) Any of the master regulatory genes that control placement and spatial organization of body parts in animals, plants, and fungi by controlling the developmental fate of groups of cells.

hominin (ho′-mi-nin) A species on the human branch of the evolutionary tree. Hominins include *Homo sapiens* and our ancestors, a group of extinct species that are more closely related to us than to chimpanzees.

homologous chromosomes (hō-mol′-uh-gus) A pair of chromosomes of the same length, centromere position, and staining pattern that possess genes for the same characters at corresponding loci. One homologous chromosome is inherited from the organism's father, the other from the mother. Also called homologs, or a homologous pair.

homologous structures Structures in different species that are similar because of common ancestry.

homology (hō-mol′-uh-jē) Similarity in characteristics resulting from a shared ancestry.

homoplasy (hō′-muh-play′-zē) Similar (analogous) structure or molecular sequence that has evolved independently in two species.

homosporous (hō-mos′-puh-rus) Referring to a plant species that has a single kind of spore, which typically develops into a bisexual gametophyte.

homozygous (hō′-mō-zī′-gus) Having two identical alleles for a given gene.

horizontal cell A neuron of the retina that helps integrate information before it is sent to the brain.

horizontal gene transfer The transfer of genes from one genome to another through mechanisms such as transposable elements, plasmid exchange, viral activity, and perhaps fusions of different organisms.

hormone In multicellular organisms, one of many types of secreted chemicals that are formed in specialized cells, travel in body fluids, and act on specific target cells in other parts of the body to change their functioning.

hornwort A small, herbaceous nonvascular plant that is a member of the phylum Anthocerophyta.

host The larger participant in a symbiotic relationship, serving as home and food source for the smaller symbiont.

host range The limited range of host cells that each type of virus can infect.

human chorionic gonadotropin (hCG) (kōr′-ē-on′-ik gō-na′-dō-trō′-pin) A hormone secreted by the chorion that maintains the corpus luteum of the ovary during the first three months of pregnancy.

Human Genome Project An international collaborative effort to map and sequence the DNA of the entire human genome.

humoral immune response (hyū′-mer-ul) The branch of acquired immunity that involves the activation of B cells and that leads to the production of antibodies, which defend against bacteria and viruses in body fluids.

humus (hyū′-mus) Decomposing organic material that is a component of topsoil.

Huntington's disease A human genetic disease caused by a dominant allele; characterized by uncontrollable body movements and degeneration of the nervous system; usually fatal 10 to 20 years after the onset of symptoms.

hybrid Offspring that results from the mating of individuals from two different species or two true-breeding varieties of the same species.

hybrid zone A geographic region in which members of different species meet and mate, producing at least some offspring of mixed ancestry.

hybridization In genetics, the mating, or crossing, of two true-breeding varieties.

hydration shell The sphere of water molecules around a dissolved ion.

hydrocarbon An organic molecule consisting only of carbon and hydrogen.

hydrogen bond A type of weak chemical bond that is formed when the slightly positive hydrogen atom of a polar covalent bond in one molecule is attracted to the slightly negative atom of a polar covalent bond in another molecule.

hydrogen ion A single proton with a charge of 1+. The dissociation of a water molecule (H_2O) leads to the generation of a hydroxide ion (OH^-) and a hydrogen ion (H^+).

hydrolysis (hī-drol′-uh-sis) A chemical process that lyses, or splits, molecules by the addition of water, functioning in disassembly of polymers to monomers.

hydronium ion A water molecule that has an extra proton bound to it; H_3O^+.

hydrophilic (hī′-drō-fil′-ik) Having an affinity for water.

hydrophobic (hī′-drō-fō′-bik) Having an aversion to water; tending to coalesce and form droplets in water.

hydrophobic interaction A type of weak chemical bond formed when molecules that do not mix with water coalesce to exclude water.

hydroponic culture A method in which plants are grown in mineral solutions rather than in soil.

hydrostatic skeleton A skeletal system composed of fluid held under pressure in a closed body compartment; the main skeleton of most cnidarians, flatworms, nematodes, and annelids.

hydroxide ion A water molecule that has lost a proton; OH^-.

hydroxyl group (hī-drok′-sil) A chemical group consisting of an oxygen atom joined to a hydrogen atom. Molecules possessing this group are soluble in water and are called alcohols.

hymen A thin membrane that partly covers the vaginal opening in the human female. The hymen is ruptured by sexual intercourse or other vigorous activity.

hyperpolarization A change in a cell's membrane potential such that the inside of the membrane becomes more negative relative to the outside. Hyperpolarization reduces the chance that a neuron will transmit a nerve impulse.

hypersensitive response A plant's localized defense response to a pathogen, involving the death of cells around the site of infection.

hypertension A disorder in which blood pressure remains abnormally high.

hypertonic Referring to a solution that, when surrounding a cell, will cause the cell to lose water.

hypha (plural, **hyphae**) (hī′-fuh, hī′-fē) One of many connected filaments that collectively make up the mycelium of a fungus.

hypocotyl (hī′-puh-cot′-ul) In an angiosperm embryo, the embryonic axis below the point of attachment of the cotyledon(s) and above the radicle.

hypothalamus (hī′-pō-thal′-uh-mus) The ventral part of the vertebrate forebrain; functions in maintaining homeostasis, especially in coordinating the endocrine and nervous systems; secretes hormones of the posterior pituitary and releasing factors that regulate the anterior pituitary.

hypothesis (hī-poth′-uh-sis) A tentative answer to a well-framed question, narrower in scope than a theory and subject to testing.

hypotonic Referring to a solution that, when surrounding a cell, will cause the cell to take up water.

imbibition The physical adsorption of water onto the internal surfaces of structures.

immigration The influx of new individuals into a population from other areas.

immune system An animal body's system of defenses against agents that cause disease.

immunization The process of generating a state of immunity by artificial means. In active immunization, also called vaccination, an inactive or weakened form of a pathogen is administered, inducing B and T cell responses and immunological memory. In passive immunization, antibodies specific for a particular microbe are administered, conferring immediate but temporary protection.

immunodeficiency A disorder in which the ability of an immune system to protect against pathogens is defective or absent.

immunoglobulin (Ig) (im′-yū-nō-glob′-yū-lin) Any of the class of proteins that function as antibodies. Immunoglobulins are divided into five major classes that differ in their distribution in the body and antigen disposal activities.

imprinting In animal behavior, the formation at a specific stage in life of a long-lasting behavioral response to a specific individual or object. (*See also* genomic imprinting.)

***in situ* hybridization** A technique used to detect the location of a specific mRNA using nucleic acid hybridization with a labeled probe in an intact organism.

***in vitro* fertilization (IVF)** (vē′-trō) Fertilization of oocytes in laboratory containers followed by artificial implantation of the early embryo in the mother's uterus.

***in vitro* mutagenesis** A technique used to discover the function of a gene by cloning it, introducing specific changes into the cloned gene's sequence, reinserting the mutated gene into a cell, and studying the phenotype of the mutant.

inclusive fitness The total effect an individual has on proliferating its genes by producing its own offspring and by providing aid that enables other close relatives to increase the production of their offspring.

incomplete dominance The situation in which the phenotype of heterozygotes is intermediate between the phenotypes of individuals homozygous for either allele.

incomplete flower A flower in which one or more of the four basic floral organs (sepals, petals, stamens, or carpels) are either absent or nonfunctional.

incomplete metamorphosis A type of development in certain insects, such as grasshoppers, in which the young (called nymphs) resemble adults but are smaller and have different body proportions. The nymph goes through a series of molts, each time looking more like an adult, until it reaches full size.

incus The second of three bones in the middle ear of mammals; also called the anvil.

indeterminate cleavage A type of embryonic development in deuterostomes in which each cell produced by early cleavage divisions retains the capacity to develop into a complete embryo.

indeterminate growth A type of growth characteristic of plants, in which the organism continues to grow as long as it lives.

induced fit Induced by entry of the substrate, the change in shape of the active site of an enzyme so that it binds more snugly to the substrate.

inducer A specific small molecule that binds to a bacterial repressor protein and changes the repressor's shape so that it cannot bind to an operator, thus switching an operon on.

induction The process in which one group of embryonic cells influences the development of another, usually by causing changes in gene expression.

inductive reasoning A type of logic in which generalizations are based on a large number of specific observations.

inflammatory response An innate immune defense triggered by physical injury or infection of tissue involving the release of substances that promote swelling, enhance the infiltration of white blood cells, and aid in tissue repair and destruction of invading pathogens.

inflorescence A group of flowers tightly clustered together.

ingestion The first stage of food processing in animals: the act of eating.

ingroup A species or group of species whose evolutionary relationships we seek to determine.

inhibin A hormone produced in the male and female gonads that functions in part by regulating the function of the anterior pituitary by negative feedback.

inhibitory postsynaptic potential (IPSP) An electrical change (usually hyperpolarization) in the membrane of a postsynaptic neuron caused by the binding of an inhibitory neurotransmitter from a presynaptic cell to a postsynaptic receptor; makes it more difficult for a postsynaptic neuron to generate an action potential.

innate behavior Animal behavior that is developmentally fixed and under strong genetic control. Innate behavior is exhibited in virtually the same form by all individuals in a population despite internal and external environmental differences during development and throughout their lifetimes.

innate immunity A form of defense common to all animals that is active immediately upon exposure to pathogens and that is the same whether or not the pathogen has been encountered previously.

inner cell mass An inner cluster of cells at one end of a mammalian blastocyst that subsequently develops into the embryo proper and some of the extraembryonic membranes.

inner ear One of three main regions of the vertebrate ear; includes the cochlea (which in turn contains the organ of Corti) and the semicircular canals.

inositol trisphosphate (IP_3) (in-ō′-suh-tol) A second messenger that functions as an intermediate between certain nonsteroid hormones and a third messenger, a rise in cytoplasmic Ca^{2+} concentration.

inquiry The search for information and explanation, often focused by specific questions.

insertion A mutation involving the addition of one or more nucleotide pairs to a gene.

insulin (in′-suh-lin) A hormone secreted by pancreatic beta cells that lowers blood glucose levels. It promotes the uptake of glucose by most body cells and the synthesis and storage of glycogen in the liver and also stimulates protein and fat synthesis.

integral protein Typically a transmembrane protein with hydrophobic regions that extend into and often completely span the hydrophobic interior of the membrane and with hydrophilic regions in contact with the aqueous solution on either side of the membrane (or lining the channel in the case of a channel protein).

integrin In animal cells, a transmembrane receptor protein that interconnects the extracellular matrix and the cytoskeleton.

integument (in-teg′-yū-ment) Layer of sporophyte tissue that contributes to the structure of an ovule of a seed plant.

integumentary system The outer covering of a mammal's body, including skin, hair, and nails.

intercalated disk (in-ter′-kuh-lā′-ted) A special junction between cardiac muscle cells that provides direct electrical coupling between the cells.

interferon (in′-ter-fēr′-on) A protein that has antiviral or immune regulatory functions. Interferon-α and interferon-β, secreted by virus-infected cells, help nearby cells resist viral infection; interferon-γ, secreted by T cells, helps activate macrophages.

intermediate disturbance hypothesis The concept that moderate levels of disturbance can foster greater species diversity than low or high levels of disturbance.

intermediate filament A component of the cytoskeleton that includes filaments intermediate in size between microtubules and microfilaments.

internal fertilization The fusion of eggs and sperm within the female reproductive tract. The sperm are typically deposited in or near the tract.

interneuron An association neuron; a nerve cell within the central nervous system that forms synapses with sensory and/or motor neurons and integrates sensory input and motor output.

internode A segment of a plant stem between the points where leaves are attached.

interphase The period in the cell cycle when the cell is not dividing. During interphase, cellular metabolic activity is high, chromosomes and organelles are duplicated, and cell size may increase. Interphase accounts for 90% of the cell cycle.

intersexual selection Selection whereby individuals of one sex (usually females) are choosy in selecting their mates from individuals of the other sex; also called mate choice.

interspecific competition Competition for resources between individuals of two or more species when resources are in short supply.

interspecific interaction A relationship between individuals of two or more species in a community.

interstitial fluid The fluid filling the spaces between cells in an animal.

intertidal zone The shallow zone of the ocean adjacent to land and between the high- and low-tide lines.

intracellular digestion The hydrolysis of food inside vacuoles.

intracytoplasmic sperm injection (ICSI) The fertilization of an egg in the laboratory by the direct injection of a single sperm.

intrasexual selection A direct competition among individuals of one sex (usually the males in vertebrates) for mates of the opposite sex.

introduced species A species moved by humans, either intentionally or accidentally, from its native location to a new geographic region; also called non-native or exotic species.

intron (in′-tron) A noncoding, intervening sequence within a primary transcript that is removed from the transcript during RNA processing; also refers to the region of DNA from which this sequence was transcribed.

invagination The infolding, or pushing inward, of cells due to changes in cell shape.

invasive species A species, often introduced by humans, that takes hold outside its native range.

inversion An aberration in chromosome structure resulting from reattachment of a chromosomal fragment in a reverse orientation to the chromosome from which it originated.

invertebrate An animal without a backbone. Invertebrates make up 95% of animal species.

involution The process by which sheets of cells roll over the edge of the lip of the blastopore into the interior of the embryo during gastrulation.

ion (ī′-on) An atom or group of atoms that has gained or lost one or more electrons, thus acquiring a charge.

ion channel A transmembrane protein channel that allows a specific ion to flow across the membrane down its concentration gradient.

ionic bond (ī-on′-ik) A chemical bond resulting from the attraction between oppositely charged ions.

ionic compound A compound resulting from the formation of an ionic bond; also called a salt.

IPSP *See* inhibitory postsynaptic potential.

iris The colored part of the vertebrate eye, formed by the anterior portion of the choroid.

islets of Langerhans Clusters of endocrine cells within the pancreas that produce and secrete the hormones glucagon (from alpha cells) and insulin (from beta cells).

isomer (ī′-sō-mer) One of several compounds with the same molecular formula but different structures and therefore different properties. The three types of isomers are structural isomers, geometric isomers, and enantiomers.

isomorphic Referring to alternating generations in plants and certain algae in which the sporophytes and gametophytes look alike, although they differ in chromosome number.

isopod A member of one of the largest groups of crustaceans, which includes terrestrial, freshwater, and marine species. Among the terrestrial isopods are the pill bugs, or wood lice.

isotonic (ī′-sō-ton′-ik) Referring to a solution that, when surrounding a cell, has no effect on the passage of water into or out of the cell.

isotope (ī′-sō-tōp′) One of several atomic forms of an element, each with the same number of protons but a different number of neutrons, thus differing in atomic mass.

iteroparity Reproduction in which adults produce offspring over many years; also known as repeated reproduction.

joule (J) A unit of energy: $1\ \text{J} = 0.239\ \text{cal}$; $1\ \text{cal} = 4.184\ \text{J}$.

juvenile hormone A hormone in arthropods, secreted by the corpora allata (a pair of glands), that promotes the retention of larval characteristics.

juxtaglomerular apparatus (JGA) (juks′-tuh-gluh-mār′-yū-ler) A specialized tissue in nephrons that releases the enzyme renin in response to a drop in blood pressure or volume.

juxtamedullary nephron In mammals and birds, a nephron with a loop of Henle that extends far into the renal medulla.

karyogamy (kār′-ē-og′-uh-mē) The fusion of two nuclei, as part of syngamy (fertilization).

karyotype (kār′-ē-ō-tīp) A display of the chromosome pairs of a cell arranged by size and shape.

keystone species A species that is not necessarily abundant in a community yet exerts strong control on community structure by the nature of its ecological role or niche.

kilocalorie (kcal) A thousand calories; the amount of heat energy required to raise the temperature of 1 kg of water by 1°C.

kin selection Natural selection that favors altruistic behavior by enhancing the reproductive success of relatives.

kinesis (kuh-nē′-sis) A change in activity or turning rate in response to a stimulus.

kinetic energy (kuh-net′-ik) The energy associated with the relative motion of objects. Moving matter can perform work by imparting motion to other matter.

kinetochore (kuh-net′-uh-kōr) A structure of proteins attached to the centromere that links each sister chromatid to the mitotic spindle.

kinetoplastid A protist, such as a trypanosome, that has a single large mitochondrion that houses an organized mass of DNA.

kingdom A taxonomic category, the second broadest after domain.

***K*-selection** Selection for life history traits that are sensitive to population density; also called density-dependent selection.

labia majora A pair of thick, fatty ridges that encloses and protects the rest of the vulva.

labia minora A pair of slender skin folds that surrounds the openings of the vagina and urethra.

labor A series of strong, rhythmic contractions of the uterus that expel a baby out of the uterus and vagina during childbirth.

lactation The continued production of milk from the mammary glands.

lacteal (lak′-tē-ul) A tiny lymph vessel extending into the core of an intestinal villus and serving as the destination for absorbed chylomicrons.

lactic acid fermentation Glycolysis followed by the conversion of pyruvate to lactate, with no release of carbon dioxide.

lagging strand A discontinuously synthesized DNA strand that elongates by means of Okazaki fragments, each synthesized in a $5' \rightarrow 3'$ direction away from the replication fork.

lancelet Member of the subphylum Cephalochordata, small blade-shaped marine chordates that lack a backbone.

landmark A location indicator—a point of reference for orientation during navigation.

landscape An area containing several different ecosystems linked by exchanges of energy, materials, and organisms.

landscape ecology The study of how the spatial arrangement of habitat types affects the distribution and abundance of organisms and ecosystem processes.

large intestine The tubular portion of the vertebrate alimentary canal between the small intestine and the anus; functions mainly in water absorption and the formation of feces.

larva (lar′-vuh) (plural, **larvae**) A free-living, sexually immature form in some animal life cycles that may differ from the adult animal in morphology, nutrition, and habitat.

larynx (lār′-inks) The portion of the respiratory tract containing the vocal cords; also called the voice box.

lateral geniculate nucleus One of a pair of structures in the brain that are the destination for most of the ganglion cell axons that form the optic nerves.

lateral inhibition A process that sharpens the edges and enhances the contrast of a perceived image by inhibiting receptors lateral to those that have responded to light.

lateral line system A mechanoreceptor system consisting of a series of pores and receptor units along the sides of the body in fishes and aquatic amphibians; detects water movements made by the animal itself and by other moving objects.

lateral meristem (mār′-uh-stem) A meristem that thickens the roots and shoots of woody plants. The vascular cambium and cork cambium are lateral meristems.

lateral root A root that arises from the pericycle of an established root.

lateralization Segregation of functions in the cortex of the left and right hemispheres of the brain.

law of conservation of mass A physical law stating that matter can change form but cannot be created or destroyed. In a closed system, the mass of the system is constant.

law of independent assortment Mendel's second law, stating that each pair of alleles segregates, or assorts, independently of each other pair during gamete formation; applies when genes for two characters are located on different pairs of homologous chromosomes.

law of segregation Mendel's first law, stating that the two alleles in a pair segregate (separate) into different gametes during gamete formation.

leading strand The new complementary DNA strand synthesized continuously along the template strand toward the replication fork in the mandatory 5′→3′ direction.

leaf The main photosynthetic organ of vascular plants.

leaf primordium A finger-like projection along the flank of a shoot apical meristem, from which a leaf arises.

learning The modification of behavior based on specific experiences.

lens The structure in an eye that focuses light rays onto the photoreceptors.

lenticel (len′-ti-sel) A small raised area in the bark of stems and roots that enables gas exchange between living cells and the outside air.

lepidosaur (leh-pid′-uh-sōr) Member of the reptilian group that includes lizards, snakes, and two species of New Zealand animals called tuataras.

leukocyte (lū′-kō-sīt′) A blood cell that functions in fighting infections; also called a white blood cell.

Leydig cell (lī′-dig) A cell that produces testosterone and other androgens and is located between the seminiferous tubules of the testes.

lichen The symbiotic collective formed by the mutualistic association between a fungus and a photosynthetic alga or cyanobacterium.

life cycle The generation-to-generation sequence of stages in the reproductive history of an organism.

life history The traits that affect an organism's schedule of reproduction and survival.

life table A table of data summarizing mortality in a population.

ligament A fibrous connective tissue that joins bones together at joints.

ligand (lig′-und) A molecule that binds specifically to another molecule, usually a larger one.

ligand-gated ion channel A protein pore in cellular membranes that opens or closes in response to a signaling chemical (its ligand), allowing or blocking the flow of specific ions.

light chain One of the two types of polypeptide chains that make up an antibody molecule and B cell receptor; consists of a variable region, which contributes to the antigen-binding site, and a constant region.

light microscope (LM) An optical instrument with lenses that refract (bend) visible light to magnify images of specimens.

light reactions The first of two major stages in photosynthesis (preceding the Calvin cycle). These reactions, which occur on the thylakoid membranes of the chloroplast or on membranes of certain prokaryotes, convert solar energy to the chemical energy of ATP and NADPH, releasing oxygen in the process.

light-harvesting complex A complex of proteins associated with pigment molecules (including chlorophyll *a*, chlorophyll *b*, and carotenoids) that captures light energy and transfers it to reaction-center pigments in a photosystem.

lignin (lig′-nin) A hard material embedded in the cellulose matrix of vascular plant cell walls that provides structural support in terrestrial species.

limiting nutrient An element that must be added for production to increase in a particular area.

limnetic zone In a lake, the well-lit, open surface waters farther from shore.

linear electron flow A route of electron flow during the light reactions of photosynthesis that involves both photosystems (I and II) and produces ATP, NADPH, and O_2. The net electron flow is from H_2O to $NADP^+$.

linkage map A genetic map based on the frequencies of recombination between markers during crossing over of homologous chromosomes.

linked genes Genes located close enough together on a chromosome that they tend to be inherited together.

lipid (lip′-id) One of a group of compounds, including fats, phospholipids, and steroids, that mix poorly, if at all, with water.

littoral zone In a lake, the shallow, well-lit waters close to shore.

liver The largest internal organ in the vertebrate body. The liver performs diverse functions, such as producing bile, preparing nitrogenous wastes for disposal, and detoxifying poisonous chemicals in the blood.

liverwort A small, herbaceous nonvascular plant that is a member of the phylum Hepatophyta.

loam The most fertile soil type, made up of roughly equal amounts of sand, silt, and clay.

lobe-fin Member of the vertebrate subgroup Sarcopterygii, osteichthyans with rod-shaped muscular fins, including coelacanths and lungfishes as well as the lineage that gave rise to tetrapods.

local regulator A secreted molecule that influences cells near where it is secreted.

locomotion Active motion from place to place.

locus (lō′-kus) (plural, **loci**) A specific place along the length of a chromosome where a given gene is located.

logistic population growth Population growth that levels off as population size approaches carrying capacity.

long-day plant A plant that flowers (usually in late spring or early summer) only when the light period is longer than a critical length.

long-term memory The ability to hold, associate, and recall information over one's lifetime.

long-term potentiation (LTP) An enhanced responsiveness to an action potential (nerve signal) by a receiving neuron.

loop of Henle The hairpin turn, with a descending and ascending limb, between the proximal and distal tubules of the vertebrate kidney; functions in water and salt reabsorption.

lophophore (lof′-uh-fōr) In some lophotrochozoan animals, including brachiopods, a crown of ciliated tentacles that surround the mouth and function in feeding.

lophotrochozoan Member of a group of animal phyla identified as a clade by molecular evidence. Lophotrochozoans include organisms that have lophophores or trochophore larvae.

low-density lipoprotein (LDL) A particle in the blood made up of cholesterol and other lipids surrounded by a single layer of phospholipids in which proteins are embedded. LDL carries more cholesterol than a related lipoprotein, HDL, and high LDL levels in the blood correlate with a tendency to develop blocked blood vessels and heart disease.

lung An infolded respiratory surface of a terrestrial vertebrate, land snail, or spider that connects to the atmosphere by narrow tubes.

luteal phase That portion of the ovarian cycle during which endocrine cells of the corpus luteum secrete female hormones.

luteinizing hormone (LH) (lū′-tē-uh-nī′-zing) A tropic hormone that is produced and secreted by the anterior pituitary and that stimulates ovulation in females and androgen production in males.

lycophyte (lī′-kuh-fīt) An informal name for a member of the phylum Lycophyta, which includes club mosses, spike mosses, and quillworts.

lymph The colorless fluid, derived from interstitial fluid, in the lymphatic system of vertebrates.

lymph node An organ located along a lymph vessel. Lymph nodes filter lymph and contain cells that attack viruses and bacteria.

lymphatic system A system of vessels and nodes, separate from the circulatory system, that returns fluid, proteins, and cells to the blood.

lymphocyte A type of white blood cell that mediates acquired immunity. The two main classes are B cells and T cells.

lysogenic cycle (lī′-sō-jen′-ik) A type of phage reproductive cycle in which the viral genome becomes incorporated into the bacterial host chromosome as a prophage and does not kill the host.

lysosome (lī′-suh-sōm) A membrane-enclosed sac of hydrolytic enzymes found in the cytoplasm of animal cells and some protists.

lysozyme (lī′-sō-zīm) An enzyme that destroys bacterial cell walls; in mammals, found in sweat, tears, and saliva.

lytic cycle (lit′-ik) A type of phage reproductive cycle resulting in the release of new phages by lysis (and death) of the host cell.

macroclimate Large-scale patterns in climate; the climate of an entire region.

macroevolution Evolutionary change above the species level, including the origin of a new group of organisms or a shift in the broad pattern of evolutionary change over a long period of time. Examples of macroevolutionary change include the appearance of major new features of organisms and the impact of mass extinctions on the diversity of life and its subsequent recovery.

macromolecule A giant molecule formed by the joining of smaller molecules, usually by a condensation reaction. Polysaccharides, proteins, and nucleic acids are macromolecules.

macronutrient A chemical substance that an organism must obtain in relatively large amounts. *See also* micronutrient.

macrophage (mak′-rō-fāj) A phagocytic cell present in many tissues that functions in innate immunity by destroying microbes and in acquired immunity as an antigen-presenting cell.

magnoliid Member of the angiosperm clade most closely related to eudicots. Extant examples are magnolias, laurels, and black pepper plants.

major depressive disorder A mood disorder characterized by feelings of sadness, lack of self-worth, emptiness, or loss of interest in nearly all things.

major histocompatibility complex (MHC) A family of genes that encode a large set of cell-surface proteins that function in antigen presentation. Foreign MHC molecules on transplanted tissue can trigger T cell responses that may lead to rejection of the transplant.

malignant tumor A cancerous tumor that is invasive enough to impair the functions of one or more organs.

malleus The first of three bones in the middle ear of mammals; also called the hammer.

malnourishment The long-term absence from the diet of one or more essential nutrients.

Malpighian tubule (mal-pig′-ē-un) A unique excretory organ of insects that empties into the digestive tract, removes nitrogenous wastes from the hemolymph, and functions in osmoregulation.

mammal Member of the class Mammalia, amniotes with mammary glands—glands that produce milk.

mammary glands Exocrine glands that secrete milk to nourish the young. These glands are characteristic of mammals.

mandible One of a pair of jaw-like feeding appendages found in myriapods, hexapods, and crustaceans.

mantle One of the three main parts of a mollusc; a fold of tissue that drapes over the mollusc's visceral mass and may secrete a shell. *See also* foot, visceral mass.

mantle cavity A water-filled chamber that houses the gills, anus, and excretory pores of a mollusc.

map unit A unit of measurement of the distance between genes. One map unit is equivalent to a 1% recombination frequency.

marine benthic zone The ocean floor.

mark-recapture method A sampling technique used to estimate the size of animal populations.

marsupial (mar-sū′-pē-ul) A mammal, such as a koala, kangaroo, or opossum, whose young complete their embryonic development inside a maternal pouch called the marsupium.

mass extinction Period of time when global environmental changes lead to the elimination of a large number of species throughout Earth.

mass number The sum of the number of protons and neutrons in an atom's nucleus.

mast cell A vertebrate body cell that produces histamine and other molecules that trigger inflammation in response to infection and in allergic reactions.

mate choice copying Behavior in which individuals in a population copy the mate choice of others, apparently as a result of social learning.

maternal effect gene A gene that, when mutant in the mother, results in a mutant phenotype in the offspring, regardless of the offspring's genotype. Maternal effect genes were first identified in *Drosophila*.

matter Anything that takes up space and has mass.

maximum likelihood As applied to systematics, a principle that states that when considering multiple phylogenetic hypotheses, one should take into account the hypothesis that reflects the most likely sequence of evolutionary events, given certain rules about how DNA changes over time.

maximum parsimony A principle that states that when considering multiple explanations for an observation, one should first investigate the simplest explanation that is consistent with the facts.

mechanoreceptor A sensory receptor that detects physical deformation in the body's environment associated with pressure, touch, stretch, motion, or sound.

medulla oblongata (meh-dul′-uh ob′-long-go′-tuh) The lowest part of the vertebrate brain, commonly called the medulla; a swelling of the hindbrain anterior to the spinal cord that controls autonomic, homeostatic functions, including breathing, heart and blood vessel activity, swallowing, digestion, and vomiting.

medusa (muh-dū′-suh) The floating, flattened, mouth-down version of the cnidarian body plan. The alternate form is the polyp.

megapascal (MPa) (meg′-uh-pas-kal′) A unit of pressure equivalent to about 10 atmospheres of pressure.

megaphyll (meh′-guh-fil) A leaf with a highly branched vascular system, characteristic of the vast majority of vascular plants.

megaspore A spore from a heterosporous plant species that develops into a female gametophyte.

meiosis (mī-ō′-sis) A modified type of cell division in sexually reproducing organisms consisting of two rounds of cell division but only one round of DNA replication. It results in cells with half the number of chromosome sets as the original cell.

meiosis I The first division of a two-stage process of cell division in sexually reproducing organisms that results in cells with half the number of chromosome sets as the original cell.

meiosis II The second division of a two-stage process of cell division in sexually reproducing organisms that results in cells with half the number of chromosome sets as the original cell.

melanocyte-stimulating hormone (MSH) A hormone produced and secreted by the anterior pituitary that regulates the activity of pigment-containing cells in the skin of some vertebrates.

melatonin A hormone secreted by the pineal gland that regulates body functions related to seasonal day length.

membrane potential The difference in electrical charge (voltage) across a cell's plasma membrane, due to the differential distribution of ions. Membrane potential affects the activity of excitable cells and the transmembrane movement of all charged substances.

memory cell One of a clone of long-lived lymphocytes, formed during the primary immune response, that remains in a lymphoid organ until activated by exposure to the same antigen that triggered its formation. Activated memory cells mount the secondary immune response.

menopause The cessation of ovulation and menstruation marking the end of a human female's reproductive years.

menstrual cycle (men′-strū-ul) In humans and certain other primates, a type of reproductive cycle in which the nonpregnant endometrium is shed through the cervix into the vagina.

menstrual flow phase That portion of the uterine (menstrual) cycle when menstrual bleeding occurs.

menstruation The shedding of portions of the endometrium during a uterine (menstrual) cycle.

meristem (mār′-uh-stem) Plant tissue that remains embryonic as long as the plant lives, allowing for indeterminate growth.

meristem identity gene A plant gene that promotes the switch from vegetative growth to flowering.

meroblastic cleavage (mār′-ō-blas′-tik) A type of cleavage in which there is incomplete division of a yolk-rich egg, characteristic of avian development.

mesoderm (mez′-ō-derm) The middle primary germ layer in an animal embryo; develops into the notochord, the lining of the coelom, muscles, skeleton, gonads, kidneys, and most of the circulatory system in species that have these structures.

mesohyl (mez′-ō-hīl) A gelatinous region between the two layers of cells of a sponge.

mesophyll (mez′-ō-fil) The ground tissue of a leaf, sandwiched between the upper and lower epidermis and specialized for photosynthesis.

mesophyll cell In C_4 plants, a type of loosely arranged photosynthetic cell located between the bundle sheath and the leaf surface.

messenger RNA (mRNA) A type of RNA, synthesized using a DNA template, that attaches to ribosomes in the cytoplasm and specifies the primary structure of a protein.

metabolic pathway A series of chemical reactions that either builds a complex molecule (anabolic pathway) or breaks down a complex molecule into simpler compounds (catabolic pathway).

metabolic rate The total amount of energy an animal uses in a unit of time.

metabolism (muh-tab′-uh-lizm) The totality of an organism's chemical reactions, consisting of catabolic and anabolic pathways, which manage the material and energy resources of the organism.

metamorphosis (met′-uh-mōr′-fuh-sis) A developmental transformation that turns an animal larva into either an adult or an adult-like stage that is not yet sexually mature.

metanephridium (met′-uh-nuh-frid′-ē-um) (plural, **metanephridia**) An excretory organ found in many invertebrates that typically consists of tubules connecting ciliated internal openings to external openings.

metaphase The third stage of mitosis, in which the spindle is complete and the chromosomes, attached to microtubules at their kinetochores, are all aligned at the metaphase plate.

metaphase plate An imaginary plane midway between the two poles of a cell in metaphase on which the centromeres of all the duplicated chromosomes are located.

metapopulation A group of spatially separated populations of one species that interact through immigration and emigration.

metastasis (muh-tas′-tuh-sis) The spread of cancer cells to locations distant from their original site.

methanogen (meth-an′-ō-jen) An organism that obtains energy by using carbon dioxide to oxidize hydrogen, producing methane as a waste product; all known methanogens are in domain Archaea.

methyl group A chemical group consisting of a carbon bonded to three hydrogen atoms. The methyl group may be attached to a carbon or to a different atom.

microclimate Very fine scale patterns of climate, such as the specific climatic conditions underneath a log.

microevolution Evolutionary change below the species level; change in the allele frequencies in a population over generations.

microfilament A cable composed of actin proteins in the cytoplasm of almost every eukaryotic cell, making up part of the cytoskeleton and acting alone or with myosin to cause cell contraction; also known as an actin filament.

micronutrient An element that an organism needs in very small amounts and that functions as a component or cofactor of enzymes. *See also* macronutrient.

microphyll (mī′-krō-fil) In lycophytes, a small leaf with a single unbranched vein.

micropyle A pore in the integument(s) of an ovule.

microRNA (miRNA) A small, single-stranded RNA molecule, generated from a hairpin structure on a precursor RNA transcribed from a particular gene. The miRNA associates with one or more proteins in a complex that can degrade or prevent translation of an mRNA with a complementary sequence.

microspore A spore from a heterosporous plant species that develops into a male gametophyte.

microtubule A hollow rod composed of tubulin proteins that make up part of the cytoskeleton in all eukaryotic cells and is found in cilia and flagella.

microvillus (plural, **microvilli**) One of many fine, finger-like projections of the epithelial cells in the lumen of the small intestine that increase its surface area.

midbrain One of three ancestral and embryonic regions of the vertebrate brain; develops into sensory integrating and relay centers that send sensory information to the cerebrum.

middle ear One of three main regions of the vertebrate ear; in mammals, a chamber containing three small bones (the malleus, incus, and stapes) that convey vibrations from the eardrum to the oval window.

middle lamella (luh-mel′-uh) In plants, a thin layer of adhesive extracellular material, primarily pectins, found between the primary walls of adjacent young cells.

migration A regular, long-distance change in location.

mineral In nutrition, a simple nutrient that is inorganic and therefore cannot be synthesized.

mineralocorticoid A steroid hormone secreted by the adrenal cortex that regulates salt and water homeostasis.

minimum viable population (MVP) The smallest population size at which a species is able to sustain its numbers and survive.

mismatch repair The cellular process that uses specific enzymes to remove and replace incorrectly paired nucleotides.

missense mutation A base-pair substitution that results in a codon that codes for a different amino acid.

mitochondrial matrix The compartment of the mitochondrion enclosed by the inner membrane and containing enzymes and substrates for the citric acid cycle.

mitochondrion (mī′-tō-kon′-drē-un) (plural, **mitochondria**) An organelle in eukaryotic cells that serves as the site of cellular respiration.

mitosis (mī-tō′-sis) A process of nuclear division in eukaryotic cells conventionally divided into five stages: prophase, prometaphase, metaphase, anaphase, and telophase. Mitosis conserves chromosome number by allocating replicated chromosomes equally to each of the daughter nuclei.

mitotic (M) phase The phase of the cell cycle that includes mitosis and cytokinesis.

mitotic spindle An assemblage of microtubules and associated proteins that is involved in the movements of chromosomes during mitosis.

mixotroph An organism that is capable of both photosynthesis and heterotrophy.

model A representation of a theory or process.

model organism A particular species chosen for research into broad biological principles because it is representative of a larger group and usually easy to grow in a lab.

molarity A common measure of solute concentration, referring to the number of moles of solute per liter of solution.

mold Informal term for a fungus that grows as a filamentous fungus, producing haploid spores by mitosis and forming a visible mycelium.

mole (mol) The number of grams of a substance that equals its molecular weight in daltons and contains Avogadro's number of molecules.

molecular clock A method for estimating the time required for a given amount of evolutionary change, based on the observation that some regions of genomes appear to evolve at constant rates.

molecular formula A type of molecular notation representing the quantity of constituent atoms, but not the nature of the bonds that join them.

molecular mass The sum of the masses of all the atoms in a molecule; sometimes called molecular weight.

molecular systematics A scientific discipline that uses nucleic acids or other molecules in different species to infer evolutionary relationships.

molecule Two or more atoms held together by covalent bonds.

molting A process in ecdysozoans in which the exoskeleton is shed at intervals, allowing growth by the production of a larger exoskeleton.

monoclonal antibody (mon′-ō-klōn′-ul) Any of a preparation of antibodies that have been produced by a single clone of cultured cells and thus are all specific for the same epitope.

monocot Member of a clade consisting of flowering plants that have one embryonic seed leaf, or cotyledon.

monogamous (muh-nog′-uh-mus) Referring to a type of relationship in which one male mates with just one female.

monohybrid An organism that is heterozygous with respect to a single gene of interest. All the offspring from a cross between parents homozygous for different alleles are monohybrids. For example, parents of genotypes *AA* and *aa* produce a monohybrid of genotype *Aa*.

monomer (mon′-uh-mer) The subunit that serves as the building block of a polymer.

monophyletic (mon′-ō-fī-let′-ik) Pertaining to a group of taxa that consists of a common ancestor and all its descendants. A monophyletic taxon is equivalent to a clade.

monosaccharide (mon′-ō-sak′-uh-rīd) The simplest carbohydrate, active alone or serving as a monomer for disaccharides and polysaccharides. Also known as simple sugars, monosaccharides have molecular formulas that are generally some multiple of CH_2O.

monosomic Referring to a cell that has only one copy of a particular chromosome instead of the normal two.

morphogen A substance, such as Bicoid protein in *Drosophila*, that provides positional information in the form of a concentration gradient along an embryonic axis.

morphogenesis (mōr′-fō-jen′-uh-sis) The development of body shape and organization.

morphological species concept A definition of species in terms of measurable anatomical criteria.

morphology An organism's external form.

moss A small, herbaceous nonvascular plant that is a member of the phylum Bryophyta.

motor neuron A nerve cell that transmits signals from the brain or spinal cord to muscles or glands.

motor protein A protein that interacts with cytoskeletal elements and other cell components, producing movement of the whole cell or parts of the cell.

motor system An efferent branch of the vertebrate peripheral nervous system composed of motor neurons that carry signals to skeletal muscles in response to external stimuli.

motor unit A single motor neuron and all the muscle fibers it controls.

movement corridor A series of small clumps or a narrow strip of quality habitat (usable by organisms) that connects otherwise isolated patches of quality habitat.

MPF Maturation-promoting factor (M-phase-promoting factor); a protein complex required for a cell to progress from late interphase to mitosis. The active form consists of cyclin and a protein kinase.

mucus A viscous and slippery mixture of glycoproteins, cells, salts, and water that moistens and protects the membranes lining body cavities that open to the exterior.

Müllerian mimicry (myū-lār′-ē-un) A mutual mimicry by two unpalatable species.

multifactorial Referring to a phenotypic character that is influenced by multiple genes and environmental factors.

multigene family A collection of genes with similar or identical sequences, presumably of common origin.

multiple fruit A fruit derived from an inflorescence, a group of flowers tightly clustered together.

muscle tissue Tissue consisting of long muscle cells that can contract, either on its own or when stimulated by nerve impulses.

mutagen (myū′-tuh-jen) A chemical or physical agent that interacts with DNA and causes a mutation.

mutation (myū-tā′-shun) A change in the nucleotide sequence of an organism's DNA, ultimately creating genetic diversity. Mutations also can occur in the DNA or RNA of a virus.

mutualism (myū′-chū-ul-izm) A symbiotic relationship in which both participants benefit.

mycelium (mī-sē′-lē-um) The densely branched network of hyphae in a fungus.

mycorrhiza (mī′-kō-rī′-zuh) (plural, **mycorrhizae**) A mutualistic association of plant roots and fungus.

mycosis (mī-kō′-sis) General term for a fungal infection.

myelin sheath (mī′-uh-lin) Around the axon of a neuron, an insulating coat of cell membranes from Schwann cells or oligodendrocytes. It is interrupted by nodes of Ranvier, where action potentials are generated.

myofibril (mī′-ō-fī′-bril) A fibril collectively arranged in longitudinal bundles in muscle cells (fibers); composed of thin filaments of actin and a regulatory protein and thick filaments of myosin.

myoglobin (mī′-uh-glō′-bin) An oxygen-storing, pigmented protein in muscle cells.

myosin (mī′-uh-sin) A type of protein filament that acts as a motor protein with actin filaments to cause cell contraction.

myotonia (mī′-uh-tō′-nī-uh) Increased muscle tension, characteristic of sexual arousal in certain human tissues.

myriapod (mir′-ē-uh-pod′) A terrestrial arthropod with many body segments and one or two pairs of legs per segment. Millipedes and centipedes comprise the two classes of living myriapods.

NAD^+ Nicotinamide adenine dinucleotide, a coenzyme that can accept an electron and acts as an electron carrier in the electron transport chain.

$NADP^+$ Nicotinamide adenine dinucleotide phosphate, an electron acceptor that, as NADPH, temporarily stores energized electrons produced during the light reactions.

natural family planning A form of contraception that relies on refraining from sexual intercourse when conception is most likely to occur; also called the rhythm method.

natural killer (NK) cell A type of white blood cell that can kill tumor cells and virus-infected cells as part of innate immunity.

natural selection A process in which organisms with certain inherited characteristics are more likely to survive and reproduce than are organisms with other characteristics.

negative feedback A primary mechanism of homeostasis, whereby a change in a physiological variable triggers a response that counteracts the initial change.

negative pressure breathing A breathing system in which air is pulled into the lungs.

nematocyst (nem′-uh-tuh-sist′) In a cnidocyte of a cnidarian, a specialized capsule-like organelle containing a coiled thread that when discharged can penetrate the body wall of the prey.

nephron (nef′-ron) The tubular excretory unit of the vertebrate kidney.

neritic zone The shallow region of the ocean overlying the continental shelf.

nerve A ropelike bundle of neuron fibers (axons) tightly wrapped in connective tissue.

nerve net A weblike system of neurons, characteristic of radially symmetrical animals, such as hydra.

nervous system The fast-acting internal system of communication involving sensory receptors, networks of nerve cells, and connections to muscles and glands that respond to nerve signals; functions in concert with the endocrine system to effect internal regulation and maintain homeostasis.

nervous tissue Tissue made up of neurons and supportive cells.

net primary production (NPP) The gross primary production of an ecosystem minus the energy used by the producers for respiration.

neural crest cells In vertebrates, groups of cells along the sides of the neural tube where it pinches off from the ectoderm. The cells migrate to various parts of the embryo and form pigment cells in the skin and parts of the skull, teeth, adrenal glands, and peripheral nervous system.

neural plasticity The capacity of a nervous system to change with experience.

neural tube A tube of infolded ectodermal cells that runs along the anterior-posterior axis of a vertebrate, just dorsal to the notochord. It will give rise to the central nervous system.

neurohormone A molecule that is secreted by a neuron, travels in body fluids, and acts on specific target cells to change their functioning.

neuron (nyūr′-on) A nerve cell; the fundamental unit of the nervous system, having structure and properties that allow it to conduct signals by taking advantage of the electrical charge across its plasma membrane.

neuropeptide A relatively short chain of amino acids that serves as a neurotransmitter.

neurotransmitter A molecule that is released from the synaptic terminal of a neuron at a chemical synapse, diffuses across the synaptic cleft, and binds to the postsynaptic cell, triggering a response.

neutral theory The hypothesis that much evolutionary change in genes and proteins has no effect on fitness and therefore is not influenced by Darwinian natural selection.

neutral variation Genetic variation that does not appear to provide a selective advantage or disadvantage.

neutron A subatomic particle having no electrical charge (electrically neutral), with a mass of about 1.7×10^{-24} g, found in the nucleus of an atom.

neutrophil The most abundant type of white blood cell. Neutrophils are phagocytic and tend to self-destruct as they destroy foreign invaders, limiting their life span to a few days.

nitric oxide (NO) A gas produced by many types of cells that functions as a local regulator and as a neurotransmitter.

nitrogen cycle The natural process by which nitrogen, either from the atmosphere or from decomposed organic material, is converted by soil bacteria to compounds that can be assimilated by plants. This incorporated nitrogen is then taken in by other organisms and subsequently released, acted on by bacteria, and made available again to the nonliving environment.

nitrogen fixation The conversion of atmospheric nitrogen (N_2) to ammonia (NH_3). Biological nitrogen fixation is carried out by certain prokaryotes, some of which have mutualistic relationships with plants.

nociceptor (nō′-si-sep′-tur) A sensory receptor that responds to noxious or painful stimuli; also called a pain receptor.

node A point along the stem of a plant at which leaves are attached.

node of Ranvier (ron′-vē-ā′) Gap in the myelin sheath of certain axons where an action potential may be generated. In saltatory conduction, an action potential is regenerated at each node, appearing to "jump" along the axon from node to node.

nodule A swelling on the root of a legume. Nodules are composed of plant cells that contain nitrogen-fixing bacteria of the genus *Rhizobium*.

noncompetitive inhibitor A substance that reduces the activity of an enzyme by binding to a location remote from the active site, changing the enzyme's shape so that the active site no longer functions effectively.

nondisjunction An error in meiosis or mitosis in which members of a pair of homologous chromosomes or a pair of sister chromatids fail to separate properly from each other.

nonequilibrium model A model that maintains that communities change constantly after being buffeted by disturbances.

nonpolar covalent bond A type of covalent bond in which electrons are shared equally between two atoms of similar electronegativity.

nonsense mutation A mutation that changes an amino acid codon to one of the three stop codons, resulting in a shorter and usually nonfunctional protein.

norepinephrine A catecholamine that is chemically and functionally similar to epinephrine and acts as a hormone or neurotransmitter; also known as noradrenaline.

norm of reaction The range of phenotypes produced by a single genotype, due to environmental influences.

normal range An upper and lower limit of a variable.

Northern blotting A technique that enables specific nucleotide sequences to be detected in a sample of mRNA. It involves gel electrophoresis of RNA molecules and their transfer to a membrane (blotting), followed by nucleic acid hybridization with a labeled probe.

northern coniferous forest A terrestrial biome characterized by long, cold winters and dominated by cone-bearing trees.

no-till agriculture A plowing technique that involves creating furrows, resulting in minimal disturbance of the soil.

notochord (nō′-tuh-kord′) A longitudinal, flexible rod made of tightly packed mesodermal cells that runs along the anterior-posterior axis of a chordate in the dorsal part of the body.

nuclear envelope The double membrane in a eukaryotic cell that encloses the nucleus, separating it from the cytoplasm.

nuclear lamina A netlike array of protein filaments lining the inner surface of the nuclear envelope; it helps maintain the shape of the nucleus.

nucleariid Member of a group of unicellular, amoeboid protists that are more closely related to fungi than they are to other protists.

nuclease An enzyme that cuts DNA or RNA, either removing one or a few bases or hydrolyzing the DNA or RNA completely into its component nucleotides.

nucleic acid (nū-klā′-ik) A polymer (polynucleotide) consisting of many nucleotide monomers; serves as a blueprint for proteins and, through the actions of proteins, for all cellular activities. The two types are DNA and RNA.

nucleic acid hybridization The process of base pairing between a gene and a complementary sequence on another nucleic acid molecule.

nucleic acid probe In DNA technology, a labeled single-stranded nucleic acid molecule used to locate a specific nucleotide sequence in a nucleic acid sample. Molecules of the probe hydrogen-bond to the complementary sequence wherever it occurs; radioactive or other labeling of the probe allows its location to be detected.

nucleoid (nū′-klē-oyd) A dense region of DNA in a prokaryotic cell.

nucleolus (nū-klē′-ō-lus) (plural, **nucleoli**) A specialized structure in the nucleus, consisting of chromatin regions containing ribosomal RNA genes along with ribosomal proteins imported from the cytoplasmic site of rRNA synthesis and ribosomal subunit assembly. *See also* ribosome.

nucleosome (nū′-klē-ō-sōm′) The basic, bead-like unit of DNA packing in eukaryotes, consisting of a segment of DNA wound around a protein core composed of two copies of each of four types of histone.

nucleotide (nū′-klē-ō-tīd′) The building block of a nucleic acid, consisting of a five-carbon sugar covalently bonded to a nitrogenous base and a phosphate group.

nucleotide excision repair A repair system that removes and then correctly replaces a damaged segment of DNA using the undamaged strand as a guide.

nucleus (1) An atom's central core, containing protons and neutrons. (2) The chromosome-containing organelle of a eukaryotic cell. (3) A cluster of neurons.

nutrition The process by which an organism takes in and makes use of food substances.

obligate aerobe (ob′-lig-et ār′-ōb) An organism that requires oxygen for cellular respiration and cannot live without it.

obligate anaerobe (ob′-lig-et an′-uh-rōb) An organism that only carries out fermentation or anaerobic respiration. Such organisms cannot use oxygen and in fact may be poisoned by it.

oceanic pelagic zone Most of the ocean's waters far from shore, constantly mixed by ocean currents.

odorant A molecule that can be detected by sensory receptors of the olfactory system.

Okazaki fragment (ō′-kah-zah′-kē) A short segment of DNA synthesized away from the replication fork on a template strand during DNA replication, many of which are joined together to make up the lagging strand of newly synthesized DNA.

olfaction The sense of smell.

oligodendrocyte A type of glial cell that forms insulating myelin sheaths around the axons of neurons in the central nervous system.

oligotrophic lake A nutrient-poor, clear lake with few phytoplankton.

ommatidium (ōm′-uh-tid′-ē-um) (plural, **ommatidia**) One of the facets of the compound eye of arthropods and some polychaete worms.

omnivore An animal that regularly eats animals as well as plants or algae.

oncogene (on′-kō-jēn) A gene found in viral or cellular genomes that is involved in triggering molecular events that can lead to cancer.

oocyte A cell in the female reproductive system that differentiates to form an egg.

oogenesis (ō′-uh-jen′-uh-sis) The process in the ovary that results in the production of female gametes.

oogonium (ō′-uh- gō′-nē-em) A cell that divides mitotically to form oocytes.

oomycete (ō′-uh-mī′- sēt) A protist with flagellated cells, such as a water mold, white rust, or downy mildew, that acquires nutrition mainly as a decomposer or plant parasite.

open circulatory system A circulatory system in which fluid called hemolymph bathes the tissues and organs directly and there is no distinction between the circulating fluid and the interstitial fluid.

operant conditioning (op′-er-ent) A type of associative learning in which an animal learns to associate one of its own behaviors with a reward or punishment and then tends to repeat or avoid that behavior; also called trial-and-error learning.

operator In bacterial DNA, a sequence of nucleotides near the start of an operon to which an active repressor can attach. The binding of the repressor prevents RNA polymerase from attaching to the promoter and transcribing the genes of the operon.

operculum (ō-per′-kyuh-lum) In aquatic osteichthyans, a protective bony flap that covers and protects the gills.

operon (op′-er-on) A unit of genetic function found in bacteria and phages, consisting of a promoter, an operator, and a coordinately regulated cluster of genes whose products function in a common pathway.

opisthokont (uh-pis′-thuh-kont′) Member of the diverse clade Opisthokonta, organisms that descended from an ancestor with a posterior flagellum, including fungi, animals, and certain protists.

opposable thumb A thumb that can touch the ventral surface of the fingertips of all four fingers.

opsin A membrane protein bound to a light-absorbing pigment molecule.

optic chiasm The place where the two optic nerves meet and where the sensations from the left visual field of both eyes are transmitted to the right side of the brain and the sensations from the right visual field of both eyes are transmitted to the left side of the brain.

optimal foraging model The basis for analyzing behavior as a compromise between feeding costs and feeding benefits.

oral cavity The mouth of an animal.

orbital The three-dimensional space where an electron is found 90% of the time.

order In classification, the taxonomic category above the level of family.

organ A specialized center of body function composed of several different types of tissues.

organ identity gene A plant homeotic gene that uses positional information to determine which emerging leaves develop into which types of floral organs.

organ of Corti The actual hearing organ of the vertebrate ear, located in the floor of the cochlear duct in the inner ear; contains the receptor cells (hair cells) of the ear.

organ system A group of organs that work together in performing vital body functions.

organelle (ōr-guh-nel′) Any of several membrane-enclosed structures with specialized functions, suspended in the cytosol of eukaryotic cells.

organic chemistry The study of carbon compounds (organic compounds).

organismal ecology The branch of ecology concerned with the morphological, physiological, and behavioral ways in which individual organisms meet the challenges posed by their biotic and abiotic environments.

organogenesis (ōr-gan′-ō-jen′-uh-sis) The process in which organ rudiments develop from the three germ layers after gastrulation.

orgasm Rhythmic, involuntary contractions of certain reproductive structures in both sexes during the human sexual response cycle.

origin of replication Site where the replication of a DNA molecule begins, consisting of a specific sequence of nucleotides.

orthologous genes Homologous genes that are found in different species because of speciation.

osculum (os′-kyuh-lum) A large opening in a sponge that connects the spongocoel to the environment.

osmoconformer An animal that is isoosmotic with its environment.

osmolarity (oz′-mō-lār′-uh-tē) Solute concentration expressed as molarity.

osmoregulation Regulation of solute concentrations and water balance by a cell or organism.

osmoregulator An animal that controls its internal osmolarity independent of the external environment.

osmosis (oz-mō′-sis) The diffusion of water across a selectively permeable membrane.

osmotic potential A component of water potential that is proportional to the osmolarity of a solution and that measures the effect of solutes on the direction of water movement; also called solute potential, it can be either zero or negative.

osteichthyan (os′-tē-ik′-thē-an) Member of a vertebrate subgroup with jaws and mostly bony skeletons.

outer ear One of three main regions of the ear in reptiles (including birds) and mammals; made up of the auditory canal and, in many birds and mammals, the pinna.

outgroup A species or group of species from an evolutionary lineage that is known to have diverged before the lineage that contains the group of species being studied. An outgroup is selected so that its members are closely related to the group of species being studied, but not as closely related as any study-group members are to each other.

oval window In the vertebrate ear, a membrane-covered gap in the skull bone, through which sound waves pass from the middle ear to the inner ear.

ovarian cycle (ō-vār′-ē-un) The cyclic recurrence of the follicular phase, ovulation, and the luteal phase in the mammalian ovary, regulated by hormones.

ovary (ō′-vuh-rē) (1) In flowers, the portion of a carpel in which the egg-containing ovules develop. (2) In animals, the structure that produces female gametes and reproductive hormones.

overnourishment The consumption of more calories than the body needs for normal metabolism.

oviduct (ō′-vuh-duct) A tube passing from the ovary to the vagina in invertebrates or to the uterus in vertebrates, where it is also known as a fallopian tube.

oviparous (ō-vip′-uh-rus) Referring to a type of development in which young hatch from eggs laid outside the mother's body.

ovoviviparous (ō′-vō-vī-vip′-uh-rus) Referring to a type of development in which young hatch from eggs that are retained in the mother's uterus.

ovulation The release of an egg from an ovary. In humans, an ovarian follicle releases an egg during each uterine (menstrual) cycle.

ovule (o′-vyūl) A structure that develops within the ovary of a seed plant and contains the female gametophyte.

oxidation The loss of electrons from a substance involved in a redox reaction.

oxidative phosphorylation (fos′-fōr-uh-lā′-shun) The production of ATP using energy derived from the redox reactions of an electron transport chain; the third major stage of cellular respiration.

oxidizing agent The electron acceptor in a redox reaction.

oxytocin (ok′-si-tō′-sen) A hormone produced by the hypothalamus and released from the posterior pituitary. It induces contractions of the uterine muscles during labor and causes the mammary glands to eject milk during nursing.

P generation The parent individuals from which offspring are derived in studies of inheritance; P stands for "parental."

P site One of a ribosome's three binding sites for tRNA during translation. The P site holds the tRNA carrying the growing polypeptide chain. (P stands for peptidyl tRNA.)

***p53* gene** A tumor-suppressor gene that codes for a specific transcription factor that promotes the synthesis of cell cycle–inhibiting proteins.

paedomorphosis (pē′-duh-mōr′-fuh-sis) The retention in an adult organism of the juvenile features of its evolutionary ancestors.

pain receptor A sensory receptor that responds to noxious or painful stimuli; also called a nociceptor.

paleoanthropology The study of human origins and evolution.

paleontology (pā′-lē-un-tol′-ō-jē) The scientific study of fossils.

pancreas (pan′-krē-us) A gland with the following dual functions: The nonendocrine portion functions in digestion, secreting enzymes and an alkaline solution into the small intestine via a duct; the ductless endocrine portion functions in homeostasis, secreting the hormones insulin and glucagon into the blood.

pandemic A global epidemic.

Pangaea (pan-jē′-uh) The supercontinent that formed near the end of the Paleozoic era, when plate movements brought all the landmasses of Earth together.

parabasalid A protist, such as a trichomonad, with modified mitochondria.

paracrine Referring to a secreted molecule that acts on a neighboring cell.

paralogous genes Homologous genes that are found in the same genome as a result of gene duplication.

paraphyletic (pār′-uh-fī-let′-ik) Pertaining to a group of taxa that consists of a common ancestor and some, but not all, of its descendants.

parareptile First major group of reptiles to emerge, consisting mostly of large, stocky quadrupedal herbivores; died out in the late Triassic period.

parasite (pār′-uh-sīt) An organism that feeds on the cell contents, tissues, or body fluids of another species (the host) while in or on the host organism. Parasites harm but usually do not kill their host.

parasitism (pār′-uh-sit-izm) A symbiotic relationship in which one organism, the parasite, benefits at the expense of another, the host, by living either within or on the host.

parasympathetic division One of three divisions of the autonomic nervous system; generally enhances body activities that gain and conserve energy, such as digestion and reduced heart rate.

parathyroid gland Any of four small endocrine glands, embedded in the surface of the thyroid gland, that secrete parathyroid hormone.

parathyroid hormone (PTH) A hormone secreted by the parathyroid glands that raises blood calcium level by promoting calcium release from bone and calcium retention by the kidneys.

parenchyma cell (puh-ren′-ki-muh) A relatively unspecialized plant cell type that carries out most of the metabolism, synthesizes and stores organic products, and develops into a more differentiated cell type.

parental type An offspring with a phenotype that matches one of the parental phenotypes; also refers to the phenotype itself.

Parkinson's disease A progressive brain disease characterized by difficulty in initiating movements, slowness of movement, and rigidity.

parthenogenesis (par′-thuh-nō′-jen′-uh-sis) Asexual reproduction in which females produce offspring from unfertilized eggs.

partial pressure The pressure exerted by a particular gas in a mixture of gases (for instance, the pressure exerted by oxygen in air).

passive immunity Short-term immunity conferred by the transfer of antibodies, as occurs in the transfer of maternal antibodies to a fetus or nursing infant.

passive transport The diffusion of a substance across a biological membrane with no expenditure of energy.

pathogen An organism or virus that causes disease.

pattern formation The development of a multicellular organism's spatial organization, the arrangement of organs and tissues in their characteristic places in three-dimensional space.

peat Extensive deposits of partially decayed organic material formed primarily from the wetland moss *Sphagnum*.

pedigree A diagram of a family tree showing the occurrence of heritable characters in parents and offspring over multiple generations.

penis The copulatory structure of male mammals.

PEP carboxylase An enzyme that adds CO_2 to phosphoenolpyruvate (PEP) to form oxaloacetate in C_4 plants. It acts prior to photosynthesis.

pepsin An enzyme present in gastric juice that begins the hydrolysis of proteins.

pepsinogen The inactive form of pepsin that is first secreted by chief cells located in gastric pits of the stomach.

peptide bond The covalent bond between the carboxyl group on one amino acid and the amino group on another, formed by a dehydration reaction.

peptidoglycan (pep′-tid-ō-glī′-kan) A type of polymer in bacterial cell walls consisting of modified sugars cross-linked by short polypeptides.

perception The interpretation of sensory system input by the brain.

perennial (puh-ren′-ē-ul) A flowering plant that lives for many years.

pericycle The outermost layer in the vascular cylinder from which lateral roots arise.

periderm (pār′-uh-derm′) The protective coat that replaces the epidermis in woody plants during secondary growth, formed of the cork and cork cambium.

peripheral nervous system (PNS) The sensory and motor neurons that connect to the central nervous system.

peripheral protein A protein loosely bound to the surface of a membrane or to part of an integral protein and not embedded in the lipid bilayer.

peristalsis (pār′-uh-stal′-sis) (1) Alternating waves of contraction and relaxation in the smooth muscles lining the alimentary canal that push food along the canal. (2) A type of movement on land produced by rhythmic waves of muscle contractions passing from front to back, as in many annelids.

peristome A ring of interlocking, tooth-like structures on the upper part of a moss capsule (sporangium), often specialized for gradual spore discharge.

peritubular capillary One of the tiny blood vessels that form a network surrounding the proximal and distal tubules in the kidney.

permafrost A permanently frozen soil layer.

peroxisome (puh-rok′-suh-sōm′) An organelle containing enzymes that transfer hydrogen (H_2) from various substrates to oxygen (O_2), producing and then degrading hydrogen peroxide (H_2O_2).

petal A modified leaf of a flowering plant. Petals are the often colorful parts of a flower that advertise it to insects and other pollinators.

petiole (pet′-ē-ōl) The stalk of a leaf, which joins the leaf to a node of the stem.

pH A measure of hydrogen ion concentration equal to $-\log [H^+]$ and ranging in value from 0 to 14.

phage (fāj) A virus that infects bacteria; also called a bacteriophage.

phagocytosis (fag′-ō-sī-tō′-sis) A type of endocytosis in which large particulate substances are taken up by a cell. It is carried out by some protists and by certain immune cells of animals (in mammals, mainly macrophages, neutrophils, and dendritic cells).

pharyngeal cleft (fuh-rin′-jē-ul) In chordate embryos, one of the grooves that separate a series of pouches along the sides of the pharynx and may develop into a pharyngeal slit.

pharyngeal slit (fuh-rin′-jē-ul) In chordate embryos, one of the slits that form from the pharyngeal clefts and communicate to the outside, later developing into gill slits in many vertebrates.

pharynx (fār′-inks) (1) An area in the vertebrate throat where air and food passages cross. (2) In flatworms, the muscular tube that protrudes from the ventral side of the worm and ends in the mouth.

phase change A shift from one developmental phase to another.

phenotype (fē′-nō-tīp) The physical and physiological traits of an organism, which are determined by its genetic makeup.

pheromone (fār′-uh-mōn) In animals and fungi, a small molecule released into the environment that functions in communication between members of the same species. In animals, it acts much like a hormone in influencing physiology and behavior.

phloem (flō′-em) Vascular plant tissue consisting of living cells arranged into elongated tubes that transport sugar and other organic nutrients throughout the plant.

phloem sap The sugar-rich solution carried through sieve tubes.

phosphate group A chemical group consisting of a phosphorus atom bonded to four oxygen atoms; important in energy transfer.

phospholipid (fos′-fō-lip′-id) A lipid made up of glycerol joined to two fatty acids and a phosphate group. The hydrocarbon chains of the fatty acids act as nonpolar, hydrophobic tails, while the rest of the molecule acts as a polar, hydrophilic head. Phospholipids form bilayers that function as biological membranes.

phosphorylated Referring to a molecule that is covalently bonded to a phosphate group.

photic zone (fō′-tic) The narrow top layer of an ocean or lake, where light penetrates sufficiently for photosynthesis to occur.

photoautotroph (fō′-tō-ot′-ō-trōf) An organism that harnesses light energy to drive the synthesis of organic compounds from carbon dioxide.

photoheterotroph (fō′-tō-het′-er-ō-trōf) An organism that uses light to generate ATP but must obtain carbon in organic form.

photomorphogenesis Effects of light on plant morphology.

photon (fō′-ton) A quantum, or discrete quantity, of light energy that behaves as if it were a particle.

photoperiodism (fō′-tō-pēr′-ē-ō-dizm) A physiological response to photoperiod, the relative lengths of night and day. An example of photoperiodism is flowering.

photophosphorylation (fō′-tō-fos′-fōr-uh-lā′-shun) The process of generating ATP from ADP and phosphate by means of a proton-motive force generated across the thylakoid membrane of the chloroplast or the membrane of certain prokaryotes during the light reactions of photosynthesis.

photoreceptor An electromagnetic receptor that detects the radiation known as visible light.

photorespiration A metabolic pathway that consumes oxygen and ATP, releases carbon dioxide, and decreases photosynthetic output. Photorespiration generally occurs on hot, dry, bright days, when stomata close and the oxygen concentration in the leaf exceeds that of carbon dioxide.

photosynthesis (fō′-tō-sin′-thi-sis) The conversion of light energy to chemical energy that is stored in sugars or other organic compounds; occurs in plants, algae, and certain prokaryotes.

photosystem A light-capturing unit located in the thylakoid membrane of the chloroplast or in the membrane of some prokaryotes, consisting of a reaction-center complex surrounded by numerous light-harvesting complexes. There are two types of photosystems, I and II; they absorb light best at different wavelengths.

photosystem I (PS I) One of two light-capturing units in a chloroplast's thylakoid membrane or in the membrane of some prokaryotes; it has two molecules of P700 chlorophyll *a* at its reaction center.

photosystem II (PS II) One of two light-capturing units in a chloroplast's thylakoid membrane or in the membrane of some prokaryotes; it has two molecules of P680 chlorophyll *a* at its reaction center.

phototropism (fō′-tō-trō′-pizm) Growth of a plant shoot toward or away from light.

phragmoplast (frag′-mō-plast′) An alignment of cytoskeletal elements and Golgi-derived vesicles that forms across the midline of a dividing plant cell.

phyllotaxy (fil′-uh-tak′-sē) The arrangement of leaves on the shoot of a plant.

PhyloCode System of classification of organisms based on evolutionary relationships: Only groups that include a common ancestor and all of its descendents are named.

phylogenetic bracketing An approach in which features shared by two groups of organisms are predicted (by parsimony) to be present in their common ancestor and all of its descendants.

phylogenetic species concept A definition of species as the smallest group of individuals that share a common ancestor, forming one branch on the tree of life.

phylogenetic tree A branching diagram that represents a hypothesis about the evolutionary history of a group of organisms.

phylogeny (fī-loj′-uh-nē) The evolutionary history of a species or group of related species.

phylum (fī′-lum) (plural, **phyla**) In classification, the taxonomic category above class.

physical map A genetic map in which the actual physical distances between genes or other genetic markers are expressed, usually as the number of base pairs along the DNA.

physiology The processes and functions of an organism and their study.

phytochrome (fī′-tuh-krōm) A type of light receptor in plants that mostly absorbs red light and regulates many plant responses, such as seed germination and shade avoidance.

phytoremediation An emerging nondestructive biotechnology that seeks to cheaply reclaim contaminated areas by taking advantage of some plant species' ability to extract heavy metals and other pollutants from the soil and to concentrate them in easily harvested portions of the plant.

pineal gland (pī′-nē-ul) A small gland on the dorsal surface of the vertebrate forebrain that secretes the hormone melatonin.

pinocytosis (pī′-nō-sī-tō′-sis) A type of endocytosis in which the cell ingests extracellular fluid and its dissolved solutes.

pistil A single carpel or a group of fused carpels.

pith Ground tissue that is internal to the vascular tissue in a stem; in many monocot roots, parenchyma cells that form the central core of the vascular cylinder.

pituitary gland (puh-tū′-uh-tār′-ē) An endocrine gland at the base of the hypothalamus; consists of a posterior lobe (neurohypophysis), which stores and releases two hormones produced by the hypothalamus, and an anterior lobe (adenohypophysis), which produces and secretes many hormones that regulate diverse body functions.

placenta (pluh-sen′-tuh) A structure in the pregnant uterus for nourishing a viviparous fetus with the mother's blood supply; formed from the uterine lining and embryonic membranes.

placental transfer cell A plant cell that enhances the transfer of nutrients from parent to embryo.

placoderm A member of an extinct class of fishlike vertebrates that had jaws and were enclosed in a tough outer armor.

planarian A free-living flatworm found in unpolluted ponds and streams.

Plantae (plan′-tā) The kingdom that consists of multicellular eukaryotes that carry out photosynthesis.

plasma (plaz′-muh) The liquid matrix of blood in which the cells are suspended.

plasma cell The antibody-secreting effector cell of humoral immunity; arises from antigen-stimulated B cells.

plasma membrane The membrane at the boundary of every cell that acts as a selective barrier, regulating the cell's chemical composition.

plasmid (plaz′-mid) A small, circular, double-stranded DNA molecule that carries accessory genes separate from those of a bacterial chromosome. Plasmids are also found in some eukaryotes, such as yeasts.

plasmodesma (plaz′-mō-dez′-muh) (plural, **plasmodesmata**) An open channel in the cell wall of a plant through which strands of cytosol connect from an adjacent cell.

plasmodial slime mold (plaz-mō′-dē-ul) A type of protist that has amoeboid cells, flagellated cells, and a plasmodial feeding stage in its life cycle.

plasmodium A single mass of cytoplasm containing many diploid nuclei that forms during the life cycle of some slime molds.

plasmogamy (plaz-moh′-guh-mē) The fusion of the cytoplasm of cells from two individuals; occurs as one stage of syngamy (fertilization).

plasmolysis (plaz-mol′-uh-sis) A phenomenon in walled cells in which the cytoplasm shrivels and the plasma membrane pulls away from the cell wall; occurs when the cell loses water to a hypertonic environment.

plastid One of a family of closely related organelles that includes chloroplasts, chromoplasts, and amyloplasts (leucoplasts). Plastids are found in cells of photosynthetic organisms.

platelet A pinched-off cytoplasmic fragment of a specialized bone marrow cell. Platelets circulate in the blood and are important in blood clotting.

pleiotropy (plī-o′-truh-pē) The ability of a single gene to have multiple effects.

pluripotent Describing a cell that can give rise to many, but not all, parts of an organism.

point mutation A change in a gene at a single nucleotide pair.

polar covalent bond A covalent bond between atoms that differ in electronegativity. The shared electrons are pulled closer to the more electronegative atom, making it slightly negative and the other atom slightly positive.

polar molecule A molecule (such as water) with opposite charges on different ends of the molecule.

polarity A lack of symmetry; structural differences in opposite ends of an organism or structure, such as the root end and shoot end of a plant.

pollen grain In seed plants, a structure consisting of the male gametophyte enclosed within a pollen wall.

pollen tube A tube formed after germination of the pollen grain that functions in the delivery of sperm to the ovule.

pollination (pol′-uh-nā′-shun) The transfer of pollen to the part of a seed plant containing the ovules, a process required for fertilization.

poly-A tail A sequence of 50 to 250 adenine nucleotides added onto the 3′ end of a pre-mRNA molecule.

polyandry (pol′-ē-an′-drē) A polygamous mating system involving one female and many males.

polygamous Referring to a type of relationship in which an individual of one sex mates with several of the other.

polygenic inheritance (pol′-ē-jen′-ik) An additive effect of two or more genes on a single phenotypic character.

polygyny (puh-lij′-en-ē) A polygamous mating system involving one male and many females.

polymer (pol′-uh-mer) A long molecule consisting of many similar or identical monomers linked together.

polymerase chain reaction (PCR) (puh-lim′-uh-rās) A technique for amplifying DNA *in vitro* by incubating it with specific primers, a heat-resistant DNA polymerase, and nucleotides.

polynucleotide (pol′-ē-nū′-klē-ō-tīd) A polymer consisting of many nucleotide monomers in a chain; nucleotides can be those of DNA or RNA.

polyp The sessile variant of the cnidarian body plan. The alternate form is the medusa.

polypeptide (pol′-ē-pep′-tīd) A polymer (chain) of many amino acids linked together by peptide bonds.

polyphyletic (pol′-ē-fī-let′-ik) Pertaining to a group of taxa derived from two or more different ancestors.

polyploidy (pol′-ē-ploy′-dē) A chromosomal alteration in which the organism possesses more than two complete chromosome sets. It is the result of an accident of cell division.

polyribosome (polysome) (pol′-ē-rī′-bō-sōm′) A group of several ribosomes attached to, and translating, the same messenger RNA molecule.

polysaccharide (pol′-ē-sak′-uh-rīd) A polymer of many monosaccharides, formed by dehydration reactions.

polytomy (puh-lit′-uh-mē) In a phylogenetic tree, a branch point from which more than two descendant taxa emerge. A polytomy indicates that the evolutionary relationships among the descendant taxa are not yet clear.

pons Portion of the brain that participates in certain automatic, homeostatic functions, such as regulating the breathing centers in the medulla.

population A localized group of individuals of the same species that can interbreed, producing fertile offspring.

population dynamics The study of how complex interactions between biotic and abiotic factors influence variations in population size.

population ecology The study of populations in relation to their environment, including environmental influences on population density and distribution, age structure, and variations in population size.

positional information Molecular cues that control pattern formation in an animal or plant embryonic structure by indicating a cell's location relative to the organism's body axes. These cues elicit a response by genes that regulate development.

positive feedback A physiological control mechanism in which a change in a variable triggers mechanisms that amplify the change.

positive pressure breathing A breathing system in which air is forced into the lungs.

posterior Pertaining to the rear, or tail end, of a bilaterally symmetrical animal.

posterior pituitary Also called the neurohypophysis; an extension of the hypothalamus composed of nervous tissue that secretes oxytocin and antidiuretic hormone made in the hypothalamus; a temporary storage site for these hormones.

postsynaptic cell The target cell at a synapse.

postzygotic barrier (pōst′-zī-got′-ik) A reproductive barrier that prevent hybrid zygotes produced by two different species from developing into viable, fertile adults.

potential energy The energy that matter possesses as a result of its location or spatial arrangement (structure).

predation An interaction between species in which one species, the predator, eats the other, the prey.

pregnancy The condition of carrying one or more embryos in the uterus.

preprophase band Microtubules in the cortex (outer cytoplasm) of a cell that are concentrated into a ring.

prepuce (prē′-pyūs) A fold of skin covering the head of the clitoris or penis.

pressure potential (Ψ_P) A component of water potential that consists of the physical pressure on a solution, which can be positive, zero, or negative.

presynaptic cell The transmitting cell at a synapse.

prezygotic barrier (prē′-zī-got′-ik) A reproductive barrier that impedes mating between species or hinders fertilization if interspecific mating is attempted.

primary cell wall In plants, a relatively thin and flexible layer first secreted by a young cell.

primary consumer An herbivore; an organism that eats plants or other autotrophs.

primary electron acceptor In the thylakoid membrane of a chloroplast or in the membrane of some prokaryotes, a specialized molecule that shares the reaction-center complex with a pair of chlorophyll *a* molecules and that accepts an electron from them.

primary growth Growth produced by apical meristems, lengthening stems and roots.

primary immune response The initial acquired immune response to an antigen, which appears after a lag of about 10 to 17 days.

primary oocyte (ō′-uh-sīt) An oocyte prior to completion of meiosis I.

primary plant body The tissues produced by apical meristems, which lengthen stems and roots.

primary producer An autotroph, usually a photosynthetic organism. Collectively, autotrophs make up the trophic level of an ecosystem that ultimately supports all other levels.

primary production The amount of light energy converted to chemical energy (organic compounds) by autotrophs in an ecosystem during a given time period.

primary structure The level of protein structure referring to the specific sequence of amino acids.

primary succession A type of ecological succession that occurs in an area where there were originally no organisms present and where soil has not yet formed.

primary transcript An initial RNA transcript; also called pre-mRNA when transcribed from a protein-coding gene.

primary visual cortex The destination in the occipital lobe of the cerebrum for most of the axons from the lateral geniculate nuclei.

primase An enzyme that joins RNA nucleotides to make the primer using the parental DNA strand as a template.

primer A short stretch of RNA with a free 3′ end, bound by complementary base pairing to the template strand, that is elongated with DNA nucleotides during DNA replication.

primitive streak A thickening along the future anterior-posterior axis on the surface of an early avian or mammalian embryo, caused by a piling up of cells as they congregate at the midline before moving into the embryo.

prion An infectious agent that is a misfolded version of a normal cellular protein. Prions appear to increase in number by converting correctly folded versions of the protein to more prions.

problem solving The cognitive activity of devising a method to proceed from one state to another in the face of real or apparent obstacles.

producer An organism that produces organic compounds from CO_2 by harnessing light energy (in photosynthesis) or by oxidizing inorganic chemicals (in chemosynthetic reactions carried out by some prokaryotes).

product A material resulting from a chemical reaction.

production efficiency The percentage of energy stored in food that is not used for respiration or eliminated as waste.

progesterone A steroid hormone that prepares the uterus for pregnancy; the major progestin in mammals.

progestin Any steroid hormone with progesterone-like activity.

progymnosperm (prō′-jim′-nō-sperm) An extinct seedless vascular plant that may be ancestral to seed plants.

prokaryotic cell (prō′-kār′-ē-ot′-ik) A type of cell lacking a membrane-enclosed nucleus and membrane-enclosed organelles. Organisms with prokaryotic cells (bacteria and archaea) are called prokaryotes.

prolactin (PRL) A hormone produced and secreted by the anterior pituitary with a great diversity of effects in different vertebrate species. In mammals, it stimulates growth of and milk production by the mammary glands.

proliferative phase That portion of the uterine (menstrual) cycle when the endometrium regenerates and thickens.

prometaphase The second stage of mitosis, in which discrete chromosomes consisting of identical sister chromatids appear, the nuclear envelope fragments, and the spindle microtubules attach to the kinetochores of the chromosomes.

promiscuous Referring to a type of relationship in which mating occurs with no strong pair-bonds or lasting relationships.

promoter A specific nucleotide sequence in DNA that binds RNA polymerase, positioning it to start transcribing RNA at the appropriate place.

prophage (prō′-fāj) A phage genome that has been inserted into a specific site on a bacterial chromosome.

prophase The first stage of mitosis, in which the chromatin condenses, the mitotic spindle begins to form, and the nucleolus disappears, but the nucleus remains intact.

prostaglandin (PG) (pros′-tuh-glan′-din) One of a group of modified fatty acids secreted by virtually all tissues and performing a wide variety of functions as local regulators.

prostate gland (pros′-tāt) A gland in human males that secretes an acid-neutralizing component of semen.

protease An enzyme that digests proteins by hydrolysis.

proteasome A giant protein complex that recognizes and destroys proteins tagged for elimination by the small protein ubiquitin.

protein (prō′-tēn) A functional biological molecule consisting of one or more polypeptides folded and coiled into a specific three-dimensional structure.

protein kinase An enzyme that transfers phosphate groups from ATP to a protein, thus phosphorylating the protein.

protein phosphatase An enzyme that removes phosphate groups from (dephosphorylates) proteins, often functioning to reverse the effect of a protein kinase.

proteoglycan (prō′-tē-ō-glī′-kan) A glycoprotein consisting of a small core protein with many carbohydrate chains attached, found in the extracellular matrix of animal cells. A proteoglycan may consist of up to 95% carbohydrate.

proteomics (prō′-tē-ō′-miks) The systematic study of the full protein sets (proteomes) encoded by genomes.

protist An informal term applied to any eukaryote that is not a plant, animal, or fungus. Most protists are unicellular, though some are colonial or multicellular.

protobiont A collection of abiotically produced molecules surrounded by a membrane or membrane-like structure.

proton (prō′-ton) A subatomic particle with a single positive electrical charge, with a mass of about 1.7×10^{-24} g, found in the nucleus of an atom.

proton pump An active transport protein in a cell membrane that uses ATP to transport hydrogen ions out of a cell against their concentration gradient, generating a membrane potential in the process.

protonema (plural, **protonemata**) A mass of green, branched, one-cell-thick filaments produced by germinating moss spores.

protonephridia (prō′-tō-nuh-frid′-ē-uh) (singular, **protonephridium**) An excretory system, such as the flame bulb system of flatworms, consisting of a network of tubules lacking internal openings.

proton-motive force The potential energy stored in the form of an electrochemical gradient, generated by the pumping of hydrogen ions across a biological membrane during chemiosmosis.

proto-oncogene (prō′-tō-on′-kō-jēn) A normal cellular gene that has the potential to become an oncogene.

protoplast fusion The fusing of two protoplasts from different plant species that would otherwise be reproductively incompatible.

protostome development In animals, a developmental mode distinguished by the development of the mouth from the blastopore; often also characterized by spiral cleavage and by the body cavity forming when solid masses of mesoderm split.

provirus A viral genome that is permanently inserted into a host genome.

proximal tubule In the vertebrate kidney, the portion of a nephron immediately downstream from Bowman's capsule that conveys and helps refine filtrate.

proximate causation The mechanistic explanation of "how" a behavior (or other aspect of an organism's biology) occurs or is modified; that is, how a stimulus elicits a behavior, what physiological mechanisms mediate the response, and how experience influences the response.

pseudocoelomate (sū′-dō-sē′-lō-māt) An animal whose body cavity is lined by tissue derived from mesoderm and endoderm.

pseudogene (sū′-dō-jēn) A DNA segment very similar to a real gene but which does not yield a functional product; a DNA segment that formerly functioned as a gene but has become inactivated in a particular species because of mutation.

pseudopodium (sū′-dō-pō′-dē-um) (plural, **pseudopodia**) A cellular extension of amoeboid cells used in moving and feeding.

pterophyte (ter′-uh-fīt) An informal name for a member of the phylum Pterophyta, which includes ferns, horsetails, and whisk ferns and their relatives.

pterosaur Winged reptile that lived during the Mesozoic era.

pulmocutaneous circuit A branch of the circulatory system in many amphibians that supplies the lungs and skin.

pulmonary circuit The branch of the circulatory system that supplies the lungs.

pulse The rhythmic bulging of the artery walls with each heartbeat.

punctuated equilibria In the fossil record, long periods of apparent stasis, in which a species undergoes little or no morphological change, interrupted by relatively brief periods of sudden change.

Punnett square A diagram used in the study of inheritance to show the predicted results of random fertilization in genetic crosses.

pupil The opening in the iris, which admits light into the interior of the vertebrate eye. Muscles in the iris regulate its size.

purine (pyū′-rēn) One of two types of nitrogenous bases found in nucleotides, characterized by a six-membered ring fused to a five-membered ring. Adenine (A) and guanine (G) are purines.

pyrimidine (puh-rim′-uh-dēn) One of two types of nitrogenous bases found in nucleotides, characterized by a six-membered ring. Cytosine (C), thymine (T), and uracil (U) are pyrimidines.

quantitative character A heritable feature that varies continuously over a range rather than in an either-or fashion.

quaternary structure (kwot′-er-nār-ē) The particular shape of a complex, aggregate protein, defined by the characteristic three-dimensional arrangement of its constituent subunits, each a polypeptide.

R plasmid A bacterial plasmid carrying genes that confer resistance to certain antibiotics.

radial cleavage A type of embryonic development in deuterostomes in which the planes of cell division that transform the zygote into a ball of cells are either parallel or perpendicular to the vertical axis of the embryo, thereby aligning tiers of cells one above the other.

radial glia In an embryo, supporting cells that form tracks along which newly formed neurons migrate from the neural tube; can also act as stem cells that give rise to other glia and neurons.

radial symmetry Symmetry in which the body is shaped like a pie or barrel (lacking a left side and a right side) and can be divided into mirror-image halves by any plane through its central axis.

radiation The emission of electromagnetic waves by all objects warmer than absolute zero.

radicle An embryonic root of a plant.

radioactive isotope An isotope (an atomic form of a chemical element) that is unstable; the nucleus decays spontaneously, giving off detectable particles and energy.

radiolarian A protist, usually marine, with a shell generally made of silica and pseudopodia that radiate from the central body.

radiometric dating A method for determining the absolute ages of rocks and fossils, based on the half-life of radioactive isotopes.

radula A straplike rasping organ used by many molluscs during feeding.

***ras* gene** A gene that codes for Ras, a G protein that relays a growth signal from a growth factor receptor on the plasma membrane to a cascade of protein kinases, ultimately resulting in stimulation of the cell cycle.

ratite (rat′-īt) Member of the group of flightless birds.

ray-finned fish Member of the class Actinopterygii, aquatic osteichthyans with fins supported by long, flexible rays, including tuna, bass, and herring.

reabsorption In excretory systems, the recovery of solutes and water from filtrate.

reactant A starting material in a chemical reaction.

reaction-center complex A complex of proteins associated with a special pair of chlorophyll *a* molecules and a primary electron acceptor. Located centrally in a photosystem, this complex triggers the light reactions of photosynthesis. Excited by light energy, the pair of chlorophylls donates an electron to the primary electron acceptor, which passes an electron to an electron transport chain.

reading frame On an mRNA, the triplet grouping of ribonucleotides used by the translation machinery during polypeptide synthesis.

receptacle The base of a flower; the part of the stem that is the site of attachment of the floral organs.

receptor potential An initial response of a receptor cell to a stimulus, consisting of a change in voltage across the receptor membrane proportional to the stimulus strength. The intensity of the receptor potential determines the frequency of action potentials traveling to the nervous system.

receptor tyrosine kinase A receptor protein in the plasma membrane, the cytoplasmic (intracellular) part of which can catalyze the transfer of a phosphate group from ATP to a tyrosine on another protein. Receptor tyrosine kinases often respond to the binding of a signaling molecule by dimerizing and then phosphorylating a tyrosine on the cytoplasmic portion of the other receptor in the dimer. The phosphorylated tyrosines on the receptors then activate other signal transduction proteins within the cell.

receptor-mediated endocytosis (en′-dō-sī-tō′-sis) The movement of specific molecules into a cell by the inward budding of membranous vesicles containing proteins with receptor sites specific to the molecules being taken in; enables a cell to acquire bulk quantities of specific substances.

recessive allele An allele whose phenotypic effect is not observed in a heterozygote.

reciprocal altruism Altruistic behavior between unrelated individuals, whereby the altruistic individual benefits in the future when the beneficiary reciprocates.

recombinant chromosome A chromosome created when crossing over combines the DNA from two parents into a single chromosome.

recombinant DNA A DNA molecule made *in vitro* with segments from different sources.

recombinant type (recombinant) An offspring whose phenotype differs from that of the parents; also refers to the phenotype itself.

recruitment The process of progressively increasing the tension of a muscle by activating more and more of the motor neurons controlling the muscle.

rectum The terminal portion of the large intestine where the feces are stored until they are eliminated.

red alga A photosynthetic protist, named for its color, which results from a red pigment that masks the green of chlorophyll. Most red algae are multicellular and marine.

redox reaction (rē′-doks) A chemical reaction involving the complete or partial transfer of one or more electrons from one reactant to another; short for oxidation-reduction reaction.

reducing agent The electron donor in a redox reaction.

reduction The addition of electrons to a substance involved in a redox reaction.

reflex An automatic reaction to a stimulus, mediated by the spinal cord or lower brain.

refractory period (rē-frakt′-ōr-ē) The short time immediately after an action potential in which the neuron cannot respond to another stimulus, owing to the inactivation of voltage-gated sodium channels.

regulator An animal for which mechanisms of homeostasis moderate internal changes in the face of external fluctuations.

regulatory gene A gene that codes for a protein, such as a repressor, that controls the transcription of another gene or group of genes.

reinforcement A process in which natural selection strengthens prezygotic barriers to reproduction, thus reducing the chances of hybrid formation. Such a process is likely to occur only if hybrid offspring are less fit than members of the parent species.

relative abundance The proportional abundance of different species in a community.

relative fitness The contribution an individual makes to the gene pool of the next generation, relative to the contributions of other individuals in the population.

renal artery The blood vessel bringing blood to the kidney.

renal cortex The outer portion of the vertebrate kidney.

renal medulla The inner portion of the vertebrate kidney, beneath the renal cortex.

renal pelvis The funnel-shaped chamber that receives processed filtrate from the vertebrate kidney's collecting ducts and is drained by the ureter.

renal vein The blood vessel that carries blood away from the kidney.

renin-angiotensin-aldosterone system (RAAS) A hormone cascade pathway that helps regulate blood pressure and blood volume.

repeated reproduction Reproduction in which adults produce offspring over many years; also known as iteroparity.

repetitive DNA Nucleotide sequences, usually noncoding, that are present in many copies in a eukaryotic genome. The repeated units may be short and arranged tandemly (in series) or long and dispersed in the genome.

replication fork A Y-shaped region on a replicating DNA molecule where the parental strands are being unwound and new strands are growing.

repressor A protein that inhibits gene transcription. In prokaryotes, repressors bind to the DNA in or near the promoter. In eukaryotes, repressors may bind to control elements within enhancers, to activators, or to other proteins in a way that blocks activators from binding to DNA.

reproductive isolation The existence of biological factors (barriers) that impede members of two species from producing viable, fertile offspring.

reproductive table An age-specific summary of the reproductive rates in a population.

reptile Member of the clade of amniotes that includes tuataras, lizards, snakes, turtles, crocodilians, and birds.

residual volume The amount of air that remains in the lungs after forceful exhalation.

resource partitioning The division of environmental resources by coexisting species such that the niche of each species differs by one or more significant factors from the niches of all coexisting species.

respiratory pigment A protein that transports oxygen in blood or hemolymph.

response (1) In cellular communication, the change in a specific cellular activity brought about by a transduced signal from outside the cell. (2) In homeostasis, a physiological activity that helps return a variable to a set point.

resting potential The membrane potential characteristic of a nonconducting excitable cell, with the inside of the cell more negative than the outside.

restoration ecology Applying ecological principles in an effort to return ecosystems that have been disturbed by human activity to a condition as similar as possible to their natural state.

restriction enzyme An endonuclease (type of enzyme) that recognizes and cuts DNA molecules foreign to a bacterium (such as phage genomes). The enzyme cuts at specific nucleotide sequences (restriction sites).

restriction fragment A DNA segment that results from the cutting of DNA by a restriction enzyme.

restriction fragment length polymorphism (RFLP) A single nucleotide polymorphism (SNP) that exists in the restriction site for a particular enzyme, thus making the site unrecognizable by that enzyme and changing the lengths of the restriction fragments formed by digestion with that enzyme. A RFLP can be in coding or noncoding DNA.

restriction site A specific sequence on a DNA strand that is recognized and cut by a restriction enzyme.

reticular formation (re-tik′-yū-ler) A diffuse network of neurons in the core of the brainstem that filters information traveling to the cerebral cortex.

retina (ret′-i-nuh) The innermost layer of the vertebrate eye, containing photoreceptor cells (rods and cones) and neurons; transmits images formed by the lens to the brain via the optic nerve.

retinal The light-absorbing pigment in rods and cones of the vertebrate eye.

retrotransposon (re′-trō-trans-pō′-zon) A transposable element that moves within a genome by means of an RNA intermediate, a transcript of the retrotransposon DNA.

retrovirus (re′-trō-vī′-rus) An RNA virus that reproduces by transcribing its RNA into DNA and then inserting the DNA into a cellular chromosome; an important class of cancer-causing viruses.

reverse transcriptase (tran-skrip′-tās) An enzyme encoded by certain viruses (retroviruses) that uses RNA as a template for DNA synthesis.

reverse transcriptase–polymerase chain reaction (RT-PCR) A technique for determining expression of a particular gene. It uses reverse transcriptase and DNA polymerase to synthesize cDNA from all the mRNA in a sample and then subjects the cDNA to PCR amplification using primers specific for the gene of interest.

Rhizaria (rī-za′-rē-uh) One of five supergroups of eukaryotes proposed in a current hypothesis of the evolutionary history of eukaryotes; a morphologically diverse protist clade that is defined by DNA similarities. *See also* Excavata, Chromalveolata, Archaeplastida, and Unikonta.

rhizobacterium A soil bacterium whose population size is much enhanced in the rhizosphere, the soil region close to a plant's roots.

rhizoid (rī′-zoyd) A long, tubular single cell or filament of cells that anchors bryophytes to the ground. Unlike roots, rhizoids are not composed of tissues, lack specialized conducting cells, and do not play a primary role in water and mineral absorption.

rhizosphere The soil region close to plant roots and characterized by a high level of microbiological activity.

rhodopsin (rō-dop′-sin) A visual pigment consisting of retinal and opsin. When rhodopsin absorbs light, the retinal changes shape and dissociates from the opsin, after which it is converted back to its original form.

rhythm method A form of contraception that relies on refraining from sexual intercourse when conception is most likely to occur; also called natural family planning.

ribonucleic acid (RNA) (rī′-bō-nū-klā′-ik) A type of nucleic acid consisting of nucleotide monomers with a ribose sugar and the nitrogenous bases adenine (A), cytosine (C), guanine (G), and uracil (U); usually single-stranded; functions in protein synthesis, gene regulation, and as the genome of some viruses.

ribose The sugar component of RNA nucleotides.

ribosomal RNA (rRNA) (rī′-buh-sō′-mul) The most abundant type of RNA, which together with proteins makes up ribosomes.

ribosome (rī′-buh-sōm′) A complex of rRNA and protein molecules that functions as a site of protein synthesis in the cytoplasm; consists of a large and a small subunit. In eukaryotic cells, each subunit is assembled in the nucleolus. *See also* nucleolus.

ribozyme (rī′-bō-zīm) An RNA molecule that functions as an enzyme, catalyzing reactions during RNA splicing.

RNA interference (RNAi) A technique used to silence the expression of selected genes. RNAi uses synthetic double-stranded RNA molecules that match the sequence of a particular gene to trigger the breakdown of the gene's messenger RNA.

RNA polymerase An enzyme that links ribonucleotides into a growing RNA chain during transcription.

RNA processing Modification of RNA transcripts, including splicing out of introns, joining together of exons, and alteration of the 5′ and 3′ ends.

RNA splicing After synthesis of a eukaryotic primary RNA transcript, the removal of portions (introns) of the transcript that will not be included in the mRNA.

rod A rodlike cell in the retina of the vertebrate eye, sensitive to low light intensity.

root An organ in vascular plants that anchors the plant and enables it to absorb water and minerals from the soil.

root cap A cone of cells at the tip of a plant root that protects the apical meristem.

root hair A tiny extension of a root epidermal cell, growing just behind the root tip and increasing surface area for absorption of water and minerals.

root pressure The upward push of xylem sap in the vascular tissue of roots.

root system All of a plant's roots, which anchor it in the soil, absorb and transport minerals and water, and store food.

rooted Describing a phylogenetic tree that contains a branch point (typically, the one farthest to the left) representing the last common ancestor of all taxa in the tree.

rough ER That portion of the endoplasmic reticulum studded with ribosomes.

round window In the mammalian ear, the point of contact between the stapes and the cochlea, where vibrations of the stapes create a traveling series of pressure waves in the fluid of the cochlea.

***r*-selection** Selection for life history traits that maximize reproductive success in uncrowded environments; also called density-independent selection.

rubisco (rū-bis′-kō) Ribulose bisphosphate (RuBP) carboxylase, the enzyme that catalyzes the first step of the Calvin cycle (the addition of CO_2 to RuBP).

ruminant (rū′-muh-nent) An animal, such as a cow or a sheep, with an elaborate, multicompartmentalized stomach specialized for an herbivorous diet.

S phase The synthesis phase of the cell cycle; the portion of interphase during which DNA is replicated.

sac fungus *See* ascomycete.

saccule In the vertebrate ear, a chamber in the vestibule behind the oval window that participates in the sense of balance.

salicylic acid (sal′-i-sil′-ik) A signaling molecule in plants that may be partially responsible for activating systemic acquired resistance to pathogens.

salivary gland A gland associated with the oral cavity that secretes substances to lubricate food and begin the process of chemical digestion.

salt A compound resulting from the formation of an ionic bond; also called an ionic compound.

saltatory conduction (sol′-tuh-tōr′-ē) Rapid transmission of a nerve impulse along an axon, resulting from the action potential jumping from one node of Ranvier to another, skipping the myelin-sheathed regions of membrane.

sarcomere (sar′-kō-mēr) The fundamental, repeating unit of striated muscle, delimited by the Z lines.

sarcoplasmic reticulum (SR) (sar′-kō-plaz′-mik ruh-tik′-yū-lum) A specialized endoplasmic reticulum that regulates the calcium concentration in the cytosol of muscle cells.

saturated fatty acid A fatty acid in which all carbons in the hydrocarbon tail are connected by single bonds, thus maximizing the number of hydrogen atoms that are attached to the carbon skeleton.

savanna A tropical grassland biome with scattered individual trees and large herbivores and maintained by occasional fires and drought.

scaffolding protein A type of large relay protein to which several other relay proteins are simultaneously attached, increasing the efficiency of signal transduction.

scanning electron microscope (SEM) A microscope that uses an electron beam to scan the surface of a sample to study details of its topography.

schizophrenia (skit′-suh-frē′-nē-uh) Severe mental disturbance characterized by psychotic episodes in which patients lose the ability to distinguish reality from hallucination.

Schwann cell A type of glial cell that forms insulating myelin sheaths around the axons of neurons in the peripheral nervous system.

scion (sī′-un) The twig grafted onto the stock when making a graft.

sclera (sklār′-uh) A tough, white outer layer of connective tissue that forms the globe of the vertebrate eye.

sclereid (sklār′-ē-id) A short, irregular sclerenchyma cell in nutshells and seed coats. Sclereids are scattered throughout the parenchyma of some plants.

sclerenchyma cell (skluh-ren′-kim-uh) A rigid, supportive plant cell type usually lacking a protoplast and possessing thick secondary walls strengthened by lignin at maturity.

scrotum A pouch of skin outside the abdomen that houses the testes; functions in maintaining the testes at the lower temperature required for spermatogenesis.

second law of thermodynamics The principle stating that every energy transfer or transformation increases the entropy of the universe. Ordered forms of energy are at least partly converted to heat.

second messenger A small, nonprotein, water-soluble molecule or ion, such as a calcium ion (Ca^{2+}) or cyclic AMP, that relays a signal to a cell's interior in response to a signaling molecule bound by a signal receptor protein.

secondary cell wall In plants, a strong and durable matrix often deposited in several laminated layers for cell protection and support.

secondary consumer A carnivore that eats herbivores.

secondary endosymbiosis A process in eukaryotic evolution in which a heterotrophic eukaryotic cell engulfed a photosynthetic eukaryotic cell, which survived in a symbiotic relationship inside the heterotrophic cell.

secondary growth Growth produced by lateral meristems, thickening the roots and shoots of woody plants.

secondary immune response The acquired immune response elicited on second or subsequent exposures to a particular antigen. The secondary immune response is more rapid, of greater magnitude, and of longer duration than the primary immune response.

secondary oocyte (ō′-uh-sīt) An oocyte that has completed the first of the two meiotic divisions.

secondary plant body The tissues produced by the vascular cambium and cork cambium, which thicken the stems and roots of woody plants.

secondary production The amount of chemical energy in consumers' food that is converted to their own new biomass during a given time period.

secondary structure The localized, repetitive coiling or folding of the polypeptide backbone of a protein due to hydrogen bond formation between constituents of the backbone.

secondary succession A type of succession that occurs where an existing community has been cleared by some disturbance that leaves the soil or substrate intact.

secretion (1) The discharge of molecules synthesized by a cell. (2) The discharge of wastes from the body fluid into the filtrate.

secretory phase That portion of the uterine (menstrual) cycle when the endometrium continues to thicken, becomes more vascularized, and develops glands that secrete a fluid rich in glycogen.

seed An adaptation of some terrestrial plants consisting of an embryo packaged along with a store of food within a protective coat.

seed coat A tough outer covering of a seed, formed from the outer coat of an ovule. In a flowering plant, the seed coat encloses and protects the embryo and endosperm.

seedless vascular plant An informal name for a plant that has vascular tissue but lacks seeds. Seedless vascular plants form a paraphyletic group that includes the phyla Lycophyta (club mosses and their relatives) and Pterophyta (ferns and their relatives).

selective permeability A property of biological membranes that allows them to regulate the passage of substances.

self-incompatibility The ability of a seed plant to reject its own pollen and sometimes the pollen of closely related individuals.

semelparity Reproduction in which an organism produces all of its offspring in a single event; also known as big-bang reproduction.

semen (sē′-mun) The fluid that is ejaculated by the male during orgasm; contains sperm and secretions from several glands of the male reproductive tract.

semicircular canals A three-part chamber of the inner ear that functions in maintaining equilibrium.

semiconservative model Type of DNA replication in which the replicated double helix consists of one old strand, derived from the old molecule, and one newly made strand.

semilunar valve A valve located at each exit of the heart, where the aorta leaves the left ventricle and the pulmonary artery leaves the right ventricle.

seminal vesicle (sem′-i-nul ves′-i-kul) A gland in males that secretes a fluid component of semen that lubricates and nourishes sperm.

seminiferous tubule (sem′-i-nif′-er-us) A highly coiled tube in the testis in which sperm are produced.

senescence (se-nes′-ens) The growth phase in a plant or plant part (as a leaf) from full maturity to death.

sensitive period A limited phase in an individual animal's development when learning of particular behaviors can take place; also called a critical period.

sensor In homeostasis, a receptor that detects a stimulus.

sensory adaptation The tendency of sensory neurons to become less sensitive when they are stimulated repeatedly.

sensory neuron A nerve cell that receives information from the internal or external environment and transmits signals to the central nervous system.

sensory reception The detection of the energy of a stimulus by sensory cells.

sensory receptor An organ, cell, or structure within a cell that responds to specific stimuli from an organism's external or internal environment.

sensory transduction The conversion of stimulus energy to a change in the membrane potential of a sensory receptor cell.

sepal (sē′-pul) A modified leaf in angiosperms that helps enclose and protect a flower bud before it opens.

septum (plural, **septa**) One of the cross-walls that divide a fungal hypha into cells. Septa generally have pores large enough to allow ribosomes, mitochondria, and even nuclei to flow from cell to cell.

serial endosymbiosis A hypothesis for the origin of eukaryotes consisting of a sequence of endosymbiotic events in which mitochondria, chloroplasts, and perhaps other cellular structures were derived from small prokaryotes that had been engulfed by larger cells.

serotonin (ser′-uh-tō′-nin) A neurotransmitter, synthesized from the amino acid tryptophan, that functions in the central nervous system.

set point In animal bodies, a value maintained for a particular variable, such as body temperature or solute concentration, to achieve homeostasis.

seta (sē′-tuh) (plural, **setae**) The elongated stalk of a bryophyte sporophyte.

sex chromosome A chromosome responsible for determining the sex of an individual.

sex pilus (plural, **sex pili**) (pī′-lus, pī′-lī) In bacteria, a structure that links one cell to another at the start of conjugation; also known as a conjugation pilus.

sex-linked gene A gene located on a sex chromosome (usually the X chromosome), resulting in a distinctive pattern of inheritance.

sexual dimorphism (dī-mōr′-fizm) Marked differences between the secondary sex characteristics of males and females.

sexual reproduction A type of reproduction in which two parents give rise to offspring that have unique combinations of genes inherited from the gametes of the parents.

sexual selection A form of natural selection in which individuals with certain inherited characteristics are more likely than other individuals to obtain mates.

Shannon diversity An index of community diversity symbolized by H and represented by the equation $H = [(p_A \ln p_A) + (p_B \ln p_B) + (p_C \ln p_C) + \ldots]$, where A, B, C . . . are the species in the community, p is the relative abundance of each species, and ln is the natural logarithm.

shared ancestral character A character, shared by members of a particular clade, that originated in an ancestor that is not a member of that clade.

shared derived character An evolutionary novelty that is unique to a particular clade.

shoot system The aerial portion of a plant body, consisting of stems, leaves, and (in angiosperms) flowers.

short tandem repeat (STR) Simple sequence DNA containing multiple tandemly repeated units of two to five nucleotides. Variations in STRs act as genetic markers in STR analysis, used to prepare genetic profiles.

short-day plant A plant that flowers (usually in late summer, fall, or winter) only when the light period is shorter than a critical length.

short-term memory The ability to hold information, anticipations, or goals for a time and then release them if they become irrelevant.

sickle-cell disease A human genetic disease caused by a recessive allele that results in the substitution of a single amino acid in a globin polypeptide that is part of the hemoglobin protein; characterized by deformed red blood cells (due to protein aggregation) that can lead to numerous symptoms.

sieve plate An end wall in a sieve-tube element, which facilitates the flow of phloem sap in angiosperm sieve tubes.

sieve-tube element A living cell that conducts sugars and other organic nutrients in the phloem of angiosperms; also called a sieve-tube member. Connected end to end, they form sieve tubes.

sign stimulus An external sensory cue that triggers a fixed action pattern by an animal.

signal In animal behavior, transmission of a stimulus from one animal to another. The term is also used in the context of communication in other kinds of organisms and in cell-to-cell communication in all multicellular organisms.

signal peptide A sequence of about 20 amino acids at or near the leading (amino) end of a polypeptide that targets it to the endoplasmic reticulum or other organelles in a eukaryotic cell.

signal transduction The linkage of a mechanical, chemical, or electromagnetic stimulus to a specific cellular response.

signal transduction pathway A series of steps linking a mechanical or chemical stimulus to a specific cellular response.

signal-recognition particle (SRP) A protein-RNA complex that recognizes a signal peptide as it emerges from a ribosome and helps direct the ribosome to the endoplasmic reticulum (ER) by binding to a receptor protein on the ER.

simple fruit A fruit derived from a single carpel or several fused carpels.

simple sequence DNA A DNA sequence that contains many copies of tandemly repeated short sequences.

single bond A single covalent bond; the sharing of a pair of valence electrons by two atoms.

single circulation A circulatory system consisting of a single pump and circuit, in which blood passes from the sites of gas exchange to the rest of the body before returning to the heart.

single nucleotide polymorphism (SNP) A single base-pair site in a genome where nucleotide variation is found in at least 1% of the population.

single-lens eye The camera-like eye found in some jellies, polychaetes, spiders, and many molluscs.

single-strand binding protein A protein that binds to the unpaired DNA strands during DNA replication, stabilizing them and holding them apart while they serve as templates for the synthesis of complementary strands of DNA.

sinoatrial (SA) node A region in the right atrium of the heart that sets the rate and timing at which all cardiac muscle cells contract; the pacemaker.

sister chromatid Either of two copies of a duplicated chromosome attached to each other by proteins at the centromere and, sometimes, along the arms. While joined, two sister chromatids make up one chromosome; chromatids are eventually separated during mitosis or meiosis II.

sister taxa Groups of organisms that share an immediate common ancestor and hence are each other's closest relatives.

skeletal muscle Muscle that is generally responsible for the voluntary movements of the body; one type of striated muscle.

sliding-filament model The theory explaining how muscle contracts, based on change within a sarcomere, the basic unit of muscle organization. According to this model, thin (actin) filaments slide across thick (myosin) filaments, shortening the sarcomere. The shortening of all sarcomeres in a myofibril shortens the entire myofibril.

slow block to polyspermy The formation of the fertilization envelope and other changes in an egg's surface that prevent fusion of the egg with more than one sperm. The slow block begins about 1 minute after fertilization.

slow-twitch fiber A muscle fiber that can sustain long contractions.

small interfering RNA (siRNA) A small, single-stranded RNA molecule generated by cellular machinery from a long, double-stranded RNA molecule. The siRNA associates with one or more proteins in a complex that can degrade or prevent translation of an mRNA with a complementary sequence. In some cases, siRNA can also block transcription by promoting chromatin modification.

small intestine The longest section of the alimentary canal, so named because of its small diameter compared with that of the large intestine; the principal site of the enzymatic hydrolysis of food macromolecules and the absorption of nutrients.

smooth ER That portion of the endoplasmic reticulum that is free of ribosomes.

smooth muscle A type of muscle lacking the striations of skeletal and cardiac muscle because of the uniform distribution of myosin filaments in the cell; responsible for involuntary body activities.

social learning Modification of behavior through the observation of other individuals.

sociobiology The study of social behavior based on evolutionary theory.

sodium-potassium pump A transport protein in the plasma membrane of animal cells that actively transports sodium out of the cell and potassium into the cell.

soil horizon A soil layer that parallels the land surface and has physical characteristics that differ from those of the layers above and beneath.

solute (sol′-yūt) A substance that is dissolved in a solution.

solute potential (Ψ_S) A component of water potential that is proportional to the osmolarity of a solution and that measures the effect of solutes on the direction of water movement; also called osmotic potential, it can be either zero or negative.

solution A liquid that is a homogeneous mixture of two or more substances.

solvent The dissolving agent of a solution. Water is the most versatile solvent known.

somatic cell (sō-mat′-ik) Any cell in a multicellular organism except a sperm or egg.

somite One of a series of blocks of mesoderm that exist in pairs just lateral to the notochord in a vertebrate embryo.

soredium (plural, **soredia**) In lichens, a small cluster of fungal hyphae with embedded algae.

sorus (plural, **sori**) A cluster of sporangia on a fern sporophyll. Sori may be arranged in various patterns, such as parallel lines or dots, which are useful in fern identification.

Southern blotting A technique that enables specific nucleotide sequences to be detected in a sample of DNA. It involves gel electrophoresis of DNA molecules and their transfer to a membrane (blotting), followed by nucleic acid hybridization with a labeled probe.

spatial learning The establishment of a memory that reflects the environment's spatial structure.

spatial summation A phenomenon of neural integration in which the membrane potential of the postsynaptic cell is determined by the combined effect of EPSPs or IPSPs produced nearly simultaneously by different synapses.

speciation (spē′-sē-ā′-shun) An evolutionary process in which one species splits into two or more species.

species (spē′-sēz) A population or group of populations whose members have the potential to interbreed in nature and produce viable, fertile offspring, but do not produce viable, fertile offspring with members of other such groups.

species diversity The number and relative abundance of species in a biological community.

species richness The number of species in a biological community.

species-area curve The biodiversity pattern, first noted by Alexander von Humboldt, that shows that the larger the geographic area of a community is, the more species it has.

specific heat The amount of heat that must be absorbed or lost for 1 g of a substance to change its temperature by 1°C.

spectrophotometer An instrument that measures the proportions of light of different wavelengths absorbed and transmitted by a pigment solution.

sperm The male gamete.

spermatheca (sper′-muh-thē′-kuh) In many insects, a sac in the female reproductive system where sperm are stored.

spermatogenesis The continuous and prolific production of mature sperm cells in the testis.

spermatogonium A cell that divides mitotically to form spermatocytes.

sphincter (sfink′-ter) A ringlike valve, consisting of modified muscles in a muscular tube, that regulates passage between some compartments of the alimentary canal.

spinal nerve In the vertebrate peripheral nervous system, a nerve that carries signals to or from the spinal cord.

spiral cleavage A type of embryonic development in protostomes in which the planes of cell division that transform the zygote into a ball of cells are diagonal to the vertical axis of the embryo. As a result, the cells of each tier sit in the grooves between cells of adjacent tiers.

spliceosome (splī′-sē-ō-sōm) A large complex made up of proteins and RNA molecules that splices RNA by interacting with the ends of an RNA intron, releasing the intron and joining the two adjacent exons.

spongocoel (spon′-jō-sēl) The central cavity of a sponge.

sporangium (spōr-an′-jē-um) (plural, **sporangia**) A multicellular organ in fungi and plants in which meiosis occurs and haploid cells develop.

spore (1) In the life cycle of a plant or alga undergoing alternation of generations, a haploid cell produced in the sporophyte by meiosis. A spore can divide by mitosis to develop into a multicellular haploid individual, the gametophyte, without fusing with another cell. (2) In fungi, a haploid cell, produced either sexually or asexually, that produces a mycelium after germination.

sporocyte A diploid cell, also known as a spore mother cell, that undergoes meiosis and generates haploid spores.

sporophyll (spō′-ruh-fil) A modified leaf that bears sporangia and hence is specialized for reproduction.

sporophyte (spō-ruh-fīt′) In organisms (plants and some algae) that have alternation of generations, the multicellular diploid form that results from the union of gametes. The sporophyte produces haploid spores by meiosis that develop into gametophytes.

sporopollenin (spōr-uh-pol′-eh-nin) A durable polymer that covers exposed zygotes of charophyte algae and forms the walls of plant spores, preventing them from drying out.

stabilizing selection Natural selection in which intermediate phenotypes survive or reproduce more successfully than do extreme phenotypes.

stamen (stā′-men) The pollen-producing reproductive organ of a flower, consisting of an anther and a filament.

standard metabolic rate (SMR) The metabolic rate of a resting, fasting, and nonstressed ectotherm at a particular temperature.

stapes The third of three bones in the middle ear of mammals; also called the stirrup.

starch A storage polysaccharide in plants, consisting entirely of glucose monomers joined by α glycosidic linkages.
statocyst (stat′-uh-sist′) A type of mechanoreceptor that functions in equilibrium in invertebrates by use of statoliths, which stimulate hair cells in relation to gravity.
statolith (stat′-uh-lith′) (1) In plants, a specialized plastid that contains dense starch grains and may play a role in detecting gravity. (2) In invertebrates, a grain or other dense granule that settles in response to gravity and is found in sensory organs that function in equilibrium.
stele (stēl)The vascular tissue of a stem or root.
stem A vascular plant organ consisting of an alternating system of nodes and internodes that support the leaves and reproductive structures.
stem cell Any relatively unspecialized cell that can produce, during a single division, one identical daughter cell and one more specialized daughter cell that can undergo further differentiation.
stenohaline (sten′-ō-hā′-līn) Referring to organisms that cannot tolerate substantial changes in external osmolarity.
steroid A type of lipid characterized by a carbon skeleton consisting of four rings with various chemical groups attached.
sticky end A single-stranded end of a double-stranded restriction fragment.
stigma (plural, **stigmata**) The sticky part of a flower's carpel, which traps pollen grains.
stimulus In homeostasis, a fluctuation in a variable that triggers a return to a set point.
stipe A stemlike structure of a seaweed.
stock The plant that provides the root system when making a graft.
stoma (stō′-muh) (plural, **stomata**) A microscopic pore surrounded by guard cells in the epidermis of leaves and stems that allows gas exchange between the environment and the interior of the plant.
stomach An organ of the digestive system that stores food and performs preliminary steps of digestion.
stramenopile A protist in which a "hairy" flagellum (one covered with fine, hairlike projections) is paired with a shorter, smooth flagellum.
stratum (strah′-tum) (plural, **strata**) A rock layer formed when new layers of sediment cover older ones and compress them.
striated muscle Muscle in which the regular arrangement of filaments creates a pattern of light and dark bands.
strobilus (strō-bī′-lus) (plural, **strobili**) The technical term for a cluster of sporophylls known commonly as a cone, found in most gymnosperms and some seedless vascular plants.
stroke The death of nervous tissue in the brain, usually resulting from rupture or blockage of arteries in the head.
stroke volume The volume of blood pumped by a heart ventricle in a single contraction.
stroma (strō′-muh) Within the chloroplast, the dense fluid of the chloroplast surrounding the thylakoid membrane; involved in the synthesis of organic molecules from carbon dioxide and water.
stromatolite Layered rock that results from the activities of prokaryotes that bind thin films of sediment together.
structural formula A type of molecular notation in which the constituent atoms are joined by lines representing covalent bonds.
structural isomer One of several compounds that have the same molecular formula but differ in the covalent arrangements of their atoms.
style The stalk of a flower's carpel, with the ovary at the base and the stigma at the top.
substance P A neuropeptide that is a key excitatory neurotransmitter that mediates the perception of pain.
substrate The reactant on which an enzyme works.
substrate feeder An animal that lives in or on its food source, eating its way through the food.
substrate-level phosphorylation The formation of ATP by an enzyme directly transferring a phosphate group to ADP from an intermediate substrate in catabolism.
sugar sink A plant organ that is a net consumer or storer of sugar. Growing roots, shoot tips, stems, and fruits are sugar sinks supplied by phloem.
sugar source A plant organ in which sugar is being produced by either photosynthesis or the breakdown of starch. Mature leaves are the primary sugar sources of plants.
sulfhydryl group A chemical group consisting of a sulfur atom bonded to a hydrogen atom.
suprachiasmatic nucleus (SCN) A group of neurons in the hypothalamus of mammals that functions as a biological clock.
surface tension A measure of how difficult it is to stretch or break the surface of a liquid. Water has a high surface tension because of the hydrogen bonding of surface molecules.
surfactant A substance secreted by alveoli that decreases surface tension in the fluid that coats the alveoli.
survivorship curve A plot of the number of members of a cohort that are still alive at each age; one way to represent age-specific mortality.
suspension feeder An aquatic animal, such as a sponge, clam, or baleen whale, that feeds by sifting small food particles from the water.
sustainable agriculture Long-term productive farming methods that are environmentally safe.
sustainable development Development that meets the needs of people today without limiting the ability of future generations to meet their needs.
swim bladder In aquatic osteichthyans, an air sac that enables the animal to control its buoyancy in the water.
symbiont (sim′-bē-ont) The smaller participant in a symbiotic relationship, living in or on the host.
symbiosis An ecological relationship between organisms of two different species that live together in direct and intimate contact.
sympathetic division One of three divisions of the autonomic nervous system of vertebrates; generally increases energy expenditure and prepares the body for action.
sympatric speciation (sim-pat′-rik) The formation of new species in populations that live in the same geographic area.
symplast In plants, the continuum of cytoplasm connected by plasmodesmata between cells.
synapse (sin′-aps) The junction where one neuron communicates with another cell across a narrow gap. Neurotransmitter molecules released by the neuron diffuse across the synapse, relaying messages to the other cell.
synapsid Member of an amniote clade distinguished by a single hole on each side of the skull. Synapsids include the mammals.
synapsis (si-nap′-sis) The pairing and physical connection of replicated homologous chromosomes during prophase I of meiosis.
synaptic cleft (sin-ap′-tik) A narrow gap separating the synaptic terminal of a transmitting neuron from a receiving neuron or an effector cell.
synaptic terminal A bulb at the end of an axon in which neurotransmitter molecules are stored and from which they are released.
synaptic vesicle Membranous sac containing neurotransmitter molecules at the tip of an axon.
systematics A scientific discipline focused on classifying organisms and determining their evolutionary relationships.
systemic Occurring throughout the body and affecting many or all body systems or organs.
systemic acquired resistance A defensive response in infected plants that helps protect healthy tissue from pathogenic invasion.
systemic circuit The branch of the circulatory system that supplies all body organs except those involved in gas exchange.
systems biology An approach to studying biology that aims to model the dynamic behavior of whole biological systems.
systole (sis′-tō-lē) The stage of the cardiac cycle in which a heart chamber contracts and pumps blood.
systolic pressure Blood pressure in the arteries during contraction of the ventricles.
T cell receptor The antigen receptor on T cells; a membrane-bound molecule consisting of one α chain and one β chain linked by a disulfide bridge and containing one antigen-binding site.
T cells The class of lymphocytes that mature in the thymus and that includes both effector cells for the cell-mediated immune response and helper cells required for both branches of adaptive immunity.
taproot A main vertical root that develops from an embryonic root and gives rise to lateral (branch) roots.
tastant Any chemical that stimulates the sensory receptors in a taste bud.
taste bud A collection of modified epithelial cells on the tongue or in the mouth that are receptors for taste in mammals.

TATA box A DNA sequence in eukaryotic promoters crucial in forming the transcription initiation complex.

taxis (tak′-sis) An oriented movement toward or away from a stimulus.

taxon (plural, **taxa**) A named taxonomic unit at any given level of classification.

taxonomy (tak-son′-uh-mē) A scientific discipline concerned with naming and classifying the diverse forms of life.

Tay-Sachs disease A human genetic disease caused by a recessive allele for a dysfunctional enzyme, leading to accumulation of certain lipids in the brain. Seizures, blindness, and degeneration of motor and mental performance usually become manifest a few months after birth, followed by death within a few years.

technology The application of scientific knowledge for a specific purpose, often involving industry or commerce but also including uses in basic research.

telomerase An enzyme that catalyzes the lengthening of telomeres in eukaryotic germ cells.

telomere (tel′-uh-mēr) The tandemly repetitive DNA at the end of a eukaryotic chromosome's DNA molecule that protects the organism's genes from being eroded during successive rounds of replication. *See also* repetitive DNA.

telophase The fifth and final stage of mitosis, in which daughter nuclei are forming and cytokinesis has typically begun.

temperate broadleaf forest A biome located throughout midlatitude regions where there is sufficient moisture to support the growth of large, broadleaf deciduous trees.

temperate grassland A terrestrial biome dominated by grasses and forbs.

temperate phage A phage that is capable of reproducing by either a lytic or lysogenic cycle.

temperature A measure of the intensity of heat in degrees, reflecting the average kinetic energy of the molecules.

template strand The DNA strand that provides the pattern, or template, for ordering the sequence of nucleotides in an RNA transcript.

temporal summation A phenomenon of neural integration in which the membrane potential of the postsynaptic cell in a chemical synapse is determined by the combined effect of EPSPs or IPSPs produced in rapid succession.

tendon A fibrous connective tissue that attaches muscle to bone.

terminator In bacteria, a sequence of nucleotides in DNA that marks the end of a gene and signals RNA polymerase to release the newly made RNA molecule and detach from the DNA.

territoriality A behavior in which an animal defends a bounded physical space against encroachment by other individuals, usually of its own species.

tertiary consumer (ter′-shē-ār′-ē) A carnivore that eats other carnivores.

tertiary structure Irregular contortions of a protein molecule due to interactions of side chains involved in hydrophobic interactions, ionic bonds, hydrogen bonds, and disulfide bridges.

testcross Breeding an organism of unknown genotype with a homozygous recessive individual to determine the unknown genotype. The ratio of phenotypes in the offspring reveals the unknown genotype.

testis (plural, **testes**) The male reproductive organ, or gonad, in which sperm and reproductive hormones are produced.

testosterone A steroid hormone required for development of the male reproductive system, spermatogenesis, and male secondary sex characteristics; the major androgen in mammals.

tetanus (tet′-uh-nus) The maximal, sustained contraction of a skeletal muscle, caused by a very high frequency of action potentials elicited by continual stimulation.

tetrapod A vertebrate with two pairs of limbs. Tetrapods include mammals, amphibians, and birds and other reptiles.

thalamus (thal′-uh-mus) One of two integrating centers of the vertebrate forebrain. Neurons with cell bodies in the thalamus relay neural input to specific areas in the cerebral cortex and regulate what information goes to the cerebral cortex.

thallus (plural, **thalli**) A seaweed body that is plantlike, consisting of a holdfast, stipe, and blades, yet lacks true roots, stems, and leaves.

theory An explanation that is broad in scope, generates new hypotheses, and is supported by a large body of evidence.

thermal energy *See* heat.

thermocline A narrow stratum of rapid temperature change in the ocean and in many temperate-zone lakes.

thermodynamics (ther′-mō-dī-nam′-iks) The study of energy transformations that occur in a collection of matter. *See* first law of thermodynamics; second law of thermodynamics.

thermoreceptor A receptor stimulated by either heat or cold.

thermoregulation The maintenance of internal body temperature within a tolerable range.

theropod Member of an ancient group of dinosaurs that were bipedal carnivores.

thick filament A filament composed of staggered arrays of myosin molecules; a component of myofibrils in muscle fibers.

thigmomorphogenesis A response in plants to chronic mechanical stimulation, resulting from increased ethylene production. An example is thickening stems in response to strong winds.

thigmotropism (thig-mo′-truh-pizm) A directional growth of a plant in response to touch.

thin filament A filament consisting of two strands of actin and two strands of regulatory protein coiled around one another; a component of myofibrils in muscle fibers.

threatened species A species that is considered likely to become endangered in the foreseeable future.

threshold The potential that an excitable cell membrane must reach for an action potential to be initiated.

thrombus A fibrin-containing clot that forms in a blood vessel and blocks the flow of blood.

thylakoid (thī′-luh-koyd) A flattened membranous sac inside a chloroplast. Thylakoids exist in an interconnected system in the chloroplast and contain the molecular "machinery" used to convert light energy to chemical energy.

thymus (thī′-mus) A small organ in the thoracic cavity of vertebrates where maturation of T cells is completed.

thyroid gland An endocrine gland, located on the ventral surface of the trachea, that secretes two iodine-containing hormones, triiodothyronine (T_3) and thyroxine (T_4), as well as calcitonin.

thyroxine (T_4) One of two iodine-containing hormones that are secreted by the thyroid gland and that help regulate metabolism, development, and maturation in vertebrates.

Ti plasmid A plasmid of a tumor-inducing bacterium (the plant pathogen *Agrobacterium*) that integrates a segment of its DNA (T DNA) into a chromosome of a host plant. The Ti plasmid is frequently used as a vector for genetic engineering in plants.

tidal volume The volume of air a mammal inhales and exhales with each breath.

tight junction A type of intercellular junction in animal cells that prevents the leakage of material between cells.

tissue An integrated group of cells with a common function, structure, or both.

tissue system One or more tissues organized into a functional unit connecting the organs of a plant.

TLR Toll-like receptor. A membrane receptor on a phagocytic white blood cell that recognizes fragments of molecules common to a set of pathogens.

tonicity The ability of a solution surrounding a cell to cause that cell to gain or lose water.

top-down model A model of community organization in which predation influences community organization by controlling herbivore numbers, which in turn control plant or phytoplankton numbers, which in turn control nutrient levels; also called the trophic cascade model.

topoisomerase A protein that breaks, swivels, and rejoins DNA strands. During DNA replication, topoisomerase helps to relieve strain in the double helix ahead of the replication fork.

topsoil A mixture of particles derived from rock, living organisms, and decaying organic material (humus).

torpor A physiological state in which activity is low and metabolism decreases.

torsion In gastropods, a developmental process in which the visceral mass rotates up to 180°, causing the animal's anus and mantle cavity to be positioned above its head.

totipotent (tō′-tuh-pōt′-ent) Describing a cell that can give rise to all parts of the embryo and adult, as well as extraembryonic membranes in species that have them.
trace element An element indispensable for life but required in extremely minute amounts.
trachea (trā′-kē-uh) The portion of the respiratory tract that passes from the larynx to the bronchi; also called the windpipe.
tracheal system In insects, a system of branched, air-filled tubes that extends throughout the body and carries oxygen directly to cells.
tracheid (trā′-kē-id) A long, tapered water-conducting cell found in the xylem of nearly all vascular plants. Functioning tracheids are no longer living.
trait Any detectable variant in a genetic character.
trans fat An unsaturated fat containing one or more *trans* double bonds.
transcription The synthesis of RNA using a DNA template.
transcription factor A regulatory protein that binds to DNA and affects transcription of specific genes.
transcription initiation complex The completed assembly of transcription factors and RNA polymerase bound to a promoter.
transcription unit A region of DNA that is transcribed into an RNA molecule.
transduction (1) A type of horizontal gene transfer in which phages (viruses) carry bacterial DNA from one host cell to another. (2) In cellular communication, the conversion of a signal from outside the cell to a form that can bring about a specific cellular response.
transfer cell In a plant, a companion cell with numerous ingrowths of its wall, which increase the cell's surface area and enhance the transfer of solutes between apoplast and symplast.
transfer RNA (tRNA) An RNA molecule that functions as an interpreter between nucleic acid and protein language by picking up specific amino acids and recognizing the appropriate codons in the mRNA.
transformation (1) The conversion of a normal animal cell to a cancerous cell. (2) A change in genotype and phenotype due to the assimilation of external DNA by a cell.
transgenic Pertaining to an organism whose genome contains a gene introduced from another organism of the same or a different species.
translation The synthesis of a polypeptide using the genetic information encoded in an mRNA molecule. There is a change of "language" from nucleotides to amino acids.
translocation (1) An aberration in chromosome structure resulting from attachment of a chromosomal fragment to a nonhomologous chromosome. (2) During protein synthesis, the third stage in the elongation cycle when the RNA carrying the growing polypeptide moves from the A site to the P site on the ribosome. (3) The transport of organic nutrients in the phloem of vascular plants.
transmission The passage of a nerve impulse along axons.
transmission electron microscope (TEM) A microscope that passes an electron beam through very thin sections and is primarily used to study the internal ultrastructure of cells.
transpiration The evaporative loss of water from a plant.
transport epithelium One or more layers of specialized epithelial cells that regulate solute movements.
transport protein A transmembrane protein that helps a certain substance or class of closely related substances to cross the membrane.
transport vesicle A tiny membranous sac in a cell's cytoplasm carrying molecules produced by the cell.
transposable element A segment of DNA that can move within the genome of a cell by means of a DNA or RNA intermediate; also called a transposable genetic element.
transposon A transposable element that moves within a genome by means of a DNA intermediate.
transverse (T) tubule An infolding of the plasma membrane of skeletal muscle cells.
triacylglycerol (trī-as′-ul-glis′-uh-rol) Three fatty acids linked to one glycerol molecule; also called a fat or a triglyceride.
triiodothyronine (T_3) (trī′-ī-ō′-dō-thī′-rō-nēn) One of two iodine-containing hormones that are secreted by the thyroid gland and that help regulate metabolism, development, and maturation in vertebrates.
trimester In human development, one of three 3-month-long periods of pregnancy.
triple response A plant growth maneuver in response to mechanical stress, involving slowing of stem elongation, a thickening of the stem, and a curvature that causes the stem to start growing horizontally.
triplet code A set of three-nucleotide-long words that specify the amino acids for polypeptide chains.
triploblastic Possessing three germ layers: the endoderm, mesoderm, and ectoderm. Most eumetazoans are triploblastic.
trisomic Referring to a diploid cell that has three copies of a particular chromosome instead of the normal two.
trochophore larva (trō′-kuh-fōr) Distinctive larval stage observed in some lophotrochozoan animals, including some annelids and molluscs.
trophic efficiency The percentage of production transferred from one trophic level to the next.
trophic structure The different feeding relationships in an ecosystem, which determine the route of energy flow and the pattern of chemical cycling.
trophoblast The outer epithelium of a mammalian blastocyst. It forms the fetal part of the placenta, supporting embryonic development but not forming part of the embryo proper.
tropic hormone A hormone that has another endocrine gland as a target.
tropical rain forest A terrestrial biome characterized by high levels of precipitation and high temperatures year-round.
tropics Latitudes between 23.5° north and south.
tropism A growth response that results in the curvature of whole plant organs toward or away from stimuli due to differential rates of cell elongation.
tropomyosin The regulatory protein that blocks the myosin-binding sites on actin molecules.
troponin complex The regulatory proteins that control the position of tropomyosin on the thin filament.
true-breeding Referring to plants that produce offspring of the same variety when they self-pollinate.
tubal ligation A means of sterilization in which a woman's two oviducts (fallopian tubes) are tied closed to prevent eggs from reaching the uterus. A segment of each oviduct is removed.
tube foot One of numerous extensions of an echinoderm's water vascular system. Tube feet function in locomotion, feeding, and gas exchange.
tumor-suppressor gene A gene whose protein product inhibits cell division, thereby preventing the uncontrolled cell growth that contributes to cancer.
tundra A terrestrial biome at the extreme limits of plant growth. At the northernmost limits, it is called arctic tundra, and at high altitudes, where plant forms are limited to low shrubby or matlike vegetation, it is called alpine tundra.
tunicate Member of the subphylum Urochordata, sessile marine chordates that lack a backbone.
turgid (ter′-jid) Swollen or distended, as in plant cells. (A walled cell becomes turgid if it has a greater solute concentration than its surroundings, resulting in entry of water.)
turgor pressure The force directed against a plant cell wall after the influx of water and swelling of the cell due to osmosis.
turnover The mixing of waters as a result of changing water-temperature profiles in a lake.
turnover time The time required to replace the standing crop of a population or group of populations (for example, of phytoplankton), calculated as the ratio of standing crop to production.
twin study A behavioral study in which researchers compare the behavior of identical twins raised apart with that of identical twins raised in the same household.
tympanic membrane Another name for the eardrum, the membrane between the outer and middle ear.
ultimate causation The evolutionary explanation of "why" a behavior (or other aspect of an organism's biology) occurs, that is, the benefit to survival and reproduction or the

evolutionary significance of the behavioral act.

undernourishment A condition that results from a diet that consistently supplies less chemical energy than the body requires.

uniformitarianism The principle stating that mechanisms of change are constant over time. *See* catastrophism.

Unikonta (yū′-ni-kon′-tuh) One of five supergroups of eukaryotes proposed in a current hypothesis of the evolutionary history of eukaryotes. This clade, which is supported by studies of myosin proteins and DNA, consists of amoebozoans and opisthokonts. *See also* Excavata, Chromalveolata, Rhizaria, and Archaeplastida.

unsaturated fatty acid A fatty acid possessing one or more double bonds between the carbons in the hydrocarbon tail. Such bonding reduces the number of hydrogen atoms attached to the carbon skeleton.

urea A soluble nitrogenous waste produced in the liver by a metabolic cycle that combines ammonia with carbon dioxide.

ureter (yū-rē′-ter) A duct leading from the kidney to the urinary bladder.

urethra (yū-rē′-thruh) A tube that releases urine from the mammalian body near the vagina in females and through the penis in males; also serves in males as the exit tube for the reproductive system.

uric acid A product of protein and purine metabolism and the major nitrogenous waste product of insects, land snails, and many reptiles. Uric acid is relatively nontoxic and largely insoluble.

urinary bladder The pouch where urine is stored prior to elimination.

uterine cycle The changes that occur in the uterus during the reproductive cycle of the human female; also called the menstrual cycle.

uterus A female organ where eggs are fertilized and/or development of the young occurs.

utricle In the vertebrate ear, a chamber in the vestibule behind the oval window that opens into the three semicircular canals.

vaccination *See* immunization.

vaccine A harmless variant or derivative of a pathogen that stimulates a host's immune system to mount defenses against the pathogen.

vacuole (vak′-yū-ōl′) A membrane-bounded vesicle whose function varies in different kinds of cells.

vagina Part of the female reproductive system between the uterus and the outside opening; the birth canal in mammals. During copulation, the vagina accommodates the male's penis and receives sperm.

valence The bonding capacity of a given atom; usually equals the number of unpaired electrons required to complete the atom's outermost (valence) shell.

valence electron An electron in the outermost electron shell.

valence shell The outermost energy shell of an atom, containing the valence electrons involved in the chemical reactions of that atom.

van der Waals interactions Weak attractions between molecules or parts of molecules that result from localized charge fluctuations.

variation Differences between members of the same species.

vas deferens In mammals, the tube in the male reproductive system in which sperm travel from the epididymis to the urethra.

vasa recta The capillary system in the kidney that serves the loop of Henle.

vascular cambium A cylinder of meristematic tissue in woody plants that adds layers of secondary vascular tissue called secondary xylem (wood) and secondary phloem.

vascular plant A plant with vascular tissue. Vascular plants include all living plant species except mosses, liverworts, and hornworts.

vascular tissue Plant tissue consisting of cells joined into tubes that transport water and nutrients throughout the plant body.

vascular tissue system A transport system formed by xylem and phloem throughout a vascular plant. Xylem transports water and minerals; phloem transports sugars, the products of photosynthesis.

vasectomy The cutting and sealing of each vas deferens to prevent sperm from entering the urethra.

vasocongestion The filling of a tissue with blood, caused by increased blood flow through the arteries of that tissue.

vasoconstriction A decrease in the diameter of blood vessels caused by contraction of smooth muscles in the vessel walls.

vasodilation An increase in the diameter of blood vessels caused by relaxation of smooth muscles in the vessel walls.

vector An organism that transmits pathogens from one host to another.

vegetal pole The point at the end of an egg in the hemisphere where most yolk is concentrated; opposite of animal pole.

vegetative reproduction Cloning of plants by asexual means.

vein (1) In animals, a vessel that carries blood toward the heart. (2) In plants, a vascular bundle in a leaf.

ventilation The flow of air or water over a respiratory surface.

ventral Pertaining to the underside, or bottom, of an animal with bilateral symmetry.

ventricle (ven′-tri-kul) (1) A heart chamber that pumps blood out of the heart. (2) A space in the vertebrate brain, filled with cerebrospinal fluid.

venule (ven′-yūl) A vessel that conveys blood between a capillary bed and a vein.

vernalization The use of cold treatment to induce a plant to flower.

vertebrate A chordate animal with a backbone: the mammals, reptiles (including birds), amphibians, sharks and rays, ray-finned fishes, and lobe-fins.

vesicle (ves′-i-kul) A sac made of membrane in the cytoplasm.

vessel A continuous water-conducting micropipe found in most angiosperms and a few nonflowering vascular plants.

vessel element A short, wide water-conducting cell found in the xylem of most angiosperms and a few nonflowering vascular plants. Dead at maturity, vessel elements are aligned end to end to form micropipes called vessels.

vestigial structure A structure of marginal, if any, importance to an organism. Vestigial structures are historical remnants of structures that had important functions in ancestors.

villus (plural, **villi**) (1) A finger-like projection of the inner surface of the small intestine. (2) A finger-like projection of the chorion of the mammalian placenta. Large numbers of villi increase the surface areas of these organs.

viral envelope A membrane that cloaks the capsid that in turn encloses a viral genome.

viroid (vī′-royd) A plant pathogen consisting of a molecule of naked, circular RNA a few hundred nucleotides long.

virulent Describing a pathogen against which an organism has little specific defense.

virulent phage A phage that reproduces only by a lytic cycle.

visceral mass One of the three main parts of a mollusc; the part containing most of the internal organs. *See also* foot, mantle.

visible light That portion of the electromagnetic spectrum that can be detected as various colors by the human eye, ranging in wavelength from about 380 nm to about 750 nm.

vital capacity The maximum volume of air that a mammal can inhale and exhale with each breath.

vitamin An organic molecule required in the diet in very small amounts. Vitamins serve primarily as coenzymes or as parts of coenzymes.

vitreous humor The jellylike material that fills the posterior cavity of the vertebrate eye.

viviparous (vī-vip′-uh-rus) Referring to a type of development in which the young are born alive after having been nourished in the uterus by blood from the placenta.

vocal cord One of a pair of bands of elastic tissue in the larynx. Air rushing past the tensed vocal cords makes them vibrate, producing sounds.

voltage-gated ion channel A specialized ion channel that opens or closes in response to changes in membrane potential.

vulva Collective term for the female external genitalia.

water potential (Ψ) The physical property predicting the direction in which water will flow, governed by solute concentration and applied pressure.

water vascular system A network of hydraulic canals unique to echinoderms that branches into extensions called tube feet, which function in locomotion, feeding, and gas exchange.

wavelength The distance between crests of waves, such as those of the electromagnetic spectrum.

wetland A habitat that is inundated by water at least some of the time and that supports plants adapted to water-saturated soil.
white matter Tracts of axons within the CNS.
wild type An individual with the phenotype most commonly observed in natural populations; also refers to the phenotype itself.
wilting The drooping of leaves and stems as a result of plant cells becoming flaccid.
wobble Flexibility in the base-pairing rules in which the nucleotide at the 5′ end of a tRNA anticodon can form hydrogen bonds with more than one kind of base in the third position (3′ end) of a codon.
xerophyte A plant adapted to an arid climate.
X-ray crystallography A technique that depends on the diffraction of an X-ray beam by the individual atoms of a crystallized molecule to study the three-dimensional structure of the molecule.
xylem (zī′-lum) Vascular plant tissue consisting mainly of tubular dead cells that conduct most of the water and minerals upward from the roots to the rest of the plant.
xylem sap The dilute solution of water and dissolved minerals carried through vessels and tracheids.
yeast Single-celled fungus that reproduces asexually by binary fission or by the pinching of small buds off a parent cell; some species exhibit cell fusion between different mating types.
yeast artificial chromosome (YAC) A cloning vector that combines the essentials of a eukaryotic chromosome—an origin for DNA replication, a centromere, and two telomeres—with foreign DNA.
yolk Nutrients stored in an egg.
yolk plug A group of large, nutrient-laden endodermal cells surrounded by the completed blastopore in an amphibian gastrula. These cells will be covered by ectoderm and end up inside the embryo.
yolk sac One of four extraembryonic membranes. It encloses the yolk in reptiles and is the first site of blood cell and circulatory system function.
zero population growth (ZPG) A period of stability in population size, when the per capita birth rate and death rate are equal.
zona pellucida The extracellular matrix surrounding a mammalian egg.
zone of polarizing activity (ZPA) A block of mesoderm located just under the ectoderm where the posterior side of a limb bud is attached to the body; required for proper pattern formation along the anterior-posterior axis of the limb.
zoned reserve An extensive region that includes areas relatively undisturbed by humans surrounded by areas that have been changed by human activity and are used for economic gain.
zoonotic pathogen A disease-causing agent that is transmitted to humans from other animals.
zoospore Flagellated spore found in chytrid fungi and some protists.
zygomycete (zī′-guh-mī′-sēt) Member of the fungal phylum Zygomycota, characterized by the formation of a sturdy structure called a zygosporangium during sexual reproduction.
zygosporangium (zī′-guh-spōr-an′-jē-um) In zygomycete fungi, a sturdy multinucleate structure in which karyogamy and meiosis occur.
zygote (zī′-gōt) The diploid product of the union of haploid gametes during fertilization; a fertilized egg.

Index

NOTE: A page number in regular type indicates where a topic is discussed in text (topic may also be in a figure on that page); a **bold** page number indicates where a term is bold and defined; an *f* following a page number indicates a figure (topic may also be discussed in text on that page); a *t* following a page number indicates a table (topic may also be discussed in text on that page).

Index

Index

E

Index

Index

F

G

Index

H

Index

Index

Index

Index

Index

Index

Index

Index

P

Index

Index

Index

Index

S

Index

Index

Index

Index

W